Lecture Notes in Computer Science 1123
Edited by G. Goos, J. Hartmanis and J. van Leeuwen

Advisory Board: W. Brauer   D. Gries   J. Stoer

Springer
*Berlin
Heidelberg
New York
Barcelona
Budapest
Hong Kong
London
Milan
Paris
Santa Clara
Singapore
Tokyo*

Luc Bougé   Pierre Fraigniaud
Anne Mignotte   Yves Robert   (Eds.)

Euro-Par'96
Parallel Processing

Second International Euro-Par Conference
Lyon, France, August 26-29, 1996
Proceedings, Volume I

 Springer

Series Editors

Gerhard Goos, Karlsruhe University, Germany
Juris Hartmanis, Cornell University, NY, USA
Jan van Leeuwen, Utrecht University, The Netherlands

Volume Editors

Luc Bougé
Pierre Fraigniaud
Anne Mignotte
Yves Robert
École Normale Supérieure de Lyon, Laboratoire LIP
46 allée d'Italie, F-69364 Lyon Cedex 07, France
Pierre.Fraigniaud@lip.ens-lyon.fr

Cataloging-in-Publication data applied for

Die Deutsche Bibliothek - CIP-Einheitsaufnahme

Parallel processing : proceedings / EURO-PAR '96. Second
International EURO-PAR Conference, Lyon, France, August 26
- 29, 1996. Luc Bougé ... (ed.). - Berlin ; Heidelberg ; New
York ; Barcelona ; Budapest ; Hong Kong ; London ; Milan ;
Paris ; Santa Clara ; Singapore ; Tokyo : Springer
NE: Bougé, Luc [Hrsg.]; EURO-PAR <2, 1996, Lyon>
Vol. 1 (1996)
  (Lecture notes in computer science ; Vol. 1123)
  ISBN 3-540-61626-8
NE: GT

CR Subject Classification (1991): C.1-4, D.1-4, F.1-2, G.1-2, E.1, H.2

ISSN 0302-9743
ISBN 3-540-61626-8 Springer-Verlag Berlin Heidelberg New York

This work is subject to copyright. All rights are reserved, whether the whole or part of the material is
concerned, specifically the rights of translation, reprinting, re-use of illustrations, recitation, broadcasting,
reproduction on microfilms or in any other way, and storage in data banks. Duplication of this publication
or parts thereof is permitted only under the provisions of the German Copyright Law of September 9, 1965,
in its current version, and permission for use must always be obtained from Springer-Verlag. Violations are
liable for prosecution under the German Copyright Law.

© Springer-Verlag Berlin Heidelberg 1996
Printed in Germany

Typesetting: Camera-ready by author
SPIN 10513487   06/3142 – 5 4 3 2 1 0   Printed on acid-free paper

Preface

Euro-Par is the annual European Conference on Parallel Processing. It merges the former CONPAR-VAPP and PARLE conferences, already two major events in the field. The goal of Euro-Par is to gather people interested in Parallel Computing and Architectures.

Euro-Par'96 consists of a large panel of workshops on all aspects of parallel processing, from theory to practice and from academy to industry. These workshops are expected to present the latest advances in their respective domains and are chaired by leading researchers in the field.

Euro-Par'96 is a workshop-based conference. The idea of organizing the conference in workshops stemmed from the following observation. While general-purpose conferences provide opportunities to meet a large number of people and to listen to talks on many different topics, attendees (that is, all of us) often feel frustrated as only a few talks are in direct connection with their research interests. On the other hand, highly focused workshops often present the opposite features. Specialists meet among themselves, and it is difficult to open up new horizons. Our objective in proposing an original format for this conference is to preserve the nice properties of focused workshops while pushing on cross-fertilization between communities.

Euro-Par'96 is a fork/join conference. We originally planned 22 independent workshops, whose topics hopefully cover most of the subjects related to parallel computing. Workshops are introduced by several high-level tutorials on general interest subjects (organized on a single day), and invited conferences are scheduled at the beginning and at the end of the conference.

The story of Euro-Par'96

The responsibility for each of the 22 workshops has been assigned to a workshop chairperson, who had to select two additional program committee members and a local chairperson to assist him or her. The Program Committee of Euro-Par'96 is made of all these persons. It so happened that 4 out of the 22 original workshops had to be cancelled as they did not get enough submitted and/or accepted papers. The accepted papers, if any, were moved to workshops on connected topics. Furthermore, small workshops covering related research domains were merged so that each workshop lasts at least one day. The resulting structure is a 15-workshop package organized on four half-days.

Papers were reviewed in a distributed manner, and the refereeing process was handled electronically. Each paper submitted to a workshop was expected to be reviewed by four referees, one referee per PC member. The average number of referee reports collected per submitted paper was 3.27. We are deeply grateful to all the referees, and especially to those who spent time to write additional comments to be forwarded to the authors. Each workshop chairperson was responsible for the selection of the papers submitted to his or her workshop, in collaboration with the other PC members.

The role of the PC meeting was then to control the selection of each workshop, check the overall fairness, and possibly move papers between workshops to improve their focus. Papers authored by program committee members have been processed separately, in order to preserve fairness and confidentiality during the PC meeting. We feel this point is essential. We are grateful to the workshop chairpersons and to the PC members who (mostly!) did a tremendous job. It has been a great pleasure to work with them.

Selection results

We wish to thank all the authors who submitted papers to Euro-Par'96. We received 383 submissions from 40 different countries. All but 11 were submitted electronically (PostScript file). More than 50 submissions came from France, from Germany, and from the USA. About 40 submissions came from the UK, about 20 from Spain, and more than 10 from Italy. Moreover, more than 10 submissions were received from Eastern Europe, and about 10 from Japan, from Taiwan, and from Korea.

We selected 4% of the submissions as *distinguished papers* to appear as 12-page papers in the proceedings, and to be presented in 30 minutes. Another 27% of the submitted papers were selected as *regular papers* to appear as 8-page papers in the proceedings, and to be presented in 30 minutes. Finally, we also selected 25% of the submitted papers as *short papers*. Short papers present on-going research and/or interesting contributions for which further polishing is needed. They appear as 4-page papers in the proceedings and are given 15-minute presentation time. All authors of accepted papers were offered the opportunity to buy 2 extra pages.

Acknowledgments

Euro-Par has been supported by the following French organizations: CNET, INRIA, LIP, *Ministère des Affaires Etrangères*, *Progammes de Recherches Coordonnées* ANM and PRS, and *Région Rhône-Alpes*. We gratefully acknowledge their support. Without them, Euro-Par'96 would not even have existed!

And, last but not least, we would like to thank Valérie Roger, the Euro-Par'96 secretary, whose great experience in organizing large events at ENS Lyon was invaluable. We also wish to thank all the staff of LIP, in particular Sylvie Boyer and Jocelyne Richerd. We have deeply appreciated the help of many "volunteer students". Their contribution was decisive in solving so many practical details. Jean-Christophe Dubacq, who was in charge of the Euro-Par Web server, deserves a special mention.

We hope you will enjoy the conference and/or these proceedings as much as we have enjoyed organizing Euro-Par'96.

July 1996   Luc Bougé, Pierre Fraigniaud,
                                              Anne Mignotte and Yves Robert

Steering Committee

Chair: Chris Jesshope (Univ of Surrey, UK), c.jesshope@ee.surrey.ac.uk

Steering committee members:
- Luc Bougé (ENS Lyon, F), luc.bouge@lip.ens-lyon.fr
- Agnès Bradier (EU), abra@dg13.cec.be
- Michel Cosnard (ENS Lyon, F), michel.cosnard@ens-lyon.fr
- Lucio Grandinetti (Univ della Calabria, I) lugran@ccusc1.unical.it
- Constantine Halatsis (Univ of Athens, GR),
    halatsis@uranus.di.uoa.ariadne-t.gr
- Seif Haridi (Swedish Insitute, Kista, S), seif@sics.se
- Peter Kacsuk (Central Research Institute for Physics, Budapest, H),
    kacsuk@sunserv.kfki.hu
- Ron Perrott (Queen's Univ, Belfast, UK), r.perrott@v2.qub.ac.uk
- Ivan Plander (Slovak Academy, Bratislava, S.K), upsycai@savba.savba.sk
- Dieter Reinartz (Univ Erlangen, D), reinartz@immd7.informatik.uni-erlangen.de
- Richard Wait (Univ of Liverpool, UK), wait@supr.scm.liv.ac.uk
- Emilio L. Zapata (Univ of Malaga, E), ezapata@atc.ctima.uma.es

Program Committee

Workshop 01: Programming environment and tools.
   Chair: Jack Dongarra (Knoxville, USA), dongarra@cs.utk.edu
   Program Committee:
   - Arndt Bode (TU Munchen, D), bode@informatik.tu-muenchen.de
   - Dennis Gannon, (Indiana Univ, Bloomington, USA), gannon@iuvax.cs.indiana.edu
   - Bernard Tourancheau (ENS Lyon, F), bernard.tourancheau@ens-lyon.fr

Workshop 02: Routing and communication in interconnection networks.
   Chair: Robert Cypher (Baltimore, USA), cypher@cs.jhu.edu
   Program Committee:
   - Jose Duato (Univ Valencia, SP), jduato@pleiades.upv.es
   - Pierre Fraigniaud (ENS Lyon, F), pierre.fraigniaud@ens-lyon.fr
   - Eli Upfal (Weizmann Institute, Rehovot,Israel), eli@wisdom.weizmann.ac.il

Workshop 03: Automatic parallelization and high performance compilers.
   Chair: Chris Lengauer (Passau, D), lengauer@fmi.uni-passau.de
   Program Committee:
   - Francois Irigoin (Ecole des Mines, Paris, F), francois.irigoin@ensmp.fr
   - Yves Robert (ENS Lyon, F), yves.robert@ens-lyon.fr
   - Mateo Valero (Univ Barcelone, SP), mateo@ac.upc.es

Workshop 04: Distributed systems and algorithms.
   Chair: Friedemann Mattern (Darmstadt, D), mattern@isa.informatik.th-darmstadt.de

Program Committee:
- Claude Jard (Irisa-Rennes), claude.jard@inria.fr
- Sacha Krakoviak (IMAG, Grenoble, F), sacha.krakowiak@imag.fr
- Santosh Shrivastava (Univ of Newcastle, UK), santosh.shrivastava@newcastle.ac.uk

Workshop 05+21: Parallel languages, programming, and high-level control.
Chair: Ian Foster (Argonne, USA), itf@mcs.anl.gov
Co-chair: Jean-Pierre Briot (Univ Paris, F), jpbriot@litp.ibp.fr
Program Committee:
- Gul Agha (Univ of Illinois, Urbana, USA), agha@cs.uiuc.edu
- Luc Bougé (ENS Lyon, F), luc.bouge@ens-lyon.fr
- Marc Gengler (ENS Lyon, F), marc.gengler@ens-lyon.fr
- Steve Gregory (Univ of Bristol, UK), steve@cs.bris.ac.uk
- Suresh Jagannathan (NEC, Princeton, USA), suresh@research.nj.nec.com
- Ron Perrot (Queen's Univ, Belfast, UK), perrott@queens-belfast.ac.uk

Workshop 06: Parallel discrete algorithms.
Chair: Burkhard Monien (Paderborn, D), bm@pbinfo.uni-paderborn.de
Program Committee:
- Afonso Ferreira (ENS Lyon, F), afonso.ferreira@ens-lyon.fr
- Joaquim Gabarro (Univ Politecnica de Catalunya, Barcelona, SP), gabarro@lsi.upc.es
- Giles Villard (IMAG Grenoble, F), gvillard@mistral.imag.fr

Workshop 07: Parallel numerical algorithms
Chair: Iain Duff (Didcot, UK), i.duff@letterbox.rl.ac.uk
Program Committee:
- Jean Roman (Univ Bordeaux, F), jean.roman@labri.u-bordeaux.fr
- Dirk Roose (KU Leuven, B), dirk.roose@cs.kuleuven.ac.be
- Marian Vajtersic (Slovak Academy, Bratislava, SK), marian@par.univie.ac.at

Workshop 08+09+10: Parallel image/video processing and computer arithmetic.
Chairs: Larry Davis (Univ of Maryland, College park, USA), lsd@umiacs.umd.edu
and Jean-Marc Delosme (Yale, USA), delosme@cs.yale.edu
Co-chairs: Francky Catthoor (IMEC, Belgium), catthoor@imec.be
and Joseph Jaja (Univ of Maryland, College park, USA), joseph@umiacs.umd.edu
Program Committee:
- Virginio Cantoni (Univ di Pavia, I), cantoni@ipvvis.unipv.it
- Luigi Dadda (Univ di Milano, I), dadda@elet.polimi.it
- Peter Kornerup (Univ of Odense, D), kornerup@imada.ou.dk
- Fadi Kurdahi (Univ of California, Irvine, USA),kurdahi@ece.uci.edu
- Anne Mignotte (ENS Lyon, F), anne.mignotte@ens-lyon.fr
- Serge Miguet (ENS Lyon, F), serge.miguet@ens-lyon.fr
- Jean-Michel Muller (ENS Lyon, F), jean-michel.mmuller@ens-lyon.fr
- Norbert Wehn (Siemens, Munchen, D), wehn@vincent.hl.siemens.de
- Bertrand Zavidovique (ETCA, Arcueil, F), zavidovique@etca.fr

Workshop 11: High performance computing and application.
  Chair: Wolfgang Gentzsch (Genias, D), gentzsch@gimli.genias.de
  Program Committee:
  - Frédéric Desprez (ENS Lyon and INRIA, F), frederic.desprez@ens-lyon.fr
  - Pierre Kuonen (EPF, Lausanne, CH), pierre.kuonen@di.epfl.ch
  - Nikolay Petkov (Univ of Groningen,D), petkov@cs.rug.nl

Workshop 12: Theory and models for parallel computing.
  Chair: Bill McColl (Oxford, UK), bill.mccoll@comlab.oxford.ac.uk
  Program Committee:
  - Rob Bisseling (Utrecht Univ, NL), rob.bisseling@math.ruu.nl
  - Alan Gibbons (Univ of Warwick, Conventry, UK), amg@dcs.warwick.ac.fr
  - Jacques Mazoyer (ENS Lyon, F), jacques.mazoyer@ens-lyon.fr

Workshop 13: Parallel computer architecture.
  Chair: Chris Jesshope (Univ of Surrey, UK), c.jesshope@ee.surrey.ac.fr
  Program Committee:
  - Daniel Litaize (IRIT, Toulouse, F), litaize@irit.fr
  - Denis Nicole (Univ of Southampton, UK), dan@ecs.soton.ac.uk
  - Wolfgang Paul (Univ of Saarbrucken, D), wjp@wurzelausix.cs.uni-sb.de

Workshop 17: Scheduling and load balancing.
  Chair: Apostolos Gerasoulis (Rutgers Univ, USA), gerasoul@cs.rutgers.edu
  Program Committee:
  - Philippe Chretienne (Univ Paris, F), philippe.chretienne@litp.ibp.fr
  - Constantine Polychronopoulos (USA), cdp@csrd.uiuc.edu
  - Tao Yang (Univ of California, Santa Barbara, USA), tyang@cs.ucsb.edu

Workshop 19: Performance evaluation.
  Chair: Francois Baccelli (INRIA, Sophia Antipolis, F), francois.baccelli@inria.fr
  Program Committee:
  - Gianfranco Balbo (Univ di Torino, I), balbo@di.unito.it
  - Brigitte Plateau (IMAG Grenoble, F), brigitte.plateau@imag.fr
  - Ken Sevcick (Univ of Toronto, CA), kcs@cs.toronto.edu

Workshop 20: Instruction level parallelism.
  Chair: Guang Gao (McGill, Montreal, Canada), gao@andy.cs.mcgill.ca
  Program Committee:
  - Christine Eisenbeis (INRIA Rocquencourt, F), christine.eisenbeis@inria.fr
  - Jesus Labarta (UPC, Barcelone, SP), jesus@ac.upc.es
  - Andre Seznec (INRIA Rocquencourt, F), andre.seznec@inria.fr

Workshop 22: Parallel and distributed databases.
Chair: Erhard Rahm (Univ of Leipzig, D), rahm@informatik.uni-leipzig.de
Program Committee:
- Peter Apers (Univ. of Twente, NL), apers@cs.utwente.nl
- Lionel Brunie (ENS Lyon, F), lionel.brunie@ens-lyon.fr
- Theo Haerder (Univ of Kaiserslautern, D), haerder@informatik.uni-kl.de

Other program commitee members.
- Pascal Berthomé (ENS Lyon, F), pascal.berthome@ens-lyon.fr
- François Blayo (Lyon, F), blayo@babel.asi.fr
- Karl-Heinz Brenner (Univ Mannheim, D), brenner@rummelplatz.uni-mannheim.de
- Pierre Chavel (Univ Paris 11, Orsay, F), pierre.chavel@iota.u-psud.fr
- Marie Cottrell (Univ. Paris, F), ecottrell@obelix.univ-paris1.fr
- Jean-Claude Fernandez (IMAG Grenoble, F), jean-claude.fernandez@imag.fr
- Paul Kuehn (Univ Stuttgart, D), kuehn@ind.uni-stuttgart.d400.de
- Ahmed Louri (Univ of Arizona, Tucson, USA), louri@ece.arizona.edu
- Helene Paugam-Moisy (ENS Lyon, F), hpaugam@ens-lyon.fr
- Amir Pnueli (Weizman Institute, Rehovot, Israel), amir@wisdom.weizmann.ac.il
- Michel Verleysen (U.C. Louvain, B), verleysen@dice.ucl.ac.be

Euro-Par'96 Referees

Abdelaziz, *Mzoughi*
Abdelrahman, *Tarek*
Agha, *Gul*
Albanesi, *Maria Grazia*
Alexandrov, *Albert*
Alouini, *Ilies*
Altman, *Erik*
Amestoy, *Patrick*
Anastasiadis, *Stergios*
Andresen, *Daniel*
André, *Françoise*
Anglano, *Cosimo*
Apers, *Peter*
Arvind, *Damal K.*
Astley, *Mark*
Auguin, *Michel*
Auletta, *Vincenzo*
Authié, *Gérard*
Ayguade, *Eduard*
Aykanat, *Cevdet*
Azéma, *Pierre*
Badia, *Rosa M.*
Bagherzadeh, *Nader*
Balbo, *Gianfranco*
Balla, *Katalin*
Bampis, *Euripide*
Baron, *Richard*
Barrado, *Cristina*
Barth, *Dominique*
Barthou, *Denis*
Basu, *Sujoy*
Baude, *Françoise*
Benkner, *Siegfried*
Benzi, *Michèle*
Berenbrink, *Petra*
Bernard, *Pierre-Eric*
Berthomieu, *Bernard*
Berthomé, *Pascal*
Bezrukov, *Sergej*
Biancardi, *Alberto*
Bisseling, *Rob*
Blanc-Talon, *Jacques*

Blanken, *Henk*
Bode, *Arndt*
Bodin, *François*
Bond, *Johny*
Bonnin, *Patrick*
Borgeest, *Rolf*
Bouaziz, *Samir*
Bouchitté, *Vincent*
Bougé, *Luc*
Boulet, *Pierre*
Brandes, *Thomas*
Branger, *Vincent*
Brenner, *Karl-Heinz*
Briggs, *Bill*
Briot, *Jean-Pierre*
Brorsson, *Mats*
Bruell, *Steve*
Brunie, *Lionel*
Burdakov, *Oleg*
Burger, *Doug*
Bäumker, *Armin*
Bétourné, *Claude*
Caceres, *Edson Norberto*
Cachera, *David*
Calland, *Pierre-Yves*
Calvin, *Christophe*
Calégari, *Patrice*
Cantoni, *Virginio*
Cao, *Xuejun*
Capobianco, *Fabrizio*
Carlson, *Bradley*
Caromel, *Denis*
Cartier, *Sylvain*
Castaneda, *Martha*
Castro-Alves, *Vladimir*
Catthoor, *Francky*
Chalmers, *Alan*
Charot, *François*
Charrier, *Pierre*
Charron-Bost, *Bernadette*
Chassin de Kergommeaux, *Jacques*
Chaumette, *Serge*

Chavel, *Pierre*
Cheng, *Chung-Ta*
Chich, *Thierry*
Chrétienne, *Philippe*
Clint, *Maurice*
Clérot, *Fabrice*
Coelho, *Fabien*
Colin, *Jean-Noël*
Collard, *Jean-François*
Collin, *Bertrand*
Colombet, *Laurent*
Cornu, *Thierry*
Correa, *Ricardo*
Cosnard, *Michel*
Cottrell, *Marie*
Craig, *David*
Cremonesi, *Paolo*
Creusillet, *Béatrice*
Cypher, *Bob*
Cypher, *Robert*
Dadda, *L.*
Dadda, *Luigi*
Dai, *H. K.*
Dallery, *Yves*
Daniel, *Frédéric*
Darte, *Alain*
Daumas, *Marc*
Dayde, *Michel*
Decker, *Karsten*
Decker, *Thomas*
Dehne, *Frank*
Dekeyser, *Jean-Luc*
Delaplace, *Franck*
Delmas, *Olivier*
Delorme, *Charles*
Delorme, *Marianne*
Delosme, *Jean-Marc*
Desprez, *Frédéric*
Diderich, *Claude*
Dimitriou, *Georgios*
Dimopoulos, *Nikitas*
Dion, *Michèle*
Dittrich, *Wolfgang*
Domas, *Stéphane*
Donaldson, *Val*

Donatelli, *Susanna*
Doreille, *Mathias*
Drach, *Nathalie*
Duato, *Jose*
Duff, *Iain*
Dulac, *Didier*
Dunne, *Paul*
Dupont de Dinechin, *Benoît*
Durand, *Bruno*
Eager, *Derek*
Eisenbeis, *Christine*
Elkihel, *Moussa*
Ezhilchelvan, *Paul*
Faltings, *Boi*
Farley, *Art*
Feautrier, *Paul*
Feldmann, *Rainer*
Ferguson, *Warren*
Fernandez, *Agustin*
Ferreira, *Afonso*
Ferreira Rezende, *Fernando de*
Filali, *Mamoun*
Finta, *Lucian*
Fitzpatrick, *Stephen*
Flammini, *Michèle*
Fleury, *Eric*
Folliot, *Bertil*
Fortes, *Jose*
Foster, *Ian*
Fraigniaud, *Pierre*
Franceschinis, *Giuliana*
Frolund, *Svend*
Fu, *Cong*
Fujita, *Satoshi*
Fulgham, *Melanie*
Gabarr, *Joaquim*
Gaeta, *Rossano*
Gannon, *Dennis*
Gao, *Hum Guang R.*
Gao, *Lixin*
Garcia, *Jordi*
Garcia, *Miguel Angel*
Gargano, *Luisa*
Gaujal, *Bruno*
Gavoille, *Cyril*

Gehring, *Joern*
Geib, *Jean-Marc*
Gengler, *Marc*
Gentzsch, *Wolfgang*
Gerbessiotis, *Alex*
Germain, *Cécile*
Gesmann, *Michael*
Gibbons, *Alan*
Gibbons, *Richard*
Ginzburg, *Ilan*
Girau, *Bernard*
Girkar, *Milind*
Gonzalez, *Antonio*
Goodman, *James*
Gorlatch, *Sergei*
Govindarajan, *R.*
Grammatikakis, *Miltos*
Gransart, *Christophe*
Gravano, *Luis*
Grefen, *Paul*
Gregory, *Steve*
Griebl, *Martin*
Grout, *John*
Guidec, *Frédéric*
Guinand, *Frédéric*
Gurd, *John*
Guyot, *Alain*
Haerder, *Theo*
Hagimont, *Daniel*
Hahn, *Gena*
Hains, *Gaetan*
Hameurlain, *Abdelkader*
Hanen, *Claire*
Harmer, *Terence*
Hasenfuss, *Sven*
Hasselbring, *Wilhelm*
Hellwagner, *Hermann*
Herrmann, *Christoph*
Heudin, *Jean-Claude*
Heun, *Volker*
Heydemann, *Marie-Claude*
Hill, *Jonathan*
Hily, *Sebastien*
Ho, *C. T.*
Holliday, *Mark*

Holyer, *Ian*
Houzet, *Dominique*
Hum, *Herbert*
Iannello, *Giulio*
Di Ianni, *Miriam*
Ibel, *Maximilian*
Irigoin, *François*
Issarny, *Valerie*
Jacques, *Jorda*
Jaffray, *Jean-Yves*
Jagannathan, *Suresh*
Jalby, *William*
Jamali, *Nadeem*
Jansen, *Klaus*
Jaray, *Jacques*
Jard, *Claude*
Jayasimha, *D.*
Jegou, *Yvon*
Jesshope, *Chris*
Jolion, *Jean-Michel*
Jouvelot, *Pierre*
Joux, *Antoine*
Juanole, *Guy*
Julliand, *Jacques*
Juurlink, *Ben*
Jéron, *Thierry*
Karl, *Wolfgang*
Kenyon, *Claire*
Keryell, *Ronan*
Kienzle, *Martin*
Kim, *Ji-Yun*
Kim, *Jin S.*
Kim, *M. K.*
Kim, *Woo Young*
Kirchner, *Claude*
Knoop, *Jens*
Koiran, *Pascal*
Konig, *Jean-Claude*
Kordon, *Fabrice*
Kornerup, *Peter*
Kosch, *Harald*
Kotlyar, *Vladimir*
Kraetzl, *Miro*
Krakowiak, *Sacha*
Kremer, *Ulrich*

Krieger, *Orran*
Krumme, *David*
Kuonen, *Pierre*
Kurdahi, *Fadi*
Labahn, *Roger*
Labarta, *Jesus*
Lachaud, *Jacques-Olivier*
Laforest, *Christian*
Lamberts, *Stefan*
Landmann, *Joerg*
Lang, *Tomas*
Larcher, *Gerhard*
Laure, *Erwin*
Le Gall, *Françoise*
Leberecht, *Markus*
Lecoupanec, *Jacques*
Lee, *Hyuk-Jae*
Lee, *Jaejin*
Lefèvre, *Laurent*
Lelait, *Sylvain*
Lengauer, *Christian*
Leservot, *Arnauld*
Levaire, *Jean-Luc*
Leveugle, *Régis*
Lichnewsky, *Alain*
Liestman, *Arthur*
Litaize, *Daniel*
Little, *Mark*
Liu, *Hung-Kang*
Liu, *Jason*
Llosa, *Josep*
Lo, *Adley*
Lo, *Virginia*
Lombardi, *Luca*
Lopez, *Pedro*
Lorts, *Daniel*
Louri, *Ahmed*
Lozano, *Luis*
Ludwig, *Thomas*
Luksch, *Peter*
Macedo, *Raimundo*
Maheo, *Yves*
Maier, *Ursula*
Maillet, *Eric*
Malumbres, *Perez*

Manneback, *Pierre*
Maquelin, *Olivier*
Maranget, *Luc*
Marchetti-Spaccamela, *Alberto*
Margenstern, *Maurice*
Marie, *Raymond*
Marques, *Osni*
Marquet, *Philippe*
Marsolf, *Bret*
Martin, *Bruno*
Martinez, *Jesus*
Martorell, *Xavier*
Marzullo, *Keith*
Mattern, *Friedemann*
May, *Michael*
Mayr, *Ernst*
Mazoyer, *Jacques*
McColl, *Bill*
Mehofer, *Eduard*
Merigot, *Alain*
Michallon, *Philippe*
Mignotte, *Anne*
Miguet, *Serge*
Miller, *Quentin*
Mitrevski, *Jovan*
Mitschang, *Bernhard*
Mohapatra, *Prasant*
Monien, *Burkart*
Montanvert, *Annick*
Montuschi, *Paolo*
Mosconi, *Mauro*
Muench, *Michael*
Muller, *Gilles*
Muller, *Henk*
Muller, *Jean-Michel*
Munier, *Alix*
Munoz, *Xavier*
Mussi, *Philippe*
Méry, *Dominique*
Namyst, *Raymond*
Nemawarkar, *Shashank*
Neri, *Filippo*
Newman, *Ilan*
Nguyen, *Anthony-Trung*
Ni, *Lionel*

Nicod, *Jean-Marc*
Nicola, *Mazzocca*
Nicole, *Denis*
Nielsen, *Asger Munk*
Ning, *Qi*
Nink, *Udo*
O'Boyle, *Michael*
Oliveira Stein, *Benhur de*
Opatrny, *Jaroslav*
Pacherie, *Jean-Lin*
Panaite, *Petrisor*
Paugam-Moisy, *Hélène*
Paul, *Wolfgang*
Paulo, *Fernandes*
Pazat, *Jean-Louis*
Pekergin, *Fehran*
Pelc, *Andrzej*
Pellegrini, *François*
Perennes, *Stéphane*
Perrin, *Guy-René*
Perroton, *Laurent*
Perrott, *Ron*
Peters, *Joseph*
Petiton, *Serge*
Petkov, *Nikolay*
Petrini, *Fabrizio*
Peyrat, *Claudine*
Picouleau, *Christophe*
Pierson, *Jean-Marc*
Rousset de Pina, *Xavier*
Piuri, *Vincenzo*
Plateau, *Brigitte*
Polychronopoulos, *Constantine*
Pottier, *Bernard*
Pouzet, *Marc*
Pradin-Chezalviel, *Brigitte*
Preis, *Robert*
Prins, *Christian*
Priol, *Thierry*
Proskurowski, *Andrzej*
Prylli, *Loïc*
Puaut, *Isabelle*
Puglisi, *Chiara*
Puzenat, *Didier*
Queinnec, *Christian*

Quenot, *Georges*
Quinton, *Patrice*
Rahm, *Erhard*
Rajopadhye, *Sanjay*
Ramaswamy, *Shankar*
Rao, *Sreenivasa*
Rapine, *Christophe*
Raspaud, *Andre*
Rathmayer, *Sabine*
Rau-Chaplin, *Andrew*
Raynal, *Michel*
Rebreyend, *Pascal*
Redon, *Xavier*
Rehrmann, *Ralf*
Reinert, *Joachim*
Ren, *Shangping*
Renaud, *Christophe*
Reymann, *Olivier*
Reynaud, *Roger*
Richter, *Harald*
Risset, *Tanguy*
Ritter, *Norbert*
Riveill, *Michel*
Robert, *Yves*
Roch, *Jean-Louis*
Rochange, *Christine*
Roeder, *Christian*
Rohou, *Erven*
Roka, *Zsuzanna*
Rolim, *Jose*
Roman, *Jean*
Roose, *Dirk*
Rosti, *Emilia*
Rottmann, *Valentin*
Roudier, *Yves*
Roux, *Olivier*
Rozoy, *Brigitte*
Ruiz, *Daniel*
Russo, *Stefano*
Sainrat, *Pascal*
Saito, *Hideki*
Sakho, *Ibrahima*
Sampels, *Michael*
Sanchez, *Fermin*
Sandhu, *Harjinder*

Saouter, *Yannick*
Sawaya, *Antoine*
Schacht, *Susanne*
Schlichting, *Rick*
Schmidt, *Olaf*
Schott, *René*
Schuster, *Assaf*
Schwarz, *Reinhard*
Schwiegelshohn, *Uwe*
Sebban, *Marc*
Serazzi, *Giuseppe*
Sereno, *Matteo*
Setti, *Alessandra*
Sevcick, *Ken*
Seznec, *Andre*
Shang, *Weijia*
Sharma, *Arun*
Shen, *Hong*
Shen, *Kish*
Shrivastava, *Santosh*
Sibeyn, *Jop*
Simon, *Jens*
Singh, *Ambuj*
Siniolakis, *Constantinos*
Sivasubramaniam, *Anand*
Skillicorn, *David*
Smith, *Jim*
Sotteau, *Dominique*
Stallard, *Paul*
Stavrakos, *Nicholas*
Stefanescu, *Dan*
Stellner, *Georg*
Stewart, *Alan*
Stoutchinin, *Artour*
Strothmann, *Willy-B.*
Sturm, *Peter*
Su, *Bogong*
Suh, *Young-Joo*
Sun, *Yong*
Sykora, *Ondrej*
Syska, *Michel*
Sérot, *Jocelyn*
Takagi, *Naofumi*
Tarlescu, *Dana*
Tel, *Gerard*

Temam, *Olivier*
Theobald, *Kevin*
Thiele, *Lothar*
Thomas, *Joachim*
Thomasset, *François*
Thoraval, *René*
Théron, *Eric*
Tian, *Xin-Min*
Timsit, *Claude*
Tiskin, *Alexandre*
Tisserand, *Arnaud*
Topham, *Nigel*
Torres, *Edgar L.*
Tougne, *Laure*
Tourancheau, *Bernard*
Truong, *Dan*
Trystram, *Denis*
Tschoeke, *Stefan*
Tseng, *Yu-Chee*
Ubeda, *Stéphane*
Unger, *Walter*
Upfal, *Eli*
Utard, *Gil*
Vajtersic, *Marian*
Valero, *Mateo*
Valero-Garcia, *Miguel*
Varavithya, *Vara*
Vauquelin, *Bernard*
Venkatasubramanian, *Nalini*
Verleysen, *Michel*
Villard, *Gilles*
Vincent, *Jean-Marc*
Violard, *Eric*
Vivien, *Frédéric*
Vladimirova, *Tanya*
Waldby, *James*
Wang, *Jian*
Wang, *Wei*
Wang, *Weining*
Wanka, *Rolf*
Watson, *Paul*
Wedler, *Christoph*
Wehn, *Norbert*
Weidmann, *Matthias*
Welsh, *Matt*

Wilschut, *Annita*
Wilson, *Gregory*
Wismueller, *Roland*
Wolfe, *Michael*
Wolper, *Pierre*
Wool, *Avishai*
Wray, *Paul*
Yalamanchili, *Sudhakar*
Yang, *Rong*
Yang, *Tao*
Yoo, *Joonkyoo*
Yu, *Yizhou*
Zapata, *Emilio L.*
Zavidovique, *Bertrand*
Zerovnik, *Janez*
Zerrouk, *Belkacem*
Zerrouki, *L.*
Zhang, *Xiaodong*
Zhang, *Ye*
Zhu, *Yingchun*

Table of Contents: Volume 1

Invited Talks 1

High-Performance Distributed Computing: The I-WAY Experiment and Beyond
    Ian Foster 3
Design and Implementation of a Parallel Architecture for Biological Sequence Comparison
    Pascale Guerdoux-Jamet, Dominique Lavenier, Charles Wagner, Patrice Quinton 11
Universal Computing
    W. F. McColl 25
Dynamic Load Balancing in Parallel Database Systems
    Erhard Rahm 37

Workshop 01
Programming Environment and Tools 53

Distributed Array Query and Visualization for High Performance Fortran
    Steven T. Hackstadt, Allen D. Malony 55
Annai Scalable Run-Time Support for Interactive Debugging and Performance Analysis of Large-Scale Parallel Programs
    Christian Clémençon, Akiyoshi Endo, Josef Fritscher, Andreas Müller, Brian J. N. Wylie 64
On the Implementation of a Replay Mechanism
    M. A. Ronsse, L. J. Levrouw 70
Concepts and Functionalities of the DOSMOS-Trace Monitoring Tool
    Lionel Brunie, Olivier Reymann 74
An Open Monitoring System for Parallel and Distributed Programs
    Thomas Ludwig, Michael Oberhuber, Roland Wismueller ... 78
Millipede: Easy Parallel Programming in Available Distributed Environments
    Roy Friedman, Maxim Goldin, Ayal Itzkovitz, Assaf Schuster .. 84
An Adaptive Cost System for Parallel Program Instrumentation
    Jeffrey K. Hollingsworth, Barton P. Miller 88
SVMview: A Performance Tuning Tool for DSM-Based Parallel Computers
    Didier Badouel, Thierry Priol, Luc Renambot 98
Cautious, Machine-Independent Performance Tuning for Shared-Memory Multiprocessors
    Sarah A. M. Talbot, Andrew J. Bennett, Paul H. J. Kelly 106
Dealing with Heterogeneity in Stardust: An Environment for Parallel Programming on Networks of Heterogeneous Workstations
    Gilbert Cabillic, Isabelle Puaut 114

An Integrated Environment to Design Parallel Object-Oriented Applications
    Klaus Wolf, Ottmar Kraemer-Fuhrmann 120
MPI-2: Extending the Message-Passing Interface
    Al Geist, William Gropp, Steve Huss-Lederman, Andrew Lumsdaine, Ewing Lusk, William Saphir, Tony Skjellum, Marc Snir 128
Optimizing Sisal Programs: A Formal Approach
    Isabelle Attali, Denis Caromel, Romain Guider, Andrew L. Wendelborn 136
A Refinement Methodology for Developing Data-Parallel Applications
    Lars Nyland, Jan Prins, Allen Goldberg, Peter Mills, John Reif, Robert Wagner 145
Task Parallelism: What a Tool Can Provide and What Should Be Left to the User
    Silvia A. Crivelli, Elizabeth R. Jessup 151
Efficient Block Cyclic Data Redistribution
    Loïc Prylli, Bernard Tourancheau 155
Optimal Grain Size Computation for Pipelined Algorithms
    Frédéric Desprez, Pierre Ramet, Jean Roman 165
Dynamic Redistribution on Heterogeneous Parallel Computers
    Dominique Sueur, Jean-Luc Dekeyser 173
Supporting Distributed Sparse Matrix Objects
    C. Addison, T. Oliver, A. Sunderland 178

Workshop 02
Routing and Communication in Interconnection Networks   185

Low-Latency Communication over Fast Ethernet
    Matt Welsh, Anindya Basu, Thorsten von Eicken 187
A Comparison of Input and Output Driven Routers
    Melanie L. Fulgham, Lawrence Snyder 195
Optimal Topology for Distributed Shared Memory Multiprocessors: Hypercubes Again?
    José Duato, M. P. Malumbres 205
A Pattern-Associative Router for Interconnection Network Adaptive Algorithms
    Daniel G. Rice, José G. Delgado-Frias, Douglas H. Summerville   213
On Stack-Graph OPS-Based Lightwave Networks
    H. Bourdin, A. Ferreira, K. Marcus 218
A General Method for Efficient Embeddings of Graphs into Optimal Hypercubes
    Volker Heun, Ernst W. Mayr 222
The Size Complexity of Strictly Non-blocking Fixed Ratio Concentrators with Constant Depth
    H. K. Dai 234

Bandwidth and Cutwidth of the Mesh of d-ary Trees
  Dominique Barth 243
Variable-Dilation Embeddings of Hypercubes into Star Graphs: Performance Metrics, Mapping Functions, and Routing
  Marcelo Moraes de Azevedo, Nader Bagherzadeh, Shahram Latifi 247
Overlapping Communication and Computation in Hypercubes
  Luis Díaz de Cerio, Miguel Valero-García, Antonio González .. 253
Efficient Delay Routing
  Miriam Di Ianni 258
Multipacket Hot-Potato Routing on Processor Arrays
  Christos Kaklamanis, Danny Krizanc 270
A Necessary and Sufficient Condition for Proper Routing in Omega-Omega Network
  Myung-Kyun Kim, Hyunsoo Yoon, Seung-Ryoul Maeng 278
Rubik Routing Permutations on Graphs
  Charles Delorme, Petrişor Panaite 283
The Effect of Flow Control and Routing Adaptivity on Priority-Driven Traffic in Multiprocessor Networks
  Shobana Balakrishnan, Füsun Özgüner 287
Routing on Networks of Optical Crossbars
  Friedhelm Meyer auf der Heide, Klaus Schroeder, Frank Schwarze 299
Latency and Bandwidth Requirements of Massively Parallel Programs: FFT as a Case Study
  Fabrizio Petrini, Marco Vanneschi 307
Induced Broadcasting Algorithms in Iterated Line Digraphs
  Jean-Claude Bermond, Xavier Muñoz, Alberto Marchetti-Spaccamela 313
Lower Bounds on Broadcasting Time of de Bruijn Networks
  Stéphane Perennes 325
Gossip in Trees under Line-Communication Mode
  Christian Laforest 333
Total Exchange in Cayley Networks
  Vassilios V. Dimakopoulos, Nikitas J. Dimopoulos ... 341
Leaf Communications in Complete Trees
  Vassilios V. Dimakopoulos, Nikitas J. Dimopoulos ... 347
A Gossip Algorithm for Bus Networks with Buses of Limited Length
  Satoshi Fujita, Christian Laforest, Stéphane Perennes 353
Worm-Hole Gossiping on Meshes
  Ben H. H. Juurlink, P. S. Rao, Jop F. Sibeyn 361
Circuit-Switched Gossiping in 3-Dimensional Torus Networks
  Olivier Delmas, Stéphane Perennes 370

Workshop 03
Automatic Parallelization and High Performance Compilers   375

Automatic Parallelization and High Performance Compilers
   Christian Lengauer 377
On the Optimality of Allen and Kennedy's Algorithm for Parallelism Extraction in Nested Loops
   Alain Darte, Frédéric Vivien 379
Memory Reuse Analysis in the Polyhedral Model
   Doran Wilde, Sanjay Rajopadhye 389
Cycle Shrinking by Dependence Reduction
   Kunio Okuda 398
A Unified Transformation Technique for Multilevel Blocking
   Marta Jiménez, José M. Llabería, Agustin Fernández, Enric Morancho 402
Array Dataflow Analysis for Explicitly Parallel Programs
   Jean-François Collard, Martin Griebl 406
Semantic Foundations of Commutativity Analysis
   Martin C. Rinard, Pedro C. Diniz 414
Applications of Fuzzy Array Dataflow Analysis
   Denis Barthou, Jean-François Collard, Paul Feautrier 424
Simplifying Communication Induced by Operations on Block-Distributed Arrays
   Andreas Eberhart, Jingke Li 428
Compiler Reduction of Invalidation Traffic in Virtual Shared Memory Systems
   Michael F. P. O'Boyle, Rupert W. Ford, Andrew P. Nisbet ... 432
Partial Dead Code Elimination for Parallel Programs
   Jens Knoop 441
Solving the Constant-Degree Parallelism Alignment Problem
   Claude G. Diderich, Marc Gengler 451
Topographic Data Mapping by Balanced Hypersphere Tesselation
   Matthias Besch, Hans Werner Pohl 455
Implementing Pipelined Computation and Communication in an HPF Compiler
   Thomas Brandes, Frédéric Desprez 459
Efficient Mapping of Interdependent Scans
   Michel Barreteau, Paul Feautrier 463
Classifying Loops for Space-Time Mapping
   Martin Griebl, Christian Langauer 467

Workshop 04
Distributed Systems and Algorithms 475

PACA: A Cooperative File System Cache for Parallel Machines
   Toni Cortes, Sergi Girona, Jesús Labarta 477
A System for Fault-Tolerant Execution of Data and Compute Intensive Programs over a Network of Workstations
   J. A. Smith, S. K. Shrivastava 487
A Framework for Viewing Atomic Events in Distributed Computations
   Ajay D. Kshemkalyani 496
Worker-Based Parallel Computing on PVM
   Dae-Kyun Yoon, Jean-Luc Gaudiot 506
An Efficient Distributed Tuple Space Implementation for Networks of Workstations
   Antony Rowstron, Alan Wood 510
A Highly Available Partition-Processing Protocol for Distributed Shared Memory Systems
   Jenn-Wei Lin, Sy-Yen Kuo 514
I/O Data Mapping in ParFiSys: Support for High-Performance I/O in Parallel and Distributed Systems
   Jesus Carretero, Fernando Pérez, Pedro de Miguel, Felix García, Luis Alonso 522
Correctness Proof for a Distributed Memory System
   Vicente Cholvi-Juan, José M. Bernabéu-Aubán 526
Distributed Shared Memory Based on Group Large Causality
   José M. Piquer 532

Workshop 05+21
Parallel Languages, Programming, and High-Level Control 539

A Framework for Integrated Communication and I/O Placement
   Rajesh Bordawekar, Alok Choudhary, J. Ramanujam 541
Formal Derivation of Parallel Program for 2-Dimensional Maximum Segment Sum Problem
   Zhenjiang Hu, Hideya Iwasaki, Masato Takeichi 553
The Migrating Tasks: An Execution Model for Irregular Codes
   Yvon Jégou 563
Discussing HPF Design Issues
   Fabien Coelho 571
Parallelizing Conditional Recurrences
   Wei-Ngan Chin, John Darlington, Yike Guo 579
Adaptive Data Parallel Computation in the Parallel Object-Oriented Language OCore
   Hiroki Konaka, Yoshiaki Itoh, Takashi Tomokiyo, Munenori Maeda, Yutaka Ishikawa, Atsushi Hori 587

The Use of Interpreted Languages for Implementing Parallel Algorithms on Distributed Systems
    Noemi Rodriguez, Cristina Ururahy, Roberto Ierusalimschy, Renato Cerqueira 597

Co-ordinating Heterogeneous Parallel Computation
    Peter Au, John Darlington, Moustafa Ghanem, Yi-ke Guo, Hing Wing To, Jin Yang 601

Correctness of a Distributed-Memory Model for Scheme
    Luc Moreau 615

Partial Evaluation Scheme for Concurrent Languages and Its Correctness
    Haruo Hosoya, Naoki Kobayashi, Akinori Yonezawa 625

Support for Implementation of Evolutionary Concurrent Systems in Concurrent Programming Languages
    Raju Pandey, J. C. Browne 633

Structured Dagger: A Coordination Language for Message-Driven Programming
    Laxmikant V. Kalé, Milind A. Bhandarkar 646

TPascal - A Language for Task Parallel Programming
    Ansgar Brüll, Herbert Kuchen 654

OB(PN)2: An Object Based Petri Net Programming Notation
    Johan Lilius 660

Reusable Coordinator Modules for Massively Concurrent Applications
    F. Arbab, C. L. Blom, F. J. Burger, C. T. H. Everaars 664

Introducing Dynamicity in the Data-Parallel Language $8_{1/2}$
    Olivier Michel 678

Astro-Gofer: Parallel Functional Programming with Co-ordinating Processes
    Andrew Douglas, Niklas Röjemo, Colin Runciman, Alan Wood 686

Multiple OR-Parallel Resolution: Meta-Level Control of Parallel Logic Programs
    Petros Kefalas, Ioannis Vlahavas 694

High Level Parallel Programming Based on Automatic Coordination
    Juergen Knopp 704

Eden - The Paradise of Functional Concurrent Programming
    S. Breitinger, R. Loogen, Y. Ortega-Mallén, R. Peña-Marí 710

A Straightforward Translation of DOL Systems in the Declarative Data-Parallel Language $8_{1/2}$
    Olivier Michel 714

Efficient Parallel Programming with Algorithmic Skeletons
    George Horatiu Botorog, Herbert Kuchen 718

A Loosely Synchronized Execution Model for a Simple Data-Parallel Language
    Yann Le Guyadec, Emmanuel Melin, Bruno Raffin, Xavier Rebeuf, Bernard Virot 732

A Nonannotative Approach to Distributed Data-Parallel Computing
    A. Shafarenko 742

Petri Net Modelling of PARSE Designs
    Stefano Russo, Carlo Savy, Innes Jelly, Peter Collingwood   752
Synchronization Expressed in Types of Communication Channels
    Franz Puntigam   762
Laws of Data Parallel Assignment
    J. P. Wray   770
Proving Progress Properties of non Terminating Programs under Fairness Assumptions
    Ricardo Peña, Luis A. Galán   775

Workshop 06
Parallel Discrete Algorithms   779

A Simple Parallel Dictionary Matching Algorithm
    Paolo Ferragina   781
Scalability and Granularity Issues of the Hierarchical Radiosity Method
    Axel Podehl, Thomas Rauber, Gudula Rünger   789
List Ranking on Interconnection Networks
    Jop F. Sibeyn   799
Parallel Algorithm for Computing the Fragment Vector in Steiner Triple Systems
    Erik Urland   809
Representation of the Gabow Algorithm for Finding Smallest Spanning Trees with a Degree Constraint on Associative Parallel Processors
    Ann S. Nepomniaschaya   813
Runtime Support for Replicated Parallel Simulators of an ATM Network on Workstation Clusters
    Kam Hong Shum, Shuo-Yen Robert Li   818
Shared-Memory Implementation of an Irregular Particle Simulation Method
    Thomas Rauber, Gudula Rünger, Carsten Scholtes   822
A Parallel Algorithm for the Technology Mapping of LUT-Based FPGAs
    Vamsi Boppana, Prashant Saxena, Prithviraj Banerjee, W. Kent Fuchs, C. L. Liu   828
Distributed String Matching Algorithm on the N-cube
    Fouzia Moussouni, Christian Lavault   832

Index of Authors   837

Table of Contents: Volume 2

Workshop 07
Parallel Numerical Algorithms   1

Optimization of the ScaLAPACK LU Factorization Routine Using Communication/Computation Overlap
    Frédéric Desprez, Stéphane Domas, Bernard Tourancheau 3

On Experiments with a Parallel Direct Solver for Diagonally Dominant Banded Linear Systems
    Peter Arbenz 11

The Computation of Partial Eigensolutions on a Distributed Memory Machine Using a Modified Lanczos Method
    Kieran Murphy, Maurice Clint, Marek Szularz, Jim Weston ... 22

The Parallel Computation of Partial Eigensolutions of Large Matrices on a Massively Parallel Processor
    James Weston, Marek Szularz, Maurice Clint, Kieran Murphy . 26

Preprocessing of Sparse Unassembled Linear Systems for Efficient Solution Using Element-by-element Preconditioners
    Michel J. Daydé, Jean-Yves L'Excellent, Nicholas I. M. Gould . 34

Implementing The Parallel Quasi-Laguerre's Algorithm for Symmetric Tridiagonal Eigenproblems
    T. Y. Li, Xiulin Zou 44

Comparing Task and Data Parallel Excecution Schemes for the DIIRK Method
    Thomas Rauber, Gudula Rünger 52

Numerical Turbulence Simulation on Different Parallel Computers Using the Sparse Grid Combination Method
    Walter Huber 62

Parallel Fourier-Motzkin Elimination
    Christoph W. Kessler 66

Comparison of Three Monte Carlo Methods for Matrix Inversion
    Vassil N. Alexandrov, Spyridoula Lakka 72

Parallel Solution of the Volume Integral Equation of Electromagnetic Scattering
    Jussi Rahola 81

Optimization of Parallel Multilevel-Newton Algorithms on Workstation Clusters
    Robert Graeb, Michael Guenther, Utz Wever, Qinghua Zheng .. 91

A Time and Space Parallel Algorithm for the Heat Equation: The Implicit Collocation Method
    Fabienne Jézéquel 97

Workshop 08+09+10
Parallel Image/Video Processing and Computer Arithmetic 101

Parallel Image/Video Processing and Computer Arithmetic
   Larry Davis, Jean-Marc Delosme, Francky Catthoor 103
A High Performance Image Database System for Remotely Sensed Imagery
   Carter T. Shock, Chialin Chang, Larry Davis, Samuel Goward, Joel Saltz, Alan Sussman 109
An Asynchronous Parallel Algorithm for Symbolic Grouping Operations in Vision
   Yongwha Chung, Viktor K. Prasanna 123
A Parallel Pipelined Hough Transform
   Nicolas Guil, Emilio L. Zapata 131
High-Performance SAR-Image Formation and Post-Processing
   Enrico Appiani, Marco Corvi, Giovanni Garibotto, C. Coelho .. 139
A Parallel Implementation of Image Coding Using Linear Prediction and Iterated Functions Systems
   Gennaro Della Vecchia, Riccardo Distasi, Michele Nappi, Domenico Vitulano 147
Parallel Algorithms for Using Non-Stationary MRA in Image Compression
   Andreas Uhl 151
High Radix Cordic Rotation Based on Selection by Rounding
   Elisardo Antelo, Javier D. Bruguera, Tomas Lang, Julio Villalba, Emilio L. Zapata 155
On-Line Algorithms for Computing Exponentials and Logarithms
   Asger Munk Nielsen, Jean-Michel Muller 165
Parallel and On-Line Addition in Negative Base and some Complex Number Systems
   Christiane Frougny 175
A Variable Latency Pipelined Floating-Point Adder
   Stuart F. Oberman, Michael J. Flynn 183
Basic Linear Algebra Operations in SLI Arithmetic
   Michael A. Anuta, Daniel W. Lozier, Nicolas Schabanel, Peter R. Turner 193
CAM2: A Highly-Parallel 2-D Cellular Automata Architecture for Real-Time and Palm-Top Pixel-Level Image Processing
   Takeshi Ikenaga, Takeshi Ogura 203
A Self-Optimising Coprocessor Model for Portable Parallel Image Processing
   D. Crookes, T. J. Brown, Y. Dong, G. McAleese, P. J. Morrow, D. K. Roantree, I. T. A. Spence 213
System-Level Memory Management for Weakly Parallel Image Processing
   Koen Danckaert, Francky Catthoor, Hugo De Man 217
Multidimensional Periodic Scheduling: Model and Complexity
   W. F. J. Verhaegh, P. E. R. Lippens, E. H. L. Aarts, J. L. van Meerbergen, A. van der Werf 226

Global Approach for Compiled Bit-True Simulation of DSP Systems
  Luc De Coster, Marc Engels, Rudy Lauwereins, J. A. Peperstraete   236

Workshop 11
High Performance Computing and Application 241

Parallel Implementation of RBF Neural Networks
  Vladimir Demian, Frédéric Desprez, Hélène Paugam-Moisy, Makan Pourzandi   243
Selected Results from the ParkBench Benchmark
  Jack J. Dongarra, Tony Hey, Erich Strohmaier   251
Exploiting Symmetry in Parallel Computations for Structural Biology
  Ioana M. Boier Martin, Dan C. Marinescu   255
An Object-Oriented and Parallel Simulation of a Power-Plant
  Klaus Wolf, António Mano, Sérgio Prata dos Santos, Jean-Marc Letteron   259
A Planning System for Aircraft Production with Parallel Constraint Logic Programming
  Patrick Albers, Jacques Bellone   266
Modelling and Optimising Flows Using Parallel Spatial Interaction Models
  Ian Turton, Stan Openshaw   270

Workshop 12 (16)
Theory and Models for Parallel Computing 277

The Queue-Read Queue-Write Asynchronous PRAM Model
  Phillip B. Gibbons, Yossi Matias, Vijaya Ramachandran   279
ERCW PRAMs and Optical Communication
  Philip D. MacKenzie, Vijaya Ramachandran   293
Goodness of Time-Processor Optimal PRAM Simulations
  Ville Leppänen   303
Simulations of PRAM on Complete Optical Networks
  Anssi Kautonen, Ville Leppänen, Martti Penttonen   307
Adaptive Parallelism in the Bulk-Synchronous Parallel Model
  Mohan V. Nibhanupudi, Boleslaw K. Szymanski   311
Implementation Issues Relating to the WPRAM Model for Scalable Computing
  Jonathan M. Nash, Peter M. Dew, John R. Davy, Martin E. Dyer   319
The Bulk-Synchronous Parallel Random Access Machine
  Alexandre Tiskin   327
The E-BSP Model: Incorporating General Locality and Unbalanced Communication into the BSP Model
  Ben H. H. Juurlink, Harry A. G. Wijshoff   339

Communication Efficient Data Structures on the BSP Model with Applications in Computational Geometry
    Alexandros V. Gerbessiotis, Constantinos J. Siniolakis 348
Submachine Locality in the Bulk Synchronous Setting
    Pilar de la Torre, Clyde P. Kruskal 352
Algebraic Laws for BSP Programming
    Jifeng He, Quentin Miller, Lei Chen 359
Realistic Parallel Algorithms: Priority Queue Operations and Selection for the BSP* Model
    Armin Baeumker, Wolfgang Dittrich, Friedhelm Meyer auf der Heide, Ingo Rieping 369
Multilayer Perceptron Learning Control
    Gilles Verley, Jean-Pierre Asselin de Beauville 377
Evaluating the Hyperbolic Model on a Variety of Architectures
    Ion Stoica, Florin Sultan, David Keyes 387
SPC: A Model of Parallel Computation
    Arjan J. C. van Gemund 397
Systematic Efficient Parallelization of Scan and Other List Homomorphisms
    Sergei Gorlatch 401
Array Structures and Data-Parallel Algorithms
    Gaétan Hains, John Mullins 409
Compile-Time Cost Analysis for Parallel Programming
    Roopa Rangaswami 417

Workshop 13 (15)
Parallel Computer Architecture   423

HPP: A High Performance PRAM
    Arno Formella, Joerg Keller, Thomas Walle 425
Relaxing the Inclusion Property in Cache Only Memory Architecture
    Jinseok Kong, Gyungho Lee 435
Using Proxies to Reduce Controller Contention in Large Shared-Memory Multiprocessors
    Andrew J. Bennett, Paul H. J. Kelly, Jacob G. Refstrup, Sarah A. M. Talbot 445
A RISC Approach to Weak Cache Coherence
    Juergen Risau, Alfred Mikschl, Werner Damm 453
3D Optoelectronic Computer Architectures for the Conjugate Gradient and Multigrid Benchmark Algorithms
    George A. Betzos, Pericles A. Mitkas 457
MSparc: A Multithreaded Sparc
    Alfred Mikschl, Werner Damm 461
A New Concept for Parallel Neurocomputer Architectures
    Alfred Strey, Narcís Avellana 470

Transformation of a 2-D VLSI Systolic Adder Circuit in 3-D Circuits Using Optical Interconnections
    Dietmar Fey 478
Scalable Software Latency Hiding Schemes: Evaluation of the Poststore and Prefetch Options
    Chaitanya Tumuluri, Alok N. Choudhary 486
Reducing Coherence Overhead in Shared-Bus Multiprocessors
    Sangyeun Cho, Gyungho Lee 492

Workshop 17
Scheduling and Load Balancing   499

An Asymptotically Optimal Affine Schedule on Bounded Convex Polyhedric Domains
    Patrick Le Gouëslier d'Argence 501
List Scheduling in the Presence of Branches - A Theoretical Evaluation
    Franco Gasperoni, Uwe Schwiegelshohn 515
Iterative Approach for the Clustering Problem
    Christophe Rapine, Denis Trystram 527
Compile-Time Task Scheduling for Multi-Phase Programming
    Abdelhamid Benaini, David Laiymani 535
Scheduling with Unit Processing and Communication Times on a Ring Network: Approximation Results
    Chams Lahlou 539
Efficient Parallel Algorithms for Scheduling with Tree Precedence Constraints
    Ernst W. Mayr, Hans Stadtherr 543
Bulk Synchronous Parallel Scheduling of Uniform Dags
    Radu Calinescu 555
Generalized Multiprocessor Scheduling
    Zhonghua Li, Chris C. Kirkham 563
A Static Scheduling Heuristic for Heterogeneous Processors
    Hyunok Oh, Soonhoi Ha 573
On the Cyclic Scheduling Problem with Small Communication Delays
    Aristotelis Giannakos, Jean-Claude König, Alix Munier 578
A New Scheduling Method for Parallel Discrete-Event Simulation
    Edwin Naroska, Uwe Schwiegelshohn 582
Minimum Length Scheduling of Precedence Constrained Messages in Distributed Systems
    Piera Barcaccia, Maurizio A. Bonuccelli, Miriam Di Ianni 594
Optimization of Parallel Programs on Machines with Expensive Communication
    Welf Löwe, Jörn Eisenbiegler, Wolf Zimmermann 602
Eager Scheduling with Lazy Retry for Dynamic Task Scheduling
    Huey-Ling Chen, Chung-Ta King 611

Load Management for Load Balancing on Heterogeneous Platforms: A
Comparison of Traditional and Neural Network Based Approaches
    Bettina Schnor, Stefan Petri, Horst Langendoerfer   615
Application-Assisted Dynamic Scheduling on Large-Scale Multi-Computer
Systems
    Ravi B. Konuru, José E. Moreira, Vijay K. Naik   621
Economic-Based Dynamic Load Distribution in Large Workstation
Networks
    Martin Backschat, Alexander Pfaffinger, Christoph Zenger . . .   631
Flexible Scheduling for Non-Deterministic, And-parallel Execution of Logic
Programs
    Kish Shen, Manuel Hermenegildo   635
A Load Balancing Task Allocation Scheme in a Hard Real Time System
    Jean-Louis Lanet   640
A Library Implementation of the Nano-Threads Programming Model
    Xavier Martorell, Jesús Labarta, Nacho Navarro, Eduard Ayguade   644

Workshop 19
Performance Evaluation   651

Speedup and Efficiency of Large Size Applications on Heterogeneous
Networks
    Laurent Colombet, Laurent Desbat   653
DiP: A Parallel Program Development Environment
    *Jesus Labarta, Sergi Girona, Vincent Pillet, Toni Cortes, Luis
    Gregoris*   665
Accurate Performance Prediction for Massively Parallel Systems and Its
Applications
    Jens Simon, Jens-Michael Wierum   675
Contention in the Cray T3D Communication Network
    Thierry Cornu, Michel Pahud   689
Theory, Practice, and a Tool for BSP Performance Prediction
    Jonathan M. D. Hill, Paul I. Crumpton, David A. Burgess . . .   697
Applying the Semi-Markov Memory and Cache Coherence Interference
Model to an Updating Based Cache Coherence Protocol
    Kazuki Joe, Akira Fukuda   706
Analytical Models of Multithreading with Data Prefetching
    Vladimir Vlassov, Lars-Erik Thorelli   714
IDRA (IDeal Resource Allocation): Computing Ideal Speedups in Parallel
Logic Programming
    Maria José Fernández, Manuel Carro, Manuel Hermenegildo . .   724
Estimation of the Throughput for some Stochastic Resources Sharing
Systems
    Matthieu Brilman, Jean-Marc Vincent   734
Some Closed Form Results for Circuit Switching in a Hypercube Network
    Vishal Sharma, Emmanouel A. Varvarigos   738

Workshop 20
Instruction Level Parallelism 743

Introduction to ILP Workshop
  Guang R. Gao, Christine Eisenbeis, Jian Wang 745
Instruction Fetching Mechanisms for Superscalar Microprocessors
  Steven Wallace, Nader Bagherzadeh 747
Designing Dynamic Two-Level Branch Predictors Based on Pattern Locality
  Chien-Ming Chen, Chung-Ta King 757
Streaming Prefetch
  Olivier Temam 765
Functionality Distribution on a Superscalar Architecture
  Eliseu M. C. Filho, Edil S. T. Fernandes, Andrew Wolfe 773
Investigating the Limits of Fine-Grained Parallelim in a Statically Scheduled Superscalar Architecture
  Richard Potter, Gordon Steven 779
On-Chip Multiprocessing
  Bernard Goossens, Duc Thang Vu 789
Identifying Bottlenecks in a Multithreaded Superscalar Microprocessor
  Ulrich Sigmund, Theo Ungerer 797
Aggregate Operation Movement: A Min-Cut Approach to Global Code Motion
  Raymond Lo, Sun Chan, Jim Dehnert, Ross Towle 801
Global Instruction Scheduling - A Practical Approach
  Sebastian Schmidt 815
Génération de micro-code parallèle pour la carte coprocesseur Rapid-2
  Laurent Winckel 819
RESIS: A New Methodology for Register Optimization in Software Pipelining
  Fermín Sánchez, Jordi Cortadella 824
Optimal Software Pipelining Through Enumeration of Schedules
  Erik R. Altman, Guang R. Gao 833

Workshop 22
Parallel and Distributed Databases 841

Triangular Grid Protocol: An Efficient Scheme for Replica Control with Uniform Access Quorums
  Cheng-Hong Cho, Jer-Tsang Wang 843
Mapping a Parallel Complex-Object DBMS to Operating System Processes
  Michael Gesmann 852
A Transaction Model for Multidatabase Systems
  Timuçin Devirmiş, Özgür Ulusoy 862

Multi-dimensional Declustering Methods for Parallel Database Systems
  *Manuel Barrena, Juan Hernández, José M. Martínez, Antonio Polo,
  Pedro de Miguel, M. Nieto*   866
Modelling Resource Utilization in Pipelined Query Execution
  Myra Spiliopoulou, Johann Christoph Freytag   872
On Transforming a Sequential SQL-DBMS into a Parallel One: First
Results and Experiences of the MIDAS Project
  *Giannis Bozas, Michael Jaedicke, Andreas Listl, Bernhard Mitschang,
  Angelika Reiser, Stephan Zimmermann*   881
DPLGraphs - A powerful Representation of Parallel Relational Query
Execution Plans
  Lionel Brunie, Harald Kosch   887
BLOCKER: A Variable and Multiattribute Declustering for Parallel
Database Machines
  Oduz Dikenelli, M. Osman Ünalýr, Esen Ozkarahan   892

Industrial Session   897

The PALLAS Portable Parallel Programming Environment
  Werner Krotz-Vogel, Hans-Christian Hoppe   899

Late Papers   905

On Optimal Parallel Algorithm for Gaussian Elimination
  Mounir Marrakchi   907
Synthesis of Massively Pipelined Algorithms for List Manipulation
  Ali E. Abdallah   911

Index of Authors   921

Invited Talks

High-Performance Distributed Computing: The I-WAY Experiment and Beyond

Ian Foster[1]

Mathematics and Computer Science Division
Argonne National Laboratory
Argonne, IL 60439
foster@mcs.anl.gov
http://www.mcs.anl.gov/globus/

Recent developments in networking are enabling innovative applications that integrate geographically distributed high-performance computing, database, display, and networking resources. However, there is as yet little understanding of the higher-level services needed to support these applications, or of the techniques required to implement these services in a scalable, secure manner. In this brief paper, I describe the I-WAY networking experiment, a large-scale wide-area computing testbed that has been used to investigate these issues. I also introduce the Globus project, a multi-institutional effort that is developing key technologies for I-WAY–like systems, including mechanisms for resource location, scheduling, authentication, and automatic configuration of high-performance distributed computations.

1 High-Performance Distributed Computing

High-performance distributing computing, or *metacomputing* as it is sometimes called, refers to the use of high-speed networks to connect supercomputers, databases, scientific instruments, and advanced display devices located at geographically remote sites [3]. In principle, metacomputing can both increase accessibility to supercomputing capabilities and enable the assembly of unique capabilities that could not otherwise be created in a cost-effective manner.

Experience with the I-WAY and other high-speed networking testbeds has provided convincing demonstrations that there are indeed applications of considerable scientific and economic importance that can benefit from access to high-performance distributed computing capabilities. Many of these applications fall into the following four general classes.

1. *Desktop supercomputing.* These applications couple high-end graphics capabilities with remote supercomputers and/or databases. This coupling connects users more tightly with computing capabilities, while at the same time achieving distance independence between resources, developers, and users.
2. *Smart instruments.* These applications connect users to instruments such as microscopes, telescopes, or satellite downlinks [17] that are themselves coupled with remote supercomputers. By allowing both quasi-realtime processing of instrument output and interactive steering, the utility of the instrument can be increased significantly.

3. *Distributed supercomputing.* More traditional supercomputing applications couple multiple, geographically distributed supercomputers in order to tackle problems that are too large for a single supercomputer or that can benefit from executing different problem components on different computer architectures [22, 19, 23].
4. *Collaborative environments.* A fourth set of applications couple multiple virtual environments so that users at different locations can interact with each other and with supercomputer simulations [8, 7].

Applications in the first and second classes are prototypes for future "network-enabled tools" that enhance local computational environments with remote compute and information resources; applications in the fourth class are prototypes of future collaborative environments.

2 The Globus Project

High-performance, geographically distributed computing requires tools for requesting, locating, scheduling, and programming diverse computational and network resources; for authenticating users, authorizing access to resources, and protecting the security of user computations; and for accessing both shared data and user file systems from geographically remote locations. These tools must scale to meet the requirements of computations that link tens or hundreds of resources located in multiple administrative domains and connected using networks of widely varying capabilities.

The Globus project is a multi-institutional research and development activity that is addressing these problems. Its goal is to provide software technologies that support the dynamic identification and composition of resources available on national-scale internets, and that provide mechanisms for authentication, authorization, and delegation of trust within environments of this scale.

In order to support the dynamic composition of computational and information resources, we are investigating the following topics.

- *Resource location*: uniform and scalable mechanisms for naming and locating computational and communication resources on remote systems, and for incorporating these resources into parallel and distributed computations.
- *Protocol and resource management*: scalable techniques for locating available network connections, making choices between alternatives according to their service type and security level, integrating chosen networks into running computations.
- *Resource-aware programming tools*: versions of high-level libraries and languages, such as MPI [16, 12], CC++ [4], Fortran M [9], and HPF, that allow programmers to specify high-performance distributed computations in a portable manner, while also providing access to low-level information when this is required for performance.

In the security area, we are focusing on two problems.

- *Data Access*: techniques for providing computations with uniform, efficient, and secure access to files. In particular, new protocols and algorithms that allow secure distributed file systems [5] to function efficiently in large-scale internetworked environments.
- *Authentication and Authorization*: authorization and access control mechanisms that provide fine-grain control over access to communication, computational, and information resources. Also, techniques based on delegation of trust for managing trust relationships and access control in large and dynamically-changing user communities across multiple administrative domains.

These basic techniques are being incorporated in a prototype software system for constructing high-performance distributed computations in national-scale internetworked environments. Preliminary versions of this software were used in the I-WAY networking experiment to support extensive experiments in wide area supercomputing.

3 I-WAY and I-Soft

The I-WAY, or Information Wide Area Year [6], was a wide-area computing experiment conducted throughout 1995 with the goal of providing a large-scale testbed in which innovative high-performance and geographically-distributed applications could be deployed. The I-WAY linked 11 existing national testbeds based on ATM (asynchronous transfer mode) technology to interconnect supercomputer centers, virtual reality research locations, and applications development sites across North America (Figure 1). When demonstrated at the Supercomputing conference in San Diego in December 1995, it connected supercomputers, mass storage systems, and advanced visualization devices at 17 different sites. This distributed supercomputing environment was used by over 60 application groups for experiments in high-performance computing (e.g., [22, 23]), collaborative design [7], and the coupling of remote supercomputers and databases into local environments (e.g., [17]). A primary thrust was applications that use multiple supercomputers and virtual reality devices to explore collaborative technologies in which shared virtual spaces are used to perform computational science. For the purposes of this experiment, all communication was performed by using standard IP protocols running over ATM Adaptation Layer 5 (AAL5).

The I-WAY experiment was intended not only as an opportunity for large-scale application experiments, but also as a testbed within which solutions to various software infrastructure problems could be deployed and studied in a somewhat controlled environment. Because the number of users (few hundred) and sites (around 20) were moderate, issues of scalability could, to a large extent, be ignored. However, issues of security, usability, and generality were of critical concern. Important secondary requirements were to minimize development and maintenance effort, both for the I-WAY development team and the participating sites and users.

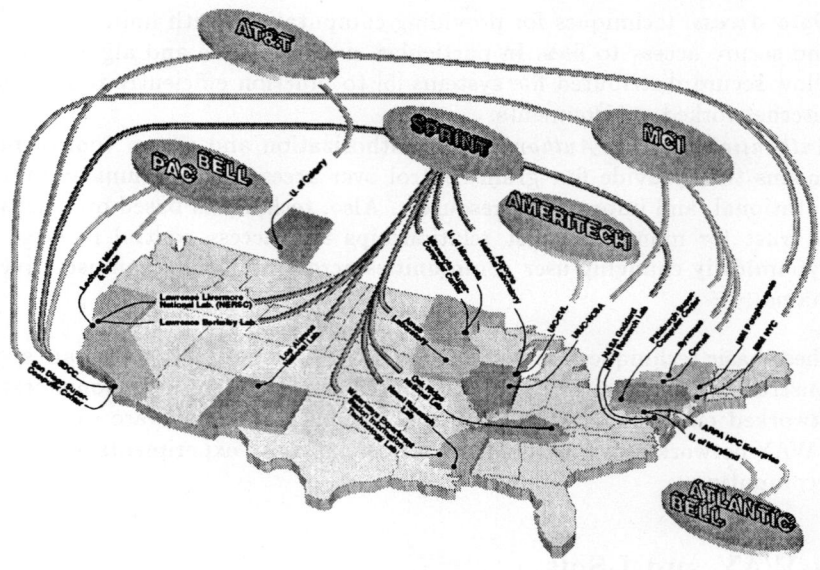

Fig. 1. The I-WAY network. Figure produced by Linda Winkler and Richard Foster.

To this end, members of the Globus project worked in collaboration with researchers and programmers at various I-WAY sites to develop a system management and application programming environment called I-Soft [11] that provided uniform authentication, resource reservation, process creation, and communication functions across I-WAY resources. A novel aspect of our approach was the deployment of a dedicated I-WAY Point of Presence, or I-POP, machine at each participating site (Figure 2). These machines provided a uniform environment for deployment of management software, and also simplified validation of system management and security solutions by serving as a "neutral" zone under the joint control of I-WAY developers and local authorities.

The task of developing software systems for environments such as the I-WAY is complicated by the fact that resources and users exist at different sites and in different administrative domains. Different sites have different access mechanisms for their resources, and cannot be expected to relinquish control to an external authority. Hence, the problem of developing management systems is in large part one of defining protocols and interfaces that support a negotiation process between users (or brokers acting on their behalf) and the sites that control the resources that users want to access. I-Soft addressed this issue by providing a simple computational resource broker that used scheduler proxies to provide a uniform scheduling environment integrating diverse local schedulers, and by

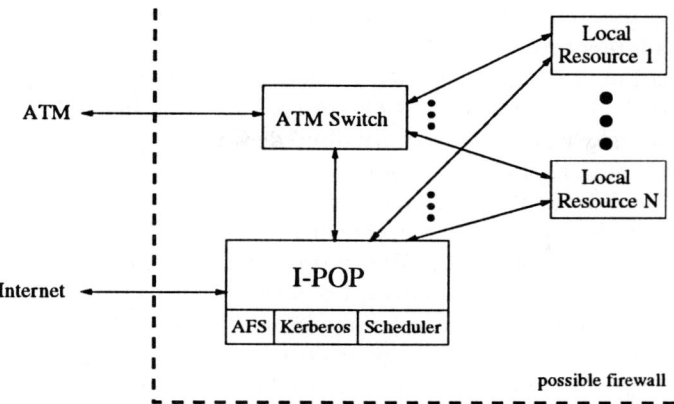

Fig. 2. An I-WAY Point of Presence (I-POP) machine

using authorization proxies to construct a uniform authentication environment and define trust relationships across multiple administrative domains [11]. I-Soft also addressed issues of system heterogeneity by providing resource-aware parallel programming tools based on the Nexus runtime system [13]. These tools used configuration information regarding topology, network interfaces, startup mechanisms, and node naming to provide a uniform view of heterogeneous systems and to optimize communication performance [10]. These various I-Soft services allowed a user to log on to any I-POP and then schedule resources on heterogeneous collections of resources, initiate computations, and communicate between computers and with graphics devices—all without being aware of where these resources were located or how they were connected.

The I-WAY experiment proved extremely useful as a means of identitying truly important issues in wide-area high-performance computing. In particular, we learned that system components that are typically developed in isolation must be more tightly integrated if performance, reliability, and usability goals are to be achieved. For example, resource location services in future I-WAY–like systems will need low-level information on network characteristics; schedulers will need to be able to schedule network bandwidth as well as computers; and parallel programming tools will need up-to-date information on network status.

4 Related Work

The Globus and I-WAY projects build on the results of considerable previous work in distributed computing, parallel computing, and high-speed networking. To name just a few examples, Condor [18], Nimrod [1], and Prospero [21] address the problem of locating and/or accessing distributed resources; AFS [20] and

DFS address problems of sharing distributed data; and MPI [16], PVM [14], and Isis [2] address problems of coupling distributed computational resources.

The Distributed Computing Environment (DCE) and Common Object Request Broker Architecture (CORBA) are two major industry-led attempts to provide a unifying framework for distributed computing. Both define (or will soon define) a standard directory service, remote procedure call (RPC), security service, and so forth; DCE also defines a Distributed File Service (DFS) derived from AFS. Some DCE mechanisms (RPC, DFS) may well prove to be appropriate for implementing I-POP services; CORBA directory services may be useful for resource location. However, both DCE and CORBA appear to have significant deficiencies as a basis for application programming in I-WAY–like systems. In particular, the remote procedure call is not well-suited to applications in which performance requirements demand asynchronous communication, multiple outstanding requests, and/or efficient collective operations.

Two other research projects are addressing similar themes and goals. The Legion project at the University of Virginia [15] and the Globe project at the Vrije Universiteit in Amsterdam [24] both seek to develop a universal, object-based software infrastructure for computing in wide area environments; research topics include scheduling, file systems, security, fault tolerance, and network protocols. The Globus and I-WAY efforts are distinguished by their focus on high-performance systems and applications, in which the efficient management and scheduling of high-speed networks are central concerns.

5 Future Directions

The Globus project is now addressing various issues identified in the course of I-Soft development. Particular focus areas include resource location, scheduling, automatic configuration, scalable trust management, and resource-aware tools and applications. In addition, we are working with colleagues to define and construct a "persistent I-WAY" that will provide further opportunities for application experiments and evaluation of wide area management and application programming tools. We hope to extend the scope of the I-WAY to include international connections in the near future.

Acknowledgments

The Globus project is led by the author and Carl Kesselman of the California Institute of Technology. The I-WAY project is centered at Argonne, the National Center for Supercomputer Applications, and the Electronic Visualization Laboratory. This work was supported in part by the Mathematical, Information, and Computational Sciences Division subprogram of the Office of Computational and Technology Research, U.S. Department of Energy, under Contract W-31-109-Eng-38. Future work on Globus will be supported by DARPA.

References

1. D. Abramson, R. Sosic, J. Giddy, and B. Hall. Nimrod: A tool for performing parameterised simulations using distributed workstations. In *Proc. 4th IEEE Symp. on High Performance Distributed Computing*. IEEE Press, 1995.
2. K. Birman. The process group approach to reliable distributed computing. *Communications of the ACM*, 36(12):37–53, 1993.
3. C. Catlett and L. Smarr. Metacomputing. *Communications of the ACM*, 35(6):44–52, 1992.
4. K. M. Chandy and C. Kesselman. CC++: A declarative concurrent object oriented programming notation. In *Research Directions in Object Oriented Programming*. The MIT Press, 1993.
5. J. Cook, S.D. Crocker, Jr. T. Page, G. Popek, and P. Reiher. Truffles: Secure file sharing with minimal system administratore intervention. In *Proc. SANS-II, The World Conference On Tools and Techniques For System Administration, Networking, and Security*. 1993.
6. T. DeFanti, I. Foster, M. Papka, R. Stevens, and T. Kuhfuss. Overview of the I-WAY: Wide area visual supercomputing. *International Journal of Supercomputer Applications*, 1996. in press.
7. D. Diachin, L. Freitag, D. Heath, J. Herzog, W. Michels, and P. Plassmann. Remote engineering tools for the design of pollution control systems for commercial boilers. *International Journal of Supercomputer Applications*, 1996. To appear.
8. T. L. Disz, M. E. Papka, M. Pellegrino, and R. Stevens. Sharing visualization experiences among remote virtual environments. In *International Workshop on High Performance Computing for Computer Graphics and Visualization*, pages 217–237. Springer-Verlag, 1995.
9. I. Foster and K. M. Chandy. Fortran M: A language for modular parallel programming. *Journal of Parallel and Distributed Computing*, 26(1):24–35, 1995.
10. I. Foster, J. Geisler, C. Kesselman, and S. Tuecke. Multimethod communication for high-performance metacomputing applications. Preprint, Mathematics and Computer Science Division, Argonne National Laboratory, Argonne, Ill., 1996.
11. I. Foster, J. Geisler, W. Nickless, W. Smith, and S. Tuecke. Software infrastructure for the I-WAY high-performance distributed computing experiment. In *Proc. 5th IEEE Symp. on High Performance Distributed Computing*. IEEE Computer Society Press, 1996.
12. I. Foster, J. Geisler, and S. Tuecke. MPI on the I-WAY: A wide-area, multimethod implementation of the Message Passing Interface. In *Proceedings of the 1996 MPI Developers Conference*. IEEE Computer Society Press, 1996.
13. I. Foster, C. Kesselman, and S. Tuecke. The Nexus approach to integrating multithreading and communication. *Journal of Parallel and Distributed Computing*, 1996. To appear.
14. A. Geist, A. Beguelin, J. Dongarra, W. Jiang, B. Manchek, and V. Sunderam. *PVM: Parallel Virtual Machine—A User's Guide and Tutorial for Network Parallel Computing*. MIT Press, 1994.
15. A. Grimshaw, W. Wulf, J. French, A. Weaver, and P. Reynolds, Jr. Legion: The next logical step toward a nationwide virtual computer. Technical Report CS-94-21, Department of Computer Science, University of Virginia, 1994.
16. W. Gropp, E. Lusk, and A. Skjellum. *Using MPI: Portable Parallel Programming with the Message Passing Interface*. MIT Press, 1995.

17. C. Lee, C. Kesselman, and S. Schwab. Near-real-time satellite image processing: Metacomputing in CC++. *Computer Graphics and Applications*, 1996. to appear.
18. M. Litzkow, M. Livney, and M. Mutka. Condor - a hunter of idle workstations. In *Proc. 8th Intl Conf. on Distributed Computing Systems*, pages 104–111, 1988.
19. C. Mechoso et al. Distribution of a Coupled-ocean General Circulation Model across high-speed networks. In *Proceedings of the 4th International Symposium on Computational Fluid Dynamics*, 1991.
20. J.H. Morris et al. Andrew: A distributed personal computing environment. *CACM*, 29(3), 1986.
21. B. Clifford Neumann and Santosh Rao. The Prospero resource manager: A scalable framework for processor allocation in distributed systems. *Concurrency: Practice and Experience*, June 1994.
22. J. Nieplocha and R. Harrison. Shared memory NUMA programming on I-WAY. In *Proc. 5th IEEE Symp. on High Performance Distributed Computing*. IEEE Computer Society Press, 1996.
23. M. Norman et al. Galaxies collide on the I-WAY: An example of heterogeneous wide-area collaborative supercomputing. *International Journal of Supercomputer Applications*, 1996. in press.
24. M. van Steen, P. Homburg, L. van Doorn, A. Tanenbaum, and W. de Jonge. Towards object-based wide area distributed systems. In *Proc. International Workshop on Object Orientation in Operating Systems*, pages 224–227, 1995.

Design and Implementation of a Parallel Architecture for Biological Sequence Comparison

Pascale Guerdoux–Jamet[2], Dominique Lavenier[2], Charles Wagner[2] and Patrice Quinton[2]

Irisa, Campus de Beaulieu, 35042 Rennes Cedex, France

Abstract. SAMBA is a full custom parallel hardware accelerator dedicated to the comparison of biological sequences. It implements a parameterized version of the Smith and Waterman algorithm allowing the computation of local or global alignments with or without gap penalty. The speed-up provided by SAMBA over standard workstations ranges from 50 to 500, depending on the application. SAMBA was designed with an effort of less than one person/year. This includes the design, fabrication and test of a full-custom VLSI chip which is used as a building block for the 128 processor systolic array implementing the string alignment algorithm. We describe the SAMBA architecture, its performance characteristics, and we detail its design steps, from the initial specification to its full implementation. We report a first application of SAMBA to the study of yeast orphan sequences.

1 Introduction

SAMBA – Systolic Accelerator for Molecular Biological Applications – is a hardware accelerator designed for speeding up biological sequence comparisons. Computations which require several hours on standard workstations are performed in a few tens of seconds on SAMBA.

SAMBA implements a parameterized version of the Smith and Waterman algorithm [15] [5]. By setting a few parameters, local or global comparisons can be performed, with or without gap penalty. Thus, a variety of softwares, such as BLAST [1], FASTA [12] or SSEARCH [13], may be implemented on SAMBA.

The complete SAMBA system comprises a workstation, a systolic array of 128 full custom hardwired 12-bit processors, and a FPGA-based interface (see Fig. 1). The FPGA interface is the PeRLe-1 board developed by Vuillemin et al. [2]; it acts as a hardware programmable driver for the systolic array.

SAMBA may be compared with two other hardware prototypes – BISP [3] and BIOSCAN [18] – also designed to speed up biological sequence comparisons with full-custom chips. Both systems include a linear systolic array made out of dedicated processors. BISP and BIOSCAN contain respectively 256 16-bit processors and 12 000 1-bit processors.

The architecture of BISP and SAMBA are similar. However, they differ on the way the systolic array is supplied with data: the BISP array is driven by a Motorola 68020 programmable processor while SAMBA uses FPGA technology.

This last approach is simpler and ensures that the high data throughput required by the systolic array is sustained.

BIOSCAN does not support dynamic programming algorithms. Like BLAST, it has been designed to detect similar segments of identical length. Hence, the algorithm is simpler and this enables a very high density of processors (812 per chips) to fit onto silicon. The speed of SAMBA depends on the application. The

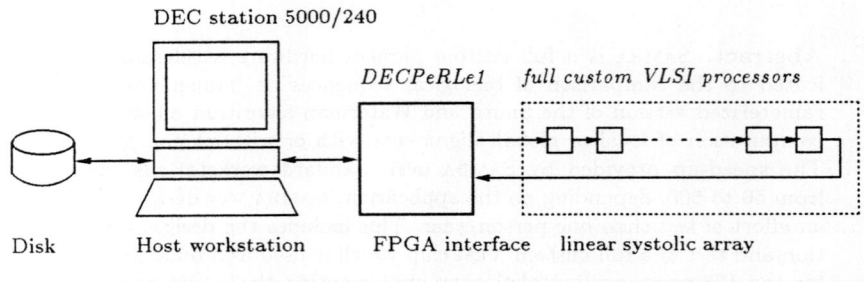

Fig. 1. SAMBA comprises a workstation (with its local disk), a systolic array made out of 128 VLSI full custom processors and a FPGA interface which fills the gap between a complete hardwired array of processors and a programmable Von Neuman machine

best speed is achieved for bank to bank comparison, that is to say, when two sequence data bases are matched against one another. In this case, the systolic array is continuously supplied with data, leading to a speed-up higher than two orders of magnitude over standard workstations.

The scan of biological banks – comparison of one query sequence against a bank, – provides also very interesting speed-ups which, however, are limited by disk access and depend on the length of the query sequence. SAMBA is thus better suited to intensive comparison tasks than to scanning biological banks, even if this latter task provides a noticeable speed-up.

The design of SAMBA took less than one person/year. It was started in September 1994, and the architecture was ready in December 1995. Since then, SAMBA has been used by biologists for sequence comparisons. The first application was to study so-called "orphan sequences" of the yeast genome, in order to compare the sensitivity of the Smith and Waterman algorithm with that of other less time consuming algorithms.

This paper is organized as follows: section 2 presents the class of algorithms supported by SAMBA. Section 3 explains how these algorithms are implemented on a linear systolic array. Section 4 gives more details on the hardware, particularly on the FPGA-based interface and the ASIC developed for the array. Finally, section 5 presents the results of a biological study that SAMBA made possible and analyses the speed of SAMBA.

2 The problem and its algorithmic solution

The algorithm executed by SAMBA is a parameterized version of the Smith and Waterman algorithm [15]. This algorithm was introduced in 1981 and extended by Gotoh [5] in 1982. It allows biological sequences to be aligned using two basic *edit operations*:

- *substitution* of characters;
- *insertion* or *omission* of characters; these last operations are called a *gap*.

By using a series of such edit operations, two sequences may be transformed into one another. The smallest number of edit operations required to change the first sequence into the second one is thus a measure of the distance between them, called the edit distance. The computation of the edit distance is achieved by dynamic programming, which consists of computing recursively an $N \times N$ matrix, where N is the length of the sequences, as we shall see in more details now. In subsection 2.1, we recall the principles of the Smith and Waterman algorithm, while subsection 2.2 explains how this algorithm was parameterized.

2.1 The Smith and Waterman algorithm

Consider two strings S1 and S2 of length respectively $l1$ and $l2$. For example,

```
S1 = AGTCCGAGGGCTACTCTACTGAAC
S2 = CCAATCTACTACTGCTTGCAGTAC
```

A comparison of these sequences will provide the alignments shown in figure 2.

```
S1      AGTCCGAGGG CTACTCTACT GAAC
                   |:|:||||||
S2                 CCAATCTACT ACTGCTTGCAGTAC

S1      AGTCCGAGGG CTACT.CTACT GAAC
                   ||||| ||:||
S2                 CCAAT CTACTACTGCT TGCAGTAC
```

Fig. 2. Examples of sequence alignments. Bar represent character matchs, and dotted lines represent gaps

To identify common subsequences, the Smith and Waterman algorithm computes the similarity $H(i,j)$ of two segments ending at position S1$_i$ and S2$_j$ of the two sequences S1 and S2.

The computation of $H(i,j)$ is given by the following recurrences:

$$H(i,j) = \max \begin{cases} 0 & 0 < i \leq l1 \quad 0 < j \leq l2 \\ E(i,j) \\ F(i,j) \\ H(i-1, j-1) + \text{Sbt}(S1_i, S2_j) \end{cases} \quad (1)$$

where

$$E(i,j) = Max \begin{cases} H(i,j-1) - \alpha \\ E(i,j-1) - \beta \end{cases} \quad 0 < i \leq l1 \quad 0 < j \leq l2$$

$$F(i,j) = Max \begin{cases} H(i-1,j) - \alpha \\ F(i-1,j) - \beta \end{cases} \quad 0 < i \leq l1 \quad 0 < j \leq l2$$

Sbt is a character substitution cost table.
Initialization of these values are given by:

$$\forall i, 0 \leq i \leq l1 : H(i,0) = E(i,0) = 0$$
$$\forall j 0 \leq j \leq l2 : H(0,j) = F(0,j) = 0$$

Multiple gap costs are taken into consideration as follows: α is the cost of the first gap; β is the cost of the following gaps. The total gap cost function $G(k)$ is given by: $G(k) = \alpha + \beta \times (k-1)$ with $\beta \leq \alpha$.

In the example of figure 2, with gap costs $\alpha = 20$ and $\beta = 10$ and **Sbt** function defined as:

$$\text{Sbt}(x,y) = \begin{cases} 10 \text{ if } (x = y) \\ -9 \text{ otherwise} \end{cases}$$

the result of this algorithm is the matrix shown in figure 3. Each position ($S1_i, S2_j$) of the matrix is a similarity value $H(i,j)$. The two segments of S1 and S2 producing this value can be determined by a backtracking procedure (see [17]). In that example, the best score is 62 and the two segments detected as similar are CTACTCTACT and CCAATCTACT.

2.2 Parameterized version

SAMBA implements a modified version of the Smith and Waterman algorithm in order to cover a set of useful sequence alignment algorithms. The reader is referred to [9] for details on the parameterized equations.

Using a slight modification of equation (1), one can solve the following problems:

- find the alignment of S1 and S2 which maximizes a similarity measure with simple gap cost. This is the equation introduced by Needleman and Wunsch in [11]: the cost of k gaps is a simple function $G(k) = g \times k$ of the cost of one gap.
- locate similar subsequences between two sequences. This is probably the most useful algorithm for current research, since two biological sequences which present a small overall similarity may share surprising relationships on short segments.

In such a way, the most commonly used softwares, such as BLAST [1], FASTA [12] or SSEARCH [13], can be executed by SAMBA.

	C	C	A	A	T	C	T	A	C	T	A	C	T	G	C	T	T	G	C	A	G	T	A	C
A	0	0	10	10	0	0	0	10	0	0	10	0	0	0	0	0	0	0	0	10	0	0	10	0
G	0	0	0	1	1	0	0	0	1	0	0	1	0	10	0	0	0	10	0	0	20	0	0	1
T	0	0	0	0	11	0	10	0	0	11	0	0	11	0	1	10	10	0	1	0	0	30	10	0
C	10	10	0	0	0	21	1	1	10	0	2	10	0	2	10	0	1	1	10	0	0	10	21	20
C	10	20	1	0	0	10	12	0	11	1	0	12	1	0	12	1	0	0	11	1	0	0	1	31
G	0	1	11	0	0	0	1	3	0	2	0	0	3	11	0	3	0	10	0	2	11	0	0	11
A	0	0	11	21	1	0	0	11	0	0	12	0	0	0	2	0	0	0	1	10	0	2	10	1
G	0	0	0	2	12	0	0	0	2	0	0	3	0	10	0	0	0	10	0	0	20	0	0	1
G	0	0	0	0	0	3	0	0	0	0	0	0	10	1	0	0	10	1	0	10	11	0	0	
G	0	0	0	0	0	0	0	0	0	0	0	0	10	1	0	0	10	1	0	10	1	2	0	
C	.10	.10	0	0	0	10	0	0	10	0	0	10	0	0	20	0	0	0	20	0	0	1	0	12
T	0	.1	1	0	10	0	20	0	0	20	0	0	20	0	0	30	10	0	0	11	0	10	0	0
A	0	0	.11	.11	0	1	0	30	10	0	30	10	0	11	0	10	21	1	0	10	2	0	20	0
C	10	10	0	.2	.2	10	0	10	40	20	10	40	20	10	21	1	1	12	11	0	1	0	0	30
T	0	1	1	0	.12	.0	20	0	20	50	30	20	50	30	20	31	11	1	3	2	0	11	0	10
C	10	10	0	0	0	.22	.2	11	10	30	41	40	30	41	40	20	22	2	11	0	0	0	2	10
T	0	1	1	0	10	2	.32	.12	2	20	21	32	50	30	32	50	30	20	10	2	0	10	0	0
A	0	0	11	11	0	1	12	.42	.22	12	30	12	30	41	21	30	41	21	11	20	0	0	20	0
C	10	10	0	2	2	10	2	22	.52	.32	22	40	20	21	51	31	21	32	31	11	11	0	0	30
T	0	1	1	0	12	0	20	12	32	.62	.42	32	50	30	31	61	41	31	23	22	2	21	1	10
G	0	0	0	0	0	3	0	11	22	42	53	33	30	60	40	41	52	51	31	21	32	12	12	0
A	0	0	10	10	0	0	0	10	12	32	52	44	24	40	51	31	32	43	42	41	21	23	22	3
A	0	0	10	20	1	0	0	10	2	22	42	43	35	30	31	42	22	23	34	52	32	22	33	13
C	10	10	0	1	11	11	0	0	20	12	22	52	34	26	40	22	33	13	33	32	43	23	13	43

Fig. 3. An example of alignment matrix. Numbers followed by a dot belong to the best alignment

3 From the algorithm to a parallel architecture

In this section, we describe the parallel implementation of the Smith and Waterman algorithm, we explain how the results can be sent to the border processors, and we give the complexity of the algorithm.

3.1 Basics

The parallelization of string comparison algorithms on linear systolic arrays has been abundantly described in the literature. Readers interested by details are refered to [8] [6] [3] [10] [18].

The architecture of SAMBA (see Figure 4) is composed of identical processors, linearly arranged, which perform one step of the matrix computation described in the previous section. Data move from the left to the right. Typically, a parallel comparison of two sequences is performed as follows:

1. the array is initialized with one sequence (usually called the query sequence) at the rate of one character per processor;
2. the other sequence (bank sequence) is flushed to the left of the array and progresses every systolic cycle;
3. the result is collected on the rightmost processor when the last character of the second sequence is output.

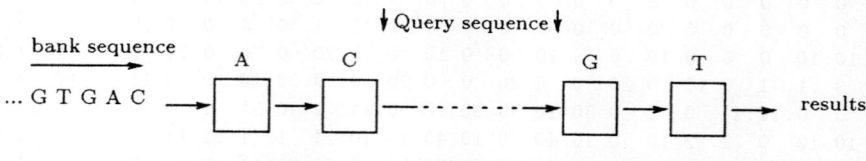

Fig. 4. Sequence comparison on a linear systolic array: one sequence (the query sequence) is stored into the array (one character per processor) and the other sequence flows from the left to right through the array. During each systolic step, one elementary matrix computation is performed on each processor. The result is available on the rightmost processor when the last character of the flowing sequence is output.

3.2 Getting the results

In a systolic array, only the border processors communicate with the "outside world". Hence, the results must necessarily be forwarded to the border to be output. In the present case, the computation consists of calculating a matrix in which the interesting values can appear anywhere: for example, the identification of similar segments requires the full matrix to be analyzed in order to detect high scores.

Forwarding values to the array extremities requires additional computations. The SAMBA algorithm is completed with the computation of two terms:

- MaxRow(i, j), the maximum score $H(i, j)$ computed on the ith row of matrix H;
- IdxCol(i, j), an index specifying the column where MaxRow is modified.

These two terms are computed using the following recurrence equations:

$$\text{MaxRow}(i, j) = \max \begin{cases} \text{MaxRow}(i, j-1) \\ H(i, j) \end{cases}, \tag{2}$$

$$\text{IdxCol}(i, j-1) = \begin{cases} \text{IdxCol}(i, j-1) \text{ if MaxRow}(i, j-1) > H(i, j)) \\ j \text{ otherwise.} \end{cases} \tag{3}$$

Let N be the size of the array and i the systolic step number. During each systolic step, the rightmost processor of the array delivers MaxRow(i, N) and IdxCol(i, N). These informations enable the coordinates of local maxima to be found in matrix H.

3.3 Complexity

Let l_q be the length of the query sequence and l_b is the length of the bank sequence. The computation is performed in $l_q + l_b - 1$ systolic steps, providing a speed-up S_1 over sequential machines given by:

$$S_1 = \frac{l_q \times l_b}{l_q + l_b - 1} \ .$$

When comparing one sequence against a bank of sequences, the speed-up may be increased by pipelining the bank sequences flowing through the array: when the last character of a sequence enters the array, the first character of the following sequence can be input during the next systolic cycle. For k sequences, $l_q + k \times l_b - 1$ systolic steps are required. The speed-up (S_k) is then given by:

$$S_k = \frac{k \times l_q \times l_b}{l_q + k \times l_b - 1} \ .$$

When scanning a biological database, the number k of sequences is generally high and the speed-up S_k can be approximated by l_q (the length of the query sequence). This last case represents the application domain where SAMBA excels.

4 From the parallel architecture to the hardware

The design of SAMBA was decided in september 1994. The main design decision was to choose the technology of the systolic array. The choice was between a programmable or a fully dedicated architecture. It was decided to build a systolic array with fully dedicated processors, as the result would be about 10 times faster than a programmable version, and there would be no need to design a programming environment – a task which usually requires a lot of effort. Moreover, the algorithms to be implemented were very stables, and a parameterized architecture would make it possible to execute all the commonly used algorithms.

We now present the architecture of SAMBA (subsection 4.1), then we describe the design trajectory of its implementation in subsection 4.2.

4.1 Architecture of SAMBA

As mentioned previously, SAMBA is composed of a host workstation, a FPGA-interface and a full custom systolic array. This section describes in details each one of these elements.

Host workstation and I/O

The host of SAMBA is a DECstation DS5000/240 model, which uses a 40 MHz MIPS R3400 Risc processor. The I/O system is based on the TURBOchannel open interconnect developed by Digital and provides three slots for connecting optional peripherals. It runs on the 25 MHz memory system clock and has a peak bandwidth of 100 MB/s. It is connected to a custom I/O controller which interfaces the TURBOchannel on one side to several different devices on the other side, mostly controllers for particular types of peripherals or data communication interfaces, such as serial ports, SCSI controller, Ethernet, etc.

The SCSI controller can perform asynchronous or synchronous data transfers at up to 5 MB/s through DMA access. The hard disk drive available on the system has a capacity of 426 MB with a maximum bus bandwidth of 4 MB/s. This local disk is used to store the banks of biological sequences.

Host/array interface

The host/array interface uses the concept of PAM (Programmable Active Memories) introduced by Vuillemin et al. [2]. As mentioned by the authors, the purpose of a PAM is to implement a virtual machine which can be dynamically configured as a large number of specific hardware devices. The PAM is connected through two links to a host processor. A third connection allows a configuration bit stream to be downloaded into the PAM; after the configuration phase, the PAM behaves like an ASIC. The PAM may operate in different modes:

Stand-alone mode: the PAM is hooked to external systems through the external links;

Co-processor mode: the PAM is controlled by the host and specialized to speed-up some crucial computations;

Mixed mode: the PAM is both controlled by the host and connected to some specific hardware.

SAMBA uses the mixed mode. The external links are connected to the systolic array. The major role of the PAM is to provide the VLSI systolic array with data, and to filter the results on the fly.

The PAM used by SAMBA is the PeRLe-1 prototype board, a configurable co-processor organized around a central computational matrix of 16 Xilinx XC3090 FPGAs, surrounded by 4 memory banks for local storage, and 7 other FPGAs to implement switch and control functions. PeRLe-1 is connected to the workstation through one of the TURBOchannel extension slots, providing a very fast communication channel between the PAM and the CPU.

VLSI Systolic array

The array is a linear systolic array composed of 32 full custom chips – named API256 – distributed over two printed boards. One chip (designed in 1 micron CMOS technology, see figure 5) integrates 4 processors, leading to a systolic array of 128 processors. Each chip contains about 100.000 transitors, and has a cycle of 100 ns. The datapath of the chip is 12 bits wide; this may be insufficient for some applications but is enough to validate the SAMBA prototype.

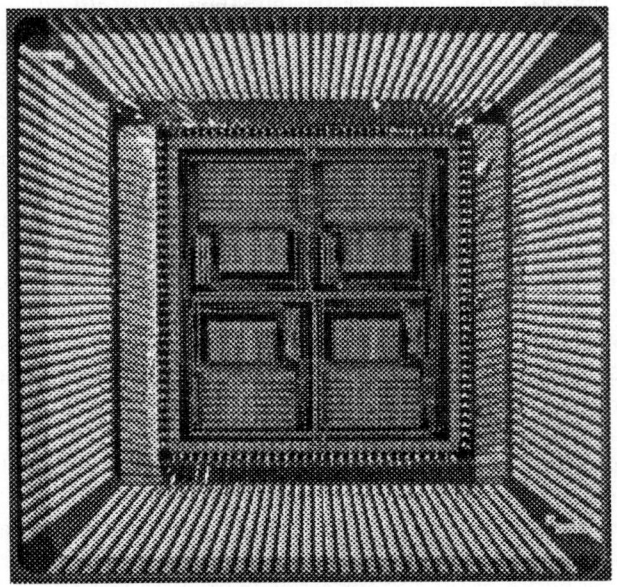

Fig. 5. Layout of the API256 chip

Each processor computes the recurrence equations presented in section 2 as well as the two auxiliary equations used for locating the results presented in section 3.2. A performance measure commonly used in computational biology is the number of *million of cell operation per second* (MCOPS), that is, the complete computation of one entry of matrix $H(i,j)$, including all comparisons, additions and maxima calculations. SAMBA performs such a computation in one systolic step. The peak speed of the 128 processor array is thus

$$P_p = Nb_{proc} \times freq = 128 \times 10^7 = 1280 \text{ MCOPS} \quad .$$

Furthermore, each processor contains a 32 word internal static memory for storing the substitution costs. This memory size is sufficient since a processor has to know a maximum of 20 substitution costs (they are 20 different animo acids) which refer to one character of the query sequence.

The systolic array is interfaced to the workstation through the direct custom links available on the PeRLe-1 board.

4.2 Design of SAMBA

SAMBA was designed as follows:

Functional specification: the specification of the sequence alignment application, including the parallel version of the algorithms, was written using the C-STOLIC language [14]. This language is a slight extension of C allowing parallel systolic algorithms to be simulated on Unix workstations. The C-stolic compiler provided three outputs: a C program emulating one processor of the systolic array, a C program emulating the interface between the host and the systolic array, and a C program for the host workstation.

Functional simulations: the functional specification was simulated on test data obtained from the application.

Architectural specification: the program run by the processors of the VLSI systolic array was translated "manually" into VHDL. Only the subset of VHDL accepted by the COMPASS VLSI synthesis tool was used, in order to be able to obtain a hardware specification.

Validation of the architecture: the architectural specification was run on the same data as the functional simulation, and validated by comparison.

Synthesis of the VLSI chip: the VLSI chip API256 was synthesized using the COMPASS hardware synthesis software, and sent to fabrication. The chips were fabricated by ES2 through the french multi-chip project. The 32 chips needed for the architecture were obtained in one run.

Board design: the systolic array was designed by assembling the 32 chips in two boards.

Interface design: thanks to the flexibility of the PAM, the interface was designed in two steps. First, the PAM was configured as a very simple link between the host and the systolic array, in order to test the full functionality of the architecture. In this configuration, the interface program produced by the C-STOLIC compiler was run by the host. Then, in order to reach the full speed of the systolic array, the PAM was configured to implement the interface program.

Application integration: finally, the whole application was integrated, by installing the application program on the workstation.

The design effort was 11 person/months: 1 month for functional specification and simulation, one month for architectural specification and simulation, 2 months for architectural synthesis, 2 month for chip testing, 0.5 month for board design, 4 month for interface design, and 0.5 month for final integration.

5 Application and speed of SAMBA

In this section we give some results obtained using SAMBA on a real intensive protein bank to bank comparison example. We first describe briefly the application, then we present and discuss the performance characteristics of SAMBA.

5.1 Application

SAMBA has been used continuously since December 1995. Several research projects are currently using this architecture. The most advanced one is described in [7], and can be summarized as follows.

When determining the sequence of some genome, some of the frames which are analyzed do not share significant similarities with protein sequences already known and stored in the databanks. These sequences are called "orphan sequences", and one problem for biologists is to find out why orphan sequences do not show similarity with databanks. Indeed, a sequence is said to be "orphan" if the computer program used for the comparison does not detect a significant similarity. Thus, this notion certainly depends on the program used.

The Smith and Waterman algorithm is known to be much more complex, and supposed to be more sensitive, than the programs commonly used on workstations, such as Fasta or Blast. Due to the complexity of this algorithm, no significant study has been performed to prove, or disprove, that it is worth using this technique. Using SAMBA, it was possible to make such a study.

The purpose was to compare a set of 814 proteins against release 31 of the databank SWISS-PROT using Smith and Waterman algorithm with different substitution matrices and different gap penalties. More precisely, the sequences to test belong to yeast protein sequences considered as orphan (i.e. with no similarities with other sequences) when compared to SWISS-PROT with software such as BLAST or FASTA. The idea was to evaluate the capability of the Smith and Waterman algorithm to find *parents* for some orphan sequences.

The study has shown that some of the orphan sequences were found to share probably significant similarities with sequences of SWISS-PROT, while these similarities were undetected by Blastp or Fasta. One can conclude that the Smith and Waterman algorithm is complementary to the other algorithms.

5.2 Performance characteristics of SAMBA

Bank to bank comparison

A sequential version of the Smith and Waterman algorithm is available in the the package FASTA as the program SSEARCH. To evaluate the speed of SAMBA, we compared the time needed by SSEARCH on the orphan sequence application.

SSEARCH was run on a DEC-Alpha workstation with a 21064 150MHz microprocessor. Measurements indicate that the computation of a one matrix cell is achieved in $0.25\,\mu s$ – which corresponds to 4 MCOPS – when run as follows

```
ssearch -Q -b 20 -d 0 query_seq sprot31.fasta
```

Release 31 of SWISS-PROT contains exactly 43 470 sequences distributed on 15 335 248 amino acids. The 814 yeast sequences represent a total of 307 400 amino acids. The time t_{seq} for comparing this set of sequences against the bank is thus given by:

$$t_{seq} = 15\,335\,248 \times 307\,400 \times 0.25 \times 10^{-6} = 1\,178\,514\,s \quad .$$

This is equivalent to 327 hours (or 13 days and 15 hours) of non-stop computation. Knowing that this task should be repeated, at least, with three different substitution matrices and two different gap costs, such an experiment would have required nearly three months of intensive computation.

On SAMBA, the comparison of the 814 sequences against the bank is achieved in 1 hour and 45 minutes. The speed-up is thus 190 for this particular application. The time was measured using the UNIX command **time**. Thus, it is the total elapsed time, as it directly affects the user, including the time for reading the database from the disk and the time for re-computing on the workstation the exact score which could not be found by the array due to the 12-bit processor arithmetic (in that case the array indicates an overflow and the comparison is performed by the workstation). The bank to bank comparison, as illustrated by this example, fits completely the SAMBA functionalities, and the speed-up is therefore very high.

Scanning data bases

Other applications such as the scan of databases, that is to say, the comparison of one query sequence against one database, may also be performed very quickly, as we shall see.

Table 1 indicates the SSEARCH scan time (in second) on different workstations of the bank SWISS-PROT (release 31) with proteins of different length. It also indicates the speed-up when SAMBA is used instead. Obviously, the longer the

Query sequence length	10	30	100	300	1000	3000	10000
SAMBA	25	25	26	30	40	77	210
DEC-Alpha - 150 MHZ CPU	57	120	350	1041	3468	11510	38450
speed-up	*2.3*	*4.8*	*13.5*	*34.7*	*86.7*	*150*	*183*
SUN-SPARC 5 - 110 MHZ CPU	95	239	746	2215	7300	24269	80300
speed-up	*3.8*	*9.5*	*28.6*	*74*	*183*	*315*	*382*
DEC 5000/250 - 40 MHZ CPU	182	548	1407	4054	12920	41169	131193
speed-up	*7.3*	*22*	*54*	*135*	*323*	*534*	*625*

Table 1. SSEARCH scan time (in seconds) of SWISS-PROT 31 for various length of the query sequence on SAMBA and on different workstations. The speed-up compared to SAMBA is also reported. These results show clearly that the speed-up increases with the length of the query sequence.

query sequence, the better the speed-up. This is mainly due to the restricted bandwidth of the I/O disk system which prevents the array from being fed at its maximum rate. Indeed, a short sequence does not require, on the array, the computation to be split into several passes. Consequently, the array is fed at the disk access rate, which is generally much slower than the array feeding rate. On the other hand, the comparison of a long query sequence – i.e., a sequence whose length is much higher than the number of processors of the systolic array – requires the computation to be partitioned into several passes which re-use the data coming from the bank. The array average feeding rate is then substantially decreased and does not become a limitation factor.

6 Conclusion

The SAMBA prototype demonstrates that a dedicated full custom systolic array is very well suited for biological sequence comparison. This is a low cost solution compared to the implementations on massively parallel machines such as MPSEARCH [16]) or commercial available systems such as BIOCCELERATOR [4]).

The speed of SAMBA comes from the association of full custom components and FPGA devices. The 128 processors of the systolic array deliver a high computational power, while the FPGA components of the interface board manage efficiently the host/array communications, computation partitioning, data supply and result filtering.

The design effort for building SAMBA was less than one person/year, and is therefore very moderate. This was possible thanks to the use of synthesis tools for the design of the custom VLSI systolic array, and to the use of reconfigurable logic for the interface.

SAMBA is well suited to sequence analyses that require intensive computations, such as bank to bank comparison or scanning banks with long query sequences.

The SAMBA prototype is currently split into three distinct printed boards: the FPGA-interface board and two 64-processor boards. Today's integration density makes possible to fit SAMBA onto a single printed board. The chip could easily contains twice as many processors running at double speed, while the interface could be reduced to a few state of the art FPGA components associated with a few Mbytes of local memory. We estimate that the cost of SAMBA would roughly the same as the cost of a middle range workstation. In other words, the computational power of SAMBA is accessible to every laboratory which has to face a daily need of power for sequence analysis.

Acknowledgements

This work was partially funded by the French Research Group GREG (Groupement de Recherches et d'Etudes sur les Génomes) and the French Coordinated Research Program ANM (Architectures Nouvelles de Machines)

References

1. S.F. Altschul, W. Gish, W. Miller, E.W. Myers, and D.J. Lipman. – Basic local alignment search tool. – *J. Mol. Biol*, 215:403–410, 1990.
2. P. Bertin, D. Roncin, and J. Vuillemin. – Programmable active memories: a performance assessment. – In F. Meyer, B. Monien, and A.L. Rosenberg, editors, *Parallel Architectures and their efficient use*, pages 119–130. LNCS, Springer-Verlag, oct 1992.

3. E. Chow, T. Hunkapiller, and J. Peterson. – Biological Information Signal Processor. – In *ASAP*, pages 144–160, sep 1991.
4. Compugen. – The bioccelerator machine. – Israel, 1993.
5. O. Gotoh. – An Improved Algorithm for Matching Biological Sequences. – *J. Mol. Biol.*, 162:705–708, 1982.
6. P. Guerdoux-Jamet and D. Lavenier. – Systolic filter for fast DNA similarity search. – In *ASAP'95*, Strasbourg, July 1995.
7. P. Guerdoux-Jamet and J.L. Risler. – Searching for a family to orphan sequences with SAMBA, a parallel hardware dedicated to biological applications. – *Biochemistry*, 1996.
8. D. Lavenier. – An Integrated 2D Systolic Array for Spelling Correction. – *Integration : the VLSI journal*, 15:97–111, August 1993.
9. D. Lavenier. – Samba. Systolic Accelerators for Molecular Biological Applications. – Internal Publication 988, IRISA, Irisa, Campus de Beaulieu, 35042, Rennes-Cedex, France, April 1996.
10. D. Lopresty and al. – Building and using a highly parallel programmable logic array. – *Computers*, pages 81–89, jan 1991.
11. S.B. Needleman and C.D. Wunsh. – A General Method Applicable to the Search of Similarities in the Amino Acid Sequence of Two Proteins. – *J. Mol. Biol*, 48:443–453, 1970.
12. W. R. Pearson and D.J. Lipman. – Improved tools for biological sequence comparison. – *Proc. Natl. Acad. Sci.*, 85:3244–3248, 1988.
13. W.R. Pearson. – Searching protein sequence libraries: comparison of the sensitivity and selectivity of the smith and waterman and fasta algorithms. – *Genomics*, 11:635–650, 1991.
14. F. Raimbault and D. Lavenier. – ReLaCS for Systolic Programming. – In *ASAP'93*, October 1993.
15. T.F. Smith and M.S. Waterman. – Identification of common molecular subsequences. – *J. Mol. Biol*, 147:195–197, 1981.
16. S.S. Sturrock and J.F. Collins. – Mpsearch version 1.3. – Technical report, University of Edinburgh, Biocomputing Research Unit, 1993.
17. M.S. Waterman. – *Mathematical Methods for DNA Sequences.* – CRC Press, Inc, 1989.
18. C.T. White, R.K. Singh, P.B. Reintjes, J. Lampe, B.W. Erickson, W.D. Dettloff, V.L. Chi, and S.F. Altschul. – BioSCAN: A VLSI-Based System for Biosequence Analysis. – In *IEEE Int. Conf on Computer Design: VSLI in Computer and Processors*, pages 504–509. IEEE Computer Society Press, oct 1991.

Universal Computing

W F McColl

Oxford University

Abstract. The field of scalable computing is being redefined by the emergence of low cost parallel servers based on standard commodity microprocessors and off-the-shelf networking technologies. Many high performance applications will, in future, be carried out on such systems. Other applications, with very demanding communications requirements, will continue to be run on more specialised and expensive supercomputer systems. Over the next few years we will see the growth of a large and diverse global parallel software industry similar to that which currently exists for sequential computing. The main goal of that industry will be to produce scalable programs which, in addition to being fully portable, will offer high performance, in a predictable way, on any general purpose parallel architecture. The BSP model provides a discipline for the design of universal programs of this kind.

1 Introduction

Over the last thirty years or so we have seen the emergence and development of a huge global software industry. A large part of that industry is concerned with the production of portable applications software for the wide variety of sequential computers which are currently in use, from personal computers to large mainframes. The roots of this success story can be traced back to the work of von Neumann in the 1940s and to the developments in the areas of high level languages and compilers which followed on from that work in the 1950s and early 1960s. The model of a general purpose sequential computer proposed by von Neumann has served as the basic framework for almost all of the sequential computers which have been produced from the late 1940s to the present time. The stability which the model has provided has been crucial to the growth of the software industry over the years.

Since the earliest days of computing it has been clear that, sooner or later, sequential computing would be superseded by parallel computing. This has not yet happened, despite the availability of numerous parallel machines and the insatiable demand for increased computing power. For parallel computing to become the normal form of computing we require a model which can play a similar role to that which the von Neumann model has played in sequential

---

* This work was supported in part by ESPRIT Basic Research Project 9072 - GEPP-COM (Foundations of General Purpose Parallel Computing). Address: Programming Research Group, Oxford University Computing Laboratory, Wolfson Building, Parks Road, Oxford OX1 3QD, England. Email: mccoll@comlab.ox.ac.uk

computing. The emergence of such a model would stimulate the development of a new parallel software industry, and provide a clear focus for future hardware developments. For a model to succeed in this role it must offer three fundamental properties:

scalability - the performance of software and hardware must be scalable from a single processor to several hundreds or thousands of processors.
portability - software must be able to run unchanged, with high performance, on any general purpose parallel architecture.
predictability - the performance of software on different architectures must be predictable in a straightforward way.

It should also, ideally, permit the correctness of parallel programs to be determined in a way which is not much more difficult than for sequential programs.

During the last thirty years, a large number of parallel models have been proposed. They include: SIMD parallelism, synchronous message passing, logic programming, graph reduction, dataflow, and various forms of cache-based virtual shared memory. None of these approaches has, however, achieved all three main requirements.

The BSP model, which we describe in this paper, is quite different from many of the approaches which have been proposed in the past. It decouples the two fundamental aspects of parallel computation - communication and synchronisation. This decoupling is the key to achieving universal applicability across the whole range of parallel architectures. Recent research on BSP algorithms, architectures and languages has demonstrated convincingly that the BSP model can achieve all three of the requirements mentioned above.

2 Parallel Architectures

In recent years, the pursuit of higher performance from computers has forced the introduction of parallel computing in many areas, particularly in scientific and engineering applications and in database systems and transaction processing. The transition from sequential to parallel has not, however, been without its problems. On most of the early commercial parallel machines produced in the 1980s, scalable parallel performance could only be achieved by carefully exploiting the particular architectural details of the machine. Besides being extremely tedious and time consuming in many cases, this usually resulted in parallel applications software which could not be easily adapted to run on other machines. In a world of rapidly changing and diverse parallel architectures, this architecture dependence of the parallel software was a major weakness and seriously inhibited the growth of the field.

Over the last few years the situation has improved somewhat. For a variety of technological and economic reasons, the various classes of parallel computer in use (distributed memory machines, shared memory multiprocessors, clusters of workstations) have been steadily becoming more and more alike. The economic advantages of using standard commodity components has been a major factor in this convergence. Other influential factors have been the need to efficiently support a single address space on distributed memory machines for ease of programming, and the need to replace buses by networks to achieve scalability in shared memory multiprocessors. These various pressures have acted to produce a rapid and significant evolutionary restructuring of the parallel computing industry.

There is now a growing consensus that for a combination of technological, commercial, and software reasons, we will see a steady evolution over the next few years towards a "standard" architectural model for scalable parallel computing. It is likely to consist of a collection of (workstation-like) processor-memory pairs connected by a communications network which can be used to efficiently support a global address space. As with all such successful models, there will be plenty of scope for the use of different designs and technologies to realise such systems in different forms depending on the cost and performance requirements sought.

The simplest, cheapest, and probably the most common architectures will be based on clusters of personal computers. The Intel Pentium Pro microprocessor is an example of a high volume, commodity microprocessor for the personal computer market which provides hardware support for multiprocessing. Using such commodity microprocessor components, various companies are now producing very low cost multiprocessor systems for use as parallel servers. Commodity networking technologies such as ATM and new scalable software systems for communications will allow these to be assembled into clusters which will provide a very low cost option for many high performance commercial, scientific and internet applications.

At the other end of the spectrum, there will continue to be a small group of companies producing very large, very powerful and very expensive parallel supercomputer systems for those with applications which require those computing resources. The CRAY T3D is a good example of such a system. It is a very powerful distributed memory architecture based on the DEC Alpha microprocessor. In addition to offering high bandwidth global communications, it has several specialised hardware mechanisms which enable it to efficiently support parallel programs which execute in a global address space. The mechanisms include hardware barrier synchronisation and direct remote memory access. The latter permits each processor to get a value directly from any remote memory location in the machine, and to put a value directly in any remote memory location. This is done in a way which avoids the performance penalties normally incurred in executing such operations on a distributed memory architecture, due to processor synchronisation and other unnecessary activities in the low level systems software.

The process of architectural convergence which has been described brings with it the hope that we can, over the next few years, establish parallel computing as the standard method of computing, and begin to see the growth of a large and diverse global parallel software industry similar to that which currently exists for sequential computing. The main goal of that industry will be to produce scalable programs which, in addition to being fully portable, will offer high performance, in a predictable way, on any general purpose parallel architecture. The BSP model provides a discipline for the design of universal programs of this kind.

3 The BSP Model

3.1 Supersteps

In architectural terms, the BSP model is essentially the standard model described above. A *bulk synchronous parallel (BSP) computer* [9, 12] consists of a set of processor-memory pairs, a global communications network, and a mechanism for the efficient barrier synchronisation of the processors. A BSP computer operates in the following way. A computation consists of a sequence of parallel *supersteps*, where each superstep consists of a sequence of steps, followed by a barrier synchronisation at which point all data communications will be completed. During a superstep, each processor can perform a number of computation steps on values held locally at the start of the superstep, send and receive a number of messages, and handle various remote read and write requests.

Although we have described the BSP computer as an architectural model, one can also view bulk synchrony as a programming model or, indeed, as a kind of programming methodology. The essence of the BSP approach to parallel programming is the notion of the superstep, in which communication and synchronisation are completely decoupled. A "BSP program" is simply one which proceeds in phases, with the necessary global communications taking place between the phases. This approach to parallel programming is applicable to all kinds of parallel architecture: distributed memory architectures, shared memory multiprocessors, and networks of workstations. It provides a consistent, and very general, framework within which to develop portable parallel software for scalable parallel architectures.

Since communication and synchronisation are decoupled in a BSP program, the programmer does not have to worry about problems such as deadlock, which can occur with synchronous message passing. Debugging a BSP program is also made much easier by this decoupling. The barrier at the end of a superstep provides an appropriate breakpoint at which the global state of the parallel computation is well defined and can be interrogated. Debugging and reasoning about the correctness of a BSP program are, therefore, not much more difficult than for a sequential program.

3.2 Cost Modelling

If we define a time step to be the time required for a single local operation, i.e. a basic operation (such as addition or multiplication) on locally held data values, then the performance of any BSP computer can be characterised by three parameters: p = number of processors; l = number of time steps for barrier synchronisation; g = (total number of local operations performed by all processors in one second)/(total number of words delivered by the communications network in one second, in a situation of continuous traffic). There is also, of course, a fourth parameter s, the number of time steps per second. However, since the other parameters are normalised with respect to that one, it can be ignored in the design of algorithms and programs. The parameter g corresponds to the frequency with which non-local memory accesses can be made; in a machine with a higher value of g one must make non-local memory accesses less frequently. Any parallel computing system can be regarded as a BSP computer, and can be benchmarked accordingly to determine its BSP parameters l and g. The BSP model is therefore not prescriptive in terms of the physical architectures to which it applies. Every general purpose parallel architecture can be viewed by an algorithm designer or programmer as simply a point (p, l, g) in the space of all BSP machines.

The time for a superstep S is determined as follows. Let the work w be the maximum number of local computation steps executed by any processor during S. Let h_s be the maximum number of messages sent by any processor during S, and h_r be the maximum number of messages received by any processor during S. The time for S is then at most $w + g \cdot \max\{h_s, h_r\} + l$ steps. The total time required for a BSP computation is easily obtained by adding the times for each superstep to obtain an expression of the form $W + H \cdot g + S \cdot l$, where W, H, S will typically be functions of n and p. Analysing and predicting the cost of a BSP program is, therefore, no more difficult than analysing and predicting the cost of a sequential program.

4 Networks and Routing Methods

The use of the parameters l and g to characterise the communications performance of the BSP computer contrasts sharply with the way in which communications performance is described for most distributed memory architectures on the market today. A major feature of the BSP model is that it lifts considerations of network performance from the local level to the global level. We are thus no longer particularly interested in whether the network is a 2D array, a butterfly or a hypercube, or whether it is implemented in VLSI or in some optical technology. Our interest is in global parameters of the network, such as l and g, which describe its ability to support data communications in a uniformly efficient manner.

In the design and implementation of a BSP computer, the values of l and g which can be achieved will depend on the capabilities of the available technology and the amount of money that one is willing to spend on the communications network. In asymptotic terms, the values of l and g one might expect for various p processor networks are: ring $[l = O(p), g = O(p)]$, 2D array $[l = O(p^{1/2}), g = O(p^{1/2})]$, butterfly $[l = O(\log p), g = O(\log p)]$, hypercube $[l = O(\log p), g = O(1)]$. These asymptotic estimates are based entirely on the degree and diameter properties of the corresponding graph. In a practical setting, the channel capacities, routing methods used, VLSI implementation etc. would also have a significant impact on the actual values of l and g which could be achieved on a given machine. New optical technologies may offer the prospect of further reductions in the values of l and g which can be achieved, by providing a more efficient means of non-local communication than is possible with VLSI.

If we are interested in the problem of designing improved networks and routing methods which reduce g then perhaps the most obvious approach is to concentrate instead on the alternative problem of reducing l. This strategy is suggested by the following simple reasoning: If messages are in the network for a shorter period of time then, given that the network capacity is fixed, it will be possible to insert messages into the network more frequently. In studying BSP algorithms we see that, in many cases, the performance of a BSP computation is limited much more by g than by l. This suggests that in future, when designing networks and routing methods it may be advantageous to accept a significant increase in l in order to secure even a modest decrease in g. This raises a number of interesting architectural questions which have not yet been fully explored. It is also interesting to note that the characteristics of many modern communication systems (slow switching, very high bandwidth) may be very compatible with this alternative approach.

5 BSP Algorithms

In designing an efficient BSP algorithm or program for a problem which can be solved sequentially in time $T(n)$ our goal will, in general, be to produce an algorithm requiring total time $W + H \cdot g + S \cdot l$ where $W(n,p) = T(n)/p$, $H(n,p)$ and $S(n,p)$ are as small as possible, and the range of values for p is as large as possible. In many cases, this will require that we carefully arrange the data distribution so as to minimise the frequency of remote memory references. Another property of interest in BSP algorithm design is the space (or memory) efficiency of the computation. We will use $M(n,p)$ to denote the maximum number of values which any one processor has to store at any point during the computation. In this section we describe some BSP algorithms for matrix multiplication and discuss their time and space complexity.

Many static computations can be conveniently modelled by directed acyclic graphs, where each node corresponds to some simple operation, and the arcs correspond to inputs and outputs. Let C_n denote the directed acyclic graph which has n^3 nodes $v_{i,j,k}$, $0 \leq i,j,k < n$, and arcs from $v_{i,j,k}$ to $v_{i+1,j,k}$, $v_{i,j+1,k}$ and $v_{i,j,k+1}$ where those nodes exist. The graph C_n can be scheduled for a p processor BSP computer by partitioning C_n into $p^{3/2}$ subgraphs, each of which is isomorphic to $C_{n/p^{1/2}}$. Let $s = n/p^{1/2}$ and $C^{\hat{i},\hat{j},\hat{k}}$, $0 \leq \hat{i},\hat{j},\hat{k} < p^{1/2}$, denote the subset of s^3 nodes $v_{i,j,k}$ in C_n where $i \ div \ s = \hat{i}$, $j \ div \ s = \hat{j}$ and $k \ div \ s = \hat{k}$. The following simple schedule for C_n requires $3p^{1/2} - 2$ supersteps: During superstep $s(t)$, each $C^{\hat{i},\hat{j},\hat{k}}$ for which $\hat{i} + \hat{j} + \hat{k} = t$ is computed by one of the p processors, with no two of them computed by the same processor. ¿From the structure of C_n it is clear that during a superstep, each processor will receive n^2/p values, send n^2/p values, and perform $n^3/p^{3/2}$ computation steps. The total time required for the BSP implementation of any computation which can be modelled by C_n

is therefore at most $n^3/p + (n^2/p^{1/2}) \cdot g + p^{1/2} \cdot l$. [Throughout this paper we will omit the various small constant factors in such formulae.]

Consider the problem of multiplying two $n \times n$ dense matrices A, B to produce C. In [8] it is shown that the product C can be computed using the following set of definitions: For all $0 < i, j, k \leq n$,

$$a_{i,j,k} = a_{i,j-1,k}$$
$$b_{i,j,k} = b_{i-1,j,k}$$
$$c_{i,j,k} = c_{i,j,k-1} + (a_{i,j,k} \cdot b_{i,j,k})$$

where $a_{i,0,k} = a_{i,k}$, $b_{0,j,k} = b_{k,j}$ and $c_{i,j,0} = 0$. These definitions can be directly translated into a labelled version of the directed acyclic graph C_n. The BSP time complexity of matrix multiplication is therefore at most $n^3/p + (n^2/p^{1/2}) \cdot g + p^{1/2} \cdot l$. For the standard n^3 sequential matrix multiplication algorithm, this BSP schedule is optimal in terms of its computation cost $W(n,p) = n^3/p$. It is also optimal in terms of its space complexity. The matrices A and B can be uniformly distributed across the p processors, with each one holding an $n/p^{1/2} \times n/p^{1/2}$ block of the matrix. Given this uniform input distribution, we can schedule the reuse of memory locations in a straightforward way to ensure that no processor will be required to store more than n^2/p values in any superstep. Therefore we have $M(n,p) = n^2/p$.

In [9] a more efficient BSP realisation of the standard n^3 algorithm, due to McColl and Valiant, was described. Its BSP time complexity is $n^3/p + (n^2/p^{2/3}) \cdot g + l$. As in the previous schedule we begin with A, B distributed uniformly but arbitrarily across the p processors. At the end of the computation, the n^2 elements of C should also be distributed uniformly across the p processors. Let $s = n/p^{1/3}$ and $A[\hat{\imath}, \hat{\jmath}]$ denote the $s \times s$ submatrix of A consisting of the elements $a_{i,j}$ where $i \; div \; s = \hat{\imath}$ and $j \; div \; s = \hat{\jmath}$. Define $B[\hat{\imath}, \hat{\jmath}]$ and $C[\hat{\imath}, \hat{\jmath}]$ similarly. Then we have $C[\hat{\imath}, \hat{\jmath}] = \sum_{0 \leq \hat{k} < p^{1/3}} A[\hat{\imath}, \hat{k}] \cdot B[\hat{k}, \hat{\jmath}]$. Let $PROC(\hat{\imath}, \hat{\jmath}, \hat{k})$, $0 \leq \hat{\imath}, \hat{\jmath}, \hat{k} < p^{1/3}$, denote the p processors. In the first superstep each processor $PROC(\hat{\imath}, \hat{\jmath}, \hat{k})$ gets the set of elements in $A[\hat{\imath}, \hat{k}]$ and those in $B[\hat{k}, \hat{\jmath}]$. The cost of this step is $(n^2/p^{2/3}) \cdot g + l$. In the second superstep $PROC(\hat{\imath}, \hat{\jmath}, \hat{k})$ computes $A[\hat{\imath}, \hat{k}] \cdot B[\hat{k}, \hat{\jmath}]$ and sends each one of the $n^2/p^{2/3}$ resulting values to the unique processor which is responsible for computing the corresponding value in C. The cost of this step is $n^3/p + (n^2/p^{2/3}) \cdot g + l$. In the final superstep, each processor computes each of its n^2/p elements of C by adding the $p^{1/3}$ values received for that element. The cost of this step is $n^2/p^{2/3} + l$.

An input-output complexity argument can be used to show that for any BSP implementation of the standard n^3 sequential algorithm, if $W(n,p) = n^3/p$ then $H(n,p) \geq n^2/p^{2/3}$. This second BSP schedule for matrix multiplication therefore provides a realisation of the standard n^3 method which simultaneously achieves the optimal values for computation cost $W(n,p)$, communication cost $H(n,p)$ and synchronisation cost $S(n,p)$. The memory requirement of this algorithm is, however, inferior to the first algorithm. Its memory complexity $M(n,p)$ is $n^2/p^{2/3}$.

6 BSP Programming

In this section we briefly describe the main characteristics of BSP programming. We also compare the BSP approach with two other approaches to parallel programming - data parallelism and message passing.

As was noted earlier, the essence of the BSP approach to parallel programming is the notion of the superstep, in which communication and synchronisation are completely decoupled. A "BSP program" is simply one which proceeds in phases, with the necessary global communications taking place between the phases. One simple way of specifying the data communications in a BSP program is to use remote memory access primitives. The operation *put* deposits locally held data into a remote memory area on another process. The *get* operation reaches into the local memory of another process to copy values held there into a data structure in its own local memory. The put and get operations are both one-sided communication primitives. They do not require the active participation of the other process. In accordance with BSP superstep semantics, they are also both non-blocking. All put and get operations initiated during a superstep will be completed before the start of the next superstep.

Bulk synchronous remote memory access is a very convenient style of programming for BSP computations which can be statically analysed in a straightforward way. It is less convenient for computations where the volumes of data being communicated between supersteps is irregular and data dependent, and where the computation to be performed in a superstep depends on the quantity and form of data received at the start of that superstep. A more appropriate style of programming in such cases is bulk synchronous message passing. In bulk synchronous message passing, a non-blocking *send* operation is used to transfer values held locally into a buffer on the destination process. The values are guaranteed to be in the remote buffer before the start of the next superstep, and can be safely inspected and manipulated by the receiving process at that time.

6.1 Data Parallelism

Data parallelism is an important niche within the field of scalable parallel computing. A number of interesting programming languages and elegant theories have been developed in support of the data parallel style of programming. The BSP approach, as outlined in this paper, aims to offer a more flexible and general style of programming than is provided by data parallelism. The two approaches are not, however, incompatible in any fundamental way. For some applications, the increased flexibility provided by the BSP approach may not be required and the more limited data parallel style may offer a more attractive and productive setting for parallel software development, since it frees the programmer from having to provide an explicit specification of the various processor scheduling, communication and memory management aspects of the parallel computation. In such a situation, the BSP cost model can still play an extremely important role in terms of providing an analytic framework for performance prediction of the data parallel program.

6.2 Message Passing

Since the early 1980s, message passing has been the dominant programming approach in the area of parallel computing. In recent years, the PVM message passing library [3] has been widely implemented and widely used. In that respect, the goal of source code portability in parallel computing has already been achieved by PVM. What then, are the advantages of BSP programming, if any, over a message passing framework such as PVM? On shared memory architectures and on modern distributed memory architectures with powerful global communications, message passing models such as PVM are likely to be less efficient than the BSP model, where communication and synchronisation are decoupled. This will be especially true on those modern distributed memory architectures which have hardware support for direct remote memory access (or one-sided communications). PVM and all other message passing systems based on pairwise, rather than barrier, synchronisation also suffer from having no simple analytic cost model for performance prediction, and no simple means of examining the global state of a computation for debugging.

MPI [6] has been proposed as a new standard for those who want to write portable message passing programs in Fortran and C. At the level of point-to-point communications (send, receive etc.), MPI is similar to PVM, and the same comparisons apply. [The MPI standard is very general and appears to be very complex relative to the BSP model. However, one could use some carefully chosen combination of the various non-blocking communication primitives available in MPI, together with its barrier synchronisation primitive, to produce an MPI based BSP programming model.] At the higher level of collective communications, MPI provides support for various specialised communication patterns which arise frequently in message passing programs. These include broadcast, scatter, gather, total exchange, reduction, scan etc. These standard communication patterns also arise frequently in the design of BSP algorithms. It is important that such structured patterns can be conveniently expressed and efficiently implemented in any BSP programming language, in addition to the more primitive operations such as put and get which generate arbitrary and unstructured communication patterns.

Comparing it to PVM and MPI, it might be argued that the BSP approach offers (a) a simple programming discipline (based on supersteps) which makes it easier to determine the correctness of programs, (b) a cost model for performance analysis and prediction which is simpler and compositional, and (c) more efficient implementations on many machines.

7 BSP Programming Libraries

The Cray T3D SHMEM library provides primitives for direct remote memory access which can be used for BSP programming. The Oxford BSP Library [10] and the Oxford BSP Toolset [7] both provide a similar set of programming primitives for bulk synchronous remote memory access. The Green BSP Library

[4] provides a set of bulk synchronous message passing primitives based on fixed sized packets. Considerable experience of BSP programming has been gained through the use of these libraries. A number of major projects in universities and in industry are now using them to develop parallel applications, see e.g. [7, 11]. The experience gained in these practical projects would appear to confirm the various claims made above, regarding BSP and its advantages over message passing.

In December 1995, the inaugural meeting of BSP Worldwide was held in Oxford. BSP Worldwide is a new global organisation to coordinate research and development activities in the area of BSP computing, and to work on the standardisation of programming tools for the growing number of software developers who are now adopting this approach. It has recently launched an initiative to produce a standard low level BSP programming library for use with sequential languages such as Fortran and C. An initial proposal for this library is given in [5]. Its main characteristics are as follows:

– Single Program Multiple Data (SPMD) parallelism.
– Primitives for buffered and unbuffered bulk synchronous remote memory access.
– Primitives for buffered bulk synchronous message passing with tagged messages.
– Primitives for address registration to (a) support data communications into static, stack and heap allocated data structures, and (b) support BSP programming in heterogeneous environments.

A number of features which are semantically well defined, such as nested parallelism and subset synchronisation, have been excluded from the initial version of the library since they can have an adverse effect on the predictability of performance.

8 BSP and other models

In the 1980s we had a large number of different types of parallel architecture. With hindsight we now see that this variety was both unnecessary and unhelpful. It stifled the commercial development of parallel applications software by requiring that, to achieve acceptable performance, any such software had to be tailored to the specific architectural properties of the machine.

The BSP model provides software developers with an attractive escape route from the world of architecture dependent parallel software. The emergence of the model has also, as we have seen, coincided with the convergence of commercial parallel machine designs to a standard architectural form which is very compatible with the model. These developments have been enthusiastically welcomed by a rapidly growing community of software engineers charged with the task of producing scalable and portable parallel applications. However, while the parallel applications community has welcomed the approach, there is still a surprising degree of skepticism amongst parts of the computer science research community.

Many people seem to regard some of the claims made in support of the BSP approach as "too good to be true". This has led to a new proliferation, this time of models, in the 1990s.

Over the last few years a large number of variants of the BSP model, and alternatives to the BSP model, have been proposed for consideration. The number of such models probably greatly exceeds the number of different architectures that the parallel programmer had to contend with ten years ago! Most of the variants and alternatives have been developed in response to one or both of the following perceptions:

- Barrier synchronisation is an inflexible mechanism for structuring parallel programs.
- Some network characteristics other than l and g have to be taken into account in designing an efficient parallel program.

The only one of these alternative models which has generated any serious interest is the LogP model [2]. LogP differs from BSP in two ways:

- It uses a form of message passing based on pairwise synchronisation.
- It adds a third parameter representing the overhead involved in sending a message.

Over the last few years:

- Experience in developing software using the LogP model has shown that to analyse the correctness and efficiency of LogP programs it is often necessary, or at least convenient, to use barriers.
- Major improvements in network hardware and in communications software have greatly reduced the overhead associated with sending messages.

Given that LogP + barriers − overhead = BSP, the above points would suggest that the LogP model does not improve upon BSP in any significant way. However, it is natural to ask whether or not the more "flexible" LogP model can enable a designer to produce a more efficient algorithm or program for some particular problem, at the expense of a more complex style of programming. Recent results show that this is not the case. In [1] it is shown that the BSP and LogP models can efficiently simulate one another, and that there is therefore no loss of performance in using the more structured BSP programming style.

It is an interesting and important activity to look for alternative models of parallel computation which improve upon what we already have. In encouraging researchers to contribute to our understanding in this area, I would however make the following point. The only sensible way to evaluate an architecture independent model of parallel computation such as BSP, LogP, or the PRAM model, is to consider it in terms of *all* of its properties, i.e. (a) its usefulness as a basis for the design and analysis of algorithms, (b) its universal applicability across the whole range of general purpose architectures and its ability to provide efficient scalable performance on them, and (c) its support for the design of fully portable programs with analytically predictable performance. If we focus on only one of these at a time, then we will simply be replacing the zoo of parallel architectures which we had in the 1980s by a new zoo of parallel models in the 1990s. It seems likely that this viewpoint on the nature and role of models will gain more and more support as we move from the straightforward world of parallel algorithms to the much more complex world of parallel software systems.

References

1. G Bilardi, K T Herley, A Pietracaprina, G Pucci, and P Spirakis. BSP vs LogP. In *Proc. 8th Annual ACM Symposium on Parallel Algorithms and Architectures*, 1996. (to appear).
2. D Culler, R M Karp, D A Patterson, A Sahay, K E Schauser, E Santos, R Subramonian, and T von Eicken. LogP: Towards a realistic model of parallel computation. In *Proc. 4th ACM SIGPLAN Symposium on Principles and Practice of Parallel Programming*, pages 1–12, May 1993.
3. A Geist, A Beguelin, J Dongarra, W Jiang, R Manchek, and V Sunderam. *PVM: Parallel Virtual Machine - A Users' Guide and Tutorial for Networked Parallel Computing*. MIT Press, Cambridge, MA, 1994.
4. M Goudreau, K Lang, S Rao, T Suel, and T Thanasis. Towards efficiency and portability: Programming with the BSP model. In *Proc. 8th Annual ACM Symposium on Parallel Algorithms and Architectures*, 1996. (to appear).
5. M W Goudreau, J M D Hill, K Lang, W F McColl, S B Rao, D C Stefanescu, T Suel, and T Thanasis. A proposal for the BSP Worldwide Standard Library (preliminary version). Technical report, available via BSP Worldwide home page http://www.bsp-worldwide.org/, April 1996.
6. W Gropp, E Lusk, and A Skjellum. *Using MPI: Portable Parallel Programming with the Message-Passing Interface*. MIT Press, Cambridge, MA, 1994.
7. J M D Hill, P I Crumpton, and D A Burgess. The theory, practice, and a tool for BSP performance prediction applied to a CFD application. Technical Report PRG-TR-4-1996, Oxford University Computing Laboratory, 1996. To appear in Proc. Euro-Par '96.
8. W F McColl. Special purpose parallel computing. In A M Gibbons and P Spirakis, editors, *Lectures on Parallel Computation. Proc. 1991 ALCOM Spring School on Parallel Computation*, volume 4 of *Cambridge International Series on Parallel Computation*, pages 261–336. Cambridge University Press, Cambridge, UK, 1993.
9. W F McColl. Scalable computing. In J van Leeuwen, editor, *Computer Science Today: Recent Trends and Developments. LNCS Volume 1000*, pages 46–61. Springer-Verlag, 1995.
10. R Miller. A library for bulk-synchronous parallel programming. In *Proc. British Computer Society Parallel Processing Specialist Group workshop on General Purpose Parallel Computing*, December 1993.
11. M Nibhanupudi, C Norton, and B Szymanski. Plasma simulation on networks of workstations using the bulk synchronous parallel model. In *Proceedings of the International Conference on Parallel and Distributed Processing Techniques and Applications*, Athens, GA, November 1995.
12. L G Valiant. A bridging model for parallel computation. *Communications of the ACM*, 33(8):103–111, 1990.

Dynamic Load Balancing in Parallel Database Systems

Erhard Rahm

University of Leipzig, Institute of Computer Science
E-mail: rahm@informatik.uni-leipzig.de

Abstract

Dynamic load balancing is a prerequisite for effectively utilizing large parallel database systems. Load balancing at different levels is required in particular for assigning transactions and queries as well as subqueries to nodes. Special problems are posed by the need to support both inter-transaction/query as well as intra-transaction/query parallelism due to conflicting performance requirements. We compare the major architectures for parallel database systems, Shared Nothing and Shared Disk, with respect to their load balancing potential. For this purpose, we focus on parallel scan and join processing in multi-user mode. It turns out that both the degree of query parallelism as well as the processor allocation should be determined in a coordinated way and based on the current utilization of critical resource types, in particular CPU and memory.

1 Introduction

A significant trend in the commercial database field is the increasing support for parallel database processing [DG92, Va93]. This trend is both technology-driven and application-driven. Technology supports large amounts of inexpensive processing capacity by providing "super servers" [Gr95] consisting of tens to hundreds of fast standard microprocessors interconnected by a scalable high-speed interconnection network. The aggregate memory is in the order of tens to hundreds of gigabytes, while databases of multiple terabytes are kept online within a parallel disk subsystem. New application areas requiring parallel database systems for processing massive amounts of data and complex queries include data mining and warehousing, digital libraries, new multimedia services like video on demand, geographic information systems, etc.. Even traditional DBMS applications increasingly face the need of parallel query processing due to growing database sizes and query complexity [MPTW94]. In addition, high transaction rates must be supported for standard OLTP applications.

The effective use of super-servers for database processing poses many implementation challenges that are largely unsolved in current products [Se93, Gr95]. One key problem is the effective use of intra-query parallelism in multi-user mode, i.e., when complex queries are executed concurrently with other complex queries and OLTP transactions. Multi-user mode (inter-query/inter-transaction parallelism) is mandatory to achieve acceptable throughput and cost-effectiveness, in particular for super-servers where a high number of processors must effectively be utilized. While proposed algorithms for parallel query processing also work in multi-user mode, their performance may be substantially lower than in single-user mode. This is because multi-user mode inevitably leads to data and resource contention that can significantly limit the attainable response time improvements due to intra-query parallelism.

Data contention problems may be solved by a multiversion concurrency control scheme which guarantees that read-only queries do not suffer from or cause any lock conflicts [CM86, BC92]. Increased resource contention, on the other hand, is unavoidable since complex queries pose high CPU, memory and disk bandwidth requirements which can result in significant delays for concurrently executing transactions. Furthermore, resource contention can be aggravated by the communication overhead associated with parallel query processing. In order to limit and control resource contention in multi-user mode, dynamic strategies for resource allocation and load balancing become necessary. In particular, the workload must be allocated among the processing nodes such that the capacity of different processing nodes be evenly utilized.

We first discuss the major forms of workload allocation and dynamic load balancing for database processing. Section 3 introduces the major architectures for parallel database processing, in particular Shared Nothing and Shared Disk systems. Their potential for dynamic load balancing is then evaluated for parallel relational database processing, in particular with respect to the two most important operators: scan (Section 4) and join (Section 5). In Section 6 we discuss additional considerations for supporting mixed OLTP/query workloads, in particular transaction routing.

2 Workload allocation

The general term "workload allocation" refers to the assignment of workload requests (processing steps) to physical or logical resources (processors, processes, memory, etc.). In this sense it corresponds to the term "resource allocation" which only expresses another perspective of the allocation problem. Depending on the workload or resource type special allocation problems can be considered, e.g., transaction and query allocation or processor and memory allocation. *Load balancing* refers to workload allocation in distributed systems where workload requests must be distributed among several processing nodes.

Heterogeneous database workloads consisting of OLTP transactions of different types as well as complex decision support queries pose special resource management problems even in the central case. One problem is to find a memory allocation that avoids that large queries monopolize the available buffer space thus causing unacceptable hit ratios for concurrent OLTP transactions. This problem can be addressed by giving higher priority to OLTP transactions and by using disjoint buffer areas for OLTP and large queries where the relative buffer sizes are dynamically controlled depending on the current workload. Such schemes have been proposed in [ZG90, PCL93, DG94] with respect to hash join queries. In [MD93, BMCL94], heuristics for dynamically controlling the number of concurrent queries are proposed in order to limit memory contention. Some commercial DBMS already support such dynamic memory allocation schemes, e.g., Tandem NonStop SQL and Informix.

For parallel database processing, load balancing is the major resource allocation problem in order to effectively utilize all available resources. Load balancing can be applied for different workload granularities depending on the level of parallelism. At the highest level, we have inter-transaction and *inter-query parallelism* with a concurrent execution of independent transactions and queries (multi-user mode). The corresponding load balancing is concerned with distributing transactions and queries among processing nodes (transaction and query routing). *Intra-query parallelism* requires additional forms of load balancing for assigning subqueries to nodes. Several forms of intra-query parallelism can be distinguished in this context, namely inter-operator and intra-operator as well as pipeline and data parallelism [DG92]. Correspondingly, load balancing is necessary for operators (e.g., scan, join, sort) and sub-operators. In all cases, load balancing should be dynamic, that is the assignment decisions should be based on the current system utilization at runtime. Otherwise an even utilization of all nodes cannot be achieved due to typically high variations in the load composition (load surges, etc.) and system state.

Pipeline parallelism is typically used for inter-operator parallelism in order to overlap the execution of several operators within a query. Data parallelism, on the other hand, is applicable for both inter- and intra-operator parallelism and requires a data partitioning so that different (sub) operators can concurrently process disjoint sets of data. While both data and pipeline parallelism are needed, pipeline parallelism is generally considered less effective for reducing query response times [DG92]. This is because typically only comparatively few (<= 10) operators can be used within a pipeline because the total number of operators is mostly small and because there are blocking operators like sort that require the total input data before they can produce their output.

Load balancing is difficult to achieve for pipeline parallelism due to precedence dependencies between operators and because individual operators can substantially vary in their resource requirements and execution times. Furthermore, the size of temporary results cannot be predicted very well [Gr93]. A careful flow control must be exercised at runtime in order to avoid that the input data generated by producer operators cannot be processed fast enough by consumer operators. Otherwise, the consumers' input data would have to be stored within temporary files introducing a potentially high amount of disk I/O.

For these reasons our further discussions on load balancing will concentrate on data parallelism allowing much higher degrees of intra-query parallelism than pipeline parallelism. In order to support both high throughput for OLTP and short response times for complex queries it is important to dynamically determine the degree of intra-query parallelism as well as which processing nodes should process the subqueries. As we will see, the implementation of such an approach largely depends on the respective architecture and query type.

3 Architecture of Parallel Database Systems

Parallel database systems are typically based on multiple standard microprocessors interconnected by a local high-speed network. Effective support for inter- and intra-transaction parallelism requires both adequate use of I/O parallelism and processing parallelism. I/O parallelism must be supported by an allocation of the database across multiple disks (declustering), either within conventional disk farms or disk arrays [PGK88, CLG94]. Declustering supports intra-query parallelism by reading and writing large amounts of data processed by a single query in parallel from or to multiple disks. Inter-transaction parallelism is supported because independent I/O requests on different disks can be served in parallel.

With respect to processing parallelism, there are three major architectures for parallel database systems [DG92, Va93]:

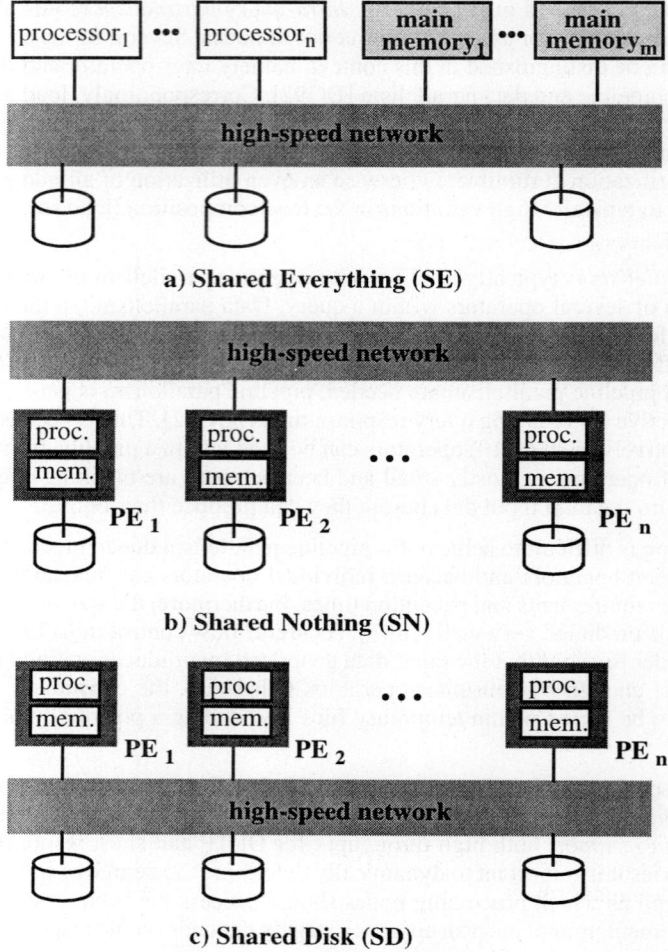

Figure 1: Architecture of parallel database systems

- *Shared Everything* (SE, Fig. 1a) refers to the use of multiprocessors for database processing. In this case, we have a tightly coupled system where all processors share a common main memory as well as peripheral devices (terminals, disks). There is only a single copy of the DBMS code that can be executed in multiple processes to utilize all processors. This approach is also referred to as Symmetric Multiprocessing (SMP).
- *Shared Nothing (SN,* Fig. 1b) systems consist of multiple autonomous processing elements (PE) each owning a private main memory and running separate copies of the operating system, DBMS and other software. Inter-processor communication takes place by means of message passing (loose coupling). A PE can consist of 1 or more processors, i.e., each node in a SN system may be a multiprocessor. The database is partitioned among the PEs so that each DBMS instance can di-

rectly access only data from the local partition. Access to non-local data requires a distributed query and transaction execution.
- Similar to SN, *Shared Disk (SD,* Fig. 1c) systems consist of multiple loosely coupled PE. However, the database is not partitioned but shared among the PE so that each DBMS instance has direct access to any data object. This assumes that each node can access any disk.

All three architectures are supported by commercial DBMS for both inter- and intra-transaction parallelism. Virtually all commercial DBMS are able to utilize multiprocessors (SE systems) for inter-transaction parallelism; support for intra-query processing is being added to most DBMS. Well-known SN systems supporting intra-query parallelism include Tandem NonStop SQL and ATT/Teradata's database machine; newer implementations are Sybase MPP, DB2/6000 Parallel Edition and Informix XPS (eXtended Parallel Server). Parallel SD implementations include Oracle Parallel Server and IBM's database systems (IMS, DB2) for parallel sysplex configurations. Oracle Parallel Server is available on many platforms, in particular on parallel computers (e.g., nCUBE) as well as on most "cluster" architectures (VaxCluster, Sequent, Pyramid, Encore, Sun, etc.).

SE systems have the advantage that shared memory supports efficient cooperation and synchronization between DBMS processes. Furthermore, effective load balancing is supported by the operating system that automatically assigns the next ready process/subquery to the next free CPU. These advantages are especially valuable for parallel query processing leading to increased communication and load balancing requirements to start/terminate and coordinate multiple subqueries. Furthermore, large intermediate results can efficiently be exchanged between subqueries. Several studies addressed dynamic load balancing for parallel query processing in SE systems [HSIT91, Om91, Ho92, LT92].

On the other hand, there are significant availability problems since the shared memory reduces failure isolation between processors, and since there is only a single copy of system software like the operating system or the DBMS [Ki84]. Furthermore, scalability is limited because the shared memory can introduce performance bottlenecks. Consequently, the number of processors is quite low in current SE systems (≤ 30). Due to these problems, SN and SD are generally considered as more appropriate to meet high-performance and high-availability requirements [DG92, MPTW94].

From a hardware point of view, SN systems appear particularly attractive. They allocate each disk drive to one particular PE and interconnect all PE by a local network, which is feasible with standard hardware at little cost. Furthermore, a large number of PE can be interconnected in this way because there are no shared resources (other than the network). SD systems, on the other hand, require an interconnection between all PE and disk drives (Fig. 1c). Because all I/O requests (page transfers) have to go over this network, an extremely fast and scalable (multi-stage) interconnection network is needed that may be much more expensive than the network of SN systems. Furthermore, SD may face an increased potential of performance bottlenecks in the network and disk subsystem.

On the other hand, with current fiber-optic interconnection technology it appears unlikely that high bandwidth requirements pose a major problem. Furthermore, high-performance SN systems also need a very fast and scalable network in particular for a larger number of PE and for parallel processing of complex queries that often requires a dynamic redistribution of large amounts of data. In addition, even in SN systems it is typically necessary to interconnect each disk drive to at least two PE for fault tolerance reasons thus increasing hardware cost.

Hence, we conclude that hardware-related aspects are less significant when comparing SN and SD than software-related aspects, in particular with respect to database processing[1]. This is also because providing powerful hardware, e.g. large numbers of PE, does by no means imply that this hardware can effectively be utilized for database processing. A key prerequisite for achieving this goal is dynamic load balancing. As we will see, the potential for such a load balancing differs significantly between SN and SD DBMS. Apart from the feasible approaches for parallel query processing and dynamic load balancing, SN and SD systems differ in additional areas like transaction management (global concurrency control and global logging for SD; distributed commit for SN) and the need for coherency control (for SD). Efficient solutions for these problems have been proposed [ÖV91, GR93, MN91, Ra93] but are beyond the scope of this paper.

4 Parallel scan processing

Scan is the simplest and most common relational operator. If predicate evaluation cannot be supported by an index, a complete *relation scan* is necessary where each tuple (record) of the relation must be read and processed. An *index scan* accesses tuples via an index (typically a B+ tree) and restricts processing to a subset of the tuples; in the extreme case, no tuple or only one tuple needs to be accessed (e.g., exact-match query on unique attribute). Parallel scan processing requires a declustering of the relation and index structures across several disks in order to allow for I/O parallelism.

We first analyze parallel scan processing for SN; SD scan processing is discussed afterwards.

Shared Nothing

In SN systems, the database partitioning among PE implies a corresponding data allocation to disks because each disk is exclusively assigned to one PE[2]. Database partitioning is typically based on a horizontal (tuple-wise) declustering of relations defined by a hash or range function on a *partitioning attribute* (e.g., primary key) [DG92]. Indices are also partitioned with the relation so that each PE holds a (sub-)index for the local records. Parallelizing a scan operation is straight-forward and determined by the database allocation. For hash and range partitioning, exact-match queries on the partitioning attribute can be restricted to a single processor; range partitioning also allows restricting the number of nodes for range queries on the partitioning attribute. However, all other scan queries must be processed by all data processors, i.e. all PE holding a partition of the respective relation.

The performance of parallel scan processing is thus very dependent on the degree of declustering as it coincides with the degree of scan parallelism in many cases. To evaluate

---

1. This is also underlined by the fact that there are SN database systems (e.g., Teradata) running on SD hardware platforms and vice versa. The former case is easily feasible by not utilizing the accessibility of all disks during normal processing but restricting each DBMS/node to a subset of the disks. The latter case is used by Oracle on SN platforms like IBM SP2 and is made feasible by the operating system implementing a "virtual shared disk" environment where the distinction between local and remote I/O is transparent to the DBMS.
2. There may be multiple disks per PE.

the impact of different degrees of parallelism in both single-user and multi-user mode, we have performed several simulation experiments with a detailed simulator of a SN database system[3]. In Fig. 2, we show the average response times and speedup values of different scan queries on a relation of 1 million tuples in single-user mode. The degree of declustering and scan parallelism (P) is varied between 1 and 64.

Fig. 2 shows that parallel processing of a relation scan is very effective in single-user mode and that a linear speedup could be obtained (the sequential processing time of about 30 minutes is reduced by a factor 60 for 64 PE). Still, response time for 64 PE was higher than in the case of a selective index scan for which only 0.1% of the tuples qualify. This illustrates that the use of an index may be more effective than employing parallel processing in order to improve response time (of course, not all queries can be supported by an adequate index). Parallel processing of index scans also improves response time but to a much lower degree than for relation scans. This is because the number of records to be processed for selective index scans is much lower than for a relation scans. Furthermore, the actual work (number of records) per processor is reduced for growing degrees of parallelism while the overhead for starting and terminating the subqueries increases proportionally to P.

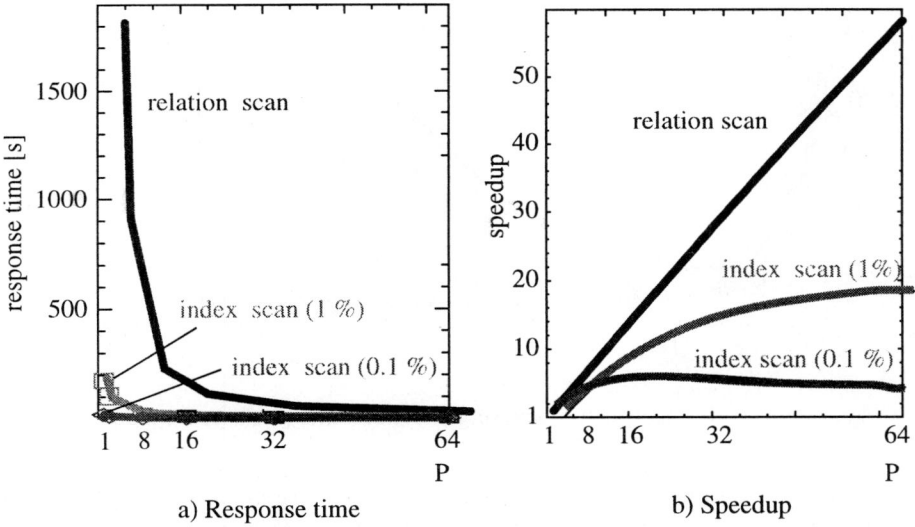

Figure 2:   Parallel scan processing in SN systems (single-user mode)

A major implication of this observation is that different scan queries on the same relation have their response time minimum for different degrees of parallelism. As can be seen from Fig. 2b, a relation scan may be best processed by 64 PE while the index scans have their optimum for smaller degrees of parallelism. The optimal degree of scan parallelism may be computed by an analytical model that considers the tradeoffs between the actual amount of work (determined by factors like relation size, query selectivity, index usage etc.) and the overhead introduced by intra-query parallelism [Gh90, WFA92, Ma95]. Unfortunately, SN requires to statically choose the degree of declus-

---

3. Details on the simulation system can be found in [MR92, RM93].

tering so that there is no way to choose the degree of scan parallelism dependent on the scan type. As a result, the actual degree of declustering must be a compromise value for an average load profile, e.g., 30 for our example. However, this implies suboptimal performance for the different scan types, in particular a overly high communication overhead for index scans and an insufficient degree of parallelism for relation scans[4].

For scan processing in multi-user mode, we consider a homogeneous workload with multiple index scan queries. Because we want to linearly increase throughput with the number of PE we increase the arrival rate of our scan query proportionally with the system size. Fig. 3 shows the resulting response times for parallel scan processing and different arrival rates (in Queries Per Second per PE, QPS/PE). The main observation is that the optimal degree of scan parallelism depends on the arrival rate and thus on the current system (CPU) utilization; it becomes the lower the higher the system is utilized. This is because the communication overhead associated with a higher degree of parallelism is less affordable under high CPU utilization (high arrival rates).

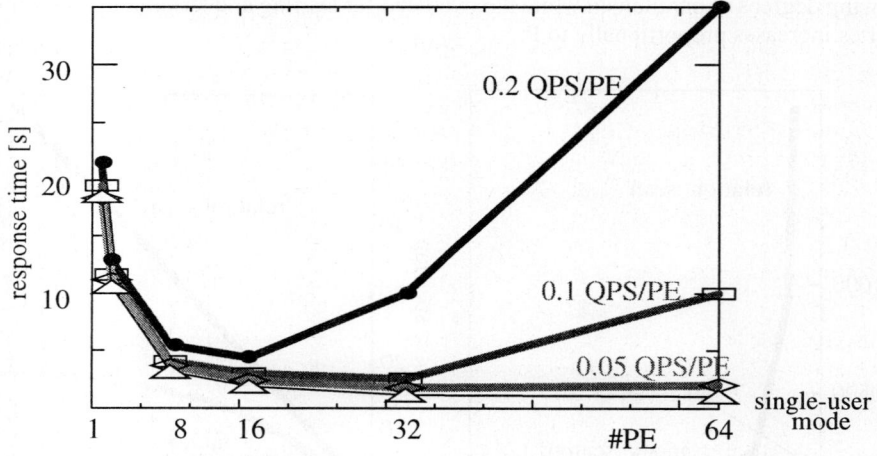

Figure 3: Parallel scan processing in SN systems (multi-user mode)

Hence, it would be desirable to choose the degree of scan parallelism according to the current system state. Unfortunately, this is not possible for SN as the data allocation statically determines the degree of scan parallelism. Furthermore, the scan processors themselves are also determined by the database allocation so that there is no support at all for dynamic load balancing.

Shared Disk

In SD systems each node has access to the entire database on disk. Hence, scan operations on a relation can be performed by any number of nodes. This gives the required flexibility to choose the degree of scan parallelism according to the query type and the

---

4. Some SN systems (e.g. Teradata) simply use a *full declustering* so that each relation is partitioned across all PE. Such an approach is clearly too simple and introduces a high communication overhead, in particular for small relations and index scans [CABK88]. The high overhead is especially harmful in multi-user mode [RM93].

current system utilization. For example, index scans on any attribute may be performed by a single processor thereby minimizing communication overhead. This would especially be appropriate for exact-match and selective range queries, and supports high throughput. For relation scans, on the other hand, a high degree of processing parallelism can be employed to utilize intra-query parallelism to reduce response time. Furthermore, under high system utilization a smaller degree of scan parallelism can be chosen to limit the communication overhead and support high throughput. A high potential for dynamic load balancing is also supported by the fact that all PE are eligible for scan processing allowing a dynamic decision about where a scan should be processed. For instance, a scan may be allocated to a set of processors with low CPU utilization in order to avoid interference with concurrent transactions on other nodes.

The load balancing potential of the SD architecture must be supported by an adequate data allocation that avoids disk contention between parallel subqueries of the same query. While in SN systems the data is partitioned among PE, SD (and SE) only requires a data declustering across multiple disks. Such a data allocation merely prescribes the maximal degree of I/O parallelism while the degree of scan parallelism can be chosen smaller. Parallel processing of relation scans is easily supported by choosing a degree of declustering D that is high enough for providing sufficiently short response times in single-user mode. Parallel relation scans are possible without disk contention for different degrees of parallelism P by choosing P such that $P * k = D$, where k is the number of disks to be processed per subquery. For instance, if we have D=100 we may process a relation scan with P = 1, 2, 4, 5, 10, 20, 25, 50 or 100 subqueries without disk contention between subqueries. Furthermore, each subquery processes the same number of disks (k) so that data skew can largely be avoided for equally sized partitions. CPU contention between subqueries is also avoided if each subquery is assigned to a different processor which is feasible as long as P does not exceed the number of processors.

Selective index scans returning only a few records are best processed sequentially which is feasible for SD with minimal communication overhead. Parallel index scans, on the other hand, may lead to disk contention between subqueries on both index and actual data, even when the shared index is declustered across several disks. The performance study [RS95] showed that this problem primarily exists for clustered index scans which should therefore be processed sequentially (unless the selectivity is so high that the data of multiple disks needs to be accessed). Parallel non-clustered index scans, on the other hand, did not suffer from a significant disk contention in the case of larger degrees of declustering.

Multi-user mode typically not only leads to CPU but also to disk contention for both SN and SD architectures. In [RS95] it was found that SD is able to reduce the level of disk contention by using smaller degrees of scan parallelism under high disk utilization. SN does not provide such a flexibility.

5 Parallel join processing

Parallel (equi-)join processing typically consists of a parallel scan phase and a parallel join phase. During the scan phase, the scan processors read the input relations from disk and perform selections on them. The scan output is then redistributed among multiple join processors performing the join phase using any sequential algorithm (e.g., hash join or sort-merge). Finally, the local join results are merged at a designated node. Data redistribution between scan and join processors is performed by applying a partitioning function (hash or range) on the join attribute. This ensures that matching tuples of both input relations arrive at the same join processor. The advantage of such a scheme is that there is a high potential for dynamic load balancing even for SN. This is because the number of join processors as well as the choice of these processors can be based on the current load situation, similarly as for parallel scan processing in SD systems. On the other hand, the communication overhead for data redistribution can be substantial.

In the following, we outline some general tradeoffs to consider for dynamic load balancing and such a parallel join processing. Afterwards, we briefly discuss load balancing in the presence of data skew as well as some other join strategies with reduced redistribution overhead.

General tradeoffs

Similar to parallel scan processing, it is possible to determine the optimal degree of intra-query processing (i.e. the optimal number of join processors) in single-user mode by means of an analytical model [Ma95]. In addition to the mentioned tradeoff between actual (CPU) work per subquery and communication overhead there are additional factors that need to be considered for determining the optimal number of join processors. In particular, the overhead for redistributing the data between scan and join processors increases with the number of join processors. Furthermore, the I/O overhead for join processing is very much dependent on the aggregate memory of the join processors that increases with the degree of join parallelism. For instance, if a hash join is used for local join processing an optimal I/O performance is achieved if the smaller join input can be completely kept in the join processors' memory [Gr93]. If less memory is available, additional I/O is necessary to keep the join input in temporary disk files at the join processors leading to a substantial response time degradation. Hence, the optimal degree of join parallelism in single-user mode is at least as high as required to avoid temporary file I/O (or, if this is unachievable, the total number of PE). Since all PE are lightly loaded in single-user mode, selection of the join processors is no problem (e.g., random selection is sufficient).

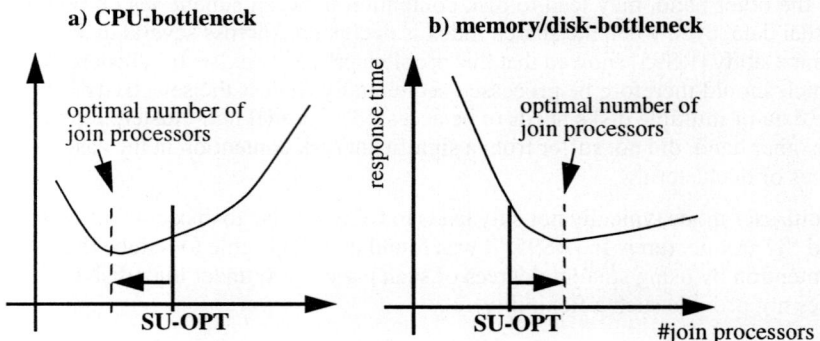

Figure 4:   Optimal number of join processors in multi-user mode

The studies [RM93, RM95] showed however, that this changes significantly in multi-user mode. Similar to parallel scan processing, the optimal number of join processors is lower than in single-user mode under high CPU utilization (Fig. 4a). Moreover, the optimal degree of join parallelism is generally the lower the higher the system is utilized

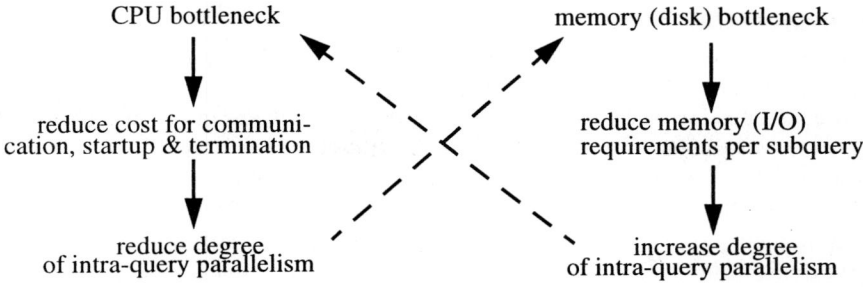

Figure 5:   Tradeoffs in dynamic load balancing with multiple bottlenecks

due to the high communication overhead associated with many join processors. Furthermore, the least utilized CPUs should be selected for join processing [RM93]. However, if response times are largely dominated by temporary file I/O, i.e., if we have memory or disk bottlenecks, it is generally advisable to choose more join processors to obtain more memory and thus to reduce the amount of temporary I/O [RM95]. As a result, under high memory (disk) utilization the optimal degree of join parallelism is typically higher than in single-user mode (Fig. 4b). These tradeoffs are summarized in Fig. 5 which shows that the degree of join parallelism in multi-user mode must be chosen dynamically based on the current memory, disk and CPU utilization. Similarly, selection of the join processors must consider the current CPU and memory utilization of the PE.

Policies

Specific approaches for such a dynamic load balancing considering multiple bottleneck types have been proposed and evaluated in [RM95]. A major finding is that the dynamic decisions should be based on the actual resource utilization at the individual nodes rather than on average utilization levels (since there may be large differences in resource utilization). Furthermore, it is important to use integrated policies for drawing the two control decisions (degree of parallelism, processor allocation) in a coordinated way. For example, a policy called MIN-IO-SUOPT proved to be effective in the presence of memory bottlenecks. Based on the current memory availability it determines the degree of join parallelism p so that the amount of temporary file I/O is minimized. In the case of multiple configurations avoiding temporary file I/O the number of processors closest to the single-user optimum is selected (in order to ensure a sufficiently high degree of parallelism). The subjoins are assigned to the p nodes with the most available memory.

In [MD95] an algorithm called RateMatch for dynamically determining the number of join processors is presented. This scheme is based on the observation that the size of the join input is less significant for finding the optimal number of join processors than the rate at which the scan processors generate the join input. Thus the scheme tries to determine the number of join processors such that their aggregate join processing rate matches the rate at which the join input is provided by the scan processors. However, RateMatch is an isolated scheme that uses an independent algorithm for selecting the join processors. Furthermore, it only considers average values for the current resource utilization rather than node-specific utilization information.

Treatment of Data Skew

For data parallelism to be effective the underlying data partitioning must ensure about equally sized input partitions for subqueries. In general, this is difficult to achieve due to non-uniform value distributions for the partitioning attributes or because of nonuniform query selectivities. The resulting data skew can result in large differences in the execution times of subqueries thereby reducing the effectiveness of intra-query parallelism (since response times are determined by the slowest subqueries).

For parallel join processing based on a dynamic redistribution of the join input it is feasible to dynamically check the size of the scan output and to extend the redistribution scheme in order to generate about equally sized join inputs in the presence of data skew. Approaches for such a dynamic load balancing have been proposed in [WDJ91, DNSS92, HLY93, WDYT94, HLH95]. However, these studies assumed single-user mode corresponding to a best-case situation with little or no resource contention. Hence, only intra-query load balancing is supported and the effectiveness of the proposals in multi-user mode must be questioned. In fact, the overhead associated with the proposed load balancing schemes is likely to be a major problem in multi-user mode. A more appropriate approach to reduce the skew problem in multi-user mode may be to avoid the expensive generation of equally-sized subjoins but to use information on the current system utilization to select the join processors dependent on the size of the subjoins (by assigning larger subjoins to less loaded nodes, etc.).

Limiting the redistribution overhead

Dynamically redistributing both input relations supports dynamic load balancing, but incurs a high communication overhead for large relations. In SN systems, this overhead is reduced if at least one of the join inputs is already declustered on the join attribute. In this case, the data processors of this relation act as join processors and only the second relation needs to be redistributed. Data redistribution is completely avoided if both relations are equally partitioned on the join attributes and allocated to the same set of PE. Unfortunately, these special cases leave no room for dynamic load balancing because the degree of join parallelism and the join processors are statically determined by the database allocation, similar as for scan processing. Still, in many cases the communication savings are more significant than the lost load balancing potential [RM93].

SD systems can also avoid the redistribution overhead if one or both join inputs are physically declustered across several disks by a logical partitioning on the join attribute. In this case, however, a large potential for dynamic load balancing remains because each PE may be selected as join processor. Furthermore, the declustering still leaves several choices for the degree of join parallelism without causing disk contention between subqueries (similar to parallel scan processing). For instance assume two relations declustered across 50 disks by using the same value partitioning (range or hash) on the join attributes. Then there may be 1, 2, 5, 10, 25 or 50 scan and join processors working in parallel and accessing disjoint sets of disks. In this case, the redistribution overhead is completely avoided while a high potential for dynamic load balancing is preserved.

6 Mixed workloads

In the previous sections we have already shown the need for dynamic load balancing for parallel query processing in multi-user mode, i.e. with both intra-query and inter-query parallelism. Similar requirements are posed for mixed workloads consisting of simple OLTP transactions and complex queries. Since efficient processing of OLTP transactions has typically highest priority, the need for limiting resource contention due to parallel query processing is aggravated. As we have seen SD offers significant advantages over SN to achieve this goal:

- SN requires definition of a (static) database allocation for an "average" load profile that must be a compromise between the different requirements for OLTP and complex queries. This inevitably leads to sub-optimal performance for both workload types and does not support dynamic load balancing. In particular, complex queries have to be restricted to fewer nodes than desirable to limit the communication overhead so that response times may not sufficiently be reduced. On the other hand, OLTP transactions cannot be confined to a single node in many cases thereby causing extra communication overhead and lowering throughput. In both cases, the sub-optimal performance must be accepted even if only one of the two workload types is temporarily active.

- In SD systems, declustering of data across multiple disks does not increase the communication overhead for OLTP. In general, OLTP transactions are completely executed on one node to avoid the communication overhead for intra-transaction parallelism and distributed commit. On the other hand, the degree of processing parallelism and thus the communication overhead for complex queries can be adapted to the current load situation. Furthermore, resource contention for CPU and memory between OLTP transactions and complex queries may largely be avoided by assigning these workload types to disjoint sets of processors which is not possible for SN, in general.

SD also offers an increased flexibility for *transaction routing*, i.e. for determining the PE where a transaction request should be routed to. This decision is particularly important for OLTP transactions where it can largely influence the number of remote requests. In SN systems, the best load assignment is primarily determined by the static database allocation. This is because a transaction should be assigned to the PE where most of the needed data is locally available. Assigning the transaction to another PE would still require that the data-owning PE has to process the operations on the required data. Hence, only little work would be saved for this node but additional communication overhead would be introduced for starting the subqueries, returning the results and for commit processing.

In SD systems, on the other hand, each PE can process any database operation and thus any transaction leaving a high potential for dynamic transaction routing and load balancing. However, sequentially executing a transaction on one PE can also require inter-PE communication for SD, in particular for global concurrency and coherency control [Ra93]. To limit the communication overhead for these functions it is generally advisable to support locality of reference by means of a so-called affinity-based transaction routing [YCDI87, Ra92, Ra93]. It assigns transactions with an affinity to the same database portions to the same processing nodes which supports a local concurrency and coherency control (depending on the chosen protocol) and good I/O performance. A survey of transaction routing schemes can be found in [Ra92].

7 Summary and Outlook

Dynamic load balancing is a prerequisite for effective utilization of parallel database systems consisting of many processing elements. Intra-query parallelism must effectively be supported in multi-user mode, i.e., in combination with inter-query and inter-transaction parallelism. The major control decisions to draw dynamically include determining the degree of intra-query parallelism and selecting the processors for executing subqueries. Load balancing must also be supported by a dynamic transaction routing for the initial assignment of queries and transactions.

Dynamic load balancing is most easily achieved for Shared Everything, but these systems suffer from availability and scalability limitations introduced by the shared memory. Shared Nothing and Shared Disk DBMS, on the other hand, require a comparatively large communication overhead for parallel query processing, in particular if intermediate query results are dynamically distributed in the system. We found out that Shared Nothing only offers a limited flexibility for dynamic load balancing because the database allocation determines in many cases where operations have to be processed, in particular for scan operations. Shared Disk systems, on the other hand, provide a higher load balancing potential since each PE can access any data and thus process any transaction, query or subquery which is especially valuable for query processing in multi-user mode and for mixed workloads. SD also requires an appropriate declustering of the data across multiple disks. But this data allocation only determines the maximal degree of intra-query parallelism, while the actual degree of parallelism can be chosen smaller (depending on the query type and current system state) without causing disk contention between subqueries.

Both SD and SN can employ dynamic load balancing for parallel join processing if the join inputs are dynamically redistributed among several join processors. We have discussed basic performance tradeoffs to consider for determining the optimal degree of join parallelism and selecting join processors in multi-user mode. Under high CPU utilization we found it necessary to reduce the degree of join parallelism in order to limit CPU contention (communication overhead for startup/termination and data redistribution). Under disk and memory bottlenecks, on the other hand, the degree of join parallelism should be increased in order to reduce the memory and I/O requirements per subquery. Furthermore, both control decisions should be drawn in a coordinated way and based on node-specific utilization information rather than system-wide averages. The redistribution overhead is reduced or avoided if the join input is already physically declustered on the join attribute. In this case, only SD preserves a potential for dynamic load balancing while for SN the data processors have to perform the join.

More work is needed on dynamic load balancing in several areas, in particular for inter-operator parallelism, to deal with data skew in multi-user mode, and to integrate local resource allocation policies for mixed workloads. Moreover, parallel database processing for object-oriented databases needs further investigation.

8 References

BC92 Bober, P.M., Carey, M.J.: On Mixing Queries and Transactions via Multiversion Locking. *Proc. 8th IEEE Data Engineering Conf.*, 535-545, 1992

BMCL94 Brown, K.P.; Mehta, M.; Carey, M.J.; Livny, M.: Towards Automated Performance Tuning for Complex Workloads. *Proc. 20th VLDB Conf.*, 72-84, 1994

CABK88 Copeland, G., Alexander, W., Boughter, E., Keller, T.: Data Placement in Bubba. *Proc. ACM SIGMOD Conf.*, 99-108, 1988

CLG94 Chen, P.M., Lee, E.K., Gibson, G.: RAID: High-Performance, Reliable Secondary Storage. *ACM Computing Surveys* 26 (2), 145-185, 1994

CM86 Carey, M.J., Muhanna, W.A.: The Performance of Multiversion Concurrency Control Algorithms. *ACM Trans. on Computer Systems* 4 (4), 338-378, 1986

DG92 DeWitt, D.J., Gray, J.: Parallel Database Systems: The Future of High Performance Database Systems. *Comm. ACM* 35 (6), 85-98, 1992

DG94 Davison, D.L.; Graefe, G.: Memory-Contention Responsive Hash Joins. *Proc. 20th VLDB Conf.*, 379-390, 1994.

DG95 Davison, D.L.; Graefe, G.: Dynamic Resource Brokering for Multi-User Query Execution. *Proc. ACM SIGMOD Conf.*, 281-292, 1995

DNSS92 DeWitt, D.J., Naughton, J.F., Schneider, D.A., Seshadri, S.: Practical Skew Handling in Parallel Joins. *Proc. 18th VLDB Conf.*, 27-40, 1992

Gh90 Ghandeharizadeh, S.: Physical Database Design in Multiprocessor Systems. Ph.D. Thesis, Univ. of Wisconsin-Madison, 1990

Gr93 Graefe, G.: Query Evaluation Techniques for Large Databases. *ACM Comput. Surveys* 25 (2), 73-170, 1993

Gr95 Gray, J.: Super-Servers: Commodity Computer Clusters Pose a Software Challenge. *Proc. German Database Conf. BTW*, March 1995

GR93 Gray, J., Reuter, A.: *Transaction Processing.* Morgan Kaufmann, 1993

Ho92 Hong, W.: Exploiting Inter-Operation Parallelism in XPRS. *Proc. ACM SIGMOD Conf.*, 19-28, 1992

HLH95 Hua, K.A., Lee, C.; Hua, C.M.: Dynamic Load Balancing in Multicomputer Database Systems Using Partition Tuning. *IEEE Trans. on Knowledge and Data Engineering* 7(6), 968-983, 1995

HLY93 Hua, K.A., Lo, Y., Young, H.C.: Considering Data Skew Factor in Multi-Way Join Query Optimization for Parallel Execution. *VLDB Journal* 2(3), 303-330, 1993

HSIT91 Hirano, Y., Satoh, T., Inoue, U., Teranaka, K.: Load Balancing Algorithms for Parallel Database Processing on Shared Memory Multiprocessors. *Proc. 1st Int. Conf. on Parallel and Distributed Information Systems*, 210-217, 1991

Ki84 Kim, W.: Highly Available Systems for Database Applications. *ACM Computing Surveys* 16 (1), 71-98, 1984

LT92 Lu, H., Tan, K.: Dynamic and Load-Balanced Task-Oriented Database Query Processing in Parallel Systems. *Proc. EDBT*, LNCS 580, 357-372, 1992

Ma95 Marek, R.: A Cost Model for Parallel Query Processing in Shared Nothing DBS (in German). *Proc. German Database Conf. BTW*, March 1995

MD93 Mehta, M., DeWitt, D.J.: Dynamic Memory Allocation for Multiple-Query Workloads. *Proc 19th VLDB Conf.*, 354-367, 1993

MD95 Mehta, M., DeWitt, D.J.: Managing Intra-Operator Parallelism in Parallel Database Systems. *Proc 21th VLDB Conf.*, 382-394, 1995

MN91 Mohan, C., Narang, I.: Recovery and Coherency-control Protocols for Fast Intersystem Page Transfer and Fine-Granularity Locking in a Shared Disks Transaction Environment. *Proc. 17th VLDB Conf.,* 193-207, 1991

MPTW94 Mohan, C., Pirahesh, H., Tang, W.G., Wang, Y.: Parallelism in Relational Database Management Systems. *IBM Systems Journal 33 (2), 1994*

MR92 Marek, R., Rahm, E.: Performance Evaluation of Parallel Transaction Processing in Shared Nothing Database Systems, *Proc. 4th Int. PARLE Conf.,* LNCS 605, 295-310, 1992

Om91 Omiecinski, E.: Performance Analysis of a Load-Balancing Hash-Join Algorithm for a Shared-Memory Multiprocessor. *Proc 17th VLDB Conf.,* 375-385, 1991

ÖV91 Özsu, M.T., Valduriez, P.: *Principles of Distributed Database Systems.* Prentice Hall, 1991

PCL93 Pang, H., Carey, M.J., Livny, M.: Partially Preemptible Hash Joins. *Proc. ACM SIGMOD Conf.,* 59-68, 1993

PGK88 Patterson, D.A., Gibson, G., Katz, R.H.: A Case for Redundant Arrays of Inexpensive Disks (RAID). *Proc. ACM SIGMOD Conf.,* 109-116, 1988

Ra92 Rahm, E.: A Framework for Workload Allocation in Distributed Transaction Processing Systems. *Journal of Systems and Software* 18, 171-190, 1992

Ra93 Rahm, E.: Empirical Performance Evaluation of Concurrency and Coherency Control for Database Sharing Systems. *ACM Trans. on Database Systems* 18 (2), 333-377, 1993

RM93 Rahm, E., Marek, R.: Analysis of Dynamic Load Balancing Strategies for Parallel Shared Nothing Database Systems. *Proc 19th VLDB Conf.,* 182-193, 1993

RM95 Rahm, E., Marek, R.: Dynamic Multi-Resource Load Balancing in Parallel Database Systems. *Proc 21th VLDB Conf.,* 395-406, 1995

RS95 Rahm, E., Stöhr, T.: Analysis of Parallel Scan Processing in Shared Disk Database Systems. *Proc. Euro-PAR95,* LNCS 966, 485-500, 1995

Se93 Selinger, P.: Predictions and Challenges for Database Systems in the Year 2000. *Proc 19th VLDB Conf.,* 667-675, 1993

Va93 Valduriez, P.: Parallel Database Systems: Open Problems and New Issues. *Distr. and Parallel Databases* 1 (2), 137-165, 1993

WDJ91 Walton, C.B; Dale A.G.; Jenevein, R.M.: A Taxonomy and Performance Model of Data Skew Effects in Parallel Joins. *Proc. 17th VLDB Conf.,* 537-548, 1991

WDYT94 Wolf, J.L., Dias, D.M., Yu, P.S., Turek, J.: New Algorithms for Parallelizing Relational Database Joins in the Presence of Data Skew. *IEEE Trans. on Knowledge and Data Engineering* 6(6), 990-997, 1994

WFA92 Wilschut, A.; Flokstra, J.; Apers, P.: Parallelism in a Main-Memory DBMS: The performance of PRISMA/DB. *Proc. 18th VLDB Conf.,* 521-532, 1992

YCDI87 Yu, P.S., Cornell, D.W., Dias, D.M., Iyer, B.R.: Analysis of Affinity-based Routing in Multi-system Data Sharing. *Performance Evaluation* 7 (2), 87-109, 1987

ZG90 Zeller, H., Gray, J.: An Adaptive Hash Join Algorithm for Multiuser Environments. *Proc. 16th VLDB Conf.,* 186-197, 1990

Workshop 01

Programming Environment and Tools

Distributed Array Query and Visualization for High Performance Fortran

Steven T. Hackstadt and Allen D. Malony

Computer & Information Science Department
University of Oregon, Eugene, OR 97403 USA

Abstract. This paper describes the design and implementation of the Distributed Array Query and Visualization (DAQV) system for High Performance Fortran, a project sponsored by the Parallel Tools Consortium. DAQV's implementation utilizes the HPF language, compiler, and runtime system to address the general problem of providing high-level access to distributed data structures. DAQV supports a framework in which visualization and analysis clients connect to a distributed array server (*i.e.*, the HPF application with DAQV control) for program-level access to array values. Implementing key components of DAQV in HPF itself has led to a robust and portable solution.

1 Introduction

During the execution of a parallel program it is often necessary for the user to inspect the values of parallel data structures that are distributed across the processing nodes of a parallel machine. Similarly, to interact with parallel applications (*e.g.*, for computational steering), it may be necessary to access and analyze distributed data at runtime. Any tool that addresses a particular user need where distributed data access capabilities are required, whether it be debugging, performance analysis, or application interaction, has either to implement these capabilities itself or rely on some other (open) infrastructure to provide them. In the former case, tool implementations tend to become specific to low-level system details, the tool implements only what is necessary for its purposes, and abstractions of distributed data access tend not to be developed. The common anecdote from users that PRINT is the only good tool is indicative of these problems: tools for parallel systems are often complex, not integrated, and vary widely across machines. PRINT is the only thing that they can rely on.

The case of relying on an existing infrastructure for distributed data access also suffers because no such infrastructure currently exists. One might think that PRINT provides suitable functionality. But the problem is really the level of the tool's interface to the infrastructure; PRINT would require significant scaffolding to give it a high-level interface to an external tool. In a similar vein, many parallel debuggers are not always aware of how parallel data structures are distributed. Whereas they provide low-level mechanisms to get the data, the higher-level semantics about what the data is (*i.e.*, the type and characteristics of the distributed data object) are generally not utilized in the infrastructure and are not apparent in its external interface. However, it is exactly these semantics that are helpful in developing open frameworks and portable tools.

The DAQV project highlights the dilemma between the two alternatives described above and how it faces tool designers. As a project sponsored by the Parallel Tools (Ptools) Consortium, DAQV set out to address the user requirement of being able to access and visualize distributed array data coming from a parallel program. Initially, it placed more emphasis on the environment for array interaction (query specifications, array operations, visualization types, etc.) than the underlying infrastructure for interfacing with the running program [14]. However, based on feedback from the Ptools community [16], it quickly became clear that DAQV's mission was too broad and that it should be more focused. The most interesting feedback was from users who helped refine the focus of the project to create a robust and well-defined infrastructure rather than a multitude of user interface features. Users made it clear that the most important contribution that the project could make was to provide a technology that lets them "get the data" (*i.e.*, a high-level PRINT capability) and use other tools to "work with the data." The goal then became to develop interfaces for both (1) low-level extraction of data from the program, and (2) the higher-level request/delivery of data to an external client (*e.g.*, for visualization). Users also acknowledged the importance of HPF as the primary language target for the DAQV project.

The result is the DAQV tool as described in this paper. We present this background because we think that the Ptools project evaluation process was instrumental in forcing the design to seriously consider users' needs. This gave DAQV its focus, but how DAQV is integrated with the HPF language is what makes it unique. One final comment is that DAQV is more characteristic of a framework or meta-tool than of a tool. It provides the intended capabilities in a form that can be utilized in other tools that may want to do something with the data that DAQV provides; a visualization tool is but one example. The rest of this paper describes the DAQV framework for HPF. We proceed by discussing related work, design, functionality, and implementation. Examples of DAQV's application are then followed by a discussion of future work.

2 Related Work

We review the related research from four perspectives: language-level parallel tools, distributed data visualization, client-server tool models, and program interaction frameworks.

There has been a strong advocation in the last few years for parallel tools to be integrated in parallel language systems and to support the language semantics in their operation. For instance, the research work integrating Pablo with the Fortran D compiler [1] and Tau with the pC++ compiler [2] demonstrates the importance of providing high-level semantic context. This is also clearly seen in the Prism [18] environment, which is, perhaps, the best example of the ease of use that can come from an integrated tool system. DAQV clearly follows in this spirit and goes one step further by actually using the language system itself for part of its implementation. This was also a feature of Breezy, a forerunner of DAQV that provided high-level program interaction for the pC++ system [3].

One of the key programming abstractions found in parallel language systems is data parallelism. Because distribution of parallel data is an important factor in the performance behavior of a program, viewing data and performance information in relation to the distribution aids the user in tuning endeavors. The GDDT tool [13] provides a static depiction of how parallel array data gets allocated on processors under different distribution methods and also supports an external interface by which runtime information can be collected. Kimelman *et al.* show how a variety of runtime information can be correlated to data distribution to better visualize the execution of HPF programs [11]. The IVD tool [10] uses a data distribution specification provided by the user to reconstruct a distributed data array that has been saved in partitioned form. In DAQV's case, array reconstruction is done implicitly (and portably) by array access functions implemented in HPF.

The increasing importance of portability and extendability in parallel tools has evoked designs following client/server models. The Panorama debugger [15] demonstrates how the use of interoperating modules can lead to increased functionality and generality in debugging systems. The p2d2 debugger [4] extends this concept considerably by proposing a full client/server debugging framework with comprehensive abstractions of operating system, language, library interfaces, and protocols for distributed object interaction. DAQV is clearly adopting the client/server approach for similar reasons, but in contrast to these two particular tools, the functionality is at a higher level, which actually affords the possibility of layering DAQV on top of systems like Panorama and p2d2.

The final perspective is one of dynamic program interaction. There has been a growing interest in runtime visualization of parallel programs and computational steering. Implementing such support raises interesting user and implementation issues. On the one hand, runtime visualization for a particular application domain might be able to utilize domain knowledge to implement a system meeting certain performance constraints. The pV3 system [6] is a good example. However, such specialized implementations may be limited when considering the general runtime visualization problem. DAQV, in many respects, is a direct descendant of the Vista research [17] since it embodies many of the same design goals: client/server operation, automated data access support, runtime operation, and structured interaction. The improvement DAQV offers is in language-level implementation to increase portability. We believe that this will also improve DAQV's ability for computational steering. Although Vista was extended for interactive steering in the VASE tool [5], the steering operations were still very much dependent on the target implementation. Supporting steering at the application language level is a more robust, general solution and is something we intend to investigate in DAQV.

3 Design

DAQV is somewhat different from many of the systems and tools discussed in the previous section. Its goal is to "expose" the distributed data structures of a HPF program to external tools — and to do so in a way that does not require external tools to have any knowledge about data decompositions, symbol tables, or the number of processors involved. The problem, then, is to provide access at a meaningful and portable level — a level at which the user is able to interpret program data and at which external tools need only know logical structures. DAQV's design has been motivated by a range of factors, including input from the Ptools user group, the need to support semantic-based access to distributed arrays, and a desire to maximize DAQV's portability. The key characteristics of this design are discussed below.

Global View of Data. High Performance Fortran (HPF) supports a global name space [12]; the programmer views distributed arrays at a global level, often performing operations on whole arrays and referring to array elements with respect to the entire array, not some local array section on a particular processor. Distribution directives allow programmers to use their knowledge about the application and advise the compiler on data distribution schemes. But this concern for data distribution does not affect how the programmer references the data. HPF syntax insulates the user from ever dealing with an array in an explicitly distributed manner. For this reason, DAQV supports a similar perspective of program data when interacting with the user (through external tools).

Client/Server Model. DAQV is a software infrastructure that enables runtime visualization and analysis of distributed arrays. It is not a stand-alone application or tool that performs these tasks itself; rather, it must interoperate with external tools. The goal in this design is to allow the continued use of existing visualization and analysis tools. The feedback from the Ptools user group during design discussions was clear: they did not need another fancy, self-contained visualization tool; they just wanted improved facilities for querying and extracting distributed arrays such that their existing tools could be easily used with the system. To this end, we logically view the entire HPF program (not individual processes or processors) as a *distributed array server* to which external *client tools* connect and then interact with the program and its data.

Portability. Another goal in the design of DAQV is to minimize the degree to which DAQV is machine- or compiler-dependent. By targeting HPF in the first place, machine portability is inherent to the extent that a given HPF compiler is. As for compiler portability, DAQV primarily accomplishes this in three ways: key components of DAQV are implemented in HPF; compiler and runtime systems are utilized; and compiler-dependent code is minimized and isolated.

4 Functionality

This section describes, in general terms, the two levels of functionality supported by DAQV. Called *push* and *pull*, these two models differ in the degree of interactivity available with the HPF program at runtime.

Push Model. The *push model* forms the basis of DAQV and constitutes the simplest and least intrusive way to access distributed arrays from an external tool, or *data client*. The push model is implemented by inserting simple DAQV subroutine calls into the HPF source code. These calls allow the programmer to (1) indicate the DAQV model (*i.e.*, push or pull) to be used in the program, (2) register distributed arrays with DAQV, (3) set parameters for communicating with a data client, (4) make DAQV connect with a data client, and (5) send the data values of a distributed array to the data client.

The functionality of the push model is the practical solution to the feedback from the Ptools user group. That is, with minimal additions to the HPF source code, users can extract the data values of distributed arrays and visualize them with other tools, never having to worry about array reconstruction. The push model can be used to spot check the state of an array or to create animations of data values over the iterations of a loop. Multiple arrays can be pushed out of the program to multiple data clients.

Pull Model. The push model is adequate if the programmer knows at compile-time exactly which arrays they want to visualize and when they want to view them. However, to support a more exploratory and flexible approach to array visualization, DAQV supports the *pull model* which enables controlling program execution and selecting arrays for visualization through an external interface.

The pull model allows the user to repeatedly run an HPF code for a period of time and then extract data values from the distributed arrays of interest. Two types of clients are used in the pull model. As in the push model, data clients process data values from arrays. In fact, any data client that works in the push model also works in the pull model because the pull model is layered on top of the push model. In addition, though, the pull model requires a single *control client* to direct program execution and configure and initiate array transfers to data clients.

Thus, the primary conceptual difference between the push and pull models is where the decision to extract an array originates. The names "push" and "pull" are meant to reflect this difference in perspective. In the push model, the HPF program "pushes" data out, while in the pull model, an external client reaches in and "pulls" data out. However, the implementation of these very different conceptual models is built upon a common infrastructure supported by DAQV.

5 Implementation

Our goal with DAQV is to facilitate simple and useful conceptual models for distributed array collection and extraction, and to implement those models in a high-level, portable manner. This section briefly explains the two key components of the DAQV implementation: procedural interface and client/server interface.

Procedural Interface. DAQV's core requirement is to support interaction with HPF programs. In part, this interaction is similar to what might be provided by a HPF debugger. However, a HPF debugger may not provide a high-level interface for distributed array query, relying instead on low-level support for gathering array data on a node-by-node basis. Such system dependencies defeat DAQV's goal of portability, making it too reliant on the target compiler or machine. Instead, key components of DAQV are implemented as HPF subroutines, allowing array access and other functions to utilize the HPF compiler and runtime system implicitly and automatically. In this respect, we view DAQV as a HPF language-level tool design and implementation.

Thus, the high-level operation of DAQV demands a different method for interacting with the HPF program than what a HPF debugger can provide. Our solution is to implement DAQV as a library that is linked with a HPF object file, creating an executable that supports a procedural interface between the original HPF program and the

DAQV distributed array server. The routines in this interface handle initialization, registration of arrays, configuration of data clients, and data extraction.

Client/Server Interface. Conceptually, the client/server interface supported by DAQV exists between the entire HPF program (*i.e.*, all SPMD processes representing the HPF program) and external data and control clients. DAQV uses a bidirectional, string-based event protocol between the HPF server and its clients. In the push model, data clients respond to array data transmissions with a confirmation that signals DAQV to let the HPF code continue executing. The pull model uses a more complete event protocol that allows the control client to direct program execution, configure client communications, and extract array values.

One HPF process is designated as a "manager" by DAQV. The manager is responsible for handling all client communications, and the other HPF "worker" processes cooperate with the manager to effect DAQV operations. This distinction is only evident to DAQV; it has no effect on the regular execution of the HPF program. So, events sent to the server are actually received only by the manager process and must be "shared" with the other HPF processes by special mechanisms implemented in HPF.

As an open framework that is intended to work with a variety of different data clients, efforts have been made to allow existing visualization and analysis tools to be easily ported for use with DAQV. First, the current implementation uses only sockets for communication. Second, a client must be able to parse data sent by DAQV. And third, a client must be able to respond over the socket with a simple string-based confirmation event after receiving data.

Control clients, on the other hand, require somewhat more sophistication. They have the same requirements as the push model, but in addition, they must handle control events. A C-library has been developed for supporting event processing, and support for other languages (*e.g.*, Tcl/Tk) will be forthcoming. Furthermore, a control client usually interacts with the user. The DAQV pull model and event specification establish a framework for other tool developers by providing a simple protocol that can be incorporated into other control clients to gain high-level access to distributed data.

Language, Compiler and Runtime System Requirements. With a primary goal being portability, DAQV does not require any modifications to the HPF language implementation, however certain assumptions have been made in the DAQV reference implementation. Currently, DAQV assumes that HPF data parallelism is achieved with a SPMD execution model. (In many respects this is a more difficult problem when compared to, say, a multithreaded implementation.) An HPF program must be able to invoke C routines, and vice-versa. The process of array registration requires knowledge about the transformations applied to function and subroutine arguments by a given compiler. However, the information required is minimal and so far has not required vendors to divulge proprietary information. DAQV can optionally take advantage of compiler-specific tracing/profiling support, though this is not required by the reference implementation. Finally, in two instances, DAQV *requires* HPF distribution directives to be carried out by the compiler. That is, DAQV will not work if the directives are ignored. With these minimal requirements and assumptions, we feel DAQV will be able to achieve a high degree of portability across both machines and compilers.

6 Application: Laplace Heat Equation

To illustrate what the current implementation of DAQV is capable of, we consider a HPF program that implements an iterative finite difference method for solving the Laplace Heat Equation [7]. The code continues until a steady state is reached to within some tolerance. DAQV is used in the push mode to visualize the heat flow through the two-dimensional surface at each iteration of the main loop. The sequence of images in Figure 1 was generated by a data client called Dandy. Implemented in Tcl/Tk, Dandy was easily ported from the pC++ Tau Tools [2] for use with DAQV. Under this scheme, when the program reaches the DAQV procedure that establishes communication with a data client, the Dandy client connects by a socket to the HPF program. At this point, execution resumes and the animation of the data values begins. At each loop iteration, the new array values are sent to Dandy and displayed. The Dandy interface allows the user to pause/resume the animation and redraw the display. Dandy automatically determines the range of the data values and maps them onto a fixed colormap. However, if the viewer wishes to fix the color range across several iterations, the automatic scaling feature can be disabled, allowing minimum and maximum data values to be set. This

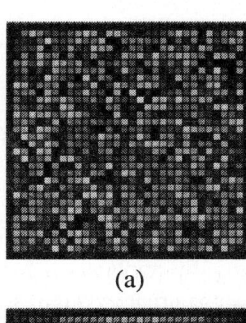

(a)

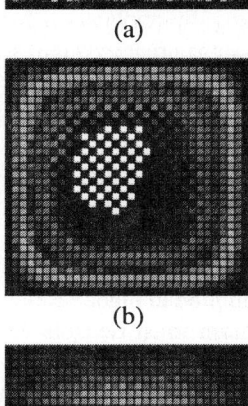

(b)

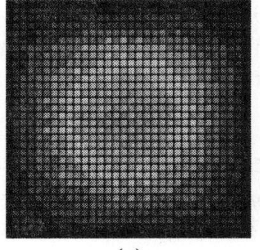

(c)

Fig. 1. Convergence of the finite method for solving the Laplace Heat Equation can be seen by visualizing the two-dimensional array representing the heated surface.

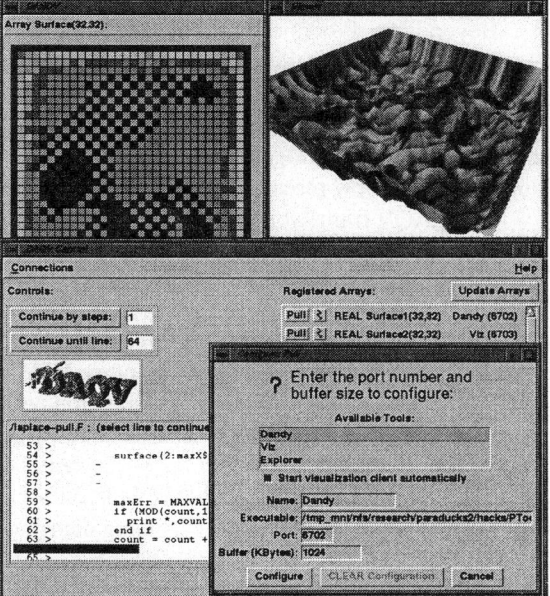

Fig. 2. A DAQV session with a control client. A simple two-dimensional data client and a three-dimensional data client both represent the 2D array from the Laplace Heat Equation code.

feature can also be used to identify outliers or to identify values beyond a certain threshold. For example, Figure 1(b) shows values above the specified range as white. As the algorithm nears convergence (Figure 1(c)), these values move into the data range.

The same application can also be used with DAQV's pull model. Figure 2 shows several windows from a DAQV session. A prototype control client interface (bottom, partially obscured) supports several DAQV features, including a source code browser, program controls, a list of HPF arrays that have been registered with DAQV. In this case, Surface, a 32x32 real-valued array, has been registered twice so that it may be viewed with two different data clients. Other controls allow the user to select a data client for an array and to send the values of an array to the appropriate data client. The first registered array, Surface1, has been mapped to the Dandy client in the upper left part of the screen, while Surface2 is being sent to a more complex, three-dimensional display created in the Viz visualization environment builder [8].

7 Future Work

In its current form, DAQV should be regarded as a reference implementation that demonstrates functionality, specifies components and their system requirements, and defines APIs and transport protocols. In fact, being the primary derivative of a Ptools project, the reference implementation serves as a prototype for vendors to evaluate and potentially adopt. If the Ptools reference platform were to be adopted by a HPF compiler vendor, several opportunities for improvement exist. In particular, DAQV lends itself nicely to compiler/preprocessor support for instrumentation.

Beyond the Ptools portion of the DAQV project, there are several interesting research directions for this work. One of the most important short-term goals is to evaluate the benefit of DAQV in a significant scientific application. At the University of Oregon, we are developing an HPF-version of a seismic tomography application for modeling mid-ocean ridges. The DAQV framework is being used to create visualizations that will allow researchers to better understand sea floor geology.

The DAQV framework is also being extended to allow distributed data to be altered as well as accessed. As a result, DAQV can be applied to application scenarios where changing runtime array data or program control variables is beneficial (*e.g.*, for computational steering in the seismic tomography application). In the near future, DAQV will be tested on other parallel machine platforms supported by PGI's HPF compiler. In addition, we hope to port DAQV to other HPF compilers. We also believe that the DAQV model (and parts of its implementation perhaps) can be applied in other parallel language systems where distributed data is involved, and we intend to retarget the DAQV reference code to the HPC++ [9] system. More generally, we see no major intellectual difficulties with porting DAQV to SPMD environments where some runtime means (language or library) for accessing distributed data exists.

Acknowledgments

Thanks to the Parallel Tools Consortium, The Portland Group Inc., David Presberg, Craig Lee, Kurt Windisch, Harold Hersey, Bernd Mohr, and Chris Harrop for their contributions to the design and implementation of DAQV. The full technical report describing DAQV can be found at http://www.cs.uoregon.edu/~hacks/research/daqv/.

References

[1] V. Adve, et al., *An Integrated Compilation and Performance Analysis Environment for Data-Parallel Programs*, Supercomputing '95, Dec. 1995.

[2] D. Brown, S. Hackstadt, A. Malony, & B. Mohr, *Program Analysis Environments for Parallel Language Systems: The TAU Environment*, Proc. of the Workshop on Environments & Tools For Parallel Scientific Computing, Townsend, TN, May 1994, pp. 162-171.

[3] D. Brown, A. Malony, B. Mohr, *Language Based Parallel Program Interaction: The Breezy Approach*, Proc. of Intl. High Performance Computing Conference (HiPC'95), New Delhi, India, Dec. 1995.

[4] D. Cheng & R. Hood, *A Portable Debugger for Parallel and Distributed Programs*, Proc. of IEEE Supercomputing '94, Washington, D.C., Nov. 1994.

[5] R. Haber, et al., *A Distributed Environment for Runtime Visualization and Application Steering in Computational Mechanics*, Univ. of Illinois at Urbana-Champaigne, CSRD Technical Report 1235, Jun. 1992.

[6] R. Haimes, *pV3: A Distributed System for Large-Scale Unsteady CFD Visualization*, American Institute of Aeronautics & Astronautics Paper 94-0321, Reno NV, Jan. 1994.

[7] P. B. Hansen, *Studies in Computational Science: Parallel Programming Paradigms*. Prentice Hall, Inc., Englewood Cliffs, NJ, 1995.

[8] H. Hersey, S. Hackstadt, L. Hansen, & A. Malony, *Viz: A Visualization Programming System*, Univ. of Oregon, Dept of Computer & Information Science, Technical Report CIS-TR-96-05, Apr. 1996.

[9] HPC++ Working Group, *HPC++ Whitepapers and Draft Working Documents*, Supercomputing '95 Workshop, URL:http://www.extreme.indiana.edu/hpc++/hpc++wp/, Feb. 1996.

[10] A. Karp & M. Hao, *Ad Hoc Visualization of Distributed Arrays*, Hewlett-Packard Technical Report HPL-93-72, Sep. 1993.

[11] D. Kimelman, P. Mittal, E. Schonberg, P. Sweeney, K. Wang, D. Zernik, *Visualizing the Execution of High Performance Fortran (HPF) Programs*, Proc. of the 9th Intl. Parallel Processing Symposium (IPPS), IEEE Computer Society Press, Apr. 1995, pp. 750-757.

[12] C. Koelbel, D. Loveman, R. Schreiber, G. Steele Jr., & M. Zosel, *The High Performance Fortran Handbook*, MIT Press, Cambridge, MA, 1994.

[13] R. Koppler, S. Grabner, & J. Volkert, *Visualization of Distributed Data Structures for HPF-like Languages*, to appear in Scientific Programming, spec. issue on implementations of High-Performance-Fortran, 1995.

[14] A. Malony, J. May, A. Karp, D. Presberg, & V. Schuster, *Distributed Array Query and Visualization*, Parallel Tools Consortium Working Document, Jan. 1995.

[15] J. May & F. Berman, Panorama: *A Portable, Extensible Parallel Debugger*, Proceedings of the ACM/ONR Workshop on Parallel & Distributed Debugging, May 1993, pp. 96-106.

[16] Parallel Tools (Ptools) Consortium DAQV Working Group, *Ptools Meeting Summary: Distributed Array Query and Visualization (DAQV) Project*, available at URL:http://www.cs.uoregon.edu/~hacks/research/daqv/, May 1995.

[17] A. Tuchman, D. Jablonowski, G. Cybenko, *Runtime Visualization of Program Data*, Proceedings of IEEE Visualization '91, Oct. 1991, pp. 255-261.

[18] Thinking Machines Corp., *Prism*, available at URL:http://www.think.com/tmhtml/ProdServ/Products/prism.html, Feb. 1996.

Annai Scalable Run-Time Support for Interactive Debugging and Performance Analysis of Large-Scale Parallel Programs

Christian Clémençon[1], Akiyoshi Endo[2], Josef Fritscher[1], Andreas Müller[1], Brian J. N. Wylie[1]
(wylie@cscs.ch)

[1] Centro Svizzero di Calcolo Scientifico, CH-6928 Manno, Switzerland
[2] NEC European Supercomputer Systems, Swiss Branch

Abstract. The Annai tool environment helps exploit distributed-memory parallel computers with High Performance Fortran and/or explicit communication, using MPI as a portable machine interface. Integration within a unified environment allows the component parallelization and compilation support, debugging and performance tools to synergetically use common facilities. Additionally, massive quantities of partitioned data and execution information, from large-scale applications on multiple processors, needs to be effectively managed and presented during program engineering. This has been achieved by scalable design and cooperative integration of tool component run-time libraries, supporting flexible interactive debugging and performance analysis.

1 Introduction

The main justification for investing development effort in producing a parallel program is usually the desire to handle large, or even huge, problems. Managing the sheer quantity of data itself is a major challenge for the program developer, especially when it must be partitioned and distributed in a manner which computers can handle efficiently. Users therefore expect appropriate tools which will help them to take care of their data and program development needs.

Compiler technology, however, remains a long way from being able to automatically and effectively parallelize sequential programs. Current technology requires programmers to explicitly write parallel programs, or at least provide appropriate directives informing the compiler of potentially exploitable parallelism or the best-suited data layouts. Tools are required which efficiently present data in forms which are easily and quickly interpreted, by being presented in a manner which closely relates to the programmer's conception of their program and data objects. The fact that large amounts of data, and possibly large numbers of processors, are involved shouldn't interfere with this process, at least not until it becomes necessary for the programmer to consider these 'details.' Effective data reduction has to be incorporated within the run-time support system of a parallel program, and much of the processing has to be done in parallel on the same processors where the parallel program resides.

The Annai integrated tool environment [1] provides a comprehensive parallel program engineering environment for distributed-memory computer systems. After briefly introducing Annai and its tool components, for which detailed accounts are separately available, attention is focussed on the run-time libraries which do much of the 'behind-the-scenes' processing, and which together make interactive high-level debugging and performance analysis possible.

2 Annai integrated tool environment

The Annai environment offers integrated tools for parallelization, debugging, and performance monitoring and analysis, with common user and machine interfaces. Building on emerging standards, 'high-level' extended High Performance Fortran (HPF) and 'low-level' explicit message-passing programs (based on MPI) are handled by all of the tools, supporting application flexibility and portability.

PST, the **Parallelization Support Tool** [2], extends the current HPF definition (v1.1) by providing language constructs and extensive run-time support for the parallelization of irregular computations. PST extensions support dynamic data distributions and run-time preprocessing of critical code sections, and PST also provides comprehensive compilation support for mixed-language program sources and the other tools.

PDT, the **Parallel Debugging Tool** [3], is a conventional source-level symbolic debugger, enhanced to support different levels of abstraction. At the data-parallel level, PDT provides coherent graphical representations of large, distributed data-sets, and control- and data-breakpoints with global break conditions. At the message-passing level, PDT assists programmers with deadlock and race detection, and deterministic execution replay.

PMA, the **Performance Monitor and Analyzer** [4], exploits profile summary and trace information from interactively specified source code regions where instrumentation is activated and data collected during the execution of a parallel program. PMA then assists with the performance tuning and interpretation of program execution through visualization and analysis of this information. Different levels of abstraction are supported, from execution summary profiles and global views of time-varying behavior down to individual processes and analysis of communication events and memory utilization.

PDT and PMA have a common interface to the parallel computing platform via the *Tool Services Agent* (TSA), which provides basic, low-level functions for controlling parallel program execution. An optional *User Interface* (UI) to the run-time environment and Annai tool components also provides synchronized source listing and program structure browsers.

Within the parallel machine user programs execute utilizing the MPI and Annai tool libraries. Optimized implementations have been developed for the NEC Cenju-3, along with versions for Unix workstations and clusters. Applications developed with Annai, based on both **PST**/HPF and MPI, have also run readily on Intel Paragon and Cray T3D systems, verifying their portability.

3 Annai run-time libraries

The modular construction of the Annai interactive user environment from a number of component tools also applies to the Annai run-time system of libraries linked with the user program and running on the parallel computer system.

The most basic configuration, for fully-optimized explicitly-parallelized programs without detailed debugging and performance analysis support, only uses the standard run-time and system libraries, including MPI. HPF programs need run-time support to manage distributed data, and perform automatic message routing to maintain data consistency. In the Annai environment all communication within the HPF and PST libraries is MPI-based. When requested by the programmer, the Annai compilation system incorporates additional run-time libraries, which assist with detailed program debugging and performance analysis.

3.1 Functionality

The main Annai run-time libraries are those associated with the different tool components. Portable versions of these libraries are often complemented by optimized implementations for particular parallel systems. In each case, the library is based on a scalable design with clean functional interfaces.

An additional support library, known as TSAlib, provides implementations of (or interfaces to) system-specific functions which interact with the operating system. Examples include timing and clock synchronization functions, and support for special 'lightweight' breakpoints which are notified directly to the run-time library and processed locally on the respective processor. These breakpoints are set like normal breakpoints, but don't halt all of the processors nor interact with Annai via TSA. This avoids unnecessary synchronization and interference with other processors, providing a convenient mechanism for modifying instrumentation and performing other reconfiguration actions during execution.

PSTlib The core of the PST run-time library provides run-time analysis support for PST/HPF programs, automatically managing shared and distributed variables by constructing and controlling essential message transfers, and handling the execution of parallel loops. Fundamental support is also provided for debugging and performance analysis of such programs.

Support for dynamic array allocation, dynamic array re-alignment and redistribution, assumed-size array arguments, user-defined and inherited data distributions all mean that representations of distributed arrays and symbol tables are complex structures which have to be set up at run-time. Functions are therefore provided by the PST run-time library which allow access to distributed arrays and their structure and distribution information.

Information about the static structure of the source program, and how this relates to the transformed code actually running on the nodes of the parallel machine, allows PDT and PMA to relate low-level events to the corresponding high-level source statements. Examples where such information is essential are matching addresses of breakpoint locations or MPI message-passing events to the appropriate line (or lines) of PST/HPF program source.

A specification of the static program structure is also incorporated within source objects during compilation, and after linking constituted into a complete specification added as an extra 'section' of the executable. This mechanism ensures that the specification is always consistent with the corresponding source files, and that it is readily accessible to Annai interactive components.

PDTlib The PDT debugger is able to reconstruct a distributed array from the fragments on each processor using the array distribution information available from the PST run-time library. This 'raw' data is most conveniently processed within the parallel machine to extract relevant parts for subsequent presentation.

Data breakpoints, or 'watchpoints,' provide an indispensable debugging service when tracking down run-time violations when they first occur. The PDT run-time library includes a scalable mechanism where all memory updates performed by the program are checked locally on each processor. Store operations are instrumented to determine whether the address is within the range of an array section to which a predicate has been specified, and then, if appropriate, the value itself is also examined.

Messages in distributed systems are considered to race if they are simultaneously in transit and the order of receipt at a single point is not completely defined. Re-running a program which has races complicates debugging, since execution is non-deterministic. The PDT run-time library uses vector timestamps sent with each message to determine on receipt whether a race has occured. On subsequent re-execution, the trace of messages which raced can be replayed.

PMAlib During program execution, the PMA library is responsible for managing profile summary and event trace collection from instrumented MPI communication library functions and instrumentation inserted in the program by the compilation system. Users can also choose to add extra instrumentation of their own. Initial library configuration determines what classes of information are gathered, and how this should be processed. This 'base' instrumentation can then be flexibly modified by specifying different instrumentation for selected program regions.

The state and utilization of the message-passing system is determined from the instrumented communication functions. Similarly, the instrumentation inserted by the compilation system makes it possible to determine the program state in terms directly related to the user's view of the program structure, i.e., routines, loops, etc. Additional information is also provided by the compilation system about parallelization overheads, such as run-time analysis, (explicit and implied) data re-distributions, and extra storage for communication buffers. Memory utilization provided by these means, and other system information, including the message-passing events, can always be related back to the program.

Two complementary performance information formats are supported by the PMA run-time library. The simplest is an event trace log which is buffered on each processor. A more convenient and scalable alternative is a profile summary, accumulated as the program runs, which keeps track of essential execution statistics for each entity of the program structure on each processor. Profiles of the current execution status can be presented on demand, or summarize the complete execution when the program finishes. Instrumentation processing overheads are also available as an integral part of the analysis.

3.2 Interaction and Operation

PDTlib uses PSTlib to access the relevant parts of distributed arrays, for reduction and presentation to users as single consistent objects. Similarly, communication events can be checked by PDTlib to determine whether a race has occured and/or noted by PMAlib as part of its execution record. Low-level events such as this are efficiently related to the appropriate high-level source statements via PSTlib program structure mapping information.

If all of their functionality were always enabled, PDTlib and PMAlib would have significantly intrusive effects, on each other as well as the user program. Their functionality is, however, only *latent*, requiring appropriate run-time configuration. When debugging, PDT can activate functionality to enable message-race checking, or to configure watched memory regions. Alternatively, during performance analysis, PMA will activate profiling and/or tracing and configure the various types of instrumentation.

Both PDT and PMA use the common facilities provided by TSA, the only part of Annai which interacts directly with the parallel program on the target platform. In addition to controlling program execution, TSA is able to read from and write into the program's address space, enabling it to configure the Annai run-time libraries. TSA also allows PDT and PMA to specify breakpoints and (where appropriate) subsequently informs the high-level tools of breakpoint hits which they can then process as they wish. At any time program execution has been halted, the run-time library configurations can be further modified by TSA: tables of memory regions corresponding to arrays which should be watched with associated predicates, and instrumentation actions and configurations to be applied at 'lightweight' breakpoint hits, can be updated.

4 Conclusion

Integrated environments for parallel systems have recently become a matter of considerable user and developer interest. Thinking Machines Corporation's commercial *Prism* environment [5] successfully demonstrated the principle, and motivated research groups to investigate similar environments for *Fortran D* [6] and *Vienna Fortran* [7], as well as other programming languages. Numerous more restricted efforts have also partially-integrated various editors, compilers, debuggers and analysis tools in research and commercial contexts, some now coordinated under the auspices of the Parallel Tools Consortium [8].

Annai is the practical realization of the further development of these ideas in a consistent, easy-to-use environment. Advanced parallelization, debugging, performance monitoring and analysis techniques have been incorporated in integrated modular tools, providing comprehensive scalable support for flexible data-parallel and explicit message-passing application development with standardized languages and portable machine interfaces. The design of the Annai run-time libraries, such that they can handle distributed events and process huge amounts of data — from both the user program and the system itself — in a scalable fashion, ensures effective support for interactive parallel program engineering.

Acknowledgements Roland Rühl played an integral part in the design and implementation of Annai and its run-time libraries. NEC C&CRL (Tokyo) provided the PST compiler front-end, the code generator and run-time library for standard HPF subroutines, as well as specific support for the optimized Cenju-3 run-time libraries.

References

1. C. Clémençon, K. M. Decker, V. R. Deshpande, A. Endo, J. Fritscher, P. A. R. Lorenzo, N. Masuda, A. Müller, R. Rühl, W. Sawyer, B. J. N. Wylie, and F. Zimmermann, "Tool-supported development of parallel application kernels," in *Proc. IPCCC'96 (Phoenix, USA)*, pp. 294–302, IEEE Comp. Soc. Press, Mar. 1996. [ISBN: 0-7803-3255-5]. Further details in CSCS-TR-95-03.
2. A. Müller and R. Rühl, "Extending High Performance Fortran for the support of unstructured computations," in *Proc. 9th Intl. Conf. on Supercomputing (ICS'95, Barcelona, Spain)*, pp. 127–136, ACM, July 1995. [ISBN: 0-89791-728-6]. Further details in CSCS-TR-94-08 and CSCS-TR-95-04.
3. C. Clémençon, J. Fritscher, and R. Rühl, "Visualization, execution control and replay of massively parallel programs within Annai's debugging tool," in *Proc. High Performance Computing Symp., (HPCS'95, Montréal, Canada)*, pp. 393–404, Centre de recherche informatique de Montréal, July 1995. [ISBN: 2-921316-12-9]. Further details in CSCS-TR-94-09 and CSCS-TR-95-01.
4. B. J. N. Wylie and A. Endo, "Annai/PMA multi-level hierarchical parallel program performance engineering," in *Proc. 1st Intl. Work. on High-Level Programming Models and Supportive Environments (HIPS'96, Honolulu, USA)*, pp. 58–67, IEEE Comp. Soc. Press, Apr. 1996. [ISBN: 0-8186-7567-5]. Further details in CSCS-TR-94-07 and CSCS-TR-95-05.
5. S. Sistare, D. Allen, R. Bowker, K. Jourdenais, J. Simons, and R. Title, "Data visualization and performance analysis in the *Prism* programming environment," in *Programming Environments for Parallel Computing*, pp. 37–52, North-Holland, 1992.
6. S. Hiranandani, K. Kennedy, C.-W. Tseng, and S. Warren, "The D Editor: A new interactive parallel programming tool," in *Proc. Supercomputing'94 (Washington DC)*, IEEE Comp. Soc. Press, Nov. 1994.
7. M. Pantano and H. P. Zima, "An integrated environment for the support of automatic compilation," in *High Performance Computing: Technology, Methods and Applications*, pp. 159–176, Elsevier Science B.V., 1995.
8. The Parallel Tools Consortium (Ptools). http://www-ptools.llnl.gov/.

[*The complete version of this paper, including a comprehensive example of an* Annai *debugging and performance analysis session, is available as CSCS-TR-96-04. For further details on the WWW see*: http://www.cscs.ch/Official/Project_CSCS-NEC.html]

On the Implementation of a Replay Mechanism

Michiel Ronsse* and Luk Levrouw

Department of Electronics and Information Systems
Universiteit Gent, Sint-Pietersnieuwstraat 41, B-9000 Gent, Belgium

Abstract. Parallel programs can be nondeterministic: consecutive runs with the same input can result in different executions. Therefore we cannot use cyclic debugging techniques. In order to be able to use those techniques we need a tool that traces information about an execution so it can be replayed for debugging. Because the recording interferes with the program, possibly perturbating the execution, we must limit the amount of information and keep the algorithm simple. This paper presents an implementation of the ROLT replay mechanism for a multi-threaded operating system (Solaris).

1 Introduction

The debugging of most parallel programs is a time-consuming task due to the complex nature of parallel programs. Moreover, most parallel programs are non-deterministic, limiting the use of cyclic debugging techniques. These techniques are based on the fact that re-executions of a program will result in the same program flow if we supply the same input.[2] During those re-executions we can analyze the program execution by setting breakpoints and watching variables until we find the error.

2 Replay Mechanisms

If we can force re-executions to be 'equivalent' to the execution that contains an error, we can still use cyclic debugging techniques. This can be accomplished using a replay mechanism: we trace a program execution (record phase), and use those traces to force subsequent executions (replay phase) to be 'equivalent' to the traced one. As these forced re-executions will be deterministic, we can use intrusive debugging techniques (breakpoints, collecting data, ...). To be practical, it is important that the trace mechanism produces small trace files and has a small overhead.

Recently, a new replay method called ROLT (Reconstruction Of Lamport Timestamps) was introduced [LAV94]. The mechanism produces smaller trace

---

* Michiel Ronsse is supported by a grant from the Flemish Institute for the Promotion of the Scientific-Technological Research in the Industry (IWT).
[2] For the remainder of this paper, we will assume that the user input, file input, system calls, ... return the same result during subsequent executions.

files and is much less intrusive than comparable mechanisms [Net93, LM87]. The implementation described in this paper traces the order of the synchronisation operations. Forcing these operations to occur in the same order during replay will yield the same execution as long as the program is data race free. A data race is an unsynchronized access to a shared variable by two processors when at least one modifies the variable. This is the caused by a lack of synchronization (or the wrong synchronization). As there is no synchronization, no information will be stored during the record phase. So, during replay, we will be unable to force the accesses to occur in the same order as in the record phase. This doesn't mean that the mechanism is totally unusable in the presence of a data race. A replay of an execution that contains a data race will yield an equivalent re-execution up to the point were the data race occured. Therefore, an (intrusive) data race detection method can be used during replay to find the race.

The ROLT mechanism uses Lamport clocks [Lam78] to attach logical timestamps to synchronization operations. As logical timestamps may not reflect the actual real-time order of the operations and because it is required that operations on the same synchronization variable have consistent timestamps (the timestamps of all operations on the same object should reflect their execution order) a simple update scheme is used at every synchronisation operation [LAV94].

During replay, the Lamport timestamps of the original execution are used to add sufficient synchronization for a correct replay: operations with a lower Lamport timestamp are executed first. To be able to do this, the same Lamport timestamps as in the original execution must be attached to the corresponding operations. Some of these timestamps can be deterministically recomputed during replay, the others were traced.

3 Implementation

Solaris offers the parallel programmer different synchronization types. These facilities are build using a layered approach. The lowest layer consists of a synchronization facility directly supported by the Sparc processors: `ldstub` (load-store unsigned byte). This is a simple *read-modify-write* instruction. The next layer (offered by `libc`) consists of two functions: `_lock_try` and `_lock_clear`. The first one tries to lock a byte and returns the result (succeeded or not), the last one clears a lock. The next layer consists of the four different synchronization types offered by the thread library (`libthread`). These are: mutual exclusion (`mutex`) locks, condition variables (`condvar`), counting semaphores (`sema`) and multiple readers, single writer locks (`rwlock`).

The replay mechanism was implemented using the dynamic linking facility of Solaris: we wrote two new `libthread` libraries, one for the record phase, and one for the replay phase. By using the dynamic linking facilities, the user doesn't have to do anything to use the replay mechanism. He doesn't have to recompile or relink his program, he can use whatever interactive debugger he likes (`gdb`, `dbx`, `debugger`, ...). Moreover, a user can add his own synchronization primitives (e.g. barriers) and use the instrumented Solaris functions in the record library to implement them.

As mentioned before, we will generate trace information about a particular execution during the tracing. The information is stored in a buffer in memory, and written to disk when the buffer is full. We have to consider four different cases :

The program exits normally, and the result is correct. This means that this particular program execution is correct. We can discard the traces, or we can use them to force a deterministic replay and collect more information about the program execution.

The program exits normally, but the result is not correct. In that case, we can use the traces to force a replay. Using an interactive debugger we can find the error using watchpoints, breakpoints, ...

The program ends, but at a wrong exit point (it crashes). As the tracing is performed by the library, the tracing mechanism will crash together with the application. This means that the information that is still in the memory buffer is lost. To tackle this problem, we use a Unix-daemon that controls and owns the memory used by the library. It checks on a regular base if one of the programs it provides with memory has crashed, and saves the memory to disk if necessary.

The program never ends (deadlock, infinitive loop). In this case, we have to force a program crash (i.e. sending a kill signal using ^C).

Record Phase During the record phase, we have to trace the execution order of the synchronization operations. As Solaris provides different levels of synchronization primitives, we can trace at different levels:

- we can trace at the lowest level (lock level). This will force all levels above this level to be replayed in the correct order;
- we can trace at the mutex level. All levels above this level will be replayed in the correct order, the level beneath it (lock level) won't. As it is possible that mutex functions have to call _lock_try several times before the lock is grabbed, this will diminish the number of operations to be traced;
- we can trace at the highest level: the synchronization operations (mutex, rwlock, condvar, sema) performed by the application are traced, the synchronization operations called by the rwlock, condvar and sema operations aren't.

Replay Phase Every thread recomputes its Lamport timestamps during replay. When a thread wants to perform a synchronization operation, the thread waits until all other threads have executed the operations with smaller Lamport clocks. This adds the extra synchronization needed to yield an equivalent execution.

4 Experimental Evaluation

Up to now, only limited experiments were performed with a parallel implementation of the Maximum Likelihood Expectation Maximization algorithm. The program was executed on a Sun with 4 processors. The following table shows the number of synchronization operations performed, the size of the trace files and the execution time.

level	#operations	logsize (b)	real time (s)
sema	2406	11824	44.98
mutex	7140	12424	45.11
lock	7791	17056	45.51

It is clear that, as expected, we have to log less, and that the overhead is smaller if we trace at a higher abstraction level. The total execution time for a normal execution was 44.23 seconds, for a traced execution 44.98 seconds and for a re-execution 48.42 seconds (mean values for 100 program runs; sema level). As we will be using an interactive debugger during the program re-executions, the increase in execution time causes no harm.

5 Conclusions

This paper showed that it is necessary to use a replay mechanism if one wishes to use cyclic debugging techniques for the debugging of non-deterministic parallel programs. An implementation of the ROLT mechanism for Solaris was proposed. It generates small logfiles and has a low overhead.

References

[Lam78] Leslie Lamport. Time, clocks, and the ordering of events in a distributed system. *Communications of the ACM*, 21(7):558–565, July 1978.

[LAV94] Luk J. Levrouw, Koenraad M. Audenaert, and Jan M. Van Campenhout. A new trace and replay system for shared memory programs based on Lamport Clocks. In *Proceedings of the Second Euromicro Workshop on Parallel and Distributed Processing*, pages 471–478. IEEE Computer Society Press, January 1994.

[LM87] Thomas J. LeBlanc and John M. Mellor-Crummey. Debugging parallel programs with Instant Replay. *IEEE Transactions on Computers*, C-36(4):471–482, April 1987.

[Net93] Robert H.B. Netzer. Optimal tracing and replay for debugging shared-memory parallel programs. In *Proceedings ACM/ONR Workshop on Parallel and Distributed Debugging*, pages 1–11, May 1993.

[RLB95] M.A. Ronsse, L.J. Levrouw, and K. Bastiaens. Efficient coding of execution-traces of parallel programs. In J. P. Veen, editor, *Proceedings of the ProRISC / IEEE Benelux Workshop on Circuits, Systems and Signal Processing*, pages 251–258. STW, Utrecht, March 1995.

Concepts and Functionalities of the DOSMOS-Trace Monitoring Tool

Lionel Brunie and Olivier Reymann

Laboratoire de l'Informatique du Parallélisme
Ecole Normale Supérieure de Lyon
69364 LYON Cedex 07, France

Abstract. This paper presents a description of the concepts and functionalities of the DOSMOS-Trace monitoring tool. The designing choices lead us to introduce the concepts of Event Manager Processes and Meta-Objects. It allows to keep a good scalability of the system but also to get a low intrusive monitoring tool.

1 Introduction

By gathering the advantages of shared memory systems and distributed memory systems, Distributed Shared Memory (DSM) systems offer an intuitive and easy-to-use programming environment and therefore receive increasing attention. The total transparency of shared data accesses hides the behaviour of its application (and of the system) to the programmer. A monitoring tool would be a great help for him to better understand how the whole system (*i.e.* the application plus the DSM system) reacts. There are many monitoring tools but are essentially designed for message passing systems. In that framework, the relevant information is related to the sequence of "send" and "receive", but for DSM systems, it concerns the operations performed on shared data. Thus, this paper presents the concepts and the functionalities of our DSM monitoring tool, DOSMOS-Trace (section 3). This tool was developed to monitor the DOSMOS DSM system which is quickly described in section 2. Section 4 concludes this paper by giving the main direction of our future work.

2 The DSM System Model

Developed on top of PVM, DOSMOS (Distributed Objects Shared MemOry System) [1] already benefits from the portability property.
   The four main features of DOSMOS are:

Variable sharing: The system allows the sharing of user defined variables in a totally transparent way. Basic types of data (*i.e.* integers, reals) but also static arrays can be declared.

The array splitting up: This feature is provided in order to reduce the contention on variable accesses in case of false-sharing.

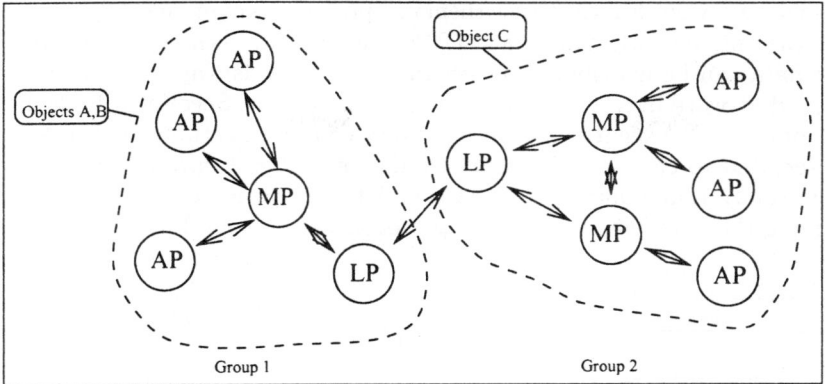

Fig. 1. DOSMOS system : an example of software configuration

Weak memory consistency model: The system allows variable replication. Thus, to ensure a consistency between the variable copies, we have implemented a weak consistency model (the release consistency) which permits to avoid expensive synchronizations and useless communications.

Hierarchic groups: The idea is to gather the processes that access regularly to a common set of variables. Only processes belonging to the group of a variable are allowed to get a copy. Due to the process clustering, the number of variable copies is limited and the cost of the consistency maintenance is bounded. This allows a good system scalability.

Figure 1 presents the software configuration of a DOSMOS application. It is composed by three types of processes:

Application Processes (AP) which contain the code of the user application (in C) plus code for communications with the Memory Processes.

Memory Processes (MP) which are in charge of the management of the shared objects. It answers the access requests sent by the Application Processes.

Link Processes (LP) are gateways for a Memory Process which wants to access a variable not present in the same group.

3 The Monitoring Tool: DOSMOS-Trace

3.1 The Concepts

In order to provide a low intrusive and scalable monitoring tool, we choose to add a new kind of process in the system called *Event Manager Processes (EMP)*. These processes receive information from the Memory Processes about variable accesses. Depending on the further use of the received trace data, they can be stored in two ways. Firstly, to use them in a post-mortem manner, they are stored in a trace file on disk. Since each EMP generates its own trace file, the whole trace data can be distributed on several disks. By this way, DOSMOS-Trace is designed to work in a scalable and little intrusive way. Secondly, if the data are used during the execution, we use a *Meta-Object* structure. Such a structure is associated to each variable and contains only useful information to perform on-line analysis and optimization. A complete description of the monitoring tool including the detailed concepts and protocols can be found in [2].

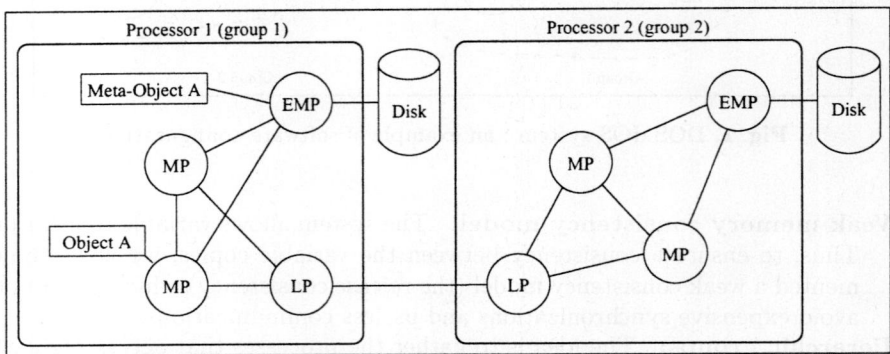

Fig. 2. DOSMOS-Trace: example of a monitoring environment architecture

3.2 The Functionalities

Visualization The user needs to see how its application runs by displaying various views of the trace data. Our tool provides two kinds of views. The first one displays access statistics on a variable (*cf.* fig 3). The second one displays an "history" of the operations performed either on a shared variable or by an Application Process. For instance, figure 4 shows the variable activity versus the execution time. One can see that the variable is shared by three processes.

Analysis and Optimization The trace information can also be analysed to detect some critical situations (*i.e.* bottlenecks, ping-pong effect, *etc...*) or a bad group structure which lead a performance lowering. Such analysis can as easily be done in a post-mortem way as on-line [3].

The result of the trace analysis can be used to optimize the application execution. By an on-line analysis, the system would be able to automatically modify some application parameters (*i.e.* the group structure). Nevertheless, it could also restrict its work to just give the programmer some advice.

Debugging We already have developed an extension to the system that allows us to retrieve the state of each shared variable in the system when the application is hanging.

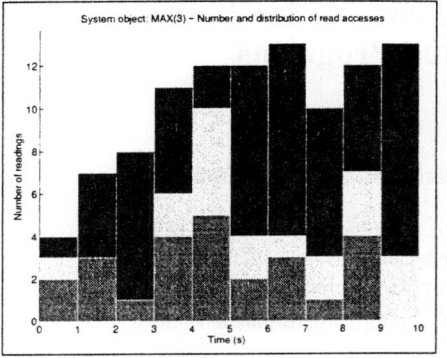 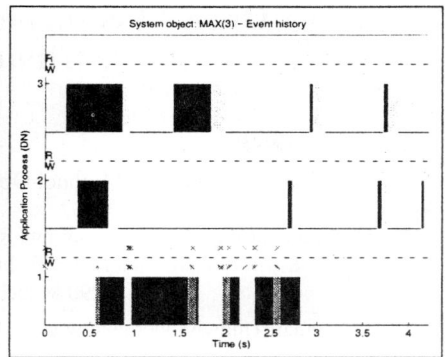

Fig. 3. Number and origin of the read accesses performed on an object vs execution time

Fig. 4. Variable's activity vs execution time

4 Conclusion and Future Work

This paper has presented the concepts of the DOSMOS DSM system and its monitoring tool DOSMOS-Trace. The implementation and our experiments have proved their scalability, efficiency and robustness. The introduction of Event Manager Processes and Meta-Object allows to get a small monitoring overhead and a data storage structure suited for an on-line use. Further developments are mainly focus on the definition and the implementation of an analysis and optimization tool. Moreover, we currently have a project with people developing OMIS [4] to extend their tool in order to be able to debug DSM applications and especially DOSMOS ones.

References

1. Lionel Brunie and Laurent Lefèvre. New propositions to improve the efficiency and scalability of DSM systems. June 1996. to be published in the proceedings of the IEEE ICA3PP'96 conference (Singapore).
2. Lionel Brunie, Laurent Lefèvre, and Olivier Reymann. Execution Analysis of DSM Applications: A Distributed and Scalable Approach. In ACM Press, editor, *SPDT'96 : SIGMETRICS Symposium on Parallel and Distributed Tools*, pages 51–60, Philadelphia, Pennsylvania, USA, May 1996. FCRC.
3. Lionel Brunie, Laurent Lefèvre, and Olivier Reymann. Monitoring and performance evaluation of distributed shared memory applications. In *Second International Conference on Massively Parallel Computing Systems*, pages 382–389, Ischia, Italy, May 1996. Euromicro.
4. Thomas Ludwig, Roland Wismüeller, Vaidy Sunderam, and Arndt Bode. OMIS : On-line Monitoring Interface Specification. Technical Report 342/05/96A, Technische Universität München, Institut für Informatik, February 1996.

An Open Monitoring System for Parallel and Distributed Programs

Thomas Ludwig, Michael Oberhuber, Roland Wismüller

Lehrstuhl für Rechnertechnik und Rechnerorganisation (LRR-TUM)
Institut für Informatik der Technischen Universität München
D-80290 München, Germany
email: {ludwig|oberhube|wismuell}@informatik.tu-muenchen.de

Abstract. The on-line monitoring interface specification OMIS provides means for developing more powerful tool environments for parallel and distributed systems. It specifies the interaction between any tool and a monitoring system which is responsible for observing and manipulating the programs' execution. By having this well defined interface it is now possible to concurrently use several tools of possibly different developers with the same program run. The research group at LRR-TUM currently designs an OMIS compliant monitoring system for PVM running on workstation clusters. Tool developers can use this implementation to attach their own on-line tools to the system.

1 Motivation

The available support for parallel programming in environments with distributed memory varies considerably in quality and quantity. Results of the Second Pasadena Workshop on System Software and Tools for High Performance Computing Environments [?] show that main problems are a lack of sophisticated tools and the almost complete absence of any interoperability of tools of different origin. Also there are no uniform environments at all, where the developer could use the same tools for different types of hardware or parallel programming libraries.

Tools which support an investigation of the running parallel program can be divided into on-line and off-line tools. In case of off-line tools, the task of a monitoring system is restricted to only collecting information about the system behavior and forwarding it to a file for storage. Since we do not know in advance which information the user wants to evaluate, enormous amounts of data have to be recorded, thus increasing the probability for a *probe effect*. With on-line tools the user evaluates this data immediately, which requires additional capabilities to be added to the monitor: programmability and adaptability.

Caused by its intermediate position, a monitoring system has two interfaces: one to the tools and one to the running program. Up to now, there are no approaches undertaken to standardize at least the tool/monitor-interface. Without a reasonable standard, however, new tools must also have new monitoring systems even if their functionality is not completely disjunct to already existing tools. Finally, the missing standard makes the integral use of a set of tools impossible as they are currently always based on differently implemented monitoring concepts which are incompatible to each other.

Having such an agreed on on-line monitoring interface, the amount of effort for having n tools on m systems (being composed of hardware, operating system, and runtime library of the programming environment) would be reduced from $n \times m$ to $n + m$, thus bringing more tools onto more parallel systems which finally would ease software development on these architectures. A further implication of having OMIS compliant monitoring systems is that finally we will reach the goal of having uniform environments, i.e. tools which are identical for a variety of target architectures.

The paper will describe the approach followed by OMIS. It will give an outline of the concepts which are incorporated into the specification. At the end we will show how to use the interface in a short example.

2 Related Work

Although methods and strategies for monitoring parallel and distributed systems seem to converge, no standard interface has yet been developed. This is especially true for the area of heterogeneous on-line monitoring. The reasons are additional demands like extendibility, programmability and asynchronous behavior. Several articles about monitoring methodologies have been published ([?], [?], [?]), but no comprehensive approach summing up state-of-the-art techniques to a common on-line monitoring system has been proposed. Comparable efforts are undertaken by the Ptools Consortium related to other areas of tool development.

Already in 1991 B. Bruegge was proposing a heterogeneous on-line monitor, called BEE (basis for distributed event environments)[?], albeit its capabilities are restricted to observe events. This constraint is also true for Xab [?]. Independence of the program environment and heterogeneity is proposed in the monitoring systems of the parallel debugger p2d2 [?], which is supposed to work with MPI as well as with PVM. But the integral use of different tools that are served by a common on-line monitor was realized a few times only [?, ?, ?] with the restriction to debuggers and performance analyzers.

The event–action paradigm is a common way to monitor parallel programs. It convinces through its intuitive use and simplicity, as shown in [?], but it lacks efficiency. There are two feasible ways to reduce the overhead without auxiliary hardware, filtering and programmability. But for an all purpose monitoring system it is not enough to recognize events and trigger actions. For example, Performance analysis requires a time driven monitoring to sample the current position of an application. Therefore, Petrenko is talking about *passive* and *active* monitoring systems [?], i.e. the monitor not does not only react, but it is acting itself.

Regarding event based monitoring systems and attempts to deal with heterogeneity the client/server approach has proven to be a feasible model for the architecture of a monitoring system [?, ?, ?]. To decouple tools and monitoring systems results in two major advantages, high portability and reusability.

The standard for on-line monitoring systems we want to propose in this article should involve the essence of the discussion above. OMIS is specified with respect to the event–action paradigm and the effort to minimize the overhead of event processing. Since a monitoring system is embedded between the tools and the object of observation it should have both, properties of a client and a server.

3 Requirements to a Monitoring Interface

The design of the monitoring interface imposes several requirements which the specification will have to meet. These requirements can be divided into three categories: (1) functional requirements, (2) conceptional requirements, and (3) efficiency requirements. We will give a short overview on major issues within these categories.

Functional requirements can be summarized as follows: the monitoring interface should be versatile enough to allow all possible tools to observe and manipulate all objects of the running program (e.g. processes, messages, variables etc.) and to watch hardware objects like the amount of available memory. In order to achieve the desired degree of versatility the monitoring interface should support the composition of more complex service requests from the basic ones.

As we neither know the complete functionality of the tools in advance nor the types of tools themselves we have to require that the monitoring interface satisfies the conceptional requirement of extendibility for future developments.

Finally, we have efficiency requirements. In order to keep the communication overhead minimal it is necessary to have powerful basic services and a possibility of composing service requests into a single request. Furthermore, the monitoring system should also be able to handle certain kinds of events occurring in the application program without interacting with the tool.

4 The Interface Structure

4.1 The System Model

An abstract view of the system's architecture is presented in figure 1. The application programs consist of a certain number of tasks which communicate by mechanisms provided by the parallel programming environment. The part which joins the tools to the running program is just the monitoring system. Its role is to establish the tool/program-interaction.

In contrast to usual interactive tools, like tool A, there may be some tools that are spread all over the target nodes, like e.g. a load balancer. We will call them distributed tools, like tool B. Regarding the middle layer we see two types of extensions (ME and DTE) that will be described in more detail in section ??.

How do the individual layers of figure 1 inter-operate? The monitoring system has two interfaces: one for interaction with the different tools (tool/monitor-interface), a second one for interaction with the program and all underlying layers which keep the program running. For simplicity reasons we will call the latter the monitor/program-interface.

The subject of OMIS is the specification of semantics for the on-line monitoring interface and the design of means to extend it. The tool/monitor-interface is therefore the union of the on-line monitoring interface and possible extensions.

4.2 Services

Conceptually the tool/monitor-interface consists of only a single procedure that can be invoked by both centralized tools and distributed tools. This procedure receives a string

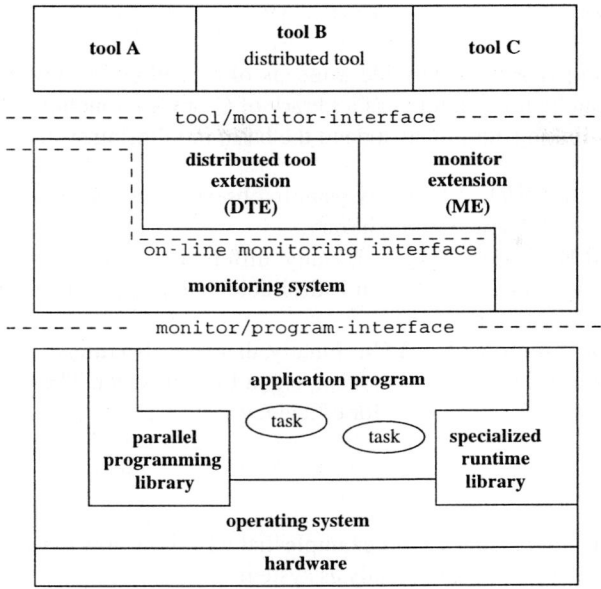

Fig. 1. System model: embedding an OMIS compliant monitoring system into an environment with tools and a parallel programming library

as an input parameter, interprets this string, and returns a result. The tool may either wait for this result from the monitoring system or may specify a call-back function that is invoked whenever the monitoring system returns a reply. The individual monitoring functions available by invoking the interface are called *services*; the string that is passed to the procedure (requesting the activation of a service) is called *service request*.

In order to meet the efficiency requirements the structure of a service request follows the event-action-paradigm, allowing the monitoring system to quickly react on state changes in the monitored system without having to communicate with the tool. If no event definition is present, the service request is unconditional and all actions are invoked immediately.The monitoring system automatically takes care of forwarding the requests to the proper nodes. In addition, we allow services that are global, i.e. that involve more than one monitor.

We classify services according to their input/output behavior:

1. **Manipulation services**. These services receive some input parameters from the caller, but will not have a result, e.g. "stop processes".
2. **Synchronous services**. These services receive some input parameters and immediately return a result exactly once, e.g. "show all tasks on node 0".
3. **Asynchronous services**. These services can have an arbitrary number of replies. Usually, neither the number of replies nor the time of the replies is known. This happens in case of an action that is triggered by an event.

4.3 Extensions

It is of utmost importance to provide a means of extending the interface. The additional services can be implemented as a library of C or C++ functions. There are three situations where linking additional code to the basic monitoring system is profitable:

Monitor Extension (ME): a tool may want to observe new objects within an application, e.g. those of a new runtime library.

Distributed Tool Extension (DTE): Usually, different tools use very different methods to process events. Therefore, it is desirable to define new services for this kind of tool-specific data processing.

Distributed Tool Component (DTC): Finally, there are also fully distributed tools without a central user interface component, e.g. a load balancer. These tools should be linked to the monitoring system for efficiency reasons.

4.4 Examples

In this section we will present a short example that will show how to use the monitoring interface in case of a performance analysis system.

Assume that a performance analysis tool wants to measure the time spent by task 4 (located on node 1) in the **pvm_send** call. In addition, the tool may want to know the total amount of data sent by this task, and it may want to store a trace of all barrier events. Then it may send the following service requests in form of strings to the monitoring system:

```
10 [1]StartLibcall([4],"pvm_send"): \
              11 [1]StartIntegrator(1),\
              12 [1]AddCounter(2,$4)
13 [1] EndLibcall([4],"pvm_send"):\
              14 [1]stop_integrator(1)
15 [] StartLibcall([],"pvm_barrier"):\
              16 [$0]Trace("barrier0",$0,$1,$2,$3)
17 [] EndLibcall([],"pvm_barrier"):\
              18 [$0]Trace("barrier1",$0,$1,$2)
19 [1] Enable(10),20 [1]Enable(13),\
              21 []Enable(15),22 []Enable(17)
```

Each request (except the last one) is composed of an event specification followed by a colon and an action specification. Several actions can be specified in one request, but only one event is allowed. Each individual specification is composed of an identifier (provided by the tool), a list of node numbers in square brackets to which the request refers, and the actual name of the service to be invoked followed by a list of parameters. Event specifications imply a list of well defined output parameters which can be referred to in the action list by using a $-notation.

The numbers 10 ... 22 are the request identifiers that will be included in the monitoring system's replies. However, in this example there will be no replies, except on error conditions, because only manipulation services are used as actions.

5 Current and Future Work

OMIS version 1.0 [?]is now available for the research community since February 1996. You can get it via WWW at http://wwwbode.informatik.tu-muenchen.de/~omis. We are in the progress of discussing specific aspects of OMIS with groups being interested in using and extending OMIS.

We expect to have a detailed specification of the compliant monitoring system by middle of this year and will then start with an implementation. Currently with the implementation of the monitoring system we will adapt existing tools of LRR-TUM to PVM.

References

1. T. Sterling, P. Messina, and J. Pool. Findings of the Second Pasadena Workshop on System Software and Tools for High Performance Computing Environments. Technical report, Center of Excellence in Space Data and Information Sciences, NASA Goddard Space Flight Center, Greenbelt, Maryland, 1995.
2. Bernd Bruegge. A Portable Platform for Distributed Event Environments. In *Proceedings of the ACM/ONR Workshop on Parallel and Distributed Debugging*, volume 26 of *ACM SIGPLAN Notices*, pages 184–193, December 1991.
3. James E. Lumpp, Howard Jay Siegel, and Dan C. Marinescu. Specification and Identification of Events for Debugging and Performance Monitoring of Distributed Multiprocessor Systems. In *Proceedings of the 10th International Conference on Distributed Computing Systems*, pages 476–483, Paris, Mai 1990.
4. C. Clemencon, J. Fritscher, and R. Rühl. Visualization, Execution Control and Replay of Massively Parallel Programs within Annai's Tool. Technical Report CSCS TR-94-09, CSCS, Manno, 1994.
5. Adam Louis Beguelin. Xab: A Tool for Monitoring PVM Programs. In *Workshop on Heterogeneous Processing*, pages 92–97, Los Alamos, April 1993.
6. Doreen Cheng and Robert Hood. A Portable Debugger for Parallel and Distributed Programs. In *Proc. of Supercomputing'94*, pages 723–732. IEEE, November 1994.
7. O. Hansen, J. Krammer, M. Oberhuber, and R. Wismüller. A Scalable Tool Environment for Observing the Runtime Behavior of Massively Parallel Applications. *To appear in Parallel Computing*, 1996.
8. Dan C. Marinescu, James E. Lumpp, Thomas L. Casavant, and Howard J. Siegel. Models for Monitoring and Debugging Tools for Parallel and Distributed Software. *Journal of Parallel and Distributed Computing*, 9:171–182, 1990.
9. A. K. Petrenko. Methods for debugging and monitoring parallel programs: A survey. *Programming and Computer Software*, 20(3):113–129, may–june 1994.
10. Suresh K. Damodaran-Kamal and Jeffrey S. Brown. Towards Heterogeneous Distributed Debugging. Technical Report LA-UR-95-906, Los Alamos National Laboratory, 1995.
11. T. Ludwig, R. Wismüller, V. Sunderam, and A. Bode. OMIS — On-line Monitoring Interface Specification (Version 1.0). Technical Report TUM-I9609, SFB-Bericht Nr. 342/05/96 A, Technische Universität München, Munich, Germany, February 1996.

Millipede: Easy Parallel Programming in Available Distributed Environments* **
(Extended Abstract)

Roy Friedman[1], Maxim Goldin[2], Ayal Itzkovitz[2], and Assaf Schuster[2]

[1] Department of Computer Science, Cornell University, Ithaca, NY 14850.***
[2] Department of Computer Science, The Technion, Haifa, 32000, Israel.

Abstract. MILLIPEDE is a generic run-time system for executing parallel programming languages in distributed environments. In this project, a set of basic constructs which are sufficient for most parallel programming languages is identified. These constructs are implemented on top of a cluster of workstations such that in order to run a specific parallel programming language in this distributed environment, all that is needed is a compiler, or a preprocessor, that maps the source language parallel code to the MILLIPEDE constructs. Some performance measurements of parallel programs on MILLIPEDE are also presented.

1 Introduction

The idea of using a cluster of workstations as a cost/effective high-performance parallel computer is becoming feasible as local area networks become faster and faster. This has led to the development of many run-time systems that support parallel programming languages on cluster environments in recent years [1, 3, 4, 5, 7, 8, 9, 10, 11, 12]. (A detailed comparison with these systems and others appears in the long version of this paper.) These run-time systems mainly differ by the memory consistency protocols they support, the load balancing schemes they employ, and the programming language they are targeted for. However, each of these run-time systems supports only one parallel programming language.

This means that in order to use an existing run-time system, one needs to adapt himself/herself to the one and only programming language which is supported by it. In this project, we have developed MILLIPEDE, a generic run-time system for parallel applications, which can support a variety of parallel programming languages. MILLIPEDE provides a flexible interface for creating parallel activities in the system, a distributed shared memory management, load balancing, and synchronization methods in the form of library routines. These

---

* The MILLIPEDE project is supported by Intel Academic Relations Grant, by Microsoft R&D group, and by a grant from the Israeli Ministry of Industry and Commerce.
** The MILLIPEDE URL: http://www.cs.technion.ac.il/~assaf/millipede.html
*** This author is currently supported by ARPA/ONR grant N00014-92-J-1866. Much of this work was done while this author was still with the Department of Computer Science at the Technion.

can be used by a compiler, or a preprocessor, to map the primitives of the language to the constructs provided by MILLIPEDE. Thus, the only thing that is required in order to implement most of the parallel programming languages on MILLIPEDE is to make the appropriate changes in its compiler, adjusting it to MILLIPEDE's interface, which is much easier than developing a new run-time system from scratch.

The goal of MILLIPEDE is therefore to support a large variety of parallel programming languages using commodity parts and standard operating systems so that most people can continue using their "favorite" parallel programming language on the same run-time environment which works with their existing equipment. Also, since it is relatively easy to incorporate new programming languages to the system, languages which are not supported at this point can be added on demand. For people that have no experience with parallel programming, MILLIPEDE currently supports the PARC programming language, which is a simple, yet powerfull, extension of C to parallel programming [2]. (We elaborate on PARC in the full version of this paper [6].) Other languages that we are currently supporting include a parallel version of C++ (CParPar), Java, and Par-Fortran 90. We intend to support Cilk, SPLASH macros, and Split-C soon.

For further flexibility and in order to be able to adapt to the particular needs of different applications, MILLIPEDE supports both *strong* and *weak consistency* memory management, which can be chosen by the application, and a variant of *release consistency* is being implemented. Also, several load balancing schemes have been implemented, and they can also be picked by the application. These protocols and schemes are described in more detail in the full version of this paper [6]. MILLIPEDE is fully implemented on MACH, and a more advanced version is under final stages of implementation on Windows-NT^{TM}.

We have conducted several performance measurements on the MACH implementation, and initial results are very encouraging. These results prove that at least for a large class of programs, it is possible to provide reasonable performance while running on a cluster of off-the-shelf workstations with an unmodified operating system, connected by an off-the-shelf network, and using a generic run-time system and a friendly programming language. Most programs achieve good speedups, while the speedups for programs that have natural parallelization is close to linear. Also, there seems to be a correlation between the problem sizes and the speedups, indicating that when the problem is large enough, the benefits of using our system outweighs the overhead imposed by it. We believe that by further optimizing the implementation, we will be able to achieve even better performance.

2 Millipede's support in ParC's constructs

MILLIPEDE provides powerfull interface that support PARC's constructs. PARC provides four constructs for creating parallel activities in the system:

pparblock creates a given number of activities, each being specified separately. In particular, this can be used to implement **fork**-like primitives.

lparblock similar to pparblock, but the number of parallel activities created is limited to some multiplication of the number of processor, so that if more blocks are specified, some of them will be executed within the same activity. This is useful for improved performance.

pparfor creates a given number of similar parallel activities which differ only by an index which is passed to them. Mainly useful for creating parallel loops.

lparfor similar to pparfor, but the number of actual parallel activities which are created is limited to some multiplication of the number of processor in the system.

The the following synchronization and atomic access constructs are provided: **sync** which is a barrier synchronization, **faa** for atomic fetch-and-add operations, **tss** as an atomic test-and-set operation, **rst** as an atomic reset operation, and **semaphore_wait** and **semaphore_signal** for mutual exclusion in critical sections. PARC also supports two main calls for early termination of parallel activities: **pcontinue**, terminates the current parallel activity and **pbreak**, terminates the current parallel activity and its siblings.

In order to run a PARC program in MILLIPEDE, a compiler for the source code programming language must first translate it into object code, with the appropriate calls to MILLIPEDE library functions, and must add a call to MILLIPEDE initialization routine. Then, the code can be linked together with the MILLIPEDE library to create the executable which runs both the MILLIPEDE run-time system and the application. When the application is executed, typically one activity is first created on one of the nodes. Then, as the execution develops, more parallel activities may be created or terminated on various nodes of the system, and parallel activities may migrate from one node to the other, depending on the load balancing policy chosen. This process is explained in more detail in the full version of this paper [6].

3 Performance

We have conducted several performance measurements which include matrix multiplication, finding the shortest distance in graphs, the traveling salesperson problem, and a linear equation solver. Due to space limitations, we only present here the high-lights of these measurements, but the full version of this paper contains a detailed description of these results [6].

In the case of matrix multiplication and finding the shortest distance in graphs, our system achieved close to linear speed-ups with problem size (matrix size and number of nodes respectively) of 400 elements or more and 4 nodes. For the traveling salesperson problem we achieved a speedup of 60% with 3 nodes, but only marginal speedup for the linear equation solver.

4 Conclusions and Further Research

In this work we have demonstrated that building a generic run-time system that can support a variety of programming languages and provide reasonable performance while using only commodity parts can be done. In our approach, adding support for a new parallel programming language means only slightly modifying its compiler, which increases the accessibility of distributed environments to parallel applications. This way all programming languages ported to our system can enjoy the same set of memory consistency and load balancing protocols, which can match the performance needs of their applications.

We are currently experimenting with porting more parallel programming languages to MILLIPEDE, supporting more memory consistency protocols, and developing new load balancing schemes. We are also investigating ways of incorporating fault-tolerance into our system.

References

1. H. Assenmacher, T. Breitbach, P. Buhler, V. Hubsch, and R. Schwarz. PANDA - Supporting Distributed Programming in C++. In *Proc. 7th European Conf. on Object-Oriented Programming (ECOOP)*, Kaiserslautern, July 1993.
2. Y. Ben-Asher, D.G. Feitelson, and L. Rudolph. ParC: An Extension of C for Shared Memory Parallel Processing. *Software Practice & Experience*, 1995.
3. B.N. Bershad, M.J. Zekauskas, and W.A. Sawdon. The Midway Distributed Shared Memory System. In *COMPCON*, pages 528–537, February 1993.
4. Robert D. Blumofe and David S. Park. Scheduling Large-Scale Parallel Computations on Networks of Workstations. In *Proc. 3rd Symp. on High-Performance Distributed Computing*, pages 96–105, August 1994.
5. J.B. Carter, J.K. Bennett, and W. Zwaenepoel. Implementation and Performance of Munin. In *In proc. 13th Symp. on Operating Systems Principles*, pages 152–164, October 1991.
6. R. Friedman, M. Goldin, A. Itzkovitz, and A. Schuster. Millipede: Easy Parallel Programming in Available Distributed Environments. Technical Report LPCR-9506, Department of Computer Science, The Technion – Israel Institue of Technology, November 1995.
7. A.M. Vahdat D.P. Ghormley and T.E. Anderson. Efficient, Portable, and Robust Extension of Operating System Functionality. Technical Report CS-94-842, University of California - Berkeley, 1994.
8. P. Keleher, S. Dwarkadas, A. Cox, and W. Zwaenepoel. TreadMarks: Distributed Shared Memory On Standard Workstations and Operating Systems. In *USENIX*, pages 115–131, January 1994.
9. D. Lenoski, J. Laudon, K. Gharachorloo, A. Gupta, and J. Hennessy. The directory-based cache coherence protocol for the DASH multiprocessor. In *In Proceedings of the 17th Annual International Symposium on Computer Architecture*, pages 148–159, May 1990.
10. H. Mehl. Distributed Shared Memory: A Survey. Technical Report SFB124-33/92, University of Kaiserslautern, Department of Computer Science, P.O.Box 3049, D-6750 Kaiserslautern, Germany, 1992.
11. A. Mohindra and U. Ramachandran. A Survey of Distributed Shared Memory in Loosely-Coupled Systems. Technical Report GIT-CC-91/01, College of Computing, Georgia Institute of Technology, Atlanta, GA 30332, January 1991.
12. B. Nitzberg and V. Lo. Distributed Shared Memory: A Survey of Issues and Algorithms. *COMPUTER*, pages 52–59, August 1991.

An Adaptive Cost System for Parallel Program Instrumentation *

Jeffrey K. Hollingsworth[1] and Barton P. Miller[2]

[1] University of Maryland, hollings@cs.umd.edu
[2] University of Wisconsin, bart@cs.wisc.edu

Abstract. We present a new data collection cost system that provides programmers with feedback about the impact data collection is having on their application. We allow programmers to define the level of perturbation their application can tolerate and then we regulate the amount of instrumentation to ensure that threshold is not exceeded. Our approach is unique in that the type of data gathered remains constant; instead we regulate when it is collected. This permits programmers to trade speed of isolation of a performance problem for less application perturbation. We implemented this cost system in the Paradyn Performance Tools and present case studies demonstrating the accuracy of the cost system.

1 Introduction

We present a new way to manage the perturbation caused by software data collection. Our approach is based on an instrumentation cost system that ensures that data collection and analysis can be accomplished while controlling the performance overhead of the instrumentation. The unique feature of our approach is that it lets the programmer see and control the overhead introduced by monitoring rather than simply being subjected to it.

The best way to handle instrumentation overhead is to avoid introducing it in the first place. In a previous paper [4], we described a new approach to performance monitoring called *Dynamic Instrumentation*. Dynamic Instrumentation delays instrumenting an application until it is in execution, permitting dynamic insertion and alteration of the instrumentation during program execution. Enabling instrumentation only when it is needed greatly reduces the amount of data collected, and thus the overhead due to the instrumentation system.

We have developed an instrumentation cost system to ensure that data collection and analysis does not excessively alter the performance of the application being studied. The system associates a cost with different resources. Possible resources include processors, interconnection networks, disks, and data analysis workstations. The cost system is divided into two parts: predicted cost and

---

* Supported in part by Wright Laboratory Avionics Directorate (WLAD), Air Force Material Command, USAF, grant F33615-94-1-1525 (ARPA order B550), NSF Grants CCR-9100968 and CDA-9024618, DOE Grant DE-FG02-93ER25176, and ONR Grant N00014-89-J-1222. The U.S. Government is authorized to reproduce and distribute reprints for Governmental purposes notwithstanding any copyright notation thereon. The views and conclusions contained herein are those of the authors and should not be interpreted as necessarily representing the official policies or endorsements, either expressed or implied, of WLAD or the U.S. Government.

observed cost. Predicted cost is computed when an instrumentation request is received, and observed cost while the instrumentation is enabled.

By computing the predicted cost of instrumentation before data collection starts, it is possible to decide if the requested data is worth the cost of collection. This predictive information can be used as feedback to reduce or defer an instrumentation request. Our higher-level performance analysis tools use the cost prediction to control how aggressively they instrument a program in search of performance bottlenecks. In many cases, control of instrumentation overhead permits our tools to more quickly isolate a performance problem (see Section 6).

Although predicting the cost of data collection prior to instrumentation execution provides useful data, it is important to make sure that the actual cost of data collection matches the predicted cost. The observed cost tracks the impact the currently enabled instrumentation has on the application. To be useful, our observed cost system needs to be both cheap to compute and accurately reflect the true impact of data collection. If the observed cost exceeds predefined limits, feedback is provided to the user or higher-level tool; this feedback allows us to dynamically maintain (approximately) a fixed level of instrumentation overhead.

2 Dynamic Instrumentation and W^3 Search Model

Our recent work in performance monitoring tools has focused on two areas. First, how can we efficiently collect performance data for large, long running applications? Second, how can we help programmers to understand the source of their performance problems rather than providing them raw performance data.

Our approach, called Dynamic Instrumentation, defers instrumenting the program until it is in execution. This approach permits dynamic insertion and alteration of the instrumentation during program execution. At any time during a program's execution, a consumer of performance data can start collecting a metric for a particular combination of resources. To satisfy this request, instrumentation code is generated and inserted into the program. When performance data is no longer required, its instrumentation code is removed from the program.

Dynamic instrumentation is designed to be usable for a variety of high level tools, and so it has a simple interface. The interface is based on two abstractions: *resources* and *metrics*. Resources are the objects about which we gather performance information. Resources include processors, interconnection networks, processes, procedures, and synchronization objects. Metrics are time varying functions that characterize a program's performance; they can be computed for any subset of the resources in the system. For example, CPU time can be computed for a single procedure executing on one processor or for the entire application.

We have also been investigating how to help programmers interpret the collected performance data. The W^3 Search Model[6] is a structured methodology for programmers to quickly and precisely isolate a performance problem without having to examine a large amount of extraneous information. It is based on answering three separate questions: *why* is the application performing poorly, *where* is the bottleneck, and *when* does the problem occur. By iteratively refining the answer to these three questions, we can precisely describe to programmers the reason their program is not performing as expected. Refining the answer to these questions requires testing different hypotheses about the source of performance problems. To deliver answers rather than asking the questions, we automate this search process. In an automated search, the tool refines the answers to these three questions by enabling and disabling the collection of performance data. The module that implements our model is called the *Performance Consultant*.

3 Cost System

With Dynamic Instrumentation, the data collected at a particular point in the program no longer remains fixed for the entire program's execution. Each time a new request for instrumentation is received, the instrumentation overhead for that point can change. Our system associates an instrumentation cost with different resources. Possible resources include processors, interconnection networks, disks, and data analysis workstations. The cost system is divided into two parts: *predicted cost* and *observed cost*. Predicted cost is computed when an instrumentation request is received, and can be used to estimate the overhead for the desired instrumentation. Observed cost is computed while the instrumentation executes and provides notification if the overhead has exceeded the user's expectations. In essence, observed cost is just another performance metric; we use the W^3 Search Model to implement much of the observed cost system.

3.1 Predicted Cost

Predicted cost is the expected overhead of collecting the data necessary to compute a metric for a particular combination of resources. We compute the predicted cost when an instrumentation request arrives, but before the instrumentation is inserted into the application. The predicted cost is expressed as the utilization of each measured resource required to collect the desired data.

In Dynamic Instrumentation, CPU time overhead is due to the insertion of instrumentation primitives at various points in the program's executable. To predict this overhead at a single point in the program we need to know: what instrumentation will be inserted, the cost of executing that instrumentation, and the frequency of execution of that point. We multiply the instrumentation overhead by the point's expected execution frequency to compute the predicted overhead. The sum of the overhead for all points is the predicted cost for an instrumentation request. Based on measurements, we know the cost of each instrumentation primitive; the difficult part is estimating their execution frequency.

Data about the frequency of instrumentation execution comes from a static estimate of procedure call frequency. This approach is at best a wild guess, but since we adjust this value based on runtime data the initial value does not need to be accurate. We associate with every point in the program an expected frequency. The initial value for each point is based on the point's type (e.g., user vs. system call). We denote the predicted cost for the application as C_{pred}.

3.2 Observed Cost

The observed cost monitors the effect of data collection on the application. Its purpose is to check that the overhead of data collection does not exceed pre-defined levels, and if it exceeds these levels, report it to the higher level consumers of the data. If the predicted and observed costs differ significantly, we can adjust the amount of instrumentation enabled.

Actual cost also might differ from predicted cost because of resource contention between the application and the data collection. For example, the predicted cost system does not include the memory hierarchy (e.g., caches and TLB). If the application is constrained by that resource then the impact on the performance could be significant.

Our current implementation includes only the direct CPU and cache pollution costs. Conceptually, computing the cost of executing the instrumentation is easy: we record the time spent executing the primitives. However, computing the cost of cache pollution is problematic. The difficulty is that it is impossible, without sophisticated hardware instrumentation, to know if cache lines used by the instrumentation will cause subsequent cache misses for the application. However, we can compute a bounds for the impact of cache pollution. The lower bound is that there was no cache pollution. This happens when none of cache items replaced due to instrumentation were subsequently used by the application. An upper bound is that every cache item loaded by the instrumentation code will result in a subsequent cache miss for the application[3].

Our observed cost has two values reflecting the lower and upper bound of the cache pollution. The actual cost of instrumentation should lie within this range. To compute the observed cost range we use two values:

C_{obs_direct}: The measured time spent executing the instrumentation. To efficiently compute this value we use statistical sampling.

C_{obs_cache}: The waiting time for cache misses during instrumentation code.

To compute C_{obs_cache}, we use C_{obs_ideal}. C_{obs_ideal} is the time required to execute the instrumentation assuming an ideal memory model (i.e., all memory requests are satisfied by the cache). Differences between the measured and ideal times are due to the memory hierarchy. So: $C_{obs\_cache} = C_{obs\_direct} - C_{ideal}$. To compute C_{obs_ideal}, we added an additional instruction at each instrumentation point to record the number of machine cycles required for the primitives at that point. The cycle count provides a precise measure of the instrumentation instructions executed. However, we still need to convert instruction counts to time. For the processors used in this case study, we divide the cycle times of the instrumentation instruction sequences by the clock frequency of the machine[4].

We then compute the lower and upper bounds for the observed cost as: $C_{obs\_low} = C_{obs\_direct}$, and $C_{obs\_high} = C_{obs\_direct} + C_{obs\_cache}$.

---

[3] While this one-to-one ratio is the worse case for direct mapped caches, for set-associative caches, the cache pollution penalty is also a function of the associativity.

[4] For super-scalar processors an approach similar to [8] will be required.

Observed cost can be viewed as just another a performance metric to characterize a type of bottleneck in a parallel program. The only difference is that the bottleneck in which we are interested was created by the data collection system rather than the programmer. We treat instrumentation as a potential bottleneck like an application bottleneck (such as too much synchronization blocking time) and use the W^3 Search Model to look for it. In the W^3 Search Model, the observed cost is expressed as additional hypotheses along the "Why" axis, that can be isolated to specific resources.

4 Evaluation of the Cost System

We ran three sequential and three parallel applications, and then compared the predicted and observed costs to the actual perturbation. For each program, we measured its performance with four different levels of instrumentation enabled:

Base: The minimum dynamic instrumentation is inserted (recording the start and end of the application).

Procedure: CPU time metrics for each procedure computed for the lifetime of the application (similar to the UNIX prof utility).

PC Base: The initial instrumentation used by the Performance Consultant to search for a bottleneck in the application is enabled.

PC Full: The Performance Consultant run in automated mode, turning on and off instrumentation as needed.

Since we were interested in assessing the accuracy of the cost system, we did not want to use the cost system to control the number of refinements being considered. However, we also did not want to overwhelm the application with instrumentation by enabling all refinements at once. As a compromise, we configured the Performance Consultant to consider ten refinements at once.

For each of the four levels of instrumentation, we recorded C_{obs_low}, C_{obs_high}, and C_{pred}. In addition we recorded two additional values:

T_{obs}: the user CPU time of the application program with the dynamic instrumentation, as measured by UNIX timing commands.

$C_{observed}$: the *timed* cost of the instrumentation, calculated as difference between T_{obs} for the current and base levels of instrumentation.

The range between C_{obs_low} and C_{obs_high} represents the bounds on the instrumentation overhead. If $C_{observed}$ is inside this range, our system accurately computed the instrumentation overhead. Differences between $C_{observed}$ and C_{pred} represent the inaccuracies in our calculations of the predicted cost. Accurate calculation of observed cost is crucial; accurate calculation of predicted cost is less critical since it can be corrected by feedback from the observed cost system.

The three sequential applications we measured (Ear, Fpppp, and Doduc) are from the floating point SPEC92 benchmark suite. Since instrumentation is currently inserted at procedure boundaries, we wanted a cross section of procedure size and procedure call frequency. The programs were run on an otherwise idle

SPARCstation 5 running at 85Mhz. PVM[2] versions of three parallel Computational Fluid Dynamics (CFD) kernels from the NAS parallel benchmarks[1] were also run.

4.1 Observed Cost

The results for the first sequential application, Ear, are shown at the top of Figure 1. The base time for this program is about 11.5 minutes, and averages 11,000 procedure calls per second during its execution. The values in the table show that the measured observed cost is within the range between C_{obs_low} and C_{obs_high} as is true for all three benchmarks. The total instrumentation overhead ($C_{observed}$) ranged from 9% to just over 42% of the CPU time of the base time to run the program in Paradyn. The Performance Consultant overhead was larger than we expected due to our naive code generator inserting duplicate copies of instrumentation to satisfy different metric requests.

Program Version	Time	$C_{observed}$ Time	Percent	C_{obs_low} Time	Percent	Delta	C_{obs_high} Time	Percent	Delta
Ear	687.9								
Procedure	753.4	65.5	9.5%	52.1	7.6%	(2.0)	87.9	12.8%	(-3.3)
PC Base	838.6	150.7	21.9%	124.0	18.0%	(3.9)	213.6	31.0%	(-9.1)
PC Full	978.2	290.2	42.2%	261.2	38.0%	(4.2)	453.2	65.9%	(-23.7)
Fpppp	293.8								
Procedure	309.4	15.6	5.3%	11.1	3.8%	(1.5)	19.5	6.6%	(-1.3)
PC Base	308.4	14.6	5.0%	11.9	4.1%	(0.9)	20.7	7.0%	(-2.1)
PC Full	314.8	21.0	7.1%	14.6	5.0%	(2.1)	24.2	8.2%	(-1.1)
Doduc	58.0								
Procedure	109.7	51.7	89.2%	48.1	82.9%	(6.3)	74.2	128.0%	(-38.8)
PC Base	61.2	3.2	5.6%	2.1	3.6%	(2.0)	3.6	6.2%	(-0.7)
PC Full	67.3	9.3	16.1%	7.1	12.2%	(3.8)	10.2	17.5%	(-1.4)

Fig. 1. Observed vs. Timed Overhead (seconds).

The second program was Fpppp, a quantum chemistry benchmark that does electron integral derivatives. It averaged 2,100 procedure calls per second.

The third program is Doduc, a Monte Carlo simulation of the time evolution of a thermo-hydraulical model of a nuclear reactor. This program averages 107,000 procedure calls per second. The timed observed cost for this program ranged from 5 to 89 percent of the base time to run the program using Paradyn. This program has the largest difference between C_{obs_low} and C_{obs_high}, indicating that the program is sensitive to cache perturbation.

Next, we tested our cost system with three NAS CFD benchmarks running on a network of SPARCstations 5's connected by Ethernet. The applications were configured with one master and four worker processes. We ran each program with the same same four levels of instrumentation that we used before.

A comparison of the low and high values of the observed cost and measured perturbation for these program appears in Figure 2. The Time column shows the total time of all processes. $C_{observed}$ ranged from 1.8% to 12.7% for these three programs. For all of the programs at all levels of instrumentation, the value of $C_{observed}$ was in the range from C_{obs_low} to C_{obs_high}.

Program Version	Time	$C_{observed}$ Time	Percent	C_{obs_low} Time	Percent	Delta	C_{obs_high} Time	Percent	Delta
PVM_BT	892.5								
Procedure	1005.8	113.3	12.7%	96.3	10.8%	(1.9)	161.2	18.1%	(-5.4)
PC Base	914.4	21.9	2.5%	13.8	1.6%	(0.9)	26.2	2.9%	(-0.5)
PC Full	925.5	33.0	3.7%	18.6	2.1%	(1.6)	33.4	3.7%	(0.0)
PVM_LU	99.0								
Procedure	105.2	6.2	6.3%	5.2	5.3%	(1.0)	9.0	9.1%	(-2.8)
PC Base	103.4	4.4	4.4%	3.3	3.3%	(1.1)	6.3	6.3%	(-1.9)
PC Full	104.6	5.6	5.7%	3.2	3.2%	(2.4)	5.9	6.0%	(-0.3)
PVM_SP	474.5								
Procedure	483.1	8.6	1.8%	7.4	1.6%	(0.3)	9.1	1.9%	(-0.1)
PC Base	501.7	27.2	5.7%	15.7	3.3%	(2.4)	29.6	6.2%	(-0.5)
PC Full	509.2	35.0	7.4%	20.2	4.3%	(3.2)	35.0	7.4%	(0.0)

Fig. 2. Timed vs. Observed Overhead for NAS CFD Kernels (seconds).

4.2 Predicted Cost

To gauge the effectiveness of the Predicted Cost, we ran the same applications we used to study the observed cost metric, and measured the predicted cost metric. These results are based on using the static predicted cost information and do not include any compensation based on the observed cost.

The predicted cost for the Ear program is shown at the top of Figure 3. The value for the Procedure and PC Base cases are each within 6% of the observed cost. The value for the PC Full case had an error of almost 40%. This is due the Performance Consultant inserting instrumentation into a single procedure that is called thousands of times a second. The middle section of table shows the predicted cost for the Fpppp application. For all three cases, the estimated cost was within 6% of base time to run the application using Paradyn. The last part of table shows the predicted cost for the Doduc application. The errors in the Predicted Cost ranged from 4.7% to 73.1% in this case. The largest error occurred instrumenting the CPU time for all procedures.

We also compared the predicted cost data for the three PVM applications. The results are shown in Figure 3. The difference between the actual predicted running time for all three applications was within 6% for all three levels of instrumentation.

5 Using Cost to Control Searching

We now describe how the predicted cost can be used with the W^3 Search Model. First the programmer sets the tolerable instrumentation overhead. We use this

Program Version	$C_{observed}$		C_{pred}			Program Version	$C_{observed}$		C_{pred}		
	Time	Percent	Time	Percent	Delta		Time	Percent	Time	Percent	Delta
Ear						**PVM_BT**					
Procedure	65.5	9.5%	97.3	14.4%	-4.6	Procedure	113.3	12.7%	112.7	12.6%	0.1
PC Base	150.7	21.9%	111.6	16.2%	5.7	PC Base	21.9	2.5%	1.6	0.2%	2.3
PC Full	290.2	42.2%	18.8	2.7%	39.5	PC Full	33.0	3.7%	4.1	0.5%	3.2
Fpppp						**PVM_LU**					
Procedure	15.6	5.3%	7.3	2.5%	2.8	Procedure	6.2	6.3%	9.6	9.7%	-3.4
PC Base	14.6	9.4%	9.4	3.2%	1.8	PC Base	4.4	4.4%	0.0	0.0%	4.4
PC Full	21.0	7.1%	5.6	1.9%	5.2	PC Full	5.6	5.7%	0.2	0.2%	5.4
Doduc						**PVM_SP**					
Procedure	51.7	89.2%	9.3	16.1%	73.1	Procedure	8.6	1.8%	0.9	0.2%	1.6
PC Base	3.2	5.6%	0.6	0.9%	4.7	PC Base	27.2	5.7%	2.5	0.5%	5.2
PC Full	9.3	16.1%	0.5	0.9%	15.2	PC Full	35.0	7.4%	16.2	3.4%	4.0

Fig. 3. Observed Cost vs. Predicted Cost for Six Applications (in seconds).

value to moderate how much instrumentation gets inserted. In manual search mode, the predicted cost acts as a check to see if the request associated with a hypothesis can be satisfied without undue perturbation. In automated search mode, we enumerate possible refinements, and then work down that list adding instrumentation and evaluating the results. When a test request pushes the instrumentation overhead too high, we delay requesting new instrumentation. Thus the perturbation threshold regulates how many hypotheses (potential performance problems) are considered simultaneously. Raising the threshold permits the search system to try more tests at once, but with a higher overhead. However, changing the threshold does not change what hypotheses get tested; it simply changes when they get tested.

We quantified how well our cost system regulated the perturbation of the Performance Consultant. For each application, we ran the Performance Consultant three times. The first time was with an overhead limit of 10% perturbation, the second time for a fixed limit of three refinements to the current hypothesis, and third with no limit on the refinements to the current hypothesis. The limit of three refinements provides a comparison to an alternative strategy for controlling the cost of data collection. The unlimited case measures the worst case impact of instrumentation for each application.

We evaluated two criteria about the effectiveness of our search system. First, we verified that the instrumentation cost was held within the limit set by the user (10% in this case). Second, we compared how quickly a performance problem could be isolated using each method.

For each run of an application, we compared the bottlenecks identified by the Performance Consultant. For Ear, the same performance bottleneck was found for all three cases, but the order was different. For the Doduc application, the same performance bottleneck was found in the 10% limit and three hypothesis limit, but the perturbation was so high in the unlimited case that no bottleneck was identified[5]. For Fpppp, the hypothesis limit and unlimited cases identified one procedure as the bottleneck and the cost limit identified another procedure. A CPU time profile for the application showed that both procedures consumed

---

[5] The application finished execution before the search was completed

enough CPU time to be flagged as bottlenecks according the thresholds.

The results for Doduc, Ear, and Fpppp are shown in Figure 4. The Search Time column reports the amount of elapsed time required for the Performance Consultant to execute its search. For all the applications, the time required for the search was least when cost was used to regulate hypothesis evaluation. The improvement in search time ranged from 29% (Fpppp) to 71% (Ear) compared to the limit of three hypotheses. The cost-based limit was able to evaluate the available hypotheses faster because different hypotheses have different costs and the cost-based limit permits evaluation of more hypotheses simultaneously, while keeping the overhead within the limit. The cost-based limit was able to identify a problem faster than the unlimited search case because it saved time by not inserting instrumentation for all possible refinements.

Program Control Method	Search Time	C_{obs_ideal} Avg	Max	Std. Dev.
Doduc				
10%	48.2	3.1%	8.6%	2.3%
3 Hypotheses	77.1	3.1%	5.6%	1.7%
unlimited	258.7	6.2%	41.6%	9.3%
Fpppp				
10%	52.0	1.2%	3.7%	0.9%
3 Hypotheses	73.2	0.5%	1.3%	0.3%
unlimited	227.2	1.9%	3.7%	0.4%
Ear				
10%	37.9	2.7%	8.1%	3.3%
3 Hypotheses	130.8	5.1%	17.5%	6.3%
unlimited	226.5	15.4%	17.2%	3.5%

Fig. 4. Summary of fixed vs. cost-based hypothesis evaluation.

6 Related Work

Perturbation compensation[5] reconstructs the performance of an un-perturbed execution from a perturbed one. These techniques require a trace based instrumentation system and post-mortem analysis to reconstruct the correct ordering of events. Our approach does not try to factor out perturbation; instead we try to avoid it using the predicted cost, and quantify it using the observed cost.

Pablo[7] uses an adaptive instrumentation system. The programmer specifies events to log for post mortem analysis. If the volume of data collected exceeds certain thresholds, the system will stop producing event logs and instead produce summary information. Pablo leaves the underlying instrumentation in place and controls the logging of data. However, our technique has the advantage that disabling data collection completely removes the instrumentation code and so there is no latent perturbation due to instrumentation code that is disabled but must execute code to learn that it is disabled.

Goldberg and Hennessy[3] used the difference between the measured and predicted time of a code region to quantify the affects of the memory hierarchy. Our approach differs in two ways from theirs. First, since we need to be able to characterize the impact of small, but (potentially) frequently accessed instrumentation code blocks, we use statistical sampling instead of timers. Second, our goal is to compute the impact of the instrumentation on the original code rather than the impact of the cache on a single block.

7 Conclusion

Our cost system controls software instrumentation overhead based on feedback. We predict the amount of overhead we will cause and then use our instrumentation facility to provide information about the actual costs. The mechanisms that we have built give the programmer direct control over their instrumentation. We expose the overhead of data collection as a first class metric in Paradyn. The programmer is also given explicit control of the overhead, which controls the rate at which the performance tool searches for bottlenecks. We evaluated our model with six programs, and demonstrated that the actual instrumentation overhead was within the range of our observed cost model.

References

1. D. H. Bailey, E. Barszcz, J. T. Barton, and D. S. Browning. The NAS parallel benchmarks. *Journal of Supercomputer Applications*, 5(3):63–73, Fall 1991.
2. J. Dongarra, A. Geist, R. Manchek, and V. S. Sunderam. Integrated PVM framework supports heterogeneous network computing. *Computers in Physics*, 7(2):166–174, March-April 1993.
3. A. J. Goldberg and J. L. Hennessy. Performance debugging shared memory multiprocessor programs with MTOOL. *Supercomputing 1991*, pages 481–490, Nov. 18-22 1991.
4. J. K. Hollingsworth, B. P. Miller, and J. Cargille. Dynamic program instrumentation for scalable performance tools. *1994 Scalable High-Performance Computing Conf.*, pages 841–850, May 1994.
5. A. Malony. *Performance Observability*. PhD Dissertation, Department of Computer Science, University of Illinois, Oct. 1990.
6. B. P. Miller, M. D. Callaghan, J. M. Cargille, J. K. Hollingsworth, R. B. Irvin, K. L. Karavanic, K. Kunchithapadam, and T. Newhall. The paradyn parallel performance measurement tools. *IEEE Computer*, 28(11), Nov. 1995.
7. D. A. Reed, R. A. Aydt, R. J. Noe, P. C. Roth, K. A. Shields, B. W. Schwartz, and L. F. Tavera. Scalable performance analysis: The pablo performance analysis environment. In A. Skjellum, editor, *Scalable Parallel Libraries Conference*. IEEE Computer Society, 1993.
8. K.-Y. Wang. Precise compile-time performance prediction of superscalar-based computers. *ACM SIGPLAN'94 Conf. on Programming Language Design and Implementation*, pages 73–84, June 20-24 1994.

SVMview: A Performance Tuning Tool for DSM-Based Parallel Computers*

Didier Badouel[1], Thierry Priol[2], Luc Renambot[2]

[1] Ecole des Mines de Nantes, 4 rue Alfred Kastler, La Chantrerie, 44070 Nantes
[2] IRISA/INRIA - Campus de Beaulieu - 35042 Rennes, France
badouel@emn.fr, [priol | renambot]@irisa.fr

Abstract. This paper describes a performance tuning tool for DSM-based parallel computers. SVMview is a tool for post-mortem analysis of page movements caused by the execution of Fortran-S programs. This tool is able to analyze particular phenomena such as false-sharing which occurs when several processors write into the same page simultaneously.

1 Introduction

Since 1985, the design of High Performance Parallel Processing Systems (HPPPS) is mainly based on distributed memory parallel computers (DMPCs). Exploiting the performance of such computers is not an easy task since the programming model is often seen as a break through process as compared to the one associated with sequential computers. Programmers have to manage explicitly message-passing between processors as well as data distribution over the local memories which is considered as some sort of low level programming. Several alternatives have already been investigated to program DMPCs using a high level programming model. Most of them aim at providing a global address space implemented either by a compiler (HPF-like approach) or by an operating system layer (Distributed Shared Memory)

This latter idea arose from the K. Li's PhD thesis [11]. Since then, a great deal of work has been carried out in that area for both distributed systems (networks of workstations) and parallel computers (DMPCs). However, at the present time, none of the existing DMPCs provides a shared virtual memory layer although the usefulness of such a concept has clearly been demonstrated for some applications [2]. There are various reasons to explain such limited impact of innovating technology. Among them, the availability of a shared memory abstraction makes the programmer not pay attention to data locality thus entailing bad performance. Contrary to the message passing, communications are implicit through page movements. Therefore, a programmer does not easily figure out where communications occur in his program. Two other reasons have been recently pointed out by J. Carter in [6]. The first one is that *"existing DSM systems are not well integrated with the rest of the software environment such as the compiler"*. The second one is because *"there is a dearth of tools for debugging and tuning the performance of DSM programs"*. These two reasons are closely related since tools for tuning performance of DSM systems often depend on the availability of programming environments. However, there are several on-going research efforts dealing with the design of programming environments for DSM systems: among them Fortran-S [4] and SVM-Fortran [7].

---

* This work is supported by the Esprit BRA APPARC and an Intel ERDP

This paper describes a performance tuning tool for DSM systems called SVMview [3]. SVMview has been designed to analyze page movements caused by the execution of Fortran-S programs. It has been targeted to various parallel computers such as the iPSC/2 running KOAN DSM [10] and the Paragon XP/S running either ASVM [14] or MYOAN[5].

2 A DSM-based Programming Environment

This section gives an overview of a programming environment we designed for the Paragon XP/S. It consists of a DSM system (MYOAN) for the OSF/1 MK-AD operating system and a parallel Fortran code generator (Fortran-S).

MYOAN [5] is a shared virtual memory system designed for the Intel Paragon supercomputer that is implemented as an external pager of the MACH microkernel. It provides the same functionalities as KOAN which was originally designed for the Intel iPSC/2 [10]. MYOAN offers several consistency protocols: strong (atomic) consistency and a relaxed form of consistency, that allows multiple writers into the same page. In addition, MYOAN provides a page broadcast mechanism to avoid contention when several processors request the same pages (producer-consumer scheme). A monitoring support has been added to provide the user with a detailed feedback (counters and events) of the execution of parallel programs. Counters give a global view of the DSM activity while not consuming too much memory. However, such information is inadequate whenever a detailed analysis is required. In this case, it is better to gather DSM events for further processing. Unfortunately, storing all events that occur during a parallel execution is a highly memory demanding operation. Nevertheless, MYOAN is able to trace events related to page management, like the activation of the page fault handler and servers that process incoming page requests and page invalidations. These events are stored in a temporary user buffer on each node.

Fortran-S [4] is a parallel extension to Fortran targeted for DSM parallel architectures. A set of directives provides the programmer with a simple programming model based on shared array variables and parallel loops. Shared variables are mapped into distributed shared memory regions provided by a DSM system. Parallel loops can be distributed among the available processors using several predefined scheduling policies. In addition, Fortran-S features directives to handle specific behavior such as false-sharing or producer-consumer scheme. Similarly to other control-oriented parallel fortran languages, Fortran-S offers several directives to specify reduction operations or synchronization such as critical section, atomic update and user events signals. Some other directives in Fortran-S are meant for execution monitoring. Among them, a pair of directives define a code section where a detailed analysis is in order.

3 Analyzing Parallel Programs Running on a DSM

The combination of a DSM system with a "high-level" programming model, such as MYOAN and Fortran-S, allows an incremental approach to parallelization. Since the parallel programming model is based on the use of directives, the programmer can simply modify or insert directives either to parallelize a new part of the code or to optimize a part which has already been parallelized. The optimization is a loop-based analysis process: the programmer modifies his source code, runs it on a parallel computer, and then pinpoints the bottlenecks using a performance tuning tool. During program execution, events are gathered in order to be visualized using a dedicated tool in a post-mortem process. To make program optimization easier, a performance tuning tool has to deal with three aspects: communication, load balancing and synchronization.

Communication Bottlenecks in DSM Systems Several data access patterns may increase the amount of communication generated by the DSM system. Memory management may decide to remove a copy of a page that is still needed for further computations. This well-known problem in virtual memory management is called page thrashing. Another phenomenon is false-sharing. It occurs when several processors are writing simultaneously to the same page. Due to the strong cache coherence protocol, such a page will move back and forth between processors. Several solutions can be applied to solve this problem such as array padding, specific loop scheduling strategy, compiler optimizations or the use of specific cache coherence protocols. Depending on the degree of false-sharing and the way the program has been parallelized, one of these optimizations is probably better than the others. In order to decide which of these optimizations has to be used, a fine-grained analysis of this phenomenon is necessary.

Another severe drawback of a DSM system is its lack of support for efficient management of a producer-consumer scheme. Our experiments have shown that such data access patterns can entail very bad performance when the number of processors crosses over a given threshold. The actual threshold depends on both the target architecture and the implementation of the DSM system. A performance tuning tool has to offer a way to monitor such a problem.

Load Balancing Some of the loop scheduling strategies may cause a heavier communication traffic than others. For instance, a scheduling strategy based on cyclic distribution of the iteration space, is sometimes well suited to distribute the load when the amount of computations depends on the iteration value. On the other hand, such a strategy may have the drawback of increasing the false-sharing effect. A block distribution strategy may badly distribute the load whereas it allows keeping the amount of false-sharing low. Another strategy based on affinity scheduling may have the advantage of reducing the page movements by exploiting the underlying temporal locality of the code. However, it may lead to a poor load balance. A trade-off has often to be found by users to select the scheduling strategy offering the maximum performance (i.e. a few page movements with a good balance of the load). A performance tuning tool has to provide the programmer with the possibility to analyze page movements and load balance in connection with a scheduling strategy.

Synchronization When a SPMD execution model is considered, barrier synchronizations play an important role in synchronizing processor activities. The Fortran-S compiler automatically inserts barrier synchronizations before and after a parallel loop to ensure the correctness of the program execution. Removing useless barrier synchronizations is part of the optimization process since some of them are not really needed depending on data dependencies. Therefore, a performance tuning tool must inform the programmer about the synchronization cost. The programmer may then take the decision to remove some barriers. For this purpose, Fortran-S provides several directives to inform the compiler to suppress barrier synchronization before or after a parallel loop or both.

3 Event Data Collection

In our programming environment, three software components are involved during the execution of a parallel program: MYOAN, a runtime library which implements some services for the SPMD code generated by Fortran-S and the SPMD code itself. Each one of these software components has to be instrumented to log specific events. Such events provide their contribution to the analysis of communication bottlenecks, load balancing or synchronization issues as discussed in the previous sections.

DSM Events Among the different software components, the DSM system is the one that will generate most of the communication during the execution of the program to be analyzed. Our approach focuses on the logging of events which are meaningful for the programmer. Such events correspond either to a page fault when reading from or writing into a page or to a modification of the access right. As for the page fault event, several pieces of information are recorded: the identifier of the processor on which the event occurs, the identifier of the shared memory region that contains the faulting address, two time stamps to specify when the page fault occurs and when it is processed, the faulting address, the faulting page number and the identifier of the processor that sends the page.

To make the analysis of any section of the program possible, it is necessary to take a snapshot of the page distribution right before the execution of that code section. A specific event is generated on request on every processor to know exactly the page distribution for a particular shared memory region.

Despite the fact that we decided not to generate an event for every communication, the execution of a parallel program may generate a huge number of events. This is especially true when an inexperienced programmer is parallelizing his sequential code for the first time. Events have to be managed carefully in order not to alter the program behavior and not to lose the collected events. Our design choice consists in letting the programmer allocate a buffer in the user space of the local memory associated with each processor. Buffer address is then communicated to the DSM to store in the events. Buffers are managed using a round-robin strategy. In case of a buffer overflow, a specific event is generated.

Runtime Library Events The SPMD code generated by Fortran-S calls a runtime library for several services such as shared memory allocation, message-passing communication, synchronization, loop distribution, etc... The runtime library has been instrumented to collect information needed to know the mapping of Fortran shared variables to shared memory regions, to estimate the cost of synchronization as well as to provide information about load balancing. One of the main motivations in designing a performance tuning tool for DSM was to integrate it to a programming environment. In other words, a relationship between DSM events and program elements (shared variable, parallel loop) has to be highlighted to the programmer. Considering this objective, we included in the trace an event which specifies the memory mapping of a shared variable to a shared memory region managed by the DSM. Such event has the following parameters: variable name, variable type, virtual address, number of dimensions and the parameters for each dimension. With such information, a DSM event, like a page fault, can be linked to the access to an element of the shared variable that caused the page fault. Moreover, events are generated when entering/leaving a sequential code section or a parallel loop. As for the MYOAN events, they are stored into buffers which are managed using a round-robin strategy. In case of buffer overflow, a special event is stored in the buffer.

SPMD code instrumentation The compiler makes very few modifications to the SPMD code. It basically inserts function calls to open and close trace files, and instruct both the runtime and the DSM system to start or quit logging events. Since only the compiler is able to provide variable names and types, several function calls are also inserted at each shared variable allocation. A function call is also inserted before the code section to be analyzed, to take a snapshot of the DSM status. An event is generated in the trace indicating, for each shared memory region, the access rights for every page.

4 SVMview: an DSM-based performance tuning tool

Tuning the performance of a parallel code requires a precise knowledge of its execution. Such knowledge can be extracted from the trace files which contain specific events like the ones described in the previous section. Unfortunately, the number of events is so large that the user is usually not able to extract useful information from the traces and to take proper actions. Visualization techniques can help the programmer better understanding his code by mapping the events into graphical views. To be pertinent, such graphical views cannot be independent from the programming model.

Several performance tuning tools have been designed in the past, but most of them focused mainly on performance tuning for message-passing parallel codes like Paragraph [9] or AIMS [13]. None of these tools can provide graphical views adapted to our requirements (*i.e.* able to show the DSM activity with a link to the high-level programming source code). Moreover, few of them are actually available. Instead of modifying an existing tool, another approach consists in using a generic tool which can be targeted for a specific need. Pablo [12] is one of these tools. Contrary to other tools, Pablo is easily extensible. It provides a set of standard graphical views which can be used through a graphical language. This language describes the processing of the trace file through a graph where each node represents a specific operation or visualization to be applied to the input data coming from ancestors in the graph. Pablo reads trace files using the Self-Defining Data Format (SDDF) [1] which is a trace description language. Despite its programmability, we found Pablo difficult to use since the processing of our trace files require often complex computations which cannot be easily described with its graphical language. For these reasons, we decided to build a new tool, called SVMview.

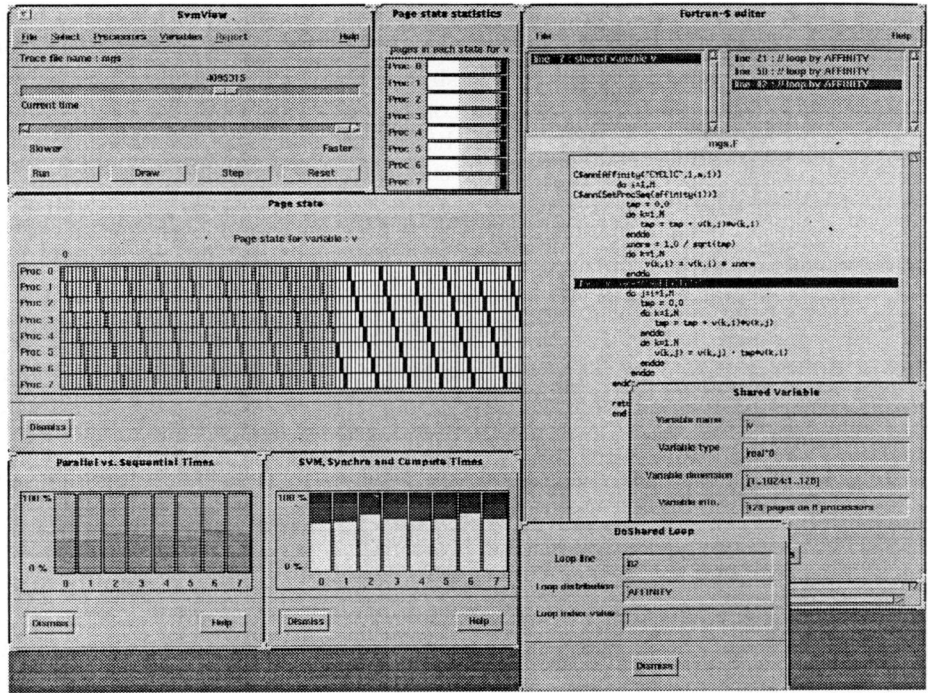

Fig. 1. SVMview and some of its graphical views

Overview of SVMview SVMview has been designed as a platform to experiment with *post-mortem* visualization tools for DSM-based architectures. SVMview takes as input SDDF trace files. Events taken out from the traces are passed to several graphical displays according to their types. SVMview allows the programmer to analyze DSM events linked to Fortran-S programming objects such as shared variables, sequential code section and parallel loops. As for instance, at any time, SVMview is able to inform the programmer that an event related to a DSM activity is related to a shared variable and where these events occurred in the code.

Figure 1 shows different graphical views from SVMview. A *monitor* window (at the upper left) allows to play back the execution and dispatches the events to the graphical views. SVMview has a specific graphical view to display the Fortran-S source code. This graphical view (at the upper right) links execution steps to Fortran-S source code. It allows the programmer to know exactly from what part of the Fortran-S source code the events, which are displayed in other graphical views, have been generated. Within the editor window, several areas are active, which means that the programmer can select a directive specifying a shared variable or a parallel loop to get some detailed information such as, for instance, the number of pages to map a given shared variable.

The *pages state* window shows the access rights (read-write, read-only, invalid) of every pages in the local memory of each processor using specific colors. During the execution playback, the programmer can see the access right modifications and page movements in an animated way. Several communication bottleneck patterns can be discovered using this view. However, specific bottlenecks such as false-sharing or producer-consumer can be identified using proper graphical views which are described in the following sections.

Many other possible views exist. Generally, the views are based on classical graphical displays such as bar-graphs (to display information related to synchronization and load balancing issues), Gantt charts, Kiviat diagrams and spreadsheet tables. Most of the graphical components we used in SVMview come from the Pablo environment [12] or from the Xbae toolkit. SVMview is also able to supply data to external applications such as MAPLE for specific processing.

False-sharing detection SVMview can both quantify and graphically exhibit the false-sharing phenomenon [3]. The graphical views correspond to different levels of interpretation to analyze traces related to a parallel loop:

1. a low-level view selectively highlights code sections where the number of pages exchanged between nodes is important;
2. another view allows to tell which pages are particularly involved in the traffic;
3. finally, a third view quantifies the false sharing effect for a given page.

The false sharing effect is measured by means of a two-dimensional array of counters. These counters are updated within a parallel loop to record the number of write page faults for a given page and for every processor involved in the communication. Each row corresponds to the number of times the selected page was sent; each column corresponds to the number of times the selected page was received. Due to the Fortran-S programming model, the write page faults, occuring during the execution of a parallel loop, are mostly due to write operations into different locations within a page. For this reason, these counters are well-suited to characterize false-sharing effects.

Producer-consumer phase detection The producer-consumer effect is a different phenomenon characterized by two phases separated by a barrier synchronization. During the production phase (a sequential code section), one particular processor produces values by means of one or more write operations. During the consumption phase (in a parallel loop), several processors attempt to read the produced value(s). As a result, several messages are sent requesting a copy of a particular page to the producer processor. The intensity of the phenomenon can be characterized by the number of page replicas produced during the consumption phase. SVMview offers a graphical view to show the producer-consumer effect on a variable [3]. SVMview characterizes a producer-consumer effect by counting for each page the number of consumers. This is done by counting the read requests for each page when stepping into a parallel loop.

5 Related works

There is few works dealing with the design of performance tuning tools for DSM systems. Another similar project to our own is currently in progress at KFA in Juelich. They designed a programming environment based on the ASVM DSM system [14] designed at Intel ESDC in Munich. One component of this environment is the SVM-Fortran language which has some similarities with Fortran-S. A source-code-based optimizer and a locality analyzer tool, called OPAL [8], have been designed. Trace files can be visualized using the ParVIS tool.

6 Conclusion

SVMview has been developed to allow the analysis of Fortran-S codes. It focuses mainly on data locality management to provide useful information to the programmer to optimize his parallel code. When programming with a DSM architecture, the main objective is to prevent page movements by exploiting the underlying locality. SVMview is also a testbed to carry out experiments on the usefulness of graphical displays in monitoring the behavior of parallel codes. Most of the displays we used came from existing tools such as Pablo or Maple. Adding new graphical displays is very easy as they can be either internal or external to SVMview.

References

1. R. A. Aydt. The Pablo Self-Defining Data Format. TR, U. of Illinois, July 1994.
2. D. Badouel, K. Bouatouch, and T. Priol. Ray tracing on distributed memory parallel computers: Strategies for distributing computation and data. *IEEE Computer Graphics and Application*, 14(4):69–77, July 1994.
3. D. Badouel, T. Priol, and L. Renambot. Svmview: a performance tuning tool for dsm-based parallel computer. TR 966, IRISA, November 1995 1995.
4. F. Bodin, L. Kervella, and T. Priol. Fortran-S: A fortran interface for shared virtual memory architectures. In *Supercomputing'93*. IEEE CSP, November 1993.
5. G. Cabillic, T. Priol, and I. Puaut. MYOAN: an implementation of the KOAN shared virtuel memory on the Intel Paragon. TR 812, IRISA, March 1994.
6. J.B. Carter, D. Khandekar, and L. Kamb. Distributed Shared Memory: Where we are and where we should be headed. In *Proc. of HOTOS'95*, 1995.
7. M. Gerndt and R. Berrendorf. Parallelizing applications with SVM-Fortran. In *HPCN, LNCS*, volume 919. Springer-Verlag, May 1995.
8. M. Gerndt, A. Krumme, and S. Özmen. Performance analysis for SVM-Fortran with OPAL. TR KFA-ZAM-IB-9519, KFA-ZAM, August 1995.
9. M. Heath and J. Etheridge. Visualizing the performance of parallel programs. *IEEE Software*, 8(5):29–39, August 1991.
10. Z. Lahjomri and T. Priol. KOAN: a shared virtual memory for the iPSC/2 hypercube. In *CONPAR/VAPP92*, September 1992.
11. K. Li. *Shared Virtual Memory on Loosely Coupled Multiprocessors*. PhD thesis, Yale University, September 1986.
12. D.A. Reed, R.A. Aydt, R.J. Noe, P.C. Roth, K.A. Shields, B.W. Schwartz, and L.F. Tavera. Scalable performance analysis: The PABLO performance analysis environment. In *SPLC*, pages 104–113. IEEE Computer Society, 1993.
13. J. Yan, S. Sarukkai, and P. Mehra. Performance measurement visualization and modeling of parallel and distributed programs using the AIMS toolkit. *Software, Practice and Experience*, 25(5):429–461, April 1995.
14. S. Zeisset. Evaluation and enhancement of the Paragon multiprocessor's shared virtual memory system. Master's thesis, TU Munich, November 1993.

Cautious, Machine-Independent Performance Tuning for Shared-Memory Multiprocessors

Sarah A. M. Talbot, Andrew J. Bennett, Paul H. J. Kelly

Department of Computing
Imperial College of Science, Technology and Medicine
London SW7 2BZ

Abstract. Coherent-cache shared-memory architectures often give disappointing performance which can be alleviated by manual tuning. We describe a new trace analysis tool, CLARISSA, which helps diagnose problems and pinpoint their causes. Unusually, CLARISSA works by analysing potential contention, instead of measuring predicted contention by simulating a specific memory system design. This is important because, after tuning, the software will be executed on different inputs and different configurations. The goal is to produce a program with robustly good performance. This paper explains the principle behind cautious trace analysis, describes our implementation, and presents our experience of using the tool.

1 Introduction

There has been considerable recent interest in developing tools to support manual performance optimisation of applications running on coherent-cache shared-memory multiprocessors (e.g. [1, 3]). The purpose of a performance tuning tool is to direct the programmer's attention to where a program is spending its time and to give as much guidance as possible into how to reduce the performance bottlenecks.

Existing performance tools measure (using special monitoring circuitry) or predict (using a simulation of the shared memory architecture) the behaviour of the machine for which the program is being developed. Although this is very useful in understanding the factors influencing performance, there are two fundamental problems in using such tools for producing high-quality software:

1. In the field, the software will be run with many different inputs, leading to behaviour different from that seen during tuning, and
2. The software will be installed on hardware with different characteristics from that used during tuning.

In this paper we present an alternative approach, *cautious trace analysis*, aimed at addressing these problems. The key idea is to identify behaviour which might lead to lost performance on some reasonable architecture, or with different timing assumptions. If we can diagnose and eliminate, or at least minimise, these characteristics, the program should behave well in service.

2 Cache Line Contention in Shared-Memory Systems

A shared-memory multiprocessor consists of several CPUs with associated caches linked to memory units via an interconnection network. A cache coherency protocol is required to ensure that CPUs do not use stale cached data. In addition to the overheads of maintaining coherence, such architectures can suffer from three problems: contention for nodes, cache lines and communication links. These all conspire to increase memory access times, and hence slow down the execution time of tasks running on the processors. The challenge is to minimise the causes of contention, i.e. to keep data in the local cache whenever possible and to avoid using the network. We use the following definitions of cache line sharing:

Active sharing: a data item is accessed by more than one processor during the execution of a program.
False sharing: this is where processors share a cache line without sharing data items within the cache line. With invalidation, a write to an item in a shared cache line requires all copies of that cache line to be invalidated, even if the other processors never use the data item which was changed.
Passive sharing: this occurs where shared data still remains in a processor's cache even though no objects on the cache line will be accessed by that processor again [2]. Since a write by another processor to any item in that cache line will require all other copies of the cache line to be invalidated, it is desirable that the redundant cache line is ejected after its last use.

These characteristics interact, and are affected by the memory access characteristics of a particular program and the shared-memory architecture. Cache line size is particularly relevant: larger cache lines would allow many objects to be allocated on each cache line, which could be helpful if an application has locality of access. However, the larger line size can lead to contention for cache lines, especially if false sharing plays a significant role in the behaviour of an application.

3 Cautious Trace Analysis and CLARISSA

The analysis process operates as a sequence of phases, in order to reflect barrier synchronisations (which prevent events occurring on different sides of a barrier from overlapping), limit the amount of analysis time and space required, and prune overlaps which are unlikely because they appear at widely-differing times in the trace (overlap is possible, since no synchronisation prevents them, but are unlikely). Essentially, what needs to be considered is which events could possibly occur in the same phase. Whatever the hardware configuration, the events for a particular CPU will always occur in the same order, but the order in which events occur between different CPUs can vary. In the example, Fig. 1, it is possible that an event in CPU_0 may occur before or after any event in, say, CPU_1 from the start of the program up to the first barrier synchronisation. However, it is not

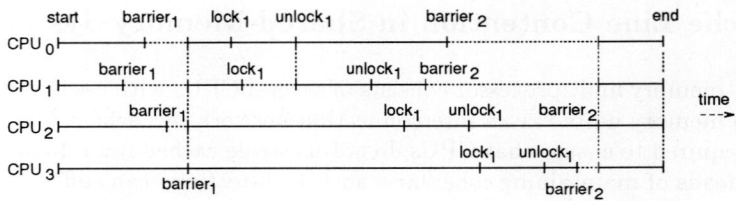

Fig. 1. An example execution path

possible for events occurring before the first barrier to overlap with events after that barrier.

Considering only barriers, Fig. 1 has three phases. Sharing can occur across barriers, but it is not currently part of our analysis. In programs where barriers are used regularly to ensure synchronisation, such as MP3D (Sect. 4), cautious trace analysis between barriers shows the sharing effects that may occur within the weak ordering programming model.

In applications where there are few or no barriers, the analysis becomes so broad that it is likely that CLARISSA will over-report the potential sharing, and the volume of data and phase end processing will be problematic. In such cases, we introduce fixed time-slots. The length of the time-slots depends on the overall length of the program. Too short a time-slot will result in some effects being missed and give too fine a level of summary information, whereas too long a time-slot will tend to over-report sharing and contention, and give a summary which is too coarse. In addition, edge effects have to be allowed for, i.e. the analysis must take into consideration sharing effects which cross a time-slot boundary. In Sect. 5 an example is given of using time-slot analysis (with overlaps) to tune the performance of an application which makes little use of barrier synchronisations.

Cautious analysis also has to allow for the use of locks. For example, in Fig. 1, if each CPU only reads and writes a particular data item when the processor has obtained $lock_1$ then, although there is still active sharing of the item between the barriers, the programmer has protected the data item from the possibility of simultaneous update by two or more processors.

3.1 Using CLARISSA

The CLARISSA tool is based on [5]. Input parameters include cache line size, class threshold (the N value in Table 1), phase type (barrier or time-slot), time-slot length and overlap. A classification system is needed for summarising the wealth of data. Table 1 gives the classification used in the SM-prof performance debugging tool, which reports cache line access for fixed time-slots in terms of read or write accesses and the number of CPUs involved [1]. In CLARISSA, an enhanced version of this categorisation is used, where the sharing categories (ending in E/F/M) are further split according to active or false sharing. The

Table 1. SM-prof classification of cache line accesses [1]

class	degree of sharing	access mode	comments
UNR	none	none	no processor referenced the cache line
ROE	exclusive	read only	one processor has done a read operation, but no write operation to the cache line
ROF	shared by few	read only	i processors have done read operations, but no write operations, to the cache line[a]
ROM	shared by many	read only	N or more processors have done read operations, but no write operations, to the cache line
RWE	exclusive	read/write	one processor has performed a read-modify-write sequence on the cache line
RWF	shared by few	read/write	i processors have performed read-write-modify sequences to the cache line.
RWM	shared by many	read/write	N or more processors have performed read-write-modify sequences to the cache line

[a] where $1 < i < N$

results are used to provide histograms of sharing activity for each phase during the execution of the program.

4 MP3D

MP3D is a particle-based wind tunnel simulation, from SPLASH [4]. It is used to study the shock waves created as an object flies at high speed through the upper atmosphere. It was run for 30000 molecules, using the supplied geometry file *test.geom*, for 10 time-steps. Two large arrays of structures account for more than 99% of the static data space used by MP3D; the first structure stores the state information for each *particle* and the second structure stores the properties of each *cell* in the active space.

The way MP3D uses barriers for synchronisation within each step meant that CLARISSA barrier analysis was the most appropriate, and the resulting phase level graphs are shown in Fig. 2. These graphs, in conjunction with the summary and detail level sharing information generated by CLARISSA, showed that the dominant sharing was active sharing of the *cells* array. For many data items within the *cells* array, more than one processor updates the same data item between barriers, and this generates a high number of coherency protocol invalidation messages.

To avoid the active sharing of *cells*, MP3D was modified so that the scheduling of work for the CPUs was driven by cells rather than particles. When the trace output for MP3D-NEW was analysed by CLARISSA, there was a substantial

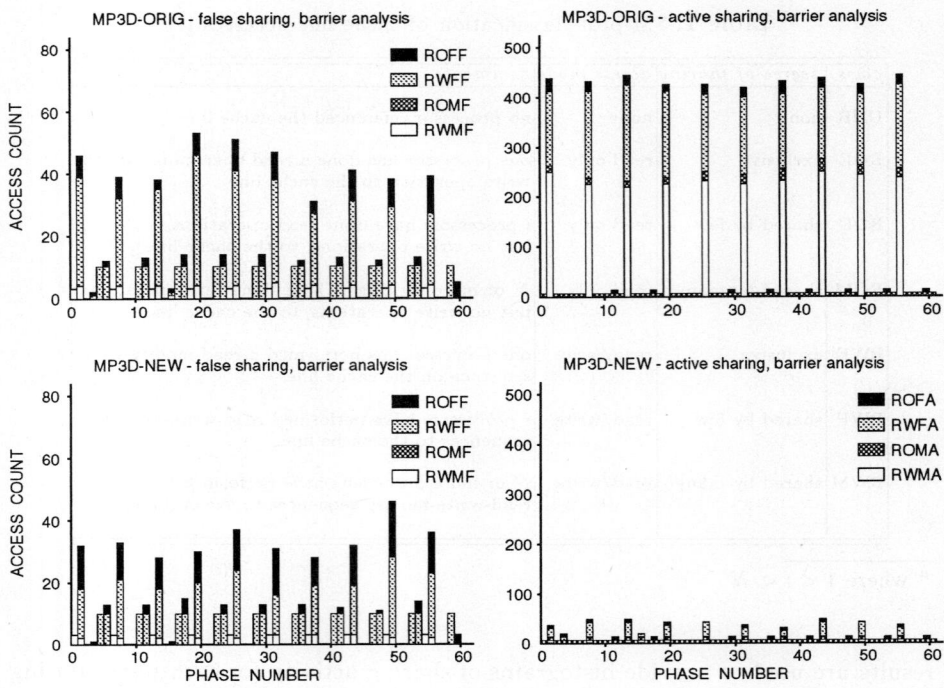

Fig. 2. Phase level graphs for MP3D-ORIG and MP3D-NEW

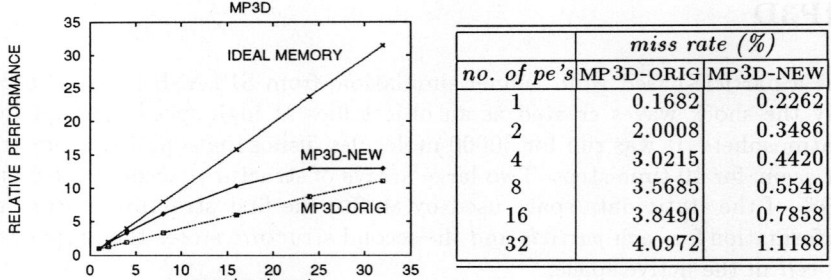

	miss rate (%)	
no. of pe's	MP3D-ORIG	MP3D-NEW
1	0.1682	0.2262
2	2.0008	0.3486
4	3.0215	0.4420
8	3.5685	0.5549
16	3.8490	0.7858
32	4.0972	1.1188

Fig. 3. Relative performance and cache miss rates for MP3D

reduction in the active sharing of the *cells* array, and false sharing had also been reduced[1]. This is illustrated by the active and false sharing phase level graphs shown in Fig. 2.

The two different versions, MP3D-ORIG and MP3D-NEW, were run on an execution-driven simulator to obtain CC-NUMA execution timings, cache miss

---

[1] Similar changes have been made by other researchers e.g. [1], but we made the change because it was specifically indicated by the active sharing information from CLARISSA.

rates and relative performance. The results are shown in Fig. 3. The best speedup is achieved by MP3D-NEW. Simulation statistics confirm that the miss rate is substantially lower for MP3D-NEW, so the strategy of reducing sharing has been successful. However, this is starting to lose its "edge" at 32 CPUs: the drop in performance is believed to be because of poor load balancing given the relatively small problem size (i.e. 30000 molecules) used for these tests.

5 Computational Fluid Dynamics

CFD is a major application area of high performance computing. The system modelled in our CFD application is a laminar flow in a square cavity with a lid which slides across the cavity introducing a zone of re-circulatory fluid. CFD uses barriers to ensure that CPU_1 has updated global variables before all the processors move on to the next stage, but there are long periods without a barrier [6]. Time-slot analysis was therefore appropriate, and time-slot length was chosen to give around 100 slots over the execution time, i.e. to give a reasonably detailed profile without being swamped by too much information.

The phase level graphs generated by CLARISSA are shown in Fig. 4. The graphs, in conjunction with the summary and detail level sharing information, indicated that the most significant sharing was false sharing within the *var* data structure. The 2-D arrays in *var* were originally distributed to the CPUs columnwise, and the program was modified to create CFD-NEW, which uses square block distribution. Without the help of CLARISSA, it would only have been possible to pinpoint the performance problem by gaining a thorough knowledge of the application program. When the trace output for CFD-NEW was analysed by CLARISSA, there was a substantial reduction in false sharing messages relating to the *var* data structure, reflected in the improved false sharing phase level graph shown in Fig. 4. Active sharing was increased by the change but, as shown by the performance results below, any coherence overhead incurred by this increase is more than compensated for by the reduction in false sharing. CFD-ORIG and CFD-NEW were run on the simulator to obtain timings, cache miss rates and relative performance, shown in Fig. 5. The best speedup running under real memory is achieved by CFD-NEW. In addition, the simulations showed that the cache miss rate was always lower for CFD-NEW in comparison with CFD-ORIG. The reduction in false sharing lead to a significant improvement in performance, even though active sharing increased slightly.

6 Related Shared-Memory Tools

MemSpy [3] assists in locating bottlenecks by providing detailed information that focuses the programmer's attention on the problem areas in the application. SM-prof [1], is similar to that presented here, but has the drawback that it does not distinguish between active and false sharing of cache lines. It also splits a program's execution up into time-slots, but does not allow for boundary effects

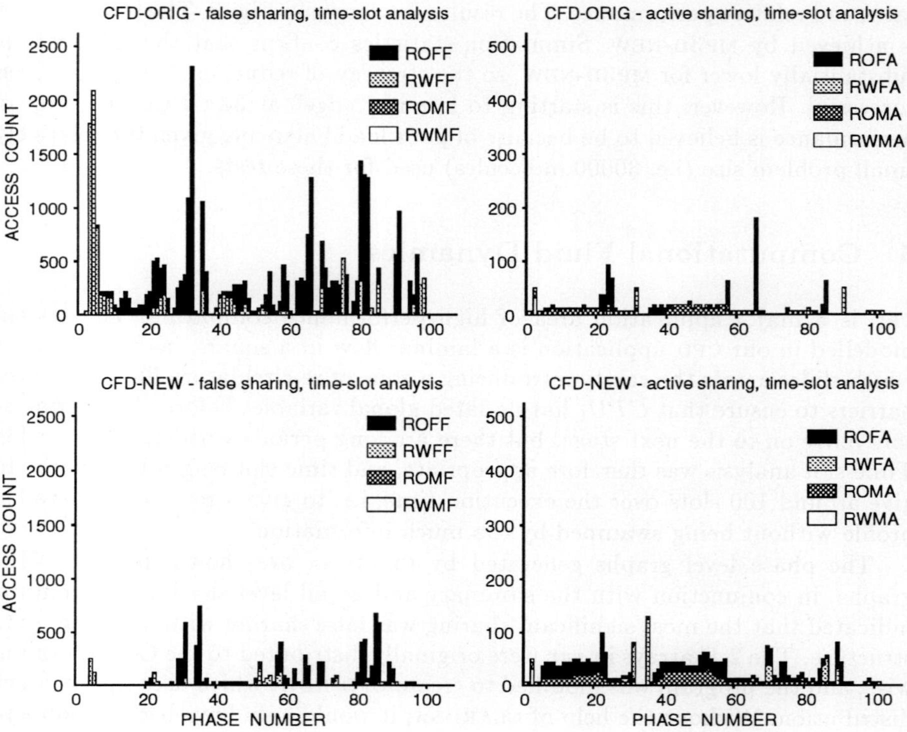

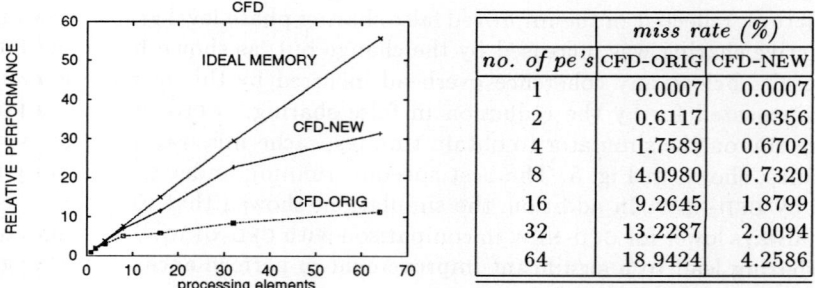

Fig. 4. Phase level graphs for CFD-ORIG and CFD-NEW

	miss rate (%)	
no. of pe's	CFD-ORIG	CFD-NEW
1	0.0007	0.0007
2	0.6117	0.0356
4	1.7589	0.6702
8	4.0980	0.7320
16	9.2645	1.8799
32	13.2287	2.0094
64	18.9424	4.2586

Fig. 5. Relative performance and cache miss rates for CFD

between adjacent slots; consequently analysis has to be performed multiple times with different time-slot lengths.

The information given by CLARISSA differs from that of existing performance analysis tools because it diagnoses *potential* contention rather than problems arising from a particular architecture. It also distinguishes between different types of sharing, i.e. active, passive and false sharing. This is important as the action to be taken depends on the type of sharing that needs to be eliminated. For example, in [1], false sharing is "suspected" in MP3D, but CLARISSA reports false sharing precisely. Similarly MemSpy can report that cache misses are high for a particular variable, but cannot say whether this is due to active, false or passive sharing. Finally, CLARISSA allows the analysis to be carried out on a barrier or time-slot basis, so that the timing of phases is appropriate for a particular application.

7 Conclusions

CLARISSA is a new tool which has been shown to be effective in analysing the cache-line sharing effects in shared-memory parallel applications. It uses a novel approach, *cautious trace analysis*, to locate potential cache line contention rather than measuring actual contention in a specific memory system design. We have shown how it was used to greatly improve the performance and cache behaviour of two scientific programs.

As further work, we plan to enhance CLARISSA to provide boundary analysis of cache lines, i.e. to cater for false sharing effects that may not show up for a particular problem size due to data structures happening to align with cache line boundaries. In addition, the performance of CLARISSA should be improved by closer integration with the simulator.

Acknowledgements. This work was funded by the U.K. Engineering and Physical Sciences Research Council through an Advanced Course Studentship and project GR/J 99117. Enormous thanks are due to Ashley Saulsbury for allowing us to use his simulator.

References

1. Mats Brorsson. SM-prof: A tool to visualise and find cache coherence performance bottlenecks in multiprocessor programs. In *Proceedings of the ACM SIGMETRICS and Performance '95*, pages 178–187, May 1995.
2. Susan J. Eggers and Randy H. Katz. A characterisation of sharing in parallel programs and its application to coherency protocol evaluation. *15th Annual International Symposium on Computer Architecture, Honolulu, May, in Computer Architecture News*, 16(2):373–382, May 1988.
3. M. Martonosi, A. Gupta, and T. Anderson. Tuning memory performance of sequential and parallel programs. *IEEE Computer*, 28(4):32–40, April 1995.
4. Jaswinder Pal Singh, Wolf-Dietrich Weber, and Anoop Gupta. SPLASH: Stanford parallel applications for shared-memory. *Computer Architecture News*, 20(1):5–44, March 1992.
5. Sarah A. M. Talbot. Performance tuning of programs for shared-memory multiprocessors. Master's thesis, Department of Computing, Imperial College, London, U.K., 1995.
6. B. A. Tanyi. *Iterative Solution of the Incompressible Navier-Stokes Equations on a Distributed Memory Parallel Computer*. PhD thesis, UMIST, 1993.

Dealing with Heterogeneity in Stardust: An Environment for Parallel Programming on Networks of Heterogeneous Workstations

Gilbert Cabillic and Isabelle Puaut

IRISA, Campus Universitaire de Beaulieu, 35042 Rennes Cédex, FRANCE
e-mail cabillic/puaut@irisa.fr

Abstract. This paper describes the management of heterogeneity in Stardust, an environment for parallel programming above networks of heterogeneous machines, which can include distributed memory multi-computers and networks of workstations. Applications using Stardust can communicate both through message passing and distributed shared memory. Stardust is currently implemented on an heterogeneous system including an Intel Paragon running Mach/OSF1 and an ATM network of Pentiums running Chorus/classiX.

1 Introduction

The proliferation of inexpensive and powerful workstations has continued to increase at a rapid rate in the last few years. This increase of machine performance is likely to continue for several years, with faster processors and multiprocessor machines. Studies have shown that for a large percentage of their lifetime, the machines are used for small tasks, thus demonstrating an average idle percentage of at least 90% even during peak hours. One possible use of these idle cycles is to run parallel applications. A number of research activities have tried to exploit the computing power of networks of workstations, like PVM [1] and MPI [2], based on the message-passing paradigm, and Mairmaid [3] based on the shared-memory (DSM) paradigm. To be fully usable, an environment for parallel computing above networks of workstations should: (i) support hardware and software heterogeneity, (ii) provide multiple programming paradigms, (iii) include mechanisms for load balancing and application reconfiguration, and (iv) tolerate machine failures.

Stardust is an environment that provides such facilities. Stardust allows the execution of parallel applications based both on the message-passing and page-based DSM paradigm. It executes on an heterogeneous computing environment composed of a parallel machine (the Intel Paragon) and an ATM network of workstations (PCs). Stardust includes a support for load balancing and check-pointing (see [4, 5] for more details). This paper focuses on Stardust support for heterogeneity. Section 2 gives an overview of the management of heterogeneity in Stardust. Section 3 then gives some performance data. Concluding remarks are given in section 4.

2 Dealing with Heterogeneity in Stardust

Stardust includes transparent mechanisms for converting data between different types of hosts. These mechanisms are used for converting both the contents of buffers exchanged in messages and the contents of shared virtual memory regions; they are described in paragraph 2.1. Furthermore, in order to cope with the processors different instruction formats, each program developed on top of Stardust is compiled for every type of architecture. Issues other than data conversion have to be dealt with, for processes to communicate through DSM (see [3, 4] for an overview of these issues). The main design decision for addressing these issues in Stardust consists in choosing a common unit for data conversion and data transfer between hosts, called *heterogeneous page* (see paragraph 2.2).

2.1 Mechanisms for data conversion and data typing

Stardust uses a standard architecture independent format for communications between different types of hosts. When sending a message, its contents is first converted into the standard data format; the receiver of the message then converts the message contents into its own physical format. The standard data format used for data transfer is the SUN eXternal Data Representation (XDR) format [6]; it was chosen because of its availability on a wide range of architectures and operating systems. The XDR library, linked with the Stardust environment, provides routines for converting basic data types to/from the XDR format.

In order to apply the XDR conversion routines when a piece of data is transferred between hosts of different types, it must be possible to have the type of each data structure. One approach consists in analyzing the source language (for instance by modifying the C compiler). Another approach, taken in [3] is to exploit information generated by compilers in object files, and intended to be used by symbolic debuggers (Symbol TABle, or stab). In Stardust, we did not want to modify the compiler or make assumptions on the kind of run-time information it generates in object files. Consequently, the type of every data structure is given explicitly by the application programmer when the data structure is allocated (region creation, allocation of a communication buffer). This is done by using a simple language, taken from [7] which is analyzed when the data structure is transferred between different types of hosts. A type is of the form *(TypeString, N_1, N_2, ... , N_n)*, where *TypeString* is the string identifying the data type (see syntax below), and N_i is the number of elements of the ith nested subtype of string *TypeString*.

Type : '{' (BasicType | Type)+ '}'
BasicType : 'C' | 'I' | 'L' | 'F' | 'D'

For example, the type *("{C{D}}", 10,20)* corresponds to a data structure which is an array of 10 structures of type *t_str*, each structure being composed of a character and 20 double precision floating point values.

Let us call *base element* the first nested subtype of a data structure (in the above example, the base element is the type *t_str*). Note that due to the method used to associate a type to a data structure, the conversion algorithm can only apply to a multiple number of base elements.

2.2 Mechanisms for sharing virtual memory regions

Mainly two issues have to be addressed when building an heterogeneous DSM. Firstly, the different hosts can have distinct page sizes. Secondly, depending on the architecture, the number of base elements in a page can vary, because of the differences in the physical representation of data. Let us consider for example an application running on an Intel Paragon and on a Dec Alpha-based architecture. Assume that the application is sharing a vector whose type is *("{LC}",8192)*. Due to the architectures' characteristics and to the alignment constraints, the size of the base element on the Paragon is 6 bytes (4 for the long integer, 1 for the character and an extra-byte for the alignment constraints), while its size is 10 bytes on the DEC alpha, where long integers take 8 bytes. Even if the page sizes on the two architectures are identical (8 Kbytes), they do not contain the same number of base elements. On this example, they do not even contain an integral number of base elements. To our knowledge, no heterogeneous DSM fully addresses all the issues coming from heterogeneity. In [3], it is assumed that all basic data types have the same size on all architectures; in addition, the problem of a base element crossing a page boundary is solved by adding an additional alignment constraint so that this problem does not occur.

In Stardust, the issues of different data representations and page sizes are addressed by choosing a common unit for data conversion and data transfer between different types of hosts, called *heterogeneous page*. The size of an heterogeneous page is a multiple of the size of a base element, so that the same data item is never on the memory of the two hosts. In addition, the size of an heterogeneous page is a multiple of the page size of every architecture, so that an action on an heterogeneous page can always be mapped on a set of actions on pages of every architecture. In the above example, an heterogeneous page contains 4096 base elements, and corresponds to 5 Paragon pages and 3 Dec pages. Note that the decomposition of a shared region into heterogeneous pages not only depends on the page sizes of all the machines hosting the application. It is also dependent on the type of data structure contained in the region. If the region taken as example above had contained characters instead of a more complex data structure, the size of the heterogeneous page would have been different (a single virtual memory page for both architectures).

Fragmentation into heterogeneous pages: A region is fragmented into heterogeneous pages at region creation time. At this time, all the types of architectures on which the application is running are known, as well as the type of the data structure contained into the region. The fragmentation algorithm finds out, for every architecture, a page size for which no base element crosses a page boundary, and then takes the least common multiple (LCM) of theses values, thus obtaining the size of the heterogeneous page for the region.

Consistency protocol: Consistency of shared data is managed by a 2-level system. The *intra-architecture* level manages data consistency for a subset of homogeneous hosts. It uses a *homogeneous memory manager* (HoMM) per node. HoMMs manage data consistency within the associated architecture. They use the architecture virtual memory page as a unit of data transfer between machines. Pages are transferred using the most efficient communication protocol that exists on the architecture. In addition, pages are transferred in the architecture physical data representation, without conversion into an architecture-independent format.

In the *inter-architecture* level, data consistency is managed by an *heterogeneous memory manager* (HeMM) per group of nodes of the same type. The unit of data transfer between HeMMs for a given region is the heterogeneous page that corresponds to the region's type. Transfers are achieved via a common communication protocol (TCP) and data is transferred in the XDR architecture independent format. In the current implementation, both HoMMs and HeMMs implement sequential consistency through a write-invalidate consistency protocol, and use K Li's static distributed scheme for locating shared pages [8].

Note that with such a structure of heterogeneous DSM there is no time overhead due to the management of heterogeneity when an application is running on hosts of the same type.

3 Implementation and Performance of Stardust

Stardust is currently implemented on a 56 nodes Intel Paragon [9] running a Mach/OSF1 kernel and on an 155 ATM network of Pentium PC machines[1] running the Chorus/classiX operation system [10]. Each Paragon node is equipped with 16 Mbytes of memory, of which nearly 8 Mbytes are consumed by the operating system. The size of pages on the paragon is 8 Kbytes. The measured transfer rate between nodes is 60 Mbytes/s. Each PC has 32 Mbytes of memory, of which only 8 Mbytes are left free for the paging activity. The size of pages on the Pentiums is 4 Kbytes. For all the performance measures given in this paragraph, the size of heterogeneous pages is 8 Kbytes.

Stardust is made up of a set of software modules, which have an OS independant interface in order to ease their portability. The *consistency manager* maintains data consistency using a write-invalidate protocol ; the same code is used by HoMMs and HeMMs. The *network manager* offers primitives for exchanging messages. It uses for intra-architecture communications are running the most efficient communication protocol of the architecture (NX communication library on the Intel Paragon, Chorus IPC on the Pentium PCs). If the message sender and receiver run on different types of hosts, TCP is used via sockets, and the message is converted into the XDR format before being sent. Finally, the OS

---

[1] The ATM driver being currently under debug, this version of the paper includes figures measured on a 10Mb/s Ethernet.

interaction manager is responsible of the communication with the underlying operating system concerning memory management and thread management.

Figure 1 shows the performance of the *MGS* (Modified Gram-Schmidt) application, which produces from a set of vectors an orthonormal basis of the space generated by these vectors. The problem size for this application is 256x1024 double precision floats. The application only uses DSM and does not exhibit false sharing. The figure shows the elapsed time for the application when it runs on (i) a set of Paragon nodes, (ii) a set of Pentium nodes, and (iii) an heterogeneous system with the same number of nodes of both architectures. Measurements were done in a system with 1 up to 8 nodes.

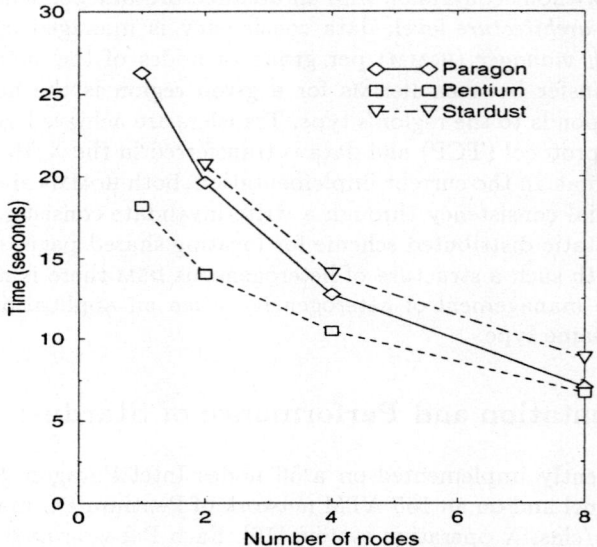

Fig. 1. Execution time of MGS application

The curves show that the application launched on an homogeneous set of Pentium nodes always exhibits better performance compared to the same application running on Paragon nodes or on heterogeneous nodes. In addition, the applications still exhibits a good speedup when the number of nodes increases. Finally, only a small percentage of the total execution time is needed to deal with heterogeneity compared to the execution time for the Paragon machines.

4 Concluding Remarks

Many environments for parallel programming on networks of heterogeneous workstations have been designed and implemented in the last ten years. A key difference between Stardust and most of these environments is that Stardust runs both on parallel machines and networks of workstations, and includes mechanisms for load balancing and checkpointing [4, 5]. In addition, unlike most environments, Stardust supports both the shared-memory and message-passing paradigm. From an implementation point of view, we have tried in Stardust to use the most efficient communication protocol of a given architecture for communications between hosts of this architecture. In comparison, most implementations of PVM use TCP for all communications, which can be less efficient than architecture-specific communication protocols. Few DSM systems have been designed for heterogeneous systems. Marmaid [3] differs from Stardust by the way types are associated to shared memory regions. Mairmaid requires less effort from the programmer, but its implementation depends on the format of the object files. Unlike Mairmaid, it is not assumed in the Stardust prototype that basic data types have the same size on all architectures.

References

1. A. Geist, A. Beguelin, J. Dongarra, W. Jiang, R. Manchek, and V. Sunderam. *PVM: Parallel Virtual Machine - A Users' Guide and Tutorial for Networked Parallel Computing.* MIT Press, 1994.
2. W. Grop, E. Lusk, and A. Skjellum. *Using MPI: Portable Parallel Programming with the Message-Passing Interface.* MIT Press, 1994.
3. S. Zhou, M. Stumm, K. Li, and D. Wortman. Heterogeneous distributed shared memory. *IEEE Transactions on Parallel and Distributed Systems*, 3(5):540–554, September 1992.
4. G. Cabillic and I. Puaut. Stardust: an environment for parallel programming on networks of heterogeneous workstations. Technical Report 1006, IRISA, April 1996. Available by anonymous ftp at ftp.irisa.fr.
5. G. Cabillic, G. Muller, and I. Puaut. The performance of consistent checkpointing in distributed shared memory systems. In *Proc. of the 14th Symposium on Reliable Distributed Systems*, pages 96–105, Bad Neuenahr, Germany, September 1995.
6. Sun Microsystems Inc. *Network Programming Guide - External Data Representation Standard: Protocol Specification*, 1990.
7. C. Pinkerton. A heterogeneous distributed file system. In *Proc. of 10th International Conference on Distributed Computing Systems*, May 1990.
8. K. Li and P. Hudak. Memory coherence in shared virtual memory systems. *ACM Transactions on Computer Systems*, 7(4):321–357, November 1989.
9. Intel. *Paragon User's Guide*, 1993.
10. M. Rozier, V. Abrossimov, F. Armand, I. Boule, M. Gien, M. Guillemont, F. Herrmann, P. Léonard, S. Langlois, and W. Neuhauser. The Chorus distributed operating system. *Computing Systems*, 1(4):305–370, 1988.

An Integrated Environment to Design Parallel Object-Oriented Applications

Klaus Wolf, Ottmar Krämer-Fuhrmann

German National Research Center for Computer Science (GMD)
Institute for Algorithms and Scientific Computing (SCAI)
Schloss Birlinghoven, D-53754 Sankt Augustin, GERMANY
Phone: +49-2241-14 2557 Fax: +49-2241-14 2181 Email: klaus.wolf@gmd.de

Abstract. The TRAPPER–paradise C++ toolset is an integrated programming environment well equiped for the design of object-oriented parallel applications. A variety of graphical tools and a parallel extension of C++ are provided to form a modern environment for all phases of the software development cycle.

1 Introduction

TRAPPER is a hardware independent programming environment being developed in cooperation between Daimler-Benz Research, GMD and GENIAS GmbH mainly for the use in industrial automotive applications. TRAPPER is a graphical tool with the ability to design and configure software for MIMD type parallel computers. Furthermore TRAPPER features to monitor the process and communication load of a running parallel program and animates application data.

paradise C++ [HPC++] is a parallel object-oriented library realised on top of sequential C++ and PVM resp. MPI. It is dedicated to distributed memory parallel computers. Portability and the facility to run on heterogeneous networks of (parallel) systems are ensured. paradise C++ was specified and implemented by GMD. The work was funded by the C.E.C project 8451 SOFTPAR [Softpar].

This paper describes the work being done to expand the functionality of both systems in order to build an integrated programming environment. Its key-features are *object-orientation, integration, parallelism* and *portability*.

2 Features of an Integrated Parallel Design Environment

TRAPPER [TRAPPER] is objected to support a modular specification of parallel applications. Such an application consists of processes, ports and links. *Process structures* can be clustered to allow the *hierarchical design* of large process nets. paradise C++ is very well suited to be a target-language for code generation from TRAPPER-design because of the one-to-one correspondence between the class structure and the graphical representation. User code only has to be inserted in predefined frames representing class local method definitions.

paradise C++ offers modularization and structurization facilities. *Hierarchies* of parallel components allow the integration of independent parallel algorithms in complex application. *Active objects* act sequentially and are provided with a local scope. *Interfaces* separate local process behavior from any global runtime-configuration. Structured object-networks may be encapsulated in *subnet classes* representing prototypes of parallel schemes. By inheritance and modification those prototypes can be adapted to specific problem solutions.

The instrumentation of paradise C++ library supports an automatic monitoring of active-object states in the user-defined design layout. Events of active objects being executed on the parallel machine are attached to their related graphical objects and displayed.

3 TRAPPER

The strength of TRAPPER is the integration of different tools for various phases of the software development cycle.

The *DesignTool* supports a hybrid approach. The object structure of the parallel application is designed graphically while sequential code or local class methods has to be integrated in a paradise C++ program frame generated by TRAPPER.

The *ConfigTool* allows the user to specify the configuration of the hardware system and allows a semi-automatic mapping of the software net to the actual hardware configuration.

The *Monitor* accompanies the application to collect run-time information about software and hardware events. Software events being monitored are object construction and destruction, interprocess-communication and remote procedure calls (RPCs). Hardware events are computation and communication loads of the processors. Furthermore the paradise C++ -programmer can define his own events which are written to the tracefile together with the automatically generated events.

3.1 Program Design

The design of parallel applications with TRAPPER is based on the model of communicating sequential processes (CSP). The programmer has to specify the parallelism on the level of independent tasks which are able to cooperate by message passing resp. RPC mechanisms.

Figure 1 shows the graphical design of a small example application containing three client and two serving objects on its highest level called *"EuroPar96-Network"* (frame on the left side). The object-interaction graph consists of nodes and edges. Nodes represent active objects which are instances of process or subnet classes written in paradise C++ . Edges may define two different types of interaction-protocols: either message-passing or remote procedure calls.

Nodes may be refined into more specific and independent subsystems which themselves consist of a network of nodes and edges. While the objects inside the subsystem *GridMaster* communicate via message passing those in the subnetwork *ServSub* invoke partner-methods by remote procedure calls. The types of the interfaces are specified by related graphical port symbols.

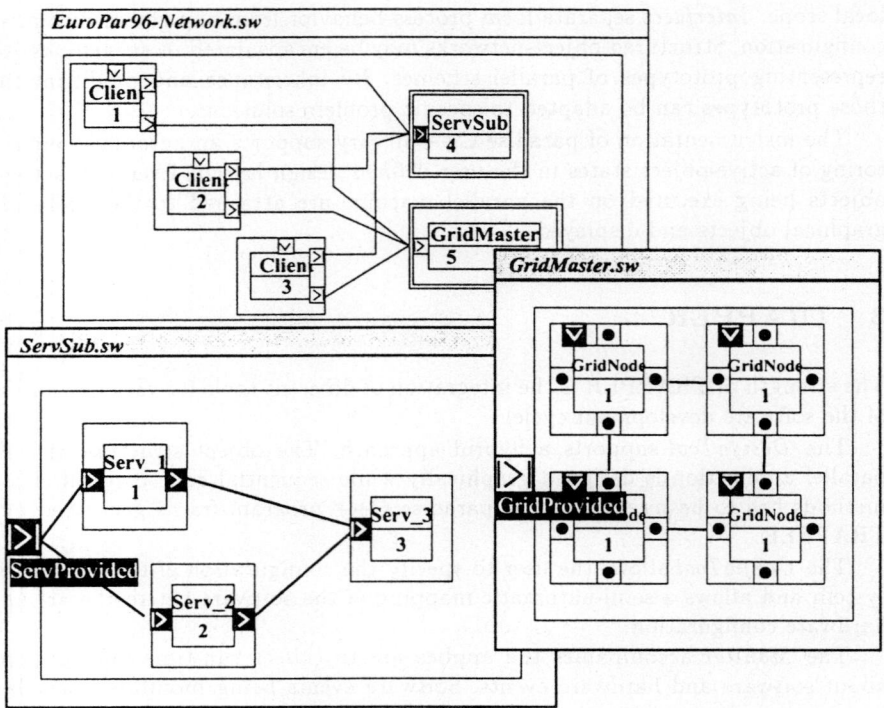

Fig. 1. Top Level of Example Design

Processes and Subnets Each designed box consists of an unique type identifier – the class denoted by its name – and dedicated interfaces called ports. Additional attributes are to be set to specify the correct system base-class, i.e. process resp. subnet. During design-phase a class is defined only once; if its name is used in several boxes a related number of object-instances of this class will be created at runtime. Instance numbers during design-phase are unique only inside the actual subnetwork - a global unique numbering on all activated objects will be defined **after** design is complete and the size of the whole network is fixed. Large object graphs can be designed hierarchically as composition of sub-systems. A subsystem is a graphical entity (i.e. it is displayed in a window of its own) and can be considered as a black box, which contains a subgraph. A subsystem itself can contain other subsystems, thus specifying a dedicated *Master-Slave* relationship. A master-object is at runtime responsible for creation, configuration and - at last - deletion of its internal slave objects; the code for those actions will be generated automatically by TRAPPER.

Interfaces As each object-instance is defined to execute in its local scope it has no access to global information or data. Any remote interaction between different objects has to be defined as an action on *interfaces*. An interface is *declared local* to a process/subnet class and *instantiated* at the same time when the process/subnet object is constructed. A *connection* has to be defined explicitly: i.e. a link between any two interfaces has to be specified in the design. Calling a method on an unconnected interface however results in a *no-operation* call resp. returns an error at runtime. During design drawing links is checked to suppress invalid connections, for example links between a message-passing and a server interface.

Interfaces (ports) may be attached with different attributes determining the protocol and potential link modes. The protocol may be of type RPC (remote procedure call) where there is a client-object with a *requiring interface* and a server-object with a *providing* one. The port protocol defines a symmetrical message-passing connection. The connection-rate of an interface may range from 1-1 to n-n.

Local Fine-Coding While the graphical design defines object instantiation and interface connection-graphs the sequential behavior inside an object has still to be described by textual representations, the traditional program code. But there is already a rough functional skeleton defined in the graphics by specifying *providing* server interfaces. In contrast to 'Fortran' or 'C' target language a design-box represents an object-instance instead of a main-routine. An object is not running all the time but is waiting for requests to start one of its provided (via interface) methods. Thus TRAPPER is able to generate a paradise C++ source-code frame out of the graphical specification. This frame is a class definition with method skeletons for each provided RPC-interface.

4 paradise C++

4.1 Standardization of Parallel C++

In 1994 a group of researchers [CC++] [CHARM++] [pC++] [Mentat] et.al. began a series of discussions about building a common framework for a parallel programming with C++. The goal was to identify common language and library mechanisms that would allow a variety of parallel programming systems and libraries to work together and to make it easy to port applications across a wide range of computing platforms. To give it a name this framework and the programming language which might emerge from this process were called **HPC++** [USHPC++]. At the same time european groups [UC++] [Eiffel] started investigations with similar objectives calling their target **EC++** [EC++]. First meetings to bring both lines into one started at the end of 1995.

In the SOFTPAR project [Softpar] which was defined and set up in 1993 the same name *HPC++* [HPC++] [SwEng] [HOOD] was used as a working title for the parallel C++ library to be developed. During the project runtime it became clear, that the SOFTPAR approach would not lead to a *standard parallel C++ language*, but to an object-oriented and parallel library with strong emphasis on design facilities. This library is actually built on top of sequential C++ and PVM resp. MPI but may easily be ported on a forthcoming standard of parallel

C++. To avoid further name-clashings and being aware of the fact that there should be only one environment called *HPC++* representing the parallel C++ standard (comparable to HPF) we changed our name into **paradise C++** – **parallel distributed C++**.

4.2 Concepts

paradise C++ is a class library defined on top of standard C++ and PVM or MPI. By this, general portability is guaranteed. The library offers basic classes for active objects, communication and operation interfaces and resource management. Describing a parallel application in paradise C++ means specifying a tree-like hierarchy of process and subnet objects. In that hierarchy higher leveled objects are intended for management and synchronization, while lower leveled objects perform the intensive computations. In a **functional decomposition** subnet- or process-objects of different classes have to be combined. Data or results are passed from one object to another by *remote procedure calls* or by *explicit message passing*. The connection of process objects with independent control flows in paradise C++ is organized in *hierarchical trees or encapsulated sub-systems*.

The visibility of remote object references is defined by the hierarchical tree of objects. A process object has access and control only on those remote objects created by itself; i.e. a parent-node only sees its own childs. *'References'* to child process objects may not be passed to other contexts.

Interaction structures between process objects having the same parent are expressed by explicit connections of *interfaces*. Where a graphical design-box in TRAPPER represents an active process object, process interfaces are to be used by methods of the process-object itself. However, interfaces of subnet-boxes are only used as transparent or virtual entries to interfaces of the subnet's slave objects. Interfaces are declared as (data) member fields of process classes before compile time and connected at runtime by the joint parent object.

Based on the interface idea described above paradise C++ supports explicit *message-passing* and *remote operation calls*. Various synchronization modes are offered to enable concurrent execution without blocking of partners.

Creating new process objects with independent control flows (i.e. with a context of their own) in paradise C++ is coordinated by **resource management** to enable deterministic load balancing. To avoid racing conditions when mapping new process objects, access to any computing resources is exclusive, i.e. each process object has a domain of computing resources for its exclusive usage.

5 Code Generation

Generating paradise C++ source code from the TRAPPER design and configuration tools profits from the fact that there is a 1-1 mapping from nearly all design symbols into paradise C++ source code.

- Terminal process boxes of one type result in one class hierarchy in paradise C++ . Interfaces and internal data structures are identical for all object instances.

- Non-terminal boxes additionally generate configuration statements for the creation and configuration of their slave objects represented by inner boxes and links in TRAPPER.
- Ports in TRAPPER have attributes specifying them as paradise C++ communication ports or as required resp. provided operation interfaces. Those interfaces are built as member fields of the related class.
- As the TRAPPER design is static in terms of paradise C++ notion resource management is easy to be realised just by counting the number of nodes of a sub-system and all its descendants.

5.1 Basic Generation Rules

In the design phase, TRAPPER is able to generate program frames out of the graphical object specification. For each box-type a hierarchy of three classes is generated:

- **Interface:** The base class is derived directly from one of the paradise C++ process resp. subnet classes. This interface class encapsulates the declaration and initialization of the interfaces.
- **Configuration:** The second class is derived from the interface class and defines creation and configuration of slave objects which are lower leveled in the design hierarchy.
- **Application:** The application class is derived from the configuration class and contains method frames for each defined interface of RPC-type. User code to define the application has to be inserted in those frames.

The base class generated for a design node is responsible for **interface declaration** and initialization. Depending on the node attribute the generated class will directly be derived from the paradise C++ system classes *HPC_Process* or *HPC_Subnet*.

Non-Terminal Boxes in TRAPPER represent **subsystems** containing internal hierarchies of object graphs. Thus non-terminal boxes generate a config-class with declarations for reference-objects to all child paradise C++ processes and subnet objects created in the subsystem. This class describes the code to configure the subsystem (object creation and related interface connections). Resource management (utilization of available processor-resources) is defined by the total number of nodes represented by the slave boxes.

Interfaces of internal child boxes which are mapped to (virtual) interfaces of the non-terminal parent box result in a macro call to define a transparent and direct connection from remote of parent to its child object interfaces.

While for the two base classes discussed above – for interface-description and for slave-configuration – all code will be generated automatically by TRAPPER, the following class has to be edited by the user.

For each interface with attribute *provided operation* a corresponding declaration for a virtual method was generated in the base-class describing the class-interfaces. A frame of method-headers defines those empty areas where **the user** has to insert his sequential and local code.

The Design Root of the TRAPPER design tree has to be marked explicitly. The total amount of computing resources needed is given by the number of boxes in the root and all subsequent child sub-systems. This number and the top-view of the process graph define the contents of the paradise C++ software-configuration file. This file describes the process objects to be created automatically at program start.

6 Monitoring with TRAPPER

As each graphic object in TRAPPER is mapped 1-1 onto an object in paradise C++ code animation of user data or monitoring of system loads is obvious. Each component in paradise C++ (process & subnet & interface objects) has its unique identification in the TRAPPER layout. So each event can be displayed in a proper relation to its graphical counterpart. The way the information of a running process is collected and provided to the monitor tool is adopted from the standard TRAPPER environment which already supports PVM and MPI directly – and both are used as implementation layer for paradise C++ . TRAPPER provides two very different modes for observing running parallel programs: the first is an automatic monitoring of paradise C++ system states, the second is allows the visualization of user-data.

The object-oriented target language allows to distinguish between more states than only 'IDLE' and 'ACTIVE'. Each time a process or subnet object is activated by a request coming from any client one dedicated method of its class is invoked and can be monitored. As paradise C++ supports different kinds of synchronization for RPC (from complete asynchronous through semi-committed to fully synchronous calls) those synchronization modes also will be displayed. On the other side an object may act as a server as well as a client at the same time. So while being activated from remote a server itself may call other objects whose provided interfaces are connected to one of its own required RPC-interface.

Data-structures defined in the user's code, can be made visible on the TRAPPER display. Each process resp. subnet-box provides a (usually 60 by 60) pixel display, which is under full and local control of the running process object. A box-display allows the colored animation of two-dimensional arrays or histograms and bar charts of one scalars.

7 Conclusion

paradise C++ proved to be well suited as target environment for the TRAPPER toolset. Dedicated interfaces for remote-procedure-calls as well as for explicit message-passing are provided. Transparent virtual interfaces for non-terminal design-boxes map to subnet-interfaces and deterministic mapping of active objects to processors is supported by resource-management. Common to the design method in TRAPPER and to paradise C++ is the idea of hierarchical encapsulation resp. ordering of computational entities.

By combining TRAPPER design tool, paradise C++ library and the TRAPPER monitor a complete and integrated parallel environment covering most of the software development phases from design to code and runtime support is available.

TRAPPER is under commercial exploitation by GENIAS GmbH, a prototyped adaptation to paradise C++ was realised in 1995. Beta versions of paradise C++ were distributed at the end of 1995. A commercial exploitation will begin 1996.

References

[CC++] K. M. Chandy, C. Kesselman. *Compositional C++: Compositional Parallel Programming*, Technical Report Caltech-CS-TR-92-13, Department of Computer Science, California Institute of Technology, 1992

[CHARM++] L. V. Kale, S. Krishnan. *CHARM++: A Portable Concurrent Object Oriented System Based On C++*, OOPSLA'93, 1993

[EC++] The EC++ Working Group on Parallel C++ Architecture SIG *EC++ - EUROPA Parallel C++ Draft Definition 0.9* November 1995, WWW: http://www.lpac.ac.uk/europa

[Eiffel] I. Attali, D. Caromel, M. Oudshoorn, *A Formal Definition of the Dynamic Semantics of the Eiffel Language* Gopta Gupta, George Mohay, and Rodney Topor, editors, Sixtenth Australian Computer Science Conference (ACSC-16), Griffith University, Feb 93

[HPC++] Wolf, Lang, Holtz (GMD), *HPC++ - High Performance C++*, HPCN95 Proceedings May 1995, Springer LNCS 919

[HOOD] Letteron, Bancroft (Sema Grp), Gerlich, Debus (DORNIER) Wolf, Lang, Holtz (GMD) *HOOD and Parallelism in the Softpar project*, HPCN95 Proceedings May 1995, Springer LNCS 919

[Mentat] A. Grimshaw, et all. *Mentat 2.5 Tutorial*, Department of Computer Science, University of Virginia, Charlottesville, 1993

[pC++] F. Bodin, P. Beckman, D. Gannon, S. Narayana, S. Yang. *Distributed pC++: Basic Ideas for an Object Parallel Language*, Department of Computer Science, Indiana University

[Softpar] CEC ESPRIT PROJECT 8451. *Softpar A Software Factory for the development of Parallel Applications*, Softpar Project Report, SOFTPAR-S-29.2, January 1996

[Split-C] D. E. Culler, A. Dusseau, S. C. Goldstein, A. Krishnamurthy, S. Lumetta, T. von Eicken, K. Yelick. *Introduction to Split-C Version 1.0*, Computer Science Division - EECS, University of California, Berkley, 1993

[SwEng] Lang, Wolf (GMD), Letteron (SEMA Group), *Software-Engineering and Parallel Object-Oriented Programming* Parallel Computing: State-of-the-Art Perspective, Proceedings of the International Conference ParCo95, Elsevier Science Editor

[TRAPPER] Lorenz Schäfers, Christian Scheidler, and Ottmar Krämer-Fuhrmann, *Trapper: A graphical programming environment for parallel systems*, Future Generations Computer Systems, 11(4-5):351–361, August 95

[UC++] T. O'Brien, G. Roberts, M. Wei, R. Winder. *UC++ v.1.0 Language and Compiler Documentation*, The London Parallel Applications Centre (LPAC), 1994

[USHPC++] The HPC++ Working Group *HPC++ Whitepapers and Draft Working Documents* December 1996, WWW: http://www.extreme.indiana.edu

MPI-2: Extending the Message-Passing Interface

Al Geist[1] and William Gropp[2] and Steve Huss-Lederman[3] and Andrew Lumsdaine[4] and Ewing Lusk[2] and William Saphir[5] and Tony Skjellum[6] and Marc Snir[7]

[1] Oak Ridge National Laboratory
[2] Argonne National Laboratory
[3] University of Wisconsin
[4] University of Notre Dame
[5] NASA Ames
[6] Mississippi State University
[7] IBM Research Yorktown

Abstract. This paper describes current activities of the MPI-2 Forum. The MPI-2 Forum is a group of parallel computer vendors, library writers, and application specialists working together to define a set of extensions to MPI (Message Passing Interface). MPI was defined by the same process and now has many implementations, both vendor-proprietary and publicly available, for a wide variety of parallel computing environments. In this paper we present the salient aspects of the evolving MPI-2 document as it now stands. We discuss proposed extensions and enhancements to MPI in the areas of dynamic process management, one-sided operations, collective operations, new language binding, real-time computing, external interfaces, and miscellaneous topics.

1 Introduction

During 1993 and 1994, a group of parallel computer vendors, library writers, and application scientists met regularly to define a standard interface for message-passing libraries. The result of this effort was MPI (Message-Passing Interface) [7]. Implementations of MPI are now widely available, including portable and freely available implementations [2, 3, 8] and specialized versions from vendors. General information on MPI is available at [1]. For the purposes of this paper, it will be useful to refer to the result of the initial MPI standardization effort as "MPI-1."

MPI-1 defined an interface for a specific *message-passing model* of parallel computation, in which a fixed number of processes with disjoint address spaces communicate through a cooperative mechanism (when two processes communicate, one sends and the other receives). MPI provides many types of point-to-point communication, to incorporate requirements for robustness, expressivity, and performance. Messages are strictly typed and scoped, allowing for communication in a heterogeneous environment. MPI also contains an extensive set of collective operations, process topology functions, and a profiling interface.

The most distinctive feature of the current MPI-2 proposals described in this paper is that they go beyond the strict message-passing model defined above. In MPI-2, processes may create other processes, so that the number of processes in an MPI computation can change dynamically (Section 2). Processes can interact directly with the memory of other processes (Section 3). Extensions, semantic modifications, and subset definitions in support of real-time and embedded systems (Section 4) also represent changes to the computational model.

Other topics being discussed in MPI-2 include extending MPI-1's collective operations to intercommunicators and nonblocking operations (Section 5), bindings for C++ and Fortran 90 (Section 6), and interface definitions for some of MPI's opaque objects so that they can be used more effectively in support of profiling and other libraries (Section 7). Finally, a number of issues, such as interlanguage communication, a portable startup mechanism, and minor repairs to the MPI-1 specification (Section 8), are under consideration in MPI-2.

In the rest of this paper, we present an overview of each of these areas. We assume familiarity with the current MPI Standard. In the Conclusion we describe the current status of these proposals and prospects for their early appearance in implementations.

2 Dynamic Process Management

MPI-1 describes how a group of processes can communicate with one another. It does not specify how those processes are created, nor does it allow processes to enter or leave a parallel application after the application has started. This static process model enables the specification of deterministic semantics and facilitates efficient implementations of MPI.

Nevertheless, a number of important applications cannot use MPI-1 because of the constraints imposed by its static process model. These include manager-worker applications, where the number and type of workers are not known until the manager has started, task farms, applications that can adapt to changing resources, applications with varying resource requirements, and client/server applications. Much of the impetus for relaxing the static process model comes from the PVM community, which is familiar with PVM's relatively rich support for dynamism.

2.1 The Interface

A fundamental concept in MPI-1 is MPI_COMM_WORLD, which defines the communication space containing all processes in an MPI application. With MPI-2's ability to add more processes to an application, the definition is modified to be the communication space containing all processes started together. Groups of newly started processes each have their own unique MPI_COMM_WORLD, but they also have an intercommunicator that allows them to merge with their parent group, forming a single bigger communicator. MPI-2 also provides an attribute,

MPI_UNIVERSE_SIZE, that suggests how many new processes might usefully be spawned in the environment.

A powerful new functionality being added to MPI-2 is the ability to establish contact between two groups of processes that initially do not share a communicator and may have been started independently. This functionality would be useful, for example, in enabling a visualization tool to start up and attach to a running simulation, or in enabling two parts of a large application, started separately at two different sites to communicate with each other. The collective functions MPI_CONNECT and MPI_IACCEPT create an intercommunicator that allows the two groups to communicate.

3 Remote Memory Access

The message-passing communication paradigm requires explicit involvement of two processes (sender and receiver), in order to transfer data from the memory of one to the memory of another. Remote Memory Access (RMA) extends the communication mechanisms of MPI by allowing the transfer to occur with the explicit involvement of only one of the two processes.

3.1 Motivation

Remote memory access facilitates the coding of some applications with irregular communication patterns. One situation occurs when a distributed-memory application needs some randomly accessed read-only shared memory (for large shared tables). Some of the processes can be used as "memory servers", while the other processes access the data by using get calls. Another situation occurs with a distributed-memory code where the data distribution is fixed or slowly changing, but where the pattern of use changes dynamically. Each process can compute what data it needs from remote processes and generate the required receives. To generate the matching sends, one needs to compute the inverse of the receive mapping, a time-consuming process that requires all processes to coordinate the data exchange. The use of get calls avoids the need for sends. A generic example is the execution of an assignment of the form A = B(map), where map is a permutation vector, and A, B, and map are distributed in the same manner.

RMA can be supported on distributed memory systems by an "RMA agent" at the target node that accepts RMA requests and performs the required read or write accesses in the memory of the target process. A portable implementation might use an asynchronous receive handler to implement this RMA agent. Systems with dedicated put/get hardware (for example, the Cray T3D) could take advantage of that hardware, at least for simple transfers. Systems with communication coprocessors can take advantage of that coprocessor in order to run the RMA agent without interfering with the application processor at the target node. On shared-memory systems, if the caller can directly access the memory of the target process, RMA can be implemented without an RMA agent: the caller process can directly copy data to or from the memory of the target process.

3.2 Interface Summary

The current MPI-2 draft proposes the following RMA operations:

Put: transfer data from caller memory to target memory
Get: transfer data from target memory to caller memory
Accumulate: update variables in target memory by values from the caller memory. The update operation is an associative operation such as addition or minimum.
Read-Modify-Write: update variables in target memory by values from the caller memory, and return the initial value of the target memory variables. With a suitable choice of the update operation, one obtains synchronization operations such as test-and-set, fetch-and-add, or compare-and-swap.

In addition, a generic asynchronous handler mechanism is provided. This mechanism can be used for a software implementation of remote memory access, as well as for implementing many other communication paradigms. However, the very generality of this mechanism prevents many implementation optimizations that are possible for the more specific RMA operations.

4 Real-Time Extensions to MPI

MPI has helped to promote performance-portable programming of traditional high-performance computing and cluster systems. It has also proven desirable to leverage the success of MPI on parallel applications in the real-time community.

Taking advantage of this opportunity, a number of new organizations and the existing MPI Forum participants initiated an effort to explore what "real-time MPI" might look like. It is not expected that real-time MPI will be a required part of the MPI-2 Standard or that all HPC and cluster MPI implementations will support the real-time profiles.

Time-Based Profile For the time-based profile, it has been tacitly accepted that an outside calendar must be provided, in addition to the MPI services, in order to schedule the computations associated with this profile of MPI/RT. The calendar will specify when to start MPI communication. The anticipated strategy is to extend the MPI interface by using persistent communications that support this timed startup of communication. Timeout-based communication also will be supported in this way.

Priority-Based Profile Priority-based messaging and threading are commonly occurring strategies in real-time and non-real-time systems. Priority levels are supported by various operating systems and by certain message-passing networks, though not widely. Furthermore, some network systems support virtual channels, which themselves may provide a mechanism of reservation, if not priority, for given "flows" of data.

5 Collective Communication Extensions

MPI-1 has a rich set of collective operations, but they are subject to a number of restrictions. MPI-2 is considering generalizing them is a number of directions.

Asynchronous Operations In the current draft, each collective operation specified by MPI-2 has an asynchronous analog. A wide variety of MPI-2 features use asynchronous collective operations on both intracommunicators and intercommunicators.

Intercommunicator Collective Operations The purpose of intercommunicator collective operations is to support broadcast, reductions, and other operations, extended to include the two-group model of parallel processing offered in MPI-1 by intercommunicators.

Original proposals for extending intercommunicators to support collective operations, in addition to their MPI-1 point-to-point facilities, were first based on [10], which included model implementations.

The additional functionality came in three forms: more collective constructors and manipulators, what is now called "half-duplex" intercommunicator operations that extend intracommmunicator collective operations, and virtual topology-oriented versions of both the constructors and the communication procedures.

6 Language Bindings

6.1 C++ Bindings

The design of MPI itself is very much object-based, and the C++ bindings are based on the underlying object-based design principles. The bindings define a small set of classes corresponding to the fundamental object types in MPI with the functionality of MPI provided as member functions of these objects. This interface is fairly lightweight and seeks to meet the requirements of a language binding while still using advanced features of the target language. For instance, MPI error codes are still returned by function calls, no new types of objects are introduced, and the type arguments to function calls must be explicitly provided. Thus, only minimal use of advanced features of C++ such as polymorphism would be available to MPI programmers. This is an approach similar to that taken in [6]. A full-fledged class library that uses such advanced features has been developed in conjunction with the bindings and can be found at [9].

6.2 Fortran 90 Interface

Fortran 90 adds a wide range of features to Fortran 77. These include the module facility, derived types, array syntax, dynamic memory allocation, "pointers", the ability to do strict type checking, and function overloading. At first glance, it seems that MPI-2 should be able to make wide use of these new features. Unfortunately, most of them are too "high level" for MPI to use, and many in fact cause more problems than they solve. The MPI-2 approach to Fortran 90 bindings therefore focuses more on trying to avoid introducing new problems than on trying to solve old ones.

7 External Interfaces

MPI-1 has a number of features that allow users to layer various capabilities on top of MPI. For example, user-defined reduction operations allow the programmer to use MPI for all communication requirements but still perform specialized reduction operations.

Generalized Requests MPI-1 had nonblocking operations for basic point-to-point send and receive calls. MPI-2 is proposing nonblocking calls for all collective operations, many one-sided operations, and dynamic spawning. Although these significantly expand the areas covered by nonblocking operations, users still may want additional nonblocking operations. For example, in the current MPI-IO effort [4, 5], nonblocking read and write operations are proposed. It would be advantageous to offer a standard MPI mechanism to perform these additional nonblocking operations. This would allow the use of other MPI features such as MPI_WAIT, reducing the effort in creating such requests and allowing one to control both types of nonblocking operations together.

Access to Opaque Objects One area that has caused difficulties in writing portable tools is the information stored with opaque objects. MPI-1 was deliberately designed with opaque objects. These allow flexibility in implementations and allow for future enhancements without changing the user's view of objects already present in MPI. To allow users to gain access to needed information in opaque objects, MPI has a number of accessor functions. For example, MPI_GET_COUNT will return the number of entries received as stored in the opaque part of the status object. One drawback to this approach is that only information with explicit accessor functions can be obtained in an easy and portable way from an MPI implementation. In MPI-1, the MPI Forum included all the accessor functions that seemed to be needed by users. However, tool writers have noted that they need access to information not typically needed by users. For example, a profiling library often needs the length of a message begun by MPI_START for a persistent request. To enable these tools to be truly portable, MPI-2 includes a number of functions to expose information stored in opaque objects.

Finally, the external interface definition in MPI-2 allows a generalization of the MPI-1 caching mechanism to allow caching on additional handles. The same calls are used but in MPI-2 apply to MPI_COMM, MPI_DATATYPE, and MPI_GROUP.

8 Miscellaneous

A number of topics are being considered by the MPI-2 Forum that do not fall into the categories above.

In MPI-1, although both C and Fortran-77 bindings were defined, nothing was specified regarding the interoperability of these two languages. Interoperability comprises at least three subareas: initialization, passing of MPI opaque objects from one language to another, and sending a message from one language and having it received in the other.

Only one form of MPI_INIT need be called. After the call, the MPI library will be completely initialized for all supported languages.

In order to deal with the portability of MPI opaque objects, such as datatypes, communicators, and requests, conversion functions will be provided that convert the language-dependent "handles" to 32-bit integers and back again. These integers will be portable (among languages) versions of the objects they reference.

Sending a message from a Fortran program to a C program or vice versa will be explicitly allowed, as long as the signatures of the datatypes match. Here we are aided by the fact that the elementary datatypes defined in MPI-1 are distinct in the two languages, and no equivalence (such as one that might exist between the C datatype int and the Fortran datatype INTEGER on some machines) is assumed. Thus, in sending messages between programs written in different languages, one sends data of a given MPI datatype; no automatic conversion takes place.

9 Conclusion

We have described the current state (February, 1996) of MPI-2 discussions. The precise content of MPI-2 remains to be decided in the coming months. Although a few implementors are beginning to experiment with some of the notions described here, most are waiting to see what the final specification will look like. The MPI-2 features will be more difficult to implement than those of MPI-1. Nonetheless, enough discussion has taken place that it is possible to discern the likely scope of the functionality that MPI-2 will add to MPI-1. In this paper we have described that functionality.

References

1. World Wide Web MPI home page.
   http://www.mcs.anl.gov/mpi/standard.html.
2. R. Alasdair, A. Bruce, James G. Mills, and A. Gordon Smith. CHIMP/MPI user guide. Technical Report EPCC-KTP-CHIMP-V2-USER 1.2, Edinburgh Parallel Computing Centre, June 1994.
3. Greg Burns, Raja Daoud, and James Vaigl. LAM: An open cluster environment for MPI. In John W. Ross, editor, *Proceedings of Supercomputing Symposium '94*, pages 379–386. University of Toronto, 1994.

4. Peter Corbett, Dror Feitelson, Yarsun Hsu, Jean-Pierre Prost, Marc Snir, Sam Fineberg, Bill Nitzberg, Bernard Traversat, and Parkson Wong. MPI-IO: A parallel file I/O interface for MPI, version 0.3. Technical Report NAS-95-002, NAS, January 1995.
5. Peter Corbett, Yarsun Hsu, Jean-Pierre Prost, Marc Snir, Sam Fineberg, Bill Nitzberg, Bernard Traversat, Parkson Wong, and Dror Feitelson. MPI-IO: A parallel file I/O interface for MPI, version 0.4. http://lovelace.nas.nasa.gov/MPI-IO, December 1995.
6. Nathan E. Doss, Purushotam V. Bangalore, and Anthony Skjellum. MPI++ : Issues and Features. In *Proceedings of OONSKI '94*, January 1994.
7. The MPI Forum. The MPI message-passing interface standard. http://www.mcs.anl.gov/mpi/standard.html, May 1995.
8. William Gropp and Ewing Lusk. *User's Guide for mpich, a Portable Implementation of MPI*. Argonne National Laboratory, 1994.
9. Andrew Lumsdaine, Brian M. McCandless, and Jeffrey M. Squyres. Object-oriented MPI, 1996. http://www.cse.nd.edu/lsc/research/oompi/.
10. Anthony Skjellum, Nathan E. Doss, and Kishore Viswanathan. Intercommunicator extensions to MPI in the MPIX (MPI eXtension) Library. Technical report, Mississippi State University — Dept. of Computer Science, April 1994. Draft version.

Optimizing Sisal Programs: A Formal Approach

I. Attali[1] and D. Caromel[1] and R. Guider[1] and A. L. Wendelborn[2]

[1] INRIA Sophia Antipolis, CNRS - I3S - Univ. Nice Sophia Antipolis,
BP 93, 06902 Sophia Antipolis Cedex - France
{ia, caromel}@sophia.inria.fr
[2] Dept. Computer Science, University of Adelaide 5005 - Australia
andrew@cs.adelaide.edu.au

Abstract. We formally describe optimization techniques for the compilation of the language Sisal 2.0. More precisely, we translate Sisal programs into data-flow IF1 graphs and optimize these graphs. An interactive visualization environment for IF1 graphs is also provided.

1 Introduction

Software engineering of parallel programming is becoming an important issue in computer science; one focus has been to design compiler schemes that will provide efficient parallel code running on various parallel architectures. Sometimes complementary, sometimes orthogonal, the increasing importance of tools and environments dedicated to parallel computing is another significant aspect (see [9, 1, 18] for recent developments).

The trade-off between the programmer's and compiler's part in designing parallel programs has been debated at least for a decade. One solution to ensure portability and reusability is a well-specified language with a high-level of abstraction; moreover, a compiler for such language must automatically exploit the parallelism of programs, with a clean model of execution, independent, as much as possible, of the architecture of the computer.

Our goal is to investigate this trade-off, namely on the Sisal programming language [5], a strongly typed, applicative, single assignment language in use on a variety of parallel processors (see [5] for a description of Sisal and associated research). We intend to establish a base for the definition and implementation of validated compilers for Sisal, using formal specifications and verifiable transformations. We have observed indeed that the description of Sisal and its implementation call for a more formal approach. The semantics of Sisal is typically given in English and although the various authors have attempted to be precise, we have found ambiguities in the extant definitions of Sisal. We have already given in [3] a formal definition of the dynamic semantics of a significant part of the language Sisal 2.0 in Natural Semantics [11], using the Typol formalism [7], within the Centaur system [6], a generic tool for designing languages. The Sisal 1.2 compiler[1], (OSC, Optimizing Sisal Compiler, OSC [10]), has proved remarkably successful through exploitation of a variety of sophisticated optimizations.

---
[1] At the moment, there is no compiler for the version 2.0 of the language.

Unfortunately, the most precise definition of the optimizations currently resides in the compiler itself, making it difficult to extend either the language or the catalogue of optimizations. Two intermediate formats IF1 [14] and IF2 [17] are used in Sisal 1.2 implementation. IF1 is a textual description of data-flow graphs which allows traditional code optimizations and IF2 allows optimizations to be applied in terms of abstract storage.

We give a syntactic definition of the IF1 format in Section 2 and describe an interactive visualization environment for IF1 graphs in Section 3. Section 4 describes the translation scheme from Sisal to IF1 and points out some deficiencies in the IF1 formalism, as it was originally defined for Sisal 1.2. In Section 5, we give an example of transformation, the elimination of common subexpressions. Finally, Section 6 concludes with some directions for future work.

2 The IF1 intermediate format

IF1 is a hierarchical, single assignment language defining data-flow graphs. An IF1 program consists of one or more acyclic graphs made up of simple and compound nodes, edges, and types. Nodes correspond to computations: simple nodes represent arithmetic operations and structure manipulation. Compound nodes represent structured expressions, using subgraphs for the components of conditionals and loops. Edges show the transmission of data between nodes; types are associated with the data transmitted on edges. An IF1 program expresses data dependencies, with control left implicit; for example, an iteration is represented as a compound node with subgraphs describing generation of index values, the body of the loop, and the packaging of results. However, the control relationships between these components are left unspecified, and can be

```
file -> GRAPH_TYPE * ...;                    FILE ::= file;
type -> INT INT PARAMS PRAGMA;               GRAPH_TYPE ::= GRAPH type;
global_graph -> GRAPH_ELT_S STR INT PRAGMA;  GRAPH ::= global_graph imported_graph
imported_graph -> STR INT PRAGMA;                        local_graph subgraph;
local_graph -> GRAPH_ELT_S STR INT PRAGMA;
subgraph -> INT GRAPH_ELT_S PRAGMA;
graph_elt_s -> GRAPH_ELT * ...;              GRAPH_ELT_S ::= graph_elt_s;
snode -> NODE_LABEL INT PRAGMA;              GRAPH_ELT ::= snode cnode literal edge;
edge -> INT INT INT INT PRAGMA;
cnode -> GRAPH_LIST INT INT INT
         ASSOC_LIST PRAGMA;
literal -> INT INT INT VALUE PRAGMA;
params -> INT * ...;                         PARAMS ::= params;
int -> implemented as INTEGER;               INT ::= int;
pragma -> PRAGM * ...;                       PRAGMA ::= pragma;
pragm -> STR VAL;                            PRAGM ::= pragm;
str -> implemented as STRING;                STR ::= str;
no_val -> implemented as SINGLETON;          VAL ::= no_val id INT STR CHAR complex;
error_value -> implemented as STRING;        VALUE ::= error_value VAL;
graph_list -> GRAPH * ...;                   GRAPH_LIST ::= graph_list;
assoc_list -> INT * ...;                     ASSOC_LIST ::= assoc_list;
```

Figure 1. Abstract Syntax for IF1

interpreted as, for example, a data flow machine graph, or a loop structure for a conventional multiprocessor.

We specify syntactic aspects — concrete and abstract syntax of the IF1 language. From this specification, one can derive a parser that transforms the textual form of a program into a structural representation, an abstract syntax tree (AST), well-typed with respect to an abstract syntax[2] (see Figure 1). From syntactic specifications, we derive a structure editor in order to parse IF1 graphs (coming from OSC or our own translation from Sisal), edit and visualize them in a textual representation, as shown in Figure 2.

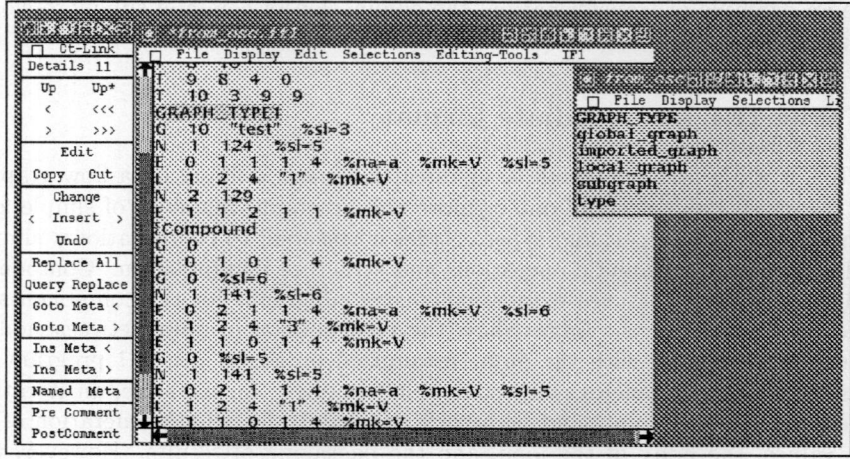

Figure 2. Textual Presentation of IF1 Graphs

3 Graphical Visualization of IF1 Graphs

This textual representation of an IF1 program describing a graph is not appropriate for understanding of the complete graph structure. Using the Centaur system and a specific package for graph display [12], we provide an automatic graphical representation of IF1 graphs (see Figure 3).

An IF1 graph is a hierarchical graph made up with subgraphs, each component with its own inputs and outputs denoted by ports. Two types of relations are represented: (1) a graph hierarchy representing the program structure, and (2) data dependencies (between nodes of the graphs).
Instead of visualizing the whole IF1 graph in a single window, since the number of nodes can be huge, we separate the relations as follows:

- the graph itself appears in a main window which is a synthesized view of the hierarchy of the program (a loop in a selection, etc);
- data dependencies appear in separate windows.

---

[2] Lower-case (resp. upper-case) identifiers define operators (resp. sorts).

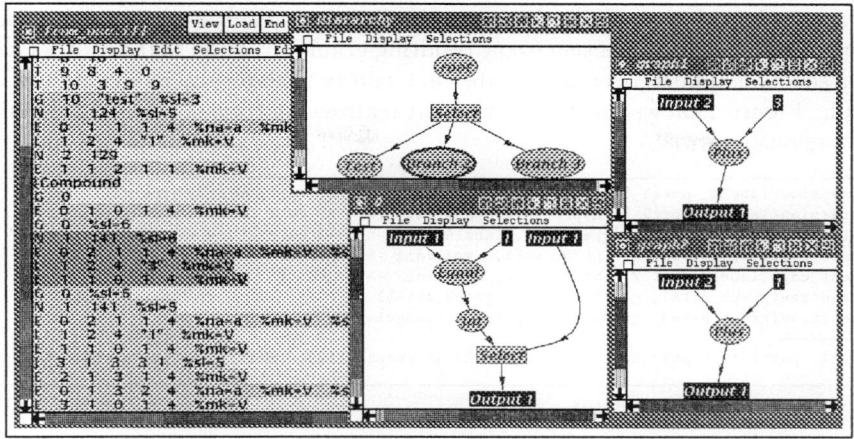

Figure 3. Graphical Visualization of IF1 graphs

We use different colors, fonts, and shapes to distinguish nodes and edges (compound nodes have a rectangular shape and graphs are denoted with oval shapes). The graph server takes care of the layout of the graph according to some specific heuristic (planarity for instance). This visualization is interactive since the graph server reacts to selection or move of a node.

We provide a zoom mechanism which expands on demand graph nodes with dependencies. Then, we have (in a main window named Hierarchy) a synthesized view of the graph, and we can also visualize data dependencies on a given graph in a new window, when zooming. In Figure 3, the root graph in the Hierarchy window is zoomed in the 0 window.

Because the whole graph (hierarchy and dependencies) is complex, we maintain the coordination of multiple views of graph objects, with a selection mechanism: clicking in windows displaying data dependencies shows the corresponding node in the hierarchical graph and highlights the textual fragment in the IF1 source, as illustrated in Figure 3 with a dark selection in the textual presentation as well as on the Branch2 node. We think that this type of representation is mandatory in the context of a large number of nodes in the data-flow graph.

4 Translating Sisal to IF1

In this section, we outline the methods used in translating Sisal 2.0 to IF1, especially ordinary expressions and loops; we also consider possible extensions to IF1 in order to reflect the powerful array operations of Sisal 2.0.

To translate expressions and function calls, we write Typol inference rules which take as input Sisal expressions and construct an IF1 AST. IF1 expressions, and associated subgraph edges and ports, can be produced from the operands and operators of the expression. Each Sisal construction can be thought of as an operator $Oper(N_1, \ldots, N_n)$; the corresponding IF1 graph is obtained by translating each N_i and connecting every resulting graph to a node that denotes $Oper$. This scheme, applied over a Sisal abstract syntax tree, results in a complete IF1 graph. Figure 4 shows the Typol rule that realizes the translation of Sisal binary expressions:

```
expression(label1, port1 |- Expr1 -> graph_elts1, label2, val1, port2, type_val1) &
expression(label2, port2 |- Expr2 -> graph_elts2, label3, val2, port3, type_val2) &
operator(label3 |- Oper -> graph_elt, label4) &
create_edge(label3, int 1, int 0, val1, type_val1 -> edge1) &
create_edge(label3, int 2, int 0, val2, type_val2 -> edge2) &
appendtree(graph_elts1, graph_elts2 -> graph_elts3) &
appendtree(graph_elts3, graph_elt_s[graph_elt,edge1,edge2]-> graph_elts4)
-----------
label1, port1 |- binary(Expr1, Oper, Expr2) -> graph_elts4, label3, label4, port3;
```

Figure 4. Translating Sisal expressions into IF1

The IF1 graph produced for a Sisal binary expression given as the subject `binary(Expr1, Oper, Expr2)`, is the composition (predicate `appendtree`) of the subgraphs respectively produced from `Expr1` and `Expr2` with recursive calls to the predicate `expression`. From the operator `Oper`, the predicate `operator` produces a node and the calls to the predicate `create_edge` build two edges, which will augment the IF1 graph. It is also necessary to maintain specific information such as the current node label (`label1`), or the current available port number (`port1`), in a left-to-right manner.

4.1 Translating loops

In translating loops, one problem that must be addressed is the classification of loops in two categories: parallel or sequential. Sisal 2.0's loop syntax is quite powerful, allowing expression of a variety of loop structures over index ranges and array structures. Unlike Sisal 1.2, it is not apparent from the loop syntax in Sisal 2.0 whether or not it is a parallel loop. This must be determined from analysis of variable usage within the loop—if the value at one iteration is dependent in any way upon the value at a previous iteration, then the loop is sequential, and iterations must be executed sequentially, otherwise all the executions of the loop body are notionally independent, and can be executed in parallel. This choice affects the IF1 produced from a loop, as IF1 provides three compound nodes with which loops can be expressed: for sequential loops with post-test (*LoopA*) and pre-test termination conditions (*LoopB*), and the *ForAll* node for parallel loops. Let us focus on distribution control loops (iteration control loops are easily translated into *LoopA* or *LoopB* nodes, depending on the presence of a pre- or post-test). The syntax for such loops is the following:

        for in-exp-list   % top part
            [decl-def-part]
        do [decl-def-part]   % body part
        returns return-clause   % returns part

A loop expression has three parts: the *top* part establishes indices (`in-exp-list`) and initialization of the loop (`decl-def-part` is evaluated only once, so it does not affect the dependency analysis); the *body* part defines the actions to be carried out in a distributed manner (`decl-def-part`); the *returns* part packages the produced values. In the body part, carried values are transmitted from one iteration to the following. We study dependencies between variable definition and usage to detect if this loop can be evaluated in a sequential or in a parallel manner. If the body part contains references of one variable which is defined in the previous iteration of the body part, then the loop must be executed in a sequential manner and will be translated into a *LoopB* node. Otherwise, the loop is a parallel one and can be translated into a *ForAll* node. Such an analysis of dependencies is facilitated by the functional nature of Sisal: side-effects and aliasing problems are banished by definition.

4.2 Discussion

IF1 is a powerful intermediate form capable of expressing data dependencies in expressions, iterative control flow, and recursive and higher-order functions, as well as standard data structures; there are, however, some aspects in which it lack expressiveness and this leads to somewhat clumsy translations of Sisal 2.0. This is principally a consequence of a much more sophisticated model of array structures employed in Sisal 2.0, whereby arrays are multidimensional values, can be constructed monolithically, and sophisticated sub-array selection operations can be expressed. We firstly examined how translation could be achieved with existing nodes. However, we discovered that this option leads to a loss of efficiency and abstraction in terms of evaluation policy for those parts of the language. This allowed us to identify those aspects of IF1 which needed modification and provided insight into how to do it. A forthcoming document will describe in detail our proposed extensions to IF1, especially concerning arrays.

5 Optimizing IF1 graphs

Optimizations on the IF1 graph constitute an early phase of the OSC compiler [10] (inline expansion, common subexpression elimination, loop invariant removal, dead code elimination, see [15] for an overview). To demonstrate our approach using IF1, we formally describe common subexpression elimination.
The principle of this optimization is to eliminate redundant computations. Redundant computation occurs in an IF1 graph when the same sub-expression is computed more than once. To perform this transformation, for each node (i.e. sub-expression), we search for a sub-graph which matches the one whose root is the node currently visited (although the structure is a DAG, it is similar to a tree when traversed from nodes to their predecessors). When such a sub-graph is detected, all its "clients" (nodes using its results) are connected to the visited node in the same way they were to the removed node.

Visualization of IF1 graphs uses a representation with an unordered list of nodes and edges but this is inadequate for the transformation. We define an alternative representation which uncovers actual data dependencies between graph nodes and encodes nodes predecessors. This representation permits to identify any two equal subgraphs and to restructure the whole graph. This alternative structure (named PGRAPH) comprises a list of nodes, and for each of them a list of predecessors (see definition Figure 5); the graph is then traversed from output to input: a subexpression can be identified with a node and all its predecessors, with the transitive closure of the relation "is predecessor of".

```
pgraph -> GRAPH_ELT_S NODE_S                          PGRAPH  ::= pgraph ;
     - - GRAPH_ELT_S are only cnodes and snodes
node_s -> NODE * ...;                                 NODE    ::= node;
node -> INT PRED_S;                                   - - <node, list of predecessors>
pred_s -> PRED * ...;                                 PRED_S  ::= pred_s;
pred -> INT GRAPH_ELT;                                PRED    ::= pred GRAPH_ELT;
     - - a predecessor can be a literal or an output port number + a predecessor node.
```

Figure 5. Representation of IF1 graphs by predecessors

The common subexpression elimination can be expressed in three phases: change the representation, detect and eliminate common subexpressions, and convert back to the primary representation (for visualization purpose).

Modularity makes it possible to reuse various sets (especially graph traversals) for the specification of other transformations.

The first step of the transformation is made of two traversals of the list of nodes and edges composing the primary graph representation (GRAPH) to build the PGRAPH structure. Information contained in the node description (label for the node's operation, name, etc) is needed during the detection of redundant computations, and when returning to the primary format. A first traversal constructs the list of predecessors from the edge description for each node. At this step, information on nodes themselves has not been retained. So a second traversal is needed to construct from a list describing predecessors with a pair of integers (node number, input port), a new predecessor list containing full informations on nodes.

Equality on graphs is expressed in the two following rules: equality is first checked on nodes (predicate node_equal) and then recursively on their predecessors (predicate pred_equal). A termination rule applies when nodes with no predecessors have been reached.

```
graph_equal(pgraph |- literal(_, _, _, P4, _), literal(_, _, _, P4, _) -> bool true ;

pred(node1 |- pgraph -> pred1) &                      - predecessors of node1
pred(node2 |- pgraph -> pred2) &                      - predecessors of node2
node_equal(|- node1, node2 -> bool true) &            - equality of nodes
pred_equal(pgraph |- pred1, pred2 -> bool true)       - equality of predecessors
_____
graph_equal(pgraph |- node1, node2 -> bool true) ;
           provided diff(node1, literal);
```

After identification of redundant computation, the next step in the transformation consists in their elimination. Given a redundant subgraph (w.r.t. an original subgraph), identified by its root node, its elimination comprises a search for all nodes whose predecessors contain this root. When detected, such a node is replaced by the original subgraph's root properly connected (input ports are connected to the appropriate output ports). Returning to the primary IF1 format is done in a straightforward manner with a traversal on the PGRAPH structure.

Due to lack of space, we do not detail the whole specification of graph manipulation in either form. They are composed of 150 rules or axioms and make extensive use of pattern-matching and unification.

6 Conclusion and Future Work

From the specifications, using the Centaur system, we have derived a program development environment for Sisal, illustrated in Figure 6. This environment provides a sequential execution of Sisal programs, the translation of Sisal programs into IF1 graphs, their textual and graphical presentations, and transformations on IF1 format. We first intend to complete the formal specification of the suite of transformations as used in the OSC compiler, contributing not only to a formal definition of transformation techniques for parallelization of Sisal programs, but also to a wide availability. Moreover, thanks to formal specifications, we intend to prove the correctness of the transformations, using proof assistant systems. Experiments have been done in this sense to specify semantic definitions and program transformations, and to study and prove their properties [4].

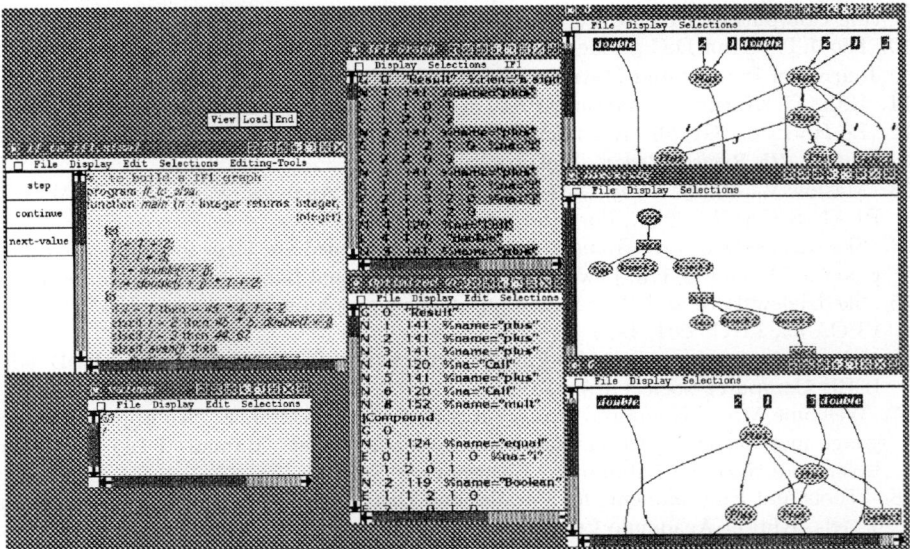

Figure 6. A Graphical Interactive Environment for Sisal and IF1.

Finally, our environment could be used for other high-level languages (logic, data-flow, object-oriented [16, 2]), or imperative languages where array operations are critical (e.g. Fortran, where parallelization is an active research area [13, 8]). Once a comprehensive set of transformation have been formally specified in a modular manner, it should be possible to take advantage of it for other paradigms, possibly through common intermediate formats.

References

1. "Parallel and Distributed Technology - Systems and Applications", Agha G. editor, 3 (4), 1995.
2. Attali I., Caromel D., Ehmety S. O., Lippi S. Semantic-based visualization for parallel object-oriented programming, To appear in OOPSLA'96, ACM Press, Sigplan Notices, San Jose, CA, 1996.
3. Attali I., Caromel D. and Wendelborn A. "A Formal Semantics and an Interactive Environment for Sisal", pp 231-258, in [18].
4. Bertot Y. and Fraer R. "Reasoning with Executable Specifications", Proc. of TAPSOFT, LNCS 915, Aarhus, Denmark, 1995.
5. Böhm A. P. W., Cann D.C., Feo J.T., Oldehoeft R.R., "Sisal Reference Manual (language version 2.0)" Draft Report, 1992.
6. Borras P. et al., "CENTAUR: the system", Third Annual Symposium on Software Development Environments, Boston, 1988.
7. Despeyroux T. "Typol: a formalism to implement Natural Semantics" INRIA research report 94, 1988.
8. Detert U. and Gerndt M., "TOP^2 - Tool Suite for the Development and Testing of Parallel Applications", CONPAR'94, Linz, Austria, LNCS 854, 1994.
9. Proceedings of the Second Workshop on Environments and Tools for Parallel Scientific Computing, Dongarra J.J. & Tourancheau B. eds, SIAM, Townsend, 1994.
10. Feo J.T., Cann D.C., Oldehoeft R.R., "A Report on the Sisal Language Project" Journal of Parallel and Distributed Computing, 1990.
11. Kahn G. "Natural Semantics", Proc. of STACS, Passau, Germany, LNCS 247, 1987.
12. Le Hors A., "Graph: A Directed Graph Displaying Server", in *GIPE 2 ESPRIT project, 4th Review Report*, Workpackage 4, 1992.
13. Maslov V., "Lazy Array Data-Flow Dependence Analysis" Proc. 21st ACM SIGPLAN-SIGACT POPL, Portland, Oregon, 1994.
14. Skedzielewski S. and Glauert J. "IF1 - An intermediate form for applicative languages" Manual M-170, Lawrence Livermore National Laboratory, Livermore, 1985.
15. Skedzielewski S. and Welcome M. "Data-flow graph optimization in IF1" Proc. of FPCA'85, LNCS 201, 1985.
16. "Programming Languages for Parallel Processing", Skillicorn D. B. & Talia D. eds, IEEE Computer Society Press, 1995.
17. Welcome M.L., Szymanski B.K., Yates R.K., Ranelletti J. E. "An applicative language intermediate form explicit memory management" Manual M-195, Lawrence Livermore National Laboratory, Livermore, 1986.
18. "Tools and Environments for Parallel and Distributed Systems", Zaky A. & Lewis T. eds, Kluwer Academic Publishers, 1996.

A Refinement Methodology for Developing Data-Parallel Applications[*]

Lars Nyland,[1] Jan Prins,[1] Allen Goldberg,[2] Peter Mills,[3] John Reif[3] and Robert Wagner[3]

[1] Dept. of Computer Science, University of North Carolina
Chapel Hill, NC 27599-3175 USA
[2] Kestrel Institute, 3260 Hillview Ave., Palo Alto, CA USA
[3] Dept. of Computer Science, Duke University, Durham, NC 27708 USA

Abstract. Data-parallelism is a relatively well-understood form of parallel computation, yet developing simple applications can involve substantial efforts to express the problem in low-level data-parallel notations. We describe a process of software development for data-parallel applications starting from high-level specifications, generating repeated refinements of designs to match different architectural models and performance constraints, supporting a development activity with cost-benefit analysis. Primary issues are algorithm choice, correctness and efficiency, followed by data decomposition, load balancing and message-passing coordination. Development of a data-parallel multitarget tracking application is used as a case study, showing the progression from high to low-level refinements. We conclude by describing tool support for the process.

1 Introduction

Data-parallelism can be generally defined as the concurrent application of an arbitrary function to all items in a collection of data, yielding a degree of parallelism that typically scales with problem size. This definition permits many computationally intensive problems to be expressed in a data-parallel fashion, but is far more general than the data-parallel constructs found in typical parallel programming notations. As a result, the development of a data-parallel application can involve substantial effort to recast the problem to meet the limitations of the programming notation and target architecture. From a methodological point of view, the problems faced developing data-parallel applications are:

- **Target architecture.** Different target parallel architectures may require substantially different algorithms to achieve good performance, hence the target architecture has an early and pervasive effect on application development.
- **Multiplicity of target architectures.** For the same reasons just cited, the frequent requirement that an application must operate on a variety of *different* architectures (parallel or sequential) substantially complicates the development.
- **Changes in problem specification or target architecture(s).** Changes in problem specification and/or target environment must be accommodated in a systematic fashion, because of the large impact that either causes for parallel applications.

We consider a refinement methodology that generates a tree of data-parallel applications whose level of development effort varies with the parallel architecture(s) targeted and the required level of performance. The methodology explicates activities whose time and expense can be regulated via a cost-benefit analysis. The main features of the methodology are:

- **High-level design capture**, where an executable description of the design can be evaluated for its complexity and scalability.
- **A data-parallel design notation** that supports a fully generalized definition of data-parallelism and eliminates dependence on specific architectures.

---
[*] This work supported by Rome Laboratory under contract #F30602-94-C-0037.

- **Analysis and refinement** of the design based on fundamental considerations of complexity, communication, locality, and target architecture.
- **Prototyping.** Disciplined experimentation with the design at various levels of abstraction to gain information used to further direct the refinement process.

Our methodology is based on the spiral model of software development with prototyping, a model favored by the software engineering community. During the risk analysis phase of development, development relies on concise high-level, executable notations for data parallel problems, such as FP, Sisal, Nesl and Proteus [3, 5, 4, 9]. The next phase is migration and integration (perhaps automatic) to achieve an efficient application that can be tested and deployed. As new requirements or target architectures are introduced, the development repeats the cycle.

The target architecture will have an influence on many aspects of the design, and here we classify parallel architectures into four classes – two shared memory classes and two distributed memory classes. The effects of each of the architectural classes is considered with regard to its impact on application development.

- **UMA shared memory.** In this class, global memory has uniform memory access (UMA) time. This challenging requirement is currently met only by cacheless machines with memory bandwidth matched to processor speed. Examples: Cray T90, NEC SX-4, and the Tera Computer.
- **NUMA shared memory.** A non-uniform memory access architecture has hierarchically organized memory, often viewed as multiple levels of cache. Examples are SGI Challenge machines and the Convex Exemplar.
- **Small-grain distributed memory.** No shared memory abstraction is provided in hardware, but fine-grain message passing and low-latency synchronization are supported. We've labeled this class SIMD, since it describes machines such as the MasPar MP-2, but in also describes MIMD machines with fine-grain support, such as the Cray T3D.
- **Large-grain distributed memory.** Message-passing (MP) machines where large messages must be used to achieve reasonable performance of processor interconnect. Examples are the Intel Paragon and IBM SP/2.

2 Design Refinement Methodology

Effective software development involves prototyping at several different design levels to discover the characteristics of a problem. Figure 1 describes the stages along the refinement paths from high-level designs to architecture-specific implementations. A description of each of the stages follows.

2.1 Problem Definition and Validation

At the *specification* stage (top of figure 1), we seek an initial description of the problem and validation of the problem against functional requirements. The focus is on determining the feasibility and practicality of obtaining a high-performance algorithmic solution for a succinctly-described problem.

Algorithm selection and parallel complexity analysis. Solutions to complex problems are often expressed as highly algorithmic, mathematically sophisticated descriptions. This makes the high-level analysis both necessary, because the algorithms are often computationally intensive, and feasible, because of their succinct description. Asymptotic analysis, prototyping, and exploration of algorithm variants can all be performed quickly at this level.

One set of measures for selecting parallel algorithms are *work* and *step* complexities. The *work* complexity is a measure of all operations performed, while the *step* complexity measures the number of parallel steps (minimum number of sequential steps). While these measures are not completely realistic, we seek to separate concerns and formulate a tractable, staged analysis methodology. Finding a work-efficient algorithms is the goal, even though they may lead to irregular solutions with data-dependent behavior. For large problems on fixed size machines, work-efficient algorithms give the best performance.

Analysis for parallel architectures. The analysis techniques at this stage are based on medium-level parallel computing cost models such as LogP [6], BSP [12], and HMM [1] that can be used to estimate communication and memory performance. The model parameters of the algorithm are obtained analytically (when possible, or by instrumenting a prototype of the algorithm). One of the 4 architectural classes must be supplied to complete the analysis.

2.2 Design Refinement

The refinement of programs, shown in figure 1 between the dashed lines, occurs in a high-level design notation. We've chosen a data-parallel subset of Proteus [9] as our design notation. An initial implementation of specification provides a starting point that provides a foundation for correctness, analysis, information-conveyance and measurement purposes. At a high level, programming for data-parallel execution is only slightly more complicated than programming for serial execution in a language where collections of data are fully supported.

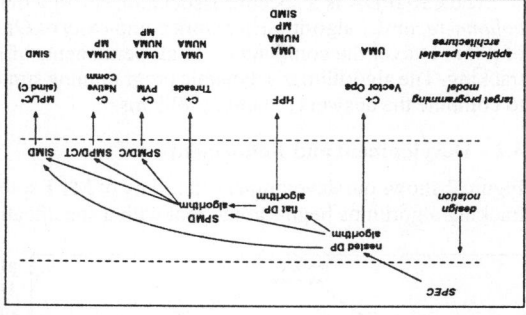

Fig. 1. Refinement and translation steps in the development of data-parallel applications

The different refinement paths are:

- Nested data-parallel design refined to flat data-parallel design by converting nested iterators and nested sequences to loops with integer iterators and rectangular arrays. Flat data parallel programs are suitably expressed using languages such as HPF, Fortran90, C* and MPL.
- Nested (or flat) data-parallel refined to SPMD. SPMD programs have multiple threads of control using a single shared-memory, where decomposition is under program control, varying from simple memory-decomposition models to complex, adaptive, load-balancing methods.
- SPMD refined to SPMD/C. This model adds explicit communications to an SPMD program, eliminating the shared-memory constructs.
- Refining to an SIMD Model. SIMD machines can be viewed as fine-grain message-passing machines, or as vector architectures. Refinement to an SIMD model can begin with either the nested or flat data-parallel versions, or it can be derived as similar to other message-passing versions.

2.3 Translation to Target Programming Models

Translation of high-level data-parallel designs can target a variety languages as shown by the vertical arrows in figure 1. Different designs are matched to different languages and different parallel architectures. Translation from the design notation may be manual or automatic; the automatic translation of nested data-parallel programs to vector models is an active area of research [11, 4].

3 Case Studies: Multitarget Tracking

In this section, we turn our attention to an example, demonstrating the methodology and the reliance on particular tools used for the process. The problem chosen to solve here is that of multitarget tracking, and through the development process, we will develop a tree of refined algorithms and show the need for additional automated support.

Multitarget tracking (MTT) can be described simply: a set of N targets are being tracked when new location data arrives as a set of M positions. Prediction models compute the expected locations of the tracked targets, but the new data may not coincide with the estimated positions. The problem is to find the joint probability that a target t, $1 \leq t \leq N$, is represented by a measurement j, $1 \leq j \leq M$. Any algorithm that computes this result falls in the category of multitarget tracking algorithms that we are studying.

3.1 Solutions to the Multitarget Tracking Problem

We have explored two published algorithms that solve the multitarget tracking problem. The first is the *column-recursive joint probability data association* (CR-JPDA) algorithm, since it has been the subject of parallel implementation studies [8, 2]. Zhou and Bose also present a parallel MTT algorithm, the *tree-search joint probability data association filter* (ZB-JPDAF) algorithm, that performs much worse than the CR-JPDA in the worst case, but has claims of better performance in average and highly likely cases [14].

The CR-JPDA is a specific association strategy that uses a weighted average of returns. The *column-recursive* algorithm has work complexity of $O(NM^2 2^N)$, but this is improved by a factor of $N(M+1)$ over the computation of all permanents of a matrix, the direct method of multitarget tracking. The algorithm is a dynamic programming strategy, relying on solutions for sub-problems to compute the answers to larger problems.

3.2 Development and Refinement

Figure 2 shows our development hierarchy of MTT solutions. The development of our multitarget tracking algorithms begins at the root with a specification of the problem to be solved.

Below the specification are the two algorithms under consideration, the CR-JPDA and the ZB-JPDAF. These two implementations are written in Proteus, and serve the purpose of achieving a baseline implementation with no initial concern for parallelism. The two implementations are concise, requiring about 40 lines of Proteus each. They were validated against one another prior to further development (details in [10]).

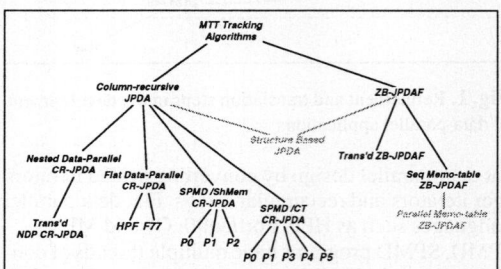

Fig. 2. Our development tree of the MTT algorithms. The grayed instances have yet to be developed.

The first descendent of the CR-JPDA targets parallelism, it is a nested data-parallel implementation in Proteus, which is suitable for automatic translation to C with vector operations. The initial version of the ZB-JPDAF, written in Proteus, is also suitable for translation to C with vector operations. At this level, all implementations are architecture-independent.

Additional studies were performed, as outlined by the design refinements described in section 2. A flat data-parallel version of the CR-JPDA was developed and translated manually into HPF and Fortran77, where memory decomposition and vectorization could be examined. An SPMD version of the CR-JPDA was developed, to explore assignment of work to processors, and alternate memory decompositions that improved performance. And finally, a paper study was performed to estimate the performance of a variety of message-passing implementations of the CR-JPDA [13].

3.3 Results of the Implementation Study

A brief description of the variants of multitarget tracking programs follows. For a complete description, including code and analysis, see [10].

Nested Data-parallel Implementations. Prototyping the CR-JPDA and the ZB-JPDAF with a high-level language yielded quick development and concise descriptions on which further implementations can be based. Not only were the two prototypes validated against one another, but their work complexity was examined by running both on a variety of input data. As suggested by the authors, a *memoized* implementation of the ZB-JPDAF drastically improved the performance. At this high level, it is possible to recognize that the CR-JPDA is a dynamic programming variation of the ZB-JPDAF, where the tabular nature of the implementation removes all variance from the execution time. The CR-JDPA always has the same performance, where the ZB-JPDAF is highly sensitive due to data-dependent branching factors.

Flat Data-parallel Implementations. The next results were from conversion to flat data-parallel implementations. The (manual) transformation removes nested structures, such as nested iterators, nested function calls, and nested summations, turning them into loops populated with assignment statements. The nested data structures are also made regular (same length at in each dimension), where space is wasted to provide regularity. These versions, written in HPF and Fortran77, gave us the opportunity to explore regular data decompositions for parallel execution. Unfortunately, the work in the CR-JPDA is unevenly distributed over the data, so regular data decompositions will never balance the work evenly, necessitating further refinement for higher performance.

SPMD Implementations. In an SPMD model, memory decomposition is under program control, as is work decomposition. Three versions were explored at this level: one similar to the flat data-parallel version (as a benchmark); an attempt to balance memory requests with a modified block-cyclic layout; and uneven data decomposition to balance the work. Both of the attempts to improve performance were successful, with good load-balancing leading to good scaling behavior.

Message-passing Studies. Our final study looked at the CR-JPDA as it might be implemented on a message-passing architecture (described fully in [13]). A simple model was developed to describe the computation and communication costs of different decompositions of the CR-JPDA. The initial implementation was based on the simple observation that M processors could be used to compute results with almost no communication. Since this provides only limited parallelism, several more implementations were studied in an attempt to use more processors (64-2048). It took substantial algorithm enhancement and complex message coordination to achieve any solutions that could outperform the initial solution.

Summary of Implementations. The refinement strategy used to develop the MTT algorithms cleanly separates development issues. In the prototypes, the primary concern was one of correctness and agreement between the two algorithms. The flat data-parallel implementations explored regular memory decompositions. The SPMD version focused on load-balancing the work and data with irregular decompositions. Finally, the message-passing study looked at several variants of the algorithm in an effort to achieve a scalable program on a given architecture. Developing correct implementations early and quickly allowed the later developments to focus mainly on performance issues.

4 Tools to support parallel software development

The work performed in the case study used a collection of software tools that only minimally support the task. What follows is a suggested set of tools to support the design methodology described here.

- **Parallel programming language for prototyping.** The brevity, clarity, and analyzability of our Proteus implementations along with other implementations (Sisal implementation in [8]) make a strong argument for using high-level languages for prototyping. High-level programming languages help the developer gain intuition and insight about the inner workings of complex algorithms prior to exploring optimizations for high-performance.
- **Repository version manager.** Developing a tree of programs is not a one-way process, rather there is constant flow of information among program instances up and down the version tree. It must be possible to develop new versions on the development tree by taking parts of other versions, or perhaps new parts, and composing them. Currently, no such tool exists, so the development of the versions in this report were managed manually.
- **Powerful transformation tools.** The translation and optimization strategies developed by us, the Scandal group at CMU and the Sisal group at LLNL permit the developer to express an parallel solution in a concise, clear notation that can be translated to code that runs competitively with hand-written, hard-to-understand, low-level implementations.
- **Multi-lingual programs.** If prototypes are refined into reliable applications, often performance can be increased with multi-lingual capabilities. In one instance, a small part might be much more efficient if rewritten in a low-level language. Rather than commit the entire prototype to the lower-level language, support for multi-lingual programs could drastically reduce development time. As an alternative, consider the integration of a prototype into a much larger existing system, such as a larger tracking system.

- **Information gathering tools.** The main goal of developing parallel programs is high performance. As such, it is important to know where the program is performing and consuming resources as expected and where it is not. Performance analysis tools that provide information about highly optimized programs are key to achieving this goal.
- **Portable compilation targets.** Architecture-independent compilation targets ensure the portability and longevity of parallel applications, as is demonstrated by the Nesl project [4], the Sisal project [5], and the Fortran-M project [7].

5 Conclusions

We have proposed a tree-based refinement strategy for developing data-parallel applications. As development progresses down the tree, algorithms are specialized, perhaps to meet architectural and performance considerations. Branches represent different algorithms or different specializations. To accommodate new target architectures, we add new branches; to accommodate specification changes, we retrace the refinements steps through the tree. The application is represented by the tree of refined programs, not one particular refined version in the tree.

We have used the refinement methodology to develop parallel multitarget tracking algorithms. The separation of concerns allowed us first to concentrate on the correctness and high-level performance of 2 competing algorithms (the CR-JPDA and the ZB-JPDAF). Further refinement allowed the explorations of regular and irregular decomposition to achieve a balanced work-load. Additional models were used to determine the costs of a family of message-passing implementations, and what was required to achieve scalable performance over a wide range of parallelism.

References

1. B. Alpern, L. Carter, and E. Feig. Uniform memory hierarchies. In *Proc. Foundations of Computer Science*, 1990.
2. John K. Antonio. Architectural influences on task scheduling: A case study implementation of the jpda algorithm. Technical Report RL-TR-94-200, Rome Laboratory, Nov. 1994.
3. J. Backus. Can programming be liberated from the von neumann style? a functional style and its algebra of programs. *Comm. of the ACM*, 21(8):613–641, 1978.
4. Guy E. Blelloch. Programming parallel algorithms. *CACM*, 39(3), Mar. 1996.
5. David C. Cann. SISAL 1.2: A brief introduction and tutorial. Technical report, Lawrence Livermore National Laboratory, 1993.
6. D. Culler, R. Karp, D. Patterson, A. Sahay, K. E. Schauser, E. Santos, R. Subramonian, and T. von Eicken. LogP: Towards a realistic model of parallel computation. In *Proc. Symposium on Principles and Practice of Parallel Programming*, 1993.
7. Ian Foster. *Designing and building parallel programs*. Addison Wesley, 1995.
8. Richard A. Games, John D. Ramsdell, and Joseph J. Rushanan. Techniques for real-time parallel processing: Sensor processing case studies. Technical Report MTR 93B0000186, MITRE, April 1994.
9. Allen Goldberg, Peter Mills, Lars Nyland, Jan Prins, John Reif, and James Riely. Specification and development of parallel algorithms with the proteus system. In G. Blelloch, M. Chandy, and S. Jagannathan, editors, *Specification of Parallel Algorithms*. American Mathematical Society, 1994.
10. Lars S. Nyland, Jan F. Prins, Allen T. Goldberg, Peter H. Mills, John H. Reif, and Robert A. Wagner. A design methodology for data-parallel applications. Technical report, Univ. of N. Carolina, 1995. Available as http://www.cs.unc.edu/Research/aipdesign.
11. Jan Prins and Daniel Palmer. Transforming high-level data-parallel programs into vector operations. In *Proceedings of Principles and Practice of Parallel Programming*, pages 119–128, San Diego, CA, 1993.
12. Leslie G Valiant. A bridging model for parallel computation. *CACM*, 33(8):103, August 1990.
13. Robert A. Wagner. Task parallel implementation of the jpda algorithm. Technical report, Department of Computer Science, Duke University, Durham, NC 27708-0129, June 1995.
14. B. Zhou and N. K. Bose. An efficient algorithm for data association in multitarget tracking. *IEEE Trans. on Aerospace and Electronic Systems*, 31(1):458–468, 1995.

Task Parallelism: What a Tool Can Provide and What Should Be Left to the User

Silvia A. Crivelli and Elizabeth R. Jessup

Dept. of Comp. Science, Univ. of Colorado, Boulder, CO 80309-0430, USA

Abstract. This paper discusses some programming issues involved in the implementation of task parallelism on distributed-memory MIMD computers. In particular, we separate those issues that are application-independent and so can be part of a library from those that should be controlled by the user to maximize the performance.

1 Introduction

Task parallelism represents an approach for program implementation that concentrates on the decomposition of the computation to be performed rather than of the data domain. It is directed at solving problems presenting a complex task that can be decomposed into asynchronous subtasks to be assigned to the processors. Thus, it offers a valuable alternative to data parallelism which focus on the partition of the data structure.

Task parallelism is difficult to implement on distributed-memory MIMD computers because the number of the tasks and their execution time usually vary dynamically and unpredictably. Therefore, it requires a dynamic, asynchronous approach that lets processors work independently as much as possible and that only adds enough coordination for a more efficient use of the resources.

In this paper, we discuss the aspects to be considered in the development of a programming library to support task-parallel computation. In particular, we emphasize that, although a great deal of programming support can be provided for task parallelization, ample decision must be given to users as the only way to maximize the performance. The paper is organized as follows. In section 2, we identify the different issues involved in a task-parallel implementation. In section 3, we present an example of task-parallel problem. In sections 4 and 5, we discuss those issues that can be covered by a library and those that should be left to the users. Finally, in section 6, we provide a reference for further reading.

2 An Approach for Task Parallelism

In this section, we identify the main features of an efficient task-parallel implementation. Such an implementation begins with the identification of the tasks. Tasks are placed in a queue that can be maintained by a single processor or split into local queues maintained by all the processors. Processors select a task from the queue for computation, create more tasks along the way, and place them in the queue. This process repeats until there are no more tasks in the queue or until a transfer of tasks becomes necessary to keep the workload distributed among the processors. Thus, load distribution is an important ingredient of achieving good performance. Observe that, to avoid unnecessary transfers of work, the load balancing mechanisms should not attempt to keep the number of tasks *evenly* distributed as the tasks sizes are mostly uneven and usually unknown.

Although an efficient implementation of task-parallel problems must keep the load distributed, it also must ensure that processors do not waste their time exploring unproductive tasks. Unnecessary work is avoided by allowing processors to share information and by assigning different priorities to the tasks.

A final ingredient is termination detection. In task parallelism, load balancing is asynchronously checked and transfer of work accordingly made from the heavily loaded processors to the lightly loaded ones. In this context, idle processors waiting for heavily loaded processors to send some work away may wait forever if not informed that all of the work has been completed. Because it is impossible for an idle processor to correctly decide whether or not to quit based on its local information, a global check for termination becomes necessary.

3 An Example: The Traveling Salesman Problem (TSP)

TSP is a discrete optimization problem in which a salesman must visit N cities in such a way as to minimize the cost of the trip. All cities, except for the starting city, must be visited only once. The TSP can be solved by using a branch-and-bound algorithm. This algorithm finds the solutions by searching through a tree of partial solutions that are created dynamically. The root of the tree corresponds to the given problem and the leaves represent the solutions.

The branch-and-bound procedure consists of two phases: a branching phase that generates new partial solutions by branching out from the current ones and a bounding phase that computes the cost associated with any partial solution. The cost of a partial solution is considered a lower bound on the cost of a complete solution deriving from it. Thus, partial solutions whose cost is larger than the least cost of the complete solutions computed so far are discarded. The cost associated with the partial solutions is also used to direct the search towards the most promising branches of the tree. Tasks with less cost are assigned higher priorities because they are more likely to produce a solution.

TSP is task-parallel because it can be associated with a tree that can be searched in parallel by partitioning it into subtrees. A task may consist of finding a new partial solution (i.e, a node of the tree) by applying the branch-and-bound procedure to a given one. An efficient parallel implementation of this problem needs to be asynchronous to exploit the fact that different subtrees can be searched independently and without involving synchronous communication. It also needs to be adaptive to dynamically redistribute the tree among the processors as new branches are created and others are pruned.

4 What a Library Can Provide

Of the components of a task-parallel implementation discussed in section 2, some are completely independent of the application. As these components are hard to implement, they discourage many potential users from trying task parallelism. Thus, they should be supported in a library. Charm [3] and PMESC [2] offer ample support for implementing task-parallel programs using the Single Program Multiple Data (SPMD) approach. However, as both tools are still evolving and there still is a great margin for improvement, we believe that it is important to discuss these issues. The main topics that should be supported by a library are:

- *Handling the task queue structure:* This process includes retrieving tasks from and storing tasks in the queue using different queueing mechanisms such as LIFO, FIFO, and other strategies for priority queues.
- *Load balancing:* It is particularly hard to implement in task parallelism due to the asynchronicity of the target problems. Load balancing strategies can be centralized, hierarchical, distributed or hybrid. They can also be sender- or receiver-initiated. Refer to [1] for a description of these methods.
- *Sharing of Information:* The trick is how to share global information without having to synchronize the processors or to interrupt them many times. An efficient approach is the implementation of pseudo-global variables. To implement a pseudo-global variable each processor maintains its own copy and propagates its value to the other processors by periodically activating an asynchronous communication mechanism.
- *Termination Detection:* Another aspect concerning task parallelism is termination detection. Its efficient implementation is sometimes crucial to achieving overall performance. Different approaches are discussed in [1].
- *Embedding:* It is important not only to allow users to choose the virtual machine to use for implementing their codes but also to provide them with the algorithms for efficiently embedding those virtual topologies into the real machine architectures.

There are different strategies proposed in the literature to deal with all these issues. However, a strategy that works well for one problem type may not work for another. Therefore, a high degree of customization must be provided to users so they can finely tune their codes. In the next section, we discuss this subject.

5 What Should Be Left to Users

The issues and decisions that should be left to the user so as to maximize the performance are:

- *Should tasks be prioritized?:* Some problems can be split into subproblems that can be executed in any order without affecting either the correctness of the results or the overall performance. There are however, other cases (like TSP) where the assignment of priorities to the tasks may have a tremendous

impact in the performance (and even the feasibility) of the problem. Because this issue is so important to achieving good performance, the user should be given the alternative of assigning priorities to the tasks.
- *How processors share some information?:* In task-parallel computation, processors share information through pseudo-global variables. Although maintaining a pseudo-global variable involves a communication overhead, keeping it updated allows processors to make a more efficient use of the information shared. In many problems, like TSP, a frequent update of the pseudo-global variable may reduce the amount of computation to be performed. The frequency of the updatings of the pseudo-global variable then becomes a subject of trade-off. If the frequency is too low, many wasted computations may result. If the frequency is too high, the updating procedure introduces a large overhead. The user should control this frequency.
- *Which load balancing strategy is the most efficient?:* Although there are many different strategies for load balancing, there is no single one that can be efficiently applied to all the problems on all the computers. For that reason, the best approach to use in a library is to provide different strategies and let the user decide which is the most appropriate for the particular problem. The user does not need to know the strategy to use in advance, and so she or he should be provided with modules that can be easily changed.
- *When should the load balancer be invoked?:* There is a compromise between the frequency of the load balancing calls and the overhead introduced by the transfer procedure. Too frequent calls may incur high overhead, while less frequent calls may reduce the overhead but also increase idle processors. Thus, the frequency of these calls should be controlled by the users.

6 Final Comments

The conclusions discussed in this paper are based on our experience designing and testing the PMESC library. Our analyses showed us that even naive users could improve the performance of their implementations by changing parameters and strategies. Refer to [1] for a thorough discussion of this topic.

References

1. Crivelli S.A.:
   A Programming Paradigm and Library for Distributed-Memory Computers.
   Ph.D. Thesis, Tech. Report CU-CS-787-95, Dept. of Comp. Science,
   Univ. of Colorado, Boulder, (1995).
2. Crivelli S.A. and Jessup E.R.:
   A User's Manual for the PMESC Library.
   Dept. of Comp. Science, Univ. of Colorado, Boulder, (1995).
3. Gursoy A. and Kale L.:
   Charm 4.3 Programming Language Manual.
   Dept. of Comp. Science, Univ. of Illinois, Urbana-Champaign (1994).

Efficient Block Cyclic Data Redistribution

Loïc Prylli and Bernard Tourancheau*

LIP, ENS-Lyon, 46 allée d'Italie, 69364 Lyon - France

Abstract. Block cyclic distribution seems to suit well for most linear algebra algorithms and this type of data distribution was chosen for the ScaLAPACK library as well as for the HPF language. But one has to choose a good compromise for the size of the blocks (to achieve a good computation and communication efficiency and a good load balancing). This choice heavily depends on each operation, so it is essential to be able to go from one block cyclic distribution to another very quickly. Moreover, it is also essential to be able to choose the right number of processors and the best grid shape for a given operation. We present here the data redistribution algorithms we implemented in the ScaLAPACK library in order to go from one block cyclic distribution on a grid to another one on another grid. A complexity study is made that proves the efficiency of our solution. Timing results on the Intel Paragon and the Cray T3D corroborate our results.

1 Introduction

This paper describes the solution of the data redistribution problem arising when implementing linear algebra in a distributed system. We point out that the paper is not addressing the problem of how to determine a relevant data distribution (even if experimental results are given for 3 numerical kernels), but how to implement a given redistribution.

The problem of data redistribution occurs as soon as you deal with arrays (from vectors to multi-dimensional ones) on parallel distributed memory computer. It applies both to data-parallel languages such as HPF and to SPMD programs with message-passing. In the first case the redistribution is implicit in array statements like $A = B$ where A and B are two matrices with different distributions. In the second case a library function has to be called explicitly.

In HPF, redistribution of data is twofold. The first level correspond to an affine mapping of the data on a template. The second level is a block cyclic distribution of the template on the processors. In the ScaLAPACK library ([5, 4]) a block cyclic distribution is used that is equivalent to an HPF distribution with a unit stride mapping on the template.

The efficiency of redistribution is crucial because it should be negligible or at least smaller than the elapse time gain it was done for. This is especially

---

* This work has been supported by CNRS contract PICS, GDR-PRC PRS action EXEC and CEE-EUREKA contract EUROTOPS

difficult since the redistribution operation has to be done dynamically, with no compile-time or static information.

We present here the algorithm and implementation of a redistribution of data from a given block cyclic distribution on a given virtual grid of processors to another block cyclic distribution on another grid of processors. Our solution is a runtime approach to construct the communication data sets and then efficiently communicate them. In order to be very fast, our algorithm uses several strategies depending on the amount of data to be communicated and on the target architecture capabilities. It runs for any number of processors, making available the possibility of loading and down-loading from/to one processor to/from many others.

2 Related works

For a long time, redistribution of block cyclically distributed data was considered very difficult in the general case. A lot of work was done for the HPF compilers, exploring the general case with all parameters known at compile time and a fixed number of processors (see [2, 3] for instance). In the compile time solutions, the communication sets are generally computed by solving linear Diophantine equations.

The runtime approach we are investigating for block-cyclic distribution is much more critical in term of elapse time and was investigated more recently. To our knowledge, the first efficient approaches were proposed in [7]²,[12] and [10]. Most other works are restricting the possible data distributions to block or cyclic distributions, or to block cyclic with block-sizes multiple of each others or treat the general case with a brute force algorithm.

A direct solution for block distribution with linear and spiral mapping on the processors is given in [16], a runtime solution for the redistribution of block distributions in an application is presented in [1], but these two papers do not target the block cyclic general case.

A compile time solution for the block cyclic general case where the data communication sets are computed as a union of intersection of cyclic patterns is presented in [13]. This solution could also be use at runtime but it is rapidly inefficient when the block size is not very small (see [11]).

A runtime solution for the redistribution with routines in a library for a HPF compiler is proposed in [14]. Analytical efficient solutions are given only for special cases. The general case is treated with a algorithm that requires a costly operation for each item to be sent.

In [17] and [15], the redistribution is studied in the case of block cyclic distribution where blocks are multiple of each other, and using gcd or lcm algorithms with two redistribution for the general case.

An approach close to ours, but only element-wise, is proposed in [10] in order to built the communication sets for block-cyclic distributions of HPF/Fortran-D

---

² Our redistribution routines in the ScaLAPACK library were released in the end of 1994 on the netlib W^3 server.

compilers. In a recent paper, [12], a syntactical representation of family of line segments (i.e., contiguous elements) the FALLS is introduced. A redistribution algorithm is presented, it has the same complexity as our but a different strategy that needs a more expensive basic operation.

3 Block cyclic data distribution and redistribution

The ScaLAPACK library and the HPF language use the block cyclic data distribution on a virtual grid of processors. Arrays are wrapped by blocks in all dimensions corresponding to the processor grid (if other dimensions exist they are not wrapped and stay associated to the wrapped elements). The Figure 1 illustrates the organization of the block cyclic distribution of a 2D array on a 2D grid of P processors.

The distribution of a matrix is defined by four main parameters: a block width size, r_1; a block height size, r_2; the number of processors in a row, P_1; the number of processors in a column, P_2, plus other parameters to determine, when a sub-matrix is used, where it is located in the global matrix.

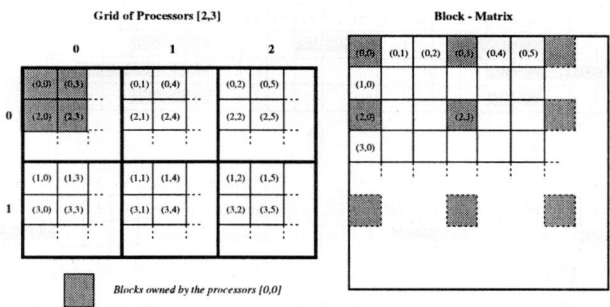

Fig. 1. The block cyclic data distribution of a 2D array on a 2×3 grid of processors.

Our dynamic approach implies that we deal with the most general case of redistribution allowed by our constraints, namely block cyclic with blocks of size (r_1, r_2) on a $P_1 \times P_2$ virtual grid to block cyclic with blocks of size (r'_1, r'_2) on a $P'_1 \times P'_2$ virtual grid.

For the sake of simplicity, we have presented here the block cyclic distribution for a 2D array, but it extends straightforwardly to multi-dimensional array of any dimension.

4 Redistribution algorithm

The whole problem of array redistribution is, for each processor, to find which data stored locally has to be sent to the others, respectively, how much data it will receive from the others, and where it will store it locally. In the following a "data set" will design all the items of an array that are transferred between a pair of processors.

We will considered in this section only the one-dimension case, thus a distribution can be described by two parameters, the block size r, and the number of processors P. Hence, in the following, we assume we want to redistribute from block-cyclic(r,P) to block-cyclic(r',P'). The problem is to find which data items stored on processor P_i in the initial distribution must go to processor P_j in the final distribution. These data items have to be packed in one message before being sent to P_j in order to avoid start-up delays. In our case, the set of elements to be sent (or received) on a processor will be described by a list of intervals in local memory.

Our algorithm scans at the same time the matrix indices of the data blocks stored on P_i and those that will be stored on P_j. More precisely, it maintains two counters, corresponding to the start of P_i's blocks in the global matrix and to those of P_j's. We increment them progressively by block as in a merge sort in order to determine the overlap areas. In figure 2 an example from a distribution (3,2) to a distribution (5,3) shows which data the processor 1 in the first distribution should send to the processor 0 in the second distribution.

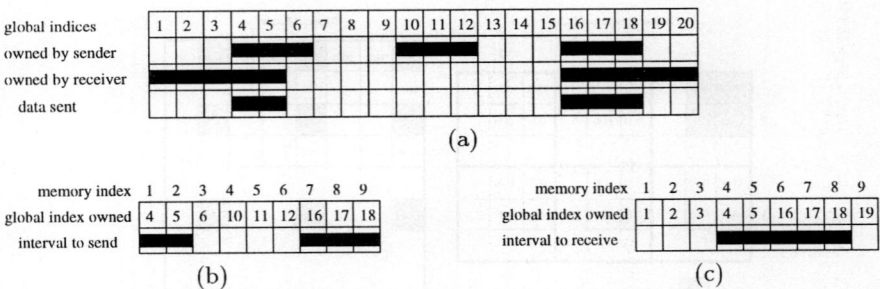

Fig. 2. Example of redistribution from (3,2) to (5,3), data sent from processor 1 and processor 0, (a) shows the data items to be sent in a global index view, (b) shows the data to be sent with the source processor memory index, and (c) shows the data with destination memory index

For the multi-dimensional case, the block scanning is done dimension by dimension. The data sets corresponding are the Cartesian product of the intervals computed in each dimension [11].

In order to improve the algorithm, we show how we reduce the computation of a data set, by using the periodic structure of such set. This periodicity has been noted in other works [13, 10, 12].

Proposition 1 *[11] The intersection of two block cyclic patterns (r, P) and (r', P') is periodic of period $lcm(rP, r'P')$.*

Thanks to the proposition 1, we restrict the scan from the whole range of indices to a range of length lcm(rP,r'P').

5 Communication

In the general case, the communications that occur between processors correspond to a personalized "all-to-all", with message sizes that depend on the processors. Each emission buffer filling and the computation of the size of each reception buffer are done thanks to the computation of the data sets. But often in practice, some pairs of processors do not communicate (the set of processors involved in the distribution may be partially disjoint or there may be no data to exchange between certain pair of processors).

In the two approaches below, we sent (and receive) only meaningful messages between processor pairs, i.e., not of null size.

Asynchronous (receive) communication: In this first approach, the sizes of the messages to be received are computed first. Then, asynchronous receives are posted. Then each data sets is sent.

Synchronous communication : In some contexts, asynchronous receives may not be available[3]. Moreover in order to avoid OS deadlocks due for instance to the fixed number of communication buffers, we designed an algorithm that is using a blocking receive protocol. At each step of this algorithm processors are grouped into pairs, messages are exchanged inside each pair. At the end, of the algorithm, we ensure that every processor has been paired with every other processor. This strategy can be compared to a rolling caterpillar of all the involved processors where, at a given step d, each processor P_i ($0 \leq i < P$) exchanges its data with processor $P_{((P-i-d) mod P)}$.

Communication pipelining: Instead of waiting for all the information from another processor, each processor P_i receives a small packet of elements, and in the same time packs a small packet of elements to be sent and unpacks the elements it just received. This is an overlapping strategy as in [6] implemented within the caterpillar method when the data size is large enough.

6 Complexity Study

This complexity study aims to give an insight on the algorithm and also shows what is the dominant part in the execution time. We consider in the following the complexities in term of basic elementary operations like comparison, addition, move, ...

Data set computation complexity: The indices scanning is done for each processor pair in order to send the intersecting data, and respectively to receive and store the data at the right place in local memory.

In the following, let M be the size of the vector to be redistributed and let B be $\max(M, \text{lcm}(rP, r'P'))$. As described in section 4, we can stop the scanning as

---

[3] This is the case of the BLACS message-passing library used for SCALAPACK.

soon as we reach the global index B while constructing the intersection patterns. In practice this optimization is interesting when the redistribution deals with small block sizes or large matrices[4]. Although there are extreme cases where it is more interesting to analytically determine the pattern (see [13] and [11]), generally for blocks and matrices of reasonable size our strategy is better because its cost is independent of the block sizes.

Proposition 2 *[11] The optimized scanning complexity is obtained by counting the number of blocks present in the interval $[1..B]$ for each distributions :*

$$T_{scan}^2 = O\left(\frac{B}{rP} + \frac{B}{r'P'}\right)$$

The result can be straightforwardly extended to multidimensional arrays by adding the cost in each dimension.

Comparison between scanning and packing/communication complexities: The complexity of the determination of the data sets augments "linearly" with the number of dimensions. On the contrary, the volume of each data set augments "exponentially" with the number of dimensions, as it is the Cartesian products of one-dimension element sets. So, in practise the total redistribution cost is roughly equal to two memory copies (one when the data is sent and one when it is received) plus the transfer time of the items owned locally.

7 Timing results

In this section, we want not only to show the efficiency of our algorithms, but also to demonstrate the real gain that is obtained, in practice, for linear algebra kernel computation using our redistribution routines. For that, we chose 3 routines of the ScaLAPACK library[5]: the matrix multiply (MM), LU factorization (LU) and triangular solve (TS). In the ScaLAPACK library, the routines run for any virtual grid of processors and any size of square blocks.

Our experiments were ran on 2 parallel computers: the Paragon, with a compiled version of the ScaLAPACK and BLACS libraries on top of the vendor optimized BLAS and the T3D, with a compiled version of the redistribution routine with a vendor implementation of the LU, MM, TS, BLACS and BLAS routines.

For a given number of processors (16) and a given matrix size (1024 × 1024), we first determine the best block size and grid shape for each computation kernel [6]. We measured timings of each routine as a function of the block sizes for virtual

---

[4] Depending of the target computer communication and computation speeds.
[5] Notice that our aim is not to prove that these routines are often chained together but only to use them as a numerical application testbed.
[6] A similar work was done for several ScaLAPACK routines in [4].

processors grids of all possible shapes : 1×16, 2×8, 4×4, 8×2, 16×1. For clarity, the block sizes are chosen power of 2, but our redistribution routines run for any sizes. Figure 3 shows, for instance, the performance of the LU routine on the Intel Paragon. The summary results are given in table 1 (see [11] for details).

Knowing these best configurations, we ran the tests for the 3 numerical kernels. We compare on Figure 4 the timings of the LU routine with and without redistributions to the best block size and grid shape and back.

The results are extremely good because there is only slight differences between the best and worst cases with the addition of our redistribution routines. Hence this performance stays close to the optimal. On the contrary, a bad distribution choice can lead to double the execution time for LU.

With that result, we then study an example with the 3 linear algebra operations (MM, LU and SOLVE) executed sequentially in the same parallel program[7]. The best grid shapes and block sizes are very different for these 3 routines (see Table 1) and the timings differences are not negligible. Hence the redistribution improvement can lead to an important gain in elapse time.

Remark that if several arrays are implied in the computation kernel, then several data redistributions must be called (in the Matrix Multiply for instance). Hence, the redistribution solution includes 6 calls to the redistribution routine (i.e., 2 for the MM kernel, 1 for the LU kernel, 2 for the TS kernel, plus one to go back to the initial data distribution).

Kernel	Paragon		T3D	
	BBS	BGS	BBS	BGS
MM	64	4×4	64	4×4
LU	8	1×16	16	1×16
TS	64	16×1	64	16×1

Table 1. Best Block Size (BBS) and Best Grid Shape (BGS) for the 2 parallel computers of our tests while running the Matrix Multiply (MM), LU decomposition (LU) and Triangular Solves (TS) computation kernels.

The results in figure 5 and 6 proves that in all cases, it is better to use the redistributions than to stay with the initial distribution.

The solid-line curve of Figure 5 and 6 are the worst case, where we assume that none of the array is in the good distribution and we have to redistribute them at the beginning and end of the algorithm. These 2 steps are not mandatory and the dotted-line curve represent this optimized case.

---

[7] the number of systems solved is set up to one fourth of the matrix size, so that the solve routine execution time is comparable to the multiply and LU routine

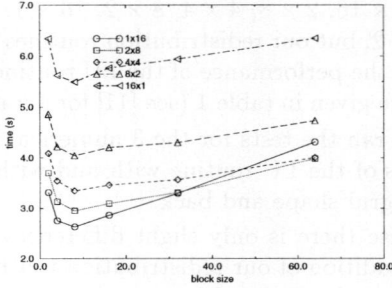
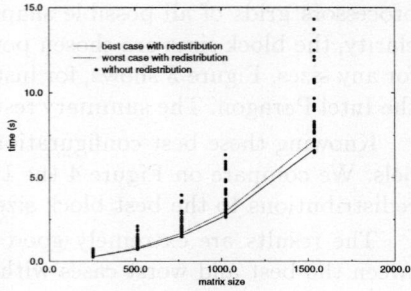

Fig. 3. LU decomposition elapse time on the Intel Paragon with 16 processors in different grid shapes and with different square block sizes.

Fig. 4. LU decomposition elapse time on the Intel Paragon with 16 processors for all grid shapes and block sizes without data redistributions (dots) compared with redistribution to the best configuration and back (curves).

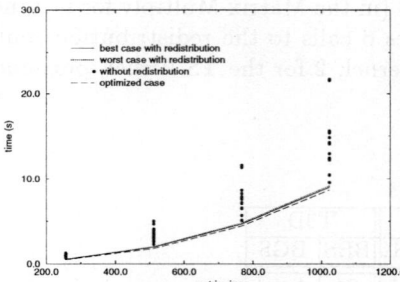
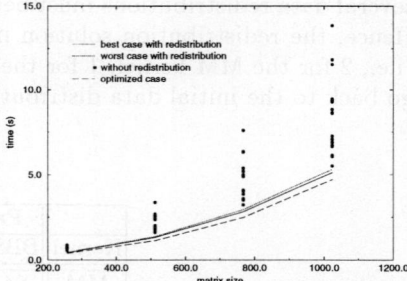

Fig. 5. Comparison of the succession of the 3 numerical kernels on the Intel Paragon with 16 processors for all grid shapes and block sizes with and without data redistributions and in the optimized case.

Fig. 6. Comparison of the succession of the 3 numerical kernels on the Cray T3D with 16 processors for all grid shapes and block sizes with and without data redistributions and in the optimized case.

8 Conclusion

We propose a new very fast merge algorithm with the addition of a very short data set computation using the property of periodicity of block cyclic intersections. Thus with these optimizations for the scanning and good communications strategies, it is sufficient to know the distribution parameters only at runtime and there is no more constraint about providing them at compile-time.

We show that this strategy is efficient with a complexity study of our algorithm and a comparison with the data transfers cost. Moreover, in the general case, our solution is performing better in practise than all the literature experiments of our knowledge. Our routines are robust, well tested and available on any of the parallel computer running the BLACS [8, 9] communication library (i.e., most of the multicomputers) or BLACS-PVM library (i.e., NOW[8] and multicomputers running PVM).

The experiments on a Cray T3D and on an Intel Paragon corroborate very well our expectations, the computation of the data sets is negligible compared to the communication and packing, and the global redistribution routine execution time is very good.

Our results on linear algebra kernels show that the redistribution of data with our algorithms leads to efficiency in practice: basically, the redistribution is useless only when the matrix is already in the optimal configuration. Moreover, it also assures that on the average the computation time stays very close to the optimal even with bad initial data distribution choices (for both block sizes and grid shape).

From our results, these redistribution routines should be integrated in the linear algebra kernel themselves and also used in compiled codes generated by HPF.

A more detailed version of this paper is available in [11] or on http://www.ens-lyon.fr/~btouranc/PAPERS/INRIA-RR-2766.ps

Our algorithms are implemented within the ScaLAPACK library and are accessible for test and use on the W^3 site
http://www.netlib.org/ScaLAPack/index.html.

References

1. G. Agrawal, A. Sussman, and J. Saltz. Compiler and runtime support for structured and block structured applications. In *Supercomputing*, pages 578–587, Portland, 1993. ACM Press.
2. S. P. Amarasinghe and M. S. Lam. Communication optimization and code generation for distributed memory machines. In *Conference on programming language design and implementation*, Albuquerque, NM, June 1993. ACM SIGPLAN.
3. S. Chatterjee, J. R. Gilbert, F. J. E. Long, R. Schreiber, and S. H. Teng. Generating local addresses and communication sets for data-parallel programs. In *Symposium on Principles and Practice of Parallel Programming*, San Diego, CA, May 1993. ACM SIGPLAN.
4. J. Choi, J. Demmel, J. Dhillon, J. Dongarra, S. Ostrouchov, A. Petitet, K. Stanley, D. Walker, and R.C. Whaley. ScaLAPACK: A portable Linear Algebra Library for Distributed Memory Computers: Design Issues and Performance. Technical Report LAWN95, Oak Ridge National Laboratory, Oak Ridge, TN 37831-6367, 1995.
5. J. Choi, J.J. Dongarra, and D.W. Walker. The Design of Scalable Software Libraries for Distributed Memory Concurrent Computers. In J.J. Dongarra and B. Tourancheau, editors, *Environments and Tools for Parallel Scientific Computing*, pages 3–15. Elsevier, 1992.

---

[8] Network Of Workstations.

6. F. Desprez and B. Tourancheau. LOCCS: Low Overhead Communication and Computation Subroutines. *Future Generation Computer Systems*, 10:279–284, 1994.
7. J. Dongarra, C. Randriamaro, L. Prylli, and B. Tourancheau. Array Redistribution in ScaLAPACK using PVM. In *EuroPVM users' group*. Hermes, 1995.
8. J.J. Dongarra, R. Van De Geijn, and D.W. Walker. A Look at Dense Linear Algebra Libraries. Technical Report ORNL/TM-12126, Oak Ridge National Laboratory, July 1992.
9. J.J. Dongarra, R. Van De Geijn, and R.C. Whaley. Two Dimensional Basic Linear Algebra Communication Subprograms. In J.J. Dongarra and B. Tourancheau, editors, *Environments and Tools for Parallel Scientific Computing*, pages 17–29. Elsevier, September 1992.
10. S. Hiranandani, K. Kennedy, J. Mellor-Crummey, and A. Sethi. Compilation techniques for block-cyclic distributions. Technical Report TR95521-S, CRPC, Rice Univ., Houston, TX 77005, March 1995.
11. L. Prylli and B. Tourancheau. Efficient block cyclic data redistribution. Technical Report 2766, INRIA Rhone-Alpes, 1995.
12. Shankar Ramaswamy and Prithviraj Banerjee. Automatic Generation of Efficient Array Redistribution Routines for Distributed Memory Multicomputers. In *The Fifth Symposium on the Frontiers of Massively Parallel Computation*, pages 342–349, McLean, VA, February 1995.
13. J. M. Stichnoth, D. O'Hallaron, and T. R. Gross. Generating communications for array statements: design, implementation, and evaluation. *JPDC*, 21:150–159, 1994.
14. R. Thakur, A. Choudhary, and G. Fox. Runtime array redistribution in HPF programs. In *Scalable High-Performance Computing Conference*, pages 309–316. IEEE, May 1994.
15. R. Thakur, A. Choudhary, and J. Ramanujam. Efficient Algorithm for Array Redistribution. *to appear in IEEE TPDS*, 1996.
16. A. Wakatani and M. Wolfe. Optimization of Array Redistribution for Distributed Memory Multicomputers. *Parallel Computing*, 21(9), 1995.
17. D. Walker and S. Otto. Redistribution of Block-Cyclic Data Distributions Using MPI. Technical Report ORNL/TM-12999, Oak Ridge National Laboratory, Oak Ridge, TN 37831-6367, June 1995.

Optimal Grain Size Computation for Pipelined Algorithms *

Frédéric Desprez[1]
Pierre Ramet[2]
Jean Roman[2]

[1] LIP (URA CNRS 1398), INRIA Rhône-Alpes et Ecole Normale Supérieure de Lyon
69 364 Lyon Cedex, France
desprez@lip.ens-lyon.fr
[2] LaBRI (URA CNRS 1304), ENSERB et Université Bordeaux I
33 405 Talence Cedex, France
(ramet,roman)@labri.u-bordeaux.fr

Abstract. In this paper, we present a method for overlapping communications on parallel computers for pipelined algorithms. We first introduce a general theoretical model which leads to a generic computation scheme for the optimal packet size. Then, we use the OPIUM [3] library, which provides an easy-to-use and efficient way to compute, in the general case, this optimal packet size, on the column LU factorization; the implementation and performance measures are made on an Intel Paragon.

Keywords : Communications overlap, pipelined algorithms, optimal packet size computation.

1 Introduction

Parallel distributed memory machines improve performances and memory capacity but their use adds an overhead due to the communications. To obtain programs that perform and scale well, this overhead must be hidden. Several solutions exist. The choice of a good data distribution is the first step that can be done to lower the number and the size of communications. Depending of the dependences within the code, asynchronous communications can be used to overlap computations and communications. The call to the communication routine (send or receive) will be put as soon as possible in the code. A wait routine will then be used to check for the completion of the communication. Unfortunately, this is not always legal due to the dependences between computations and communications. Pipeline schemes are also sometimes found like the one on Figure 1 (A). There is a sequentiality within the code, and the total execution time is higher than the sequential one because of the overhead of the communications. One first solution is to start the communications as soon as possible, i.e. as soon as one processor has computed one data; this is called a fine-grain pipeline ((B) on Figure 1). This solution adds an higher overhead because of the communication startup time. A trade-off has to be found which minimizes the execution time. This is a coarse grain pipeline ((C) on Figure 1).

---

* This work is supported by the INRIA Rhône-Alpes, the french GDR-PRS and the Eureka-Eurotops project.
[3] Optimal Packet sIze compUtation Methods.

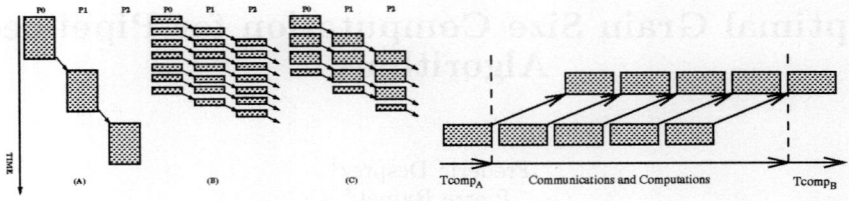

Fig. 1. Overlap of communications using a pipeline method.

Fig. 2. Communication/Computation Overlap.

The linear case is an easy one [Des94b]. This is no longer the case as soon as the complexity of the computations and communications is not a linear function of the packet size. The aim of this paper is to present a library which can be used to improve pipeline computations in the general case. This library, called OPIUM (Optimal Packet sIze compUtation Methods) will be used in another library which implements coarse grain pipelines, the LOCCS [4] [Des94a, DT94].

Some previous studies exist, especially around the compilation of data-parallel languages like High Performance Fortran. The pipelining of computation is a classical optimization of such codes [Tse93, BD96]. The linear case from which this study has started is described in [Des94b]. The results are also used within the LOCCS library. In [OSKO95], the optimal size of packets is computed for general DOACROSS loop nests. The optimizations of the Fortran D compiler are presented in [Tse93]. Here again, only the simple cases that arise in the compilation of simple loop nests are given. The author admits that a more accurate evaluation of the optimal size of packets should be given to get the best performances. Finally in [SS95], the case of a loop nest with two loops is presented on a network of workstations, using a computation of the optimal size done at run-time, which allows some kind of load-balancing.

In the general case presented in this paper, the computation of the optimal grain size is no longer trivial. It can require complicated methods and a very accurate model to obtain correct results; a library is necessary to compute the optimal size without a knowledge of the methods, to do run-time optimizations, to avoid to re-compute the size if it is not necessary, and to include this kind of optimization in a data-parallel compiler.

The rest of the paper is organized as follows: the section 2 introduces the problem formulation and the theoretical model. In section 3, we describe experiments on the Intel Paragon, based on the OPIUM library, including performance results and analysis. Then we conclude with remarks on the benefits of this study and on our future work.

2 Computation/Communication Overlap

2.1 Problem formulation and theoretical model

In this section, we first describe a theoretical model for the overlap of communications by computations. It will be used as the reference model until the

---
[4] Low Overhead Communication and Computation Subroutines.

end of the paper. Then, the way to resolve this model is presented and illustrated with some examples. Finally, a method to avoid the re-computation of the optimal packet sizes as far as possible is described.

Let L be the Amount of Data to be Pipelined (ADP). We split the data to be sent in μ packets of size ν (see Figure 2). Moreover, in the following, ν_{opt} denotes the Optimal Packet Size (OPS).

Definition 1. The "forward computation time" denotes the time spent in the forward computation on a packet of size ν. It can be defined by

$$Tcomp_A(\nu, L) = R_A(L)\, f_A(\nu)\, \tau_A \quad \text{where}$$

- $R_A(L)$ is the forward complexity constants that may depend on the ADP,
- $f_A(\nu)$ is the complexity function of the forward computation *implementation*,
- and τ_A denotes the elementary forward computation time.

Definition 2. In the same way, the "backward computation time" denotes the time spent in the backward computation on a packet of size ν with

$$Tcomp_B(\nu, L) = R_B(L)\, f_B(\nu)\, \tau_B.$$

Definition 3. In the following, the notation $\theta(f_A(\nu))/\theta(f_B(\nu))$ will be used to represent a pipeline algorithm with a "forward" complexity growing as $\theta(f_A(\nu))$ and a "backward" complexity growing as $\theta(f_B(\nu))$.

Example 1. Let us study the column LU factorization. A column cyclic distribution is used to balance the computation load all over the execution [Saa86]. With this distribution, we can extract three phases for each factorization step

- the "local scale": the processor holding the pivot column of the current step computes its contribution, and sends it to the other processors,
- the "local update": the same processor updates its local columns,
- the "remote update": the other processors receive the contribution, and update their columns.

We clearly have a pipeline scheme. The "local change" must be in the "forward computation" and the "remote update" is the "backward computation". The "local update" is put in the "forward computation" to well balance the "forward" and the "backward" computations such as we can hope to obtain an optimal overlap gain. For this implementation, the complexities of the "forward" and "backward" computations are $\theta(\Delta\,\nu)$, where Δ represents the number of local columns to update. It is important to notice that these complexities depend on the implementation of the pipeline and on the data distribution. The column LU factorization is typically a $\theta(\nu)/\theta(\nu^2)$ algorithm, whereas we have a $\theta(\nu)/\theta(\nu)$ pipelined implementation.

Definition 4. $Tcomm(\nu)$ denotes the time spent to send a packet of size ν.

In general, the expression of $Tcomm(\nu)$ depends on the communication scheme (broadcast, all-to-all, ...), and on the computer architecture. In the following we will consider a linear communication model $Tcomm(\nu) = \beta + \nu\,\tau$ where β is the startup time and τ the per-word transfer time.

Proposition 5. *The OPS computation problem can be rewritten into the minimization problem of*

$$Tcomp_A(\nu, L) + \frac{L}{\nu} Tcomm(\nu) + Tcomp_B(\nu, L) \tag{1}$$

$$under\ the\ constraints \quad \begin{cases} Tcomp_A(\nu, L) < Tcomm(\nu) \\ Tcomp_B(\nu, L) < Tcomm(\nu) \\ 1 \leq \nu \leq L \end{cases} \tag{2}$$

Proof. See [DRR96]. □

If the computation time is always greater than the communication time then we split the ADP with the minimal packet size and thus $\nu_{opt} = 1$.

We now want to solve the reference model. There are three steps:

1. First we compute the extremum ν_{min} of (1),
2. then we look for the validity domain of the constraints; we compute ν_{bal} the highest packet size which verify the constraints (2),
3. and finally we deduce the optimal size ν_{opt}.

• We focus in the following on polynomial complexity functions. If the constraints are true then ν_{min} is the solution of $\frac{\partial}{\partial \nu} Ttotal(\nu, L) = 0$.

Lemma 6. *Consider that* $\begin{cases} Tcomp_A(\nu) = R_A\ \nu^{\alpha_A}\ \tau_A \\ Tcomp_B(\nu) = R_B\ \nu^{\alpha_B}\ \tau_B\ \ with\ \alpha_A \geq 1,\ \alpha_B \geq 1. \\ Tcomm(\nu) = \beta + \nu\ \tau \end{cases}$
ν_{min} *is the unique solution of:* $\alpha_A\ R_A\ \nu^{\alpha_A - 1}\ \tau_A + \alpha_B\ R_B\ \nu^{\alpha_B - 1}\ \tau_B - \frac{L\beta}{\nu^2} = 0$.

Proof. See [DRR96]. □

Remark. For the most usual complexity functions that are over-linear and communication types, we can verify that the second order derivate of $Ttotal(\nu)$ is always positive on the validity domain of ν. For all these cases, the problem admits one and only one solution.

• Now that we have the minimum of equation (1), we must take care of the constraints. An other approach to optimize the overlap of computations by communications consists in finding, when possible, the packet size such that the communication time for one packet balances exactly the computation time for this packet [MP92]. As the "forward" and "backward" computations play a symmetrical part in the minimization of equation (1), we can assume that $Tcomp_A > Tcomp_B$. Then the constraint balance is given by

$$Tcomp_A(\nu, L) = Tcomm(\nu). \tag{3}$$

Let ν_{bal} be the solution of this equation. Often, ν_{bal} is not the OPS; for a packet size upper than ν_{bal} we can rewrite the expression of $Ttotal(\nu, L)$ as

$$Ttotal(\nu, L) = Tcomp_A(\nu, L) + \frac{L}{\nu} Tcomp_A(\nu, L) + Tcomp_B(\nu, L). \tag{4}$$

If $Tcomp_A(\nu, L)$ is over-linear on ν then $Ttotal(\nu, L)$ increases with ν and ν_{bal} minimizes $Ttotal(\nu, L)$ for $\nu \geq \nu_{bal}$. Otherwise, we have to find ν_{bal} which minimizes (4).

• Finally, the following proposition leads to the computation of ν_{opt}.

Proposition 7. *If $1 \leq \nu_{bal} \leq L$ then two cases are possible:*
- *if $\nu_{bal} \leq \nu_{min}$ then the constraint occurs and $\nu_{opt} = \nu_{bal}$,*
- *otherwise the constraint does not occur and $\nu_{opt} = \nu_{min}$.*

The end of this section is dedicated to some examples.

Lemma 8. *Consider that* $\begin{cases} Tcomp_A(\nu) = R_A\,\nu\,\tau_A \\ Tcomp_B(\nu) = R_B\,\nu\,\tau_B \\ Tcomm(\nu) = \beta + \nu\,\tau \end{cases}$
Two cases are possible:

- *if $R_A\,\tau_A - \tau \leq 0$ then $\nu_{opt} = \sqrt{\frac{L\,\beta}{R_A\,\tau_A + R_B\,\tau_B}}$.*
- *if $R_A\,\tau_A - \tau > 0$ then*
  - *if $L \leq \frac{\beta\,(R_A\,\tau_A + R_B\,\tau_B)}{(R_A\,\tau_A - \tau)^2}$ then $\nu_{opt} = \sqrt{\frac{L\,\beta}{R_A\,\tau_A + R_B\,\tau_B}}$,*
  - *else $\nu_{opt} = \frac{\beta}{R_A\,\tau_A - \tau}$.*

Proof. See [DRR96]. □

Remark. In the case of the column LU factorization, R_A and R_B depend on the ADP ($R_A = R_B = \frac{L}{NPROC}$ within one, where $NPROC$ is the number of processors). So $\nu_{min} = \sqrt{\frac{NPROC\,\beta}{\tau_A + \tau_B}}$ and the optimal packet size is independent of the ADP.

Lemma 9. *Consider that* $\begin{cases} Tcomp_A(\nu) = R_A\,\nu^2\,\tau_A \\ Tcomp_B(\nu) = R_B\,\nu\,\tau_B \\ Tcomm(\nu) = \beta + \nu\,\tau \end{cases}$

The validity domain for the OPS is $\nu \in \left[0, \frac{\tau + \sqrt{\tau^2 + 4\,R_A\,\tau_A\,\beta}}{2\,R_A\,\tau_A}\right]$ and there is one and only one real positive solution for ν_{min} among the 3 Jordan solutions:

$\begin{cases} \nu_1 = \sqrt[3]{\frac{-q+\delta}{2}} + \sqrt[3]{\frac{-q-\delta}{2}} - \frac{A}{3} \\ \nu_2 = \sqrt[3]{\frac{-q+\delta}{2}} \cdot j + \sqrt[3]{\frac{-q-\delta}{2}} \cdot j^2 - \frac{A}{3} \\ \nu_3 = \sqrt[3]{\frac{-q+\delta}{2}} \cdot j^2 + \sqrt[3]{\frac{-q-\delta}{2}} \cdot j - \frac{A}{3} \end{cases}$ *with* $\begin{cases} p = -\frac{A^2}{3} \\ q = \frac{2\,A^3}{27} - B \\ \delta^2 = \frac{4\,p^3 + 27\,q^2}{27} \end{cases}$

Proof. See [DRR96] □

Lemma 10. *Consider that* $\begin{cases} Tcomp_A(\nu) = R_A\,\ln(\nu)\,\tau_A \\ Tcomp_B(\nu) = R_B\,\ln(\nu)\,\tau_B \\ Tcomm(\nu) = \beta + \nu\,\tau \end{cases}$
then $\nu_{min} = \frac{L\,\beta}{R_A\,\tau_A + R_B\,\tau_B}$.

The proof is straightforward. However, in order to compute ν_{bal}, we have to solve $R_A\,ln(\nu)\,\tau_A = \beta + \nu\,\tau$ and we cannot explicitly extract the solution. The same problem occurs when extracting ν_{min} with, for instance, a $\theta(\nu ln(\nu))/\theta(\nu ln(\nu))$ pipeline scheme or a $\theta(\nu^n)/\theta(\nu^m)$ pipeline scheme with $n, m > 4$.

We can see with this last examples that it is not always possible to explicitly extract the OPS. For more complex computations or for non linear communication types, we will have to use numerical schemes such as Newton or Brent methods.

2.2 Avoid re-computation of optimal packet sizes

In practical applications, pipeline are chained and only the ADP differs. So it is not always useful to re-compute the OPS for each pipeline. In fact, it is sometimes possible to find the gap between problem sizes such that optimal packet sizes are equal.

Proposition 11. *Assuming that the analytic expression of $\nu_{opt}(L)$ is known, the gap of problem sizes corresponding to equal optimal packet sizes is the largest integer l_{max} over all $l \geq 0$ such that*

$$|\nu_{opt}(L+l) - \nu_{opt}(L)| < 1. \tag{5}$$

Lemma 12. *If $\nu_{opt}(L)$ can be rewritten into $C L^\gamma$, with $C > 0$ and $\gamma \geq 1$, then $l_{max} = (L^\gamma + 1/C)^{1/\gamma} - L$.*

Proof. See [DRR96]. □

Remark. We can formulate three remarks:

- Consider the linear example (lemma 8). We have $C = \sqrt{\frac{\beta}{R_A \tau_A + R_B \tau_B}}$ and $\gamma = 1/2$, and so $l_{max} = \frac{R_A \tau_A + R_B \tau_B}{\beta} + 2\sqrt{\frac{L(R_A \tau_A + R_B \tau_B)}{\beta}}$.
- For the column LU factorization (see example (1)), the optimal size is independent of the ADP, so we compute ν_{opt} only one time.
- If the analytic expression of ν_{opt} can be explicitly extract but cannot be rewritten into $C L^\gamma$ (see for instance lemma 9), the l_{max} value will be computed by a numerical scheme.

3 Experiments

In this section, we discuss the experiments done to evaluate the validity of the theoretical model and the benefits of the overlap of communications by computations with the optimal packet size. Our experiments are based on the LOCCS library [Des94b, DT94] which brings some useful overlap management primitives. The performance measures are made on a 32 nodes Intel Paragon.

The main example is the column LU factorization (see example 1) with a good balance between the "forward" and the "backward" computations. We use both the general and linear specific OPIUM methods. A description of the OPIUM library is given in [DRR96].

The optimal packet size predicted and computed with our model is confirmed by the experimental measures; on Figure 3, the vertical line corresponds to the OPS computed by the OPIUM routine. Moreover, according to the theory, this size is constant for each pipeline of the factorization until the constraint occurs. So it is computed only one time. On Figure 4, we see that the more important is the ADP, the more efficient is the communication/computation overlap. The gain brought by the overlap decreases when the number of processors increases (see Figure 5). This is due to the fact that local computations, for each processor, are not large enough to have an efficient overlap . But we can see that the ratio between the two algorithms (with or without overlap) is quite independent of

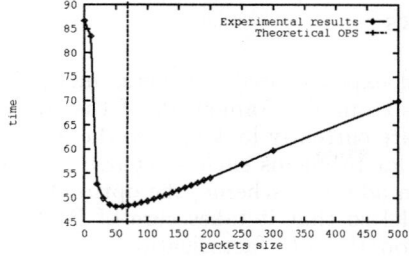 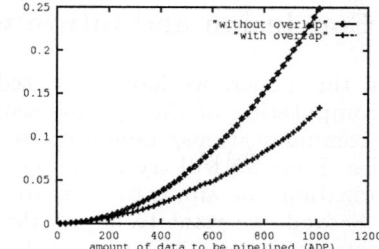

Fig. 3. Factorization time of a matrix of size 1024 as a function of the packet size.

Fig. 4. Time spent in each pipeline of the factorization of a matrix of size 1024 (with and without overlap).

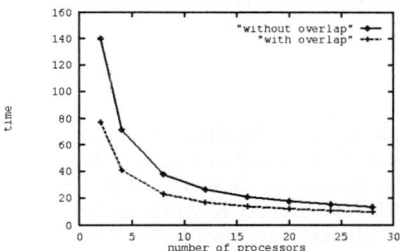 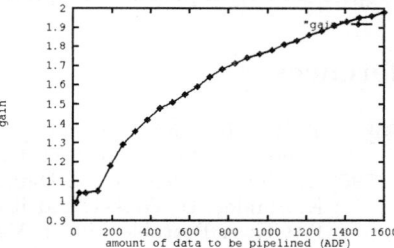

Fig. 5. Factorization time (with and without overlap) of a matrix of size 960 as a function of the number of processors.

Fig. 6. Ratio between the factorization time, with and without overlap, as a function of the matrix size.

the number of processors which shows the scalability of this optimization. The gain curve (see Figure 6) shows that for a large enough matrix, we quickly reach a good overlap (the factorization time with overlap is reduced by half, compared to the factorization without overlap) and we also see that, whatever the matrix size may be, the overhead time generated by the management of the overlap is negligible. We can remark that for less than 100 data, no splitting is done ($\nu_{opt} > L$). Figure 7 represents the activity of the 4 processors allocated for the factorization of a matrix of size 512, with and without overlap. This diagram shows that processors are almost always busy using pipeline optimizations.

We made some experiments with pipeline schemes for different "forward" and "backward" complexities to compare the generic method with specifics pipeline methods. The OPS are identical, and if the ADP increases slowly, the overhead due to the iterative scheme of the generic method is negligible. Indeed, the precomputed OPS leads to fewer iterations.

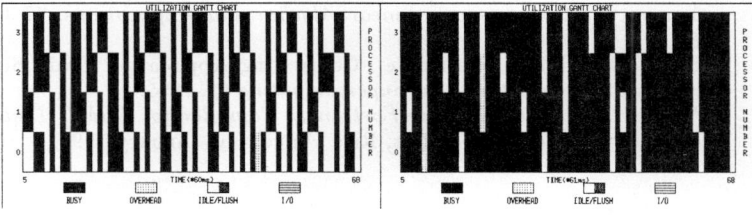

Fig. 7. Paragraph Gantt diagram (in white the inactivity and in black the activity of processors).

4 Conclusion and future works

In this paper, we have presented and experimented a general method for the computation of the optimal packet size in the framework of the overlapping communications/computations. We are currently looking into the following studies. First we will try to tackle irregular problems such as sparse Cholesky factorization; the aim is to compute, in an adaptive scheme, the optimal packet size. Second, we want to adjust the model so that it takes care of BLAS optimizations; indeed, the gains obtained on the LU factorization using a column algorithm may not be obtained for a block algorithm like the one used in ScaLAPACK [CDD+95]. There is a trade-off to be found between the computation grain size and the pipeline optimization. Finally, we wish to integrate those techniques into data parallel compilers like High Performance Fortran ones.

References

[BD96] T. Brandes and F. Desprez. Implementing Pipelined Computation and Communication in an HPF Compiler. Submitted to Europar'96, 1996.

[CDD+95] J. Choi, J. Demmel, I. Dhillon, J. Dongarra, S. Ostrouchov, A. Petitet, K. Stanley, D. Walker, and R.C. Whaley. LAPACK Working Note: ScaLAPACK: A Portable Linear Algebra Library for Distributed Memory Computers - Design Issues and Performances. Technical Report 95, Department of Computer Science - University of Tennessee, 1995.

[Des94a] F. Desprez. A Library for Coarse Grain Macro-Pipelining in Distributed Memory Architectures. In *IFIP 10.3 Conference on Programming Environments for Massively Parallel Distributed Systems*, pages 365–371. Birkhaeuser Verlag AG, Basel, Switzerland, 1994.

[Des94b] F. Desprez. *Procédures de Base pour le Calcul Scientifique sur Machines Parallèles à Mémoire Distribuée*. PhD thesis, Institut National Polytechnique de Grenoble, January 1994. LIP ENS-Lyon.

[DRR96] F. Desprez, P. Ramet, and J. Roman. Optimal grain size computation for pipelined algorithms. Technical report, Laboratoire Bordelais de Recherche en Informatique, 1996.

[DT94] F. Desprez and B. Tourancheau. LOCCS: Low Overhead Communication and Computation Subroutines. *Future Generation Computer Systems*, 10(2&3):279–284, June 1994.

[MP92] B. Tourancheau M. Pourzandi. Recouvrement Calcul/Communication dans l'Elimination de Gauss sur un iPSC/960. Technical report, Ecole Normal Supérieure de Lyon, 1992.

[OSKO95] H. Ohta, Y. Saito, M. Kainaga, and H. Ono. Optimal Tile Size Adjustement in Compiling General DOACROSS Loop Nests. In ACM Press, editor, *International Conference on Supercomputing*, pages 270–279, Barcelona, Spain, July 1995. ACM SIGARCH.

[Saa86] Y. Saad. Communication Complexity of the Gaussian Elimination Algorithm on Multiprocessors. *Linear Algebra and Applications*, 77:315–340, 1986.

[SS95] B.S. Siegel and P.A. Steenkiste. Controlling Application Grain Size on a Network of Workstations. In *Supercomputing'95*, 1995.

[Tse93] C.W. Tseng. *An Optimizing Fortran D Compiler for MIMD Distributed-Memory Machines*. PhD thesis, Rice University, January 1993.

Dynamic Redistribution on Heterogeneous Parallel Computers

Dominique Sueur and Jean Luc-Dekeyser

Laboratoire d'Informatique Fondamentale de Lille
Université des Sciences et Technologies de Lille

Abstract. New generations of scientific codes trend to mix different types of parallelism. Algorithms are defined as a set of modules, with data parallelism inside modules and task parallelism between them. With high speed networks, tasks running on a heterogeneous computing environment can exchange data in a reasonable delay. Therefore data-parallel tasks distributed on different parallel computers can interact efficiently by reading or writing Data Parallel Objects. These objects are distributed on the physical nodes according to the mapping directives. Migrations of data parallel objects from one parallel computer to another lead us to define efficient algorithms for runtime array redistribution.
In this work, we have specially cared about the ability to handle distinct source and target processor sets while performing redistribution and the ability to overlap communications and computations. Performance results on a farm of *ALPHA* processors are discussed.

1 Introduction

Thanks to data parallelism, we can take advantage of parallel machines keeping an unique instruction stream to express parallelism. This programming model takes advantage of fine grain parallelism which can be found in numerous algorithms (matrix computations, image processing, ...). However, large applications often have an heterogeneous structure. A large grain parallelism insures the cooperation between few tasks, each task refers the data parallelism model to express the local parallel algorithm. With efficient networks like FDDI, HIPPI or ATM, it becomes possible to structure heterogeneous applications as a set of communicating data parallel tasks and to use the best architecture for each of them[2, 3].

To insure data parallel object migrations, we developed a command language called **Dpshell** [4]. Its goal is to build an heterogeneous application with data parallel program libraries. Programs are data parallel tasks which communicate by reading or writing Data Parallel Objects (DPO) by the way of **Dpshell** commands. Dependencies between data parallel modules are explicitly expressed at the command language level. The user can write his local data parallel algorithm on any machine and then combines the running programs to perform his entire application. Modules can then be reused to interactively create new applications. One of the characteristics of the **Dpshell** is to move the DPO from one parallel machine to another. From the processor point of view, this migration triggers a parallel Input/Output. The input (resp output) subroutine only needs to know the source (resp target) data distribution. It is the `Dpshell` responsibility to provide the elementary data migration according to the required distributions.

Implementation of parallel Input/Output between data-parallel tasks required dynamic redistribution over heterogeneous sets of processors. In this paper, we present a mathematical representation of $Cyclic(K)$ to $Cyclic(K')$ redistributions from a set of P processors to a distinct set of P' processors. Note that $Cylic(K)$ distributions include usual $Block(K)$ distributions. Based on this representation, we study special cases for which efficient algorithms can be identified. For each case we are able to overlap communications and computations. Performance results show that specialized algorithms are up to ten times faster than general algorithm.

2 Mathematical representation

Let T be an array of size N distributed $cyclic(K)$ over P processors. We visualize the layout of array elements in processor memories as sequences of blocks. Three values describe the location of an array element $T(i)$: $p = (i/K) \bmod P$ is the processor holding $T(i)$, $n = i/PK$ is the block index on processor p and $x = i \bmod K$ is the offset within the block. Therefore, we can write :

$$i = pK + nPK + x \qquad (1)$$

with $x \in [0, K[$, $n \in [0, L(p)/K[$ and $p \in [0, P[$.

An heterogeneous redistribution involves two distinct sets of processor. The problem is to move an array T from one distribution D, $Cyclic(K)$ over P processors towards an other distribution D', $Cyclic(K')$ over P' processors. Let $E_D(p)$ be the set of indices of array elements hold by processor p with the source distribution D, and $E'_{D'}(p')$ the set of indices of array elements hold by processor p' with the target distribution D'. A redistribution routine needs to figure out exactly which data need to be sent between each processor pair. The processor p have to send to the processor p' the set $I(p, p') = E_D(p) \cap E'_{D'}(p')$. By using the equation (1) we can write :

$$i \in I(p, p') \Longleftrightarrow i \in E_D(p) \cap E'_{D'}(p')$$

$$\Longleftrightarrow pK + nPK + x = p'K' + n'P'K' + x' \wedge \begin{cases} p \in [0, P[\\ n \in [0, L(p)/K[\\ x \in [0, K[\\ p' \in [0, P'[\\ n' \in [0, L'(p')/K'[\\ x' \in [0, K'[\end{cases} \qquad (2)$$

Solutions of this diophantine equation can be enumerated by using the general algorithm decribed in the next section.

3 General algorithm

In this enumeration approach, each source processor computes using equation (1) the index of each element it owns according to the source distribution. This index is then used to compute the target processor. The corresponding data item is packed into a buffer meant for that processor. When all address computations have been made data are sent to the target machine. Target processors perform the reverse job when they have received all their data. This algorithm is very simple and can be used for every $Cyclic(K)$ to $Cyclic(K')$ distributions, but it is very expensive because data migrations only takes place when all address computations have been achived. To take advantage of an heterogeneous environment and to overlap communications and computations, other algorithms have to be considered.

4 Specialized algorithms

To solve the diophantine equation (2) and to build efficient algorithms, we consider specific redistributions in which one or more variables can be fixed to zero. For these cases, we show that we can enumerate the solutions of the diophantine equation by using a set of regular sections ($first : last : stride$). By using regular sections, a given processor p can create the $I(p,p')$ set meant for target processor p' without having to compute adresses of his whole local data. At each redistribution step, p creates one message and send it to the target machine. Computations in source and target processor sets can then be done in parallel and communications are overlapped with computations. All these specialized algorithms have the following properties :

The number of message sent is minimal: there is at most one message for each processor pair ;
Only required data are sent : only array items are sent (to be fully asynchrone we also send the source processor index) ;
Computations are dynamics : it is sufficient to know distribution parameters only at runtime ;
Constant time adress computations : the preprocessing time required to compute the regular sections only depends of distribution parameters ;
Computations are overlapped with communications ;
Two levels of parallelism are used : two data parallel tasks are runing concurently on the source and target computers.

Figure (1) presents a global view of every studied case. In the next section performance results obtained with these specialized algorithms will be discussed.

Variables set to zero	Redistribution algorithms	Step of the regulars sections
n, n'	$Block(K)$ to $Block(K')$	$\Delta x = \Delta x' = \Delta p = 1$
n	$Block(K)$ to $Cyclic(K')$	$\Delta x' = \Delta n' = \Delta p' = 1$ $\Delta x = P'K'$
n'	$Cyclic(K)$ to $Block(K')$	$\Delta x = \Delta n' = \Delta p' = 1$ $\Delta x' = PK$
x'	$Cyclic(TK')$ to $Cyclic(K')$	$\Delta n = P'/GCD(PT, P')$ $\Delta n' = PT/GCD(PT, P')$ $\Delta p' = GCD(PT, P')$
x	$Cyclic(K)$ to $Cyclic(TK)$	$\Delta n = TP' * GCD(P, P'T)$ $\Delta n' = P/GCD(P, P'T)$ $\Delta p' = GCD(P, P'T)/GCD(T, D)$

Fig. 1. Specialize algorithms

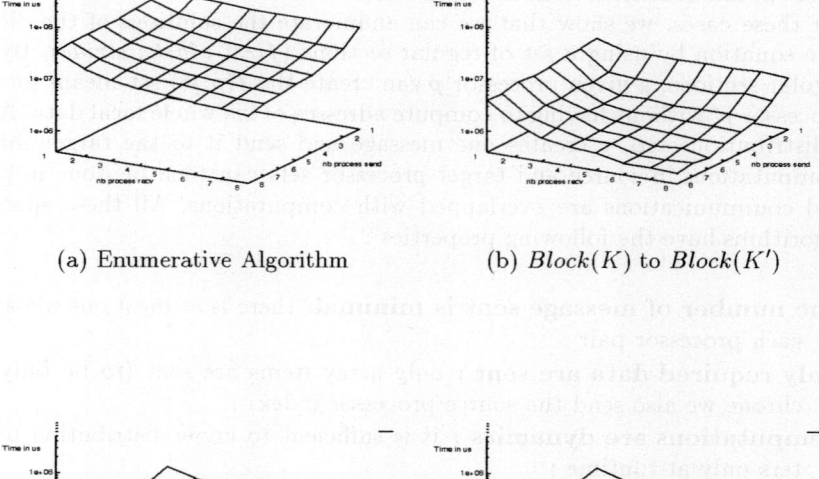

(a) Enumerative Algorithm

(b) $Block(K)$ to $Block(K')$

(c) $Block(K)$ to $Cyclic(K')$

(d) $Cyclic(TK')$ to $Cyclic(K')$

5 Experimental results

All experiments reported on this section have been done on a ALPHA farm of 16 processors interconnected by a cross-bar supporting FDDI bandwidth. We used an array of 4 Millions of integers. Algorithms were implemented over PVM3 library. In each case, we record the time necessary to redistribute the array T from a first set, S of P processors, to another set, T, of P' processors. We always keep $S \bigcap T = \emptyset$. Specialized algorithms always perform better than the enumeration algorithm because the redistribution is done in an overlapped fashion. In the $Block(K)$ to $Block(K')$ redistribution the difference is more accused as there is only one loop and all accesses are contiguous.

6 Conclusion

In this article, we proposed a mathematical representation of $Cyclic(K)$ to $Cyclic(K')$ redistributions. By using this representation, we resolved the corresponding diophantine equation and obtained algorithms which overlap communications and computations. These algorithms can be used in most usual cases. In heterogeneous environments, parallel I/O generaly required multiple redistributions. First data are moved from source compute nodes to source I/O nodes, then between source and target I/O nodes, and at last from target I/O nodes to target compute nodes. Strategies have to be developed to chose the inner distributions in order to minimized redistribution cost.

Complex scientific applications provide opportunities for exploiting multiple levels of parallelism. Thanks to our efficient algorithms, data-parallel tasks could exchange data in a reasonable delay. Note that these algorithms can also be used to implement HPF distribution and redistribution directives when there are several processor directives.

References

1. B. Avalani, A. Choudhary, I. Foster, R. Krihnaiyer, and M. Xu. A data transfer library for communicating data-parallel tasks. Technical report, Syracuse University, Argone Natinal Laboratory, NY 13244, 1994.
2. D. A. Carlson. Ultrahigh-performance FFTs for the CRAY-2 and CRAY-YMP supercomputers. *The journal of supercomputing*, 6(2):107–116, june 1992.
3. G. Edjlali, N. Emad, and S. Petiton. Hybrid methods on network of heterogeneous parallel computers. *Proceedings of the 14th Imacs world congress Atlanta, USA*, july 1994.
4. D. Sueur. Shell hétérogène à parallélisme de données. In *Renpar'7, Actes des 7 Rencontres Francophones du parallélisme*, pages 58–61, PIP-FPMs Mons, Belgique, June 1995.

Supporting Distributed Sparse Matrix Objects*

C. Addison[1] and T. Oliver[1] and A. Sunderland[2]

[1] University of Liverpool, Liverpool, U.K.
[2] Daresbury Laboratory, Daresbury, U.K.

Abstract. The distributed data library (DDL) was developed at the University of Liverpool partly to provide sparse matrix support for a parallel commercial compositional oil reservoir simulation code. The DDL is a portable combination of C and Fortran that uses MPI as its underlying message passing layer. DDL sparse data structures contain not only pointers to standard matrix formats that can be passed directly to the matrix setup routines but also support user-definable properties associated with operators applied to the sparse matrix. This paper describes the design and implementation of operator-specific properties within a distributed sparse object in more detail.

1 Introduction

The EUREKA PARSIM project, [1], aims to produce an interactive 3-D oil reservoir simulator. This requires distributed memory parallel processing to obtain the necessary computational power. The underlying sequential code is a compositional solver with several non-linear equations per grid block. The spatial discretization is based upon finite volume techniques, but the structure of the underlying matrices is fairly regular.

As is described in [2], and summarized in [3], each time step requires the solution of a system of nonlinear equations via Newton's method, which is computationally expensive. In the underlying sequential code, the associated Jacobian matrix is formed explicitly and a preconditioned iterative solver is used to solve the system of linear equations at each Newton iteration.

Therefore, it was important that the sparse matrix data structures on each process were based upon the data structures assumed in the sequential matrix set-up code. This allowed the parallel matrix-setup to reuse nearly all of the sequential code. Performanace and portability contraints also required the parallel preconditioner to scale well and be effective on a range of processor configurations.

The first point was not difficult to achieve, partly because a standard sparse matrix format was already employed. The second was (and remains) a challenge. Some preliminary results with a block preconditioner are found in [1].

Investigations into suitable parallel preconditioners, [4], suggested that the DDL needed to include the ability to solve sparse triangular systems of linear

---
* This work was funded in part by the U.K.'s Department of Trade and Industry as part of the EUREKA Project EU 638, PARSIM.

equations and support for global row and column permutations of a matrix ordering. This work also highlighted a design problem – calls to preconditioner set-up and solve routines should have a uniform calling sequence, but different preconditioners form different secondary matrices associated with the preconditioning operation.

Our solution was to allow user-defined "properties" to be associated with a matrix and the relevant operators. Therefore, a preconditioner set-up routine adds a property to a sparse matrix object that then contains all of the secondary information required by the solve routine. The inter-process communication pattern when performing an operation such as a matrix-vector multiplication can also be stored as a property and reused.

2 DDL Design Basics

The DDL was developed in response to the requirements of the PARSIM project. Its emphasis therefore is somewhat narrow, but it was designed to be extensible. Further details can be found in [5].

The DDL assumes that all objects are 1-D or 2-D matrices. These objects are distributed over a 1-dimensional process grid. Further, 2-D objects can be belong to one of two matrix classes – dense or sparse (all 1-D objects are dense). Within a class, objects may have a variety of types. Each type specifies the particular data structures that should be used to represent the object. Where possible, the choice of data structure is left to the users so that they can select the more appropriate structure for their applications. However, one of the aims of the DDL is to hide data structure details and the data manipulation required to perform an operation. Hence DDL procedures should accept objects of any type compatible with the specified operation. As examples, I/O procedures should be able to read and write any of the object types and a matrix-vector multiply procedure should accept both dense and sparse matrices of any type. At present, the variants of three sparse matrix formats are supported: compressed sparse row format, compressed sparse column format and coordinate format (see the SPARSKIT reference, [6], for more details on these formats).

An SPMD programming model is assumed, so that every process executes the same basic code. By being careful about the design of the application, it is often possible to make the sequential code a special case of the parallel code, which reduces maintenance costs further. As mentioned in the abstract, MPI is the underlying message passing layer.

All DDL objects presently share the same distribution pattern, which is consistent with the block distribution of HPF in one dimension (and no other partitionings in any other dimension). Thus, matrices can be distributed by rows or by columns across the processes. Notice that this partitioning is by index, so that a 1000 by 1000 sparse matrix, stored in the distributed equivalant of compressed row format and distributed over 4 processes, would have all of columns 1 to 250 on process 0. If it was stored in compressed column format, then rows 1 to 250 would be on process 0. The number of non-zeros per process would not usually be the same.

3 Properties

Sparse matrices are represented as a linked list of properties, as is discussed in detail in the next section. This design evolved from the requirements to keep auxillary information about a sparse matrix that was needed by a particular operator. In a sequential setting, such information is often put into real and integer work arrays that are then passed into the required routines. Alternatively, reverse communication is utilised, such as in SPARSKIT and recent parallel extensions of SPARSKIT. Reverse communication removes the need for auxillary information to be passed into a routine, but severely complicates its calling. It also means that the managment of the storage required for any auxillary information becomes partly the responsibility of the user.

The use of opaque handles for distributed objects in the DDL meant that it was relatively simple to encapsulate properties in an unobtrusive, yet easy to extend, manner. The PETSc library from Argonne also exploits encapsulation but it embeds the actual operators in the object definition. This is a good way to proceed when the operators are known in advance. However, the DDL approach of encapsulating properties is intended for the situation where the operator set cannot be predefined, where properties may be required by several operators and where different variants of an operator (such as forming a preconditioner) will have differing data requirements.

To illustrate the types of issues properties are intended to address, consider two of the base-level property classes that are supported: permutations and communication patterns.

3.1 Permutation properties

While an object has a fixed physical partitioning across processes, it is possible to associate a global permutation to the object's logical indices. Thus, matrices can have both row and column indices permuted in order to obtain a more favourable ordering for a given application. For instance, incomplete factorisations, which define upper and lower triangular matrices from the original matrix, are a common and often effective class of preconditioner. Solving triangular systems that are derived from a matrix in something like a standard finite difference ordering has poor scalability, but the solution of triangular systems derived from a matrix with a red-black ordering, potentially has extremely good scalability properties. Such an ordering is easily possible by applying a permutation to the original matrix rows and columns.

Notice that the matrix elements are not redistributed. If both row and column permutations were present, then a matrix element with index (i,j) in the original matrix would have index (row_perm(i),col_perm(j)) in the reordered system. The implied mapping of matrix elements onto processes (ie. via the inverse permutation) is nearly optimal when permuting a finite difference ordering into a red-black ordering and is reasonable in several of the other cases relevant in PARSIM. A redistribution function, which would physically move matrix elements according to the permutations, would be a desirable complement to the permutations, but has not been written yet.

3.2 Communication properties

It is imporant to coordinate the gathers, sends, receives and scatters required within many sparse operations. Consider matrix-vector multiplication, $y = \mathbf{A}x$, where $\mathbf{A}$ is distributed by rows across the processes. For row i, the owning process needs the values of x whose indices match those of the non-zero elements $a_{i,j}$. Conversely, it owns the elements x_{low} to x_{high}, so it will need to send a selection of these values to those processes where the column index of an owned non-zero value of $\mathbf{A}$ lies in the range low to $high$. Therefore, the process needs to have a list of which elements of x to send to other processes and a second list of which elements of x to receive from other processes. By storing these lists as properties, they are available for reuse.

Given a matrix distributed by rows, the matrix-vector multiply routine on each process then consists of posting non-blocking receives to obtain the required off-process elements of x, buffering and sending the owned elements of x required by other processes, performing the partial matrix-vector multiplications with the owned elements of x, completing the receive of off-process elements and then completing the matrix-vector multiplication.

Ordering the list by process number is convenient when all of the elements of the distributed vector are available at once. In other situations, such as a triangular solve, it might be advantageous to order the list by index value so that elements can be dealt with as they become available or as they are required. For each element to be sent, there is a list of processes that require that element. For each element to be received, the process sending that element is stored. Such communication is very fine grain but this can be offset to a certain extent by using groups of non-blocking receives and sends.

4 DDL data structures

As mentioned earlier, objects are represented as a nested collection of properties on each process. The structural representation for an object is identical on each process. Figure 1 provides an example showing the salient features for a sparse matrix with permutations. Each property has a header containing 4 basic fields:

- `type`: the type of the property (such as the object is a sparse matrix, a dense vector or a communications pattern).
- `tag`: used to give a unique name to a property. Current tags include a row or column permutation (of type dense vector) or a row communications pattern for matrix-vector multiply.

- `prop`: a pointer to the next property at the current nesting level.
- `data`: a pointer to the type specific record containing the actual property information.

These are shown as the linked list at the top level of Figure 1. The handle for a distributed object is a pointer to the property header record for the object, the left-most top record in the figure. In this header record, the `prop` pointer points to the first item in the list of special properties associated with the object, the process's share of row permutations in the example.

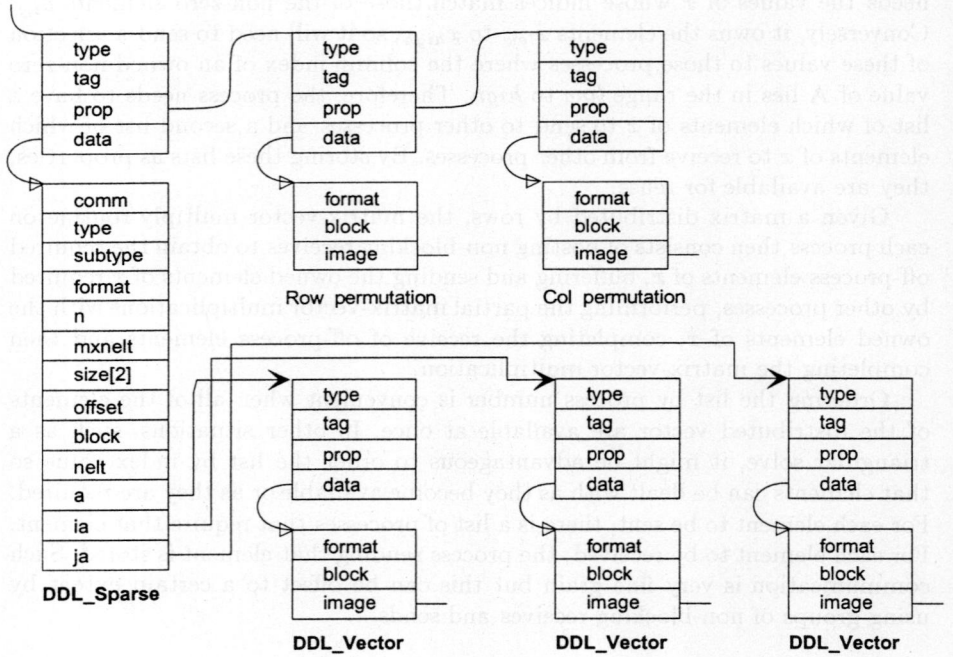

Fig. 1. Example sparse matrix data structure

The data area for a sparse matrix object is pointed to by the `data` pointer in the header record; the data area in Figure 1 lies immediately below the header record. It contains pointers to three distributed vectors, one for row indices or pointers, one for column indices or pointers and one for the actual element values. In addition, it contains the dimension of the matrix, the total number of non-zeros, the data type of the matrix elements, the data format, which determines how the three vectors should be interpreted, information on the data stored on each process and information on the communications topology. The permutations and the triple vectors that contain the actual sparse matrix data are all vector objects. The **image** pointers for each vector points to the contiguous memory area where the appropriate data values are stored.

The communciations pattern property objects are somewhat more complicated and the interested reader is referred to [7].

5 Status and future developments

The detailed documentation, [5, 7], and software for the current version of the DDL is available at http://supr.scm.liv.ac.uk/~tim/parsim/parsim.html. The software is not public domain. It is owned by the University of Liverpool, but it is freely available for non-commercial experimentation and use.

We feel that the sparse matrix DDL represents a flexible basis upon which to build a range of practical, parallel application software.

There are a number of areas where future work is desired. These include:

- performance testing on a range of parallel machines,
- adding support for dense matrices,
- providing support for a wider range of sparse matrix formats,
- investigating new high-level objects, such as a sparse preconditioner,
- investigating new object properties,
- increasing the range of high-level operations supported,
- investigating the integration of the library with HPF.

References

1. J. Larsen, L. Frellesen, J. Jansson, F. If, C. Addison, A. Sunderland, T. Oliver, *Parallel Oil Reservoir Simulation*, **Proceedings of PARA'95**, Lyngby, Denmark, August, 1995.
2. K. Aziz and A. Settari, **Petroleum Reservoir Simulation**, Applied Science Publ., London, England, 1979.
3. C. Addison, T. Christensen, J. Larsen, T. Oliver and A. Sunderland, *Supporting an oil reservoir simulator in a distributed memory environment*, **Proceedings of HPCN'94**, Munich, Germany, April, 1994, pp.340-345.
4. A. Sunderland, *Parallel solution strategies for triangular systems arising from oil reservoir simulations*, **Proceedings of HPCN'95**, Milan, Italy, May, 1995, pp.148-155.
5. T. Oliver, *Sparse DDL Version 2.1: User Guide*, IASC Technical Report, August, 1995.
6. Y. Saad, *SPARSKIT: a basic tool kit for sparse matrix computations, Version 2*, June, 1994.
7. T. Oliver, *Sparse DDL Version 2.1: Technical Guide*, IASC Technical Report, August, 1995.

Workshop 02

Routing and Communication in Interconnection Networks

Low-Latency Communication over Fast Ethernet

Matt Welsh, Anindya Basu, and Thorsten von Eicken
{mdw,basu,tve}@cs.cornell.edu

Department of Computer Science
Cornell University, Ithaca, NY 14853

http://www.cs.cornell.edu/Info/Projects/U-Net

Abstract

Fast Ethernet (100Base-TX) can provide a low-cost alternative to more esoteric network technologies for high-performance cluster computing. We use a network architecture based on the U-Net approach to implement low-latency and high-bandwidth communication over Fast Ethernet, with performance rivaling (and in some cases exceeding) that of 155 Mbps ATM. U-Net provides protected, user-level access to the network interface and enables application-level round-trip latencies of less than 60µs over Fast Ethernet.

1 Introduction

High-performance computing on clusters of workstations requires low-latency communication in order to efficiently implement parallel languages and distributed algorithms. Recent research [1, 6, 8] has demonstrated that direct application access to the network interface can provide both low-latency and high-bandwidth communication in such settings and is capable of showing performance comparable to state-of-the-art multiprocessors. Previous work in this area has concentrated on high-speed networks such as ATM, the technology for which is still emerging and somewhat costly. This paper presents U-Net/FE, a user-level network architecture employing low-cost Fast Ethernet (100Base-TX) technology.

U-Net circumvents the traditional UNIX networking architecture by providing applications with a simple mechanism to access the network device as directly as the underlying hardware permits. This shifts most of the protocol processing to user-level where it can often be specialized and better integrated into the application thus yielding higher performance. Protection is assured through the virtual memory system and kernel control of connection set-up and tear-down. A previous implementation of U-Net over ATM[6] demonstrated that this architecture is able to efficiently support low-latency communication protocols such as Active Messages[7] and parallel languages such as Split-C[2]. However, two important outstanding questions were whether the U-Net model is only feasible over connection oriented networks such as ATM and whether the use of a programmable co-processor on the network adaptor (as in the ATM implementation) is a necessary part of the design.

The U-Net/FE implementation described here demonstrates directly that U-Net can indeed be implemented efficiently over a network substrate other than ATM. The performance results show that low-latency communication over 100Mbps Fast Ethernet is possible using off-the-shelf hardware components. As a result, the cost of workstation clusters is brought down through the use of inexpensive personal computers and a commodity interconnection network.

1.1 Related Work

User-level networking issues have been studied in a number of recent projects. Several of these models propose to introduce special-purpose networking hardware. Thekkath[5] proposes to separate the control and data flow of network access using a shared-memory model; remote-memory operations are implemented as unused opcodes in the MIPS instruction set. The Illinois Fast Messages[4] implementation achieves high performance on a Myrinet network using communication primitives similar to Active Messages. The network interface is accessed directly from user-space but without providing support for simultaneous use by multiple applications. The HP Hamlyn[9] network architecture also implements a user-level communication model similar to Active Messages but uses a custom network interface where message sends and receives are implemented in hardware. Shrimp[1] allows processes to connect virtual memory pages on two nodes through the use of custom network interfaces; memory accesses to such pages on one side are automatically mirrored on the other side. The ParaStation[8] system obtains small-message (4-byte) send and receive processor overheads of about 2.5μsec using specialized hardware and user-level unprotected access to the network interface; however, this does not include the round-trip latency.

2 U-Net user-level communication architecture

The U-Net architecture[6] virtualizes the network interface in such a way that a combination of operating system and hardware mechanisms can provide every application the illusion of owning the interface to the network. The U-Net platform in itself is not dependent on the underlying hardware. Depending on the sophistication of the actual hardware, the U-Net components manipulated by a process may correspond to real hardware in the NI, to software data structures that are interpreted by the OS, or to a combination of the two. The role of U-Net is limited to multiplexing the actual NI among all processes accessing the network and enforcing protection boundaries. In particular, an application has control over both the contents of each message and the management of send and receive resources.

2.1 Sending and receiving messages

U-Net is composed of three main building blocks shown in Figure 1: *endpoints* serve as an application's handle into the network and contain a *buffer area* to hold message data as well as *message queues* to hold descriptors for messages that are to be sent or that have been received. Each process that wishes to access the network first creates one or more endpoints.

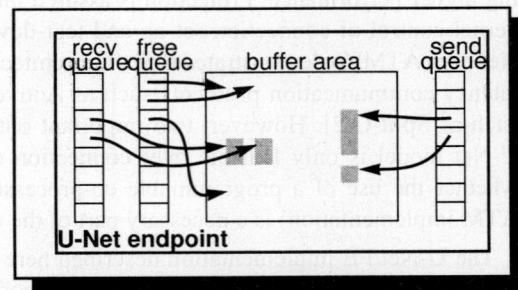

Figure 1: U-Net building blocks.

To send a message, a user process composes the data in the endpoint buffer area and pushes a descriptor for the message onto the send queue. The network interface then picks-up the message and inserts it into the network. The management of the buffers is entirely up to the application: the U-Net architecture does not place constraints on the size or number of buffers nor on the allocation policy used.

Incoming messages are demultiplexed by U-Net based on a tag in each incoming message to determine its destination endpoint and thus the appropriate buffer area for the data and message queue for the descriptor. The exact form of this message tag depends on the network substrate; for example, in an ATM network the virtual channel identifiers (VCIs) may be used. A process registers these tags with U-Net by creating communication channels: on outgoing messages the channel identifier is used to place the correct tag into the message (as well as possibly the destination address or route) and on incoming messages the tag is mapped into a channel identifier to signal the origin of the message to the application. An operating system service needs to assist the application in determining the correct tag to use based on a specification of the destination process and the route between the two nodes.

After demultiplexing, the data is transferred into one or several free buffers and a message descriptor with pointers to the buffers is pushed onto the receive queue. As an optimization for small messages—which are used heavily as control messages in protocol implementations—the receive queue may hold entire small messages in descriptors. Note that the application cannot control the order in which receive buffers are filled with incoming data.

3 Fast Ethernet Implementation of U-Net

U-Net/FE was implemented on a 133Mhz Pentium system running Linux and using the DECchip DC21140 network interface. The DC21140 is a PCI busmastering Fast Ethernet controller capable of transferring complete frames to and from host memory via DMA. The controller includes a few on-chip control and status registers, a DMA engine, and a 32-bit Ethernet CRC generator/checker. It maintains circular send and receive rings containing descriptors which point to buffers for data transmission and reception in host memory. This interface was designed for traditional in-kernel networking layers in which the network interface is controlled by a single agent on the host. As a result the DC21140 lacks any mechanisms which would allow direct user-level access to the chip without compromising protection. This means that U-Net has to be implemented in the kernel: a device driver and a special trap are used to safely multiplex the network interface among multiple applications.

The DC21140's transmit and receive descriptor rings are stored in host memory: each descriptor contains pointers to up to two buffers (also in host memory), a length field, and flags. Multiple descriptors can be chained to form a PDU out of an arbitrary number of buffers. These descriptor rings must be shared among all U-Net/FE endpoints and are therefore distinct from the U-Net transmit, free, and receive queues stored in the communication segment. Figure 2 shows the various rings, queues and buffer areas used in the U-Net/FE design.

3.1 Endpoint and Channel Creation

Creation of user endpoints and communication channels is managed by the operating system. This is necessary to enforce protection boundaries between processes and to properly manage system resources. Endpoint creation consists of issuing an *ioctl* to

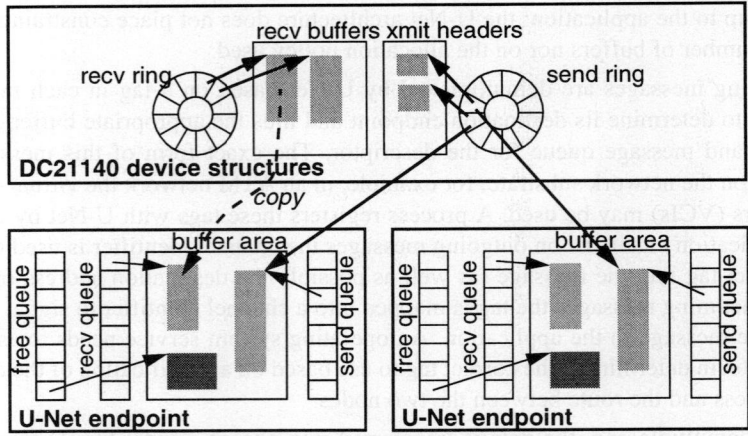

Figure 2: U-Net/FE endpoint and device data structures.

the U-Net device driver requesting space for the message queues and buffer areas. The kernel allocates a segment of pinned-down physical memory for the endpoint, which is mapped into the process address space by use of an *mmap* system call.

A communication channel in the U-Net/FE architecture is associated with a pair of endpoints, each of which is identified by a combination of a 48 bit Ethernet MAC address and a one byte U-Net port ID. A communication channel can be created by issuing an *ioctl* and specifying the two sets of Ethernet MAC addresses and port IDs. The Ethernet MAC address is used to route outgoing messages to the correct interface on the network; the port ID is used to demultiplex incoming messages to a particular endpoint. The operating system registers the requested addresses and returns a channel tag to the application. The channel tag is subsequently used by the application to specify a particular end-to-end connection when pushing entries onto the U-Net send queue. Similarly, the operating system uses the incoming channel tag when placing new entries on the receive queue for the application.

3.2 Transmit

When the user process wishes to transmit data on the network, it first constructs the message in the buffer area and then pushes an entry onto the U-Net send queue, specifying the location, size, and transmit channel tag for each buffer to send. The DC21140 device send ring is shared between all endpoints and must therefore be managed by the kernel. The user process issues a fast trap to the kernel where the U-Net driver services the user's send queue. This is implemented as an x86 trap gate into kernel space, requiring under 1μs for a null trap on a 133MHz Pentium system. This form of trap does not incur the overhead of a complete system call, and the operating system scheduler is not invoked upon return.

The kernel service routine traverses the U-Net send queue and, for each entry, pushes corresponding descriptors onto the DC21140 send ring. Each ring descriptor contains pointers to two buffers: the first one being an in-kernel buffer with the Ethernet header and packet length field, and the second being the user buffer containing the

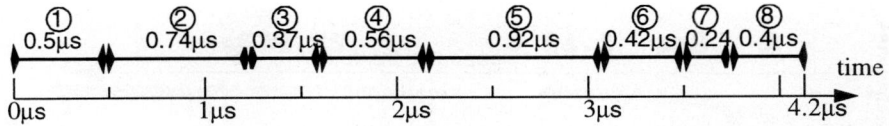

1. trap entry overhead
2. U-Net send param check
3. Ethernet header set-up
4. device send ring descr set-up
5. issue poll demand to DC21140
6. free send ring descr of prev message
7. free U-Net send queue entry of prev message
8. return from trap

Figure 3: Fast Ethernet transmission time-line for a 40 byte message (66 bytes with the Ethernet and U-Net header).

data (for multi-buffer user messages additional descriptors are used). By pointing directly to the U-Net buffer a copy is avoided and the DC21140 transfers the data directly from user-space into the network.

The in-kernel buffers for the ethernet header are preallocated and protected from user-access. Each buffer holds 16 bytes: 14 are filled by the service routine with the Ethernet header for the appropriate U-Net channel, and two contain the actual length of the user message. This last field is needed at the receiving end to identify the exact length of messages under the minimum Ethernet payload size of 46 bytes.

After all the descriptors have been pushed onto the device transmit ring the service routine issues a transmit poll demand to the DC21140. The latter processes the transmit ring, adds padding for small payloads if necessary, and adds the 32-bit packet CRC. Once the frame has been serialized onto the wire, the DC21140 sets a bit in the transmit ring descriptor to signal that the transmission is complete. The kernel service routine uses this information to mark the associated U-Net send queue entries as free.

Figure 3 shows the time-line of a kernel send trap for a 40-byte message which, with the Ethernet and U-Net headers, corresponds to a 66-byte frame. The DC21140 engages the DMA to access the device send ring descriptor after it receives the host's poll demand (i.e. towards the end of segment 5 in the figure). The transfers on the PCI bus are shown in Figure 4 which depicts an oscilloscope screen shot taken of the active-low PCI *FRAME* signal during a loop-back message transmission. The five accesses at the left show that the DC21140 rapidly fetches the message data from the two buffers and serializes it onto the wire. The transmission of the Ethernet frame takes 5.4μs after which the first data of the loop-back message is DMA-ed back into a receive buffer. A timing analysis of the U-Net trap code shows that the processor overhead required to push a message into the network is approximately 4.1μs of which about 20% are consumed by the trap overhead.

3.3 Receive

Upon packet receipt the DC21140 transfers the data into buffers in host memory pointed to by a device receive ring analogous to the transmit ring. The controller then checks the CRC of the incoming packet and interrupts the host, which consumes new entries on the device receive ring and hands them back to the DC21140.

The kernel interrupt routine determines the destination endpoint and channel tag from the U-Net port number contained in the Ethernet header, copies the data into the

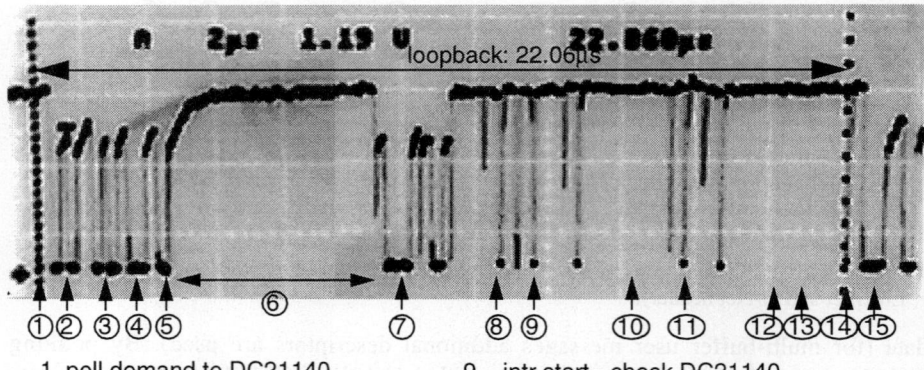

1. poll demand to DC21140
2. DMA fetch of tx descr
3. DMA fetch of header buffer
4. DMA fetch of data buffer
5. DMA prefetch of next tx descr
6. message transmission and reception
7. DMA of incoming message
8. DMA of tx descr (completion)
9. intr start - check DC21140
10. copy message
11. return from interrupt check DC21140
12. user-level recv
13. xmit of next message
14. next poll demand
15. DMA fetch of tx descr

Figure 4: Oscilloscope screen shot of a loop-back transmission and reception of a 66-byte Ethernet frame on the DC21140. The trace shows the PCI bus cycle FRAME signal (active low).

appropriate U-Net buffer area and enqueues an entry in the user receive queue. As an optimization, small messages (under 56 bytes) are copied directly into the U-Net receive descriptor itself.

Figure 5 shows the time line for 40 and 100-byte messages. The short message optimization is effective in that it saves over 15% by skipping the allocation of a receive buffer. For messages of more than 64 bytes the copy time increases by 1.42μs for each additional 100 bytes. The latency between frame data arriving in memory and invocation of the interrupt handler is roughly 2μs and the major cost of the receive interrupt handler is the additional memory copy required to place incoming data into the appropriate user buffer area.

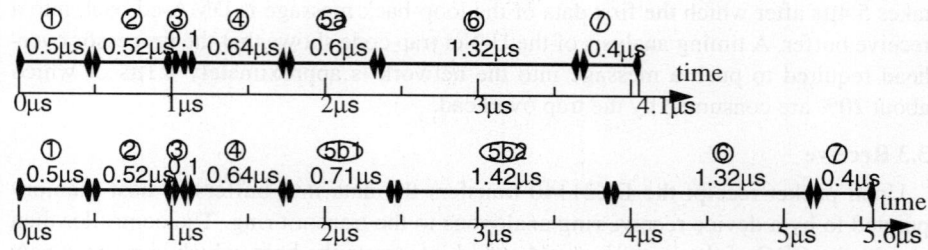

1. interrupt handler entry
2. poll device recv ring
3. demux to endpoint
4. alloc+init U-Net recv descr
5a. copy 40 byte message
5b1. allocate U-Net recv buffer
5b2. copy 100 byte msg
6. bump device recv ring
7. return from interrupt

Figure 5: Fast Ethernet reception time-line for a 40-byte and a 100-byte message.

3.4 Latency and Bandwidth Performance

Figure 6 depicts the application-to-application message round-trip time as a function of message size for U-Net/FE on the DC21140 and compares it to the ATM implementation of U-Net on the FORE Systems PCA-200 ATM interface[1]. Message sizes range from 0 to 1498 bytes for U-Net/FE and ATM. Three Fast Ethernet round-trip times are shown: with a broadcast hub, with a Bay Networks 28115 16-port switch, and with a Cabletron FastNet100 16-port switch. The round-trip time for a 40-byte message over Fast Ethernet rages from 57µsec (hub) to 91µsec (FN100), while over ATM it is 89µsec[2]. This corresponds to a single-cell send and receive which is optimized for ATM.

Increase in latency over Fast Ethernet is linear with a cost of about 25µsec per 100 bytes; over ATM, the increase is about 17µsec per 100 bytes. This can be attributed in part to the higher serialization delay over 100Mbps Fast Ethernet as opposed to 155Mbps for ATM.

Figure 7 shows the bandwidth over the raw U-Net interface for Fast Ethernet and ATM in Mbits/sec for message sizes ranging from 0 to 1498 bytes. For messages as small as 1Kbyte the bandwidth approaches the peak of about 97Mbps (taking into account Ethernet frame overhead) for Fast Ethernet. Due to SONET framing and cell-header overhead the maximum bandwidth of the ATM link is not 155Mbps, but rather 135 Mbps.

4 Summary

The U-Net/FE architecture has been presented as an efficient user-level communication mechanism for use over 100Mbit Fast Ethernet. This system provides low-latency

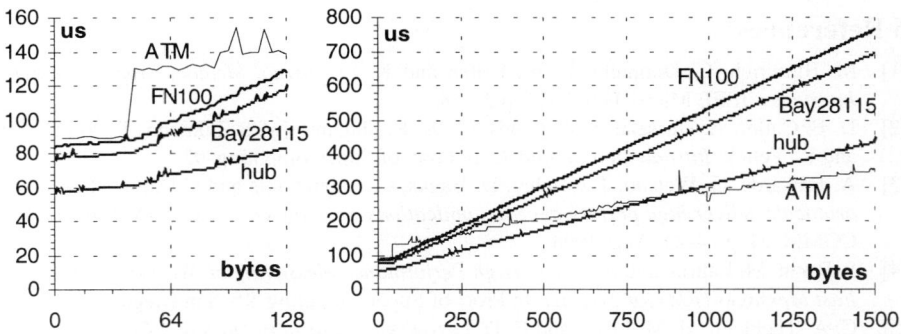

Figure 6: Round-trip message latency vs. message size for Fast Ethernet and ATM.

---

1. The Fore Systems PCA-200 ATM network interface includes an on-board i960 processor to perform the segmentation and reassembly of cells as well as DMA to/from host memory. The i960 processor is controlled by firmware (downloaded from the host) which implements U-Net directly: for message transmission user applications push descriptors directly into i960 memory and incoming messages are DMA-ed straight into user-space.
2. A previous implementation of U-Net over ATM on 140Mbps TAXI demonstrated 65msec round-trip latency; however, here the physical medium is 155Mbps OC-3 and considerable additional overhead is incurred due to SONET framing.

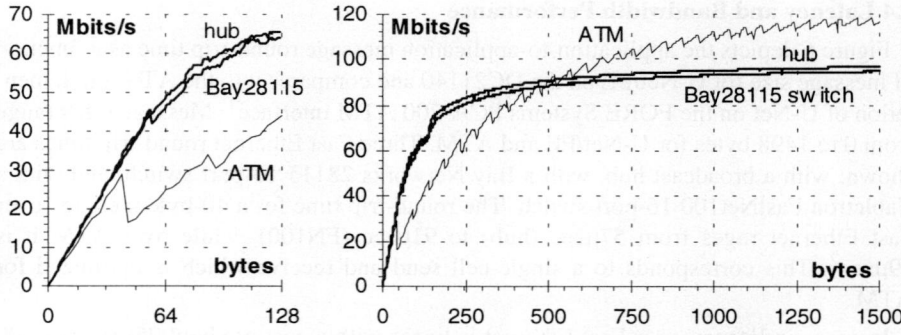

Figure 7: Bandwidth vs. message size for Fast Ethernet and ATM.

communication with performance rivaling that of ATM. For small messages the round-trip latency is in fact less than that for 155Mbps ATM which demonstrates that Fast Ethernet can be employed as a low-latency interconnect for workstation clusters.

This paper has also shown that the U-Net architecture as implemented on ATM can be extended to other networks and network interfaces. In particular, a kernel trap can be used successfully in case the network interface hardware is not capable of multiplexing/demultiplexing messages directly.

5 Acknowledgments

The authors thank Donald Becker of the Beowulf Project at CESDIS for sharing his Linux kernel driver for the DC21140. The U-Net project is supported by the Air Force Material Contract F30602-94-C-0224 and ONR contract N00014-92-J-1866.

6 References

[1] M. Blumrich, C. Dubnicki, E. W. Felten and K. Li. *Virtual-Memory-Mapped Network Interfaces*. IEEE Micro, Feb. 1995, p 21-28.
[2] D. E. Culler, A. Dusseau, S. C. Goldstein, A. Krishnamurthy, S. Lumetta, T. von Eicken, and K. Yelick. *Introduction to Split-C*. In Proc. of Supercomputing '93.
[3] A. Edwards, G. Watson, J. Lumley, D. Banks, C. Calamvokis and C.Dalton. *User-space protocols deliver high performance to applications on a low-cost Gb/s LAN*. Proc. of SIGCOMM-94, p 14-23, Aug. 1994
[4] S. Pakin, M. Lauria, and A. Chien. *High Performance Messaging on Workstations: Illinois Fast Messages (FM) for Myrinet*. In Proc. of Supercomputing '95, San Diego, California.
[5] C. A. Thekkath, H. M. Levy, and E. D. Lazowska. *Separating Data and Control Transfer in Distributed Operating Systems*. Proc. of the 6th Int'l Conf. on ASPLOS, Oct 1994.
[6] T. von Eicken, A. Basu, V. Buch, and W. Vogels. *U-Net: A User-Level Network Interface for Parallel and Distributed Computing*. Proc. of the 15th ACM SOSP, p 40-53, Dec 1995.
[7] T. von Eicken, D. E. Culler, S. C. Goldstein, and K. E. Schauser. *Active Messages: A Mechanism for Integrated Communication and Computation*. Proc. of the 19th ISCA, p 256-266, May 1992.
[8] T. M. Warschko, W. F. Tichy, and C. H. Herter. *Efficient Parallel Computing on Workstation Clusters*. http://wwwipd.ira.uka.de/~warschko/parapc/sc95.html
[9] J. Wilkes. *An interface for sender-based communication*. Tech. Rep. HPL-OSR-92-13, Hewlett-Packard Research Laboratory, Nov. 1992.

A Comparison of Input and Output Driven Routers

Melanie L. Fulgham and Lawrence Snyder

Department of Computer Science and Engineering, Box 352350
University of Washington, Seattle, WA 98195-2350 USA

Abstract. Communication in parallel computers requires a low latency router. Once a suitable routing algorithm is selected, an implementation must be designed. Issues such as whether the router should be input or output driven need to be considered. In this paper, we use simulations to compare input driven and output driven routing algorithms. Three algorithms, the Dally-Seitz oblivious router, the *-channels router, and the minimal triplex algorithm are evaluated. Each router is implemented as both an input and an output driven router. Experiments are run for each of the router implementations with seven different traffic patterns on both a 256-node two dimensional mesh and torus networks. The results show that in almost all cases, the output driven router matches or outperforms the input driven router. Furthermore, we find that randomization of output buffer selection in the input driven algorithm increases its performance and substantially reduces the performance discrepancy between the input and output driven algorithms. Although the findings apply to the routers considered, we believe the results generalize to other routers.

1 Introduction

Communication in parallel computers continues to be a very difficult problem. In this paper we consider communication in multicomputer networks, networks with point to point connections between processors. In this model, processors communicate by sending messages through the network. Messages are forwarded at each node to their destination by a hardware router that implements the routing algorithm, the rules of which specify the path a message takes to reach its destination. Initially the communication bottleneck was the large software overhead for sending and receiving messages. Nevertheless, research has reduced this overhead [20] and exposed the network interface. Designing low latency network interfaces has become a hot topic for research [2, 18]. Eventually network interfaces will no longer overshadow the routing time in the network. Then, both the routing algorithm and its implementation will have an impact on communication performance.

In this paper, we experimentally compare two methods of implementing a particular router, as an input driven or as an output driven algorithm. Although the two choices are conceptually similar, they result in noticeable performance differences when compared with each other. Three algorithms, the Dally-Seitz oblivious router, the *-channels router, and the minimal triplex router, are simulated on a 256-node two dimensional (2D)

This work was supported in part by the National Science Foundation Grant MIP-9213469 and by an ARPA Graduate Research Fellowship.

mesh and torus. Each algorithm is configured as both an input and an output driven router. The output driven versions of the algorithms are almost always superior when the networks are congested. This is the critical area of network performance, since communication is often bursty and does not always present easy, light loads to the network.

The paper is organized as follows. Section 2 defines input and output driven routers. The simulation methodology and the three routing algorithms are described in Section 3. Results follow in Section 4, related work in Section 5, and finally conclusions in Section 6.

2 Input versus Output Driven

Routers that make decisions based on local information can usually be classified as either input driven or output driven. The input driven algorithms make routing decisions from an occupied input buffer while the output driven algorithms select routes from an empty output buffer. Typically the input (output) driven algorithms service the input (output) buffers in round robin order. After routing, the pointer to the current buffer is advanced to the next dimension where a valid buffer resides. For input driven algorithms a valid (input) buffer is one that contains a message that is ready to be routed across the node. For output driven algorithms a valid (output) buffer is one that has room for a message to be routed to it.

An input driven router operates as follows. First it computes which output buffers the message in the current input buffer needs. Then it selects one of these output buffers, if available, and routes the message to that output buffer. Many of the routing algorithms in the literature fall into this category.

An output driven algorithm considers the current empty output buffer. The router tries to find a message in an input buffer that needs to be routed to the current output buffer. If messages are found, it selects one to be routed to the current output buffer. The Chaos router is a good example of an output driven router [16].

Many algorithms can be implemented as either input or output driven algorithms. For these algorithms, it would be advantageous to implement the router with the best performing routing mechanism. A later section shows that for the algorithms examined, the output driven algorithm almost always performed better than the input driven algorithm. Unfortunately some algorithms cannot easily be transformed into output driven algorithms, for example algorithms that take probabilities over the possible output choices.

3 Methodology

Three algorithms are simulated as both input and output driven routers: the Dally-Seitz oblivious router [8], the *-channels router [1], and the minimal triplex router [11] using a flit-level simulator of 256-node 2D torus and mesh networks. The number of virtual channels per channel per direction for the routers are 2 (1), 3 (2) and 3 (2), respectively, for the torus (mesh) algorithms. Each node has an injection and delivery buffer in addition to an input and an output buffer for each virtual channel in each dimension in each direction. The oblivious algorithm also has an extra set of virtual lanes [7], yielding a total of 34 (18), 26 (18), and 26 (18) buffers per node for the torus (mesh) routers, respectively. Transmission of a flit across a channel from an output buffer to input buffer takes one cycle; the pipelined logic at each node takes 3 cycles for the oblivious, and 4 cycles for the adaptive routers [3]. Routing options are computed in parallel, although a router connects at most a single message from an input buffer to an output buffer per cycle. Each buffer holds exactly one 20-flit message, and virtual cut through flow control [14] is used to avoid store-and-forward penalties. Results are normalized to the maximum theoretically sustainable load, constrained by the network bisection bandwidth, of one message every 80 (160) cycles for the torus (mesh), respectively. Messages are introduced at each node at every cycle with a probability specified by the applied load. Traffic considered is random (all nodes equally likely), permutations including bit reversal, transpose, complement, and perfect shuffle, as well as hot spot traffic (ten random nodes four times more likely to be destinations). The hot spots are listed in Appendix A. The traffic patterns are found in the literature, and are generally thought to be difficult, useful, or both. See [10] and [4] for further details.

4 Results

The results are presented in two parts. The first set of experiments compare output driven routers to input driven routers that choose deterministically the first available output buffer from the set of needed output buffers. Any deterministic order suffices, in this case the routers search in dimension order. In the second set of experiments the input driven router selects an available output buffer at random from the set of needed output buffers. For this set-up, the input driven router is very similar to the output driven router which chooses messages from the input buffers at random. In both cases the output driven routers perform better than the input driven routers for all but a few cases. As expected, the results are more dramatic for the first case.

4.1 Fixed Order Output Buffer Selection

The first set of experiments consider input driven routers with a fixed order output buffer selection policy. Table 1 specifies the first normalized applied load, in increments of .05, at which messages arrive faster then they can be injected and delivered by the 256-node torus network. For all three algorithms and all traffic patterns except one, the point of saturation for the output driven router is at least as great as the saturation point of the corresponding input driven router. The advantage ranges from zero to seventy percent of the input driven saturation point.

We conjecture that the output driven algorithms are better at utilizing the channels between nodes because the empty output buffers are filled by dimension in round-robin order which naturally tries to keep all the physical channels busy. More specifically, when a message has been routed to an empty output buffer in a dimension, the router will consider the next dimension that has an empty output buffer. However the input driven algorithm routes a message to the first available output buffer it finds. This buffer may share the same physical channel as a recently routed message. With virtual cut-through routing the buffer chosen may even be the same buffer as the previously routed message.

Table 1. Minimum normalized applied load within .05 at which saturation is detected.

Traffic	16x16 Torus Saturation, input driven is fixed order					
	oblivious		*-channels		min triplex	
	input	output	input	output	input	output
Random	0.70	0.80	0.85	0.95	0.80	0.85
Bit reversal	0.40	0.50	0.70	0.80	0.65	0.75
Complement	0.45	0.50	0.45	0.40	0.40	0.40
Perfect shuffle	0.40	0.45	0.50	0.50	0.40	0.45
Transpose	0.50	0.55	0.55	0.55	0.55	0.55
Hot Spot 1	0.60	0.65	0.70	0.90	0.65	0.85
Hot Spot 2	0.50	0.55	0.50	0.85	0.60	0.80

This cannot happen in an output driven router unless all the other output buffers are full, or no message in the input buffers needs any other output buffer. Thus, the benefit of output over input driven routers should be less pronounced in oblivious algorithms where a message only waits on one virtual channel, particularly those with fewer virtual lanes. The data supports this hypothesis for the torus and for the mesh which is presented later.

Routing the complement permutation using the *-channels algorithm is the one case where the output driven saturation point did not equal or exceed the input driven saturation load. The complement is unusual since dimension order routing helps prevent conflicts in this traffic pattern, i.e. when two messages compete for the same output buffer. See [10] for an explanation. For the *-channels algorithm the input driven router has a slight advantage since it prefers the lowest dimension output buffer needed, while the output driven algorithm selects a random message that needs the current output buffer. However, the second set of experiments show this advantage disappears when the input driven algorithm picks from the needed output buffers at random.

Table 2 presents the saturation loads for the 256-node mesh and shows that for the mesh, output driven routers are superior to input driven routers. The advantage ranges from zero to twenty percent. Again, the only exception is the complement traffic for the *-channels and minimal triplex adaptive routers.

For the mesh, the oblivious algorithm was also simulated with a single virtual channel (vc) per channel, i.e. with no extra lanes. In this case, a message waiting in an input buffer needs exactly one output buffer. Therefore, the input and output driven algorithms do not make different routing decisions. Rather, the difference in performance is due to the ordering of messages and the router's ability to keep the channels busy. As hypothesized, the difference in saturation points between input and output driven routers is very modest, non-existent in four traffic patterns and within a normalized load of .05 for the remaining ones. For the two lane oblivious routers, the difference is more distinct. The output driven router has a higher saturation point than the input driven router in all but one of the traffic patterns, and the maximum difference in saturation points is .15.

Figures 1 and 2 in the Appendix show the expected throughput and latency versus the normalized applied load for each of the traffic patterns on the 256-node torus. Results for the second hot spot case are similar to the first hot spot case and have been

Table 2. Minimum normalized applied load within .05 at which saturation is detected.

16x16 Mesh Saturation, input driven is fixed order								
Traffic	oblivious (1 vc)		oblivious		*-channels		min triplex	
	input	output	input	output	input	output	input	output
Random	0.85	0.90	0.90	0.95	0.95	0.95	0.90	0.90
Bit reversal	0.50	0.50	0.50	0.50	0.80	0.80	0.70	0.75
Complement	0.45	0.50	0.45	0.50	0.45	0.35	0.45	0.35
Perfect shuffle	0.75	0.75	0.75	0.90	0.80	0.95	0.80	0.90
Transpose	0.50	0.50	0.50	0.55	0.85	0.85	0.85	0.85
Hot Spot 1	0.75	0.75	0.75	0.80	0.80	0.85	0.75	0.85
Hot Spot 2	0.70	0.70	0.70	0.75	0.75	0.85	0.75	0.85

omitted. Throughput represents messages delivered per cycle but is normalized to reflect bandwidth limitations of the network. Latency measures time in the network and does not include source queueing, since after saturation, the source queue length is not well defined.

Initially the input and output driven routers are indistinguishable. Nevertheless, output driven routers achieve a higher or equivalent throughput than the corresponding input driven router for all the traffic patterns and routers simulated. As with the saturation data, the complement traffic is the only exception. For the minimal triplex router, the peak throughput of the output driven router under the complement traffic does not quite reach that of the input driven router.

The expected latency curves have three phases. Initially input and output schemes have equivalent latencies. Any router and routing decision will do when the router is lightly loaded. When the applied load is in the neighborhood of saturation and the network is congested, the output driven routers exhibit a lower latency (and latency variance) than the input driven router. We believe this is because the output driven router is doing a better job at keeping the physical channels utilized. Finally after saturation, the latencies are less predictable, but often tend to converge. After saturation the network cannot keep up with the message arrivals; and again, any type of routing will do. The oblivious router is an exception since the sustained higher throughput of the output driven router results in higher latencies.

4.2 Random Output Buffer Selection

In order to validate our conjecture that the disadvantage of input driven routers is not from the deterministic search for available output buffers, the following changes were made to make the input driven routers as similar as possible to the output driven routers. Each input driven router was modified so that it selects a needed output buffer at random, instead of choosing the lowest dimension output buffer available. The output driven routers were unchanged and choose messages from the input buffers at random. Experiments showed the modified input driven algorithms to have better channel utilization resulting in improved performance[1].

---

[1] It is likely that round-robin selection could approximate this improvement.

Table 3 compares the saturation points of the input and the output driven routers for the mesh and torus. The change from fixed order output buffer selection to random does not change the routing decisions in the one lane oblivious input driven algorithm. The change between the two lane oblivious input driven algorithms is small; the two routers may select a different lane to route a message in the input buffer. Nevertheless, there is no noticeable change in saturation points in either the oblivious mesh or torus routers, most likely since the lanes share the same underlying physical channel.

Table 3. Minimum normalized applied load within .05 at which saturation is detected.

Traffic	16x16 Network Saturation, input driven is random											
	Mesh						Torus					
	oblivious		*-channels		min triplex		oblivious		*-channels		min triplex	
	input	output	input	output	input	output	input	output	input	output	input	output
Random	0.90	0.95	0.95	0.95	0.90	0.90	0.70	0.80	0.95	0.95	0.85	0.85
Bit reversal	0.50	0.50	0.80	0.80	0.75	0.75	0.40	0.50	0.80	0.80	0.70	0.75
Complement	0.45	0.50	0.35	0.35	0.35	0.35	0.45	0.50	0.40	0.40	0.35	0.40
Perfect shuffle	0.75	0.90	0.95	0.95	0.90	0.90	0.40	0.45	0.50	0.50	0.45	0.45
Transpose	0.50	0.55	0.80	0.85	0.75	0.85	0.50	0.55	0.55	0.55	0.55	0.55
Hot Spot 1	0.75	0.80	0.90	0.85	0.85	0.85	0.60	0.65	0.85	0.90	0.80	0.85
Hot Spot 2	0.70	0.75	0.90	0.85	0.85	0.85	0.50	0.55	0.75	0.85	0.75	0.80

For the adaptive routers the difference between the saturation points of the input and output driven algorithms on the mesh and torus has nearly been eliminated for almost all the traffic patterns. The input driven routers with random output buffer selection are superior to those with fixed order output selection, since removing the bias of fixed order selection improves the utilization of the physical channels of a node. The only exception is the complement. The complement, as mentioned earlier, prefers dimension order routing, and hence saturates at a lower load than previously for both adaptive routers on the mesh and torus.

For the hot spot traffic, the input driven router with random selection has higher saturation point than the output driven router, even though the two routers achieve the same throughput. The input driven router is able to prevent congestion around the hot spots from spreading as quickly into the whole network compared to the output driven router. The reasons for this are unclear, but most likely related to the lack of wrap edges in the mesh since the torus does not exhibit this behavior.

Due to space limitations, the expected throughput and latency of the input and output driven algorithms on the torus are not presented, but can be found in [10]. The oblivious input driven algorithm does not exhibit significant change between the fixed order and random selection of output buffers, except for a decrease in latency for some of the traffic patterns including the complement, perfect shuffle, and transpose.

The input driven adaptive routers improve their maximum achieved throughput, and in some cases match the throughput of the corresponding output driven router, as with *-channels under bit reversal and random traffic and triplex under random traffic. For most cases there is also a substantial decrease in the after saturation latency of the input driven algorithms for all the traffic patterns. In addition, the steep increase in latency is delayed and occurs at a higher applied load for random, bit reversal, perfect shuffle, and hot spot traffic.

5 Related Work

Most of the work in routing algorithms has been in developing deadlock-free algorithms. Numerous frameworks have been presented for developing deadlock-free algorithms with varying complexity, resource requirements, and switching techniques [8, 17, 9, 13, 6, 19]. Each of these factors influences the overall performance of the router. Nevertheless, only a few studies have been devoted to improving performance or comparing various implementations of a routing algorithm. Dally increased throughput of wormhole algorithms by adding virtual channels to separate the buffering resources from the transmission resources of the router [7]. Konstantinidou reduced overall message latency in bimodal length traffic by introducing segment routing [15]. Segment routing provides a separate buffer for large messages, allowing small messages to pass larger ones. Cherkasova and Rokicki replaced FIFO injection, the traditional method of introducing messages into the network, with alpha scheduling [5]. With variable length messages, alpha scheduling approximates the optimal average message latency of shortest first scheduling without introducing starvation. Finally, a study by Glass and Ni compared the performance of various policies for selecting input and output buffers for two different routing algorithms on the mesh [12].

6 Conclusions

We have experimentally compared the performance of input and output driven algorithms on the mesh and torus. Although the two are conceptually similar, for almost all the cases examined, the performance of the output driven algorithms is equivalent or superior to that of the input driven algorithms. The difference is diminished when randomization is added to the output buffer selection of the input driven algorithm. Although the findings presented only apply to the routers considered, we believe that the results can be generalized to routers where the designer is indifferent to which approach to use. Future work may compare input and output driven routers using longer messages or a non-minimal router.

A Appendix

Assuming the nodes are labeled in row major order from 0 to 255, the hot spot nodes for case 1 are 158, 186, 216, 236, 121, 86, 6, 152, 201, and 123. For case 2 they are 51, 92, 254, 140, 51, 70, 201, 155, 124, and 245.

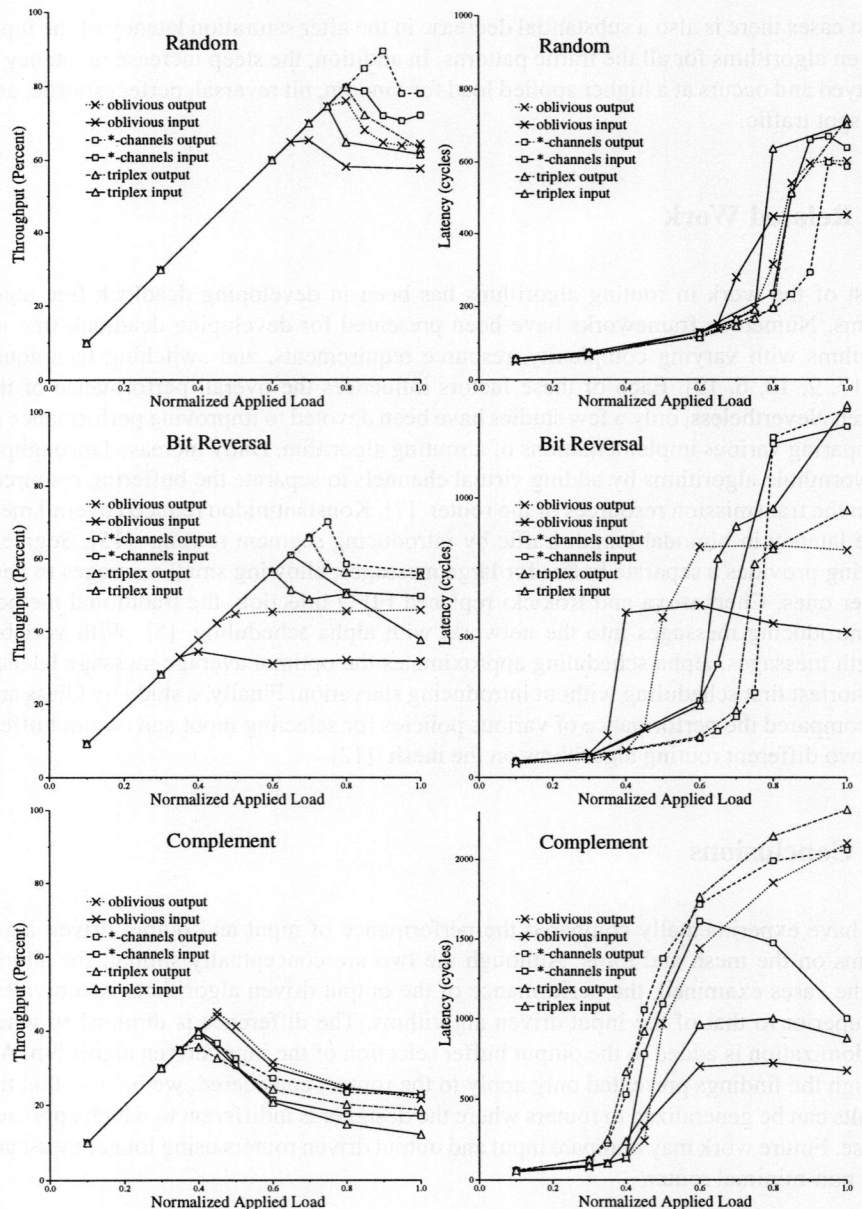

Fig. 1. Throughput and latency on a 256-node 2D torus with fixed output buffer selection.

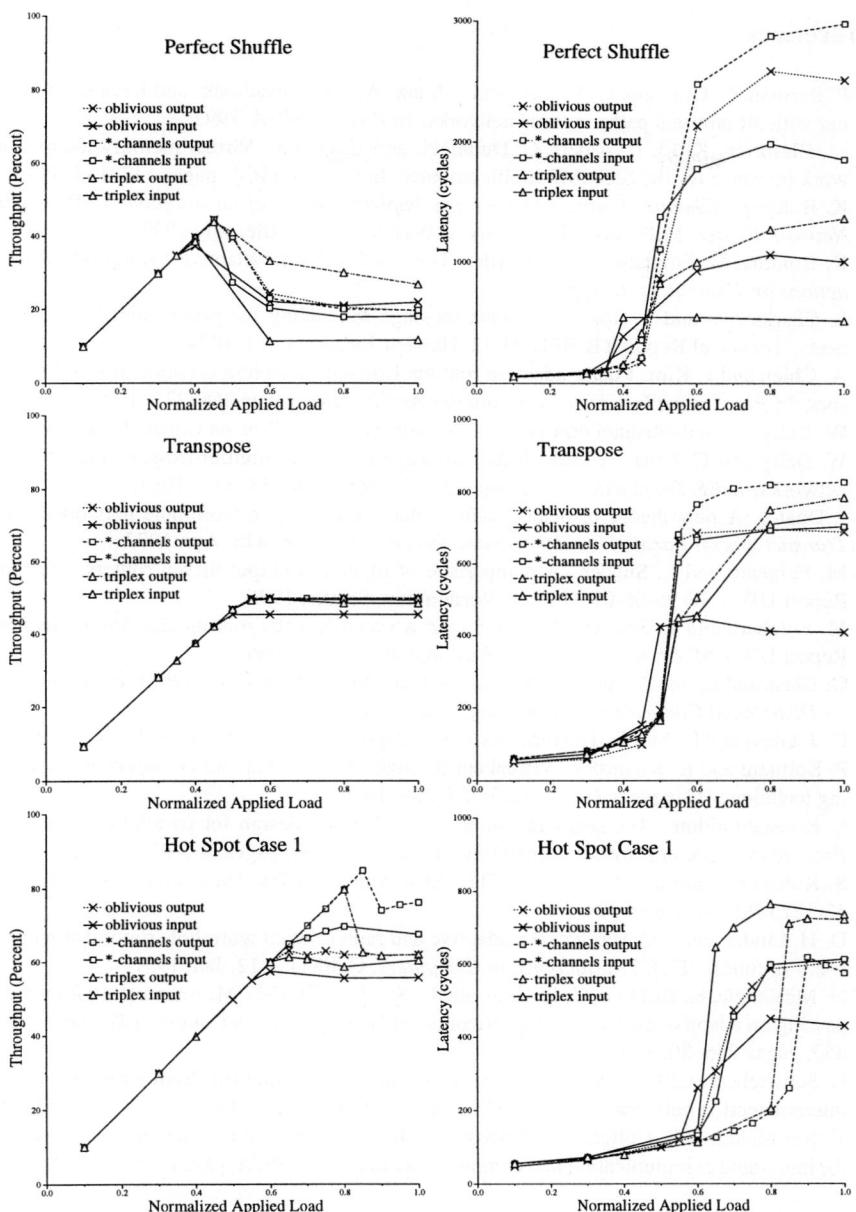

Fig. 2. Throughput and latency on a 256-node 2D torus with fixed output buffer selection.

References

1. P. Berman, L. Gravano, G. Pifarré, and J. Sanz. Adaptive deadlock- and livelock-free routing with all minimal paths in torus networks. In *Proc. of SPAA*, 1992.
2. M. Blumrich, K. Li, R. Alpert, C. Dubnicki, and E. Felten. Virtual memory mapped network interface for the SHRIMP multicomputer. In *Proc. of ISCA*, pages 142–153, 1994.
3. K. Bolding. *Chaotic Routing: Design and Implementation of an Adaptive Multicomputer Network Router*. PhD thesis, University of Washington, Seattle, July 1993.
4. K. Bolding, M. Fulgham, and L. Snyder. The case for chaotic adaptive routing. *IEEE Transactions on Computers*, to appear.
5. L. Cherkasova and T. Rokicki. Alpha message scheduling for packet-switched interconnects. Technical Report TR HPL-94-72, Hewlett-Packard Labs, 1994.
6. A. Chien and J. Kim. Planar-adaptive routing: Low-cost adaptive networks for multiprocessors. In *Proc. of the Intl. Sym. on Computer Architecture*, pages 268–277, 1992.
7. W. Dally. Virtual–channel flow control. In *Proc. of the Intl. Sym. on Comp. Arch.*, May 1990.
8. W. Dally and C. Seitz. Deadlock-free message routing in multiprocessor interconnection networks. *IEEE Transactions on Computers*, C-36(5):547–553, May 1987.
9. J. Duato. A new theory of deadlock-free adaptive routing in wormhole networks. *IEEE Transactions on Parallel and Distributed Systems*, 4(4):466–475, Apr. 1993.
10. M. Fulgham and L. Snyder. A comparison of input and output driven routers. Technical Report UW-CSE-96-06-01, Univ. of Washington, Seattle, 1996.
11. M. Fulgham and L. Snyder. Triplex router: a versatile torus routing algorithm. Technical Report UW-CSE-96-01-11, Univ. of Washington, Seattle, 1996.
12. C. Glass and L. Ni. Adaptive routing in mesh-connected networks. In *Proc. of the Intl. Conf. on Distributed Computing Systems*, pages 12–19, 1992.
13. C. J. Glass and L. M. Ni. The turn model for adaptive routing. *JACM*, 41(5):874–902, 1994.
14. P. Kermani and L. Kleinrock. Virtual cut-through: A new computer communication switching technique. *Computer Networks*, 3:267–286, 1979.
15. S. Konstantinidou. The segment router: a novel router design for parallel computers. In *Proc. of the Sym. of Parallel Algorithms and Architectures*, pages 364–373, 1994.
16. S. Konstantinidou and L. Snyder. The chaos router. *IEEE Transactions on Computers*, 43(12):1386–97, Dec. 1994.
17. D. H. Linder and J. C. Harden. An adaptive and fault tolerant wormhole routing strategy for k-ary n-cubes. *IEEE Transactions on Computers*, C-40(1):2–12, Jan. 1991.
18. N. McKenzie, K. Bolding, C. Ebeling, and L. Snyder. CRANIUM: An interface for message passing on adaptive packet routing networks. In *Lecture Notes in Computer Science*, volume 853, pages 266–80, 1994.
19. L. Schwiebert and D. Jayasimha. A universal proof technique for deadlock-free routing in interconnection networks. In *Proc. of the Sym. on Par. Alg. and Arch.*, pages 175–184, 1995.
20. T. von Eicken D.E. Culler, S. Goldstein, and K. Schauser. Active messages: a mechanism for integrated communication and computation. In *Proc. of ISCA*, pages 256–266, May 1992.

Optimal Topology for Distributed Shared-Memory Multiprocessors: Hypercubes Again?

José Duato and M.P. Malumbres

Facultad de Informática, Universidad Politécnica de Valencia
P.O.B. 22012, 46071 - Valencia, SPAIN. E-mail: {jduato,mperez}@gap.upv.es

Abstract. Many distributed shared-memory multiprocessors (DSM) use a direct interconnection network to implement a cache coherence protocol. An interesting characteristic of the message traffic produced by coherence protocols is that all the messages are very short. Most current multicomputers use low dimensional meshes or tori because these topologies usually achieve a higher performance. However, when messages are very short, latency is mainly dominated by the distance traveled in the network. As a consequence, higher dimensional topologies may achieve a lower latency than low-dimensional topologies. In this paper, we compare the 2D-mesh and the hypercube topologies assuming a very detailed router model. Network load has been modeled taking into account the traffic produced by cache coherence protocols. Performance results show that average latency for hypercubes is slightly lower than for meshes. Moreover, hypercubes achieve a much higher throughput than meshes, making them suitable for DSMs.

1 Introduction

Distributed shared-memory multiprocessors (DSM) are gaining acceptance because they are easier to program than multicomputers. Recently proposed DSMs use a direct interconnection network to access remote memory locations, making these architectures scalable [12, 11]. Most DSM implement a cache coherence protocol. An interesting characteristic of the message traffic produced by coherence protocols is that all the messages are very short. In particular, invalidations and acknowledgments require a single flit. Messages containing a cache line usually range from four to sixteen flits depending on flit width and line size.

Most current multicomputers like Intel Paragon and Cray T3D use wormhole switching [4] and low dimensional meshes or tori [10, 9]. Low-dimensional topologies allow the use of wider channels, therefore increasing channel bandwidth [1]. This is important when messages are long. However, when messages are very short, latency is mainly dominated by the distance traveled in the network. As a consequence, higher dimensional topologies may achieve a lower latency than low-dimensional topologies. Taking into account that invalidation protocols require a very low message latency to work efficiently, the optimal topology for DSMs may differ from the one for multicomputers. Throughput is also important for cache

---

This work was supported by the Spanish CICYT under Grant TIC94-0510-C02-01

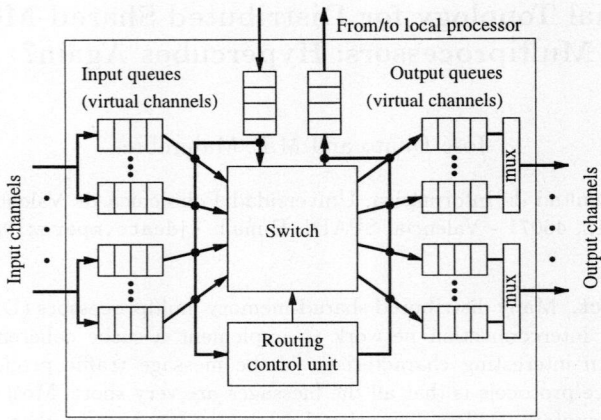

Fig. 1. Router model used in our study

coherence protocols, especially for update protocols. Again, high-dimensional topologies may achieve a higher throughput for short messages because they offer more alternative paths, especially when adaptive routing is used.

In this paper, we present a performance evaluation of deterministic and fully adaptive routing algorithms for meshes and hypercubes. Performance evaluation uses a very detailed router model. The design complexity of the router has been considered while computing internal delays. Also, wire length has been taken into account to compute channel delay and bandwidth. Network load has been modeled taking into account the traffic produced by cache coherence protocols in DSMs. In section 2, we describe the router model that we have considered. In section 3, we present the network traffic pattern. In section 4 we present the simulation results for different approaches. Finally, conclusions are drawn.

2 Router Model

Each router consists of a routing control unit, a switch, and several channels. Figure 1 shows the router model used in our simulator. The routing control unit selects the output channel for a message as a function of its destination node, the current node and the output channel status. The routing control unit can only process one message header at a time. We also considered an alternative router design in which each virtual channel has an independent routing control unit, allowing a concurrent processing of message headers that arrive at the node. The switch is a crossbar. So, it allows multiple messages traversing it simultaneously without interference.

Physical channels can be split into several virtual channels. Virtual channels are assigned to the physical link using a demand-slotted round-robin arbitration scheme. Each virtual channel has an associated buffer. This buffer is divided into two halves, one associated with the output port of the switch (m flits), and another one associated with the input to the next node's switch (n flits). We will use the $m + n$ notation to specify this buffer size. We have used a variable buffer size to study the impact on performance.

We assume that all operations inside each router are synchronized by its local clock signal. To compute the clock frequency of each router, we use the delay model proposed in [3]. It assumes 0.8 micron CMOS gate array technology for the implementation. The delay of each component is computed as follows:

- Routing control unit. Routing a message involves the following operations: Address decoding, routing decision, and header selection. According to [3], the address decoder delay is equal to 2.7 ns. The routing decision logic has a delay that grows logarithmically with the number of alternatives, or degree of freedom, offered by the routing algorithm. Denoting by F the degree of freedom, this circuit has a delay given by $0.6 + 0.6 \log F$ ns. Finally, the routing control unit must compute the new header, depending on the output channel selected. This operation has a delay given by $1.4 + 0.6 \log F$ ns. Therefore, total routing time will be the sum of all the delays, yielding:

$$T_r = 2.7 + 0.6 + 0.6 \log F + 1.4 + 0.6 \log F = 4.7 + 1.2 \log F \text{ ns.}$$

- Switch. The time required to transfer a flit from one input channel to the corresponding output channel is the sum of the delay involved in the internal flow control unit, the delay of the crossbar, and the set-up time of the output channel latch. The flow control unit has a constant delay equal to 2.2 ns. The crossbar delay grows logarithmically with the number of ports. Assuming that P is the number of ports of the crossbar, its delay is given by $0.4 + 0.6 \log P$ ns. Finally, the set-up time of a latch is 0.8 ns. Therefore, switch time is:

$$T_s = 2.2 + 0.4 + 0.6 \log P + 0.8 = 3.4 + 0.6 \log P \text{ ns.}$$

- Channels. The time required to transfer a flit across a physical channel includes the off-chip delay across the wires, and the time required to latch it onto the destination. The latter time is the sum of input buffer, input latch and synchronizer delays. Typical values for the technology used are 0.6, 0.8, and 1.0 ns, respectively, yielding 2.4 ns per flit. The off-chip delay across the wires depends on their length. In particular, topologies like 2D-meshes have constant wire length. For the technology used, assuming 25 pF load, typical propagation delay across wires is 1.5 ns. However, hypercubes have wires with different lengths. So, wire delay must be computed for hypercubes taking into account wire length in order to make a fair comparison with mesh topologies. For example, if we have an 8D-hypercube, we can assemble the topology in three dimensions as shown in Figure 2. As can be seen, the shortest wires have the same length that they would have in a 2D-mesh. Also, there are some wires twice as long as the shortest ones. Finally, the longest wire is four times the size of the shortest one. Thus, the off-chip delay in an 8D-hypercube will depend on wire length. For the shortest wires, a typical value will be 1.5 ns (the same as for 2D-meshes). For the remaining wire lengths, the off-chip delay will be 3 ns and 6 ns, respectively.

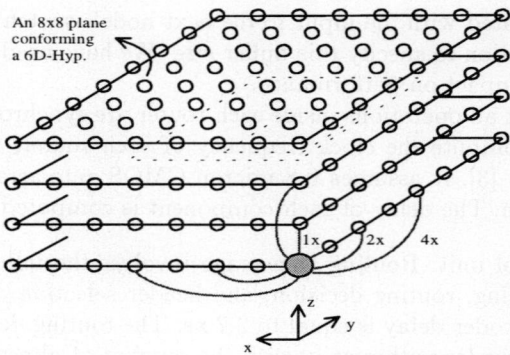

Fig. 2. A three-dimensional implementation of a 8D-hypercube

If virtual channels are used, the time required to arbitrate and select one of the ready flits must be added. The virtual channel controller has a delay logarithmic in the number of virtual channels per physical channel. If V is the number of virtual channels per physical channel, virtual channel controller delay is $1.24 + 0.6 \log V$ ns. If we denote the off-chip delay for the shortest wires as T_{c1}, and the off-chip delays for wires that are twice and four times longer as T_{c2} and T_{c4}, respectively, total channel delay yields:

$T_{c1} = 2.4 + 1.5 + 1.24 + 0.6 \log V = 5.14 + 0.6 \log V$ ns.
$T_{c2} = 2.4 + 1.5 * 2 + 1.24 + 0.6 \log V = 6.64 + 0.6 \log V$ ns.
$T_{c4} = 2.4 + 1.5 * 4 + 1.24 + 0.6 \log V = 9.64 + 0.6 \log V$ ns.

In what follows, we compute the value of F and P as a function of the topology and the number of virtual channels for deterministic and adaptive routing algorithms. The deterministic routing algorithm is the dimension-order routing algorithm. The adaptive routing algorithm is based on Duato's theory [7]. In this routing algorithm, all the virtual channels but one are used for fully adaptive minimal routing. The remaining virtual channel is used to avoid deadlocks by routing in dimension-order.

1. 2D-mesh with deterministic routing. A deterministic routing algorithm has a single routing choice. However, messages can use any virtual channel. If there are V virtual channels per physical channel then $F = V$. Also, each crossbar has $P = 4V + 1$ ports, including the port connecting to the local processor.

2. 2D-mesh with adaptive routing. The number of routing choices is equal to $F = 2(V - 1) + 1$, because we have $V - 1$ virtual channels in each dimension that can be used to cross the dimensions in any order plus one additional virtual channel to avoid deadlock [7]. As above, the number of ports is given by $P = 4V + 1$.

3. 8D-hypercube with deterministic routing. In this case, we have $F = V$ and $P = 8V + 1$.

4. 8D-hypercube with adaptive routing. The number of routing choices is $F = 8(V-1) + 1$ and the number of ports is $P = 8V + 1$.

3 Network Traffic Pattern Description

We assume a distributed shared-memory multiprocessor architecture. So, traffic consists of two kinds of messages: control messages containing invalidations and acknowledgments (1 data flit), and data messages that transport cache lines (8 data flits). Control messages represent a 60% of the total number of messages and may have more than one destination (multicast messages). The rest of messages will be unicast data messages, following the usual distribution in DSMs.

For each simulation run, we have considered that message generation rate is constant and the same for all the nodes. Once the network has reached a steady state, the flit generation rate is equal to the flit reception rate (traffic). We have evaluated the full range of traffic, from low load to saturation. The message destination is randomly chosen among all the nodes. This pattern has been widely used in other performance evaluation studies [6, 2].

4 Simulation Results

We evaluated 2D-mesh and hypercube topologies with 256 nodes, considering different routing algorithms and network parameters, and assuming the specific traffic pattern described above. We ran simulations using the minimum number of virtual channels per physical channel and the minimum plus one. The evaluation methodology used is based on the one proposed in [7]. The most important performance measure in DSMs is latency (time required to deliver a message). So, plots will represent latency versus network traffic. Traffic is the flit reception rate. Latency is measured in nanoseconds. Traffic is measured in flits per node per microsecond. Note that we have used absolute measurement units because we are going to compare routing algorithms that involve different implementations, and consequently, different clock frequencies. When simulation results do not consider the router implementation, the latency and traffic will be measured in clock cycles and flits per node per cycle, respectively.

4.1 Effect of queue size and multiple routing control units

We analyzed the effect of flit queues with capacity for $2 + 2$ and $4 + 5$ flits. Note that a queue with capacity for $4 + 5$ flits is able to store a whole message, including its header. Also we considered two router models: one of them has the typical routing control unit in which header flits are routed sequentially, and the other one is able to make routing operations in parallel. So, we can determine if the routing control unit is a bottleneck when short messages are used.

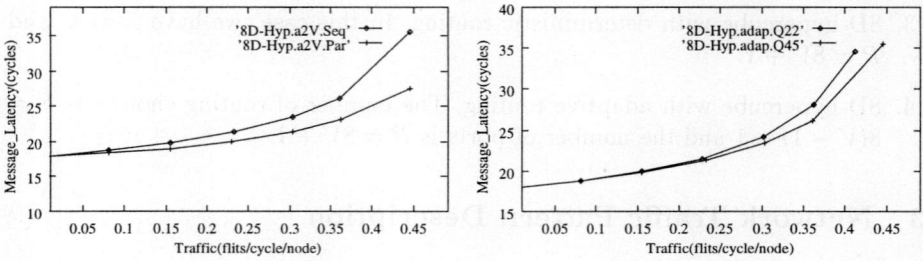

Fig. 3. Single vs multiple routing control units, and effect of queue size in 8D-hypercubes using an adaptive routing algorithm

Figure 3a shows the difference between using single and multiple routing control units in an 8D-hypercube with adaptive routing. As can be seen, latency is almost identical when traffic load is low. When traffic increases, the difference between both approaches also increases, achieving an improvement of up to 25% in average latency. In general, the latency improvement is not worth the additional cost of replicating the routing control units. Similar results are obtained for 2D-mesh topologies. So, in what follows, a single routing control unit will be used. Figure 3b shows the effect of using flit queues of different capacities. This effect is very small. So, in what follows, we will use a 4+5 queue size per channel.

4.2 Considering router delays and implementation constraints

The most important implementation constraints proposed in the literature are pin count and network bisection width [1]. For small and medium size networks, channel width is only limited by pin count, existing proposals to increase the utilization of the available bisection bandwidth without increasing pin count [5].

In order to make a fair comparison between the 2D-mesh and the hypercube, both topologies should have the same pin count. So, physical channels are twice as wide in the 2D-mesh than in the 8D-hypercube. Thus, the 8D-hypercube will spend 2 clock cycles to transfer one flit across a short physical channel, while the 2D-mesh will only require one cycle. Also, depending on parameters like number of virtual channels, routing algorithm, network dimensions and physical layout, we can compute the delays for all the router components. Using the expressions obtained in the router model section, we present in table 1 the delays and clock periods for 2D-mesh and 8D-hypercube topologies.

2D-mesh	T_r	T_s	T_{c1}	T_{clk}
Det-1V	4.7	4.79	5.14	5.14
Det-2V	5.9	5.3	5.74	5.9
Adap-2V	6.6	5.3	5.74	6.6
Adap-3V	7.49	5.62	6.09	7.49

8D-hyp	T_r	T_s	T_{c1}	T_{c2}	T_{c4}	T_{clk}
Det-1V	4.7	5.3	5.14	6.64	9.64	5.3
Det-2V	5.9	5.85	5.74	7.24	10.24	5.9
Adap-2V	8.5	5.85	5.74	7.24	10.24	5.85
Adap-3V	9.6	6.2	6.09	7.59	10.59	6.2

Table 1. Delays for routing control unit (T_r), switch (T_s) and channels of different lengths (T_{ci}), and router clock period (T_{clk}) in 8D-hypercube and 2D-mesh topologies.

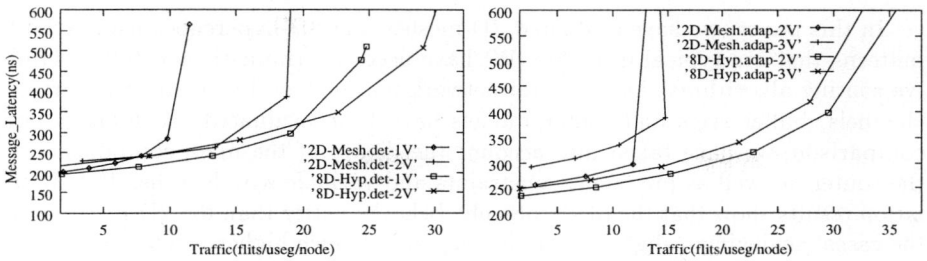

Fig. 4. Performance of deterministic and adaptive routing algorithms for 2D-mesh and 8D-hypercube when constant pin count and variable channel length are considered

In the 2D-mesh, the delays for the routing control unit, switch and channels are very similar. So, the clock period is determined by the slowest component.

However, in 8D-hypercubes the delay for long channels can be up to twice the switch delay. Delays for the switch and routing control unit are very similar when deterministic routing is used. However, these delays considerably differ for adaptive routing. So, we have chosen clock period to be equal to $\max(T_r, T_s)$ for deterministic routing, and equal to T_s for adaptive routing. As a consequence, the routing operation will require two clock cycles when adaptive routing is used. Taking into account that transferring one flit across a short physical channel requires 2 clock cycles (due to pin count constraints), transferring one flit across the rest of the channels can be done in four clock cycles.

Figure 4a shows simulation results for deterministic routing algorithms with one and two virtual channels, assuming the timing considerations mentioned above. Both topologies achieve similar average message latency when network load is low, being slightly better for the hypercube with one virtual channel. However, the hypercube increases throughput with respect to the 2D-mesh by a factor of 2 and 1.5 for one and two virtual channels, respectively.

Figure 4b shows simulation results for adaptive routing algorithms. As can be seen, the hypercube offers better performance than 2D-meshes, achieving a 25% reduction in message latency for three virtual channels, and a 10% reduction for two virtual channels. Moreover, in the latter case, the hypercube improves throughput by a factor of 2.7 over the 2D-mesh. On the other hand, performance does not improve when the number of virtual channels is increased from two to three. This result was already presented in [8].

5 Conclusions

Previous research has shown that low-dimensional topologies achieve higher performance than high-dimensional topologies. While this is true for multicomputers, previous research efforts did not consider the particular characteristics of network traffic in DSMs. When cache coherence protocols are implemented in hardware, messages are very short. In this case, latency heavily depends on the distance between nodes, therefore favoring the use of high-dimensional topologies.

In this paper, we have evaluated 2D-meshes and 8D-hypercubes using traffic patterns that are typical in DSMs. We have used deterministic and fully adaptive routing algorithms. Also, several network parameters, like number of virtual channels, buffer sizes and router models have been evaluated. To make a fair comparison, we have taken into account the delays of the main components of the router, as well as pin count constraints and variable wire lengths. The simulation results show that the 8D-hypercube behaves better than the 2D-mesh in all the cases, achieving a slightly lower latency and a much higher throughput. Also, the maximum performance is obtained for both topologies when the minimum number of virtual channels is used.

In summary, hypercubes perform better than 2D-meshes when traffic consists of very short messages, and router designs are optimized for each topology. Note however that we did not analyze very large networks. Bisection width may be the main constraint for channel width in networks with thousands of processors. So, results may change for large networks. Anyway, current DSMs have a few hundreds of nodes at most.

References

1. A. Agarwal, "Limits on interconnection network performance", *IEEE Trans. Parallel Distributed Syst.*, vol. 2, no. 4, pp. 398–412, Oct. 1991.
2. R.V. Boppana, and S. Chalasani, "A comparison of adaptive wormhole routing algorithms," in *Proc. 20th Annu. Int. Symp. Comput. Architecture*, May 1993.
3. A.A. Chien, "A cost and speed model for k-ary n-cube wormhole routers," in *Proc. Hot Interconnects'93*, Aug. 1993.
4. W.J. Dally and C.L. Seitz, "Deadlock-free message routing in multiprocessor interconnection networks," *IEEE Trans. Comput.*, vol. 36, no. 5, pp. 547–553, May 1987.
5. W.J. Dally, "Express cubes: Improving the performance of k-ary n-cube interconnection networks," *IEEE Trans. Comput.*, vol. 40, no. 9, pp. 1016–1023, Sept. 1991.
6. W.J. Dally, "Virtual-channel flow control," *IEEE Trans. Parallel Distributed Syst.*, vol. 3, no. 2, pp. 194–205, Mar. 1992.
7. J. Duato, "A new theory of deadlock-free adaptive routing in wormhole networks," *IEEE Trans. Parallel Distributed Syst.*, vol. 4, no. 12, pp. 1320–1331, Dec. 1993.
8. J. Duato and P. López, "Performance evaluation of adaptive routing algorithms for k-ary n-cubes," in *Proc. Workshop on Parallel Computer Routing and Communication*, May 1994.
9. Intel Scalable Systems Division, "*Intel Paragon Systems Manual*," Intel Corporation.
10. R.E. Kessler and J.L. Schwarzmeier, "CRAY T3D: A new dimension for Cray Research," in *Compcon*, pp. 176–182, Spring 1993.
11. J. Kuskin et al., "The Stanford FLASH multiprocessor," in *Proc. 21st Annu. Int. Symp. Comput. Architecture*, April 1994.
12. D. Lenoski, J. Laudon, K. Gharachorloo, W. Weber, A. Gupta, J. Hennessy, M. Horowitz and M. Lam, "The Stanford DASH multiprocessor," *IEEE Computer Magazine*, vol. 25, no. 3, pp. 63–79, March 1992.

A Pattern-Associative Router for Interconnection Network Adaptive Algorithms

Daniel G. Rice[1,2]   Jose G. Delgado-Frias[1]   and   Douglas H. Summerville[1]

[1]Electrical Eng., SUNY, Binghamton, NY 13902   [2]Lockheed Martin, Owego, NY 13827

Abstract. In this paper, we propose a high-performance router approach to handle adaptive, deadlock-free, wormhole routing algorithms in a number of interconnection network topologies. The router allows programmable support for a wide range of networks and routing algorithms. Routing algorithms are mapped to a set of bit-patterns which are matched in parallel. The number of bit patterns required depends on the network topology and adaptive routing algorithm; in general this number is of O(network degree). To show the applicability of this adaptive router, we have studied a number of adaptive routing algorithms.

1 Introduction

A number of reconfigurable routers have been proposed to provide the flexibility to support various network topologies and routing algorithms. Two main approaches have been used: dedicated processors and look-up tables. The dedicated processor approach allows the router to perform with great flexibility the computations necessary to support multiple configurations of the network [7]. A drawback of this type of router is that the routing algorithm program must be executed in a sequential fashion. The look-up table approach requires that each routing alternative for any given destination address from a given node be stored in memory. The look-up table approach requires only one memory access delay to determine the output channel. However, large tables are required to store the routing information. Interval routing, a derivative of look-up table routing, assigns output ports to a range of destination node addresses, thus reducing the amount of memory required to store the look-up table [8]. This type of router might not be well suited to adaptive routing algorithms because the look-up table provides only one possible solution to the routing problem for a given set of inputs.

In this paper, we propose a new approach to adaptive routing algorithm execution. We will use wormhole routing algorithms in this paper as a means to introduce the approach; however the approach is applicable to any flow control technique. Our proposed router scheme provides the flexibility to support oblivious and adaptive routing algorithms for a variety of interconnection network topologies. The proposed router utilizes a pattern matching array that contains a relatively small number of bit patterns that are used to compare in parallel a number of potential routing alternatives. The adaptive router is composed of pattern match units to support routing, selection, and channel dependency enforcement functions. This approach allows the same scheme to be utilized for both oblivious and adaptive algorithms on a large number of interconnection networks.

2 Pattern-Associative Adaptive Router

An adaptive router approach uses two main functions, namely a routing function (R) and a selection function (ρ). The routing function defines a set of channels (physical/virtual channel pairs) that may be used for the current message. The selection function then determines which of these channels to use based on network status information (α_{pv}). This adaptive router approach has been used by Dally and Aoki [1] and Duato [2]. Our proposed adaptive router approach separates the selection of physical channel from the selection of virtual channel, according to Theorem 1, below. This separation allows simplification of the routing and selection functions without sacrificing an algorithm's deadlock-freedom. A proof for this theorem is presented in [10].

Theorem 1: A connected and adaptive routing function R for an interconnection network I may be divided into selection of physical channel followed by enforcement of virtual channel dependencies without sacrificing deadlock-freedom.

The adaptive router proposed by Theorem 1 is shown in Figure 1. The three main functions are routing (R), selection (ρ), and dependency enforcement (E). The routing function (R) determines a set of alternative physical channels which may be used by the message to reach its destination. The selection function (ρ) then selects one of these physical channels from the alternatives presented. The dependency enforcement function (E) ensures that the proper virtual channel is used to meet the requirements of the channel dependency graph for the routing algorithm. The proposed adaptive router approach divides the adaptive routing problem into smaller subtasks. First, a set of valid alternative physical channels are determined. Second, the best physical channel is chosen from this set of alternatives based on information about these channels. Finally, a valid virtual channel is determined for the physical channel selected, and the message is routed on the resulting physical channel/virtual channel pair. Adaptive routing algorithms that can be partitioned into physical and virtual channels should not be affected by the proposed adaptive router scheme.

The proposed adaptive router scheme offers four major advantages over the basic router. These advantages are: *small subtasks*, the approach divides the routing function into smaller subtasks, which can be optimized to reduce the number of bit-pattern entries; *small set of alternative physical paths*, the routing function determines alternative physical channels from the domain of physical channels available; *simpler selection function*, the selection function is simplified by reducing the domain of possible inputs; and *simplified dependency enforcement*, the dependency enforcement function can be optimized for the particular routing algorithm to be implemented.

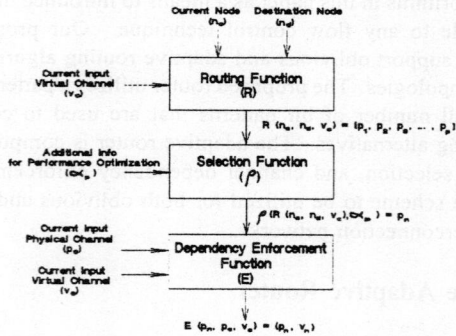

Figure 1. Proposed Adaptive Router Scheme

The bit-pattern associative router relies on network topological information inherent in the network addressing scheme. The router uses sets of bits from the destination node address to determine an output port assignment for the current message. For adaptive algorithms, the router allows multiple paths to be considered, and selects the "best" path based on network status information available to the router. The router's basic constructs allow implementation of a large number of oblivious and adaptive routing algorithms for a large variety of interconnection networks. A block diagram representation of the bit pattern matching unit scheme is shown in Figure 2(a). The bit pattern match section utilizes a matching mechanism which allows multiple bit-pattern alternatives to be considered in parallel and each viable alternative is presented to the priority selection section. [6] The matching unit allows "don't care" bits to used as part of the patterns to be tested, to reduce the number of patterns required in the array. This technique allows only those bits which are important to the routing decision to be considered for each routing alternative. The priority selection section selects a single alternative from the bit pattern match "hits" based on a fixed priority scheme. The port assignment memory contains physical or virtual port assignments for each bit pattern match.

The three main functions (routing, selection and dependency enforcement) of this approach are implemented using two pattern matching units as shown in Figure 2(b). The first pattern matching unit performs the routing and selection functions, and the second matching unit performs the dependency enforcement function. Each pattern matching unit consists of a bit pattern match section, a priority selector section, and a port assignment memory. The routing function uses the destination address from the message header, and the "free/busy" bits for the corresponding physical output ports. This approach is well-suited to use a single "free/busy" bit for each output port, although more complex channel status reporting mechanisms are possible. The selection function chooses the most attractive output port from the options available. For general adaptive algorithm implementations, a miscellaneous output port assignment field is provided in addition to the physical channel output port field. This miscellaneous field allows simpler implementation of some adaptive routing algorithms. Both the selected physical channel port and the miscellaneous field data are passed to the dependency enforcement matching unit (virtual channel selection). The dependency enforcement function also uses a pattern matching unit. The bit pattern fields that are used to determine which virtual channel will be utilized are the selected physical channel, the virtual channel status for the selected physical channel, and in some instances, the

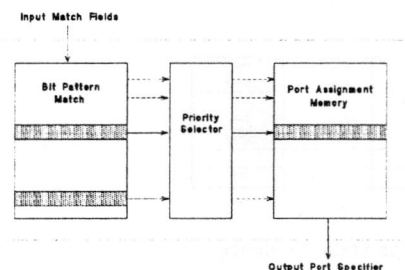

Figure 2(a). Bit Pattern Matching Unit Block Diagram

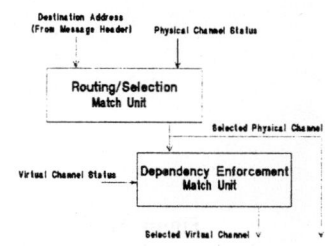

Figure 2(b). Proposed Adaptive Router Approach

miscellaneous field from the router/physical channel selection function, or the current input physical and virtual channels for the message. The output port memory for the dependency enforcement matching unit contains virtual channel alternatives for the selected physical channel.

3 Mapping of Adaptive Routing Algorithms to Proposed Scheme

We have chosen an example algorithm to demonstrate the bit-pattern associative router concept. Our example is a deadlock-free static dimension-reversal algorithm for mesh-connected k-ary n-cube networks [1] applied to the hypercube. This adaptive algorithm utilizes a count of the number of times a packet has been routed from a channel in one dimension to channel in a lower dimension (the dimension-reversal or DR number) to select the virtual channel to be used for routing to avoid deadlock. Each physical channel is divided into non-empty classes of virtual channels numbered zero to r, where r is the maximum number of dimension reversals permitted. Packets with DR $< r$ may be routed in any direction, but must use virtual channels of class DR. Once a packet's DR $= r$, it must be routed in ascending dimension order towards its destination, and may not re-enter the adaptive channels. The bit patterns for this algorithm are shown in Figure 3.

The selection function for the algorithm chooses an unoccupied, productive channel if possible. Only when no adaptive channels are available (ie. packet DR $= r$) does the packet resort to deterministic dimension-order routing. The selection function may utilize a variety of schemes to select the output channel from those available.

Destination Address				DR	Output Channel Status [Free / Busy]				
					l_{-1}	l_{-2}	....	l_i	l_j
a_{-1}	a_{-2}	a_i	a_j	x	x	x	....	x	x	This Node
x	x	x	$\bar{a}_j$	r	x	x	....	x	x	Link 0
x	x	$\bar{a}_i$	x	r	x	x	....	x	x	Link 1
....					....					....
$\bar{a}_{-1}$	x	x	x	r	x	x	....	x	x	Link n-1
x	x	x	a_j	x	x	x	....	x	1	Link 0
....					....					....
$\bar{a}_{-1}$	x	x	x	x	1	x	....	x	x	Link n-1
x	x	x	$\bar{a}_j$	x	x	x	....	x	x	Link 0
....					....					....
$\bar{a}_{-1}$	x	x	x	x	x	x	....	x	x	Link n-1
x	x	x	x	x	x	x	....	x	x	Default

X = Don't Care P_{out}

Virtual Channel Status		P_o		
V_c				
x	x	Link 0	Link 0	DR
x	x	x	Link 0	DR+1
x	x	Link 0	Link 1	DR
x	x	Link 1	Link 1	DR
x	x	x	Link 1	DR+1
....				
x	x	x	Link n-1	DR

V_{out}

Figure 3. Bit Pattern Mapping for Static Dimension Reversal Algorithm

4 Concluding Remarks

The proposed scheme to adaptive routing presented in this paper is applicable to a wide range of interconnection networks and adaptive routing algorithms. The mapping of some of these algorithms have been provided to demonstrate the applicability of this approach. Additional adaptive routing algorithms including the Fully Adaptive Algorithm [2,5], Planar Adaptive Routing (PAR) [3], Linder and Harden's virtual networks concepts [4] and turn-model based algorithms [9] have been mapped to this approach. The adaptive router uses a novel bit-pattern match unit to create a flexible router for a variety of network implementations and routing algorithms. The performance of the pattern-associative adaptive router is based on the performance of each of the two stages of the router. If no stage pipelining is used, a total of two comparison delays and two read delays are needed for the full adaptive router. Compared to other flexible adaptive router schemes, the proposed bit-pattern adaptive router is a high-performance approach. The adaptive router can be utilized with no changes to hardware or bit patterns as the network expands, so long as sufficient address bits are included in the initial implementation to support the additional node addresses.

References

[1] W. Dally and H. Aoki, "Deadlock-free adaptive routing in multicomputer networks using virtual channels," *IEEE Trans. Parallel Dist. Syst.*, vol. 4, no. 4, Apr. 1993, pp. 466- 475.
[2] J. Duato, "A new theory of deadlock-free adaptive routing in wormhole networks," *IEEE Trans. Parallel Dist. Syst.*, vol. 4, no. 12, Dec. 1993, pp. 1320-1331.
[3] J. Kim and A. Chien, "An evaluation of planar adaptive routing (PAR)", *Proc. 4th IEEE Symp. Parallel Distrib. Processing*, Dec. 1992, pp. 470-478.
[4] D. Linder and J. Harden, "An adaptive and fault tolerant wormhole routing strategy for k-ary n-cubes," *IEEE Trans. Comput.*, vol. 40, no. 1, Jan. 1991, pp. 2-12.
[5] J. Duato, "Deadlock-free adaptive routing algorithms for multicomputers: Evaluation of a new algorithm," *Proc. 3rd IEEE Symp. Parallel Distrib. Processing*, 1991, pp. 840-847.
[6] D. Summerville, J. Delgado-Frias and S. Vassiliadis, "A flexible bit-associative router for interconnection networks," *IEEE Trans. Parallel Dist. Syst.*, vol. 6, no. 5, May 1996.
[7] J. Park, S. Vassiliadis, and J.G. Delgado-Frias, "Flexible oblivious router architecture," *IBM Journal of Research and Development*, vol. 39, no. 3, pp. 315-329, May 1995.
[8] J. Van Leeuwen and R.B. Tan, "Interval routing," *The Computer Journal*, vol. 30, no. 4, pp. 298-307, 1987.
[9] C.J. Glass and L.M. Ni, "The turn model for adaptive routing," *Journal of the ACM*, vol. 41, no. 5, pp. 874-902, Sept. 1994.
[10] D.G. Rice, J.G. Delgado-Frias and D.H. Summerville, "A Pattern-Associate Router for Adaptive Algorithms in Hypercube Networks," *IASTED Int. Conf. Parallel and Dist. Systems*, Washington, DC, October, 1995.

On Stack-Graph OPS-Based Lightwave Networks*

H. Bourdin**, A. Ferreira*** and K. Marcus[†]

Abstract. Most of the proposed architectures for interconnecting nodes in processor networks are based on graph topologies. Recently, stack-graph topologies based on the hypergraph theory have emerged, taking advantage of the fact that the huge bandwidth of optical fiber can be divided into several high-speed channels in networks using Optical Passive Star (OPS) couplers. We show through simulation that these stack-graph networks compare very well against graph-based ones, in terms of stochastic behavior for routing.

1 Introduction

Optical technologies such as tunable optical transmitters and receivers, and wavelength division multiplexing (WDM) allow the construction of very efficient local and metropolitan area networks (LAN and MAN, respectively), and could be used as interconnection networks in parallel computers. Using Passive Star (OPS) couplers, one can build *single-hop* systems, where every processor is able to directly communicate with one another. Clearly, however, this could represent a severe drawback when building very large networks.

Alternatively, the same kind of couplers can be used in the construction of *multihop* networks, where a node is assigned to a small and static set of predefined channels. Pairwise communication may then need to hop through intermediate nodes [7]. Thus, in multihop systems, communications take longer, but nodes are simpler, cheaper, and more reliable than in single-hop systems.

Several topologies were proposed as *point-to-point* logical architectures for WDM networks, based on graphs [6, 8, 9]. Unfortunately, these topologies either use too many channels, have too many transceivers, or suffer of a large diameter. Furthermore, an intrinsic feature of optical communications is that the channels induced by WDM can span a large number of nodes in the network. Hence, point-to-point logical topologies do not efficiently use optical technology and new avenues have to be explored.

On the other hand, *one-to-many* topologies are best represented by hypergraphs [1], that can be seen as a generalization of graphs in which edges are replaced by hyperedges [10] joining sets of nodes, instead of only two nodes. As

---

* This work has been partially supported by the PRC PRS of the French CNRS, by the European HCM project MAP, and by the Brazilian CAPES.
** LIP ENS Lyon, 69364 Lyon Cédex 07, France.
*** CNRS - LIP ENS Lyon, 69364 Lyon Cédex 07, France.
[†] CSI, University of Ottawa, P.O. Box 450 Stn. A, K1N 6N5, Ottawa, Canada.

for graphs, a hyperedge represents a communication means; a message sent on a hyperedge can be read by all nodes in that hyperedge.

Our work is focused on new one-to-many architectures for multihop systems, whose main performance characteristics are their regularity and modularity [2, 4]. Positive points are that they can be incrementally expanded and the number of links between two processors is fixed, thus allowing the transmitters and receivers to be tuned to some pre-determined frequencies. The high fault tolerance ensures a reliable communication and network performance. Finally, only a small number of hops is required in any pairwise communication, and the network allows easy schemes for global routing operations [2, 4].

In this paper we present a comparative study of three hypergraph–based networks, namely the *stack-ring*, the *stack-torus*, and the *hypertorus* [2], and two very well known graph–based topologies, namely the hypercube and the two–dimensional torus [3]. In Section 3 we discuss their routing–related stochastic characteristics that were obtained through simulation.

2 Emerging networks

Many interconnection topologies have been proposed for the design of LAN's and MAN's using multihop lightwave techniques as the Manhattan Street Network[6], de Bruijn graphs based network [9] and the Supercube [9]. These networks are based on graph models, and the possibility of a node to communicate with another one is represented by an edge joining the two nodes. Thus, in any communication step, only pairs of nodes are involved. In [2, 4], it was proposed to grow the number of nodes involved in each step, profiting mostly from WDM that allows the huge bandwidth of optical fiber to be divided into a set of high-speed logical channels. Such channels can be seen as logical attributes shared among many processors. The network topologies are based on hypergraph models, and thus called hypertopologies.

Stack-rings and stack-tori can be seen as a generalization of rings and tori from graphs to hypergraphs using the concept of *stack-graphs*. Briefly, stack-graphs are obtained by piling up copies of the original graph and subsequently replacing the edges by hyperedges. Hypertori, on the other hand, are defined as a Cartesian product of stack-rings. Because of severe space limitations, we refer the interested reader to the definitions appearing in [2, 5].

3 Simulation & Results

Sen and Maitra presented in [8] a simulator used to evaluate the dynamic quality of point-to-point lightwave networks. Using the same framework, we implemented a similar simulator for the hypergraph–based lightwave networks studied in this paper. There are two control parameters. The *offered load* (exponentially distributed), with channel speed equal to 100 M-bits/s, and the *channel speed* with offered load equal to 100 packets/s.

We simulated, whenever possible, (hyper) networks with 120 nodes such that all hypernetworks have diameter 3. Table 1 shows the simulated (hyper) networks, giving the chosen configuration with number of channels (hyperedges), number of fixed transceivers per node (degree), and number of processors per channel (rank).

topology	configuration	nodes	hyperedges	degree	rank
Hyperring	$\mathcal{R}_{6,20}$		6	2	40
Hypertorus	$\mathcal{R}_{5,2} \times \mathcal{R}_{3,4}$		90	4	8
Stack-torus	$\varsigma(T(5,3),8)$	120	30	4	16
Torus	$T(10,12)$		240	4	2
Hypercube	$H(7)$	128	448	7	2

Table 1. Hypertopologies with 120 nodes and a diameter of 3.

Figure 1 shows that, with respect to the packet delivery time, the hypertopologies have a better performance than the graph–based ones, although the asymptotic behavior is quite the same for all networks. Concerning the average wait time / delivery time shown in Figure 1, we recall that both parameters depend on the number of channels. Therefore, the stack-ring behaves poorly in comparison to the others, that have all a similar behavior.

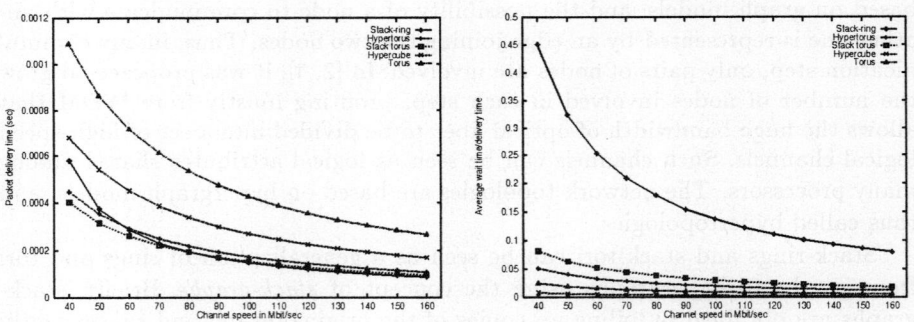

Fig. 1. Packet delivery time and wait ratio versus channel speed.

In Figure 2 one can see that the small number of channels of the stacking-ring troubles its performance. With respect to the packet delivery time, the hypertopologies have the same behavior, better than both the torus and the hypercube.

From our experiments we can conclude that, as expected, because of their good (hyper)graph-theoretic properties discussed in [4, 5], the hypernetworks outperform graph–based networks in almost all aspects. However, an adequate balance among channels, degree, and diameter should be obtained for stack-graph-based LAN's, MAN's, and optical interconnection networks for parallel computers.

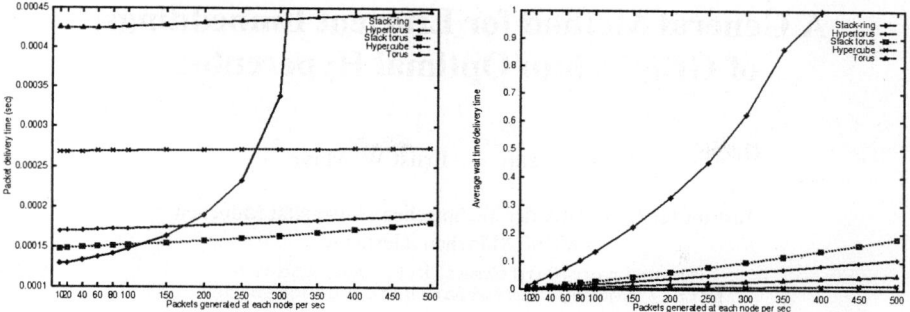

Fig. 2. Packet delivery time and wait / delivery time versus load.

References

1. C. Berge. *Hypergraphs*. North Holland, 1989.
2. H. Bourdin, A. Ferreira, and K. Marcus. A comparative study of one-to-many WDM lightwave interconnection networks for multiprocessors. In E. Schenfeld, editor, *Proceedings of the 2nd IEEE International Workshop on Massively Parallel Processing using Optical Interconnections*, San Antonio (USA), October 1995. IEEE Press.
3. A. Ferreira. *Handbook of Parallel and Distributed Computing*, chapter Hypercubes. McGraw-Hill, New York (USA), 1995.
4. A. Ferreira and K. Marcus. Modular multihop WDM–based lightwave networks, and routing. In S. I Najafi and H. Porte, editors, *Proceedings of The European Symposium on Advanced Networks and Services, Conference on Receivers, transmitters, and WDMs for fibre optic networks*, volume 2449 of *Proc. SPIE*, pages 78–86, Amsterdam, March 1995. SPIE – The International Society for Optical Engineering.
5. A. Ferreira and K. Marcus. A theoretical framework for the design of lightwave networks. In J. Chrostowski, editor, *Proceedings of The 10th Annual International Symposium on High Performance Computers*, Ottawa, June 1996. IEEE Press.
6. N.F. Maxemchuk. Regular mesh topologies in local and metropolitan area networks. *AT&T Tech. J*, 64(7):1659–1685, 1985.
7. B. Mukherjee. WDM-Based Local Lightwave Networks Part II: Multi-hop Systems. *IEEE Networks*, pages 20–32, jul 1992.
8. A. Sen and P. Maitra. A comparative study of Shuffle-Exhange, Manhattan Street and Supercube network for lightwave applications. *Computer Networks and ISDN Systems*, 26:1007–1022, 1994.
9. K.N. Sivarajan and R. Ramaswami. Lightwave Networks Based on de Bruijn Graphs. *IEEE/ACM Transactions on Networking*, 2(1):70–79, apr 1994.
10. T. Szymanski. Hypermeshes: Optical Interconnection Networks for Parallel Computing. *Journal of Parallel and Distributed Computing*, 26:1–23, 1995.

A General Method for Efficient Embeddings of Graphs into Optimal Hypercubes

Volker Heun   Ernst W. Mayr

Institut für Informatik der Technischen Universität München
D-80290 München, Germany
{heun|mayr}@informatik.tu-muenchen.de
http://wwwmayr.informatik.tu-muenchen.de/

Abstract. Embeddings of several graph classes into hypercubes have been widely studied. Unfortunately, almost all investigated graph classes are regular graphs such as meshes, complete trees, pyramids. In this paper, we present a general method for one-to-one embedding irregular graphs into their optimal hypercubes based on extended-edge-bisectors of graphs. An extended-edge-bisector is an edge-bisector with the additional property that a subset of the vertices is distributed more or less evenly among the two halves of the bisected graph. The dilation and congestion of the embedding depends on the quality of the extended-edge-bisector. Moreover, if the extended bisection can be efficiently computed on the hypercube, so can the embedding.

1 Introduction

Hypercubes are a very popular model for parallel computation because of their regularity and their relatively small number of interprocessor connections. Another important property of an interconnection network is its ability to simulate efficiently the communication of parallel algorithms. Thus, it is desirable to find suitable embeddings of graphs representing the communication structure of parallel algorithms into hypercubes representing the interconnection network of a parallel computer.

Embeddings of graphs with a regular structure, like rings, (multidimensional) grids, complete trees, binomial trees, pyramids, X-trees, meshes of trees and so on, have been investigated by many researcher, see, e.g., [4, 5, 6, 7, 8, 11, 15, 20, 21, 22, 25]. Unfortunately, the communication structure of a parallel algorithm can often be very irregular.

For graphs whose structure is less regular, it is in general hard to decide whether there is a good embedding into a given host graph. Here, an embedding is considered good if it has small dilation, load, and expansion. Given a graph G and a positive integer d, it is $\mathcal{NP}$-complete to decide whether G is a subgraph of the d-dimensional hypercube, even if G is a tree [23]. Given a graph G and a positive integer d, it is also $\mathcal{NP}$-complete to decide whether G has a dilation 2 embedding into the d-dimensional hypercube [24].

For arbitrary binary trees, one-to-one embeddings into their optimal hypercubes with constant dilation and constant node-congestion have been constructed in [3]. Another one-to-one embedding of an arbitrary binary tree into its optimal hypercube with constant dilation is given in [18], but in this paper neither node-congestion nor edge-congestion is considered. The embedding given in [13] yields dilation 8 and constant node-congestion. This is the best known bound on the dilation. Furthermore, this embedding can be efficiently computed on the hypercube itself. In [10], Havel has conjectured that every binary tree has a one-to-one embedding with dilation 2 into its optimal hypercube. This conjecture is still open. A simple parity argument shows that the complete binary tree is not a subgraph of its optimal hypercube [4, 20, 25]. In [12], it was shown that balanced caterpillars with legs of unit length are subgraphs of their optimal hypercubes. This was generalized in [2]; it was shown that balanced caterpillars with the additional property that the lengths of the legs have the same parity are subgraphs of their optimal hypercubes. Embeddings of graphs with bounded treewidth into optimal hypercubes have been studied in [14]. A one-to-one embedding with dilation $3\lceil \log((d+1)(t+1)) \rceil + 8$ was obtained which can be efficiently constructed on the optimal hypercube itself.

In this paper, a general technique for embedding irregular graphs based on extended-edge-bisectors is presented. An extended-edge-bisector is an edge-bisector with the additional property that a subset of the vertices is distributed more or less evenly among the two halves of the bisected graph. Depending on the size of the extended-edge-bisector, roughly the number of vertices incident to the edges cut by the bisector, the quality of distributing marked vertices, and the maximal degree of the given graph, we compute bounds on the dilation and node-congestion for our embedding method. Moreover, if the extended bisection can be efficiently computed on the hypercube, so can the embedding.

The remainder of this paper is organized as follows. First, we recall some basic definitions and notations which we will use later. In the third section, we present the main tool for our embedding, called the (k, h, o, λ)-tree, and we discuss the quality of the embedding achieved. We define extended-edge-bisectors in the fourth section and describe how to efficiently obtain extended-edge-separators from extended-edge-bisectors. We also determine the quality of the embedding in terms of the quality of the extended-edge-bisector for a given family of graphs. In the fifth section, we present some applications of our method. Finally, we give some concluding remarks.

2 Preliminaries

An *embedding* of a graph $G=(V_G, E_G)$, called *guest graph*, into a graph $H=(V_H, E_H)$, called *host graph*, is a mapping $\phi:G\to H$ consisting of two mappings $\phi_V:V_G\to V_H$ and $\phi_E:E_G\to \mathcal{P}(H)$. Here, $\mathcal{P}(G)$ denotes the set of paths in the graph $G=(V, E)$. The mapping ϕ_E maps each edge $\{v, w\}\in E_G$ to a path $p\in\mathcal{P}(H)$ connecting $\phi_V(v)$ and $\phi_V(w)$. We call an embedding *one-to-one* if the mapping ϕ_V is 1-1.

The *dilation* of an edge $e\in E_G$ under an embedding ϕ is the length of the path $\phi_E(e)$. Here, the length of a path p is the number of its edges. The *dilation of an embedding* ϕ is the maximal dilation of an edge in G. The number of vertices of a guest graph which are mapped onto a vertex v in the host graph, is called the *load* of the vertex v. The *load of an embedding* ϕ is the maximal load of a vertex in the host graph. In this paper, unless noted otherwise, we only consider embeddings with load one. The ratio $|V_H|/|V_G|$ is called the *expansion* of the embedding ϕ. The *congestion* of an edge $e'\in E_H$ is the number of paths in $\{\phi_E(e) \mid e\in E_G\}$ that contain e'. The *edge-congestion* is the maximal congestion over all edges in H. The *congestion* of a vertex $v\in V_H$ is the number of paths in $\{\phi_E(e) \mid e\in E_G\}$ containing v. Again, the *node-congestion* is the maximal congestion over all vertices in H. In the following, we initially restrict our attention to finding a suitable mapping ϕ_V, and we will use shortest paths in the hypercube for the mapping ϕ_E. Nevertheless, it is still important to decide which paths we choose, since we are interested in obtaining an embedding with small node-congestion.

A *hypercube of dimension* d is a graph with 2^d vertices, labeled 1-1 with the strings in $\{0,1\}^d$. Two vertices are connected iff their labels differ in exactly one position. The smallest hypercube into which we can embed a given graph $G=(V,E)$ with load one is called its *optimal* hypercube. Thus, its dimension is $\lceil \log(|V|) \rceil$. Hence, an embedding of a graph G into its optimal hypercube has expansion less than 2.

The *level* of a vertex v in a tree is the number of vertices on the path from the root to v. Hence, the level of the root is 1. The *height* of a tree T is the maximum level of a vertex in T. A *complete* d-ary tree T of height h is a tree such that each internal vertex has exactly d children, and such that all leaves have the same level. Given a vertex v in a tree, a vertex w is called a descendant of v if it lies on a path from v to a leaf of the tree. A subtree rooted at a vertex v is the induced subgraph of all descendants of v in the tree. We call a vertex of a tree an *internal* vertex if it is not a leaf of the tree.

3 The (k, h, o, λ)-Tree

3.1 Definition of a (k, h, o, λ)-Tree

To construct our embedding, we use the data structure of a (k, h, o, λ)-tree. The (k, h, o, λ)-*tree* is a complete 2^k-ary tree of height h with integer node weights, also called the *capacities* of the nodes. The capacity of a node depends on the level of the node and additional integer parameters $o \geq 0$ and $\lambda \in [0:k-1]$. We will call o the order and λ the type of a (k, h, o, λ)-tree. The capacity of a node at level ℓ of a (k, h, o, λ)-tree is defined as follows ($\delta_{i,j}$ is the Kronecker-delta):

$$c(\ell) = 2^o \left[\sum_{i=0}^{k-1} 2^i \binom{k(h-l+1)-i-1}{\lambda} + \delta_{\ell,1} \sum_{i=0}^{\lambda} \binom{kh}{i} \right],$$

In the sequel, we call vertices of a (k, h, o, λ)-tree *nodes*, and we denote the capacity of a node at level ℓ by $c(\ell)$.

Lemma 1. *The total capacity of a (k, h, o, λ)-tree is 2^{kh+o}.*

Proof. By definition of the capacities we get:

$$\sum_{\ell=1}^{h} 2^{k(\ell-1)} \cdot c(\ell) = \sum_{\ell=1}^{h} 2^{k(\ell-1)} \cdot 2^o \sum_{i=0}^{k-1} 2^i \binom{k(h-\ell+1)-i-1}{\lambda} + 2^o \sum_{i=0}^{\lambda} \binom{kh}{i}$$

$$= 2^o \left[\sum_{\ell=1}^{h} \sum_{i=0}^{k-1} 2^{kh-1-(k\ell-i-1)} \binom{k\ell-i-1}{\lambda} + \sum_{i=0}^{\lambda} \binom{kh}{i} \right]$$

$$= 2^o \left[\sum_{i=0}^{kh-1} 2^{kh-1-i} \binom{i}{\lambda} + \sum_{i=0}^{\lambda} \binom{kh}{i} \right]$$

using $\sum_{i=\lambda}^{n-1} 2^{n-1-i} \binom{i}{\lambda} = \sum_{i=\lambda+1}^{n} \binom{n}{i}$

$$= 2^o \left[\sum_{i=\lambda+1}^{kh} \binom{kh}{i} + \sum_{i=0}^{\lambda} \binom{kh}{i} \right] = 2^{kh+o} \qquad \square$$

We state without proof the following inequalities for $c(\ell)$ which we will use later.

Lemma 2. *For $\lambda \leq \frac{k}{2}$ the following inequalities are valid:*

i) $\frac{1}{2} \cdot 2^{k+o} \left(\frac{k(h-\ell)}{\lambda}\right)^\lambda \leq \frac{3}{4} c(\ell)$, ii) $c(\ell-1) \leq 3^{\lambda+1} c(\ell)$, iii) $2^o(2^k - 1) \leq c(\ell)$.

For $\lambda=0$ the term in parentheses on the left hand side of the first inequality is set to be 1.

3.2 Embedding the (k, h, o, λ)-Tree into the Hypercube

We now describe a mapping of a (k, h, o, λ)-tree into its optimal hypercube such that each node of the (k, h, o, λ)-tree occupies as many vertices of the hypercube as given by its capacity. Each node in a (k, h, o, λ)-tree can be represented by a string in $(\{0,1\}^k)^*$ as follows. The empty string ϵ represents the root of the (k, h, o, λ)-tree. If α represents a node v of the (k, h, o, λ)-tree, then the strings $\alpha\beta$ for $\beta \in \{0,1\}^k$ represent the 2^k children of v from left to right. Note that the string representing a node at level ℓ has length $k(\ell - 1)$. For a string $v \in \{0,1\}^*$, we denote by $|v|_1$ the number of 1's in v and by $|v|$ the length of v. We define the following sets of hypercube locations, where α represents some arbitrary node in a (k, h, o, λ)-tree:

$$S_\alpha := \{\alpha\beta\gamma\delta \in \{0,1\}^{kh+o} : \beta \in (0+1)^*1 \wedge |\beta| \leq k \wedge |\gamma|_1 = \lambda \wedge |\delta| = o\}$$
$$T := \{\gamma\delta \in \{0,1\}^{kh+o} : |\gamma|_1 \leq \lambda \wedge |\delta| = o\}$$

We also define the set $S := \bigcup_\alpha S_\alpha$. The vertices of the given graph mapped to the node of a (k, h, o, λ)-tree represented by α, will finally be mapped to hypercube locations in the set $L_\alpha := S_\alpha$ if $\alpha \neq \epsilon$, and $L_\epsilon := S_\epsilon \cup T$ otherwise. We will now show that the capacity of a node in the (k, h, o, λ)-tree is equal to the cardinality of the set of vertices in the hypercube to which it is mapped. Let α represent a node in the (k, h, o, λ)-tree at level ℓ; therefore, $|\alpha| = k(\ell - 1)$. Hence we get:

$$|T| = 2^o \sum_{i=0}^{\lambda} \binom{kh}{i}, \quad |S_\alpha| = 2^o \sum_{\substack{\beta \in (0+1)^*1 \\ |\beta| \leq k}} \binom{kh - |\alpha| - |\beta|}{\lambda} = 2^o \sum_{i=0}^{k-1} 2^i \binom{k(h-l+1) - i - 1}{\lambda}.$$

Thus, $|L_\alpha| = c(\ell)$. Furthermore, it can easily be verified that for every $\alpha \neq \alpha' \in (\{0,1\}^k)^*$ $S_\alpha \cap S_{\alpha'} = \emptyset$ and $S_\alpha \cap T = \emptyset$, and hence $L_\alpha \cap L_{\alpha'} = \emptyset$ for every $\alpha \neq \alpha' \in (\{0,1\}^k)^*$. Hence, for each string $s \in S \cup T$ there is a unique decomposition $s = \alpha\beta\gamma\delta$ as used in the definition of S_α and T. For a given hypercube location v, we will call the (k, h, o, λ)-tree node represented by α such that $v \in L_\alpha$ the *corresponding* (k, h, o, λ)-tree node.

Lemma 3. *Let v and w be representations of two nodes in a (k, h, o, λ)-tree and let $u = \text{lca}(v, w)$ be their lowest common ancestor in the (k, h, o, λ)-tree. If the lengths of the paths from u to v and from u to w are at most Δ then any pair r, s of hypercube locations, where $r \in L_v$ and $s \in L_w$, differ in at most $k(\Delta+1) + o + 2\lambda$ positions.*

Proof. We first consider the case where both vertices r and s belong to the set S. The diagram in Figure 1 shows the labels of the two hypercube locations r and s to which v and w are mapped. In this picture, α represents the lowest common ancestor of v and w and $\alpha\alpha'$ (resp., $\alpha\alpha''$) represents the vertex v (resp., w). Without loss of generality, we

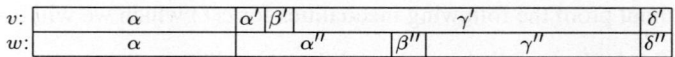

Fig. 1. Hypercube Locations of Adjacent Graph Vertices

assume $|\alpha'|\leq|\alpha''|$. Since the lowest common ancestor of v and w is at distance $\leq\Delta$ from both vertices, we get $|\alpha''|\leq k\Delta$. The definition of the mapping from the (k,h,o,λ)-tree to the hypercube implies that $|\beta''|\leq k$, $|\delta''|=o$, and that γ' and γ'' contain exactly λ 1's each. Hence, the labels r and s differ in at most $k\Delta+k+\lambda+\lambda+o=k(\Delta+1)+o+2\lambda$ positions.

We now consider the case that the hypercube location r belongs to the set T, implying that v represents the root of the (k,h,o,λ)-tree. The hypercube location $s=\alpha'\beta'\gamma'\delta'$, belongs again to S. Since the distance between v and w is Δ, we have $|\alpha'|\leq k\Delta$. Hence, there are at most $k\Delta+k+\lambda=k(\Delta+1)+\lambda$ 1's in the first kh positions of s. By the definition of the set T, r has at most λ 1's in the first kh positions. Thus, the hypercube locations r and s differ in at most $k(\Delta+1)+\lambda+\lambda+o=k(\Delta+1)+o+2\lambda$ positions.

Finally, if both vertices r and s belong to the set T, they obviously differ in at most $o+2\lambda$ positions. □

In the following, we will bound the node-congestion of the given embedding. We restrict our attention to bound the node-congestion, since the edge-congestion is less than or equal to the node-congestion. Moreover, for a better understandable representation we consider a $(k,h,o,1)$-tree. The argument for (k,h,o,λ)-trees is similar.

Consider two adjacent vertices of the given graph which are mapped to hypercube locations labeled v and w. We decompose the label into four segments A, B, C, and D. Segment D consists of the last o bits. The lengths of the segments A, B, and C are multiples of k; Segment C is the longest suffix before segment D such that it contains at most one 1 in both v and w and its length is a multiple of k. Segment B consists of the $(\Delta+1)k$ positions before segment C, and segment A is the remainder. See Figure 2 for an illustration of this decomposition. Recall that the hypercube locations of the vertices v and w can differ only in segments B and D, and in at most 2 positions of segment C (cf. Figure 2 for the case $v,w\in S$, the positions where the labels can differ are indicated by shading). Also note that segment A can be empty. For a path $p_{v,w}$ from v to w, we call v the *lower* endpoint of $p_{v,w}$, if $v=\alpha_v\beta_v\gamma_v\delta_v\in S$, $w=\alpha_w\beta_w\gamma_w\delta_w\in S$ (cf. Figure 2) and $|\gamma_v|<|\gamma_w|$, or if $v\in S$ and $w\in T$. Otherwise, if $v,w\in S$ and $|\gamma_v|=|\gamma_w|$ or if $v,w\in T$, we arbitrarily select one endpoint of the path $p_{v,w}$ to be the lower endpoint. The endpoint of the path $p_{v,w}$ which is not the lower endpoint is called the *upper* endpoint.

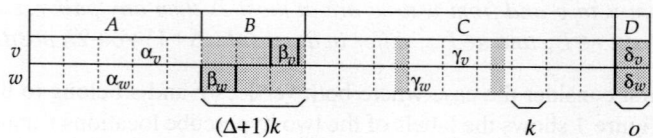

Fig. 2. Hypercube Locations of Two Adjacent Vertices

To construct a shortest path $p_{v,w}$ from v to w in the hypercube, we proceed in four phases. Without loss of generality, we assume that v is the lower endpoint; otherwise, we execute the routing about to be described in reverse order. In the first phase, we flip the bit position in segment C which has to be changed from 0 to 1, if it exists. In the second phase, we flip those bit positions in segment D that need to be changed. In the third phase, we first flip in segment B 0's to 1's from right to left that have to be changed; then, we flip 1's to 0's from left to right whenever necessary. Finally, we flip the bit position in segment C which has to be changed from 1 to 0, if it exists.

To obtain an upper bound on the node-congestion, we consider a fixed hypercube location u and bound the number of hypercube locations which can be an upper or lower endpoint of a path hitting u. The details are omitted due to space limitations.

Theorem 4. *Let G be a graph of size n and maximal degree d. Let an embedding of G into a (k, h, o, λ)-tree be given such that the load of no (k, h, o, λ)-tree node exceeds its capacity. If a pair of adjacent graph vertices is mapped to a pair of (k, h, o, λ)-tree nodes such that their lowest common ancestor is at most at distance Δ from either node, we obtain an embedding of G into a hypercube with unit load, dilation at most $(\Delta+1)k+o+2\lambda$, and expansion at most $2^{kh+o}/n$. Choosing the shortest paths between each pair of such hypercube locations as described above, the node-congestion of this embedding is at most $O(d2^{(\Delta+1)k+o+2\lambda})$.*

4 The Embedding

In this section, we will construct embeddings of arbitrary graphs into (k, h, o, λ)-trees provided the graphs have small bisectors. Further, if the bisectors can be efficiently constructed on the hypercube then the embedding can also be efficiently implemented on the hypercube.

4.1 Edge-Bisectors and Edge-Separators

Let $G=(V, E)$ be a graph containing marked vertices and let $\mu(V) \subseteq V$ be the set of marked vertices in G. Let $\alpha, \beta : \mathbb{R} \to \mathbb{R}$ be two functions. Let $S \subseteq E$ be a subset of edges. We denote by $G(E')$ the subgraph $G'=(V, E')$ of G for some $E' \subseteq E$. Further, let $C_1=(V_1, E_1), \ldots, C_\ell=(V_\ell, E_\ell)$ be the connected components of the graph $G(E \setminus S)$. We call S an (α, β)-*extended-edge-bisector* if there exists a partition of the connected components of the graph $G(E \setminus S)$ into sets S_1 and S_2 such that the following conditions hold for $i \in \{1, 2\}$:

i) $\left\lfloor \dfrac{n}{2} \right\rfloor \leq \displaystyle\sum_{C_j \in S_i} |V_j| \leq \left\lceil \dfrac{n}{2} \right\rceil$, ii) $|S| \leq \alpha(|V|)$, iii) $\displaystyle\sum_{C_j \in S_i} |\mu(V_j)| \leq \lceil \beta(|\mu(V)|) \rceil$.

For $\beta(n)=\frac{n}{2}$, such a separator is known in the literature as a 2-color-bisector [1, 9]. We call an edge belonging to the set S a *separator edge*. Vertices incident to a separator edge will be called *separator vertices*. We will denote the set of separator vertices in a connected component $C=(V, E)$ (or more generally in a graph $G=(V, E)$) by $\sigma(V)$.

Again, let $G=(V, E)$ be a graph containing marked vertices and let $\mu(V) \subseteq V$ be the set of marked vertices in G. Let $\gamma : \mathbb{R} \times \mathbb{R} \to \mathbb{R}$ be a real valued function. Let $S \subseteq E$ be a subset of the edges. We denote by $C_1=(V_1, E_1), \ldots, C_\ell=(V_\ell, E_\ell)$ the connected components of the graph $G(E \setminus S)$. We call S a (γ, κ)-*extended-edge-separator* if there exists a partition of the connected components of the graph $G(E \setminus S)$ into sets $\mathcal{S}_i$, for $i \in [1:\kappa]$, such that the following conditions hold:

$$\forall i \in [1:\kappa] : \left\lfloor \frac{n}{\kappa} \right\rfloor \leq \sum_{C_j \in \mathcal{S}_i} |V_j| \leq \left\lceil \frac{n}{\kappa} \right\rceil \wedge \sum_{C_j \in \mathcal{S}_i} |\mu(V_j) \cup \sigma(V_j)| \leq \gamma(|V|, |\mu(V)|).$$

A set $\mathcal{F}$ of graphs is called a *family* of graphs if each subgraph of a graph in $\mathcal{F}$ is also in $\mathcal{F}$. More formally, for each $G=(V,E) \in \mathcal{F}$, for each subset $V' \subseteq V$, and for each subset $E' \subset (E \cap (V' \times V'))$, it follows that $G'=(V',E') \in \mathcal{F}$. A family $\mathcal{F}$ of graphs is called (α,β)-*bisectable* if each graph $G \in \mathcal{F}$ has an (α,β)-extended-edge-bisector. A family $\mathcal{F}$ of graphs is called (γ,κ)-*separable* if each graph $G \in \mathcal{F}$ has a (γ,κ)-extended-edge-separator.

4.2 Constructing Edge-Separators from Edge-Bisectors

Using a (α,β)-extended-edge-bisector, we are able to construct a $(\gamma,2^k)$-extended-edge-separator.

Theorem 5. *Let $\mathcal{F}$ be a family of (α,β)-bisectable graphs such that β is monotonically increasing, i.e., $x \leq y \Rightarrow \beta(x) \leq \beta(y)$, and β satisfies the triangle inequality, i.e., $\beta(x+y) \leq \beta(x) + \beta(y)$. Then $\mathcal{F}$ is also $(\gamma,2^k)$-separable for every $k \in \mathbb{N}$, where*

$$\gamma(x,y) = \sum_{i=0}^{k-1} \left[(\beta^{(i)} \circ \alpha) \left(\left\lceil \frac{x}{2^{k-1-i}} \right\rceil \right) \right] + \beta^{(k)}(y) + \sum_{i=0}^{k-1} \beta^{(i)}(1).$$

If $\beta(x) = \frac{x}{2}$ then we get for γ: $\quad \gamma(x,y) = \sum_{i=0}^{k-1} \left[\frac{1}{2^i} \cdot \alpha \left(\left\lceil \frac{x}{2^{k-1-i}} \right\rceil \right) \right] + \left\lceil \frac{y}{2^k} \right\rceil.$

Proof. We will prove the theorem by induction on k.
Induction basis ($k=1$): Let $G=(V,E) \in \mathcal{F}$. Since $\mathcal{F}$ is a family of (α,β)-bisectable graphs, the first condition in the definition of an extended-edge-separator follows immediately from condition i) in the definition of an extended-edge-bisector. By definition, the number of marked and separator vertices in each half is at most

$$\alpha(|V|) + \lceil \beta(|\mu(V)|) \rceil \leq \alpha(|V|) + \beta(|\mu(V)|) + 1 \leq \alpha(|V|) + \beta(|\mu(V)|) + \beta^{(0)}(1),$$

as claimed in the theorem. If $\beta(x) = \frac{x}{2}$ then the number of marked and separator vertices in each half is at most $\alpha(|V|) + \lceil |\mu(V)|/2 \rceil$.
Induction step ($k \to k+1$): First, we apply the induction hypothesis for $k=1$ to the graph G. Hence, we get a partition of the connected components $C_1, \ldots, C_\ell$ of the graph $G(E \setminus S)$ into two sets $\mathcal{S}_1$ and $\mathcal{S}_2$ satisfying the required conditions. Let $G^i = (V^i, E^i)$, for $i \in \{1,2\}$, be the union of the connected components in $\mathcal{S}_i$, i.e., $V^i = \bigcup_{C_j \in \mathcal{S}_i} V(C_j)$ and $E^i = \bigcup_{C_j \in \mathcal{S}_i} E(C_j)$. Since $\mathcal{F}$ is a family of graphs, it follows that $G^i \in \mathcal{F}$ and, therefore, the graphs G^i are (α,β)-bisectable. Now we mark all separator vertices and apply the induction hypothesis for $k-1$ to each graph $G^i = (V^i, E^i)$ in parallel. Thus, we obtain 2^k graphs $G_i = (V_i, E_i)$. Thus, $\lfloor \frac{n}{2^k} \rfloor = \lfloor \frac{\lfloor n/2 \rfloor}{2^{k-1}} \rfloor \leq |V_i| \leq \lceil \frac{\lceil n/2 \rceil}{2^{k-1}} \rceil = \lceil \frac{n}{2^k} \rceil$. Moreover, the number of marked and separator vertices in each G_i is $|\mu(V_i) \cup \sigma(V_i)|$

$$\leq \sum_{i=0}^{k-2}\left[(\beta^{(i)}\circ\alpha)\left(\left\lceil\frac{\lceil|V_i|/2\rceil}{2^{k-2-i}}\right\rceil\right)\right] + \beta^{(k-1)}\Big(\alpha(|V_i|) + \lceil\beta(|\mu(V_i)|)\rceil\Big) + \sum_{i=0}^{k-2}\beta^{(i)}(1)$$

using the triangle inequality for $\beta(\cdot)$ and the monotonicity of $\beta(\cdot)$

$$\leq \sum_{i=0}^{k-1}\left[(\beta^{(i)}\circ\alpha)\left(\left\lceil\frac{|V_i|}{2^{k-1-i}}\right\rceil\right)\right] + \beta^{(k-1)}\big(\beta(|\mu(V_i)|) + 1\big) + \sum_{i=0}^{k-2}\beta^{(i)}(1)$$

using the triangle inequality for $\beta(\cdot)$

$$\leq \sum_{i=0}^{k-1}\left[(\beta^{(i)}\circ\alpha)\left(\left\lceil\frac{|V_i|}{2^{k-1-i}}\right\rceil\right)\right] + \beta^{(k)}(|\mu(V_i)|) + \sum_{i=0}^{k-1}\beta^{(i)}(1) \; = \; \gamma(|V_i|,|\mu(V_i)|)$$

If $\beta(x)=\frac{x}{2}$ then the number of marked and separator vertices in each G_i is

$$|\mu(V_i)\cup\sigma(V_i)| \leq \sum_{i=0}^{k-2}\left[\frac{1}{2^i}\cdot\alpha\left(\left\lceil\frac{\lceil|V_i|/2\rceil}{2^{k-2-i}}\right\rceil\right)\right] + \left\lceil\frac{\alpha(|V_i|)+\lceil|\mu(V_i)|/2\rceil}{2^{k-1}}\right\rceil$$

$$\leq \sum_{i=0}^{k-1}\left[\frac{1}{2^i}\cdot\alpha\left(\left\lceil\frac{|V_i|}{2^{k-1-i}}\right\rceil\right)\right] + \left\lceil\frac{|\mu(V_i)|}{2^k}\right\rceil \; = \; \gamma(|V_i|,|\mu(V_i)|)$$

The claim on the computation times follows immediately from the observation that the bisections in each of the k stages can be done in parallel in an optimal subcube. Further, routing of the subgraphs created by the bisection to their own subcubes can be done in logarithmic time using concentration routing [19]. □

We should mention here that the hypercube algorithm for the bisection must satisfy some special requirements. Whenever the hypercube algorithm requires some conditions on how the graph is stored in the hypercube before the bisection, then the algorithm must ensure that these conditions are also true for the subgraphs resulting from the bisection.

4.3 The Embedding Algorithm

Given a family $\mathcal{F}$ of (α,β)-bisectable graphs, we now consider the embedding of a graph $G=(V,E)\in\mathcal{F}$ into the hypercube. We denote the number of vertices in G by n, and the maximal degree of a vertex in G by d. In the sequel, we assume that $d\geq 3$, since all graphs with degree ≤ 2 can easily be embedded one-to-one into its optimal hypercube.

Our embedding of (α,β)-bisectable graphs into optimal hypercubes is achieved in two steps. First, we embed the graph into a (k,h,o,λ)-tree. This will be explained in detail in the following. Then, we use the mapping presented in the previous section to complete the embedding. To obtain small dilation, adjacent vertices of the graph should be mapped to nodes which are close in the (k,h,o,λ)-tree. Our goal is to obtain an embedding of the graph into a (k,h,o,λ)-tree such that adjacent vertices are mapped to two nodes of the (k,h,o,λ)-tree with distance at most 1 from their lowest common ancestor of the (k,h,o,λ)-tree. Our method leads to an embedding of the graph G into the hypercube with dilation $2k+o+2\lambda$.

The embedding of the graph G into the (k,h,o,λ)-tree will be achieved as follows. First, we fill up the root of the (k,h,o,λ)-tree with arbitrarily chosen vertices of G and remove these vertices from G, obtaining G'. Then we mark the unmapped neighbors of the mapped vertices in the resulting graph G'. We associate this graph G' with the root of the (k, h, o, λ)-tree. We decompose G' into 2^k parts using Theorem 5 and associate to each of the children of the root one of the parts of the decomposed graph G'. In the next step, we fill up the children of the root with the marked vertices in the associated subgraph of G'. Additionally, we map the separator vertices of the previous decomposition to the 2^k children of the root. Finally, we fill up the nodes of the (k, h, o, λ)-tree with arbitrarily chosen vertices of the associated subgraph until the capacity of these nodes is reached. We repeat this process until we reach the leaves of the (k, h, o, λ)-tree.

4.4 Quality of the General Embedding Strategy

In this subsection, we will show how the parameters k, o, and λ can be chosen depending on the extended-edge-separator and the maximal degree d of the given graph to obtain an embedding with small dilation. Since we are looking for an embedding into the optimal hypercube, we have to choose the parameters of the (k, h, o, λ)-tree such that $kh+o=\lceil\log(n)\rceil$ implying that the height h is determined by k and o.

We make the following assumptions on our (α, β)-bisectable family of graphs. We recall from Theorem 5 that β is function satisfying the triangle inequality such that $\beta(x) \in [\frac{x}{2}, x]$. We first assume that $\beta : \mathbb{R} \to \mathbb{R}$ is a linear function. For our convenience, we will write $\beta(n)=\beta \cdot n$. Further, we assume that $\alpha : \mathbb{R} \to \mathbb{R}$ is a poly-logarithmic function, i.e., $\alpha(x)=\alpha \cdot \log^{\lambda}(x)$ for some $\alpha \in \mathbb{R}^+$.

In order to compute the number of graph vertices mapped to a single node in the (k, h, o, λ)-tree, let $n(\ell)$ be the maximum number of marked vertices and separator vertices mapped to a single node of the (k, h, o, λ)-tree at level ℓ, and let $f(\ell)$ be the size of the associated graph which is partitioned at a node at level ℓ of the (k, h, o, λ)-tree. An obvious upper bound for $f(\ell)$ is $\lceil \frac{2^{kh+o}-c(1)}{2^{k(\ell-1)}} \rceil$. Recall that $c(\ell)$ denotes the capacity of a node in the (k, h, o, λ)-tree at level ℓ. The number of marked vertices in a forest corresponding to a node at level ℓ before the partitioning is at most $(d-1)c(\ell)$, since each mapped vertex has at most $d-1$ unmapped neighbors. It what follows, we will show that $n(\ell) \leq c(\ell)$. Note that by construction of the embedding $n(1) \leq c(1)$ and $n(h) \leq c(h)$. Now we can establish the following inequality for $2 \leq \ell \leq h-1$ using Theorem 5.

$$n(\ell) \leq \sum_{i=0}^{k-1} \left[\beta^{(i)} \left(\alpha \left(\frac{f(\ell-1)}{2^{k-1-i}} \right) \right) \right] + \beta^{(k)} \left((d-1) \cdot c(\ell-1) \right) + \sum_{i=0}^{k-1} \beta^{(i)}$$

$$\leq \alpha \frac{1-\beta^k}{1-\beta} \log^{\lambda} \left(\frac{2^{kh+o}}{2^{k(\ell-2)}} \right) + k + (d-1)\beta^k c(\ell-1) + k$$

using Lemma 2 $ii)$ for $\lambda \leq \frac{k}{2}$

$$\leq \alpha \frac{1-\beta^k}{1-\beta} (k(h-\ell+2)+o)^{\lambda} + \beta^k (d-1) 3^{\lambda+1} c(\ell) + 2k \stackrel{!}{\leq} c(\ell)$$

Thus, it is sufficient to satisfy the following three inequalities.

$$2k \stackrel{!}{\leq} \frac{1}{8} c(\ell) \qquad (1)$$

$$\beta^k 3^{\lambda+1}(d-1)c(\ell) \stackrel{!}{\leq} \frac{1}{8}c(\ell) \qquad (2)$$

$$\alpha \frac{1-\beta^k}{1-\beta}(k(h-\ell+2)+o)^\lambda \stackrel{!}{\leq} \frac{1}{2} \cdot 2^{k+o}\left(\frac{k(h-\ell)}{\lambda}\right)^\lambda \leq \frac{3}{4}c(\ell) \qquad (3)$$

For inequality (3) we have used Lemma 2 *i)*. Inequality (1) is satisfied if we choose $o > k$ and $k \geq 3$, since using Lemma 2 *iii)*, we get:

$$2k \leq 2^{2k-3} \leq 2^{k+o-4} \leq \frac{1}{8}2^o(2^k-1) \leq \frac{1}{8}c(\ell).$$

From inequality (2) we may conclude for $\lambda > 0$ that

$$k \geq \frac{\log(d) + \lambda \log(3) + 4}{-\log(\beta)}.$$

Clearly, inequality (3) is satisfied if the following inequality holds, assuming that $o \leq 2k$.

$$2\alpha \frac{1-\beta^k}{1-\beta}\lambda^\lambda\left(1+\frac{2k+o}{k(h-\ell)}\right)^\lambda \leq 2\alpha\frac{1}{1-\beta}2^{\lambda \log(5\lambda)} \stackrel{!}{\leq} 2^{k+o}$$

Hence, the following inequalities should be satisfied, if we choose $o \in [k+1:2k]$:

$$k+o \stackrel{!}{\geq} \lambda \log(5\lambda) + \log\left(\frac{2\alpha}{1-\beta}\right) \Rightarrow k \stackrel{!}{\geq} \frac{1}{2}\left(\lambda \log(5\lambda) + \log\left(\frac{2\alpha}{1-\beta}\right)\right)$$

Altogether, if we choose k and o as described above, we never map more vertices to a (k, h, o, λ)-tree node at level $\ell \leq h-1$ than its capacity allows. Using Lemma 3, we can conclude that the dilation of the embedding is at most $2k + o + 2\lambda \leq 4k + 2\lambda$.

Theorem 6. *Let $\mathcal{G}$ be a family of (α, β)-bisectable graphs, where $\alpha(n) = \alpha' \cdot \log^\lambda(n)$ is a poly-logarithmic function and $\beta(x) \leq \beta' \cdot x$. Then a graph $G \in \mathcal{G}$ can be one-to-one embedded into its optimal hypercube with dilation at most $4k + 2\lambda$ and node-congestion at most $O(d2^{4k+2\lambda})$, where*

$$k = \lceil \max\{\tfrac{1}{2}(\lambda \log(5\lambda) + \log\left(\tfrac{2\alpha}{1-\beta}\right)), -(\log(d) + \lambda \log(3) + 4)/\log(\beta)\}\rceil$$

Moreover, if a (α, β)-bisection can be computed in time $\mathcal{B}(n) = \Omega(\log(n))$ for any graph of size n on its optimal hypercube, this embedding can computed in time $O(\log(n) \cdot \mathcal{B}(n))$ on the optimal hypercube.

Obviously, the dilation and node-congestion can be improved for special graph classes by a more sophisticated analysis of the inequality $n(\ell) \leq c(\ell)$.

5 Applications

In this section, we will apply our results of the previous sections to some special families of graphs. Some of the results are well known, but others are new. It has been shown in [13] that a binary tree has a $(4\log(n), \frac{n}{2})$-extended-edge-bisector. As shown in [13], this extended-edge-bisector can be computed on the optimal hypercube in logarithmic time after some preprocessing requiring time $O(\log^2(n) \log\log(n) \log^*(n))$.

Theorem 7. *An arbitrary binary tree can be embedded into its optimal hypercube with constant dilation, constant node-congestion, and unit load. The embedding can be computed in time $O(\log^2(n) \log\log(n) \log^*(n))$ on the optimal hypercube itself.*

The graphs with treewidth t and maximal degree d have a $(2(t+1)(d+1)\log(n), \frac{n}{2})$-extended-edge-bisector, as shown in [14].

Theorem 8. *A graph with treewidth t and maximal degree d can be embedded into its optimal hypercube with dilation $\max\{14+\lceil 2\log((d+1)(t+1))\rceil, 24+4\lceil\log(d)\rceil\}$. The embedding can be computed in time $O(\log^2(n)\log\log\log(n)\log^*(n))$ on the optimal hypercube itself if the input graph is given by its tree decomposition.*

It can also be shown that graphs with pathwidth t and maximal degree d have a $((t+1)(d+1), \frac{n}{2})$-extended-edge-bisector.

Theorem 9. *A graph with pathwidth t and maximal degree d can be embedded into its optimal hypercube with dilation $\max\{4+\lceil 2\log((d+1)(t+1))\rceil, 16+4\lceil\log(d)\rceil\}$. The embedding can be computed in time $O(\log^2(n)\log\log\log(n)\log^*(n))$ on the optimal hypercube itself if the input graph is given by its tree decomposition.*

Using the results in [1, 9], it can be shown that the family of circular-arc graphs have a $(d(d+1), \frac{n}{2})$-extended-edge-bisector, with d the maximal degree of a vertex in the graph.

Theorem 10. *A circular-arc graph with maximal degree d can be embedded into its optimal hypercube with dilation $4\lceil\log(d)\rceil+4$. The embedding can be computed in time $O(\log^2(n)\log\log\log(n)\log^*(n))$ on the optimal hypercube itself if the input graph is given by its circular-arc representation.*

Since interval graphs are a subclass of circular-arc graphs, it follows:

Corollary 11. *An interval graph with maximal degree d can be embedded into its optimal hypercube with dilation $4\lceil\log(d)\rceil+4$. The embedding can be computed in time $O(\log^2(n)\log\log\log(n)\log^*(n))$ on the optimal hypercube itself if the input graph is given by its interval representation.*

The family of k-outerplanar graphs has a $(6k(d+1)\log(n), \frac{n}{2})$-extended-edge-bisector.

Theorem 12. *A k-outerplanar graph can be one-to-one embedded into its optimal hypercube with dilation at most $\max\{2\lceil\log(k(d+1))\rceil+8, 26+4\lceil\log(d)\rceil\}$.*

6 Conclusion

We have presented a general technique for embedding irregular graphs based on extended-edge-bisectors. Depending on the size of the extended-edge-bisector, the quality of distributing marked vertices, and the maximal degree of the given graph, we have given bounds on the dilation and node-congestion for our embedding. It remains as an open problem whether an embedding can be computed dynamically as the graph grows. It has been proven in [17] that any deterministic algorithm for dynamically embedding caterpillars into hypercubes cannot simultaneously achieve constant dilation, constant load, and constant expansion. Thus, either randomized techniques (as in [17]) or migration, i.e., remapping of vertices, has to be used for dynamic embeddings. In [16], a deterministic algorithm for dynamic one-to-one embeddings of arbitrary binary trees into hypercubes with small dilation, expansion, and node-congestionusing migration is presented. Furthermore, this embedding can be efficiently computed on the hypercube. It seems that this method can at least be extended for k-trees.

References

1. N. Alon, D. West: The Borsuk-Ulam Theorem and Bisection of Necklaces, *Proc. Am. Math. Soc.*, **98**(1986), 623–628.
2. S. Bezrukov, B. Monien, W. Unger, G. Wechsung, Embedding Ladders and Caterpillars into the Hypercube, Preprint, GH-Univ. Paderborn, 1995, (to appear in *Disc. Appl. Math.*).
3. S. Bhatt, F. Chung, T. Leighton, A. Rosenberg: Efficient Embeddings of Trees in Hypercubes, *SIAM J. Comput.*, **21**(1992), 151–162.
4. S. Bhatt, I. Ipsen: How to Embed Trees in Hypercubes, *Yale Univ.*, RR-443, 1985.
5. M. Chan: Embedding of d-Dimensional Grids into Optimal Hypercubes, *Proc. of the 1989 Symp. on Parallel Algorithms and Architectures*, 52–57.
6. M. Chan: Embedding of Grids into Optimal Hypercubes, *SIAM J. Comput*, **20**(1991), 834–864.
7. M. Chan, F. Chin, C. Chu, W. Mak: Dilation-5 Embedding of 3-Dimensional Grids into Hypercubes, *J. Parallel Distrib. Comput.*, **33**(1996), 98-106.
8. K. Efe: Embedding Mesh of Trees in the Hypercube, *J. Parallel Distrib. Comput.*, **11**(1991), 222–230.
9. C. Goldberg, D. West: Bisection of Circle Colorings, *SIAM J. Alg. Disc. Meth.*, **6**(1985), 93–106.
10. I. Havel: On Hamiltonian Circuits and Spanning Trees of Hypercubes (in Czech.), *Časopis. Pěst. Mat.*, **109**(1984), 145–152.
11. I. Havel, P. Liebl: Embedding the Polytomic Tree into the n-Cube, *Časopis. Pěst. Mat.*, **98**(1973), 307–314.
12. I. Havel, P. Liebl: One-Legged Caterpillars Span Hypercubes, *J. Graph Theory*, **10**(1986), 69–76.
13. V. Heun, E. Mayr: A New Efficient Algorithm for Embedding an Arbitrary Binary Tree into Its Optimal Hypercube, *J. Algorithms*, **20**(1996), 375–199.
14. V. Heun, E. Mayr: Embedding Graphs with Bounded Treewidth into Optimal Hypercubes, *Proc. of the 13th Symp. on Theoretical Aspects of Computer Science*, LNCS 1046, 157–168.
15. V. Heun, E. Mayr: Optimal Dynamic Edge-Disjoint Embeddings of Complete Binary Trees into Hypercubes, (to appear in *Proc. of the 4th Workshop on Parallel Systems and Algorithms*).
16. V. Heun, E. Mayr: Efficient Dynamic Embeddings of Arbitrary Binary Trees into Hypercubes, (to appear in *Proc. of the IRREGULAR'96*).
17. T. Leighton, M. Newman, A. Ranade, W. Schwabe: Dynamic Tree Embeddings in Butterflies and Hypercubes, *SIAM J. Comput.*, **21**(1992), 639–654.
18. B. Monien, H. Sudborough: Simulating Binary Trees on Hypercubes, *Proc. of the Aegean Workshop on Computing*, LNCS 319, 170–180.
19. D. Nassimi, S. Sahni: Data Broadcasting in SIMD Computers, *IEEE Trans. Comput.*, **C-30**(1981), 101-107.
20. Y. Saad, M. Schulz: Topological Properties of the Hypercube, *Yale Univ.*, RR-389, 1985.
21. X. Sheen, Q. Hu, W. Liang: Embedding k-ary Complete Trees into Hypercubes, *J. Parallel Distrib. Comput.*, **24**(1995), 100–106.
22. Q. Stout: Hypercubes and Pyramids, *Proc. of the NATO Advanced Research Workshop on Pyramidal Systems for Computer Vision 1986*, 75–89.
23. A. Wagner, D. Corneil: Embedding Tree in a Hypercube is NP-Complete, *SIAM J. Comput.*, **19**(1990), 570–590.
24. A. Wagner, D. Corneil: On the Complexity of the Embedding Problem for Hypercube Related Graphs, *Discrete Appl. Math.*, **43**(1993), 75-95.
25. A. Wu: Embedding of Tree Networks into Hypercubes, *J. Parallel Distrib. Comput.*, **2**(1985), 238-249.

The Size Complexity of Strictly Non-blocking Fixed Ratio Concentrators with Constant Depth

H. K. Dai

Department of Computer Science, University of North Dakota
Grand Forks, North Dakota 58202, U. S. A.

Abstract. Concentrators are interconnection networks that provide vertex-disjoint directed paths to satisfy interconnection requests. An interconnection network is non-blocking in the strict sense if every compatible interconnection request can be satisfied regardless of any existing interconnections. We show a size optimal bound of $\Theta(n^{1+\frac{1}{k}})$ for synchronous strictly non-blocking γn-limited $(\alpha n, \beta n)$-concentrators with non-full capacity and constant depth k, and present a size upper bound of $O(n^{1+\frac{1}{\lceil k/2 \rceil}})$ for synchronous strictly non-blocking βn-limited $(\alpha n, \beta n)$-concentrators with full capacity and constant depth k.

1 Introduction

The interconnection network of an information transmission system provides mediation between a set of inputs (sources of information) and a set of outputs (sinks of information). The graph-theoretic design and analysis of interconnection patterns can idealize such networks. Interconnection networks are classified according to the type of interconnections that they provide. Several generic types of interconnections that have been studied extensively in the literature (see [Clo53], [Ben65], [DDPW83], [FFP88], and [Dai91]) are concentration, superconcentration, connection, expansion, and partition. An orthogonal classification of interconnection networks, based upon the network capability of allowing an interconnection to be destroyed and created dynamically without disturbing any of the other existing interconnections in the network, results in three genera of interconnection capability: rearrangeability, wide-sense non-blockingness, and strictly non-blockingness.

The study of interconnection networks is relevant to theoretical computer science in areas such as parallel computations, graph pebbling, oblivious computations for many naturally occurring functions, modeling circuits with limited depth and unbounded fan-in, and implementation on parallel computers of algorithms for sorting.

In this paper, we study the size-depth complexity tradeoff for strictly non-blocking fixed ratio concentrators. More specifically, we show a size optimal bound of $\Theta(n^{1+\frac{1}{k}})$ for synchronous strictly non-blocking γn-limited $(\alpha n, \beta n)$-concentrators with non-full capacity and constant depth k, and present a size upper bound of $O(n^{1+\frac{1}{\lceil k/2 \rceil}})$ for synchronous strictly non-blocking βn-limited $(\alpha n, \beta n)$-concentrators with full capacity and constant depth k. Our motivation for studying the size-depth tradeoffs for concentrators and generalized-concentrators derives from considering the gap between the lower and upper size bounds versus depth for connectors and generalized-connectors ([Fri88], [FFP88], and [Clo53]) in the strictly non-blocking context.

1.1 Preliminaries

We shall abbreviate "directed graph" to "digraph". For a graph or digraph G, denote by $V(G)$ the vertex set of G, and by $E(G)$ the edge set of G. The out-neighborhood of a vertex v in G is denoted by $\Gamma_{out}(v)$. For a directed path P in a digraph G, denote by $term(P)$ the terminus of P.

For positive integers n and m, an (n, m)-network $G = (V, E, I, O)$ is an acyclic digraph with vertex set V and edge set E, a set I of n distinguished vertices called inputs, and a set O of m other distinguished vertices called outputs. We shall use the same denotation G for a network and its underlying acyclic digraph when the context is clear. The size of a network is the size of its underlying digraph.

Two directed paths in a digraph are compatible if their intersection is a common initial directed subpath of them, which may possibly be empty. A route in a network $G = (V, E, I, O)$ is a directed path from an input to an output. The depth of G is the maximum number of edges in a route in G. A state S of G is a set of pairwise compatible routes, or equivalently, S is the set of all routes of a directed forest with its roots at I and leaves at O. A state S saturates a vertex v in G if v appears in a directed path of S.

A concentration request in a network is an input. A concentration request is satisfied by a route if the route is directed from the concentration request. A generalized-concentration assignment is a multi-set of concentration requests. A generalized-concentration assignment A is satisfied by a state S if each concentration request $u \in A$ is satisfied by a number of routes of S equal to the multiplicity of u in A.

Two network parameters c and r that govern respectively the network capacity (an achievable lower bound on the number of possible simultaneous concentration requests allowed in the network) and request multiplicity (an achievable lower bound on the number of possible simultaneous concentration requests with common input allowed in the network) are associated with generalized-concentration. For positive integers c and r with $c \geq r$, a generalized-concentration assignment A is (c, r)-limited if each concentration request in A has multiplicity at most r and the sum of all multiplicities in A is at most c. A state S is (c, r)-limited if the unique generalized-concentration assignment $A(S)$ satisfied by S is (c, r)-limited. A concentration request u is (c, r)-limited in a state S if the multi-set obtained by adjoining the concentration request u to $A(S)$ is a (c, r)-limited generalized-concentration assignment.

A strictly non-blocking (c, r)-limited (n, m)-generalized-concentrator is an (n, m)-network for which the set of all (c, r)-limited states is closed under the (c, r)-limited generalized-concentration extension: for every (c, r)-limited state S_1 and every (c, r)-limited concentration request u in S_1, there exists a (c, r)-limited state S_2 such that $S_1 \subseteq S_2$ and S_2 contains an additional route satisfying the concentration request u.

The notion of generalized-connection is defined analogously (see [Dai91]). For (c, r)-limited (n, m)-generalized-concentrators and (n, m)-generalized-connectors, the case when $c = m$ is referred to as full capacity. The adjective "$(c, 1)$-limited (n, m)-generalized-" refers to a non-generalized context and is thus abbreviated to "c-limited (n, m)-" in this paper.

A network is synchronous provided that all the routes in the network have the same length. In a synchronous network $G = (V, E, I, O)$ with depth k, the vertex set V can be partitioned into $k + 1$ disjoint ranks $(V_i)_{i=0}^{k}$ and the edge set E can be partitioned into k disjoint stages $(E_i)_{i=1}^{k}$ in an obvious manner such that $V_0 = I$, $V_k = O$, and E_i is the set of edges directed from V_{i-1} into V_i for $i = 1, 2, \ldots, k$.

A special type of concentrators with their parameters, input and output cardinalities and network capacity, in a fixed ratio are called fixed ratio concentrators, that is, they are represented as γn-limited $(\alpha n, \beta n)$-concentrators for some positive integer constants α, β, and γ such that $\alpha > \beta \geq \gamma$.

1.2 Previous Work

Pippenger [Pip74] showed that strictly non-blocking concentrators with capacity c must have size at least $3c \log_3 c - O(c)$. Dai ([Dai93] and [Dai94]; see Theorem 3 below) obtained the size optimal bounds for strictly non-blocking c-limited (n, m)-concentrators and (c, r)-limited (n, m)-generalized-concentrators with depth 1, and lower bound size-depth tradeoffs for their synchronous versions with arbitrary depth k.

For non-blocking connectors, Bassalygo and Pinsker [BP74] proved that strictly non-blocking (n, n)-connectors exist with size $O(n \log n)$; and an explicit construction was obtained through the work of Margulis [Mar75] and Gabber and Galil [GG81]. Friedman [Fri88] showed a lower bound size-depth tradeoff, n^2 and $\Omega(n^{1+\frac{1}{k-1}})$ for synchronous strictly non-blocking (n, n)-connectors with depth 2 and depth $k \geq 3$, respectively. Clos [Clo53] gave a size upper bound versus depth for these networks, which is $O(n^{1+\frac{1}{j}})$ for depth $2j - 1$. For strictly non-blocking (n, n)-generalized-connectors, Bassalygo and Pinsker [BP80] proved that such networks must have size $\Omega(n^2)$ for any depth.

2 Construction of Expanding Networks

We present an upper bound on the size of synchronous strictly non-blocking fixed ratio concentrators with non-full capacity versus constant depth, which turns out to be optimal. The size bound is achieved via an explicit construction of interconnection networks with strong expanding property and appropriate size.

Suppose that G is a synchronous network with depth k, and rank partition $(V_i)_{i=0}^{k}$ and stage partition $(E_i)_{i=1}^{k}$. Construct a synchronous network G' with depth $k+1$, and rank partition $(V_i')_{i=0}^{k+1}$ and stage partition $(E_i')_{i=1}^{k+1}$ such that:

1. $V_i' = V_i$ for $i = 0, 1, \ldots, k$, V_{k+1}' is disjoint from $V(G)$ with $|V_{k+1}'| \geq |V_k'|$, and V_{k+1}' is partitioned into $|V_k'|$ pairwise disjoint subsets of vertices with cardinalities either $\lfloor \frac{|V_{k+1}'|}{|V_k'|} \rfloor$ or $\lceil \frac{|V_{k+1}'|}{|V_k'|} \rceil$; and these $|V_k'|$ subsets are indexed by vertices in V_k' as $\{S_v \mid v \in V_k'\}$, and

2. $E_i' = E_i$ for $i = 1, 2, \ldots, k$, and $E_{k+1}' = \cup \{\{v\} \times S_v \mid v \in V_k'\}$.

That is, the network G' evolves from G with its last stage E_{k+1}' being the uniform "projection" of V_k' into V_{k+1}'. We call G' the projection of G onto V_{k+1}', denoted by $proj(G, V_{k+1}')$.

Let $\mathcal{G}$ be a set of pairwise vertex-disjoint synchronous networks with depth k. Denote by $proj(\mathcal{G}, V_{k+1})$ the synchronous network with depth $k+1$ that is the graph-theoretic union of the projections $proj(G, V_{k+1})$ for $G \in \mathcal{G}$. The projections created in this progressive manner are magnified as the depth increases, and this allows each vertex in the input rank to access a sufficiently large number of vertices in the projected ranks. The following theorem shows an evolution of a synchronous network with depth $k+1$ from a set of pairwise vertex-disjoint synchronous networks with depth k and expanding capability via projection, which results in a network with stronger expansion by properly manipulating the cardinalities of inputs and outputs.

The expansion capability of a network is measured by the quantity of accessible output vertices from an input vertex regardless of the existing interconnections. Formally, for a network $G = (V, E, I, O)$ and a positive integer Δ, G is said to satisfy the

Δ-expanding property provided that for every set S of pairwise vertex-disjoint routes of G and every input vertex v that is not saturated by S, there exist at least Δ output vertices that are not saturated by S to which v can be directed, via routes that are vertex-disjoint from each route in S.

Theorem 1. *For integer constants k, α_i and β_i, $i = 1, 2, \ldots, k$, let*

$$\Delta_i = \beta_i - \alpha_i - \left(\sum_{\eta=1}^{i-1} \frac{\alpha_\eta}{\beta_\eta}\right)\beta_i - (i-1)\frac{\beta_i}{\beta_1}$$

for $i = 1, 2, \ldots, k$ (note that $\sum_{\eta=1}^{i-1} \frac{\alpha_\eta}{\beta_\eta} = 0$ if $i = 1$).

If $\Delta_i > 0$ for $i = 1, 2, \ldots, k$, then, for sufficiently large n ($\geq N_k$, a constant), there exists a synchronous $(\alpha_k n, \beta_k n)$-network G_k with depth k and size $O(n^{1+\frac{1}{k}})$ that satisfies the $\Delta_k n$-expanding property (the implied constants depend on k, α_i and β_i for $i = 1, 2, \ldots, k$).

Proof. We prove the theorem by induction on the depth k. For the basis of induction that $k = 1$, suppose that α_1 and β_1 are positive integer constants such that $\Delta_1 = \beta_1 - \alpha_1 > 0$. Let G_1 be the digraph whose underlying graph is the complete bipartite graph $K_{\alpha_1 n, \beta_1 n}$ with bipartition (V_0, V_1) where $|V_0| = \alpha_1 n$ and $|V_1| = \beta_1 n$, and $V(G_1) = V(K_{\alpha_1 n, \beta_1 n})$ and $E(G_1) = V_0 \times V_1$, that is, the rank and stage partitions of G_1 are respectively (V_0, V_1) and $(V_0 \times V_1)$. For every set S of pairwise vertex-disjoint directed edges from V_0 into V_1, and for every input vertex $v \in V_0 - S(V_0)$, $|\Gamma_{out}(v) - S(V_1)| = |V_1 - S(V_1)| = |V_1| - |S(V_1)| \geq \beta_1 n - (\alpha_1 n - 1)$ since $|S(V_1)| \leq \alpha_1 n - 1$. Thus, the digraph G_1 satisfies the desired $\Delta_1 n$-expanding property.

For the induction step, assume that the statement in the theorem is true for all depths less than k where $k > 1$. Suppose that α_i and β_i, $i = 1, 2, \ldots, k$, are positive integer constants such that $\Delta_i > 0$ for $i = 1, 2, \ldots, k$. The positiveness of $\Delta_1, \Delta_2, \ldots, \Delta_{k-1}$ enables us to apply the induction hypothesis in the case of depth $k - 1$. Let n be sufficiently large that $\lfloor n^{\frac{k-1}{k}} \rfloor \geq N_{k-1}$. Then there exists a synchronous $(\alpha_{k-1} \lfloor n^{\frac{k-1}{k}} \rfloor, \beta_{k-1} \lfloor n^{\frac{k-1}{k}} \rfloor)$-network G_{k-1} with depth $k-1$ and size $O((\lfloor n^{\frac{k-1}{k}} \rfloor)^{1+\frac{1}{k-1}})$ that satisfies the corresponding $\Delta_{k-1} \lfloor n^{\frac{k-1}{k}} \rfloor$-expanding property. For sufficiently large n, let $\tau = \lceil \frac{\alpha_k n}{\alpha_{k-1} \lfloor n^{\frac{k-1}{k}} \rfloor} \rceil$ and $\mathcal{G}$ be a set of τ pairwise vertex-disjoint copies of G_{k-1}, say $\mathcal{G} = \{G_{k-1}^{(1)}, G_{k-1}^{(2)}, \ldots, G_{k-1}^{(\tau)}\}$ where $G_{k-1}^{(i)}$ has rank partition $(V_j^{(i)})_{j=0}^{k-1}$ and stage partition $(E_j^{(i)})_{j=1}^{k-1}$, and consider the synchronous network $G = proj(\mathcal{G}, V_k)$ where $|V_k| = \beta_k n$.

Clearly, G is a synchronous $(\tau \alpha_{k-1} \lfloor n^{\frac{k-1}{k}} \rfloor, \beta_k n)$-network (note that $\tau \alpha_{k-1} \lfloor n^{\frac{k-1}{k}} \rfloor = \lceil \frac{\alpha_k n}{\alpha_{k-1} \lfloor n^{\frac{k-1}{k}} \rfloor} \rceil \alpha_{k-1} \lfloor n^{\frac{k-1}{k}} \rfloor \geq \alpha_k n$) with depth k. We can obtain the size of G by noting that the projection of $G_{k-1}^{(i)}$ onto V_k, $proj(G_{k-1}^{(i)}, V_k)$, has size $|E(G_{k-1}^{(i)})| + \beta_k n = |E(G_{k-1})| + \beta_k n$, and therefore $|E(G)| = \sum_{\eta=1}^{\tau}(|E(G_{k-1}^{(\eta)})| + \beta_k n) = \tau(|E(G_{k-1})| + \beta_k n)$. Observe that $|E(G_{k-1})|$ is $O((\lfloor n^{\frac{k-1}{k}} \rfloor)^{1+\frac{1}{k-1}})$, and $(\lfloor n^{\frac{k-1}{k}} \rfloor)^{1+\frac{1}{k-1}} \leq (n^{\frac{k-1}{k}})^{\frac{k}{k-1}} = n$, therefore $|E(G)|$ is $O(\tau n)$. We note that $\tau n = \lceil \frac{\alpha_k n}{\alpha_{k-1} \lfloor n^{\frac{k-1}{k}} \rfloor} \rceil n$, and thus $|E(G)|$ is $O(n^{1+\frac{1}{k}})$.

We now show that the synchronous network G satisfies the $\Delta_k n$-expanding property. Let S be a set of pairwise vertex-disjoint routes of G, and v be an input vertex of G that is not saturated by S (i.e., $v \in (\cup_{\eta=1}^{\tau} V_0^{(\eta)}) - S(\cup_{\eta=1}^{\tau} V_0^{(\eta)}))$. Observe that,

$$(\cup_{\eta=1}^{\tau} V_0^{(\eta)}) - S(\cup_{\eta=1}^{\tau} V_0^{(\eta)}) = \cup_{\eta=1}^{\tau}(V_0^{(\eta)} - (\cup_{\xi=1}^{\tau} S(V_0^{(\xi)}))) = \cup_{\eta=1}^{\tau}(V_0^{(\eta)} - S(V_0^{(\eta)}))$$

by the pairwise disjointedness of $V_0^{(i)}$, $i = 1, 2, \ldots, \tau$. Therefore, $v \in V_0^{(i)} - S(V_0^{(i)})$ for some $i \in \{1, 2, \ldots, \tau\}$. Note that, by the pairwise vertex-disjointedness in $\mathcal{G} = \{G_{k-1}^{(i)} \mid i = 1, 2, \ldots, \tau\}$ and the definition of $proj(\mathcal{G}, V_k)$, the set $S^- = \{P - \{term(P)\} \mid P \in S\}$ can be partitioned into $S^{(1)}, S^{(2)}, \ldots, S^{(\tau)}$ where $S^{(i)}$ is a set of pairwise vertex-disjoint directed paths in $G_{k-1}^{(i)}$ that are directed from $V_0^{(i)}$ into $V_{k-1}^{(i)}$ for $i = 1, 2, \ldots, \tau$.

Since $v \in V_0^{(i)} - S(V_0^{(i)}) \; (= V_0^{(i)} - S^{(i)}(V_0^{(i)}))$, then by the $\Delta_{k-1} \lfloor n^{\frac{k-1}{k}} \rfloor$-expanding property of $G_{k-1}^{(i)}$, there exist at least $\Delta_{k-1} \lfloor n^{\frac{k-1}{k}} \rfloor$ vertices in $V_{k-1}^{(i)} - S^{(i)}(V_{k-1}^{(i)})$ to which v can be directed via directed paths that are vertex-disjoint from each directed path in $S^{(i)}$, and therefore from each directed path in S^-. Thus, by the definition of projection in $proj(\mathcal{G}, V_k)$, there exist at least $\Delta_{k-1} \lfloor n^{\frac{k-1}{k}} \rfloor \lfloor \frac{|V_k|}{|V_{k-1}^{(i)}|} \rfloor$ vertices in V_k to which v can be directed via directed paths that are vertex-disjoint from each directed path in S^-, that is, there exist at least $\Delta_{k-1} \lfloor n^{\frac{k-1}{k}} \rfloor \lfloor \frac{|V_k|}{|V_{k-1}^{(i)}|} \rfloor - |S^- - S^{(i)}|$ vertices in V_k to which v can be directed via routes of G that are vertex-disjoint from each route in S. Consider that

$$\delta_k = \Delta_{k-1} \lfloor n^{\frac{k-1}{k}} \rfloor \lfloor \frac{|V_k|}{|V_{k-1}^{(i)}|} \rfloor - |S^- - S^{(i)}| \geq \Delta_{k-1} \lfloor n^{\frac{k-1}{k}} \rfloor \lfloor \frac{|V_k|}{|V_{k-1}^{(i)}|} \rfloor - |S^-|$$

$$\geq \Delta_{k-1} \lfloor n^{\frac{k-1}{k}} \rfloor (\frac{|V_k|}{|V_{k-1}^{(i)}|} - 1) - (\tau \alpha_{k-1} \lfloor n^{\frac{k-1}{k}} \rfloor - 1)$$

$$\geq (\Delta_{k-1} \frac{\beta_k}{\beta_{k-1}} - \alpha_k) n - (\Delta_{k-1} + \alpha_{k-1}) \lfloor n^{\frac{k-1}{k}} \rfloor + 1.$$

The coefficient of n in δ_k is

$$\Delta_{k-1} \frac{\beta_k}{\beta_{k-1}} - \alpha_k = (\beta_{k-1} - \alpha_{k-1} - (\sum_{\eta=1}^{k-2} \frac{\alpha_\eta}{\beta_\eta}) \beta_{k-1} - (k-2) \frac{\beta_{k-1}}{\beta_1}) \frac{\beta_k}{\beta_{k-1}} - \alpha_k$$

$$= \beta_k - \alpha_k - (\sum_{\eta=1}^{k-1} \frac{\alpha_\eta}{\beta_\eta}) \beta_k - (k-1) \frac{\beta_k}{\beta_1} + \frac{\beta_k}{\beta_1} = \Delta_k + \frac{\beta_k}{\beta_1}$$

(note that $\sum_{\eta=1}^{k-2} \frac{\alpha_\eta}{\beta_\eta} = 0$ if $k \leq 2$). Thus, $\delta_k \geq (\Delta_k + \frac{\beta_k}{\beta_1}) n - (\Delta_{k-1} + \alpha_k) \lfloor n^{\frac{k-1}{k}} \rfloor + 1 \geq \Delta_k n$ for sufficiently large n.

We observe that the $\Delta_k n$-expanding property of G is preserved if any set of input vertices together with the edges incident with them are deleted. Therefore, the digraph G_k, obtained by deleting a set of input vertices from G such that the input rank has cardinality $\alpha_k n$, satisfies the statement in the theorem. This completes the induction step, and the proof of the theorem. □

3 Size Optimal Bound versus Constant Depth

Our scheme for constructing synchronous strictly non-blocking fixed ratio concentrators with non-full capacity and constant depth k is to employ the above-mentioned expanding networks as principal components. The construction is composed of two facets of interconnection, in which the first $k - 1$ stages yield a set of pairwise vertex-disjoint synchronous expanding networks and the last stage unites the projections of moderate expansion from the components onto the output rank. Hence the concern is the actual accessibility of any input of a given component to the outputs, which is limited by the existing concentration effect created by the other components.

Theorem 2. *For positive integer constants α, β, γ, and k with $\alpha > \beta > \gamma$, and for sufficiently large n, there exists a synchronous strictly non-blocking γn-limited $(\alpha n, \beta n)$-concentrator with non-full capacity γn and depth k, and size $O(n^{1+\frac{1}{k}})$ (the implied constants depend on α, β, γ, and k).*

Proof. Since $\beta > \gamma > 0$, we can see that there exists a sufficiently large constant t such that $t \geq 2k - 3$ and $1 - \frac{\gamma}{\beta} \geq \frac{2k-2}{2^t}$. Let α_i and β_i be positive integer constants such that $\beta_1 = 2^t$ and $\frac{\alpha_i}{\beta_i} = \frac{1}{2^t}$ for $i = 1, 2, \ldots, k - 1$.

We show that the positive integer parameters α_i and β_i, $i = 1, 2, \ldots, k - 1$ satisfy the hypothesis of Theorem 1. For $i = 1, 2, \ldots, k - 1$,

$$\Delta_i = \beta_i - \alpha_i - (\sum_{\eta=1}^{i-1} \frac{\alpha_\eta}{\beta_\eta})\beta_i - (i - 1)\frac{\beta_i}{\beta_1} = 2^t \alpha_i - \alpha_i - (\sum_{\eta=1}^{i-1} \frac{1}{2^t})2^t \alpha_i - (i - 1)\frac{2^t \alpha_i}{2^t}$$

$$= (2^t - 2i + 1)\alpha_i \geq (2^t - 2(k - 1) + 1)\alpha_i > (t - (2k - 3))\alpha_i \geq 0.$$

Thus, for sufficiently large n such that $\lfloor n^{\frac{k-1}{k}} \rfloor \geq N_{k-1}$ (a constant), there exists a synchronous $(\alpha_{k-1}\lfloor n^{\frac{k-1}{k}} \rfloor, \beta_{k-1}\lfloor n^{\frac{k-1}{k}} \rfloor)$-network G_{k-1} with depth $k - 1$ and size $O((\lfloor n^{\frac{k-1}{k}} \rfloor)^{1+\frac{1}{k-1}})$ that satisfies the $\Delta_{k-1}\lfloor n^{\frac{k-1}{k}} \rfloor$-expanding property stated in Theorem 1.

For sufficiently large n, let $\tau = \lceil \frac{\alpha n}{\alpha_{k-1}\lfloor n^{\frac{k-1}{k}} \rfloor} \rceil$. Define $\mathcal{G}$ as in the proof of Theorem 1, and $G = proj(\mathcal{G}, V_k)$ where $|V_k| = \beta n$.

The derivation in the proof of Theorem 1 can also show that G is a synchronous $(\tau \alpha_{k-1}\lfloor n^{\frac{k-1}{k}} \rfloor, \beta n)$-network (note that $\tau \alpha_{k-1}\lfloor n^{\frac{k-1}{k}} \rfloor = \lceil \frac{\alpha n}{\alpha_{k-1}\lfloor n^{\frac{k-1}{k}} \rfloor} \rceil \alpha_{k-1}\lfloor n^{\frac{k-1}{k}} \rfloor \geq \alpha n$) with depth k and size $O(n^{1+\frac{1}{k}})$. It must be shown that the network G is a synchronous strictly non-blocking γn-limited $(\tau \alpha_{k-1}\lfloor n^{\frac{k-1}{k}} \rfloor, \beta n)$-concentrator with depth k, that is, for a state S of G and a γn-limited concentration request v in S, there exists a state S_1 of G that contains S and an additional route satisfying the concentration request v. We proceed as in showing the $\Delta_k n$-expanding property of G in Theorem 1, and we can see that the desired state S_1 exists provided that $\delta = \Delta_{k-1}\lfloor n^{\frac{k-1}{k}} \rfloor \lfloor \frac{|V_k|}{|V_{k-1}^{(i)}|} \rfloor - (|S^-| - |S^{(i)}|) > 0$ where $|S^-| = |S| \leq \gamma n - 1$. Noting that $\Delta_{k-1} = (2^t - (2k - 3))\alpha_{k-1}$, we have

$$\delta \geq \Delta_{k-1}\lfloor n^{\frac{k-1}{k}} \rfloor (\frac{|V_k|}{|V_{k-1}^{(i)}|} - 1) - |S^-| \geq \Delta_{k-1}\lfloor n^{\frac{k-1}{k}} \rfloor (\frac{\beta n}{\beta_{k-1}\lfloor n^{\frac{k-1}{k}} \rfloor} - 1) - (\gamma n - 1)$$

$$= (\frac{\Delta_{k-1}}{\beta_{k-1}}\beta - \gamma)n - \Delta_{k-1}\lfloor n^{\frac{k-1}{k}} \rfloor + 1 = ((2^t - (2k - 3))\frac{\alpha_{k-1}}{\beta_{k-1}}\beta - \gamma)n - \Delta_{k-1}\lfloor n^{\frac{k-1}{k}} \rfloor + 1$$

$$= (1 - \frac{\gamma}{\beta} - \frac{2k-2}{2^t})\beta n + \frac{1}{2^t}\beta n - \Delta_{k-1}\lfloor n^{\frac{k-1}{k}} \rfloor + 1 \geq \frac{1}{2^t}\beta n - \Delta_{k-1}\lfloor n^{\frac{k-1}{k}} \rfloor + 1 > 0$$

for sufficiently large n.

Since the strictly non-blocking concentration property and its capacity in a network are preserved if any set of input vertices together with the edges incident with them are deleted, the digraph obtained by deleting a set of input vertices from G such that the input rank has cardinality αn is the desired synchronous strictly non-blocking γn-limited $(\alpha n, \beta n)$-concentrator with depth k and size $O(n^{1+\frac{1}{k}})$. This completes the proof of the theorem. □

The following theorem, which gives lower bound size-depth tradeoffs for synchronous strictly non-blocking concentrators and generalized-concentrators with arbitrary depth, together with the size upper bound in Theorem 2, provide the size optimal bound for synchronous strictly non-blocking fixed ratio concentrators with non-full capacity and constant depth.

Theorem 3.

1. *[Dai93] For positive integers n, m, and c with $n > m \geq c$, the optimal size of strictly non-blocking c-limited (n,m)-concentrators with depth 1 is $(n - m + c)c + (m - c)$. For positive integers n, m, c, and k such that $n > m \geq c$, a synchronous strictly non-blocking c-limited (n,m)-concentrator with depth k has size at least $k(n - m + \frac{k}{k+1}c)c^{\frac{1}{k}} - (k-1)(n-c)$.*

2. *[Dai94] For positive integers n, m, c, and r with $n > m \geq c \geq r$, the optimal size of strictly non-blocking (c,r)-limited (n,m)-generalized-concentrators with depth 1 is $nc - \lfloor \frac{m-c}{r} \rfloor (c - r)$.*
For positive integers n, m, c, r, and k such that $n > m \geq c \geq r$, a synchronous strictly non-blocking (c,r)-limited (n,m)-generalized-concentrator with depth $k \geq 2$ has size at least $k(n - m + \frac{k}{k+1}c)c^{\frac{1}{k}} - (k-1)(n-c)$ if $r < k - 1$, and $\alpha_k(n - \frac{m}{r} + \beta_k \frac{k}{k+1} \frac{c}{r})r^{\frac{k-1}{k}}c^{\frac{1}{k}} - \frac{1}{2}(n - \frac{m}{r})r$ otherwise, where $\alpha_k = \frac{1}{2} \frac{k}{(k-1)^{\frac{k-1}{k}}}$ ($> \frac{1}{2}$) and $\beta_k = 1 - \frac{1}{2^{1+\frac{1}{k}}}$ ($> \frac{1}{2}$).

Theorem 4. *For positive integer constants α, β, γ, and k with $\alpha > \beta > \gamma$, the optimal bound on the size of synchronous strictly non-blocking γn-limited $(\alpha n, \beta n)$-concentrator with non-full capacity γn and depth k is $\Theta(n^{1+\frac{1}{k}})$ (the implied constants depend on α, β, γ, and k).*

4 Size Upper Bound for Full Capacity

Dai [Dai91] showed a folklore result how the capacity parameter of networks promotes the interconnection property possessed by the networks from concentration to connection in both wide-sense and strictly non-blocking contexts, and thus increases the size complexity of the networks.

Theorem 5. *For positive integers n and m with $n > m$, if G is a strictly (wide-sense) non-blocking (n,m)-concentrator with full capacity, then G is a strictly (respectively, wide-sense) non-blocking (n,m)-connector with full capacity.*

The best known upper bound on the size of synchronous strictly non-blocking (n,n)-connectors with full capacity and depth k is $O(n^{1+\frac{1}{\lceil k/2 \rceil}})$, via an explicit recursive construction due to Clos [Clo53]. We show below the same size upper bound for these concentrators versus constant depth, but it employs the explicit construction of expanding networks detailed in Theorem 1.

Theorem 6. *For positive integer constants α, β, and k with $\alpha \geq \beta$, there exists a synchronous strictly non-blocking βn-limited $(\alpha n, \beta n)$-concentrator with full capacity and depth k (that is, connector in the same context), and size $O(n^{1+\frac{1}{\lceil k/2 \rceil}})$ (the implied constants depend on α, β, and k).*

Proof. It suffices to prove the theorem for the case of $k = 2j - 1$. Let t be a sufficiently large integer constant such that $2^{t-1} > 2j-3$. Consider the sequences of positive integer constants α_i and β_i such that $\beta_1 = 2^t$ and $\frac{\alpha_i}{\beta_i} = \frac{1}{2^t}$ for $i = 1, 2, \ldots, j-1$, as in the proof of Theorem 2, which satisfy the hypothesis of Theorem 1, i.e., $\Delta_i = (2^t - 2i + 1)\alpha_i > 0$ for $i = 1, 2, \ldots, j - 1$. Thus, for sufficiently large n, there exists a synchronous $(\alpha_{j-1}\lfloor n^{\frac{i-1}{j}} \rfloor, \beta_{j-1}\lfloor n^{\frac{i-1}{j}} \rfloor)$-network G_0 with depth $j - 1$ and size $O((\lfloor n^{\frac{i-1}{j}} \rfloor)^{1+\frac{1}{j-1}})$, which satisfies the $\Delta_{j-1}\lfloor n^{\frac{i-1}{j}} \rfloor$-expanding property stated in Theorem 1. We also note that in this case, $\Delta_{j-1} > \frac{1}{2}\beta_{j-1}$ since

$$\Delta_{j-1} = (2^t - (2j-3))\alpha_{j-1} = 2^{t-1}\alpha_{j-1} + (2^{t-1} - (2j-3))\alpha_{j-1}$$
$$= \frac{1}{2}\beta_{j-1} + (2^{t-1} - (2j-3))\alpha_{j-1} > \frac{1}{2}\beta_{j-1}.$$

For a digraph $H = (V, E)$, its mirror image denoted by H^r is defined to be the digraph (V, E^r) where $E^r = \{(u, v) \mid (v, u) \in E\}$. For sufficiently large n, let $a = \lceil \frac{\alpha n}{\alpha_{j-1}\lfloor n^{\frac{i-1}{j}} \rfloor} \rceil$ and $b = \lceil \frac{\beta n}{\alpha_{j-1}\lfloor n^{\frac{i-1}{j}} \rfloor} \rceil$. Let G be a synchronous $(a\alpha_{j-1}\lfloor n^{\frac{i-1}{j}} \rfloor, b\alpha_{j-1}\lfloor n^{\frac{i-1}{j}} \rfloor)$-network with depth k, together with rank partition $(V_i)_{i=0}^k$ and stage partition $(E_i)_{i=1}^k$, defined as follows (note that $a\alpha_{j-1}\lfloor n^{\frac{i-1}{j}} \rfloor = \lceil \frac{\alpha n}{\alpha_{j-1}\lfloor n^{\frac{i-1}{j}} \rfloor} \rceil \alpha_{j-1}\lfloor n^{\frac{i-1}{j}} \rfloor \geq \alpha n$, and similarly, $b\alpha_{j-1}\lfloor n^{\frac{i-1}{j}} \rfloor \geq \beta n$). The subdigraph G_{in} of G induced by its first j ranks $V_0, V_1, \ldots, V_{j-1}$ is the graph-theoretic union of $G_1, G_2, \ldots, G_a$, which is a sequence of a pairwise vertex-disjoint copies of above-mentioned synchronous network G_0, while the subdigraph G_{out} of G induced by its last j ranks $V_j, V_{j+1}, \ldots, V_k$ is the graph-theoretic union of $G_1^r, G_2^r, \ldots, G_b^r$, which is a sequence of b pairwise vertex-disjoint copies of G_0^r; and the center stage E_j is composed of all the edges directed from the output rank of G_p into the input rank of G_q^r in any one-to-one correspondence manner for $p = 1, 2, \ldots, a$, and $q = 1, 2, \ldots, b$.

Computation can show that $|E(G_{in})|$ and $|E(G_{out})|$ are $O(n^{1+\frac{1}{j}})$, and $|E_j| = ab\beta_{j-1}\lfloor n^{\frac{i-1}{j}} \rfloor$ that is $O(n^{1+\frac{1}{j}})$. Thus $|E(G)|$ is $O(n^{1+\frac{1}{j}})$. To show that the network G is a synchronous strictly non-blocking $(a\alpha_{j-1}\lfloor n^{\frac{i-1}{j}} \rfloor, b\alpha_{j-1}\lfloor n^{\frac{i-1}{j}} \rfloor)$-connector with depth $k = 2j - 1$, let S be a state of G and (u, v) be a $b\alpha_{j-1}\lfloor n^{\frac{i-1}{j}} \rfloor$-limited connection request in S. Therefore there exist $p \in \{1, 2, \ldots, a\}$ and $q \in \{1, 2, \ldots, b\}$ such that u is an input of G_p and v is an output of G_q^r. Since G_p satisfies the $\Delta_{j-1}\lfloor n^{\frac{i-1}{j}} \rfloor$-expanding property stated in Theorem 1, there exist at least $\Delta_{j-1}\lfloor n^{\frac{i-1}{j}} \rfloor$ $(> \frac{1}{2}\beta_{j-1}\lfloor n^{\frac{i-1}{j}} \rfloor)$ vertices in the output rank of G_p to which u can be directed via directed paths that are vertex-disjoint from each route in S. By symmetry, a similar argument gives that there exist $\Delta_{j-1}\lfloor n^{\frac{i-1}{j}} \rfloor$ vertices in the input rank of G_q^r from which v can be directed via directed paths that are vertex-disjoint from each route in S. Then, since the directed edges in E_j provide an one-to-one correspondence between the outputs of G_p and the inputs of G_q^r, and $2\Delta_{j-1} > \beta_{j-1}$, there exists a route that satisfies the connection request (u, v) and is vertex-disjoint from each route in S. This shows the strictly non-blocking connection property of G. A desired synchronous strictly non-blocking fixed ratio $(\alpha n, \beta n)$-connector can be obtained by deleting all but αn inputs and all but βn outputs of G; and the theorem is proved. □

5 Concluding Remarks

The equivalences between the full-capacity generalized-concentration (concentration) and the full-capacity generalized-connection (respectively, connection) in the strictly and wide-sense non-blocking contexts allow us to apply the known results on size lower bound and upper bound versus depth ($\Omega(n^{1+\frac{1}{k-1}})$ and $O(n^{1+\frac{1}{\lceil k/2 \rceil}})$, respectively) for synchronous strictly non-blocking (n,n)-connectors with full capacity and depth $k \geq 2$ to synchronous strictly non-blocking concentrators with full capacity. These two size bounds show an optimality result for depth $k \leq 3$, but there is a considerable gap between them for depth $k \geq 4$. An improvement in narrowing this gap is desirable.

References

[Ben65] V. E. Beneš. *Mathematical Theory of Connecting Networks and Telephone Traffic*. Academic Press, New York, 1965.

[BP74] L. A. Bassalygo and M. S. Pinsker. Complexity of an optimum nonblocking switching network without reconnections. *Problems of Information Transmission*, 9:64–66, 1974.

[BP80] L. A. Bassalygo and M. S. Pinsker. Asymptotically optimal networks for generalized rearrangeable switching and generalized switching without rearrangement. *Problemy Peredachi Informatsii*, 16:94–98, 1980.

[Clo53] C. Clos. A study of non-blocking switching networks. *Bell System Technical Journal*, 32:406–424, 1953.

[Dai91] H. K. Dai. Complexity issues in strictly non-blocking networks. Ph. D. Dissertation, Department of Computer Science and Engineering, University of Washington. 1991.

[Dai93] H. K. Dai. On synchronous strictly non-blocking concentrators and generalized-concentrators. In *Proceedings of the 7th International Parallel Processing Symposium*, pages 406–412, April 1993.

[Dai94] H. K. Dai. An improvement in the size-depth tradeoff for strictly non-blocking generalized-concentration networks. In C. Halatsis, D. Maritsas, G. Philokyprou, and S. Theodoridis, editors, *Lecture Notes in Computer Science (817): PARLE'94 Parallel Architectures and Languages Europe*, pages 214–225, Springer-Verlag, Berlin Heidelberg, 1994.

[DDPW83] D. Dolev, C. Dwork, N. Pippenger, and A. Wigderson. Superconcentrators, generalizers and generalized connectors with limited depth. In *Proceedings of the Fifteenth ACM Symposium on the Theory of Computing*, pages 42–51. Association for Computing Machinery, May 1983.

[FFP88] P. Feldman, J. Friedman, and N. Pippenger. Wide-sense nonblocking networks. *SIAM Journal on Discrete Mathematics*, 1(2):158–173, 1988.

[Fri88] J. Friedman. A lower bound on strictly non-blocking networks. *Combinatorica*, 8(2):185–188, 1988.

[GG81] O. Gabber and Z. Galil. Explicit construction of linear-sized superconcentrators. *Journal of Computer and System Sciences*, 22:407–420, 1981.

[Mar75] G. A. Margulis. Explicit constructions of concentrators. *Problems of Information Transmission*, 9:325–332, 1975.

[Pip74] N. Pippenger. On the complexity of strictly non-blocking concentration networks. *IEEE Transactions on Communications*, COM-22(11):1890–1892, November 1974.

Bandwidth and Cutwidth of the Mesh of d-Ary Trees*

Dominique BARTH

LRI U.R.A. CNRS 410, Université de Paris Sud, Centre d'Orsay, Bât. 490
91405 Orsay, France
e-mail:barth@lri.fr

Abstract. We mainly show that the cutwidth of the mesh of d-ary trees $MT(d,n)$ satisfies $\frac{d^{n-2}(d+1)^2}{8} - 1 \leq c(MT(d,n)) \leq \frac{d^{n+3}}{d-1}$; if $d > 2$, we also show that the bandwidth of this graph $b(MT(d,n))$ is in $\theta\left(d^{n+1}\frac{d^n-1}{n(d-1)}\right)$.

1 Introduction and definitions

In this paper, we focus on the bandwidth and the cutwidth of a graph, two well known graph parameters. They are defined as follows. We use graph theory notations of [3]; let G be a graph, $V(G)$ (resp. $E(G)$) the set of vertices (resp. edges) of G. Consider $\mathcal{L}(G)$ the set of all the labelings of $V(G)$; a *labeling* of $V(G)$ is a bijection l between $V(G)$ and $\{0, \ldots, |V(G)| - 1\}$. The bandwidth of G is defined by $b(G) = \min_{l \in \mathcal{L}(G)} \left(\max_{[X,X'] \in E(G)} |l(X) - l(X')| \right)$.

The cutwidth of G is $c(G) = \min_{l \in \mathcal{L}(G)} \left(\max_{X \in V(G)} c_l(X) \right)$, where

$$c_l(X) = |\{[Y,Y'] \in E(G) : l(Y) \leq l(X) < l(Y')\}|.$$

Finding the bandwidth and the cutwidth of a graph are known to be NP-complete problems [5, 7]. These parameters are useful to determine good VLSI designs for interconnection networks, by considering the *Thompson grid model* of VLSI layout [8]. We deal here with the mesh of d-ary trees. This graph is an interesting interconnection network for parallelism, since it uses both tree and grid structures (see section 2). Good parallel algorithms have been developed in it [1, 6], and it has been proposed as a good parallel computer topology for some applications to images analysis [4].

Let us first precise that in all the following, we use some language theory notations. Let $\{0, \ldots, d-1\}$ be an alphabet, with $d \geq 2$, and v be a word on it. We denote by $|v|$ the *length* of v, i.e. the number of letters in v; we denote by e the empty word, i.e. $|e| = 0$. Each letter $x \in \{0, \ldots, d-1\}$ is also considered as an element of $\mathbb{Z}_d$.

---
* This work was supported by the "Opération RUMEUR" of the French PRC PRS.

The mesh of d-ary trees, introduced in [1], is a generalization of the mesh of binary trees (i.e. $MT(2,n)$) [6]. It is defined as follows.

- The vertices of $MT(d,n)$ are all the couples $(u;v)$ of words of $\{0,..,d-1\}^*$, such that $|u|=n$ and $|v|<n$, or $|v|=n$ and $|u|\le n$.
- There is an edge between two vertices $(u;v)$ and $(w;z)$ in $MT(d,n)$ iff $|u|=n$, $u=w$ and $[v,z]\in E(T(d,n))$, or $|v|=n$, $v=z$ and $[u;w]\in E(T(d,n))$. This edge is denoted by the pair $[(u;v),(w;z)]$.

By definition, for any $u\in\{0,\ldots,d-1\}^n$, the subgraph of $MT(d,n)$ induced by the vertices of the form $(u;v)$ (resp. $(v;u)$), with $v\in\{0,\ldots,d-1\}^*$, is isomorphic to $T(d,n)$. In $MT(d,n)$, the column tree (resp. line tree), isomorphic to $T(d,n)$ with root $(e;u)$ (resp. $(u;e)$), is denoted by $T_{(e;u)}$ (resp. $T_{(u;e)}$).
The number of vertices of $MT(d,n)$ is $|V(MT(d,n))|=d^n\left(d^n+2\frac{d^n-1}{d-1}\right)$. We will also use the following recursive construction of $MT(d,n)$ from $MT(d,n-1)$.
1. Consider first d^2 disjoint copies of $MT(d,n-1)$, each one denoted by $MT_{i,j}(d,n-1)$ with $(i,j)\in\{0,\ldots,d-1\}^2$. A vertex $(u;v)$ in $MT_{i,j}(d,n-1)$ is denoted by $(u;v)_{i,j}$.
2. We add $2d^n$ new vertices : d^n vertices $(w;e)$, with $w\in\{0,\ldots,d-1\}^n$ and d^n vertices $(e;w)$. Then, we add the set of edges

$$\bigcup_{(i,j)\in\{0,\ldots,d-1\}^2}\{[(e;w),(e;u)_{i,j}]:w=ui\}\bigcup\{[(w;e),(u;e)_{i,j}]:w=ui\}.$$

By associating each vertex $(u;v)_{i,j}$ to the vertex $(ui;vj)$ in $MT(d,n)$, it is easy to see that the graph we obtain is isomorphic to $MT(d,n)$.

We also give the notations and the definitions we use in the next section, similar to the ones of [2]. Let G be a graph and Π be a partition of $V(G)$. For each element $\pi\in\Pi$, we define $\omega(\pi)$ as the *cocycle* of π, i.e. the set $\{[X,X']\in E(G):X\in\pi,X'\notin\pi\}$. We note $max_\Pi=\max_{\pi\in\Pi}|\pi|$ and $max_\omega=\max_{\pi\in\Pi}|\omega(\pi)|$. We also denote by $G[\pi]$ the subgraph of G induced by π.
- The *quotient graph of G by Π*, denoted by $Q=G/_\Pi$, is defined by
  - $V(Q)=\Pi$,
  - $[\pi,\pi']\in E(Q)\Leftrightarrow(\pi\ne\pi',\exists X\in\pi,\exists X'\in\pi':[X,X']\in E(G))$.
- Let l_Q be a labeling of $V(Q)$ and l_G be a labeling of $V(G)$. For each $\pi\in\Pi$, we represent by $l_G[\pi]$ the set of all the labels of vertices in π by l_G. We say that l_G is *compatible* with l_Q if for any $\pi\in\Pi$, we have $l_G[\pi]$ is an interval $[m_\pi,..,M_\pi]$ and if for each $\pi'\in\Pi$ such that $l_Q(\pi')<l_Q(\pi)$ (resp. $l_Q(\pi')<l_Q(\pi)$), $M_{\pi'}<m_\pi$ (resp. $m_{\pi'}>M_\pi$).
- For each $\pi\in\Pi$,

$$\delta_\omega^+(\pi)=\min_{X\in\pi}|\{[X,X']\in E(G):X'\in\pi',\pi' \text{ such that } l_Q(\pi')>l_Q(\pi)\}|$$
$$\delta_\omega^-(\pi)=\min_{X\in\pi}|\{[X,X']\in E(G):X'\in\pi',\pi' \text{ such that } l_Q(\pi')<l_Q(\pi)\}|.$$

The *edge-bissection* of G is denoted by $bis_e(G)$ (see [3]).
With a general result of [2] and with an original construction, we show in [1] the next result.

Proposition 1. $b(MT(d,n))$ is in $\theta\left(d^{n+1}\frac{d^n-1}{n(d-1)}\right)$

In the next section, we also use the following result from [2].

Theorem 2. *Let l_Q be a labeling of $V(Q)$ achieving the cutwidth of Q, and let l_G be a labeling of $V(G)$ compatible with l_Q.*

(1.) $c(G) \leq \left(c(Q) - \left\lceil\frac{\delta(Q)}{2}\right\rceil\right) \cdot max_\omega$
$\quad\quad + \max_{\pi \in \Pi}\left(c(G[\pi]) + (|\omega(\pi)| - |\pi| \cdot \min\left(\delta_\omega^+(\pi), \delta_\omega^-(\pi)\right))\right)$

(2.) $c(G) \geq bis_e(G)$

2   The cutwidth of $MT(d,n)$

Proposition 3. *If $n \geq 2$, then $c(MT(2,n))$ is in $\theta(2^n)$ and if $d \geq 3$,*
- $c(MT(d,n)) \geq \frac{d^{n+2}(d^n(d+1)-2)^2}{4(d^{2n}((d^3-d)(d+2)+1)-d^n(d^4+2d^3+3d^2-4d+1)+d^3+3d^2-d))}$
- $c(MT(d,n)) \leq \frac{d^n(d^3+d^2+4)-(d^4-3d+2)}{2(d-1)}$

Corollary 4. *If $n \geq 2$, then if $d \geq 3$, $\frac{d^{n-2}(d+1)^2}{8} - 1 \leq c(MT(d,n)) \leq \frac{d^{n+3}}{d-1}$.*

Proof of the proposition.
1. *Let us show the upper bound.*
a. Let us first define a partition Π of $V(MT(d,n))$, with $d \geq 2$ and $n > 1$. Π contains $d^2 + 2d$ parts : d^2 parts $\pi_{i,j}$ with $(i,j) \in \{0,\ldots,d-1\}^2$; d other parts, each one denoted by $\pi_{i,e}$, and d last parts $\pi_{e,j}$. They are defined by

$$\begin{cases} -\ \pi_{i,j} \text{ is the set of all the vertices } (m;m') \in V(MT(d,n)) \text{ such that } i \text{ is the} \\ \quad \text{first letter of } m \text{ and } j \text{ the first letter of } m'. \\ -\ \pi_{i,e} \text{ (resp } \pi_{e;j}) \text{ is the set of all the vertices } (m;e) \text{ (resp } (e;m')) \text{ where } i \\ \quad (\text{resp. } j) \text{ is the first letter of } m \text{ (resp. } m') \end{cases}$$

By definition, $\pi_{i;e}$ and $\pi_{e;j}$ are two independant sets of vertices of $MT(d,1)$. Moreover, it is easy to see that Π is a partition of $V(MT(d,n))$. Let us denote by Q the graph G/Π. We now consider a couple $(i,j) \in \{0,\ldots,d-1\}^2$.
• Consider $(iu;jv)$ a vertex in a part $\pi_{i,j}$. If X is a vertex in $MT(d,n)$, with $X \notin \pi_{i,j}$, and if $[(iu;jv),X] \in E(MT(d,n))$, then $v = e$ and $X = (iu;e) \in \pi_{i,e}$, or $u = e$ and $X = (e;jv) \in \pi_{e,j}$.
Hence, the edges of Q are pairs $[\pi_{i,j}, \pi_{i;e}]$ and $[\pi_{i,j}, \pi_{e;j}]$. Then, by associating to each part $\pi_{i,j}$ the couple $(i;j)$, and to each part $\pi_{i,e}$ (resp. $\pi_{e,j}$) the couple $(i;e)$ (resp. $(e;j)$), we can conclude that Q is isomorphic to $MT(d,1)$.
• The subgraph of $MT(d,n)$ induced by $\pi_{i,j}$ is isomorphic to $MT(d,n-1)$. This can be directly deduced from the definition of Π and by following the recursive construction of $MT(d,n)$ from $MT(d,n-1)$ given in section 1: we associate to each vertex $(iu;jv)$ in $\pi_{i,j}$ the vertex $(u;v) \in V(MT(d,n-1))$. Since $|\pi_{i,e}| = |\pi_{e,j}| = d^n$, then $max_\Pi = |V(MT(d,n-1))|$.
• $|\omega(\pi_{i,j})| = |\{(iu;j) \in \pi_{i,j}\} \cup \{(i;vj) \in \pi_{i,j}\}| = 2d^{n-1}$. Moreover, in $MT(d,n)$

the degree of each vertex from $\pi_{i;e}$ and from $\pi_{e,j}$ is equal to d. Hence, since $|\pi_{i,e}| = |\pi_{e,j}| = d^{n-1}$, then $|\omega(\pi_{i,e})| = |\omega(\pi_{e,j})| = d|\pi_{i,e}| = d^n$. So $max_\omega = d^n$.

b. To use Theorem 2, we give an upper bound for $c(MT(d,1))$ by using a labelling of $T(d,1)$ (see [1]), for $d > 2$. Hence, we show that if $d > 2$, $c(MT(d,1)) \leq (d+2)\lceil \frac{d}{2} \rceil - d$. If $d = 2$, $MT(2,1)$ is a cycle of length 8 and so $c(MT(2,1)) = 2$.

c. We can now apply Theorem 2. Assume $d > 2$ and $n > 1$,

$$c(MT(d,n)) \leq \left((d+2)\lceil \tfrac{d}{2} \rceil - d - 1\right) d^n + \\ max\left(c(MT(d,n-1)) + 2d^{n-1};\ d^n - (d^{n-1} \cdot \lfloor \tfrac{d}{2} \rfloor)\right)$$

We then deduce from this inequality an upper bound for $c(MT(d,n))$, i.e. $c(MT(d,n)) \leq \left\lceil \frac{d^n(d^3+d^2+4)-(d^4+3d+2)}{2(d-1)} \right\rceil$. If $d = 2$, we show by the same way that $c(MT(2,n)) \leq 2^{n+2} - 6$.

2. *We now deal with the lower bound. We give a detailed sketch of the proof.* We know that $bis_e(MT(2,n))$ is in $\theta(2^n)$ [6]. To determine $bis_e(MT(d,n))$ with $d > 2$, we give a routing function R in $MT(d,n)$. Then, $bis_e(MT(d,n)) \geq \frac{|V(MT(d,n))|^2+1}{2 \cdot cg(R)}$, with $cg(R)$ the congestion of R (see [1]). Thus, we show that $bis_e(MT(d,n)) \geq$

$$\frac{d^{n+2}(d^n(d+1)-2)^2}{4(d^{2n}((d^3-d)(d+2)+1) - d^n(d^4+2d^3+3d^2-4d+1) + d^3+3d^2-d))}$$

We conclude with Theorem 2.2 . □

References

1. D. Barth, Réseaux d'interconnexion : structures et communications, Thesis, Université Bordeaux I, 1994.
2. D. Barth, F. Pellegrini, J. Roman, A. Raspaud, On bandwidth, cutwidth and quotient graphs, LaBRI intern report, RAIRO Theor. Comp. Science, vol. 29, n. 6, pp. 487-508, 1995.
3. C. Berge, Graphes et Hypergraphes, North Holland Publishing, 1971.
4. M.M. Eshaghian, V.K. Prasanna Kumar, Parallel Geometric Algorithms for Digitized Pictures on Mesh of Trees, *Proc. of IEEE FOCS*, pp 270-273, 1986.
5. F. Gavril, Some NP-complete problems on graphs, Proc. 11th Conf. on Inf. sciences and systems, W.H Freeman pub., San Francisco, 1979.
6. F.T. Leighton, Introduction to Parallel Algorithms and Architecture: arrays, trees, hypercubes, Morgan Kaufman Publisher, 1992.
7. C.H. Papadimitriou, The NP-completeness of the bandwidth minimization problem, Computing, vol. 16, pp. 263-270, 1976.
8. C.D. Thompson, Area-time complexity for VLSI, Proc. 11th STOCS, ACM, pp. 81-88, 1979.

Variable-Dilation Embeddings of Hypercubes into Star Graphs: Performance Metrics, Mapping Functions, and Routing

Marcelo Moraes de Azevedo*[1], Nader Bagherzadeh[1], and Shahram Latifi[2]

[1] Dept. of Elec. & Comp. Engr. – Univ. of Calif. – Irvine, CA 92717
[2] Dept. of Elec. & Comp. Engr. – Univ. of Nevada – Las Vegas, NV 89154

Abstract. We present load 1 embeddings of a k-dimensional hypercube Q_k into an n-dimensional star graph S_n. Dimension i links of Q_k are mapped into paths of length at most d_i in S_n, where d_i varies with i rather than being fixed. Our embeddings are an attractive alternative to previously known techniques, producing small *average dilation* and small *average congestion* without sacrificing expansion. We provide a thorough characterization of our embeddings, which spans several combinations of node mapping functions and routing algorithms in S_n.

1 Introduction

The star graph [1] is regarded as an attractive network for parallel processing, featuring smaller degree and diameter than a hypercube [5] of comparable size. However, the repertory of star graphs algorithms is still small compared to that of the hypercube. In this paper, we investigate load 1 embeddings of Q_k into S_n, which can be used to port algorithms developed for the hypercube to a star graph. Previous work on this problem sought to minimize *dilation* and *expansion*, but has produced embeddings for which a trade-off between these metrics exists [7]. The difficulty in minimizing dilation and expansion is due to topological differences between the two networks (e.g., degree and minimum cycle length) [7]. Moreover, previous research on embeddings of Q_k into S_n has not addressed some important performance metrics, which are discussed in this paper. These include *average dilation*, *congestion*, and *average congestion*. The average dilation and average congestion metrics are good approximations for the communication slowdown induced by an embedding, and are often correlated. In particular, average dilation has been used as a standard performance metric in practical evaluations of embedding heuristics into hypercubes [4].

We present *variable-dilation embeddings* (*VDEs*) which consistently achieve small average dilation and small average congestion (e.g., one of our VDE techniques produces values for these metrics respectively in the ranges $[1.50, 3.24]$ and $[1.00, 3.21]$, for $n : 4 \rightarrow 10$). Simultaneously, the expansion of our VDEs matches that achieved by dilation 4 embeddings of [7]. Another advantage which stems

---

* This research is supported in part by CNPq, Brazil, under the grant No. 200392/92-1.

from our techniques include the capability of employing unused nodes in S_n (up to 100% capacity) to host additional VDEs (see [2] for more on this topic).

Using several performance metrics which are defined in Sec. 2, and a combination of mathematical analysis and computer simulation, we provide a thorough characterization of our VDEs. Metrics which are derived analytically include expansion, dilation, and dilation along each of the hypercube dimensions. Average dilation, average congestion, and congestion are computed with a custom simulation program, which supports all of the VDEs presented in this paper. Measures for these last three metrics were computed over a selection of four different *node mapping functions* (*NMFs*), and four different routing algorithms in S_n. The paper illustrates some of the measures we obtained, pointing out the most promising combinations of NMFs and routing algorithms.

2 Variable-Dilation Embeddings (VDEs)

Performance Metrics: Definitions. Let $G_k = \{V(G_k), E(G_k)\}$ be a hierarchical k-dimensional graph, such that G_{k+1} is obtained recursively from $c(k)$ many copies of G_k. We refer to the links connecting the $c(k)$ copies of G_k that exist within G_{k+1} as *dimension $(k + 1)$ links*.

An *embedding* of G_k into H_n, which we denote by $f : G_k \mapsto H_n$, is a mapping of $V(G_k)$ into $V(H_n)$ and of $E(G_k)$ into paths of H_n. In this paper, f is uniquely specified by a *node mapping function* (*NMF*) $f_V : V(G_k) \mapsto V(H_n)$ and a *deterministic routing algorithm* r_H of H_n. Thus, a link (u, v) of G_k is mapped to a path $f(u, v) = r_H(f_V(u), f_V(v))$.

The *node image of f* is $f(V(G_k)) = \{f_V(u) : u \in V(G_k)\}$. The *load of f*, denoted by $\lambda(f)$, is the maximum number of nodes of G_k that are mapped to any single node of H_n. The *dilation of f* is $d(f) = \max\{dist_H(f_V(u), f_V(v)) : (u,v) \in E(G_k)\}$, where $dist_H(x, y)$ is the distance in H_n between two vertices x and y of H_n. The *expansion of f* is $X(f) = |V(H_n)|/|V(G_k)|$.

Let $E_i(G_k)$ denote the subset of dimension i links in $E(G_k)$. The *dilation of f along the ith dimension of G_k* is $d_i(f) = \max\{dist_H(f_V(u), f_V(v)) : (u, v) \in E_i(G_k)\}$. Hence, $d(f) = \max\{d_i(f) : 1 \leq i \leq k\}$. f is a *variable-dilation embedding* (*VDE*) if $d_i(f) < d(f)$, for at least one dimension i of G_k. Accordingly, f is a *fixed-dilation embedding* if $d_i(f) = d(f)$, $\forall i$, $1 \leq i \leq k$. The *dilation vector of f* is $\overline{d(f)} = [d_1(f), d_2(f), \ldots, d_k(f)]$. The *average dilation of f* is $d_{avr}(f) = \left(\sum_{(u,v) \in E(G_k)} dist_H(f_V(u), f_V(v)) \right) \cdot (|E(G_k)|)^{-1}$.

Let (u, v) and (x, y) be links of G_k and of H_n, respectively. The congestion induced by (u, v) into (x, y), denoted by $cg_{(x,y)}(f(u, v))$, is 1 if $f(u, v)$ traverses (x, y), and 0 otherwise. The congestion induced by f into (x, y) is $cg_{(x,y)}(f) = \sum_{(u,v) \in E(G_k)} cg_{(x,y)}(f(u, v))$. The *congestion of f* is $cg(f) = \max\{cg_{(x,y)}(f) : (x, y) \in E(H_n)\}$. The *congestion induced by dimension i links* of G_k into H_n is $cg(f(E_i(G_k)) = \max\{\sum_{(u,v) \in E_i(G_k)} cg_{(x,y)}(f(u, v)) : (x, y) \in E(H_n)\}$. The *link image of f* is $f(E(G_k)) = \{(x, y) \in E(H_n) : cg_{(x,y)}(f) \geq 1\}$. The *average congestion* of f is $cg_{avr}(f) = \left(\sum_{(x,y) \in f(E(G_k))} cg_{(x,y)}(f) \right) \cdot (|f(E(G_k))|)^{-1}$.

The Guest Graph. A k-dimensional hypercube graph Q_k contains 2^k nodes, which are labeled with binary strings of length k. A node $\phi = q_1 \ldots q_i \ldots q_k$ is connected to k distinct nodes, respectively labeled with strings $\phi_i = q_1 \ldots \overline{q_i} \ldots q_k$, $1 \leq i \leq k$, where $\overline{q_i}$ denotes the binary negation of bit q_i [5]. The link connecting ϕ and ϕ_i is a *dimension i link* of Q_k.

The Host Graph. An n-dimensional star graph S_n contains $n!$ nodes which are labeled with the $n!$ possible permutations of n distinct symbols. In this paper, we use the integers $\{1, 2, \ldots, n\}$ to label the nodes of S_n. A node $\pi = p_1 \ldots p_i \ldots p_n$ is connected to $(n-1)$ distinct nodes, respectively labeled with permutations $\pi_i = p_i \ldots p_{i-1} p_1 p_{i+1} \ldots p_n$, $2 \leq i \leq n$ [1]. The link connecting π and π_i is a *dimension i link* of S_n.

The Intermediary Graph. Our embeddings of Q_k into S_n use an $(n-1)$-dimensional mesh of size $2 \times 3 \times \ldots \times n$ as an intermediary reference graph. We denote such a graph by M_{n-1}, and label its nodes with $(n-1)$-integer vectors $m_1 \ldots m_i \ldots m_{n-1}$, where $0 \leq m_i \leq i$. A node $\omega = m_1 \ldots m_i \ldots m_{n-1}$ is connected to at most $2n - 3$ distinct nodes, respectively labeled with vectors $\omega_i^- = m_1 \ldots (m_i - 1) \ldots m_{n-1}$ and $\omega_i^+ = m_1 \ldots (m_i + 1) \ldots m_{n-1}$, $1 \leq i < n$. ω_i^- (ω_i^+) is a *left (right) dimension i neighbor of ω* if ω_i^- (ω_i^+) exists.

NMFs $g_V : V(M_{n-1}) \mapsto V(S_n)$. Our embeddings of Q_k into S_n use two-step NMFs. Initially, we employ an NMF $h_V : V(Q_k) \mapsto V(M_{n-1})$. The second step uses an NMF $g_V : V(M_{n-1}) \mapsto V(S_n)$. The composite NMF $f_V : V(Q_k) \mapsto V(S_n)$ is denoted by $f_V = h_V \odot g_V$.

In what follows, we describe four different NMFs $g_V : V(M_{n-1}) \mapsto V(S_n)$, which we denote by g_V^{nonh}, g_V^{mnonh}, g_V^{hier}, and g_V^{qhier}. These NMFs are respectively referred to as the *non-hierarchical NMF* [6, 8], the *modified non-hierarchical NMF* [2], the *hierarchical NMF*, and the *quasi-hierarchical NMF*. All four NMFs embed M_{n-1} into S_n with load 1, expansion 1, and dilation 3. The dilation vectors of the embeddings produced by g_V^{nonh}, g_V^{mnonh}, g_V^{hier}, and g_V^{qhier} are respectively $[3, \underbrace{\ldots, 3}_{n-2}, 1]$, $[3, \underbrace{\ldots, 3}_{n-1}]$, $[1, 2, 3, \underbrace{\ldots, 3}_{n-3}]$, and $[1, 3, \underbrace{\ldots, 3}_{n-2}]$.

Let π be a permutation of n symbols. We denote the transposition of *symbols* i and j in π by $(i\ j)_s$. Similarly, we denote the transposition of the symbols occupying the ith and the jth *positions* in π by $(i\ j)_p$. We define an operator $\circ$ which applies transpositions to permutations. Hence, $2413 \circ (2\ 4)_s = 4213$, and $2413 \circ (2\ 4)_p = 2314$. NMFs g_V^{nonh}, g_V^{mnonh}, g_V^{hier}, and g_V^{qhier} can be generically described by the algorithm depicted in Table 1. For example, $g_V^{mnonh}(102) = 2134 \circ (4\ 3)_p \circ (3\ 2)_p = 2413$.

NMF $h_V : V(Q_k) \mapsto V(M_{n-1})$. In what follows, we present an NMF $h_V : V(Q_k) \mapsto V(M_{n-1})$, which is common to all of our VDEs. We denote the corresponding composite NMFs $f_V : V(Q_k) \mapsto V(S_n)$ by $f_V^{nonh} = h_V \odot g_V^{nonh}$, $f_V^{mnonh} = h_V \odot g_V^{mnonh}$, $f_V^{hier} = h_V \odot g_V^{hier}$, and $f_V^{qhier} = h_V \odot g_V^{qhier}$.

Let $F(x, y) = x(y+1) - 2^{x+1} + 2$, and let n, ℓ, and k be integers such that $n \geq 4$, $2 \leq \ell \leq \lfloor \log_2 n \rfloor$, and $F(\ell-1, n) < k \leq F(\ell, n)$. Let $\phi[\ ]$ and $\omega[\ ]$ be nodes of Q_k and M_{n-1}, respectively. An algorithmic description of h_V follows:

Table 1. Algorithmic description for NMFs $g_V : V(M_{n-1}) \mapsto V(S_n)$

a. Choose an initial permutation π_0 from Table 1a, according to the type of NMF g_V and mesh coordinates $m_1 m_2$.						
NMF	$m_1 m_2 = 00$	$m_1 m_2 = 10$	$m_1 m_2 = 01$	$m_1 m_2 = 11$	$m_1 m_2 = 02$	$m_1 m_2 = 12$
g_V^{nonh}	$12345\ldots n$	$12435\ldots n$	$13245\ldots n$	$13425\ldots n$	$14235\ldots n$	$14325\ldots n$
g_V^{mnonh}	$12345\ldots n$	$21345\ldots n$	$13245\ldots n$	$23145\ldots n$	$31245\ldots n$	$32145\ldots n$
g_V^{hier}	$12345\ldots n$	$21345\ldots n$	$31245\ldots n$	$13245\ldots n$	$23145\ldots n$	$32145\ldots n$
g_V^{qhier}	$12345\ldots n$	$21345\ldots n$	$23145\ldots n$	$13245\ldots n$	$32145\ldots n$	$31245\ldots n$

b. For $i : 3 \to (n-1)$, apply the first m_i transpositions specified for dimension i and NMF g_V in Table 1b to π_0.				
NMF	$i = 3$	$i = 4$	$\cdots$	$i = n - 1$
g_V^{nonh}	$(1\ 2)_s \circ (2\ 3)_s \circ (3\ 4)_s$	$(1\ 2)_s \circ \cdots \circ (4\ 5)_s$	$\cdots$	$(1\ 2)_s \circ \cdots \circ (n-1\ n)_s$
g_V^{mnonh}	$(4\ 3)_p \circ (3\ 2)_p \circ (2\ 1)_p$	$(5\ 4)_p \circ \ldots \circ (2\ 1)_p$	$\cdots$	$(n\ n-1)_p \circ \cdots \circ (2\ 1)_p$
g_V^{hier}, g_V^{qhier}	$(4\ 3)_s \circ (3\ 2)_s \circ (2\ 1)_s$	$(5\ 4)_s \circ \cdots \circ (2\ 1)_s$	$\cdots$	$(n\ n-1)_s \circ \cdots \circ (2\ 1)_s$

```
for (i = 1; i < n; i++) ω[i] = 0;
for (e = 1; F(e-1,n) < k; e = e + 1)
  for (i = F(e-1,n) + 1; i ≤ min(F(e,n),k); i = i + 1)
    ω[i - F(e-1,n) + 2^e - 2] = ω[i - F(e-1,n) + 2^e - 2] + 2^{e-1} · φ[i];
```

Table 2 depicts the VDE $h : Q(4) \mapsto M(3)$ produced by NMF h_V. Properties for h are $X(h) = 1.5$, $\lambda(h) = 1$, $d(h) = 2$, $\overline{d(h)} = [1, 1, 1, 2]$, and $d_{avr}(h) = 1.25$. Also shown in Table 2 are VDEs $f : Q(4) \mapsto S(4)$ produced by f_V^{nonh}, f_V^{mnonh}, f_V^{hier}, and f_V^{qhier}. The properties of f^{hier}, for example, are $X(f^{hier}) = 1.5$, $\lambda(f^{hier}) = 1$, $d(f^{hier}) = 4$, $\overline{d(f^{hier})} = [1, 2, 3, 4]$, and $d_{avr}(f^{hier}) = 2$.

Table 2. Image nodes for NMFs h_V, f_V^{nonh}, f_V^{mnonh}, f_V^{hier}, and f_V^{qhier} ($n = k = 4$)

NMF	Hypercube node (ϕ)															
	0000	1000	0100	1100	0010	1010	0110	1110	0001	1001	0101	1101	0011	1011	0111	1111
h_V	000	100	010	110	001	101	011	111	002	102	012	112	003	103	013	113
f_V^{nonh}	1234	1243	1324	1342	2134	2143	2314	2341	3124	3142	3214	3241	4123	4132	4213	4231
f_V^{mnonh}	1234	2134	1324	2314	1243	2143	1342	2341	1423	2413	1432	2431	4123	4213	4132	4231
f_V^{hier}	1234	2134	3124	1324	1243	2143	4123	1423	1342	3142	4132	2341	3241	4231	2431
f_V^{qhier}	1234	2134	2314	1324	1243	2143	2413	1423	1342	3142	3412	1432	2341	3241	3421	2431

Properties of VDEs $f : Q_k \mapsto S_n$. (see [3] for a proof).

Theorem 1. Let $F(x, y) = x(y + 1) - 2^{x+1} + 2$, and let n, ℓ, and k be integers such that $n \geq 4$, $2 \leq \ell \leq \lfloor \log_2 n \rfloor$, and $F(\ell - 1, n) < k \leq F(\ell, n)$. Let f_V be one of the node mapping functions f_V^{nonh}, f_V^{mnonh}, f_V^{hier}, and f_V^{qhier}, and let $f : Q_k \mapsto S_n$ be one of the corresponding embeddings f^{nonh}, f^{mnonh}, f^{hier}, and f^{qhier} generated by f_V. For each f, we define integers $\gamma_i(f)$ as follows:

- $\gamma_i(f^{nonh}) = 0$ if $i = F(e,n)$, $\forall e \in [1,\ell]$, and $\gamma_i(f^{nonh}) = 2$ otherwise.
- $\gamma_i(f^{mnonh}) = 2$, $\forall i$.
- $\gamma_1(f^{hier}) = 0$, $\gamma_2(f^{hier}) = 1$, and $\gamma_i(f^{hier}) = 2$, for all $i > 2$.
- $\gamma_1(f^{qhier}) = 0$, $\gamma_2(f^{qhier}) = 1$, and $\gamma_i(f^{qhier}) = 2$, for all $i > 2$.

Then, $\lambda(f) = 1$, $X(f) = n!/2^k$, $d(f) = \max_i\{d_i(f)\}$, and:

$$\overline{d(f)} = [\underbrace{1 + \gamma_i(f), \ldots}_{n-1}, \underbrace{2 + \gamma_i(f), \ldots}_{n-3}, \ldots, \underbrace{2^{e-1} + \gamma_i(f), \ldots}_{F(e,n)-F(e-1,n)}, \ldots, \underbrace{2^{\ell-1} + \gamma_i(f), \ldots}_{k-F(\ell-1,n)}].$$

Routing and Simulation Results. We used simulation to characterize other important metrics of our VDEs, such as average dilation, average congestion, congestion, and congestion induced by dimension i links of Q_k. The tool supports all of the NMFs f_V^{nonh}, f_V^{mnonh}, f_V^{hier}, and f_V^{qhier}, and four deterministic routing algorithms in S_n. These algorithms are denoted by r_S^{can}, r_S^{rcan}, r_S^{ecan}, and r_S^{ocan}.

Recall that a link $(u,v) \in E(Q_k)$ is mapped to a path $f(u,v)$ in S_n with endpoints $f_V(u)$ and $f_V(v)$. Let $\pi_{u,v}$ denote the permutation which sorts $f_V(u)$ into $f_V(v)$. Each routing algorithm employs a particular format for the *cyclic representation of $\pi_{u,v}$*, and executes $\pi_{u,v}$ accordingly [3]. r_S^{can}, r_S^{rcan}, r_S^{ecan}, and r_S^{ocan} represent $\pi_{u,v}$ in *canonical format, reverse canonical format, even-only canonical format*, and *odd-only canonical format*, respectively [3].

Figure 1 depicts the average dilation and the average congestion produced by VDEs employing the combination $f_V^{qhier} + r_S^{rcan}$. Similar plots for other combinations of NMF and routing algorithm can be found in [3]. We consider the cases $k : 2 \to 19$ and $n : 4 \to 10$, which correspond respectively to hypercubes of sizes $4 \to 524,288$, and star graphs of sizes $24 \to 3,628,800$. f_V^{nonh}, f_V^{mnonh}, f_V^{hier}, and f_V^{qhier} achieve average dilation which lie in the ranges $[2.25, 3.53], [2.00, 3.60]$, $[1.50, 3.21]$, and $[1.50, 3.24]$, respectively. Accordingly, measures for average congestion produced by the combinations $f_V^{nonh} + r_S^{can}$, $f_V^{mnonh} + r_S^{can}$, $f_V^{hier} + r_S^{rcan}$, and $f_V^{qhier} + r_S^{rcan}$ lie in the ranges $[1.00, 5.47]$, $[1.19, 2.96]$, $[1.07, 3.81]$, and $[1.00, 3.21]$, respectively. Note that the average dilation computed for a VDE does not depend on the choice of routing algorithm.

Hierarchical NMFs achieve smaller average dilation than non-hierarchical NMFs, and produce smaller average congestion when used in combination with r_S^{rcan}. Conversely, r_S^{can} minimizes the average congestion produced by non-hierarchical NMFs. The combinations $f_V^{mnonh} + r_S^{can}$ and $f_V^{qhier} + r_S^{rcan}$ seem to be the most appropriate for star graphs employing canonical and reverse canonical routing, respectively. r_S^{ecan} and r_S^{ocan} produce congestion metrics which lie between the minima and maxima obtained with r_S^{rcan} (r_S^{can}) and r_S^{can} (r_S^{rcan}), when used in combination with hierarchical and non-hierarchical NMFs, respectively.

Several of our VDEs produce congestion 1 or 2 on the links of S_n when a single dimension of Q_k is used [3]. This is particularly important for algorithms which employ only a fraction of the links of Q_k at any point of their execution (e.g., SIMD algorithms). From the viewpoint of congestion, some interesting results are: 1) $f_V^{mnonh} + r_S^{can}$ produces VDEs whose congestion is less than k, over the ranges $k : 2 \to 19$ and $n : 4 \to 10$, and 2) $f_V^{qhier} + r_S^{rcan}$ produces VDEs with congestion 1 when $k \leq n - 1$ [3].

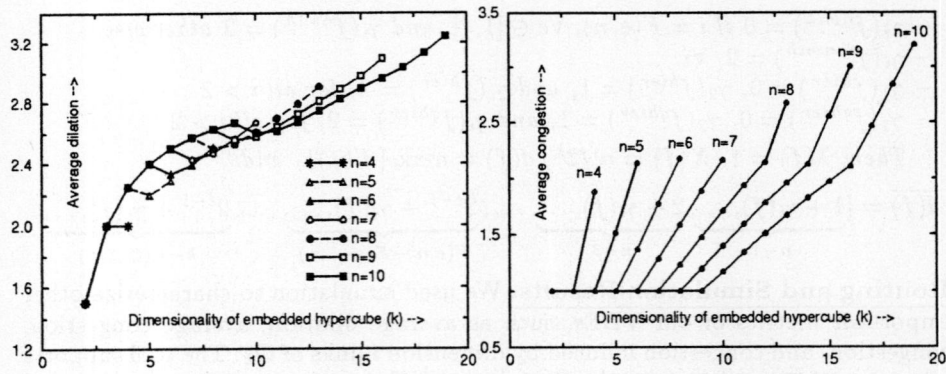

Fig. 1. Average dilation and average congestion produced by f_V^{qhier} and r_S^{rcan}

3 Conclusion

This paper presented novel techniques for embedding a hypercube into a star graph. Our embeddings are designed for performance, and consistently produce small average dilation and small average congestion. We achieve these goals by employing variable-dilation embeddings, and a careful selection of node mapping functions and routing algorithms. Our techniques demonstrated the possibility of embedding large hypercubes into the star graph, with corresponding small expansion. On continued research, we are expanding our investigation on congestion metrics to a related technique introduced by the authors. Such a technique is referred to as *packing*, and can produce optimal expansion (i.e., 1) [2].

References

1. Akers, S. B., Harel, D., Krishnamurthy, B.: The star graph: an attractive alternative to the n-cube. Proc. Int. Conf. Par. Proc. (1987) 393-400
2. Azevedo, M. M., Latifi, S., Bagherzadeh, N.: Low expansion packings and embeddings of hypercubes into star graphs. Proc. IEEE 15th Phoenix Conf. Comp. Comm. (1996) 115-122
3. Azevedo, M. M., Bagherzadeh, N., Latifi, S.: Variable-dilation embeddings of hypercubes into star graphs: performance metrics, mapping functions, and routing. Dept. Elec. & Comp. Engr. U. Cal., Irvine. Tech. Rep. ECE 96-05-01 (1996)
4. Chen, W.-K., Stallmann, M. F. M., Gehringer, E. F.: Hypercube embedding heuristics: an evaluation. Int. J. Par. Prog. **18** No. 6 (1989) 505-549
5. Hayes, J. P., Mudge, T.: Hypercube supercomputers. Proc. IEEE. **77** No. 12 (1989) 1829-1841
6. Jwo, J. S., Lakshmivarahan, S., Dhall, S. K.: Embedding of cycles and grids in star graphs. J. Circ., Sys., and Comp. **1** No. 1 (1991) 43-74
7. Nigam, M., Sahni, S., Krishnamurthy, B.: Embedding Hamiltonians and hypercubes in star interconnection networks. Proc. Int. Conf. Par. Proc. (1990) 340-343
8. Ranka, S., Wang, J.-C., Yeh, N.: Embedding meshes on the star graph. J. Par. Dist. Comp. **19** (1993) 131-135

Overlapping Communication and Computation in Hypercubes*

Luis Díaz de Cerio, Miguel Valero-García and Antonio González

Dept. d'Arquitectura de Computadors - Univ. Polit. de Catalunya
c/ Gran Capitán s/n, Campus Nord - D6, E-08071 Barcelona, Spain
E-mail: {ldiaz,miguel,antonio}@ac.upc.es

Abstract. This paper presents a method to derive efficient algorithms for hypercubes. The method exploits two features of the underlying hardware: a) the parallelism provided by the multiple communication links of each node and b) the possibility of overlapping computations and communications, which is a feature of machines supporting an asynchronous communication protocol. The method can be applied to a generic class of hypercube algorithms. Many examples of this class of algorithms are found in the literature for different problems. The paper shows the efficiency of the method using two of these problems as an example: FFT and Vector Add. The results show that the reduction in communication overhead is very significant in many cases and the algorithms produced by our method are always very close to the optimum in terms of execution time.

1 Introduction

Hypercube multicomputers are interesting because there are many problems for which parallel algorithms with a hypercube communication topology are obtained in a natural way [7], [14].

Communication overhead is a crucial issue when considering the performance of a multicomputer. In some cases, communication overhead is the most significant factor in the execution time. To help in reducing communication overhead a) the nodes of a multicomputer can send several messages in parallel through different links (communication parallelism), and/or b) communication through one or several links can be overlapped with computation in the node (communication/computation overlapping)

Designing parallel algorithms which are able to exploit features (a) and/or (b) is not an easy task. Many of those natural hypercube algorithms found in [7], [14] cannot exploit these features efficiently. In any case, some papers can be found in the literature which propose hypercube algorithms for particular problems which are efficient for particular machine configurations. Examples are [1],[2], [4], [9], [12], and [13], just to mention a few.

In this paper we propose a method to derive hypercube algorithms which are able to exploit features (a) and (b). The method takes as an starting point a hypercube algorithm to solve the problem and transforms it in a systematic way, using a technique that we call communication pipelining. The starting algorithm must belong to a class of hypercube algorithms which we call CC-cube algorithms. Many numerical and symbolic computation problems can be solved using a CC-cube algorithm. FFT [9], Hartley transform [3], All-to-All personalized communications [8] and Jacobi methods for singular value decomposition and eigenvalue computation [11] are just some examples.

---

* This work has been supported by the Ministry of Education and Science of Spain (CICYT TIC-92/880 and TIC-91/1036) and the European Center for Parallelism in Barcelona (CEPBA).

In this paper, performance figures are given for two concrete application examples: FFT and Vector Add. Hypercube algorithms for FFT have been extensively studied in the literature. Among others, [9] and [2] are two concrete examples of algorithms which try to exploit communication parallelism ([9]) and communication/computation overlapping ([2]). In both cases the degree of communication parallelism and overlapping is fixed and, therefore, the results are only efficient for some machine configurations.

The communication pipelining technique, which is the basis of the proposed method, has been used in a previous paper [5], in which only communication parallelism is considered. The main contribution of this paper is the extension of the method in order to include the overlapping of communications and computations.

The rest of the paper is organized as follows. Next section describes the architecture assumptions. Section 3 defines the characteristics of a CC-cube algorithm. Section 4 reviews the communication pipelining technique. Section 5 establish the overlapping method that is based on the previously described concept of communication pipelining. Section 6 shows the performance figures that the method provides. Finally, the conclusions are found in section 7.

2 Target architecture

This study assumes a distributed memory multicomputer consisting of 2^d processors connected by bidirectional point-to-point links in a d-dimensional hypercube topology. Every node can send messages to any of its d neighbors following an asynchronous protocol. This means that, after initiating a communication operation through one or several of its links, a processor can continue performing computations in parallel with the transmission of the data. It is also assumed that every node can send messages in parallel along different links of the hypercube. However, the start-up times for the different communications cannot be overlapped (we assume that the start-up time corresponds mostly to time spent by the processor to initiate each transmission). Therefore, the cost of sending c messages in parallel along c different dimensions of the hypercube and performing afterwards C computations is considered to be:

$$cT_{sup} + \text{Max}\,(CT_a, L_{max}T_e)$$

where T_a is the time to perform a computation, T_{sup} is the communication start-up, T_e is the communication time per size unit and L_{max} is the size of the longest message to be sent. This model is in fact an upper bound that will be exact only in the case that the longest message is the last one to be sent. Notice that this model is valid for any control flow method (store-and-forward, wormhole, circuit switching) since the communications always take place between neighbor nodes.

3 CC-cube algorithms

A CC-cube algorithm consists of 2^d processes that perform some computation and exchange data among them. Each process communicates only with other d processes following a hypercube communication topology. Every process executes the same code, which has the following structure:

```
do i = 1, K
    compute x_i[1:N] plus some local data
    exchange x_i with neighbor in dimension d_i
enddo
```

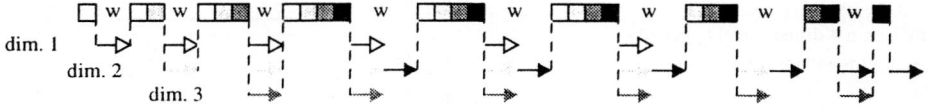
w: Node is waiting for data.

Fig. 1. An example of communication pipelining. The packets with the same gray level belong to the same iteration of the original CC-cube algorithm.

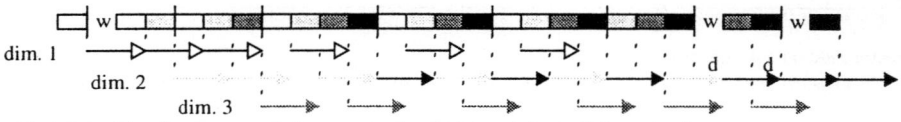
w: Node is waiting for data.   d: Communication is delayed due to link contention.

Fig. 2. Pipelining and overlapping of communications and computations.

where d_i is one of the dimensions of the hypercube ($d_i \in [1, d]$). Note that it is not necessary that each iteration uses a different dimension. The computation of $x_i[j]$ is a function of $x_{i-1}[j]$ (which was computed in iteration $i-1$ by the neighbor in dimension d_{i-1}) and possibly some local data.

4 Communication pipelining

The communication pipelining technique is inspired in the software pipelining approach used to generate code for VLIW processors [10]. Communication pipelining is based on the fact that, in order to compute $x_i[j]$ it is not necessary to have received the whole vector x_{i-1} from the neighbor in dimension d_{i-1} but simply element $x_{i-1}[j]$. In this situation, the algorithm is rewritten in as follows. Every vector x_i is decomposed into Q packets. As Fig. 1 shows, in the first iteration every node computes the first packet of x_1 and sends the result to neighbor through dimension d_1. In the second iteration, every node computes the second packet of x_1 and the first packet of x_2 (it has all the data required to perform these computations). At the end of this second iteration, each node sends two messages, one of them to neighbor through dimension d_1 containing the second packet of x_1, and the other one to neighbor through dimension d_2, containing the first packet of x_2. If $d_1 \neq d_2$, both packets can be sent in parallel; otherwise, they are combined into a single message and sent to its destination. Proceeding in this way, at the end of the third iteration every node can send three messages in parallel (if the involved dimensions are different). Following this approach, a parallel algorithm that makes use of all the links of the hypercube at the same time can be designed.

5 Overlapping communication and computation

After a packet is computed it can be sent to its destination at the same time that the following packet is computed. As Fig. 2 shows, the nodes do not send the data at the end of every iteration but they send the data after the computation of every packet (if the link is busy the communication is delayed until it becomes idle). In parallel with the transmission, the nodes compute consecutive packets if they have the necessary data.

The complete study of the execution time and the corresponding analytical models can be found in [6]. These models have been used to obtain the performance figures presented in the next section.

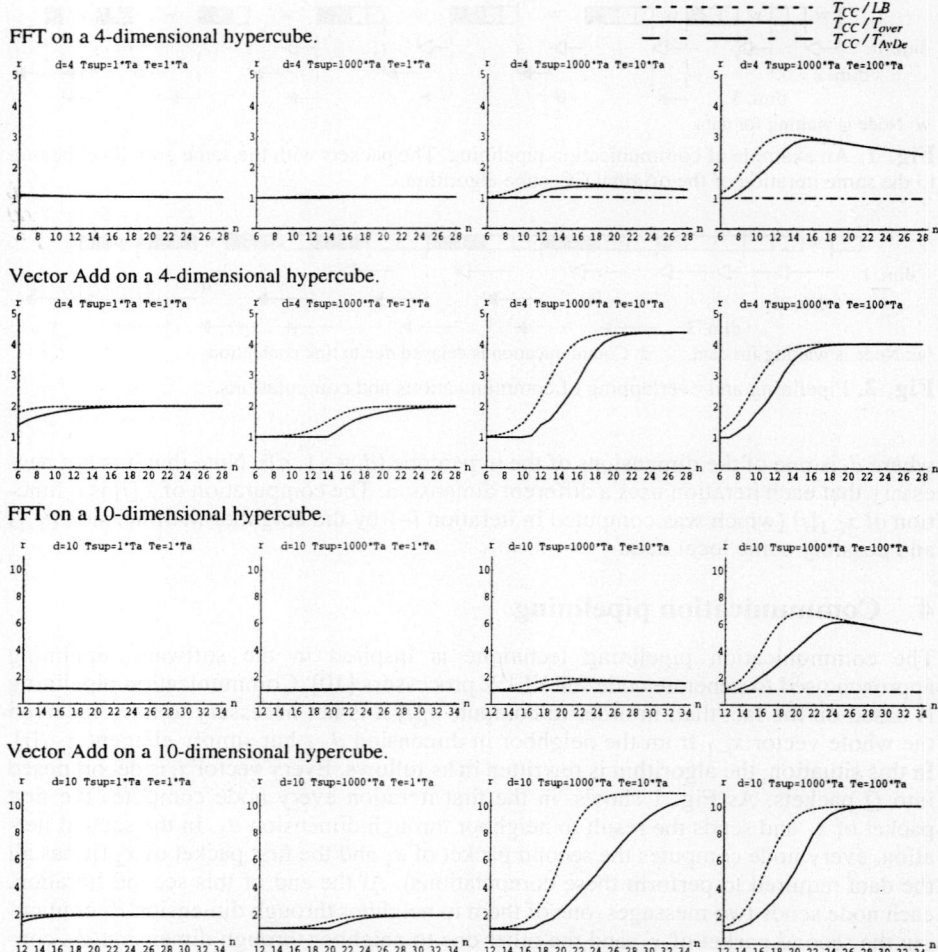

Fig. 3. Performance improvement of the overlapping scheme in relation to the CC-cube algorithm.

6 Performance figures

The plots in figure 3 show the performance improvement of the overlapping scheme in relation to the CC-cube algorithm (T_{CC}/T_{ov}, where T_{ov} is the execution time when using the overlapping method). In addition, they also show the performance improvement of a hypothetical algorithm that achieves an execution time equal to a lower bound (T_{CC}/LB). This lower bound represents the execution time under the assumption that an ideal overlapping of computations and communications can be achieved. The objectives of these plots are twofold: it is intended to show that overlapping provides a very important improvement and that it is very close to the optimum. These figures combine both problems (FFT and Vector Add) for two different sizes of the hypercube ($d = 4, 10$), varying the size of problems ($n \in [6, 34]$) and the communication parameters ($T_e/T_a = 1$, 10, 100 and $T_{sup}/T_a = 1$, 1000). In case of the FFT problem, the figures also show the relative improvement of the method proposed by Aykanat and Dervis (T_{CC}/T_{AyDe}) [2].

7 Conclusions

The main conclusion of the performance figures is that the proposed overlapping scheme gives results very close to the lower bound for all machine configurations and problem sizes. In addition, they show that a significant improvement is achieved in many cases (more than a factor 10 in some of the examples).

The main features of the proposed overlapping method are: a) It is a method that can be applied to a wide range of algorithms for hypercubes. In the literature, there are particular solutions to specific problems that are difficult to adapt to other problems. We have illustrated the method by means of two real problems: FFT and Vector Add. There are many other problems on which the method can be applied. b) The method permits to tune the degree of overlapping for a given problem and a given architecture (due to lack of space we have omitted the derivation of this parameter). Other proposals, as the one of Aykanat and Dervis, besides being a particular solution to a specific problem, they cannot vary the amount of overlapping and therefore the performance of the algorithm is much lower than that of our scheme, for some machine configurations.

8 References

1. Agarwal, R. C., Gustavson, F. G., Zubair, M.: An Efficient Algorithm for the 3-D FFT NAS Parallel Benchmark. Scalable High-Performance Computing Conf. (1994) 129-133
2. Aykanat, C., Dervis, A.: An Overlapped FFT Algorithm for Hypercube Multicomputers. ICPP (1991) III-316 - III-317
3. Aykanat, C., Dervis, A.: Efficient Fast Hartley Transform Algorithms for Hypercube - Connected Multicomputers. IEEE Transactions on Parallel and Distributed Systems, vol. 6, no. 6 (1995) 561-577
4. Clement, M. J., Quinn, M. J.: Overlapping Computations, Communications and I/O in Parallel Sorting. Journal of Parallel and Distributed Computing 28 (1995) 162-172
5. Díaz de Cerio, L., González, A., Valero-García, M.: Communication Pipelining in Hypercubes (submitted for publishing)
6. Díaz de Cerio, L., Valero-García, M., González, A.: Overlapping Communication and Computation in Hypercubes. DAC/UPC Research Report No. RR-96/02 (1996)
7. Fox, G. et al.: Solving Problems on Concurrent Processors. Englewood Cliffs, N. J. Prentice - Hall (1988)
8. Johnsson, S. L., Ho, C. T.: Optimum broadcasting and Personalized Communication in Hypercubes. IEEE Trans. Comput. 38 (1989) 1249-1268
9. Johnsson, S. L., Krawitz, R. L.: Cooley-Tukey FFT on the Connection Machine. Parallel Computing 18 (1992) 1201-1221
10. Lam, M.: Software Pipelining: An Effective Scheduling Technique for VLIW machines. Conf. on Programming Language Design and Implementation (1988) 318-328
11. Mantharam, M., Eberlein, P. J.: Block Recursive Algorithm to Generate Jacobi-sets. Parallel Computing 19 (1993) 481-496
12. Sahay, A.: Hiding Communication Costs in Bandwidth-Limited Parallel FFT Computation. Report: UCB/CSD 93/722, University of California (1993)
13. Suarez A., Ojeda-Guerra, C.: Overlapping Computations and Communications in Tours Networks. 4th Euromicro Workshop on Parallel and Distributed Processing (1996) 163-169
14. Thomson Leighton, F.: Introduction to Parallel Algorithms and Architectures: Arrays, Trees and Hypercubes. Morgan Kaufmann Publishers (1992)

Efficient delay routing*
(extended abstract)

Miriam Di Ianni[1]

Dipartimento di Scienze dell'Informazione, University of Rome "La Sapienza", via Salaria 113, I-00198 Rome, Italy. E-mail: diianni@dsi.uniroma1.it

Abstract. In this paper the computational complexity of finding minimum end-to-end delay packet routing schemes is studied. The existence of polynomial-time algorithms able to optimize both the end-to-end delay achievable when the number of packets in the network increases, and the number of packets that can be accepted in the network in order to keep the end-to-end delay within a constant value is investigated. In particular, it is proved the hardness of approximating in polynomial time both the minimum end-to-end delay and the maximum number of accepted packets even within a sublinear error in the number of packets.

1 Introduction

Efficient routing of messages is a fundamental task in parallel and distributed systems. Many packet routing algorithms trying to minimize the completion time of delivering packets have been proposed in the past [9, 11, 12, 17] but they were mainly devised for particular topologies. The first step towards the design of topology independent routing algorithms was the randomized technique proposed by Valiant and Brebner [24, 25], even if the authors first used it to route on hypercubes. Successively, a series of fundamental papers [1, 10, 14, 20, 21, 22, 23] showed the effective advantage of randomization in the design of efficient routing strategies. Universal deterministic packet routing was significantly approached by Leighton, Maggs Ranade and Rao [13, 14, 15]. Their solution to routing consists of two steps: during the first one the paths to be followed by packets are selected, while the second step, usually called *scheduling*, is used for the timing of packet movements is decided in order to minimize the total delivery time without violating network constraints (limited channel bandwidth and queue size). In [13], the authors proved the existence of a schedule bringing all packets to their respective destinations in $O(C + D)$ steps for any set P of paths used to route the packets, where C is the *congestion* (maximal number of paths in P using the same channel) and D the *dilation* (length of the longest path in P) of P. In [15] a polynomial time algorithm able to find such a schedule has been shown.

Usually, the delivery time of a packet is expressed as the sum of *network latency* and *end-to-end delay*. The first quantity is strictly dependent on the

---

* Work supported by the "Human Capital and Mobility" MAP project.

architectural choices for the network (topology, switching technique) and it is not affected by network congestion; instead, the end-to-end delay measures the number of times a packet must wait for traversing the links because of their limited bandwidth. In order to study the effects of network congestion on its performance, in this paper the attention is focused on the end-to-end delay: given a network and a set of packets, route them in order to minimize the maximum end-to-end delay. The interest in minimum end-to-end delay schedules, is also motivated by other reasons. The first one relies on some switching techniques yielding network latencies which can be considered almost independent from the lengths of the paths (virtual cut-through and wormhole routing). In such cases, the end-to-end delay is the main factor which strongly affects the delivery time of packets. The second reason is more theoretical and is related to the results in [13, 15]: even if the bound $\mathbf{O}(C + D)$ on the delivery time of a schedule is asymptotically optimal, it was still unknown if a schedule that completes the delivery of packets within $D+\mathbf{O}(C)$ steps can be found in in polynomial time. Notice that, while D is a "physical" constraint on the delivery time of any schedule, the number of steps required after the first D ones is a measure of the goodness of a schedule with respect to the congestion.

1.1 Results and paper organization

The minimum end-to-end delay problem has already been considered in [4, 5], where the authors proved the hardness of optimally routing and of approximating the minimum end-to-end delay in a network model in which the main resource packets must share is storage inside nodes. It will be called *buffer quarrel model*. In this paper, the results of [4, 5] are extended to a model in which an unbounded amount of buffers is associated to every node while the bandwith is kept bounded. This model will be called *channel quarrel model*. In this case, the main resource packets must share is edge bandwidth.

Both the routing paradigm proposed in [13], of first choosing the paths and then scheduling channel assignments, and the more general one, in which the choice of the paths and the scheduling are interleaved, are considered. The former will be called *fixed paths routing*, the latter *arbitrary paths routing*.

In section 2.1 it is proved that, in the fixed paths routing case, the minimum end-to-end delay cannot be approximated within a relative error in $\mathbf{O}(k^{\frac{1}{13}-\delta})$ for any $\delta > 0$, where k is the number of packets, unless P=NP. It follows that it is impossible to find in polynomial time schedules whose delivery time is $C+\mathbf{O}(D)$. Such result can be easily extended (via generalization) to the arbitrary paths routing case.

Assuming knowledge of the whole network available at each node is somehow unrealistic in large distributed systems since this requires a great amount of information to be exchanged between nodes. A relevant issue is thus to investigate the performance of *local strategies*, i.e. routing algorithms in which nodes send packets according to the knowledge of their neighbors' state only. A first contribution to this aim has been given in [5] by introducing a "local optimum" criterium: if two or more packets simultaneously require a buffer in a node and one of them can alternatively choose another node belonging to a different shortest path towards its destination and which is not requested by any other packet then it will use such node. In fact, if each node has the knowledge of the occupancy state of its neighbors only, it has not sufficient information for delaying a packet which can advance along another shortest path. The behavior of similar greedy strategies has already been investigated in [3, 19] with respect to the minimum delivery time and, as remarked in the second paper, such policies are widely used. While the hardness result for fixed paths routing can be easily extended to local strategies, a stronger result has been proved for arbitrary paths routing. In this case, approximating the minimum end-to-end delay with respect to local strategies within a relative error of $f(k)$ is NP-hard for *any* sublinear function f in the number k of packets. In section 2.2, the idea of "local greedy schedule" is considered in the channel quarrel model and the above result is extended to it.

Next, a flow control mechanism is considered: a call admission algorithm selects a subset of communication requests which can be satisfied within a fixed end-to-end delay. In section 3 it is proved that it is NP-hard to approximate the maximum number of packets which can be accepted in the network in order to schedule them with no end-to-end delay within a relative error in $O(r^{\frac{1}{6}-\delta})$, for any $\delta > 0$, where r is the number of communication requests.

In the last section conclusive remarks and open questions are discussed.

1.2 Preliminary definitions

A network is usually represented as a graph, where nodes stand for sites containing the processing elements and edges for communication links. A *bandwidth* is associated to each edge, representing the maximum number of packets which can be simultaneously transmitted on it. Each packet in the network follows a route starting at its *source node* and ending at its *destination*, and each transmission along one edge requires one unit of its bandwidth. If a packet cannot be transmitted along one link at a given time, it is stored in a buffer included in the output queue of that link. In this paper, it is always assumed that edge bandwidth is 1 and that transmission along one edge takes one time unit.

A network is in the *initial configuration* if all the packets are into their respective source nodes; it is in the *final configuration* if each packet has reached its own destination. The schedule ends when the network reaches the final configuration and the number of configurations met by a schedule S to end is called *delivery time* or *length* of S. If in a configuration i met by a schedule S a packet p has neither reached its destination nor is transmitted along any edge, then p is said to be *delayed* at i by S. The *delay* $d(p, S)$ of packet p denotes how many times p has been delayed by S. The end-to-end delay $d(S)$ of schedule S is the maximum of the $d(p, S)$'s.

Since the main results presented in this paper are hardness proofs, and since if a problem is intractable even under particular conditions then, a fortiori, it is intractable in the general case, some restrictions are imposed to the model in order to make the results stronger. Thius, unless differently stated, *layered networks* are always considered: nodes are partitioned in $L + 1 \geq 2$ sets or *levels* V^i, with $0 \leq i \leq L$ and edges exist only between consecutive levels. Furthermore, source nodes are always included in level 0, and destinations are always included in level L. Finally, only off-line schedules are considered in this paper, that is, all the packets are known before the schedule is started.

2 Approximating the minimum end-to-end delay

The attention in this section is focused on polynomial time algorithms able to find approximate solutions, that is, solutions whose sizes have bounded relative error with respect to sizes of the optimal ones. Here, the relative error of an algorithm A for a minimization problem Π is defined as follows:

$$\frac{m(S_A(x))}{m(S^*(x))}$$

where $S^*(x)$ is an optimum solution relative for instance x of Π, $S_A(x)$ is the approximate solution found by A, and $m(S^*(x))$ and $m(S_A(x))$ are, respectively, their sizes. A problem is said to be ϵ-approximable if a polynomial time algorithm A exists such that the relative error is never greater than ϵ.

In [4, 5] the minimum delivery time schedule and the minimum end-to-end delay schedule problems have been studied in the buffer quarrel model in which a packet cannot be transmitted from node u to node v if v does not contain a free buffer when the transmission is requested. Although the two models look quite similar, it does not seem easy to transform a buffer quarrel network N into a channel quarrel network N' in such a way that there is a strong relation between any schedule in N and the corresponding schedule in N'.

In fact, an intuitive transformation of N into N', would map each node u of N into an edge (u, u') of N' and assign to it a bandwidth equal to the number of buffers included in u. However, this transformation is not correct. Consider for instance the buffer quarrel network N in figure 1(a) in which every node contains one buffer: by making packet p_2 occupy earlier than p_1 the buffer contained in u and packet p_3 occupy earlier than p_2 the buffer contained in v, we get a schedule of length $L + 4 = 7$ in which no pair of packets reach their destinations at the same time. In figure 1(b), the network N' corresponding to N (according to the previous described transformation) is shown: in this case edges (u, u') and (v, v') have both bandwidth 1. By using the same priorities as before in order to assign edges to packets (i.e., edge (u, u') is assigned first to p_2 and edge (v, v') is assigned first to p_3), we get a schedule of length $L + 1$ in which packets p_1 and p_2 reach their destinations at the same time.

In spite of the previously remarked differences, in the model considered throughout this paper it is still possible to prove very similar results to the ones proved in [4, 5] for the buffer quarrel model.

2.1 Centralized strategies

In this section strategies that take decisions about packet transmissions according to the knowledge of the state of the entire network are considered.

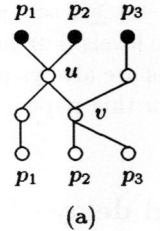
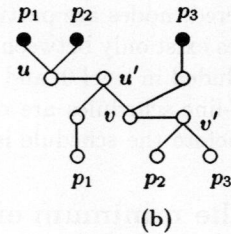

Fig. 1. Transformation of a buffer quarrel network into a channel quarrel network.

Theorem 1. *In the fixed paths routing case, the minimum end-to-end delay problem cannot be approximated with an error in $O(k^{\frac{1}{13}-\delta})$ for any $\delta > 0$, where k is the number of packets to be transmitted, unless P=NP.*

Proof. The proof is similar to the one presented in [5]. Let us consider the well-known NP-hard min-colorability problem [6]: given a graph $G = (V, E)$, find the minimum-size partition $V_1, \ldots, V_{hmin}$ of V such that no pair of nodes contained in a same V_i are adjacent in G. A reduction from min-colorability is shown such that the original graph G can be colored with h colors if and only if the corresponding network admits a schedule having end-to-end delay $h - 1$.

Let $\langle G = (V, E)\rangle$ be an instance of min-colorability with $n = |V|$ and $m = |E|$. The reduction maps $\langle G = (V, E)\rangle$ into a network N^G containing n source nodes s_i, each of them corresponding to a node of G. The packet contained in s_i will be denoted as x_i.

Basically the network is a chain of n identical *filters* such that the n outputs of one filter are connected to the n inputs of the next one (see figure 2 (c)). The goal of a filter is to create conflicts between pairs of packets representing adjacent edges in the input graph G. Each filter is a layered network with n source and n destinations. The pair of inner levels $(2i-1, 2i)$ corresponds to edge $e_i = (v, w) \in E$ of the input graph G: each of them contains $n-1$ nodes with one node in level $2i-1$ having indegree two and one node in level $2i$ having outegree two. Such nodes are connected by an edge which must be used by both of the two packets representing v and w. All the others nodes have indegree and outdegree one. Thus, there are $2m$ inner levels (see figure 2 (b)). Clearly, such a network can be constructed in polynomial time.

To prove that G can be colored with $h \leq n$ colors if and only if a schedule for the network exists with end-to-end delay $k = h - 1$ suppose first that a partition of V into h subsets $V_1, \ldots, V_h$ having the required property exists. In this case, packets are partitioned into h sets $S_1, \ldots, S_h$ where each S_i contains packets associated to the nodes in V_i: packets in S_i leave level 0 (i.e. they start) with a delay equal to $i-1$, for $i = 1, \ldots, h$. Since nodes in V_i are pairwise not adjacent, all packets included in a same set S_i, $i = 1, \ldots, h$, use distinct edges within the network and, thus, they are not furtherly delayed. Hence, all packets in S_i reach their destinations with delay i, that is, the end-to-end delay of such a schedule is $h - 1$.

Conversely, suppose the network admits a schedule with end-to-end delay $h - 1$ ($h \leq n$). Hence, the packets arrive at the end of the last filter partitioned into h sets $S_1, \ldots, S_h$. Thus, none of the S_i contains any pair of packets which are in conflict for the use of some edge. Indeed, this property is true when the set of packets having delay 0 (i.e. the set S_1) leaves the first filter. However, some pair of packets leaving the first filter with the same delay may have a conflict for the use of some edge in the

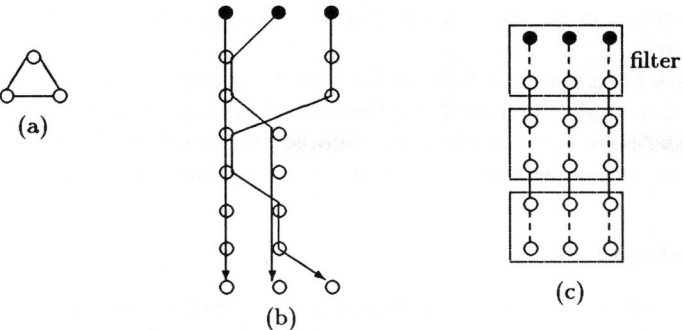

Fig. 2. An example of the reduction: (a) graph G, (b) filter, (c) the network as a chain of filters.

next levels: for instance, this happens if packet x and packet y have a conflict in level i of the first filter and both of them are delayed by some packet in S_1, respectively, in levels $j_x > i$ and $j_y > i$. But, at the end of the second filter, packets having delay 1 (i.e. the set S_2) are pairwise non conflicting. In general, at the end of the i-th filter this property is true for the set S_i. Thus, after at most h filters none of the sets S_i can contain conflicting packets. It follows that nodes corresponding to packets included in a same set S_i are pairwise not adjacent, that is, nodes of G can be partitioned into h pairwise disjoint subsets $V_1, V_2, \ldots, V_h$, where each V_i contains the nodes corresponding to packets included in S_i.

Since the delay of the schedule and the size of the coloring are linearly related, the previous reduction preserves approximation properties. Indeed, suppose that a polynomial-time $\epsilon(k)$-approximation algorithm A for the minimum end-to-end delay problem in the fixed paths routing case exists (k being the number of packets to be transmitted), that is, there exists a $\epsilon(k)$ such that for any networks N, A yields a scheduling $S(A, N)$ whose end-to-end delay $d(S(A, N))$ satisfies the following relation:

$$\frac{d(S(A,N))}{d^*(N)} < \epsilon(k)$$

where $d^*(N)$ denotes the optimum end-to-end delay for the network N. Then, A can be applied to the network N^G corresponding to some graph G with k nodes according to the reduction above: consider the partition $V_1, \ldots, V_{h(A)}$ for the nodes of graph G induced by the scheduling $S(A, N^G)$. Since the size h_{min} of an optimum coloring for G is $h_{min} = d^*(N) + 1$ then

$$\frac{h(A)}{h_{min}} = \frac{d(S(A,N)) + 1}{d^*(N) + 1} < \epsilon(k) + 1$$

This implies that any $\epsilon(k)$-approximation algorithm for the minimum end-to-end delay problem can be used also for the min-colorability one with the same performances. Since the min-colorability problem cannot be approximated with an error in $O(k^{\frac{1}{13}-\delta})$ for any $\delta > 0$ [18, 2], the assertion is proved.

The previous theorem can be easily extended to the arbitrary paths routing case by noticing that arbitrary paths routing can be viewed as a generalization of fixed paths routing, in which each entry of the routing table consists of one outgoing edge only.

Since the length l of a schedule S for an $L+1$-levels layered network N satisfies the relation $l = L + d(S)$, theorem 1 implies that it is impossible to optimally solve in polynomial time the minimum length schedule problem, unless P=NP. However, the same cannot be said concerning its approximation properties.

2.2 Local strategies

In large distributed systems centralized strategies are somehow unrealistic, since they require a great amount of information to be exchanged between nodes in the network. More frequently, the strategy to solve collisions and to decide the outgoing edge along which to forward a packet at a given time is chosen by each node according to the knowledge of the state of its neighborhood only. A relevant issue is thus to investigate the performance of such *local strategies*. As a first step towards this aim, it is assumed that if two or more packets are simultaneously requiring a transmission along the same edge and one of them can alternatively choose another edge belonging to a shortest path towards its destination and which is not requested by any other packet, then it is forced to advance along that edge. This is a sort of "local optimum" criterium: if each node has the knowledge of the occupancy state of its incident edges only then it has not sufficient information for delaying a packet which can advance along another shortest path. Due to its similarity with deflection routing, this kind of schedules will be called *deflection schedules*.

Observe that, according to the standard definition of approximability [6, 7], the performance of a local strategy should be compared to the performance of an optimal one, no matter if this optimum can be actually achieved by a local strategy. This clearly implies that all negative results obtained for global strategies can be immediately extended to the performances achieved by any local strategy. However, this criterium is often too pessimistic, that is, if global strategies are unrealistic, the performance of a local strategy should be compared to the optimum achievable by local strategies (see [8] for more discussions). In what follows, the last criterion will be adopted.

In the fixed paths routing case, just notice that, in the proof of theorem 1, the optimal strategy forces packets to advance whenever it is possible. Thus, theorem 1 can be extended to local strategies and it also holds for the new criterion. A stronger result can be proved for the arbitrary paths routing case.

Theorem 2. *In the arbitrary paths routing case, the minimum end-to-end delay deflection schedule problem is not approximable within an error $f(k)$, where f is any sublinear function and k is the number of packets, unless* P=NP.

Sketch of Proof. The proof is based on a reduction from the disjoint connecting paths problem (in short DCP) that produces a large gap between the minimum end-to-end delay in a network corresponding to a yes instance of DCP and the minimum end-to-end delay in a network corresponding to a no instance. The DCP problem is defined as follows: given a graph G and a set $\{(s_1,t_1),(s_2,t_2),\ldots(s_h,t_h)\}$ of pairs of nodes of G, decide if G contains h pairwise disjoint paths, each connecting a pair (s_i,t_i), $i = 1,\ldots,h$. DCP is a well-known NP-complete problem [6] and it has been recently proved to remain NP-complete also for instances restricted to layered graphs [4].

Given graph G with L_G levels and h pairs of nodes, the network N^G of the corresponding instance of the minimum end-to-end delay deflection schedule problem is composed by h identical subnetworks $N^1, N^2, \ldots, N^h$ plus a final 'funnel' F. Each N^i contains $2L_G + 2$ levels and is partitioned into two further subnetworks, N_1^i and N_2^i. N_1^i contains h pairs of source nodes and the two packets belonging to a pair, x_j^i and y_j^i, are forced to use the same edge. This device is used in order to avoid 0 delays that are inconsistent with the definition of relative error. The pair of levels 2 and 3 of N_1^i corresponds to the first two levels of G and, in general, the pair of levels $2j$ and $2j+1$ of N_1^i corresponds to the pair of levels j and $j+1$ of G. The remaining pairs of levels ($2j-1$ and $2j$) only contain 'vertical' edges, that is, edges connecting pairs of nodes of N_1^i that correspond to the same node in G. N_2^i does not contain any source node and starts at level 4; furthermore, each level of N_2^i contains h nodes and N_2^i consists of a set of h disjoint chains. Finally, level $2j+1$ of N_1^i and level $2j+2$ of N_2^i, $j \geq 1$, are a complete bipartite graph. F contains 3 levels and the first and the second levels contain $h^2 + 1$ nodes: all nodes of the last level of N_2^i, $i = 1,\ldots,h$, are connected with the last node of the first level of F, while node j of the last level of N_1^i is connected with node $h(i-1) + j$ of the first level of F. Node j of the first level of F is connected with node j of its second level which, in turn, is connected with the jth destination, $j = 1, 2, \ldots, h^2$. Finally, the last node of the first level of F is connected with the last node of second level which, in turn, is connected with every destination.

In figure 3 it is shown an example of the reduction. For the sake of simplicity, levels 0 and 1 of N_1^1 and N_1^3 have not been drawn and only packets x have been depicted; finally, the last node of the first two levels of F has been drawn in the center.

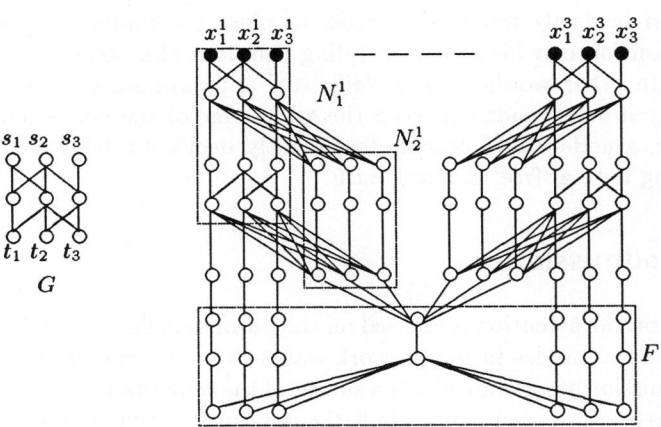

Fig. 3. Graph G and the corresponding N^G.

Notice that, if a packet x_j^i or y_j^i passes through N_2^i then there is a unique (shortest) path it can follow to reach its destination. Notice also that N_1^i contains h disjoint paths between the source and the destinations of each packet x_j^i (or y_j^i) if and only if the input graph G contains h disjoint paths between the pairs (s_j, t_j) of nodes.

Thus, if G contains h disjoint paths between the pairs (s_j, t_j) an optimum deflection schedule with delay $d^* = 1$ is easily achieved when all the x_j^i's follow the disjoint paths in N_1^i, followed by the y_j^i's in a pipeline fashion, for any $i = 1, \ldots, h$. Conversely, if G does not contain the h disjoint paths, all deflection schedules have end-to-end delay $d \geq 2h - 1$. In fact, whenever two packets $x_{j_1}^i$, $x_{j_2}^i$ (or $y_{j_1}^i$, $y_{j_2}^i$) have a conflict for using some edge of N_1^i, one of them is forced to advance in the first node of the next level of N_2^i, according to the definition of deflection schedule. This implies that all of them have to pass through the 'funnel' in the last node of the first level of F.

Suppose now an $f(h)$-approximation algorithm A for the minimum end-to-end delay schedule problem exists, where $f(h) = o(h)$. Without loss of generality, since $f(h) = o(h)$, consider an instance of DCP in which $h > \frac{f(h)+1}{2}$: apply to it the reduction above and, finally, apply A to the corresponding N^G. If the h disjoint paths exist, A finds a deflection schedule having end-to-end delay $d \leq f(h)$. Similarly, if the h disjoint paths do not exist, the algorithm finds a deflection schedule having end-to-end delay $d \geq 2h - 1 > f(h)$. This implies that the $f(h)$-approximation algorithm for the minimum end-to-end delay deflection schedule problem corresponds to a polynomial-time algorithm which decide the DCP problem, an absurd.

Since $k = 2h^2$, it has been proved that the minimum end-to-end delay deflection schedule problem cannot be approximated with an error in $O(k^{\frac{1}{2}-\delta})$ for any $\delta > 0$. Notice now that the same reasoning can be repeated for a network N^G composed by h^r identical subnetworks $N^1, N^2, \ldots, N^{h^r}$, for any $r > 0$: in this case, it can be shown that the minimum end-to-end delay deflection schedule problem cannot be approximated with an error in $O(k^{\frac{r}{r+1}-\delta})$ for any $\delta > 0$. Since this claim holds for any $r > 0$, the assertion is completely proved.

It is interesting to observe that the network N^G in the proof of the previous theorem can be easily modified in order to place the funnel F arbitrarily 'far' from the position in which the scheduling decisions that generate a 'big' delay are taken. In other words, a more 'efficient' approximating schedule could be derived only if every node can keep the knowledge of the entire network. Only in this case, a node would be able to correctly decide to delay a packet rather than sending it to a 'free' shortest path.

3 Call admission

In this section the attention is focused on call admission flow control mechanisms: whenever a set of nodes in the network wants to send messages to other nodes, the call admission algorithm selects a subset of the requests which can be satisfied within a fixed end-to-end delay. Next theorem proves that maximizing or even approximating the number of admitted requests in polynomial time is an hard problem also when the required fixed end-to-end delay is 0. Here, the relative error of an algorithm A for a maximization problem Π is defined as follows:

$$\frac{m(S^*(x))}{m(S_A(x))}$$

where $S^*(x)$, $S_A(x)$, $m(S^*(x))$ and $m(S_A(x))$ are defined similarly to section 2.

Theorem 3. *The maximum number of communication requests which can be accepted by the system in order to be satisfied with no end-to-end delay cannot be approximated with an error in $O(r^{\frac{1}{6}-\delta})$ for any $\delta > 0$, where r is the number of communication requests, unless P=NP.*

Sketch of Proof. The theorem is proved for the paths beeing fixed before the requests are submitted to the system. The assertion for the general case follows by generalization (similarly to what noticed at the end of theorem 1).

The proof is an approximation preserving reduction from the well-known NP-hard max-clique problem [6]: given a graph $G = (V, E)$ with $|V| = r$, find the maximum-size complete subgraph of G.

Let $\langle G = (V, E) \rangle$ be an instance of max-clique with $r = |V|$ and $m = |E|$. The reduction maps $\langle G = (V, E) \rangle$ into a layered network N^G containing r level 0 nodes s_i, each of them corresponding to a node of G, and r level L nodes t_i. Every pair (s_i, t_i) is a communication request submitted to the flow control procedures of the system.

N^G contains $L-1 = r(r-1) - 2m$ inner levels; it is described in terms of the paths chosen for the communication requests. Each pair of inner levels $(2i-1, 2i)$ corresponds to a pair of non adjacent nodes v, w of the input graph G: they contain $r-1$ nodes with exactly one node in level $2i-1$ having indegree two and one node in level $2i$ having outegree two. Such nodes are connected by an edge that must be used by both packets representing v and w. The other nodes have indegree and outdegree one (see figure 4). Clearly, N^G can be constructed in polynomial time.

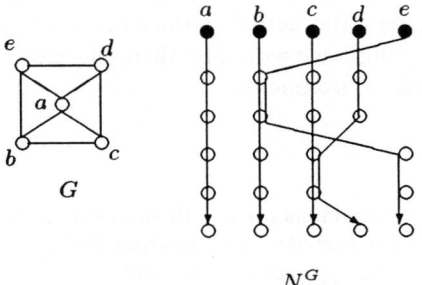

Fig. 4. Network corresponding to graph G

Let us now prove that G contains a clique of k nodes if and only if k packets exist which can be scheduled with end-to-end delay 0. Notice that, by the above construction, a set of packets reaches level L without any end-to-end delay if and only if they correspond to pairwise adjacent nodes in G. Thus, the above statement is trivially true.

Suppose now that a polynomial-time $\epsilon(r)$-approximation algorithm A for the maximum call admission problem exists (r being the number of communication requests), that is, there exists a $\epsilon(r)$ such that for any networks N, A accepts $R(A, N)$ requests that can be scheduled with end-to-end delay 0, with

$$\frac{R^*(N)}{R(A, N)} < \epsilon(r)$$

where $R^*(N)$ denotes the maximum number of requests that can be scheduled with end-to-end delay 0. Then, A can be applied to the network N^G which corresponds to some graph G with r nodes according to the reduction described above: since the size of a maximum clique in G is $k_{max} = R^*(N)$, then the above relation bounds also the ratio between the approximate and the maximum clique in G. In other words, A can be easily transformed into an $\epsilon(r)$-approximation algorithm for the maximum clique problem. Since this last problem cannot be approximated with an error in $O(r^{\frac{1}{6}-\delta})$, for any $\delta > 0$ [2], then the assertion is proved.

4 Conclusions

In this paper some computational complexity results have been shown related to the problem of finding minimum end-to-end delay packet routing schemes, trying to optimize both the end-to-end delay, when the number of packets (and, thus, the congestion) increases, and the number of packets which can be accepted in the network in order to keep the end-to-end delay low. Unfortunately, all the results are negative, even with respect to approximate solutions. This means that it is impossible to design an algorithm that is efficient (with respect to running time) and performs well (with respect to the quality of solutions) in the worst case. It is thus worth to perform an average case analysis, that is, to study if the minimum end-to-end delay schedule problem is NP-complete on the average with respect to some reasonable probability distribution in the set of instances, or even to search for heuristics that are on the average approximating.

Finally, since little is known about the actual performances of the various proposed heuristics, another interesting issue would be their effective implementation and simulation in distributed environments.

References

1. R. Aleliunas, "Randomized parallel communication", *Proc. of the ACM SIGACT-SIGOPS Symposium on Principles of Distributed Computing*, 60-72, 1982.
2. M. Bellare, M. Sudan, "Improved non-approximability results", *Proc. of the 26th Annual ACM Symposium on Theory of Computing*, 184-193, 1994.
3. I. Cidon, S. Kutten, Y. Mansour, D. Peleg, "Greedy packet scheduling", *Proc. of 4th Workshop on Distributed Algorithms*, 1990.
4. A. Clementi, M. Di Ianni, "Optimum Schedule Problem in Store and Forward networks", *Proc. of the IEEE INFOCOM'94*, 1336-1343, 1994.
5. A. Clementi, M. Di Ianni, "On the hardness of approximating optimum schedule problems in Store and Forward networks", to appear in *IEEE-ACM Transactions on Networking*, 1995.
6. M.R. Garey, D.S. Johnson, *Computers and intractability: a guide to the theory of NP-completeness*, Freeman, San Francisco, 1979.
7. D. S. Johnson, "Approximation algorithms for combinatorial problems", *Journal of Computer and System Sciences*, 9, 256-278, 1974.
8. E. Koutsoupias, C. H. Papadimitriou, "Beyond competitive analysis (on-line algorithms)", *Proc.of the 35th IEEE Annual Symposium on Foundations of Computer Science*, 394-400, 1994.

9. D. Krizanc, S. Rajasekaran and T. Tsantilis, "Optimal routing algorithms for mesh-connected processor arrays", *Lecture Notes in Computer Science*, Springer-Verlag, New York, vol. 319, 411-422, 1988.
10. T. Leighton, *Introduction to parallel algorithms and architectures: arrays, trees, hypercubes*, Morgan Kaufmann Publishers (San Mateo, Ca), 1992.
11. T. Leighton and B. Maggs, "Fast algorithms for routing around faults in multibutterflies and randomly-wired splitter networks", *IEEE Trans. on Comput.*, 41, N. 5, 578-587, 1992.
12. T. Leighton, F. Makedon and I. Tollis, "A $2N - 2$ step algorithm for routing in an $N \times N$ mesh", *Proc. of the 1989 ACM Symposium on Parallel Algorithms and Architectures*, 1989.
13. T. Leighton, B. Maggs, S. Rao, "Universal packet routing algorithms", *Proc. of the 29th IEEE Annual Symposium on Foundations of Computer Science*, 256-269, 1988.
14. T. Leighton, B. Maggs, S. Rao, A.G Ranade "Randomized routing and sorting on fixed-connection networks", *Journal of Algorithms*, 17, 157-205, 1994.
15. T. Leighton, B. Maggs, "Fast algorithms for finding O(congestion + dilation) packet routing schedules", Unpublished Manuscript, 1994.
16. T. Leighton, B. Maggs, S. Rao, "Packet routing and job-shop scheduling in O(congestion + dilation) steps", *Combinatorica*, 2, 167-186, 1994.
17. C.E. Leiserson and B. Maggs, "Communication-efficient parallel graph algorithms for distributed random-access machines", *Algorithmica* 3, 53-77, 1988.
18. C. Lund, M. Yannakakis, "On the hardness of approximating minimization problems", *Proc. of 25th Annual ACM Symposium on Theory of Computing*, pp. 286-293, 1993.
19. Y. Mansour, B. Patt-Shamir, "Greedy packet scheduling on shortest paths", *Proc. of the ACM PODC '91*, 165-175, 1991.
20. F. Meyer auf der Heide, B. Vöcking, "A packet routing protocol for arbitrary networks", *Proc. of the 12th Symposium on Theoretical Aspects of Computer Sience*, 291-302, 1995.
21. N. Pippenger, "Parallel communication with limited buffers", *Proc. of the 25th IEEE Annual Symposium on Foundations of Computer Science*, 127-136, 1984.
22. A.G. Ranade, "How to emulate shared memory", *Journal of Computer and System Sciences*, 42, 307-326, 1991.
23. E. Upfal, "Efficient schemes for parallel communication", *Proc. of the ACM SIGACT-SIGOPS 1989 ACM Symposium on Principles of Distributed Computing*, 55-59, 1982.
24. L.G. Valiant, "A schemes for fast parallel communication", *SIAM Journal on Computing*, 11/2, 350-361, 1982.
25. L.G. Valiant and G.J. Brebner, "Universal schemes for parallel communication", *Proc. 13th ACM Symposium on Theory of Computing*, 263-277, 1981

Multipacket Hot-Potato Routing on Processor Arrays

Christos Kaklamanis[1]* Danny Krizanc[2]**

[1] Computer Technology Institute, Patras University, Rio, Greece.
[2] School of Computer Science, Carleton University, Ottawa, Canada.

Abstract. In this paper, we consider the problems of multipacket batch and balanced routing on d-dimensional (constant $d \geq 2$) torus and mesh-connected processor arrays. We present new "hot-potato" routing algorithms which achieve the best known average-case and worst-case time bounds for both problems on all such networks. In particular, our results include the following:

1. Algorithms that route almost all batch routing problems where each node is the source of at most $\lfloor d/2 \rfloor - 1$ packets in $dn/2 + O(\log^2 n)$ time steps on the d-dimensional n^d-node torus and in $dn + O(\log^2 n)$ time steps on the d-dimensional n^d-node mesh.
2. Randomized algorithms that route any routing problem where each node is the source and destination of at most $\lfloor d/2 \rfloor - 1$ packets in $dn + O(\log^2 n)$ time steps on the d-dimensional n^d-node torus and in $2dn + O(\log^2 n)$ time steps on the d-dimensional n^d-node mesh, with high probability.

To achieve these bounds we introduce a number of novel techniques for constructing routing schemes for processor arrays.

1 Introduction

A number of researchers [1, 2, 3, 4, 5, 6, 7, 8, 9, 10, 12] have suggested algorithms for routing packets in a network with the property that on each step, each node in the network sends all of the packets it received on the previous step along one of its outgoing edges (with at most one packet leaving per edge). Such schemes are generally referred to as *hot-potato* or *deflection* routing schemes since the packets are always moving (i.e., they are treated as "hot-potatoes") and if two packets conflict for the use of an edge one of them may be "deflected" to a node further from its destination. They have the distinct advantage over traditional store-and-forward packet routing algorithms, that use buffers to store packets between time steps, of requiring only one buffer per incoming edge. If the choice of which edge to send the incoming packets out is a simple one, then the resulting algorithm is easy to implement with minimal time between routing steps.

While the apparent advantages of hot-potato algorithms have been borne out by numerous simulation studies [1, 6, 7, 10, 13], exact analysis of their behavior (i.e.,

---
* Partially supported by the European Union ESPRIT Basic Research Project GEPPCOM. Email: kakl@cti.gr
** Partially supported by the Natural Sciences and Engineering Research Council of Canada. Email: krizanc@scs.carleton.ca

without any independence assumptions) has proven to be difficult. Hajek [8] showed that for a natural algorithm on the N-node hypercube, k packets with worst-case destinations are delivered in $2k + \log N$ time steps. An algorithm for the hypercube suggested by Borodin and Hopcroft [4] was shown by Prager [13] to terminate in $O(\log N)$ steps on a special class of permutations. Feige and Raghavan [5] were the first to present an exact analysis of the average-case and worst-case behavior of hot-potato algorithms for two- and three-dimensional tori. Feige and Raghavan also showed that a straightforward algorithm for the hypercube terminates in $O(\log N)$ steps on average. Newman and Schuster [12] analyse the worst-case behavior of deterministic hot-potato algorithms for d-dimensional meshes ($d \geq 2$) and for the hypercube.

The model we use is the same as that in [5]. A network is modelled as a directed graph where the nodes are processors and the unidirectional edges are communication links between processors. The routing is performed in discrete, synchronous time steps. We only consider *deflection networks* where the out-degree of each node is greater or equal to its in-degree and the digraph is strongly connected. During each step, a processor receives zero or one packet along each incoming edge and must send all the packets it received out along outgoing edges with at most one packet leaving per outgoing edge. Note that no buffers are required to hold the packets between time steps (except possibly at the source or destination nodes).

In the k-*batch* routing problem all processors generate k packets simultaneously with arbitrary destinations. We assume that k is less than or equal to the outdegree of the nodes. The $k - k$ routing problem is a special case of the k-batch routing problem where each node is the destination of k packets. The time required to route a given routing problem is the maximum over all packets of the number of time steps required to deliver each packet.

In this paper, we present new hot-potato routing algorithms for the d-dimensional n^d-node torus and for the d-dimensional n^d-node mesh, for any constant d. We give algorithms that route almost all ($\lfloor d/2 \rfloor - 1$)-batch routing problems in $dn/2 + O(\log^2 n)$ time steps on the d-dimensional torus and in $dn + O(\log^2 n)$ time steps on the d-dimensional mesh. (By "for almost all" we mean all but a $1/n^c$ fraction for some $c \geq 1$.) Note that these algorithms are optimal in the sense that for almost all batch routing problems there is a packet that is, initially, $dn/2 - o(n)$ distance from its destination on the torus and $dn - o(n)$ distance from its destination on the mesh. Our bounds are the best known for all $d \geq 6$ and can be easily adjusted to match the bounds known for $d < 6$. Kaklamanis, Krizanc and Rao [9] gave algorithms that route almost all 1-batch routing problems in $dn/2 + O(\log^2 n)$ time steps on the d-dimensional torus and in $dn + O(\log^2 n)$ time steps on the d-dimensional mesh, for any constant d. Recently, Meyer auf der Heide and Westermann [11] presented an algorithm that routes almost all $d/88$-batch routing problems in $dn + O(d^3 \log n)$ steps, for any $d = O(n^\epsilon)$, with $0 < \epsilon < 1/2$. The results of Ben-Dor, Halevi and Schuster [3] imply a simple greedy algorithm can perform a d-batch routing problem in $O(n^d)$ steps on a d-dimensional mesh in the worst-case, but no bounds on the average-case are given.

Combining the above average-case algorithms with well-known randomized routing techniques [14] we give randomized algorithms that route every ($\lfloor d/2 \rfloor - 1$) − ($\lfloor d/2 \rfloor - 1$) routing problem on a d-dimensional torus in $dn + O(\log^2 n)$ time steps and on a d-dimensional mesh in $2dn + O(\log^2 n)$ time steps, with high probability. (By "with high probability" we mean with probability $1 - 1/n^c$ for some $c \geq 1$.) Again, these are the best known bounds for all $d \geq 6$ and the only bounds we are aware of for worst-case multipacket routing on multidimensional tori and meshes. Kaklamanis, Krizanc and Rao [9] give algorithms which solve any $1-1$ routing problem on a d-dimensional torus in $dn + O(\log^2 n)$ time steps and on a d-dimensional mesh in $2dn + O(\log^2 n)$ time steps, with high probability.

All of our algorithms have the same general structure: Packets are divided into groups which are routed in parallel in disjoint subtori along paths that closely approximate their natural greedy paths. Small cycles (or "snakes") that pass through all of the nodes in small regions of the subtori are used to facilitate moving from one dimension to another. Informally, the snakes are being used as a storage mechanism at the turns of a packet's greedy path.

It is possible for a deflection routing algorithm on certain inputs to route a set of packets in such a manner that they mutually deflect each other in an infinite loop from which they never recover. This situation is referred to as *livelock*. All of the above algorithms are constructed so as to guarantee they are free from livelock. The proof of this assertion is straightforward and is not presented in this paper.

In the next section we present in detail our algorithm for the average-case multipacket batch routing problem on the d-dimensional torus. This is followed by a discussion of how it can be extended to perform average-case multipacket batch routing on the d-dimensional mesh, and worst-case $(\lfloor d/2 \rfloor - 1) - (\lfloor d/2 \rfloor - 1)$ routing on the d-dimensional torus and mesh.

2 Multipacket Routing on the d-Dimensional Torus

In this section we prove our main result:

Theorem 1. *For any constant $d \geq 4$, there exists a hot-potato routing algorithm which routes almost all $(\lfloor d/2 \rfloor - 1)$-batch routing problems on the d-dimensional torus in $dn/2 + O(\log^2 n)$ steps.*

The proof of the theorem follows easily from the following lemma which shows how a set of packets can "fix" one of their dimensions using only edges in two dimensions of the torus. A packet is said to be *oscillating* in a dimension x if it is moving back and forth along the same edge in the x dimension on successive time steps.

Lemma 2. *Consider a two-dimensional subtorus of a d-dimensional torus consisting of nodes in the xy plane. Assume $(1-\epsilon)n$ packets are oscillating on edges in each row of the x dimension of the torus with the exception of every $(c \log n)$th row (c satisfying the probabilistic conditions below). Assume further that each packet is within $c \log n$ distance of its original x-coordinate in a given batch routing problem. Then, there exists a hot-potato routing algorithm using only edges in the x and y dimensions of the torus which for almost all batch routing problems routes each packet to its final destination in the x dimension and within $c \log n$ of its original position in the y dimension with the exception of packets destined for every $(c \log n)$th column. At the* *end of the routing the packets are oscillating in the y dimension. For a randomly chosen batch routing problem, the algorithm performs this task in $n/2 + O(\log^2 n)$ steps, with high probability.*

Proof: We first divide the $n \times n$ torus into $n/c \log n$ vertical strips that consist of $c \log n$ columns each. Each strip is subdivided into blocks that consist of $c \log n$ rows each, i.e., blocks of size $c \log n \times c \log n$.

In each block, for the purposes of the routing, we construct two edge disjoint directed cycles that visit every node in the block. In particular the first cycle, referred to as the xy "snake" below, goes through the nodes in column-major snake-like order "starting" at the top-left node in the block and then moving alternately down along odd columns and up along even columns until it reaches the top right node in the block (assuming $c \log n$ is even) and then returning along the top row to the top left node of the block. The second cycle, referred to as the yx snake below, goes through the nodes in row-major snake-like order starting at the top-left node in the block and then moving alternately across odd rows and back along even rows until it reaches the bottom left node in the block and then returning along the leftmost column to the top left node of the block. The "backbone" of each snake (i.e., the top row of xy snake and the leftmost column of the yx snake) correspond to the unused rows and columns mentioned in the lemma. Note also that the backbone in one block overlaps with the last row or column of the snake corresponding to its neighboring block.

We use the following algorithm to perform the above task:

1. At a fixed time, all packets stop oscillating in dimension x and instead continue to move in the direction dictated by their last oscillation. Note that this may not be on the shortest path towards the packet's destination in the x dimension.
2. Packets not travelling on their shortest path attempt to switch directions on each step by either finding a packet moving in the opposite direction that is also going in the "wrong" direction or by finding an empty position in the flow of packets moving in its desired direction.
3. Once a packet reaches the block of processors containing its destination column in the x dimension, it attempts to turn in the y dimension in the opposite direction of the flow of the xy snake in the block. It attempts this turn at every column of the block. If it fails to turn at all columns in the block it becomes a low priority packet and is routed along arbitrary free edges so as to not interfere with the movement of the other packets.
4. Once a packet has entered the xy snake it moves around the snake in the "wrong" direction until it finds an empty position in the flow of packets moving in the snake's direction and reverses its direction.
5. After $n/2 + 2(c \log n)^2$ steps from the beginning of the algorithm, all packets in xy snakes switch to moving in the yx snakes covering their same block.
6. As a packet travelling in the yx snake passes its final x dimension column it attempts to turn in the y dimension in either the positive or negative direction (with the exception of packets destined for the backbone of the snake). If it is successful it oscillates in the y dimension along the edge it turned into. If it fails to turn, it continues around the snake until the next position that intersects its final x dimension column and attempts to turn there. It continues in this manner

until it successfully turns into its destination column and is oscillating in the y dimension.

The lemma is easily obtained from the following claims concerning the algorithm above.

Claim 1 *For a random batch routing problem, all packets successfully execute step 2 (i.e., start moving in the correct direction) within $c\log n$ steps, with high probability.*

Proof: For a random batch routing problem, for each packet, the probability that its final destination column is in the positive x direction is $1/2$. It is easy to show, using Chernoff bounds, that with high probability in any segment of $c\log n$ at most $(1-\epsilon)(c\log n)$ packets have destination columns in the positive x direction. Therefore any packet with a destination column in the positive x direction that begins travelling in the opposite direction will find an empty position in the flow of packets travelling in the positive direction within $c\log n$ steps, with high probability. □

Claim 2 *With high probability, for any block and any time window of length $(c\log n)^2$, there are at most $O(1)$ packets whose destination column is inside that block and that arrive during that time window.*

Proof: For a packet to enter a block during the time window $[t, t+(c\log n)^2]$ the distance between its origin and the block must be between $t+(c\log n)^2$ and $t-c\log n$ (by claim 1). Thus for a particular block and for a particular time window there are $O(\log^3 n)$ packets that could enter the block and each has its destination in the block with probability $c\log n/n$. From this it is easy to show that at most $O(1)$ packets arrive at the block during a particular time window, with high probability. Since there are $O(n^2)$ blocks and $O(n)$ time windows the claim holds, with high probability. □

Claim 3 *With high probability, all packets that have successfully executed step 3, will successfully execute step 4 within $(c\log n)^2$ steps.*

Proof: From Chernoff bounds we know that the number of packets with destination column in a given block and origin row coinciding with the block is $(1-\epsilon)(c\log n)^2 + o(\log^2 n)$, with high probability. That is, at any time there are at most this many packets travelling in the correct direction in the snake. Since the snake is $(c\log n)^2$ nodes long, a packet travelling in the opposite direction must find an empty position in the flow of packets travelling in the correct direction. □

Claim 4 *With high probability, for any block and any time window of size $(c\log n)^2$, any packet whose destination is inside the block and arrives at that block during that time window successfully completes step 3 within $O(1)$ steps of arrival.*

Proof: The proof is by induction on successive time windows for some particular block. For the basis step consider a block and the packets that are destined for that block and arrive in the first $(c\log n)^2$ time steps. With high probability there are at most $O(1)$ such packets according to claim 2. Also there are at most another

$O(1)$ such packets that arrive in the next time window. Now consider one of the packets arriving in the first time window. After it arrives at the block containing its destination, it starts trying to turn in the y dimension and thus enter the "wrong" direction of the snake. At each attempt it fails if it encounters some other packet that is also moving in the "wrong" direction of the snake. Such a packet must have of course arrived at the destination block before the moment of the encounter. Furthermore, since the attempts are made at successive columns of the snake, any consecutive failures must be due to distinct packets. Therefore, the packet can fail at most $O(1)$ times due to packets that arrive during the first $(c \log n)^2$ time steps, and if these failures occur very close to time $(c \log n)^2$, at most another $O(1)$ times due to packets in the second time window. Thus it will successfully complete step 3 within $O(1)$ steps of its arrival at the block.

Now let us assume that the claim is true for the first m time windows of size $(c \log n)^2$. Then consider some packet that arrives at the destination block during the $(m+1)$th time window of size $(c \log n)^2$. Because of the inductive hypothesis and claim 3 all other packets destined for the block that arrived during the first $m-2$ time windows of size $(c \log n)^2$ are already moving in the correct direction of the snake. Thus the packet can only conflict with other such packets that arrive during the $(m-1)$th, the mth, the $(m+1)$th (i.e. the current), or the $(m+2)$th window of size $(c \log n)^2$. According to claim 2 with high probability there are at most $O(1)$ such packets and each one of them can be encountered only once. Therefore, every packet destined for the block that arrives during the $(m+1)$th time window of size $(c \log n)^2$ successfully completes step 3 within $O(1)$ time steps with high probability.
□

Claim 5 *For a random batch routing problem, all packets successfully complete step 6 within $(c \log n)^2$ steps, with high probability.*

Proof: From the above claims, by $n/2 + 2(c \log n)^2$ steps after the beginning of the algorithm, all packets are travelling in the correct direction in a xy snake in the block containing their destination column, with high probability. At this point, all the packets begin moving in a yx snake in the block. The yx snake crosses each column $c \log n$ times (with the exception of the first and last column of the snake) giving each packet $2c \log n$ oppurtunities to turn in the y direction and begin their oscillation. By Chernoff bounds, there are at most $c \log n$ packets with a destination in a particular column and therefore each packet must find a free edge to turn into in one of its attempts, with high probability. This completes the proof of the claim and of the lemma.
□

Sketch of Proof of Theorem 1: Randomly distribute the $\lfloor d/2 \rfloor - 1$ packets at each node to one of $\lfloor d/2 \rfloor$ different groups. The algorithm consists of d phases. In each phase, the algorithm of lemma 2 is applied to fix a single dimension of the packets in each of the groups using edge disjoint tori in parallel. During the first phase (numbered 0), packets in group i use the torus consisting of dimensions $2i$ and $2i+1$ to fix their dimension $2i$, $i = 0, 1, \ldots, \lfloor d/2 \rfloor - 1$. During the kth phase, they use the torus consisting of dimensions $(2i + k) \bmod d$ and $(2i - 1 + k) \bmod d$ to fix their dimension $(2i + k) \bmod d$. During the last phase, they use the torus consisting of dimensions $(2i - 1) \bmod d$ and $2i \bmod d$ to fix their dimension $(2i - 1) \bmod d$. At this point they are in a snake that passes through their final destination and the routing is completed.

Note that for the lemma to apply we must be certain that (1) $(1-\epsilon)n$ packets are to be routed in each row during each application of the algorithm and (2) none of the packets to be routed have origins in a snake backbone row or a destination in a snake backbone column. For a randomly chosen batch routing problem, using Chernoff bounds one can show the number of packets being routed in any row during any application of the lemma is at most $(1-2/d)n + o(n) = (1-\epsilon)n$, for an appropriately chosen ϵ. To insure that no packets are routed in backbone rows or columns, any packet that would use such a column or row according to its random group assignment and the schedule above is routed along a shifted path reached using initial and final redistribution phases of $O(\log^2 n)$ steps each. Since the number of such packets is small, this can be done in such a way that only $o(n)$ packets are added to any group.

By repeatedly applying the lemma, we get a hot-potato algorithm which routes almost all ($\lfloor d/2 \rfloor - 1$)-batch routing problems in $dn/2 + O(\log^2 n)$ steps.

As described above, the algorithm is not livelock-free. To make it livelock-free, route any packets that are unsuccessful at any stage of the algorithm along arbitrary free dimensions until time $dn/2 + k\log^2 n$, for an appropriately chosen k, at which time these packets can be routed using one large snake encompassing all nodes of the torus. □

Techniques similar to those used above and those in [14] can be used to show the following theorems. Proofs are ommitted due to space limitations.

Theorem 3. *For any constant $d \geq 4$, there exists a hot-potato routing algorithm which routes almost all ($\lfloor d/2 \rfloor - 1$)-batch routing problems on the d-dimensional mesh in $dn + O(\log^2 n)$ steps.*

□

Theorem 4. *For any constant $d \geq 4$, there exists a hot-potato routing algorithm which routes every ($\lfloor d/2 \rfloor - 1$)-($\lfloor d/2 \rfloor - 1$) routing problem on the d-dimensional torus in $dn + O(\log^2 n)$ steps, with high probability.*

□

Theorem 5. *For any constant $d \geq 4$, there exists a hot-potato routing algorithm which routes every ($\lfloor d/2 \rfloor - 1$)-($\lfloor d/2 \rfloor - 1$) routing problem on the d-dimensional mesh in $2dn + O(\log^2 n)$ steps, with high probability.*

□

References

1. Acampora, A. and Shah, S., "Multihop lightwave networks: a comparison of store and forward and hot-potato routing", IEEE INFOCOM, 1991, 10–19.
2. Bar-Noy, I., Raghavan, P., Schieber, B. and Tamaki, H., "Fast Deflection Routing for Packets and Worms", Symp. on Principles of Distributed Computing, 1993, 225-234.
3. Ben-Dor, A., Halevi, S. and Schuster, A., "Potential Function Analysis of Greedy Hot-Potato Routing", Symp. on Principles of Distributed Computing, 1994, 225-234.

4. Borodin, A. and Hopcroft, J., "Routing, merging, and sorting on parallel models of computation", Journal of Computer and System Sciences, 30, 1985, 130–145.
5. Feige, U. and Raghavan, P., "Exact analysis of hot-potato routing", Symposium on the Foundations of Computer Science, 1992, 553–562.
6. Greenberg, A. and Goodman, J., "Sharp approximate models of deflection routing in mesh networks", IEEE Transactions on Computers, to appear.
7. Greenberg, A. and Hajek, B., "Deflection routing in hypercube networks", IEEE Transactions on Computers, to appear.
8. Hajek, B., "Bounds on evacuation time for deflection routing", Distributed Computing, 5, 1991, 1–6.
9. Kaklamanis, C., Krizanc, D. and Rao, S., " Hot-Potato Routing on Processor Arrays", Symp. on Parallel Algorithms and Architectures, 1993, 273-282.
10. Maxemchuk, N., "Comparison of deflection and store and forward techniques in the Manahattan street and shuffle-exchange networks", IEEE INFOCOM, 1989, 800–809.
11. Meyer auf der Heide, F. and Westermann, M., "Hot-Potato Routing on Multi-Dimensional Tori", Workshop on Graph-Theoretic Concepts in Computer Science (WG 95), 1995.
12. Newman, I. and Schuster, A., "Hot-potato algorithms for permutation routing", Technion Technical Report, CS-LPCR 9201, 1992.
13. Prager, R., "An algorithm for routing in hypercube networks", University of Toronto Technical Report, 1986.
14. Valiant, L., "A Scheme for fast parallel communication", SIAM Journal of Computing, 11, 1982, 350–361.

A Necessary and Sufficient Condition for Proper Routing in Omega-Omega Network

Myung-Kyun Kim[1] and Hyunsoo Yoon[2] and Seung-Ryoul Maeng[2]

[1] Woosuk University, Wanju-Gun, Cheonbuk 565-701, Korea
[2] Korea Advanced Institute of Science and Technology, Daejun 305-701, Korea

Abstract. Recently, a routing algorithm, called inside-out algorithm, was proposed for routing an arbitrary permutation in the omega-based $2\log_2 N$ stage networks including the omega-omega network. This paper discusses the problems of the inside-out routing algorithm and shows that the suggested condition for proper routing in the omega-omega network is insufficient. This paper suggests an extended necessary and sufficient condition for proper routing in the omega-omega network.

1 Introduction

Multistage interconnection networks(MINs) have been widely used in multiprocessor systems to connect thousands of processors and memory modules. The rearrangeability and the routing algorithm for $2\log_2 N$(or $2\log_2 N - 1$) stage symmetric MINs, which have the same decomposition structure as the Benes network, were studied by many authors [1] [3] [5] [6] [7]. But, the rearrangeability and the routing algorithm for asymmetric MINs are not so simple as the symmetric MINs because they do not have the recursive decomposition structures. Recently, Feng and Seo [2] proposed a new routing algorithm called inside-out routing algorithm which claims to route an arbitrary permutation in the omega-based $2\log_2 N$ stage MINs including the omega-omega network.

This paper discusses the problems of the inside-out routing algorithm and presents a modified condition for proper routing in the omega-omega network. It is shown that the inside-out routing algorithm needs backtracking in the terminal number assignment procedure because of assignment conflicts. Also the conditions suggested in [2] for proper routing are insufficient. A baseline × baseline^{-1}(BL×BL^{-1}) network is shown to be topologically equivalent to the omega^{-1}×omega network [2], and a modified necessary and sufficient condition for proper routing in the BL^{-1} network is suggested.

Section 2 describes a few notations and definitions and introduces a BL×BL^{-1} network. Section 3 discusses the problems of the inside-out routing algorithm and presents an extended necessary and sufficient condition for proper routing in the BL^{-1} network. Section 4 concludes this paper.

This work was partially supported by *Center for Artificial Intelligence Research* of KAIST.

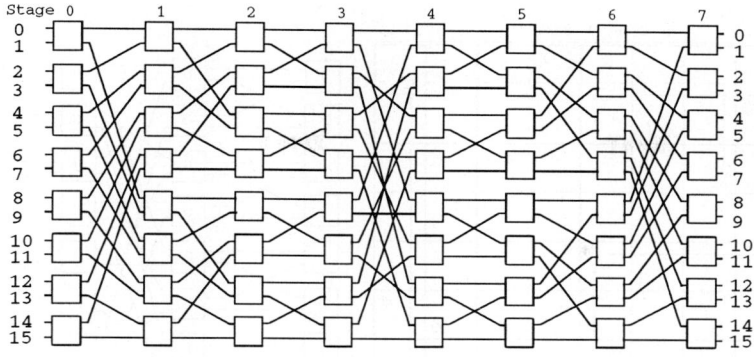

Fig. 1. 16×16 BL× BL^{-1} network.

2 BL× BL^{-1} Network

Throughout this paper, $N = 2^n$ and the binary representation of an integer x, where $0 \leq x \leq N-1$, is $x_{n-1}x_{n-2}\cdots x_0$ where x_{n-1} is the most significant bit.

Definition 1. Let S be an ordered set with N numbers in the range of $[0, N-1]$, then a segment partition of S by b is a partition of S into N/b segments with each segment having b contiguous numbers. The k-th segment of a segment partition of S by b is denoted by $SP(N, b, k)$ where $0 \leq k \leq N/b - 1$.

A similar definition to a complete residue system modulo m used in [5] is introduced here to describe the passable condition of the $m\Omega$ network.

Definition 2. Let S be an ordered set of N numbers in the range of $[0, N-1]$, then a complete quotient system of N by k, denoted by $CQS(N, k)$, is a set of N/k integers of S whose quotients divided by k are different among the others.

For the sake of easy presentation, BL× BL^{-1} network is introduced. An $N \times N$ BL× BL^{-1} network is a network obtained by concatenating serially the BL, bit-reversal, and BL^{-1} networks, and has a mirror image according to the virtual center line. An $N \times N$ BL× BL^{-1} network is topologically equivalent to an $N \times N$ omega^{-1}×omega network [4]. Fig. 1 shows 16×16 BL× BL^{-1} networks.

3 A Necessary and Sufficient Condition for Proper Routing in the BL^{-1} Network

The first problem of the inside-out routing algorithm is that the assignment conflicts of terminal numbers can occur in the the generation and propagation procedures, thus the routing algorithm needs backtracking. The fact will be shown by an example in [2]. For the given permutation P of [2] such as

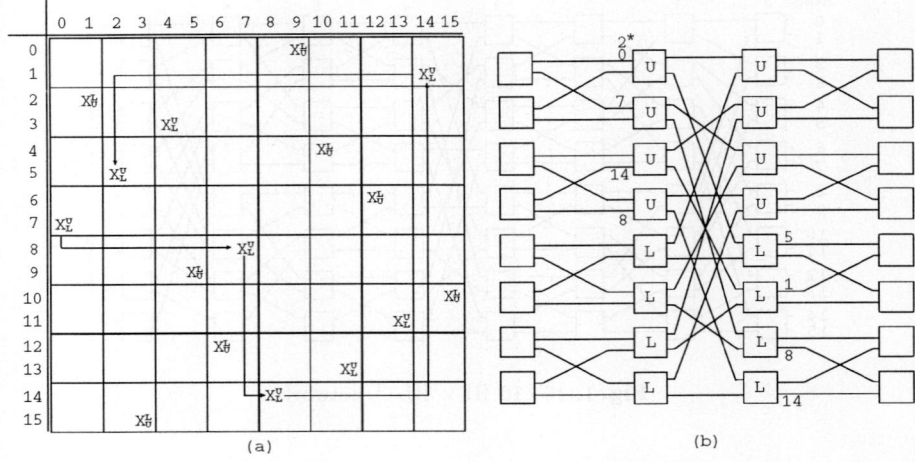

Fig. 2. An example of assignment conflicts of terminal numbers in the center stages.

$$P = \begin{pmatrix} 0 & 1 & 2 & 3 & 4 & 5 & 6 & 7 & 8 & 9 & 10 & 11 & 12 & 13 & 14 & 15 \\ 7 & 2 & 5 & 15 & 3 & 9 & 12 & 8 & 14 & 0 & 4 & 13 & 6 & 11 & 1 & 10 \end{pmatrix},$$

the inside-out routing algorithm first makes a decision chart such as shown in Fig. 2(a). Next, it determines the interchangeable groups by making complete loops by the generation and propagation procedures. In [2], a complete loop was made by the following sequence of generation and propagation procedures:

$$(0,7) \xrightarrow{i} (4 \xrightarrow{p} 3) \xrightarrow{o} (2 \xleftarrow{p} 5) \xrightarrow{i} (7 \xrightarrow{p} 8) \xrightarrow{o} (8 \xleftarrow{p} 14) \xrightarrow{i}$$
$$(13 \xrightarrow{p} 11) \xrightarrow{o} (11 \xleftarrow{p} 13) \xrightarrow{i} (14 \xrightarrow{p} 1) \xrightarrow{o} (0 \xleftarrow{p} 7).$$

In this, $\xrightarrow{i}$ and $\xrightarrow{o}$ denote the input and output generation procedures, respectively, and $\xrightarrow{p}$ and $\xleftarrow{p}$ denote the propagation procedures. Different complete loops can be constructed by changing the initial position or by selecting different terminal numbers in the generation procedures. However, all of the complete loops do not lead to the successful assignments. In the decision chart of Fig. 2, it is possible to construct a complete loop by the following sequence of generation and propagation procedures:

$$(0,7) \xrightarrow{i} (7 \xrightarrow{p} 8) \xrightarrow{o} (8 \xleftarrow{p} 14) \xrightarrow{i} (14 \xrightarrow{p} 1) \xrightarrow{o} (2^* \xleftarrow{p} 5) \xrightarrow{i} \cdots.$$

In the loop, after the output terminal 1 in the (14,1) pair is assigned by the generation procedure, the output terminal 5 in the (2,5) pair can be assigned because the pair (2,5) only has "L" at the output strip. The output terminal 5 is assigned to the upper position of the lower SE of the corresponding buddy, and 5 accompanies 2 to the upper SE of the input side center stage. The input terminal 2 must be assigned to the upper position of the upper SE of the buddy, but it is not possible because the upper position of the SE is already assigned to the input terminal 0. The assignment conflict means that it is not a proper terminal number assignment. Thus, the routing algorithm needs backtracking to find another complete loop which satisfies all the conditions suggested in

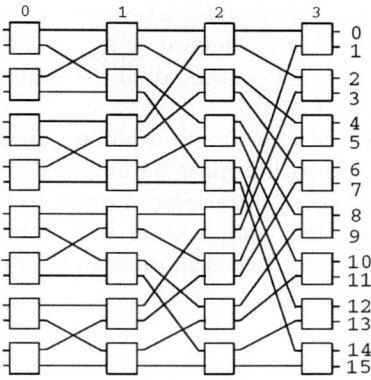

Fig. 3. 16×16 BL^{-1} Network

[2]. There are many other constructions of the complete loops which cause the assignment conflicts of terminal numbers.

The second problem is that the condition for proper routing in the $m\Omega$ network suggested in [2] is insufficient. Fig. 3 shows 16×16 BL^{-1} network. $SE(i,j)$ denotes the j-th SE of stage i where $0 \leq i \leq n-1$ and $0 \leq j \leq N/2-1$. The condition for proper routing in the $m\Omega$ network can be obtained by analyzing the connectivity of the network. The following theorem describes BL^{-1}-passable condition of a permutation.

Theorem 3. For a given permutation P, P is BL^{-1}-passable if and only if $SP(N, b, k)$ of P is $CQS(N, N/b)$ for all b and k such that $b = 2^{i+1}$ where $0 \leq i \leq n-1$ and $0 \leq k \leq N/b - 1$.

Proof. A brief sketch of the proof is the following. From the decomposition structure of the BL^{-1} network [3], at stage i where $1 \leq i \leq n-1$, the successive 2^{i+1} output links of the SEs from $SE(i, 2^i \cdot k)$ to $SE(i, 2^i \cdot k + 2^i - 1)$ can be connected to the output terminals numbered $0^{i+1}t^{n-i-1}$, $0^i 1 t^{n-i-1}$, $\cdots$, $1^i 0 t^{n-i-1}$, and $1^{i+1}t^{n-i-1}$, respectively, where 0^i(or 1^i) denotes i bits string with 0(or 1) in each bit and t^{n-i-1} denotes $(n-i-1)$ bits string with 0 or 1 in each bit. Thus, after being routed through the SEs of the stages from 0 to $i-1$, the successive 2^{i+1} elements of P assigned in the SEs from $SE(i, 2^i \cdot k)$ to $SE(i, 2^i \cdot k + 2^i - 1)$, where $0 \leq k \leq 2^{n-i-1}-1$, must be $CQS(N, N/2^{i+1})$ to be routed without a conflict in any SE of stage i. Also it can be seen easily that the set of successive 2^{i+1} elements of P assigned in the SEs of stage i from $SE(i, 2^i \cdot k)$ to $SE(i, 2^i \cdot k + 2^i - 1)$ is the k-th segment of the segment partition of P, $SP(N, 2^{i+1}, k)$. Thus, to be routed without a conflict in any SE of stage i, $SP(N, b, k)$ must be $CQS(N, N/b)$ for all b and k such that $b = 2^{i+1}$ and $0 \leq k \leq 2^{n-i-1}-1$. From the above facts, the theorem is proved. □

The conditions for proper routing suggested in [2] state only the following three conditions for a given permutation P: $SP(P, 2, k)$, where $0 \leq k \leq N/2-1$, of P must be $CQS(N, N/2)$, $SP(P, 4, k)$, where $0 \leq k \leq N/4 - 1$, of P must be $CQS(N, N/4)$, and $SP(P, N/2, k)$, where $0 \leq k \leq 1$, of P must be $CQS(N, 2)$. For $N \leq 16$, the conditions in [2] contain all of the conditions of Theorem 3, but they are insufficient for N, $N > 16$.

From the above facts, the input and output terminal assignment procedure of the inside-out routing algorithm must be modified to check more conditions in Theorem 3, and also needs backtracking in order to find the proper terminal number assignment. Thus, the computation time complexity of the modified inside-out routing algorithm may need more than $O(N)$.

4 Conclusions

This paper has addressed the problems of the inside-out routing algorithm. It has been shown that assignment conflicts can occur in the input and output terminal number assignment procedures and that the suggested conditions for proper routing in the omega-omega network are insufficient. Also, an $N \times N$ $BL \times BL^{-1}$ network has been shown to be topologically equivalent to an $N \times N$ omega$^{-1} \times$omega network, and an extended necessary and sufficient condition has been suggested for proper routing in the BL^{-1} network. The assignment conflict and insufficient conditions for the terminal number assignment mean that the inside-out routing algorithm needs checking more conditions and backtracking during the input and output terminal assignments. But, it is still unknown whether the modified inside-out routing algorithm with backtracking will always find proper assignments of terminal numbers without any conflict for all permutations or not. Thus, the rearrangeability and routing algorithm for the $2\log_2 N$ stage asymmetric networks such as the omega-omega network still remain open problems.

References

1. Benes, V.: On Rearrangeable Three-Stage Connecting Networks. Bell Syst. Tech. J. **XLI** (1962) 1481–1492
2. Feng, T., Seo, S.: A New Routing Algorithm for a Class of Rearrangeable Networks. IEEE Trans. on Computers (1994) 1270–1280
3. Kim, M., Yoon, H., Maeng, S.: Bit-Permute Multistage Interconnection Networks. Microprocessing and Microprogramming **41** (1995) 449–468
4. Kim, M.: Generalized Classes of Rearrangeable Multistage Interconnection Networks. Ph.D. Thesis Korea Advanced Institute of Science and Technology (1996)
5. Lee, K.: On the Rearrangeability of $2(log_2 N) - 1$ Stage Permutation Networks. IEEE Trans. on Computers (1985) 412–425
6. Opferman, D., Tsao-Wu, N.: On a Class of Rearrangeable Switching Networks-Part I: Control Algorithm. Bell Syst. Tech. J. **50** (1971) 1579–1600
7. Yeh, Y., Feng, T.: On a Class of Rearrangeable Networks. IEEE Trans. on Computers (1992) 1361–1379

Rubik Routing Permutations on Graphs*

Charles Delorme and Petrişor Panaite

LRI, Université de Paris-Sud, Bât. 490, 91405 Orsay Cedex, France

Abstract. We consider the permutation routing problem on undirected graphs when the routings correspond to products of elements of some fixed set. Previous work has concerned only two-way links. We introduce here the Rubik routing model in order to study the one-way case. We give a complete characterization of Rubik rearrangeable graphs and we study how the 2D grid and the hypercube behave with respect to Rubik routing of permutations.

1 Introduction

Let G be an undirected graph, π a permutation of $V(G)$, and let us assume that each vertex $v \in V(G)$ has a packet and a destination address $\pi(v)$. To *route* π means to send the packets to their destinations.

In this paper, we deal with a special model of permutation routing which is not very attractive from a practical point of view, but which could be useful in characterizing the intrinsic permutation power of a graph. At each routing step, the packets are permuted; that is, the packets – or, more suggestively, the "pebbles" – are moved so that each vertex has again exactly one pebble. Let us denote with $\text{Gen}(G)$ the set of permutations routable in one step on a graph G and with $\text{Perm}(G)$ the permutation group generated by $\text{Gen}(G)$. The routing of a permutation π on G can be regarded as one of its factorizations $\pi = \alpha_1 \alpha_2 \ldots \alpha_t$, where $\alpha_i \in \text{Gen}(G)$ for all $i \in \{1, \ldots, t\}$. Let us call such a routing model a *product-routing model*.

A product-routing model is defined by the choice of a unique set of generators $\text{Gen}(G)$ for each graph G. Thus, in [2], every permutation in $\text{Gen}(G)$ corresponds to a matching $\mathcal{M}$ of G: the pebbles at the two endpoints of each edge in $\mathcal{M}$ are interchanged. More generally, in [6], every permutation in $\text{Gen}(G)$ corresponds to a matching $\mathcal{M}$ and a set $\{C_1, \ldots, C_k\}$ of vertex-disjoint directed cycles with $V(\mathcal{M}) \cap V(C_i) = \emptyset$ for all $i \in \{1, \ldots, k\}$. This time, each pebble is interchanged with another pebble along an edge of $\mathcal{M}$, or moved along exactly one arc on the directed cycle that contains it, or keeps its position. All these patterns of communication work only if the edges are *two-way* (*i.e.*, two pebbles can cross simultaneously the same edge in opposite directions).

What happens if the edges are *one-way* (*i.e.*, only one pebble can cross a link at each routing step)? Can we route every permutation on the graph G if in the

---

* This work was supported by the "Opération RUMEUR" of the French PRC PRS.

product-routing model introduced in [6] we impose $\mathcal{M} = \emptyset$ at each step? This is the problem we study in the present paper.

Let us call *Rubik routing* (or *R-routing*) a routing in which, at each step, a pattern of vertex-disjoint directed cycles (of length ≥ 3) is chosen and each pebble is moved along exactly one arc on the directed cycle that contains it. The pebbles outside these cycles keep their positions during the step. The similarities with the movements in the Rubik's cube motivate the name of the routing model. Thus, every graph could be regarded as the basis of a new puzzle-game.

We introduce here the model of R-routing in order to capture the ideas of product-routings and of one-way edges. As far as we know, the routing algorithm given in [5], to perform the special class of bit-permute permutations on the hypercube, could be considered as the first R-routing algorithm.

A graph G is *R-rearrangeable* if Perm(G) (corresponding to the R-routing model) is equal to the whole symmetric group that acts on $V(G)$, denoted with $\mathcal{S}_{V(G)}$. Let $t_R(G) = \max\limits_{\pi \in \text{Perm}(G)} \min\{t \; : \; \pi$ is R-routable on G in t steps$\}$.

2 Rubik rearrangeability and connectivity

We refer the reader to [7] for notions and notation from the theory of groups. Clustering suitably the transpositions from the disjoint cycle decomposition of a permutation, we can prove the following result.

Proposition 1 *For $n \geq 4$, the complete graph K_n is R-rearrangeable and $t_R(K_n) = 2$.*

Note that K_n is rearrangeable via matchings in two steps, too [2].

Proposition 2 *A 2-connected graph G of order n which is not a cycle is R-rearrangeable and $t_R(G) \leq 2n^2 + n - 3$.*

PROOF. Each permutation of $V(G)$ is the product of at most $n - 1$ transpositions. We can prove that each transposition is R-routable on G in at most $2n + 3$ steps. □

It is well-known [3, Sect.9.2] that every graph G can be considered as a union of *2-connected blocks*, any two such blocks having in common at most one vertex. Let $b(G)$ denote $\max\{|H| \; : \; H$ is a 2-connected block in $G\}$.

Lemma 3 *If G is a 2-edge-connected graph, different from a cycle, then* Perm(G) *is $b(G)$-transitive.*

Theorem 4 *G is R-rearrangeable if and only if G is 2-edge-connected, different from a cycle, and contains an even cycle.*

PROOF. Clearly, if G is R-rearrangeable then it is 2-edge-connected and different from a cycle. If G has only odd length cycles then Perm(G) is at most $\mathcal{A}_{V(G)}$ (the alternating group). Therefore, G must contains at least one even cycle.

On the other hand, if G is 2-edge-connected and different from a cycle then, by Lemma 3, Perm(G) is $\mathcal{A}_{V(G)}$ or $\mathcal{S}_{V(G)}$. If in addition G has a cycle of even length then Perm(G) $\neq \mathcal{A}_{V(G)}$ and, consequently, Perm(G) = $\mathcal{S}_{V(G)}$. □

A similar study on group theoretic properties of permutations on graphs can be found in [8].

3 The 2D grid and the hypercube

We denote with $G_{m \times n}$ the *2D grid with m rows and n columns*.

Lemma 5 *For $n \geq 3$, $t_R(G_{2 \times n}) \leq 2n + O(\log_2 n)$ and $t_R(G_{3 \times n}) \leq 3n + O(\log_2 n)$.*

PROOF. Let π be a permutation of $V(G_{2 \times n})$. We use the *divide and conquer* strategy to R-route π. Let M' and M'' be the two halves of $G_{2 \times n}$, that is, the subgraphs of $G_{2 \times n}$ induced by the first $\lceil \frac{n}{2} \rceil$ columns and the last $\lfloor \frac{n}{2} \rfloor$ columns of $G_{2 \times n}$, respectively. Each pebble is moved into the half of the $G_{2 \times n}$ which contains its destination. This can be accomplished in at most $n + 2$ steps. Such a procedure is recursively applied on M' and M''. The *divide* process is stopped when the sub-grid obtained has 3, 4, or 5 columns.

The proof is analogous for $G_{3 \times n}$. □

Theorem 6 *If $n \geq 3$, then $G_{n \times n}$ is R-rearrangeable and*
$$t_R(G_{n \times n}) \leq \begin{cases} 6n + O(\log_2 n), & n \text{ even} \\ 9n + O(\log_2 n), & n \text{ odd} \end{cases}.$$

PROOF. Let n be even. Covering $G_{n \times n}$ with disjoint copies of $G_{2 \times n}$ and applying Lemma 5, we can permute in any way the pebbles on the rows and the columns of $G_{n \times n}$ within $2n + O(\log_2 n)$ steps. Hence, we can apply the general scheme of three phases given in [1] to route permutations on a product of graphs.

The proof is analogous for n odd; we cover $G_{n \times n}$ with disjoint copies of both $G_{2 \times n}$ and $G_{3 \times n}$. □

We denote with H_n the *n-dimensional hypercube*. Let $\pi \in \mathcal{S}_{V(H_n)}$ and let us suppose that each vertex x in H_n has a pebble with the destination address $\pi(x)$. In the *greedy routing* of π (see [4, pp.515-518]), each vertex x constructs a list of the dimensions in which x and $\pi(x)$ differ. The pebbles are sent along these dimensions in turn, the two-way edge hypothesis being implicitly assumed. A permutation $\pi \in \mathcal{S}_{V(H_n)}$ is said to be *easily routable* if it can be routed on H_n with two-way edges by a greedy routing without vertex contention.

Lemma 7 *If $\pi \in \mathcal{S}_{V(H_n)}$ is easily routable then π is R-routable on H_n in at most $2n + O(1)$ steps.*

Theorem 8 *If $n \geq 3$ then H_n is R-rearrangeable and $t_R(H_n) \leq 4n + O(1)$.*

PROOF. Via the Beneš network, each permutation of $V(H_n)$ can be decomposed in two easily routable permutations. □

4 Conclusions

Due to their 2-connectivity, almost all graphs used as patterns for interconnection networks are R-rearrangeable.

In the case of two-way edges, $G_{n \times n}$ and H_n are rearrangeable via matchings within $3n$ and $2n - 1$ steps, respectively [2]. If we replace each bidirectional step with two monodirectional steps, we obtain "classic" algorithms for an one-way communication model. These algorithms, which store two packets per vertex, require hence $6n$ steps for $G_{n \times n}$ and $4n - 2$ steps for H_n. In this paper, we have proved that $t_R(G_{n \times n}) \leq 6n + O(\log n)$, n even, and $t_R(H_n) \leq 4n + O(1)$. In other words, a permutation can be routed within the same number of steps (the low order additive terms ignored) by storing only one packet per vertex.

We note that lower bounds on $t_R(G)$ are most likely very hard to prove.

References

1. F. Annexstein and M. Baumslag. A unified approach to off-line permutation routing on parallel networks. In *Proc. of ACM Symposium on Parallel Algorithms and Architectures*, pp.398-406, 1990.
2. N. Alon, F. R. K. Chung, and R. L. Graham. Routing permutations on graphs via matchings. *SIAM Journal of Discrete Mathematics*, **7** (1994) 513-530.
3. C. Berge. *Graphs and hypergraphs*. North-Holland Publishing Co., Amsterdam, 1973.
4. F. T. Leighton. *Introduction to parallel algorithms and architectures: arrays, trees and hypercubes*. Morgan Kaufmann Publishers, Inc., San Mateo, 1992.
5. C. S. Raghavendra and M. A. Sridhar. Optimal routing of bit-permutes on hypercube machines. In *Proc. of International Conference on Parallel Processing*, vol.I, pp.286-290, 1990.
6. M. Ramras. Routing permutations on a graph. *Networks*, vol.23, pp.391-398, 1993.
7. J. J. Rotman. *An introduction to the theory of groups*. 3rd edition, Allyn-Bacon, 1984.
8. R. M. Wilson. Graph puzzles, homotopy, and the alternating group. *Journal of Combinatorial Theory (B)*, vol.16, pp.86-96, 1974.

The Effect of Flow Control and Routing Adaptivity on Priority-Driven Traffic in Multiprocessor Networks

Shobana Balakrishnan and Füsun Özgüner

Dept. of Electrical Engineering, The Ohio State University,
Columbus OH 43210-1272, USA

Abstract. We study the impact of two flow control schemes and routing adaptivity on the performance of priority-driven traffic produced by real-time applications. The first is wormhole routing (WR) with real-time extensions [1] and the second is preemptive pipelined circuit switching (PPCS-RT) [2]. Our simulations show that for a fixed number of virtual channels (VCs)/link, parallel lanes are more effective than adaptive routing alone, in reducing the number of messages that miss their deadlines. PPCS-RT performs better than WR due to the VC preemption protocol supported by it.

1 Introduction

Wormhole routing (WR) [3] is a flit-buffered flow control scheme increasingly used in multiprocessor point-to-point multi-hop networks, since it provides the design for low latency networks, and high link bandwidth, at low cost. The flit buffers with associated control are referred to as virtual channels (VCs). The commonly used routing algorithm in wormhole routers is the dimension order routing algorithm [3], which is an oblivious shortest path routing scheme. More recently, a number of adaptive routing algorithms have been proposed [3]. Here, the routing algorithm provides a choice of routing options and a selection function selects one of the options for routing the header. In order to prevent deadlocks, the routing algorithm uses routing classes (also called virtual networks), implemented by multiple VCs that are demand multiplexed on the link, with restrictions on the use of the classes [4]. For example, a simple minimal fully adaptive routing algorithm for meshes proposed in [4] requires two routing classes. Multiple VCs without any routing restrictions called lanes, were also proposed in [5] to improve the utilization of the physical link bandwidth. The speed advantage of WR using dimension order routing with a single lane, that requires only one VC/link, and the cost associated with multiple VCs, has resulted in the prominence of parallel systems with a modest number of VCs/link.

Real-time applications, unlike regular applications, produce tasks that must be performed before their associated deadlines. These tasks typically read sensor data and process it periodically. Multiprocessor systems are likely candidates for running real-time applications, since they offer increased computational power and fault tolerance, at good price points. Real-time applications produce messages, where a real-time message M_i is characterized by three parameters; namely period (p_i), deadline (d_i), and size (s_i). Real-time messages are classified as *hard* (also called guarantee seeking) and *soft* deadline (also called best effort) messages. For a hard deadline message, it is imperative that the communication model guarantees its timely delivery, while for a soft deadline message, timely delivery is a desirable feature, although occasional deadline misses can be tolerated. Real-time messages have an associated *global priority* that determines the importance level of the message. For example, when the message with the earliest deadline is assigned the highest priority, the assignment scheme is called Deadline Monotonic Scheduling (DMS) [6]. We shall use this priority assignment scheme in this paper.

With priority-driven real-time traffic, the priority is used to resolve the competition among messages for shared network resources. In WR, there are two types of shared resources, namely the VCs and the physical link. A WR message can compete with any of the messages that share at least one hop on the path of the message, for the use of the resources. Hence, the global priority order among all competing messages must be enforced by the communication model, for predictability. With a limited number of parallel lanes, a global priority ordering on the usage of the VCs during the course of the application is desirable. Two real-time extensions to WR have been proposed in [1], namely, priority based VC allocation and link arbitration. With priority-based VC allocation, the highest priority message amongst the competing messages *at that time* is allocated the output VC at a router. Local priority order on the usage of the link is enforced by priority based link bandwidth arbitration, instead of the conventional demand multiplexing, whereby, the physical link is in effect *locally preemptable*. A higher priority header arriving at an input VC whose routing request for an output VC is granted, can preempt a lower priority message using the link. A higher priority message, if not granted an output VC due to unavailability, cannot in any way affect lower priority messages occupying the output VCs, since a VC once allocated can only be released by the tail flit of the message. Hence, depending on the order of arrival of messages, a higher priority message may be blocked by a lower priority message using a VC, for an *unbounded* amount of time. Such a situation is referred to as *priority inversion*, which was first investigated in the context of task scheduling. The priority inversion is unbounded, because in WR, the message delivery time cannot be bounded due to the fact that the header may be blocked at each hop, depending on the contending messages. The priority inversion problem is magnified with fewer lanes. Such a WR model with v lanes, can only enforce a v level priority order, although the priority levels of the competing messages may be more fine grained. In this paper, we compare a new flit-buffered communication model that we recently proposed [2] called *Preemptive Pipelined Circuit Switching for Real-Time* (PPCS-RT) messages with WR [1]. Both models employ parallel lanes as well as deadlock-free routing. We evaluate the effectiveness of v parallel lanes under dimension order routing. We also consider the effect of routing adaptivity *i.e.*, WR using multiple VCs for adaptive routing classes as well as parallel lanes.

In terms of alternative flow control schemes to WR, Pipelined Circuit Switching (PCS) was proposed by Gaughan and Yalamanchili [7] for achieving fault tolerance. PCS does not address the problem of real-time message scheduling. Unlike the nonblocking header in PCS, which requires an extra forward virtual channel, the header is blocking in PPCS-RT, and backtracks when *any* of the VCs acquired by it are requested by a higher priority message. A unified approach to real-time task and message scheduling has been adopted in [9] for a single hard deadline periodic input using a scheduled routing method. The SPIDER [8] network adapter supports mixed mode flow control, namely both packetized store-and-forward switching and WR, to cater to the conflicting requirements of hard and soft deadline messages, by partitioning the network into two separate virtual networks, one for hard deadline, and the other for soft deadline traffic. We believe that WR is a good choice in massively parallel systems when the application traffic does not have any deadline constraints. However, for soft and hard deadline traffic, where priorities of messages are fine grained such as on the basis of deadline, we argue that WR is inadequate. The rest of this paper is organized as follows. In Section 2 we describe the basic PPCS-RT model and the protocol for preemption. In Section 3 we compare the performance of PPCS-RT and WR with varying number of virtual channels for parallel lanes and deadlock-free routing.

2 A New Flow Control Model for Real-Time Messages

For completeness, we present in this section a brief description of the PPCS-RT flow control model proposed in [2, 10]. Additional implementation details may be found in [2]. The goal of PPCS-RT is to keep the advantages of wormhole routing such as small flit buffers (that result in single chip, fast, inexpensive routers) and prevent the unbounded priority inversion problem.

2.1 PPCS-RT

In WR, the delivery time of a message is unpredictable due to the blocking faced by the header. The basic idea of PPCS-RT is to enforce the global priority ordering on the usage of the VCs by supporting preemption during this unpredictable period when the header is in transit from the source to the destination. For this purpose, the message delivery is performed in two decoupled phases; Path Establishment (PE) and Data Delivery (DD). The header flit first establishes the path similar to circuit switching in the PE phase. If a header cannot receive the requested output VC, it blocks and holds the VCs acquired so far, as in WR. All VCs reserved by the header are preemptable in this phase, by higher priority messages. When a path is successfully established, an acknowledgment is returned to the source node, that terminates the PE phase and initiates the DD phase. Data flits are pipelined in the DD phase. In the DD phase, the VCs occupied by the data flits are not preemptable. However, the duration of the DD phase is bounded and dependent on the message size, and the link bandwidth only.

For each VC that transfers the flits in the forward direction (denoted as v_f), PPCS-RT requires a reverse VC to transmit control flits in the reverse direction (denoted as v_r). The forward and reverse VCs of adjacent nodes are multiplexed on the physical link. The header, data, and tail flits use the forward virtual channel v_f as in WR. A *success* flit is returned along the reverse path using virtual channel v_r to the source router, when the header reaches the destination successfully, which then starts the DD phase. A message whose priority is higher than the owner of a VC can request preemption, if the owner is in the PE phase. This is done by generating a non-blocking *preempt* flit at the preemption point (preempted output channel) which also travels along v_f in the forward direction, until it reaches the channel containing the header to be preempted. If the header flit is not blocked, a preempt flit may or may not be able to catch up with the header depending on the preemption point and the position of the header with respect to the destination. The preempt flit has a router cycle time that is lower than the header flit since it does not go through the routing logic at each router. If the header is close to reaching its destination and successfully completes the PE phase before the preempt flit reaches it, the success flit returning kills the preempt flit, thereby aborting the preemption request. In this case, the higher priority message must face a blocking time equal to the DD phase of the lower priority message, and contend for the VC in priority order again.

When a header is preempted, a *free* flit is produced which backtracks along a reverse path using v_r and frees all forward VCs held by the preempted message until the point of preemption. Note that all VCs on the path up to the preemption point are still held by the preempted message. The header, preempt, free, and success flits will be referred to as control flits. It was shown that PCS [7] suffers from larger no-blocking message latency (minimum latency) due to the overhead of path setup, compared to WR, which is also the case with PPCS-RT. The experiments conducted in [7] assumed 16 VCs/link when the effect of blocking due to VC unavailability is negligible. With fewer VCs, by resolving VC contention based on global priority rather than local priority as in WR, we will show in the next section, that PPCS-RT results in improved mean latency over WR for high priority traffic, despite the additional overhead of the PE phase.

2.2 Control Flit Format

Figure 1 illustrates the fields in a control flit. Two bits are used to encode the control flit function. Note that the tail flit is not considered here since its function is identical to WR. The header flit carries the address of the destination. The preempt flit carries the address of the node at which the preemption occurs. This information is recorded in a writable register associated with each channel along which the preempt flit is propagated downstream. The free flit uses this information to detect when to stop its reverse path. The free flit carries the address of the destination node, which is used by the control logic to generate a new header when the free flit reaches the preemption point, so that the preempted message can resume from the preemption point. Note that the header and free flits differ only in the first two bits. The success flit does not carry any address, and simply follows the reverse path up to the source node. The header and free flits also carry the priority of the message used to determine which channels are eligible for preemption by the routing logic.

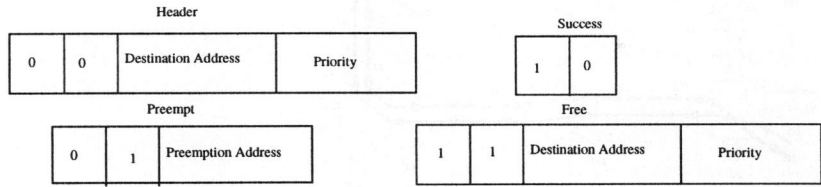

Fig. 1. Control flit format

2.3 Protocol for Virtual Channel Preemption

In order to support VC preemption, the state of the output virtual channels, and the priority of the owner must be maintained for each VC. A VC is in the PE state when the message owning it is in the PE phase. Similarly, a VC is in the DD state when the message owning it is in the DD phase. A VC is in the preempt state (PT) when a preemption request has been propagated on that VC. A VC, if available, is in the AV state. All examples are based on a 2D mesh, and one VC/link is assumed for simplicity. We first illustrate the *header-success* paths for a message originating at node A and destined to node D in Figure 2 (a), using the bidirectional links at each router. The actual flow of header and success flits is shown in Figures 2(b) and (c) respectively. At each node we show the input virtual channel, the state of the output virtual channel, and the crossbar connection made. Details of the implementation of the forward and reverse paths are discussed in [2]. The crossbar is a VC to VC crossbar. In this example, the crossbar is a 5x5 crossbar with inputs numbered xi and outputs numbered xo, where x denotes the directions $0-3$ or the node N.

The header flit shown in Figure 2(b) is responsible for switching the output virtual channel state to PE. The success flit generated at node D returns on the reverse VC path until it reaches the source node A as shown in Figure 2(c). A success flit arriving on an input channel j switches the state of output channel j to DD. At node A, the success flit is propagated to the reverse node channel No where it terminates, and the DD phase is started thereafter. Although the header and success flits are shown to exist in several VC FIFOs in the figure, only a single flit exists in any FIFO of a path, at any given time.

In Figure 3(a) we illustrate the same message from A to D, whose header is blocked at node C due to the fact that the output channel $3o$ is in DD state. While the header is blocked, a higher priority message, occupying channel $0i$ at

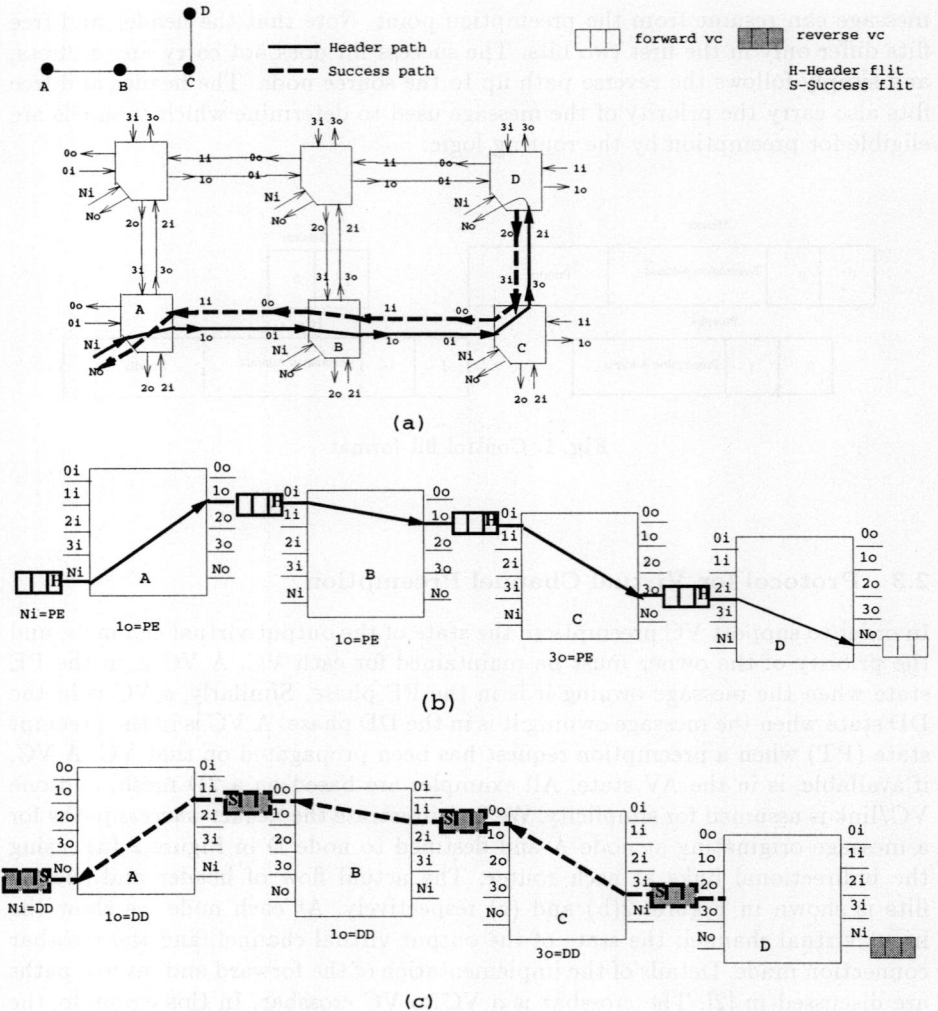

Fig. 2. PE phase (a) Header-Success paths (b) Header flit path (c) Success flit path

node A (Figure 3(b)) generates a preemption request for output channel 1*o*. A preemption request is generated if the output channel to be preempted is in the PE state, and the owner priority is lower than the requesting message priority. The preempt flit propagated on an output channel j, changes the state of the channel to PT. The node address at which the preemption request is generated, must be recorded at each preempted channel for future preemption requests, as well as for the returning free flit to determine if it has reached the preemption point. For this purpose, a writable register *Paddr[j]* associated with each output

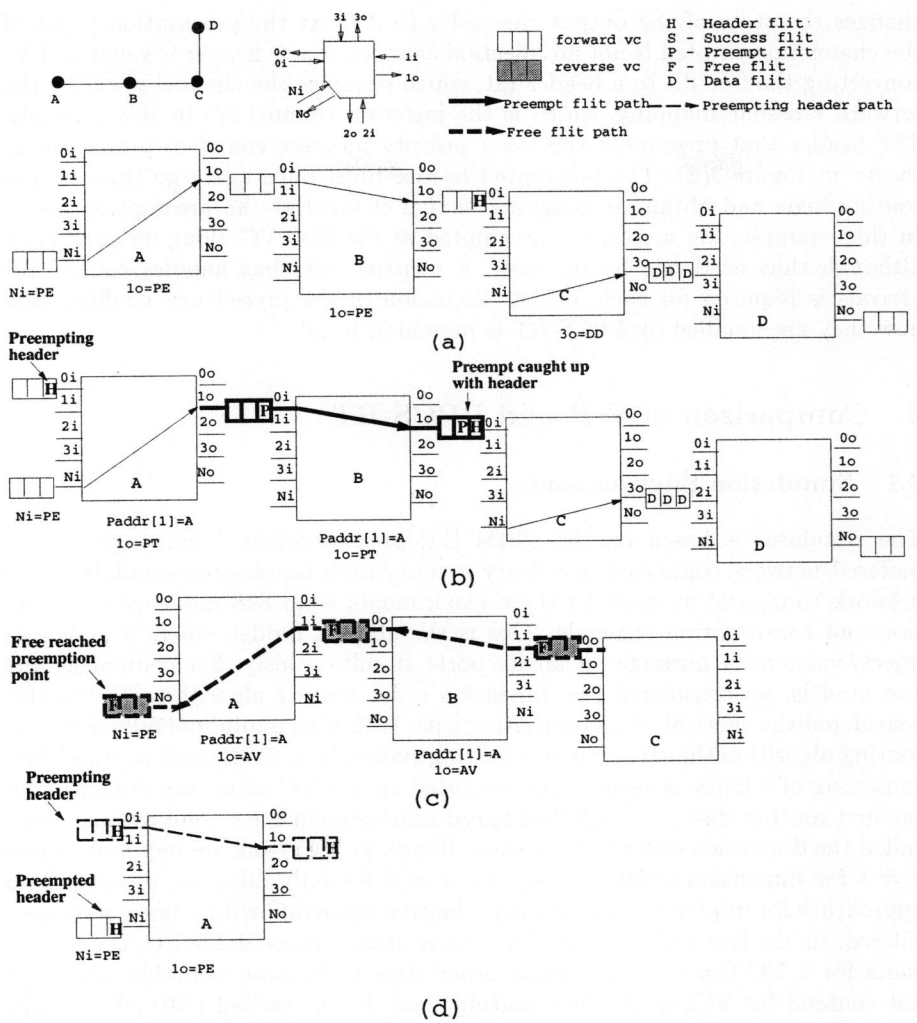

Fig. 3. Preempt-free flit paths (a) Header is blocked at node C (b) Preempt flit path (c) Free flit path (d) Preempting header path

VC j is required. A preempt flit that is propagated on an output VC j, updates the *Paddr[j]* register to the preemption address carried by it. In this example, at nodes A and B, *Paddr[1]* is set to the address of node A.

A free flit is generated when the preempt flit reaches node C, where the preempted header is blocked. The free flit shown in Figure 3(c), traverses the reverse path from node C to node A. At each hop, the free flit is propagated upstream if the preemption point is not reached. The free flit arriving at input channel 1 at node A terminates. The free flit arriving on an input channel j, changes the state of the output channel j to AV. At the preemption point, if the channel preempted is not an injection channel, a new header is generated by converting the free flit to a header flit, which occupies the channel given by the forward crossbar mapping, which is the injection channel Ni in this example. The header that preempted the lower priority message can then propagate as shown in Figure 3(d). The preempted header must once again go through the routing logic and obtain an outgoing virtual channel at the preemption point. In this example, the message is preempted at the first VC along its path ($1o$), although that need not be the case. A control unit that handles control flit arrivals is required for each VC. A discussion of the precedence conflicts and how they are handled by PPCS-RT is presented in [2].

3 Comparison of WR and PPCS-RT

3.1 Simulation Environment

The simulator is based on the CSIM [11] process oriented language. A flit-buffered network connected in a k-ary n-cube/mesh topology is simulated. The network configuration used for these experiments is an 8x8 mesh with 4 injection and consumption channels. This is the all-port model, where a node can inject/consume a message on all its ports simultaneously. For comparing the two models, we considered the dimension order routing algorithm [3]. We also considered the effect of adaptive routing with WR, with a minimal fully adaptive routing algorithm that requires two routing classes ($r = 2$) [4], each routing class consisting of v lanes. A header can be routed on any VC along any dimension in the first routing class, or along the highest dimension in the second routing class, called the dimension order routing class. Hence, rv VCs/link are required, where $r = 1$ for dimension order routing, and $r = 2$ for fully adaptive routing. Two approaches for implementing the fully adaptive algorithm with v lanes were considered. In the first (A1), if a header cannot obtain an available VC, the header waits for a VC from the dimension order class to become available, and does not contend for VCs in the first routing class. In the second (A2), if a header cannot obtain an available VC, it contends in round robin fashion until a free VC becomes available in either routing class. Approach A2 requires a centralized round robin arbitration scheme.

The timing parameters for the simulated PPCS-RT and WR models are as follows: Both models assume a physical link bandwidth (B) of 10 MB/s, *i.e.*, it takes $0.4\mu s$ to transfer a data flit across a link where a flit is assumed to be 4 bytes. The header flit in WR takes $1.2\mu s$ to propagate one hop. This includes the routing decision and the flit transfer time. The header flit in PPCS-RT takes $1.6\mu s$ to propagate one hop (t_r), due to the added complexity of the PPCS-RT router. The success flit has a cycle time per hop of $0.4\mu s$. We assume that the preempt, and free flits propagate without delay. This is a reasonable assumption since the overhead of preemption is very much smaller than the PE and DD phases, due to the fact that both the preempt and free flits are non-blocking.

The reason for this assumption, is that we used events to simulate preempt and free conditions thereby simplifying the PPCS-RT programming model. The ideal PPCS-RT latency C_i is given by:

$$C_i = h_i t_r + h_i/B + (h_i + s_i - 1)/B \tag{1}$$

where s_i is the size of message M_i, h_i is the number of hops traversed by M_i, B is the bandwidth of the link, and t_r is the header propagation time per hop. The first term in Eq. 1 is the time for the header to reach the destination, the second term is the time for the success flit to return, and the last term is the time taken by the DD phase.

A message generation program generates messages that are periodic with random source destination pairs. The message size is randomly chosen to be one of four message sizes (16, 128, 1024, 4096 flits). Depending on the message size, the period is randomly chosen from 16 values within a range of $2 - 100$ ms so that larger messages are restricted to larger periods. For example, a message of size 16 flits has a period in the range of $2 - 10$ ms while a message of size 4096 flits has a period in the range $20 - 100$ ms. This is done so as to represent real-time traffic more realistically, where smaller messages (often critical messages) occur more frequently than larger messages. Furthermore, we do not want very large messages to occur with a high frequency thereby saturating the network very quickly. The deadline of the message is chosen to be less than the period. Messages are generated until a predefined average link utilization (LU) is reached ranging from $0.1 - 0.5$. The link utilization at a link with k messages, is defined as the fraction of the link bandwidth utilized by the messages using the link, i.e., $\sum_{i=1}^{k} C_i/p_i$, where C_i is given by Equation 1. Due to the periodic nature of messages, at a link utilization of 0.5, the number of messages existing in the network is very large (over 15,000) resulting in long simulation times. The simulator injects each generated message in the message set into the network initially at time 0. Thereafter, new instances are injected at their respective periods. Each message instance is a process which creates a header process. The header process in the case of WR spawns a data process at each node which is responsible for transferring the flits. In the case of PPCS-RT, the header process simulates the PE phase and backtracking if preempted.

3.2 Results

In Figure 4(a) we compare the performance in terms of Overall Miss Ratio ($OMR\%$) defined as the ratio of missed messages to total number of messages delivered, using PPCS-RT and WR with $v = 1$ and $v = 2$ under dimension order routing. In general, PPCS-RT significantly outperforms WR for the same v, with the difference in performance being more significant at increased link utilizations. The performance difference between PPCS-RT and WR is greatest for $v = 1$. PPCS-RT($v = 1$) outperforms WR($v = 2$) at high link utilization in terms of $OMR\%$ (Figure 4(a)). Hence, at increased link utilization, a large number of messages suffer from priority inversion in WR resulting in more deadline misses than PPCS-RT even though WR uses twice the number of lanes used by PPCS-RT.

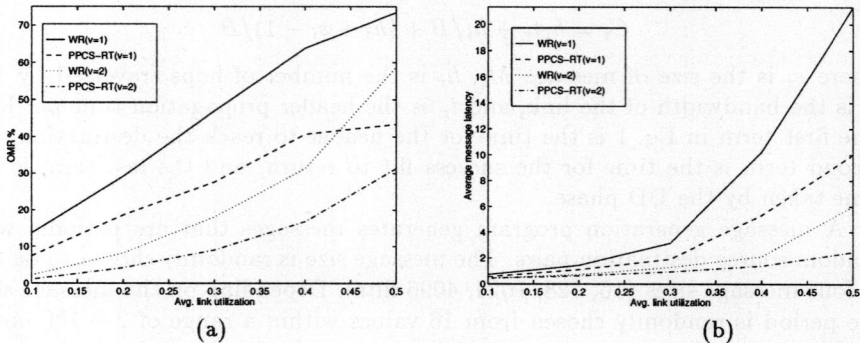

Fig. 4. WR versus PPCS-RT with dimension order routing(a) $OMR\%$ versus average link utilization (b) Average message latency versus average link utilization

Figure 4(b) shows the average latency of a message using WR and PPCS-RT for $v = 1$, and $v = 2$. PPCS-RT results in lower average message latency compared to WR for the same v despite the overhead of the PE phase. By resolving VC contention according to global priority rather than local priority, PPCS-RT achieves a lower average blocking time, and hence a lower average latency compared to WR. Although, PPCS-RT($v = 1$) has lower $OMR\%$ compared to WR($v = 2$) at increased link utilization, the mean latency of messages using PPCS-RT($v = 1$) is higher than that of WR($v = 2$) messages, as seen in Figure 4 (b). While mean latency is a good indicator of real-time performance with WR, this is not the case with PPCS-RT. With WR, most messages that miss their deadlines, do so by a small margin, while with PPCS-RT, the low priority messages miss by a large margin, contributing to the increased average message latency.

Next, we evaluated the effectiveness of adaptive routing with WR using a variable number of lanes v. Figure 5(a) shows the $OMR\%$ with the two implementations of fully adaptive minimal routing. A2 results in lower $OMR\%$ than A1 at low link utilizations with $v = 1$ and, for all link utilizations with $v = 2$. At high link utilizations and with only one lane ($v = 1$), the centralized round robin arbitration scheme performs worse than strict priority-based waiting for a channel in the dimension order routing class. This is due to the fact that with a single lane a large number of messages are blocked, and the centralized arbitration scheme becomes a bottleneck. A1 is also easier to implement in hardware. Comparing Figures 4(a) and 5(a), fully adaptive routing using A1 results in lowering the $OMR\%$ as compared to dimension order routing, for the same number of lanes v. However, the fully adaptive scheme requires twice the number of VCs compared to dimension order routing for the same v. When $v = 1$, adaptive routing alone performs marginally better (< 10% difference in $OMR\%$) than dimension order routing. This is not the case when $v = 2$, where the difference in the $OMR\%$ is over 25%. In meshes, fully adaptive routing alone with no parallel lanes is not very effective since a large number of messages have only a single shortest routing path. Fully adaptive routing with parallel lanes effectively distributes the traffic resulting in good real-time performance. Figure 5(b) compares the average latency of wormhole routed messages using dimension order routing and adaptive routing (using A1) that require the same number of total VCs per link. As also seen from the $OMR\%$, the average latency of a WR message using adaptive routing is lower than a message using dimension order routing for the same v. However if the total number of VCs/link is fixed, then parallel lanes are more effective compared to adaptive routing. For example, WR($v = 2, r = 1$) performs better than WR($v = 1, r = 2$), where both require 2 VCs/link. Note that with priority-based link bandwidth arbitration, the physical link bandwidth is not continuously time multiplexed as with demand multiplexing. Hence the overhead of multiplexing is small as compared to demand multiplexing. We conclude that parallel lanes are more effective for priority-based traffic and require less hardware compared to adaptive routing.

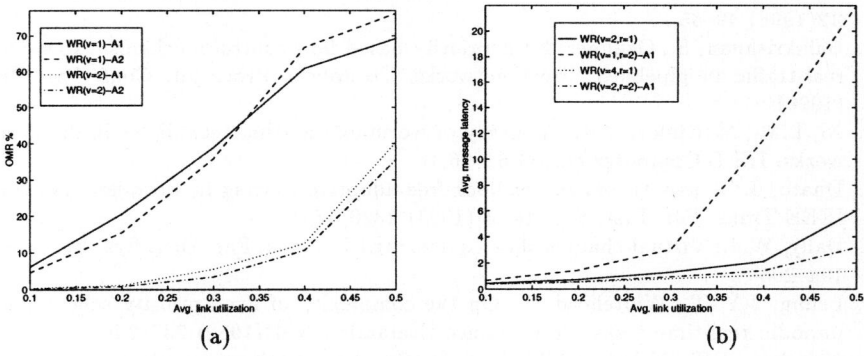

Fig. 5. Comparison of WR with minimal fully adaptive routing using A1 and A2 (a) $OMR\%$ versus average link utilization for two implementations (b) Average latency versus average link utilization for adaptive and dimension order routing using the same number of VCs

Finally, PPCS-RT is much better suited than WR for real-time traffic when only a few VCs per link exist, which is the case in most systems today. For example, PPCS-RT($v = 2$) employing dimension order routing has a lower $OMR\%$ than WR($v = 2$) using fully adaptive minimal routing at a link utilization of 0.5. Hence, we infer that complex adaptive routing schemes are not really required if a well suited priority-based flow control scheme exists for real-time messages under heavy traffic conditions. Our simulation experiments did not account for the overhead of preemption. This overhead, affects the preempting message delivery time by a small amount. Although the number of misses is expected to increase marginally with this overhead, the general trend that low priority messages will miss with large margins, the enforcement of priority order, and the lower $OMR\%$ of PPCS-RT compared to WR for small v, are expected to remain.

4 Conclusions

In this paper we compared the performance of PPCS-RT, a new flow control model that we proposed for priority-driven traffic, and WR, in terms of overall miss ratio. We also examined the effect of VCs for routing adaptivity and parallel lanes using WR. Parallel lanes provide better performance compared to adaptive routing for the same number of VCs/link. An important problem that we are investigating is a comparison of PPCS-RT and WR taking into account the increased latency due to additional flow control complexity, and routing complexity, and the transmission times of preempt and free flits.

References

1. Li, J.-P., Mutka, M.W.: Real-time virtual channel flow control. J. Par. Dist. Comp. **32**(1996) 49–65
2. Balakrishnan, S., Özgüner, F.: A priority-based flow control mechanism to support real-traffic in pipelined direct networks. To appear Proc. Int. Conf. Par. Proc. (1996)
3. Ni, L.M., McKinley, P.K.: A survey of wormhole routing techniques in direct networks. IEEE Computer (1993) 62–76
4. Duato, J.: A new theory of deadlock-free adaptive routing in wormhole networks. IEEE Trans. Par. Dist. Sys. **4(12)**(1993) 1320–1331
5. Dally, W.J.: Virtual-channel flow control. IEEE Trans. Par. Dist. Sys. **3(2)**(1992) 194–205
6. Leung, J.Y.-T., Whitehead, J.: On the complexity of fixed-priority scheduling of periodic real-time tasks. Performance Evaluation **2(4)**(1982) 237–250
7. Gaughan, P.T., Yalamanchili, S.: A family of fault-tolerant routing protocols for direct multiprocessor networks. IEEE Trans. Par. Dist. Sys. **6(5)**(1995) 482–497
8. Dolter, J. et. al.: SPIDER: Flexible and efficient communication support for point-to-point distributed systems. Proc. Int. Conf. Dist. Comp. Sys.(1995) 574–580
9. Shukla, S.B., Agrawal, D.P.: Scheduling pipelined communication in distributed memory multiprocessors for real-time applications. Proc. Int. Symp. Comp. Arch. (1991) 222–231
10. Balakrishnan, S., Özgüner, F.: Providing message delivery guarantees in pipelined flit-buffered multiprocessor networks. To appear Proc. IEEE Real-time Technology and Applications Symp. (1996)
11. Schwetman, H.: CSIM: A C-based process-oriented simulation language. Proc.Winter Simulation Conf. (1986) 387–396

Routing on Networks of Optical Crossbars*

(Extended Abstract)

Friedhelm Meyer auf der Heide, Klaus Schröder, and Frank Schwarze

Heinz Nixdorf Institute and Computer Science Department, University Paderborn,
33102 Paderborn, Germany

Abstract. We describe routing algorithms on networks composed of optical busses. Using networks with short busses and small degree we are able to give very fast routing algorithms. First, we describe a leveled optical network and routing algorithms for it. Next, we show how to simulate this network on high–dimensional meshes of optical busses (MOBs). We present algorithms, e.g., for h–relations with runtime being double–logarithmic in the size of the mesh, linear in h, and polynomial in the dimension. Previous results are exponential in the dimension. We use a novel type of protocol and analysis inspired by hashing based shared memory simulations with redundant storage representation from [MSS95].

1 Introduction

In recent years, the possibility of using optical devices to build very fast, high bandwidth communication networks has attracted many researchers, engineers as well as (theoretical) computer scientists. Anderson and Miller [AM88] were the first to consider routing algorithms for h–relations using the following model, later called the Completely Connected Optical Communication Parallel Computer (OCPC), or, as a routing device, the optical bus. It is motivated by possibilities and restrictions of optical communication technology.

An optical bus of length k connects k processors. In one step, each processor can try to send a message to an arbitrary other processor. The sending is successful, only if the receiving processor gets no other message in this step. For a discussion of this model see [AM88], [GJLR93] and [GJLR94]. Extensions are discussed in [DM93] and [MSS95].

In order to avoid too long optical busses, 2–dimensional meshes in which rows and columns are connected by optical busses are examined in [GJLR94]. These meshes are called mesh of optical busses (MOBs). Here, length k busses suffice to obtain an efficient routing device for k^2 processors.

In this paper we examine the routing capability of this and other networks of optical busses. Formally such a network can be described by a hypergraph whose nodes are the processors and whose edges are the optical busses. The following parameters characterize the quality of such a network of size N:

— the maximum length k of the busses

---

* supported by the DFG-Graduiertenkolleg "Parallele Rechnernetzwerke in der Produktionstechnik", ME 872/4-1 and by DFG–SFB 376 "Massive Parallelität"

- the maximum degree d
- the routing time for permutations, h-relations, random functions etc.

Our main results are very fast routing algorithms on certain leveled networks of optical busses, the split&hash networks. They are inspired by high-ary, low depth versions of Butterfly networks or Multibutterfly networks [Upf89].

We derive a network of busses from such a network by replacing certain subgraphs by optical busses. Furthermore we show how to simulate such split&hash networks on high-dimensional MOBs, achieving the fastest known routing algorithms for them. Our techniques for permutation routing differ significantly from those previously derived for MOBs of constant dimension as in [GJLR94]. They are inspired by hashing techniques to simulate shared memory on an OCPC as described in [MSS95].

1.1 Previous Results

Anderson and Miller [AM88] present an algorithm for realizing h-relations in $O(h + \log N)$ expected time on a single OCPC.

In [GJLR93] Goldberg et al. present an algorithm for h-relations using $O(h + \log \log N)$ time, with high probability,[2] if $h < \log N$.

The algorithm of Goldberg et al. works on a single OCPC. In [GJLR94] they give an algorithm on a 2-dimensional MOB using $O(h + \log \log N)$ time, with high probability, if $h < \log N$. In the same paper they suggest an extension of their algorithm to higher dimensions still using $O(h + \log \log N)$ time, if the dimension d is constant. Extensions to higher dimensions result in a runtime exponential in d.

1.2 New Results

We present a new Butterfly-like class of networks, the so called *split&hash networks*, and its realization using optical busses. On such networks we give algorithms for delivering packets from the N sources to the N sinks. Our algorithm for permutation routing requires $O(d^2 \cdot \log d \cdot \log \log N)$ steps on a split&hash network of depth d, with high probability. The probability space is described by some random choices done for constructing the networks. The main building block of our algorithm is an adaption of a scheme presented in [MSS95] where it is used and analyzed for obtaining fast shared memory simulations on OCPCs and similar machines.

The algorithm can be extended to route a random function. The running time is $O\left(\frac{\log N}{\log \log N}\right)$, with high probability, if $d = O(\log^\delta N)$, $\delta < \frac{1}{2}$. The last step of the algorithm makes use of an algorithm from [GJLR93]. The overall running time is optimal, since some sink gets $\Theta\left(\frac{\log N}{\log \log N}\right)$ packets, with high probability.

Combining our methods with another technique from [GJLR93] and [AM88] we derive an algorithm for h-relations. The algorithm is randomized and requires $O(d^3 \cdot \log d \cdot \log \log N + d^3 \cdot h)$ steps, with high probability, if $h \leq \log N$, $d = O(\log^\delta N)$ and $\delta < \frac{1}{2}$.

---

[2] With high probability means a probability of at least $1 - N^{-\gamma}$, where $\gamma > 0$ can be chosen arbitrary.

Simulating a split&hash network of depth d we obtain algorithms for the d–dimensional MOB.

We get an algorithm using $O(d^4 \cdot \log d \cdot \log \log N)$ steps, with high probability, to route permutations and an algorithm for random functions using $O(d^4 \cdot \log d \cdot \log \log N + \frac{\log N}{\log \log N})$ steps, with high probability. This is optimal for $d = O(\log^\delta N)$, $\delta < \frac{1}{4}$. The MOB version of our h–relation algorithm requires $O(d^5 \cdot \log d \cdot \log \log N + d^3 \cdot h)$ steps, with high probability, for any h–relation, for $d = O(\log^\delta N)$, $\delta < \frac{1}{2}$.

Our results show a tradeoff between running time for various routing problems, the size of the optical busses and the degree of the network. The size of the optical busses in a split&hash network of depth d is $O(d \cdot N^{\frac{1}{d}})$ and exactly $N^{\frac{1}{d}}$ in a d–dimensional MOB with N processors. The degree is $O(d)$, both in the split&hash network and in the MOB.

1.3 Organization of the Paper

In the next section we describe the split&hash networks. Section 3 contains the algorithm for permutation routing on split&hash networks. The result is transferred to the MOB by simulation in Section 4. In Section 5 we give the extensions to random functions and h–relations. Most proofs are only sketched. Complete proofs are contained in a full version of this paper, available soon as technical report [MSS96][3].

2 The Split&Hash Network

A (N, d)–*Butterfly–type network* has the following structure. For $d = 1$ it is the complete bipartite network with N sources and N sinks. For $d > 1$, it consists of $k = N^{\frac{1}{d}}$ many $(N^{\frac{d-1}{d}}, d-1)$–Butterfly–type networks $B_1, \ldots, B_k$ and N new nodes, the sources. These sources are connected by k identical bipartite graphs, called *funnels*, to the sources of each B_j, $j = 1, \ldots, k$. These k funnels all share the same N sources. For an illustration see the picture below. The nodes on level 0 are the sinks, those on level d the sources.

(Note: Butterfly and Multibutterfly are $(N, \log N)$–Butterfly–type networks.)

In a Butterfly–type network a packet is sent from an input node to an output node by traveling along the unique sequence of sub–Butterfly–type networks. This sequence is called the *coarse path* of the packet.

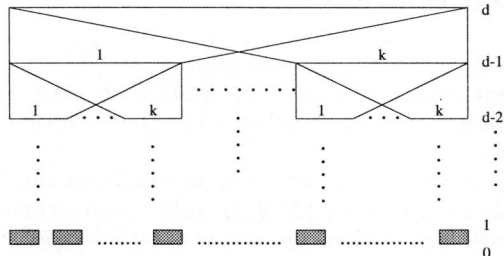

---

[3] http://www.uni-paderborn.de/cs/ellern.html or
http://www.uni-paderborn.de/cs/fmadh.html

To specify an (N,d)–Butterfly–type network we have to specify the funnels. Let $a \in \mathbb{N}$ be fixed. The (N,d,a)-split&hash network is defined as follows. Between level $i+1$ and level i the following (k^{i+1}, k^i, a_i)-funnel is used. It is defined by $a_i := \lceil a \cdot \frac{d}{i} \rceil$ functions $h^i_j : [k^{i+1}] \to \left\{ \left\lfloor \frac{jk^i}{a_i} \right\rfloor + s \,\middle|\, s \in \left[\left\lfloor \frac{k^i}{a_i} \right\rfloor\right]\right\}$, $j \in [a_i]$[4]. So the k^i bottom nodes of the funnel are split into a_i blocks of equal size, each $h^i_j, j \in [a_i]$, maps to one block. We assume that each h^i_j is uniform, i.e. $|(h^i_j)^{-1}(x)| = |(h^i_j)^{-1}(y)|$ for all $x,y \in \mathrm{image}(h^i_j)$. So the funnel has the edges $\{(l, h^i_j(l)) | \, l \in [k^{i+1}], j \in [a_i]\}$.

A random (N,d,a)-split&hash network is obtained by choosing a random permutation $\varphi : [k^{i+1}] \to [k^{i+1}]$ and putting $h^i_j(l) := \varphi(l) \bmod \left\lfloor \frac{k^i}{a_i} \right\rfloor + j \cdot \left(\left\lfloor \frac{k^i}{a_i} \right\rfloor + 1 \right)$. In the remainder of the paper we assume all split&hash networks to be *random* split&hash networks.

Since we want to deal with optical crossbars, we give the following optical crossbar version of the split&hash network. Note that every node of level $d, \ldots, 2$ is a top node of a group of k funnels. Every l–th bottom node of a funnel in the group is connected to the same set of top nodes. So we replace all edges adjacent to any l–th bottom node by one optical crossbar connecting the i-th bottom nodes with the top nodes adjacent to the l–th bottom nodes. Due to the choice of the functions h^i_j the optical crossbar has length $k + k \cdot a_i$. Further we replace the complete bipartite graphs between level 1 and 0 by optical crossbars of length $2 \cdot k$.

The picture shows the optical crossbar (thick dotted line) replacing the edges of two functions in three funnels in a split&hash network for $k = 3$.

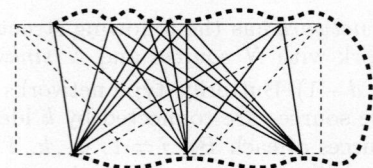

Remark A (N,d,a)–split&hash network has depth d, optical busses of length $O(N^{\frac{1}{d}})$ and degree $O(d)$.

3 Permutation Routing

Our first result is about routing a permutation $\pi : [N] \to [N]$ from the input nodes to the output nodes of a (N,d,a)-split&hash network, so the (unique) packet of the l-th source has to be delivered to the $\pi(l)$-th sink.

Theorem 1. *Let $k,d \in \mathbb{N}$, $N := k^d$. Any permutation can be routed from the sources to the sinks of a random (N,d,a)-split&hash network in $O(d^2 \cdot \log d \cdot \log \log N)$ steps, with high probability, if a is a sufficiently large constant (independent of N).*

---
[4] $[l]$ denotes the set $\{0, \ldots, l-1\}$

Proof. Fix permutation π. For $\epsilon > 0$ (to be fixed later) we show how to route a batch of $\epsilon \cdot N$ messages which contains the messages with destinations $\pi(i)$ where $\pi(i) \bmod \left(\frac{1}{\epsilon}\right)$ has some fixed value. The complete routing consists of $\frac{1}{\epsilon}$ batches.

The routing of a batch is performed in d rounds. At the beginning of round i all packets are on level $d - i + 1$ in different processors. Round i delivers them to different processors of level $d - i$ obeying their coarse path.

We have to show how a round can be done. The last round delivering the packets from level 1 to their destinations on level 0 takes at most one step, since only partial permutations on complete bipartite graphs have to be executed.

It remains to show how the other rounds can be realized. We make use of the following *Funnel Algorithm* which is a modification of an algorithm given in [MSS95]. The algorithm is executed in all (k^{i+1}, k^i, a_i)-funnels, i.e. all funnels of level i.

Funnel Algorithm:
set all processors holding a packet of the actual batch as active
while there are active processors do

    for $k := 1$ to a_i do
        if processor $l \in [k^{i+1}]$ is active: try to send the packet to $h_k^i(l)$
        if sending was successful (i.e. only processor l tried to send a packet to $h_k^i(l)$): set processor l as inactive

Every run through the 'for' loop is called a *round* of the algorithm.

Main Lemma. *Let $i \in \{2, \ldots, d\}$, $\beta > 1$, and $\epsilon > 0$ such that $\epsilon \beta a_i < 1$. The following holds for k large enough. (a) A set of $\epsilon \cdot m := \epsilon \cdot k^{i+1}$ packets given on arbitrary distinct sources of a random (k^{i+1}, k^i, a_i)-funnel can be delivered to distinct sinks of the funnel using $t_i := \frac{\log\log(k^i)}{\log(a_i - 1)} - 1$ rounds, with probability at least $1 - m^{-\gamma \cdot \lceil \frac{d}{i} \rceil}$ for a constant $\gamma > 0$ depending on a and $\frac{1}{\epsilon}$.*
(b) A constant fraction of the $\epsilon \cdot m$ packets is delivered to distinct bottom nodes after a constant number of rounds, with high probability.
(c) Each round contains a_i steps.

The *proof* is omitted due to space limitations. It is based on a proof given in [MSS95] for the running time of the so called (N, ϵ, a, b, c)-*scheme* which is similar to the Funnel Algorithm given here. Our proof is much simpler than the one given in [MSS95] and can also be used in their situation.

Using this lemma we can bound the running time of our routing algorithm. For any $\gamma > 0$ the Main Lemma yields a probability of at most $k^{-(i+1) \cdot \gamma \cdot \lceil \frac{d}{i} \rceil} \leq N^{-\gamma}$ for a packet to be undelivered after t_i rounds. So the probability for an undelivered packet in any funnel of level i is almost $k^i \cdot N^{-\gamma}$. Hence the probability for an undelivered packet in any funnel of the split&hash network is at most $N^{-\gamma} \cdot \sum_{i=1}^{d-1} k^i \leq N^{-\gamma} \cdot k^d \leq N^{-\gamma+1}$

Using the Main Lemma the running time per batch of $\epsilon \cdot N$ packets is at most

$$\sum_{i=1}^{d-1} t_i \cdot a_i \leq d \cdot \frac{\log \log N}{\log(a-1)} \cdot \log d$$
$$= O\left(d \cdot \log d \cdot \log \log N\right).$$

As it suffices to chose $\frac{1}{\epsilon} = O(d)$ Theorem 1 follows. □

4 Simulation on the Mesh of Busses

Theorem 2. *Let $k, d, a \in \mathbb{N}$, $N := k^d$. Any permutation can be routed on the d–dimensional MOB of edge-size k using $O\left(d^4 \cdot \log d \cdot \frac{\log \log N}{\log(a-1)}\right)$ steps, with high probability, if a is a sufficiently large constant.*

Proof. The proof is done by simulating the split&hash network on the MOB. We embed a (N, d, a)–split&hash network into the d–dimensional MOB of edge-size $k = N^{\frac{1}{d}}$ in the natural way, embedding the l–th node of every level into node $(l_1, \ldots, l_d)$ of the MOB where $\sum_{\nu=1}^{d} l_\nu \cdot k^{\nu-1} = l$.
Consider a level i of the split&hash network. Since we define the functions h_j^i as folded permutations we can find partial permutations $\psi_j^{i,\iota}$, $\iota \in [a_i]$, so that $\left(\psi_j^{i,\iota}(l), h_j^i(l)\right)$ is an edge of the MOB for one $\iota \in [a_i]$ if $(l, h_j^i(l))$ is an edge of the split&hash network.
In a preprocessing phase, we initialize the MOB such that all these $\sum_{i=1}^{d-1} a_i^2 = O(d^3)$ permutations can be routed in time $O(d)$. (Note that the choice of the permutations does not depend on the input permutation.) So a step of the Funnel Algorithm can be replaced by a_i phases, each routing one of the permutations described above and then sending the packet to the destination sink of the funnel. If a node gets more than one packet while simulating a step it rejects all received packets, thus we ensure that the packets are delivered to distinct nodes. Since $a_i = O(d)$, simulating a step of the Funnel Algorithm requires $O(d^2)$ steps on the MOB. □

5 Other Routing Problems

In Section 3 we showed how to realize a permutation on the split&hash network. In this section we will extend the results to h-relations and random functions. The result for random functions is a corollary of our Theorems 1 and 2:

Corollary 3. *A random function can be routed*

(a) from the input nodes to the output nodes of a (N, d, a)–split&hash network in $O\left(\frac{\log N}{\log \log N}\right)$ steps,

(b) on a d-dimensional MOB of edge-size k using $O(d^4 \cdot \log d \cdot \log \log N + \frac{\log N}{\log \log N})$ steps,

with high probability, if $d < \log^{\frac{1}{2}} N$.

Proof. We just have to assure that each funnel on each level gets no more than $\epsilon \cdot k^{i+1}$ packets.
Since even the smallest funnels have a size of $k = N^{\frac{1}{d}} = \Omega(\log N)$, a random choice of $\epsilon \frac{1}{\rho e}$ packets assures for any $\rho > 1$ that at most $\epsilon \cdot k^i$ packets have to

travel through a certain funnel, with high probability, if $N^{\frac{1}{d}}$ is large enough. So we choose batches of size $N \cdot \frac{\epsilon}{\rho e}$ and make use of the Funnel Algorithm to deliver the packets from level d to level 1.

After delivering all packets of all batches to level 1 we have to deliver the packets to their destinations on level 0. This may be done using the h–relation algorithm of [GJLR93] on the complete bipartite graphs for $h = \frac{\log N}{\log \log N}$, since each node of level 1 gets at most $\frac{\rho e}{\epsilon} = O(d) = O\left(\frac{\log N}{\log \log N}\right)$ packets, and each sink is the destination of at most $O\left(\frac{\log N}{\log \log N}\right)$ packets, with high probability. The h–relation algorithm is repeated $O(d)$ times, so the failure probability is sufficiently small. □

Theorem 4. *Any h-relation can be routed*

(a) from the input nodes to the output nodes of a (N, d, a)-split&hash network in $O(d^3 \cdot \log d \cdot \log \log N + d^3 h)$ steps,
(b) on the d–dimensional MOB of edge-size k in $OO(d^5 \cdot \log d \cdot \log \log N + d^3 h)$ steps,

with high probability, if $h \leq \log N$, $d < \log^{\frac{1}{2}} N$ and N large enough.

Proof. (SKETCH)
(a) Consider $\epsilon \cdot k^{i+1} \cdot h$ packets on the input nodes of a funnel. Each node to contain at most h packets. We deliver the main part of the packets to the output nodes of the funnel by a modified version of the so called *Thinning Phase* introduced in [AM88] and also used in [GJLR93].
A node is called *active* if it contains at least one packet.
Let $T \in \mathbb{N}$ and $t_i = O(\frac{h}{2^i} + \log h)$.
Modified Thinning Phase
for $g := 1$ to $\log h$ do

> for $j := 1$ to t_i do
> every active input node chooses a random $l \in \{1, \ldots, \frac{h}{2^{g-1}}\}$.
> if the node contains at least l packets it takes part at the Funnel Algorithm using functions $h_0^j, \ldots, h_{a-1}^j$ for T steps using the l–th packet.
> if a node contains more than $\frac{h}{2^g}$ packets it becomes inactive.

Part (b) of our Main Lemma assures that a certain packet has a constant probability to be delivered in a constant number of steps of the Funnel Algorithm. So the analysis for the Thinning Phase as given by Goldberg et al. can be modified yielding: With probability $1 - \left(\frac{1}{k^i}\right)^{\gamma \log k^i}$, for some $\gamma > 0$ depending only on the choice of the T, ϵ and β it holds: (i) there are at most k^i undelivered packets, (ii) at most $\frac{k^{i+1}}{h^2}$ nodes of the splitter become inactive in round k of the outer loop. The running time is obviously $t := O(T \cdot a_i \cdot \log^2 h + T \cdot a_i \cdot h)$. To ensure that at most h packets are on the node of the next level we use $s \cdot a_i$ functions with distinct images using each group of a_i functions at most h times.
The routing is performed for each of $\frac{1}{\epsilon}$ batches chosen as in Theorem 1. For each batch the Modified Thinning Phase is applied for d rounds. Round i delivers a

main part of the packets at level $d-i+1$ to level $d-i$. After that the packets are delivered from level 1 to their destination on level 0 using the algorithm of Goldberg et al. $O(d)$ times.

Now each funnel of level i contains at most k^i packets, with high probability. These remaining packets are sent using the random function algorithm. To apply this algorithm we distribute the packets of each level, so that each node contains at most two packets, with high probability. This may be done by partitioning the top node sets into distinct sets of size $O(\log^3 k^{i+1})$. With high probability, each set contains at most 2 · size packets, witch can be distributed using a parallel prefix algorithm on each set. After that the remaining packets are delivered to their destinations by applying the random function algorithm once for each level. To prove part (b): There are only $O(d)$ additional permutations needed to bring together the sets of the partitions used for the redistribution. □

References

[AM88]   Richard J. Anderson and Gary L. Miller. Optical communication for pointer based algorithms. Technical Report CA 90089-0782 USA, Computer Science Department, University of Southern California, 1988.

[DM93]   Martin Dietzfelbinger and Friedhelm Meyer auf der Heide. Simple efficient shared memory simulations. In *Proceedings of the 5th Ann. Symp. on Parallel Algorithms and Architectures*, 1993.

[GJLR93] Leslie Ann Goldberg, Mark Jerrum, Tom Leighton, and Satish Rao. A doubly logarithmic communication algorithm for the completely connected optical communication parallel computer. In *Proceedings of the 5th Ann. ACM Symp. on Parallel Algorithms and Architectures*, pages 300–309, 1993. Preprint.

[GJLR94] Leslie Ann Goldberg, Mark Jerrum, Tom Leighton, and Satish Rao. Doubly logarithmic communication algorithms for the completely connected optical communication parallel computers. Preprint, 1994.

[McD89] C. McDiarmid. On the method of bounded differences. *Combinatorics, London Math. Soc. Lecture Notes Series*, (141):148–188, 1989.

[MSS95] Friedhelm Meyer auf der Heide, Christian Scheideler, and Volker Stemann. Exploiting storage redundancy to speed up randomized shared memory simulations. In *Annual Symposium on Theoretical Aspects of Computer Science, to appear in Theoretical Computer Science*, 1995.

[MSS96] Friedhelm Meyer auf der Heide, Klaus Schröder, and Frank Schwarze. Routing on networks of optical crossbars (Technical Report). Technical report, Heinz Nixdorf Institute and Computer Science Department, University Paderborn, 1996. to appear.

[Upf89] Eli Upfal. An O(log N) deterministic packet routing scheme. In *Symposium on Theory of Computing*. ACM, 1989.

Latency and Bandwidth Requirements of Massively Parallel Programs: FFT as a Case Study

Fabrizio Petrini and Marco Vanneschi

Dipartimento di Informatica, Università di Pisa,
Corso Italia 40, 56125 Pisa, Italy,
tel +39 50 887248, fax +39 50 887226
e-mail: {petrini,vannesch}@di.unipi.it

Abstract. In this paper we compare three routing algorithms for massively parallel architectures, each offering an increasing degree of adaptivity: a *deterministic* algorithm, a minimal adaptive based on *Duato*'s methodology and a non-minimal adaptive, the *Chaos* routing. Rather than using a synthetic benchmark, the comparison is done with a real application, the transpose FFT algorithm. The simulation results collected on bi-dimensional tori with up to 256 processing nodes show that both adaptive algorithms suffer from post-saturation problems that degrade the network throughput.

1 Introduction

In many existing works in the literature routing algorithms are compared using synthetic benchmarks. These include uniform, transpose, complement and other communication patterns. In this paper we adopt a detailed simulation model, that describes both the interconnection network and the internal structure of the processing nodes, to compare three routing algorithms using a real application, the transpose FFT algorithm.

The remainder of this paper is organized as follows. Section 2 presents some parallel FFT algorithms. Section 3 describes the relevant details of the simulation model that we use in Section 4 to compare the routing algorithms. An overview of the experimental results and some concluding remarks are given in section 5.

2 FFT algorithms

The most commonly used parallel FFT algorithm maps each row r of an n-input butterfly on processor $\lfloor \frac{rP}{n} \rfloor$, where P is the number of processors[1]. With this mapping, processors need to communicate with each other during the computation of the first $\log P$ columns. The remaining $\log \frac{n}{P}$ columns are available on the same processor. While this mapping is effective on the hypercubes because

---
[1] We will assume that both n and P are powers of two.

generates a *normal* algorithm, low-dimensional cubes cannot make efficient use of a large number of processors [7].

An alternative formulation, often named *transpose* algorithm, divides the computation into two computational phases separated by a global communication phase [6]. Each processor computes in the first phase an $\frac{n}{P}$-input butterfly available locally, allocated according to a cyclic mapping. In the second computational phase each processor has assigned $\frac{n}{P^2}$ butterflies, whose size is P. There is a direct connection between any two butterflies belonging to different computational phases. This implies that the communication phase is an *all-to-all personalized broadcast* (AAPB) where each processor during the first computational phase has to send the values of $\frac{n}{P^2}$ output butterfly nodes to the remaining $P - 1$ processors.

3   The simulated model

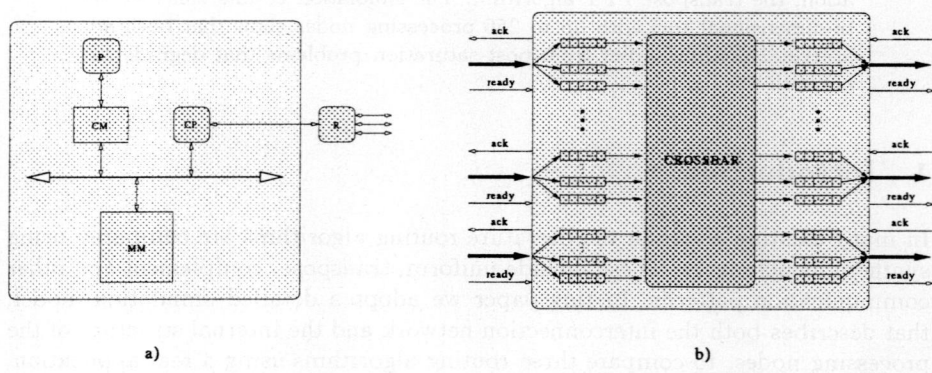

Fig. 1. a) The internal structure of the processing node. b) The internal structure of the router. P processor, CM cache memory, MM main memory, CP communication processor, R router.

This section presents the simulation model of parallel architectures that is used to analyze the behavior of the routing algorithms. The internal structure of the processing node is shown in Figure 1 a) and is centered around the main bus. We distinguish the following units: the *main processor*, its *cache memory*, the *main memory*, the *communication processor* and the *router*.

The computational part of the algorithm is executed on the main processor. The communication processor is dedicated to message handling and acts as a Message Driven Processor [3]. It fragments outgoing messages in packets that are enqueued in a FIFO directed to the router and receives incoming packets from another FIFO. The communication processor can perform DMA-like activities and store the content of incoming messages in proper memory locations. There is a simple coherency mechanism between the cache memory and the communication processor. The cache controller snoops on the main bus, and every time a memory location is overwritten by the communication processor, it invalidates the corresponding cache line, if present in the cache memory. Symmetrically, when the communication processor tries to read a memory location replicated on the cache memory, it gets the updated copy.

Figure 1 b) outlines the internal structure of a router. We can distinguish the external channels or links, the input and the output buffers or lanes that implement the buffer space of the virtual channels and an internal crossbar.

3.1 Routing algorithms

We compare three algorithms, each offering a different degree of adaptivity: deterministic, minimal adaptive based on Duato's methodology and non-minimal adaptive, the Chaos routing.

The deterministic algorithm is a dimension order routing based on a static channel dependency graph [4]. Packets are sent to their destination along a unique, predetermined and minimal path. The potential deadlocks caused by the wrap-around connections are avoided doubling the number of virtual channels and creating two distinct virtual networks. Packets enter the first virtual network at the beginning and switch to the second network upon crossing a toroidal connection. Our version uses two virtual channels for each physical link.

Rather than using a static channel dependency graph, Duato's methodology only requires the absence of cyclic dependencies on a connected channel subset. The remaining channels can be used in almost any way. We associate four virtual channels to each link: on two of these channels, called *adaptive* channels, packets are routed along any minimal path between source and destination. The remaining two channels are *escape* channels where packets are routed deterministically when the adaptive choice is limited by network contention. A static channel dependency graph is used in the channel subset [5].

The Chaos routing is a non-minimal cut-through routing algorithm. Under normal conditions, it routes packets from its input buffers to the output buffers as in minimal adaptive routing. The Chaos router has a central multiqueue to avoid blocking and deadlocks. A packet is moved into the multiqueue if it is potentially in a deadlock loop or it has stalled at an input buffer. When an output buffer becomes available, packets in the multiqueue have priority over packets in the input buffers and, if more than a packet in the multiqueue desires the channel, it is allocated using a fair policy. If a packet must be moved into the multiqueue when it is full, another packet in the queue is chosen at random and derouted to the next available output buffer along a potentially non-minimal path. As suggested in [1], we utilize a multiqueue with five packet buffers.

3.2 SMART

This model is evaluated in the SMART (Simulator of Massive ARchitectures and Topologies) environment. Implemented in C++, SMART is an object-oriented discrete-event simulation tool for evaluating massively parallel architectures. Configuring some shell scripts, it is possible to select the node internal structure, the network topology and the internal router policies. The simulator allows the definition of the network topology, routing algorithm, packet length, number of virtual channels and buffers for both input and output lanes. Also, it is possible to monitor several metrics and time-dependent events, that are gathered in trace files.

The computation of the non-input nodes of the butterfly requires 4 clock cycles when the operands are on chip. A read hit is served in 3 cycles and a read miss in 26 cycles. The message marshaling overhead in the communication processor is 40 cycles, for both incoming and outgoing packets. The flit size is 16 bits and the link delay to transmit a flit across a physical link is 4 cycles.

Adaptive algorithms have more degrees of freedom but require larger crossbars and more complex arbitration [2]. For this reason the routing delay is normalized with the router complexity. The routing delay is 4 cycles for the deterministic algorithm, 8 cycles for the Duato algorithm and 12 cycles for the Chaos routing.

4 Experimental results

In our experiments we mapped a 65536-input butterfly [2] on machines with up to 256 processors. In Figure 2 a) we can see that the speedups for the three routing algorithms are very close. It is worth noting that there is a superlinear speedup with up to 64 processors. This is due to the favorable computation/communication ratio of the transpose algorithm and to smaller working sets that lead to a better memory hierarchy utilization. The computation with 256 processors is bandwidth-limited: the Chaos routing provides a slightly better speedup of 116, followed by the Duato's algorithm with 110 and the deterministic with 107. The number of active processors is shown in Figure 2 b). We can easily distinguish the two local phases separated by the global communication.

Though similar in terms of accepted bandwidth, the three algorithms show widely different behaviors, as can be seen in Figure 2 c) and d). While the network utilization, i. e. the fraction of active links, of the deterministic algorithm is stable around 22%, in the Duato's algorithm it oscillates between 5% and 42%. In the Chaos routing the network utilization reaches 63%, even if the communication pattern is executed with a comparable amount of time.

The Duato's algorithm suffers from post-saturation problems. In a saturated network incoming packets tend to flood the adaptive channels and this limits the network throughput. The lowest network utilization is reached around 27000

---

[2] A 65536-input butterfly is the minimum data set size supported by the transpose algorithm on a 256 processors architecture.

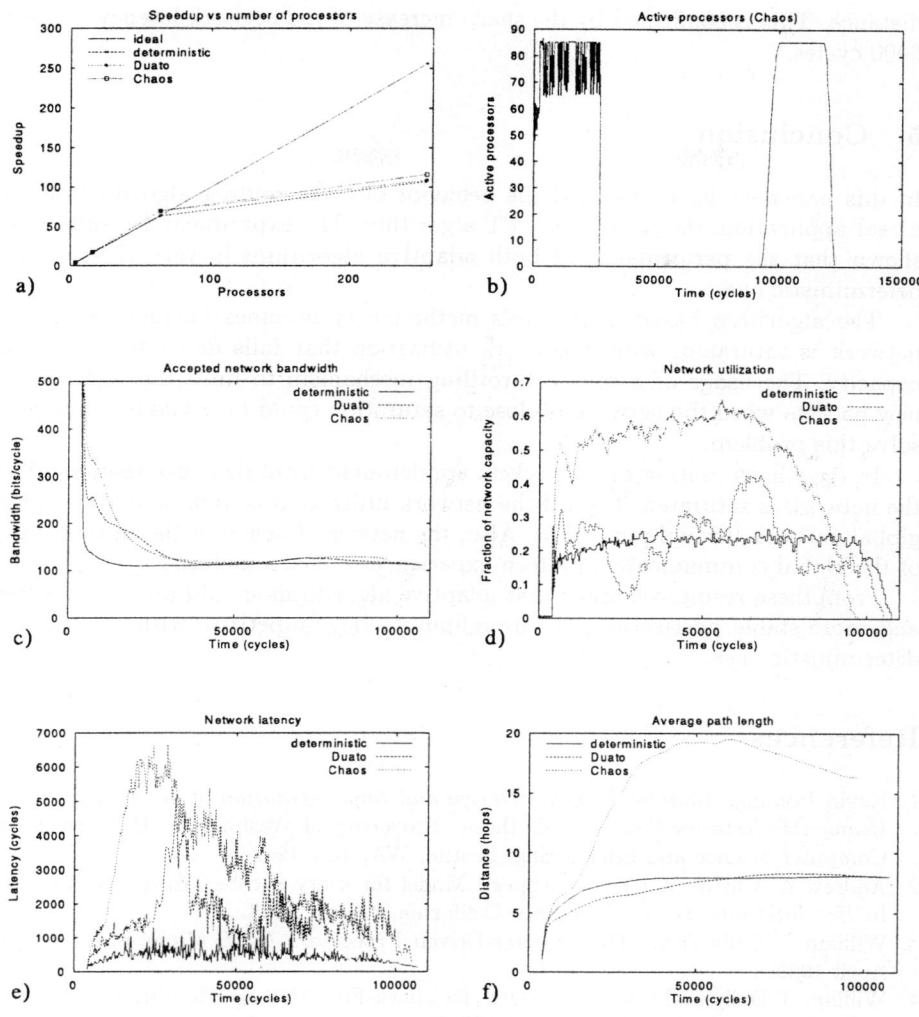

Fig. 2. Performance results on a 256 processors bi-dimensional torus.

cycles, at the end of the first computational phase. When the pressure at the source nodes begins to decrease, there is a corresponding increase in the accepted bandwidth, which reaches 42%.

The gap between the network utilization and throughput in the Chaos routing can be explained looking at the graphs in Figure 2 e) and f). At saturation, many packets are derouted and temporally sent away from their minimal path. In fact the average path length reaches 20 hops, more than twice the topological average distance. This is confirmed by the sharp increase of the network latency[3] at over 6000 cycles.

5 Conclusion

In this paper we have analyzed the behavior of three routing algorithms using a real application, the transpose FFT algorithm. The experimental results have shown that the performance of both adaptive algorithms is very close to the deterministic one.

The algorithm based on Duato's methodology becomes unstable when the network is saturated, with a network utilization that falls down to 5% of the capacity. The usage of a source throttling mechanism to limit the injection of new packets when the network is close to saturation could be a viable solution to solve this problem.

In the Chaos routing many packets are derouted from their destination when the network is saturated. Even if the network utilization is high, above 60%, the global network throughput is low. Also, the network latency in the initial phase of the global communication pattern experiences a sharp increase.

From these results we argue that adaptive algorithms should provide a better and more stable post-saturated throughput to be competitive with the simpler deterministic ones.

References

1. Kevin Bolding. *Chaotic Routing: Design and Implementation of an Adaptive Multicomputer Network Router*. PhD thesis, University of Washington, Department of Computer Science and Engineering, Seattle, WA, July 1993.
2. Andrew A. Chien. A Cost and Speed Model for k-ary n-cube Wormhole Routers. In *Hot Inteconnects '93*, Palo Alto, California, August 1993.
3. William J. Dally et al. The Message-Driven Processor. *IEEE Micro*, pages 23–39, April 1992.
4. William J. Dally and Charles L. Seitz. Deadlock-Free Message Routing in Multiprocessor Interconnection Networks. *IEEE Transactions on Computers*, C-36(5):547–553, May 1987.
5. José Duato. A Necessary and Sufficient Condition for Deadlock-Free Adaptive Routing in Wormhole Networks. *IEEE Transactions on Parallel and Distributed Systems*, 6(10):1055–1067, October 1995.
6. Anshul Gupta and Vipin Kumar. The Scalability of FFT on Parallel Computers. *IEEE Transactions on Parallel and Distributed Systems*, 4(8):922–932, August 1993.
7. F. Thomson Leighton. *Introduction to Parallel Algorithms and Architectures: Arrays, Trees, Hypercubes*. Morgan Kaufmann Publishers, San Mateo, CA, USA, 1992.

---

[3] The network latency is the time spent by a packet in the network, without including source queuing delay

Induced Broadcasting Algorithms in Iterated Line Digraphs

Jean-Claude Bermond[1], Xavier Muñoz[2] and Alberto Marchetti-Spaccamela[3]

[1] Project SLOOP I3S-CNRS/INRIA/Université de Nice–Sophia Antipolis
930 route des Colles, F–06903 Sophia-Antipolis, France. (bermond@unice.fr)
[2] Departament de Matemàtica Aplicada i Telemàtica, Universitat Politècnica de Catalunya. Campus Nord C3, c/Gran Capità s/n, 08034 Barcelona, Spain. (xml@mat.upc.es)
[3] Dipartimento di Informatica e Sistemistica, Università di Roma "La Sapienza". via Salaria 113 - 00198 - Roma, Italy. (alberto@dis.uniroma1.it)

Abstract. We propose broadcasting algorithms for line digraphs in the telephone model. The new protocols use a broadcasting protocol for a graph G to obtain a broadcasting protocol for the graph $L^k G$, the graph obtained by applying k times, the line digraph operation to G. As a consequence improved bounds for the broadcasting time in De Bruijn, Kautz and Wrapped Butterfly digraphs are obtained.

1 Introduction

Considerable attention has been recently devoted to the dissemination of information in parallel computing, since the time needed to disseminate information among the processors in a network often determines the running time of the whole algorithm. Broadcasting is the process in which a node of the network (the originator) sends one piece of information to all members of the network by placing calls over the communication lines of the network. This is to be completed as quickly as possible subjected to the constraints of the considered model. This problem has been widely studied both for its theoretical and practical interest for different network configurations and under different models (see [7, 10] for recent surveys or the book [11]).

In this paper we are concerned with the telephone model, that is, each call involves only two vertices of the network and requires one unit of time (round); moreover we require that a vertex can participate in at most one call per unit of time by either sending or receiving a call and we restrict links to be one-way, that is, information will always flow in the same direction.

---

* Work partially supported by the HCM Network of Excellence project MAP. Part of this work was developed while J-C.Bermond and X.Muñoz were visiting the University of Rome "La Sapienza". J-C. Bermond has been also supported by the French action RUMEUR of the GDR PRS, X. Muñoz by the Spanish Research Council (Comisión Interministerial de Ciencia y Tecnología, CICYT) under project TIC 92-1228-E and A. Marchetti-Spaccamela by the project MURST 40% Algoritmi e strutture di calcolo.

Following the literature we model the network topology by a digraph $G = (X, U)$; a vertex $x \in X$ corresponds to a node in the network, and an arc $(x, y) \in U$, corresponds to a link in which x is the sender and y is the receiver. $b(G, x)$ represents the minimum number of time units (number of rounds) necessary to complete broadcasting form vertex x; $b(G)$ is defined as the maximum of $b(G, x)$ taken over all vertices x of G. It is well known that for any digraph G with n vertices $b(G) \geq \lceil \log_2 n \rceil$ and that the problem of determining $b(G)$ is a NP–hard problem. Values of $b(G)$ are known for many usual interconnection networks; however for Butterfly digraphs and related networks the order is still unknown.

In this paper we propose a general methodology for designing broadcasting algorithms in line digraphs and iterated line digraphs. In Section 3.1 we give a simple protocol for broadcasting in LG (the graph obtained by applying the line digraph operation to the graph G) using a broadcast protocol in G where G is a regular digraph with degree d; if broadcast in G can be performed in time t, then the algorithm performs broadcasting in LG in time $t + \lceil \log_2 d \rceil + 1$. In section 3.2 we show that if G satisfies an additional property, that we call A, then there exists a broadcasting protocol in LG that runs in time $t + \lfloor \log_2 d \rfloor + 1$.

By iterating the algorithms of Sections 3.1 and 3.2 we obtain protocols for iterated line digraphs ($L(L^{n-1}G) = L^n G$). In Section 4.1 we show an improved algorithm for broadcasting in $L^2 G$ using a broadcasting protocol in G. For some values of d, the direct derivation of the protocol for $L^2 G$ has running time $t + 2\lfloor \log_2 d \rfloor + 1$ if G satisfies property A. In Section 5 we sketch the extension to $L^k G$. For example we obtain for $d = 2^\alpha$, $\alpha > 1$, a broadcasting algorithm in $L^3 G$ running in time $t + 3 \log_2 d + 1$.

The iteration of the line digraph operation is a good method to obtain large digraphs with fixed degree and diameter. Besides, many other good properties are observed in such digraphs when used as models of communication networks ([3, 9]). In fact some of the best known families of digraphs are indeed line digraphs: De Bruijn, Kautz,Wrapped Butterfly, among others.

Property A is satisfied in particular if G is itself a line digraph. So we obtain new and better protocols for De Bruijn, Kautz and Butterfly networks. In the case of De Bruijn digraphs $B(d, D)$ and Kautz digraphs $K(d, D)$ (the De Bruijn/Kautz digraph with degree d and diameter D) we obtain the following bounds

$$b(B(d, D)) \leq D(\log_2 d + f(d));$$

$$b(K(d, D)) \leq D(\log_2 d + f(d)) + 1.$$

where $f(d)$ satisfies: $\log_2(1 + \frac{1}{d}) \leq f(d) \leq 0.508..$

This improves the best current known bounds both for digraphs ([3], [9]) and graphs ([2]).

2 Notation and Preliminaries

We refer to [5, 11] for basic definitions concerning graphs and digraphs and we recall basic definitions and properties of line digraphs (see also [6, 8]).

Given a directed graph $G = (X, U)$, if (x, y) is an arc, then x is *adjacent to y* and y is *adjacent from x*. We also say that arc (x, y) is *incident to y* and *incident from x*, and that arc $u = (x, y)$ is adjacent to arc $v = (y, z)$. We denote by $d^+(x)$ $(d^-(x))$ the number of vertices adjacent from x (adjacent to x); a digraph is said to be *d-regular* if $d^+(x) = d^-(x) = d$ for any vertex x.

A *dipath* of *length h* from a vertex x to a vertex y is a sequence of vertices $x = x_0, x_1, \ldots, x_{h-1}, x_h = y$ where (x_i, x_{i+1}) is an arc of G. A digraph is *strongly connected* if, for any couple of vertices x, y, there exists a dipath from x to y. The length of a shortest dipath from x to y is the *distance* from x to y. Its maximum value over all couples of vertices is the *diameter* of the digraph denoted $D(G)$.

Given a digraph $G = (X, U)$ the *line digraph* operation allows to define a new digraph LG whose vertex set is in one to one correspondence with the set of arcs of G. Vertex u of LG representing the arc $u = (x, y)$ is adjacent to the vertex representing the arc v if and only if $v = (y, z)$, that is arcs of LG represent dipaths of length 2 of G. It can be easily shown that if G is d-regular with n vertices then LG is d-regular with dn vertices; furthermore if G is a strongly connected digraph different from a directed cycle and $d > 1$, then the diameter of LG is the diameter of G plus one. We denote by $L^k G$ the graph $L(L^{k-1}G)$.

Let G be a digraph with n vertices and LG the corresponding line digraph. The arcs of LG can be partitioned into n complete bipartite digraphs isomorphic to $\overrightarrow{K}_{d,d}$ that are in one to one correspondence with vertices of G. If x is a vertex of a digraph G then we denote by B_x the corresponding bipartite digraph in LG. Note also that each vertex of LG belongs to exactly to two of such bipartite digraphs: in one case with in-degree 0 and in the other with out-degree 0.

An *arc-labeling* for a digraph G is a labeling of its arcs such that any two arcs incident to the same vertex have different labels and any two arcs incident from the same vertex have also different labels. If G is d-regular then it is always possible to obtain such an arc-labeling with d labels. Since we always consider d-regular digraphs we simply say an arc-labeling instead of an arc-labeling with d labels. Given a digraph with an arc-labeling, we identify the labels with the elements of $\mathbf{Z}_d$ and we refer to the arcs incoming from or outgoing to the same vertex by their labels.

The De Bruijn digraph $B(d, D)$ is a d-regular digraph with diameter D defined as follows: vertices are the strings of length D, $z_1 z_2 \ldots z_D$, where $z_i \in \mathbf{Z}_d$, $i = 1, 2, \ldots, D$, and vertex $z_1 z_2 \ldots z_D$ is adjacent to vertices $z_2 \ldots z_D \alpha$, $\alpha = 0, 1, 2, \ldots, d-1$. In [6] it is shown that $B(d, D) = L^{D-1} K_d^+$, where K_d^+ denotes the complete symmetric digraph on d vertices with an additional loop in each vertex.

The Kautz digraph $K(d, D)$ is defined analogously: vertices are all strings of length D, $z_1 z_2 \ldots z_D$, $z_i \in \mathbf{Z}_{d+1}$, $i = 1, 2, \ldots, D$, such that two consecutive symbols cannot be equal. Vertex $z_1 z_2 \ldots z_D$ is adjacent to vertices $z_2 \ldots z_D \alpha$, $\alpha \in \mathbf{Z}_{d+1}$, $\alpha \neq z_D$. The digraph $K(d, D)$ is d-regular and has diameter D; moreover $K(d, D) = L^{D-1} K_{d+1}^*$, where K_{d+1}^* denotes the complete symmetric digraph on $d+1$ vertices [6].

The Directed Wrapped Butterfly $\overrightarrow{WBF}(d, n)$ has as vertex set the couples (x, l) where x is a string of length n $(x_{n-1} x_{n-2} \ldots x_0)$, $x_i \in \mathbf{Z}_d$, $i = 0, 1, \ldots, n-1$, and $l \in \mathbf{Z}_n$. Vertex $(x_{n-1} x_{n-2} \ldots x_0, l)$ is adjacent to vertices $(x_{n-1} \ldots x_{l+1} \alpha x_{l-1} \ldots x_0, l+1)$, $\alpha \in \mathbf{Z}_d$. In [4] it is shown that $\overrightarrow{WBF}(d, n) = L^n dC_n$, where dC_n denotes the graph obtained from the directed cycle with n vertices by replacing each arc with d parallel arcs.

3 Broadcasting in line digraphs

In this section we will show that, given a broadcasting algorithm in a digraph G running in time t, it is possible to construct a broadcasting protocol in LG that runs in time $t + \lceil \log_2 d \rceil + 1$. The running time is $t + \lfloor \log_2 d \rfloor + 1$ if G satisfies certain properties.

3.1 A broadcasting protocol in a line digraph

A broadcasting algorithm in a digraph $G = (X, U)$ with originator r can be described via the broadcast tree $T(r)$ with root r and vertex set X. $T(r)$ contains the arcs through which the information has been broadcasted; since each vertex in $T(r)$ has in-degree 1, then $T(r)$ is an arborescence.

Any broadcasting algorithm P with originator r induces a partial broadcasting in LG with originator any vertex in LG representing an arc (r', r), which informs those vertices of LG corresponding to the arcs of the broadcast tree of G. If P is a broadcasting protocol running in time t in G, then at time t, for any x, there is at least one vertex $u = (\cdot, x)$ in LG, where · stands for an undetermined symbol, that has been informed (namely vertex corresponding to (r', r) and the vertices that correspond to the arcs of the broadcast tree $T(r)$).

During round $t + 1$, for any x in G, the informed vertex $u = (\cdot, x)$ in LG sends the information, denoted by **i**, to vertex $(x, \cdot)$ of label 1. Since the labeling of the arcs is an arc–labeling in G, for all x', vertices $(\cdot, x')$ of label 1 have been informed. At round $t + 2$, all vertices $(\cdot, x)$ of label 1 send **i** to vertices $(x, \cdot)$ of label 2. Thus at the end of this round all vertices of LG with labels 1 or 2 will know **i**; so they will be able to send **i** to vertices of LG of labels 3 and 4 during round $t + 3$. Following this pattern, we claim that at the end of round $t + k$, vertices of LG with label i, $1 \le i \le 2^{k-1}$, will know **i**. The proof is by induction on k; it is true for $k = 1$ and $k = 2$. Suppose it is true at time $t + k$, then at time $t + k + 1$ for any x the vertex $(\cdot, x)$ of LG of label i informs the vertex $(x, \cdot)$ of label $i + 2^{k-1}$. In this way all vertices with label i, $2^{k-1} + 1 \le i \le 2^k$, get the information. Since G is d-regular at round $t + \lceil \log_2 d \rceil + 1$, the protocol is completed. So we can state the following theorem.

Theorem 1. *Given a regular digraph G with degree d, such that broadcast in G can be completed in t rounds, then there exists a broadcast protocol in LG that runs in time $t + \lceil \log_2 d \rceil + 1$.*

3.2 An improved protocol in a line digraph

In the previous protocol we have not used the facts that **i** arrived at vertex x of G on some arc of a specific label and that **i** might arrive before time t. For example, suppose that it arrives to an arc $(\cdot, x)$ not of label 1, then before round $t + 2$ the information **i** has reached two vertices of LG. Therefore, at round $t+2$, **i** could have been sent on two vertices of LG instead of only one; analogously if **i** reaches x at time $t-h$ in the successive h rounds **i** can be sent on h new arcs.

In the sequel we obtain an improved broadcasting protocol in LG if G admits a suitable arc-labeling. To this aim let us consider the broadcasting tree $T(r)$ given by a broadcasting algorithm P with originator vertex r and let us label its arcs in the following way: for each vertex, the last arc used to send **i** is labeled 0, the penultimate one is labeled -1 (where we perform addition modulo d), the preceding one is labeled with -2 and so on. We will call such a labeled tree a *labeled broadcasting tree*. If it is possible to make an arc-labeling of the digraph G in which the labels of the arcs of $T(r)$ are the ones in the labeled broadcasting tree, we say that this arc-labeling is *consistent* with the labeling of the broadcasting tree. We say that a vertex of G is *of kind k* (in the protocol P) if it has received **i** though an arc of label $-k$ (modulo d).

We note the following immediate lemma.

Lemma 2. *If a vertex is of kind k, then it has been informed in the protocol P at time $t-k$ or before.*

Definition 3. A d-regular digraph G satisfies property A if there exists a broadcasting protocol such that, for any vertex r, there exists an arc-labeling of G that is consistent with the labeling of the broadcasting tree $T(r)$.

Lemma 4. *If G is a line digraph then it satisfies property A.*

Proof. If G is a line digraph, due to the partition of the arcs into complete bipartite digraphs, it is possible to label its arcs by arc-labeling independently the bipartite subdigraphs.

Let $T(r)$ be the broadcasting tree associated to the broadcasting protocol P in G and for a given bipartite digraph B_x of the decomposition of G, let V_1 and V_2 be the partition of the vertices of B_x. For each vertex y of G there exists exactly one arc entering in y that belongs to $T(r)$ (the arc through which **i** has arrived). Let $\delta_T(x)$ be the out-degree of x in T (that is the number of vertices which have been informed by x). Now we rank the vertices of V_1 and V_2 as follows.

Choose a vertex x in V_1, with $\delta_T(x) = \delta \geq 1$; label this vertex x_0 and label its out-neighbors in V_2 with $y_0, y_1, \ldots, y_{\delta-1}$ in the order they have been informed, i.e. according to their kind (the out-neighbor of kind 0 is labeled y_0, the one of kind 1, y_1 and so on. In general suppose that at some step of the algorithm we have used all labels till y_i for vertices in V_2.

If $i < d-1$, then choose a vertex x in X with out-degree $\delta_T(x) = \delta_i \geq 1$; label this vertex x_{i+1} and its out-neighbors y_{i+j}, $1 \leq j \leq \delta_i$, according to their kind. If $i = d-1$, then there might be vertices in V_1 that have not been labeled; we label them in any order using the remaining labels. Note that the labeling is not unique (in fact it depends on the order in which vertices in V_1 have been chosen).

Let us now consider the following arc-labeling of G: arc (x_i, y_j) has label $(i-j) \bmod d$. It is immediate to see that it is an arc-labeling which furthermore is consistent with the labeling of $T(r)$.

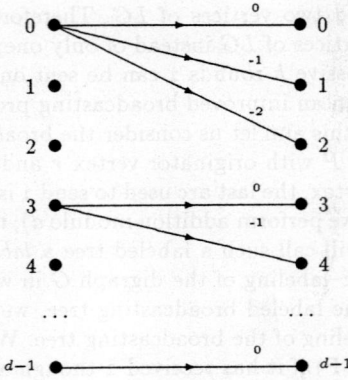

Fig. 1. Labeling of the vertices of a bipartite digraph. Drawn arcs are those in the broadcasting tree T of G

Lemma 5. *Given a d-regular digraph G that satisfies property A and a broadcasting protocol P running in G in time t, then there exists a protocol in LG such that at time $t + h$*

i) for a vertex of kind k, $-k \in \{1, 2, \ldots, 2^h - 1\}$, all its outgoing arcs have been informed;

ii) for a vertex of kind k, $-k \notin \{1, 2, \ldots, 2^h - 1\}$, all the arcs of labels j, $1 \leq j \leq 2^h - 1$, have been informed.

Proof. The proof is by induction on h and uses the following protocol Q in LG. Until time t we will use the broadcasting algorithm P in G but with the following modification. Let x be a vertex of G of kind k. By lemma 2, vertex x has received i at time less or equal to $t - k$. If we consider protocol P two possible cases arise: either x has sent i on at least k arcs or x has sent i to k' arcs, $k' < k$, and it has been idle for at least $k - k'$ rounds before t. In the modified protocol that we use until time t we will use $k - k'$ rounds to send the information to $k - k'$ arcs in such a way that, at time t, x has sent the information through arcs $0, -1, \ldots, -k + 1$.

The specification of protocol Q is completed as follows. At round $t + 1$ each vertex $(\cdot, x)$ of LG that corresponds to an arc of the broadcasting tree sends i to its neighbor of label 1. At the end of this round, if $x \in V(G)$ is of kind $d - 1$ then $(\cdot, x)$ has informed all its neighbors in LG (in rounds before t, the vertices corresponding to arcs of label $-j$, $j = 0, 1, \ldots, d - 2$, and, at round $t + 1$, arc of label $1 = -(d - 1)$). If x is of kind k different from $d - 1$ then there exist at least two vertices of LG (y_1, x) and (y_{-k}, x) that have received i (these vertices correspond to arcs with labels 1 and $-k$); in this case these two vertices can send i to the two vertices $(x, \cdot)$ with labels 2 and 3 during round $t + 2$; note that in the previous protocol at round $t + 2$, i was sent to only one vertex $(x, \cdot)$.

Now we can start the induction. The lemma is satisfied for the cases $h = 0$ and $h = 1$. Suppose the lemma is true for h_0 and let us consider vertex x of kind

k. Then, at round $t + h_0 + 1$, if $-k \in \{1, 2, \ldots, 2^{h_0} - 1\}$ the lemma is true by induction. If $-k \notin \{1, 2, \ldots, 2^{h_0} - 1\}$, x has received at the end of round $t + h_0$ i on at least 2^{h_0} arcs: arc of label $-k$ (before time $t - k$) and arcs of label j, $j = 1, 2, \ldots, 2^{h_0} - 1$ (at rounds $t, t+1, \ldots, t + h_0$). Therefore x can send i on 2^{h_0} arcs of label j, $j = 2^{h_0}, \ldots, 2^{h_0+1} - 1$. So if $-k \in \{1, 2, \ldots, 2^{(h_0+1)} - 1\}$ all outgoing arcs of x have been informed; otherwise the set of informed arcs is the claimed one.

The proof of the following theorem is an immediate consequence of the lemma and of the fact that $t + \lceil \log_2(d+1) \rceil = t + \lfloor \log_2(d) \rfloor + 1$.

Theorem 6. *Given a d-regular digraph G that satisfies property A and such that broadcast can be performed in time t, then there exists a protocol in LG that runs in time $t + \lfloor \log_2(d) \rfloor + 1$.*

4 Broadcasting in $L^2 G$

Given a d-regular digraph G with $d = 2^\alpha(1+\beta)$, $0 \le \beta < 1$, that satisfies property A, in which there exists a broadcasting algorithm running in time t, by applying twice the protocol for broadcasting in a line digraph, we obtain a broadcasting algorithm in $L^2 G$ running in time $t + 2\alpha + 2$. The aim of this section is to design a broadcasting protocol in $L^2 G$ directly from the protocol in G running in time $t + 2\alpha + 1$; this is possible if β is not too large.

Recall that, given a graph $G = (X, U)$, the vertex set of $L^2 G$ is the set of all possible dipaths of length 2 in G, and that broadcasting in $L^2 G$ is equivalent to broadcast the information through all paths of length 2, or equivalently, every arc of G informs every other arc adjacent in G.

Observe that, in the protocol for LG described in the previous section, most of the vertices are idle during the last unit of time. This happens because all their neighbors have been already informed and it should not make sense to send again the information. Nevertheless, if we iterate the process in order to obtain a broadcasting algorithm in $L^2 G$, each arc must send i to any other arc and, therefore, the fact described before represents a waste of time.

4.1 The design of a protocol from small values of d

Lemma 7. *Given a d-regular graph G that satisfies property A and a broadcast protocol in G that runs in time t, then there exists a broadcast protocol in $L^2 G$ such that, at time $t + \alpha$, $\alpha = \lfloor \log_2 d \rfloor$, the information is arrived to all vertices that correspond to pairs of arcs of G with labels $(i, i + 2^j)$, where $1 \le i \le 2^{\alpha - 1} - 1$ and $\lfloor \log_2 i \rfloor < j \le \alpha - 1$, and also to vertices that correspond to a not completely specified pair of arcs with labels $(\cdot, 2^j)$, where $0 \le j \le \alpha - 1$ and $\cdot$ denotes the undetermined label of an arc (depending on the broadcast algorithm in G).*

Proof. We use the protocol Q described in Lemma 5 in the following precise way. We know that, for any vertex x, at time $t + h$, the information i is arrived on all the arcs i, $1 \le i \le 2^h - 1$, and on the arc $-k$ (if x is a vertex of kind k). At time

$t+h+1$, we impose that, if it has not already been done, arc i sends the information to $i+2^h$ and arc $-k$ to 2^h.

Note that if $-k \in \{1,2,\ldots,2^h-1\}$ there is no conflict as arc $-k$ has already sent i to 2^h and $2^h + k'$ with $k' = -k \bmod d$, $k' \geq 0$. So the lemma follows by induction and considering the case $h = \alpha = \lfloor \log_2 d \rfloor$.

Let us call the above protocol a *weak protocol*. If, furthermore, we suppose that the undetermined · is 0 we say that the protocol is a *strong protocol*. In this case arc labeled 0 has informed arcs labeled 2^j, $0 \leq j \leq \alpha - 1$, or equivalently all the pairs of arcs$(0, 2^j)$ have been informed.

Let D_α be the set of values of d, with $\lfloor \log_2 d \rfloor = \alpha$, such that any weak protocol obtained at time $t + \alpha$ can be completed in a full protocol for L^2G in $\alpha + 1$ more steps. Since we can always find a weak protocol in time $t + \alpha$ for any d in D_α, it follows that if $d \in D_\alpha$ then there exists a protocol in L^2G running in time $t + 2\alpha + 1$.

Analogously let E_α be the set of values of d with $\lfloor \log_2 d \rfloor = \alpha$ such that any strong protocol obtained at time $t + \alpha$ can be completed as in the case of a weak protocol (G is a d-regular digraph) in $\alpha + 1$ more steps. Notice that $D_\alpha \subseteq E_\alpha$.

Lemma 8. *1. If $d \in D_\alpha$ then $2d \in D_{\alpha+1}$.*
2. If $d \in D_\alpha$ and $d + 1 \in E_\alpha$ then $2d + 1 \in D_{\alpha+1}$.
3. If $d \in E_\alpha$ then $2d \in E_{\alpha+1}$.

Proof. 1) Let $d \in D_\alpha$, thus $\alpha = \lfloor \log_2 d \rfloor$ and $\alpha + 1 = \lfloor \log_2 2d \rfloor = \lfloor \log_2(2d+1) \rfloor$. Let us recall that there exists a weak protocol for G with degree $2d$ as it is explained in the protocol for LG. Let us consider the subdigraphs G_e and G_o induced by the even and odd arcs of G respectively. The protocol for G_e (where we label arc $2i$ as i') is exactly a weak protocol for G_e (with degree d) completed in time $t + \alpha + 1$. Indeed the information is arrived on all the arcs $2i, 2 \leq 2i \leq 2^{\alpha+1} - 1$, that is, $1 \leq i' \leq 2^\alpha - 1$, and arc i' has informed arcs $i' + 2^j$ with $\lfloor \log_2 i' \rfloor + 1 < j \leq \alpha - 1$. Similarly the protocol induced in G_o (where we identify $2i + 1$ with i'') is a strong protocol for G_o.

The protocol in G_e and in G_o can be completed in $\alpha + 1$ steps more since $d \in D_\alpha$. Thus at time $t + 2\alpha + 2$ all the vertices in L^2G of the form $(2i, 2j)$ and $(2i+1, 2j+1)$, $0 \leq i, j \leq d - 1$ know the information. Using an extra step $(2i, 2j)$ can inform $(2j, 2i+1)$ and $(2i+1, 2j+1)$ can inform $(2j+1, 2i)$ completing the protocol in time $2(\alpha + 1) + 1$. Thus $2d \in D_{\alpha+1}$.

2) Let us consider subdigraphs G_e and G_o as in the case before, but labeling arc $2d$ as $2d+1$, that is considering $2d$ as if it were an odd number, and thus belonging to G_o. The protocols induced in G_e and G_o are a weak protocol and a strong protocol respectively. Since $d \in D_\alpha$ and $d + 1 \in E_\alpha$ it is possible for the vertices of the form $(2i, 2j)$, $0 \leq i, j \leq d - 1$, and $(2i+1, 2j+1)$, $0 \leq i, j \leq d$ to be informed at time $2\alpha + 2$. Nevertheless let us change a bit the protocol in G_o in order that for each j, $0 \leq j \leq d - 1$, one of the informed vertices during last step, say $(2i+1, 2j+1)$ is replaced by the vertex $(2i+1, 2j)$, that is $2i+1$ informs $2j$ instead of informing $2j+1$.

Notice that at time $2\alpha + 2$ arcs in G_e have been informed by $d + 1$ other arcs, that is $(2i, 2j)$ know the information for all $i, j \leq d - 1$ as well as $(2k+1, 2j)$ for some k. Thus $(2j, 2i+1)$ can get the information in the next step for all $i \leq d, j \leq d - 1$.

Also at time $2\alpha + 2$ the arcs in G_o have been informed at least d times, that is, for every $j \leq d$ there are at least d vertices of the form $(2i+1, 2j+1)$ which are informed, and thus they can inform the $d - 1$ remaining vertices of the form $(2j+1, 2i)$, $0 \leq i \leq d-1$ which were not informed at time $2\alpha + 2$ and the vertex $(2i+1, 2j+1)$ which was neither informed at that time. Hence the protocol can be completed in $2(\alpha + 1) + 1$ steps, and so $2d + 1 \in D_{\alpha+1}$.

3) This case is analogous to Case 1 and it is omitted.

Lemma 9. $4 \in D_2$ and $5 \in E_2$.

Proof. In the following tables we show an algorithm for completing the protocol in L^2G. In the tables columns represent successive rounds and rows represent labels of arcs incident to any vertex of G. At each round it is shown what must be done by each incoming arc. For instance in table a) at time $t + 3$ all arcs inform 0; thus 0 will have been informed by three different arcs and at time $t + 4$ it can inform 3 different arcs (with labels 1, 2, 3).

Recall that, by Lemma 7, at time less than or equal to $t + 2$ arcs 1, 2 and 3 have been informed and moreover the incoming arc 1 has informed the outgoing arc with label 3.

	$\leq t+2$	$t+3$	$t+4$	$t+5$
0			1,2,3	0
1	3	0	1	2
2		0	2	1,3
3		0	3	1,2

a) table for $4 \in D_2$

	$\leq t+2$	$t+3$	$t+4$	$t+5$
0		1,2	4	3	0
1		3	4	2	0,1
2			2	1,4	0,3
3			3	1,4	0,2
4				2,3	0,1,4

b) table for $5 \in E_2$

It is not difficult to construct analogous tables to show that $10 \in D_3$ and $11 \in E_3$, but for a lack of space we leave them to the full paper. This result and the ones shown in Lemmas 8 and 9, imply that given d, $d = 2^\alpha(1 + \beta)$ and $\alpha \geq 3$, then $d \in D_\alpha$ if $\beta < 3/8$ and $d \in E_\alpha$ if $\beta \leq 3/8$.

We use a different analysis in order to obtain a broadcast protocol for L^2G that runs in time $t + 2\alpha + 1$, when $d = 2, 5$.

Lemma 10. *Given a d-regular line digraph with $d = 2$ or $d = 5$ there exists a protocol in L^2G running in time $t + 3$, respectively $t + 5$.*

Proof. We give only the proof for $d = 2$ (the proof for $d = 5$ uses analogous ideas). Let us consider the table below:

It seems to be a problem at time $t + 3$, because the arc of label 0 incident to vertex of kind 1 might have received the information only once, and thus cannot send twice. But if we have a little more care in our analysis we can realize that in the case it has received the information only once (i.e. the incoming arc of label 0 in a vertex of kind 1 is an outgoing arc of a vertex of kind 1) it has received the information in time less or equal than t, and thus can have sent this information previous to $t + 3$. In other case, (i.e. the incoming arc of label 0 in a vertex of kind 1 is an outgoing arc of a vertex of kind 0), it has received the information twice, and so it can send twice the information. In any case it is possible to have a protocol for L^2G running in time $t+3$, which is $t+2\alpha+1$.

		$\leq t$	$t+1$	$t+2$	$t+3$
kind 0	0		1	0	
	1			0	1
kind 1	0				0,1
	1	0	1		

The following theorem is the main result of this section and is a straightforward consequence of previous lemmas.

Theorem 11. *Given a d-regular digraph G, $d = 2^\alpha(1+\beta)$ and $\beta < 3/8$, that satisfies property A and such that broadcast can be performed in time t then there exists a protocol in L^2G that runs in time $t + 2\lfloor \log_2(d) \rfloor + 1$.*

It is possible to improve the value of β that appears in Theorem 11 by providing tables analogous to the tables of Lemma 9 for larger values. We leave this extension to the full paper. For example in the full paper we prove that $44 \in D_5$ and $45 \in E_5$ and so we have a theorem analogous to Theorem 11 with $\beta = 13/32$.

We now show an upper bound for the value of β for which it possible to get this improvement by the proposed methodology. Namely it cannot be applied if $\beta > \sqrt{2} - 1$. In fact the method uses for the first t rounds a broadcast protocol in G; it follows that at time t we know that only n vertices in L^2G have got the information. Since at each subsequent round the number of vertices knowing the information at most double, it follows that, at time $t+2\alpha+1$, the number of vertices being informed is at most $n2^{2\alpha+1}$. On the other hand the number of vertices in L^2G is $n2^{2\alpha}(1+\beta)^2$. The bound is obtained by comparing both expressions. Note that $13/32 = 0.406..$ is very close to $\sqrt{2} - 1 = 0.414..$.

4.2 Broadcasting in some families of digraphs

We consider the following families of iterated line digraphs: De Bruijn $B(d, D)$, Kautz $K(d, D)$ and Directed Wrapped Butterfly $\overrightarrow{WBF}(d, n)$. Since K_d^+, K_{d+1}^* and dC_n (the first cases) satisfy property A then all these families satisfy property A as well and the following corollary of theorems 6, 11 easily follows.

Corollary 12. *Let $d = 2^\alpha(1+\beta)$, $\alpha > 4$, and let $c_d = 1/2$ if $\beta < 13/32$ and $c_d = 1$ otherwise. Broadcasting in De Bruijn, Kautz and Wrapped Butterfly digraphs can be performed in time*

1. $b(B(d, D)) \leq D(\lfloor \log_2 d \rfloor + c_d)$
2. $b(K(d, D)) \leq D(\lfloor \log_2 d \rfloor + c_d) + 1$
3. $b(\overrightarrow{WBF}(d, n)) \leq n(\lfloor \log_2 d \rfloor + c_d) + n - 1$

Remark. For the De Bruijn digraph the broadcasting time can be written as

$$b(B(d, D)) \leq D(\log_2 d + f(d))$$

where $f(d)$ is a function bounded by $\log_2(1 + (1/d)) \leq f(d) \leq 0.508..$.

The lower bound is obtained for $d = 2^\alpha - 1$ and the upper one is obtained for $\beta = 13/32$. Nevertheless this result can be improved using protocols in $L^k G$ as it will be sketched in the next section.

5 Protocol in $L^k G$

The results of the previous section can be extended by showing that, for some values of d, it is possible to construct a protocol in $L^k G$ running in time $t + k\lfloor \log_2 d \rfloor + h$, $h < k$ directly from the broadcasting algorithm in G. In the following we omit proofs. First of all we observe that if G is a d-regular line digraph G with $d = 2^\alpha(1+\beta)$, $0 \leq \beta < 1$, with a broadcasting algorithm running in time t, it is possible to construct a broadcasting algorithm in $L^k G$ running in time $t + k\alpha + h$, with this method, only if $1 + \beta \leq 2^{\frac{h}{k}}$.

Furthermore we can extend the notion of a weak protocol given in Section 4 as follows: let $D_\alpha^{(k,h)}$ be the set of values of d with $\lfloor \log_2 d \rfloor = \alpha$ such that any weak protocol can be completed in a full protocol for $L^k G$ in $(k-1)\alpha + h$ steps more. The next result is an extension of Lemma 8.

Lemma 13. *If $d \in D_\alpha^{(k,h)}$ and $b(B(2^s, k-1)) = (k-1)s$, then $2^s d \in D_{s+\alpha}^{(k,h)}$.*

In [9] it is shown that $b(B(8,2)) = 3$. Moreover we have been able to complete the weak protocols for degrees 4, 8 and 16 into full protocols for $L^3 G$ in $2\alpha + 1$ time steps more. Thus we can state the following result.

Theorem 14. *Let G be a line digraph with degree $d = 2^\alpha$, $\alpha > 1$, such that there exists a broadcasting algorithm running in time t, then there exist a broadcasting algorithm in $L^3 G$ running in time $t + 3\alpha + 1$.*

6 Conclusions

We have described a constructive method to design good broadcasting protocols in line digraphs and in iterated line digraphs. The broadcasting time given by this method is in most cases better than with any other method known up to date, even considering the underlying graph of the line digraph (see [2]).

We summarize in the following table the running time for De Bruijn digraphs. The table shows for each value of d the running time in terms of $b(B(d,D)) = s_d D$. We have used the fact that we know how to construct protocols in $L^3 G$ for $d = 9$ (resp. $d = 11$) running in time $t + 10$ (resp. $t + 11$), and in $L^4 G$ for $d = 4$ (resp. $d = 8$) running in time $t + 9$ (resp. $t + 13$).

d	2	3	4	5	6	7	8	9	10	11	12
Previous bound	1.5	2	2.5	3	3.5	4	4.5	5	5.5	6	6
Undirected case	1.5	2	2.5	2.8	3	3.28	3.5	3.67	3.8	3.9	4
New result	1.5	2	2.25	2.5	3	3	3.25	3.33	3.5	3.66	4
Lower bound	1.44	1.80	2.11	2.38	2.62	2.82	3.01	3.17	3.32	3.46	3.59

Acknowledgments The authors want to thank S. Perennes for his help.

References

1. J-C. Bermond, P. Hell, A. L. Liestman, J. G. Peters, "Broadcasting in bounded degree graphs", *SIAM J. Discrete Math.*, 5(1), (1992), pp. 10–24.
2. J-C. Bermond, S. Perennes, "Efficient broadcast protocols on the de Bruijn and similar networks". In *Proceedings of SIROCCO'95*, Olympia, (1995), International informatics series 2, Carleton U. Press, (1995), pp. 199-204.
3. J-C. Bermond, C.Peyrat, "Broadcasting in De Bruijn networks", *Proc. of the 19th S-E Conf. on Combin., Graph Theory and Comp.*, Florida, (1988), pp. 283–292.
4. J-C.Bermond, E.Darrot, O. Delmas, S. Perennes, "Hamilton circuits in directed butterfly networks", Tech. Rep. 95-47, I3S, Univ. Nice S.A., (1995), submitted to Discrete Applied Mathematics.
5. G.Chartrand and L.Lesniak, *Graphs and Digraphs* Wadsworth, Monterrey, (1986).
6. M.A. Fiol, I. Alegre, J.L.A. Yebra, "Line digraph iterations and the (d,k) problem", *IEEE Trans. Comp.*, C–33, (1984), pp.400–403.
7. P.Fraignaud, E.Lazard, "Methods and problems of communication in usual networks", *Discrete Appl. Math.*, 53, (1994), pp. 79–133.
8. R.L.Hemminger et al. *Selected topics in Graph Theory*, vol.1, Academic Press (1983).
9. M.C., Heydemann, J.Opatrny, D.Sotteau, "Broadcasting and spanning tree in De Brujin and Kautz networks", *Discrete Appl. Math.*, 37-38, (1992), pp. 297–317.
10. J.Hromkovic, R.Klasing, B.Monien, R.Peine, "Dissemination of information in interconnection networks (broadcasting and gossiping)", in *Combinatorial Network Theory*, D-Z.Du, D.F. Hsu (Eds)., Kluwer Academic Publ., (1996).
11. Jean de RUMEUR, *Communications dans les réseaux de processeurs*, Masson, Paris, 1994 (english translation to appear).

Lower Bounds on Broadcasting Time of de Bruijn Networks

Stéphane Perennes.

Laboratoire I3S U.R.A 1376
930 Route des Colles- B.P 145 F - 06903 Sophia Antipolis France

Abstract. Under the *telephone model*, the broadcasting time of most of the logarithmic networks (where the degree is fixed and the diameter logarithmic in the number of nodes) is not known, as the time of the known protocols is different from the known lower bounds. That is the case for de Bruijn. In this paper we present a technique enabling to derive better lower bounds on the broadcasting time of various networks, the technique is applied in the case of the de Bruijn graphs.

1 Introduction

The problem of dissemination of information has been investigated for most of the classical networks. We refer the reader to one of the recent surveys [8] [11] or the book [7]. Our aim is to describe a technique enabling to derive lower bounds on the broadcasting time of various networks. In this short version we apply the technique to obtain lower bounds on the broadcasting time of de Bruijn networks.

The *maximum degree* Δ of a graph is the maximum degree of the vertices. In the case of a digraph, the number of arcs going out (resp. entering) a vertex is called the out-degree (resp. in-degree) of this vertex. The out(in)-degree $d^+(d^-)$ of a digraph is the maximum out(in)-degree of the vertices.

Definition 1. The parameter d of a graph (resp. digraph) is its maximum degree minus one ($\Delta - 1$) (resp. its maximum out-degree d^+)

De Bruijn networks The **de Bruijn** digraph (resp. graph), denoted by $B(d, D)$ (resp. $UB(d, D)$), has d^D vertices, diameter D and in-degree or out-degree d (resp. degree $2d$). The vertices correspond to the words of length D over an alphabet of d symbols. The arcs (or edges) correspond to the shift operations: Given a word $x = x_1 \cdots x_D$ on an alphabet $\mathcal{A}$ of d letters, where $x_i \in \mathcal{A}$, $i = 1, 2, \cdots, D$, and given $\lambda \in \mathcal{A}$, the operations: $x_1 \cdots x_D \longrightarrow x_2 \cdots x_D \lambda$ and $x_1 \cdots x_D \longrightarrow \lambda x_1 \cdots x_{D-1}$ are called respectively left-shift and right-shift. In the de Bruijn digraph $B(d, D)$, the successors are obtained by left-shift operations, whereas in the de Bruijn graph $UB(d, D)$, the neighbors are obtained by either left or right-shift operations. The de Bruijn graph is obtained by removing the directions of the arcs. Note that, since the de Bruijn digraph contains loops and symmetric arcs, this leads to a multi-graph. Thus some authors define the undirected de Bruijn graph as obtained by removing loops and multiple edges. Here we still keep them for homogeneity purpose, and in order to have a regular graph. The symmetric de Bruijn digraph $B^*(d, D)$ is the digraph obtained by replacing in $UB(d, D)$ each edge by a pair of opposite arcs.

Broadcasting One of the main problems of information dissemination investigated in the current literature and used in parallel and distributed computing is **broadcasting**. Broadcasting (also called One to All) refers to the process of message dissemination in an interconnection network whereby a message, originated at one node x, is transmitted to all the nodes of the network. Broadcasting is accomplished by placing a series of calls over the communication lines of the network. Therefore the communication protocol is a sequence of rounds, each one being performed by a set of local calls.

Here we restrict ourselves to what is called the "telephone model" where during one round each vertex can communicate with at most one of its neighbors (out-neighbors in case of digraphs). The broadcasting time $b(x, G)$ is the minimum number of rounds to achieve in G a broadcasting protocol originated at x. The broadcast time $b(G)$ of a graph G, is the maximum of $b(x, G)$ on all vertices x. (In the notation of [8] it would be $b_{F_1}(G)$, F_1 standing for "full duplex one port model"). **In all the paper all the logarithms used are in base 2.**

2 Previous Results

Previous Upper bounds In the case of de Bruijn networks upper bounds have been given in many papers for example [4] [10] and [13] and have still be recently improved in [5, 6] . In particular this leads to:

$$b(B(2, D)) \leq 1.5(D + 1) \quad b(B(3, D)) \leq 2(D + 1)$$

Previous lower bounds Non trivial lower bounds have been derived using available results for graphs or digraphs of bounded degree [3, 12]. Here we recall the basic counting argument of these articles:

Let G be a graph (resp. digraph) of parameter d. The set of informed nodes cannot double at each round, as a node informed during some round can, in the optimal case, only forward the information during the d next rounds [1]. As example, when $d = 2$, for $t \geq 4$, the number $S(t)$ of nodes receiving an information during the round t is such that $S(t) \leq S(t - 1) + S(t - 2)$. We can then compute an upper bound $N(t)$ of $S(t)$. In fact $N(t)$ is the solution of $N(t) = N(t-1) + N(t-2)$ for $t \geq 4$ with $N(1) = S(1)$ and $N(2) = S(2)$, $N(3) = S(3)$. The difference between graphs and digraphs with the same parameter d appears in the computation of the initial values of $N(t)$. As example for graphs

---

[1] This is not true for the originator of the broadcasting in a graph, as such a node can possibly inform new vertices during $d + 1$ rounds

(digraphs) of parameter 2 we have : the values $N(2) = 2, N(3) = 4$ (resp.. . $N(2) = 2, N(3) = 3$). The lower bound is then deduced by:

$$\sum_{0 \leq t \leq b(G)} N(t) \geq \sum_{0 \leq t \leq b(G)} S(t) \geq |V|$$

The analysis of the recurrences leads to the bounds of [3, 12]:

$$d = 2, b(G) \geq 1.4404 \log(|V|)(1 + o(1)) \quad d = 3 b(G) \geq 1.1374 \log(|V|)(1 + o(1))$$

Note that $1.4404 = \frac{1}{\log(\tau)}$ where $\tau = \frac{1+\sqrt{5}}{2}$ is the **golden ratio**, that is the greatest positive root of polynomial $x^2 - x - 1$. This property can be generalized as in[3]: $b(G) \geq \frac{1}{\log(\tau_d)} \log(|V|)(1 + o(1))$, where τ_d is the largest positive root of $x^d - x^{d-1} - \ldots - 1 = 0$.

In the case of binary de Bruijn and Butterfly graphs, the lower bounds have been improved later in [11], the authors proved that $b(UB(2, D)) \geq 1.31171 D$, and that $b(\mathcal{WBF}(2, D)) \geq 1.7456 D$.

In that paper we will introduce a technique enabling us to prove that broadcasting in the undirected de Bruijn graph is not significantly faster than in the digraph. We will prove that $b(B^*(2, D)) \geq 1.4404 D, b(B^*(3, D)) \geq 1.8028 D$. For a generalization of the technique to the case of *iterated line digraph* (Kautz and Butterflies networks) we refer to the extended version.

3 Counting informed vertices in a digraph (graph) of parameter d.

For this we consider a broadcast protocol in a graph (digraph) of parameter d as an infinite directed tree of out-degree d.

Definition 2. Given a digraph, the **broadcast tree** associated to the protocol is a directed infinite tree of out-degree d whose root is the originator. The arcs outgoing from a given vertex are labeled $0, 1, 2, \ldots, d - 1$ according to the following rule *the arc (x, y) is labeled i, if y is the $i+1-th$ vertex that x calls after having itself being informed.*

So the labeling reflects the strategy used locally by each node to forward the information in the tree, each label corresponding to the **delay** introduced by using the arc. In the case of a finite graph of out-degree d, the broadcast tree is finite and in fact obtained as a quotient graph of the infinite broadcast tree.

Definition 3. – Given a graph, the **broadcast tree** is defined in the same way, except that the out-degree of the originator is $d+1$ with delays $0, 1, 2, \ldots, d$.
– If w is a path from the root to x we will say that x is informed along w. If w if of length l we will say that x is at **depth** l

- The **delay** of a path is the sum of the delays on each arc. The **cost** of a path is the sum of its length and of the delays of its arcs, it corresponds to the time at which the end vertex of the path is informed along the path.
- We will denote $I(l,t)$ the ideal [2] number of nodes informed at time exactly t on a path of length l.
- In a broadcast protocol let $S(t)$ denotes the number of nodes informed at time exactly t, note that $S(t) \leq \sum_l I(l,t)$.

Proposition 4. *In a graph (digraph) of parameter d.*

$$I(l,t) = \binom{l}{t-l}_d$$

Where the multinomial coefficient $\binom{l}{i}_d$ is defined by the generating series identities:

$$(1+z+\ldots z^{d-1})^l = \sum_{i \geq 0} \binom{l}{i}_d z^i \quad \textit{For digraphs}$$

$$(1+z+\ldots+z^{d-1}+z^d)(1+z+\ldots z^{d-1})^{l-1} = \sum_{i \geq 0} \binom{l}{i}_d z^i \quad \textit{For graphs}$$

Proof. As our bound is valid for any graph (digraph), we study a broadcasting protocol in a directed infinite tree of out-degree d whose root is the originator. Under this assumption, the number of paths of length l and cost t (or equivalently of delay $t-l$) is exactly the number of words of length l and weight $t-l$ on the alphabet $\{0,1\ldots d-1\}$ that is $\binom{l}{t-l}_d$. In the case of a (di)graph of parameter d the value of $I(l,t)$ is always smaller or equal than this value. Indeed, a vertex can appear many times in the broadcast tree, and so might be counted as informed more than once. This only means that our counting does not take into account possible additional properties of the graph, and that the bound can be refined for some (di)graph. Note that in the case of graphs as the originator (.i.e the root of the tree) have degree $d+1$ instead of d with delays $0,1,\ldots d$, the generating serie differs a bit.

As one can check, the difference between graphs and digraph of parameter d is not significant when deriving asymptotical evaluations. Consequently we will just forget it, considering for graphs the serie $(1+z+\ldots z^{d-1})^l$ instead of $(1+z+\ldots+z^{d-1}+z^d)(1+z+\ldots z^{d-1})^{l-1}$.

Case $d=2$ In the case $d=2$, we can do a detailed analysis which gives not only the former bound $\frac{1}{\log(\tau)}\log(|V|)$ but much more information on how should be a protocol reaching the lower bound if any. As $d=2$, $I(l,t)$ the estimation of $I(l,t)$ is $\binom{l}{t-l}_2$, the number of combinaisons of $t-l$ elements of a set of cardinal l: $\binom{l}{t-l}$.

For a fixed time t, we shall estimate what is the length $l_0(t)$ which maximize $I(l,t)$ at time t. Indeed, we will compute at what depth are mainly located the

---

[2] That is a generic upper bound valid for a specific class of graph

nodes counted as informed at that time. The set of nodes informed at time t at depth $l_0(t)$ will be called the *main level* at time t. Stirling approximation or usual techniques of estimation for coefficients of generating series (see [9]) allow us to claim that: $\log(\binom{n}{xn}) \sim n\phi(x)$ where ϕ is the classical entropy function $\phi(x)) = -\log((1-x)^{1-x}x^x)$. Then setting $l = \alpha t$, we have to maximize for a fixed t: $\log(I(\alpha t, (1-\alpha)t) = (1+o(1))\alpha t\phi(\frac{1-\alpha}{\alpha})$ relatively to the variable α. Computations leads to: $l_0(t) = \alpha_0 t = \frac{1+\frac{1}{\sqrt{5}}}{2}t$. Thus the main level at time t is at depth $\alpha_0 t$ and its size is $\binom{\alpha_0 t}{(1-\alpha_0)t}$. Note that $\log(\binom{l_0(t)}{t-l_0(t)})_2 \sim t\log(\tau)$ [3] this means that the cardinality of the main level is potentially growing exponentially with the time. Due to the behavior of binomial coefficients one can shows that $\log(S(t))$ is bounded above by a quantity equivalent to $\log(I(l_0(t),t)$, that is $\log(S(t)) \leq (1+o(1))t\log(\tau)$. Furthermore due to exponential growing of the value bounding $S(t)$: $2^{(1+o(1))t\log(\tau)}$, one knows that $\log(\sum_{t \leq T} S(t)) \leq (1 + o(1))T\alpha_0\phi(\frac{1-\alpha_0}{\alpha_0}) = T\log(\tau)(1+o(1))$. And the number of vertices informed at time t is bounded by $2^{t\log(\tau)(1+o(1))}$. This is a more precise way to derive the result of [3], as to achieve broadcasting at time T we must have $2^{T\log(\tau)(1+o(1))} \geq |V|$. Our refined calculus is useful, as it enables us to precise the depth at which the nodes informed at time t are mainly located. In fact one can easily check that at time t it is possible to consider that *all the vertices counted as informed are at depth $\alpha_0 t$*. As example, if a broadcast protocol finishes in a graph G of parameter 2 in $\frac{\log(|V|)}{\log(\tau)}$ rounds, then most of the informed vertices are at depth: $l = \alpha_0 \frac{\log(|V|)}{\log(\tau)} = \gamma \log(|V|)$. Computation shows that $\gamma = \frac{\alpha_0}{\log(\tau)} \sim 1.0423 > 1$. So we can assert that if a graph (digraph) of parameter 2 has a nearly optimal broadcast protocol this one informs most of the nodes along paths of length around $\gamma \log(|V|)$. For example, in the case of degree 3 graphs, we need a graph with mean eccentricity less than $\gamma \log(|V|)$. As γ is very close to 1, such a graph is nearly a Moore graph (just note that the graph has $|V|$ vertices,[4] and most of these are supposed to lie at distance at most $\gamma \log(|V|)$). Our condition is clearly more restrictive than the straight one stating that the diameter of G is not greater than $1.4404 \log(|V|)$, and gives also hints to tackle the following question:

Conjecture 1 $b(B(2,D)) \sim D\frac{1}{\log(\tau)} \sim 1.4404D$

Note that here $\log(|V|)$ is D, so a protocol proving this assumption should use paths of length γD instead of shortest paths of length of order D used in existing protocols.

---

[3] τ has been defined in section 2 as $\frac{1+\sqrt{5}}{2}$
[4] At the moment, no explicit procedure is known to construct such a graph, existence is proved by probabilistic methods

4 Counting in the de Bruijn graph $B^*(d, D)$

Any vertex in the digraph has $2d$ out going arcs. Among this $2d$ arcs, d correspond to a left-shift and lie in the de Bruijn digraph. These arcs will be called L (left) ones. The others corresponding to a right-shift are R (right) arcs. For purpose of simplicity we associate to each protocol P a derived protocol P_1 faster than P.

The derived protocol The protocol P is completely defined by a local labeling of the outgoing arcs. Suppose that at each vertex the outgoing arcs are sorted according to their labels, then the derived protocol P_1 is defined by the new labeling given below:

R arcs	$R_{a_0} < R_{a_1} \ldots < R_{a_{d-1}}$
L arcs	$L_{a_0} < L_{a_1} \ldots < L_{a_{d-1}}$

L arcs labels	L_{a_0}	L_{a_1}	$\ldots$	$L_{a_{d-1}}$
	0	1	$\ldots$	$d-1$
R arcs labels	R_{a_0}	R_{a_1}	$\ldots$	$R_{a_{d-1}}$
	0	1	$\ldots$	$d-1$

Counting in the derived protocol $\overline{I(l,t)}$ and $\overline{S(t)}$ will denote the values of the estimation in the derived protocol P_1. $I(l,t)$ and $S(t)$ still refer to a directed protocol. The counting problem is now simple; we consider once again the infinite tree associated to P_1; for which at each vertex there are $2d$ outgoing arcs. Using the labeling of P_1 they are:

$$\begin{cases} (R,0), (R,1), \ldots, (R, d-1) \\ (L,0), (L,1), \ldots, (L, d-1) \end{cases}$$

So, a path of length l and cost t, in the infinite broadcast tree is defined by the pair (a, w) where: $a = (a_1 \ldots a_l)$ with $0 \le a_i \le d-1$ and $\sum a_i = t - l$, and w is a word of length l over the alphabet $\{L, R\}$. If the $i-th$ letter of the word w is L (resp. R) that means that the $i-th$ arc used in the path is (L, a_i) (resp. (R, a_i)). **The word w will be called the (L, R) word of the path**. A first bound on $\overline{I(l,t)}$ can be obtained by multiplying $\binom{l}{t-l}_d$ by the number of words of length l over the alphabet $\{L, R\}$. So the number of vertices informed at time t along a path of length l is at first glance: $\overline{I(l,t)} \le I(l,t)2^l$. Not surprisingly, this estimation is useless as it does not take into account properties of de Bruijn graphs. Hopefully we will prove that most of the paths are always redundant (i.e useless for the broadcasting protocol).

Useful (L, R) words

Definition 5.
- A path in the broadcast tree is redundant if, there exists either another path of smaller cost, or another path of equal cost with smaller length which leads to the same vertex. In other words, if this path informs a vertex which can be informed earlier by another path, or at the same time along a shorter path.
- A word w of $\{L, R\}^l$ is useless if each walk in the broadcast tree having w as $\{L, R\}$ word is redundant. A word is useful if it is not useless.

Lemma 6. *In the undirected de Bruijn graphs $UB(2,D)$ and $UB(3,D)$, useful (L,R) words have no sub-word of kind $L^i R^i L^i$ or $R^i L^i R^i$.*

Proof. We consider a path in the tree, leading to a node z, with as (L,R) word $w_1 L^i R^i L^i w_2$. Let x (resp. y) be the vertex informed along the path corresponding to the word w_1 (resp. $w_1 L^i R^i L^i$). Suppose that x is informed at time t. Then according to our protocol y will not be informed, thanks to the path, before time $t + 3i$. Due to de Bruijn iterated line graph structure, there is a path with a (L,R) word L^i from x to y. So, once again, according to our protocol, y will be also informed at time at most $t + di$ (as $d \leq 3$, $t + di \leq t + 3i$) along a path with (L,R) word $w_1 L^i$. As z is informed from y, we can claim that any z informed along a path with (L,R) word $w_1 L^i R^i L^i w_2$ can be informed sooner (or at the same time but along a shorter path) along a path with (L,R) word $w_1 L^i w_2$. Hence $w_1 L^i R^i L^i w_2$ is useless.

Lemma 7. *A partition of an integer n is an increasing sequence of integers $i_1 \ldots i_p$ such that $\sum_{k=1}^{k=p} i_k = n$. The number $P(n)$ of the partitions of n has the following asymptotic behavior: $P(n) \sim \frac{1}{4\sqrt{n}} exp(\pi\sqrt{\frac{2n}{3}})$*

Proof. A proof of this lemma is given for example in the book [1]

Proposition 8. *Let $U(l)$ denotes the cardinality of the set of useful (L,R) words of length l, then for $UB(2,D)$ and $UB(3,D)$; $\log(U(l)) = \Theta(\sqrt{l})$*

Proof. To a (L,R) word we associate a partition as follows: if the word is of the form $L^{i_1} R^{i_2} \ldots L^{i_p}$ we associate to it the sequence $i_1, i_2, \ldots, i_p$. As w is useful, the sequence $\{i_k\}$ is a *bitonic* one. (i.e. the sequence is increasing then decreasing). To check that, just notice that if $i_k + 1 \leq i_k$ then $i_{k+2} < i_{k+1}$ (if not, one of the pattern $L^{i_{k+1}} R^{i_{k+1}} L^{i_{k+1}} R^{i_{k+1}} L^{i_{k+1}} R^{i_{k+1}}$ would appear contradicting lemma 6). Then to each useful word we can associate an increasing sequence and a decreasing one with sums respectively n_1 and n_2 with $n_1 + n_2 = l$. So a crude estimation of $U(l)$ can be $U(l) \leq \sum_{n_1+n_2=l} P(n_1)P(n_2) \leq \sum_{0 \leq i \leq l} P(i)P(n-i) \leq lP(\frac{l}{2})^2$. Lemma 7 allows us to conclude.

Proposition 9. $b(UB(2,D)) \geq \frac{1}{\log(\tau)} D + o(D), b(UB(3,D)) \geq \frac{\log(3)}{\log(\tau_3)} D + o(D)$

Proof. As $\overline{I(l,t)} \leq I(l,t)U(l)$, and as $\log(U(l)) = o(l)$, we have in fact $\log(\overline{I(l,t)}) = \log(I(l,t)) + o(l)$. As $l = \Theta(t)$ we can claim that $\log(\overline{S(t)}) = \log(\sum I(l,t)) \leq \log(\sum I(l,t)U(l)) \sim \log(S(t)) + o(t)$. So, we get the same estimation than for directed graphs of out-degree d: $\log(\sum_{t \leq T} \overline{S(t)}) \leq \log(\tau_d)T + o(T)$. So, undirected de Bruijn graphs of small degree behave like the digraphs.

5 Conclusion

In this article we have used a general technique and the structural properties of the de Bruijn networks to derive new lower bounds on the broadcasting time.

The results obtained show that the protocol is not likely to be more efficient than in the de digraphs. We conjecture that bounds obtained give the right order for the broadcasting time in de Bruijn networks. So improvements have to been done in designing faster protocols. However our results show that the diffusion will not be done on shortest paths which implies that these protocols might be complicated or difficult to imagine. Extension of the technique to iterated line digraphs (butterflies and Kautz networks) and to n-dimensional grids will be addressed in the full paper.

Acknowledgment: This work has been supported by the European HCM project MAP, the french action RUMEUR and the Canadian/French project PICS.

I thanks very much J-C Bermond, B. Martin and O. Delmas for their advice.

References

1. George E. Andrews. *The Theory of partitions, Encyclopedia of Mathematics and its applications. Volume 2.* Addison-Wesley Publishing Company, 1976. Section: Number Theory.
2. C. Berge. *Graphes.* North Holland, 1981.
3. J-C. Bermond, P. Hell, A.L. Liestman, and J.G. Peters. Broadcasting in bounded degree graphs. *SIAM journal of Discrete Mathematics*, 5:10–24, 1992.
4. J-C. Bermond and C. Peyrat. Broadcasting in de Bruijn networks. Proceedings of the 19-th Southeastern conference on combinatorics, Graph theory and Computing, Congressus Numerantium, 66, 267-282, 1988.
5. J-C. Bermond, A. Marchetti, and X. Munoz. Improved broadcasting protocols for line digraphs. submitted to *EUROPAR 96*
6. J-C. Bermond and S. Perennes. Efficient broadcasting protocols on the de Bruijn and similar networks. In *Proceedings of SIRROCO'95*, 1995.
7. J. de Rumeur. *Communications dans les Réseaux de Processeurs.* Masson, 1994. To be translated.
8. P. Fraigniaud and E. Lazard. Methods and problems of communication in usual networks. *Discrete Applied Mathematics*, 53, Special issue on "Broadcasting and gossiping":79–133, 1995.
9. D. Gardy. The asymptotic behaviour of coefficients of large power of functions. In *Serie formelles et combinatoire algébrique*, pages 237–248, 1992.
10. M-C. Heydemann, J. Opatrny, and D. Sotteau. Broadcasting and spanning trees in de Bruijn and Kautz networks. *Discrete Applied Math.*, 27-28:297–317, 1992.
11. R. Klasing, B. Monien, R. Peine, and E. Stohr. Broadcasting and gossiping in the Butterfly and de Bruijn networks. In Spinger Verlag, editor, *Proc. STACS'92*, volume Lectures Notes in Computer Science 577, pages 351–362, 1992.
12. A.L. Liestman and J.G. Peters. Broadcast networks of bounded degree. *SIAM Journal of Disc. Math.*, 1(4), 1988.
13. S. Perennes. Broadcasting and gossiping on de Bruijn, shuffle exchange and similar networks. Technical report 93-53, I3S, 1993.

Gossip in Trees under Line-Communication Mode

C. Laforest*

LRI, Université de Paris-Sud, Bât. 490, 91405 Orsay Cedex, France

Abstract. This paper addresses the problem of performing efficiently an important operation of communication in networks, namely *gossip*. We study it in the *line-communication model* that is similar to the circuit-switching technique. We construct algorithms to gossip in any tree network and prove that they are optimal or asymptotically optimal. In all the cases, we show that the difference between lower and upper bounds is constant. In addition, these results give a general method to gossip, since any (connected) network has a spanning tree.

1 Introduction

This paper deals with the problem of gossip in parallel computers. We answer the question on how to realize it efficiently even with a restricted network. In a parallel computer consisting of *nodes* (processor with distributed memory) and a *network* for communications, the *broadcast* is the operation where one node, the *originator*, has a *piece of information* (*PI* for short) that must be transmitted to all the others nodes. A *gossip* is a simultaneous broadcast from all the nodes. We also define the *accumulation* in which all the nodes have a distinct PI that must be transmitted to one distinguished node called the *accumulator*. These fundamental operations are often used in parallel programs (for scientific calculations for example). It is then interesting to write a library containing them, independently of the applications that will use them. This implementation is dependent of the network and the type of parallel computer used. These basic operations have been studied by a large number of people under many types of models (see surveys [3, 4] and the book [10] on interconnection networks). A majority of results have been obtained in local communications models i.e. in which a node can only communicate with one (or more) neighbor. However, this restriction can be replaced by a more powerful model: the *line-communications* in which two nodes can communicate even when they are not neighbors (this is a good approximation of the *circuit switching* technique).

The model and related work. A network is modeled by a graph $G = (V, E)$ where the set V of *vertices* represents the nodes (we will not distinguish now node and vertex), and the set E of *edges* represents the links of communication

---
* This work was supported by the "Opération RUMEUR" of the French PRC PRS.

(the edge $[x,y] \in E$ represents the physical link between x and y). Vocabulary and undefined terms on graphs can be found in [1]. In the line-communication model, two nodes can communicate directly via a *path* and such a communication is called a *call*. Transmissions are performed synchronously, by *step*; in each step, many calls can occur. During any step, two calls must use two *edge-disjoint paths* (i.e., any two different paths cannot share a same edge at the same step). The time of a step (i.e. to establish the communication, to send information and to cancel the path) takes one time unit, regardless of the quantity of information transported, or length of paths used during the calls. This is what we call the *constant time* hypothesis (we are only interested in the total number of synchronous steps). Another important feature of our model is that, during a step, a node can make at most one call; this is what we call the *1-port* hypothesis. The last assumption concerns the possibilities to use the paths; during a call, two nodes x and y can exchange their PI's on the communication path, we say that the path is *Full-duplex*. We denote the complete model by $\mathcal{F}$, which is edge-disjoint, constant time, 1-port, full-duplex. An example of a gossip algorithm satisfying the constraints of the model $\mathcal{F}$ is shown in Figure 1.

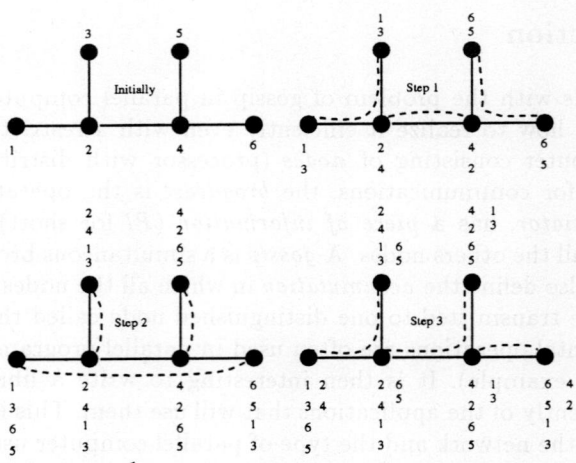

Fig. 1. A gossip algorithm in 3 steps in a tree with 6 vertices.

Let G be a graph with n vertices, r any vertex in G. Let us denote by $b(r,G)$ (resp. $a(r,G)$) the minimum time to broadcast (resp. to accumulate) with originator (resp. accumulator) r in the graph G. We denote by $b(G) = \max_{r \in V(G)} b(r,G)$ and $a(G) = \max_{r \in V(G)} a(r,G)$. In 1980, Farley [2] gave the minimum time to broadcast under the $\mathcal{F}$ model in any connected graph (a short proof of this result can be found in [9]):

Theorem 1. [2]
Under the $\mathcal{F}$ model, for any connected graph G with n vertices, for any vertex u of G,
$$b(u, G) = a(u, G) = \lceil \log_2 n \rceil$$

In [7], Kane and Peters address minimizing the sum of the lengths of paths used to broadcast in minimum time in the cycle. In [5, 6] Hromkovič et al. study similar models and prove results on the gossip time in graphs like the grid and planar graphs.

Our results. In our paper, we are interested in the minimum gossip time, $g(T)$, under the $\mathcal{F}$ model in any tree T. We choose to study the tree, because it is a minimal network that can be found to span any connected graph. Thus, our method can always be applied in any graph. For that, we only require that the network be connected. We only use parameters on trees that can be calculated in linear time in the number of vertices (we show how to calculate them). More precisely, we give some general lower bounds for $g(T)$ that lead us to find some exact (optimal) gossip times for classes of trees. We then construct gossip algorithms that give general upper bounds on $g(T)$. We note that our lower bounds differ from our upper bounds by at most seven time units. This gives us asymptotically optimal results on the gossip time in any tree.

2 Gossip

Let G be any connected graph. We denote by $g(G)$ the minimum time to perform a gossip in G under the $\mathcal{F}$ model. We first give general bounds on the function g in the following lemma. We then prove in Corollary 5 that the bounds are tight.

Lemma 2. *Let G be a connected graph with n vertices. We have:*

- *If n is even, $\lceil \log_2 n \rceil \leq g(G) \leq 2\lceil \log_2 n \rceil - 1$.*
- *If n is odd, $\lceil \log_2 n \rceil + 1 \leq g(G) \leq 2\lceil \log_2 n \rceil - 1$.*

Proof. During each step $t > 0$, the number of vertices that know a given PI can at most double and therefore $\lceil \log_2 n \rceil \leq g(G)$. In addition, if n is odd, during the first step there is a vertex which does not communicate, and thus needs at least $\lceil \log_2 n \rceil$ additional steps to broadcast its PI.

Theorem 1 demonstrates that accumulation or broadcast can be done in $\lceil \log_2 n \rceil$ steps in any tree with n vertices. A gossip algorithm in G is the following: accumulate all the PI's at a vertex r in $\lceil \log_2 n \rceil$ steps and then broadcast this accumulated message also in $\lceil \log_2 n \rceil$ steps from vertex r. As the broadcast can be done by reversing the accumulation, and because the paths are full-duplex, the last step of the accumulation can be done simultaneously with the first step of the broadcast. Thus, $g(G) \leq 2\lceil \log_2 n \rceil - 1$. □

We will now consider gossip in a tree T. We prove lower bounds for $g(T)$ and show that some of these bounds can be reached for different classes of trees. We then give upper bounds of the function $g(T)$, by constructing an algorithm that approximates the lower bounds. This gives us an asymptotically optimal expression of $g(T)$ for any tree T.

2.1 Lower bounds for trees and applications

We begin by defining important parameters that we will need in all what follows. Then, we use them to derive lower bounds.

Definition 3. Let $T = (V, E)$ be a tree with n vertices. For any edge $e \in E$ we denote by $T_{max}(e)$ and $T_{min}(e)$ the two trees obtained by deleting e in T and $n_{max}(e)$ and $n_{min}(e)$ the number of vertices in $T_{max}(e)$ and $T_{min}(e)$ respectively; we choose $n_{max}(e) \geq n_{min}(e)$. Finally, we denote by $n_{min} = \max_{e \in E}\{n_{min}(e)\}$ and $n_{max} = n - n_{min}$. The edge e such that $n_{min}(e) = n_{min}$ is called *maximal edge* and $T_{max}(e)$ is denoted by T_{max} and $T_{min}(e)$ is denoted by T_{min}.

Parameters n_{min} and n_{max} can be calculated by a polynomial algorithm. For example, for each of the $n - 1$ edge e, calculate n_{min} with a breath first search and then take the maximum of all these values (a fastest algorithm can be described, see [9]).

Lemma 4. *Let T be a tree with n vertices:*

$$g(T) \geq \begin{cases} \lceil \log_2 n \rceil + \lceil \log_2(n_{min} + 1) \rceil - 1 & \text{if } n = 2^d \\ \lceil \log_2 n \rceil + \lceil \log_2(n_{min} + 1) \rceil - 2 & \text{if } n \neq 2^d \end{cases}$$

Proof. Let $d = \lceil \log_2 n \rceil$ and let T_{max} and T_{min} associated to a maximal edge e. Our argument for the lower bound is based on the fact that all the PI's in T_{max} must be sent in T_{min}. The fact that the paths used for calls must be edge-disjoint implies that at most one call per step can occur between a vertex in T_{max} and a vertex in T_{min}. Moreover, a vertex can at most double the number of PI's that it knows at each step (because of the 1-port assumption). Thus, at the end of step 1, there is at most one PI of T_{max} in T_{min}, at the end of step 2, there are at most three PI's of T_{max} in T_{min} etc. More generally at the end of step t, $1 \leq t \leq d - 1$, there are at most $2^t - 1$ PI's of T_{max} in T_{min}: a maximum of $2^{t-1} - 1$ are known at the end of step $t - 1$ and at most 2^{t-1} new PI's can be received in T_{min} during step t because the sender in T_{max} knows at most this amount of PI's at the end of step $t - 1$.

We now calculate the greatest t_0 such that at the end of step t_0 not all the n_{max} PI's of T_{max} are in T_{min}. We know that $n_{max} \geq n_{min}$ and that $n = n_{max} + n_{min}$, $n_{max} \geq \frac{n}{2}$.

- If $n = 2^d$, $n_{max} \geq 2^{d-1}$ thus, at the end of step $d - 1$, there are at most $2^{d-1} - 1 < n_{max}$ PI's of T_{max} in T_{min} i.e. $t_0 \geq d - 1$.

- If $2^{d-1} < n < 2^d$, at the end of step $d-2$ there are at most $2^{d-2} - 1 < n_{max}$ PI's of T_{max} in T_{min} i.e. $t_0 \geq d - 2$.

Let us denote by S the non empty set of PI's from T_{max} that are not yet known in T_{min} at the end of step t_0. This set S must be known by all the vertices in T_{min} at the end of the gossip. Let t_1 be the minimum number of steps to inform of S all the vertices in T_{min}. This broadcast is done in T_{min} with, at each step, an additional source from T_{max}. Thus, by Theorem 1, we know that: $t_1 \geq \lceil \log_2(n_{min} + 1) \rceil$.
The lemma is proved by noting that $g(T) \geq t_0 + t_1$. □

From these results we can derive exact bounds for the gossip time in some special trees, which demonstrate that both lower and upper bounds from Lemmas 2 and 4 can be reached.
Let $P_n = (\{1, \ldots, n\}, \{[i, i+1] : 1 \leq i < n\})$ be the path with n vertices, and $S_n = (\{1, \ldots, n\}, \{[1, i] : 1 < i \leq n\})$ be the star with n vertices and let BT_{2^d} be the binomial tree with 2^d vertices (see figure 2 for examples). The binomial tree is defined recursively: $BT_1 = (\{r\}, \emptyset)$ and the unique vertex r is called the root; BT_{2^d} is constructed by connecting the roots of two copies of $BT_{2^{d-1}}$; the root of the new graph is one of the two previous roots.

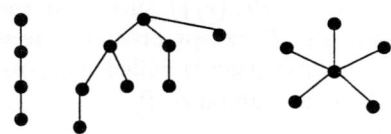

Fig. 2. P_4, BT_8 and S_6.

Corollary 5. *We have:*

- *For any even integer n, $g(S_n) = \lceil \log_2 n \rceil$.*
- *For any odd integer n, $g(S_n) = \lceil \log_2 n \rceil + 1$.*
- *For any tree T with 2^d vertices in which $n_{min} = 2^{d-1}$, $g(T) = 2d - 1 = 2\lceil \log_2 n \rceil - 1$.*

In particular this is true for P_{2^d} and BT_{2^d}.

Proof. We know from Lemma 2 that $g(S_n) \geq \lceil \log_2 n \rceil$ if n is even and $g(S_n) \geq \lceil \log_2 n \rceil + 1$ steps if n is odd.
Let us consider any set $A = \{(x, y) : x \neq y, x \text{ and } y \in V(S_n)\}$ of pairs of vertices such that each x in $V(S_n)$ is in at most one pair. In S_n, all the paths (of length at most two) connecting each pair of vertices of A are edge-disjoint. This strong property will help us to construct a gossip algorithm in S_n. Indeed, in [8], Knödel shows that in the complete graph K_n (where there is an edge between each pair of vertices) the gossip takes $\lceil \log_2 n \rceil$ steps if n is even and $\lceil \log_2 n \rceil + 1$ steps if

n is odd under the $\mathcal{F}$ model (in fact Knödel only uses paths of length 1). His algorithm consists in steps in which each vertex participates to at most one call and sends all the PI's that it knows. These calls at each step represent pairs of vertices of K_n and with the previous observation we can use edge-disjoint paths to do the same calls in S_n between the same pairs of vertices.

Let now T be a tree with 2^d vertices in which $n_{min}(e) = 2^{d-1}$. Lemma 4 shows that $g(T) \geq 2d-1$ and Lemma 2 shows that $g(T) \leq 2d-1$. In particular, these properties are satisfied for P_{2^d} and BT_{2^d}. □

2.2 Upper bounds for trees and algorithms

In what follows we construct a gossip algorithm to obtain upper bounds for the function g. Let T be a tree with n vertices and $e = [r_1, r_2]$ be a maximal edge of T. According to Definition 3 we denote by T_{max} (resp. T_{min}) the tree with n_{max} (resp. n_{min}) vertices such that $r_1 \in V(T_{max})$ (resp. $r_2 \in V(T_{min})$) with $n_{max} \geq n_{min}$. Let δ be the degree of r_1 in T. Let $B_1, \ldots, B_\delta$ be the set of vertices in the subtrees of r_1 (the B_i's do not contain r_1, so are disjoint). From the definition of the maximal edge it is easy to see that for all i, $1 \leq i \leq \delta$, $|B_i| \leq n_{min}$.

Definition 6. We say that a tree T is n_{min}-partitioned if there exists a partition of the family of subsets $\{B_1, \ldots, B_\delta, \{r_1\}\}$ such that each set of the partition contains exactly n_{min} vertices of T, except possibly one set (which may contain less than n_{min} vertices). This partition is called n_{min}-partition and is denoted by $A_1, \ldots, A_\alpha$ (where each A_i is a union of B_j).

Remark: it is easy to show that if T is n_{min}-partitioned then $\alpha = \lceil \frac{n}{n_{min}} \rceil$ and that one of the A_i is $V(T_{min})$.

Lemma 7. *Let T be a tree with n vertices:*

$$g(T) \leq \begin{cases} \lceil \log_2 n \rceil + \lceil \log_2 n_{min} \rceil + 2 & \text{if } T \text{ is } n_{min}\text{-partitioned} \\ \lceil \log_2 n \rceil + \lceil \log_2 n_{min} \rceil + 1 & \text{if } T \text{ is } n_{min}\text{-partitioned and if } n_{min} = 2^k \\ & \text{or } \alpha \text{ is even} \\ \lceil \log_2 n \rceil + \lceil \log_2 n_{min} \rceil & \text{if } T \text{ is } n_{min}\text{-partitioned and if } n_{min} = 2^k \\ & \text{and } \alpha \text{ is even} \\ \lceil \log_2 n \rceil + \lceil \log_2 n_{min} \rceil + 4 & \text{if } T \text{ is not } n_{min}\text{-partitioned} \end{cases}$$

Proof. We first suppose that T is n_{min}-partitioned. Let $A_1, \ldots, A_\alpha$ with $\alpha = \lceil \frac{n}{n_{min}} \rceil$ be an n_{min}-partition of T. We denote by A_1 the set that contains r_1. For all i, $2 \leq i \leq \alpha$, we choose v_i in A_i such that v_i is a neighbor of r_1 in T. For A_1 we choose $v_1 = r_1$.
The algorithm proceeds in 3 phases:

1. For each i, accumulate in v_i all the PI's contained in A_i.
2. Gossip in the star $S_\alpha = (\{r_1, v_2, \ldots, v_\alpha\}, \{[r_1, v_i] : i = 2, \ldots, \alpha\})$.
3. For each i, broadcast the global information from v_i to all the vertices of A_i.

Phase 1 and phase 3 have the same complexity; each phase can be done in $\lceil \log_2 n_{min} \rceil$ steps. Indeed, each set of vertices A_i can independently proceed communications in T (possibly using paths passing through r_1) without conflict. Thus we can apply Theorem 1 to obtain the result.

Concerning the intermediate phase 2, we use the first result of Corollary 5 showing that in S_α a gossip can be done in $\lceil \log_2 \alpha \rceil$ steps if α is even and in $\lceil \log_2 \alpha \rceil + 1$ steps if α is odd.

Thus the total time of our algorithm is at most:

$$2\lceil \log_2 n_{min} \rceil + \lceil \log_2 \lceil \tfrac{n}{n_{min}} \rceil \rceil + 1$$
$$\leq 2\lceil \log_2 n_{min} \rceil + \lceil \lceil \log_2 n \rceil - \log_2 n_{min} \rceil + 1$$
$$\leq \lceil \log_2 n \rceil + \lceil \log_2 n_{min} \rceil + (\lceil \log_2 n_{min} \rceil - \lfloor \log_2 n_{min} \rfloor) + 1$$
$$\leq \lceil \log_2 n \rceil + \lceil \log_2 n_{min} \rceil + 2$$

We can see now that if $\alpha = \lceil \tfrac{n}{n_{min}} \rceil$ is even then the gossip time in S_α is $\lceil \log_2 \alpha \rceil$ instead of $\lceil \log_2 \alpha \rceil + 1$ and if n_{min} if a power of 2, $\lceil \log_2 n_{min} \rceil - \lfloor \log_2 n_{min} \rfloor = 0$.

The second case is when T is not n_{min}-partitionable or (when the n_{min}-partitionability of T is not known). We will apply the same kind of technique as in the first part of the proof but without n_{min}-partition. However, we construct a partition $A_1, \ldots, A_\alpha$ of the set $\{B_1, \ldots, B_\delta, \{r_1\}\}$ with the following algorithm:

$i := 1;$
$A_i := \{r_1\};$
$F := \{B_1, \ldots, B_\delta\}$
While $|F| \neq 0$ Do
Begin
   If $|A_i| < n_{min}$ Then
     Begin
       $A_i := A_i \cup B_j;$ {with $B_j \in F$}
       $F := F - B_j$
     End
   Else Begin $i := i + 1;\ A_i := \emptyset;$ End
End

It is easy to prove that each $|A_i| \leq 2n_{min} - 1$ (because each $|B_j| \leq n_{min}$), and that $\alpha \leq \lceil \tfrac{n}{n_{min}} \rceil$. Thus, as in the previous case we choose in each A_i a vertex v_i and we apply the same 3-phase algorithm. The global time is now at most:

$$2\lceil \log_2 \max_i |A_i| \rceil + g(S_\alpha)$$
$$\leq 2\lceil \log_2 2n_{min} \rceil + \lceil \log_2 \alpha \rceil + 1$$
$$\leq \lceil \log_2 n \rceil + \lceil \log_2 n_{min} \rceil + 4.$$

In all the cases, it is always possible to apply this polynomial algorithm to decompose the tree, even if T is n_{min}-partitionable. □

To summarize the results, we have the two extreme bounds:

Theorem 8. *Let T be a tree with n vertices and let n_{min} be defined by Definition 3:*

$$\lceil \log_2 n \rceil + \lceil \log_2(n_{min} + 1) \rceil - 2 \leq g(T) \leq \lceil \log_2 n \rceil + \lceil \log_2(n_{min}) \rceil + 4.$$

This theorem gives an asymptotically optimal bound for the gossip time in any tree.

3 Conclusion

In this paper, we have given lower and upper bounds for the gossip time in any tree under the line-communication model. The upper bounds are obtained by algorithms that can be constructed in polynomial time in the number of vertices of the considered tree. The lower bounds are obtained by structural arguments on the considered tree and on the operation. In all the cases, the difference between the lower and upper bounds is constant. A general method to construct gossip algorithm is simply to do this operation in any spanning tree of the graph. In addition, in [9], we give a precise idea of the shape of trees having a gossip time equal to $\lceil \log_2 n \rceil$ steps.

Acknowledgments. I thank the third anonymous referee for his helpful comments and to have indicated a way to simplify the proof of Lemma 4. I also thank my supervisor, D. Sotteau, for her attention during this work.

References

1. A. Bondy, U. Murty, Graph theory with applications, Mac Millan Press LTD, 1981.
2. A. Farley, Minimum-time broadcast networks, Networks **10** (1980) 59-70.
3. P. Fraigniaud, E. Lazard, Methods and problems of communication in usual networks, Discrete Applied Mathematics **53** (1994) 79-133.
4. S. Hedetniemi, S. Hedetniemi, A. Liestman, A survey of gossiping and broadcasting in communication networks, Networks **18** (1988) 319-349.
5. J. Hromkovič, R. Klasing, E. Stöhr, H. Wagener, Gossiping in vertex-disjoint paths mode in d-dimensional grids and planar graphs, Proc. First Annual European Symposium on Algorithms (ESA'93), LNCS 726, Springer Verlag (1993) 200-211.
6. J. Hromkovič, K. Klasing, W Unger, H. Wagener, Optimal algorithm for broadcast and gossip in the edge-disjoint path mode. Proc. 4th Scandinavian Workshop on Algorithm Theory (SWAT'94), LNCS 824, Springer-Verlag (1994) 219-230.
7. J. Kane, J. Peters, Line broadcasting in cycles, Technical repport number CMPT TR 94-11, School of computing science, Simon Fraser University, Burnaby, Canada.
8. W. Knödel, New gossips and telephones, Discrete Mathematics **13** (1975) 95.
9. C. Laforest, Broadcast and gossip in line-communication mode, Technical repport number 1005, LRI, Bât. 490, Université de Paris-Sud, 91405 Orsay Cedex France (1995).
10. J. de Rumeur, Communications dans les réseaux de processeurs. *Edition Masson*, Etudes et recherches en informatique (1994).

Total Exchange in Cayley Networks *

Vassilios V. Dimakopoulos and Nikitas J. Dimopoulos

Department of Electrical and Computer Engineering,
University of Victoria
P.O. Box 3055, Victoria, B.C., CANADA, V8W 3P6.

Abstract. We present a time-optimal solution to the *total exchange* problem, under the *single-port* assumption, for any Cayley graph. We exploit symmetries inherent in Cayley graphs to devise what we call *node-invariant* algorithms which behave uniformly across the network.

1 Introduction

Collective communications for distributed-memory multiprocessors have received considerable attention, as evidenced from their inclusion in the Message Passing Interface standard and in supporting various constructs in High Performance Fortran. In *total exchange* each node in a network has distinct messages to send to all the other nodes. Algorithms to solve the problem for a number of networks under a variety of models/assumptions have already appeared, mostly concentrating in hypercubes and tori (e.g. [13, 3, 14]). Here we follow the so-called *single-port* model in a store-and-forward network where:
- only adjacent nodes can exchange messages,
- a message is transferred in one time unit between two nodes,
- a node can send *and/or* receive at most one message in each step.

Under this model, time-optimal total exchange algorithms have been given in [4, pp. 81–83] for hypercubes and in [11] for star graphs. In this work we present algorithms that solve the total exchange problem in the minimum time for any Cayley network.

We note that this paper provides a short report of the results, avoiding formal proofs. A more detailed exposition of the material can be found in [9].

2 Graph-theoretic and group-theoretic notions

An (undirected) graph G consists of a set V of *nodes* (or *vertices*) interconnected by a set E of (undirected) *edges*. This is the usual model of representing a multiprocessor interconnection network. Thus the terms 'graph' and 'network' will be considered synonymous here.

The *eccentricity* of v, $e(v)$, is the distance to a node farthest from v, i.e. $e(v) = \max_{u \in V}\{dist(v, u)\}$.

An *automorphism* of a graph G is a permutation σ of V such that $(\sigma(v), \sigma(u)) \in E$ iff $(v, u) \in E$. If for any pair of vertices v, u there exists an automorphism mapping v to u, the graph is *node symmetric*. The set of automorphisms $\Pi(G)$ of a graph G is a group with respect to composition.

---

* This research was supported in part through grants from NSERC and the University of Victoria.

Given a set $\Gamma = \{\gamma_1, \gamma_2, \ldots, \gamma_d\}$ of generators for a group $\mathcal{G}$, a *Cayley graph* [5, 1] has vertices corresponding to the elements of $\mathcal{G}$ and edges corresponding to the action of the generators. That is, if $v, u \in \mathcal{G}$, the edge (v, u) exists in G iff there is a $\gamma \in \Gamma$ such that $v \cdot \gamma = u$. Usually the identity element of $\mathcal{G}$ does not belong to Γ (in order to avoid edges from a node to itself) and Γ is closed under inverses (so that the graph is in effect undirected).

Cayley graphs include the *hypercube*, the (wrapped) *butterfly*, the *cube connected cycles* [2, 12, 8] and connected *circulant graphs* [6] (which include the rings).

3 An automorphism property of Cayley graphs

Let G be node symmetric with node set $V = \{v_0, v_1, \ldots, v_{n-1}\}$, and let $\Pi(G)$ be its automorphism group. Denote by $\Pi_{v_i}(G)$ the subset of $\Pi(G)$ consisting of all automorphisms that map v_0 to v_i: $\Pi_{v_i}(G) = \{\sigma \mid \sigma(v_0) = v_i, \sigma \in \Pi(G)\}$. Since G is node symmetric $\Pi_{v_i}(G)$ is nonempty. From each set $\Pi_{v_i}(G)$ we select one automorphism σ_{v_i} and form the set $\Sigma(G) = \{\sigma_{v_i} \mid \sigma_{v_i} \in \Pi_{v_i}(G), i = 0, 1, \ldots, n-1\}$. In particular, we select σ_{v_0} to be the *identity* mapping. Let $\sigma\sigma'$ be the composition of σ and σ'. The selected mappings have the following property: for every neighbor v_a of node v_0 and for every $i = 0, 1, \ldots, n-1$,

$$\sigma_{\sigma_{v_i}(v_a)} = \sigma_{v_i}\sigma_{v_a}. \tag{1}$$

Lemma 1. *Every Cayley graph has a set $\Sigma(G)$ of automorphisms satisfying (1).*

Proof. The mapping $\sigma_{v_i}(v_x) = v_i \cdot v_0^{-1} \cdot v_x$, is an automorphism of the graph [1] that maps v_0 to v_i. Then $\sigma_{\sigma_{v_i}(v_a)}(v_x) = \sigma_{v_i}(\sigma_{v_a}(v_x))$, which satisfies (1).

Example 1. Consider a ring R_n with n nodes. Node v_i is adjacent to nodes $v_{i\oplus 1}$ and $v_{i\ominus 1}$ where $\oplus$ and $\ominus$ denote addition and subtraction modulo n. A set $\Sigma(G)$ of automorphisms with the desired properties consists of the following mappings: $\sigma_{v_i}(v_x) = v_{i\oplus x}$, $i = 0, 1, \ldots, n-1$. Clearly, $\sigma_{\sigma_{v_i}(v_a)}(v_x) = \sigma_{v_{i\oplus a}}(v_x) = v_{i\oplus a\oplus x} = \sigma_{v_i}(v_{a\oplus x}) = \sigma_{v_i}(\sigma_{v_a}(v_x))$, satisfies (1).

During total exchange nodes send messages to various destinations. A node must then pick one of its neighbors through which the message will be routed to its destination. We have the following lemma (see [9] for the proof).

Lemma 2. *Let $\Sigma(G)$ be a set of automorphisms satisfying (1). If v_0 "picks" one of its neighbors, v_a, and every node v_i, $i = 1, 2, \ldots, n-1$, "picks" neighbor $\sigma_{v_i}(v_a)$ then* **(a)** *every node is picked by exactly one other node and* **(b)** *if v_b is the node that picks v_0 then $\sigma_{v_i}(v_b)$ is the node that picks v_i.*

4 Lower bound on total exchange time

During total exchange, a node v sends $n-1$ distinct messages, one for each of the other nodes in an n-node network. If there exist n_d nodes in distance d from v, where $d = 1, 2, \ldots, e(v)$, then the messages sent by v must cross $s(v) = \sum_{d=1}^{e(v)} dn_d$ links. For all messages to be exchanged, the total number of link traversals is $S_G = \sum_{v \in V} s(v)$. The quantity $s(v)$ is known as the the *status* [7] of node v.

Every time a message passes between adjacent nodes one link traversal occurs. If nodes are allowed to transmit only one message per step, at most n links are traversed per step. Consequently, we can at best subtract n units from S_G in each step, so that a lower bound on the total exchange time is $T \geq \frac{S_G}{n}$. Because all nodes in a node symmetric graph have the same status [7], it is seen that for such networks the lower bound is simply $T \geq s(v)$, where v is any node.

Based on the above discussion we have the following sufficient conditions in order for a total exchange scheme to achieve the lower bound:

all nodes are busy all the time, and, (2)

every transmitted messages gets closer to its destination. (3)

The conditions guarantee that n units are subtracted from S_G at every step. Notice that we require that transmitted messages are not *derouted*.

5 Optimal algorithms

Every node v_i in the network maintains a *message queue*, Q_{v_i}, where incoming messages are deposited except of the ones destined for v_i which are forwarded to the local processor immediately. At each node v_i a local algorithm $\mathcal{A}_{v_i}$ selects a message from Q_{v_i} and a neighbor of v_i to which the message will be sent.

Definition 3. A distributed total exchange algorithm $\mathcal{A} = (\mathcal{A}_{v_0}, \mathcal{A}_{v_1}, \ldots, \mathcal{A}_{v_{n-1}})$ is a collection of local algorithms $\mathcal{A}_{v_i}$ running on node v_i, $i = 0, 1, \ldots, n-1$. Algorithm $\mathcal{A}_{v_i}$ is written as $\mathcal{A}_{v_i} = (f_{v_i}, w_{v_i})$, where, given a message queue Q_{v_i}, f_{v_i} selects a message $f_{v_i}(Q_{v_i}) = m$ and w_{v_i} selects a neighbor $w_{v_i}(m)$ of v_i.

The idea now is to let every node v_i select a message "corresponding" to the message selected by node v_0 and to send it to a neighbor "corresponding" to the neighbor selected by v_0, ensuring that the algorithm behaves uniformly across the network. This implies that all nodes have "corresponding" message queues at each step; hence queues that have the same size. We will then be able to guarantee that all queues become empty at the same time. This is exactly the time when total exchange completes, and condition (2) is satisfied.

In order to describe algorithms with a uniform behavior, we need the following notation. Let $m_{v_x}(v_y)$ be the message of node v_x (source) meant for node v_y (destination). For an automorphism $\sigma \in \Pi(G)$, let $\sigma(m_{v_x}(v_y))$ be the message of node $\sigma(v_x)$ destined for node $\sigma(v_y)$, i.e. $\sigma(m_{v_x}(v_y)) \stackrel{\text{def}}{=} m_{\sigma(v_x)}(\sigma(v_y))$. Finally, let Q be a set of messages. We define $\sigma(Q) \stackrel{\text{def}}{=} \{\sigma(m_{v_x}(v_y)) \mid m_{v_x}(v_y) \in Q\}$.

Definition 4. Let G be a Cayley graph and let $\Sigma(G)$ be a set of automorphisms that satisfy (1). A total exchange algorithm $\mathcal{A} = (\mathcal{A}_{v_0}, \ldots, \mathcal{A}_{v_{n-1}})$ where $\mathcal{A}_{v_i} = (f_{v_i}, w_{v_i})$, $i = 0, 1, \ldots, n-1$, will be called *node-invariant* if for any message queue Q and any message m it satisfies

$$f_{v_i}(\sigma_{v_i}(Q)) = \sigma_{v_i}(f_{v_0}(Q))$$
$$w_{v_i}(\sigma_{v_i}(m)) = \sigma_{v_i}(w_{v_0}(m)).$$

The proofs of the following two lemmas can be found in [9].

Lemma 5. *If $Q_{v_i}(t)$ is the queue of node v_i at time t, $i = 0, 1, \ldots, n-1$, then any node-invariant algorithm guarantees that $Q_{v_i}(t) = \sigma_{v_i}(Q_{v_0}(t))$, for all $t \geq 0$.*

Lemma 6. *If node v_0 never deroutes a message then the same is true for every other node v_i, $i = 1, 2, \ldots, n-1$.*

Theorem 7. *Any node-invariant algorithm for which w_0 selects shortest paths is an optimal total exchange algorithm for Cayley graphs.*

Proof. From Lemma 5 it is seen that all nodes have the same queue size at any step. Thus all nodes become idle (all queues are empty, hence total exchange is completed) at the same time. From Lemma 6 no message is derouted if w_0 selects shortest paths. Consequently, both conditions (2) and (3) are satisfied and the algorithm solves the problem optimally.

Summarizing, we showed that there exists a class of algorithms, called node-invariant algorithms, which are able to solve the total exchange problem optimally in any Cayley network. Most reasonable algorithms, such as furthest-first, closest-first, etc. are valid candidates. In the next section we provide a simple node-invariant algorithm and we give an example in the context of hypercubes.

6 A simple node-invariant algorithm

Assume that we have an algorithm $\mathcal{W}$ which takes a message, looks at its destination and picks a neighbor of v_0 which lies on a shortest path from v_0 to the destination of the message.

Example 2. In a ring R_n we can have

$$\mathcal{W}(m_{v_x}(v_y)) = \begin{cases} v_1 & \text{if } y \leq n/2 \\ v_{n-1} & \text{otherwise} \end{cases}$$

(nodes v_1 and v_{n-1} are the two neighbors of node v_0).

Let us treat a message queue as a set of messages that behaves as a FIFO queue. At node v_0 we initially sort destinations in any desired order. For instance,

$$Q_{v_0}(0) = \{m_{v_0}(v_1), m_{v_0}(v_2), \ldots, m_{v_0}(v_{n-1})\}.$$

Suppose that the right end is the head of the FIFO queue and the left end is its tail. Departing messages will leave from the head of the queue. Arriving messages will join at the tail of the queue as long as they are not destined for the current node; otherwise they are immediately forwarded to the local processor. We have to guarantee that initially $Q_{v_i}(0)$ is equal to $\sigma_{v_i}(Q_{v_0}(0))$, so we let

$$Q_{v_i}(0) = \{m_{v_i}(\sigma_{v_i}(v_1)), m_{v_i}(\sigma_{v_i}(v_2)), \ldots, m_{v_i}(\sigma_{v_i}(v_{n-1}))\}.$$

The local algorithm $\mathcal{A}_{v_i} = (f_{v_i}, w_{v_i})$ is defined as follows:

$f_{v_i}(Q)$: select the message at the head of the queue Q,

and it is trivial to see that $f_{v_i}(\sigma_{v_i}(Q)) = \sigma_{v_i}(f_{v_0}(Q))$.

$$\mathcal{A}_{v_i}: \quad (i = 0, 1, \ldots, n-1)$$
At $t = 0$ set
$$Q_{v_i} = \left\{ m_{v_i}(\sigma_{v_i}(v_1)), m_{v_i}(\sigma_{v_i}(v_2)), \ldots, m_{v_i}(\sigma_{v_i}(v_{n-1})) \right\},$$
and let
$f_{v_i}(Q_{v_i})$: select the message at the head of the queue Q_{v_i},
$w_{v_i}(m)$: if $m = f_{v_i}(Q_{v_i})$, select neighbor $\sigma_{v_i}\left(\mathcal{W}(\sigma_{v_i}^{-1}(m))\right)$,

Fig. 1. An optimal total exchange algorithm for Cayley networks. The queues are FIFO. Messages join at the left end and depart from the right.

Finally, let σ^{-1} be the inverse of $\sigma \in \Pi(G)$. The existence and the uniqueness of σ^{-1} is guaranteed by the fact the $\Pi(G)$ is a group. Given $\mathcal{W}$ we define

$w_{v_i}(m)$: for message m select neighbor $\sigma_{v_i}\left(\mathcal{W}(\sigma_{v_i}^{-1}(m))\right)$.

It can be shown that $w_{v_i}(\sigma_{v_i}(m)) = \sigma_{v_i}(w_{v_0}(m))$, for any message m.

In summary, the algorithm shown in Fig. 1 is, based on Definition 4, node-invariant. Therefore, it is an optimal total exchange algorithm for any Cayley network, according to Theorem 7.

6.1 An example in hypercubes

To illustrate the theory developed in the previous sections we will construct an algorithm for hypercubes, based on the algorithm in Fig. 1. An optimal algorithm was given in [4, pp. 81–83] but is not in explicit form.

Let $\oplus$ be the exclusive-or operation. If the binary representation of x is $(x_{d-1}, \ldots, x_1, x_0)$ then the bitwise exclusive-or operation, $\oplus_b$, is defined as

$$x \oplus_b y = (x_{d-1} \oplus y_{d-1}, \ldots, x_1 \oplus y_1, x_0 \oplus y_0).$$

Dropping 'v' from the name of node v_i, a hypercube Q_d has node set $V = \{0, 1, \ldots, 2^d - 1\}$. A node i has neighbors $i \oplus_b 2^0$, $i \oplus_b 2^1$, ..., $i \oplus_b 2^{d-1}$. The following is an automorphism of the hypercube [10] that maps node 0 to node i:

$$\sigma_i(x) = i \oplus_b x. \tag{4}$$

Because of associativity, $\sigma_{\sigma_i(a)}(x) = i \oplus_b a \oplus_b x = \sigma_i(\sigma_a(x))$, for any node a, so that the set of automorphisms given by (4) for $i = 0, 1, \ldots, 2^d - 1$ satisfy (1). Because $i \oplus_b i = 0$, it is seen that $\sigma_i^{-1} = \sigma_i$. Finally, it is known that if in the binary representation of y, $y_k = 1$ for some k then neighbor 2^k of node 0 lies on a shortest path from 0 to y, that is $\mathcal{W}(m_x(y)) = 2^k$. To comply with the standard e-cube routing, k is selected to be the leftmost non-zero bit position of y. Thus, the algorithm of the last section takes the form shown in Fig. 2.

7 Discussion

We considered the total exchange problem under the single-port model in the setting of Cayley graphs. It was shown that as long as every node sends a message at every step and the message is not derouted, the optimal completion time is

$\mathcal{A}_i$: $(i = 0, 1, \ldots, n - 1)$
At $t = 0$ set
$$Q_i = \Big\{ m_i(i \oplus_b 1), m_i(i \oplus_b 2), \ldots, m_i(i \oplus_b (n - 1)) \Big\}.$$
At any step $t \geq 0$,
- select the message at the head of Q_i (say $m_x(y)$)
- send it to node $i \oplus_b 2^k$ where k is the leftmost non-zero bit position of $i \oplus_b y$.

Fig. 2. An optimal total exchange algorithm for d-dimensional hypercubes. The standard e-cube routing paths are followed at every transmission.

guaranteed. A particular type of algorithms, which we named node-invariant algorithms, always satisfy these optimality conditions.

The only requirement for our arguments to work was that the network possesses a set of isomorphisms that satisfy (1).

A detailed exposition of this material is available in [9] and can be obtained through the World Wide Web at http://www-lapis.uvic.ca.

References

1. Akers, S. B., Krishnamurthy, B.: A group-theoretic model for symmetric interconnection networks. IEEE Trans. Comput. **38** (Apr. 1989) 555–566
2. Annexstein, F., Baumslag, M., Rosenberg, A. L.: Group action graphs and parallel architectures. SIAM J. Comput. **19** (June 1990) 544–569
3. Bertsekas, D. P., et. al. Optimal communication algorithms for hypercubes. J. Parallel Distrib. Comput. **11** (1991) 263–275
4. Bertsekas, D. P., Tsitsiklis, J. N.: Parallel and Distributed Computation: Numerical Methods. Englewoods Cliffs, N.J.: Prentice - Hall 1989
5. Biggs, N.: Algebraic Graph Theory (2nd edition). Cambridge, G.B.: Cambridge University Press 1993
6. Boesch, F., Tindell, R.: Circulants and their connectivities. Journal of Graph Theory **8** (1984) 487–499
7. Buckley, F., Harary, F.: Distance in Graphs. Reading, Mass.: Addison - Wesley 1990
8. Carlsson, G. E., Cruthirds, J. E., Sexton, H. B., Wright, C. G.: Interconnection networks based on a generalization of cube-connected cycles. IEEE Trans. Comput. **C-34** (Aug. 1985) 769–772
9. Dimakopoulos, V. V., Dimopoulos, N. J.: Optimal total exchange in Cayley graphs. Technical Report ECE-96-1, University of Victoria. (Jan. 1996)
10. Leighton, F. T.: Introduction to Parallel Algorithms and Architectures: Arrays, Trees, Hypercubes. San Diego, CA: Morgan Kaufmann 1992
11. Misŝić, J., Jovanović, Z.: Communication aspects of the star graph interconnection network. IEEE Trans. Parall. Distrib. Syst. **5** (July 1994) 678–687
12. Preparata, F. P., Vuillemin, J.: The cube-connected cycles: a versatile network for parallel computation. Commun. ACM **24** (May 1981) 300-309
13. Saad, Y., Schultz, M. H.: Data communications in hypercubes. J. Parallel Distrib. Comput. **6** (1989) 115–135
14. Varvarigos, E. A., Bertsekas, D. P.: Communication algorithms for isotropic tasks in hypercubes and wraparound meshes. Parallel Comput. **18** (1992) 1233–1257

Leaf Communications in Complete Trees *

Vassilios V. Dimakopoulos and Nikitas J. Dimopoulos

Department of Electrical and Computer Engineering,
University of Victoria
P.O. Box 3055, Victoria, B.C., CANADA, V8W 3P6.

Abstract. In this work we consider tree-based interconnects where the processing nodes are confined to the leaves of the tree. These types of interconnects include fat tree networks. We present and analyze algorithms for collective communications problems that include broadcasting, scattering/gathering, multinode broadcasting and total exchange.

1 Introduction

Distributed memory multiprocessors are based on a collection of independent processing nodes integrated through an interconnection network. *Collective communications* problems include: *broadcasting, multinode broadcasting, scattering (gathering)* and *total exchange*. The need for their efficient solution was realized quite early, especially in the context of parallel numerical algorithms [2].

Studies in collective communication problems for hypercube-networks appear in [11, 8, 1] and for non-hypercube ones in [12]. Two excellent surveys on the subject can be also found in [7, 6].

In this work we consider the complete tree topology where processors are confined to the leaf level. *Fat trees* [9, 10], are also complete trees but with branch capacities increasing towards the root. Here we consider k-ary trees where k is any integer greater than one. We shall study the above communication problems and determine the time requirements for solving them in the optimum time. Related but not exactly applicable to fat trees are the works in [5, 3].

We note that this paper provides only a short report of the results, avoiding formal proofs. A more detailed exposition of the material can be found in [4].

1.1 Preliminaries

In a complete k-ary tree each node has k children, except for the leaves. The tree consists of $\log_k n + 1$ levels. Level 0 is the leaf level and level $h = \log_k n$ is the level of the root; n denotes the number of leaves, which will be numbered from left to right as $0, 1, \ldots, n-1$. The leaves correspond to *processing* nodes while all the other levels include only *routing* nodes. A *fat tree* is based on the same topology only branches increase in capacity as one moves towards the root. Branches between levels $i-1$ and i have capacity $c_i \geq 1$ corresponding to the number of physical links included in the branch. Two capacity patterns are of interest: constant ($c_i = 1$; $i = 1, 2, \ldots, h$) and exponential ($c_i = k^{i-1}$).

---

* This research was supported in part through grants from NSERC and the University of Victoria.

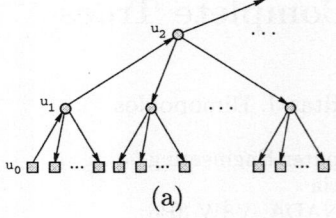

Source Node:
1 Send message to parent;
Every Routing Node:
1 Send message to parent (if any);
2 Send message to each child in turn, except to the one the message came from;

(a) (b)

Fig. 1. Broadcasting under the single-port model (a) labeling (b) algorithm

We assume packet-switching. Messages consist of a single packet. Transferring a message between two neighbors occurs in one time unit (or step). We consider two models of communication: in the first one, nodes will be able to utilize only one of their output links at a time (*single-port* model). In the *multiport* model, all links incident to a node can be utilized simultaneously.

2 Communications under the single-port model

In this section we shall derive lower bounds for the communication problems we consider and shall provide algorithms that achieve the lower bounds. Notice that we allow any node to receive messages from *all* its neighbors simultaneously but it can only send one message at a time; thus branch capacities have no effect.

2.1 Broadcasting

Let $B_r(\ell)$ denote the broadcast time from a node at level ℓ to the leaf nodes.

$$B_r(\ell) = B_r(\ell-1) + k \Rightarrow B_r(\ell) = k\ell \tag{1}$$

Our goal though is to broadcast from a leaf of a complete k-ary tree. Let the leaf, labeled u_0 in Fig. 1(a), be the source node. Also, let $b(T_{u_i})$ be the broadcast time remaining after u_i receives the message. Node u_h is the root of the tree. When u_h receives the message it must inform its $k-1$ remaining children and for this it needs $B_r(h-1)+k-1$ steps. Consequently, $b(T_{u_h}) = B_r(h-1)+k-1$.

Now consider u_{h-1}. This node has to inform $k-1$ subtrees lying below plus its parent, u_h. For the subtrees it needs $B_r(h-2)+k-1$ steps, while broadcasting from u_h, needs $B_r(h-1)+k-1$ steps. Observe that the best plan is to first send the message to the subtree with the largest broadcast time, (i.e. to u_h first). Consequently, $b(T_{u_{h-1}}) = \max\{b(T_{u_h})+1, B_r(h-2)+k\} = b(T_{u_h})+1$.

Proceeding downwards in this manner it is seen that any node u_i must first inform its parent (u_{i+1}) and then its $k-1$ remaining children, and

$$b(T_{u_i}) = \max\{b(T_{u_{i+1}})+1, B_r(i-1)+k\} = b(T_{u_h})+h-i,$$

giving $b(T_{u_0}) = b(T_{u_h}) + h = B_r(h-1)+k-1+h$. Since from (1) $B_r(h-1) = k(h-1)$, we have the following result (recall that $h = \log_k n$):

Theorem 1. *Broadcasting in fat trees under the single-port model requires* $(k+1)\log_k n - 1$ *steps.*

Every Leaf Node:
  Send message to parent;

Every Routing Node: (except the root)
  A Receive all messages from children, sending a copy of them upwards, one-by-one;
  B Until the last message arrives from parent
     Send one message to each child in turn;

Root Node:
  B While receive all messages from children, keep sending one message to each child in turn;

(a)

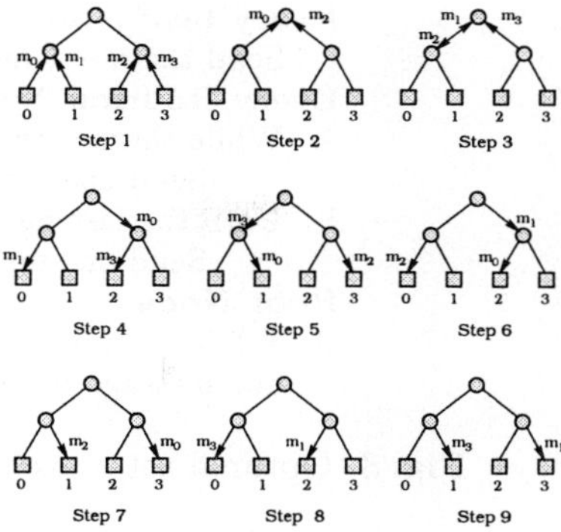

(b)

Fig. 2. Multinode broadcasting under the single-port model (a) optimal algorithm (b) example in 4 leaves

2.2 Scattering

The observation that the source node has to send or receive $n-1$ different messages over its incident link leads to a lower bound of $S(n) = G(n) \geq n-1$ steps. In reality, the exact bound is n or $n+1$ steps. In gathering $n-1$ messages at processor 0, in the first two steps we can at most receive one message since the closest leaf is at distance two. Therefore, there will be at least one step with no message reception by node 0, i.e. $S(n) = G(n) \geq n$. If the tree is binary there will be one extra step of no reception since the next closest leaves (2 and 3) are both at distance four from node 0, i.e. $S(n) = G(n) \geq n+1$ if $k=2$.

Gathering algorithms can be had from scattering ones by reversing the data paths. The lower bound is achieved using *furthest-first* scheduling, whereby the source node gives priority to messages that are destined the furthest.

Theorem 2. *The furthest-first discipline results in an optimal scattering algorithm.*

2.3 Multinode broadcasting

In multinode broadcasting every node broadcasts its own message. An optimal multinode algorithm schedules the traffic in each routing node so that all leaves receive all messages at the minimum possible time. The subsequent two theorems (the proofs of which can be found in [4]) establish the lower bound and the optimality of a multinode broadcasting algorithm.

Theorem 3. *Any multinode broadcasting algorithm under the single-port model requires at least $kn + (k+1)(\log_k n - 2) + 1$ steps.*

Theorem 4. *The algorithm presented in Fig. 2(a) is optimal.*

Every Leaf Node:
  Send the $n-1$ messages in a furthest-first order;
Every Routing Node: (except the root)
  A   While there exist upward messages from children
       Send one message to parent;
  B   Until the last message arrives from parent
       Send one message to any (appropriate) child;
Root Node:
  B   While receiving all messages from children, keep sending
       one message to any (appropriate) child;

Fig. 3. Optimal total exchange algorithm under the single-port model

2.4 Total exchange

Under the single-port model we may easily determine the lower bound of total exchange algorithms, in a similar fashion to the proof of Theorem 3.

Theorem 5. *Any total exchange algorithm under the single-port model requires at least $n^2(2k+1)(k-1)/k^3 + 2\log_k n - 3$ steps.*

An algorithm that achieves the lower bound of Theorem 5 is given in Fig. 3. (For the proof of its optimality see [4].)

Theorem 6. *The algorithm presented in Fig. 3 is optimal.*

3 Communications under the multiport model

We shall now assume that a node is able to utilize all its incident links simultaneously. This model of communication is affected by the capacity arrangement on the branches of the tree. We shall further assume that the capacity of level-1 branches is $c_1 = 1$.

3.1 Single-source communications

In any network, broadcasting from a node under the multiport model takes time equal to d, where d is the distance between the source node and a node farthest from the source. In our case, broadcasting will require $B(n) = 2h = 2\log_k n$ steps. It is easily accomplished by setting the routing nodes to a broadcast mode whereby the received message is replicated towards all directions.

Scattering and gathering, under our assumptions, is governed by the same bounds as in the single-port case; there is only one link available from a leaf, forcing only one message to be sent or received at a time. Consequently, $S(n) = G(n) \geq n$ (or $n+1$ if $k = 2$). Had we allowed $c_1 > 1$, different lower bounds would have been derived.

3.2 Multinode broadcasting

The same argument used for deriving bounds for scattering/gathering algorithms can be used to determine lower bounds for multinode broadcasting in the multiport model since every leaf has to receive $n-1$ different broadcast messages. The exact bounds are $MB(n) \geq n$ if $k \geq 3$ or $MB(n) \geq n+1$ if $k = 2$.

```
1   For all i = 0 to h − 1
      /* Phase i */
2     Do in parallel for all level h − i nodes
3       Transfer all messages from each of the k subtrees
        to the other k − 1 subtrees;
```

Fig. 4. A total exchange algorithm with no contention

Theorem 7. *Multinode broadcasting can be performed in time equal to the lower bound.*

Proof. The lower bound can be achieved by following a "flooding" procedure: each node replicates every received message to all possible directions (expect the one the massage came from). The proof can be found in [4].

This flooding algorithm achieves the lower bound but it requires large queues. In [4] we give an algorithm which is suboptimal by $2\log_k n - 2$ steps but it eliminates the queues.

3.3 Total exchange

A simple lower bound for the total exchange problem is $TE(n) \geq n$ (or $TE(n) \geq n+1$ if the tree is binary), since multinode broadcasting can be performed in at most as many steps as total exchange [2]

A total exchange algorithm that works for any capacity pattern and induces no queueing is as follows. Initially, the k subtrees of the root node exchange their $n(n - k^{h-1})$ messages meant for each other. Then, the k subtrees perform internally a total exchange in parallel. The algorithm is stated in Fig. 4.

During the ith phase a node at level $h - i$ has to pass $k^{h-i}(k^{h-i} - k^{h-i-1})$ messages over its k incident branches which have capacity c_{h-i}. This means that a maximum of kc_{h-i} messages can cross towards the node at a time. To avoid contention while maintaining maximum speed, exactly kc_{h_i} leaves should dispatch messages at a single step, and the messages should be appropriately chosen so that their destinations are distinct. By calculating the time needed for each phase, it can be seen that the algorithm needs time

$$T = \sum_{i=1}^{h} \left\lceil \frac{(k-1)k^{2i-2}}{c_i} \right\rceil + h^2.$$

Instead of executing the phases serially, it is possible to pipeline the phases appropriately and save in total $h^2 - 2h + 1$ steps (see [4]). The pipelined algorithm has a final number of steps

$$T = \sum_{i=1}^{h} \left\lceil \frac{(k-1)k^{2i-2}}{c_i} \right\rceil + 2h - 1. \quad (2)$$

For trees with exponential capacities, $c_i = k^{i-1}$, we see, from (2), that our algorithm requires $T = k^h + 2h - 2 = n + 2h - 2$ steps. Consequently, the algorithm we presented above is suboptimal by $2\log_k n - 2$ steps. It is interesting to see whether the last bound is tight or not. We have shown [4] that for the case of trees with exponential capacities arrangement, a bound tighter than $TE(n) \geq n$ exists which guarantees that the total exchange algorithm we presented is very close to optimal (within $2\log_k \log_k n$ steps) for such trees.

Theorem 8. *An optimal total exchange algorithm for exponential capacity trees needs at least* $n + 2\log_k n - 2\log_k \log_k n - 2$ *steps.*

4 Conclusion

We studied the implementation and performance of communication operations in complete k-ary trees where the processing nodes are confined to the leaf level. The results can be easily generalized to the case where nodes in level i have k_i children, where $n = k_1 k_2 \cdots k_h$. However, there are still a number of issues to be considered. Multinode broadcasting has excessive queueing requirements if it is to be performed in the minimum number of steps. It would be interesting to see what is the lower time bound under the constraint of no queueing. Another issue for consideration is total exchange under the multiport model. The algorithm we presented is close to optimal especially in the case of exponential capacities. Improved bounds though for the total exchange problem need to be found as the straightforward one does not seem to be tight.

A detailed exposition of this material is available in [4] and can be obtained through the World Wide Web at http://www-lapis.uvic.ca.

References

1. D.P. Bertsekas, et. al. "Optimal communication algorithms for hypercubes," *J. Parallel Distrib. Comput.*, Vol. 11, pp. 263–275, 1991.
2. D. P. Bertsekas and J. N. Tsitsiklis, *Parallel and Distributed Computation: Numerical Methods*. Englewoods Cliffs, N.J.: Prentice - Hall, 1989.
3. S. N. Bhatt, et. al. "Scattering and gathering messages in networks of processors," *IEEE Trans. Comput.*, Vol. 42, No. 8, pp. 938–949, Aug. 1993.
4. V. V. Dimakopoulos and N. J. Dimopoulos, "Leaf communications in complete trees," Technical Report ECE-95-6, University of Victoria, Oct. 1995.
5. R. Feldmann, et. al. "Optimal algorithms for dissemination of information in generalized communication modes," in *Proc. 4th PARLE, Parallel Architectures and Languages Europe*, Paris, France, June 1992, pp. 115–130.
6. P. Fraigniaud and E. Lazard, "Methods and problems of communication in usual networks," *Discrete Appl. Math.*, Vol. 53, pp. 79–133, 1994.
7. S. M. Hedetniemi, et. al. "A survey of gossiping and broadcasting in communication networks," *Networks*, Vol. 18, pp. 319–349, 1988.
8. S. L. Johnsson and C. - T. Ho, "Optimum broadcasting and personalized communication in hypercubes," *IEEE Trans. Comput.*, Vol. 38, pp. 1249–1268, 1989.
9. C. E. Leiserson, "Fat-trees: universal networks for hardware-efficient supercomputing," *IEEE Trans. Comput.*, Vol. C-34, No. 10, pp. 892–901, Oct. 1985.
10. C. E. Leiserson and et al, "The network architecture of the Connection Machine CM-5," in *Proc. 4th ACM Symp. Parall. Algor. Arch.*, June 1992, pp. 272–285.
11. Y. Saad and M. H. Schultz, "Data communications in hypercubes," *J. Parallel Distrib. Comput.*, Vol. 6, pp. 115–135, 1989.
12. Y. Saad and M. H. Schultz, "Data communications in parallel architectures," *Parallel Comput.*, Vol. 11, pp. 131–150, 1989.

A Gossip Algorithm for Bus Networks with Buses of Limited Length

Satoshi Fujita[1], Christian Laforest[2], and Stephane Perennes[3]

[1] Faculty of Engineering, Hiroshima University
Kagamiyama 1-4-1, Higashi-Hiroshima, 739, Japan
[2] LRI, Université de Paris-Sud, Bât. 490, 91405 Orsay Cedex, France
[3] Project SLOOP I3S-CNRS/INRIA/Université de Nice–Sophia Antipolis, 930 route des Colles, F–06903 Sophia-Antipolis Cedex, France

Abstract. In this paper we consider the **gossiping problem** in bus networks in which any subset of vertices (i.e., nodes) of size $l' \leq l$, where l is a given constant greater than or equal to 2, is connected by a shared bus of length l'. We call such a network **complete bus network**, since it is a generalization of complete point-to-point network K_n. The proposed algorithm finishes a gossip in complete bus networks in a short time under a feasible and practical communication model. For example, it finishes a gossip only in $1.137 \log_2 n + O(1)$ steps for $l = 3$.

1 Introduction

A **bus network** with vertex set V is a communication network in which vertices in V (representing nodes of a physical network) are connected by a set of shared buses. The number of vertices connected to a bus is called the **length** of the bus. In recent years, bus networks have received considerable attention from many researchers as a versatile communication topology for parallel processors, since it is inherently more powerful than usual point-to-point networks. In spite of a lot of important works in this area, there has not been found a general bus network topology which is recognized as the "best" one. A criterion of the goodness of communication topology is the performance for executing several communication patterns which are commonly used in parallel processing. In this paper, we evaluate the performance of bus networks by focusing on a communication pattern known as a **gossiping** in the literature. Note that it requires a lot of messages to be transmitted for finishing a gossiping.

In this paper, the communication in bus networks is assumed to proceed as follows [8]:

- Communications proceed step by step, and the transmission of a message along a bus takes one time unit (called a **step**) regardless of the length of the transmitted message and the length of the bus. (That is, we are interested here in the total number of steps taken by a gossiping.)
- A vertex can either send or receive a message to/from at most one bus at a given step, even if the vertex is connected to several buses.

- Given a bus, only one vertex connected with the bus can send a message through it at a given step. In addition, all vertices connected with the bus can simultaneously receive the transmitted message (i.e., each bus is accessed in CREW manner).

During a step, several message transmissions can occur simultaneously as long as all of the above three constraints are satisfied. Note that if all buses have length two, it coincides with the communication in point-to-point networks (i.e., graphs) under the single-port, half-duplex model, so-called "telegraph model" (surveys on communications in graphs can be found in [9, 12]).

In this paper, we focus on a special class of bus networks defined as follows.

Definition 1. **Complete bus network** $cbn(n, l)$, where $2 \leq l \leq n$, is a bus network consisting of n vertices and each bus is of length at most l. For any subset of vertices of size $l' \leq l$, there is a bus of length l' which connects all vertices in the subset.

For each $u \in V$, let $m(u)$ be the piece of information initially held by vertex u. The **gossiping problem** is the problem of disseminating $m(u)$ to all the other vertices in V, for all $u \in V$. This problem in bus networks has been studied in the literature [8, 10, 11, 13, 14]; e.g., Fujita and Yamashita [10, 11] considered the problem in *mesh-bus networks* in which all vertices are arranged on a two-dimensional array and vertices in each row and vertices in each column, respectively, are connected with a bus; Hily and Sotteau [13, 14] extended the result on two-dimensional arrays to the case of d-dimensional arrays for $d \geq 3$; and Fraigniaud and Laforest [8] studied bus networks of minimal size in which a gossip finishes in a minimum time. Some other results have been obtained in [1, 2, 6, 7] concerning the design of particular bus networks and the broadcasting operation in them.

Let $g_l(n)$ denote the minimum number of steps necessary to finish a gossip in $cbn(n, l)$ under our model[4]. In [8], it is shown that $g_n(n) \geq \lceil \log_2 n \rceil + 1$ holds for any $n \geq 2$. Since $g_l(n) \geq g_n(n)$ obviously holds (indeed, one may use only buses of length at most l), we have the following theorem.

Theorem 2. (*Lower Bound*) *For all* $2 \leq l \leq n$, $g_l(n) \geq \lceil \log_2 n \rceil + 1$.

The upper bound we will prove on $g_l(n)$ is related to some numbers τ_l that are defined as follows:

Definition 3. τ_l is the root of greatest modulus of the polynomial $X^l - X^{l-1} - \cdots - X - 1$. It has been proved that τ_l is a real and belongs to $[1, 2]$ (see [3] and [15], section 5.4.2, ex. 7).

When $l = 2$, buses of length 2 are indeed "edges" and our model is equivalent to the model so-called *half-duplex model* for point-to-point networks. In that case, a tight bound has been derived:

---

[4] $g_l(n)$ represents a fundamental lower bound on the number of steps for finishing a gossip in bus networks with n vertices and buses of length at most l under our model.

Theorem 4. $g_2(n) = \frac{1}{\log_2(\tau_2)} \log_2 n + O(1)$.

A proof of the upper bound was given by Entringer and Slater [4], and the proof of the lower bound was independently given by Krumme et al. [16], Sunderam and Winkler [18], Even and Monien [5], and Labahn and Warnke [17].

The objective of our paper is to find an exact value of $g_l(n)$ in terms of n for any $l \geq 2$. In Section 2, we propose a gossip algorithm for $cbn(n, l)$ to give an upper bound on $g_l(n)$. It is shown that this upper bound is very close to the lower bound given in Theorem 2, and we conjecture that the bound is optimal within a constant number of steps. Section 3 concludes the paper with future directions of this research.

2 An Upper Bound on $g_l(n)$

The time of our algorithm will be denoted $T_l(n)$.

2.1 When $n = 0 \pmod{l}$

For simplicity we describe first an algorithm valid when $n = ql$ for some integer q. The set of vertices in those networks can be viewed as a rectangle of q columns and l lines. Vertex of line i and column j will be consequently labeled (i, j) with $i \in Z_l$ and $j \in Z_q$. The set of vertices in line i will be denoted $L_i \ (= \{(i,j) \mid j \in Z_q\})$; in the same way, the set of vertices in column j will be denoted $C_j \ (= \{(i,j) \mid i \in Z_l\})$.

The algorithm runs in two phases.

Phase 1. A description of Phase 1 is given as follows.

Description of Phase 1 *In parallel, in each column C_j for $j \in Z_q$, perform a gossip.*

Phase 1 takes exactly $\lceil \log_2 l \rceil + 1$ steps according to the results obtained in [8].

Remark. Note that at the end of Phase 1, any vertex (i, j) in column C_j knows the whole information originally contained in column j; i.e., $m_j = \bigcup_{u \in C_j} m(u)$. For simplicity of notations we will say that *a vertex v knows the information of columns $[j, j']$, if v knows the information originally contained in the consecutives columns $j, j+1, j+2, \cdots, j'$* (remind that addition is performed modulo q).

Phase 2 The second phase is executed only when $q \geq 2$. We can refer it abusively as a "gossip between columns". In what follows, we shall denote by "time t" the time after the tth step of Phase 2, that is phase 2 starts at time 0. At the beginning of Phase 2, any vertex knows the information of at least one column. In particular, vertex (i, j) knows the information of columns $[j, j]$.

We will describe an algorithm such that at time t each vertex (i,j) of line L_i, knows the information of the $F_i(t)$ consecutive columns $[j, j+F_i(t)-1]$, where $F_i(t)$ depends only on i and t. This situation can be represented by the vector:

$$F(t) = \begin{pmatrix} F_0(t) \\ F_1(t) \\ \cdots \\ \cdots \\ F_{l-1}(t) \end{pmatrix}$$

where $F_i(t)$ will be called the **amount of information** known by line L_i.

Description of Phase 2 At time 0, to perform step 1 we choose a line L_{i_0} and any vertex (i_0, j) of this line sends its information to the vertices of the set $\{(i, j-1), i \neq i_0\}$. So at time 1, each vertex (i, j) of a line $L_i (i \neq i_0)$ will know the information of columns $[j, j+1]$. A vertex (i_0, j) will still know only the information of column j. At time 1, we choose a line $L_{i_1}(i_1 \neq i_0)$; then during step 2, a vertex (i_1, j) sends its information to $\{(i, j - F_i(1)), i \neq i_1\}$. Explicitly (i_1, j) sends its information to $(i_0, j-1)$ and to $(i, j-2)$ for $i \neq i_0, i_1$.

Let $i_t \in Z_l$ be such that $F_{i_t}(t)$ is maximum, that is $F_{i_t}(t) = max_{i \in Z_l}(F_i(t))$. Then, at step $t+1$ each vertex (i_t, j) sends its information to the vertices $\{(i, j - F_i(t)), i \neq i_t\}$.

For example (see table 1), in the case $l = 4$, at time 3 if we have chosen $i_0 = 0, i_1 = 1, i_2 = 2$, then the maximum of $F_i(3)$ is 8 which is attained for line 3; so the senders are in line 3. During step 4 a vertex $(3, j)$ sends its information (namely that of columns $[j, j+7]$) to $\{(0, j-7), (1, j-6), (2, j-4)\}$.

At time $t+1$ a vertex $(i, j), i \neq i_t$ will know all the information of columns $[j, j + F_i(t) - 1]$ (known at time t) plus that of columns $[j + F_i(t), j + F_i(t) + F_{i_t}(t) - 1]$ that is of columns $[j, j + F_i(t) + F_{i_t}(t) - 1]$. Vertices of line i_t will receive no new information. Consequently we obtain the following recurrence relations:

- $\forall i, F_i(0) = 1$;
- $F_{i_t}(t+1) = F_{i_t}(t)$; and
- for $i \neq i_t, F_i(t+1) = F_{i_t}(t) + F_i(t)$.

As during phase 2, at time t, any vertex (i, j) knows the information of $F_i(t)$ consecutive columns; phase 2 completes the gossip at the first time T for which $\forall i, F_i(T) \geq q$. Note that according to the previous description the algorithm can be implemented under our model in $cbn(n, l)$.

Lemma 5. *The time of Phase 2 is at most* $\lceil \log_{\tau_l}(q) \rceil + 1$.

Proof. Let $G(t)$ be the vector of values $F_i(t)$ which are sorted in a decreasing order. We will denote $G_i(t)$ the ith coordinate of $G(t)$, as example $max_{i \in Z_l}(F_i(t)) =$

Table 1. The first values of $F_i(t)$ for $l = 4$.

t	0	1	2	3	4	5	6	7	8
$F_0(t)$	1	1	3	7	15	15	44	100	208
$F_1(t)$	1	2	2	6	14	29	29	85	193
$F_2(t)$	1	2	4	4	12	27	56	56	164
$F_3(t)$	1	2	4	8	8	23	52	108	108

$F_{i_t}(t) = G_0(t)$. Then, $G(t)$ clearly satisfies:

$$G(0) = \begin{pmatrix} 1 \\ 1 \\ 1 \\ \cdots \\ 1 \\ 1 \end{pmatrix} \quad \text{and} \quad G(t+1) = \begin{pmatrix} G_0(t) + G_1(t) \\ G_0(t) + G_2(t) \\ G_0(t) + G_3(t) \\ \cdots \\ G_0(t) + G_{l-1}(t) \\ G_0(t) \end{pmatrix}$$

The relation between $G(t+1)$ and $G(t)$ can be represented by a usual linear recursion $G(t+1) = QG(t)$, where

$$Q = \begin{pmatrix} 1 & 1 & 0 & \cdots & 0 & 0 \\ 1 & 0 & 1 & \cdots & 0 & 0 \\ & & & \cdots & & \\ 1 & 0 & 0 & \cdots & 0 & 1 \\ 1 & 0 & 0 & \cdots & 0 & 0 \end{pmatrix}$$

One can check that the characteristic polynomial of Q is up to a sign of $X^l - X^{l-1} - X^{l-2} - \cdots - X - 1$. Consequently, $Q^l - Q^{l-1} - Q^{l-2} - \cdots - Q - I = 0$, and as for $t \geq 0$, $G(t+l) = Q^l G(t)$ we have:

$$t \geq 0, \quad G(l+t) = G(l+t-1) + G(l+t-2) + \cdots + G(t). \tag{1}$$

Note that the smallest coordinate of $G(t)$ is $G_{l-1}(t) = G_0(t-1)$, though coordinates of $G(t)$ lie in the interval $[G_0(t-1), G_0(t)]$. Consequently, we will only analyze the behavior of $G_0(t)$ value which will be denoted $M(t)$. Phase 2 will be completed at time T if and only if $M(T-1) \geq q$.

We will not derive an exact evaluation [5] of $M(t)$. Instead we just point out that $M(t) = 2^t$ for $t \leq l-1$, moreover for $l \leq t$, equation (1) implies that:

$$M(t) = M(t-1) + M(t-2) + \cdots + M(t-l) \tag{2}$$

Now note that the two sequences $M(t)$ and $U(t) = (\tau_l)^t$ are the solutions of the same linear recursion (2). As all the coefficients of the recursion are positive and as $\forall t \in \{0, 1, \cdots, l-1\}$, $M(t) = 2^t \geq (\tau_l)^t$ (recall that $\tau_l < 2$), we can claim that for any time t, $M(t)$ is greater than $(\tau_l)^t$. Consequently we have $T \leq \lceil \log_{\tau_l}(q) \rceil + 1$. □

---

[5] Actually one can shows that the generating function of $M(t)$ is $\frac{1-z^l}{1-2z+z^{l+1}}$, and prove then that $M(t) = c_l(\tau_l)^t + o(t)$ with $c_l = \frac{1-(\frac{1}{\tau_l})^l}{2-(l+1)(\frac{1}{\tau_l})^l} \tau_l > 1$.

Corollary 6. $g_l(ql) \leq T_l(ql) \leq \lceil \log_2 l \rceil + \lceil \log_{\tau_l}(q) \rceil + 2$

Proof. By summing the times of Phases 1 and 2. □

2.2 When $n \not\equiv 0 \pmod{l}$

Let $n = ql + r$ for some $1 \leq r < l$. If $q = 0$, since we have $n < l$, the gossiping can be achieved in an optimal time $\lceil \log_2(n) \rceil + 1$ (see [8]). If $q \geq 1$, we have to slightly modify the algorithm in such a way to take into account those r remaining vertices. In our algorithm, before starting Phase 1, each of those r vertices send their information to a vertex in column C_0. Note that it takes only one step since $r < l$. At that point, the whole information of V has been concentrated on the first q columns and we can apply the previous algorithm to the ql vertices. One additional step is enough for completing the whole gossip operation because at the end of Phase 2, ql ($\geq r$) vertices have known the whole information. So when $n = ql + r$, the total time to complete the gossiping is at most $T_l(ql) + 2$.

2.3 Total Time

By the previous subsections, the total execution time $T_l(n)$ of our gossip algorithm is bounded above by

$$T_l(n) \leq \log_2 l + \left(\frac{1}{\log_2 \tau_l}\right) \log_2 \left(\frac{n}{l}\right) + O(1) \tag{3}$$

It is worth noting that for any fixed l, an upper bound on the gossip time by the algorithm is asymptotically given by $T_l(n) \leq (1/\log_2 \tau_l) \log_2 n + O(1)$. For comparison, let us consider the performance of the following naive gossip algorithm: it first accumulates all the pieces of information into a vertex (i.e., an expert) in $\lceil \log_2 n \rceil$ steps, then the expert vertex broadcasts the accumulated information to all the other vertices in $\lceil \log_l n \rceil$ steps. The asymptotic behavior of the algorithm is given by $\{1 + (1/\log_2 l)\} \log_2 n$.

Table 2. Coefficients of the $\log_2 n$ term.

l	x	$1/\log_2 x$	$1 + (1/\log_2 l)$
3	1.839286755	1.137466951	1.630929753
4	1.927561975	1.056214652	1.500000000
5	1.965948237	1.025404040	1.430676558
6	1.983582843	1.012034454	1.386852807
7	1.991964197	1.005842216	1.356207187
8	1.996031180	1.002873979	1.333333333

Table 2 shows a numerical result on the asymptotic behavior of the above two algorithms; i.e., it compares the numerical approximation of the coefficient of the $\log_2 n$ term of two algorithms for small l's. Note that the performance of the naive algorithm is worse than that of an optimal algorithm for complete graphs. The readers can verify that $1/\log_2 \tau_l$, the coefficient of our algorithm, is smaller than $1 + (1/\log_2 l)$, the coefficient of a naive algorithm. According to Table 2, $\frac{1}{\log_2 \tau_l}$ is quickly converging to 1. Indeed, this fact is proved in [3] by the evaluation $\frac{1}{\log_2(\tau_l)} \sim 1 + \frac{\log_2(e)}{2^l}$, where e is the base of the exponential.

3 Concluding Remarks

In this paper, we have proposed a new gossip algorithm for complete bus networks. This implies that

$$g_l(n) \leq T_l(n) = \frac{\log_2(n)}{\log_2(\tau_l)} + O(1).$$

We conjecture that it is optimal within an additive constant number of steps, that is $g_l(n) = \frac{\log_2(n)}{\log_2(\tau_l)} + O(1)$. Note that this conjecture is proved in the case $l = 2$ (see theorem 4), but we have not been able to extend the proof in the general case. It would be interesting to prove that the bound is tight at least for its order: i.e., $g_l(n) \sim \frac{\log_2(n)}{\log_2(\tau_l)}$. This problem should be essential for analyzing the time complexity of the gossiping in bus networks. Indeed, it would gives a generic lower bound valid for any bus network with buses of limited length.

Another problem we want to solve is to construct bus networks with the smallest number of buses of length at most l, for given l, in which a gossip takes $g_l(n)$ steps. It is the problem of finding *minimum gossip bus networks* with a given parameter l. For that question one can already remark that the algorithm presented is valid for a bus network of degree $\Theta(\log n)$ obtained by considering the network containing only the buses used during the algorithm.

Acknowledgment: This work is supported in part by Scientific Research Grants-in-Aid from the Ministry of Education, Science and Culture of Japan 08680372. The two last authors are partially supported by the opération "Rumeur" of the French GDR/PRC PRS and also by the French/Canadian PICS of the CNRS.

We are grateful to J-C. Bermond and J. Yu for their advice.

References

1. J.-C. Bermond, J. Bond, and S. Djelloul. Dense bus networks of Diameter 2. In *Interconnection Networks and Mapping and Scheduling Parallel Computation*, DIMACS Workshop, Volume 21, D. F. Hsu, A. L. Rosenberg, D. Sotteau Editors, American Mathematical Society (1994) 9–18.

2. J.-C. Bermond and F.O. Ergincan. Bus Interconnection Networks. To appear in *Discrete Applied Mathematics* (1996).
3. J.-C. Bermond, P. Hell, A. L. Liestman, and J. G. Peters. Broadcasting in Bounded Degree Graphs. SIAM J. Disc. Math. **5(1)** (1992) 10–24.
4. R. C. Entringer and P. J. Slater. Gossips and telegraphs. *J. Franklin Inst.* **307** (1979) 353–360.
5. S. Even and B. Monien. On the number of rounds necessary to disseminate information. In *Proc. 1st ACM Symp. on Parallel Algorithms and Architectures*, (1989) 318–327.
6. A. Ferreira, A. Goldman vel Lejbman, and S. Song. Bus based parallel computers: a viable way for massive parallelism. *PARLE'94* (Lecture Notes in Computer Science 817), Springer Verlag (1994) 553–564.
7. A. Ferreira, A. Goldman Vel Lejbman, and S.W. Song. Broadcasting in bus interconnection networks. In *Proc. Parallel Processing: CONPAR 94 - VAPP VI* (Lecture Notes in Computer Science 854). Springer Verlag (1994) 797–807.
8. P. Fraigniaud and C. Laforest. Minimum gossip bus networks. *Networks*, **27** (1996) 239–251.
9. P. Fraigniaud and E. Lazard. Methods and Problems of Communication in Usual Networks. *Discrete Applied Mathematics* **53** (1994) 79–133.
10. S. Fujita and M. Yamashita. Optimal gossiping in mesh-bus computers. *Parallel Processing Letters* **3(4)** (1993) 357–361.
11. S. Fujita, M. Yamashita, and T. Ae. Gossiping on mesh-bus computers by packets. In *Proc. of Int'l Symp. on Parallel Architectures, Algorithms and Networks (ISPAN)* (1994) 222–229.
12. S. M. Hedetniemi, S. T. Hedetniemi, and A. L. Liestman. A survey of gossiping and broadcasting in communication networks. *Networks* **18** (1986) 319–349.
13. A. Hily and D. Sotteau. Communication in bus networks. *Parallel and Distributed Computing*, (Lecture Notes in Computer Science 805). Springer Verlag (1994) 197–206.
14. A. Hily and D. Sotteau. Gossiping in d-dimensional mesh-bus networks. *Parallel Processing Letters* **6(1)** (1996) 101–113.
15. D. E. Knuth. The Art of Computer Programming, Vol.3 / Sorting and Searching, Addison-Wesley Publishing Company (1973).
16. D. W. Krumme, G. Cybenko, and K. N. Venkataraman. Gossiping in minimal time. *SIAM J. Comput.*, **21(1)** (1992) 111–139.
17. R. Labahn and I. Warnke. Quick gossiping by multi-telegraphs. In Bodendiek and R. Henn, editors, *Topics in Combinatorics and Graph Theory* Phisica-Verlag Heidelberg (1990) 451–458.
18. V. S. Sunderam and P. Winkler. Fast information sharing in a complete network. *Discrete Applied Math.* **42** (1993) 75–86.

Worm-Hole Gossiping on Meshes*

Ben H.H. Juurlink[†]   P.S. Rao[‡]   Jop F. Sibeyn[§]

Abstract

Several algorithms for performing gossiping on one- and higher dimensional meshes are presented. As a routing model, we assume the practically important worm-hole routing. For one-dimensional arrays, we give a novel lower bound and an asymptotically optimal gossiping algorithm. For two-dimensional meshes, we present a simple algorithm composed of one-dimensional phases. For an important range of packet and mesh sizes, it gives clear improvements. The algorithm is analyzed theoretically, but, the achieved improvements are also convincingly demonstrated by simulations and by an implementation on the Paragon. For higher dimensional meshes, we give algorithms which are based on a generalized notion of a diagonal.

1 Introduction

Meshes. One of the most thoroughly investigated interconnection schemes for parallel computation is the $n \times n$ *mesh*, in which n^2 processing units, **PUs**, are connected by a two-dimensional grid of communication links. Its immediate generalizations are d-dimensional $n \times \cdots \times n$ meshes. Numerous parallel machines with mesh topologies have been built.

Gossiping. Gossiping is a fundamental communication problem. It appears in many contexts, both theoretical and practical. Gossiping is the problem in which each of the N PUs needs to send data to every other PU. Finally, all PUs must know the complete data of size $N \cdot L$. This is a very communication intensive operation.

Gossiping appears as a subroutine in many important problems. For example, if M numbers are to be sorted on N PUs, then a good approach is to select a set of m splitters [8, 5] which must be made available in every PU. This means that we have to perform a gossip in which every PU contributes m/N numbers. In this case the amount of data is small and, hence, the gossiping time can be made negligible with efficient gossiping algorithms. A second application of gossiping appears in algorithms for solving ordinary differential equations using parallel block predictor-corrector methods [9]. In each application of the block method, computations corresponding to the prediction are carried out by different PUs and these values are needed by all other PUs.

Earlier Work. A substantial amount of research has been performed on (variants of) the gossiping problem [1, 2, 7]. In some sense, we turn back to basics. Rather than to design an even more sophisticated algorithm, along the lines of [7], we present a fairly simple algorithm and show that it actually works in practice. An essential point is that we achieve an optimal trade-off between start-up and routing time. For relatively large messages, it is not enough to focus on the number of start-ups only. A non-trivial lower bound shows that our algorithms are close to optimal for *all* values of the involved parameters. On two-dimensional meshes, the information is concentrated on diagonals. For higher dimensional meshes we give an interesting generalization of the notion of a diagonal, which may be of independent interest.

---

*Part of this research was performed during a stay of the second author at the Max-Planck-Institute
[†]Dept. of Computer Science, Leiden University, P.O. Box 9512, 2300 RA Leiden, The Netherlands. Email: benj@cs.LeidenUniv.nl
[‡]Dept. of Computer Science, University of Hyderabad, Hyderabad - 500 046, India. Email: psraocs@uohyd.ernet.in
[§]Max-Planck-Institut für Informatik, Im Stadtwald, 66123 Saarbrücken, Germany. Email: jopsi@mpi-sb.mpg.de

2 Preliminaries and Lower Bounds

A d-dimensional **mesh** consists of $N = n^d$ processing units, PUs, laid out in a d-dimensional grid of side length n. Every PU is connected to each of its (at most) $2 \cdot d$ immediate neighbors by a bidirectional communication link. We assume the full-port model in which a PU can transmit data to all of its neighbors simultaneously.

For the communication we assume the much considered worm-hole routing model (see [6, 3] for some recent surveys). In this model a packet consists of **flits** and has a header which contains the necessary routing information. The other flits just follow the header. Initially all flits reside in the source PU. Finally all flits should reside in the destination PU. Furthermore, two or more flits may reside in the same PU only at the source and the destination. The reasons to consider worm-hole routing instead of the more traditional store-and-forward routing are of a practical nature. On modern MIMD computers, the time to issue a packet is considerably larger than the time needed to traverse a connection. The time to send a packet consisting of l flits over a distance of c connections is given by

$$t(d, l) = t_s + c \cdot t_d + l \cdot t_l. \tag{1}$$

We refer to t_s as the **start-up time**, t_d as the **hop time**, and t_l as the **flit-transfer time**. (1) is correct as long as the paths of various packets do not overlap. Our algorithms are overlap-free.

We start with a trivial but general lower bound.

Lemma 1 *For any network of N PUs with degree deg and diameter D, the time $T_{con}(N, deg, D)$ for concentrating all information in a single node satisfies:*

$$T_{con}(N, deg, D) \geq \max\{N/deg \cdot l \cdot t_l, D \cdot t_d, \log N / \log(deg + 1) \cdot t_s\}.$$

Of course, T_{con} immediately gives a lower bound for the gossiping problem. A stronger lower bound is given in the following theorem. The proof of this theorem is given in [4].

Theorem 1 *Let $r = t_s / (l \cdot t_l)$ where $r \leq n/e^2$. The time for gossiping on a linear array with n PUs satisfies*

$$T_{gos} = \Omega(n \cdot \ln n / \ln(n/r) \cdot l \cdot t_l).$$

3 Linear and Circular Arrays

We analyze gossiping on one-dimensional processor arrays. We assume that the time for routing a packet is given by (1), as long as the paths of the packets do not overlap. We only present the algorithms for circular arrays. With minor modifications, all of them carry on for linear arrays.

For gossiping on a circular array consisting of n PUs, there are two trivial approaches. Each of them is good in an extreme case.

1. Every PU sends a packet containing its data to the left and right. The packets are sent on for $\lfloor n/2 \rfloor$ steps.

2. Recursively concentrate the data into a selected PU. Then, reverse the process to disseminate the information to all other PUs.

Lemma 2 *If the packets consist of l flits each, then Approach 1 takes $T_1(n, l) = \lfloor n/2 \rfloor \cdot (t_s + t_d + l \cdot t_l)$ time.*

Lemma 3 *If the packets consist of l flits each, then the time consumption of Approach 2 can be estimated on $T_2(n, l) \simeq \log_3 n \cdot (2 \cdot t_s + n \cdot l \cdot t_l)$.*

Proof: During the concentration phase, the number of 'active' PUs is reduced by a factor of three in every step, the packets get three times as heavy and the distance over which the packets have to be sent increases by a factor of three. This gives $T_{conc} = \sum_{i=0}^{\log_3 n - 1}(t_s + 3^i \cdot (t_d + l \cdot t_l)) < \log_3 n \cdot t_s + n/2 \cdot (t_d + l \cdot t_l)$. In all steps of the dissemination phase, the packets consist of $n \cdot l$ flits each, giving $T_{dis} = \sum_{i=0}^{\log_3 n - 1}(t_s + 3^i \cdot t_d + l \cdot n \cdot t_l) < \log_3 n \cdot (t_s + n \cdot l \cdot t_l) + n/2 \cdot t_d$. Since t_d is of the same order as t_l, the term $n/2 \cdot t_d$ can be ignored. □

$n \setminus t_s/t_l'$	2		10		50		250	
	40		144		664		3264	
27	64		104		304		1304	
	40	(27, -)	**100**	(3, 1)	318	(3, 1)	1318	(3, 1)
	120		440		2040		10040	
81	257		319		593		1993	
	120	(81, -)	**239**	(5, 4)	594	(3, 1)	2013	(3, 1)
	364		1332		6172		30372	
243	990		1062		1422		3222	
	337	(4, 8)	**565**	(7, 7)	**1251**	(3, 2)	**3248**	(3, 1)
	1092		4004		18564		91364	
729	3667		3755		4195		6395	
	936	(10, 20)	**1377**	(13, 17)	**2707**	(7, 7)	**6264**	(4, 2)

Table 1: Comparison of the results obtained for gossiping on a circular arrays with n PUs applying Approach 1 (top), Approach 2 (middle) and CIRCGOS (bottom). The instances for which CIRCGOS is better are printed bold. Behind the results for CIRCGOS, the values of the parameters a and b for which the result was obtained are indicated. The cost unit is t_l'.

If $t_s \gg l \cdot t_l$, then for all reasonable values of n, the result of Lemma 3 cannot be improved. However, note that *any ratio $t_s/(l \cdot t_l)$ is possible*. For example, if a large sorting problem is solved on a relatively small system, then the packets consist of many flits. In that case it may even happen that $l \cdot t_l > t_s$. For such instances, we propose an approach which has features of both basic approaches.

We henceforth neglect the distance term, which is of minor importance anyway, and write $t_l' = l \cdot t_l$. The algorithm consists of three phases, and works with parameters a and b.

Algorithm CIRCGOS(a, b)

1. Concentrate n/a data in a evenly interspaced PUs, called *bridgeheads* or *concentration points*.

2. For $\lfloor a/2 \rfloor$ steps, send packets of size n/a among the concentration points in both directions, such that afterwards all data are known in every concentration point.

3. In $\lceil \log_a n - 1 \rceil$ further rounds, repeatedly increase the number of bridgeheads by a factor of a until n. The information is passed to the $a - 1$ new points between any two existing bridgeheads in $b \geq \lfloor a/2 \rfloor$ steps with packets of size $n/(2 \cdot b - a + 2)$.

In Phase 2, the packets are circulated around. The description is pleasant because of the circular structure. Notice that the algorithm becomes equal to Approach 1 for $a = n$.

Lemma 4 *The three phases of* CIRCGOS(a, b) *take*

$$T_{cg, 1} = \log_3(n/a) \cdot t_s + n/(2 \cdot a) \cdot t_l',$$
$$T_{cg, 2} = \lfloor a/2 \rfloor \cdot (t_s + n/a \cdot t_l'),$$
$$T_{cg, 3} = (\log_a n - 1) \cdot b \cdot (t_s + \lceil n/(2 \cdot b - a + 2) \rceil \cdot t_l').$$

The best choices for a and b have been found by a simple computer program. Table 1 lists some typical results. There are several interesting conclusions that can be derived:

- For realistic values of n and t_s/t_l', CIRCGOS may be several times faster than Approach 1 and Approach 2. At worst, CIRCGOS is hardly slower than Approach 2 (actually, for $a = 3$ and $b = 1$, it becomes equal to Approach 2, except that this knowledge is not exploited).

- The range of t_s/t_l' values for which CIRCGOS is the best increases with n. The best choices of a and b increase with n and decrease with t_s/t_l'.

Theorem 2 *Let $r = t_s/t'_l$ where $r < n$. The time consumption of CIRCGOS($n/r, n/r$) is given by*

$$T_{cg}(n/r, n/r) = \mathcal{O}(n \cdot \ln n / \ln(n/r) \cdot t'_l).$$

Thus, CIRCGOS($n/r, n/r$) is asymptotically optimal (cf. Theorem 1), and gives a natural continuous transition from gossiping times $\mathcal{O}(n)$, as achieved by Approach 1 for $r = \mathcal{O}(1)$, to gossiping times $\mathcal{O}(n \cdot \log n)$, as achieved by Approach 2 for $r = n$.

Corollary 1 *Let $r = t_s/t'_l$ and $0 < \epsilon < 1$. For all r, $\log n \leq r \leq n^\epsilon$, CIRCGOS is about a factor of $\log n$ faster than Approach 1 and 2.*

Proof: For $r \geq \log n$, Approach 1 and Approach 2 both take $\Omega(n \cdot \log n \cdot t'_l)$ time. On the other hand, for $r = n^\epsilon$, CIRCGOS takes $\mathcal{O}(n \cdot \log n / \log(n^{1-\epsilon}) \cdot t'_l) = \mathcal{O}(n \cdot t'_l)$ time. □

4 Two-Dimensional Arrays

The simplest idea for gossiping on two-dimensional (2D) tori is to send the packets first along the rows and then along the columns, choosing the best of Approach 1 and Approach 2 in each phase. A factor of two is gained when the packets in PUs (x, y) with $x + y$ even are colored 'white' and 'black' otherwise, and by routing the black packets orthogonally to the white ones. Let Approach i-j denote the algorithm in which first Approach i is applied and then Approach j, and let $T_{i,j}$ denote the time taken by Approach i-j. Approach 1-2 can be excluded.

Lemma 5 *The time consumption of the three gossiping algorithms is given by*

$$T_{1,1} \simeq 3/4 \cdot n \cdot t_s + n/4 \cdot (n+1) \cdot t'_l,$$
$$T_{2,1} \simeq (2 \cdot \log_3(n/2) + n/2) \cdot t_s + n/2 \cdot (\log_3(n/2) + n/2) \cdot t'_l,$$
$$T_{2,2} \simeq (4 \cdot \log_3 n - 3) \cdot t_s + (2 \cdot \log_3 n - 2) \cdot n/2 \cdot (n/2 + 1) \cdot t'_l.$$

The described approaches are competitive for many choices of n, t_s and t'_l, but a more truly 2D approach gives considerably better results for intermediate r values. The algorithm is a 2D analogue of CIRCGOS. We may concentrate on the white packets. The black packets are routed orthogonally to the white ones.

Algorithm TORGOS(a, b, x)

1. Concentrate all white packets in a concentration points of their rows; in row i, the PUs (i, j) with $(j - i) \bmod (n/a) = 0$. After this phase, each concentration point holds $n/(2 \cdot a)$ white packets.

2. Route the data in each concentration point in $\lfloor a/2 \rfloor$ steps to all other concentration points in the same row. Now every concentration point holds $n/2$ white packets.

3. Route the data in each concentration point in $\lfloor a/2 \rfloor$ steps to all other concentration points in the same column. Now every concentration point holds $a \cdot n/2$ white packets.

4. Determine suitable b, x and t such that $b^t = n/a$ and $x \geq \lfloor b/2 \rfloor$. Perform t rounds of further concentration. At the beginning of round j, $0 \leq j < t$, the concentration points contain $S_j = a \cdot b^j \cdot n/2$ white packets.

  a. Divide the data into packets of size $S_j/(2 \cdot x - b + 2)$. Route these for x steps along the rows, to $b - 1$ points equally interspaced between any two concentration points.

  b. Perform $\lfloor b/2 \rfloor$ steps of vertical routing with packets of size S_j.

Phase 1 is performed by a repeated concentration in $\log_3(n/a)$ steps. After this phase, all data are present on each of a diagonals. After Phase 3, all data are present on each section of length n/a of these diagonals. In Phase 4, new diagonals are created. First the data are copied to them (4.a), then they are made available in all sections (4.b).

n	$t_s/t_l = 8$		$t_s/t_l = 30$		$t_s/t_l = 100$		$t_s/t_l = 250$	
	351		796		2214		5252	
	360		761		2038		4774	
27	855		1053		1683		3033	
	444		606		1122		2227	
	363	(3, 9, 7)	**605**	(3, 3, 2)	**1122**	(3, 3, 1)	**2227**	(3, 3, 1)
	2146		3483		7736		16848	
	2155		3194		6501		13568	
81	10188		10474		11384		13334	
	3424		3652		4378		5934	
	2162	(3, 27, 22)	**2828**	(9, 9, 8)	**3982**	(3, 5, 3)	5934	(3, 3, 1)
	16281		20290		33048		60386	
	16335		19200		28317		47853	
243	119206		119580		120770		123320	
	29814		30108		31044		33049	
	16288	(5, 49, 54)	**17808**	(7, 35, 32)	**21101**	(3, 9, 9)	**25477**	(3, 9, 7)
	137416		149445		187718		269730	
	137819		146074		172341		228627	
729	1332416		1332878		1334348		1337498	
	266398		266758		267904		270360	
	137398	(3, 243, 211)	**141693**	(9, 81, 86)	**149888**	(15, 49, 49)	**162239**	(17, 43, 37)

Table 2: Comparison of four gossiping algorithms on $n \times n$ tori. Results are given for Approach 1-1 (top rows), Approach 2-1 (second rows) and Approach 2-2 (third rows). In the fourth rows we give the results for TORGOS(3, 3, 1). In the last rows, we give the results for TORGOS, with the corresponding optimal choices of a, b and x indicated in brackets. Where these results are better than any of the others, they are printed bold. The cost unit is t'_l.

Lemma 6 *The phases of* TORGOS(a, b, x) *take time*

$$T_{tg.1} = \log_3(n/(2 \cdot a)) \cdot t_s + n/(4 \cdot a) \cdot t'_l,$$
$$T_{tg.2} = \lfloor a/2 \rfloor \cdot (t_s + n/(2 \cdot a) \cdot t'_l),$$
$$T_{tg.3} = \lfloor a/2 \rfloor \cdot (t_s + n/2 \cdot t'_l),$$
$$T_{tg.4.a} \simeq x \cdot (\log_b(n/a) \cdot t_s + n^2/(2 \cdot (b-1) \cdot (2 \cdot x - b + 2)) \cdot t'_l),$$
$$T_{tg.4.b} = \lfloor b/2 \rfloor \cdot (\log_b(n/a) \cdot t_s + n^2/(2 \cdot (b-1)) \cdot t'_l).$$

In Table 2 we give some numerical results. Looking at the complete list of results, we see that Approach 2-1 and Approach 2-2 have become obsolete: in all cases TORGOS performs better. Only for small $r = t_s/t'_l$, it may happen that Approach 1-1 is the best of all.

5 Higher Dimensions

For the success of the two-dimensional algorithm it was essential that the packets were concentrated on diagonals at all times. The main problem in the construction of a gossiping algorithm for d-dimensional meshes is, that it is not clear how to generalize the concept of a diagonal. Once we have such a 'diagonal', we can perform an analogue of TORGOS.

The property of a two-dimensional diagonal that must be generalized, is the possibility of 'seeing' a full hyperplane, when looking along any of the coordinate axes. We will try to explain what this means. In $[0, 1] \times [0, 1] \subset \mathbf{R}^2$, when projecting its diagonal, the set $\{x + y = 1 | 0 \leq x, y \leq 1\}$, perpendicularly on the y-axes, we obtain the set $0 \times [0, 1]$. When projecting on the x-axes, we obtain $[0, 1] \times 0$. For algorithm TORGOS, this means that the information from diagonals in adjacent submeshes can be copied without problem onto each other. Not only in one direction, but in *both* directions. This requirement of problem-free copying between diagonals in adjacent submeshes along all coordinate axes leads us to the following:

Definition 1 *A subset of a d-dimensional cube is called a d-dimensional diagonal, if the following two conditions are satisfied:*

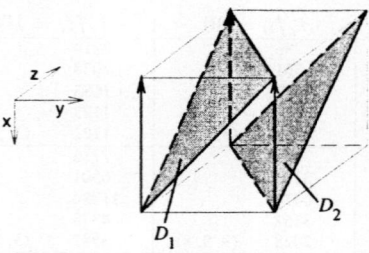

Figure 1: Diagonal of the unit cube; the projection along the x-axis is the set $0 \times [0,1] \times [0,1]$.

1. *The perpendicular projection of the diagonal of a subcube onto any of the bounding hyperplanes of this cube is surjective.*

2. *The perpendicular projection of the diagonal of a subcube onto any of the bounding hyperplanes of this cube is injective except for subsets of total measure zero.*

For $i = 1, 2, \ldots, d-1$, let $\mathcal{D}_i = \{x_0, x_1, \ldots, x_{d-1} | x_0 + x_1 + \cdots + x_{d-1} = i\}$. For the **unit-cube** $I^d = [0,1] \times \cdots \times [0,1]$, the union of the intersections of the $\mathcal{D}_i$-hyperplanes with I^d gives a diagonal:

Lemma 7 *A diagonal of I^d is given by $\bigcup_{i=1}^{d-1} \mathcal{D}_i \cap I^d$.*

Proof: As the D_i are completely symmetric, we can concentrate on the projection along the x_0-axis. Denote the projection of a set S along the x_0-axis by $\Pi_0(S)$. Then, for all $1 \leq i \leq d$,

$$\Pi_0(\mathcal{D}_i) = \{x_0, x_1, \ldots, x_{d-1} | x_0 = 0, i-1 \leq x_1 + \cdots + x_{d-1} \leq i\}. \tag{2}$$

So, for any $i < j$,

$$\Pi_0(\mathcal{D}_i) \cap \Pi_0(\mathcal{D}_j) = \begin{cases} \{x_0, x_1, \ldots, x_{d-1} | x_0 = 0, x_1 + \cdots x_{d-1} = i\} & \text{if } j = i+1, \\ \emptyset & \text{if } j > i+1. \end{cases}$$

Hence, the projection is almost injective, as required by point 2 of Definition 1. On the other hand, (2) gives that

$$\bigcup_{i=1}^{d-1} \Pi_0(\mathcal{D}_i) = \bigcup_{i=1}^{d-1} \{x_0, x_1, \ldots, x_{d-1} | x_0 = 0, i-1 \leq x_1 + \cdots + x_{d-1} \leq i\}$$
$$= \{x_0, x_1, \ldots, x_{d-1} | x_0 = 0, 0 \leq x_1 + \cdots + x_{d-1} \leq d-1\}$$

Hence, the projection is surjective as well. □

So, we successfully defined d-dimensional diagonals. The reader is advised to obtain a full understanding of the case $d = 3$, as illustrated in Figure 1. For us it was helpful to construct a model of paper (cardboard would have been better). Such a model makes it easy to convince oneself that the required property that looking along a coordinate axis indeed gives a full but non-overlapping view of the hyperplanes. Though we are not aware of any result in this direction, we are not sure that we are the first to define this concept. Still, we are very pleased with the utmost simplicity of the defined diagonal and the elegance of the proof of Lemma 7.

We now describe the algorithm for d-dimensional tori. Each packet is given a color from $\{0, 1, \ldots, d-1\}$; the packets in PU $(x_0, x_1, \ldots, x_{d-1})$ are given color $(\sum_i x_i) \bmod d$. In this way, there are n/d packets with the same color on any 1-dimensional submesh. The packets of each color are treated independently by d orthogonally operating gossiping procedures.

Algorithm CUBGOS(n, d)

1. In every PU, compute the nearest intersection of its row with the set of diagonals of all $n/3 \times \cdots \times n/3$ subtori. In each row, concentrate all color 0 packets in $\log_3 n - 1$ steps in the three computed concentration points. Each concentration point now holds $n/(3 \cdot d)$ color 0 packets.

2. Perform d steps of data spreading: in Step i, $0 \leq i \leq d - 1$, the color 0 data residing in a concentration point are routed along axis i to both other concentration points on this same axis. After Step i, every concentration point holds $n/d \cdot 3^i$ color 0 data.

3. Perform $\log_3 n - 1$ rounds of further concentration. At the beginning of round j, $1 \leq j \leq \log_3 n - 1$, the concentration points all hold $n/d \cdot 3^{j \cdot (d-1)}$ color 0 data. All data are known within every $n/3^j \times \cdots \times n/3^j$ subtorus. In round j, two diagonals are added between any two existing diagonals. Then we perform

   a. Route a copy of the color 0 data in each direction along axis 0, from every PU on an old diagonal to the PUs on the two adjacent new diagonals.

   b. Perform $d - 1$ steps of data spreading: in Step i, $1 \leq i \leq d - 1$, the color 0 data residing in a concentration point are routed along axis i to the concentration points on adjacent diagonals. After Step i, every concentration point holds $n/d \cdot 3^{j \cdot (d-1)+i}$ color 0 data.

Notice that the situation after Phase 1 is similar to the situation after Phase 3.a, and that Phase 2 is analogous to Phase 3.b. For $d = 2$, CUBGOS is identical to TORGOS(3, 3, 1).

Lemma 8 *The phases of* CUBGOS(n, d) *take time*

$$T_{cg.1} = (\log_3 n - 1) \cdot t_s + n/(6 \cdot d) \cdot t'_l,$$
$$T_{cg.3.a} = (\log_3 n - 1) \cdot t_s + n^d/(d \cdot (3^{d-1} - 1)) \cdot t'_l,$$
$$T_{cg.2} + T_{cg.3.b} = \log_3 n \cdot (d - 1) \cdot t_s + n^d/(2 \cdot d \cdot (1 - 1/3^{d-1})) \cdot t'_l.$$

Proof: We consider the last equation. Clearly, in Phase 2 and Phase 3.b, $\log_3 n$ rounds of $d - 1$ steps each are performed. In Step i, $0 \leq i \leq d - 1$, of Round j, $0 \leq j \leq \log_3 n - 1$ (where Round 0 corresponds to Phase 2), the packets have weight $n/d \cdot 3^{j \cdot (d-1)+i-1}$. Summing over i and j and using the estimate $\sum_{i=1}^{d-1} 3^{i-1} \leq 3^{d-1}/2$, gives the result. □

Adding all important contributions together gives

Theorem 3 *The time consumption of* CUBGOS *is given by*

$$(d + 1) \cdot \log_3 n \cdot t_s + \frac{(3^{d-1} + 2) \cdot n^d}{2 \cdot d \cdot (3^{d-1} - 1)} \cdot t'_l.$$

This is a very strong and general result. The algorithm is close to optimal for all n, t_s/t'_l and $d \geq 2$:

Corollary 2 *For gossiping on d-dimensional tori,* CUBGOS *is* $\max\{1 + 1/d, (1 + 2/3^{d-1})/(1 - 1/3^{d-1})\}$-*optimal.*

Proof: Because, for given d and n, $d \cdot \log_3 n \cdot t_s + n^d/(2 \cdot d) \cdot t'_l$, is a trivial lower bound for just concentrating all data in a single PU, the worst performance ratio is the maximum over all $r = t_s/t'_l > 0$ of

$$\frac{(d + 1) \cdot \log_3 n \cdot r + (1 + 2/3^{d-1})/(1 - 1/3^{d-1}) \cdot n^d/(2 \cdot d)}{d \cdot \log_3 n \cdot r + n^d/(2 \cdot d)}.$$

Differentiating for r gives that the extremal values are assumed for $r = 0$ and $r = \infty$. Substituting these values gives the stated result. □

For $d = 2$, we have 5/2-optimality, and for large d, CUBGOS almost achieves one-optimality.

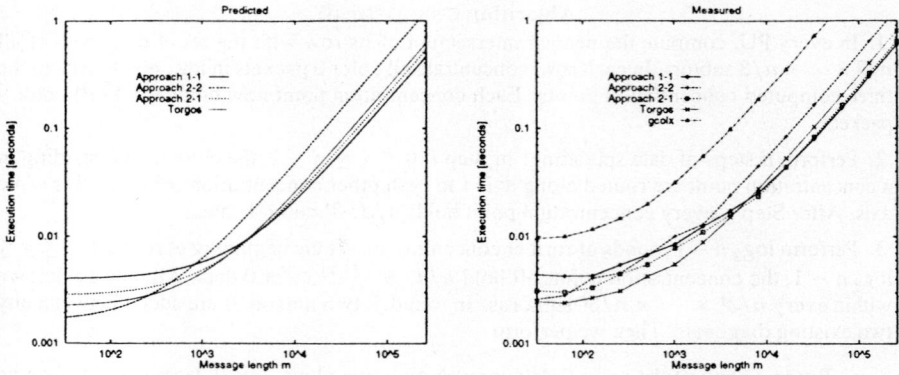

Figure 2: Predicted and measured execution times of the four one-dimensional gossiping algorithms on a 9×9 mesh.

6 Experiments

In this section, we present experimental results collected on an Intel Paragon XP/S 10 configuration of 9 nodes high and 9 nodes wide. A fuller account of the experimental data presented in this section is provided in [4].

In total eight algorithms were implemented: four one-dimensional (1D) gossiping algorithms and four two-dimensional (2D) variations. The 1D algorithms are obtained by first performing Approach 1 or 2 in the rows, then in the columns. The 2D variants divide the packet initially residing in each PU into a white packet and a black packet, and route the white packets orthogonally to the black ones.

Note that because the experimental platform is a mesh, not a torus, the runtime analysis slightly changes. For example, the time taken by Approach 1-1 for the 1D case is given by $T_{1,1} = (n-1) \cdot (t_s + t'_l) + (n-1) \cdot (t_s + n \cdot t'_l)$. Furthermore, it is necessary to refine the performance model used in the previous sections. In particular, we need to distinguish between several basic steps. The reason is that in some steps, a node needs to send or receive one message, whereas in others it may be required to send or receive multiple messages, introducing bus conflicts. For example, a *1D-send* step in which a node sends a message to its east or south neighbor (but not simultaneously), takes about $2.5 \cdot 10^{-4} + 2.2 \cdot 10^{-8} \cdot m$ seconds, where m is the message length in bytes. On the other hand, a *2D-concentrate* step in which a node receives data from all its neighbors simultaneously, requires approximately $3.0 \cdot 10^{-4} + 4.8 \cdot 10^{-8} \cdot m$ seconds.

Figure 2 plots the measured and predicted performance of the 1D implementations. The curves on the left show the predicted execution times. The curves on the right plot the measured times. For large messages, we find that the implementations run within 20% of the expected time. The error is probably due to the fact that processors operate asynchronously. However, the relative performance is quite accurately predicted.

The model predicts that for very small messages Approach 2-2 is the fastest. However, this is not observed in practice. For messages up to 4 KB, TORGOS(3, 3, 1) is faster than any other algorithm. For very long vectors (> 16 KB), Approach 1-1 is the best and Approach 2-2 performs significantly worse than the other approaches, which is in agreement with the predictions. For messages of moderate size (4–16 KB), Approach 2-1 is faster.

Figure 2 also compares the performance of our algorithms with the performance of an implementation that uses the global communication routine gcolx supplied by Intel. Clearly, our implementations are significantly faster than the gcolx communication routine. For example, for messages of 32 KB, the gcolx routine requires 716 milliseconds, while Approach 1-1 requires only 79 milliseconds.

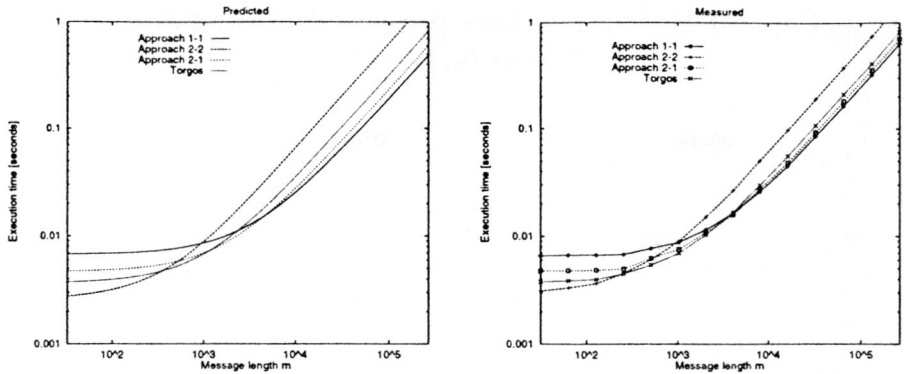

Figure 3: Predicted and measured execution times of the two-dimensional gossiping algorithms on a 9×9 mesh.

The predicted and measured execution times of the 2D variants are shown in Figure 3. As expected, these implementations do not outperform the 1D algorithms due to bus conflicts. Under ideal circumstances, the time by each 1D basic step and its corresponding 2D basic step would be about the same, but this is not observed in practice.

7 Conclusion

We presented gossiping algorithms for meshes of arbitrary dimensions. We optimized the trade-off between contributions due to start-ups and those due to the bounded capacity of the connections. This enabled us to reduce the time for gossiping in theory *and* practice for an important range of the involved parameters.

Acknowledgment. Computational support was provided by KFA Jülich, Germany.

References

[1] Barnett, M., R. Littlefield, D.G. Payne, R. van de Geijn, 'Global Combine on Mesh Architectures with Wormhole Routing,' *Proc. 7th IPPS*, pp. 13–16, IEEE, 1993.

[2] Fraignaud, P., J.G. Peters, 'Structured Communication in Torus Networks,' *Proc. 28th Hawai Conference on System Science*, 1995.

[3] Huang, Y., Ph. K. McKinley, 'An Adaptive Global Reduction Algorithm for Wormhole-Routed 2D Mesh Networks,' *Proc. 7th SPDP*, IEEE, 1995.

[4] Juurlink, B., P.S. Rao, J.F. Sibeyn, 'Gossiping on Meshes and Tori,' *Techn. Rep. MPI-I-96-1018*, Max-Planck Institut für Informatik, Saarbrücken, Germany, 1996.

[5] Kaufmann, M., S. Rajasekaran, J.F. Sibeyn, 'Matching the Bisection Bound for Routing and Sorting on the Mesh,' *Proc. 4th SPAA*, pp. 31–40, ACM, 1992.

[6] Ni, L.M., Ph.K. McKinley, 'A Survey of Wormhole Routing Techniques in Direct Networks,' *IEEE Computer*, 26(2), pp. 62–76, 1993.

[7] Peters, J.G., M. Syska, 'Circuit-Switched Broadcasting in Torus Networks,' *IEEE Transactions on Parallel and Distributed Systems*, to appear.

[8] Reif, J., L.G. Valiant, 'A Logarithmic Time Sort for Linear Size Networks,' *Journal of the ACM*, 34(1), pp. 68–76, 1987.

[9] Rao, P.S., Mouney, G. 'Data Communications in Parallel Block Predictor-Corrector Methods for solving ODEs,' *Techn. Rep.*, LAAS-CNRS, France, 1995.

Circuit-Switched Gossiping in 3-Dimensional Torus Networks

Olivier Delmas[1] and Stéphane Perennes[1]

SLOOP Project (CNRS-INRIA-UNSA)
Laboratoire I3S - CNRS URA 1376
930 Route des Colles, B.P. 145 - 06903 Sophia Antipolis Cedex
E-mail: {delmas, sp}@unice.fr

Abstract. In this paper we describe an efficient gossiping algorithm for short messages into the 3−dimensional torus networks (wrap-around or toroidal meshes) that uses synchronous circuit-switched routing. The algorithm is based on a recursive decomposition of a torus. It requires an optimal number of rounds and a quasi-optimal number of intermediate switch settings to gossip in a $7^i \times 7^i \times 7^i$ torus network.

1 Introduction

Distributed memory multicomputer architectures in which the processors communicate by exchanging messages over an interconnection network are very useful techniques for achieving cost-effective high-performance computing. At present the "circuit-switched" like routing (wormhole, direct connect, virtual cut-through, ...) is used in many recent multicomputer systems such as the Intel Paragon, IBM SP2, Cray-T3D or more recently the new Cray-T3E.

In this paper, we study the circuit-switched gossiping in 3−dimensional torus networks. After the description of the model of communications, we recall some classical definitions used in this study. In section 4 we establish a new non-trivial lower bound for the number of rounds from a circuit-switched gossiping protocol. In the last section we present a new circuit-switched gossiping algorithm for 3−dimensional torus networks which uses linear cost model. We prove that our algorithm is optimal in terms of number of rounds and quasi-optimal in terms of number of intermediate switch settings. This algorithm gives an efficient protocol when the messages are short or when the time to initiate a message transmission is much larger than the unit propagation time of a message along a link. This is the situation in many current multiprocessor networks.

2 Models of communication

In this paper we will consider the circuit-switched routing model. We will use the linear cost model in which the transmission time for a message of length L to be sent at distance d is $\alpha + d\delta + L\tau$, where α is the time to initiate a new message transmission, δ is the time to switch an intermediate node, and $1/\tau$

is the bandwidth of the communication links. We will use the all-port model of communication in which a processor can use all of its communications links simultaneously. We also assume that the communication links are full-duplex so that messages can travel in both directions simultaneously. Finally, we assume that each node has an initial distinct message, but all these messages have the same length L and we allow messages to be concatenated with negligible cost.

3 Definitions

In this article, $\mathbb{Z}_q$ will denote the set of integers modulo q. G will denote a **digraph** of order N with vertex set $V(G)$ and arc set $A(G)$. The **distance** $d_G(x,y)$ will denote the length of the shortest dipath from a vertex x to a vertex y. $D(G)$ will denote the **diameter** of a digraph G (i.e. $D(G) = max_{(x,y) \in V^2(G)} d_G(x,y)$). In a symmetric digraph G, $\Delta(G)$ (or shortly Δ) will denote **maximum in-degree** of G, that is the maximum over the in-degrees of all vertices $V(G)$.

Definition 1 [5]. The k-**dimensional torus** is the cartesian sum of k symmetric circuits of orders $p_1, p_2, \ldots, p_k$ and is denoted by $TM(p_1, p_2, \ldots, p_k) = C_{p_1} \Box C_{p_2} \Box \cdots \Box C_{p_k}$, where C_{p_i} denote the symmetric circuit of order p_i.

Remark. When $p_1 = p_2 = \cdots = p_k$, we will use the abbreviated notation $TM(p)^k$, We will assume that $p \geq 3$.

Definition 2. The total time necessary to achieve a gossiping protocol in a digraph G will be denoted by $g(G) = g_\alpha(G)\alpha + g_\delta(G)\delta + g_\tau(G)\tau$ where $g_\alpha(G)$ is the number of rounds, $g_\delta(G)$ the sum of the maximum communication distances of each couple of processors implicated in each round of the gossiping protocol and $g_\tau(G)$ measure the flow of information.

As said before, here we consider only short messages (or equivalently suppose $\tau \ll \alpha$ and $\tau \ll \delta$). So we are mainly interested in determining the optimal $g_\alpha(G)$ and $g_\delta(G)$. For $g_\delta(G)$ a trivial lower-bound is the diameter $D(G)$. In the next section we give a new non-trivial lower-bound for $g_\alpha(G)$.

4 Lower bounds

First let $\pi(G)$ be the **arc-forwarding index of the digraph** G (see [2]). For any digraph we have establised the following theorem.

Theorem 3. *Let G be a digraph with maximum degree Δ and order N and $t_0 = \lceil \log_{\Delta+1}(N) \rceil$. If $g_\alpha(G) \leq 2t_0$ then $g_\alpha(G) \geq t_0 + \log_{\Delta+1}(\frac{\pi(G)}{N}) - O(\log_{\Delta+1} \log_{\Delta+1}(N))$.*

The idea of the proof is based on a precise enumeration of the load of dipaths which can be used in a gossiping protocol. This notion is similar to the arc-forwarding index which uses the load of the arcs. Now, with this theorem we are able to state the following corollary.

Corollary 4. *Given a gossip protocol in the digraph $TM(n)^k$, the number of rounds necessary to achieve this protocol is $g_\alpha(G) \geq (k+1)\log_{2k+1}(n) - O(\log_{2k+1}\log_{2k+1}(n))$.*

Proof. This corollary is correct as in [2] it has been proved that $\pi(TM(n)^k) = \frac{n^{k-1}}{2}\lfloor\frac{n^2}{4}\rfloor$ and in [1] it has been shown that the total number of rounds to achieve a broadcasting protocol in the $TM(n)^k$ digraph is t_0. Then a trivial gossiping protocol will be the concatenation of 2 broadcasting protocols and $g_\alpha(G) \leq 2t_0$. Therefore the result follows immediatly.

5 Gossiping in the 3−dimensional torus $TM(7^i)^3$

5.1 Case of $TM(7)^3$

The idea of this section come from the original study of J.G. Peters and M. Syska [4] for the circuit-switched broadcast in the 2-dimensional torus. Here G denote the $TM(7)^3$ symmetric digraph for which we have established the following proposition.

Proposition 5. *There exists a gossiping protocol on the symmetric digraph $G = TM(7)^3$ with time $g(G) = 4\alpha + 12\delta + (1 + 7 + 7^2 + 7^3)L\tau$.*

The proof is based on the description of the gossiping protocol. But before describing it, we need some additional notations and definitions.

We will consider the vertices of G as elements of the 3−dimensional vector-space $\mathbb{Z}_7^3$, with canonical base $\{e_1, e_2, e_3\}$ [1]. If M is a 3−dimensional matrix and U is a set of vectors, MU will denote the image of U by $M : \{Mx \mid x \in U\}$. The sum of two sets of vectors U_1 and U_2 will be $U_1 + U_2 = \{x \mid x = u_1 + u_2, u_1 \in U_1, u_2 \in U_2\}$. We will denote by B_1 the set $\{e_1, e_2, e_3, 0, -e_1, -e_2, -e_3\}$ of vectors whose norm is less than or equal to 1. Note that B_1 is the sphere of radius 1 centered at zero of $\mathbb{Z}_7^3$ for the Lee distance (see [3]). $x + B_1$ is the sphere of radius 1 centered at x and it also contains the neighbours of x in G union x.

Definition 6 [3]. *The code $\mathcal{C}$ is the set of vertices such that $\mathcal{C} = \{(x_1, x_2, x_3) \in \mathbb{Z}_7^3 \mid x_1 + 2x_2 + 3x_3 = 0\}$.*

Remark [3]. $\mathcal{C}$ is a linear code of length 3 defined over $\mathbb{Z}_7$. As $\mathbb{Z}_7^3 = \mathcal{C} + B_1$, then $\mathcal{C}$ is a perfect Lee code. Note that $\mathcal{C}$ has 49 elements like $(0, 0, 0), (2, -3, -1), \cdots$

To describe the gossiping protocol of proposition 5 we introduce an additional notation.

Let x be a vertex of a digraph G and $\mathcal{A} \subset V(G)$. The notation $x \rightarrow \mathcal{A}$ *(resp. $x \leftarrow \mathcal{A}$) is used when the vertex x sends its message towards all the vertices of $\mathcal{A}$ (resp. all the vertices of $\mathcal{A}$ send their own message towards the vertex x).*

---

[1] Vertex $(x_1 e_1 + x_2 e_2 + x_3 e_3)$ will be denoted as the vector $\begin{pmatrix} x_1 \\ x_2 \\ x_3 \end{pmatrix}$.

We have found (see [1]) a 3–dimensional matrix M_0 which performs the following algorithm as a gossiping protocol.
Begin ─────────── *Gossiping Algorithm* ─────────── *in* $TM(7)^3$.
- **Round 1 - Concentration:** $\forall x \in \mathcal{C}$, $x \leftarrow \{x + B_1\}$ .... [$Cost: \alpha + \delta + L\tau$].
- **Step 2 - Gossiping between the vertices of** $\mathcal{C}$
  - **Round 2-a:** $\forall x \in \mathcal{C}$, $x \rightarrow \{x + M_0 B_1\}$ .... [$Cost: \alpha + 5\delta + 7L\tau$].
  - **Round 2-b:** $\forall x \in \mathcal{C}$, $x \rightarrow \{x + M_0^2 B_1\}$ .... [$Cost: \alpha + 5\delta + 7^2 L\tau$].
- **Round 3 - Final broadcasting:** $\forall x \in \mathcal{C}$, $x \rightarrow \{x + B_1\}$. [$Cost: \alpha + \delta + 7^3 L\tau$].

End ─────────── *Gossiping Algorithm* ─────────── *in* $TM(7)^3$.

With a precise analyse of the algorithm we have been able to exhibit a set of dipaths in G realizing each round of communication of the algorithm.

5.2 Generalization for torus $TM(7^i)^3$

Here G_i denotes the symmetric digraph $TM(7^i)^3$. We have generalized the previous result to the $TM(7^i)^3$ torus digraph.

Proposition 7. *There exists a gossiping protocol on the symmetric digraph* $G_i = TM(7^i)^3$ *with time* $g(G_i) = 4i\alpha + \frac{12}{9}D(G_i)\delta + [\frac{57}{49}(7^{i-1} - 1) + \frac{7^3}{7^{2i}} - \frac{1}{7^{3i}}]\frac{NL}{6}\tau$.

The main idea of this section is to apply recursively the gossiping protocol designed for the torus $TM(7)^3$ to the torus $TM(7^i)^3$. For this we use the **code** $\mathcal{C}_i$ which is the subset of the vertices of G_i defined as $\mathcal{C}_i = \{(x_1, x_2, x_3) \in \mathbb{Z}_{7^i}^3 | x_1 + 2x_2 + 3x_3 \equiv 0 \pmod{7}\}$. The recursion is possible because this code is once again a perfect code for the Lee distance. Indeed, spheres of radius 1 centered at each vertex of the code $\mathcal{C}_i$ cover completely the digraph G_i. That is $V(G_i) = \mathbb{Z}_{7^i}^3 = B_1 + \mathcal{C}_i$. The code $\mathcal{C}_i$ has 7^{3i-1} elements.

Acknowledgments

The authors are grateful to J-C. Bermond, M. Syska and J. Yu for helpful discussions and remarks.

References

1. O. Delmas and S. Perennes. Diffusion en mode commutation de circuits. In *RenPar'8*, pages 53–56, Bordeaux, France, 20-24 May 1996. Edition spéciale, présentation des activités du GDR-PRC Parallélisme, Réseaux et Systèmes, Richard Castanet and Jean Roman.
2. M.C. Heydemann, J.C. Meyer, and D. Sotteau. On forwarding indices of networks. *Discrete Applied Mathematics*, 23:103–123, 1989.
3. F.J. MacWilliams and N.J.A. Sloane. *The theory of Error-Correcting Codes*. North-Holland, 1977.
4. J.G. Peters and M. Syska. Circuit-switched broadcasting in torus networks. *IEEE Transactions on Parallel and Distributed Systems*, 7(3):246–255, March 1996.
5. Jean de Rumeur. *Communication dans les réseaux de processeurs*. Collection Etudes et Recherches en Informatique. Masson, 1994. (English version to appear).

Workshop 03

Automatic Parallelization and High Performance Compilers

Automatic Parallelization and High Performance Compilers

Christian Lengauer

Fakultät für Mathematik und Informatik
Universität Passau, D–94030 Passau, Germany
email: lengauer@fmi.uni-passau.de
WWW: http://www.uni-passau.de/~lengauer

1 The Call for Papers

The call for papers for this workshop covered all topics concerning the modelling of programs for an automatic parallelization and the production of parallel programs using high-performance compilers. Languages explicitly mentioned were the parallel versions of FORTRAN and C, but also of Haskell and ML. There are commonalities with other Euro-Par workshops (notably #05, #12, and #17) but, in contrast to them, workshop #03 focussed on issues of compilation and generation of efficient code. At the time of the call, we were hoping to address the following issues:

1. Data dependence analysis
2. Parallelization models & methods
3. Optimization algorithms & methods
4. Loop parallelization
5. Static vs. dynamic parallelization
6. Software Production
7. High performance languages
8. Generating efficient parallel code
9. Generating efficient communications
10. Allocation (processor, memory,...)
11. Performance studies

2 The Statistics

Of the 27 submissions 7 were accepted as regular papers, 8 as short papers, and 12 were rejected. One of the regular papers was moved to workshop #05.

3 The Topics

A brief review of the titles should convince the reader that the workshop represented the desired topics in fair balance. The only topics not addressed explicitly are at the application end: software production and performance studies. This is probably not a random occurrence but indicative of a current focus on methods in the community which submitted to the workshop. Also, there were several other workshops dedicated to applications.

One other notable lack of representation is that of the functional programming community (only representative: ALPHA). We were hoping to make some progress towards unifying—or at least comparing—the work on automatic parallelization in the imperative and the functional world. My suspicion is that there are more commonalities than meet the eye. After all, parallelization is based on dependence graphs, and functional programs might be viewed as representations of slightly sparser dependence graphs than imperative programs (no control or anti dependences). Maybe the focus on recursive data structures in functional programming languages in place of the array in imperative programming languages is the culprit—how to analyze the dependences in recursive data structures statically seems to be an unsolved problem. Another contributing factor may be that the workshop committee reflected a bias towards imperative programming.

Looking at the imperative world (since this turns out to be the world represented in the workshop), it can be said that the technology of the generation, analysis and optimization of parallel programs is clearly maturing. On the one hand, parallelization methods which started out rather restrictive are being generalized to cope with an increasing range of programs (like programs with unpredictable data dependences), and new methods are being added (like commutativity analysis). On the other hand, optimization methods which had originally been developed for sequential programs are being generalized for parallel programs (like dead code elimination and memory allocation). Also, methodologists and compiler builders are increasingly taking note of each other [1].

The call for papers mentioned one other aim: to foster a dialogue between researchers in abstract parallel models and researchers in concrete parallel architectures. This dialogue must go on at the conference level, not at the workshop level, since it involves also the attendants of workshops #12 and #13.

References

1. C. Lengauer, L. Thiele, M. Wolfe, and H. Zima, editors. Loop parallelization. Technical Report 142, Schloß Dagstuhl, April 1996.

On the Optimality of Allen and Kennedy's Algorithm for Parallelism Extraction in Nested Loops

Alain Darte and Frédéric Vivien*

LIP, URA CNRS 1398, ENS-Lyon, F - 69364 LYON Cedex 07, France
e-mail: [Alain.Darte,Frederic.Vivien]@lip.ens-lyon.fr

Abstract. We explore the link between dependence abstractions and maximal parallelism extraction in nested loops. Our goal is to find, for each dependence abstraction, the minimal transformations needed for maximal parallelism extraction. The result of this paper is that Allen and Kennedy's algorithm is optimal when dependences are approximated by dependence levels. This means that even the most sophisticated algorithm cannot detect more parallelism than found by Allen and Kennedy's algorithm, as long as dependence level is the only information available.

1 Introduction

Many automatic loop parallelization techniques have been introduced over the last 30 years, starting from the early work of Karp, Miller and Winograd [12] in 1967 who studied the structure of computations in repetitive codes called systems of uniform recurrence equations. This work defined the foundation of today's loop compilation techniques. It has been widely exploited and extended in the systolic array community, as well as in the compiler-parallelizer community: Lamport [14] proposed a parallel scheme - the hyperplane method - in 1974, then several loop transformations were introduced (loop distribution/fusion, loop skewing, loop reversal, loop interchange, ...) for vectorizing computations, maximizing parallelism, maximizing locality and/or minimizing synchronizations. These techniques have been used as basic tools for optimizing algorithms, the most two famous being certainly Allen and Kennedy's algorithm [1], designed at Rice in the Fortran D compiler, and Wolf and Lam's algorithm [18], designed at Stanford in the SUIF compiler.

At the same time, dependence analysis has been developed so as to provide sufficient information for checking the legality of these loop transformations, in the sense that they do not change the final result of the program. Different abstractions of dependences have been defined (among others dependence distance [16], dependence level [1], dependence direction vector [19], dependence polyhedron or cone [11], ...), and more and more accurate tests for dependence analysis have been designed (among others Banerjee's tests [2], I test [13], Δ test [9], λ test [15], PIP test [7], PIPS test [10], Omega test [17], ...).

---
* Supported by the CNRS-INRIA project *ReMaP*.

In general, dependence abstractions and dependence tests have been introduced with some particular loop transformations in mind. For example, the dependence level was designed for Allen and Kennedy's algorithm, whereas the PIP test is the main tool for Feautrier's method for array expansion [7] and parallelism extraction by affine schedules [8]. However, very few authors have studied, in a general manner, the links between both theories, dependence analysis and loop restructurations, and have tried to answer the following two dual questions:

- What is the minimal dependence abstraction needed for checking the legality of a given transformation?
- What is the simplest algorithm that exploits all information provided by a given dependence abstraction at best?

With the answer to the first question, we can adapt the dependence analysis to the parallelization algorithm, and avoid implementing an expensive dependence test if it is not needed. This question has been deeply studied in Yang's thesis [21], and summarized in Yang, Ancourt and Irigoin's paper [20].

Conversely, with the answer to the second question, we can adapt the parallelization algorithm to the dependence analysis, and avoid using an expensive parallelization algorithm, if a simpler algorithm extracts the same degree of parallelism. This question has been addressed by Darte and Vivien in [6] for dependence abstractions based on a polyhedral approximation of distance vectors.

Completing this previous work, we propose a more precise study of the link between *dependence abstractions* and *parallelism extraction* in the particular case of *dependence levels*. Our main result is that, in this context, Allen and Kennedy's parallelization algorithm is optimal for parallelism extraction, which means that even the most sophisticated algorithm cannot detect more parallel loops than Allen and Kennedy's algorithm does, as long as dependence level is the only information available. In other words, loop distribution is sufficient for detecting maximal parallelism in dependence graphs with dependence levels. There is no need to use more complicated transformations such as loop interchange, loop skewing, or any other transformations that could be invented, because there is an intrinsic limitation in the dependence level abstraction that prevents detecting more parallelism.

The paper is organized as follows. In Section 2, we explain what we call maximal parallelism extraction for a given dependence abstraction. In Section 3, we recall the definition of dependence levels and we present Allen and Kennedy's algorithm in its simplest form, which is sufficient for what we want to prove. In Section 4, we build a set of loops that are equivalent to the loops to be parallelized, in the sense that they have the same dependence graph. These loops contain exactly the degree of parallelism found by Allen and Kennedy's algorithm which proves the optimality (however, as the proof is quite long, we refer to [5] for an extended version). Finally, Section 5 summarizes the paper.

2 Theoretical framework

2.1 Notations

The notations used in the following sections are:

- $f(N) = O(N)$ if $\exists\, k > 0$ such that $f(N) \leq kN$ for all sufficiently large N.
- $f(N) = \Omega(N)$ if $\exists\, k > 0$ such that $f(N) \geq kN$ for all sufficiently large N.
- $f(N) = \Theta(N)$ if $f(N) = O(N)$ and $f(N) = \Omega(N)$.
- If X is a finite set, $|X|$ denotes the number of elements in X.
- $G = (V, E)$ denotes a directed graph with vertices V and edges E.
- $e = (x, y)$ denotes an edge from vertex x to vertex y.

2.2 Dependence graphs

We restrict to the case of perfectly nested loops for making the discussion simpler. The structure of perfectly nested loops can be captured by an ordered set of statements $S_1, \ldots, S_s$ (with S_i textually before S_j if $i < j$) and an iteration domain $\mathcal{D} \subset \mathbb{Z}^n$ that describes the values of the loops counters (n is the number of nested loops). Given a statement S, to each n-dimensional vector $I \in \mathcal{D}$ corresponds a particular execution (called instance) of S, denoted by $S(I)$.

Dependences between instances of statements define the **expanded dependence graph (EDG)**. The vertices of the EDG are all possible instances $\{(S_i, I) \mid 1 \leq i \leq s \text{ and } I \in \mathcal{D}\}$. There is an edge from (S_i, I) to (S_j, J) (denoted by $S_i(I) \Longrightarrow S_j(J)$) if executing instance $S_j(J)$ before instance $S_i(I)$ may change the result of the program. For all $1 \leq i, j \leq s$, one defines the **distance set** $E_{i,j}$ as follows:

$$E_{i,j} = \{(J - I) \mid S_i(I) \Longrightarrow S_j(J)\} \qquad (E_{i,j} \subset \mathbb{Z}^n) \qquad (1)$$

In general, the EDG (and the distance sets) cannot be computed at compile-time, either because some information is missing (such as the values of size parameters or the exact accesses to memory), or because generating the whole graph is too expensive. Instead, dependences are captured through a smaller, (in general) cyclic directed (multi) graph, with s vertices, called the **reduced dependence graph (RDG)**. Each edge e has a label $w(e)$. This label has a different meaning depending upon the dependence abstraction that is used: it represents [2] a set $D_e \subset \mathbb{Z}^n$ such that:

$$\forall i, j,\ 1 \leq i, j \leq s,\ E_{i,j} \subset \left(\cup_{e=(S_i, S_j)} D_e \right) \qquad (2)$$

In other words, the RDG describes, in a condensed manner, a superset of the EDG, called the **apparent dependence graph (ADG)**. The ADG and the EDG have the same vertices, but the ADG has more edges, defined by:

$$(S_i, I) \Longrightarrow (S_j, J) \text{ in the ADG} \Leftrightarrow \exists\, e = (S_i, S_j) \text{ in the RDG} \mid (J - I) \in D_e \qquad (3)$$

Equations 1 and 2 ensure that the EDG is a subset of the ADG.

---

[2] except for exact dependence analysis where it defines a subset of $\mathbb{Z}^n \times \mathbb{Z}^n$.

2.3 Maximal degree of parallelism

We now define what we call maximal parallelism extraction in reduced dependence graphs. We consider that the only information available for extracting parallelism in a set of loops L is the RDG associated to L. Any parallelization algorithm that transforms L into an equivalent code L_t has to preserve all dependences summarized in the RDG, i.e. all dependences described in the ADG (and not in the EDG which is not known!). If $(S_i, I) \Longrightarrow (S_j, J)$ in the ADG then (S_i, I) must be computed before (S_j, J) in the transformed code L_t.

Definition 1. For a given statement S in L, we define the S-**latency** of the transformed code L_t as the minimal number of clock cycles needed to execute L_t if: 1) an unbounded number of processors is available; 2) executing an instance of S requires one clock cycle; 3) any other operation requires zero clock cycle. The latency of L_t is defined as the sum of all S-latencies.

Since two instances linked by an edge in the ADG cannot be computed at the same clock cycle in the transformed code L_t, the latency of L_t, whatever the parallelization algorithm, is larger than the length of the longest path in the ADG. With this simple remark, we can define a theoretical framework in which the optimality of parallelization algorithms, with respect to a given dependence abstraction, can be discussed. In the following definitions, the RDGs are supposed to be labeled with a fixed dependence abstraction and the optimality is defined with respect to this abstraction.

Definition 2. Let G be a RDG and $\mathcal{D}$ be the n-dimensional cube of size N. Let d be the smallest non negative integer such that the length of the longest path in the ADG, is $O(N^d)$. Then, we say that the **degree of (intrinsic) parallelism** in G is $(n-d)$ or that G contains $(n-d)$ degrees of parallelism.

Definition 3. Let L be a set of n nested loops and G its RDG. Apply a parallelization algorithm A to G and suppose that $\mathcal{D}$ is the n-dimensional cube of size N. The **degree of parallelism extraction** for A is $(n-d)$ if d is the smallest non negative integer such that the latency of the transformed code is $O(N^d)$.

Definition 4. An algorithm A performs maximal parallelism extraction (or is said **optimal for parallelism extraction**) if for each RDG G, the degree of parallelism extraction is equal to the degree of intrinsic parallelism.

Then the optimality of a parallelization algorithm A can be proved as follows. Let L be a set of n nested loops, G its RDG, and G_a the corresponding ADG. Let $(n-d)$ be the degree of parallelism extraction for A in G. Then, we have at least two ways for proving the optimality of A:

1. Build in G_a a dependence path whose length is not $O(N^{d-1})$.
2. Build a set of loops L' whose RDG is also G and whose EDG contains a dependence path whose length is not $O(N^{d-1})$.

Note that (2) implies (1) since the EDG of L' is included in G_a (L and L' have the same RDG). Therefore, proving (2) is more powerful. It reveals the intrinsic limitations due to the dependence abstraction itself: even if their EDGs may be different, L and L' cannot be distinguished by the parallelization algorithm, since they have the same RDG. Therefore, since L' is parallelized optimally, the algorithm is considered optimal with respect to the dependence abstraction that is used. Figure 1 recalls the links between L, L' and their EDG, RDG and ADG. The loops L' are called **apparent nested loops**.

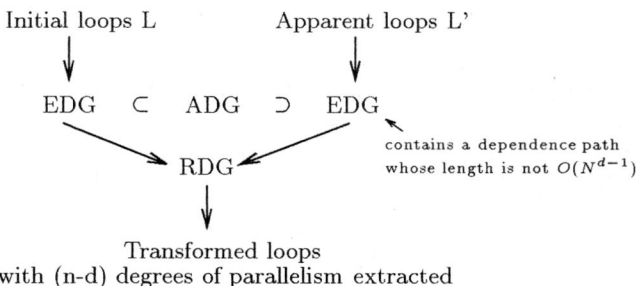

Fig. 1. Links between L, L' and their EDG, ADG and RDG

One can define a more precise notion of optimality by using the notion of S-latency instead of latency. The S-latency is related to the S-length of the longest path in the ADG, where the S-length of a path P is the number of vertices in P that are instances of S. Similarly, we define the S-degree of parallelism extraction in L_t, and the S-degree of intrinsic parallelism in a RDG. Finally, we have the following notion of optimality:

Definition 5. An algorithm A performs maximal parallelism extraction (or is said **optimal for parallelism extraction**) if for each RDG G and for each statement S of G, the S-degree of parallelism extraction for A in G is equal to the S-degree of intrinsic parallelism in G.

With this definition, we can discuss the quality of parallelizing algorithms even for statements that do not belong to the most sequential part of the code. Note that this definition of optimality is more precise than Definition 4 since the degree of intrinsic parallelism (resp. of parallelism extraction) in G is the minimal S-degree of intrinsic parallelism (resp. of parallelism extraction) in G.

One could argue that the latency and the S-latency of a transformed code are not easy to compute. Indeed, in the general case, the latency can be computed only by executing the transformed code with a fixed value of N. However, for most known parallelizing algorithms, the degree of parallelism extraction (but not necessarily the latency) can be computed simply by examining the structure of the transformed code, as shown by Lemma 6.

Lemma 6. *In addition to the hypotheses of Definition 3, assume that each statement S of the initial code L appears only once in the transformed code L_t and is surrounded by exactly n loops. Furthermore, assume that the iteration domain $\mathcal{D}_t$ described by these n loops contains a n-cube of size $\Omega(N)$ and is contained in a n-cube of size $O(N)$. Then, the number of parallel loops that surrounds S is the S-degree of parallelism extraction and the minimal S-degree of parallelism extraction is the degree of parallelism extraction.*

We now discuss, within this theoretical framework, the optimality of Allen and Kennedy's algorithm with respect to the dependence level abstraction. A similar study for other dependence abstractions can be found in [4].

3 Allen and Kennedy's algorithm

Allen and Kennedy's algorithm has first been designed for vectorizing loops. Then, it has been extended so as to maximize the number of parallel loops and to minimize the number of synchronizations in the transformed code. It has been shown (see details in [3, 22]) that for each statement of the initial code, as many surrounding loops as possible are detected as parallel loops. Therefore, one could think that what we want to prove in this paper has been already proved!

However, looking precisely into the details of Allen and Kennedy's proof reveals that what has actually been proved is the following: consider a statement S of the initial code and L_i one of the surrounding loops. Then L_i is marked as parallel if and only if there is no dependence at level i between two instances of S. This result proves that the algorithm is optimal among all parallelization algorithms that describe, in the transformed code, the instances of S with exactly the same loops as in the initial code. This does not prove a general optimality property. In particular, this does not prove that it is not possible to detect more parallelism with more sophisticated techniques than loop distribution and loop fusion. This paper gives an answer to this question.

We first recall the definition of dependence level, the dependence approximation used by Allen and Kennedy's algorithm. Then, we recall Allen and Kennedy's algorithm, in its simplest form, which is sufficient for discussing the optimality for parallelism extraction.

The **dependence level** associated to a dependence distance $J - I$ where $S_i(I) \Longrightarrow S_j(J)$ is either ∞ if $J - I = 0$ or the smallest integer l, $1 \leq l \leq n$, such that the l-th component of $J - I$ is non zero (and thus positive). A **reduced leveled dependence graph (RLDG)** is a reduced dependence graph whose edges are labeled by dependence levels. The **level** $l(G)$ of a RLDG G is the minimal level of an edge of G: $l(G) = \min\{l(e) \mid e \in G\}$.

We need to recall some graphs definitions: a **strongly connected component** of a directed graph G is a maximal subgraph of G in which for any vertices p and q ($p \neq q$) there is a path from p to q. The **acyclic condensation** of a graph G is the acyclic graph whose nodes are the strongly connected components $\mathcal{V}_1, \ldots, \mathcal{V}_c$ of G and there is an edge from $\mathcal{V}_i$ to $\mathcal{V}_j$ if there is an edge $e = (x_i, y_j)$ in G such that $x_i \in \mathcal{V}_i$ and $y_j \in \mathcal{V}_j$.

To apply Allen and Kennedy's algorithm to a RLDG G, call $AK(G, 1)$ where:

$AK(H, l)$
- $H' = H \setminus \{e \mid l(e) < l\}$
- Build H'' the acyclic condensation of H', and number its vertices $\mathcal{V}_1, \ldots, \mathcal{V}_c$ in a topological sort order.
- **For** $i = 1$ **to** c **do**
  1. If $\mathcal{V}_i$ is reduced to a single statement S, with no edge, **then** generate parallel DO loops (DOALL) in all remaining dimensions (i.e. for levels l to n) and generate code for S.
  2. Otherwise, let $k = l(\mathcal{V}_i)$. Generate parallel DO loops (DOALL) for levels from l to $k - 1$, and a sequential DO loop (DOSEQ) for level k. Call $AK(\mathcal{V}_i, k + 1)$.

4 Loop nest generation algorithm

In this section, we present a systematic procedure, called LGA (for Loops Generation Algorithm), that builds, from a RLDG G, a perfect loop nest L' whose RLDG is exactly G. L' are the desired apparent loops (see Section 2) that we use to prove the optimality of Algorithm AK (see Theorem 7). The construction of L' is based on the notion of critical edges. These edges are built in Section 4.1. The exact formulation of Procedure LGA is given in Section 4.2. Finally, in Section 4.3, we show that the RLDG associated to L' is G, as desired.

4.1 Critical edges

The procedure Critical given below defines a set of edges E_c, called **critical edges**, that we need for defining the apparent loops L'. We call Critical(G_i) for each strongly connected component G_i of the RLDG G which contains at least one edge.

$Critical(H)$
1. $l \leftarrow l(H)$.
2. Select an edge f of H with level l. Call f the **critical** edge of H. $E_c \leftarrow E_c \cup \{f\}$.
3. $H' = H \setminus \{e \mid l(e) \le l\}$.
4. Let $H_1, \ldots, H_c$ be the strongly connected components of H'.
5. Call Critical(H_i) for each H_i that has at least one edge.

4.2 Generation of apparent nested loops

Let $G = (E, V)$ be a RLDG. We assume that G has been built in a consistent way from some nested loops. Therefore, vertices can be numbered according to the topological order defined by the edges whose level is ∞: $v_i \xrightarrow{\infty} v_j \Rightarrow i < j$. We denote by d the **dimension** of G: $d = \max\{l(e) \mid e \in E \text{ and } l(e) < \infty\}$. The apparent loops L' corresponding to G will consist of d perfectly nested loops, with $|V|$ statements. Each statement will be of the form $a_i[I] = \text{rhs}(i)$ where a_i is a d-dimensional array and $\text{rhs}(i)$ is the right-hand side that defines array a_i. In the following, E_c is the set of critical edges of G (defined in Section 4.1) and "@" denotes the operator of expression concatenation.

LGA(G)

Initialization:
  For $i = 1$ to $|V|$ do rhs$(i) \leftarrow$ "1"

Computation of the statements of L':
  For each $e = (v_i, v_j) \in E$ do
    if $l(e) = \infty$ then rhs$(j) \leftarrow$ rhs(j) @ "$+a_i[I_1, \ldots, I_d]$"
    if $l(e) < \infty$ and $e \notin E_c$
        then rhs$(j) \leftarrow$ rhs(j) @ "$+a_i[I_1, \ldots, I_{l(e)-1}, I_{l(e)} - 1, I_{l(e)+1}, \ldots, I_d]$"
    if $e \in E_c$ then rhs$(j) \leftarrow$ rhs(j) @ "$+a_i[I_1, \ldots, I_{l(e)-1}, I_{l(e)} - 1, \underbrace{N, \ldots, N}_{d-l(e)}]$"

Code generation for L':
  For $i = 1$ to d do generate ("For $I_i = 1$ to N do")
  For $i = 1$ to $|V|$ do generate ("$a_i[I_1, \ldots, I_d] :=$" @ rhs(i))

4.3 Optimality result

The following results are detailed in [5]. Let L be a set of n perfectly nested loops whose RLDG is G. Use Algorithm LGA to generate the apparent loops L'. The RLDG of L' is G by construction. Let d_S be the number of calls to Procedure Critical that concern Statement S: $n - d_S$ is also equal to the degree of parallelism extraction for Allen and Kennedy's algorithm. Furthermore:

Theorem 7. *For each strongly connected component G_i of G, there is a path in the EDG of L' which visits, $\Omega(N^{d_s})$ times, each statement S in G_i.*

The proof is long, technical and painful. The fundamental corollary is:

Corollary 8. *Allen and Kennedy's algorithm is optimal for parallelism extraction in reduced leveled dependence graphs (optimal in the sense of Definition 5).*

We illustrate the optimality theorem through the following example:

For $i = 1$ to N do
  For $j = 1$ to N do
    $S_1 : \text{a}(i,j) := 1 + \text{a}(i, j-1) + \text{b}(i-1, j)$
    $S_2 : \text{b}(i,j) := 1 + \text{a}(i,j) + \text{b}(i-1, j)$

(a) Original loop nest L   *(b) RLDG*

The RLDG is strongly connected and contains one edge at level 1. Thus, the first loop is marked sequential. At level 2, each statement is in a separate strongly connected component and, after loop distribution, the second loop is marked sequential for S_1 and parallel for S_2. Indeed, one can check that L and the apparent loops L' have the same RLDG and that L' contains an exact dependence path including $\Theta(N^2)$ instances of S_1 and $\Theta(N)$ instances of S_2. This proves the optimality of Algorithm AK in our example even if the original program contains only uniform dependences, and, therefore, can be scheduled with a single sequential loop using the hyperplane method.

For $i = 1$ to N do
    For $j = 1$ to N do
        $S_1 : \mathrm{a}(i,j) := 1 + \mathrm{a}(i, j-1) + \mathrm{b}(i-1, N)$
        $S_2 : \mathrm{b}(i,j) := 1 + \mathrm{a}(i,j) + \mathrm{b}(i-1,j)$

Apparent loops L'

5 Conclusion

We have introduced a theoretical framework in which the optimality of algorithms that detect parallelism in nested loops can be discussed. We have formalized the notions of degree of parallelism extraction (with respect to a given dependence abstraction) and of degree of intrinsic parallelism (contained in a reduced dependence graph). This study explains the impact of a given dependence abstraction on the maximal parallelism that can be detected: it determines whether the limitations of a parallelization algorithm are due to the algorithm itself or are due to the weaknesses of the dependence abstraction.

In this framework, we have studied more precisely the link between dependence abstractions and parallelism extraction in the particular case of *dependence level*. Our main result is the optimality of Allen and Kennedy's algorithm for parallelism extraction in reduced leveled dependence graphs. This means that even the most sophisticated algorithm cannot detect more parallelism, as long as dependence level is the only information available. In other words, loop distribution is sufficient for detecting maximal parallelism in dependence graphs with levels.

The proof is based on the following fact: given a set of loops L whose dependences are specified by levels, we are able to systematically build a set of loops L' that cannot be distinguished from L (i.e. they have the same reduced dependence graph) and that contains exactly the degree of parallelism found by Allen and Kennedy's algorithm. We call these loops the apparent loops. We believe this construction is of interest because it better explains why some loops appear sequential when considering the reduced dependence graph while they actually may contain some parallelism.

References

1. J.R. Allen and K. Kennedy. Automatic translations of Fortran programs to vector form. *ACM Toplas*, 9:491–542, 1987.
2. U. Banerjee. *Dependence Analysis for Supercomputing*. Kluwer Academic Publishers, Norwell, MA, 1988.
3. D. Callahan. *A Global Approach to Detection of Parallelism*. PhD thesis, Dept. of Computer Science, Rice University, Houston, TX, 1987.
4. Alain Darte and Frédéric Vivien. A classification of nested loops parallelization algorithms. In *INRIA-IEEE Symposium on Emerging Technologies and Factory Automation*, pages 217–224. IEEE Computer Society Press, 1995.

5. Alain Darte and Frédéric Vivien. On the optimality of Allen and Kennedy's algorithm for parallelism extraction in nested loops. Technical Report 96-05, LIP, ENS-Lyon, France, February 1996. Extended version of Europar'96.
6. Alain Darte and Frédéric Vivien. Optimal fine and medium grain parallelism in polyhedral reduced dependence graphs. In *Proceedings of PACT'96*, Boston, MA, October 1996. IEEE Computer Society Press. To appear.
7. Paul Feautrier. Dataflow analysis of array and scalar references. *Int. J. Parallel Programming*, 20(1):23–51, 1991.
8. Paul Feautrier. Some efficient solutions to the affine scheduling problem, part II, multi-dimensional time. *Int. J. Parallel Programming*, 21(6):389–420, December 1992.
9. G. Goff, K. Kennedy, and C.W. Tseng. Practical dependence testing. In *Proceedings of ACM SIGPLAN'91 Conference on Programming Language Design and Implementation*, Toronto, Canada, June 1991.
10. F. Irigoin, P. Jouvelot, and R. Triolet. Semantical interprocedural parallelization: an overview of the PIPS project. In *Proceedings of the 1991 ACM International Conference on Supercomputing*, Cologne, Germany, June 1991.
11. F. Irigoin and R. Triolet. Computing dependence direction vectors and dependence cones with linear systems. Technical Report ENSMP-CAI-87-E94, Ecole des Mines de Paris, Fontainebleau (France), 1987.
12. R.M. Karp, R.E. Miller, and S. Winograd. The organization of computations for uniform recurrence equations. *Journal of the ACM*, 14(3):563–590, July 1967.
13. X.Y. Kong, D. Klappholz, and K. Psarris. The I test: a new test for subscript data dependence. In Padua, editor, *Proceedings of 1990 International Conference of Parallel Processing*, August 1990.
14. Leslie Lamport. The parallel execution of DO loops. *Communications of the ACM*, 17(2):83–93, February 1974.
15. Z.Y. Li, P.-C. Yew, and C.Q. Zhu. Data dependence analysis on multi-dimensional array references. In *Proceedings of the 1989 ACM International Conference on Supercomputing*, pages 215–224, Crete, Greece, June 1989.
16. Y. Muraoka. *Parallelism exposure and exploitation in programs.* PhD thesis, Dept. of Computer Science, University of Illinois at Urbana-Champaign, February 1971.
17. William Pugh. The Omega test: a fast and practical integer programming algorithm for dependence analysis. *Communications of the ACM*, 8:102–114, August 1992.
18. Michael E. Wolf and Monica S. Lam. A loop transformation theory and an algorithm to maximize parallelism. *IEEE Trans. Parallel Distributed Systems*, 2(4):452–471, October 1991.
19. Michael Wolfe. *Optimizing Supercompilers for Supercomputers.* MIT Press, Cambridge MA, 1989.
20. Y.-Q. Yang, C. Ancourt, and F. Irigoin. Minimal data dependence abstractions for loop transformations. *International Journal of Parallel Programming*, 23(4):359–388, August 1995.
21. Yi-Qing Yang. *Tests des dépendances et transformations de programme.* PhD thesis, Ecole Nationale Supérieure des Mines de Paris, Fontainebleau, France, 1993.
22. Hans Zima and Barbara Chapman. *Supercompilers for Parallel and Vector Computers.* ACM Press, 1990.

Memory Reuse Analysis in the Polyhedral Model

Doran Wilde[1] and Sanjay Rajopadhye[2]

[1] EE Department, BYU, Provo UT (`wilde@ee.byu.edu`)
[2] IRISA, Rennes, France (`rajopadh@irisa.fr`)

Abstract. In the context of developing a compiler for a ALPHA, a functional data-parallel language based on *systems of affine recurrence equations* (SAREs), we address the problem of transforming scheduled single-assignment code to multiple assignment code. We show how the polyhedral model allows us to statically compute the lifetimes of program variables, and thus enables us to derive necessary and sufficient conditions for reusing memory.

1 Introduction

The methodology of automatic systolic array synthesis from *Systems of Affine Recurrence Equations* (SAREs) has a close bearing on parallelizing compilers and on efficient implementation of functional languages. To study this relationship, we are currently developing a compiler for ALPHA [11], a functional, data parallel language based on SAREs defined over *polyhedral index domains*. The language semantics directly lead to sequential code based on demand driven evaluation. However, the resulting context switches can be avoided if the program is transformed into (sequential) imperative single assignment code (SAC) [15]. This is currently done semi-automatically—the user chooses the transformation, and the system generates the final code automatically.

In the imperative SAC, memory size is of the same order as the iteration space (for example, matrix multiplication will take $\Theta(N^3)$ memory). This is clearly unacceptable, and can be ameliorated by producing *multiple assignment code* (MAC). The idea is that memory is reused by allocating multiple index points to the same memory location. This necessitates a *lifetime analysis*, and in this paper we develop the constraints that the memory allocation must satisfy. They are based on information from the program and the schedule chosen. For the purposes of this paper, we only consider sequential code, but the technique can be adapted for parallel code.

The paper is organized as follows. In the next section we review the ALPHA language, system and the compiler. We then present the memory reuse analysis (Sec 3), discuss related work (Sec 4) and conclude (Sec 5). All proofs and additional details are available in a technical report [20].

2 The ALPHA Language, System and Compiler

ALPHA is a functional data-parallel language based on SAREs. Variables are statically declared over a *polyhedral domain*, and represent multidimensional arrays whose shapes are polyhedra. For example, "`A:{i,j| 0<j<i<=N} of real`", declares a (strictly) lower triangular, real matrix. A program is a "system", with a number of size parameters, also declared over polyhedral domains. Consider the problem of solving a lower triangular (unit diagonal) system of equations, $Ax = b$, using forward substitution. Each element x_i, of the solution (for $i = 1 \ldots n$) is given by $x_i = b_i - \sum_{j=1}^{i-1} A_{i,j} x_j$. The corresponding ALPHA program (not shown due to space constraints) is almost identical, except for the syntax. The summation (written as a *reduction* in ALPHA) allows a very high level specification, but is not practical in terms of implementation. We therefore *serialize* it by using a temporary variable (say f) which accumulates the partial sums (say, in the increasing order of j). The resulting ALPHA program is shown in Figure 1, and illustrates the main syntactic constructs of the language (more details may be found in [11]).

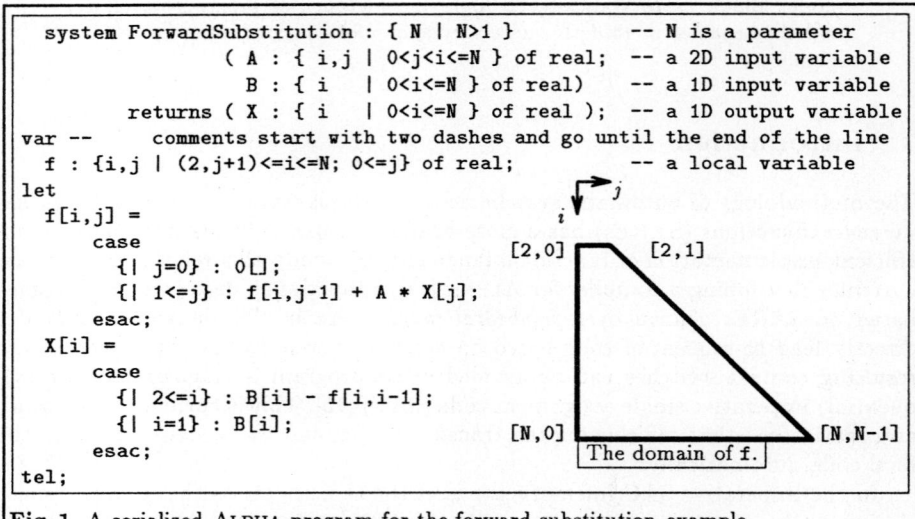

Fig. 1. A serialized ALPHA program for the forward substitution example

2.1 Change of Basis

An important transformation in the ALPHA system is the **change of basis** (COB). It is a generalization of unimodular loop transformations and includes array alignment, data distribution, etc. in a unified framework. The intuition behind it is as follows. An ALPHA variable denotes a multidimensional array defined over a polyhedral domain, and we want to change the "shape" of its domain and construct an equivalent program. When a variable, V is so transformed, the system must determine: (i) its new domain, (ii) the new case structure of its equation, (iii) the new dependencies for the uses of *all* variables in the equation for V, and (iv) the new dependencies for *uses* of V (in all other equations). All of this is done automatically using a polyhedral library [19], and relies on the fact that ALPHA is founded on well defined closure properties of domains, dependencies and transformations.

2.2 Compiling ALPHA to (Sequential) Single Assignment Code

The ALPHA compiler is semi-automatic and transformational. Currently, the user chooses the transformations based on the analyses enabled by the polyhedral model, specifies it as a COB for each variable, and the system takes care of the rest. Our long term goal is to perform the analyses and choose the transformations automatically, using the now mature research on systolic synthesis and its generalization.

Scheduling: There has been much research on the SARE scheduling problem [9, 17, 8, 18, 4], formulated as follows. For each variable V in a SARE, determine a function $t_V(z)$ that gives us the time instant, represented by a k-dimensional *time vector*, at which $V[z]$ can be computed. For example, $t_f(i,j) = i+j$ and $t_X(i) = 2i$ are valid (one dimensional) affine schedules for the forward substitution program.

Alignment: Among the f and X variables of Figure 1, there are actually three *independent* indices—it was mere coincidence that we used only two *names*, i and j. In order to generate code however, we must place all variables in a common index space. This is similar to HPF's distribution directives, but in ALPHA, the mappings are affine transformations. For parallel code, the variable domains are also mapped to processors, by **affine allocation functions**.

There is a close interplay between the schedule, the alignment and the allocation functions, and the choice of one may affect the others. The schedule achieves a partial alignment (the time indices are common to all variables). Thus each of these three functions gives us partial information about the COB to apply to each variable.

Change of Basis: In constructing the COB, there may still be freedom. The subject of which transformation to apply is currently the subject of much intense research, but we will not address it here. After applying the transformation, we have a program where all variables share the same iteration space, certain indices (in a specific order) denote time, and the others (if any) are guaranteed to carry no dependencies.

(Sequential) Code generation: A special preamble allocates memory for the (bounding boxes of the) domains of all variables. The compiler generates code that visits each point of the domain of each variable in the correct order (each visit consisting of an update of the corresponding memory location using the RHS of the equation; see [15] for details). The loop nests to visit each point are produced by first "separating" the domains according to their time indices [10],, and "sorting" the equations so that whenever an equation group textually precedes another, it is also temporally before the other. This form is called **imperative normal form**, and the final code is generated by a special pretty printer.

3 Memory Reuse

We can generate MAC as follows. For each (n_V-dimensional) variable, V, with domain, D_V, we define a **memory allocation function**, $\text{Mem}_V(z)$ which assigns a memory address to each point $z \in D_V \subseteq \mathcal{Z}^{n_V}$. In order to make this function injective (so that many points are allocated to the same address) we consider functions that are (multi) projections, i.e., $\text{Mem}_V(z) = \Pi_V z$, where Π_V is a $(n_V - m_V) \times n_V$ matrix. The total memory required will correspond to an $(n_V - m_V)$-dimensional array (a m_V-dimensional subspace of D_V is mapped to the same memory location).

Suppose that, for each variable, a k dimensional affine schedule is given. Hence, the "time" at which $Y[z']$ is computed is the k dimensional vector, $\Lambda_Y z' + \alpha_Y$, where Λ_Y is a $k \times n_Y$ matrix. Similarly, Λ_X, α_X specifies the k-dimensional schedule for X. Note that these are parallel schedules (an $(n_Y - k)$-dimensional subspace of Y and an $(n_X - k)$-dimensional subspace of X are scheduled simultaneously).

Π_Y is characterized by its null space: two points $z'_1, z'_2 \in D_Y$ are mapped to the same memory location iff $(z'_1 - z'_2) \in \text{Null}(\Pi_Y)$. Let $\rho_1, \ldots, \rho_{m_Y}$, be a basis for Null($\Pi_Y$). Note that ρ_i cannot be in the null space of Λ_Y, otherwise two points will be written to the same memory location *at the same time*, leading to a write conflict. In other words, the matrix $\begin{bmatrix} \Lambda_Y \\ \Pi_Y \end{bmatrix}$ must be of full column rank, and hence $k + (n_Y - m_Y) \geq n_Y$. Thus, $m_Y \leq k$, and no more than k dimensions of the domain of Y can be "removed" (there are examples where one cannot achieve even this).

Now, without loss of generality, let us choose the signs of the ρ_i's such that $\Lambda_Y \rho_i$ is lexicographically strictly positive $\Lambda_Y \rho_i \succ 0$. It follows that $\forall z' \in D_Y$, the write into $\text{Mem}_Y(z')$ *immediately following* that of $Y[z']$, must be one of $z' + \rho_i$.

3.1 Constraints on the Memory Allocation Function

For the final code to be correct, we must ensure that no value is overwritten before it has been used by all the computations that need it. We will now show how the resulting constraints can be expressed concisely in the polyhedral model, using the knowledge of the schedule, the variable domains and the dependency information.

Definition 1. A memory allocation function, $\text{Mem}_Y(z')$ is said to be **valid** iff, for any $Y[z']$ written into $\text{Mem}_Y(z')$, the **next** write into the same memory location occurs **after all** uses of $Y[z']$ have been executed.

Thus, for any point $z' \in D_Y$, we need to know the set of points that use $Y[z']$. This information is succinctly described by a *usage table*. It is derived from the program by inverting the dependencies (actually a relational inverse is computed, since a dependency may be many-to-one; details may be found in [20]). For each dependency in the program, the corresponding entry in the usage table has the form, $D' : Y[z'] \Rightarrow D''(z') : X[M_1 z' + M_2 r]$ (read as, "for all $z' \in D'$, each Y value at z' is used for the computation of X at the set of points, $M_1 z' + M_2 r$, where r belongs to the polyhedron, D'', parameterized by z'"). Here, $D' \subseteq D_Y$ is a polyhedron specifying the *producers* of the Y values that are used; r are auxiliary indices to describe the (sub) space where the *same* value is used; $D''(z')$ gives the range of the r indices, which may be different for different z' and hence, is a polyhedron *parameterized by* z'; and $M_1 z' + M_2 r$ gives a point in D_X which uses $Y[z']$. Substituting this into Definition 1, we have

Proposition 1 $\text{Mem}_Y(z') = \Pi_Y z'$ *is a valid memory allocation function for Y iff for each usage table entry, $D' : Y[z'] \Rightarrow D''(z') : X[M_1 z' + M_2 r]$, the following condition holds for each ρ_i*

$$\forall z' \in D', \forall r \in D''(z') : [\Lambda_X(M_1 z' + M_2 r) + \alpha_X \preceq \Lambda_Y(z' + \rho_i) + \alpha_Y$$

which can be simplified to (if we set $\bar{z} = \begin{bmatrix} z' \\ r \end{bmatrix}$, and $\bar{M} = [M_1\ M_2]$)

$$\forall \bar{z} \in D'' : (\Lambda_X \bar{M} - \Lambda_Y[Id\ 0])\bar{z} + \alpha_X - \alpha_Y \preceq \Lambda_Y \rho_i \quad (1)$$

Observe that we now have constraints that must hold for *all* points $\bar{z}$ in D''. But since the size of D'' may be arbitrary, there may be an unbounded number of constraints to be satisfied. However, the fact that the D'' is a polyhedron allows us to exploit the power of the polyhedral model:

- First, note that Eqn. 1 holds at *all* $\bar{z} \in D''$ iff it is satisfied by the point(s) in D'' that maximize, in the lexicographic order, the (multi) linear cost function, $[\Lambda_X \bar{M} - \Lambda_Y[Id\ 0]]\bar{z}$. Hence, we have a valid memory allocation function, iff for each ρ_i,

$$\text{Lmax}_{\bar{z} \in D''} \left([\Lambda_X \bar{M} - \Lambda_Y[Id\ 0]]\bar{z} + \alpha_X - \alpha_Y \right) \preceq \Lambda_Y \rho_i \quad (2)$$

- Second, The LHS of Eqn 2 can be written as a classic integer linear programming problem (actually k ILPs) whose solution is attained at a vertex of D'', and can be determined easily, thanks to the dual representation of the polyhedral library. Let d be this solution. This leads to the following.

Theorem 2. *ρ is a valid memory projection vector for Y with respect to a dependency if and only if it satisfies a finite number of linear constraints:*

$$d \preceq \Lambda_Y \rho \quad (3)$$

3.2 An Alternative Formulation

Although we needed the usage table in order to formulated our constraints they can be considerably simplified. Indeed, the usage table does not even appear in the final constraint. For any point $z' \in D'$, its *maximum lifetime* with respect to a given dependency is defined as the time between its computation and its last usage, i.e.,

$$d(z') = \text{Lmax}_{r \in D''(z')}[\Lambda_X(M_1 z' + M_2 r) + \alpha_X - \Lambda_Y z' + \alpha_Y]$$

Then, the *maximum lifetime* of the variable Y (with respect to a given dependency) is $\text{Lmax}_{z' \in D'}\ d(z')$, which can be simplified to d, the LHS of Eqn 2 above. An alternative formulation of the maximum lifetime of Y (as always, with respect to a particular dependency) is as follows[3]

$$d' = \text{Lmax}_{z \in D_0}[\Lambda_X z + \alpha_X - \Lambda_Y M z - \alpha_Y] \quad (4)$$

where $D_0 = D_X \cap \text{Preimage}((D_Y \cap \text{Closure}(\text{Image}(D_X, M))), M)$. Since ALPHA domains are closed under the preimage operation, D_0 is a valid ALPHA domain, and this generalizes and simplifies the previous formulation.

To generate multiple assignment code (MAC) from this, we find the largest number of linearly independent vectors ρ, such that $\Lambda_Y \rho \succeq d$. Such an analysis is done for each variable and memory allocation functions are chosen. Then the MAC is generated by two trivial modifications of the current SAC generator: memory is allocated for the (bounding box of the) *projection* of the domain of each variable; and all accesses (read as well as write) to $Y[Az]$ are replaced by $Y[\Pi_Y Az]$.

---

[3] The authors would like to thank Paul Feautrier for suggesting that the lifetime could be formulated independently of the usage table.

3.3 Extensions and Variations

Many related problems can be addressed in this framework depending on the different criteria that we seek to optimize. The obvious one is to minimize the memory by (i) maximizing the *number* of linearly independent ρ_i's (this gives us order of magnitude reduction), and (ii) choosing the ρ_i's so that the "footprint" of the projected domain is minimized. This is similar to the optimal allocation function problem in systolic synthesis, and we expect that the methods can be extended and adapted.

If we are seeking a sequential implementation, then we know that *all* indices are ultimately going to be interpreted as time. The schedule, Λ constrains only k of them. So we could to minimize memory by seeking the unimodular extension of Λ that maximizes the *number* of linearly independent bases that satisfy Eqn. 2.

It may also be possible to seek the memory allocation, not necessarily to minimize the memory, but to optimize other performance criteria such as cache locality, etc.

3.4 Communication Analysis

Although the usage table does not appear in the final formulation of memory reuse constrains, it has nevertheless, other applications, notably for communication analysis and optimization. In fact, the syntax of the usage table is motivated by the LACS notation [16] proposed by Rajopadhye for *specifying* a wide range of communication activity within the framework of polyhedra and linear/affine index functions. A LACS specification can also be *analyzed* to first determine if it is well formed (no write conflicts, etc.) and then to infer many communication patterns (scatters/ gathers, broadcasts, reductions, scans, etc.) This enables optimization such as message vectorization, broadcast elimination and latency hiding techniques such as "message prefetching", etc.

If we determine the usage table *after* applying a space-time COB, so that some indices are interpreted as processors, and some as time, we have a "sender centric" view of the communication—a LACS specification (actually, LACS allows reductions too, so if we could obtain the "usage table" of an ALPHA program *before* serialization, we could derive the complete LACS specification). Hence, the communication analysis of LACS can be incorporated into the ALPHA compiler.

4 Related Work

ALPHA is a specialized functional language, similar to (actually a proper subset of) Crystal [3]. The Crystal compiler is similar in spirit to ours, but cannot use the analysis presented here. SISAL was developed as an effort to use functional programming for numeric and scientific computing [12]. Being functional, all dependencies are inherently flow dependencies, although the fact that arrays can be declared and used in the language implies that memory is being reused. I-structures were introduced originally in the language Id (and currently incorporated in Haskell) for explicit support for parallelism and arrays in functional programming [1]. They can achieve very significant performance because of hardware support for fast context switching, threaded compilation, etc. Neither Sisal, Haskell/Id nor Crystal do static analyses enabled by the polyhedral model. These efforts are complementary to ours: with ALPHA, we have made a deliberate choice to sacrifice generality for complete static analyzability.

Parallelizing compliers face a dual problem to ours. In a loop program, only the flow dependencies are true dependencies, output and anti dependencies arise due to memory reuse, and can be eliminated at the price of more memory. Since the fewer the dependencies, the higher the parallelism in general, there has been considerable work on the problem of eliminating false dependencies by introducing temporary variables [14]. Usually this increases the memory space by a constant factor (the number of temporaries), but techniques such as array expansion [21] may cause orders of magnitude increases. In most parallelizing compilers, such expansion is often essential in order to obtain parallelism, and is considered worth the price.

The work closest to our effort is that of Feautrier's PAF which also uses the polyhedral model. It is known [6] that a nested loop program where the loop bounds are affine functions of outer indices (and possibly parameters), and all arrays are accessed through affine functions of the indices is equivalent to an SARE. This translation may involve array expansion, and hence results in SAC (or at least an intermediate single assignment representation). Hence they face a similar problem to that addressed in this paper. Their solution was developed independently of ours and is a bit different [7]. They use the SARE form for performing the analysis (of the choice of COB transformations to apply, etc.) but the code generation phase uses the original source code. Once all the transformations are chosen, the code generator asks itself whether it is safe to ignore a given index (introduced during array expansion). In other words their analysis seeks to verify whether a particular memory projection vector ρ is valid. Their constraints become equivalent to a particular case of ours. This has practical benefits in terms of simplicity of the tests needed and may impact the compilation time, but is otherwise restrictive. For instance, if the original program was already in single assignment form or otherwise memory inefficient, their method would not be able to (does not even seek to) improve its memory efficiency.

Chamski was the first to tackle the problem of memory reuse and lifetime analysis for an earlier version of ALPHA compiler [2]. His model was restrictive in a number of ways. It assumed that the schedule was full dimensional (not just k-dimensional), and this unnecessarily constrains the space of the memory projections. The COB was supposed to have already been applied, and only self-dependencies were considered (these are not really limitations but explain the notation below); the definition of the lifetime of a variable was as follows.

$$d = \max\{z - z' \mid z' \in D_Y, z \in M^{-1}z'\}$$

The M^{-1} here implies that the dependency is bijective (or at least admits a left inverse). Details on how to compute d were not given (all examples had only uniform dependencies). Finally, only orthogonal projections were considered.

The first notion of the usage information is probably due to Mongenet, who called it the *utilization set* [13], but did not give a method to derive it. She states constraints similar to Eqn. 1. Recently, De Greef and Catthoor have also addressed the memory reuse problem in (an extension of) the polyhedral model [5]. They too stop at the formulation of the constraints to be satisfied. Neither paper gives a procedure where the number of constraints is bounded and independent of problem parameters, nor show how this can be formulated in an ILP framework.

5 Conclusion

We addressed the problem of producing memory efficient (multiple assignment) code when compiling ALPHA, a language based on systems of affine recurrence equations over polyhedral domains. Our principal contribution has been to show that the constraints on the memory projection functions can be expressed as a *finite* number of linear inequalities, and can thus be formulated as one or more ILP problems. Different cost criteria may be used, depending on on what one seeks to optimize.

Our second contribution (described only briefly due to space constraints, see [20]) was the notion of the usage table to succinctly capture the relational inverse of a program dependency. Although our final formulation does not make use of this, the usage table has applications in communication analysis and optimization.

In the short term, perhaps the biggest beneficiaries of our work will be functional languages, since the compiler can now tap into the optimizations that seemed very far removed. It enables them to compete with imperative languages as their parallelizing compilers, albeit for a fairly narrow, but important class of programs, while retaining declarative semantics.

There are two main criticisms that one can make about the polyhedral model. The first is that one cannot (without contortions) express algorithms that have dynamic dependencies (iterative methods, pivoting algorithms, etc.) When parallelizing compilers make safe assumptions about the extent of dependencies, the size of the iteration space, etc., they impose very similar restrictions. However, their restrictions are on the analysis methods not on the programmer, who is shielded: the compiler may not be able to effectively analyze a program with pointers, but this does not mean that the program won't run, just that the best performance will not be attained. In ALPHA, such a program can't even be written without contortions. Nevertheless, we contend that this is a reasonable choice since we are exploring the limits of the analysis techniques enabled by the polyhedral model. ALPHA is a research language and is not intended to replace existing languages.

The second criticism is that these methods seem to be overkill. After all, most parallelizing compilers do detect 90% of the common communication patterns, and can generate efficient code. Is the price of the sophisticated analysis worth the (seemingly minimal) returns? Our conviction is that it will be worthwhile in the long run: even if "all" we have is a firm theoretical foundation for the optimizations that work in 90% of the cases.

Acknowledgments We would like to thank Paul Feautrier for valuable feedback, and for suggesting that the maximum lifetime may be expressed independently of the usage table.

References

1. Arvind, R. S. Nikhil, and K. K. Pingali. – I-structures: Data structures for parallel computing. – *ACM Transactions on Programming Languages and Systems*, 11(4):598–632, October 1989.

2. Z. Chamski. – Generating memory efficient imperative data structures from systolic programs. – Technical Report PI-621, IRISA, Rennes, France, December 1991.
3. Marina C. Chen. – A parallel language and its compilation to multiprocessor machines for VLSI. – In *Principles of Programming Languages*. ACM, 1986.
4. A. Darte and Y. Robert. – Constructive methods for scheduling uniform loop nests. – *IEEE Transactions on Parallel and Distributed Systems*, 5(8):814–822, Aug 1994.
5. E. De Greef, F. Catthoor, and H. De Man. – Reducing storage size for static control programs mapped to parallel architectures. – presented at Dagstuhl Seminar on Loop Parallelization, April 1996.
6. P. Feautrier. – Dataflow analysis of array and scalar references. – *International Journal of Parallel Programming*, 20(1):23–53, Feb 1991.
7. P. Feautrier and V. Le Fevre, 1996. – Unpublished manuscript on the PAF code generator (a preliminary French version was presented at RenPar 7, July 1995).
8. Paul Feautrier. – Some efficient solutions to the affine scheduling problem, Part II, multidimensional time. – Technical Report 78, Labaratoire MASI, Institut Blaise Pascal, October 1992.
9. R. M. Karp, R. E. Miller, and S. V. Winograd. – The organization of computations for uniform recurrence equations. – *JACM*, 14(3):563–590, July 1967.
10. H. Le Verge, V. Van Dongen, and D. Wilde. – La synthèse de nids de boucles avec la bibliothèque polyédrique. – In *RenPar'6*, Lyon, France, Juin 1994. – English version "Loop Nest Synthesis Using the Polyhedral Library"in IRISA TR 830, May 1994.
11. Christophe Mauras. – *ALPHA: un langage équationnel pour la conception et la programmation d'architectures parallèles synchrones.* – PhD thesis, L'Université de Rennes I, IRISA, Campus de Beaulieu, Rennes, France, December 1989.
12. J.R. McGraw, S.K. Skedzielewski, S. Allan, and D. Grit. – Sislal—streams and iteration in a single–assignment language. – *Language Reference Manual, Version 1.2*, Jan 1985.
13. Catherine Mongenet. – Data compiling for systems of uniform recurrence equations. – *Parallel Processing Letters*, 4(3):245–257, 1994.
14. D. A. Padua and M. J. Wolfe. – Advanced compiler optimizations for supercomputers. – *Communications of the ACM*, 29(12):1184–1201, December 1986.
15. P. Quinton, S. Rajopadhye, and D. Wilde. – Deriving imperative code from functional programs. – In *7th Conference on Functional Programming Languages and Computer Architecture*, pages 36–44, La Jolla, CA, Jun 1995. ACM.
16. S. V. Rajopadhye. – LACS: A language for affine communication structures. – Technical Report 712, IRISA, 35042, Rennes Cedex, April 1993.
17. S. V. Rajopadhye and R. M. Fujimoto. – Synthesizing systolic arrays from recurrence equations. – *Parallel Computing*, 14:163–189, June 1990.
18. W. Shang and J. A. B. Fortes. – On the optimality of linear schedules. – *Journal of VLSI Signal Processing*, 1:209–220, 1989.
19. D. Wilde. – A library for doing polyhedral operations. – Technical Report PI 785, IRISA, Rennes, France, Dec 1993. – an extended version of the author's MS Thesis, Computer Science Dept, Oregon State University, Corvallis, OR. Dec 1993.
20. D. Wilde and S. Rajopadhye. – The power of polyhedra. – Technical Report 95-80-8, Oregon State University, Computer Science Dept, Corvallis OR 97331, August 1995.
21. M. J. Wolfe. – *High Performance Compilers for Parallel Computing.* – Addison-Wesley, 1996.

Cycle Shrinking by Dependence Reduction*

Kunio Okuda
Universidade de São Paulo, Brasil, email: kunio@ime.usp.br

Abstract. We present a new simple cycle shrinking technique called *dependence reduction* for regular uniform nest algorithm. It consists of a transformation of the dependence graph to reduce the number of execution steps as well as the communication between processors. We gave simple and sufficient conditions for the dependence reduction method to be better than other methods.

1 Explicit domain and dependence reduction

Consider regular uniform nest algorithm with commands $S_1, \ldots, S_k$ in the loop body. To show the dependences in Dom (set of loop indexes in Z^n) explicitly we will use the following definition: $ED=$ explicit domain $=\{S_1, \ldots, S_k\} \times Dom$. For its representation ED will always be identified as a subset of R^{n+1}. ED will illustrate the idea of dependence reduction.

Example 1 *(See Dom and the dependence graph in Figure 1)*

for $i = 0$ to N do
  for $j = 0$ to N do
    $S_1 : a(i,j) = f(b(i-1, j-3) + d(i-1, j+2))$
    $S_2 : b(i,j) = g(c(i-1, j+1))$
    $S_3 : c(i,j) = h(a(i-1, j-1))$
    $S_4 : d(i,j) = k(a(i, j-1))$

We construct ED and project the four planes to R^2 (See Figure 2). Figure 2 is decomposed into Figure 3 that shows the dependences of the cycles C_1 and C_2. Figure 4.old shows the dependences in relation to $S_1(i,j)$ in particular.

In both sides of Figure 3, we have independent "zig-zags" showing dependence between commands. The scheduling of cycle 1 (one zig-zag) must be compatible to the scheduling of cycle 2 (other zig-zag) because of the intersection point (Figure 4.old). The positions of the S_1's (intersection points of the two cycles) are essential for scheduling. We thus want to analyze the dependences only in terms of S_1. We join the computations of $S_2(i-1, j-3), S_3(i-2, j-2)$ and $S_4(i-1, j+2)$ into $S_1(i,j)$. Thus we construct S as a macro command (composed by these 4

---

* Supported by FAPESP- Proc. No. 95/0767-0, and CNPq - PROTEM-2-TCPAC Proc. No. 680060/94-4, and Commission of the European Communities through Project ITDC-207.

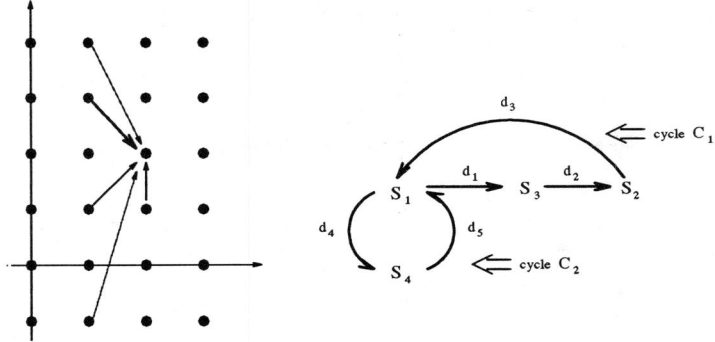

Fig. 1. *Dom* and the dependence graph of Example 1

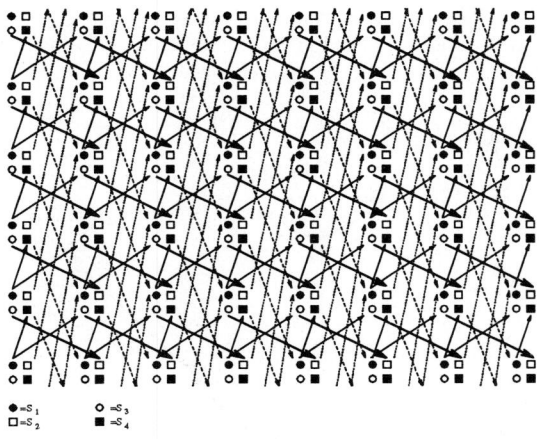

Fig. 2. *DE* for Example 1

commands) for a point of *Dom*. Figure 4.new is obtained from the transformation of Figure 4.old with this consideration. We denote this transformation by the name of *dependence reduction*.

Example 2 *(well-known example of Peir and Cytron [4])*

Let $Comm=$ unit of time for communication between processors and $Comp=$ unit of time to compute a command. The following table summarizes the results of the several methods using this example.

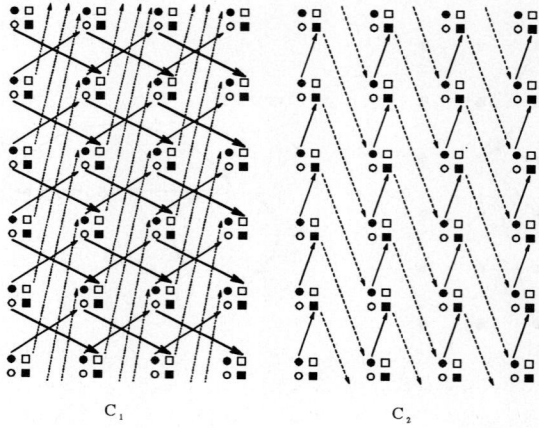

Fig. 3. Dependences of cycles C_1 and C_2

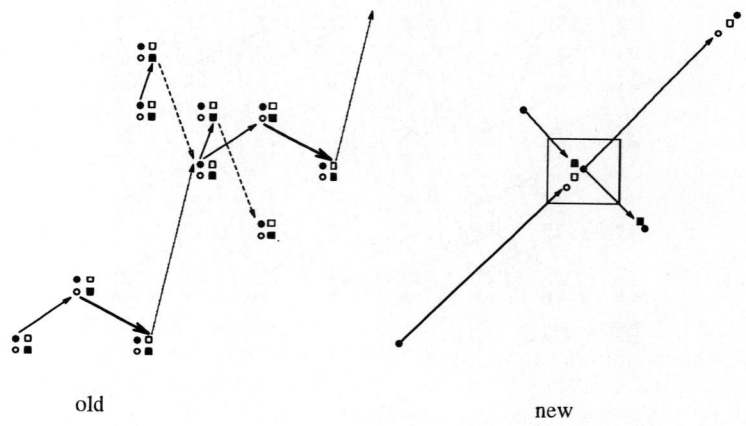

old   new

Fig. 4. Dependences for $S_1(i,j)$

	Computation	Communication
GSS ([7, 6])	$8NComp$	$8NComm$
GSS combined with ISM ([6, 2])	$2NComp$	$2NComm$
Affine by Statement ([1])	$\frac{12}{7}NComp$	$\frac{12}{7}NComm$
Dependence reduction	$\frac{20}{7}NComp$	$\frac{5}{7}NComm$

We conclude that, if $Comm > \frac{8}{7}Comp$, then *dependence reduction* presents the best time of all.

2 Formalization

Let $G = (V, E)$ be the dependence graph for regular uniform nest algorithm where V and E correspond to the set of commands $S_1, \ldots S_m$ and to the dependence between commands, respectively. Each edge of E is labeled by its dependence vector. We use the notation $u \xrightarrow{d} v$ to denote the edge from node u to node v with dependence vector d. We will divide V into two sets: the set of secondary nodes (VS) and the set of principal nodes (VP).
$VS = \{v \in V \mid v$ has exactly one edge entering the node and one edge leaving it $\}$
$VP = V - VS$ Let $\overline{G} = (\overline{V}, \overline{E})$ be a transformed graph in which $\overline{V} = VP$ and $\overline{E}$ is defined as follows:
$\overline{E} = \{ v \xrightarrow{d} v' \mid v, v' \in VP$ and in G there exists a path between v and v' whose intermediate nodes all belong to VS and d is the sum of the dependence vectors of this path $\}$. The commands corresponding to the secondary nodes in this path will be incorporated into a macro command of v'.

Applicability: We compare a *"floor"* of the total time spent in other methods with a *"ceiling"* of the total time of the dependence reduction method. If the length of the longest path in ED is T_f, then, except the case with no communication, the *"floor"* is $T_f(Comm + Comp)$. As for the *"ceiling"*, the following fact ensures that we have a good chance of computing such a value easily.

Fact: The dependence vectors before reduction are all lexicographically positive. When we perform dependence reduction, we increase the possibility of having a direction along which all the resulting dependence vectors present positive elements. If such a direction exists, then we can parallelize along this direction.
We emphasize that the comparison gives only a sufficient condition.

References

1. Darte, A. and Robert, Y. Mapping uniform loop nests onto distributed memory architectures. *Parallel Computing* 20(1994) 679-710.
2. Liu, L. S., Ho, C. W. and Sheu, J. P. On the parallelism of nested for-loops using index shift method. *Proc. Internat. Conf. on Parallel Processing* (Aug. 1990) II-119-II-123.
3. Okuda, K., Cycle shrinking by dependence reduction. Technical Report, Universidade de São Paulo, 1995.
4. Peir, J. K. and Cytron, R., Minimum distance: a method for partitioning recurrence for multiprocessors. *IEEE Trans. Comput.* 38(8) (Aug. 1989) 1203-1211.
5. Polychronopoulos, C. D., Compiler optimization for enhancing parallelism and their impact on architecture design. *IEEE Trans. Comput.* 37(8) (Aug. 1988) 991-1004.
6. Robert, Y. and Song, S.W., Revisiting cycle shrinking. *Parallel Computing*, 18(1992) 481-496.
7. Shang, W., O'Keefe, M. T. and Fortes, J. A. B. Generalized cycle shrinking. *Parallel Algorithms and VLSI Architecture II*. P. Quinton and Y. Robert (editors), North Holland, 1991.

A Unified Transformation Technique for Multilevel Blocking

M. Jiménez, J. M. Llabería, A. Fernández and E. Morancho

Departamento de Arquitectura de Computadores, Universidad Politécnica de Cataluña
Gran Capitán s/n, Módulo D6, E-08034 Barcelona, (Spain), e-mail: marta@ac.upc.es

Abstract. This paper presents a new unified method for simultaneously tiling the register and cache levels of the memory hierarchy. We will only focus on the code transformation phase of tiling. Our algorithm uses strip-mining and loop interchange on all memory hierarchy levels to determine the tiles as usual, and, afterwards, and due to the special characteristics of the register level, we apply index set splitting, unrolling and scalar replacement to this level. After applying strip-mining, the iteration space is non-convex. To perform in a single step the loop interchange in non-convex iteration spaces, we use non-unimodular matrices. The order proposed to perform index set splitting to the loops guarantees that each loop in the nest has to be processed only once and also avoids code explosion.

1 Introduction

To hide the details of the memory hierarchy from the user, several code transformation techniques have been developed. These techniques aim at exploiting the temporal and spatial locality properties of a program.

Iteration space tiling is a code restructuration technique used to reduce a program's data working set. Several proposals in the literature consider loop tiling on one, two or more levels of the hierarchy (registers, cache levels, TLB, virtual memory).Tiling an iteration space can be implemented using unroll & jam and scalar replacement at the register level[1][2], and applying strip-mining and loop interchange at all other levels of the hierarchy[3][6][9].

Tiling complex iteration spaces presents several problems. Applying strip-mining to a loop nest produces a non-convex iteration space since some of the loops in the nest will end up with a step different from 1. Therefore, the loop interchange needed after strip-mining can not be done using techniques based on unimodular transformation matrices [10]. Previous work on tiling, such as Wolf and Lam [10], uses ad-hoc methods to implement the loop interchange after strip-mining, and does not give a general algorithm. At the register level, S. Carr [2] uses pattern recognition techniques on the loop bounds to apply unroll and jam. Nevertheless, when the iteration space is complex, the loop bounds can not be matched by the patterns and no general algorithm to partition these complex iteration spaces into simpler ones, that could be recognized through patterns, is presented in [2].

Commercial compilers such as KAP from Kuck & Associates are not always able to produce an iteration space tiling when the loop bounds are affine functions of the surrounding loops iteration variables. These types of bounds are commonly found in linear algebra algorithms.

This paper presents a new unified method for simultaneously tiling the register and cache levels of the memory hierarchy. Because we do not use pattern recognition techniques, our method works for any iteration space having loop bounds defined as affine functions of the surrounding loops iteration variables. We will focus on the code transformation phase of tiling and we will assume that dependency analysis and locality analysis have already been performed [10][9]. Our algorithm uses strip-mining and loop interchange on all memory hierarchy levels (cache and registers) to determine the tiles as usual, and, afterwards, and due to the special characteristics of the register level, we also apply index set splitting, unrolling and scalar replacement to this level. The loop interchange in non-convex iteration spaces (spaces resulting after strip-mining) is done in a single step using non-unimodular matrices. We also use an order to perform index set splitting that guarantees that each loop in the nest has to be processed only once and also avoids code explosion.

2 Transformation Steps

We use the Multilevel Orthogonal Block (MOB) forms [7] to compute the tiles in each level of the memory hierarchy. The MOB forms provide maximum reuse of data in all levels simultaneously and their orthogonality property provides a simple method to optimize the form and the size of the tiles at each level. We will assume that the loops to be transformed are perfectly nested, fully permutable[8] and the loops bounds are affine functions of the surrounding loops iteration variables.
The steps we follow are:
   Tiling: To all levels of the memory hierarchy (cache and registers):
   1. Apply strip-mining to all loops selected using the MOB forms.
   2. Interchange loops as determined by the MOB forms.
   At the register level:
   3. Apply index set splitting repeatedly until being able to unroll those loops that provide data reuse at the register level.
   4. Distribute the surrounding loop of the loops we want to fully unroll.
   5. Fully unroll all innermost loops.
   6. Apply scalar replacement to all innermost loops [2].

Due to the lack of space, in this paper we will focus only at the register level. In [5] we explain how to perform strip-mining and interchange in non-convex iteration spaces in a single step.

2.1 Register Level

To exploit data locality at the register level after applying the loop interchange transformation, we need to fully unroll the most internal loops in the nest (loops that provide data reuse at this level) and to apply scalar replacement. We will refer to the loops that we want to unroll as Unroll Candidates Loops (UCLs).

Scalar replacement [2] finds opportunities for reuse of subscripted variables and replaces the references involved by references to temporal scalar variables; we use scalar replacement for determining array elements that are loop invariant with respect to the iteration variable of the innermost loop.

The loop i of Fig.1(a) is an UCL and can be fully unrolled using conditional statements in the loop body as shown in Fig.1(b). Loop i inside the if-part can be fully unrolled since it always executes the same number of iterations. The choice between the unrolled and non-unrolled version is performed at runtime by the if-test. The overhead of the if-test seems to be equivalent to the overhead of the max function in Fig.1(a). Nevertheless, the problem of using conditional statements is that, in the general case, the conditional statement cannot be moved outside the loop that surround the UCLs (loop k in Fig.1(b)) and it is not possible to apply scalar replacement to this loop.

In general, the bounds of the UCLs are affine functions of the surrounding loops iteration variables, and it is not possible to apply scalar replacement to the innermost loop, and therefore there is no data reuse at the register level.To overcome

```
do 10 iᴮ=Lᵢ, Uᵢ, Bᵢ              do 10 iᴮ=Lᵢ, Uᵢ, Bᵢ              do 10 iᴮ=Lᵢ, Uᵢ, Bᵢ
   ...                              ...                              ...
   do 10 k =Lₖ, Uₖ, Bₖ               do 10 k =Lₖ, Uₖ, Bₖ               do 20 k =Lₖ, min(iᴮ, Uₖ), Bₖ
      do 10 i = max(iᴮ, k), iᴮ+Bᵢ-1     if (iᴮ.ge.k) then                  do 20 i = iᴮ, iᴮ+Bᵢ-1
         ···A(i)···                      do 20 i= iᴮ, iᴮ+Bᵢ-1                  ···A(i)···
10 continue                                 ···A(i)···                 20 continue
                          (a)        20  continue                         do 10 k = k, Uₖ, Bₖ
                                        else                                 do 10 i = k, iᴮ+Bᵢ-1
                                           do 30 i = k, iᴮ+Bᵢ-1                 ···A(i)···
                                              ···A(i)···                10 continue
                                     30  enddo
                                        endif                                                    (c)
                                     10 continue         (b)
```

Fig. 1. (a) Example of loop nest. (b) Code using conditional statements. (c) Loop nest after applying ISS.

this problem, we use Index Set Splitting (ISS) on the external loops in order to simplify the bounds of the UCLs, so that they can be fully unrolled.

For example, to unroll loop i in Fig.1(a), it is necessary that i always executes the same number of iterations. Isolating the bounds i^B and i^B+B_i-1 from the other bounds, we will achieve this goal.

In the example we split loop k into two new loops in such a way that in every iteration of one of the new loops the constraint $k \leq i^B$ holds, and in every iteration of the other loop the constraint $k > i^B$ holds. Now we can simplify the bounds of loop i in both new loop nests. In the loop nest where $k \leq i^B$ holds, the lower complex bound of i can be simplified to i^B ($\max(i^B, k) = i^B$). Similarly, in the loop nest where $k > i^B$ holds, the lower complex bound of i can be simplified to k ($\max(i^B, k) = k$). The resulting code after index set splitting is shown in Fig.1(c). The UCL i of the first loop nest always executes B_i iterations and can be fully unrolled. The loop i of the second loop nest never executes a constant number of iteration and cannot be unrolled.

The order in which we apply ISS is very important to avoid processing a loop more than once and to avoid code explosion. We first apply ISS to the loops that we want to unroll (UCLs) from innermost to outermost (this ordering makes possible processing each loop only once). Then, ISS is applied to the rest of the loops from outermost to innermost (this second ordering avoids code explosion). See [5] for further details on how we apply ISS. Due to this order, it can happen that the UCLs are not directly surrounded by a loop. In this case it is necessary to apply loop distribution after ISS to be able to apply scalar replacement later on. In particular, we will distribute the loop that surrounds the UCLs. Figure 2(a) shows code of Fig.1(a) after applying unroll and scalar replacement.

3 Results and Conclusions

We used a triangular matrix product (MxM) algorithm and a rank 2k update (SYR2K) algorithm to evaluate the effects of the new method proposed. The matrices in the SYR2K algorithm are banded and stored in compact form, therefore the complexity of the loop bounds in SYR2K is higher than in the triangular matrix product algorithm.

The two algorithms are compiled first without applying any restructuring transformation, then using the KAP compiler to restructure the code, and finally applying our new method. We compare the performance (Mflops) obtained by each of these three versions of the algorithms on an ALPHA AXP 21064 (direct-mapped 8Kb cache memory and 32 floating point registers). In the results presented in this paper, we considered registers and first level cache.

In the triangular MxM algorithm the KAP compiler generates different code depending on the initial loop ordering so we have selected the ordering that yields the

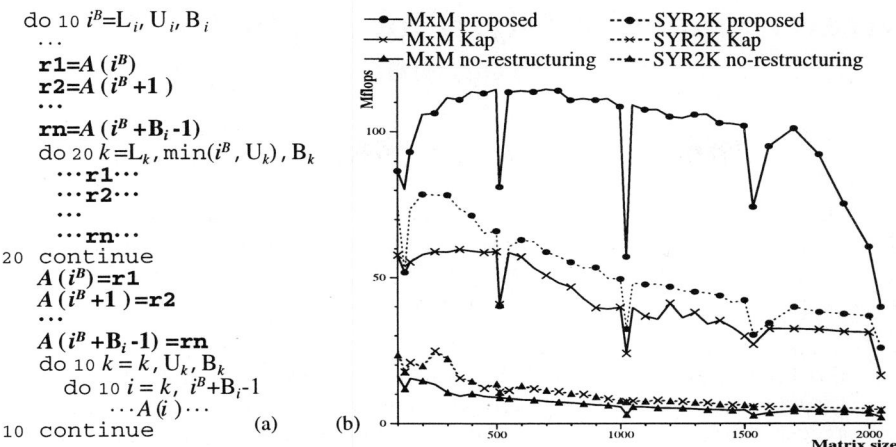

Fig. 2. (a) Loop nest after applying unroll and scalar replacement.
(b) Performance obtained by each method for different matrix sizes.

best performance. Figure 2(b) (solid lines) shows how our proposal doubles the Mflops obtained by the KAP compiler. For SYR2K the KAP compiler is only able to apply scalar replacement without unrolling. Figure 2(b) (dotted lines) shows the Mflops obtained with a fixed band of 50 and different matrix sizes for the three versions of the SYR2K algorithm. The proposed method obtains a speedup of around 4 over the KAP compiler.

In this paper we have presented a new unified method for code restructuring that aims at exploiting all levels of the memory hierarchy. The performance obtained with the method presented in this paper is substantially higher than the performance obtained by commercial compilers. Moreover, the relative improvement obtained with our method increases as the complexity of the loop bounds (such as in SYR2K) also increases.

Acknowledgments

This work was supported by the Ministry of Education and Science of Spain (CICYT TIC-0429/95).

References

1. D. Callahan, S. Carr, K. Kennedy. Improving Register Allocation for Subscripted Variables. Int. Conf. on Programming Language Design and Implementation, June 1990, pp. 53-65
2. S. Carr. Memory-Hierarchy Management. Ph.D. Dissertation, Rice University, Feb 1993.
3. S. Carr, K. McKinley, C-W. Tseng. Compiler Optimizations for Improving Data Locality. Int. Conf. on Architectural Support for Programming Languages and Operating Systems, Aug 1994, pp.252-262
4. A. Fernández, J.M. Llabería, M. Valero-García. Loop Transformation using non-unimodular matrices. IEEE Transactions on Parallel and Distributed Systems, Vol. 6, No. 8, Aug 1995, pp. 832-840
5. M. Jiménez, J.M. Llabería, A. Fernández, E. Morancho. A Unified Transformation Technique for Multilevel Blocking. TR. UPC-DAC-1995-51, Dept. of Computer Architecture, Polytechnic University of Catalonia, Dec 1995.
6. M. Lam, E.Rothberg, M. Wolf. The Cache Performance and Optimizations of Blocked Algorithms. Int. Conf. on Architectural Support for Programming Languages and Operating Systems, 1991, pp. 63-74
7. J.J Navarro, T. Juan, T. Lang. MOB Forms: A Class of Multilevel Block Algorithms for Dense Linear Algebra Operations. Int. Conf. on Supercomputing, July 1994, pp. 354-363
8. M. Wolf. Improving Locality and Parallelism in Nested Loops. Technical Report CSL-TR-92-538, Stanford University, Aug 1992.
9. M. Wolf, M. Lam. A Data Optimizing Algorithm. Int. Conf. on Programming Language Design and Implementation, June 1991, pp. 30-44
10. M. Wolf, M. Lam. A Loop Transformation Theory and an Algorithm to Maximize Parallelism. IEEE Trans. on Parallel and Distributed System, Vol. 2, No. 4, October 1991, pp. 452-471
11. M. Wolfe. More Iteration Space Tiling. Int. Conf. on Supercomputing, 1989, pp. 655-664

Array Dataflow Analysis for Explicitly Parallel Programs

Jean-François Collard[1] and Martin Griebl[2]

[1] Laboratoire PRISM, Université de Versailles St-Quentin, 45 Avenue des Etats-Unis, 78035 Versailles, FRANCE, Jean-Francois.Collard@prism.uvsq.fr
[2] FMI, Universität Passau, Innstraße 33, 94032 Passau, GERMANY, Martin.Griebl@fmi.uni-passau.de

Abstract. This paper describes a dataflow analysis of array data structures for data-parallel and/or control- (or task-) parallel imperative languages. This analysis departs from previous work because it 1) simultaneously handles both parallel programming paradigms, and 2) does not rely on the usual iterative solving process of a set of data flow equations but extends array dataflow analysis based on integer linear programming, thus improving the precision of results.

1 Introduction

After decades of parallel processing, few programming languages can claim to be used on a wide range of parallel architectures. Probably, one of the reasons lies in the difficulty of *efficiently* compiling a general language on an ever-widening spectrum of machines. Among useful analyses, *dataflow analysis* derives information about the definition and the subsequent use(s) of data values in a program. Its applications include dead-code elimination, strength reduction, array expansion [1] or equivalently conversion into single-assignment form.

Unfortunately, very few data-flow analyses have been proposed for parallel languages (but see [7]). This paper presents an analysis for data-parallel languages, e.g. HPF [2], control-parallel (also called task-parallel) languages, such as PCF [3], or a mixture of both [4, 5, 6].

2 Motivating Examples

Several languages [4, 6] have indexed parallel constructs whose semantics correspond to what we call `doall` in this paper: a statement instance is spawned for each possible value of the index variable (e.g., from 0 to $2n$ by step 1 in Program ExD, hence $2n + 1$ instances or tasks are spawned). Each instance has its own copy of shared data structures, and all reads and writes are applied to this copy. Shared data structures are updated only when the instances of all statements in the loop body have completed. Thus, within one iteration of a `doall` loop, all statements work on the same copy.

For example in Program ExD, if $n = 2$, the read of `a(1)` in Instance number 1 of Statement R may return the value produced by the instance number 1 of

	program ExD		program ExF
P_1	doall($i = 0 : 2*n : 1$)		forall($i = 0 : 2*n : 1$)
	where (...)		where (...)
S_1 :	a(i) = ...	S_1 :	a(i) = ...
S_2 :	else a(2*n-i) = ...	S_2 :	else a(2*n-i) = ...
	endwhere		endwhere
R :	... = a(i)	R :	... = a(i)

Fig. 1. Two examples

S_1, but not from Instance 3 of S_2. Consequently, the information we would like to automatically derive is that the source of a(i) in Instance i of R is either S_1 in Instance i or undefined (written as $\bot$) if $i \neq n$; or: S_1 or S_2 in Instance i if $i = n$ (because in this case, either S_1 or S_2 writes into a(n)).

Notice that we consider that all instances of both arms of the parallel conditional structure **where** are executed in parallel.

Similarly, a **forall** construct spawns as many instances as there are possible values for the index variable (e.g., $2n+1$ instances are spawned in Program **ExF**). The semantics of **forall** we consider is reminiscent from the HPF semantics[3] [2]: in a multi-statement **forall**, the array assignment semantics are applied to each statement in turn. Each instance of a statement has its own copy of shared data structures, but these shared data structures are updated before the instances of following statements begin.

Consider the read of a(1) in Program **ExF**: it may be the case that both S_1 and S_2 simultaneously wrote into this cell. Such an *over-determined source* will be denoted by top ($\top$). To sum up, our analysis derives that the source of a(i) in Instance i of R in Program **ExF** is either S_1 in Instance i or S_2 in Instance $2n - i$ or $\bot$ or $\top$, if $i \neq n$, or S_1 or S_2 in Instance i, if $i = n$ (without $\bot$ nor $\top$).

3 Definitions

The input language includes the following sequential control structures: **do**, **while**, **if**, the following parallel structures: **forall**, **doall**, **where**, and parallel sections **parsection**.

The dimension of a vector **x** is denoted by $|\mathbf{x}|$. The k-th entry of vector **x**, $k \geq 0$, is denoted by $\mathbf{x}[k]$. The sub-vector built from components k to l is written as: $\mathbf{x}[k..l]$. If $k > l$, then this vector is by convention the vector of dimension 0. Furthermore, $\leqslant$ ($\ll$) denotes the (strict) lexicographical order on such vectors.

The *index vector* of a statement S is the vector built from the counters of surrounding **do**, **forall**, **doall** and **while** constructs. An *operation* is an instance of some Statement S, and will be denoted by $\langle S, \mathbf{x} \rangle$, where **x** is some value of the index vector of S.

---
[3] Note that in HPF the only parallel constructs inside **forall**s are **forall**s and **where**s (Rules H404 and H406 of [2]).

The *depth* of a statement or construct is the number of surrounding do, forall, doall or while constructs. So, the depth of S equals to $|\mathbf{x}|$. If $\mathbf{x}[p]$, $p \geq 0$, is a counter of a do, forall or doall construct, then lower and upper affine bounds are known: $l_p(\mathbf{x}[0..p-1]) \leq \mathbf{x}[p] \leq u_p(\mathbf{x}[0..p-1])$ where l_p and u_p are syntactically given by the loop bounds. In the case of while-loops, we have by convention $1 \leq \mathbf{x}[p]$. The index domain of a statement S is denoted by $\mathbf{D}(S)$ and is given by the conjunction of all inequalities on surrounding loop bounds. We define $C_p(S)$ as the iterative (do, while, forall or doall) construct surrounding S at depth p. (When clear from the context, S will be omitted.) For example, let us consider Statement S_1 in Program ExD: $C_0 = P_1$ (the doall).

We define $\mathcal{P}(S, R)$ to be the par construct surrounding both S and R such that S and R appear in distinct sections of the par construct. (Notice that there is at most one such construct.) Moreover, let M_{SR} be the depth of $\mathcal{P}(S, R)$. (If $\mathcal{P}(S, R)$ does not exist, then M_{SR} is the number of do, forall, doall or while constructs surrounding both S and R.) In Program ExF, $\mathcal{P}(S_1, R) = \emptyset$ and $M_{S_1 R} = 1$. Predicate Δ_{SR} is true if S and R appear in opposite arms of an if..then..else or where..elsewhere construct, or in distinct sections of a par construct. Predicate T_{SR} is true if S textually precedes R and Δ_{SR} is false. We denote by $\mathcal{W}(u)$ (resp. $\mathcal{R}(u)$) the memory cell written into (resp. read) by operation u, so for instance $\mathcal{W}(\langle S_2, i \rangle) =$ a($2n - i$) and $\mathcal{R}(\langle R, i \rangle) =$ a(i) in Programs ExD and ExF.

4 The Semantics: Execution Order

The purpose of this section is not to give a complete semantical description of a parallel language. As far as dataflows are concerned, we are mainly interested in the order in which computations, and their corresponding writes and reads to memory, occur. The fact that $\langle S, \mathbf{x} \rangle$ is executed before $\langle R, \mathbf{y} \rangle$ in the parallel program will be denoted by $\langle S, \mathbf{x} \rangle \prec \langle R, \mathbf{y} \rangle$. In the case of sequential programs, $\prec$ is a total order which can be expressed as the lexicographical order on index vectors. In turn, the lexicographical order can be expressed as a disjunction of linear inequalities. Expressing the execution order $\prec$ of parallel programs is more intricate. For instance, Section 2 showed that two semantics can be chosen for a data-parallel construct: the "synchronous" semantics of the construct we call forall, where the memory is updated between the execution of two successive statements inside a forall; and the "asynchronous" semantics of the construct we call doall, where none of the spawned tasks sees the effects produced by other tasks. However, thanks to the semantics of forall and doall constructs, $\prec$ can still be expressed in a linear way:

$$\langle S, \mathbf{x} \rangle \prec \langle R, \mathbf{y} \rangle \Leftrightarrow \left(\bigvee_{\substack{p=0..M_{SR}-1, \\ C_p=\text{do} \vee C_p=\text{while}}} Pred(p, \mathbf{x}, \mathbf{y}) \right)$$
$$\vee \left(\left(\bigwedge_{p=0..M_{SR}-1} Equ(p, C_p, \mathbf{x}, \mathbf{y}) \right) \wedge T_{SR} \right) \quad (1)$$

where:
$$Pred(p, \mathbf{x}, \mathbf{y}) \equiv \left(\bigwedge_{i=0..p-1} Equ(i, C_i, \mathbf{x}, \mathbf{y}) \right) \wedge \mathbf{x}[p] < \mathbf{y}[p] \qquad (2)$$

$$Equ(p, \mathtt{do}, \mathbf{x}, \mathbf{y}) \equiv Equ(p, \mathtt{while}, \mathbf{x}, \mathbf{y}) \equiv \mathbf{x}[p] = \mathbf{y}[p] \qquad (3)$$

$$Equ(p, \mathtt{forall}, \mathbf{x}, \mathbf{y}) \equiv \mathbf{true} \qquad (4)$$

$$Equ(p, \mathtt{doall}, \mathbf{x}, \mathbf{y}) \equiv \mathbf{x}[p] = \mathbf{y}[p] \qquad (5)$$

Obviously, (1) is a partial order on operations. (In particular, it has no cycle.) Intuitively, Predicate $Pred$ in (1) formalizes the sequential order of a given $\mathtt{do}$ or $\mathtt{while}$ loop at depth p. Such a loop enforces an order up to the first $\mathtt{par}$ construct encountered at depth M_{SR} while traversing the nest of control structures, from the outermost level to the innermost. The order of sequential loops is given by the strict inequality in (2), under the condition that the two operations at hand are not ordered up to level $p-1$; hence the conjunction on the p outer predicates Equ. Notice that the instances of two successive statements inside a $\mathtt{forall}$ at depth p are always ordered at depth p due to (4), but are ordered inside a $\mathtt{doall}$ only if they belong to the same task (i.e., the values of the $\mathtt{doall}$ index are equal, cf (5)).

Note that $\forall P, \bigvee_{i \in \emptyset} P(i) = \mathbf{false}$ and $\bigwedge_{i \in \emptyset} P(i) = \mathbf{true}$. Note also that the lexicographical order on nests of $\mathtt{do}$ loops [1] and in (sequential) dynamic control programs [8] comes as a special case of (1).

Example As far as programs $\mathtt{ExD}$ and $\mathtt{ExF}$ are concerned, the order between S_1 (or S_2) and R is given by:

ExF: $\mathcal{P}(S_1, R) = \emptyset$, so $M_{SR} = 1$. $C_0 = \mathtt{forall}$ and $T_{S_1 R} = \mathbf{true}$. Thus:

$$\langle S_1, i' \rangle \prec \langle R, i \rangle \Leftrightarrow \left(\bigvee_{p \in \emptyset} Pred(p, i', i) \right) \vee (Equ(0, C_0, i', i) \wedge T_{S_1 R})$$
$$\Leftrightarrow \mathbf{false} \vee (\mathbf{true} \wedge \mathbf{true}) \Leftrightarrow \mathbf{true} \qquad (6)$$

This is a formal restatement of a semantical property of $\mathtt{forall}$s, cf Section 2. Notice that $\langle S_2, i' \rangle \prec \langle R, i \rangle$ is also always true.

ExD: $\mathcal{P}(S_1, R) = \emptyset$, so $M_{SR} = 1$. $C_0 = \mathtt{doall}$ and $T_{S_1 R} = \mathbf{true}$. Thus:

$$\langle S_1, i' \rangle \prec \langle R, i \rangle \Leftrightarrow \left(\bigvee_{p \in \emptyset} Pred(p, i', i) \right) \vee (Equ(0, C_0, i', i) \wedge T_{S_1 R})$$
$$\Leftrightarrow \mathbf{false} \vee (i' = i \wedge \mathbf{true}) \Leftrightarrow i' = i \qquad (7)$$

5 Dataflow Analysis

Our dataflow analysis first builds the set of possible sources of the flow of a given data, and then selects the latest element, i.e., the maximal element according to the original sequential order. Selecting the maximal element is then done using Integer Linear Programming techniques.

5.1 Method Overview

Let us consider two statements S and R. Suppose that S writes into an array a and that R reads that same array:

$$S : \quad \mathtt{a}(f(\mathbf{x})) = \ldots$$
$$R : \quad \ldots = \mathtt{a}(g(\mathbf{y}))$$

The aim of array dataflow analysis is to find the source of the value $\mathtt{a}(g(\mathbf{y}))$ read in R for a given $\mathbf{y}$. This source is denoted by $\sigma(\langle R, \mathbf{y} \rangle)$[4]. To be a source candidate, an operation $\langle S, \mathbf{x} \rangle$ has to satisfy the following constraints:

Existence predicate: $\langle S, \mathbf{x} \rangle$ is a valid operation: $e(\langle S, \mathbf{x} \rangle)$ evaluates to **true**. (See Section 5.2.)
Conflicting accesses: $\langle S, \mathbf{x} \rangle$ and $\langle R, \mathbf{y} \rangle$ access the same array cell: $f(\mathbf{x}) = g(\mathbf{y})$. f and g are possibly multi-dimensional affine functions.
Sequencing condition: $\langle S, \mathbf{x} \rangle$ is executed before $\langle R, \mathbf{y} \rangle$ in the parallel program: $\langle S, \mathbf{x} \rangle \prec \langle R, \mathbf{y} \rangle$. Notice that this leads to a formal definition of $\bot$: $\bot$ is the name of the operation that is executed once before all other operations of the program, i.e., $\forall S, \mathbf{x} : \bot \prec \langle S, \mathbf{x} \rangle$.
Environment: The set of source candidates is computed under the hypothesis that $\langle R, \mathbf{y} \rangle$ is a valid operation, i.e., $\mathbf{y} \in \mathbf{D}(R)$.

The set of candidate sources is thus:

$$\mathbf{Q}_{SR}(\mathbf{y}) = \{ \langle S, \mathbf{x} \rangle \mid e(\langle S, \mathbf{x} \rangle), \qquad (Existence) \\ f(\mathbf{x}) = g(\mathbf{y}), \quad (Conflicting\ accesses) \qquad (8) \\ \langle S, \mathbf{x} \rangle \prec \langle R, \mathbf{y} \rangle \} \qquad (Order)$$

The direct dependence from S to R is then $K_{SR}(\mathbf{y}) = \max_\prec \mathbf{Q}_{SR}(\mathbf{y})$. Obviously, S may not be the only statement writing into a. Let $\mathcal{S}(R)$ be the set of statements writing into the array read by R. Then, the set of candidate sources is $\mathbf{Q}_R(\mathbf{y}) = \bigcup_{S \in \mathcal{S}(R)} \mathbf{Q}_{SR}(\mathbf{y})$. The source $K_R(\mathbf{y})$ of the flow of datum $\mathtt{a}(\ g(\mathbf{y})\ )$ is the maximal element in $\mathbf{Q}_R(\mathbf{y})$ according to $\prec$:

$$K_R(\mathbf{y}) = \max_\prec \mathbf{Q}_R(\mathbf{y}). \qquad (9)$$

Clearly, three problems occur:

- We have to express Predicate e. This issue is addressed in Section 5.2.
- Maxima according to $\prec$ among sets of operations have to be computed. Section 5.3 explains how to compute $\max_\prec$ using the Omega tool.
- $K_R(\mathbf{y})$ may not be uniquely defined, since $\prec$ is not a total order. Intuitively, a non-unique maximum means that two operations wrote in an undefined order into the same memory cell. The over-determined source is denoted by $\top$ (cf Section 5.4).

---
[4] For the sake of clarity, we assume that an operation executes at most one read.

5.2 Existence of Operations

We say that an operation *exists* if this operation executes. In the case of static-control sequential programs, the only loops are do loops, and the existence predicate boils down to $e(\langle S, \mathbf{x} \rangle) \Leftrightarrow \mathbf{x} \in \mathbf{D}(S)$.

When arbitrary while, if and where constructs appear, the control flow is dynamic, and existence of operations cannot in general be predicted. The problem then boils down to finding a suitable coding of Existence predicate e. The reader is referred to [8, 10] for details. In the sequel, we will use the Omega package[9] and phrase this paper in the corresponding framework[5]. For instance, the case of if and where constructs is handled as follows: If $\mathbf{x}$ is the index vector of a conditional statement, the execution of the then branch is coded by postulating that some uninterpreted function f evaluates to, say, a nonnegative value: $f(\mathbf{x}) \geq 0$. The execution of the else or elsewhere branch is then denoted by $f(\mathbf{x}') < 0$. Since both branches cannot execute in the same instance of the construct, $f(\mathbf{x}) \geq 0 \wedge f(\mathbf{x}') < 0 \Rightarrow \mathbf{x} \neq \mathbf{x}'$.

Example Let us find possible bottoms in the source of the read $\langle R, k \rangle$ in Program **ExF**. (k is here a parameter.) The set of possible writes from S_1 is:

```
# Omega Calculator [v1.00, Mar 96]:
# symbolic n, f(1), k;
# W1 := { [iw] : 0 <= iw <= 2n && f(Set) >= 0 && iw = k } ;
```

where f(Set) means that f is applied to the bounded variable(s) that define the set (here, iw). Then, for S_2:

```
# W2 := { [iw] : 0 <= iw <= 2n && f(Set) < 0  && 2n - iw = k } ;
```

We are interested in finding all reads that do not have a corresponding write from either S_1 or S_2. We thus take the union of the two sets of writes:

```
# W1 union W2 ;
{[iw]: k = iw && 0 <= iw <= 2n && 0 <= f(iw)} union
 {[iw]: k+iw = 2n && 0 <= iw <= 2n && f(iw) <= -1}
```

We then subtract the obtained set to the set R of all reads:

```
# R := { [ir] : 0 <= ir <= 2n && ir = k } ;
# Bottoms := R intersection complement (W1 union W2);
# Bottoms ;
{[In_1]: k = In_1 && n < In_1 <= 2n && f(In_1) <= -1} union
 {[In_1]: k = In_1 && 0 <= In_1 < n && f(In_1) <= -1}
```

Since nothing is known about f, we have to take a conservative approach and assume that any predicate involving f is true. The set of possibly undefined reads (bottoms) is thus given by $\{k : 0 \leq k < n\} \cup \{k : n < k \leq 2n\}$. Notice that, as expected, the case $k = n$ does not occur, i.e., $\langle S, n \rangle$ is always defined by S_1 and/or S_2.

---

[5] When called with an input file, the Omega calculator (v1.00 dated March 1996) copies the input equations on the standard output, prefixed by the # sign.

5.3 Computing Maxima

Solving (9), i.e., computing the maximum (maxima), is equivalent to finding the element(s) $K_{SR}(\mathbf{y})$ such that $\neg\exists \mathbf{x} \in \mathbf{Q}_{SR}(\mathbf{y}), \mathbf{x} \neq K_{SR}(\mathbf{y}) : K_{SR}(\mathbf{y}) \prec \mathbf{x}$. This expression may have several solutions, since parallel programs have partial execution orders. (See Section 5.4.) Similarly, $K_R(\mathbf{y})$ in (9) can be computed by:

$$\neg\exists \langle S, \mathbf{x}\rangle, e(\langle S, \mathbf{x}\rangle), \mathcal{W}(\langle S, \mathbf{x}\rangle) = \mathcal{R}(\langle R, \mathbf{y}\rangle), \langle S, \mathbf{x}\rangle \neq K_R(\mathbf{y}), K_R(\mathbf{y}) \prec \langle S, \mathbf{x}\rangle \tag{10}$$

5.4 Detecting Over-Determined Sources

Detection of over-determined sources is done by checking that no two distinct operations satisfy their respective existence predicates, write into the same memory cell, and are not related by $\prec$. Formally: $Error(u,v) \equiv e(u) \wedge e(v) \wedge u \neq v \wedge \mathcal{W}(u) = \mathcal{W}(v) \wedge \neg(u \prec v) \wedge \neg(v \prec u)$, where u and v are operations. Notice that, in the general case, checking that no dependence is carried by a `doall`- or a `forall`-loop is not sufficient.

$\boxed{\text{Example}}$ Let us find the over-determined cases in the source of operation $\langle R, k\rangle$ in Program **ExF**.

```
# Symbolic f(1), n , k ;
# Error :=
# { [iw1] -> [iw2] : 0 <= iw1, iw2 <= 2n && f(In) >= 0 && f(Out) < 0
#     && iw1 = 2n - iw2 = k }
# union
# { [iw1] -> [iw1'] : 0 <= iw1, iw1' <= 2n && f(In) >= 0 && f(Out) >= 0
#     && iw1 != iw1' && iw1 = iw1' = k }
# union
# { [iw2] -> [iw2'] : 0 <= iw2, iw2' <= 2n && f(In) < 0 && f(Out) < 0
#     && iw2 != iw2' && 2n - iw2 = 2n - iw2' = k } ;
```

In and Out are collective names of the input (the tuple preceding the -> and output (the tuple following the ->) variables of a relation, respectively.

```
# Error ;
{[In_1] -> [iw2]  : In_1+iw2 = 2n && k = In_1 && n < In_1 <= 2n
                    && f(iw2) <= -1 && 0 <= f(In_1)} union
 {[In_1] -> [iw2] : In_1+iw2 = 2n && k = In_1 && 0 <= In_1 < n
                    && f(iw2) <= -1 && 0 <= f(In_1)}
```

As in the case of bottoms, f is unknown. The set **Error** of possibly over-determined reads ($\top$) is thus $\{k : 0 \leq k < n\} \cup \{k : n < k \leq 2n\}$. Notice that this is the result expected from Section 2.

6 Conclusion

This paper presented a dataflow analysis that can be applied to data- and/or task- parallel programs, with dynamic flows of control. Li and Wolfe proposed [11] a simple and precise framework to express the interaction of arbitrarily nested parallel control structures, together with an appropriate data dependence analysis. They did not, however, extended this work to data flows.

The main results of this paper are: a general affine expression for the execution order of programs written in a structured imperative parallel language (this expression subsumes the lexicographical execution order in sequential programs), and a general affine expression for the over-determined cases (*Error*).

Acknowledgments We would like to thank L. Bougé, T. Brandes, W. Pugh, Y. Robert, G. Utard and M. Wolfe for their helpful comments, and D. Wannacott for his technical help with Omega. Authors are supported by the CNRS and the DFG project RecuR, respectively, and in addition by a German-French programme Procope.

References

1. P. Feautrier. Dataflow analysis of scalar and array references. *Int. Journal of Parallel Programming*, 20(1):23–53, February 1991.
2. C.H. Koelbel, D.B. Loveman, R.S. Schreiber, G.L. Stelle Jr, and M.E. Zosel. *The High Performance Fortran Handbook*. The MIT Press, 1994.
3. Parallel Computing Forum. PCF fortran extensions. *Fortran Forum*, 10(3), 1991.
4. M. Gerndt and R. Berrendorf. SVM-Fortran, Reference Manual, Version 1.4. KFA-ZAM-IB-9510, Research Center Jülich May 1995.
5. I. Foster et al. A Compilation System That Integrates High Performance Fortran and Fort ran M. In *Proc. of the Scalable High-Performance Computing Conf (SHPCC'9 4)*. pages 293–300, Knoxville, TN, May 1994.
6. B. Chapman et al. Extending Vienna Fortran with Task Parallelism. In *Proc. of the 1994 Intl Conf on Parallel and Distributed System s*. pages 258–263, Hsinchu, Taiwan, December 1994.
7. J. Ferrante, D. Grunwald, and H. Srinivasan. Computing communication sets for control parallel programs. In K. Pingali et al., editor, *Proc. of 7th Int. W. on Lang. and Compilers for Parallel Comp.*, volume 892 of *LNCS*, pages 316–330, Ithaca, NY, August 1994.
8. J.-F. Collard, D. Barthou, and P. Feautrier. Fuzzy array dataflow analysis. In *Proc. of 5th ACM SIGPLAN Symp. on Principles and Practice of Parallel Programming*, pages 92–102, Santa Barbara, CA, July 1995.
9. W. Pugh. A practical algorithm for exact array dependence analysis. *Communications of the ACM*, 35(8):27–47, August 1992.
10. W. Pugh and D. Wonnacott. Nonlinear Array Dependence Analysis. In *Third Workshop on Languages, Compilers and Run-Time Systems for Sc alable Computers*, Troy, NY, May 1992.
11. J. Li and M. Wolfe. Defining, Analyzing, and Transforming Program Constructs. *IEEE Parallel and Distributed Technology*, 32–39, Spring 1994

Semantic Foundations of Commutativity Analysis

Martin C. Rinard[†] and Pedro C. Diniz[‡]

Department of Computer Science
University of California, Santa Barbara
Santa Barbara, CA 93106
{martin,pedro}@cs.ucsb.edu
http://www.cs.ucsb.edu/~{martin,pedro}

Abstract. This paper presents the semantic foundations of commutativity analysis, an analysis technique for automatically parallelizing programs written in a sequential, imperative programming language. Commutativity analysis views the computation as composed of operations on objects. It then analyzes the program at this granularity to discover when operations commute (i.e. generate the same result regardless of the order in which they execute). If all of the operations required to perform a given computation commute, the compiler can automatically generate parallel code. This paper shows that the basic analysis technique is sound. We have implemented a parallelizing compiler that uses commutativity analysis as its basic analysis technique; this paper also presents performance results from two automatically parallelized applications.

1 Introduction

Current parallelizing compilers preserve the semantics of the original serial program by preserving the data dependences [1]. They analyze the program to identify independent pieces of computation (two pieces of computation are independent if neither writes a piece of memory that the other accesses), then generate code that executes independent pieces concurrently.

This paper presents the semantic foundations of a new analysis technique called commutativity analysis. Instead of preserving the relative order of individual reads and writes to single words of memory, commutativity analysis views the computation as composed of operations on objects. It then analyzes the computation at this granularity to discover when pieces of the computation commute (i.e. generate the same result regardless of the order in which they execute). If all of the operations required to perform a given computation commute, the compiler can automatically generate parallel code. While the resulting parallel program may violate the data dependences of the original serial program, it is still guaranteed to generate the same result.

We expect commutativity analysis to eliminate many of the limitations of existing approaches. Compilers that use commutativity analysis will be able

---

[†] Supported in part by an Alfred P. Sloan Research Fellowship.
[‡] Sponsored by the PRAXIS XXI program administrated by Portugal's JNICT – Junta Nacional de Investigação Científica e Tecnológica, and holds a Fulbright travel grant.

to automatically generate parallel code for many applications that periodically update shared data structures using commuting operations and/or manipulate recursive, pointer-based data structures such as lists, trees and graphs. Commutativity analysis allows compilers to automatically parallelize computations even though they have no information about the global topology of the manipulated data structure. It may therefore be especially useful for parallelizing computations that manipulate the persistent data in object-oriented databases. In this context the code that originally created the data structure may be unavailable, negating any approach (including data-dependence based approaches) that analyzes the code to verify or discover global properties of the data structure topology.

The rest of the paper is structured as follows. In Section 2 we present an example that illustrates how commutativity analysis can automatically parallelize graph traversals. In Section 3 we describe the basic approach and informally state the conditions that the compiler uses to recognize commuting operations. In Section 4 we provide a formal model of computation and define what it means for two operations to commute in this model. This section contains a key theorem establishing that if all of the operations in a computation commute, than all parallel executions are equivalent to the serial execution. In Section 5 we state that we have developed the analysis algorithms required for the practical application of commutativity analysis and provide a reference to a technical report that presents the analysis algorithms. In Section 6 we present experimental results for two complete scientific applications parallelized by a prototype compiler that uses commutativity analysis as its basic analysis technique. In Section 7 we conclude.

2 An Example

In this section we present a simple example that shows how recognizing commuting operations can enable the automatic generation of parallel code. The visit method in Figure 1 serially traverses a graph. When the traversal completes, each node's sum instance variable contains the sum of its original value and the values of the val instance variables in all of the nodes that directly point to that node. The example is written in C++.

The traversal generates one invocation of the visit method for each edge in the graph. We call each method invocation an *operation*. The receiver of each visit operation is the node to traverse. Each visit operation takes as a parameter p the value of the instance variable val of the node that points to the receiver. The visit operation first adds p into the running sum stored in the receiver's sum instance variable. It then checks the receiver's mark instance variable to see if the traversal has already visited the receiver. If not, the operation marks the receiver, then recursively invokes the visit method for all of the nodes that the receiver points to.

The way to parallelize the traversal is to execute the two recursive visit operations concurrently. But this parallelization may violate the data dependences.

```
class graph {
  boolean mark;
  int val, sum;
  graph *left; graph *right;
};
graph::visit(int p) {
  graph *l = left;
  graph *r = right;
  int v = val;
  sum = sum + p;
  if (!mark) {
    mark = TRUE;
    if (l != NULL) l->visit(v);
    if (r != NULL) r->visit(v);
  }
}
```

Fig. 1. Serial Graph Traversal

```
class graph {
  lock mutex;
  boolean mark;
  int val, sum;
  graph *left; graph *right;
};
graph::visit(int s) {
  this->parallel_visit(s);
  wait();
}
graph::parallel_visit(int p) {
  mutex.acquire();
  graph *l = left;
  graph *r = right;
  int v = val;
  sum = sum + p;
  if (!mark) {
    mark = TRUE;
    mutex.release();
    if (l != NULL)
      spawn(l->parallel_visit(v));
    if (r != NULL)
      spawn(r->parallel_visit(v));
  } else
    mutex.release();
}
```

Fig. 2. Parallel Graph Traversal

The serial computation executes all of the accesses generated by the left traversal before all of the accesses generated by the right traversal. If the two traversals visit the same node, in the parallel execution the right traversal may visit the node before the left traversal, changing the order of reads and writes to that node. This violation of the data dependences may generate cascading changes in the overall execution of the computation. Because of the marking algorithm, a node only executes the recursive calls the first time it is visited. If the right traversal reaches a node before the left traversal, the parallel execution may also change the order in which the overall traversal is generated.

In fact, none of these changes affects the overall result of the computation. It is possible to automatically parallelize the computation even though the resulting parallel program may generate computations that differ substantially from the original serial computation. The key property that enables the parallelization is that the parallel computation generates the same set of visit operations as the serial computation and the generated visit operations can execute in any order without affecting the overall behavior of the traversal.

Given this commutativity information, the compiler can automatically generate the parallel `visit` method in Figure 2. The top level `visit` method first invokes the `parallel_visit` method, then invokes the `wait` construct, which blocks until the entire parallel computation completes. The `parallel_visit` method executes the recursive calls concurrently using the `spawn` construct, which creates a new task for the execution of each method invocation. The compiler also augments the `graph` data structure with a mutual exclusion lock `mutex`. The `parallel_visit` method uses this lock to ensure that all of its invocations execute atomically.

3 The Basic Approach

Commutativity analysis is designed for programs written using a pure object-based paradigm. Such programs structure the computation as a sequence of operations on objects. Each operation consists of a receiver object, an operation name and several parameters. Each operation name identifies a method that defines the behavior of the operation; when the operation executes, it executes the code in that method. Each object implements its state using a set of instance variables. When an operation executes it can recursively invoke other operations and/or use primitive operators (such as addition and multiplication) to perform computations involving the parameters and the instance variables of the receiver.

Commutativity analysis is designed to work with *separable* methods, or methods whose execution can be decomposed into an object section and an invocation section. The object section performs all accesses to the receiver. The invocation section invokes other operations and does not access the receiver. It is of course possible for local variables to carry values computed in the object section into the invocation section, and both sections can access the parameters. Separability imposes no expressibility limitations — although the current compiler does not do so, it is possible to automatically convert any method into a collection of separable methods via the introduction of auxiliary methods.

The following conditions, which the compiler can use to test if two operations A and B commute, form the foundation of commutativity analysis.

1. (**Instance Variables**) The new value of each instance variable of the receiver objects of A and B must be the same after the execution of the object section of A followed by the object section of B as after the execution of the object section of B followed by the object section of A.
2. (**Invoked Operations**) The multiset of operations directly invoked by either A or B under the execution order A followed by B must be the same as the multiset of operations directly invoked by either A or B under the execution order B followed by A. [1]

Note that these conditions do not deal with the entire recursively invoked computation that each operation generates — they only deal with the object and

---
[1] Two operations are the same if they execute the same method and have the same parameters.

invocation sections of the two operations. Furthermore, they are not designed to test that the entire computations of the two operations commute. They only test that the object sections of the two operations commute and that the operations together invoke the same multiset of operations regardless of the order in which they execute. As we argue below, if all pairs of operations in the computation satisfy the conditions, then all parallel executions generate the same result as the serial execution.

The instance variables condition ensures that if the parallel execution invokes the same multiset of operations as the serial execution, the values of the instance variables will be the same at the end of the parallel execution as at the end of the serial execution. The basic reasoning is that for each object, the parallel execution will execute the object sections of the operations on that object in some arbitrary order. The instance variables condition ensures that all orders yield the same final result.

The invoked operations condition provides the foundation for the application of the instance variables condition: it ensures that all parallel executions invoke the same multiset of operations (and therefore execute the same object sections) as the serial execution.

Both commutativity testing conditions are trivially satisfied if the two operations have different receivers — in this case their executions are independent because they access disjoint pieces of data. We therefore focus on the case when the two operations have the same receiver.

It is possible to determine if each of the receiver's instance variables has the same new value in both execution orders by analyzing the invoked methods to extract two symbolic expressions. One of the symbolic expressions denotes the new value of the instance variable under the execution order A followed by B. The other denotes the new value under the execution order B followed by A. Given these two expressions, a compiler may be able to use algebraic reasoning to discover that they denote the same value. The compiler uses a similar approach to determine if A and B together invoke the same multiset of operations in both execution orders.

To use the commutativity testing conditions, the compiler first computes a conservative approximation to the set of methods invoked as a result of executing a given computation. The compiler then applies the commutativity testing conditions to all pairs of potentially invoked methods that may have the same receiver. If all of the pairs commute, the compiler can legally generate parallel code.

4 Formal Treatment

We next provide a formal treatment of how commutativity enables parallel execution. We first present a formal model of computation for separable operations. We define parallel and serial execution in this model, and define what it means for two operations to commute. The key theorem establishes that if all of the operations in the parallel executions commute, then all parallel executions are equivalent to the serial execution.

4.1 The Model of Computation

We next describe a formal model of computation for separable programs. We assume a set of objects $r, o \in O$, a set of constants $c \in C \subseteq O$, a set of instance variable names $v \in IV$ and a set of operation names $op \in OP$. The operational semantics uses several functions. We model instance variable values with functions $i : I = IV \to O$. We say that an object r is in state i if the value of each instance variable v of r is $i(v)$. We model object memories with functions $m : M = O \to I$. We also have operations $a \in A = O \times OP \times O$, and write an element a in the form $r\text{->}op(o)$. Each operation has a receiver r, an operation name op and a parameter o. To simplify the presentation we assume that each operation takes only one parameter, but the framework generalizes in a straightforward way to include operations with multiple parameters.

We next describe how we model computation. The execution of each invoked operation first updates the receiver, then invokes several other operations. We model the effect on the receiver with a receiver effect function $R : A \times I \to I$. $R(r\text{->}op(o), i)$ is the new state of r when operation $r\text{->}op(o)$ executes with the receiver r in state i; it models the effect of the operation's object section on the receiver.

We model the invoked operations using sequences $s \in S = seq(A)$ of the form $a_1 \circ \cdots \circ a_k$. ϵ is the empty sequence, and $\epsilon \circ s = s \circ \epsilon = s$. We use an invoked operation function $N : A \times I \to seq(A)$ to model the sequence of operations invoked as a result of executing an operation. $N(r\text{->}op(o), i)$ is the sequence of operations directly invoked when $r\text{->}op(o)$ executes with the receiver r in state i; it models the multiset of operations invoked in the invocation section of the operation.

The serial operational semantics of the program uses a transition function $\to$ on serial states $\langle m, s \rangle \in M \times seq(A)$, where m is the current object memory and s is the sequence of operations left to invoke. $\to$ models the execution of the next operation in s. As Definition 1 shows, $\to$ updates the memory to reflect the new state of the receiver of the executed operation. It also removes the executed operation from the current sequence of operations left to invoke and prepends the multiset of operations that the executed operation invokes. The operations therefore execute in the standard depth-first execution order. Strictly speaking, $\to$ depends on R and N, but we do not represent this dependence explicitly as the correct R and N are always obvious from context.

Definition 1. $\to$ is the function on $M \times seq(A)$ defined by:

$$\frac{m' = m[r \mapsto R(r\text{->}op(o), m(r))], s' = N(r\text{->}op(o), m(r)) \circ s}{\langle m, r\text{->}op(o) \circ s \rangle \to \langle m', s' \rangle}$$

The states $\langle m, p \rangle$ in the parallel operational semantics are similar to the states in the serial operational semantics, but have a multiset p, rather than a sequence s, of operations left to invoke. Such multisets $p \in mst(A)$ are of the form $\{a_1\} \uplus \cdots \uplus \{a_k\}$. We use the sequence to multiset function $s2p : seq(A) \to mst(A)$ defined by $s2p(a_1 \circ \cdots \circ a_k) = \{a_1\} \uplus \cdots \uplus \{a_k\}$ to map sequences of operations into the corresponding multisets.

We model parallel execution by generating all possible interleavings of the execution of the operations. In the parallel execution any of the current operations may execute next, with each operation taking the current state to a potentially different state than the other operations. The parallel operational semantics therefore models execution using a transition relation $\Rightarrow$ rather than a transition function. As the following definition illustrates, $\Rightarrow$ relates a state to all of the states reachable via the execution of any of the current multiset of operations left to invoke. The definition of $\Rightarrow$ completes the semantic framework required to precisely state when two operations commute.

Definition 2. $\Rightarrow$ is the smallest relation [2] on $M \times mst(A)$ satisfying the following condition:
$$\frac{\langle m, r\text{->op}(o)\rangle \to \langle m', s'\rangle, p' = s2p(s') \uplus p}{\langle m, \{r\text{->op}(o)\} \uplus p\rangle \Rightarrow \langle m', p'\rangle}$$

For two operations to commute, they must leave their receivers in identical states and invoke the same multiset of operations regardless of the order in which they execute.

Definition 3. $r_1\text{->op}_1(o_1)$ and $r_2\text{->op}_2(o_2)$ commute in m if
$\langle m, \{r_1\text{->op}_1(o_1)\}\rangle \Rightarrow \langle m_1, p_1\rangle$ and $\langle m_1, \{r_2\text{->op}_2(o_2)\}\rangle \Rightarrow \langle m_{12}, p_{12}\rangle$ and
$\langle m, \{r_2\text{->op}_2(o_2)\}\rangle \Rightarrow \langle m_2, p_2\rangle$ and $\langle m_2, \{r_1\text{->op}_1(o_1)\}\rangle \Rightarrow \langle m_{21}, p_{21}\rangle$
implies $m_{12} = m_{21}$ and $p_1 \uplus p_{12} = p_2 \uplus p_{21}$

Definition 4. $r_1\text{->op}_1(o_1)$ and $r_2\text{->op}_2(o_2)$ commute if $\forall m : r_1\text{->op}_1(o_1)$ and $r_2\text{->op}_2(o_2)$ commute in m.

Theorem 5. *If $r_1 \neq r_2$ then $r_1\text{->op}_1(o_1)$ and $r_2\text{->op}_2(o_2)$ commute.*

Proof Sketch: Intuitively, if the receivers are different, the operations are independent because they access disjoint pieces of data.

4.2 Correspondence Between Parallel and Serial Semantics

We next present several lemmas that characterize the relationship between the parallel and serial operational semantics. Lemma 6 says that for each partial serial execution, there exists a partial parallel execution that generates an equivalent state.

Lemma 6.
If $\langle m, r\text{->op}(o)\rangle \to \cdots \to \langle m', s\rangle$ then $\langle m, \{r\text{->op}(o)\}\rangle \Rightarrow \cdots \Rightarrow \langle m', s2p(s)\rangle$.

---

[2] Under the subset ordering on relations considered as sets of pairs defined as follows. Consider two relations $\Rightarrow_1, \Rightarrow_2 \subseteq (M \times mst(A)) \times (M \times mst(A))$. $\Rightarrow_1$ is less than $\Rightarrow_2$ if $\Rightarrow_1 \subset \Rightarrow_2$.

Proof Sketch: At each step the parallel execution may execute the same operation as the serial execution.

We next establish the impact that commutativity has on the different parallel executions. Lemma 8 states that if all of the invoked operations in given computation commute and one of the parallel executions terminates, then all parallel executions terminate with the same memory. This lemma uses the function $gen : M \times A \to 2^A$, which tells which operations can be invoked as a result of invoking a given operation.

Definition 7. $gen(m, r\text{->op}(o)) = \cup \{p : \langle m, \{r\text{->op}(o)\}\rangle \Rightarrow \cdots \Rightarrow \langle m', p\rangle\}$

Lemma 8. *If* $\forall r_1\text{->op}_1(o_1), r_2\text{->op}_2(o_2) \in gen(m, r\text{->op}(o)) : r_1\text{->op}_1(o_1)$ *and* $r_2\text{->op}_2(o_2)$ *commute, then* $\langle m, \{r\text{->op}(o)\}\rangle \Rightarrow \cdots \Rightarrow \langle m_1, \emptyset\rangle$ *implies*

- *not* $\langle m, \{r\text{->op}(o)\}\rangle \Rightarrow \cdots$ *(i.e. there is no infinite parallel execution), and*
- $\langle m, \{r\text{->op}(o)\}\rangle \Rightarrow \cdots \Rightarrow \langle m_2, \emptyset\rangle$ *implies* $m_1 = m_2$

Proof Sketch: If all of the invoked operations commute then the transition system is confluent, which guarantees deterministic execution [3].

Lemma 9 characterizes the situation when the computation may not terminate. It says that if all of the operations invoked in the parallel executions commute, then it is possible to take any two partial parallel executions and extend them to identical states.

Lemma 9. *If* $\forall r_1\text{->op}_1(o_1), r_2\text{->op}_2(o_2) \in gen(m, r\text{->op}(o)) : r_1\text{->op}_1(o_1)$ *and* $r_2\text{->op}_2(o_2)$ *commute, then* $\langle m, \{r\text{->op}(o)\}\rangle \Rightarrow \cdots \Rightarrow \langle m_1, p_1\rangle$ *and* $\langle m, \{r\text{->op}(o)\}\rangle \Rightarrow \cdots \Rightarrow \langle m_2, p_2\rangle$ *implies* $\exists m' \in M, p \in mst(A) : \langle m_1, p_1\rangle \Rightarrow \cdots \Rightarrow \langle m', p\rangle$ *and* $\langle m_2, p_2\rangle \Rightarrow \cdots \Rightarrow \langle m', p\rangle$

Proof Sketch: If all of the invoked operations commute then the transition system is confluent, which guarantees deterministic execution [3].

An immediate corollary of these two lemmas is that if the serial computation terminates, then all parallel computations terminate with identical memories. Conversely, if a parallel computation terminates, then the serial computation also terminates with an identical memory.

5 Analysis

We have developed a formal semantics that, given a program, defines the receiver effect and invoked operation functions for that program [6]. We have also developed a static analysis algorithm that analyzes pairs of methods to determine if they meet the commutativity testing conditions in Section 3. The foundation of this analysis algorithm is symbolic execution [4]. Symbolic execution simply executes the methods, computing with expressions instead of values. It maintains a set of bindings that map variables to the expressions that denote their values and updates the bindings as it executes the methods. To test if the executions of two methods commute, the compiler first uses symbolic execution to extract expressions that denote the new values of the instance variables and the multiset of invoked operations for both execution orders. It then simplifies the expressions and compares corresponding expressions for equality. If the expressions denote the same value, the operations commute. We have proved a correspondence between the static analysis algorithm and the formal semantics, and used the correspondence to prove that the algorithms used in the compiler correctly identify parallelizable computations [6].

6 Experimental Results

We have implemented a prototype parallelizing compiler that uses commutativity analysis as its basic analysis technique. The compiler also uses several other analysis techniques to extend the model of computation significantly beyond the basic model of computation presented in Section 3 [5].

We used the compiler to automatically parallelize two applications: the Barnes-Hut hierarchical N-body code [2] and Water, which evaluates forces and potentials in a system of water molecules in the liquid state. We briefly present several performance results; we provide a more complete description of the applications and the experimental methodology elsewhere [5].

Figure 3 presents the speedup curve for the Barnes-Hut on two input data sets; this graph plots the running time of the sequential version running with no parallelization overhead divided by the running time of the automatically parallel version as a function of the number of processors executing the parallel computation. The primary limiting factor on the speedup is the fact that the compiler does not parallelize one of the phases of the computation; as the number of processors grows that phase becomes the limiting factor on the performance [5].

Figure 4 presents the speedup curve for Water running on two input data sets. The limiting factor on the speedup is contention for shared objects updated by multiple operations [5].

7 Conclusion

Existing parallelizing compilers all preserve the data dependences of the original serial program. We believe that this strategy is too conservative: compilers must recognize and exploit commuting operations if they are to effectively parallelize a range of applications. This paper presents the semantic foundations of commutativity analysis and shows that the basic analysis technique is sound. It also presents experimental results from two complete scientific applications that were successfully and automatically parallelized by a prototype compiler that uses commutativity analysis as its basic analysis technique. Both applications exhibit respectable parallel performance.

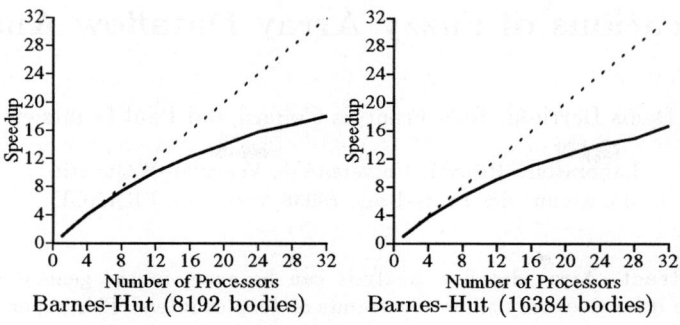

Fig. 3. Speedup for Barnes-Hut

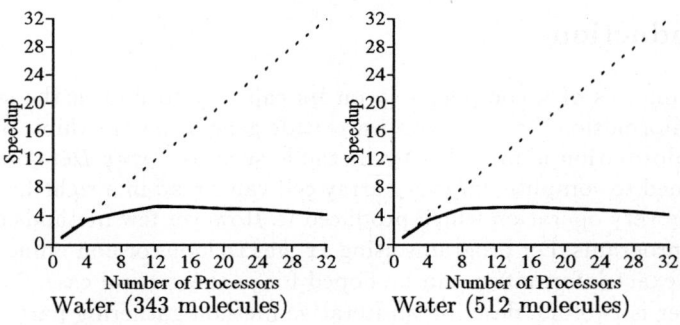

Fig. 4. Speedup for Water

References

1. U. Banerjee, R. Eigenmann, A. Nicolau, and D. Padua. Automatic program parallelization. *Proceedings of the IEEE*, 81(2):211–243, February 1993.
2. J. Barnes and P. Hut. A hierarchical O(NlogN) force-calculation algorithm. *Nature*, pages 446–449, December 1976.
3. G. Huet. Confluent reductions: Abstract properties and applications to term rewriting systems. *Journal of the ACM*, 27(4):797–821, 1980.
4. R. Kemmerer and S. Eckmann. UNISEX: a UNIx-based Symbolic EXecutor for pascal. *Software—Practice and Experience*, 15(5):439–458, May 1985.
5. M. Rinard and P. Diniz. Commutativity analysis: A new analysis framework for parallelizing compilers. Technical Report TRCS96-08, Dept. of Computer Science, University of California at Santa Barbara, May 1996.
6. M. Rinard and P. Diniz. Semantic foundations of commutativity analysis. Technical Report TRCS96-09, Dept. of Computer Science, University of California at Santa Barbara, May 1996.

Applications of Fuzzy Array Dataflow Analysis

Denis Barthou, Jean-François Collard and Paul Feautrier

Laboratoire PRiSM, Université de Versailles-S^{t}Quentin
45, avenue des États-Unis, 78035 Versailles, FRANCE

Abstract. Array dataflow analysis can be exact in the general case when it involves only affine constraints on loop counters. This paper first presents an iterative method in the framework of Fuzzy Array Dataflow Analysis and then describes applications of fuzzy analysis on some usual techniques in compilation and parallelization.

1 Introduction

The performances of a compiler rely on its capacity to find in the source program the information it needs to optimize code generation or exhibit parallelism. Detailed information is provided by methods such as *Array Dataflow Analysis* [4, 7] designed to compute, for every array cell value read in a right-hand side expression the very operation which produced it. However few methods can handle non-static programs. For programs using if, while loops or non-affine array subscripts, no exact information can be hoped for in the general case. The purpose of this paper is twofold: describe an iterative method gathering partial information that can be used in the framework of the *Fuzzy Array Dataflow Analysis* (FADA)[3] and present some applications of this technique such as program checking, parallelization and minimal memory expansion.

2 From Exact to Fuzzy Array Dataflow Analysis

The basic problem of array dataflow analysis is, given an operation $\langle R, y \rangle$ called the "sink", which is an iteration of a statement R whose iteration domain is $I(R)$, and an element $a(g(y))$ of an array a which is read by $\langle R, y \rangle$ to find the "source" of $a(g(y))$ in $\langle R, y \rangle$. The source is an operation $\sigma(\langle R, y \rangle)$ which writes into $a(g(y))$, which is executed before $\langle R, y \rangle$ and such that no operation which executes between $\sigma(\langle R, y \rangle)$ and $\langle R, y \rangle$ also writes into $a(g(y))$. The computation of the source is in two steps: first compute the source for each statement, known as the direct dependence since [2], then combine these sources in the expression of $\sigma(\langle R, y \rangle)$, as detailed in [4]. Suppose that we are investigating source candidates from a statement S: $\langle S, x \rangle$, writing into array a at subscripts $f(x)$. The candidate source has to verify the following constraints:

- Existence predicate: $\langle S, x \rangle$ is a valid operation: $x \in I(S)$.
- Subscript equation: $\langle S, x \rangle$ and $\langle R, y \rangle$, access the same array cell: $f(x) = g(y)$.
- Sequencing condition: $\langle S, x \rangle$ is executed before $\langle R, y \rangle$: $\langle S, x \rangle \prec \langle R, y \rangle$,

- Environment: sources have to be computed under the hypothesis that $\langle R, y \rangle$ is a valid operation, i.e. $y \in I(R)$.

The direct dependence is then given by $\langle S, K_S(y) \rangle$ where $K_S(y) = \max_{\ll} \{x \mid x \in I(S), f(x) = g(y), \langle S, x \rangle \prec \langle R, y \rangle\}$ and where $\ll$ represents the lexicographic order.

As soon as the program model includes conditionals, while loops or non-affine do loop bounds or subscripts, the existence predicate and subscript equation may contain non-linear terms and the exact computation of K_S cannot be achieved in the general case. However, linear relations may be found between constraints in order to compute the smallest set of all the exact sources for any shape of the non-linear constraints verifying these relations. To reach this goal, a solution is to make the source depend on parameters representing the non-linear terms. Pugh and Wonnacott [7] proposed to keep the parametric expression of the non-linear functions in the source when they depend only on y. Given a statement S, they may be represented by the set of vectors $D_S(y)$ for which they are verified, called *parameter domain*[1]. Note that the dimension M_S of the vectors of $D_S(y)$ is lower or equal to the dimension of the iteration vector of S. The expression of $K_S(y)$ is $\max L_S(y) \cap \{x \mid x[1..M_S] \in D_S(y)\}$ where L_S is the set of vectors verifying all linear constraints. If $K_S(y)$ is defined, there exists a vector $\beta_S(y)$ called parameter of the maximum such that $K_S(y) = \max L_S \cap \{x \mid x[1..M_S] = \beta_S(y)\}$. Hence the source can be computed as a function of the parameters of the maximum of all direct dependences. We have shown that for any property $\mathcal{P}$ that is a relation of inclusion between union or intersection of parameter domains and linearly defined sets, the set of the parameters of the maximum corresponding to all the parameter domains verifying $\mathcal{P}$ is defined by linear constraints and is therefore computable [1]. The aim then is to find some properties on the parameter domains. This can be done by an algorithm based on the abstract symbolic tree of the program [3] and more precise relations may be found by analyzing the expressions of the non-linear constraints.

3 Iterative Analysis

The purpose of the iterative analysis is to find relations between the non-linear constraints coming from different statements so as to compare parameter domains. Given two constraints that are the same function but appear at different places in the program, we can say that they have the same value if the variables they use are the same and have the same values. As a variable has the same value in two operations if it has the same source, the equality of the values of constraints may be proved in some cases by a dataflow analysis. Since this dataflow analysis can be fuzzy, the method can then be applied once more and eventually the fuzziness will be reduced by successive analyses. More formally, given two statements S and S' writing into array a, we will suppose that only one non-linear constraint appears in the computation of $K_S(y)$ and $K_{S'}(y)$. Let c and c' be the non-linear constraints respectively involved in $K_S(y)$ and $K_{S'}(y)$, appearing in statements T and T' .

- Partial equality: the constraints c and c' are the same, use the same variables and a dataflow analysis shows that these variables have the same sources in both operations in a context C that is defined by linear inequalities. The relation is $D_S \cap C = D_{S'} \cap C$.
- Image of a parameter domain: the constraints c and c' are the same, use the same variables and the sources of the variables of c at operation $\langle T, x \rangle$ are the same as the sources of the variables of c' at operation $\langle T', f(x) \rangle$, with f an affine function w.r.t. the iteration vector. The relation is $f(D_S) = D_{S'}$.

These relations can be generalized to any number of statements and non-linear constraints. The reader is referred to [1] for technical details.

4 Applications

We present thereafter the application of FADA to variable initialization checking and code parallelization.

4.1 Variable Initialization Checking

In a correct program, all variables are initialized before they are used. Verifying this by a dataflow analysis can help to check the correctness of the program or validate some properties on non-linear constraints. When the analysis is fuzzy, the condition for which the source of the value of a does not come from S is a conjunction of affine constraints on y and β_S. Let $q(y)$ and $r(y, \beta_S)$ be the predicates forming this condition. When the source comes from S, $\forall y \in \mathbf{I}(R)$ s.t. $q(y)$ then $r(y, \beta_S) = false$. According to the definition of the parameter of the maximum, this is equivalent to: $\forall y \in \mathbf{I}(R)$ s.t. $q(y) = true, \exists x$ s.t. $(r(y, x) = false) \land (c(y, x) = true)$ where c is the non-linear constraint involved. This condition can be generalized to any number of direct dependences and non-linear constraints. Checking the condition can be left to the programmer or submitted to an assertion generator.

4.2 Code Parallelization

There are two basic techniques for extracting parallelism from a dependence graph: one consists in computing a schedule, the other one in computing a placement.

Fuzzy Scheduling We must guarantee that: $\theta(S, x) + 1 \leq \theta(R, y)$. In the result of the corresponding FADA, x is an affine function ϕ of x and of parameters β_S which must satisfy a set of affine predicates $P(x)$. We may refine the above inequality into $y \in \mathcal{I}(R), \alpha \in P(y) \Rightarrow \theta(S, \phi(y, \alpha)) + 1 \leq \theta(R, y)$. Suppose we have expressed the schedule θ as an affine form with unknown coefficients. Since everything is affine, we are in a position to apply Farkas lemma; the result is a set of linear equations in the coefficients of the schedule and new positive unknowns, the *Farkas multipliers*. These equations may be solved as in [5].

Memory Expansion In order to take into account memory based dependences in the above schedule, a solution is to find the minimal memory expansion which is consistent with this schedule. The method presented by Lefebvre [6] can be used in the present case with little or no modification. Indeed, it is obvious that, even in the case of dynamic control structures and non-linear arrays, we may still compute an ordinary dependence graph. In the case of FADA, the shape of the source is exactly the same as in the exact analysis case, hence the same algorithms apply. In some cases, parameters will disappear, for instance, when expansion of a scalar has been deemed unnecessary. When a parameter is actually needed, its value must be recorded when the corresponding control operations are executed. If speculation has been used, this means that a read operation may not be executed before the results of the controlling operations are known. This is a new constraint which has to be taken into account when computing the schedule.

5 Conclusion

Many applications in the compilation and parallelization field take advantage of our technique, with little change in their algorithms. The Fuzzy Array Dataflow Analysis extends the scope of variable initialization checking, code parallelization to some programs with dynamic control structures. Moreover, even a fuzzy result can give enough information for a significant improvement of the output of these techniques. Further developments on the combination of compilation and parallelization methods with fuzzy analysis will be the subject of future work.

References

1. Denis Barthou, Jean-François Collard, and Paul Feautrier. Fuzzy array dataflow analysis. Technical Report 95/33, PRiSM Laboratory, 1995.
2. Thomas Brandes. The importance of direct dependences for automatic parallelization. In *ACM Int. Conf. on Supercomputing*, St Malo, France, July 1988.
3. J.-F. Collard, D. Barthou, and P. Feautrier. Fuzzy array dataflow analysis. In *Proc. of 5th ACM SIGPLAN Symp. on Principles and Practice of Parallel Programming*, Santa Barbara, CA, July 1995.
4. Paul Feautrier. Dataflow analysis of scalar and array references. *Int. J. of Parallel Programming*, 20(1):23–53, February 1991.
5. Paul Feautrier. Some efficient solutions to the affine scheduling problem, I, one dimensional time. *Int. J. of Parallel Programming*, 21(5):313–348, October 1992.
6. Vincent Lefebvre. Gestion de la mémoire dans les programmes parallèles. In *8eme rencontres francophones du parallélisme*, pages 149–152, May 1996.
7. William Pugh and David Wonnacott. An exact method for analysis of value-based array data dependences. In *Lecture Notes in Computer Science 768: Sixth Annual Workshop on Programming Languages and Compilers*, Portland, OR, August 1993. Springer-Verlag.

Simplifying Communication Induced by Operations on Block-Distributed Arrays

Andreas Eberhart and Jingke Li

Portland State University, Portland OR 97207, USA

Abstract. This paper analyzes communication patterns induced by operations on arrays with arbitrary block-distributions. A simple method is introduced that transforms the resulting pattern into a collection of independent gathers and scatters, which can be implemented efficiently.

1 Introduction

Whole array operations and array section operations are important features of many data-parallel languages. With these operations, matrix-based numerical algorithms can be expressed clearly and concisely, making them easier to read and to maintain. An *array section* $A(l:h:s)$ denotes a subarray of A: $[A(l), A(l+s), A(l+2s), ..., A(h)]$. The *array section operation* $A(l:h:s) \circ B(l':h':s')$ performs the binary arithmetic or logic operation $\circ$ in an element-wise fashion on the array sections of A and B.

Data Distribution: As in the case of other high-level programming constructs, efficient implementation of array operations on distributed-memory multicomputers is a non-trivial task. When a data-parallel program is compiled to a massively parallel processing system, the data arrays in the program are generally decomposed and distributed over the system's memory modules. *Block-distributions* are a common way of doing this. The two parameters (p, k) describe the distribution, where p is the number of nodes and k is the number of array elements on each node.[1] Consequently, each node holds either $\lfloor k/|s| \rfloor$ or $\lceil k/|s| \rceil$ elements of an array section with stride s (nodes at the beginning or the end of an array section can be an exception).

Induced Communication: Figure 1 shows the distribution and array sections of the arrays A (top) and B (bottom). In order to perform an element-wise operation on the array sections, corresponding elements must be at the same location. Without loss of generality, we assume that the computation takes place at the location of array B. Thus, the nodes holding $A(l:h:s)$ become *source nodes* transferring to the *destination nodes* where $B(l':h':s')$ is located. Since the source nodes hold more data, each source node *scatters* its data to a cluster of destination nodes. The individual scatters overlap, and one destination can receive messages from two sources.

---

[1] Two-level mappings, where arrays are aligned via a template with offset a and stride b, can be handled as well since $A(l:h:s)$ has the same distribution pattern as $A(a+lb:a+hb:sb)$, where A is aligned with a one-level mapping.

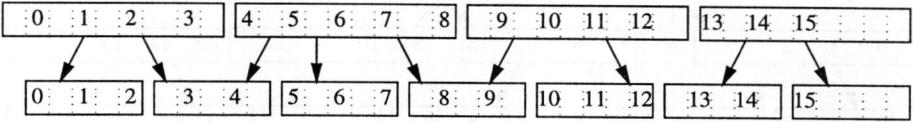

Fig. 1. Communication pattern for the transfer of a 1D array section with blocksizes $k = 9, k' = 5$ and section-strides $s = s' = 2$. The nodes are symbolized by the solid boxes. A number i indicates the ith array section element. The source nodes hold either 4 or 5 array section elements, and the destination nodes hold either 2 or 3. The arrows show the resulting communication pattern.

Approach: A simple approach for handling communication induced by an array operation is the *unscheduled transfer* that identifies sources and destinations and then generates messages for bringing the data from the sources to their destinations. However, as observed by many users of massively parallel systems, a large number of unorchestrated concurrent messages can cause high levels of resource contention, message blockage, buffer overflow, and even deadlock [3].

A better approach is to use synchronization to organize concurrent messages. Optimized collective communication routines have been developed for some simple message patterns, such as broadcast, multicast, and reduction [1, 3]. These routines have been successfully incorporated into several data-parallel language compilers [4]. This paper uses a similar approach.

2 Initial and Final Shifts

The idea is to avoid message path overlaps among concurrent gathers and scatters and to obtain regionally independent patterns that can be integrated in an overall scheduling approach.

In a scatter situation, as shown in Figure 1, this goal can be achieved by combining array segments that must be sent to a common destination from two adjacent source nodes. By convention, the two array segments are combined to the left node. Figure 2 shows the new communication pattern where any node participates exclusively in one scatter. This *initial shift* involves only neighbor-communication; therefore, the messages can all be sent in parallel inducing only a constant overhead. Each source node n can detect if it has to shift data by checking whether the destinations of its first and the preceding node's last section elements are the same. If the source nodes hold less data than the destination nodes,[2] then the pattern is reversed (*final shift right*) to obtain independent gathers.

Two-Dimensional Mappings: The communication pattern induced by a 2D array operation is basically a composition of two 1D communication patterns, one

---

[2] For this comparison $k|s'|$ and $k'|s|$ must be used.

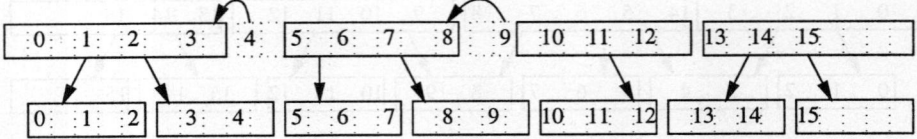

Fig. 2. Example from Figure 1 after nodes that provide only a partial data set for their leftmost destination shifted the data to their left neighbor. We obtain non-overlapping scatters.

for each dimension. Consequently, there are four *base patterns*: *scatter/scatter*, *scatter/gather*, *gather/scatter*, and *gather/gather*. One horizontal and one vertical shift must be performed before or after the main data transfer.

3 Scheduling the Base Patterns

In order to optimize the message-flow, we can develop a distributed dynamic scheduling algorithm that avoids network conflicts and uses all available resources. Obviously, this task depends on both the network topology and the routing algorithm. On a mesh or torus topology with X-Y routing, we can easily identify non-conflicting sets of patterns that can be sent concurrently: since we obtained independent patterns through the initial and final shifts, scatter messages from nodes on a diagonal line do not conflict in the network (Figure 3a). The scheduling algorithm is called from the node program with the parameters of the array sections. It identifies the base pattern in which the local node participates. Diagonals of base patterns are enumerated and sent sequentially. Further optimizations lead to the diagonal scheduling scheme which was proven to yield optimal schedules with respect to the resource usage [2].

We conducted a simulation study for wormhole-routed mesh networks which shows that link and node-contention cause unscheduled messages to get stuck in the network, blocking channels without actually using them for data transmission (Figure 3b). Scheduled transfers yield a much better performance as demonstrated in Figure 3c.

Efficient implementations of gather and scatter routines are described in [1]. The algorithms use recursive halving and doubling techniques in order to save message start-ups ($\log_2(p)$ instead of p). In addition to the benefits obtained through scheduling, the initial and final shifts also reduce the number of messages by up to 75% in the 2D case. Since only a fraction of those messages can be sent concurrently, this also saves start-up time.

4 Conclusion and Future Work

We presented the concept of initial and final shifts that simplify communication induced by operations on block-distributed arrays. Inducing only a small over-

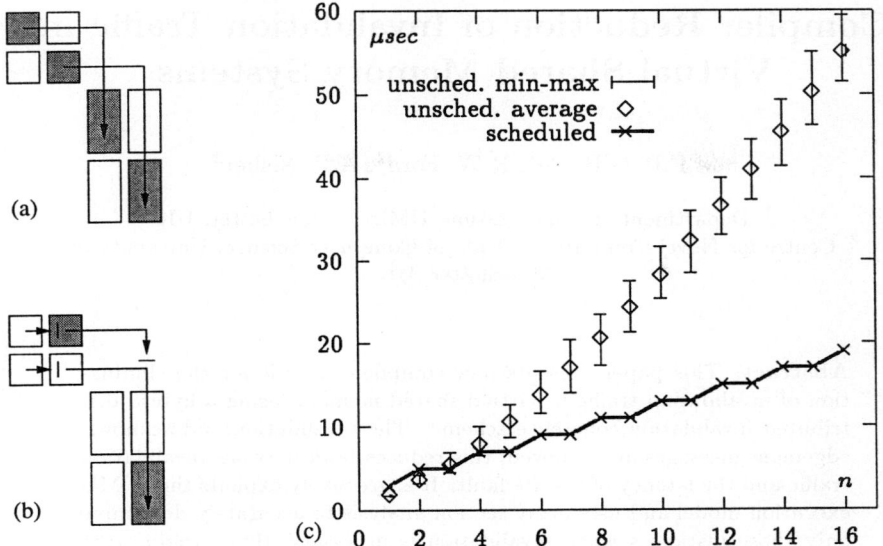

Fig. 3. Possible transfers of four base patterns with X-Y routing (Figures a, b) and simulation results for the transfer of an array section, located on $n \times n$ nodes, to its destination, located on $2n+1 \times 2n+1$ nodes (Figure c).

head, the overall communication can be partitioned into independent gathers and scatters that can be scheduled in order to avoid both link and node-contention and guarantee high resource usage. Since the schedule can be generated in a distributed manner, only the shifts cause additional communication that can be handled in two steps in which messages are sent between neighboring nodes. We are currently working on extending our scheduling algorithms to adaptive routing strategies. Furthermore we are analyzing the communication induced by cyclic array distributions.

References

1. M. Barnett, S. Gupta, D. G. Payne, L. Shuler, R. A. van de Geijn, and J. Watts. Building a high-performance collective communication library. In *Proceedings of Supercomputing '94*, pages 107–116, 1994.
2. A. Eberhart and J. Li. Optimizing communication for array operations on message-passing multicomputers. In *Proceedings of the 1996 ISPANN*, 1996.
3. G. Fox, M. Johnson, G. Lyzenga, S. Otto, J. Salmon, and D. Walker. *Solving Problems on Concurrent Processors*. Prentice Hall, 1988.
4. J. Li and M. Chen. Compiling communication-efficient programs for massively-parallel machines. *IEEE Trans. on Parallel and Dist. Systems*, 2(3), July 1991.

Compiler Reduction of Invalidation Traffic in Virtual Shared Memory Systems

M.F.P. O'Boyle[1], R.W. Ford[2], A.P. Nisbet[2]

[1] Department of Computation, UMIST, Manchester, UK
[2] Centre for Novel Computing, Dept. of Computer Science, University of Manchester, UK

Abstract. This paper presents new compiler analysis for the elimination of invalidation traffic in virtual shared memory, using a hybrid distributed invalidation coherence scheme. The invalidation and acknowledgement messages are removed; this reduces both network invalidation traffic and the latency of a write fault. It aggressively exploits the SPMD execution model and uses array section analysis to accurately determine only those instances when invalidation is necessary, thus avoiding the additional read misses of previous schemes. Equations determining precisely what data should be invalidated are presented and translated into a form amenable to compiler analysis. Preliminary experimental results on a 30 node prototype architecture demonstrate the performance attainable using this scheme.

1 Introduction

Virtual shared memory (VSM) systems provide a shared address space on distributed memory architectures. A shared memory programming model is attractive because it is simple to program, thus speeding the implementation and porting of parallel programs, and enabling the parallelisation of complex adaptive programs which may be difficult to implement in message-passing. However, for program segments with static communication patterns, sequentially consistent VSM fares significantly worse as, although it provides coherence mechanisms to support general access patterns, it seldom provides facilities to allow exploitation of compile-time knowledge. Scalable VSM systems use a directory structure (which typically records the location and the read/write access permissions of copies of data) and invalidation/update based protocols to maintain consistency. In the sequential consistency model with an invalidation protocol, an attempt to write a new value is delayed until all remote copies are invalidated. Performance can be degraded both by the delay on the writing node, and by the resulting network traffic.

This paper presents new compiler analysis to exploit a hybrid Distributed Invalidation (DI) scheme in the context of invalidation based sequential consistency. The DI technique [5, 7] removes the responsibility of invalidation from the writing processor and distributes it to the processors with the remote copies. The writing processor can now proceed without stalling and the invalidation of copies is done locally. The invalidation and acknowledgement messages are removed, reducing both network invalidation traffic and the latency of a write fault. In the case of compile-time unknowns, the directory structure guarantees correctness. Thus, the compiler may be aggressive in its analysis as it is does not have to make the conservative assumptions of compiler directed schemes where the compiler is entirely responsible for coherence.

We have implemented [5] a call to our VSM system *LocalExclusive(addr)* which sets local access for the location *addr* of a variable to exclusive and, by default, deletes the directory pointers that refer to copies of data. This allows a write to proceed without stalling (as in weak coherency) and removes the associated invalidation traffic. The corresponding call *LocalInvalidate(addr)* sets a read-only copy of the *addr* to invalid. Notice that a write can now occur before the remote copies have been invalidated, however compiler placement of barriers ensures that no read to incorrect data can take place. An important feature of these calls is that after they have completed, the system is left in a coherent state. Thus in the absence of further compiler optimisation, the directory will continue to maintain sequential consistency.

Previous Work All existing analysis for compiler directed coherence assumes a fork/join model. Parallel regions are separated into *epochs* by a synchronisation, even though there may be no cross-processor dependence between regions. Scheduling within an epoch is considered to be dynamic so it is impossible to determine the location of accesses between epochs. This has the effect of making worst case assumptions about the location of previous reads and writes, generating unnecessary invalidation or misses.

Compiler directed coherence places the entire burden of maintaining cache coherence on the compiler. Some schemes use a compiler controlled directory to help in run-time dependence analysis, whilst others remove the need for a directory altogether [2, 4]. Early work invalidated all cached data at each epoch. More recent schemes have used tags or timestamps [3] to maintain cached data across epochs. This method relies on the compiler analysis for correctness and is limited in the amount of inter-epoch locality that can be exploited by the size of the tag and epoch counters, as well as by the conservative nature of the compiler analysis. Other approaches include a directory based method, Dynamic Self Invalidation [7] and hybrid techniques, [6, 9] each of which is developed for CC-NUMAs. In [9] they note the critical role of (but do not develop) compiler analysis.

We use an SPMD execution model, exploiting data parallelism with owner-computes scheduling. Owner-computes guarantees that the write to any page will be performed by the same processor, and hence two or more write actions cannot make data stale on the processor that issued the first write. Using data layout and alignment we have implemented a scheme that determines whether two accesses can cause a cross-processor dependence. We have used this to optimise barrier placement, drastically reducing synchronisation overhead. This same analysis can be used to detect if an anti-dependence is cross-processor, thereby requiring coherence action. By using array section analysis rather than

Original Program	Local Program

Do s = 1,n
 1: b(s) = s
 Enddo
 Do i = 2,n-1
   Do j = 1,i
 2:     a(i) = a(i) + b(j) + b(i)
   Enddo
 Enddo
 Do t = 4,n-5
 3: b(t) = t
 Enddo
	Do s = max(1,lo),min(hi,n)
	 1: b(s) = s
	 Enddo
	 barrier()
	 Do i = max(2,lo),min(hi,n-1)
	   Do j = 1,i
	 2: a(i) = a(i) + b(j) + b(i)
	   Enddo
	 Enddo
	 barrier()
	 Do t = max(4,lo),min(hi,n-5)
	 3: b(t) = t
	 Enddo

Figure 1. Example to show coherence state

variable names we can accurately determine the data to invalidate, whereas previous schemes require the invalidation of entire arrays, or an additional bit mask per write instruction [8].

2 Analysis

A processor must invalidate all its copies of pages that will be written by others and ensure exclusive access to pages it will write. This section is concerned with precisely determining *which* pages should be modified, when using owner-computes scheduling. It develops a set of equations that determine the local actions each processor must take in order to maintain coherence, without incurring invalidation traffic.

2.1 Coherence Equations

Before deriving equations to mark the data invalid and exclusive, it is necessary to define a number of primitives.

- Let an action i be a read or write operation occurring after action $i-1$ and before $i+1$. This ordering must be ensured by program synchronisations or local sequential ordering.
- Let W_i, R_i be the set of a pages (global array data) written/read in action i.
- Let $\underline{W}_i, \underline{R}_i$ be the set of local pages written/read in action i.
- Let $\mathcal{E}x_i, \mathcal{R}o_i$ be the set of local pages in exclusive/read state after action i.

The state of a page after a particular action is dependent on its previous state, and on the action. After a write, the local pages in exclusive mode will be increased by any pages written locally that were not previously marked as exclusive:

$$\underline{\mathcal{E}x}_i = \underline{\mathcal{E}x}_{i-1} \cup \underline{W}_i. \tag{1}$$

The set of local pages in read state will be decreased by writes occurring across the processors if they refer to local pages previously in read state:

$$\underline{\mathcal{R}o}_i = \underline{\mathcal{R}o}_{i-1} - (\underline{\mathcal{R}o}_{i-1} \cap W_i). \tag{2}$$

Let p be the number of processors in the target machine, and z the processor id of the local processor where $z \in \{1, ..., p\}$. After a read action, the set of local pages in read state will be increased by other processors reading processor z's local pages previously in exclusive state, and by the remote pages read by the local processor. As it is necessary to distinguish between the local pages of distinct processors, the processor id will be used as a super-script.

$$\underline{\mathcal{R}o}_i = \underline{\mathcal{R}o}_{i-1} \cup ((\bigcup_{\substack{k \in \{1,...,p\} \\ k \neq z}} \underline{R}_i^k) \cap \underline{\mathcal{E}x}_{i-1}) \cup (\underline{R}_i^z - (\underline{\mathcal{E}x}_{i-1} \cap \underline{R}_i^z)) \tag{3}$$

Note that, from this equation, reads by a processor to its own local data do not change its state from exclusive to read state, only reads by other processors will do this. The local pages in exclusive state are decreased by reads from other processors to local pages previously in exclusive state.

$$\underline{\mathcal{E}x}_i = \underline{\mathcal{E}x}_{i-1} - (\underline{\mathcal{E}x}_{i-1} \cap (\bigcup_{\substack{k \in \{1,...,p\} \\ k \neq z}} \underline{R}_i^k)) \tag{4}$$

Once again the pages in exclusive state are unaffected by local reads to local data. We now introduce the equations which are used to define coherence state transition. Let $\underline{LE}_i, \underline{LI}_i$ be the local pages to be set to exclusive/invalid state due to action i. The local pages to be made exclusive are those which will be written locally but are not already exclusive and those local pages which are in a read state should be invalidated if written to by remote processors.

$$\underline{LE}_i = \underline{W}_i - (\underline{\mathcal{E}x}_{i-1} \cap \underline{W}_i) \tag{5}$$

$$\underline{LI}_i = (W_i - \underline{W}_i) \cap \underline{\mathcal{R}o}_{i-1} \tag{6}$$

2.2 Sets $\mapsto$ Constraints

So far pages that need to be invalidated, or made exclusive, have been defined in terms of set algebra. To be amenable to compiler analysis, they have to be expressed as array sections. We restrict our attention to programs where the set based description may be interpreted in terms of array sections using affine constraints and access functions.

To denote the elements or array section accessed within a loop nest of an array v we use the (iteration space, access) pair:

$$[AJ \leq b], [v(f(J))] \tag{7}$$

where $AJ \leq b$ denote the constraints on the iterators J from the bounds in the enclosing loop nest and $v(f(J))$ represents the access to variable v which is an affine function f of the iterators. For example, the elements of a accessed in the second loop of Figure 1 are:

$$[(2 \leq i \leq n-1) \wedge (1 \leq j \leq i)], [a(i)] \tag{8}$$

Let X be the *set* of elements accessed:

$$X = [A_x J_x \leq b_x], [x(f_x(J_x))] \tag{9}$$

In order to derive the relation between the set expressions of Section 2.1 and the compiler's constraint representation, we need to define the effect of set operators. The intersection of two sets is equivalent to the conjunction of the two systems of constraints from the iteration space and the requirement that the access functions are equal.

$$X \cap Y \ : \ [(A_x J_x \leq b_x) \wedge (A_y J_y \leq b_y) \wedge (f_x(J_x) = f_y(J_y))], [x(f_x(J_x))] \tag{10}$$

The addition of two non-intersecting sets corresponds simply to two systems of constraints.

$$X + Y \ : \ ([A_x J_x \leq b_x], [x(f_x(J_x))]) + ([A_y J_y \leq b_y], [y(f_y(J_y))]) \tag{11}$$

The difference of two sets, given one is the subset of the other is similar to that of intersection except that one set of constraints is negated.

$$X - Y : [(A_x J_x \leq b_x) \wedge (\neg A_y J_y \leq b_y) \wedge (x(f_x(J_x)) = y(f_y(J_y)))], [x(f_x(J_x))] \tag{12}$$

The main difficulty with the constraint equivalences of set algebra is that negation is not directly defined for polyhedra as it can lead to non-convexity. This is overcome by distributing the negation through each set of inequalities using DeMorgan's Law to give a new set of inequalities connected by $\vee$ (OR) rather than $\wedge$ (AND). This is returned to a set of inequalities connected by $\wedge$ by using the following identity.

Let x, y be inequalities then

$$(x \vee y) \Leftrightarrow (\neg x \wedge y) + (x \wedge \neg y) + (x \wedge y) \tag{13}$$

Fortunately negation is defined for a single integer inequality:

$$\neg(a \leq b) \Leftrightarrow (a > b) \Leftrightarrow (b \leq a - 1). \tag{14}$$

Thus negation and hence union can be expressed. Unfortunately expanding $\vee$ may lead to an explosion of terms. For two constraints there are only three terms, but this grows exponentially as $2^C - 1$ where C is the number of constraints. However, many of the new terms can be eliminated as they are contradictory (or more strictly their solution space is empty) which is readily determined using Fourier-Motzkin elimination. The final terms to consider are unions. The simple union of two sets can be expressed in terms of previously defined operators

$$X \cup Y = X + Y - (X \cap Y). \tag{15}$$

The indexed union, as used in equations (10) and (11) cause more difficulties. We first trivially split it:

$$\bigcup_{k \in \{1,\ldots,p\}}^{k \neq z} X^k = (\bigcup_{k \in \{1,\ldots,z-1\}} X^k) \cup (\bigcup_{k \in \{z+1,\ldots,p\}} X^k) \tag{16}$$

and then store each union as a parameterised set of inequalities

$$\bigcup_k X^k \Leftrightarrow [A^k J^k \leq b^k \ \forall k], [x(f_x(J_x))]. \tag{17}$$

2.3 Example

When the iterations and data are mapped to several processors, each processor will have its own local iteration and data spaces. For example, in Figure 1 the second column shows the local program generated by the MARS compiler [1] after partitioning; lo, hi are the introduced lower and upper bounds of the local data space. In MARS each dependence is marked as being either an internal or a cross-processor dependence. The dependences b(s) δ^f b(i) and b(i) δ^a b(t) are within a processor while b(s) δ^f b(j) and b(j) δ^a b(t) are cross-processor dependences requiring synchronisation. In this section we will derive the data regions of array b to be invalidated for the program shown in Figure 1. Before calculating the regions to be marked as exclusive and invalid, a number of terms $\mathcal{R}o_1, \mathcal{E}x_1, \mathcal{R}o_2, \mathcal{E}x_2$ must be derived. For the sake of illustration, assume that there are local read copies of b (b[2..n-1]), before the execution of the first loop, so $\mathcal{R}o_0 : [2 \leq i_0 \leq n-1], [b(i_0)]$. The write access in the first loop is $W_1 : [1 \leq s \leq n], [b(s)]$ and we wish to calculate the read state $\mathcal{R}o_1$ after this write, based on equation (2). First, we must determine $\mathcal{R}o_0 \cap W_1$:

$$[(2 \leq i_0 \leq n-1) \wedge (1 \leq s \leq n) \wedge (i_0 = s)], [b(s)] \tag{18}$$

i_0 is redundant as it does not occur in the access function. Removing i_0 by Fourier-Motzkin elimination gives the constraints on s:
$[2 \leq s \leq n-1], [b(s)]$. We now calculate $\mathcal{R}o_0 - (\mathcal{R}o_0 \cap W_1)$:
$[(2 \leq i_0 \leq n-1) \wedge \neg(2 \leq s \leq n-1) \wedge (i_0 = s)], [b(s)]$ giving

$$[(2 \leq i_0 \leq n-1) \wedge ((2 > s) \vee (s > n-1)) \wedge (i_0 = s)], [b(s)] \tag{19}$$

On expanding the $\vee$ each summand is found to be inconsistent; that is there does not exist a solution and hence $\mathcal{R}o_1 = \emptyset$. We will now derive the remaining

terms. As all of array b is written in the first loop, all local data is in exclusive state and there are no local copies of remote data,
$\mathcal{E}x_1 = \underline{W}_1 : [(\max(1, lo) \leq s \leq \min(n, hi))], [b(s)]$.

Determining $\mathcal{R}o_2$ based on equation (3) is more complex and will be dealt with piecewise. One of the simpler terms is $(\underline{R}_2 - (\mathcal{E}x_1 \cap \underline{R}_2))$, which reduces to $[1 \leq j \leq lo - 1], [b(j)]$. The indexed union term in equation (3) can be expressed as:

$$\left(\bigcup_{\substack{k \in \{1,...,p\} \\ k \neq z}} R_2^k \right) \cap \underline{\mathcal{E}x}_1 : [(1 \leq j \leq \min(hi^k, n-1)) \wedge$$
$$((\max(1, lo^z) \leq s \leq \min(n, hi^z)) \wedge (s = j), \forall k, k \neq z], [b(j)] \qquad (20)$$

We have introduced superscripts to distinguish the local constraints. On eliminating j and splitting the range of k we have:

$$[\max(1, lo^z) \leq s \leq \min(hi^z, hi^k, n-1) \, \forall k \in \{1, ..., z-1\}], [b(s)] \qquad (21)$$

and

$$[\max(1, lo^z) \leq s \leq \min(hi^z, hi^k, n-1) \, \forall k \in \{z+1, ..., p\}], [b(s)] \qquad (22)$$

. Equation (21) is inconsistent for all values of k as $lo^z > hi^{z-1}$ while equation (22) reduces to

$$[\max(1, lo^z) \leq s \leq hi^z], [b(s)] \qquad (23)$$

Combining the subexpressions gives

$$\mathcal{R}o_2 : [(1 \leq j \leq hi)], [b(j)] \qquad (24)$$

The final term can easily be found to be $\mathcal{E}x_2 = \emptyset$. Given $W_3 : [4 \leq t \leq n-5], [b(t)]$ and $\underline{W}_3 : [\max(4, lo) \leq t \leq \min(hi, n-5)], [b(t)]$ it is possible to define the data to be marked invalid and exclusive before the write W_3.

$$LE_i : [\max(4, lo) \leq t \leq \min(n-5, hi)], [b(t)] \qquad (25)$$

$$LI_i : ([4 \leq t \leq \min(\max(4, lo-1), hi)], [b(t)]) +$$
$$([\max(1, \min(n-5, hi+1)) \leq t \leq \min(n-5, hi)], [b(t)]) \qquad (26)$$

3 Experiments

In this section we apply the above techniques to two applications, an in house version of the NAS Parallel benchmark CG and the core communication structure of Shallow on a 30 node EDS prototype [12] supporting a prototype VSM implementation. For further details of the implementation see [11].

Figures 2 and 3 present counts of read/write misses and invalidation traffic with and without distributed invalidation for the two applications. Figures 4 (a)

Nodes	2	4	8	16
Read Misses	100	600	2800	12000
Write Misses	100	200	400	800
Invalidations	100	600	2800	12000

Nodes	2	4	8	16
Read Misses	100	600	2800	12000
Write Misses	0	0	0	0
Invalidations	0	0	0	0

Figure 2. Counts of misses for in-house CG (a) standard (b) with distributed invalidation.

Nodes	2	4	8	16
Read Misses	100	300	700	1500
Write Misses	100	300	700	1500
Invalidations	100	300	700	1500

Nodes	2	4	8	16
Read Misses	100	300	700	1500
Write Misses	0	0	0	0
Invalidations	0	0	0	0

Figure 3. Counts of misses for core communication of Shallow (a) standard, (b) with distributed invalidation.

and (b) show the temporal performance where the *Naive Ideal* line assumes linear speedup from the standard sequential version. The *Optimised* line presents performance improvement from the use of distributed invalidation. Note that counts and timings are taken from the second to last iteration, thereby eliminating the effect of initial (cold start) faults from the results. Our approach gives a dramatic reduction in invalidation messages without the potential increase in unnecessary misses of previous compiler-controlled schemes.

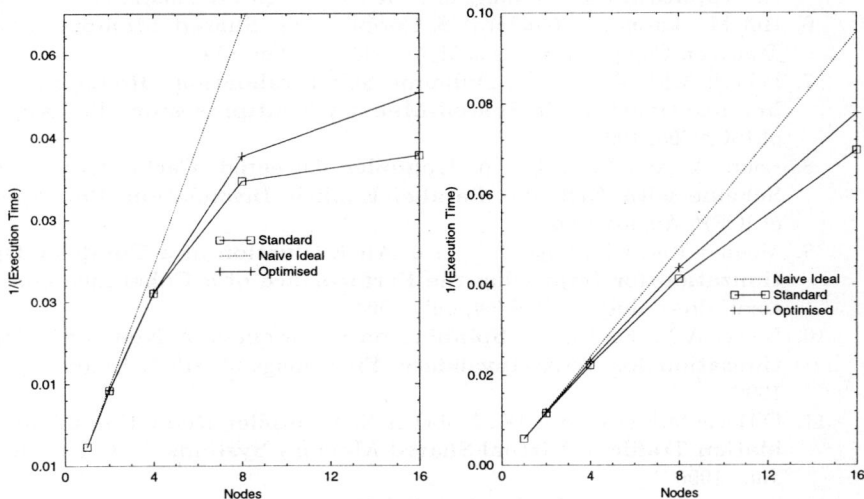

Figure 4. Temporal Performance of (a) in-house CG and (b) core communication of Shallow.

4 Conclusions and Further Work

This paper has described compiler analysis for a hybrid distributed invalidation scheme making explicit use of compile-time knowledge of scheduling and data layout. Future work will focus on completing implementation of the algorithm in our compiler and investigating the trade-off between invalidation traffic and unnecessary misses. We intend to investigate compiler analysis for other VSM optimisations such as latency-hiding post-storing [10] and to develop a unifying approach to compiling for VSM.

Acknowledgements This work was partly funded by the ESPRIT SODA project (9124). We would like to thank Thomas Fahringer and Mark Bull for their helpful comments on drafts of this paper.

References

1. Bodin F., O'Boyle M.F.P. **A Compiler Strategy for SVM**, Third Workshop on Languages, Compilers and Runtime Systems for Scalable Computing, New York, May 1995.
2. Cheong H., Veidenbaum A.V., **Compiler Directed Cache Management in Multiprocessors**, IEEE Computer, 23(6):39-48, June 1990.
3. Cheong H., **Life-Span Strategy - A Compiler-Based Approach to Cache Coherence**, Proceedings of ICS, July 1992.
4. Darnell E., Kennedy K., **Cache Coherence Using Local Knowledge**, Proceedings of Supercomputing 1993, pages 720-729, Nov 1993.
5. Ford R.W., Nisbet A.N., Bull J.M. **User-level VSM Optimisation and its Application** Proceedings of PARA95, Lyngby, Denmark, LNCS, 1995.
6. Hill M., Larus J., Reinhardt S.,**Cooperative Shared Memory**, ACM Trans. on Computer Systems 11(4): 300-318, Nov 1993.
7. Lebeck A.R., Wood D.A. **Dynamic Self Invalidation: Reducing Coherence Overhead in Shared-Memory Multiprocessors**, Proceedings of ISCA '95, 1995.
8. Louri A. and Sung H., **A Compiler Directed Cache Coherence Scheme with Fast and Parallel Explicit Invalidation**, Proceedings of ICPP, August 1992.
9. Mounes-Toussi F., Lilja D.J., Li Z. **An Evaluation of a Compiler Optimization for Improving the Performance of a Coherence Directory**, Proceedings of ICS '94, July 1994.
10. Nisbet A.N., Ford R.W., **Spinning on Coherency: A New VSM Optimisation for Write-Invalidate**, Proceedings of HPCN Europe, April 1996.
11. O'Boyle M.F.P.,Ford R.W., Nisbet A.N., **Compiler Reduction of Invalidation Traffic in Virtual Shared Memory Systems**, CNC Tech. Rep. June 1996.
12. Skelton, C.J. et al., **EDS a Parallel Computer System for Advanced Information Processing**, Parallel Architectures and Languages Europe, PARLE 92, July 1992.

Partial Dead Code Elimination for Parallel Programs

Jens Knoop

University of Passau, Germany[**]

Abstract. Eliminating *partially dead code* has proved to be a powerful technique for the runtime optimization of sequential programs. Here, we show how this technique can be adapted to *parallel* programs with shared memory and interleaving semantics on the basis of a recently presented framework for efficient and precise bitvector analyses for parallel programs. Whereas the framework allows a straightforward adaptation of the required data flow analyses to the parallel case, the transformation part of the optimization requires special care in order to preserve parallelism. This preservation is an absolute must in order to guarantee that the optimization does never impair efficiency. The introduction of an appropiate natural side condition suffices to lift even the optimality result known from the sequential setting to the parallel setting.

1 Motivation

Partial dead code elimination (*PDE*) improves the runtime efficiency of a program by avoiding the execution of unnecessary statements at runtime. This is illustrated in Figure 1, where the assignment of node **1** of Figure 1(a) is *partially dead* because it is dead on the left branch but alive on the right one. This assignment can be eliminated by moving it to the entries of node **3** and node **4** as shown in Figure 1(b), where the occurrence of node **3** is (totally) dead, and can be eliminated as shown in Figure 1(c).

As illustrated above, the effect of PDE depends in essence on the combined effects of two separate transformations: (1) *sinking of assignments*, and (2) *eliminating dead assignments*.

In the *sequential setting*, it is an important observation that assignment sinkings *never* affect the execution time. Intuitively, this fact is responsible for the optimality of the following two-step procedure, which is repeated until the program stabilizes: (1) Moving assignments *as far as possible* in the direction of the control flow (which places them in a context as specific as possible, and thus maximizes the potential of dead code), and (2), eliminating all dead assignments. In

---

[**] Author's current address: Fachbereich 10 — Informatik, Carl von Ossietzky Universität Oldenburg, Uhlhornsweg 84, D-26129 Oldenburg, Germany. E-mail: jens.knoop@informatik.uni-oldenburg.de. Most of the work was done while the author was at the University of Passau. The author's work is partially funded by the Leibniz Programme of the German Research Council (DFG) under grant Ol 98/1-1.

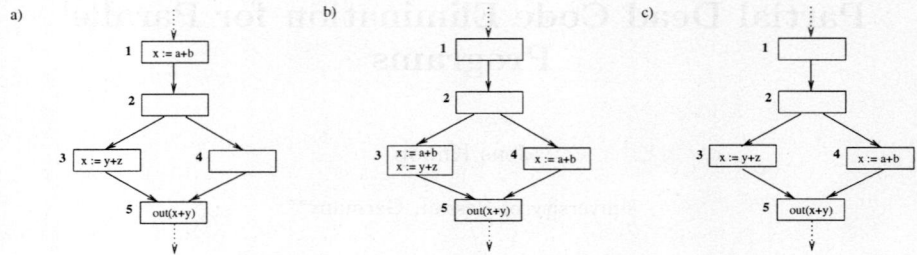

Fig. 1. Partial Dead Code Elimination

fact, this captures all *second order effects* between assignment sinkings and eliminations, and guarantees the following optimality result [KRS1]: Partially dead code remaining in the final program cannot be eliminated without changing the branching structure or the semantics of the program, or without impairing some program executions.

At first sight, the same iterative process seems to work for parallel programs as well. However, in the *parallel setting* the execution time is extremely sensitive to assignment sinkings. Intuitively, if assignments are moved from a parallel into a sequential program part, sinking can dramatically *impair* the execution time. In the worst case, for example, if the assignments of a parallel statement are all data independent and none of them is dead, a "naive" realization of the "as-far-as-possible" sinking strategy results in a completely sequentialized program without profiting from any elimination, hence drastically impairing the execution time.

In this article we show how to cure this defect and the possibly desastrous impact a "naive" transfer of the sequential transformation technique for PDE to the parallel setting can have on the execution time because there a subset of semantically correct assignment sinkings violates the basic requirement (R) that program optimization should never impair the execution time, by introducing a *natural* and *sufficient* constraint (C), which restricts assignment sinkings to semantically correct ones, which additionally satisfy (R): An occurrence of an assignment pattern $\alpha \equiv v := t$ can only be sunk out of a parallel statement if occurrences of α can be sunk out of each of its components (C).

An immediate side-effect of imposing constraint (C) is that assignment sinkings preserve the parallelism of the original program. Moreover, sinkings moving assignments from sequential to parallel program parts are further supported. In fact, under constraint (C) we can lift the optimality result known from the sequential setting to the parallel setting (cf. Theorem 3).

The power of the complete algorithm is demonstrated in the example of Figure 1(a), where it is unique to achieve the optimization result of Figure 1(b). It removes the (totally) dead assignment of node **3**, and eliminates the partially dead assignments of node **1**, **2**, **5**, and **10** by moving them to node **6**, **15**, and **14**, respectively. Moreover, this example illustrates the following three features, which are central for our approach:

(1) *Parallelism is preserved:* This is a consequence of constraint (C). Note the different treatment of the assignments to x and z in the parallel statement, which are all partially dead.

(2) *Parallelism is transparent:* Assignments can be moved across parallel statements which are "transparent" for them. See the assignment of node **2**, which is moved to node **15**.

(3) *Parallelism can be enhanced:* Assignments can be moved from sequential to parallel program parts. See the assignment of node **1**, which is moved to node **6**.

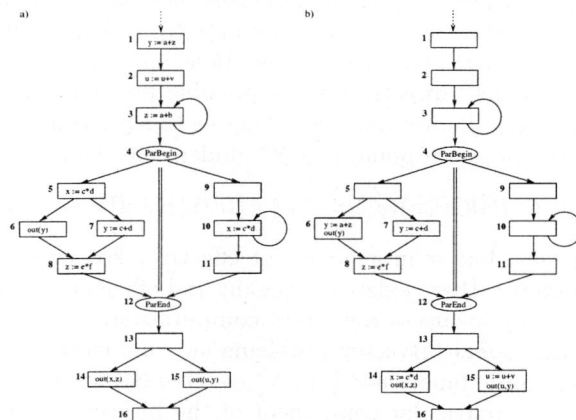

Fig. 2. PDE for *Parallel* Programs

Related Work: For sequential programs there are essentially two algorithms addressing the problem of PDE, the one of Feigen et al. [FKCX], and the one of Knoop et al. [KRS1], which is underlying the approach in this article. Both algorithms are incomparable as the one of [FKCX] considers complex program statements as sinking candidates, but is not capable of moving statements out of or across loops. For parallel programs PDE has not yet been addressed to the knowledge of the author. In fact, whereas a large number of approaches has been proposed for the *automatic parallelization* of sequential programs (cf. [BENP, GC, Le] to cite a very few), there is currently very little work on classical *optimizations* for parallel programs (cf. [KY, WS]), which may be due to the fact that naive adaptations of the sequential optimization methods fail [MP], and their straightforward correct adaptations have unacceptable costs caused by the interleavings which manifest the possible executions of a parallel program.

2 The Parallel Setting

This section sketches our setup, which has been presented in detail in [KSV]. We consider a parallel imperative programming language with interleaving semantics. Parallelism is syntactically expressed by means of a `par` statement whose components are executed in parallel on a shared memory. As usual, we assume that there are neither jumps entering a component of a `par` statement from outside nor vice versa.

Data Flow Analysis: In essence, *data flow analysis* (*DFA*) provides information about the program states that may occur at some given program points during execution. Theoretically well-founded are DFAs based on *abstract interpretation* (cf. [CC]). Usually the abstract semantics, which is tailored to deal with a specific problem, is specified by a *local semantic functional* $[\![\ ]\!] : N^* \to (\mathcal{C} \to \mathcal{C})$, which gives abstract meaning to every program statement (here: every node of a parallel flow graph G^* with node set N^*) in terms of a transformation function on a complete lattice $(\mathcal{C}, \sqcap, \sqsubseteq, \bot, \top)$, whose elements express the DFA-information of interest. In our framework this carries over to the parallel setting. A local semantic functional can easily be extended to cover *parallel paths*, i.e., paths which represent a proper interleaving execution of the program. This extension is the key for defining the solution of the parallel version of the *meet over all paths* (*MOP*) approach in the sense of Kam and Ullman [KU], which specifies the intuitively desired solution of a DFA-problem. The *MOP*-approach (in the parallel setting the *PMOP*-approach) directly mimics possible program executions in that it "meets" (intersects) all informations belonging to a program path starting in $\mathbf{s}^*$ and reaching the program point $n \in N^*$ under consideration.

The *PMOP*-**Solution:** $\forall\, c_0 \in \mathcal{C}.\ PMOP(n)(c_0) = \sqcap\, \{\, [\![\, p\,]\!](c_0)\, |\, p \in \mathbf{PP}_{G^*}[\mathbf{s}^*, n[\, \}$

This directly reflects our desires, but is in general not effective. For unidirectional bitvector problems, however, there exists an elegant and efficient way for computing the *PMOP*-solution by means of a fixpoint computation.

Bitvector Analyses: Unidirectional bitvector problems are characterized by the simplicity of their local semantic functional $[\![\ ]\!] : N^* \to (\mathcal{B} \to \mathcal{B})$, which specifies the effect of a node n on a particular component of the bitvector, where $\mathcal{B}$ is the lattice $(\{\mathit{ff}, \mathit{tt}\}, \sqcap, \sqsubseteq)$ of Boolean truth values with $\mathit{ff} \sqsubseteq \mathit{tt}$ and the logical "and" as meet operation $\sqcap$ (or its dual counterpart). Important for our efficient fixpoint approach are the following obvious facts on the semantic domain $\mathcal{F}_\mathcal{B}$ of the monotonic Boolean functions $\mathcal{B} \to \mathcal{B}$ of bitvector analyses: (1) $\mathcal{F}_\mathcal{B}$ consists of the constant functions $Const_{tt}$ and $Const_{f\!f}$, together with the identity $Id_\mathcal{B}$ on $\mathcal{B}$ only. (2) All functions of $\mathcal{F}_\mathcal{B}$ are distributive. (3) $\mathcal{F}_\mathcal{B}$ together with the orderings $\sqsubseteq_{seq}$ and $\sqsubseteq_{par}$ forms two complete lattices with least element $Const_{f\!f}$ and greatest element $Const_{tt}$, and least element $Const_{f\!f}$ and greatest element $Id_\mathcal{B}$, respectively, i.e., $Const_{f\!f} \sqsubseteq_{seq} Id_\mathcal{B} \sqsubseteq_{seq} Const_{tt}$ and $Const_{f\!f} \sqsubseteq_{par} Const_{tt} \sqsubseteq_{par} Id_\mathcal{B}$. In both cases $\mathcal{F}_\mathcal{B}$ is closed under function composition. Note that $\sqsubseteq_{seq}$ is just the pointwise ordering between functions of $\mathcal{F}_\mathcal{B}$. In the following we will usually drop the index seq. Hence, $\sqcap$ and $\sqsubseteq$ expand to $\sqcap_{seq}$ and $\sqsubseteq_{seq}$, respectively. The following lemma will be the key to the efficient computation of the "interleaving effect".

Lemma 1 Main-Lemma. *Let* $f_i : \mathcal{F}_\mathcal{B} \to \mathcal{F}_\mathcal{B}$, $1 \leq i \leq q$, $q \in I\!\!N$, *be functions on* $\mathcal{F}_\mathcal{B}$. *Then:* $\exists\, k \in \{1, ..., q\}.\ f_q \circ ... \circ f_2 \circ f_1 = f_k\ \wedge\ \forall\, j \in \{k+1, ..., q\}.\ f_j = Id_\mathcal{B}$.

Interference: The relevance of Main Lemma 1 for bitvector problems is that each possible interference in a parallel program is due to a single statement in a parallel component, i.e., due to a statement whose execution can be interleaved with the statement at the program point n under consideration, denoted as n's *interleaving predecessors* $Pred^{Itlvg}_{G^*}(n)$. This is a consequence of the fact that for each node $m \in Pred_{G^\bullet}^{Itlvg}(n)$, there exists a parallel path leading to n whose last step requires the execution of m. Together with the obvious existence of a path to n that does not require the execution of any statement of $Pred_{G^\bullet}^{Itlvg}(n)$, this implies that the only effect of interference is "destruction". This motivates the introduction of the predicate *NonDestruct* defined for each node $n \in N^*$ by

$$NonDestruct(n) \iff_{df} \forall m \in Pred_{G^\bullet}^{Itlvg}(n). [\![m]\!] \in \{Const_{tt}, Id_{\mathcal{B}}\}.$$

Only the constant function given by the precomputed value of this predicate is used below to model interference (see the definition of $[\![\ ]\!]$), and in fact, Theorem 2 guarantees that this modelling is sufficient.

Synchronization: In order to leave a parallel statement, all parallel components are required to terminate. The information that is necessary to model this effect can be computed by a hierarchical algorithm that only considers purely sequential programs. The central idea coincides with that of interprocedural DFA (cf. [KS]): we need to compute the effect of complete subgraphs, in this case of complete parallel components. This information is computed in an "innermost" fashion and then propagated to the next surrounding parallel statement. The complete three-step procedure $\mathcal{A}$ is given in [KSV]. In essence, it is a straightforward hierarchical adaptation of the functional version of the *maximal fixed point* (*MFP*) approach in the sense of [KU] to the parallel setting. Here we only consider the second step realizing the synchronization at end nodes of parallel statements in more detail. In essence, this step can be reduced to the case of parallel statements G with purely sequential components $G_1, \ldots, G_k$. Thus, the global semantics $[\![G_i]\!]^*$ of the component graphs G_i can be computed as in the sequential case. Afterwards, the global semantics $[\![G]\!]^*$ of G is given by:

$$[\![G]\!]^* = \sqcap_{par}\{[\![G_i]\!]^* \mid i \in \{1,\ldots,k\}\} \qquad (Synchronization)$$

Again, Main Lemma 1 is the key for proving the correctness of this step, i.e., $[\![G]\!]^*$ coincides with the desired *PMOP*-solution. As before the point is that a single statement is responsible for the entire effect of a path through a parallel statement. Thus, its effect is already given by the projection of this path onto the parallel component containing the vital statement, which is exploited in the synchronization step above. Adapting this step will be the key for overcoming the pitfalls of assignment sinking in Section 3. After the hierarchical preprocess, whose correctness is a consequence of the Hierarchical Coincidence Theorem of [KSV], the functional $[\![\ ]\!] : N^* \to \mathcal{F}_{\mathcal{B}}$, defined as the greatest solution of the equation system below, is the key for characterizing the *PMOP*-solution of a unidirectional bitvector problem algorithmically (refer to [KSV] for notation):

$$[\![n]\!] = \begin{cases} Id_{\mathcal{B}} & \text{if } n = \mathbf{s}^* \\ [\![pfg(n)]\!]^* \circ [\![start(pfg(n))]\!] \sqcap Const_{NonDestruct(n)} & \text{if } n \in N_X^* \\ \sqcap\{[\![m]\!] \circ [\![m]\!] \mid m \in pred_{G^\bullet}(n)\} \sqcap Const_{NonDestruct(n)} & \text{otherwise} \end{cases}$$

In analogy to the *MFP*-solution of [KU] for the sequential case, we now define the $PMFP_{BV}$-solution of unidirectional bitvector problems for the parallel setting:

The $PMFP_{BV}$-Solution: $\forall n \in N^*\ \forall b \in \mathcal{B}.\ PMFP_{BV(G^\bullet, [\![\ ]\!])}(n)(b) = [\![n]\!](b)$

As in the sequential case, the $PMFP_{BV}$-solution is practically relevant because it can efficiently be computed. Moreover, as desired we have (cf. [KSV]):

Theorem 2 The Parallel Bitvector Coincidence Theorem. *Given a parallel flow graph G^*, and a local semantic functional $[\![\ ]\!] : N^* \to \mathcal{F_B}$, the PMOP-solution and the $PMFP_{BV}$-solution coincide.*

3 Partial Dead Code Elimination

In this section we develop our algorithm for partial dead code elimination in parallel programs. Without loss of generality, we consider an arbitrary, but fixed parallel program G^*. Moreover, we assume that all edges starting in a node outside the set of start nodes of parallel statements with more than one successor have been split by inserting a synthetic node. This is typical for code motion transformations in order to avoid the blocking of the code motion process by *critical edges* (cf. [KRS2]).

As mentioned in Section 1 assignment sinkings and assignment eliminations must satisfy certain constraints in order to be semantics preserving (and profitable). For example, assignment sinkings must never move an assignment across a "blocking" instruction, i.e., an instruction which uses or modifies its left-hand side variable, or modifies an operand of its right-hand side term. Assignment eliminations may only remove *dead* assignments, i.e., assignments whose left-hand side variable is not used afterwards in the program without a preceding redefinition. These admissibility constraints can be expressed by appropiate predicates. In the following we will write $G \vdash_{(\mathcal{S},\mathcal{E})} G'$ if the parallel flow graph G' results from G by an assignment sinking or elimination to G satisfying the sinking and elimination constraints expressed by $\mathcal{S}$ and $\mathcal{E}$, respectively. Then, $\mathcal{G}_{(\mathcal{S},\mathcal{E})} =_{df} \{ G' \mid G \vdash^*_{(\mathcal{S},\mathcal{E})} G' \}$ denotes the universe of all programs being reachable from a program G by sinking and elimination steps respecting $\mathcal{S}$ and $\mathcal{E}$.

In order to compare different programs of $\mathcal{G}_{(\mathcal{S},\mathcal{E})}$, we introduce the notion of the *execution time* of a (parallel) program path as a standard of comparison. For simplicity, we assume that all assignments have unit costs. The *execution time* of a parallel program path is then given (structurally) as follows: For a parallel statement it is the maximum of the execution times of its components for the considered execution, and for a parallel program path, i.e., the sequential composition of elementary and parallel statements, it is the sum of the execution times of its components.

Based on this definition, a program $G' \in \mathcal{G}_{(\mathcal{S},\mathcal{E})}$ is *better* than a program $G'' \in \mathcal{G}_{(\mathcal{S},\mathcal{E})}$, if and only if for all paths p from the start node to the end node in G' the execution time is less or equal to that of the corresponding path in G''. Moreover, a program $G \in \mathcal{G}_{(\mathcal{S},\mathcal{E})}$ is *optimal* if and only if G is better than any other program in $\mathcal{G}_{(\mathcal{S},\mathcal{E})}$.

Now we can present our algorithm for the optimal elimination of partially dead assignments in a parallel program. It evolves from stepwise refining the straightforward parallel extension of the underlying sequential algorithm, which we call the "naive" parallel PDE algorithm. Like their sequential counterpart, both the naive and the refined parallel algorithm are composed of two procedures for assignment sinking and dead assignment elimination, which are controlled by a sinkability analysis for assignment sinkings, and a dead variable analysis for assignment eliminations, which are repeated until the program stabilizes in order to capture the second order effects of PDE. Whereas the sinking procedures are different for both algorithms, the elimination procedure is shared by them, which we thus present first. It directly evolves from its sequential counterpart.

The Elimination Procedure: Intuitively, an assignment is dead at a program point, if there is no program continuation on which its left-hand side variable is used without a preceding redefinition. This is a classical *backward* bitvector problem (cf. [He]).[3] Table 1 shows its specification in a form which directly fits into the framework of Section 2. Here, $Used_x$ and Mod_x are two local predicates of nodes, which are true for a node n, if x is a right-hand side variable of n, and if x is the left-hand side variable of n, respectively.

$$\forall\, n \in N^*.\ [\![\, n\,]\!]^{dd}_x =_{df} \begin{cases} Const_{tt} & \text{if } \neg Used_x(n) \wedge Mod_x(n) \\ Id_{\mathcal{B}} & \text{if } \neg(Used_x(n) \vee Mod_x(n)) \\ Const_{f\!f} & \text{otherwise} \end{cases}$$

Table 1. Dead Variable Analysis for Variable Pattern x

Note that $[\![\ ]\!]^{dd}_x$ is exactly the functional which is known from the sequential case. Indeed, the effect of interference is completely taken care of by the instantiated version of the predicate *NonDestruct* induced by $[\![\ ]\!]^{dd}_x$. "Destruction" by interference here means that the variable under consideration is forced to be alive by some statement of a parallel "relative". However, one does not have to bother about these details when applying the framework. In fact, the functional $[\![\ ]\!]^{dd}_x$ can directly be fed into the generic algorithm of the framework of Section 2 computing the *PMFP*-solution, which, as a consequence of the Parallel Bitvector Coincidence Theorem 2 coincides with the desired *PMOP*-solution. Hence, the algorithm comes up with the desired set of dead variables. The corresponding program transformation is then very simple: Eliminate every assignment occurrence, whose left-hand side variable is dead immediately after it.

The Sinking Procedure of the "Naive" PDE Algorithm: Central for the sinking procedure of the "naive" PDE algorithm is the admissibility constraint imposed on it. For the "naive" algorithm this is the natural extension $\mathcal{S}_{nv}$ of its sequential counterpart (cf. [KRS1]): It requires that an assignment is never moved across an instruction "blocking" it, nor moved to a program point which is not reached by a corresponding occurrence from all its predecessors. The "naive"

---

[3] In contrast, detecting "faint" variables required by the similar problem of "partial faint code elimination" is not a bitvector problem (cf. [KRS1]).

sinking procedure is then completely specified by the local semantic functional of Table 2, where $LocSinkable_\alpha$ and $LocBlocked_\alpha$ are two local predicates of nodes, which are true for a node n, if n represents an occurrence of the assignment pattern α, and if the statement of n blocks the sinking of α, respectively.[4]

$$\forall\, n \in N^*.\; [\![\, n\,]\!]^{sk}_\alpha =_{df} \begin{cases} Const_{tt} & \text{if } LocSinkable_\alpha(n) \\ Id_{\mathcal{B}} & \text{if } \neg(LocSinkable_\alpha(n) \vee LocBlocked_\alpha(n)) \\ Const_{f\!f} & \text{otherwise} \end{cases}$$

Table 2. Sinkability Analysis for Assignment Pattern $\alpha \equiv v := t$

For the sinkability analysis "destruction" by interference means that an assignment is blocked by the instruction of a parallel relative. Expressing the result of the sinkability analysis for α by the predicate $Sink_\alpha$, the corresponding program transformation is as follows: Remove all original occurrences of α satisfying $\bigwedge\{\, Sink_\alpha(m) \,|\, m \in succ(n)\,\}$, and insert new instances of α at the entry of each node satisfying $Sink_\alpha(n) \wedge (LocBlocked_\alpha(n) \vee \bigvee\{\, \neg Sink_\alpha(m) \,|\, m \in succ(n)\,\})$.

Whereas the ratio underlying the sinking procedure of the "naive" algorithm, namely to move assignments as far as possible, i.e., as long as the semantics is preserved, in order to maximize the potential of dead code, leads to optimal results in the sequential setting, it can be desastrous in the parallel one as the parallelism of the argument program can be destroyed without profiting from any elimination (cf. Section 1).

The Sinking Procedure of the Optimal PDE Algorithm: The point of refining the "naive" sinking procedure is to exclude assignment sinkings which possibly impair the execution time of a program execution. Conceptually, this is achieved by strengthening the admissibility constraint $\mathcal{S}_{nv}$ of the "naive" sinking procedure by additionally imposing constraint (C) introduced in Section 1. In essence, this means to handle synchronization on leaving a parallel statement like an ordinary join of the control flow in the sequential setting.

The First Refinement – Synchronization: In order to take care of constraint (C) during synchronization, it is sufficient to replace the synchronization step described in Section 2 by:

$$[\![\, G\,]\!]^* = \sqcap_{seq}\{\, [\![\, G_i\,]\!]^* \,|\, i \in \{1,\ldots,k\}\,\}.$$

In contrast to the "real" synchronization mechanism, which, instantiated with the sinkability functional $[\![\;]\!]^{sk}_\alpha$, allows us to move an assignment out of a parallel statement, if it can be moved out of *some* of its components, the modified one requires that this is possible for *all* components. Clearly, this requires that there is

---
[4] Occurrences of an assignment pattern α satisfy both the predicate $LocSinkable_\alpha$ and $LocBlocked_\alpha$, a fact, which cannot directly be modelled in our abstract domain. It is taken care of by associating all occurrences of α (in parallel components) with two semantic functions. This does not introduce any subtleties in the framework of Section 2. It just makes the twofold character of an occurrence of α explicit.

at least one occurrence in every component of the parallel statement under consideration. In this case, however, these occurrences interfere and therefore block each other. Thus, as a consequence of modifying the synchronization mechanism no assignment at all can be sunk out of a parallel statement. Of course, this is too restrictive, and is overcome by the second refinement dealing with interference.

The Second Refinement – Interference: The point of this refinement is to exclude that occurrences of the same assignment pattern prevent each other from sinking due to interference in a parallel statement. In the framework of Section 2, it suffices to replace the predicate *NonDestruct* by the weaker version

$$NonDestruct'(n) =_{df} \forall m \in Pred_{G^\bullet}^{Itlvg}(n).\ \mathcal{P}(m) \neq \mathcal{P}(n) \Rightarrow [\![m]\!] \in \{Const_{tt}, Id_\mathcal{B}\}$$

where $\mathcal{P}$ is a function returning the statement pattern of its argument node. The new version is used for all nodes except for the start nodes of parallel components. This is mandatory in order to prevent that assignments sink from a sequential program part into a parallel statement and get stuck in more than one of its components. In fact, correctness, even in the sense of *sequential consistency* (cf. [La]), gets lost in general. This is particularly obvious for "recursive" assignments, i.e., assignments whose left-hand side variables occur in their right-hand side terms. Conversely, sinking of occurrences of assignment patterns out of each component of a parallel statement as it is intended by the second refinement is always correct for non-recursive patterns; for recursive ones, however, this depends on the underlying interleaving model. In case of non-atomic assignments, sequential consistency would be preserved, and thus it would be correct. Otherwise, this cannot be guaranteed and hence, the second refinement has to be restricted to "non-recursive" assignment patterns in this case.

Main Result: Denoting the admissibility constraints of the second refinement by $\mathcal{S}_R$ and $\mathcal{E}$, respectively, and the program finally resulting from the application of the refined algorithm to an argument program G by G_{pde}, we have the following optimality result, which can be proved along the lines of [GKLRS]:

Theorem 3 Optimality. *G_{pde} is optimal in $\mathcal{G}_{(\mathcal{S}_R, \mathcal{E})}$.*

4 Conclusions

Optimizing explicitly parallel programs is desirable and possible. In this article we demonstrated this by developing an algorithm for the optimal elimination of partially dead code in a parallel program. An important observation during this development was that in contrast to the DFAs required, which could straightforward be adapted from their sequential counterparts by means of the framework of [KSV], the transformation itself required special care. In fact, the naive adaptation of the optimal sequential algorithm could destroy the parallelism of its argument programs, and thus dramatically impair their execution time. By introducing a natural and sufficient side condition this defect could be cured and the optimality result known from the sequential setting be lifted to the parallel one. Currently, we are investigating how to adapt our approach to other classical optimizations like *code motion* [KRS2] and *assignment motion* [KRS3].

References

[BENP] Banerjee, U., Eigenmann, R., Nicola, A., and Padua, D. A. Automatic program parallelization. In Proc. IEEE 81(2), 1993, 211- 243.

[CC] Cousot, P., and Cousot, R. Abstract interpretation: A unified lattice model for static analysis of programs by construction or approximation of fixpoints. In *Conf. Rec. 4th Int. Symp. Princ. Prog. Lang. (POPL'77)*, 1977, 238 - 252.

[FKCX] Feigen, L., Klappholz, D., Casazza, R., and Xue, X. The revival transformation. In *Conf. Rec. 21st Symp. Princ. Prog. Lang. (POPL'94)*, 1994, 421 - 434.

[GC] Griebl, M., and Collard, J.-F. Generation of synchronous code for automatic parallelization. In *Proc. 1st Europ. Conf. Parallel Process. (Euro-Par'95)*, Springer-V., LNCS 966 (1995), 315 - 326.

[GKLRS] Geser, A., Knoop, J., Lüttgen, G., Rüthing, O., and Steffen, B. Non-monotone fixpoint iterations to resolve second order effects. In *Proc. 6th Int. Conf. Comp. Constr. (CC'96)*, Springer-V., LNCS 1060 (1996), 106 - 120.

[He] Hecht, M. S. Flow analysis of computer programs. Elsevier, North-Holland, 1977.

[KRS1] Knoop, J., Rüthing, O., and Steffen, B. Partial dead code elimination. In *Proc. ACM SIGPLAN'94 Conf. Prog. Lang. Design Impl. (PLDI'94)*, ACM SIGPLAN Not. 29, 6 (1994), 147 - 158.

[KRS2] Knoop, J., Rüthing, O., and Steffen, B. Optimal code motion: Theory and practice. *ACM Trans. Prog. Lang. Syst. 16*, 4 (1994), 1117 - 1155.

[KRS3] Knoop, J., Rüthing, O., and Steffen, B. The power of assignment motion. In *Proc. ACM SIGPLAN'95 Conf. Prog. Lang. Design Impl. (PLDI'95)*, SIGPLAN Not. 30, 6 (1995), 233 - 245.

[KS] Knoop, J., and Steffen, B. The interprocedural coincidence theorem. In *Proc. 4th Int. Conf. Comp. Constr. (CC'92)*, Springer, LNCS 641 (1992), 125 - 140.

[KSV] Knoop, J., Steffen, B., and Vollmer, J. Parallelism for free: Efficient and optimal bitvector analyses for parallel programs. In *ACM Trans. Prog. Lang. Syst. 18*, 3 (1996), 268 - 299.

[KU] Kam, J. B., and Ullman, J. D. Monotone data flow analysis frameworks. *Acta Informatica 7*, (1977), 309 - 317.

[KY] Krishnamurthy, A., and Yelick, K. Optimizing parallel programs with explicit synchronization. In *Proc. ACM SIGPLAN'95 Conf. Prog. Lang. Design Impl. (PLDI'95)*, SIGPLAN Not. 30, 6 (1995), 196 - 204.

[La] Lamport, L. How to make a multiprocessor computer that correctly executes multiprocess programs. *IEEE Trans. Comput. C-28*, 9 (1979), 690 - 691.

[Le] Lengauer, Ch. Loop parallelization in the polytope model. In *Proc. 4th Int. Conf. Conc. Theo. (CONCUR'93)*, Springer-V., LNCS 715 (1993), 398 - 416.

[MP] Midkiff, S. P., and Padua, D. A. Issues in the optimization of parallel programs. In *Proc. Int. Conf. Parallel Processing, Volume II*, St. Charles, Illinois, (1990), 105 - 113.

[SW] Srinivasan, H., and Wolfe, M. Analyzing programs with explicit parallelism. In *Proc. 4th Int. Conf. on Lang. and Comp. for Parallel Computing (LCPC'91)*, Springer-V., LNCS 589 (1991), 405 - 419.

[WS] Wolfe, M, and Srinivasan, H. Data structures for optimizing programs with explicit parallelism. In *Proc. 1st Int. Conf. Austrian Center for Parallel Computation*, Springer-V., LNCS 591 (1991), 139 - 156.

Solving the Constant-Degree Parallelism Alignment Problem

Claude G. Diderich[1*] and Marc Gengler[2]

[1] Swiss Federal Institute of Technology – Lausanne, Computer Science Department,
CH-1015 Lausanne, Switzerland, E-mail: diderich@di.epfl.ch
[2] Ecole Normale Supérieure de Lyon, Laboratoire de l'Informatique du Parallélisme,
F-69364 Lyon, France, E-mail: Marc.Gengler@lip.ens-lyon.fr

Abstract. We describe an exact algorithm for finding a computation mapping and data distributions that minimize, for a given degree of parallelism, the number of remote data accesses in a distributed memory parallel computer (DMPC). This problem is shown to be NP-hard.

1 The alignment problem

An important problem when compiling nested loops towards DMPCs is how to map the computation and the data onto processors. This problem can be subdivided into two subproblems: 1) the *alignment problem* which assigns computation and data to a set of virtual processors, and 2) the *mapping problem* which folds the set of virtual processors onto the physical ones. In this paper we address the alignment problem. Following the linear algebra formulation of the alignment problem by Huang and Sadayappan [6] in 1991, researchers have primarily focused on finding linear or affine computation and data alignment functions requiring no remote data accesses [3] or on developing heuristics for minimizing communication [2].

The alignment problem is the problem of finding an alignment of loop iterations with the array elements accessed, that is, mappings of the loop iterations and array elements to a set of virtual processors. The alignment should address the two needs: i) maximize the degree of parallelism, i.e. use as many processors as possible, ii) minimize the number of non local data accesses, i.e. distribute the array elements such that a processor owns a maximal number of the elements it accesses. Depending on how the needs i) and ii) are verified, various subproblems can be defined. When allowing only local data accesses, we talk about the *communication-free alignment problem*. Another subproblem is defined by minimizing the number of remote data accesses for a given degree of parallelism. This subproblem is called the *constant-degree parallelism alignment problem*. We consider array access functions that are linear or affine and use the approach presented by Bau *et al.* [3] for expressing the alignment problem. Access l to array k is described by a function F_k^l. The unknown computation mappings C_j and data mappings D_k can also be written as matrix functions. $\mathcal{I}$ represents the index domain defined by the loop bounds, $\mathcal{D}_k$ the array access domain and $\mathcal{P}$ the virtual multi-dimensional grid of processors.

---

[*] Supported by a grant from the Swiss Federal Institute of Technology – Lausanne.

$$F_k^l: \mathcal{I} \longrightarrow \mathcal{D}_k : \quad \mathbf{i} \longmapsto F_k^l(\mathbf{i}) = \mathbf{F}_k^l\, \mathbf{i} + \mathbf{f}_k^l$$
$$C_j: \mathcal{I} \longrightarrow \mathcal{P} : \quad \mathbf{i} \longmapsto C_j(\mathbf{i}) = \mathbf{C}_j\, \mathbf{i} + \mathbf{c}_j$$
$$D_k: \mathcal{D}_k \longrightarrow \mathcal{P} : \quad \mathbf{a} \longmapsto D_k(\mathbf{a}) = \mathbf{D}_k\, \mathbf{a} + \mathbf{d}_k$$

$C_j(\mathbf{i})$ represents the processor on which iteration $\mathbf{i}$ of assignment instruction j is executed. Similarly, the function D_k indicates on which processors the elements of array k are located. The requirements i) and ii) can be formulated as follows. Eqns. (1) are called alignment or locality constraints.

$$\max_{\mathbf{C}_j, \mathbf{c}_j} \left(\min_j \left(\mathrm{rank}\left(\mathbf{C}_j^\mathsf{T} \right) \right) \right).$$
$$\forall \mathbf{i} \in \mathcal{I}: \ \mathbf{C}_j\, \mathbf{i} + \mathbf{c}_j = \mathbf{D}_k \left(\mathbf{F}_k^l\, \mathbf{i} + \mathbf{f}_k^l \right) + \mathbf{d}_k. \tag{1}$$

The algorithm for solving the communication-free alignment problem presented by Bau et al. [3], called LINEAR-ALIGNMENT, is defined as follows. The set of eqns. (1) is rewritten in the following equivalent form.

$$\forall \mathbf{i} \in \mathcal{I}: \left(\hat{\mathbf{C}}_j\ \hat{\mathbf{D}}_k \right) \begin{pmatrix} \mathbf{I} \\ -\hat{\mathbf{F}}_k^l \end{pmatrix} \begin{pmatrix} \mathbf{i} \\ 1 \end{pmatrix} = 0 \tag{2}$$

with $\hat{\mathbf{C}}_j = (\ \mathbf{C}_j\ \ \mathbf{c}_j\ ),\ \hat{\mathbf{D}}_k = (\ \mathbf{D}_k\ \ \mathbf{d}_k\ ),\ \hat{\mathbf{F}}_k^l = \begin{pmatrix} \mathbf{F}_k^l & \mathbf{f}_k^l \\ 0 & 1 \end{pmatrix}$

To simplify the problem we require that eqns. (2) hold for any vector $\mathbf{i}$, regardless of whether or not it belongs to the iteration domain $\mathcal{I}$. The alignment constraints then become equations of the form (3). Allowing no communication imposes that all locality constraints (3) are verified simultaneously, leading to (4).

$$\left(\hat{\mathbf{C}}_j\ \hat{\mathbf{D}}_k \right) \begin{pmatrix} \mathbf{I} \\ -\hat{\mathbf{F}}_k^l \end{pmatrix} = 0. \tag{3}$$

$$\hat{\mathbf{U}}\,\hat{\mathbf{V}} = 0 \tag{4}$$

where $\hat{\mathbf{U}} = (\ \hat{\mathbf{C}}_1 \cdots \hat{\mathbf{C}}_t\ \hat{\mathbf{D}}_1 \cdots \hat{\mathbf{D}}_s\ ),\ \hat{\mathbf{V}} = (\ \hat{\mathbf{V}}_{u,v,w} \cdots \hat{\mathbf{V}}_{x,y,z}\ )$

$$\hat{\mathbf{V}}_{j,k,l} = \begin{pmatrix} 0 & \cdots & 0 & \mathbf{I} & 0 & \cdots & 0 & -\hat{\mathbf{F}}_k^l & 0 & \cdots & 0 \end{pmatrix}^\mathsf{T}$$

The sub-matrix $\hat{\mathbf{V}}_{j,k,l}$, where $\mathbf{I}$ is the jth block and $-\hat{\mathbf{F}}_k^l$ the $(t+k)^{\text{th}}$ block, represents the alignment constraint of data access l of array k in statement j and the processor using that data. Eqn. (4) is equivalent to $\hat{\mathbf{V}}^\mathsf{T}\,\hat{\mathbf{U}}^\mathsf{T} = 0$. Therefore, the column vectors of the unknown matrix $\hat{\mathbf{U}}^\mathsf{T}$ are in the null space of the known $\hat{\mathbf{V}}^\mathsf{T}$.

2 The constant-degree parallelism alignment problem

Often, the degree of parallelism of a communication-free alignment is non-existing. This leads us to define the *constant-degree parallelism alignment problem* (CDPAP), which consists of finding communication and data mappings such that the degree of parallelism obtained is at least equal to the input parameter d and the communications are minimized. The *constant-degree parallelism alignment algorithm* (CDPAA) solves the CDPAP.

Assume that it is possible to find a communication-free alignment of parallelism degree d' for a given problem $\hat{\mathbf{V}}$. We simplify $\hat{\mathbf{V}}$ by finding a minimal set of alignment constraints from (1) to be left unsatisfied such that the simplified problem has a solution of degree of parallelism d when solved by the LINEAR-ALIGNMENT algorithm. To increase the parallelism introduced by the LINEAR-ALIGNMENT algorithm by $d'' = d - d'$, we have to construct a modified problem $\hat{\mathbf{V}}'$ such that the size of the basis of the null space of that problem is increased by d'' compared to the size of the null-space of the original problem.

We will use the notation of $\tilde{\mathbf{V}}$ to represent the vector space spanned by the column vectors of the matrix $\hat{\mathbf{V}}$. $\tilde{\mathbf{V}}$ represents the space of all the alignment constraints. In order to increase the degree of parallelism by at least d'', we need to find a subspace $\tilde{\mathbf{V}}'$ of $\tilde{\mathbf{V}}$ such that $\dim(\tilde{\mathbf{V}}) - \dim(\tilde{\mathbf{V}}') \geq d''$. Let $\tilde{d} = \dim(\tilde{\mathbf{V}}) - d''$. There exist an infinite number of such subspaces $\tilde{\mathbf{V}}'$ of dimension $\tilde{d}$, but only finitely many are of interest to us. In fact, all subspaces of $\tilde{\mathbf{V}}$ that contain less than $\tilde{d}$ vector columns of $\hat{\mathbf{V}}$ are uninteresting, because we know that there exists at least one subspace containing at least $\tilde{d}$ column vectors of $\hat{\mathbf{V}}$. Furthermore, the set of all the subspaces of degree at most $\tilde{d}$ containing at least $\tilde{d}$ column vectors can be easily enumerated. To do so, we select $\tilde{d}$ column vectors of $\hat{\mathbf{V}}$. Then, for each valid subset of column vectors of $\hat{\mathbf{V}}$, we compute a basis and count the number of alignment constraints that can be expressed in that basis. Finally, we select a subspace $\hat{\mathbf{V}}^*$ that contains the largest number of alignment constraints.

2.1 Some important aspects

Data dependences. As long as all alignment constraints are verified, data dependences are as well. As soon as alignment constraints are dropped this may no longer be true. Removing alignment constraints that represent part of data dependences may increase the degree of parallelism without reducing the execution time. In [4] we characterize the relation between the computed alignment and the iteration scheduling, that is, the relation between processor and time parallelism. Essentially, we show that a sufficient, but not necessary, condition to get a non constant number of active processors during each time step consists of imposing that there be at least two fulfilled alignment constraints that correspond to a data dependence.

Cost function. In the CDPAA we use a counting argument based on the number of non local data accesses as optimization function for computing efficient alignment functions. Another possibility would be to assign different weights to the different alignment constraints depending on their importance and then minimize the weighted sum of remote data accesses. Such a cost function even allows the user of the algorithm to require some alignment constraints to be verified by assigning to them an infinite weight. For example, the owner computes rule can be imposed by assigning a weight of $+\infty$ to the alignment constraint $\hat{\mathbf{C}}_j = \hat{\mathbf{D}}_k \hat{\mathbf{F}}_k^1$, where $\hat{\mathbf{F}}_k^1$ represents the element of array k being modified by instruction j. Furthermore, a weighting function may also be used, in principle, in order to ignore constant offsets or not, prefer non local constant accesses to non-local linear ones, etc.

Complexity and optimality. The following results are extracted from [4]. The input size of any CDPAP is characterized by four parameters, which are the number of data accesses n, the larger of the maximal dimension of any array and the maximal loop nest depth e, the number of assignment statements c and the number of arrays a. When considering all these parameters variable, we have the following theorem, obtained by reduction from the homogeneous, bipolar MAX FLS$^=$ problem [1].

Theorem 1 *The CDPAP is NP-hard.*

Theorem 2 *The CDPAA finds communication and data alignments that need a minimal number of non local data accesses for a given degree of parallelism.*

Theorem 3 *The CDPAA needs $O(n^{4+e\,(c+a)})$ time to find an optimal alignment.*

2.2 Experimental results and related work

To show the performance of the CDPAA, we apply the algorithm to various loop nests of different sizes extracted from various programs and benchmarks. In each example, which does not have a communication-free alignment, we search for alignment functions having at least one degree of parallelism. Results are given in [4].

Many techniques for solving the alignment problem proposed by different research teams [2, 3, 5] are related to the approach taken in this paper. Anderson and Lam [2] define necessary conditions for the data being local. These conditions admit a direct translation into the framework defined by Bau et al. [3] and are particular cases. The problem considered by Dion and Robert in [5] is also a particular case of [3]. Dion and Robert find an optimal solution to their problem. Our technique for constructing a maximal set of alignment constraints that can be verified while providing a given degree of parallelism is more general than [5], as we allow any access function, rather than restricting ourselves to access functions of full rank.

3 Conclusion

In this paper we presented an extension to the communication-free alignment algorithm of Bau et al. to remove a minimal number of unsatisfiable constraints to increase the degree of parallelism up to a given constant. In our future work, we are investigating the possibility to incorporate into our framework the notion of scheduling vector so as to be able to optimize a single function when solving both the scheduling and the alignment problem.

References

1. E. Amaldi and V. Kann. The complexity and approximability of finding maximum feasible subsystems of linear relations. *Theoret. Comput. Sci.*, 147(1-2):181-210, 1995.
2. J. M. Anderson and M. S. Lam. Global optimizations for parallelism and locality on scalable parallel machines. In *Proc. PLDI '93*, pages 112-125, 1993.
3. D. Bau, I. Kodukula, V. Kotlyar, K. Pingali, and P. Stodghill. Solving alignment using elementary linear algebra. In *Proc. LCPC '94*, LNCS 892, pages 46-60, Springer, 1994.
4. C. G. Diderich and M. Gengler. Solving the constant-degree parallelism alignment problem. Research Rep. DI-96/195, Swiss Fed. Inst. of Tech. – Lausanne, Switzerland, 1996.
5. Michèle Dion and Yves Robert. Mapping affine loop nests: New results. In *Proc. HPCN '95*, LNCS 919, pages 184-189, Springer, 1995.
6. C.-H. Huang and P. Sadayappan. Communication-free hyperplane partitioning of nested loops. In *Proc. LCPC '91*, LNCS 589, pages 186-200, Springer, 1991.

Topographic Data Mapping by Balanced Hypersphere Tessellation[*]

Matthias Besch and Hans Werner Pohl

RWCP Massively Parallel Systems GMD Laboratory
Rudower Chaussee 5, 12489 Berlin, Germany

[*]This work is supported by the *Real World Computing Partnership*, Japan

Abstract. Data mapping plays a crucial role in large-scale scientific computing on distributed-memory parallel machines. We present a new, fast mapping heuristic which is basically a generalization of vector quantization and is applicable to multipartitioning arbitrarily high-dimensional and irregular domains. Its efficiency derives mainly from the exploitation of topographic properties of the problem domain. We compare the quality and performance of our approach with the classical Kerninghan&Lin heuristic and spectral bisection based on the Lanczos algorithm.

1 Introduction

We assume the reader to be familiar with the problem of data mapping in the field of parallel computing. To conduct the following discussion in more formal terms, we define a mapping function $\psi : \mathcal{V} \to \mathcal{P}$ which assigns vertices of the problem graph, i.e. from the set of defined data points $i \in \mathcal{V} = \{1, ..., N\}$, to nodes of the processor graph, i.e. processors $k \in \mathcal{P}$. A given problem domain, i.e. set of points, is required to be embedded into some index space Z^n of appropriate dimension, thus obtaining regular or irregular sparse multidimensional arrays [1]. A general method to transform an arbitrary graph into a representation in the n-Euclidean index space is presented in [2]. Alternatively, geometrical information can be explicitly given by means of user-defined and problem-adequate topology declarations, thus avoiding the expensive transformation process [6].

An objective function usually consists of a term describing interprocessor-communication overhead, and one containing a load-balance criterion. Topographic mapping algorithms are based on the assumption that data points that are close together in the problem domain, i.e. index space, communicate more likely with each other than points that are far apart. Therefore, they do not have to make explicit use of communication structures, and the corresponding term vanishes from the objective function.

The objective of well-balanced workload L amoung the P processors is given by

$$\sum_i l_i \delta_{\psi_i, k} \approx \sum_i l_i / P = L/P ,$$

where δ is the *Kronecker* symbol and l_i denotes the computational load of the data point i. In contrast, we define the *load imbalance of a partition k* as

$$I_k = P \cdot L_k / L - 1 \quad \text{with} \quad L_k = \sum_i l_i \delta_{\psi_i, k}.$$

The advantage of this definition is its independence of absolute load weights. The imbalance is zero if the load is perfectly balanced, i.e. if $L_k = L/P$.

So we can now regard the mapping objective as minimizing the absolute values of the partition-load imbalances:

$$\Lambda(\psi) = \max_k \{|I_k|\}.$$

In the next section we present a new, fast mapping heuristic (BHT), which is basically a generalization of vector quantization [3] and is applicable to multi-partitioning arbitrarily high-dimensional and irregular domains. Finally, we compare BHT with the classical Kerninghan&Lin heuristic (KL) [4] and spectral bisection (SB) based on the Lanczos algorithm, e.g. [5].

2 Balanced Hypersphere Tessellation

The idea of the algorithm is to separate clusters of data points from each other by means of hyperspheres, such that the load is well-balanced. Alternatively, we could have chosen hyperplanes, but the former possesses some advantages as described in [1]. In classical vector quantization a cluster is defined as the set of data points $x \in Z^n$ that are closer to v_i than to any other v_k, where $v_i \in \mathcal{R}^n$ is called the representative or prototype of the cluster V_i, for $i = 1, 2, ..., P$:

$$V_i = \{x | \forall k \bullet \|x - v_i\|^2 \le \|x - v_k\|^2\}.$$

Clustering is performed with the help of vector quantization using a simple *winner-takes-all competitive learning* procedure. The prototype vectors v_i, compete with each other to be the one to move for each given input pattern (data point) x. The position of the winner v_{i^*} with $\|v_{i^*} - x\|^2 \le \|v_i - x\|^2$ is updated by

$$v_{i^*}^{new} = v_{i^*} + \gamma \cdot (x - v_{i^*}) \quad \text{for a } \gamma \in (0, 1),$$

and is, for example, carried out at the end of the presentation of all input vectors using the average of all changes (*batch learning*). This procedure is iterated until a stable state is reached w.r.t. some ε. With every update, the prototype vector moves towards the centre of its cluster. As a result we obtain a *Voronoi-tessellation* of the input space, as depicted in Figure 1.

Fig. 1. Voronoi tessellation and Delaunay triangulation

So far, clustering solely considers the geometric positions of the input patterns and does not at all guarantee a balanced load amoung the partitions. Therefore, we multiply both sides of the inequation of the learning rule by a *load factor*, such that the polyhedrons of the resulting *hypersphere tessellation* are given by

$$V_i = \{x | \forall k \bullet \|x - v_i\|^2 \cdot \alpha_i \leq \|x - v_k\|^2 \cdot \alpha_k\}.$$

The α_i itself depend on the current load imbalance I_k and are adapted in the training process by $\alpha_k^{new} = \alpha_k^{old} \cdot \eta_k$, where $\eta_k = exp(I_k)$. The prototype vectors are adapted to its (weighted) *mass centres* c_i by

$$v_k^{new} = v_k^{old} + \gamma \left(c_i - v_k^{old} \right), \text{ where } c_k = \left(\sum_{i, x_i \in V_k} l_i x_i \right) \bigg/ \left(\sum_{i, x_i \in V_k} l_i \right).$$

Initially, we randomly distribute the v_k around the mass centre of all data points. The process terminates if the objective function is smaller than a given ε. Figure 2 shows a 2d-tessellation of a finite-element mesh. Note that the separation is made up of circle arcs instead of straight lines.

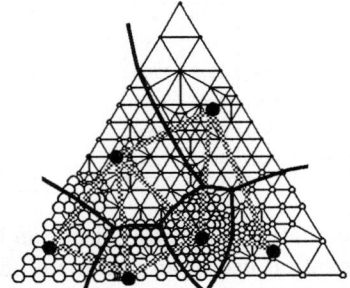

Fig. 2. Multipartitioning with hypersphere tessellation

It is easy to see that the time complexity of one iteration is $O(N \cdot P)$, in the next section we give an complexity estimation of the entire algorithm.

3 Results

The first table shows the runtime of BHT in milliseconds for various artificially generated finite-element meshes (see Figure 2 for an example) on a SUN supersparcII workstation. The runtime, i.e. the number of iterations, grows a little more than linearly with both the number of partitions and the number of mesh elements. As a rough approximation, we have found an experimental time complexity of $N^{3/2} \cdot P^{3/2}$ (see Fig. 3).

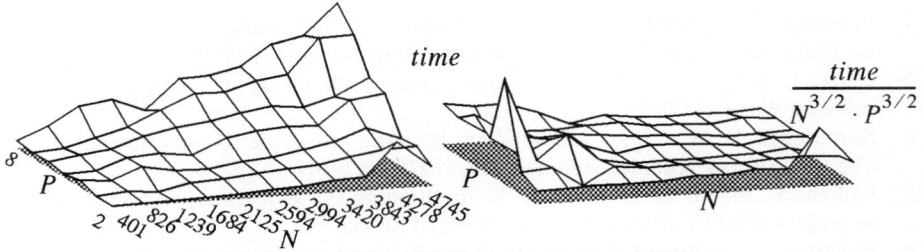

Fig. 3. Experimental time complexity of BHT

Note that the flexibility of multipartitioning has to be paid for by some decrease in efficiency. It can be shown that for a time complexity of $N^\alpha \cdot P^\beta$ with $\alpha, \beta \geq 1$ and $\alpha \cdot \beta > 1$ (sequential) recursive bipartitioning is faster than multipartitioning.

The second table compares the quality of BHT with KL and SB by means of their resulting cut-sizes. Note that all three algorithms can be combined with each other, together with multilevel graph contraction, or recursive partitioning strategies. However, we have to constraint ourselves here on comparing the kernel algorithms performing bisection.

Table 1. Runtimes for multipartitioning finite-element graphs in milliseconds; cut weights after bipartition with KL, spectral bisection, and BHT

points	partitions						
	2	3	4	5	6	7	8
401	16	50	383	183	333	316	433
1239	100	50	900	866	833	916	2150
2125	233	516	559	816	1800	2233	1700
2994	533	883	983	1916	2383	3066	3466
3843	716	1350	1616	1983	2500	3850	4966
4745	1066	1700	1583	3616	3933	5933	7050

points	KL	SB	BHT
826	67	48	65
1648	133	70	103
2594	125	78	125
3420	133	88	146
4278	265	104	157
5202	275	115	170
6087	295	119	166
6961	326	130	190

It is not surprising that the quality of spectral bisection is superior to BHT, since the latter does not make explicit use of an optimization criterion w.r.t. the connectivity. The problem with KL is that its quality largely depends on the initial bipartition, a point repeatedly made in the literature. This may lead to unpredictable results, as can be seen in the table. Still, the runtime of the BHT algorithm is convincing, ranging between a few milliseconds up to seconds on the above mentioned platform, while spectral bisection and KL require times that are one, and two, respectively, orders of magnitudes larger. BHT is thus an ideal candidate for an initial partitioning, which can than be further refined by other methods. In many cases, however, the quality of BHT should be sufficient.

References

1. Besch, M., Pohl, H. W.: Topographic Mapping of Irregular Sparse Data Structures, RWC TR, http://www.first.gmd.de/promoter/papers/topMappingTR.ps.gz, Japan (1996)
2. Fukunaga, K., Yamada, S., Stone, H.S., Kasai, T.: A Representation of Hypergraphs in the Euclidean Space, IEEE Transactions on Computers, **33**, 4 (1984)
3. Hertz, J., Krogh, A., Palmer, A.: Introduction to the theory of neural computation, Addison Wesley (1991)
4. Kerninghan, B. W., Lin, S.: An efficient heuristic procedure for partitioning graphs, The Bell System Technical Journal, Feb. (1970) 291-307
5. Pothen, A., Simon, H. D., Liou, K.-P.: Partitioning sparse matrices with eigenvectors of graphs, SIAM J. Matrix Anal., 11 (1990) 430-452
6. Promoter: Collection of Papers, http://www.first.gmd.de/promoter/papers/index.html

Implementing Pipelined Computation and Communication in an HPF Compiler

Thomas Brandes[1]*
Frédéric Desprez[2]**

[1] GMD, SCAI
Schloss Birlinghoven, P.O. Box 1319, 53754 St. Augustin, Germany
[2] LIP, ENS Lyon, CNRS URA 1398, INRIA Rhône-Alpes
46, Allée d'Italie, 69364 LYON cedex 07, France

Abstract. Many scientific applications can benefit from pipelining computation and communication. Our aim is to provide compiler and runtime support for High Performance Fortran applications that could benefit from these techniques. This paper describes the integration of a library for pipelined computations in the runtime system. Results on some application kernels are given.

1 Introduction

With the introduction of High Performance Fortran (HPF) [KLS+94], it is possible to use the data parallel programming paradigm in a very convenient way for scientific applications. With current compilation technology, these programs will execute phases of computations and communications on differents sets of data and no overlap exists between communications and computations. Moreover, communication phases are synchronous, i.e. each processor executes these phases at the same time and waits until the last processor completes his communication phase. An important task of the HPF compiler is to detect the potential of overlapping computation and communication and to take efficient use of it.

Overlapping is not always possible because of the dependences within the code. In this case the computation might be broken into smaller pieces that can be executed in a pipelined fashion. This is also called macro-pipelining [Kin88, Tse93]. Usually, the resulting code of macro-pipelining is very complicated. But we will show that it is possible to use runtime system functions that do this splitting at runtime. This does not only decrease the complexity of the HPF compiler, but also allows the optimization of overlapping computation and communication at runtime. This can be done by making some runtime measurements that determines the best size of granularity.

Though most of the techniques are already known, this paper focus on the efficient use and the integration in an existing HPF compilation system. The runtime functions are a new version of the LOCCS library (Low Overhead Communication and Computation Subroutines) that has been first presented in [DT92].

---

* This works has been supported by the Esprit-6643 project PPPE (Portable Parallel Programming Environment)
** This work has been supported by the INRIA Rhône-Alpes and the PRC-GDR PRS

2 Pipelined Computations

In pipelined computations, a processor cannot begin execution until it receives results computed by its predecessor. Though this kind of pipelined execution for its own is still a sequential execution, there are two possibilities to extract partial parallelism by overlapping computations. If one pipelined computation follows another one, all processers become busy when the pipeline is filled. The other possibility is to break the computation and to send partial results. By this way the processors may overlap their computations as shown in Figure 1.

The ADI algorithm in Figure 2 is a typical example that benefits from pipelining computation. Assume that the columns are distributed in a block fashion among the available processors. The first loop nest contains no dependence and can be executed in parallel. In the second loop nest, every processor computes the results in row order, sending the last value to the next processor as soon as it is ready. This strategy produces a pipelined effect.

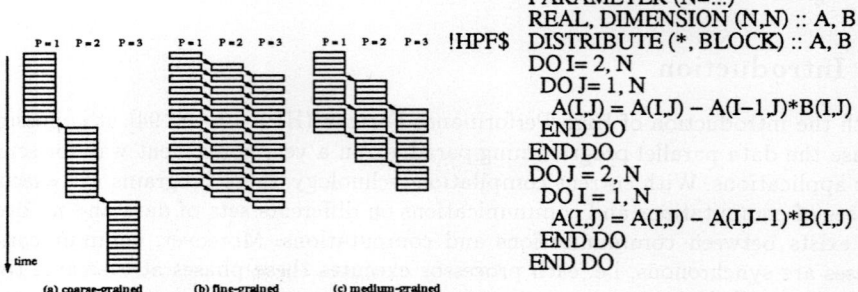

Fig. 1. Breaking up a pipelined computation.

Fig. 2. ADI algorithm in HPF.

Usually, this method is not very efficient due to the large communication startup time on MIMD message-passing machines. Therefore, a variant of the method is chosen where each processor computes a few rows before communicating the results (tiling).

3 Implementation within ADAPTOR

ADAPTOR (Automatic Data Parallelism Translator) is a public domain compilation system developed at GMD for compiling data parallel HPF programs to equivalent message passing programs [BZ94].

Instead of generating directly complex pipelining code for loop nests, ADAPTOR provides a driver routine that is implemented within the runtime system DALIB.

This routine is called with the sections belonging to the iteration space and with the corresponding data dependences. The first parameter is the name of the local subroutine that contains the code for one tile. The compiler has to find cross-processor loops and must generate the call to the routine as well as the code for the local subroutine.

```
      PARAMETER (N=...)                           EXTRINSIC (HPF_LOCAL) SUBROUTINE BLOCK (A, B)
      REAL, DIMENSION (N,N) :: A, B               REAL A(:,:), B(:,:)
!HPF$ DISTRIBUTE (*,BLOCK) :: A, B                !HPF$ DISTRIBUTE *(*,BLOCK) :: A, B
      ...                                         DO J=lbound(A,2),ubound(A,2)
      DO I = 2, N                                    DO I=lbound(A,1),ubound(A,1)
        DO J = 1, N                                    A(I,J) = A(I,J) - A(I,J-1)*B(I,J)
          A(I,J) = A(I,J) - A(I-1,J)*B(I,J)        END DO
        END DO                                    END DO
      END DO                                      END
      CALL DALIB_LOCCS_SHIFT (BLOCK, 2, 0,
        A(:,2:N), [0,1], B(:,2:N), [0,0])
```

The runtime approach allows to integrate machine dependent optimizations in the runtime system. Furthermore, the driver can deal with arbitrary distributions of the arguments. Therefore we can generate code also for cases where the distribution of the arguments is unknown at compile time. The computation of the optimal size of the tiles can be computed at run-time and dynamically adjusted.

A detailed description of the interface and of the implementation is given in our report [BD96].

4 Results

In this section, we give the results of first experiments using optimized pipelined computations within ADAPTOR.

Figure 3 shows the speedups achieved by pipelining on the IBM SP 2 (AIX 3.2.5). We compare the execution time of the parallel program against the execution time of the serial program to show the real speed-ups. The pipelined execution uses in the serial dimension the block size 16.

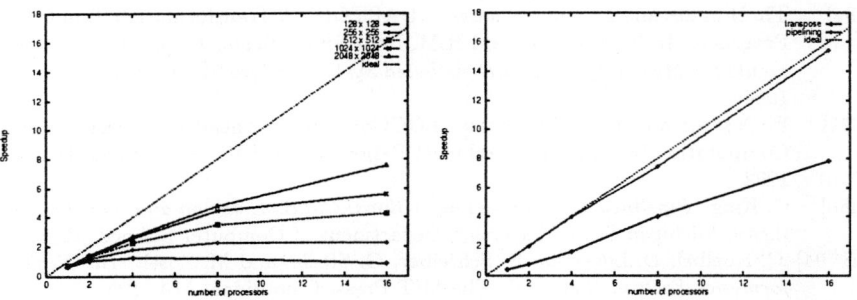

Fig. 3. Speedups for the pipelined execution of the gauss-seidel relaxation.

Fig. 4. Speedups for the ADI algorithm.

Now we discuss the parallelization of the ADI algorithm given on Figure 2. This algorithm contains dependences partly between the columns, partly between the rows. There are two strategies to solve this problem, one using a redistribution (transposition) of the array, one utilizing pipelined execution. As shown in Figure 4 the pipelined execution achieves nearly the optimal speed-up and needs no temporary data.

Our report describes other applications of pipelined executions and gives more detailled results [BD96].

5 Conclusion and Future Work

Our first impressive results show that there is no doubt about the usefulness of pipelining and about the efficient realization within ADAPTOR. Our experiments have shown that the pipelined computation using the overlap of computation and communication can be integrated successfully within an HPF compiler. Using these kind of optimizations in a message passing program is difficult and usually machine-dependent. By their integration in a compiler, the user can benefit from it a very convenient way. The interface of the LOCCS library is now well suited for the optimization of a compiler like ADAPTOR. An MPI version of the LOCCS is currently under development. Our future work concentrates on increasing the possibilities of pipelined executions and on the development of good algorithms and heuristis for choosing the order and the grain of the tiles.

Acknowledgements

We thank ZAM, Jülich for providing access to the Intel Paragon XP/S.

References

[BD96]   T. Brandes and F. Desprez. Implementing Pipelined Computation and Communication in an HPF Compiler. Technical report, LIP - ENS Lyon, 1996.

[BZ94]   Th. Brandes and F. Zimmermann. ADAPTOR - A Transformation Tool for HPF Programs. In K.M. Decker and R.M. Rehmann, editors, *Programming Environments for Massively Parallel Distributed Systems*, pages 91–96. Birkhäuser, April 1994.

[DT92]   F. Desprez and B. Tourancheau. LOCCS: Low Overhead Communication and Computation Subroutines. Technical Report 92-44, LIP - ENS Lyon, December 1992.

[Kin88]   C. King. *Pipelined data parallel algorithms: Concept, design and modeling.* PhD thesis, Michigan State University, Department of Computer Science, 1988.

[KLS+94]   C. Koelbel, D. Loveman, R. Schreiber, G. Steele, and M. Zosel. *The High Performance Fortran Handbook.* The MIT Press, Cambridge, MA, 1994.

[Tse93]   C. Tseng. *An optimizing Fortran D compiler for MIMD distributed-memory machines.* PhD thesis, Rice University, Houston, Texas, 1993.

Efficient Mapping of Interdependent Scans

Michel Barreteau and Paul Feautrier

Laboratoire PRiSM, Université de Versailles - S^{t}Quentin
45, avenue des États-Unis, 78 035 Versailles, FRANCE

Abstract. Distributed memory multiprocessors are extremely sensitive to communication costs. Some global communications such as scans and reductions are of special interest since their cost is much lower than for point to point communications. Our paper focuses on an algorithm which efficiently takes the mapping of scans into account.

1 Introduction

Communications remain the most critical aspect of performance in efficiently programming distributed memory multiprocessors. Hence minimizing communications is an indispensable task. A static placement may be obtained in two different ways: one may ask the user to insert annotations to specify data mapping à la HPF. Our approach is to leave this work to the compiler (see 2). However such a placement will not be sufficient if some particularities of the target machine are not taken into account, for instance if communication primitives with low overhead are not used. In this paper we propose a method to compute a placement which efficiently exploits interprocessor data movements such as scans and reductions. Especially we detail the Cholesky example to make the reader sensitive to the interdependences of scans. Neglecting them yields a mapping which may be incompatible with the minimization of communications.

2 Automatic parallelization

Automatic parallelization consists of extracting from the source program all restrictions on its potential parallelism. The Data Flow Graph (DFG) depicts the results of this analysis, i.e. the data movements between sets of operations [2]. A System of Affine Recurrence Equations (SARE) is built from the DFG. It is suitable for detecting recurrences [4]. We will consider the following SARE of three equations, which is extracted from a Cholesky decomposition program:

$$Ins\_2[i,k] = \begin{cases} \text{Scan}(\{i_2, k_2 \mid (3 \leq i_2 \leq n) \land (1 \leq k_2) \land (1 \leq i_2 - k_2)\}, \\ \quad ([0\ 1]),\ +,\ -\text{Ins\_6}[k_2, i_2]^2,\ \text{InitS}_2)[i, k] \\ if\ (i \leq n) \land (2 \leq k) \land (1 \leq i - k) \\ a[i,i] - (a[k,i]/\sqrt{a[k,k]})^2 \quad if\ (2 \leq i \leq n) \land (k=1) \end{cases}$$

$$Ins\_5[i,j,k] = \begin{cases} \text{Scan}(\{i_5, j_5, k_5 \mid (3 \leq i_5) \land (j_5 \leq n) \land (1 \leq k_5) \\ \quad \land (1 \leq i_5 - k_5) \land (1 \leq j_5 - i_5)\},\ ([0\ 0\ 1]),\ +, \\ \quad -(\text{Ins\_6}[k_5, j_5] * \text{Ins\_6}[k_5, i_5]),\ \text{InitS}_5)[i, j, k] \\ if\ (j \leq n) \land (2 \leq k) \land (1 \leq i - k) \land (1 \leq j - i) \\ a[i,j] - ((a[k,j]/\sqrt{a[k,k]}) * (a[k,i]/\sqrt{a[k,k]})) \\ if\ (2 \leq i) \land (j \leq n) \land (k=1) \land (1 \leq j - i) \end{cases}$$

$$Ins\_6[i,j] = \begin{cases} Ins\_5[i,j,i-1]/Ins\_2[i,i-1] \\ if\ (2 \leq i) \wedge (j \leq n) \wedge (1 \leq j-i) \end{cases}$$

3 Scans & Reductions

Since scans and reductions are particular recurrences due to their associative operator, they may be computed in a parallel way using a binary tree. A reduction is a special scan in which only the final result is used. We only consider unidirectional scans, i.e. scans whose accumulation follows a unique direction in their iteration space. Let us remember the semantics of the unidirectional operator $\text{Scan}(\mathcal{D}_S, \mathbf{v}_S, \odot, \mathbf{D}_S, \mathbf{G}_S)$: $\mathcal{D}_S$ describes the accumulation space, $\mathbf{v}_S$ is the direction vector, $\odot$ a binary and associative operator, $\mathbf{D}_S$ the expression of the data to be reduced and $\mathbf{G}_S$ are initial values. Scan computes a value for each integral point of $\mathcal{D}_S$, i.e. a multidimensional array S. Some so-called *virtual* processors are in charge of these computations. They define the virtual processors space.

Two scans S_2 and S_5 whose direction vectors are $\mathbf{v}_{S_2} = (0,1)$ and $\mathbf{v}_{S_5} = (0,0,1)$ have been detected in our SARE. S_2 computes the array S2: $\forall\ (3 \leq i_2 \leq n) \wedge (1 \leq k_2) \wedge (1 \leq i_2 - k_2)$, $\text{S2}[i_2, k_2] = \text{S2}[i_2, k_2-1] + (-Ins\_6[k_2, i_2]^2)$.

Practical implementations of the Scan operator allow only direction vectors which are parallel to the canonical axes of the virtual processors grid. A correct determination of the orientation of scans is necessary to avoid some additional communications related to non local data. But most of the time it depends on interactions between scans. Hence, the problem is to find out an efficient orientation of all the scans when placement binds them together. To this end we recall the successive stages of the mapping algorithm introduced in [3], which will be modified in order to solve our problem.

4 Placement algorithm

A placement function π_C, which is applied to a set of operations defined by an equation C of iteration vector $\mathbf{i}_C$, gives the name of the virtual processor on which C has to be executed. As regards geometry, a grid of virtual processors whose dimension g is given ($g = 2$ for Cholesky), has been elected. For each equation and for each scan, an affine placement function with g components has to be computed: for all dimension d ($1 \leq d \leq g$), $\pi_C^d(\mathbf{i}_C) = H_C^d.\mathbf{i}_C + K_C^d.\mathbf{n} + \mathbf{l}_C^d$ where unknowns are the coefficients μ_C of the $g \times |\mathbf{i}_C|$ matrix H, ν_C of the $g \times |\mathbf{n}|$ matrix K and ξ_C of the vector $\mathbf{l}$. The vector of structure parameters is $\mathbf{n}$ (n belongs to it). $|\mathbf{v}|$ gives the number of components of a vector $\mathbf{v}$.

Ex: $\pi_{Ins\_2}^d(i,k) = \mu_{Ins\_2}^{d,1}.i + \mu_{Ins\_2}^{d,2}.k + \nu_{Ins\_2}^{d,1}.n + \xi_{Ins\_2}^d$

4.1 Placement conditions

First let us consider usual placement conditions that are given by the DFG:
$$\pi_{Ins\_6}^d(i,j) = \pi_{Ins\_5}^d(i,j,i-1) \Leftrightarrow \{\mu_{Ins\_6}^{d,1} = \mu_{Ins\_5}^{d,1} + \mu_{Ins\_5}^{d,3}, \mu_{Ins\_6}^{d,2} = \mu_{Ins\_5}^{d,2}\} \quad (1)$$
$$\pi_{Ins\_6}^d(i,j) = \pi_{Ins\_2}^d(i,i-1) \Leftrightarrow \{\mu_{Ins\_6}^{d,1} = \mu_{Ins\_2}^{d,1} + \mu_{Ins\_2}^{d,2}, \mu_{Ins\_6}^{d,2} = 0\} \quad (2)$$

Then let us add the scan placement conditions which are developed in [1]:

1. Data alignment along the scan: $\forall\, C(\mathbf{i}_C) \in D_S(\mathbf{i}_S)$, $\pi_S(\mathbf{i}_S) = \pi_C(\mathbf{i}_C)$

$$\pi_{S_2}^d(i_2, k_2) = \pi_{Ins\_6}^d(k_2, i_2) \Leftrightarrow \{\mu_{S_2}^{d,1} = \mu_{Ins\_6}^{d,2}, \mu_{S_2}^{d,2} = \mu_{Ins\_6}^{d,1}\} \quad (3)$$

$$\pi_{S_5}^d(i_5, j_5, k_5) = \pi_{Ins\_6}^d(k_5, j_5) \Leftrightarrow \{\mu_{S_5}^{d,1} = 0, \mu_{S_5}^{d,2} = \mu_{Ins\_6}^{d,2}, \mu_{S_5}^{d,3} = \mu_{Ins\_6}^{d,1}\} \quad (4)$$

$$\pi_{S_5}^d(i_5, j_5, k_5) = \pi_{Ins\_6}^d(k_5, i_5) \Leftrightarrow \{\mu_{S_5}^{d,1} = \mu_{Ins\_6}^{d,2}, \mu_{S_5}^{d,2} = 0, \mu_{S_5}^{d,3} = \mu_{Ins\_6}^{d,1}\} \quad (5)$$

2. Results collection from the scan to any variable X such as $X[\mathbf{i}_X] = S[l(\mathbf{i}_X)]$ with $l(\mathbf{i}_X) = P_S.\mathbf{i}_X + Q_S$: $\pi_X(\mathbf{i}_X) = \pi_S(l(\mathbf{i}_X))$

$$\pi_{Ins\_2}^d(i, k) = \pi_{S_2}^d(i, k) \Leftrightarrow \{\mu_{Ins\_2}^{d,1} = \mu_{S_2}^{d,1}, \mu_{Ins\_2}^{d,2} = \mu_{S_2}^{d,2}\} \quad (6)$$

$$\pi_{Ins\_5}^d(i, j, k) = \pi_{S_5}^d(i, j, k) \Leftrightarrow \{\mu_{Ins\_5}^{d,1} = \mu_{S_5}^{d,1}, \mu_{Ins\_5}^{d,2} = \mu_{S_5}^{d,2}, \mu_{Ins\_5}^{d,3} = \mu_{S_5}^{d,3}\} \quad (7)$$

3. As regards the placement condition on computations which consists of finding for each scan an orientation following a j direction in the virtual processors grid: $\exists\, j\ (1 \le j \le g)$ such that $H_S^j.\mathbf{v}_S \ne 0$ and $\forall\, i \ne j$, $H_S^i.\mathbf{v}_S = 0$ (8), it should have priority to ensure an efficient computation. For each scan, g conditions $H_S^d.\mathbf{v}_S = e_S^d$ are at the top of the system $\mathcal{P}$ of placement equations in order to satisfy the strategy of the solution algorithm described below. The projection of $\mathbf{v}_S$ by H_S in the virtual processors grid is $\mathbf{e}_S$. The components of $\mathbf{e}_S$ are orientation parameters. All of them are equal to zero except one: $\forall\, S,\ \mathbf{e}_S = k.\mathbf{u}_S$ where $k \in \mathbb{Z}^*$ and $\mathbf{u}_S$ is a canonical vector (9)

$$H_{S_2}.\mathbf{v}_{S_2} = \mathbf{e}_{S_2} \Leftrightarrow \{\mu_{S_2}^{1,2} = e_{S_2}^1, \mu_{S_2}^{2,2} = e_{S_2}^2\} \quad (10)$$

$$H_{S_5}.\mathbf{v}_{S_5} = \mathbf{e}_{S_5} \Leftrightarrow \{\mu_{S_5}^{1,3} = e_{S_5}^1, \mu_{S_5}^{2,3} = e_{S_5}^2\} \quad (11)$$

Most of the placement conditions have to be satisfied in order to minimize communications. A greedy algorithm based on the Gauss-Jordan elimination takes the edges that transfer the most important data volume first into account. This is their *weight*. In this way placement equations which are associated to these edges have a high probability to be cut and taken part in the solution of $\mathcal{P}$. (The order of edges to be considered according to their weight - as computed by our prototype - is: (7),(4),(5),(1),(3),(6) and (2).)

In order to avoid the mapping of every operation on a unique processor (experience shows that it often happens when all the equations are satisfied), a heuristic is adopted. This is the *triviality test*: it only accepts an equation if the current solution of $\mathcal{P}$ (this equation included) is still able to generate g linearly independent solutions for each placement function.

The triviality test accepts for instance $\mu_{S_5}^{d,1} = 0$ from (4). But $\mu_{S_5}^{d,2} = 0$ is rejected because one cannot build for Ins_5 g (i.e. 2) linearly independent solutions any more ($\mu_{Ins\_5}^{d,2} = \mu_{S_5}^{d,2}$ from (7) and $\mu_{Ins\_5}^{d,1} = 0$ from (7) and (4)).

4.2 Interactions between scans

The pivot element of each equation is chosen to find out interactions between scans. As no assumption about relations on the coefficients μ_S can be made, we reject the problem to the relations $\mathcal{S}$ between the orientation parameters e_S.

$\mu_{Ins\_6}^{d,2} = \mu_{S_5}^{d,2}$, $\mu_{Ins\_6}^{d,1} = \mu_{S_5}^{d,3}$ according to (4) and (3) implies $\mu_{Ins\_6}^{d,2} = \mu_{S_2}^{d,1}$, $\mu_{Ins\_6}^{d,1} = \mu_{S_2}^{d,2}$. Thus we get $\mu_{S_2}^{d,1} = \mu_{S_5}^{d,2}$ and $\mu_{S_2}^{d,2} = \mu_{S_5}^{d,3}$. Finally $e_{S_2}^d = e_{S_5}^d$ from (10) and (11).

The constraint to be satisfied for building $\mathcal{S}$ comes straightforward from (8): let $(\mathbf{u}^1, \ldots, \mathbf{u}^g)$ be the canonical vectors which describe the g directions of the virtual processors space, then $\mathcal{S}$ must guarantee that each $\mathbf{e}_S$ follows any canonical axis of the processors grid, i.e. each $\mathbf{e}_S$ must be colinear to one $\mathbf{u}^j$.

Let us consider the equation $\mathcal{E}$: at first $e_{S_2}^1 = e_{S_5}^1$ since equations are processed dimension per dimension. Then a combination of canonical vectors $\mathbf{u}_S^j$ which represents the direction of reorientation vectors $\mathbf{e}_S$ is chosen. For instance $j = 1$ (cf (8)) for S_2 and S_5, which means that both scans are supposed to lie in the direction of $\mathbf{u}^1$. Rather than expressing this colinearity, it is easier to express that each $\mathbf{e}_S$ has to be orthogonal to the $g - 1$ canonical vectors $\mathbf{u}^j$ which describe the supplementary subspaces of the subspace defined by $\mathbf{u}_S^j$. In this way, one has to compute $g - 1$ dot products per scan: $\mathbf{e}_{S_2} \cdot \mathbf{u}^1 = 0$ and $\mathbf{e}_{S_5} \cdot \mathbf{u}^1 = 0$. Thus we get the following conditions $\{e_{S_2}^2 = 0, e_{S_5}^2 = 0\}$ which complete the subsystem $\mathcal{S}' = \mathcal{S} \cup \mathcal{E}$. (9) must be confirmed by the solution of $\mathcal{S}'$. $e_{S_5}^1 \neq 0$ is acceptable. This combination suits us. Otherwise we should choose another combination such as $(\mathbf{u}_{S_2}^1, \mathbf{u}_{S_5}^2)$. The second equation is: $e_{S_2}^2 = e_{S_5}^2$. The first combination $(\mathbf{u}_{S_2}^1, \mathbf{u}_{S_5}^1)$ yields: $e_{S_5}^2 = 0$. (9) is still verified. We deduce $\mathcal{S} = \mathcal{S} \cup \mathcal{E} = \{e_{S_2}^d = e_{S_5}^d\}$. On the other hand the second mentioned combination $(\mathbf{u}_{S_2}^1, \mathbf{u}_{S_5}^2)$ would yield: $\{e_{S_5}^1 = 0, e_{S_5}^2 = 0\}$. It would have been rejected because (9) is not satisfied any more. In this case $\mathcal{S}$ would remain unchanged.

When every equation has been processed, one has to replace the parameters e_S^d by their own values into the global system $\mathcal{P}$ using the last combination which satisfies $\mathcal{S}$. $\mathcal{P}$ is solved and the dth solution is applied to π_C^d. Thus we get:

$$\pi_{S_2}^1(i_2, k_2) = k_2 \qquad \pi_{S_5}^1(i_5, j_5, k_5) = k_5 \qquad \pi_{Ins\_6}^1(i, j) = i$$
$$\pi_{S_2}^2(i_2, k_2) = i_2 \qquad \pi_{S_5}^2(i_5, j_5, k_5) = j_5 \qquad \pi_{Ins\_6}^2(i, j) = j$$

5 Conclusion

We presented in this paper an algorithm which computes multidimensional placement functions. It relies on data locality but also consider some special data movements such as scans and reductions and their interdependences. It enables to optimize mapping but also to produce an efficient target code. In this way hardwired communication primitives that are provided by many distributed memory multiprocessors will be efficiently exploited.

References

1. Barreteau, Feautrier: *Automatic Mapping of Scans and Reductions*. HPCS'95. 1995.
2. Feautrier: *Dataflow Analysis of Scalar and Array References*. Int. J. of Parallel Programming, 20(1):23–53. 1991.
3. Feautrier: *Toward Automatic Partitioning of Arrays on Distributed Memory Computers*. In ACM ICS'93 Tokyo pp.175–184. 1993.
4. Redon, Feautrier: *Detection of Reductions in Sequentials Programs with Loops*. PARLE, LNCS 694. Ed. Arndt Bode and Mike Reeve and Gottfried Wolf. 1993.

Classifying Loops for Space-Time Mapping

Martin Griebl and Christian Lengauer

Fakultät für Mathematik und Informatik
Universität Passau, D–94030 Passau, Germany
email: {griebl,lengauer}@fmi.uni-passau.de
WWW: http://www.uni-passau.de/~lengauer

Abstract. We propose a class hierarchy for loops in a loop nest. Its purpose is to help identify the proper code generation methods for a space-time mapped nest. We illustrate the hierarchy and its use on a loop nest for computing the reflexive transitive closure of a graph.

1 Introduction

Traditional methods of space-time mapping apply to nests of for loops [15]. Given a for loop nest, an optimizing search can identify at compile time a space-time mapping which is minimal according to some stated metric (like the number of execution steps, processors, communication links, etc.). This is so because all information necessary for the search is static, i.e., available at compile time.

Lately researchers in loop parallelization have become interested in dynamic properties of loop nests. One useful generalization is to admit while loops [16]. The upper bound of a while loop is, in general, not known before the loop starts executing. However, it is not true that every space-time mapped for loop nest can be treated with traditional code generation methods and every nest containing while loops cannot—e.g., every for loop can be coded trivially as a while loop.

We propose a classification of loops and outline which code generation methods are necessary in each case. The crucial factors in the classification are when the bounds of the loop can be determined and which form they take. In the case of a (perceived) dynamic loop bound, different kinds of control dependences in the loop nest's dependence graph must be considered, which has repercussions on the potential parallelism and on the form of the target code. The data dependences and space-time mappings need to be (piecewise) affine for all classes. As in the Chomsky hierarchy of formal languages, the larger the class, the lower the number we give it. We comment on ways of parallelizing each class, and on the nature of the target code. In nests with loops of varying classes the general rule is: consider the biggest class (the one with the lowest number). We suspect that in many cases optimizations of this rule are possible.

2 Classification of Loops

We introduce five classes. As illustrating example imagine a double loop nest whose outer loop is on i between 0 and, say, some problem size parameter n. For each class, we shall give an example inner loop on j.

Class 4: Affine Loops. The bounds of these loops are affine expressions in the indices of the outer loops and in the structure parameters (i.e., the parameters which define the problem size). Nests with only affine loops can be treated well by traditional methods [15], which are realized in a number of systems [2, 14, 22, 23]. Inner loop: for $j := 0$ to $i + 5$ do.

Class 3: Convex Loops. If the loop, together with the loops enclosing it, enumerates a (discrete) convex set (the *execution space*), then there must be a loop nest which enumerates precisely the points of the set's image (the *target space*) under the space-time mapping; we call this fact *scannability* [11]. But there is no general mathematical framework (similar to Fourier-Motzkin elimination for Class 4 [1]) for identifying this loop nest.

The requirement that the check for convexity must be possible at compile time restricts the loop bounds to functions in the outer loop indices and structure parameters. Inner loop: for $j := 0$ to $\sqrt{i}$ do.

Class 2: Arbitrary for *Loops.* The next larger class of loops contains loops whose number of iterations is not known at compile time, but is known when the execution of the loop commences. The bounds are closed expressions in arbitrary variables and parameters. (Here, we assume that the upper bounds are evaluated once before the execution of the loop as in Pascal and Modula, not before every iteration as in C.) These loops are usually written as for loops, even though the bounds must be calculated at run time. Inner loop: for $j := 0$ to $A[i]$ do, for some array A.

If a loop of Class 2 is contained in a loop nest, then the image of the nest's index set is, in general, unscannable [11]. Therefore, we must scan a superset of the image and prevent the points which are not in the image from execution. For this purpose, we consider control dependences with dependence vector **0** from the computation of the loop bound to all statements of the loop body. These dependences reflect that the maximal number of iterations can and must be calculated before the operations of the body are executed.

For Classes 3 and 4 such control dependences need not be considered since the transformed loop bounds capture all required information. However, if the space-time mapped bounds of convex loops cannot be computed precisely but only estimated at compile time, then enumerating a superset of the image and taking explicitly care of the control dependences becomes necessary to exclude the points from execution which are not in the image.

Class 1: Static while *Loops.* In the most wide-spread case of while loops, the upper bound is also fixed when the while loop starts its execution—however, it is not given explicitly as a closed expression but as a while condition which does not hold in some iteration. Consequently, there is a while dependence, i.e., a control dependence from one iteration to the next iteration of the while loop. Obviously the target loop bounds must be computed at run time. Inner loop: for $j := 0$ while $A[i,j] > 0$ do, where array A is not modified in the body.

Class 0: Dynamic while *Loops.* In the most general case of loops, the number of iterations may be changed by the iterations of the loop body. The difference to loops of Class 1 is a data dependence from a statement in the loop body to the while condition. This has no consequences for the code generation. Inner loop: for $j := 0$ while $A[i,j] > 0$ do, where array A is modified in the body.

In the literature, a popular way of parallelizing loops of Class 1 is to execute the while loop—hopefully avoiding must of the computations in the loop body—in order to evaluate the number of iterations, and then to insert this number as the upper bound of an equivalent for loop [24]. This approach can also be applied to loops of Class 0 which means dividing them into a "control" and a "rest" part. We claim that the space-time mapping approach unifies and generalizes other approaches to the parallelization of while loops [20, 24], and that it yields the same pipelined solutions—or better ones, since one does not add unnecessary data dependences and provided one uses the fastest available by-statement scheduler [8, 9].

3 Example Problem: Transitive Closure

Our illustrating example is a loop nest which computes the reflexive transitive closure of a directed acyclic graph that is given by its adjacency list. More formally, a graph is represented by a set *node* of nodes and, for every node, by the number *nrsuc* of its successors and the set *suc* of successor nodes. *rt* of n is the adjacency list of node n in the reflexive transitive closure. Figure 1 depicts a graph and the data structure representing it.

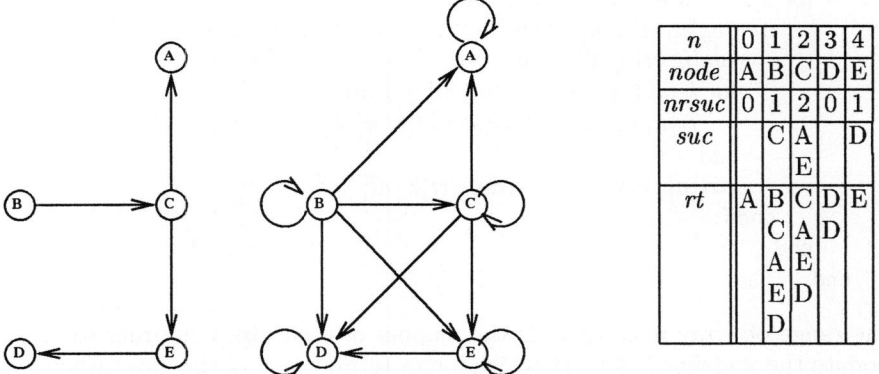

Fig. 1. A graph (left), its reflexive transitive closure (middle) and the adjacency lists representing both (right)

4 Source Program
4.1 Algorithm

The following source algorithm computes the reflexive transitive closure, under the assumption that the resulting adjacency lists rt are initially empty:

    for every node n do
        add n to rt of n
        while there is a node m not yet considered in rt of n do
            for every successor ms of m do
                add ms to rt of n

Note that this algorithm may produce adjacency lists which contain some node more than once. This is a suboptimal representation, but enforcing lists with unique elements spoils the parallelism.

4.2 Implementation

Since the polyhedron model offers no methods for dealing with sets or lists (not yet, anyway) but excels on arrays, we use arrays in our concrete representation. *node* and *nrsuc* are one-dimensional arrays, *suc* and *rt* are two-dimensional. For the computation of the reflexive transitive closure we need an auxiliary one-dimensional array *nxt* which, for every n, provides a pointer to the next free entry in the list of n's successors. Initially all undefined array elements contain the value $\bot$; rt and nxt are totally undefined. Here is the source program:

```
1:   for n := 0 while node[n] ≠ ⊥ do
2:       rt[n, 0] := node[n]
3:       nxt[n] := 1
4:       for d := 0 while rt[n, d] ≠ ⊥ do
5:           if ¬tag[n, rt[n, d]] then
6:               tag[n, rt[n, d]] := tt
7:               for k := 0 to nrsuc[rt[n, d]] − 1 do
8:                   rt[n, nxt[n]+k] := suc[rt[n, d], k]
                 end
9:               nxt[n] := nxt[n] + nrsuc[rt[n, d]]
             endif
         end
     end
```

The range of array *node* exceeds the number of nodes by 1 in order to accommodate the undefined element which forces termination of the outer while loop.

4.3 Classification of Loops

Let us classify the loops in this program.
    The outermost loop is a typical member of Class 1. If we had stored the number of nodes in some variable, we would get a loop of Class 3, and if the

number of nodes were a structure parameter known at compile time, it would even be a loop of Class 4. Target code enumerating the transformed index space precisely can be generated, since it is convex whether the outermost loop is a for or a while loop. However, if we convert this loop to Class 3 or Class 4, we can omit the unit and null control dependence vectors, which must be cited in loops of Class 1. This may result in a better schedule.

The loop on d is of Class 0 since list $rt[n]$, which determines its termination, becomes longer as execution proceeds.

The innermost loop is of Class 2 since its number of iterations is fixed when the loop starts, but is not known at compile time.

5 Space-Time Mapping

To find a valid space-time mapping we compute all data dependences and insert the necessary control dependences according to the analysis of the loops in the previous section: all loops of Classes 4, 3 and 2 get a zero dependence vector from the loop-controlling statement to every statement in the body, and all loops of Classes 1 and 0 get, additionally, a unit dependence vector from the loop-controlling statement to itself, modelling the while dependences.

Then, standard techniques can be used to determine a valid space-time mapping; techniques which do not consider the loop bounds (e.g., [6]) can be applied to any loop nest without change, whereas more precise methods considering the loop bounds (e.g., [8, 9]) must be adapted so as to deal with while loops (Classes 0 and 1).

6 Target Program

One of the most intricate problems in parallelizing general loops is to generate code for the transformed program. Even for a set of loops of Class 4 which are not perfectly nested the target code may become very complex, but there are algorithms for an automatic generation [13, 21].

In general, there may not even be a target program which enumerates precisely all transformed points of the execution space: the central problem is to find computable bounds for the loops enumerating (a superset of) the execution set.

6.1 Synchronous Parallelism

In general, a synchronous target program cannot be created with standard methods since, at compile time, there is no boundary on the outer, sequential loop without knowledge of the maximal extent of some inner loop—in general, there need not be a scannable transformation in the synchronous case [16]. Target code enhancements which deal with this typical problem in both shared and distributed memory systems are given in [4, 12]. One can also employ a speculative approach as described in [3]. However, these complex schemes are not necessary for all classes of loops.

For Class 4, the computation space is known at compile time to be a polytope. Therefore, code enumerating the target space can be generated easily with standard methods like Fourier-Motzkin elimination [1] or PIP [7]. In principle, loops of Class 3 can also be treated at compile time since their execution space is convex but, at present, no code generation methods are known.

Loops of Class 2 result in non-convex execution spaces, but the compiler can easily generate a guard for the execution of every iteration. The simplest version of such a guard for a loop of Class 2 at some level r is:

$$lb_r \leq (\mathcal{T}^{-1} * target\_coordinates)|_r \leq ub_r$$

where lb_r and ub_r are the expressions for the lower and upper bound of loop r, resp., and $|_r$ denotes the projection of the rth coordinate of a vector. The main property of this guard is that its applications in different iterations are independent of each other and can therefore be executed simultanously.

For loops of Class 1 and 0, the full scheme must be used because the guard can only be determined iteratively at run time.

6.2 Asynchronous Parallelism

For asynchronous programs one can always find a scannable transformation. This transformation yields target code without any overhead—for any class. If, for some reason, one prefers an unscannable transformation (e.g., in order to obtain space optimality), the comments of Section 6.1 apply.

However, an asynchronous program can only be written in a very abstract model, where we allow some outer loops in space to enumerate an infinite number of processors. We know that the number of processes is given by an affine function of time, i.e., the number of used processors grows affinely with time. But, since the time coordinate is not known in the outer spatial loops, one must allocate infinitely many processors initially.

In a real implementation, however, the problem of allocating an infinite number of processors at some time step is obsolete since, in general, all processors must be allocated before the parallel program starts its execution. There, the unboundedness is solved by standard partitioning or folding techniques [5, 18, 19].

This explains the non-existence of a whileall construct (a parallel while loop with an upper bound given by an arbitrary boolean expression); whileall would have to activate a set of processors in one time step (like forall) but would have to test a linearly large set of conditions, which cannot be done in constant time.

7 An Alternative Classification

Our classification is based on the question of how one can decide whether or not at some given point (in a superset of the target space) a computation must be applied. An alternative criterion, just as important for target code generation as ours, is whether target loop bounds can be determined which are functions in the outer target loop indices (if any) and structure parameters only.

This is feasible for loops of Class 4 with Fourier-Motzkin and for while loops (Classes 0 and 1), since their while dependences allow termination detection at run time [10, 12]. Thus, these classes would not change.

Classes 2 and 3, however, would in the alternative classification be divided orthogonally: in both classes there are loop nests for which target loop bounds can be found at compile time (e.g., any source loop nest under scannable transformations) and loop nests for which this is not possible—either because of a lack of mathematical methods or because of theoretically unsolvable problems at compile time (e.g., bounds depending on variables whose values vary during the program's execution). In the latter case, mathematical methods might (hopefully) provide conservative estimates of the loop bounds, (e.g., for monotonic functions); in the worst case, the loop bounds must be computed at run time.

We prefer our classification, since the classification of loops should only be based on properties of the source loop nest—but scannability is determined by the shape of the space-time matrix [11].

8 Conclusions

We hope to have demonstrated that space-time mapping methods, which in their original form from systolic design [17] can handle only a subset of Class 4 (perfect nests with uniform dependences), are becoming more and more generally applicable. Recent extensions are pushing the limit of current data dependence analysis technology: dependence tests for data structures other than static arrays are required.

Acknowledgements

The first author is grateful to Max Geigl for numerous extremely helpful discussions and careful readings of the paper. This work is part of the DFG project RecuR and received travel funds from the DAAD exchange program PROCOPE.

References

1. U. Banerjee. *Loop Transformations for Restructuring Compilers: The Foundations.* Kluwer, 1993.
2. P. Boulet, M. Dijon, E. Lequiniou, and T. Risset. Reference manual of the Bouclettes parallelizer. Technical Report 94-04, Laboratoire de l'Informatique du Parallélisme, Ecole Normale Supérieure de Lyon, October 1994.
3. J.-F. Collard. Automatic parallelization of while-loops using speculative execution. *Int. J. Parallel Programming*, 23(2):191–219, 1995.
4. J.-F. Collard and M. Griebl. Generation of synchronous code for automatic parallelization of while loops. In S. Haridi, K. Ali, and P. Magnusson, editors, *EURO-PAR '95 Parallel Processing*, Lecture Notes in Computer Science 966, pages 315–326. Springer-Verlag, August 1995.

5. A. Darte. Regular partitioning for synthesizing fixed-size systolic arrays. *INTEGRATION*, 12(3):293–304, December 1991.
6. A. Darte and F. Vivien. Automatic parallelization based on multi-dimensional scheduling. Technical Report 94-24, Laboratoire de l'Informatique du Parallélisme, Ecole Normale Supérieure de Lyon, September 1994.
7. P. Feautrier. Parametric integer programming. *Operations Research*, 22(3):243–268, 1988.
8. P. Feautrier. Some efficient solutions to the affine scheduling problem. Part I. One-dimensional time. *Int. J. Parallel Programming*, 21(5):313–348, October 1992.
9. P. Feautrier. Some efficient solutions to the affine scheduling problem. Part II. Multidimensional time. *Int. J. Parallel Programming*, 21(6):389–420, October 1992.
10. M. Griebl and J.-F. Collard. Generation of synchronous code for automatic parallelization of while loops. In S. Haridi, K. Ali, and P. Magnusson, editors, *EUROPAR '95*, Lecture Notes in Computer Science 966, pages 315–326. Springer-Verlag, 1995.
11. M. Griebl and C. Lengauer. On the space-time mapping of WHILE-loops. *Parallel Processing Letters*, 4(3):221–232, September 1994.
12. M. Griebl and C. Lengauer. A communication scheme for the distributed execution of loop nests with while loops. *Int. J. Parallel Programming*, 23(5):471–495, 1995.
13. W. Kelly, W. Pugh, and E. Rosser. Code generation for multiple mappings. Technical Report CS-TR-3317, Dept. of Computer Science, Univ. of Maryland, 1994.
14. H. Le Verge, C. Mauras, and P. Quinton. The ALPHA language and its use for the design of systolic arrays. *J. VLSI Signal Processing*, 3:173–182, 1991.
15. C. Lengauer. Loop parallelization in the polytope model. In E. Best, editor, *CONCUR'93*, Lecture Notes in Computer Science 715, pages 398–416. Springer-Verlag, 1993.
16. C. Lengauer and M. Griebl. On the parallelization of loop nests containing while loops. In N. N. Mirenkov, Q.-P. Gu, S. Peng, and S. Sedukhin, editors, *Proc. 1st Aizu Int. Symp. on Parallel Algorithm/Architecture Synthesis (pAs'95)*, pages 10–18. IEEE Computer Society Press, 1995.
17. P. Quinton and Y. Robert. *Systolic Algorithms and Architectures*. Prentice-Hall, 1990.
18. J.-P. Sheu and T.-H. Tai. Partitioning and mapping nested loops on multiprocessor systems. *IEEE Trans. on Parallel and Distributed Systems*, 2:430–439, 1991.
19. J. Teich and L. Thiele. Partitioning of processor arrays: A piecewise regular approach. *INTEGRATION*, 14(3):297–332, 1993.
20. P. P. Tirumalai, M. Lee, and M. S. Schlansker. Parallelization of while loops on pipelined architectures. *J. Supercomputing*, 5:119–136, 1991.
21. S. Wetzel. Automatic code generation in the polytope model. Diplomarbeit, Fakultät für Mathematik und Informatik, Universität Passau, 1995.
22. R. P. Wilson, R. S. French, C. S. Wilson, S. P. Amarasinghe, J. M. Anderson, S. W. K. Tjiang, S.-W. Liao, C.-W. Tseng, M. W. Hall, M. S. Lam, and J. L. Hennessy. SUIF: An infrastructure for research on parallelizing and optimizing compilers. In *Proc. Fourth ACM SIGPLAN Symp. on Principles & Practice of Parallel Programming (PPoPP)*, pages 31–37. ACM Press, 1994.
23. M. Wolfe. The Tiny loop restructuring research tool. In H. D. Schwetman, editor, *Proc. Int. Conf. on Parallel Processing*, volume II, pages 46–53. CRC Press, 1991.
24. Y. Wu and T. G. Lewis. Parallelizing while loops. In D. A. Padua, editor, *Proc. Int. Conf. on Parallel Processing*, volume II, pages 1–8. Pennsylvania State University Press, 1990.

Workshop 04

Distributed Systems and Algorithms

PACA: A Cooperative File System Cache for Parallel Machines*

Toni Cortes, Sergi Girona and Jesús Labarta

Departament d'Arquitectura de Computadors
Universitat Politècnica de Catalunya - Barcelona
E-mail: {toni, sergi, jesus}@ac.upc.es
URL: http://www.ac.upc.es/hpc

Abstract. A new cooperative caching mechanism, PACA, along with a caching algorithm, LRU-Interleaved, and an aggressive prefetching algorithm, Full-File-On-Open, are presented. The caching algorithm is especially targeted to parallel machines running a microkernel-based operating system. It avoids the cache coherence problem with no loss in performance. Comparing our algorithm with another cooperative cache one (N-Chance Forwarding), in the above environment, better results have been obtained by LRU-Interleaved. We also evaluate an aggressive prefetching algorithm that highly increases read performance taking advantage of the huge caches cooperative caching offers.

1 Introduction and Related Work

In this paper we present PACA, a specific cooperative caching mechanism built on top of a microkernel-architecture operating system. PACA defines a single parallel global cache built from the union of all the small local caches across the different nodes. As part of PACA, we have studied several caching and prefetching policies. From all the considered policies, special interest will be placed on LRU-Interleaved (caching algorithm with no cache coherence problems) and Full-File-On-Open (aggressive prefetching) algorithms.

All performance data presented in this paper is obtained through simulation. In order to understand the advantages our algorithm has, we compare it with the one presented by Dahlin et al. [4]. In addition to the simulation results presented in this paper, a working prototype has been implemented on top of the PAROS operating system microkernel [9][14].

Most of the work along this line has not been done on parallel machines but on networks of workstations running a *Unix-like* operating system which offers most services. Leff et al. studied the impact distributing cached objects over the network could have [11]. A more practical project was done by Dahlin et al. [4] as part of the xFS file system [1]. They proposed several cooperative caching algorithms and simulated their performance. In their work, N-Chance

---
* This work has been supported by the Spanish Ministry of Education (CICYT) under the TIC-94-537 and TIC-95-0429 contracts.

Forwarding was identified as the best algorithm. Our work differs from the one presented by Dahlin et al. in two main issues. First, only read operations are studied in their work while both reads and writes are covered in this paper. The second difference is the environment the cooperative caching algorithm will work on. In both previous works, the test bed was a network of workstation running a full *Unix-like* operating system. We work on a parallel machine with a microkernel-based operating system on each node. Most services, like file system operations, are implemented by a user-level server.

Similar approaches to cooperative caching have been taken in data base implementation [7], remote memory paging [13, 12] and memory management [6].

2 Target Environment

The parallel machine this file system is targeted for is made of several nodes with local memory. Full connectivity is offered by the interconnection network. Besides, each node may have none, one or even several disks.

In this work, a couple of differences between parallel machines and network of workstations are assumed. While the second one needs reasonable fault tolerance mechanisms, a parallel machine works as a unit and a node failure means whole machine failure. Besides, parallel machines have a higher interconnection network bandwidth than a network of workstations have.

In our environment, a parallel machine runs a microkernel-based operating system instead of a full *Unix-like* one. All functionalities not offered by the kernel are implemented by servers. This is also the case for the file system operations.

This work started as a file system cache prototype for the PAROS operating system microkernel [9]. This target platform defined the environment we work with. In order to be able to implement our distributed cache, the underlying microkernel should offer, besides the usual abstractions, a *memory_copy* operation. This mechanism will be used to transfer data between nodes. Our assumption is that any processor can set up a data transfer between any other two processors. The processor that invokes the copy is charged with all the overhead. When we refer to a memory copy the copy request and copy itself are all included.

3 Design

3.1 PACA (PArallel CAche)

PACA is a specific cooperative caching mechanism built on top of a microkernel-architecture operating system. PACA defines a single parallel global cache build from the union of all the nodes' local cache. This global management can lead to high performance through a high global hit ratio and good adaptability to the changing needs of the nodes. This will increase the overall system performance. When this mechanism is used we can observe two kinds of cache hits. If the requested block is kept in the local memory we will have a *local hit*. If this block is kept on a remote node we will have a *remote hit*.

As a first step, we have studied the behavior of a centralized single server. A centralized control should be able to cope with a reasonable number of processors. The simulation results show that, in most of the parallel machine installations (with less that 50 nodes), the single server should not cause a bottleneck. The causes behind this reasonable scalability are little server overhead and distributed data transfers. Regardless of these results, work on the distribution of PACA for larger systems is ongoing.

3.2 LRU-Interleaved Algorithm

LRU-Interleaved is a very simple algorithm designed to work as a caching method of PACA. It uses all the available cache in the system as a single cache. It also takes advantage of the parallelism and the high data transfer bandwidth offered by a parallel machines.

We use a set associative block placing algorithm. The number of sets equals the number of nodes and the size of each set is the size of each local cache measured in blocks. When the server has to cache a new block, it applies a hash function to the file name (or file-id) and block number. The result of this function indicates which node will cache the block. Next, a place in the local cache of that node is found using a LRU replacement algorithm. If the hash function is good enough, one of the oldest blocks in the cache will be replaced and the behavior will be an approximation to a global LRU replacement algorithm [3].

As performance is a very important issue in the design of LRU-Interleaved, we have taken several steps in order to increase it as much as possible. The first step towards it consists of designing a simple algorithm. We have eliminated all replication and cache coherence mechanisms. A second step consists of using the potential communication parallelism. When a user requests more than one cache block, these blocks, if in cache, can be sent to the client node in parallel. This parallelization decreases the overhead produced by bringing the data from a remote node.

3.3 Prefetching

Cooperative caching on a parallel machine allows the system to have huge caches. Given these cache sizes, it takes many hours to fill the cache and most of the cached data is more than several hours old. In our simulations (50 nodes and 16MB local caches) the cache needed 13 hours to be filled. This leads us to believe that huge caches should be used for aggressive prefetching. It is well known that prefetching is not always a good idea as it may end up delaying the application if many mispredictions are made [16] [8]. Nevertheless, if the cache is big enough, these mispredictions will not affect the overall cache performance as prefetched blocks will replace very old data. In this work we have studied two different prefetching algorithms: One-Block-Ahead and Full-File-On-Open.

One-Block-Ahead queues the next block to the prefetching queue after each read or write operation. As soon as the disk becomes idle, the first block in the prefetching queue will be fetched.

The best aggressive policy we have simulated so far is the one presented in this paper (Full-File-On-Open). It consists of queuing the whole file on the prefetching queue as soon as the file is opened. With this mechanism most blocks are already in the cache when the application requests them. This algorithm may be too aggressive if the files are very large. As no problems have been detected with the used workload we have not taken this into account.

4 N-Chance Forwarding Algorithm

In order to test how good our algorithm is, we compare it with N-Chance Forwarding [4] which is one of the latest cooperative cache algorithms found in the bibliography. This algorithm divides the cache a workstation has, into two parts. The first one is used to cache the local data and the second one will hold data cached by remote workstations. The size of these two parts is not fixed but dynamically adjusted depending on the node I/O activity.

In order to adapt this algorithm to our environment some changes have been made. The most significant modification is due to the microkernel architecture we work with. In the original version of N-Chance Forwarding algorithm, a *local hit* had no need to access a remote node. The workstation's operating system recognized the requested block as a *local hit* and delivered it directly to the local client. In our model, as the file system is managed by a server, all requests have to be sent to a possibly remote server. We have to notice that this implementation will increase the *local hit* access time and decrease the *remote hit* one. Each *local hit* will be increased the time needed to send the message to the server. On the other side, *remote hits* will only send one message to the server per user request instead of one message for each *remote hit*.

N-Chance Forwarding allows each node to cache the blocks its applications request. This algorithm attempts to avoid discarding unreplicated blocks (singlets) from the cache. When a client discards a block, the server checks to see if that block is the last copy in the whole cache. If the block is a singlet, rather than discarding it, it forwards the data to a random peer. The peer that receives the data adds the block to its LRU list as if it had been recently referenced. This forwarding can only take place N times before the block is referenced again. After N forwardings, if nobody references it, the block is discarded. If a client has a *remote hit*, the block is replicated from the remote cache to the local one of the requesting client.

In this paper we have used two different values for N. The first one, $N=0$, implements a cache where *remote hits* are possible but no coordination between nodes is done (Greedy policy). The second value, $N=2$, is used because it was described as the best choice in [4].

Our N-Chance Forwarding version has been implemented as a single server to avoid worrying about cache coherence. In this centralized version, the cache coherence algorithm will not need to communicate different servers decreasing the control traffic between nodes. This simplification has to be taken into account as it may speed write operations.

5 Algorithms Comparison

Before getting into performance details, it would be very useful to compare both cooperative caching algorithms: LRU-Interleaved and N-Chance Forwarding.

LRU-Interleaved is a very simple algorithm. The way blocks are distributed among the nodes makes searching a block very easy and efficient. Straight block location is achieved without using large directory tables. In N-Chance Forwarding there is a special cost on maintaining information about where the blocks are placed. They also have to keep track of which blocks are replicated in order to detect if a block is a singlet or not.

In LRU-Interleaved, much more interest is placed in obtaining full utilization of the cache and avoiding replication than minimizing the amount of data transfer between nodes. On the other hand, the N-Chance Forwarding algorithm places special interest in *local hits* trying to minimize the number of block transferences between nodes.

A third important difference is the cost *remote hits* have on both algorithms. On N-Chance Forwarding a *remote hit* implies a remote copy from the remote cache to the local cache and a local copy from the local cache to the user. On the other hand, a remote copy in LRU-Interleaved only implies a remote copy from the remote cache to the user.

Another difference between the algorithms is the way the cache coherence problem is solved. In our proposal no replication is allowed and thus no cache coherence problems appear. In N-Chance Forwarding some kind of cache coherence algorithm has to be implemented.

6 Simulation Methodology

6.1 Simulator

The file-system cache simulator used in this work is part of DIMEMAS [2] [10], a distributed memory parallel machine simulator.

We will not get into detail of the simulator functionality but the communication model should be explained in order to understand the figures presented in this work. Communications are divided into two parts: a startup and a data transfer. The startup is constant for each type of communication (port or *memory_copy*) and it is assumed to require CPU activity. The data transfer time is proportional to the size of the data sent and the intercommunication network bandwidth. In our model, all communications are synchronous. Asynchronous communication can be achieved by creating new threads.

6.2 Implementation Details

Although we don't want to get into many implementation details, some of them are very important in order to understand the results presented in this paper.

---

[2] DIMEMAS is a performance prediction simulator developed by CEPBA-UPC and it is commercially available from PALLAS

The file server is a high-priority multi-threaded application which shares the node with other applications.

It is important to note that cleaning or forwarding a block is done after the operation is finished and the user has been notified. This takes the overhead away from the critical path of the operation. From now on, we will refer to these situations as *delayed clean* and *delayed forward*. These delayed operations will be possible as long as the file server does not run out of auxiliary buffers.

As we want to optimize the critical path of the read and write operations we give them higher priority over prefetching operations.

7 Performance

7.1 Sprite Workload

In order to get the results presented in this paper, we have used the Sprite workload [2]. These traces contain the activity of 48 client machines and some servers over two day period measured in the Sprite operating system. All measures presented in this paper are taken from the 15th hour to the 48th hour in order to study the behavior of a warm cache. We have used this trace as we believe that parallel machines should not only be used for parallel applications but also for *Unix-like* ones in a time sharing manner.

7.2 General Information and Simulation Parameters

In the following subsections, most of the parameters used in the simulation are fixed. Unless otherwise specified, all runs simulated a 50 nodes machine with a 16MB of local cache, 8KB cache blocks, no prefetching and a 30 seconds sync.

We assume disk time accesses as described by Ruemmler and Wilkes [15]. Reading a 8KB disk block takes 14.7 milliseconds while writing it takes 18.3 milliseconds. All measures in this paper are taken with only one disk connected to the node where the centralized file server runs.

Unless otherwise specified, nodes are connected through a 155 Mbits/s interconnection network and local copies are done at 320 Mbits/s. We assumed a 100 microseconds port startup and a memory copy one of 50 microseconds.

In the following subsections we will use a few short expressions in order to identify several common situations. When a user requests a block that it is not in the cache it is called cache miss. This requested block has to replace another block already cached. This replaced block may have been modified since its last update to disk, or may not have. We will refer to the first situation as a *miss on dirty* and to the second one as a *miss on clean*.

7.3 Read and Write Performance

In this subsection we will study the performance of read and write operations with LRU-Interleaved, N-Chance Forwarding and Greedy policies (Fig. 1).

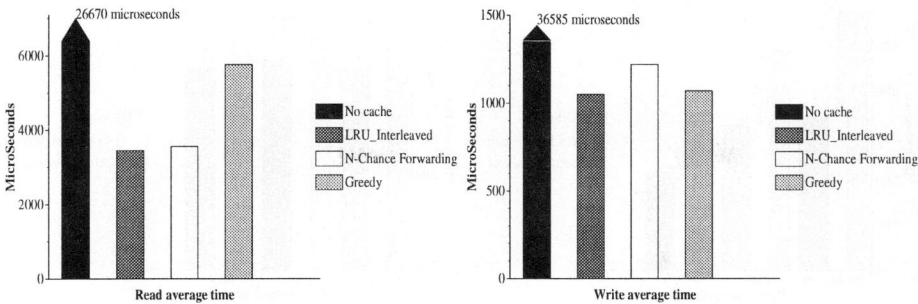

Fig. 1. Average READ and WRITE time for the various caching policies.

An important result is that both cooperative caching algorithms nearly double the read operations bandwidth if compared to non cooperative ones (Greedy and no cache).

If LRU-Interleaved and N-Chance Forwarding are compared, a 3.1% gain from the first algorithm over the second one is observed. Even though Dahlin's algorithm has a much higher *local hit* ratio no proportional gain is observed due to two main reasons: different costs of remote operations and forwardings.

A *remote hit* in N-Chance Forwarding takes longer than in LRU-Interleaved as was explained in Section 5.

There are also quite a few block forwardings that cannot be delayed due to a lack of auxiliary buffers. They have to be included in the critical path of the operation. This usually happens with large requests (100KBytes or more). We have to recall that there is a limit of 16 auxiliary buffers per node.

In Figure 1, we also observe that a write operation with LRU-Interleaved is 14% faster than with N-Chance Forwarding. In order to explain this difference we should first explain why Greedy is also faster than N-Chance Forwarding. The main difference between these two algorithms is that the first one does not forward blocks while the second one does. The overhead due to not delayed forwardings increases the write time a lot. A not delayed forwarding implies an extra remote memory copy in the write operation. As write operations are very fast, this extra time affects the overall write time significantly.

LRU-Interleaved is also faster than Greedy because of the dirty blocks. As N-Chance Forwarding and Greedy algorithms clean a dirty block just before forwarding it [5], the syncer does not have enough time to clean all blocks before being forwarded. This does not happen in LRU-Interleaved as blocks are cleaned once they are completely discarded. The overhead of block cleaning is higher than the time lost because of *remote hits*.

Finally, another important issue is the different cost *remote hits* have in LRU-Interleaved compared to N-Chance Forwarding and Greedy. The same explanation as in the read operations applies.

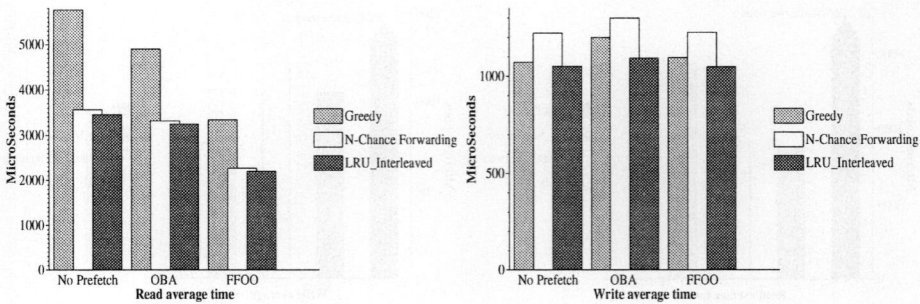

Fig. 2. Average read and write times due to the PREFETCHING policy.

7.4 Prefetching Influence on Read and Write Operations

The traditional prefetching policy One-Block-Ahead (OBA) decreases the read time a little bit (Fig. 2). With this algorithm not much gain is obtained. This is because starting to prefetch a block just after accessing the previous one is not early enough. When the user needs the prefetching block, it is still on its way from the disk.

Another problem this algorithm has is that random accesses see very little gain. If blocks are accessed in a random way, the block prefetched are not the ones the application needs.

These problems found in One-Block-Ahead have lead us to implement the Full-File-On-Open policy (FFOO). If we start prefetching the file once it is open, the probability to have finished the prefetching of a block before it is needed increases. Besides, if a random access is performed, most blocks in the first part of the file will have been prefetched before requested.

The Greedy algorithm improves much more than any other as its hit ratio is very low when no prefetching is done.

If we move to the write operations, the first thing we notice is that no real gain is obtained with either algorithm. This is because increasing the hit ratio does not increase the write performance. A *miss on clean* takes the same or even less amount of time than a cache hit. For instance, *remote hits* on N-Chance Forwarding are even more expensive than *misses on clean* [3].

Another problem appears when the block to be written is being prefetched. This write will take some of the disk read time.

On the other hand, if a block is prefetched very few *misses on dirty* will happen increasing the write-operation performance. One thing outweighs the other and not much difference is seen on write operations due to prefetching.

7.5 Interconnection Network Bandwidth Influence

In this section, we study the influence the local-remote bandwidth ratio has on the already shown results. Figure 3 presents the percentage gain obtained by LRU-Interleaved over N-Chance Forwarding when the local-remote bandwidth

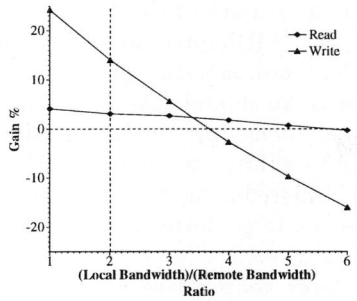
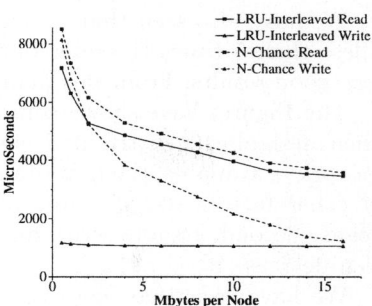

Fig. 3. Influence of the local/remote transference bandwidth.

Fig. 4. Read and write performance using different local CACHE SIZES.

ratio is modified. The most favorable case for our algorithm is where there is no difference between a local and a remote transfer. From this point we decrease the remote bandwidth until N-Chance Forwarding reads and writes become faster.

This figure shows that the results presented in this paper are valid even with lower interconnection bandwidths. It also shows that after a certain point N-Chance Forwarding is the way to go.

7.6 Cache Size Influence

In this work we were also interested in studying the effect local cache sizes had on both algorithms (Fig. 4). We can observe that LRU-Interleaved works much better than N-Chance Forwarding when small local caches are used. This is because a higher global hit ratio is obtained by our algorithm due to a better cache utilization. As no replication is allowed, the whole cache contains useful blocks. Dahlin's algorithm loses part of the cache with replicated blocks and it behaves as a smaller one. The benefit of the *local hits* cannot outweight the higher hit ratio obtained by our algorithm.

It is also important to examine the behavior of the write operations. When N-Chance Forwarding is used with small caches the probability to forward a dirty block is very high. This dirty block has to be cleaned before sending it to the new node increasing the write operation time.

8 Conclusions

In this paper we have presented a distributed-cache-oriented file system designed to work on a parallel machine running a microkernel-based operating system. Simplicity, scalability and performance have been the three main objectives in the design. While simplicity and performance have been clearly achieved, more work has to be done if a fully scalable file system is to be obtained.

We have shown that a very simple algorithm obtains similar read bandwidth and a better write performance than more complex ones.

We have also seen that in our environment, as long as the local bandwidth is less than 5 times the interconnection network one, LRU-Interleaved obtains very good results. From this point on N-Chance Forwarding is the way to go.

The Figures have shown some important aspects we should take in account when designing a distributed cache. First, if a relatively fast interconnection network is available, the importance of remote and local hits can be outweighted by other factors like avoiding block forwarding and reducing the number of cleans. Second, remote write hits do not increase write performance and may even decrease it.

We have seen that aggressive prefetching in large cooperative caches may increase the hit ratio and thus decrease the average read time. We have also shown that prefetching very rarely increases the write bandwidth.

More information may be found in the longer version of this paper [3].

Acknowledgments

We owe special thanks to Michael D. Dahlin, E. Markatos and Maite Ortega for their help and useful comments. We are also grateful to the people at Berkeley who gathered the Sprite traces used in this work.

References

1. T.E. Anderson, M.D. Dahlin, J.M Neefe, D.A Patterson et al. "Serverless Network File Systems," *15th SOSP*, December 1995, pp. 109-126
2. M.G. Baker, J.H. Hartman, M.D. Kupfer et al. "Measurements of a distributed File System," *Proc. Of the 13th SOSP*, 1991, pp. 198-212
3. T. Cortes, S. Girona and J. Labarta "PACA: A Cooperative File System Cache for Parallel Machines," *UPC-CEPBA Technical Report RR-UPC-CEPBA-1996-07*
4. M.D. Dahlin, R.Y Wang, T.E. Anderson and D.A Patterson "Cooperative Caching: Using Remote Client Memory to Improve File System Performance," *OSDI'94*, pp. 267-280
5. M.D. Dahlin *Personal communication - 1994*
6. M.J. Feeley, W.E. Morgan, F.H. Pighin et al. "Implementing Global Memory Management in a Workstation Cluster," *15th SOSP*, December 1995
7. M.J. Franklin, M.J. Carey and M. Livny. "Global Memory Management in Cliente-Server DBMS Architectures," *ICVLDB*, 1992. pp. 596-609
8. D. Kotz "Prefetching and Caching Techniques in File Systems for MIMD Multiprocessors," *PhD Thesis from Duke University, Dept. of Computer Science*, 1991
9. J. Labarta, J. Gimenez, C. Pujol, T. Jove and J.I. Navarro "PAROS: Operating System Kernel for Distributed Memory Parallel," *PACTA*, Barcelona 1992
10. J. Labarta, S. Girona, V. Pillet, T. Cortes, L. Gregoris "DiP : A Parallel Program Development," *Euro-Par'96, Lyon, August 1996*
11. A. Leff, J.L. Wolf and P.S. Yu "Replication Algorithms in a Remote Caching Architecture," *IEEE Trans. on Parallel & Distributed Systems, vol. 4, No. 11*, 1993, pp. 1185-1204
12. A. Leff, J.L. Wolf and P.S. Yu "Efficient LRU-Based Buffering ina LAN Remmote Caching Architecture," *IEEE Trans. on Parallel & Distributed Systems, vol. 7, No. 2*, 1996, pp. 191-206
13. E.P. Markatos, G. Dramitinos and K. Papachristos "Implementation and Evaluation of a Remote Memory Pager," *FORTH-ICS Technical Report TR-129*, 1995
14. M. Ortega, T. Cortes and J. Labarta "Implementation of a Cooperative File System Cache on PAROS," *UPC-DAC Technical Report UPC-DAC-96/16*, 1996
15. C. Ruemmler and J. Wilkes "UNIX Disk Access Patterns," *HP Laboratories Technical Report, HPL-92-152*, 1992
16. A.J. Smith "Disk cache - Miss Ration Analysis and Design Considerations," *ACM Transactions on Computer Systems, vol. 3, No. 3*, 1985, pp. 161-203

A System for Fault-Tolerant Execution of Data and Compute Intensive Programs over a Network of Workstations

J.A.Smith and S.K.Shrivastava

Department of Computing Science,
The University of Newcastle upon Tyne,
Newcastle upon Tyne,
NE1 7RU UK
{jim.smith,santosh.shrivastava}@newcastle.ac.uk

Abstract. The bag of tasks structure permits dynamic partitioning for a wide class of parallel applications. This paper describes a fault-tolerant implementation of this structure using atomic actions (atomic transactions) to operate on persistent objects, which are accessed in a distributed setting via a Remote Procedure Call (RPC). The system is suited to parallel execution of data and compute intensive programs that require persistent storage and fault tolerance, and runs on stock hardware and software platforms, UNIX, C++. Its suitability is examined in the context of the measured performance of three applications; ray tracing, matrix multiplication and Cholesky factorization.

1 Introduction

Many computations manipulate very large amounts of data. Matrix calculations represent one example class. In a Massively Parallel Processor (MPP) such a vast data set is typically partitioned statically between the very many distributed processing elements and moved amongst them as necessary to perform the computation. Such an approach is exemplified in Cannon's algorithm for matrix multiplication [14]. One suggestion is that a Network Of Workstations (NOW) be modelled on such an architecture [2]. However, it may be that problem size can exceed even the aggregate memory of all available machines. In such a situation, the problem cannot be statically partitioned between processors.

As the problem size increases so too does the computation time in any given configuration, and in a NOW potentially so too does the number of nodes which may be employed. As the scale of a distributed computation is increased in this way, the possibility of a failure occurring which might affect the execution of the computation must increase. If it is not possible to tolerate such an event, it is necessary to restart the entire computation.

The approach described here provides a solution for these problems by implementing a store on secondary storage which is shared between a collection of concurrent processes. A computation is organized as a bag of tasks type structure [7] where the overall computation is divided up into a number of tasks which are then scheduled dynamically between a potentially varying collection of concurrent processes. Computation data, including the bag of tasks is located in the shared store, which is organized as a repository of objects and fault tolerant access to it supported through atomic actions operating on the contained objects. It is suggested that these mechanisms provide a clear model to the user.

In this experiment, these facilities are supported through an established distributed system which runs on many versions of UNIX and C++, without alteration to either. The approach is investigated through implementation of applications of scale appropriate to parallelization and fault-tolerance in a NOW. Performance is shown to be fundamentally limited only in hardware bandwidths.

The paper continues with notes on related work in Sect. 2, a description of the applications and fault-tolerance mechanisms in Sect. 3, measured performance in Sect. 4 and summary in Sect. 5.

2 Related Work

The attraction of exploiting a readily available NOW to perform parallel computations is widely acknowledged. It is also recognized that a NOW typically has disadvantages compared to a tightly coupled multiprocessor, including a lower performance interconnect and a greater need for fault-tolerance.

Experiments have been performed to statically partition data intensive computations over a NOW, e.g. [5]. However, the size of the computation is bounded by aggregate memory of the machines. Structuring similar to the bag of tasks is often employed in practice, e.g. for seismic migration in [1], but with limited provision for fault-tolerance and for problems which are less intensive in data.

Mechanisms to support fault-tolerance may be transparent to the application programmer, e.g. [12], [15]. However, a transparent scheme is unlikely to take advantage of points in an application where data to be saved is minimum, such as when data has just been written to disk for instance.

One non transparent scheme for the static partitioning approach [17] maintains a parity copy of distributed partitions of computation state. While performance for a Cholesky factorization of 5000 element square matrix, at 1700 seconds employing 17 Sparc-2 machines, is similar to that recorded here the computation is bounded by total memory and the approach here which employs fewer machines is resilient to a greater number of failures.

An early design study [4] considered the use of atomic actions as a mechanism to support fault-tolerant parallel programming over a NOW.

Fault tolerance for a bag of tasks type structure has been considered before, e.g. [3], [8] but without providing access to large scale data on secondary storage. Plinda [11] which supports access to persistent tuple spaces and a transaction mechanism does have some similarity to this work.

The experiments described here attempt to exploit parallelism in a NOW of modest scale to perform large scale computations in a fault-tolerant way without altering operating system or language.

3 Implementation

3.1 Fault Tolerance

It is assumed that a workstation fails by crashing and that then any data in volatile storage is lost, but that held on disk remains unaffected. It is also assumed that the network does not partition.

Atomic actions operating on persistent state provide a convenient framework for introducing fault-tolerance [10] through ensuring defined concurrent behaviour and fault-tolerance. Atomic actions have the well known properties of (1) serializability, (2) failure atomicity, and (3) permanence of effect.

A convenient model is for this state to be encapsulated in the instance variables of persistent objects and accessed through member functions. Within these functions the programmer places lock requests, e.g. read or write to suit the semantics of the operation, and typically surrounds the code within the function by an atomic action, starting with *begin* and ending with *commit* or *abort*. Operations thus enclosed which can include calls on other atomic objects are then perceived as a single atomic operation. The infrastructure manages the required access from and/or to disk based state. Such objects may be distributed on separate machines, e.g. for performance, and replicated to increase availability. The applications are implemented using the Arjuna tool kit [16], an object-oriented programming system that implements in C++ this object and action model.

The following enhancements add fault-tolerance to a bag of tasks application.

1. The slave begins an atomic action before fetching a task from the bag, and commits the action after writing the corresponding result. If the slave fails the action aborts, all work pertaining to the current task is recovered and the task itself becomes available again in the bag.
2. The shared objects are replicated on at least $k+1$ machines, so that the failure of up to k of these machines may be tolerated.
3. A computation object contains a description of the computation and data objects and the computation's completion status. This object may be queried at any time to determine the status of the computation and may be replicated for availability. It is a convenient interface for a process to be started on an arbitrary machine to join in an ongoing computation.

Arjuna requires an underlying RPC to implement distribution and object server process management; accessing these services through certain interface classes. The RPC implementation employed here supports optional use of the TCP protocol with connection establishment on a per-call basis. Some optimization of this RPC mechanism has been performed to exploit homogeneity of machines. The RPC also supports reuse of an existing server process. This facility is exploited in service of the main shared data objects in order to prevent excessive contention in the shared communications medium; the common server is single threaded and therefore serializes all slave requests.

In each application, the main operands are managed as collections of smaller objects. Each task entails computation of some part of the result, which may be one or more of such objects.

At the start of the computation, the shared objects are installed in the object repository. In the fault-tolerant version, a fault-tolerant bag of tasks is created and all task descriptions stored in it. Then the chosen number of slaves is created on separate workstations. In the non fault-tolerant implementation, each slave is informed of a unique allocation of tasks to perform. In these initial experiments, a master process is employed to perform these functions and then wait for the completion of the slaves before performing any final processing to the output, such as converting to a desired file format, and finally reporting on the elapsed time. The master takes no active part during the main part of the application, so a shell script replacement is quite feasible. Also at this time the shared objects are not replicated.

The fault-tolerant bag of tasks is implemented as a recoverable queue [6] which relaxes the usual FIFO ordering to suit its use in a transactional environment. If an element is dequeued within a transaction, then it is write-locked immediately, but only actually dequeued at the time the transaction commits. Similar use of recoverable queues in asynchronous transaction processing is described in [10]. The *dequeue* operation returns a status which allows the caller to distinguish between the situation where the queue is empty and that where entries remain but are all locked by other users.

3.2 Applications

Three applications are implemented. The first is a port of a publicly available ray tracing package, *rayshade* [13]. Input data comprises only scene description and output is a two dimensional array of red-green-blue pixel values. A task is defined as computation of a number of rows of the output array. To display the output image, it is convenient to copy it to the file format used in the original package, Utah Raster RLE format. In this implementation, this operation is performed serially by the master process. A simple scene provided as an example in the package is traced for the purposes of the test. For comparison, the unaltered package is built and run as a sequential program on one of the workstations.

The remaining applications are dense matrix computations, matrix multiplication and Cholesky factorization. A preliminary description of the former was given in [19]. In linear algebra computations it is common to employ block structuring to benefit from increased locality [9]. In the implementation of both matrix computations here, matrices are composed of square blocks and a task defined as the computation of a single block of the result.

In the case of matrix multiplication, a task entails a block dot product of a row of blocks in the first and column of blocks in the second operand matrices. The implementation of Cholesky factorization employs the Pool-of-Tasks algorithm of [9], §6.3.8. The required inter task coordination is ultimately implemented through a two dimensional array of flags which indicate whether corresponding blocks in the output matrix have been written or not. Concurrent operations on the flags are controlled through locks obtained within the scope of atomic actions and are therefore recoverable. A fuller description appears in [18].

4 Performance

Each experiment is conducted during off peak time in a cluster of HP9000/710 (HP710) machines each with 32 Mbyte memory and 64 Kbyte cache, connected by 10 Mbit/s Ethernet. A small number of HP9000/730 (HP730) machines with 64 Mbyte memory and 256 Kbyte cache have sizeable temporary disk space space available. For the matrix computations a cluster containing a HP730 is used, and the shared objects located on it, but HP710 machines are used otherwise. In this way computations with data requirements of about 200 Mbyte are performed.

4.1 Cost of Queue Access

An indication of the failure free overhead cost may be obtained by comparing fault tolerant and non fault tolerant sequential computations running within a single workstation. This is done for matrix multiplication by locating a single slave and the data objects on the same host, a HP730 machine. The measured results are shown in Tab. 1 for a range of task sizes.

Table 1. Cost of employing queue in sequential multiplication of 3000 square matrices. The times in columns 3 and 4 are averages rounded to integer values.

Items of work	Block width (elements)	Execution time		Fault tolerance Cost		
		Fault-tolerant (seconds)	Non fault-tolerant (seconds)	During queue Creation (seconds)	After queue creation (seconds)	Total as % of total time
9	1000	2201	2152	6.5	41.5	2.2
16	750	2254	2224	10.3	20.3	1.4
25	600	2215	2171	15	29	2
36	500	2313	2252	22	38.3	2.7
144	250	3068	2917	93.8	58.1	5.2
225	200	3579	3352	154	73.5	6.8

The fault-tolerance costs represent the following operations:

- The cost of creating the queue and enqueueing one entry per block of the output matrix within a surrounding action, and committing that action.
- The cost incurred by the slave of binding to the queue object, essentially server creation, and then dequeuing an entry describing each piece of work.

The queue entries are simply small job descriptions and their size is independent of the data size so the cost of using the queue should be dependent on the number of tasks, rather than data size. Therefore percentage overheads should reduce for larger scale computations, but even for the size of computation performed, fault tolerance does not appear to be the significant cost.

The queue is implemented as a collection of separately lockable persistent objects, and some breakdown of the costs associated with the use of atomic actions on individual persistent objects is given in [16].

4.2 Parallel Execution

The parallel performance of the applications is shown in Fig. 1.

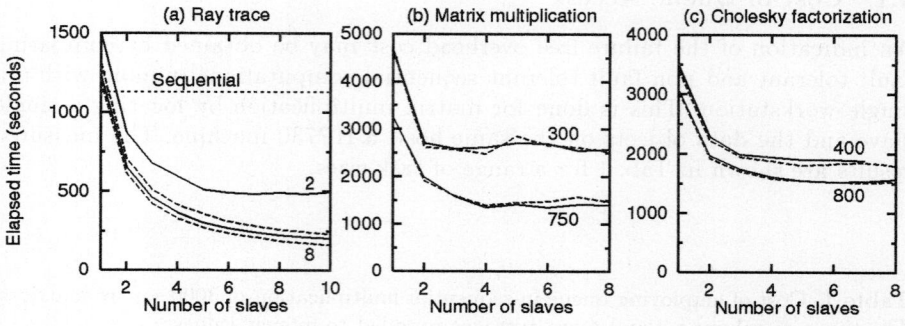

Fig. 1. Performance of parallel applications, comparing fault-tolerant (solid line) and non fault-tolerant (dashed line) versions for indicated task sizes.

In the event of slave failure and immediate resumption, or replacement by a spare, the failure free execution time is increased by a recovery time due to the loss of aborted work. This recovery time is the cost of between zero and one task executions, the *average recovery* being half of the maximum. A computation with non uniform tasks may still be characterized by a simple average recovery cost, though this may be misleading if the cost varies very considerably. If data are cached at a slave which fails, then the slave that takes over the aborted task incurs an extra cost in cache misses. If a slave fails and does not resume and there is no spare, then the increase in overall execution time depends on the exact point of failure, but may be regarded as comprising two components. First, there is the cost of redoing the failed task and secondly, the execution of the remaining tasks is slowed since there is then one less slave.

Table 2 summarizes the performance of the parallel implementations, showing for each application a measure of the performance achieved and estimate of the average recovery time. The table also indicates the total data: input (*input*), written (*put*) and read collectively by slaves during the computation (*get*).

Table 2. Fault-tolerant application parallel performance summary. The speedup shown for ray tracing is absolute, i.e. relative to that of the sequential implementation.

Application	Tasks	Task size (elements)	Data access (Mbyte)			Minimum time (seconds)	Performance (speedup or rate)	Average recovery (seconds)
			input	get	put			
Ray Trace (512^2)	256	2×512	small		6.3	483	2.3	1.5
	64	8×512				204	5.5	9.6
Matrix Multiplication (3000^2)	100	300^2	144	1440	72	2545	21 Mflop/s	24
	16	750^2		576		1353	40 Mflop/s	102
Cholesky Factorization (4800^2)	78	400^2	99	1198	99	1879	20 Mflop/s	23
	21	800^2	108	645	108	1512	24 Mflop/s	74

For all three experiments it is seen that increasing the task size improves the performance. In the matrix computations, the increase in total data read with decreasing block size seems to be the overwhelming effect. In the ray tracing example little data is read, but at 25 KByte and 98 Kbyte the task output is not so large as to be bandwidth limited and so the larger task is cheaper proportionally.

Noting that the data format conversion for ray tracing mentioned earlier takes about 23 and 13 seconds respectively for the task sizes, 2 and 8, the performance of this easy application appears promising.

The performance of the matrix computations is not exciting, though in the one case the peak performance of the memory based matrix multiplication on a single HP710, measured at 33 Mflop/s, is exceeded. Some intuition for the cost of the parallel computations may be gained by considering the cost of accessing the data. Each data access entails both a memory to memory copy between slave and server machine and a local disk, or filesystem cache access on the server machine. Some potential benefit exists both in pipelining data accesses and in caching blocks at slave machines but neither is attempted here. For block sizes above 250, the low level transfer rates for local memory to remote memory, local disk read and local disk write (new data) are found to be roughly constant at about 1, 1.6 and 0.2 Mbyte/s. Assuming no benefit is gained from caching blocks between tasks, an estimate for the total time involved in transfers for the matrix multiplication application with larger block size is 1368 seconds. This would then be a lower bound on the parallel computation time and since the implementation described almost achieves this minimum time it seems possible that bandwidth limitation is being observed. Fuller analysis [18] finds that the benefit gained in this particular situation from involuntary filesystem caching is likely to be small, strengthening the case for bandwidth limitation.

5 Summary

The work described here considers the implementation of certain large scale computations each structured as a bag of tasks over a NOW employing Persistent objects and atomic actions to support fault-tolerance. The first application is a public domain ray tracing package with moderate demands for space. Experiment suggests that respectable performance can be achieved if a suitably large granularity is chosen. The other two applications are both dense matrix computations where the space requirement can exceed available memory. In such a case a model which employs a relatively small number of machines sharing large secondary storage space has some attraction. For this type of execution, a realistic all-be-it prototype implementation has shown that the cost of introducing fault-tolerance is small and performance gain through parallelism is limited essentially by hardware bandwidths.

The system described here provides a practical solution to the question as to how to exploit commonly available clusters of workstations for running compute and data intensive programs by providing much needed support for fault-tolerance and moderate speedup. Since the toolkit developed here does not require any special hardware or software facilities other than those already available, it can readily be adapted to exploit new generations of hardware. [18] describes detailed performance analysis of applications reported here and enables prediction of the expected performance under higher network bandwidth. For example, if the communications media is replaced by fast ethernet, at 100 Mbits/s, but the configuration remains otherwise unchanged a performance of 80 Mflop/s is anticipated for matrix multiplication using 4 slaves.

The overall conclusion is that objects and actions as employed in the computations described seem to be a convenient way to express fault tolerance in parallel applications, and for appropriate scale of computation impose small cost.

Acknowledgements

The work reported here has been supported in part by research and studentship grants from the UK Ministry of Defence, Engineering and Physical Sciences Research Council (Grant Number GR/H81078) and ESPRIT project BROADCAST (Basic Research Project Number 6360). The support of the Arjuna team is acknowledged, and in particular the assistance of M. Little, G. Parrington, and S. Wheater with implementation issues relevant to this work.

References

1. George S. Almasi and Allan Gottlieb. *Highly Parallel Computing*. Benjamin/Cummings, 2nd edition, 1994. ISBN 0-8053-0443-6.

2. Thomas E. Anderson, David E. Culler, and David A. Patterson. A case for NOW (Networks of Workstations). *IEEE Micro*, 15(1):54–64, February 1995.
3. David Edward Bakken. *Supporting Fault-Tolerant Parallel Programming in Linda*. PhD thesis, Department of Computer Science, The University of Arizona, August 1994.
4. Henri E. Bal. Fault tolerant parallel programming in Argus. *Concurrency: Practice and Experience*, 4(1):37–55, February 1992.
5. A. Benzoni and M. L. Sales. Concurrent matrix factorizations on workstation networks. In A. E. Fincham and B. Ford, editors, *Parallel Computation*, pages 273–284. Clarendon Press, 1991.
6. Philip A. Bernstein, Meichun Hsu, and Bruce Mann. Implementing recoverable requests using queues. *ACM SIGMOD*, pages 112–122, 1990.
7. Nicholas Carriero and David Gelernter. *How To Write Parallel Programs: A First Course*. MIT Press, 1991. ISBN 0-262-03171-X.
8. Timothy Clark and Kenneth P. Birman. Using the ISIS resource manager for distributed, fault-tolerant computing. Technical Report 92-1289, Cornell University Computer Science Department, June 1992.
9. Gene H. Golub and Charles F. Van Loan. *Matrix Computations*. John Hopkins University Press, second edition, 1989. ISBN 0-8018-3772-3.
10. Jim Gray and Andreas Reuter. *Transaction Processing: Concepts and Techniques*. Morgan Kauffman, 1993.
11. Karpjoo Jeong. *Fault-Tolerant Parallel Processing Combining Linda, Checkpointing, and Transactions*. PhD thesis, New York University, Department of Computer Science, January 1996.
12. M. Frans Kaashoek, Raymond Michiels, Henri E. Bal, and Andrew S. Tanenbaum. Transparent fault-tolerance in parallel Orca programs. In *Proceedings of the Symposium on Experiences with Distributed and Multiprocessor Systems III*, pages 297–312, Newport Beach, CA, March 1992.
13. Craig Kolb. *rayshade*. ftp://ftp.cs.yale.edu, May 1990. version 3.0.
14. Vipin Kumar, Ananth Grama, Anshul Gupta, and George Karypis. *Introduction to Parallel Computing*. Benjamin Cummings, 1994. ISBN 0-8053-3170-0.
15. Juan Leon, Allan L. Fisher, and Peter Steenkiste. Fail-safe PVM: A portable package for distributed programming with transparent recovery. Technical Report CMU-CS-93-124, School of Computer Science, Carnegie Mellon University, Pittsburgh, PA 15213, February 1993.
16. G. D. Parrington, S. K. Shrivastava, S. M. Wheater, and M. C. Little. The design and implementation of Arjuna. *USENIX Computing Systems Journal*, 8(3):225–308, summer 1995.
17. James S. Plank, Youngbae Kim, and Jack J. Dongarra. Algorithm-based diskless checkpointing for fault tolerant matrix operations. Technical Report CS-94-268, University of Tennessee, December 1994.
18. J. Smith. *Fault Tolerant Parallel Applications Using a Network Of Workstations*. PhD thesis, University of Newcastle upon Tyne, Department of Computing Science, 1996. In Preparation.
19. J. Smith and Santosh Shrivastava. Fault-tolerant execution of computationally and storage intensive programs over a network of workstations: A case study. In ESPRIT Basic Research Project 6360 Third Year Report, July 1995.

A Framework for Viewing Atomic Events in Distributed Computations

Ajay D. Kshemkalyani

IBM Corporation, P. O. Box 12195, Research Triangle Park, NC 27709, USA
Email: ajayk@vnet.ibm.com

Abstract. We present a unifying framework for expressing and analyzing events at various levels of atomicity in distributed computations. In the framework, events at any level of atomicity are defined and composed in terms of events at a finer level of atomicity using hierarchical views. We identify and prove two properties that are satisfied by each level of atomicity. Results based on these properties that hold for any one level of atomicity apply to all levels of atomicity.

1 Introduction

In the literature on distributed system executions (also known as computations), events have been implicitly modeled in the isolated contexts of various applications, e.g., designing communication primitives [2, 3, 7], global states [5], concurrency measures [6, 9], deadlock detection [12], clock systems [10, 14, 17], termination detection [16], mutual exclusion [20], debugging [8], fault-tolerance and transactions [4, 11, 19]. The events modeled have various levels of atomicity, and there is no prior treatment of the various levels of atomicity in a unifying framework. A formal treatment of grouping events in a distributed execution is crucial in modeling distributed activities to provide different abstract views. Lamport also argued that it is useful to assume that primitive elements between which concurrency is modeled are nonatomic for studying basic questions about nonatomicity [15]. This paper provides a unifying framework for expressing and analyzing events at various levels of atomicity in distributed system executions; events at a particular level of atomicity are defined and hierarchically composed in terms of events at a finer level of atomicity. We define system executions for the various levels of atomicity by first defining a system execution dealing with the most elementary events, suitably identified. We then hierarchically compose system executions of coarser levels of atomicity by using the system executions at a finer level of atomicity.

We also prove that each level of atomicity satisfies two properties. [Property **P1:**] The events at any level of atomicity partition the events at the finer level of atomicity in terms of which this level is defined. (See Defn. 3 and Theorems 1, 3, and 5 for the four levels of atomicity considered.) [Property **P2:**] The events at any level of atomicity ordered by the corresponding ordering relation form a partially ordered set (poset) (See Defn. 3 and Theorems 2, 4, and 6 for the four levels of atomicity considered.) P1 implies that all the events at any level of atomicity are included implicitly in more abstract events at coarser levels of atomicity. Any result based on the graph property P1 or P2 that applies to any one level of atomicity applies to all levels of atomicity.

Section 2 presents the system model. Section 3 presents the events at four levels of atomicity by a hierarchical composition, and gives their applications. Section 4 concludes. The full paper [13] includes the proofs of theorems stated here.

2 System Model

A distributed system is a set of processes connected by communication channels. Depending on the level of atomicity being modeled, both processes and channels are modeled as nodes, or only processes are modeled as nodes that communicate with each other. Let E be the set of the most elementary events in a system execution, i.e., a run of a computation. We assign a semantic meaning to E later. Events of E are partitioned into local computations at a node, assuming that each event of E occurs at one node only. Each local computation is a linearly ordered set. An event e in partition i is denoted e_i. The computation at node i is a sequence of events and the system computation is the collection of computations at the various nodes. The initial event in each partition i is $\perp_i$. For finite computations, the final event in each partition i is $\top_i$.

Nodes communicate with each other by passing messages. A channel cannot generate, consume, or alter messages, but can permute the order of delivery of messages. The local action of sending (receiving) a message is a send (receive) event. The message sent at any send event is distinct from all messages sent at other send events at the level of atomicity being considered. The transfer of a message between a pair of process nodes takes finite time on a global time scale but between a process node and a channel node, it is instantaneous. The set of events that occur on any one node in a run of a computation can be decomposed into the sets $\mathcal{RC}$, $\mathcal{SD}$, and $\mathcal{IN}$, which are the sets of events of receiving a message from another node, sending a message to another node, and internal events, respectively. Individual events in the three sets are denoted by RC, SD, and IN, respectively. The sets $\mathcal{RC}$, $\mathcal{SD}$, and $\mathcal{IN}$ will be defined at multiple levels of atomicity which will be differentiated by appropriate subscripts.

Events in a computation are ordered by the causality relation $<$ on E [14]. An edge that orders two events on the same node (different nodes) is termed a *local edge* (*message edge*). A cut C is a subset of E such that if $e_i \in C$ then $\forall e'_i : e'_i < e_i$, we have $e'_i \in C$. A *consistent cut* is a downward-closed subset of E in $(E, <)$. $\mathcal{C}$, the set of cuts of a poset $(E, <)$, forms a lattice $(\mathcal{C}, \subset)$ with the operations $\bigcup$ and $\bigcap$. $\mathcal{CC}$, the set of consistent cuts of a poset $(E, <)$, forms a sublattice of $\mathcal{C}$, as shown in [17].

We use the formalism of hierarchical views of a system execution introduced by Lamport [15] to define events at various levels of atomicity in terms of elementary actions in a system. The choice of actions treated as elementary is based on the need to model sufficiently fine-grained actions for the known applications.

The set of events in the system execution at an arbitrary level of atomicity x, as well as the ordering relation among the events at that level of atomicity is represented as a tuple $\langle \mathcal{A}_x, <_x \rangle$. $\mathcal{A}_x$ and $<_x$ are different for each level of atomicity x. The term "atom" will be used interchangably with "event"; individual events (or atoms) and the set of events (or atoms) are denoted A_x and $\mathcal{A}_x$, respectively, to emphasize their atomic nature. The subscript will be dropped when the context is clear.

Consider $(\mathcal{A}_\alpha, <_\alpha)$ and $(\mathcal{A}_\beta, <_\beta)$, where $\mathcal{A}_\alpha$ and $\mathcal{A}_\beta$ are sets and $<_\alpha$ and $<_\beta$ are relations on the elements of $\mathcal{A}_\alpha$ and $\mathcal{A}_\beta$, respectively. Let mapping μ_β be a one-many surjective mapping that maps each element A_β of $\mathcal{A}_\beta$ to a non-empty subset of $\mathcal{A}_\alpha$. If μ_β^{-1} is a function then $\mathcal{A}_\beta$ defines a partition on $\mathcal{A}_\alpha$ — this means each element A_α of $\mathcal{A}_\alpha$ is contained in exactly one element A_β of $\mathcal{A}_\beta$, and an element A_β may contain multiple elements from $\mathcal{A}_\alpha$. Each element A_β in $\mathcal{A}_\beta$ is a set that is a higher level grouping of the events in $\mathcal{A}_\alpha$ that is of interest to some application. μ_β is specified so as to define meaningful events at an appropriate level of atomicity $(\mathcal{A}_\beta, <_\beta)$ in terms of the events specified at the level of finer atomicity in $(\mathcal{A}_\alpha, <_\alpha)$.

There are two cases to consider when we define a system execution $S_\beta = \langle \mathcal{A}_\beta, <_\beta \rangle$. (i) For system executions S_β at recursively higher levels of atomicity, we specify a mapping μ_β, which maps S_β to a system execution S_α at a finer level of atomicity. $\mathcal{A}_\beta$ contains events at a coarser level of atomicity than $\mathcal{A}_\alpha$. (ii) If S_β is at the level of atomicity of the most elementary actions that we choose, μ_β maps S_β to S_β and we provide a semantic model for S_β.

Definition 1 *A system execution S_β is a tuple $\langle \mathcal{A}_\beta, <_\beta \rangle$ where $\mathcal{A}_\beta$ is a set and $<_\beta$ is a ordering relation on $\mathcal{A}_\beta$.*

S_β is specified in terms of a mapping $\mu_\beta : S_\beta \longrightarrow S_\alpha$, where S_α is a system execution at a finer level of atomicity such that:
1. *μ_β maps each element in $\mathcal{A}_\beta$ to a subset of $\mathcal{A}_\alpha$.*
2. *μ_β defines $<_\beta$ in terms of $<_\alpha$.*

If S_β is at the finest level of atomicity, $S_\alpha = S_\beta$ and we give a semantic model for S_β.

At the finest level of atomicity, we will use the semantic model of E and the causality relation on E, i.e., $\langle E, < \rangle$, for the system execution.

3 Modeling Events in a Distributed Computation

In Sections 3.1, 3.2, 3.3, and 3.4, we define four levels of atomicity $S_{dist}, S_{SR}, S_{react}$, and S_{TL}, respectively, in a hierarchical manner, starting with the finest level S_{dist} to which we assign the semantic model of $\langle E, < \rangle$.

3.1 Primitive Send and Receive Events

To view the system execution at the finest level of atomicity S_{dist}, we consider primitive send and receive events that are expressed by explicitly modeling channels that connect any two processes, and the input and output buffers of the two processes. Though there are many communication constructs to send and receive messages, they are not necessarily atomic. It is shown in [7] that all such constructs can be expressed as some combination of one of the following primitive events.

1. POST-SEND, abbreviated PS, is a send event that initiates a message send to the destination process, and can complete even before the message is copied out of the sender's buffer. The set of all PS events is $\mathcal{PS}$.
2. WAIT-FOR-BUFFER-RELEASE, abbreviated WB, waits for the message to be copied out of the sender's buffer. Thus, it is a receive event at which it receives an acknowledgement from the channel that the message has been received by the channel. The set of all WB events is $\mathcal{WB}$.
3. WAIT-FOR-SEND-TO-BE-MATCHED, abbreviated WSM, is a receive event that waits for an acknowledgement from the channel that the destination process has received the message. The set of all WSM events is $\mathcal{WSM}$.
4. POST-RECEIVE, abbreviated PR, is a send event that requests the channel to deliver to it any incoming message that matches the parameters and the sender-id specified. This event can complete before the received message is stored in the receive buffer specified. The set of all PR events is $\mathcal{PR}$.

5. WAIT-FOR-RECEIVE-TO-BE-MATCHED, abbreviated WRM, is a receive event that completes only after the incoming message has been placed in the specified receive buffer. The set of all WRM events is $\mathcal{WRM}$.

The events in $\mathcal{PS}$, $\mathcal{WB}$, $\mathcal{WSM}$, $\mathcal{PR}$, and $\mathcal{WRM}$ occur on process nodes. In order that the computation can progress, we also need to model and identify events at channel nodes, by viewing each channel as an active node. For each PS and PR event (which are send events) on a process node, there exists a corresponding receive event on the channel node. For each WB, WSM and WRM event (which are receive events) on a process node, there exists a corresponding send event on the channel node. The following definition captures this relation.

Definition 2 *If e is a SD or RC event, then $match(e)$ is respectively the RC or SD event corresponding to the message that was sent at e.*

$match(e)$ exists and is unique. (Its definition can be extended to multicasts.) Blocking

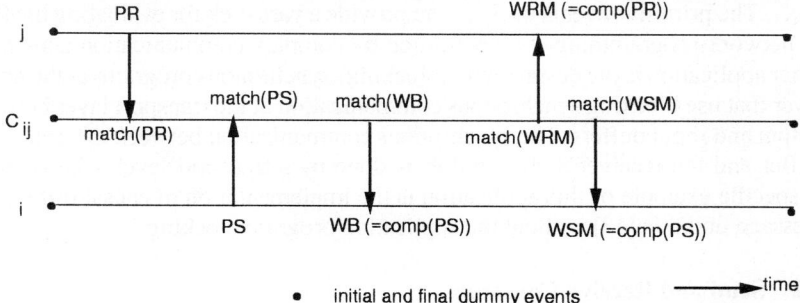

Fig. 1. Message Communication Events at the Finest Level of Atomicity.

and nonblocking, as well as synchronous and asynchronous sends and receives can be executed using the above primitive events [3, 7].

Figure 1 illustrates the effects of events PS, WB, WSM, PR and WRM, as well as Definition 2, by showing the message transfer from process i to process j on channel c_{ij}. The message send initiated by the PS event could complete by either the WB event or the WSM event. Although both WB and WSM are shown in the figure, in practice at most one of them would be used. The notation comp(PS) and comp(PR) for the events will be explained subsequently by Definition 4.

$\mathcal{A}_{dist}$, the set of elementary events in S_{dist}, can now be defined using disjoint sets.

- $\mathcal{A}_{dist} = \mathcal{PS} \bigcup \mathcal{WB} \bigcup \mathcal{WSM} \bigcup \mathcal{PR} \bigcup \mathcal{WRM} \bigcup \{match(PS) : PS \in \mathcal{PS}\}$ $\bigcup \{match(WB) : WB \in \mathcal{WB}\} \bigcup \{match(WSM) : WSM \in \mathcal{WSM}\} \bigcup$ $\{match(PR) : PR \in \mathcal{PR}\} \bigcup \{match(WRM) : WRM \in \mathcal{WRM}\} \bigcup \mathcal{IN}$.

The following decomposition of $\mathcal{A}_{dist}$ shows how the set is partitioned orthogonally to the above into internal events, send events, and receive events:

- $\mathcal{SD}_{dist} = \mathcal{PS} \bigcup \mathcal{PR} \bigcup \{match(WB) : WB \in \mathcal{WB}\} \bigcup \{match(WSM) : WSM \in \mathcal{WSM}\} \bigcup \{match(WRM) : WRM \in \mathcal{WRM}\}$

- $\mathcal{RC}_{dist} = \mathcal{WB} \bigcup \mathcal{WSM} \bigcup \mathcal{WRM} \bigcup \{match(PS) : PS \in \mathcal{PS}\} \bigcup \{match(PR) : PR \in \mathcal{PR}\}$
- $\mathcal{IN}_{dist} = \mathcal{IN}$

We can now define S_{dist}, the system execution at the finest level of atomicity in terms of the semantic model of $(E, <)$.

Definition 3 *System execution $S_{dist} = \langle \mathcal{A}_{dist}, <_{dist} \rangle$, where $\mu_{dist}(S_{dist} \longrightarrow S_{dist})$ is a 1-1 identity mapping. The semantic model of S_{dist} is $(E, <)$, where $\mathcal{A}_{dist}$ is E and $<_{dist}$ is the causality relation on $\mathcal{A}_{dist}$.*

From Definition 3, it follows that S_{dist} satisfies—[Property P1:] Atoms of $\mathcal{A}_{dist}$ partition events (atoms) in E, and [Property P2:] $(\mathcal{A}_{dist}, <_{dist})$ is a poset.

Applications: Complex communication constructs for specific communication styles, such as remote procedure calls (RPC) [2], conversations or dialogs [3], and messaging and queuing constructs, can be designed using PS, WB, WSM, PR, and WRM events of S_{dist}. The primitive events of S_{dist} can provide a yardstick for evaluating the flexibility of network programming style permitted by complex communication constructs. Another application is the design of nonblocking asynchronous programs at the application layer that use blocking synchronous communication at the transport layer between their output and input buffers. The synchronous communication between the sender's output buffer and the receiver's input buffer is done by a transport level acknowledgement. A specific example of this application is the implementation of causal ordering among message unicasts [18] without the application program blocking.

3.2 Send and Receive Constructs

Complex message send and receive events that atomically execute high-level communication constructs, e.g., constructs for various flavors of RPC [2] or the CPI-C communications programming interface [3], provide a higher level of abstraction than the primitive send and receive events of S_{dist}. A system execution at this level of atomicity, denoted S_{SR}, will be defined in terms of system execution S_{dist}. Only process nodes are considered in the S_{SR} view.

Observe that in S_{dist}, a receive initiated by a PR event completes at the corresponding WRM event. Similarly, a send initiated by a PS event completes at the corresponding WB or WSM event. Based on this observation, we define the complement, (abbreviated $comp$), of these events to define the relation between events at a process node that complement other events on the same process node. The $comp$ relation, along with the $match$ relation (Definition 2) will be used to group events in S_{dist} together at the coarser level of atomicity S_{SR}.

Definition 4 *$comp(e)$ is defined as follows [7]:*

1. *If e is a PS event, then $comp(e)$ is the corresponding WB or WSM event, and vice-versa.*
2. *If e is a PR event, then $comp(e)$ is the corresponding WRM event, and vice-versa.*

Any send or receive event e on a process node in S_{dist} identifies the set $\{e, comp(e), match(e), match(comp(e))\}$ – this set will form an atomic event in S_{SR}.

Definition 5 *System execution* $S_{SR} = \langle \mathcal{A}_{SR}, <_{SR} \rangle$ *is defined by a mapping* $\mu_{SR} : S_{SR} \longrightarrow S_{dist}$ *as follows:*

1. $\mathcal{A}_{SR} = \mathcal{IN}_{dist} \bigcup \{\{e, match(e), comp(e), match(comp(e))\} : e \in (\mathcal{PS} \bigcup \mathcal{WRM})\}$
2. For any $A_{SR} \in \mathcal{A}_{SR}$, define $key\_member(A_{SR})$ as follows:
   - $key\_member(A_{SR}) \stackrel{def}{=}$ a PS event in A_{SR},   if a PS event belongs to A_{SR}
   - $key\_member(A_{SR}) \stackrel{def}{=}$ a WRM event in A_{SR}, if a WRM event belongs to A_{SR}
   - $key\_member(A_{SR}) \stackrel{def}{=}$ a IN event in IN_{SR}, if a IN_{dist} event belongs to A_{SR}

   Then, $A_{SR} <_{SR} A'_{SR}$ iff $key\_member(A_{SR}) <_{dist} key\_member(A'_{SR})$.

It is shown in [13] that each event A_{SR} in $\mathcal{A}_{SR}$ has a uniquely defined $key\_member(A_{SR})$ which is a PS, WRM, or IN event of $\mathcal{A}_{dist}$. Note that even if $A_{SR} <_{SR} A'_{SR}$, it may be that $\exists A_{dist} \in A_{SR} \exists A'_{dist} \in A'_{SR} : A'_{dist} <_{dist} A_{dist}$.

Theorem 1 (P1:) *The atoms of $\mathcal{A}_{dist}$ are partitioned into atoms in S_{SR}.*

The proof of Theorem 1 [13] also shows that $\mathcal{A}_{SR}$ can be partitioned into $\mathcal{SD}_{SR}$, $\mathcal{RC}_{SR}$, and $\mathcal{IN}_{SR}$, where:
- $\mathcal{SD}_{SR} = \{ A_{SR} \in \mathcal{A}_{SR} : key\_member(A_{SR}) \in \mathcal{PS} \}$
- $\mathcal{RC}_{SR} = \{ A_{SR} \in \mathcal{A}_{SR} : key\_member(A_{SR}) \in \mathcal{WRM} \}$
- $\mathcal{IN}_{SR} = \{ A_{SR} \in \mathcal{A}_{SR} : key\_member(A_{SR}) \in \mathcal{IN} \}$

Theorem 2 (P2:) *The atoms in $\mathcal{A}_{SR}$ ordered by $<_{SR}$ form poset $(\mathcal{A}_{SR}, <_{SR})$.*

The following corollary is used to analyze system executions S_{TL} in Section 3.4.

Corollary 1 CC_{SR}, *the set of consistent cuts of poset* $(\mathcal{A}_{SR}, <_{SR})$, *forms a sublattice of* $\mathcal{C}_{SR}$, *the set of all cuts of* $(\mathcal{A}_{SR}, <_{SR})$. *(from Theorem 2 and [17]).*

Applications: There are many applications for which each complex send and receive construct, and internal event in the computation is explicitly modeled as a single event at the process nodes in S_{SR}. Global state and snapshot definition and computation [5], concurrency measures for a system execution [6, 9], clock systems for distributed computations [10, 14, 17], transfer of knowledge, checkpointing and recovery [4, 21], leader election, mutual exclusion algorithms [20], and distributed deadlock detection [12] all deal with send and receive events in the S_{SR} view of the system execution.

3.3 Reactive Events

A coarser atomicity of events than that of SD_{SR}, RC_{SR} or IN_{SR} events is useful for applications such as termination detection [16] and debugging [8], even though it does not reflect all the concurrency of the original execution. Events at this coarser level of atomicity are reactive because the computation in an event begins in reaction to a received message. Thus, a reactive event begins when a node receives an external message, and then it does local processing and may send messages. The reactive event is defined to end when either: (i) an application-dependent locally determinable condition ϕ becomes true at a distinguished auxiliary event $C(\phi)$, or (ii) just before a message is received after this event has sent a message, in the S_{SR} view of the execution. We define system execution S_{react} in terms of system execution S_{SR} and using regular expressions over SD_{SR}, RC_{SR} and IN_{SR} events, and the auxiliary event $C(\phi)$.

Definition 6 *System execution* $S_{react} = \langle \mathcal{A}_{react}, <_{react} \rangle$ *is defined by a mapping* $\mu_{react} : S_{react} \longrightarrow S_{SR}$ *as follows:*

1. *Reactive atoms at any node x form a sequence $\langle A_{react}^{x,1}, A_{react}^{x,2}, A_{react}^{x,3}, \ldots \rangle$ where:*
   (a) $A_{react}^{x,1}$ = *the maximal sequence of events that belong to $\mathcal{A}_{SR}$ and occur on node x, that satisfy the regular expression $\langle \perp_x (IN_{SR}|RC_{SR})^*(IN_{SR}|SD_{SR})^*(C(\phi))^* \rangle$*
   (b) $A_{react}^{x,i}, i > 1$ *is the maximal sequence of events that belong to $\mathcal{A}_{SR}$ and occur on node x, that satisfy the context-sensitive regular expression:*
   $A_{react}^{x,i-1} A_{react}^{x,i} = A_{react}^{x,i-1} \langle RC_{SR}(IN_{SR}|RC_{SR})^*(IN_{SR}|SD_{SR})^*(C(\phi))^* \rangle$
2. $\mathcal{A}_{react} <_{react} \mathcal{A}'_{react}$ *iff* $(\exists A_{SR} \in \mathcal{A}_{react}, \exists A'_{SR} \in \mathcal{A}'_{react} : A_{SR} <_{SR} A'_{SR})$.

$A_{react}^{x,i}$ is the ith reactive event on node x. The superscripts/subscript are dropped if there is no ambiguity. Figure 2 shows the reactive events in a distributed execution.

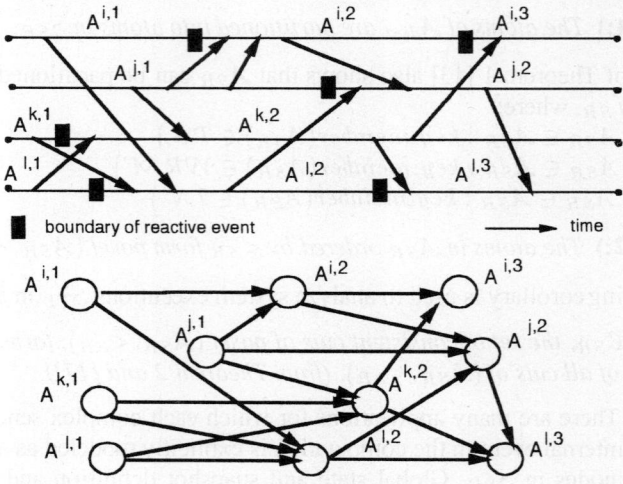

Fig. 2. Reactive Events.

Theorem 3 (P1:) *The atoms of $\mathcal{A}_{SR}$ are partitioned into atoms in S_{react}.*

Theorem 4 (P2:) *The atoms in $\mathcal{A}_{react}$ ordered by $<_{react}$ form poset $(\mathcal{A}_{react}, <_{react})$.*

It follows that no event in $\mathcal{A}_{react}$ has both an edge that goes to another event in $\mathcal{A}_{react}$ and an incoming edge from that other event.

Applications: Computation termination [16] can be modeled by reactive events as follows. Consider a system in which: (i) A process node is either idle or active. (ii) An idle process may have only a RC_{SR} event, at which time the process becomes active. (iii) An active process can become idle any time. A computation is *terminated* if each process is idle and the channels are empty. We express this as follows. Define ϕ as "there is no A_{SR} event waiting to occur." A process is idle if the reactive event has ended and presently there is no event waiting to occur, i.e., ϕ holds. A channel is empty if the number of $match(PS)$ events and $match(WRM)$ events is the same in the S_{dist} view.

A message race occurs at an RC_{SR} event if it can receive one of multiple messages. Debugging based on controlled execution of message races examines the possible executions corresponding to one space-time diagram [8]. The definition of reactive events (Defn. 6) for debugging does not use any auxiliary event $C(\phi)$, i.e., $\phi = false$. A message that could be received in a reactive event A may have been sent in a reactive event A' such that $A' < A \bigvee (A' \not< A \bigwedge A \not< A')$. For e.g., in Figure 2, if A is $A^{l,2}$, then A' is any of $A^{i,1}$, $A^{i,2}$, $A^{i,3}$, $A^{j,1}$, $A^{k,1}$, $A^{k,2}$, and $A^{l,1}$. During controlled (replay) executions for event A, such events A' are forced to complete before A begins, before permuting the order of delivery of racing messages to RC_{SR} events in A.

3.4 Events between Transitless Cuts

System executions at the next higher level of atomicity S_{TL} are defined in terms of S_{SR}. Events at this level of atomicity belong to multiple process nodes.

Definition 7 *A transitless cut TLC_{SR} is a consistent cut in $(\mathcal{A}_{SR}, <_{SR})$ such that the only ordering edges between it and the rest of $\mathcal{A}_{SR}$ are local edges at process nodes (defined in Section 2).*

The system state after the execution of events in a transitless cut is a *transitless global state*. Such states have the property that the effects of the past computation are contained in only local edges of process nodes in a S_{SR} view of the execution, viz., the process states, and no messages are in transit. We examine this level of atomicity using Corollary 1 [17] and properties of lattices, unlike previous work (see Applications).

Lemma 1 $T\mathcal{LC}_{SR}$, *the set of transitless cuts of a poset $(\mathcal{A}_{SR}, <_{SR})$, forms a sublattice of $\mathcal{CC}_{SR}$, the set of all consistent cuts of $(\mathcal{A}_{SR}, <_{SR})$, with operations $\bigcup$ and $\bigcap$.*

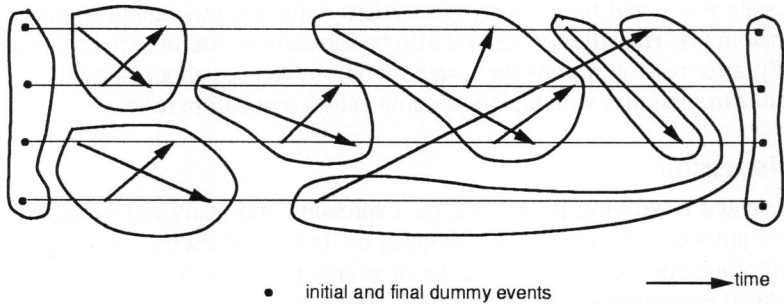

• initial and final dummy events ⟶ time

Fig. 3. Events between Transitless Global States.

From Lemma 1, note that each member of lattice $T\mathcal{LC}_{SR}$ is a set of events in $\mathcal{A}_{SR}$. Henceforth, a member of $T\mathcal{LC}_{SR}$ will be denoted by TLC. For any two comparable elements TLC^u and TLC^l of a lattice, $length[TLC^l, TLC^u]$ is the length of the longest maximal chain in the lattice between TLC^l and TLC^u. We now define the system execution S_{TL} for transitless cuts using the lattice $T\mathcal{LC}_{SR}$ and S_{SR}.

Definition 8 *System execution $S_{TL} = \langle \mathcal{A}_{TL}, <_{TL} \rangle$ is defined by a mapping $\mu_{TL} : S_{TL} \longrightarrow S_{SR}$ as follows:*

1. $\mathcal{A}_{TL} = \{ (TLC^u - TLC^l) : TLC^u, TLC^l \in \mathcal{TLC}_{SR} \bigwedge length[TLC^l, TLC^u] = 1 \}$
2. $<_{TL}$ *is the transitive closure of* $<_{tlc}$ *where* $(TLC^u - TLC^l) <_{tlc} (TLC'^u - TLC'^l)$ *iff* $(\exists e \in (TLC^u - TLC^l), \exists e' \in (TLC'^u - TLC'^l) : e <_{SR} e')$.

Events in $\mathcal{A}_{TL}$ change the system state from one transitless state to another. Events in $\mathcal{A}_{TL}$ are defined only in terms of the set difference of two elements (of the form $TLC^u - TLC^l$) of lattice $\mathcal{TLC}_{SR}$ that are separated by a length of one. The same event may be expressible as the difference of more than one pair of transitless cuts. This property is important and is used in the proof of Theorem 6. Figure 3 shows the events in S_{TL}. Each event in $\mathcal{A}_{TL}$ is marked by encircling the elements of $\mathcal{A}_{SR}$ to which μ_{TL} maps it. There is an initial dummy event, and a final dummy event for terminating computations. All the edges of $(\mathcal{A}_{SR}, <_{SR})$ entering and leaving each event $\mathcal{A}_{TL}$ in $\mathcal{A}_{TL}$ are local edges. An event $\mathcal{A}_{TL}$ signifies that the computation it represents is affected only by the incoming local edges on processes in a S_{SR} view, and it affects the rest of the computation only through outgoing local edges on processes in the S_{SR} view.

Theorem 5 (P1:) *The atoms of $\mathcal{A}_{SR}$ are partitioned into atoms by S_{TL}.*
Theorem 6 (P2:) *The atoms in $\mathcal{A}_{TL}$ ordered by $<_{TL}$ form poset $(\mathcal{A}_{TL}, <_{TL})$.*

Applications: Transitless states are used in applications like fault-tolerance, checkpointing/recovery [4, 11, 21], synchronization [19], and transactions [4, 11]. Transitless states were forced in [11, 19] for synchronization and checkpointing/recovery. Transaction systems create transitless states at the end of each transaction using commit protocols [4]. In these applications, the transitless states along the boundaries of only certain events in S_{TL} are recorded; in case of failure, the most recent recorded transitless state is restored for recovery. Transitless states and their applications were also examined in [1]. Transitless states can also be shown to be useful to reset vector clocks [10, 17]; after reset at a transitless state, wrong inferences about causality cannot be drawn due to messages with high timestamp values sent before reset.

4 Discussion

We presented a unifying framework for expressing and analysing events at various levels of atomicity in distributed computations. In the framework, events at a coarser level of atomicity are defined in terms of events at a finer level of atomicity using hierarchical composition and lattices. The global states at various levels of atomicity correspond to embedded lattices of global states. The framework was applied to four levels of atomicity here, and can be applied to parallel system executions as shown in [13]. The system model can be varied to allow message losses and multicasts as in [13].

The system execution at every level of atomicity was shown to have two properties. [Property P1]: If S_β is defined in terms of S_α, then the atoms in S_α are partitioned into atoms in S_β. [Property P2]: the atoms at any level of atomicity form a poset ordered by an ordering relation for that level of atomicity. Therefore, any result or proof that applies to one level of atomicity and is based on the above properties applies to all levels of atomicity. For example, the proof for execution S_{SR} that synchronous communication between processes guarantees causal ordering of message unicasts applies without change to the proof for execution S_{dist} that asynchronous communication between processes, with synchronous communication over channels between the (infinite) output and input process buffers, respectively, guarantees causal ordering of message unicasts [18]. A second example is the reuse of concurrency measures described in S_{SR} [6, 9] for gauging concurrency of incremental debugging in S_{react}; this latter measure is useful to determine the number of nondeterministic and deterministic replays.

References

1. Ahuja, M., Kshemkalyani, A. D., Carlson, T.: A Basic Unit of Computation in Distributed Systems. Proc. 10th IEEE Int. Conf. Distrib. Comput. Systems (1990) 12–19
2. Ananda, A., Tay, B., Koh, E.: A Survey of Asynchronous RPC. ACM Operating Systems Review (1992)
3. Arnette, W., Kshemkalyani, A.D., Riley, W., Sanders, J., Schwaller, P., Terrien, J., Walker, J.: CPI-C: An API for Distributed Applications. IBM Systems Journal **34(3)** (1995) 501–518
4. Bernstein, P.A., Hadzilacos, V., Goodman, N.: Concurrency Control and Recovery in Database Systems, Addison-Wesley (1987)
5. Chandy, K.M., Lamport, L.: Distributed Snapshots: Global States of a Distributed System. ACM Trans. Comput. Systems **3(1)** (1985) 63–75
6. Charron-Bost, B.: Measure of Parallelism of Distributed Computations. Proc. STACS 89, In LNCS 349 Springer-Verlag (1989) 434–445
7. Cypher, R., Leu, E.: Repeatable and Portable Message-Passing Programs. Proc. 13th ACM Symp. on Principles of Distributed Computing (Aug. 1994) 22–31.
8. Damodaran-Kamal, S., Francioni, J.: Nondeterminacy:Testing and Debugging in Message Passing Parallel Programs. ACM/ONR Workshop on Debugging (1993) 118–128
9. Fidge, C.A.: A Simple Run-Time Concurrency Measure. The Transputer in Australasia, Eds. T. Bossomaier, T. Hintz, J. Hulskamp, IOS Press (1990) 92–101
10. Fidge, C.A.: Timestamps in Message-Passing Systems That Preserve Partial Ordering. Australian Computer Science Communications **10(1)** (Feb. 1988) 56–66
11. Fisher, M., Griffeth, N., Lynch, N.: Global States in a Distributed System. IEEE Trans. Software Engineering **8(3)** (May 1982) 198–202
12. Kshemkalyani, A.D., Singhal, M.: On Characterization and Correctness of Distributed Deadlock Detection. Journal of Parallel and Distributed Computing **22(1)** (July 1994) 44–59
13. Kshemkalyani, A.D.: A Unifying Framework for Viewing Atomic Actions in Parallel and Distributed Systems. IBM Tech. Rep. TR29.2014 (1995)
14. Lamport, L.: Time, Clocks, and the Ordering of Events in a Distributed System. Communications of the ACM **21(7)** (July 1978) 558–565
15. L. Lamport.: On Interprocess Communication, Part I: Basic Formalism, Part II: Algorithms. Distributed Computing, **1** (1986) 77–101
16. Mattern, F.: Algorithms for Distributed Termination Detection. Distributed Computing **2** (1987) 161–175
17. F. Mattern, Virtual Time and Global States of Distributed Systems. Parallel and Distributed Algorithms, North-Holland (1989) 215-226
18. Mattern, F., Fünfrocken, S.: A Nonblocking Lightweight Implementation of Causal Order Message Delivery. In LNCS 938 Springer-Verlag (1995) 197–213
19. Randell, B.: System Structure for Software Fault Tolerance. IEEE Trans. Software Engg. **1(2)** (1975) 220–232
20. Singhal, M.: A Taxonomy of Distributed Mutual Exclusion. Journal of Parallel and Distributed Computing, **18(1)** 1993 94–101
21. Strom, R.E., Yemini, S.: Optimistic Recovery in Distributed Systems. ACM Trans. on Computer Systems **3(3)** (1985) 204–226

Worker-Based Parallel Computing on PVM *

Dae-Kyun Yoon and Jean-Luc Gaudiot

Department of Electrical Engineering - Systems
University of Southern California
Los Angeles, CA 90089-2563
USA

Abstract. Networks of workstations (NOW) are emerging as popular platforms for high performance computing. PVM (Parallel Virtual Machine) is a software package which provides basic primitives for creating asynchronous tasks on different machines as well as message passing primitives for the communication between tasks. In this paper, we present a worker (or *agent*) based run-time system on PVM to provide a simple interface between user application programs and the parallel processing subsystem using *parallel function calls*.

1 Introduction

The goal of this work is to provide a framework in which both high performance and high programmability can be achieved in network computing. Early efforts to achieve network parallel computing were centered around the construction of a layer on top of an existing operating system and the definition of interfaces between the operating system and the user program for the exploitation of parallelism. PVM (Parallel Virtual Machine) [3, 4], p4 [1], and MPI (Message Passing Interface) [5] are such packages which provide libraries and/or runtime support for network-based computing.

Our approach is to design and implement a worker-based runtime model, in which a simple conventional *function call* syntax and semantics can be applied to the *parallel function call*. [2, 6] In this system, the programmer does not have to be concerned about the correctness of the program as long as its sequential execution produces the correct result. For an acceptable performance enhancement, we also developed a runtime allocation scheme which works particularly well for medium to coarse grain parallelism. We chose to base our system on PVM since PVM is one of the most mature packages which also has been ported to a wide variety of platforms.

2 Overview of Worker-Based Runtime System

We use two basic primitives for the exploitation of parallelism:

---
* This research is supported in part by ARPA grant # DABT63-95-0093

- PARCALL : Invoke a function in parallel
- PARJOIN : Wait until the previous invoked parallel function is completed.

In order to implement these primitives, a runtime system must be built on top of the existing parallel execution sub-system, in our case, PVM. Our *Worker-based runtime system* is another abstract layer between user programs and the parallel execution sub-system. The parallel runtime primitives (*i.e.* PARCALL and PARJOIN), are used to interface between the user program and the worker based runtime system.

A worker manages several data-structures, such as the *task pool* and the *ready task queue*.[2] The *task pool* contains task blocks that are created by the *local worker* while the *ready task queue* holds task blocks that are created by *remote worker(s)* and sent to the local worker for execution. The flow from the task creation to the consumption of its result is depicted in figure 1.

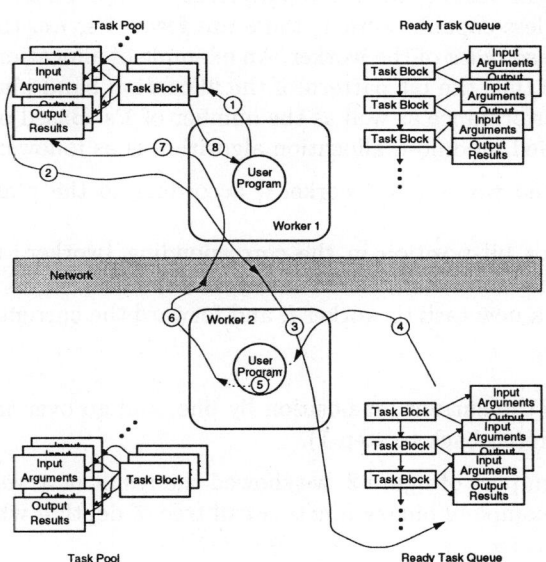

Fig. 1. Each stage of a task is numbered from *1* to *8*. For example, at stage *1* a task created by *ParCall*, and at stage *8* the result of remote execution of the task is passed to the user program.

3 Task Allocation

The important function of *ParCall* is to select the remote worker (task allocation) where the new task is to be executed. One major goal of task allocation

---
[2] Each task corresponds to a user-defined and/or compiler-generated function.

is to eliminate the message passing overhead while preserving the relative load balance among workers (hosts). First, we describe an *modulo-n allocation* scheme which allocates tasks evenly throughout the workers:

1. If the first PARCALL is encountered on a host, h, allocate a new task on $(h+1) \mod N$.
2. If the most recently allocated machine is k, then the new task is created on $l = (k+1) \mod N$.

In the network parallel computing environment, the *capability* or the *speed* is different from worker to worker. It may thus be desirable to discriminate the less-capable worker in the allocation step.

We thus introduce a *weighted modulo-N* allocation scheme in which the *less-capable* worker will have a lower chance of being assigned a new task. The key idea is to use *allocation bands* which are built based on the relative speed of each worker. An allocation band is a pattern of 0's and 1's with a certain fixed length. For the less-capable worker, there are fewer 1's, *i.e.* the number of 1's represent the *speed index* of the worker. An example of allocation bands is shown in figure 2. Note that the bit pattern of the allocation band plays an important in the overall performance as well as the number of 1's. Based on this allocation band, the modified modulo-n allocation algorithm is as follows:

1. Determine the worker, say worker i, according to the plain *modulo-N* scheme.
2. If the current bit-position in the corresponding (worker i's) allocation band is 1, then
   allocate a new task to worker i, and forward the current bit-position by one.
   Otherwise,
   Forward the current bit-position by one, and go over the allocation procedure(go back to step 1).

In the bottom part of figure 2, we showed the result of a weighted modulo-n allocation for a complete binary function call tree of depth 4 where each node is labeled A...Z, 1...5.

4 Preliminary Results and Analysis

We have run the *recursive adaptive quadrature* for a preliminary experimentation. Three machines with different relative speeds have been used for our experimentation. When the weighted modulo-N allocation scheme is applied, the execution time has been reduced from 6.23 to 4.38 compared to the straightforward modulo-N allocation scheme.

5 Conclusion

Our work intends to show that a network of workstations is a feasible platform for parallel computation both in terms of cost-effectiveness and programmability.

<Example of Allocation Band for each Worker (on Worker 0)>

For Worker 0 (Speed Index 4)	11111111 11111111 11111111 11111111
For Worker 1 (Speed Index 2)	10101010 10101010 10101010 10101010
For Worker 2 (Speed Index 1)	00100010 00100010 00100010 00100010
For Worker 3 (Speed Index 3)	11111111 11111010 10101010 10101111

<Allocation Map>

	Worker 0	Worker 1	Worker 2	Worker 3
Worker 0	A,L,R,X,4	B,S,5	V	C,M,W,Y
Worker 1	2,3		D	E
Worker 2	I,O	N		H
Worker 3	F,K,T,Z	P,1	G	J,Q,U
Speed	4	2	1	3
#of tasks	13	6	3	9

Cell$_{i,j}$ = (t_1,t_2,t_3) --> Worker i allocates tasks, (t_1,t_2,t_3), to Worker j

Fig. 2. An example of Weighted Modulo-N task allocation.

We have designed and implemented a worker-based runtime system, in which a simple *function call* can be easily replaced with the *parallel function call*. We also developed an overhead-free runtime allocation scheme and demonstrated its effectiveness in a network parallel computing environment.

References

1. R. Butler and E. Lusk. Monitors, messages, and clusters: The p4 parallel programming system. Technical Report MCS-P362-0493, Argonne National Laboratory, 1993.
2. D. Cann, C. Lee, R Oldehoeft, and S. Skedzielewski. SISAL multiprocessing support. Technical Report UCID-21115, Lawrence Livermore National Laboratory, 1987.
3. J. Dongarra, G. Geist, R. Manchek, and V. Sundaram. Integrated pvm framework supports heterogeneous network computing. *Computers in Physics*, 7(2):166–175, 1993.
4. A. Geist *et al.* . *PVM: Parallel Virtual Machine – A Users' Guide and Tutorial for Networked Parallel Computing*. The MIT Press, Cambridge, Massachusetts, 1994.
5. W. Gropp, E. Lusk, and A. Skjellum. *Using MPI: Portable Parallel Programming with the Message-Passing Interface*. The MIT Press, 1994.
6. D-K. Yoon and J-L. Gaudiot. Programming and evaluating the performance of signal processing applications in the Sisal programming environment. In *Proceedings of the Second Sisal Users' Conference*, pages 67–82. Lawrence Livermore National Laboratory, 1992.

An Efficient Distributed Tuple Space Implementation for Networks of Workstations

Antony Rowstron* and Alan Wood**

Department of Computer Science, University of York,
York, YO1 5DD, UK.

Abstract. In this paper an overview of a novel run time system for the management of tuple spaces, which utilises *implicit* information about tuple space use in Linda programs to maximise its performance. The approach is novel compared to other tuple space implementations because they either ignore the information, or expect the programmer to provide the information explicitly. A number of experimental results are given to demonstrate the advantages of using the our approach.

1 Introduction

Linda is a well known coordination model[1]. For many years implementations have followed very traditional routes based on the work at Yale[2]. However, since the initial traditional implementations the model has evolved; one of the major changes has been the addition of multiple tuple spaces. Most implementations "tack" tuple space names onto tuples in an ad hoc fashion and subsequently treating them like any other field within a tuple. Some implementations[3] use the potential locality information that tuple spaces can provide, but expect the programmer to provide *explicit* information about the "type" or "usage" of a tuple space, thus explicitly declaring the locality of a tuple space.

Our new run-time tuple space management system (kernel) uses *implicit* information to enable it to classify every tuple space as either a *local tuple space* (LTS) or a *remote tuple space* (RTS) on the fly and without extra communication or programmer guidance. A LTS is one which can only be accessed by one process and a RTS is one that many processes can access. What is novel about our approach is the use of implicit information to enable the kernel to transparently move large numbers of tuples around the system to gain optimum performance. It should be stressed that making the distinction between LTSs and RTSs does not alter the semantics of the Linda model, a user sees no distinction between a LTS and RTS.

2 The novel implementation technique

A tuple spaces classification controls where its tuples are stored. If the classification changes then the tuple space migrates to the correct storage place for its

** Contact: {ant,wood}@minster.york.ac.uk
* Funded by an EPSRC CASE grant with British Aerospace Military Aircraft.

new classification. The kernel itself is distributed and Figure 1 shows its general structure. The kernel has two distinct sections, *Tuple Space Servers* (TSS) and *Local Tuple Space Managers* (LTSM).

The TSS is a set of stand alone processes which together act as a tuple server. They receive tuples and requests for tuples. All RTSs are stored in the TSS, but usually distributed over many processes within the TSS. How tuples are distributed between processes is not important for this paper[3]. The LTSM is a set of library routines which are linked into user processes; LTSMs join and leave the kernel with the user processes they belong to. A LTSM is able to find information on whether a tuples space is a LTS or a RTS *dynamically* (with no inter-process communication), and subsequently controls the movement of tuple spaces. A tuple space will only ever migrate from a TSS to a LTSM or vice-versa (*never* from LTSM to LTSM or TSS process to TSS process). A LTS is always stored on the LTSM of the user process which know it as a LTS.

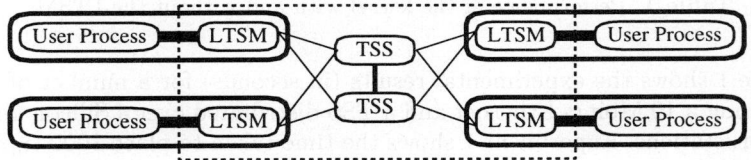

Fig. 1. Diagram showing the layout of the kernel.

When tuple space operations occur the LTSM checks internally to see if the tuple space is a LTS. If so, the LTSM updates itself accordingly, with no communication with the TSS. If it is a RTS the LTSM contacts the relevant TSS process(es). The bulk movement of tuples occurs when either a bulk movement primitive is performed (`collect`[6] or `copy-collect`[7]) or a tuple space handle for a LTS is placed in a tuple in a RTS.

The bulk tuple space primitives require a source and a destination tuple space. If the the classification of both tuple spaces are the same then they are performed in the same part of the kernel; LTS in a LTSM and RTS in the TSS and there is no migration of tuples. If the source tuple space is a RTS and destination is a LTS, then the TSS performs the duplication (if necessary) and migrates (in one or more blocks of *multiple* tuples) the tuples to the correct LTSM. If the source is a LTS and the destination a RTS then the LTSM performs the duplication and migrates the tuples to the TSS. If no movement is required there will not be any movement of tuples. When a tuple space handle for a LTS is placed in a tuple in a RTS the LTSM checks to see if the handle refers to a LTS. If so, the tuple space is migrated to the TSS (thus becoming a RTS).

In this section a very brief overview of the implementation techniques have been presented, for more information see Rowstron et al.[8]. This describes why the bulk movement of tuples is more efficient than the single movement of tuples. The most important point is that the system uses *implicit* information rather than relying on extra explicit information provided by the programmer.

---

[3] The method used is the same as for previous York kernels[4, 5].

3 Experimental results

In order to demonstrate the performance of the kernel we compare its performance with the LTSM enabled and disabled. When the LTSM is disabled the kernel degenerates into an implementation similar to many more traditional implementations[4, 5] because all tuple spaces are treated as RTSs and therefore stored in the TSS.

Experiment	1		2		3		4		5	
LTSM	Off	On	Off	On	Off	On	Off	On	Off	On
out	2.890	2.861	2.769	0.018	2.802	0.018	2.770	0.018	2.769	2.767
collect	n/a	n/a	n/a	n/a	n/a	n/a	0.007	0.038	0.007	0.007
in	3.224	3.227	3.270	0.017	3.303	3.877	3.258	3.670	3.255	0.056
Total	6.114	6.087	6.039	0.035	6.105	3.895	6.035	3.726	6.031	2.830

Table 1. Performance of the kernel with and without the LTSM.

Table 1 shows the experimental results (in seconds) for a number of experiments using a 10 Mbit/s Ethernet and a TSS distributed over 4 Silicon Graphics Indy workstations. Experiment 1 shows the time taken to place 1000 tuples in a RTS using out and retrieve them using in. This demonstrates that there is no significant difference in time between the LTSM being on or off (because in both cases the tuple space is stored on the TSS). Experiment 2 is the same except the tuple space can be classified as a LTS. The results demonstrate that when the LTSM is on it detects that the entire operation is local to the process. When the LTSM is off the tuple space is treated as a RTS and the execution times reflect this. Experiment 3 shows the time taken to place 1000 tuples in a LTS, then place in UTS[4] a tuple containing the handle of that tuple space (so the tuple space becomes a RTS) and then retrieve the tuples. This shows that the time taken to perform the operation is less when the LTSM is on. This is because, when the movement of tuple occurs (as the tuple is placed in the UTS), the tuples are packed into larger packets for dispatching to the TSS. Experiment 4 shows the time taken to place 1000 tuples in a LTS, then collect[5] them all into a RTS and then retrieve them. As one would expect the times are comparable to those of experiment 3, as essentially the same tuple space "movements" are occuring. This again demonstrates how the bulk movement of tuples creates a more efficient implementation. Experiment 5 shows the time taken to place 1000 tuples in a RTS, then collect them all into a LTS and then retrieve them. The results again demonstrate the effectiveness of the bulk movement of tuples and tuple spaces. This is an important operation, as it represents the communications behaviour of the basic operations that a process has to perform to overcome the multiple rd problem[7].

---

[4] UTS is a *universal tuple space*, which all processes have access to.

[5] This primitives moves all tuples which match a given template from a source tuple space to a destination tuple space, and returns a count of the number of tuples moved[6].

These results demonstrate the effect that the LTSM has on the execution time for a number of specific examples, and that the LTSM does not slow the kernel down, and indeed that when used it provides large speed increases for certain classes of operations. Within the scope of this paper it is not possible to demonstrate the performance of the kernel against other kernels. We have compared our kernel with a commercial version of Linda, called C-Linda[6]. For certain classes of algorithms our kernel achieves a speedup of between 10 and 70 times. For more information on the performance of the kernel (including the results for a "real-world" problem) see Rowstron et al.[8].

4 Conclusion

An overview of a new Linda kernel, which transparently uses bulk movement of tuples to achieve performance increases over traditional implementation methods has been presented. The bulk movement of tuples is achieved by using *implicit* information about tuple spaces gathered on the fly, rather than by using either compile time analysis or explicit programmer added "hints".

Currently work is focusing on the development of a multi layer hierarchical kernel, thereby altering the definition of tuple spaces from a discrete categorisation to a continuous one. This kernel uses the same implicit information to provide better locality information to the kernel and will have the potential to be used by hundreds of workstations.

References

1. N. Carriero and D. Gelernter. Linda in context. *Communications of the ACM*, 32(4):444–458, 1989.
2. R. Bjornson. *Linda on distributed memory multiprocessors*. PhD thesis, Yale University, 1992. YALEU/DCS/RR-931.
3. B. Nielsen and T. Sorensen. Implementing Linda with multiple tuple spaces. Technical report, Aalborg University, Denmark, 1993.
4. A. Douglas, A. Wood, and A. Rowstron. Linda implementation revisited. In *Transputer and occam developments*, pages 125–138. IOS Press, 1995.
5. A. Rowstron, A. Douglas, and A. Wood. A distributed Linda-like kernel for PVM. In *EuroPVM'95*, pages 107–112. Hermes, 1995.
6. P. Butcher, A. Wood, and M. Atkins. Global synchronisation in Linda. *Concurrency: Practice and Experience*, 6(6):505–516, 1994.
7. A. Rowstron and A. Wood. Solving the Linda multiple rd problem. In *Coordination Languages and Models*, volume 1061 of *LNCS*, pages 357–367. Springer-Velag, 1996.
8. A. Rowstron and A. Wood. An efficient distributed tuple space implmentation for networks of heterogenous workstations. Technical Report YCS 270, University of York, 1996.

---

[6] Available from Scientific Computing Associates, Connecticut, USA.

A Highly Available Partition-Processing Protocol for Distributed Shared Memory Systems

Jenn-Wei Lin and Sy-Yen Kuo

Department of Electrical Engineering
National Taiwan University
Taipei, Taiwan, R.O.C.

Abstract. This paper investigates the problem of network partitioning in distributed shared memory (DSM) systems. Network partitioning in a DSM system will severely reduce the availability of shared pages. We propose a partition-processing protocol for DSM systems, called *one-copy read and no-copy write* protocol, to allow that shared pages can be optimistically accessed in each partition without taking care of the consistency with other partitions. However, the proposed protocol may introduce the problem: inconsistency between partitions. To resolve the inconsistency problem, we employ a coordinated checkpointing technique to reconstruct the consistency between partitions. Finally, we perform mathematical analysis to evaluate the availability of the proposed protocol.

1 Introduction

The network partitioning problem is often neglected in the previous research on the error recovery of distributed shared memory (DSM) systems [1-3]. The communication network in a DSM system is usually assumed to be immune to failures. However, this assumption is impractical since node or communication link failures may lead to network partitioning. After network partitioning, processing nodes are divided into two or more partitions. Nodes in different partitions can not communicate with each other. Network partitioning in a DSM system will severely reduce the availability of shared pages.

A large body of literature concerning network partitioning exists, most of them dealing with distributed database systems rather than DSM systems. Based on these literature, partition-processing protocols are classified into two categories: optimistic and pessimistic [4]. Optimistic-based protocols allow each node to independently access data by sacrificing the consistency. Pessimistic-based protocols prevent inconsistency by limiting availability.

To allow a partitioned DSM system to continuously function, this paper presents a partition-processing protocol, called *one-copy read and no-copy write* protocol. This protocol permits each partition to independently access shared pages without taking

---

Acknowledgment: This research was supported by the National Science Council, Taiwan, R.O.C., under Grant NSC 85-2221-E002-018.

care of the consistency with other partitions. However, this access manner may introduce the inconsistency problem . We employ the coordinated checkpointing technique to establish recovery lines at appropriate access points. If some shared pages in different partitions are inconsistent, the recovery lines can be used to reconstruct the consistency among the shared pages.

The rest of the paper is organized as follows. Section 2 describes the system model and assumptions made. Section 3 presents the new partition-processing protocol. Section 4 evaluates availability of the proposed protocol. Finally, we give concluding remarks in Section 5.

2 System Model and Assumptions

2.1 Architecture

The distributed shared memory system considered in this paper is shown in Fig. 1. Nodes communicate with each other via a communication network. Each processing node consists of a processor (a CPU and a local memory) and a disk.

2.2 Assumptions

The system is assumed to have N processing nodes. Each node has a connection vector cv to maintain the connection status with other nodes. If node i can communicate with node j, the value of c_{ij} is set to "1"; otherwise, it is set to "0". Failures and repairs on nodes or communication links are instantly recorded in the appropriate connection vectors [5]. Each node is also assumed to be fail-stop [6]; this assumption ensures that the fault in the faulty node will not be propagated to other fault-free nodes. The page consistency is maintained by a fixed distributed manager (FDM) protocol [7].

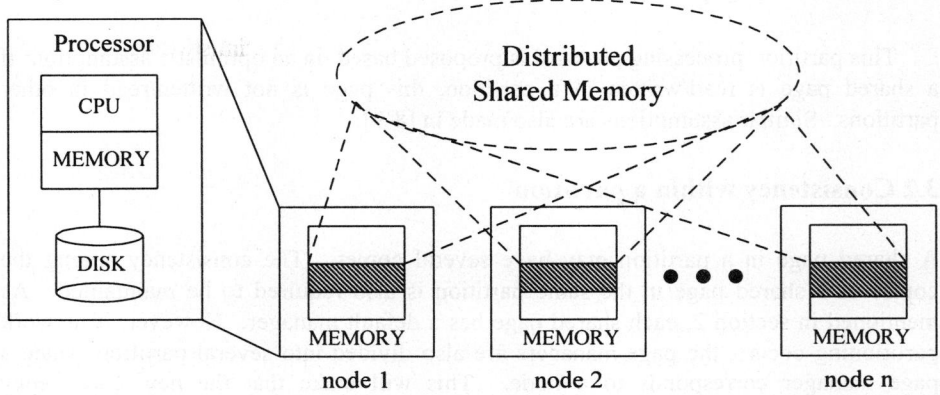

Fig. 1. System architecture.

3 Partition-Processing Protocol

In this section, we will present a new partition-processing protocol to handle the accessibility of shared pages after network partitioning. To achieve high availability, an optimistic assumption is made in our proposed protocol. However an optimistic partition-processing protocol may introduce the problem: inconsistency between partitions. To resolve the inconsistency problem, a memory-based coordinated checkpointing approach is also presented.

3.1 Availability

After network partitioning, some accesses to shared pages can not be executed normally. For example, as shown in Fig. 2, the system is partitioned into two parts: partition 1 and partition 2; partition 1 has two read-only copies of shared-page 1 and one read-only copy of shared-page 2; and partition 2 has one owner copy of shared-page 1 and one owner copy of shared-page 2. When a node in partition 1 issues a read-fault access to shared-page 1, this access can not be completed since the owner copy of shared-page 1 is not stored in partition 1. Similarly, when a node in partition 2 issues a write-fault access to shared-page 2, this access also can not be completed since a read-only copy of shared page 2 in partition 1 can not be notified. In order to allow a partitioned DSM system to continuously function, the processing protocol of shared accesses must be modified. To achieve this goal, we propose a new partition-processing protocol, called the *one-copy read and no-copy write* protocol, described as follows.

- A read access to shared-page p in partitioned i can be performed if there is at least one copy (a read-only copy or an owner copy) of the shared-page p in partition i.
- A write access to shared-page p in partition i is always performed even if none of the shared-page p's copies exist in partition i.

This partition-processing protocol is proposed based on an optimistic assumption: if a shared page is read/written in a partition, this page is not written/read in other partitions. Similar assumptions are also made in [8-9].

3.2 Consistency within a partition

A shared page in a partition may have several copies. The consistency among the copies of a shared page in the same partition is also required to be maintained. As mentioned in section 2, each shared page has a default manager. However, if network partitioning occurs, the page managers are also divided into several partitions since a page manager corresponds to a node. This will make that the new consistency information of some shared pages unable to be kept. For example, in Fig. 2, the manager of shared-page 1 is first assumed to be in partition 1. After network partitioning, if a node in partition 2 writes to shared-page 1, the new consistency

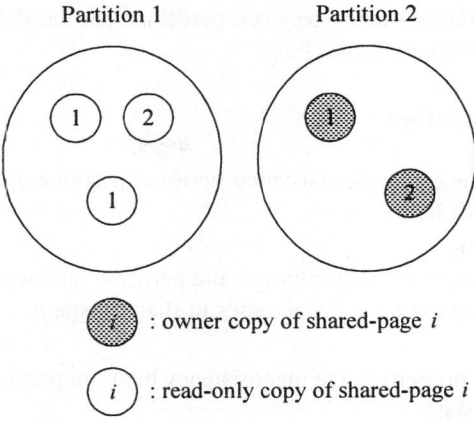

: owner copy of shared-page i

: read-only copy of shared-page i

Fig. 2. A possible partition scenario.

information of shared-page 1 can not be kept in its default manager.

For convenience in presenting how to maintain page consistency within a partition, the centralized manager algorithm is employed here. Each partition specifies a node as the page manager to keep the consistency information of the shared pages stored in it. The page manager maintains a table called *Info* which has an entry for each shared page. Each *Info* entry consists of two fields: *owner* and *copyset*. In partition i, the *owner* in the Info entry of shared-page p (*Info[p].owner*) indicates which node in partition i currently owns the shared-page p. The *copyset* in the entry of shared-page p (*Info[p].copyset*) maintains the identifiers of the nodes in partition i which have a copy of the shared-page p.

3.3 Inconsistency between partitions

The shared-write access in our partition-processing protocol is not prohibited and therefore, the page consistency between partitions is not guaranteed. For example, in Fig. 2, if shared-page 1 is written in partition 1, the owner copy of shared-page 1 in partition 2 can not be notified. Thereafter, if a node in partition 2 issues a read access to shared-page 1, there will be a write-read conflict on the accesses to shared-page 1. This is equivalent to saying that there is an inconsistency with respect to shared-page 1 between partition 1 and partition 2. Based on the above description, if each shared access issued in partition 1 or partition 2 is tracked, the inconsistency between them can be revealed by detecting conflicts on their shared access traces. The shared accesses in partition i can be tracked by the page manager of partition i, as follows. First, the *Info* of partition i's page manager is extended by adding two fields: *read* and *write*. After network partitioning, if shared-page p is first read in a partition, a *read-tracking* message is sent to the corresponding page manager. Then, the page manager makes a "read" mark in the *read* field of the shared-page p's *Info* entry (*Info[p].read*=1). Similarly, if shared-page p is first written in a node, a "write" mark is also made in the write field of the p's *Info* entry (*Info[p].write*=1). Based on the tracing information recorded in *Info*, the inconsistency between partitions can be detected and resolved, which will be discussed in later subsections.

3.4 Inconsistency detection

If shared-page p introduces an inconsistency between partition i_1 and partition i_2, the following properties must hold:

- Shared-page p is accessed by partition i_1 and partition i_2 concurrently.
- At least one conflict exists on the accesses to shared-page p.

Based on the above properties, the inconsistency between partition i_1 and partition i_2 can be detected as follows:

- Find all the shared pages that are concurrently accessed in partition i_1 and partition i_2.
- Verify each concurrently accessed shared page to determine whether there is at least one conflict on the accesses to it.

The concurrently accessed shared pages (concurrent pages) between partition i_1 and partition i_2 can be found when partition i_1 and partition i_2 are reconnected. If partition i_1 and partition i_2 are reconnected due to some node or link failures being repaired, both *Info* tables first need to be merged in order to form a new *Info* table for maintaining the consistency of the new reconnected partition. During merging *Info* tables, if shared-page p is found that it owns an *Info* entry in partition i_1 and partition i_2, this means that shared-page p is concurrently accessed in partition i_1 and partition i_2.

The next step is to determine which concurrent pages introduce inconsistency. This can be done by applying the following conflict-detection function f on each concurrent page.

$$f(Info_{i_1}[p], Info_{i_2}[p]) = (Info_{i_1}[p].read \text{ and } Info_{i_2}[p].write)$$
$$\text{or } (Info_{i_1}[p].write \text{ and } Info_{i_2}[p].read)$$
$$\text{or } (Info_{i_1}[p].write \text{ and } Info_{i_2}[p].write)$$

If function f returns the value "0", it means that there is no conflict on the accesses to shared-page p. In this case, the new *Info* entry of shared-page p can be formed by combining two entries: *Info*$_{i1}[p]$ and *Info*$_{i2}[p]$. Conversely, if function f returns the value "1", there is at least one conflict (read-write conflict, write-read conflict, or write-write conflict) on the accesses to shared-page p. This also means that shared-page p introduces inconsistency between partition i_1 and partition i_2 (shared-page p has different values in partition i_1 and partition i_2). The new Info entry of shared-page p can not be constructed.

3.5 Inconsistency reconstruction

To resolve the inconsistency between partition i_1 and partition i_2, the intuitive idea is to re-establish the consistency of each inconsistent page between partition i_1 and partition i_2. Here, an inconsistent page is the shared page which have different values in different partitions. An inconsistent page may propagate its own inconsistency. For example, in Fig 3, shared-page 1 propagates its inconsistency to shared-page 2 in partition 1 and shared-page 3 in partition 2. Although shared-page 2 and shared-page 3 are not diagnosed as inconsistent pages, the data written in shared-page 2 and shared-page 3 are incorrect. The inconsistency propagation problem needs to be considered for resolving the inconsistency between partitions Therefore, the idea to resolve the inconsistency between partition i_1 and partition i_2 is modified as: finding a consistent state between them. To achieve this goal, we employ the coordinated checkpointing technique to create checkpoints at appropriate access points. Whenever shared-page p is first accessed in partition i_1 (partition i_2), a consistent state with respect to shared-page p in partition i_1 (partition i_2) is also saved. This saved consistent state forms a recovery line with respect to shared-page p in partition i_1 (partition i_2). In the future, if shared-page p is diagnosed as an inconsistent page, all its introduced accesses (first access, second access, and etc.) and propagation in partition i_1 (partition i_2) can be eliminated by rolling back the nodes in partition i_1 (partition i_2) based on the recovery line with respect to shared-page p. Then, the original accesses to shared-page p in partition i_1 (partition i_2) can be re-executed in the new reconnected partition without any conflict since the one-copy serializability in a single partition (the consistency within a partition) is ensured.

Here, the checkpoints generated for resolving the inconsistencies can be saved in the main memory since they are not used to handle the node failures. If the memory space allocated for checkpointing is restricted, the concept of incremental memory-based checkpointing approach [10] can be applied here. This checkpointing approach is a way of increasing the concurrency of checkpointing.

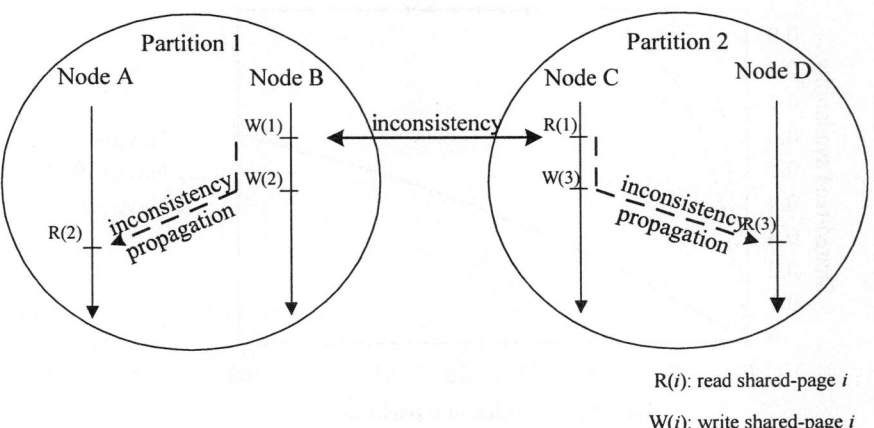

$R(i)$: read shared-page i

$W(i)$: write shared-page i

Fig. 3. An example inconsistency propagation.

4 Evaluation

In this section, we will evaluate that how high the availability can be achieved by the proposed partition-processing protocol. Two availability metrics, shared-read and shared-write, are evaluated based on the mathematical analysis. Before performing the mathematical analysis, we first assume that the system has N nodes, and the mean number of copies for each shared page is N_{avg}.

Lemma 1: The availability P_r that a read access to shared-page p is allowed to be executed in a partition i with n nodes is $1-(1-N_{avg}/N)^n$

Proof: A read access to shared-page p is allowed to be executed in partition i if there is at least one copy of shared-page p in partition i. Based on the N_{avg}, the probability that a node has a copy of shared-page p can be represented as N_{avg}/N. By further inferring, it is easy to see that $P_r = 1-(1-N_{avg}/N)^n$. □

Fig. 4 shows the shared-read availability achieved by the proposed partition-processing protocol, for $N=64$, $N_{avg}=1$, 10, and 20. As shown in Fig. 4, the curve of P_r approaches to the horizontal line with the value of 1 as N_{avg} increases. For example, when $N_{avg}= 10$, P_r increases rapidly if the number of nodes in a partition is smaller than 9 and then monotonically increases to the value of 1.

Lemma 2: The availability P_w that a write access to shared-page p is allowed to be executed in partition i with n nodes is 1.

Proof: Based on our protocol, a shared write operation is always allowed to be executed in any partition. Hence, the P_w is equal to 1. This probability is independent of the number of nodes in a partition and the mean number of copies for a shared page. □

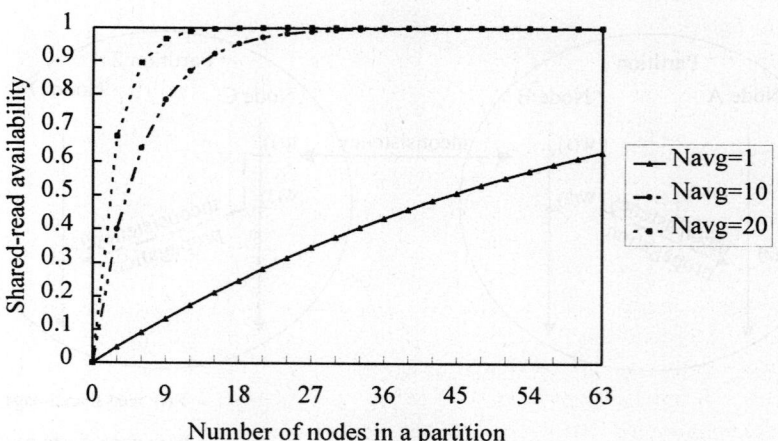

Fig. 4. Shared-read availability with different N_{avg} values.

5 Conclusions

A new partition-processing protocol has been proposed in this paper to achieve high availability of the shared pages among partitions. The proposed protocol allow shared pages in a partition can be optimistically accessed without considering the consistency without other partitions. However, this access method may introduce the inconsistency problem. To resolve the inconsistency problem, we utilize the coordinated checkpointing technique to set up consistent recovery lines at appropriate accessed points. If there is inconsistency between partition i and partition j, the recovery lines can be used to determine a consistent state between partition i and partition j. All nodes in partition i and partition j are rolled back to the determined consistent state. In addition, high availability achieved by the proposed protocol was also presented based on the mathematical analysis.

References

1. K. L. Wu and W. Fuchs, "Recoverable Distributed Shared Virtual Memory," *IEEE Trans. Comput.*, vol. 39, no. 4, pp. 460-469, April 1990.
2. G. Janakiraman and Y. Tamir, "Coordinated Checkpointing-Rollback Error Recovery for Distributed Shared Memory Multicomputers," *Proc. 24th Int'l. Symp. on Reliable Distributed Systems*, pp. 42-51, 1994.
3. G. Suri, B. Janssens and W. Kent Fuchs, "Reduced Overhead Logging for Rollback Recovery in Distributed Shared Memory," *Proc. 25th Int'l. Symp. on Fault-Tolerant Computing*, pp. 279-288, June 1995.
4. S. B. Davidson, H. Garcia-Molina, and D. Skeen, "Consistency in partitioned networks," ACM Computing Surv., vol. 17, no. 3, pp. 341-370. Sep. 1985.
5. S. Jajodia and D. Mutchler, "A Pessimistic Consistency Control Algorithm for Replicated Files which Achieves High Availability," *IEEE Trans. Software Eng.*, vol 15, pp. 39-45, Jan. 1989.
6. R. D. Schlichting and F. B. Schneider, "Fail-Stop Processors: An Approach to Designing Fault-Tolerant Computing Systems," *ACM Trans. Comput. Syst.*, vol. 1, pp. 222-238, August 1983.
7. K. Li and P. Hudak, "Memory Coherence in Shared Virtual Memory Systems," *ACM Trans. Comput. Syst.*, vol. 7, no. 4, pp. 321-359, Nov. 1989.
8. C. D. Tait and D. Duchamp, "Service InterFace and Replica Management Algorithm for Mobile File System Clients," *Proc. 1st Int'l. Conf. Parallel and Distributed Inform. Syst.*, pp. 190-197, 1991
9. P. Triantafillou and D. J. Taylor, "Multiclass Replicated Data Management: Exploiting Replication to Improve Efficiency," *IEEE Trans. on Parallel and Dist. Systems*, vol. 5, no. 2, pp. 121-138, Feb. 1994.
10. K. L. Jeffrey, F. Naughton, and J. S. Plank, "Low-Latency, Concurrent Checkpointing for Parallel Programs," *IEEE Trans. on Parallel and Dist. Systems*, vol. 5, no. 8, pp. 874-879, August 1994.

I/O Data Mapping in *ParFiSys*: Support for High-Performance I/O in Parallel and Distributed Systems

J. Carretero and F.Pérez and P. de Miguel and F. García and L. Alonso

Universidad Politécnica de Madrid (UPM)
e-28660 Madrid, España, E-mail: jcarrete@fi.upm.es

Abstract. This paper gives an overview of the I/O data mapping mechanisms of *ParFiSys*. Grouped management and parallelization are presented as relevant features. I/O data mapping mechanisms of *ParFiSys*, including all levels of the hierarchy, are described in this paper.

1 Introduction

MPPs, distributed memory systems with a high number of processors, provide a new I/O system, relying on parallel hardware. Their I/O subsystem architecture, which is distributed in nature, usually consists on several independent I/O nodes supporting one or more secondary storage devices. Using parallel I/O systems and parallel file systems seems to be a good approach to take advantage of the inherent parallelism of the MPP, as shown in some parallel file systems as Vesta [3]. In this paper we give an overview of some aspects of *ParFiSys* [2], a parallel file system developed at the UPM to provide I/O services for the GPMIMD machine, an MPP developed within the ESPRIT program P-5404. The main design goals of *ParFiSys* were to provide I/O services to scientific applications requiring high I/O bandwidth, to minimize application porting effort, and to exploit the parallelism of generic message-passing multicomputers, including processing nodes (PNs) and I/O nodes (IONs). To fully exploit all the parallel features of the I/O hardware, the architecture of *ParFiSys* is clearly divided in two levels: file services and block services. The first level is comprised into a component named *ParClient*. The second architectural level, named *ParServer* and located at the ION, deals with logical block requests interacting directly with the I/O devices located on its own ION.

2 Data Distribution in *ParFiSys*

To enhance flexibility, *ParFiSys* mapping is based on a very generic **distributed partition** represented as the tuple $(\{NODE_n\}, \{CTLR_c\}_n, \{DEV_d\}_{nc})$, which describes the set of I/O nodes, controllers per node, and devices per controller of the partition. The current implementation of *ParFiSys* supports three kinds of predefined file systems (figure 1): UNIX-like, distributed extended, and distributed cyclic. To reduce contention in ParServers and to increase fault tolerance, the distribution is applied to data and metadata of the file system. A distributed file system has a replicated superblock, and distributed bitmaps and i-node blocks. Having replicated superblocks allows *ParFiSys* to *mount* file systems with some devices damaged. Distributing bitmaps and i-nodes

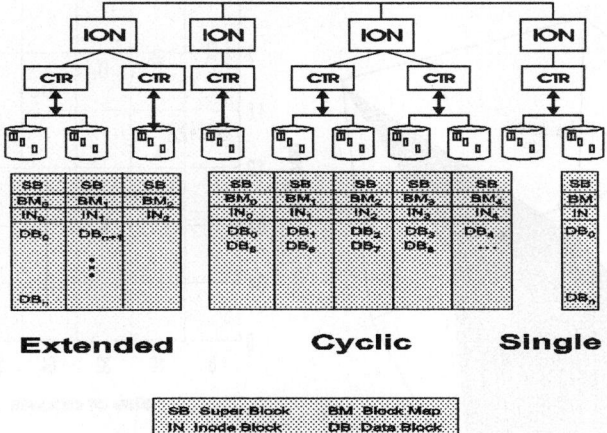

Fig. 1. *ParFiSys* Distributed File Systems

increase *ParFiSys* performance by balancing the load due to metadata management among all the ParServers of the file system.

3 Data Mapping in *ParFiSys*

Usually the user vision of the file is a byte stream, whereas the file system vision is a set of scattered logical blocks. Moreover, in parallel and distributed file systems, the blocks may be spread out among several ION and devices [1]. Thus, a parallel file system as *ParFiSys* must be able to establish some correspondence between the user and the physical image of data. To satisfy user I/O requests, each ParClient and ParServer must have some knowledge of where the data corresponding to the user image are located. Mapping functions have to be established from the user data structure to the relevant ION, controllers, devices, and disk blocks. The approach suggested for the NCube's I/O software [4] has been followed in *ParFiSys*, where data mapping is solved stepping through four correspondences: File Block to Byte (FBB), File Block to I/O Node (FBION), File Block to Controller (FBC), and File Block to Device (FBD). The two uppermost levels (FBB and FBION) are accomplished at the ParClients. The two lowermost levels are provided by the ParServers. A complete translation to *write* data is defined as $FBB^{-1} \; o \; (FBION \; o \; FBC \; o \; FBD)$.

FBB is only related to the file image used in *ParFiSys*, a UNIX-like image viewed as a string of bytes. High level I/O operations executed in each ParClient (read, write, ...) translate user I/O requests, defined in bytes, to file system block ones. The block-based approach, used on most file systems, has several drawbacks in high-performance I/O systems, where I/O requests are usually large: cache management cost is proportional to the number of blocks, there is no support to manage several user requests as a single one, and a remote access to a ParServer may be required for each requested block. To enhance FBB correspondence the whole user buffer is mapped to blocks on a single operation in *ParFiSys*, which highly reduces the number of accesses to indirect blocks. In block-based file systems, the number of accesses (na) is linear to the depth

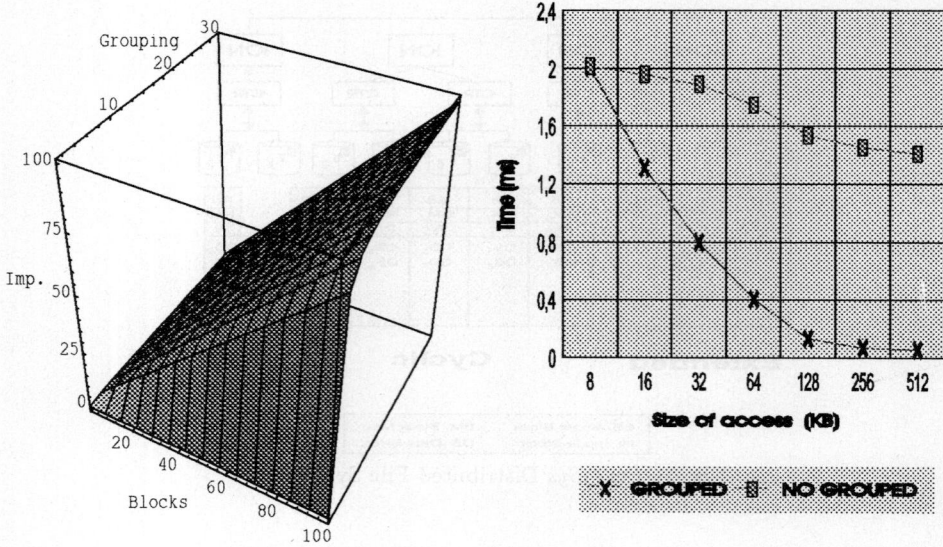

Fig. 2. FBB Improvement in ParFiSys

of the indirection level used (single, double, triple). The FBB correspondence used in *ParFiSys* allows to obtain a lists of n blocks corresponding to the whole user request with a single mapping operation. To avoid problems due to greedy processes executing many large I/O requests, a uppermost limit (*FBB factor*) is defined for the size of the user buffer that can be mapped on a single operation. It is computed using a configuration parameter (*FBB maximum*) affected by a weight factor, which depends on the process I/O behavior. The number of accesses (na) is now reduced proportionally to the *FBB factor* (k), as reflected below:

$$na = \sum_{j=1}^{i} (\frac{n}{k^j} + 1) * j \qquad (1)$$

where n is the number of blocks, and i is the indirection level. The improvement achieved with the *ParFiSys* approach is shown in figure 2, crossed section reflects the percent ratio in number of accesses to indirect blocks required to map a user buffer with several grouping factors. Even when only the first indirection level is shown in the figure, the number of accesses avoided is almost the 80% for a FBB factor of 5 blocks. To improve accesses smaller than the former threshold, 5 blocks are always premapped.

Minimizing indirect block accesses highly reduces the number of operations required to execute FBB mapping and the number of operations requested to the ParServer, which increases scalability. The improvement of the mapping time when using the *ParFiSys* FBB correspondence compared with a traditional FBB correspondence is shown in figure 2. This figure shows the experimental results obtained by reading a 10 MBytes file, using 8 Kbyte blocks, with request size varying from 8 KBytes to 512 KBytes. The FBB factor is equal to the request size. As can be seen, the improvement in time is almost exponential for *ParFiSys*, which is highly affected by the request size. For accesses larger than 128 KBytes, the improvement is very small, so it can be considered as a *maximum FBB factor*. A similar test to evaluate write operations shown an analogous behavior.

FBION, FBC, and FBD depends on the distribution strategy defined by the user or the file system to map data on I/O devices. The FBION correspondence must determine the relation between a file block x and the node where it is located. Establishing the former correspondences for extended and cyclic partitions having an non uniform layout of controllers and devices is heavy. To enhance the former computations, some information about the distributed partition and file system is computed when the file system is created and stored in the superblock. A *symmetrical* layout which greatly improves the mapping functions, as showm below for a cyclic partition with c controllers, d devices, and s blocks per device, are:

$$FBION = \frac{x \bmod pd}{cd} \qquad (2)$$

$$FBC = \frac{(x \bmod pd) \bmod cd}{d} \qquad (3)$$

$$FBD = ((x \bmod pd) \bmod cd) \bmod d \qquad (4)$$

where $pd = n*c*d$ is the number of devices in the partition, and $cd = c*d$ is the number of devices per ION.

4 Conclusion

The concurrent architecture of *ParFiSys* provides high concurrency and parallelism in ParClients and ParServers, reducing the latency of services and increasing the overall performance. Multi-level mapping is a very flexible and scalable mechanism to connect the user data image with the file system internal image. The correspondence used to map user space to file blocks provides a very good performance because of grouped algorithms and parallelization. Data distribution algorithms have shown a high flexibility to map data onto different kind of file systems, becuase of the generality of the partition structure. Finally, it is very important to remark that users may define data mapping functions to be used by *ParFiSys*, even when three mapping strategies are predefined internally.

References

1. R. Bordawekar, J. Rosario, and A. Choudhary. Design and Evaluation of Primitives for Parallel I/O. In *Supercomputing 93*, pages 452–462. IEEE, 1993.
2. J. Carretero, F. Pérez, F. De Miguel, F. García, and L. Alonso. ParFiSys: A Parallel File System for MPP. *ACM SIGOPS*, 30(2):74–80, April 1996.
3. P. Corbett, S. Johnson, and D. Feitelson. Overview of the Vesta Parallel File System. *ACM Computer Architecture News*, 21(5):7–15, December 1993.
4. E. DeBenedictis and J. M. del Rosario. nCUBE Parallel I/O Software. In *Eleventh Annual IEEE International Phoenix Conference on Computers and Communications (IPCCC)*, pages 117–124, April 1992.

Correctness Proof for a Distributed Memory System

Vicente Cholvi-Juan[1] and José M. Bernabéu-Aubán[2]

[1] University Jaume I
Castelló (Spain)
[2] Polytechnic University of Valencia
Valencia (Spain)

Abstract. Distributed shared memory systems offer the possibility of implementing parallel algorithms using the familiar operations of shared memory access. In this paper we present the proof of correctness of an algorithm implementing such a shared memory system over a distributed environment. Such proof is based in the formal model for memory systems introduced by the authors in previous works.

1 Introduction

A typical (loosely coupled) distributed systems is composed of a collection of independent computers interconnected through some type of network. In order to cooperate, applications written to span several computers on such a system need to have some mechanism to allow each one of their parts to exchange information.

The *shared memory model* (SMM) provides a shared address space which can be used by processes in the same way as local memory, even if they are executed concurrently in different processors. Thus, every process can access any address by means of the basic operations $data = read(address)$ and $write(address, data)$, where $read$ returns the $data$ in $address$, and $write$ associates $data$ with $address$.

When a SMM is built on top of a distributed system, we get what is known as a DSMM. In this paper we present the formal proof of correctness of a DSMM algorithm which implements a particular shared memory model. This model is a mixture of the *atomic* memory model [3] and the *sequential* model [4], thus presenting the developer with a familiar memory access semantics, while allowing a more efficient implementation. The proof is carried out using the formal model introduced by the authors in [1]. We base our framework on the I/O automata formalism [5] for discrete event systems. The I/O automata formalism has been successfully used to prove the correctness of algorithms as well as to specify properties of memory systems and transactional systems.

2 Correctness Proof

It is well known the inherent complexity of distributed algorithms. Indeed, there are several widely known algorithms that were considered correctly designed and that later have been found incorrect. Therefore, it is very important to be able to ensure that implementations of models work correctly.

Most of the work on shared models has not paid enough attention to the way such models are specified. The availability of an adequate formal framework makes it possible to answer questions about models and algorithms implementing them. In another work [2], we have introduced such a formal framework. Here, we show, basing on the specifications of two given models, how to we verify the correctness of a proposed DSMM implementing algorithm.

Our framework takes into account two type of operations: write(i, x, u) which associates the value u to the variable x, and read(i, x, u) which informs of the value u associated with the variable x (being i the process that executes the operation). The way these operations are executed are described as actions of that formalism. A write(i, x, u) operation can be identified as a pair of bwrite(i, x, u) − ewrite(i, x) events, and a read(i, x, u) operation can be identified as a pair of bread(i, x) − eread(i, x, u) events (the first to begin the operation and the second to finish it).

In our formalism to specify a given memory model it is enough to specify the set of executions it allows. Moreover, the sequences of the model that we are going to implement (a mixture of the atomic and sequential) are characterized by the existence of atomics and sequential "views" of those sequences. Formally, from [2], "a sequence α is *sequential* if $VIEWS(SEQ, \alpha)$ is not an empty set (ditto for the *atomic* sequences)".

The proposed DSMM implementing algorithm permits to take advantage of the larger degree of parallelism given by the sequential model, allowing at the same time the use of a stricter model (atomic) for a subset of the variables.

The way in which the DSMM implementing algorithm deals with atomic variables consists of invalidating all the replicas previously to each writing. Thus, each variable in isolation will be atomic [3]. Thereby, if we consider that the atomic model is compositional, we can assume (in the verification) that all the variables are sequential without losing generality.

Our solution contains three system's descriptions: OpSpec, the operational requirement specification; SysImpl, the hight level system implementation; and DiscSysImpl, the discrete system implementation.

The operational specification describes the system by using a single automaton. Next, we split it into two automata: one of them models the architectural platform and the other the DSMM implementing algorithm. Finally, we split the later automaton into the automata that model its distributed implementation.

Here, we present only those parts we consider necessary to understand the verification process. The details of the proofs are available in [2]. In what follows, $\mathcal{P}$ will stand for the set of processes, $\mathcal{M}$ for the set of variables (being $\mathcal{M}_0$ the initial values) and $msig(\mathcal{P}, \mathcal{M})$ for the set of memory operations.

2.1 Operational Specification

The operational specification, OpSpec, consists of a single automaton, $BLACK(\mathcal{P}, \mathcal{M}, \mathcal{M}_0)$ (Fig. 1), which models all required properties in order to solve the sequential model. Moreover, it incorporates the fairness condition that guarantees liveness of the implementation. The state transitions are described by specifying the "preconditions" under which each action can occur and the "effect" of each action. This constitutes an abstract representation harder to understand than the axiomatic specification of the sequential model. However, it is easier to use in the proofs.

Actions:
- Inputs: bread(i, x), bwrite(i, x, u)
- Outputs: eread(i, x, u), ewrite(i, x)

where $i \in \mathcal{P}$, $x \in \mathcal{M}$ and $u \in range(x)$

State: a set of elements, S, in $msig(\mathcal{P}, \mathcal{M})$, initially empty

Transitions:
- bwrite(i, x, u)

  Effect:
  $S = S \cdot$ bwrite(i, x, u)
- bread(i, x)

  Effect:
  $S = S \cdot$ bread(i, x)

- ewrite(i, x)

  Precondition:
  $last(S \mid i) =$ bwrite$(i, x, \cdot)$

  Effect:
  $S = S \cdot$ ewrite(i, x)
- eread(i, x, u)

  Precondition:
  $last(S \mid i) =$ bread(i, x)
  $VIEW(SEQ, T) \neq \emptyset$
  (where $T = S \cdot$ eread(i, x, u))

  Effect:
  $S = S \cdot$ eread(i, x, u)

Partition: $\{\{$ewrite$(i, \cdot)\}, \{$eread$(i, \cdot, \cdot)\}\}$

Fig. 1. Specification of $BLACK(\mathcal{P}, \mathcal{M}, \mathcal{M}_0)$.

2.2 System Implementation

To implement a model, memory operations have to be translated into operations of the underlying system. The nature of such operations depends on the architecture of the system on which the model is to be implemented. The architecture gives us only part of the full implementation of the model. The other part is the algorithm which, based on the architecture, implements the model (*MMS*).

In this section we provide the specification of the components that compose the system by using two automata. The first implements the architecture and the other one the MMS.

We consider a distributed architecture consisting of a set of locations $\mathcal{D}$ connected by a fault-free, ordered communication channel. We identify the set of variables at each location as $\mathcal{W}$, being $\mathcal{W}_0$ their initial values. Then, the architecture is globally modeled by using a single automaton $DST(\mathcal{D}, \mathcal{W}, \mathcal{W}_0)$

(Fig. 2). Roughly speaking, each memory operation is modeled by using three actions (the first to start it, the second to "physically perform" it and the later to finish it). Moreover, data at each location is modeled by an array of storing values. Accesses to those data are made through three FIFO operations: $HEAD$ returns the queue's bottom value, DEQ dequeues that value and ENQ puts the value on the top of the queue.

Actions:
- Inputs: $br(j,y)$, $bw(j,y,v)$
- Outputs: $er(j,y,v)$, $ew(j,y)$
- Internals: $xr(j,y,v)$, $xw(j,y,v)$

where $j \in \mathcal{D}$, $y \in \mathcal{W}$ and $v \in range(y)$

State:
- $F_{in}[j]$, $F_{out}[j]$: array of FIFO queues of elements in $in(sig(DST(\mathcal{D}, \mathcal{W}, \mathcal{W}_0)))$ and $out(sig(DST(\mathcal{D}, \mathcal{W}, \mathcal{W}_0)))$ respectively, initially empties
- Store[j,y]: array of variables in the range of y, initially y_0

Transitions:

- $bw(j,y,v)$
  Effect:
  $ENQ(F_{in}[j], bw(j,y,v))$

- $ew(j,y)$
  Precondition:
  $HEAD(F_{out}[j]) = ew(j,y)$
  Effect:
  $DEQ(F_{out}[y])$

- $xw(j,y,v)$
  Precondition:
  $HEAD(F_{in}[j]) = bw(j,y,v)$
  Effect:
  $Store[j,y] = v$
  $DEQ(F_{in}[y])$
  $ENQ(F_{out}[y], ew(j,y))$

- $br(j,y)$
  Effect:
  $ENQ(F_{in}[j], br(j,y))$

- $er(j,y,v)$
  Precondition:
  $HEAD(F_{out}[j]) = er(j,y,v)$
  Effect:
  $DEQ(F_{out}[j])$

- $xr(j,y,v)$
  Precondition:
  $HEAD(F_{in}[y]) = br(j,y)$
  $v = Store[j,y]$
  Effect:
  $DEQ(F_{in}[j])$
  $ENQ(F_{out}[j], er(j,y,v))$

Partition: $\{\{ew(j,\cdot)\}, \{er(j,\cdot,\cdot)\}, \{xw(j,\cdot,\cdot)\}, \{xr(j,\cdot,\cdot)\}\}$

Fig. 2. Specification of $DST(\mathcal{D}, \mathcal{W}, \mathcal{W}_0)$.

The MMS is modeled by using a single automaton $RED(\mathcal{P}, \mathcal{M})$ (Fig. 3). That continues being a rather abstract representation of the system which is still far away from a realistic representation of the proposed distributed memory system. However, it is intended to be used in the discrete system specification correctness proof.

At this stage it has to be proved that the interaction of the MMS with the architecture provides sequential executions.

Lemma 1. *Let* SysImpl $= hide_\Sigma(RED(\mathcal{P}, \mathcal{M}) \varPi DST(\mathcal{D}, \mathcal{W}, \mathcal{W}_0))$ *(where* $\Sigma = extsig(DST(\mathcal{D}, \mathcal{W}, \mathcal{W}_0))$*). Then* SysImpl *solves* OpSpec.

Actions:
- Inputs: bread(i, x), bwrite(i, x, u), er(j, y, v), ew(j, y)
- Outputs: eread(i, x, u), ewrite(i, x), br(j, y), bw(j, y, v)

where $i \in \mathcal{P}$, $x \in \mathcal{M}$, $u \in range(x)$, $j \in \mathcal{D}$, $y \in \mathcal{W}$ and $v \in range(y)$

State: a set of elements, S, in $sig(RED(\mathcal{P}, \mathcal{M}))$, initially empty

Transitions:
- bwrite(i, x, u)

  Effect:
  $S = S \cdot$ bwrite(i, x, u)
- bread(i, x)

  Effect:
  $S = S \cdot$ bread(i, x)
- bw(j, y, v)

  Precondition:
  $last(S \mid j) =$ bwrite(j, y, v)
  Effect:
  $S = S \cdot$ bw(j, y, v)
- br(j, y)

  Precondition:
  $last(S \mid j) =$ bread(j, y)
  Effect:
  $S = S \cdot$ br(j, y)

- ewrite(i, x)

  Precondition:
  $last(S \mid i) =$ ew(i, x)
  Effect:
  $S = S \cdot$ ewrite(i, x)
- eread(i, x, u)

  Precondition:
  $last(S \mid i) =$ er(i, x, u)
  $VIEW(SEQ, T) \neq \emptyset$
  (where $T = (S \cdot$ eread$(i, x, u)) \mid msig(\mathcal{P}, \mathcal{M}))$
  Effect:
  $S = S \cdot$ eread(i, x, u)
- ew(j, y)

  Effect:
  $S = S \cdot$ ew(j, y)
- er(j, y, v)

  Effect:
  $S = S \cdot$ er(j, y, v)

Partition: $\{\{$eread$(i, \cdot, \cdot)\}, \{$ewrite$(i, \cdot)\}, \{$br$(j, \cdot)\}, \{$bw$(j, \cdot, \cdot)\}\}$

Fig. 3. Specification of $RED(\mathcal{P}, \mathcal{M})$.

2.3 Discrete System Implementation

In the previous sections we have presented two levels of description of our solution, which, by their very nature, are quite abstract and do not correspond directly with an implementation over a distributed system.

We now proceed to describe the discrete system model of the algorithm. In this model, the distributed nature of the underlying system is made explicit by using FIFO communication channels and explicit message passing between nodes in the distributed system.

The communications system is modeled by an automaton $FIFO_M$, which, for our purposes, is in charge of sending and making receive the messages used by our protocol.

To obtain DiscSysImpl we compose the automata that will model the modules responsible of requesting variables as well as providing them ($\{WRKR_i(\mathcal{P}, \mathcal{M})\}_{i \in \mathcal{P}}$) with the module taking care of arbitrating conflicting requests for variables, ordering them and keeping track of circulating variables ($SEQ(\mathcal{P}, \mathcal{M})$) and that which models the *communication channel* module ($FIFO_M$).

At this stage it has to be proved that the interaction of the automata that model the discrete system implementation can be safely used for any task for which $RED(\mathcal{P}, \mathcal{M})$ is satisfactory.

Lemma 2.
Let DiscSysImpl = $hide_\Sigma(FIFO_M \;\Pi\; SEQ(\mathcal{P}, \mathcal{M}) \;\Pi\; \{WRKR_i(\mathcal{P}, \mathcal{M})\}_{i \in \mathcal{P}})$ (where $\Sigma = extsig(FIFO_M)$). Then DiscSysImpl solves $RED(\mathcal{P}, \mathcal{M})$.

3 Conclusion

In this paper we have shown how to model, using our formalization of memory coherency models, a particular algorithm implementing a distributed shared memory system. First we modeled the problem using an I/O automata. The possible execution sequences of this automata define the "problem" we claim our algorithm "solves". Then we propose a more detailed I/O automaton which is shown to solve the original problem. This automaton, while still abstract, and not making use of the distributed nature of the underlying system, contains some details which facilitate its further break down into a fully distributed specification.

The method used in this paper would be appropriate in general to prove the correctness of any other MMS implementation. In our approach we use the full power of the I/O automata formalism. A similar approach can also be used to prove the correctness of algorithms built upon any particular memory model.

References

1. Bernabéu-Aubán, J.M., Cholvi-Juan, V.: Formalizing memory coherency models. Journal of Computing and Information **1** (1994) 653–672
2. Cholvi-Juan, V.: Formalizing memory models. PhD thesis, Department of Computer Science, Polythechnic University of Valencia (1994)
3. Herlihy, M.P., Wing, J.M.: Linearizability: A correctness condition for concurrent objects **12** 3 (1990) 463–492
4. Lamport, L.: How to make a multiprocessor computer that correctly executes multiprocess programs, IEEE Transactions on Computers **28** 9 (1979) 690–691
5. Lynch, N.: I/O Automata: A model for discrete event system. Massachusetts Institute of Technology, Technical Report MIT/LCS/TM-351 (1988)

Distributed Shared Memory Based on Group Large Causality*

José M. Piquer

DCC - Universidad de Chile, Casilla 2777, Santiago, Chile
jpiquer@dcc.uchile.cl

Abstract. The implementation of Distributed Shared Memory (DSM) must be simple and efficient to become a general purpose tool. In general, known DSM systems use page-oriented virtual memory systems to simulate the shared address space. The usual implementation uses vector clocks or history tuples associated to pages, being very expensive and impossible to scale to large systems.
In this paper, we show a novel approach to this implementation problem, proposing an algorithm based on a relaxed causal ordering of multicasts and messages (called Group Large Causality) to provide Coherent Causal Consistency DSM in large-scale networks.

Keywords: Distributed Shared Memory, Causality, Replica Coherency

1 Introduction

In distributed systems, programming parallel applications can be very complex if only message-passing is provided. Distributed Shared Memory (DSM)[9] is a widely accepted paradigm for this environment. DSM simulates a global address space in a distributed system. The programmer can avoid message passing, or use both paradigms at the same time, easing the parallel programming task.

Most DSM systems work on virtual memory pages, but our work concentrates more on complete objects. This allow to have type information on objects, and act differently when accessing them, or sending them (eg. to avoid architecture difference problems). Eventhough, most of the discussion and algorithms can be applied both to pages and complete objects.

For efficiency reasons, local copies of the shared objects are kept on many processors. This replication introduces the problem of consistency. Causal Consistency[6] is generally accepted as a good tradeoff between efficiency and similarity with real shared memory. Many programs written for shared memory run with no modifications under Causal Consistency[1]. Another important point is to serialize write operations on the same object, which is called Coherence[4] (thus all processors see the modifications in the same order). This is called Coherent Causal Consistency DSM.

---

* This work has been partially funded by FONDECYT project 1950599.

On the other hand, DSM implementations[5, 4, 7, 3] remain complex and difficult to scale to large distributed networks (as the Internet). We strongly believe that the DSM model can be applied to large distributed computations if the implementation is efficient enough, and shared objects are grouped into shared memory segments (groups of objects) accessed by some processors only. So, a shared object belongs to a processor group. To have read or write access to it, a processor must join the group. Other processors can have references to the object, but they cannot act directly on it (read or write) and they will never have replica of it. Of course, message passing and remote procedure call are also supported, mixed with DSM.

Group Large Causality is an extension of Large Causality[11] to support efficient causal multicasts and point to point messages inside groups. In this paper we try to show that with Group Large Causality it is easy and straightforward to implement DSM.

The organization of this paper is as follows: Section 2 describes Group Large Causality, section 3 presents the DSM implementation and section 4 the conclusions.

2 Group Large Causality

In our model, a processor can belong to one or more groups of processors, it can send point-to-point messages to any other processor and it can multicast a message to any group it belongs to.

The idea is that causality will be preserved inside groups, but not between them. To represent this, we define a new "happened before"[8] relation, inside a group. Given a group g, the new relation is noted $\rightarrow^g$, which can be read "happened before inside g".

We note $\rightarrow^l$ the causal relation between two events at the same processor (local causality). $send_p(m, q)$ denotes that processor p sends message m to processor q and $mcast_p(m, g)$ that processor p sends a multicast m to group g. $deliver_p(m)$ denotes delivery of message m at processor p.

The $\rightarrow^g$ relation is defined by:

1. if a and b are two events at processor p, and $a \rightarrow^l b$, then $\forall g, p \in g, a \rightarrow^g b$.
2. if $p, q \in g$, $send_p(m, q) \rightarrow^g deliver_q(m)$.
3. $mcast_p(m, g) \rightarrow^g deliver_q(m), \forall q \in g$.
4. if $a \rightarrow^g b$ and $b \rightarrow^g c$, then $a \rightarrow^g c$.

Using this relation, we can now define the Group Large Causality delivery rules:

GLC1 Causal Group Multicast
if $mcast(m_1, g) \rightarrow^g mcast(m_2, g')$, then $deliver_q(m_1) \rightarrow^l deliver_q(m_2)$ at every site $q \in g \cap g'$.

GLC2 Multicast to Message
if $mcast(m_1, g) \rightarrow^g send(m_2, q)$ and $q \in g$,
then $deliver_q(m_1) \rightarrow^l deliver_q(m_2)$.
GLC3 Message to Multicast
if $send(m_1, q) \rightarrow^l mcast(m_2, g)$, then $deliver_q(m_1) \rightarrow^l deliver_q(m_2)$.
(Note that this is only valid when m_1 and m_2 are emitted from the same site.)

Group Large Causality can be implemented using one vector clock per group, with one component per processor member. On FIFO systems, the timestamps used at processor p are of size $O(gs)$, with g being the number of groups in which p participates, and s the average size of the groups. The main interest is that it is reasonable to suppose that $g << N$ and $s << N$ (N being the total number of hosts in the network), in other words, that each processor participates in only a few groups, and that each group is small, compared to the whole system. In fact this is the main motivation towards group-based primitives in large systems. We have measured an implementation of Group Large Causality, confirming that the execution time overhead is proportional to g and to s.

3 DSM based on Group Large Causality

Each time a shared object is created, a group is assigned to it (it can be a new group or an existing one). To share the object, a processor must join this group. Two basic operations can be performed on shared objects: read and write. Also, a send operation is provided, which allows to send a reference to the object to other processors. In general, we use one group to manage many related objects (a shared pool), but each object must belong to only one group. These primitives are used by the underlying system, the programmer is not aware of them, just using the references to objects as usual.

Processors inside the group can have references to shared objects, with a local cache to hold replica values. The object reference consists of the globally unique object identifier (obj_id), the group it belongs to (group) and a status field (status) to mark the cache as VALID or INVALID.

Processors outside the group can have references, but they cannot access their values (though they never receive replica).

The object itself has a value. To serialize the write accesses, the object also has an owner which is the only processor able to modify it. To keep trace of the write requests, the object has a write_queue where all the processors trying to become its owner are noted. The owner field is a boolean, and is only true in the owner's processor.

3.1 The read Protocol

When a read access is performed on a reference to an object, we look if the status field is VALID to return the cache contents. If it is INVALID, we multicast to the **group** a `READ` request (with the **obj_id**), and wait for a reply. (Note that many replies may be received, as noted in section 3.2.) The reply contains the object's **value**, which is copied to the cache and the **status** is changed to `VALID`.

Upon reception of a `READ` request, we check if we are the object's owner, and if yes, we reply to the sender the **value**. Otherwise, we ignore it. As we use a multicast for reading, there is no need to manage an owner hint in the object.

3.2 The write Protocol

Here we implement a Token Passing[2] algorithm to serialize the accesses, and the Token holder is the object's owner.

When a **write** access is performed on a reference to an object, we look if we are the owner. If yes, we multicast to the **group** an `INVALIDATE` message (with the **obj_id**). Otherwise, we multicast a `WRITE` message, asking to become the object's owner. There we block until our ownership is granted (and we enqueue every `READ` request received during this period for later processing). Afterwards, we modify the object's **value**.

Upon reception of an `INVALIDATE` message, we set **status** to `INVALID`.

Upon reception of a `WRITE` message, if we are the object's owner, we add the sender to the tail of **write_queue** (if it was not already in the queue). When we want to give up our ownership (and the queue is not empty), we extract the first **write_queue** element to determine the new owner, and we multicast a `MIGRATE` message (with the **obj_id** and the new owner processor). Afterwards, we send the object's **value** and the **write_queue** to the new owner in a point-to-point message. Finally, we set **status** to `INVALID` and **owner** to false.

Upon reception of a `MIGRATE` message, we compare the new owner to our identifier. If we are not the new owner, we set **status** to `INVALID`. Otherwise, we wait for a message from the sender containing the **value** and **write_queue** fields. Upon reception, we update both fields and we set **status** to `VALID` and **owner** to true. Then, we can resume all **write** operations blocked on this object, and we can process the `READ` requests received during the blocked period. These old requests may have been already answered by a previous owner, but it is not sure. This is why the `READ` multicast may have more than a reply. However, the replies will reach in order (interleaved with `INVALIDATE` messages) the requester, so the more recent value will be the last.

3.3 The send Protocol

A processor holding a reference to a remote object, can send this reference in messages or write it on another shared object. The reference consist of the **obj_id** and the **group** field. If the receiving processor participates in the group, it will create an object with an `INVALID` cache and **owner** set to false. Afterwards, it can read or write the object as usual.

If the receiving processor does not participate in the group, it can have a pointer to the object, but it cannot access it. However, it can join the group at any moment, and then access the object as usual.

The **send** operation should never send the **value**, even when the **status** is **VALID** to avoid **INVALIDATE** messages to be ignored during the message transit. Only the owner can send **value**, as it will also multicast the corresponding **INVALIDATE** or **MIGRATE** message, ensuring that the message will arrive after the **value**.

This algorithm implements a Coherent Causal Consistent DSM. The Coherency (serialization of writes on the same object) is given by the token passing protocol provided for the ownership. The Causal Consistency is provided by the **INVALIDATE/MIGRATE** multicast, based on Group Large Causality delivery order. As each write operation is signalled by a multicast, every read or write causally following this write (in the same group) will always see the new value already written in the object. Due to the relaxed rules for Group Large Causality, any message path going outside of the group looses the causal relation. However, the idea is to remain Causal Consistent inside the group sharing the object, allowing our scalability.

An advantage of this algorithm is that only the processors holding references to the object must participate in the write protocol. The other members of the group can just ignore the requests for objects unknown to them. Using the Group Large Causality rule GLC3, a new replica inside the group can be created at any time, just sending the reference and the value from the owner. As the **INVALIDATE/MIGRATE** multicast will be emitted from the same processor, it will be received after the replica, causing its correct invalidation.

The read protocol uses a multicast, to allow a simpler implementation and dynamic join/leave of new member. We preferred this scheme to keep an owner hint and to use a request forwarding to find the owner. This last mechanism is complicated for new members, and implies that all the old owners must remember the chain of forwarding pointers, complicating the Garbage Collection. A multicast can be expensive if the groups are very large, but our memory segments tend to be small and with not many members (just enough to make parallelism attractive).

4 Conclusions

In this paper we have presented a new DSM implementation, respecting Coherent Causal Consistency, based on Group Large Causality. The implementation is simple and efficient, and scalable based on processor groups sharing memory segments.

Group Large Causality implements an efficient multicast and point to point message ordering, allowing a relaxed causal order inside groups. This DSM implementation shows that Group Large Causality is a useful and powerful tool for distributed programming, and allows the programmer to use the DSM paradigm together with message passing and remote procedure call.

The DSM system and the algorithm presented in this paper were implemented as the basic memory management of a distributed parallel Lisp programming environment called TransPive[12, 13], where it has been coupled with an independent distributed Garbage Collector called Indirect Reference Count (IRC)[10]. This system has shown to be very simple to use for parallel programming running in a local area network.

We are working now on a new version running on Java to be used for Internet programming. We want to test the performance of the system when faced with really large networks. The system is not fault-tolerant at this moment, and this topic needs some work if we really want to program the Internet at large.

References

1. M. Ahamad, G. Neiger, J. Burns, P. Kohli, P. Hutto, "Causal Memory: Definitions, Implementation and Programming," Tech. Report GIT-CC-93/55, College of Computing, Georgia Institute of Technology, 1993.
2. K. P. Birman, T. A. Joseph, "Exploiting Replication in Distributed Systems," in *Distributed Systems*, edited by S. Mullender, ACM Press, N.Y., 1989.
3. A. Campos, J. Navarro, "Coherent Causal Consistency in Distributed Shared Memory," *Proc. XV International Conference of the Chilean Computer Science Society*, Arica, Chile, October 1995.
4. K. Garachorloo, K. Lenoski et al., "Memory Consistency and Event Ordering in Scalable Shared Memory Multiprocessors," *Proc. 17th Annual International Symposium on Computer Architecture*, May 1990.
5. J. Goodman, "Cache Consistency and Sequential Consistency," Tech. Report 61, IEEE Scalable Coherent Interface Working Group, 1989.
6. P. Hutto, M. Ahamad, "Slow Memory: Weakening Consistency to Enhance Concurrency in Distributed Shared Memories," *Proc. IEEE 10th International Conference on Distributed Computing Systems*, 1990.
7. R. John, M. Ahamad, "Evaluation of Causal Distributed Shared Memory for Data-Race-Free Programs", Tech. Report GIT-CC-94/34, College of Computing, Georgia Institute of Technology, 1993.
8. L. Lamport, "Time, Clocks, and the Ordering of Events in a Distributed System," *Comm. ACM*, Vol. 21, No. 7, pp. 558-565, July 1978.
9. K. Li, P. Hudak, "Memory Coherence in Shared Virtual Memory Systems", *ACM Trans. on Computer Systems*, Vol. 7, No. 4, pp. 321-359, November 1989.
10. J. M. Piquer, "Indirect Reference Counting: A Distributed GC," LNCS 505, *PARLE '91 Proceedings Vol I*, pp. 150-165, Springer Verlag, Eindhoven, The Netherlands, June 1991.
11. J. M. Piquer, "Large Causality: Ordering Broadcasts and Messages," position paper, *5th ACM SIGOPS Workshop on Models and Paradigms for Distributed Systems Structuring*, Mt Saint-Michel, France, September 1992.
12. J. M. Piquer, C. Queinnec, "TransPive: A Distributed Lisp System," *Lettre du Transputer*, Laboratoire d'Informatique de Besançon, N. 16, pp. 55-68, December 1992.
13. J. M. Piquer, "A Reimplementation of TransPive: Lessons from the Experience," Parallel Symbolic Languages and Systems (PSLS'95), Beaune, France, October 1995.

Workshop 05+21

Parallel Languages, Programming, and High-Level Control

A Framework for Integrated Communication and I/O Placement

Rajesh Bordawekar[1], Alok Choudhary[1] and J. Ramanujam[2]

[1] ECE Dept., 121, Link Hall, Syracuse University, Syracuse, NY 13244
[2] ECE. Dept., Louisiana State University, Baton Rouge, LA 70803

Abstract. This paper describes a framework for analyzing dataflow within an out-of-core parallel program. Dataflow properties of FORALL statement are analyzed and a unified I/O and communication placement framework is presented. This placement framework can be applied to many problems, which include eliminating redudant I/O incurred in communication. The framework is validated by applying it for optimizing I/O and communication in out-of-core stencil problems. Experimental performance results on an Intel Paragon show significant reduction in I/O and communication overhead.

1 Introduction

It is widely acknowledged in the high-performance computing circles that parallel input/output requires substantial improvement in order to make scalable computers truly usable. There are several reasons for a parallel application for performing input/output. These include real-time I/O, initial/final read-write, checkpointing and out-of-core computations [Bor96].

We focus on the problem of supporting *out-of-core* computations. Out-of-core computations are those computations whose primary data sets are stored on files in the secondary memory. Specifically, we concentrate on compiling out-of-core programs developed using High Performance Fortran (HPF) [Hig93].[3] HPF is a data parallel language which provides explicit language directives to partition data over processors in certain pre-defined decomposition patterns like BLOCK and CYCLIC. This data distribution results in each processor storing a *local array* associated with each array distributed in the HPF program. HPF also provides data-parallel program construsts like FORALL [Hig93].

In this paper, we describe a dataflow framework for optimizing communication in out-of-core problems. We focus on communication optimization within a single out-of-core FORALL construct. Unlike the available dataflow frameworks for optimizing inter-processor communication [KN94, KS95, GSS95], our framework takes an unified approach for placing I/O and communication calls *while preserving characteristics of these calls*. All the current frameworks focus on improving communication performance by vectorizing messages, eliminating redundant communication and overlapping communication with computation. How-

---
[3] Although the techniques are discussed with respect to HPF, they can be applied to compilation of data parallel programs in general.

ever, these frameworks do not directly extend to out-of-core problems. Another limitation of these frameworks is that they do not make efficient use of the *copy-in-copy-out* semantics of the HPF FORALL construct. We illustrate these points by applying two communication placement frameworks [KN94, KS95] to an out-of-core problem performing stencil computations (also called an *regular* problem). We then compare the results with an integrated I/O and communication placement framework which achieves substantial performance improvement by simultaneously reordering I/O and communication calls.

The paper is organized as follows: Section 2 introduces various dataflow definitions that will be used throughout the paper. In Section 3, we present an out-of-core regular problem and analyzes it's communication and I/O pattern. This problem is used as a running example throughout the paper. Section 4 presents an integrated I/O and communication framework and describes its application in eliminating extra file I/O from communication. Section 5 presents experimental performace results of optimizing out-of-core communication from stencil problems using our framework. Finally, we conclude in Section 6.

2 Background

Our program representation is based on [KS95]. Let $G=(N, E)$ be the interval flow graph representing an HPF program, with N nodes and E edges. Let s and e be the unique start and end nodes of G. Every edge in E can be classified as an entry, forward or backward edge. Let a Tarjan interval $T(h)$ represent a set of program flow nodes that correspond to a loop in the program text. $T(h)$ has a unique header h, where $h \notin T(h)$. For every node n of the interval flow graph, G, we define SUCC(n) and PRED(n) as a set of successor and predecessor nodes of n. The edges induce the following traversal order over G. Given a forward edge (m, n), a FORWARD order visits m before n and a BACKWARD order visits m after n. Let HEADER denote the header node of the interval $T(n)$. [Bor96] describes the properties of the interval flow graph.

To anlyze dataflow properties of the FORALL statement, we use the classical dataflow definitions, i.e., USE, DEF, KILL. A variable is said to be USEd if it is referred in an expression. A variable is said to be DEFed if it is *initialized* in an expression. The variable is said to be LIVE until it is *defined* again (in other words, KILLed). We can extend these definitions for objects such as arrays. An array is said be INJURED, if some elements of the array are overwritten, otherwise the array can be considered LIVE. An array is said to be ACTIVE if some of its elements are either USEd or DEFed and these elements constitute the ACTIVE set of the array.

Recall that the FORALL statement has *copy-in-copy-out* semantics [Hig93]. Consequently, during the execution of a FORALL statement, old as well as new values of an array can be LIVE. In other words, the FORALL statement satisfies the DELAYED_KILL property [Bor96]. We use variable *DKILL* to represent an array which satisfies the DELAYED_KILL property.

We now define some dataflow variables that will be used for analyzing communication and I/O access patterns in out-of-core programs. Let ACTIVE_n^p denote the set of elements that will be used in computation in processor p at a node n in the interval flow graph. Similarly INCORE_n^p denote the set of elements read by a processor p at node n. Definitions ACTIVE_n^p and INCORE_n^p are used to compute the send-recv sets for each processor, SEND_n^p and RECV_n^p. Using SEND_n^p and RECV_n^p, we can compute the set of elements communicated at a node n, COMM_n as $\bigcup_i \{\text{SEND}_n^i + \text{RECV}_n^i\}$. Similarly, we compute the set of incore elements at node n, INCORE_n, as $\bigcup_i \text{INCORE}_n^i$. For every node n, for every processor p and SEND_n^p, we define EIO_n^p as a set of elements which will be sent by p but are not members of INCORE_n^p. Formally, $\text{EIO}_n^p = \text{SEND}_n^p - (\text{INCORE}_n^p \cap \text{SEND}_n^p)$.

For any data set $d \in \text{INCORE}$ or SEND or RECV, the following predicates are defined. Bit vectors are used to represent individual data sets.

- $Used(n,d) \stackrel{df}{=} \text{TRUE}$ *iff* a subset of d is referenced at node n.
- $Kill(n,d) \stackrel{df}{=} \text{TRUE}$ *iff* a subset of d is modified at n.
- $Incore(n,d) \stackrel{df}{=} \text{TRUE}$ *iff* a subset of d is in-core at n.

3 I/O and Communication Optimization: An Example

Figure 1:2 presents an HPF example in which an out-of-core array a is distributed over 4 processors in BLOCK fashion. This example will be used as a running example throughout the paper. Our running example performs one-dimensional relaxation using 3-point stencil computations. The interior points of the array a are updated using a FORALL construct. To preserve the FORALL semantics, it is necessary to use temporaries to store initial and intermediate data. Since the primary data sets are stored in files, it is necessary to use two different files, the source local array file (LAF) for reading initial data and a temporary LAF to store the updated intermediate data. After the computation is over, the temporary LAF can be renamed as the source LAF.[4]

Figure 1:3 shows the pseudo-code for the stripmined program (assuming per processor available memory as 10). There are two stripmined iterations, each iteration reads the initial data from the source file into an in-core local array (ICLA) temp and writes the intermediate results from an ICLA temp1 to the temporary file. Each iteration, after reading the ICLA, performs communication (if required). For example, in the first iteration, processors 0,1 and 2 send elements a(16),a(32) and a(48) to processors 1,2, and 3 respectively. In the second iteration, processors 1,2, and 3 send elements a(17), a(33) and a(49) to processors 0, 1, and 2. Note that this is an example of the **Receiver-driven In-core communication** method [Bor96].

Figure 1:4 shows the initial communication and input/output placement. The communication and input/output sets for each processor are given in global name

---

[4] A more detailed description is provided in [Bor96].

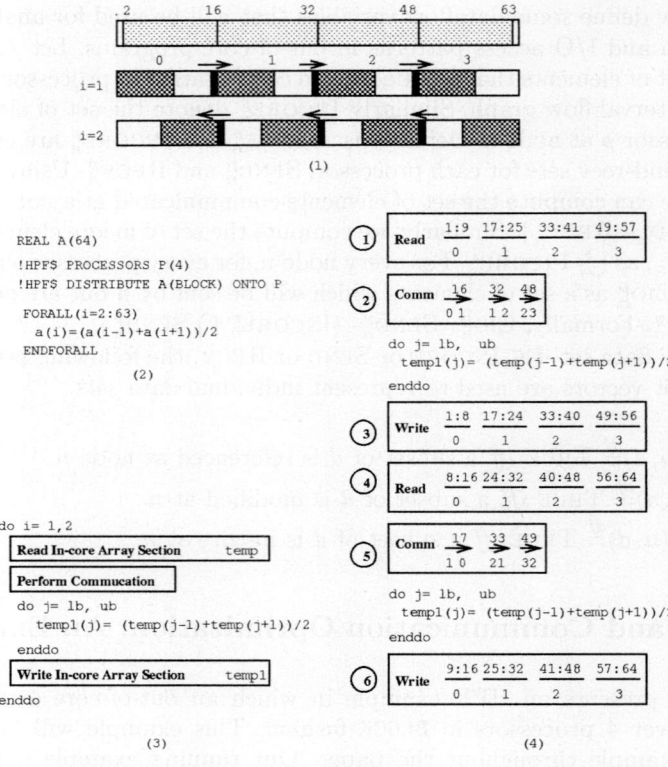

Fig. 1. Example program.

space while the bounds for the in-core computation are given in the local stripmined space (i.e., lb=1 and ub=8). For example, **Read**$\frac{1:9}{0}$ means that processor 0 is reading elements a(1) to a(9), **Comm** $0 \xrightarrow{16} 1$ represents communication of element a(16) from processor 0 to processor 1 and **Write**$\frac{1:8}{0}$ means that processor 0 writing elements a(1) to a(8).

From the computation pattern, it is easy to determine the communication pattern for each stripmined iteration [Bor96]. For example, in the first iteration, processor 0 needs to send element a(16) to processor 1. Since processor 0 does not have element a(16) in memory, however, it needs to read it from the LAF and send it to processor 1. Similarly, processors 2 and 3 need to read elements a(32) and a(48) from their LAFs and send them to their respective destinations. These file reads are termed as *extra* since the read elements are not required for computation by the owner processor. In the second iteration, processors 1, 2, and 3 perform also extra file accesses to read elements a(17), a(33) and a(49) respectively. To prevent violation of FORALL semantics, *old values* of elements, a(17), a(33), and a(49), are read from the source LAF and communicated to appropriate processors. It should be observed that elements a(17), a(33) and a(49) are brought into memory in the first iteration and could be communicated *before* or *after* they are overwritten; thus minimizing extra file accesses. The example also performs redundant reads of some elements. For example, in the first iteration, processor 0 reads elements a(1) to a(9), but writes modified values of elements a(1) to a(8) *while retaining the old set of elements*, a(1) to a(9) *in form of the temporaries*.[5] In the second iteration, processor 0 again reads the *old* values of elements a(8) and a(9). Therefore, these two reads are *partially redundant*. These partially redundant reads can be eliminated if it is possible to determine which elements can be reused across iterations.

As observed before, for our running example, communication requires both inter-processor communication (i.e., communication of in-core data) and file I/O. To improve the communication cost, it is very important to minimize the file I/O cost (or the number of file accesses). The file accesses generated by the program can be classified into: (1) *Compulsory*: These accesses are required to read and write in-core data and (2) *Extra*: These accesses are required for communicating off-processor out-of-core elements. The file I/O cost can be reduced by (1) eliminating partially redundant compulsory file accesses and (2) minimizing extra file accesses by communicating *in-core* data whenever possible. The second optimization requires reordering computation and placing the communication calls so that only in-core data is communicated [Bor96]. In an out-of-core application, the computation order is decided by the data access pattern, that is, by *placement of the read/write calls*. Therefore, to minimize overhead due to file I/O in communication, it is important that both communication and I/O calls are placed at appropriate positions.

4 A Framework for Integrated I/O and Communication Placement

In Section 3, we describe the compilation of an out-of-core FORALL statement. We observe that the implementation of out-of-core FORALL requires extra file accesses during communication and a naive implementation results in reading redundant data. In this section, we propose an integrated I/O and communication placement framework that exploits the DELAYED_KILL property of the FORALL construct and applies the array access information for improving the overall performance. Note that the indeterminacy in FORALL execution order, allows our framework to freely reorder in-core computations. Specifically, our framework reorders in-core computation such that communication would involve only inter-processor communication. Consequently, all extra file accesses will be eliminated.

4.1 The Correctness Criteria

Our integrated framework imposes the following correctness requirements:

---

[5] Note that temporaries are marked LIVE during the FORALL computation.

- *Safety*: All data either communicated or read is used *immediately*.
- *Sufficiency*: Every in-core computation is preceded by an appropriate READ call and each non-local reference is preceded by appropriate communication.
- *Balance*: For every SEND, there is exactly one matching RECV. Note that this condition does not apply for READ.[6]

In the presence of the DELAYED_KILL type of computation, the definition of Safety is considerably weakened. Hence, it is more appropriate to term it as *Weak Safety*. Note that *Weak Safety* and *Sufficiency* are applicable for both file access and communication calls, while *Balance* is applicable only to communication calls. Therefore, our framework is able to take an unified approach for placing file access and communication calls while honoring their individual characteristics.

4.2 Eliminating Extra File Accesses in Communication

It should be observed that extra file accesses are generated because an array section[7] is used several times in the stripmined FORALL iterations; once by the processor that owns the section and in remaining cases, by other processors. If it is possible for the processors to perform computation on the common array section in the same iteration, the communication will involve only inter-processor data transfer and extra file accesses could be eliminated. To satisfy this condition, we add the following constraint in the correctness criteria.

Strict Safety Constraint

- *Strict Safety*: Everything that is *read* or *communicated* (i.e., *sent* and *received*) will be used only once.

Criteria *Safety* and *Strict Safety* require that the data read by processor i at node n, INCORE_n^i, should be used *immediately* and should not be used anywhere else in the computation. Computation in any processor, j, at node n', which requires elements of INCORE_n^i (in other words, $\text{RECV}_{n'}^j \subset \text{INCORE}_n^i$), should, therefore, be placed at node n. Then, processor i needs to send only the incore data ($\text{SEND}_n^i \subset \text{INCORE}_n^i$). Applying this condition to every processor, we can observe that if node n satisfies *Strict Safety*, COMM_n is subsumed by INCORE_n and therefore, set EIO is empty and all extra I/O is eliminated.

Processors i, j satisfying the above requirements exhibit one or both of the following *inclusion* properties

- $\text{RECV}_{n'}^j \subset \text{SEND}_n^i \rightarrow \text{RECV}_{n'}^j \subset \text{INCORE}_n^i$
- $\text{RECV}_n^i \subset \text{SEND}_{n'}^j \rightarrow \text{RECV}_n^i \subset \text{INCORE}_{n'}^j$

where n and n' are nodes of the interval flow graph denoting the initial placement of the computation (in other words, placement of READ calls). To

---
[6] We currently use synchronous I/O calls.
[7] An element can be considered as a special case of section.

find i, j and n, n', it is necessary to perform both FORWARD and BACKWARD flow analysis.

Let us now define a predicate $Incl_j^i(n, n')$ as follows:

- $Incl_j^i(n, n') \stackrel{df}{=}$ TRUE if $\text{RECV}_{n'}^j \subset \text{INCORE}_n^i$ or $\text{RECV}_n^i \subset \text{INCORE}_{n'}^j$

For a processor i, the solution of the $Incl_j^i(n, n')$, for any processor j ($j \neq i$), gives the node pair (n, n') satisfying the inclusion properties. The inclusion property is then verified for every INCORE and RECV set in the program. If all the INCORE and RECV sets satisfy the *inclusion* property, then the computation is said to be *balanced*. For balanced computation, one can eliminate extra I/O by reordering computations.

We illustrate this optimization by using our running example (Figures 1). Table 1 illustrates the values of various dataflow variables corresponding to the stripmined iterations (Figure 1). There are two stripmined iterations; for each iteration, INCORE gives the set of elements that are brought in memory by each processor (ICLA). Corresponding ACTIVE, SEND and RECV sets are also shown.

Table 1. Dataflow Variables for the running example.

Iter.	Processor	Node	INCORE	ACTIVE	SENT	RECV
1	0	2	1:9	1:9	16	-
	1	2	17:25	16:25	32	16
	2	2	33:41	32:41	48	32
	3	2	49:57	48:57	-	48
2	0	6	8:16	8:17	-	17
	1	6	24:32	24:33	17	33
	2	6	40:48	40:49	33	49
	3	6	56:64	56:64	49	-

Table 2 presents the solutions for the *Incl* predicate for all processors in form of the *Inclusion* matrix. An entry (n, n') in a position $[i, j]$ denotes the pair of nodes of the interval flow graph satisfying the inclusion equations for the processors i and j. This entry is called as a *solution* entry. In other words, it defines the INCORE sections of processors i and j that satisfy the inclusion property. For example, consider the solution at position [2,1]. The solution tuple (2,6) denotes that $\text{RECV}_2^2 \subset \text{INCORE}_6^1$, i.e., the data required by the ICLA of processor 2 at node 2 (first stripmined iteration) is part of the ICLA of processor 1 at node 6 (second stripmined iteration). The entries in the positions [0,0] and [3,3] denote that processors 0 and 3 do not perform communication at nodes 2 and 6 respectively (in other words, in the first and second stripmined iteration). Such entries are called *non-solution* entries. The number of *solution* entries in

i^{th} row or j^{th} column denotes the number of times a processor i or j performs communication.

Table 2. Inclusion matrix for the running example.

Processor	Processor			
	0	1	2	3
0	(2,2)	(2,6)	-	-
1	(2,6)	-	(6,2)	-
2	-	(2,6)	-	(6,2)
3	-	-	(2,6)	(6,6)

The information provided by Table 2 can be used to reorder the computation. This reordering is an iterative procedure; every iteration tries to schedule computation such that the inclusion equations are satisfied. The iterations stop when all ICLAs represented by the solution tuples are scheduled. Let us understand the reordering procedure using our running example and its inclusion matrix.

1. In the first step, choose a random processor i. For our problem, let us choose processor 2. For this processor, select a solution entry from the second row, e.g., entry [2,1] which corresponds to the solution tuple (2,6). It states that $\text{RECV}_2^2 \subset \text{INCORE}_6^1$. Therefore, sections of local arrays of processors 2 and 1, corresponding to the nodes 2 and 6 (in the interval flow graph) should be brought in memory.
2. In the second step, using the inclusion matrix, determine if the ICLA of processor 1 requires any off-processor data. It can be easily found out by checking the first row of the inclusion matrix for solution entries containing node 6. The entry [1,2] corresponds to the solution tuple (6,2), which indicates that $\text{RECV}_6^1 \subset \text{INCORE}_2^2$. Note that the array section of processor 2, corresponding to node 2, is already in memory. Therefore, the communication between processors 1 and 2 will involve only inter-processor communication.
3. The first two steps have scheduled ICLAs of processors 1 and 2. The third step tries to schedule ICLAs of the remaining processors so that there are no extra I/O accesses. Consider processor 0. In the 0th row, the only solution entry involves processor 1 at node 2. Since ICLA of processor 1 at node 2 is already scheduled, this entry cannot be used. In this case, the non-solution entry, i.e., entry at position [0,0], [2,2], should be used. This non-solution entry suggests that the ICLA of processor 0 at node 2 does not require communication and therefore, can be scheduled along with ICLAs of processors 1 and 2. Applying the same principle to processor 3, we can see that ICLA of processor 3 at node 6 does not require communication. Hence,

this ICLA can be scheduled along with the ICLAs of processor 0, 1 and 2. For this ICLA schedule, only communication required will be inter-processor communication between processors 1 and 2.

4. Applying the same procedure, the remaining four ICLAs can be scheduled. This ICLA schedule will involve interprocessor communication between processors 0 and 1, and between processors 2 and 3. Therefore, the overall computation involves only inter-processor communication and the extra I/O accesses are eliminated. Figure 2:A illustrates the final placement of I/O and communication calls. Figure 2:B illustrates an alternative placement. This placement is obtained using a different choice of initial processor.

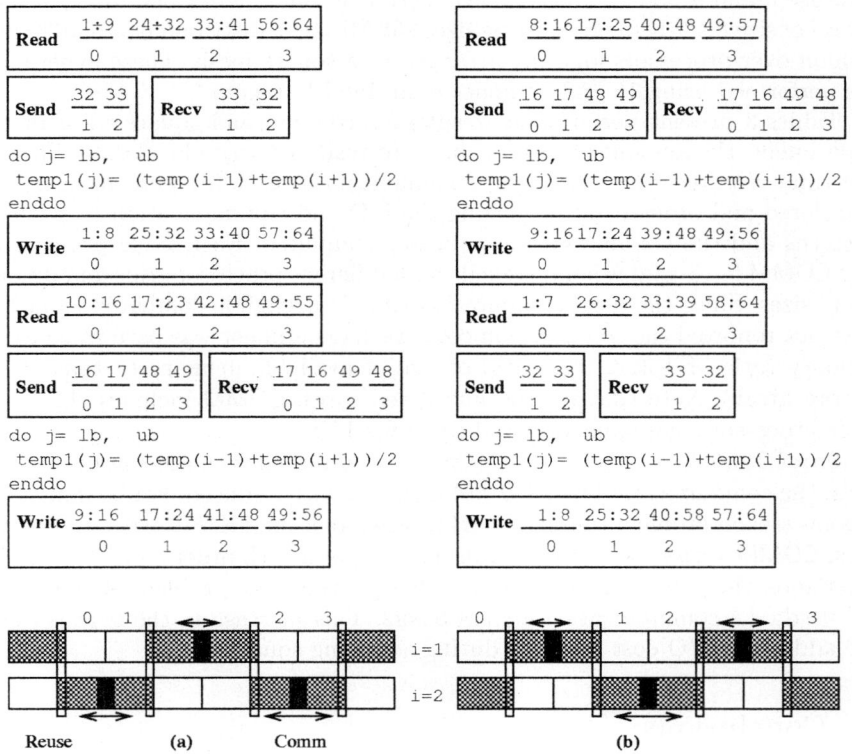

Fig. 2. Final placement of I/O and communication calls.

5 Applying Dataflow Framework to Stencil Problems

We now apply the communication and I/O placement framework to the stencil problems. We illustrate using the 5- and 9-point stencils (Figure 3 (1) and (2)).

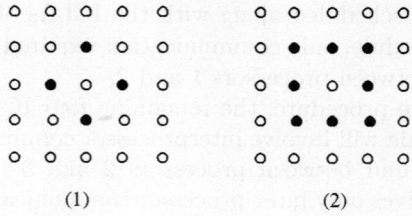

Fig. 3. 5- and 9-point Stencils

This section presents performance results of hand-coded out-of-core examples that use 5-and 9-point stencils. The experiments were performed for square real arrays of size 8K*8K (aggregate file sizes 256 Mbytes), distributed in BLOCK-BLOCK fashion over processors logically arranged as a square mesh. These experiments are performed using 16 and 64 nodes of an Intel Paragon.

Tables 3 present performance results for column, and square tiles. In each experiment, the amount of time required to read and write local data, LIO, and the time required for performing communication, COMM, were measured for unordered and ordered (after placing the I/O and communication calls) access patterns and the communication gain was computed. Each table presents LIO and COMM for 5- and 9-point stencils with different processor grids and different array sizes. Since the local computation time is negligible compared to LIO, we have not reported the computation cost. Each experiment was performed for the memory ratio of $\frac{1}{4}$ (i.e., the ratio of size of available memory to that of out-of-core array). Note that for the unordered cases, COMM includes the cost of inter-processor communication and extra file I/O.

From Table 3, we can observe that by reordering communication and I/O calls, the communication cost COMM is significantly reduced. For example, for a 9-point stencil problem running on 64 processors using 8K*8K array and column tiles, COMM without ordering is 2.06 seconds, and with ordering is 0.05 seconds (therefore, the communication gain is 39). For the same problem, if square tiles are used, the communication gain is 35992. This increase in the gain is due to the additional I/O cost incurred during accessing square tiles.

6 Conclusions

In this paper, we described a framework for optimizing communication and I/O costs in out-of-core problems. We focussed on communication and I/O optimization within a FORALL construct. We showed that existing frameworks do not extend directly to out-of-core problems and can not exploit the FORALL semantics. We presented a unified framework for the placement of I/O and communication calls and applied it for optimizing communication for stencil applications. Using the experimental results, we demonstrated that correct placement of I/O and communication calls can completely eliminate extra file I/O from communication and as a result, significant performance improvement can be obtained.

Table 3. Performance of the 5-and 9-point stencils. time in seconds.

Memory ratio	Procs.	Unordered COMM a	Unordered LIO b	Ordered COMM c	Ordered LIO d	Comm Gain e=(a/c)
5-point Stencil, Column Tiles, 8K*8K Array						
1/4	16	**1.38**	7.57	**0.04**	6.71	**35.38**
1/4	64	**1.16**	10.87	**0.05**	9.80	**25.21**
9-point Stencil, Column Tiles, 8K*8K Array						
1/4	16	**1.27**	7.70	**0.03**	7.05	**38.48**
1/4	64	**2.06**	10.20	**0.05**	10.15	**39.84**
5-point Stencil, Square Tiles, 8K*8K Array						
1/4	16	**175.11**	183.96	**0.03**	178.14	**5506.60**
1/4	64	**192.2**	197.57	**0.005**	196.40	**36061.79**
9-point Stencil, Square Tiles, 8K*8K Array						
1/4	16	**150.78**	175.88	**0.03**	182.01	**4569.09**
1/4	64	**192.2**	197.57	**0.005**	196.40	**35992.51**

Acknowledgments

The work of R. Bordawekar and A. Choudhary was supported in part by NSF Young Investigator Award CCR-9357840, grants from Intel SSD and in part by the Scalable I/O Initiative, contract number DABT63-94-C-0049 from Advanced Research Projects Agency(ARPA) administered by US Army at Fort Huachuca. R. Bordawekar is also supported by a Syracuse University Graduate Fellowship. The work of J. Ramanujam was supported in part by an NSF Young Investigator Award CCR-9457768, an NSF grant CCR-9210422 and by the Louisiana Board of Regents through contract LEQSF(1991-94)-RD-A-09. This work was performed in part using the Intel Paragon System operated by Caltech on behalf of the Center for Advanced Computing Research (CACR). Access to this facility was provided by CRPC.

References

[Bor96] Rajesh Bordawekar. *Techniques for Compiling I/O Intensive Parallel Programs*. PhD thesis, Electrical and Computer Engineering Dept., Syracuse University, April 1996.

[GSS95] Manish Gupta, Edith Schonberg, and Harini Srinivasan. A Unified Framework for Optimizing Communication in Data-Parallel Programs. *IEEE Transactions on Parallel and Distributed Systems*, 1995.

[Hig93] High Performance Fortran Forum. High Performance Fortran Language Specification. *Scientific Programming*, 2(1-2):1–170, 1993.

[KN94] Ken Kennedy and Nenad Nedeljković. Combining Dependence and Data-Flow Analyses to Optimize Communication. Technical Report CRPC-TR94484-S, CRPC, Rice University, September 1994.

[KS95] Ken Kennedy and Ajay Sethi. A Constraint Based Communication Placement Framework. Technical Report CRPC-TR95515-S, CRPC, Rice University, February 1995. Revised May 1995.

Formal Derivation of Parallel Program for 2-Dimensional Maximum Segment Sum Problem

Zhenjiang Hu[1], Hideya Iwasaki[2], Masato Takeichi[1]

[1] Department of Information Engineering, University of Tokyo
7-3-1 Hongo, Bunkyo-ku, Tokyo 113, Japan
Email: hu@ipl.t.u-tokyo.ac.jp and takeichi@u-tokyo.ac.jp
[2] Department of Computer Science, Faculty of Technology
Tokyo University of Agriculture and Technology
2-24-16 Naka-cho, Koganei, Tokyo 184, Japan
Email: hiwasaki@cc.tuat.ac.jp

Abstract. It has been attracting much attention to make use of list homomorphisms in parallel programming because they ideally suit the divide-and-conquer parallel paradigm. However, they have been usually treated rather informally and ad-hoc in the development of efficient parallel programs. This paper reports a case study on systematic and formal development of a new parallel program for the 2-dimensional maximum segment problem. We show how a straightforward, and "obviously" correct, but quite inefficient solution to the problem can be successfully turned into a semantically equivalent "almost list homomorphism" based on two transformations, namely tupling and fusion, which are defined according to the specific recursive structures of list homomorphisms.

1 Introduction

It has been attracting wide attention to make use of list homomorphisms in parallel programming. *List homomorphisms* are those functions on finite lists that *promote* through list concatenation — that is, function h for which there exists an associative binary operator $\oplus$ such that, for all finite lists xs and ys, we have $h\,(xs \mathbin{+\mkern-10mu+} ys) = h\,xs \oplus h\,ys$, where $\mathbin{+\mkern-10mu+}$ denotes list concatenation. Intuitively, the definition of list homomorphisms means that the value of h on the larger list depends in a particular way (using binary operation $\oplus$) on the values of h applied to the two pieces of the list. The computations of $h\,xs$ and $h\,ys$ are independent each other and can thus be carried out in parallel. This simple equation can be viewed as expressing the well-known divide-and-conquer paradigm of parallel programming.

Therefore, the implications for parallel program development become clear; *if the problem is a list homomorphism*, then it only remains to define a cheap $\oplus$ in order to produce a highly parallel solution. However, there are a lot of useful and interesting list functions that are not list homomorphisms and thus have no corresponding $\oplus$. One example is the function *mss* known as *(1-dimensional) maximum segment sum problem*, which finds the maximum of the sums of contiguous segments within a list of integers. For example, *mss* $[3, -4, 2, -1, 6, -3] = 7$, where the result is contributed by the segment $[2, -1, 6]$. The *mss* is not a list homomorphism, since knowing *mss xs* and *mss ys* is not enough to allow computation of *mss* $(xs \mathbin{+\!\!+} ys)$.

This paper reports a case study of a formal and systematic derivation of a new efficient and correct $O(\log^2 n)$ (n denotes the size of a matrix) parallel program for the *2-dimensional maximum segment sum problem* via construction of (almost) list homomorphisms based on the idea in [HIT96]. This problem is of interest because there are efficient but non-obvious algorithms to compute it in parallel. In [Smi87], the tuple consisting of eleven functions is used for the definition of a $O(\log^2 n)$ parallel algorithm, but the detailed derivation, which would be rather cumbersome with Smith's approach, was not given at all.

This paper is organized as follows. In Sect. 2, we review the notational conventions and some basic concepts used in this paper. After giving a specification for the 2-dimensional maximum segment sum problem in Sect. 3, we focus ourselves on deriving an efficient (almost) list homomorphism from the specification by using our two important theorems, namely the Tupling and the Almost Fusion Theorems, in Sect. 4. Concluding remarks are given in Sect. 5.

2 Preliminary

In this section, we briefly review the notational conventions known as Bird-Meertens Formalisms [Bir87] and some basic concepts which will be used in the rest of this paper.

Functions

Functional application is denoted by a space and the argument which may be written without brackets. Thus $f\,a$ means $f(a)$. Functions are curried and application associates to the left. Thus $f\,a\,b$ means $(f\,a)\,b$. Functional application is regarded as more binding than any other operator, so $f\,a \oplus b$ means $(f\,a) \oplus b$ but not $f(a \oplus b)$. Functional composition is denoted by a centralized circle $\circ$. By definition, $(f \circ g)\,a = f(g\,a)$. Functional composition is an associative operator, and the identity function is denoted by id. Infix binary operators will often be denoted by $\oplus, \otimes$ and can be *sectioned*; an infix binary operators like $\oplus$ can be turned into unary functions by: $(a\oplus)\,b = a \oplus b = (\oplus b)\,a$.

The following are some important operators (functions) used in the paper.

- The *projection* function π_i will be used to select the i-th component of tuples, e.g., $\pi_1(a, b) = a$. The $\vartriangle$ and $\times$ are two important operators related to tuples, defined by

$$(f \vartriangle g)\,a = (f\,a, g\,a), \quad (f \times g)(a, b) = (f\,a, g\,b).$$

The $\vartriangle$ can be naturally extended to functions with two arguments. So, we have $a\,(\oplus \vartriangle \otimes)\,b = (a \oplus b, a \otimes b)$.

- The *cross* operator $\mathcal{X}_\oplus$, which crosswisely combines elements in two lists with operator $\oplus$, is defined informally by:

$$[x_1, \cdots, x_n] \mathcal{X}_\oplus [y_1, \cdots, y_m] = [x_1 \oplus y_1, \cdots, x_1 \oplus y_m, \cdots, x_n \oplus y_1, \cdots, x_n \oplus y_m].$$

The cross operator enjoys many algebraic identities, e.g., $(f*) \circ \mathcal{X}_\oplus = \mathcal{X}_{f \circ \oplus}$.
- The *concat*, a function to flatten a list, is defined by:

$$concat\ [xs_1, \cdots, xs_n] = xs_1 \mathbin{+\!\!+} \cdots \mathbin{+\!\!+} xs_n.$$

- The *zip-with* operator $\Upsilon_\oplus$, a function to apply $\oplus$ pairwisely to two lists, is informally defined by

$$[x_1, \cdots, x_n] \Upsilon_\oplus [y_1, \cdots, y_n] = [x_1 \oplus y_1, \cdots, x_n \oplus y_n].$$

Lists

Lists are finite sequences of values of the same type. A list is either empty, a singleton, or the concatenation of two other lists. We write $[\,]$ for the empty list, $[a]$ for the singleton list with element a (and $[\cdot]$ for the function taking a to $[a]$), and $xs \mathbin{+\!\!+} ys$ for the concatenation of xs and ys. Concatenation is associative, and $[\,]$ is its unit. For example, the term $[1] \mathbin{+\!\!+} [2] \mathbin{+\!\!+} [3]$ denotes a list with three elements, often abbreviated to $[1, 2, 3]$.

List Homomorphisms

A function h satisfying the following three equations will be called a *list homomorphism*.

$$\begin{aligned} h\ [\,] &= \iota_\oplus \\ h\ [x] &= f\ x \\ h\ (xs \mathbin{+\!\!+} ys) &= h\ xs \oplus h\ ys \end{aligned}$$

It soon follows from this definition that $\oplus$ must be an associative binary operator with unit $\iota_\oplus$. For notational convenience, we write $(\!(f, \oplus)\!)$ for the unique function h^3, e.g., $sum = (\!(id, +)\!)$ and $max = (\!(id, \uparrow)\!)$, where $\uparrow$ denotes the binary maximum function whose unit is $-\infty$. Note when it is clear from the context, we usually abbreviate "list homomorphisms" to "homomorphism."

Two important list homomorphisms are *map* and *reduction*. Map is the operator which applies a function to every item in a list. It is written as an infix $*$. Informally, we have

$$f * [x_1, x_2, \cdots, x_n] = [f\ x_1, f\ x_2, \cdots, f\ x_n].$$

Reduction is the operator which collapses a list into a single value by repeated application of some binary operator. It is written as an infix $/$. Informally, for an associative binary operator $\oplus$, we have

$$\oplus/\ [x_1, x_2, \cdots, x_n] = x_1 \oplus x_2 \cdots \oplus x_n.$$

---

[3] Strictly speaking, we should write $(\!(\iota_\oplus, f, \oplus)\!)$ to denote the unique function h. We can omit the $\iota_\oplus$ because it is the unit of $\oplus$.

It is not difficult to see that $*$ and $/$ have simple massively parallel implementations on many architectures. For example, $\oplus/$ can be computed in parallel on a tree-like structure with the combining operator $\oplus$ applied in the nodes, whereas $f*$ is totally parallel. The relevance of list homomorphisms to parallel programming can be seen clearly from the Homomorphism Lemma [Bir87]: $(\![f, \oplus]\!) = (\oplus/) \circ (f*)$. Every list homomorphism can be written as the composition of a reduction and a map.

As stated in the introduction, quite a lot of interesting functions are not list homomorphisms. Fortunately, Cole [Col93a] argued informally that some of them can be converted into so-called *almost (list) homomorphisms* by tupling with some extra functions. To make this conversion be more formal and systematic, we proposed the idea of construction of such almost homomorphisms via tupling and fusion transformations [HIT96]. As a matter of fact, an almost list homomorphism is a composition of a projection function and a list homomorphism. Since projection functions are simple, almost homomorphisms are also suitable for parallel implementation as list homomorphisms do.

3 Specification

Before giving a specification for the 2-dimensional maximum segment sum problem, let's start with the simpler 1-dimensional maximum segment sum problem *mss* (refer [HIT96, Col93b, CS92] for more detailed discussions), an example given in the introduction. An obviously correct solution to the problem is $mss : [Int] \to Int$ defined by:

$$mss = max \circ (sum*) \circ segs$$

which is implemented by three passes; (1) computing all contiguous segments of a sequence by *segs*, (2) summing up each contiguous segment by *sum*, (3) selecting the largest value by *max*.

The only unknown function in the specification is $segs : [Int] \to [[Int]]$, computing all segments of a list. It would be likely to define it simply as

$$segs\ (xs \mathbin{+\mkern-10mu+} ys) = segs\ xs \mathbin{+\mkern-10mu+} segs\ ys \mathbin{+\mkern-10mu+} (tails\ xs\ \mathcal{X}_{+\mkern-8mu+}\ inits\ ys).$$

The equation reads that all segments in the sequence $xs \mathbin{+\mkern-10mu+} ys$ are made up of three parts: all segments in xs, all segments in ys, and all segments produced by crosswisely concatenating every *tail segment* of xs (i.e., the segment in xs ending with xs's last element) with every *initial segment* of ys (i.e., the segment in ys starting with ys's first element). Here, *inits* and *tails* are standard functions in [Bir87], though our definitions are slightly different as will be seen later. Being simple, it is a *wrong definition* for *segs*, as you may have noticed that, for example, $segs\ ([1,2] \mathbin{+\mkern-10mu+} [3]) \ne segs\ ([1] \mathbin{+\mkern-10mu+} [2,3])$ while they are expected to be equal (to $segs\ [1,2,3]$). A closer look reveals that the two resulting lists indeed consist of all segments of $[1,2,3]$ but in different order. One way to remedy this

$$
\begin{aligned}
&mss \;:\; [(Index, Int)] \to Int \\
&mss \;=\; max \circ (sum'*) \circ segs
\end{aligned}
$$

where

$$
\begin{aligned}
max &= ([id, \uparrow]) \\
sum' &= ([id, \lambda((i,x),(j,y)).x+y]) \\
segs\;[\,] &= [\,] \\
segs\;[x] &= [[x]] \\
segs\;(xs \mathbin{+\!\!+} ys) &= segs\;xs \mathbin{+\!\!+_\prec} segs\;ys \mathbin{+\!\!+_\prec} (tails\;xs\;\mathcal{X}_{+\!\!+}\;inits\;ys) \\
inits\;[\,] &= [\,] \\
inits\;[x] &= [[x]] \\
inits\;(xs \mathbin{+\!\!+} ys) &= inits\;xs \mathbin{+\!\!+} (xs \mathbin{+\!\!+}\,) * (inits\;ys) \\
tails\;[\,] &= [\,] \\
tails\;[x] &= [[x]] \\
tails\;(xs \mathbin{+\!\!+} ys) &= (\mathbin{+\!\!+}\, ys) * (tails\;xs) \mathbin{+\!\!+} tails\;ys
\end{aligned}
$$

Fig. 1. Specification for *mss* Problem

situation is to force *segs* to give result of a sorted list under a total order, say $\prec$, and thus we can define *segs* correctly as

$$segs\;(xs \mathbin{+\!\!+} ys) = segs\;xs \mathbin{+\!\!+_\prec} segs\;ys \mathbin{+\!\!+_\prec} (tails\;xs\;\mathcal{X}_{+\!\!+}\;inits\;ys)$$

where $\mathbin{+\!\!+_\prec}$ merges two sorted lists into one with respect to the order of $\prec$.

Let's see how we can define such $\prec$ in a simple way. Let $[x_{i_1}, x_{i_1+1}, \cdots, x_{j_1}]$ and $[x_{i_2}, x_{i_2+1}, \cdots, x_{j_2}]$ be two segments of the presumed list $[x_1, \cdots, x_n]$. Then, $\prec$ is defined by $[x_{i_1}, x_{i_1+1}, \cdots, x_{j_1}] \prec [x_{i_2}, x_{i_2+1}, \cdots, x_{j_2}] =_{def} (i_1, \cdots, j_1) <_D (i_2, \cdots, j_2)$, where $<_D$ stands for the lexicographic order. To capture the index information in our specification, we extend the input type of *mss* and *segs* from lists of integers, $[Int]$, to lists of pairs of index and integer, $[(Index, Int)]$.

So much for the specification of the *mss* problem, which is summarized in Fig. 1. It is a naive solution of the problem without concerning efficiency and parallelism at all, but its correctness is obvious.

Let's turn to the specification for the 2-dimensional maximum segment sum problem, *mss2*, a generalization of *mss*, which finds the maximum over the sum of all rectangular subregions of a matrix. The matrix can be naturally represented by a list of lists with the same length as shown in Fig. 2 (a), and so does its rectangular subregion as in Fig. 2 (b). Following the same thought we did for *mss*, we define *mss2* straightforwardly as in Fig. 3. Here, *segs2* computes all rectangular subregions of a matrix, then *sum2* is applied to every rectangular subregion and sums up all elements, and finally *max* returns the largest.

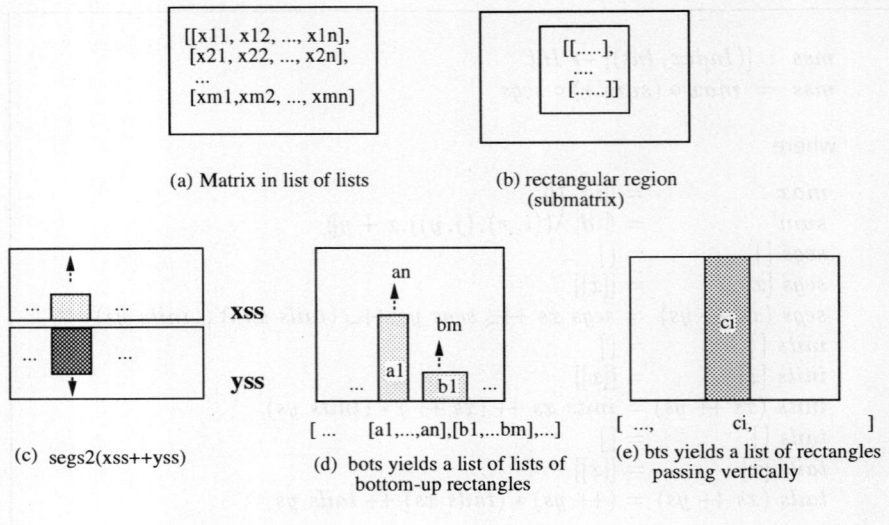

Fig. 2. The *mss2* Problem

$$mss2 \;:\; [[(Index', Int)]] \to Int$$
$$mss2 \;=\; max \circ (sum2*) \circ segs2$$

where

$$
\begin{array}{ll}
sum2 & = sum \circ sum'* \\
segs2\ [] & = [] \\
segs2\ [xs] & = [\cdot] * (segs'\ xs) \\
segs2\ (xss \mathbin{+\!\!+} yss) & = segs2\ xss \mathbin{+\!\!+_{\prec'}} segs2\ yss \mathbin{+\!\!+_{\prec'}} \\
& \quad concat\ ((bots\ xss) \Upsilon_{\mathcal{X}_{+\!\!+}} (tops\ yss)) \\
bots\ [] & = [] \\
bots\ [xs] & = [\cdot] * ([\cdot] * (segs'\ xs)) \\
bots\ (xss \mathbin{+\!\!+} yss) & = ((bots\ xss)\ \Upsilon_{\lambda(x,y).(+\!\!+ y)*x}\ (bts\ yss))\ \Upsilon_{+\!\!+}\ (bots\ yss) \\
tops\ [] & = [] \\
tops\ [xs] & = [\cdot] * ([\cdot] * (segs'\ xs)) \\
tops\ (xss \mathbin{+\!\!+} yss) & = (tops\ xss)\ \Upsilon_{+\!\!+}\ ((bts\ xss)\ \Upsilon_{\lambda(x,y).((x+\!\!+)*y)}\ (tops\ yss)) \\
bts\ [] & = [] \\
bts\ [xs] & = [\cdot] * (segs'\ xs) \\
bts\ (xss \mathbin{+\!\!+} yss) & = (bts\ xss)\ \Upsilon_{+\!\!+}\ (bts\ yss) \\
segs'\ [] & = [] \\
segs'\ [x] & = [[x]] \\
segs'\ (xs \mathbin{+\!\!+} ys) & = segs'\ xs \mathbin{+\!\!+_{\prec'}} segs'\ ys \mathbin{+\!\!+_{\prec'}} (tails\ xs\ \mathcal{X}_{+\!\!+}\ inits\ ys)
\end{array}
$$

Fig. 3. Specification for *mss2* Problem

Function *segs2* is defined in a quite similar way to *segs*. The last equation reads that all rectangular subregions of $xss \mathbin{+\!\!+} yss$, a matrix connecting xss and yss vertically (Fig. 2 (c)), are made up from those in both xss and yss, and those produced by combining every *bottom-up rectangular subregion* in xss (depicted by shallow-grey rectangle) with every *top-down rectangular subregion* in yss (depicted by dark-grey rectangle) sharing the same edge.

Let's see the definition of the total order $\prec'$ among rectangular subregions. Note that the index type *Index'* in this case should be a pair denoting the row and column of elements. So we define $\prec'$ by $[[((r_1,c_1),x_1),\cdots],\cdots,[\cdots,((r_2,c_2),x_2)]] \prec'$ $[[((r'_1,c'_1),y_1),\cdots],\cdots,[\cdots,((r'_2,c'_2),y_2)]] =_{def} ((r_1,c_1),(r_2,c_2)) <_D ((r'_1,c'_1),(r'_2,c'_2))$.

For other functions in Fig. 3, *bots* is used to calculate a list of lists each of which comprises all rectangles with the same bottom edge. Symmetrically, *tops* calculates a list of lists each of which comprises all rectangles with the same top edge. They are defined by using another function *bts* which yields a list of rectangles passing through the matrix vertically (Fig. 2 (e)). The function *segs'* is almost the same as *segs* except that it is defined under the order of $\prec'$ rather than $\prec$. Although segs and segs' could be unified into a single function by means of overloading, they are defined independently for simplicity.

4 Derivation

Our derivation of an almost homomorphism from the specification in Fig. 3 is carried out in the following procedure.

1. Derive an almost homomorphism from the recursive definition of *segs2* (Sect. 4.1);
2. Fuse $(sum2*)$ with the derived almost homomorphism to obtain another almost homomorphism and repeat this fusion for *max* (Sect. 4.2);
3. Let $\pi_1 \circ (\!|f, \oplus|\!)$ be the result obtained in (2). If f or $\oplus$ are much complicated, repeat (1) and (2) to find an efficient parallel implementation for f and $\oplus$ (Sect. 4.3).

4.1 Deriving almost homomorphisms

Our approach is based on the following theorem. For notational convenience, we define $\Delta_1^n f_i = f_1 \vartriangle f_2 \vartriangle \cdots \vartriangle f_n$.

Theorem 1 (Tupling [HIT96]). *Let $h_1, \cdots, h_n$ be mutually defined by:*

$$h_i\,[\,] = \iota_{\oplus_i}$$
$$h_i\,[x] = f_i\,x \qquad (1)$$
$$h_i\,(xs \mathbin{+\!\!+} ys) = ((\Delta_1^n h_i)\,xs) \oplus_i ((\Delta_1^n h_i)\,ys)$$

Then $\Delta_1^n h_i = (\!|\Delta_1^n f_i, \Delta_1^n \oplus_i|\!)$, and $(\iota_{\oplus_1}, \cdots, \iota_{\oplus_n})$ is the unit of $\Delta_1^n \oplus_i$. □

Theorem 1 says that if h_1 is mutually defined with other functions (i.e., $h_2, \cdots h_n$) *traversing over the same lists* in the *specific form* of (1), then tupling $h_1, \cdots, h_n$ will give a list homomorphism. Let's see how the tupling theorem is used in deriving an almost homomorphism from the definition of *segs2* given in Sect. 3.

First, we determine what functions are to be tupled, i.e., $h_1, \cdots, h_n$. As the tupling theorem suggests, the functions to be tupled are those traversing over the same lists in the mutual definitions. So, from the definition of *segs2*:

$$segs2\ (xss \mathbin{+\!\!+} yss) = \underline{segs2\ xss} \mathbin{+\!\!+}_{\prec'} \underline{segs2\ yss} \mathbin{+\!\!+}_{\prec'} concat((\underline{bots\ xss})\Upsilon_{\chi_{+\!\!+}}(\underline{tops\ yss}))$$

we know that *segs2* should be tupled with *bots* and *tops*, because *segs2* and *bots* traverse over the same list xss whereas *segs2* and *tops* traverse over the same list yss as underlined. Similarly, the definitions of *bots* and *tops* requires that *bts* be tupled with *bots* and *tops*. In summary, the functions to be tupled are *segs2*, *bots*, *tops* and *bts*, i.e., our tuple function will be:

$$segs2 \mathbin{\vartriangle} bots \mathbin{\vartriangle} tops \mathbin{\vartriangle} bts.$$

Next, we rewrite the definition of each function in the above tuple to be in the form of (1), i.e., deriving $f_1, \oplus_1$ for *segs2*, $f_2, \oplus_2$ for *bots*, $f_3, \oplus_3$ for *tops*, and $f_4, \oplus_4$ for *bts*. This is straightforward. The results are as follows. For example, from the definition of *segs2*, we can easily derive that

$$f_1\ xs = [\cdot] * (segs'\ xs)$$
$$(s_1, b_1, t_1, d_1) \oplus_1 (s_2, b_2, t_2, d_2) = s_1 \mathbin{+\!\!+}_{\prec'} s_2 \mathbin{+\!\!+}_{\prec'} concat\ (b_1 \Upsilon_{\chi_{+\!\!+}} t_2)$$
$$f_2\ xs = [\cdot] * ([\cdot] * (segs'\ xs))$$
$$(s_1, b_1, t_1, d_1) \oplus_2 (s_2, b_2, t_2, d_2) = (b_1\ \Upsilon_{\lambda(x,y).(+\!\!+\ y)*x}\ d_2)\ \Upsilon_{+\!\!+}\ b_2$$
$$f_3\ xs = [\cdot] * ([\cdot] * (segs'\ xs))$$
$$(s_1, b_1, t_1, d_1) \oplus_3 (s_2, b_2, t_2, d_2) = t_1\ \Upsilon_{+\!\!+}\ (d_1\ \Upsilon_{\lambda(x,y).((x+\!\!+)*y)}\ t_2)$$
$$f_4\ xs = [\cdot] * (segs'\ xs)$$
$$(s_1, b_1, t_1, d_1) \oplus_4 (s_2, b_2, t_2, d_2) = d_1\ \Upsilon_{+\!\!+}\ d_2$$

Finally, we apply Theorem 1 and get the following list homomorphism.

$$segs2 \mathbin{\vartriangle} bots \mathbin{\vartriangle} tops \mathbin{\vartriangle} bts = (\!| \Delta_1^4 f_i, \Delta_1^4 \oplus_i |\!)$$

And our almost homomorphism for *segs2* is thus obtained:

$$segs2 = \pi_1 \circ (\!| \Delta_1^4 f_i, \Delta_1^4 \oplus_i |\!). \tag{2}$$

4.2 Fusion with Almost Homomorphisms

In this section, we show how to fuse a function with an almost homomorphism. Our fusion theorem for this purpose is given below.

Theorem 2 (Almost Fusion [HIT96]). *Let* $(\!|\Delta_1^n f_i, \Delta_1^n \oplus_i|\!)$ *and* h *be given. If there exist* $\otimes_i\ (i = 1, \cdots, n)$ *and a map* $\Lambda h = h_1 \times \cdots \times h_n$ *where* $h_1 = h$ *such that for all* i, $\forall x, y.\ h_i\ (x \oplus_i y) = (\Lambda h)\ x \otimes_i (\Lambda h)\ y$, *then* $h \circ (\pi_1 \circ (\!|\Delta_1^n f_i, \Delta_1^n \oplus_i|\!)) = \pi_1 \circ (\!|\Delta_1^n (h_i \circ f_i), \Delta_1^n \otimes_i|\!)$. □

Returning to our example, recall that we have reached the point:

$$mss2 = max \circ (sum2*) \circ (\pi_1 \circ (\!|\Delta_1^4 f_i, \Delta_1^4 \oplus_i|\!)).$$

We can fuse $sum2*$ with $\pi_1 \circ (\!|\Delta_1^4 f_i, \Delta_1^4 \oplus_i|\!)$ by Theorem 2, and then repeat this fusion for max, giving the following result.

$$mss2 = \pi_1 \circ (\!|\Delta_1^4 f_i', \Delta_1^4 \oplus_i'|\!) \qquad (3)$$

where

$$(s_1, b_1, t_1, d_1) \oplus_1' (s_2, b_2, t_2, d_2) = s1 \uparrow s_2 \uparrow (\uparrow\!/\,(b_1 \,\Upsilon_{\mathcal{X}_+}\, t_2))$$
$$(s_1, b_1, t_1, d_1) \oplus_2' (s_2, b_2, t_2, d_2) = (b_1 \,\Upsilon_+\, d_2) \,\Upsilon_\uparrow\, b_2$$
$$(s_1, b_1, t_1, d_1) \oplus_3' (s_2, b_2, t_2, d_2) = t_1 \,\Upsilon_\uparrow\, (d_1 \,\Upsilon_+\, t_2)$$
$$(s_1, b_1, t_1, d_1) \oplus_4' (s_2, b_2, t_2, d_2) = d_1 \,\Upsilon_+\, d_2$$

and

$$f_1' = max \circ (sum'*) \circ segs'$$
$$f_2' = (sum'*) \circ segs'$$
$$f_3' = (sum'*) \circ segs'$$
$$f_4' = (sum'*) \circ segs'$$

4.3 Improving Operators in List Homomorphisms

Equation (3) has given a homomorphic solution to the 2-dimensional maximum segment sum problem. Let n be the size of the input matrix. By a simple divide-and-conquer implementation of list homomorphisms, the derived program can expect an $max(O(\Delta_1^4 f_i'), (O(\log n) * O(\Delta_1^4 \oplus_i')))$ parallel algorithm. With assumptions that $\Upsilon_\otimes$ and $\mathcal{X}_\otimes$ can be implemented fully in parallel, i.e., $O(\Upsilon_\otimes) = O(\otimes)$ and $O(\mathcal{X}_\otimes) = O(\otimes)$, we can see that $O(\Delta_1^4 \oplus_i') = O(\log n)$ due to the inherited parallelism in the reduction ($\uparrow\!/$). It follows that $mss2$ is a

$$max(O(\Delta_1^4 f_i'), \ O(\log^2 n))$$

parallel algorithm. It is, however, not so obvious about efficient parallel implementation of, e.g., f_1' (similar to 1-dimensional mss problem except for different index order). We can derive (almost) list homomorphisms for it using the above derivation strategy again, giving a $O(\log n)$ parallel algorithm. (see [HIT96] for a detailed derivation for mss).

5 Concluding Remarks

In this paper, we demonstrate our derivation of an efficient parallel algorithm for the 2-dimensional maximum segment sum problem. It is based on the *manipulation* of (almost) homomorphisms, namely the construction of almost homomorphisms from recursive definitions (Theorem 1) and the fusion of a function with almost homomorphisms (Theorem 2). After the initial solution, all the derivation are proceeded in a formal setting based on our theorems and algebraic identities of list functions. Therefore, the resulting parallel algorithm is guaranteed to be semantically equivalent to the initial naive but inefficient solution. This is in sharp contract to Cole's informal approach [Col93b, Col93a].

Smith [Smi87] applied a strategy of divide-and-conquer approach to the same problem as an application. He constructs the composing operator (analog to our associative operator $\oplus$) by employing the suitable mathematical properties of the problem. Although our initial specification is less abstract than his, our derivation is more systematic and less prone to errors (Sect. 4) In [Smi87], the tuple consisting of eleven functions is given for the 2-dimensional *mss* problem, and the corresponding manipulation with Smith's approach must be cumbersome (Smith does not present them).

References

[Bir87]   R. Bird. An introduction to the theory of lists. In M. Broy, editor, *Logic of Programming and Calculi of Discrete Design*, pages 5–42. Springer-Verlag, 1987.

[Col93a]   M. Cole. List homomorphic parallel algorithms for bracket matching. Technical report CSR-29-93, Department of Computing Science, The University of Edinburgh, August 1993.

[Col93b]   M. Cole. Parallel programming, list homomorphisms and the maximum segment sum problems. Report CSR-25-93, Department of Computing Science, The University of Edinburgh, May 1993.

[CS92]   W. Cai and D.B. Skillicorn. Calculating recurrences using the Bird-Meertens Formalism. Technical report, Department of Computing and Information Science, Queen's University, 1992.

[HIT96]   Z. Hu, H. Iwasaki, and M. Takeichi. Construction of list homomorphisms via tupling and fusion. In *21st International Symposium on Mathematical Foundation of Computer Science* (MFCS '96), Cracow, September 1996. Springer-Verlag Lecture Notes in Computer Science.

[Smi87]   D.R. Smith. Applications of a strategy for designing divide-and-conquer algorithms. *Science of Computer Programming*, (9):213–229, 1987.

The Migrating Tasks:
An Execution Model for Irregular Codes

Yvon Jégou

IRISA/INRIA, Campus de Beaulieu, 35042 RENNES Cedex, FRANCE

Abstract. We introduce the use of the task migration paradigm for compiling and executing fine grain parallel applications on distributed memory computers. With this model, the execution of each iteration of a parallel loop is a task. When a task needs to access non local data, it migrates on the processing element which owns this data, and its execution resumes on this processor. The parallel data are never moved unlike the owner compute rule paradigm where the data is explicitly sent to the processor using it or in SVM systems where the data is implicitly requested. Our experiments show that good performance on irregular codes can be obtained on a distributed memory computer even when the application exhibits poor memory locality.

1 Introduction

Distributed Memory Parallel Computers (DMPC) are expected to offer a high degree of performance on computation intensive applications in the future. Several projects such as Fortran-D [1, 2], Vienna Fortran [3] and Pandore [4] are building parallel compilers that offer a parallel programming style by means of data distribution specifications and parallel statement constructs. Other projects [5] propose more conventional compilation techniques through the use of a shared virtual memory at run-time. The bottleneck for performances with all the distributed memory systems is the data movement: each time a non local datum is referenced, messages must be exchanged, processors or threads must synchronize, and data must move. In general, the compilation methods try to optimize these data motions. When enough information can be gathered at compile-time, efficient execution becomes possible on DMPC. However, a large class of codes contain irregular accesses to data: sparse matrix or unstructured mesh computations for instance. The potential parallelism of these codes is impressive in many cases. But, because of the irregular accesses to the memories, data migrations cannot be predicted at compile time.

The data migration patterns can be analyzed during the inspector phase before the real computation using the inspector/executor paradigm ([6]). Then the executor phase alternates "gather"/"scatter" periods (the data are moved) with execution periods (all memory accesses are local).

In this paper, we explore an alternative execution paradigm for parallel computation we call migrating tasks. A task is in charge of a piece of computation. In our model, a task is not attached to a fixed processor: it migrates in order to reach its data. The execution of a parallel construct is divided into small independent tasks (for instance, each task corresponds to one iteration of a parallel loop). When the tasks are independent, as it is the case for a parallel loop execution, the system is completely asynchronous. A good overlap of computation and communication can be obtained as soon as enough parallel tasks have been created.

The remainder of the paper is structured as following. The migrating task model is introduced in Sect. 2. Section 3 describes a code generation algorithm which allows automatic code production from classical programs. Because our system is completely asynchronous, the traditional distribution of control problems must be solved: a special distributed termination algorithm has been defined in order to detect the end of the execution of the parallel constructs. The description of this algorithm is out of the scope of this paper, and can be found in [8]. An irregular mesh computation was used to experiment our model on a Paragon X/PS. This experiment is presented in Sect. 4. Our technique is shown to produce some speed-up even when poor locality is encountered in the application.

2 The Migrating Task Model

2.1 Principles

A migrating task, [7], is in charge of a chunk of a program computation. During its execution, a migrating task can access the local memory only. If the task needs to read or write a variable in the memory of some other processor, it must migrate on this processor. Each migration involves copying all the private data of the task into the local memory of the destination processor in order to restart its execution.

In order to fully exploit the parallelism of the programs, a large number of tasks can exist at the same time in the system. For instance, one task can be created for the execution of each iteration of a parallel loop. As long as the iterations of the parallel loops are independent, no synchronization scheme needs to be defined between the tasks. When the execution of a task resumes on some processor, it continues until the task migrates or until the execution of the task ends. No information on the task in kept in the processor after a migration. The specificity of the migrating tasks is that parallel data do not move, but the task needing some data migrates on the processor owning the data location.

2.2 Application Domain of the Model

As long as the iterations of a loop are independent, there is no theoretical limitation to the use of migrating tasks for the execution of a program on parallel architectures. The loop body can contain control instructions such as conditionals, gotos, sequential or parallel loops as well as subroutine calls.

The current implementation of the migrating tasks is based on the SPMD execution model in order to reduce the volume of the task contexts (duplicated variables have the same value on all the processors and need not migrate). However, with this model, all the tasks must be created inside the same context: migrations are not possible inside subroutines called by migrating tasks.

Because the distributed data are accessed only on the local processor, many loops can be executed in parallel without synchronization as long as the dependencies occur during the same iteration and can be handled between two migrations. For instance, atomic updates inside the loop body do not prevent its parallel execution. A variable update occurs when the new value of a variable depends on its previous value (`A(P(i))= A(P(i))+ ...`). Distributed data can also be accessed outside parallel loops. In this case, only one task is created and migrates on the processors. The execution of a sequential loop containing distributed memory accesses is also handled by a migrating task. In this case, only one task is created and executes the whole loop.

2.3 SPMD Implementation

Under the SPMD execution model, the program code is duplicated on all the processors. The execution of the program alternates duplicated phases and migrating phases. During a duplicated phase, all the processors execute the same instructions and compute the same values. A migrating phase is initiated when instructions containing accesses to distributed data must be executed. Upon initiation of a migrating phase, the duplicated variables contain the same value on all the processors. The SPMD execution model allows a very light-weight implementation of the migrating tasks. Because the tasks never synchronize, and because once the execution of a task is initiated on some processor, it is never interrupted until the task ends or migrates, the current state of a task is completely defined by its private data. For the compiler point of view, a task migration consists in copying all the private data on the destination processor and in renaming the private variables after this copy.

In order to manage the memory accesses to the distributed objects, each distribution template defines three functions: a locality test function, an owner computation function and a local index function. A locality test is associated to each distribution template and determines if some element element is local to the current processor. This expression is a simple interval test in the case of block distribution. The owner expression produces the number of the processor owning some array element. The local index expressions computes the local index of an array element from the global index. In the case of block distribution, this computation is limited to subtracting the lower bounds of the array section mapped on the processor from the global index.

A scalar variable can also be distributed, in which case it is stored into the memory of only one processor. This is also the case for the files which are treated as distributed objects.

2.4 Expected Runtime Behavior of Migrating Tasks

With classical techniques, the access to a non local variable involves interactions of the requesting processor with the owning processor. At least one message must be transferred into each direction. In our model, when a task migrates, no information is kept on the sending processor and no synchronization request is produced. When the execution of some task resumes on a processor, this execution always continues until the task migrates or ends. This execution is never suspended. A processor is never waiting for a message. As long as the processor finds new tasks to execute, it will remain busy. This asynchronism is expected to allow a good overlap of computation and communication. The efficiency of each execution technique depends on many factors. The number of messages exchanged and the volume of data transfered are important factors. The overlap of computation and communication is another important factor. Maintaining the coherency of all the copies of duplicated data is another important factor and can also limit the performance of some techniques. The migrating task model does not suffer from these last two factors.

3 Code Generation for Migrating Tasks

In order to be effective, the production of migrating task codes from classical programming languages must be fully automatic and integrated into a compiler. Migrating tasks can be generated directly after a simple analysis of the program instructions. But, the analysis of the high level instructions allows many optimizations: common subexpressions factorization (multiple use of the same value by the same task, for instance), code reordering (access to all data that are local before migration).

We propose an algorithm based on information present in modern compilers: the data flow graph and information on array alignments. The data flow graph is complemented with dependency edges when necessary. Each node of the graph represents one expression or one instruction. Two types of nodes are considered in the data-flow graph: task nodes and distributed nodes. A task node contains only task local or duplicated data accesses. Such a node can be generated as soon as all the nodes it depends on have been generated. A distributed nodes contains one access to a distributed variable (read, write or update). A distributed node is into one among three states: the distributed state, the local state and the non-local state. In the distributed state, the compiler has no knowledge on the locality of the node; a locality test must be generated before considering this node. In the local state, the node contains a distributed data access that has been proved to be local. Such a node can be generated as soon as all the nodes it depends on have been generated. In the non-local state, the node contains one access to a distributed data on which the locality test has failed. A migration is necessary before generating such a node.

The code generation algorithm we propose is illustrated in Fig. 1. The current state contains the partially reduced data flow graph, the sets of local, non local

```
CodeGeneration(State):
    if a task or local node can be reduced then
        reduce one node and compute NewState
        CodeGeneration(NewState)
    else if a distributed node is reducible then
        generate locality test,
        compute FailureState and SuccessState
        CodeGeneration(SuccessState)
        CodeGeneration(FailureState)
    else if non local nodes are reducible then
        generate migration request and compute MigState
        CodeGeneration(MigState)
    else should be terminated
```

Fig. 1. code generation step

and distributed nodes and informations on alignments. During each recursive call to the code generation algorithm, one action among node reduction, locality test generation and migration production is selected on the current state and the algorithm is applied to the states resulting from this action.

Each transformed code section is replaced by a while loop containing the message handling and the task computations. This loop is in charge of the creation of the initial tasks and of the execution of the migrating tasks. Its execution stops once the termination of all the tasks is detected.

4 Experiments on the Paragon XP/S

The migrating task model has been experimented on part of a right hand side assembly code for unstructured triangular meshes. This code is described in [9]. The original Fortran code is shown in Fig. 2. Our code generation algorithm has not been completely incorporated in a real compiler at this time. So the data flow graphs of these loops were coded by hand and then transformed automatically by the code generation system. Figure 3 shows the data flow graph corresponding to the first loop. The symbols <> are placeholders for the private value of the tasks. During the last phase of code generation, these placeholders are replaced by the effective value: the source expression if the computation is local, a context location if a migration occurs between the source computation and its use.

Before running the code generation algorithm, we have considered the HPF [10] directives of Fig. 2 on data alignment. These HPF directives align the arrays in such a way that all the elements me(*, t) and cq(*, t) indexed with the same triangle number are located on the same processing element. All the array elements ynm1(n), yn(n), yn1(n) and v1(n) with the same mesh node number are also aligned.

This program was experimented on two versions of the same mesh: cav200kw.d and cav200kw.r. Version cav200kw.r was obtained from cav200kw.d by a simple renumbering algorithm which improve the locality of the memory accesses.

```
      subroutine assem(n,ns,nt,me,dt,ynm1,yn,ynp1,v1,cq,aire)
      implicit real*8(a-h,o-z)
      integer ns,nt,me(3,nt),i,j
      real*8 ynm1(ns),yn(ns),ynp1(ns),v1(ns)
      real*8 cq(6,nt),aire(nt)
      real*8 gxn,gyn,dt
!HPF$ Template triangles(nt)
!HPF$ Template nodes(ns)
!HPF$ Align me(*,:) with triangles(:)
!HPF$ Align (:) with nodes(:) :: ynm1, yn, ynp1, v1
!HPF$ Align cq(*,:) with triangles(:)
!HPF$ Distribute (Block) onto Processors :: triangles, nodes
c                          time loop
c                          *********
      do l=1,n
       do j=1,nt
        gxn= yn(me(1,j))*cq(1,j)+yn(me(2,j))*cq(3,j)+yn(me(3,j))*cq(5,j)
        gyn= yn(me(1,j))*cq(2,j)+yn(me(2,j))*cq(4,j)+yn(me(3,j))*cq(6,j)
        ynp1(me(1,j))=ynp1(me(1,j))+aire(j)*(gxn*cq(1,j)+gyn*cq(2,j))
        ynp1(me(2,j))=ynp1(me(2,j))+aire(j)*(gxn*cq(3,j)+gyn*cq(4,j))
        ynp1(me(3,j))=ynp1(me(3,j))+aire(j)*(gxn*cq(5,j)+gyn*cq(6,j))
       end do
       do i=1,ns
        y=yn(i)+dt*v1(i)+0.5*dt*ynp1(i)
        ynm1(i)=yn(i)
        yn(i)=y
        ynp1(i)=0.
       end do
      end do
      end
```

Fig. 2. original Fortran program

The simulation traces on 48 processors show that, for the original mesh, between 2 and 3% of the tasks were local to the creating processor. The other 97-98% tasks migrated. In the renumbered case cav200kw.r, the execution of 93% of the tasks is completely local and does not need migration.

The mesh contains $ns = 207691$ nodes and $nt = 411380$ triangles. All the runs have been executed with twenty iterations of the time-loop l.

The code was run on a Intel Paragon X/PS, 56 processors, running Mach/OSF1/1 Release 1.0.4 Server 1.3. On the Paragon system, all the timings have been measured with the local function dclock(). The code has been run with a number of processors ranging from 8 to 56. Paging occurs under eight processors. The speedup of the parallel executions have been computed using the execution time of the sequential program (without transformations) compiled with all possible optimizations.

On the two meshes, the computation time decreases when the number of processors is increased. Figure 4 shows the speedup obtained when these runs are compared to sequential execution.

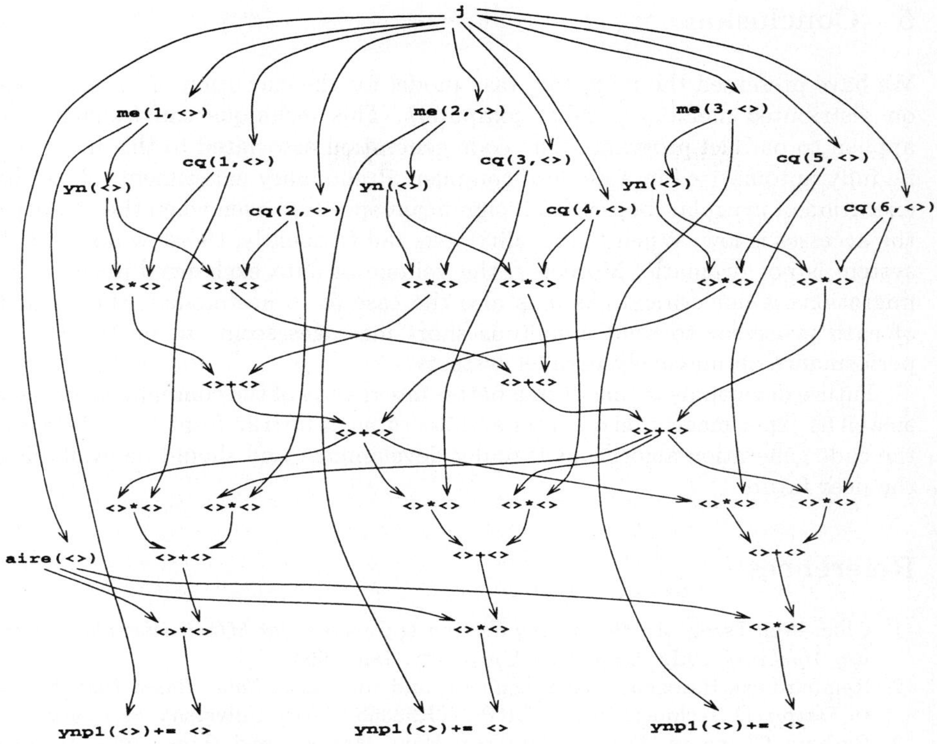

Fig. 3. data flow graph on loop j

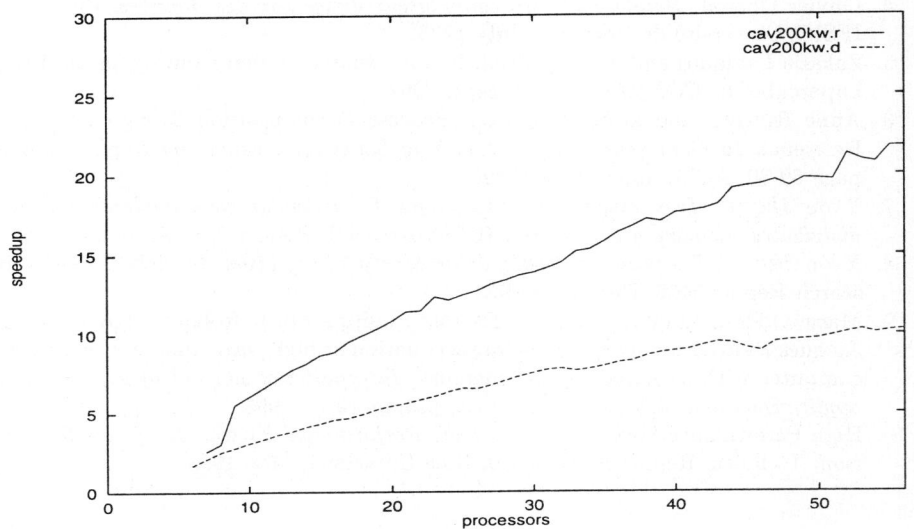

Fig. 4. speedup

5 Conclusion

We have presented the migrating task model for the execution of parallel codes on distributed memory parallel computers. This technique can frequently be applied to parallel programs. The code generation associated to this model can be fully automatized in a modern compiler. Preliminary experiments show that execution of irregular loops exhibit significant speed-up even when the locality of the accesses is low. When the locality tests fail frequently, the slow down of the system is not dramatic. Moreover, the volume of data exchanged during these migrations is not critical. As it is also the case for other models, the capacity of each processor to send numerous short messages seems to be the key for performance on massively parallel systems.

Future development consist in a better integration of the communication layer as well as experimentation on other architectures. A fortran front-end integrating the code generation algorithms is under development and should be available in the near future.

References

1. Chau-Wen Tseng. *An Optimizing Fortran D Compiler for MIMD Distributed Memory Machines*. PhD thesis, Rice University, Jan. 1993.
2. Reinhard van Hanxleden, Ken Kennedy, and Joel Saltz. *Value-Based Distributions in Fortran D*. Technical Report CRPC-TR93365-S, Rice University, February 1994.
3. Barbara Chapman, Piyush Mehrotra, Hans Moritsh, and Hans Zima. *Dynamic Data Distribution in Vienna Fortran*. Technical Report 93-92, ICASE/NASA LRC, December 1993.
4. Olivier Chéron. *Pandore II : un compilateur dirigé par des données*. PhD thesis, IFSIC/Université de Rennes I, July 1993.
5. Zakaria Lahjomri and Thierry Priol, Koan: a shared virtual memory for an iPSC/2 hupercube. In *CONPAR/VAPP*, Sept. 1992.
6. Anne ROGERS and keshav PINGALI, Process Decomposition Trough Locality of Reference, In *Conference on Programming Language Design and Implementation*, page 69-80, ACM, June 21-23 1989.
7. Yvon JÉGOU, *Task migration, a technique for irregular code implementation on distributed memory architectures*, IRISA Research Report 958, November 1995.
8. Yvon JÉGOU, *Exécution de codes irréguliers par migration de tâches*, INRIA Research Report 2436, December 1994.
9. Marie Odile BRISTEAU, Jocelyne ERHEL, Philippe FEAT, Roland GLOWINSKY and Jacques PÉRIAUX. Solving the helmotz equation at high wave numbers on a parallel computer with a shared virtual memory. *International journal of supercomputer applications and high performance computing*, (9.1), 1995.
10. High Performance Fortran Forum, *High Performance Fortran Language Specification*, Technical Report revision 1.0, Rice University, May 1993.

Discussing HPF Design Issues

Fabien COELHO (coelho@cri.ensmp.fr)

Centre de Recherche en Informatique, École des mines de Paris,
35, rue Saint-Honoré, F-77305 Fontainebleau CEDEX, FRANCE.
Phone: +33 1 64 69 48 52. URL: http://www.cri.ensmp.fr/pips

Abstract. As High Performance Fortran (HPF) is being both developed and redesigned by the HPF Forum, it is important to provide comprehensive criteria for analyzing HPF features. This paper presents such criteria related to three aspects: adequacy to applications, aesthetic and soundness in a language, and implementability. Some features already in HPF or being currently discussed are analyzed according to these criteria. They are shown as not balanced. Thus new or improved features are suggested to solve the outlined deficiencies: namely a scope provider, multiple mapping declarations and simpler remappings.

1 Introduction

Analyzing and discussing the design of a programming language [14] involves balancing issues related to targeted applications and possible implementations within a compiler. The design must also address the technical and aesthetic quality of features. HPF provides a data parallel programming model for distributed memory parallel machines. Users advise the compiler about data mapping and parallelism through directives, constructs and intrinsic functions added on top of Fortran 90 [9]. The compiler is expected to switch from the global name space and implicit communication of HPF to the distributed local name space and explicit communication of the target parallel machine. This paper studies HPF design issues, after three years of existence and as commercial compilers have been released. The language is being both developed and redesigned by the HPF Forum: new features are considered for addition and some are selected for defining an efficiently implementable kernel. There is indeed an opportunity to improve HPF on the application, language and implementation aspects, as features are questioned at light of experience.

HPF version 1.1 [6] includes directives for describing regular mappings onto multiprocessor machines, parallel constructs, and intrinsics considered as useful or expressing some parallelism. Array mappings are statically declared through **template** and **processors** objects using **align** and **distribute**. Arrays with the **dynamic** attribute may be changed their mapping at runtime with **realign** and **redistribute** *executable* directives. Subroutine arguments may be specified a *descriptive*, *prescriptive* or *transcriptive* mapping: it may be warranted, to be enforced or unknown with **inherit**. Parallel loops are tagged with **independent**. **new** specifies variables private to iterations. **forall** is a data-parallel construct which generalizes Fortran 90 array expressions. Intrinsics are available to query at runtime about data mappings and to specify reductions.

These features form the core of HPF. After three years of existence, they are reconsidered at the light of early compiler implementations, application ports or teaching sessions. The Forum is redesigning the language to allow efficient implementations. This paper presents criteria to analyze HPF features (Section 2). Some current and planed features are then analyzed with respect to these criteria (Section 3). They are shown as not balanced. This is the ground for new or improved HPF features which are suggested (Section 4) before concluding.

2 Analysis criteria

It seems natural to analyze language features with respect to the application, language and implementation aspects which are respectively the concern of users, designers and implementors. For each point, several issues are discussed, which express the interests each group might find in a feature.

2.1 Application

The preliminary goal of HPF is to suit data-parallel applications to be run efficiently on distributed memory parallel machines. HPF feature benefits for users can be seen through three perspectives.

Useful: some applications require or benefit from a feature for efficiency. Some unconstrained features such as remappings are not useful in their full generality and are difficult to handle for the compiler.

Expressive: HPF aims at providing to the user a mean to express its knowledge about the application in order to help the compiler. If some knowledge cannot be expressed while it would be useful, the language lacks something.

Elegant: notations must be concise, intuitive and as simple as possible. Thus they are easy to teach, learn and remember. Also users may misunderstood some features which are error-prone.

2.2 Language

From the designer point of view, language features must be neat, and have some technical properties which are considered as qualities, or their absence are considered as design mistakes and source of difficulties for users. It is

Homogeneous if different features share a common look and feel. This issue is illustrated in Fig. 1, where two homogeneous directive styles are presented. The first one applies on the next statement and is verbose, while the other is external and applies on the explicitly specified variables.

Orthogonal if features are not constrained one with respect to the other. HPF is built orthogonaly (with very few restrictions) on top of Fortran 90: it is up to the compiler to insure that HPF and Fortran features interact as expected. Also for instance Fortran 90 array expressions cannot specify a transposition, because dimensions are taken in the same order. HPF `forall` provides the array expression and dimension orthogonality.

Complete if all cases are expressed. The `inherit` mapping addresses this point. While this is a nice property for a language, handling any case with no motivation for efficiency can have a significant impact on the compiler costs, if not on the compiler structure.

```
!hpf$  independent                    !hpf$  independent(k,j), &
         do k=1, n                    !hpf$    new(s), reduction(p)
!hpf$    independent                           do k=1, n
           do j=1, n                             do j=1, n
!hpf$      new                                    s = 0.
             s = 0.                   !hpf$      independent(i), &
!hpf$      independent                !hpf$        new(t), reduction(s)
             do i=1, n                            do i=1, n
!hpf$        new                                    t = b(i,j,k)+c(i,j,k)
               t = b(i,j,k)+c(i,j,k)              a(i,j,k) = t*t+t
               a(i,j,k) = t*t+t                   s = s + t
!hpf$        reduction                            p = p * t
               s = s + t                        enddo
!hpf$        reduction                          p = p * s
               p = p * t                      enddo
             enddo                           enddo
!hpf$      reduction
             p = p * s
           enddo
         enddo
```

Fig. 1. Internal and external directive styles

2.3 Compiler

The last but not least issue is the implementability at a reasonable cost in a compiler. The state of the art evolves continuously because of research, but general conditions for sound and simple implementations remain. As the market for the language is not very large development costs due to useless features must be avoided. For the implementor a feature must be

Simple: features are added with *some* possible use in mind, but implementors must handle them in *all* cases, including the worst; it may also happen that some future optimizations are prevented or becomes more difficult because of complex features; this issues are illustrated below with remappings.

Efficient: to allow compilation to produce efficient code, as much static information as possible must be provided, and compile-time handling must not be prevented; see the **assume** directive suggestion in Section 4.1.

Separable: as interprocedural compiler technology [8] is not yet the common lot of software providers, features must not prevent separate compilation of subroutines; Fortran 90 explicit subroutine interfaces provide the mean to allow both maximum information about callees and separate compilation, if they are both mandatory and extended.

3 HPF feature analyses

Let us now analyze some HPF features according to these criteria. They are shown either as not simply and efficiently implementable or miss some instances. The difficulty of the design is to balance criteria for a given feature, in order to reach a sound proposal.

```
        subroutine callee(A)              !hpf$ distribute A(block)
        real A(n)                               if (...) then
!hpf$   inherit A                         !hpf$   redistribute A(cyclic)
        ! A mapping is not known!               endif
        A = ...                                 ! A is block OR cyclic
                                                A = ...
```

Fig. 2. Inherit unknowledge   **Fig. 3.** Remapping unknowledge

Inherit was added for completion purposes when specifying mappings of subroutine arguments. Some passed arguments cannot be described naturally with descriptive mappings, thus the need to be able to specify something else. Compilation of this feature is tricky because the compiler cannot assume any knowledge about the mapping (Fig. 2). Applications in which different mapping may reach a given subroutine tend to use this non-description while some knowledge could be available. Section 4.1 discusses the **assume** directive which can provide more knowledge to the compiler.

Remappings are well integrated in HPF, but not in compilers: they are definitely useful to application kernels such as ADI [10], FFT [7] and linear algebra [12]; the syntax is homogeneous to static mapping directives; remappings are orthogonal to the program control flow. This very orthogonality is the nightmare of implementors (if any): remappings may appear anywhere in the program, and may depend on runtime values. Thus for a given array reference the compiler may not know the mapping status of an array, as show in Fig. 3. This ambiguity similar to **inherit** has a strong impact on the way references can be handled, delaying everything to runtime, and preventing standard analyses and optimizations from giving their full benefice. This worst case must be handled before optimizing for programs where there is no such ambiguity. Section 4.2 discusses possible improvements in the design.

Internal style reductions (Fig. 1) were first considered by designers [13], and then changed [3]: although equivalent to the external syntax, this choice was not homogeneous with the **new** directive, which also deals with scalars. Moreover the external syntax eases the implementation: in the example the reduction on p is over the full loop nest while the one on s is only at the i level. This level is directly specified with the external syntax, while it must be extracted upward in the code checking for some condition with the internal syntax, requiring more work from the compiler.

Independent, New and Reduction are definitely not well integrated in the language: the directive styles are not homogeneous (Fig. 1): it is internal for **independent** and external for others. Moreover these directives are not orthogonal: **new** and **reduction** must be attached to an **independent**. However some non independent loops could benefit from these directives, as well as pieces of flat code. Section 4.3 develops an homogeneous and orthogonal proposal. This proposal does not change implementability issues, but allows more information to be given to the compiler.

```
assume-directive =
  ASSUME
    spec-directive-or-list
  END [ASSUME]

spec-directive-or-list =
  spec-directive-list
[ OR
  spec-directive-or-list ]

spec-directive-list =
  <nothing>
| spec-directive
[ spec-directive-list ]
```

Fig. 4. ASSUME directive proposal

```
attached-directive =
  explicit-independent-directive
| new-directive
| reduction-directive
| ...

explicit-independent-directive =
  INDEPENDENT(...)

scope-directive =
  BLOCK
    statements-and-directive-list
  END [BLOCK]
```

Fig. 5. Scope for orthogonal directives

```
      subroutine callee(A(100))
!hpf$ ASSUME
!hpf$   distribute A(block)
!hpf$ OR
!hpf$   distribute A(cyclic)
!hpf$ OR
!hpf$   ! nothing, not mapped
!hpf$ END ASSUME
```

Fig. 6. Assume example

```
!hpf$ DISTRIBUTE A(block,block)
      if (...) then
!hpf$   REDISTRIBUTE A(*, block)
        ...
      endif
! A is (*,block) or (block,*)
! thus must *not* be referenced
!hpf$ REDISTRIBUTE A(*,block)
```

Fig. 7. Remapping example

4 Improved and additional features

In the previous analysis, several deficiencies of HPF features were outlined: lacks of simple implementability, language homogeneity and orthogonality or expressiveness for some applications. In the following, several new or improved features are suggested to solve these problems.

4.1 Assume

HPF does not allow to describe several mappings that may reach subroutine arguments. Thus either **inherit** is used, warranting inefficiency, and/or subroutines are cloned by hand to build several versions to be compiled with more knowledge. There is a tradeoff between efficiency and software quality. The **assume** directive [2, 5] outlined in Fig. 4 allows to describe several mapping for arguments in the declaration part of the subroutine. Fig. 6 shows a simple example. It asserts that array A may reach the subroutine with a block or cyclic distribution, or no distribution at all (*i.e.* a local array).

It is up to the compiler to manage the cloning and selection of the right version. Some carefully designed constraints are necessary for allowing both simple and efficient implementations: (1) the **assume** directive must also appear in the explicit subroutine interface; (2) the subroutine argument mappings must appear in the same order in both declarations; These two points all together allow separate compilation and compile/link-time handling of the feature.

The expressiveness is maximum, because no constraints are needed on the mapping directives involved: descriptive and prescriptive mappings can be specified for the arguments, as well as on local variables, including partially redundant or useless specifications. Thus no special checking is required. Several and nested **assume** directives allows to describe the cross product of possible mappings. Constraints (1) and (2) allow callers and callees to agree about naming conventions for efficient and separate compile and link time implementation: when compiling a call, the list of assumed mappings is scanned for the best match and this version is called. Its name should depend on the version number, which is the same in both caller and callee declarations. From these constraints, a program with an assume specification can be translated by cloning into a standard HPF program as described in [5].

4.2 Remappings

As shown above remappings are perfectly integrated in the language and suited to some applications. It is perfect, but the implementation. Thus the current design of the HPF kernel excludes all remappings. We argue [2] that some could be kept, and especially those that are the most useful. This suggestion is supported by a full implementation in our prototype HPF compiler [4].

The following constraints are suggested: (1) all remappings must be statically known (*i.e.* as constrained as static mapping declarations); (2) array references should have at most one reaching mapping. These constraints disallow all ambiguities at the price of orthogonality: from (1) the mapping cannot depend on runtime values, and (2) implies that remappings cannot appear anywhere. Checking for (2) is similar to the the reaching definition standard data flow problem [1]. This second restriction does not apply to remapping statements: there may be some points in the program where the mapping of an array is ambiguous, provided that the array is not referenced at this point. This is illustrated in Fig. 7: between the **endif** and up to the last **redistribute** the array mapping is ambiguous to the compiler, thus the array must not be referenced. In such cases the runtime must keep track of the current status of an array in order to perform the right remapping. Remappings are much easier to compile under these constraints.

These restrictions do not prevent any use of remappings in the real application kernels cited in Section 3. Indeed, it is very hard to imagine a real application in which the desired mapping of some references would be unknown to the user. As computations are performed by the application, the user has a clear idea of the right mapping at different point and the remapping directives naturally reflect this information to the compiler without ambiguity. The pathological cases that are the nightmare of implementors are not those that would appear in any real application. Thus the suggested constraints, while allowing a much simpler implementation, does not prevent the use of remappings for efficiency.

```
!hpf$ new(time), &                  !hpf$ new(t), reduction(s)
!hpf$    reduction(s)               !hpf$ block
    do i=1, n                             t = a(1,1)+a(2,1)
       s = s + ...                        a(1,2) = t*t
       time = i*delta                     s = s + t ! out of a loop
       ! some use of time                 do i=1, n
    enddo                                    s = s + ...  ! in
    ! no use (read) of time                  ! some computation
                                          enddo
                                    !hpf$ end block
```

Fig. 8. Orthogonality of directives   **Fig. 9.** new and reduction scoping

4.3 Scope

As noted above, the **independent**, **new** and **reduction** directives are not homogeneous and orthogonal. Let us first allow for homogeneity the external syntax for **independent**. Fig. 8 shows a loop example where the non-orthogonality prevents the assertion of useful information: if iterations are performed across processors in the first example, **new** advises the compiler not to transmit and update the **time** variable between all processors, and **reduction** does the same for **s**, avoiding useless communication. These directives may also be useful outside of loops, as shown in Fig. 9 (with the syntax suggested hereafter): stencil computations typically use particular formula for corners that use temporary scalars which can be private to the processor holding the corner.

The scope proposal in Fig. 5 provides a solution to these problems. Directives are orthogonal one to the other, and apply on the next statement scope, which may be a loop or a scope-provider directive in the spirit of what is suggested for the **on** clause [11]. The **block/end block** provides scoping for flat regions of code. It would be obsoleted by the addition of such a structure to Fortran.

For consistency and homogeneity, these directives must be active for the next statement scope. Thus **independent(i)** implies that all loops with index **i** in the scope are parallel. Parallel loop indexes are necessarily and implicitly private. Moreover **new** and **reduction** do not make sense on **forall**, because no scalar can be defined within this construct. Scopes may be nested: in Fig. 1 scalar **s** is private to the full loop nest but is protected as a reduction in the inner loop. This proposal gives more homogeneity and orthogonality to HPF. Implementation issues are not changed: more information than before is available to the compiler, that can only benefit to applications.

5 Conclusion

HPF design has been discussed with respect to application, language and implementation issues. Several features have been analyzed and deficiencies outlined. Such an analysis is obviously much easier after experience. Then new features were suggested to improve the language with respect to our criteria. An **assume** directive, a new scope provider and simpler remappings were described. Possible simple and efficient implementation techniques were presented.

Other HPF features could also be discussed: replication of data through **align** seems tricky for users. The simpler and intuitive syntax suggested in Fortran D and Vienna Fortran could allow this specification in **distribute**. The **on** directive currently discussed is in the internal style spirit and includes a private scope provider. This does not improve HPF homogeneity and orthogonality.

Discussing design issues is a difficult task. Features are developed and discussed independently one of the other, thus homogeneity issues and possible interactions are easily overlooked in the difficult process of writing a sound and correct specification. Possible alternatives to design decisions are not obvious. Although a Forum may be able to select good ideas once expressed and formalized, such ideas are not always found. Some problems come only to light after experience. Teaching, implementing and using the language helps to clarify many points, to identify new problems and to offer new solutions.

Acknowledgments: Charles KOELBEL, Robert SCHREIBER, Henk SIPS and Henry ZONGARO for electronic discussions; Corinne ANCOURT, Denis BARTHOU, Béatrice CREUSILLET, Ronan KERYELL and the referees for comments.

References

1. A. V. Aho, R. Sethi, and J. D. Ullman. *Compilers Principles, Techniques, and Tools.* Addison-Wesley Publishing Company, 1986.
2. F. Coelho. Discussion about HPF kernel/HPF2 and so. hpff-distribute list, December 13, 1995.
3. F. Coelho. HPF reduce directive. hpff-task list, January 2, 1996.
4. F. Coelho and C. Ancourt. Optimal Compilation of HPF Remappings. CRI TR A 277, CRI, École des mines de Paris, Oct. 1995. To appear in JPDC, 1996.
5. F. Coelho and H. Zongaro. ASSUME directive proposal. TR A 287, CRI, École des mines de Paris, Apr. 1996.
6. H. P. F. Forum. *High Performance Fortran Language Specification.* Rice University, Houston, Texas, Nov. 1994. *Version 1.1.*
7. S. Gupta, C.-H. Huang, and P. Sadayappan. Implementing Fast Fourier Transforms on Distributed-Memory Multiprocessors using Data Redistributions. *Parallel Processing Letters,* 4(4):477–488, Dec. 1994.
8. F. Irigoin, P. Jouvelot, and R. Triolet. Semantical interprocedural parallelization: An overview of the PIPS project. In *ACM International Conference on Supercomputing,* pages 144–151, June 1991.
9. ISO/IEC. *International Standard ISO/IEC 1539:1991 (E),* second 1991-07-01 edition, 1991.
10. K. Kennedy and U. Kremer. Automatic Data Layout for High Performance Fortran. CRPC-TR94 498-S, Center for Research on Parallel Computation, Rice University, Dec. 1994.
11. C. Koelbel and R. Schreiber. ON clause proposal. hpff-task list, January 8, 1996.
12. L. Prylli and B. Tourancheau. Efficient block cyclic data redistribution. Research Report 2766, INRIA, Jan. 1996. To appear in Europar'96.
13. R. Schreiber, C. Koelbel, and J. Saltz. reductions. hpff-task list, January 2, 1996.
14. B. Stroustrup. *The Design and Evolution of C++.* Addison-Wesley Publishing Company, 1994.

Parallelizing Conditional Recurrences

W.N. CHIN[1] and J. DARLINGTON[2] and Y. GUO[2]

[1] DISCS, National University of Singapore
[2] DoC, Imperial College of Science, Technology and Medicine

Abstract. Recursive functions which use conditional constructs are common in functional (and imperative) programs. We present a collection of techniques for handling such functions for a parallel synthesis method. These techniques can help us enlarge the class of sequential functions which could be systematically transformed to parallel equivalent.

1 Introduction

Most programs are more easily expressed via sequential code. Functional programs are no exception. As a simple example, consider the following program to find the product of a list.

```
data List a = Nil | Cons(a,List a) ;
product(Nil)    = 1;
product(x:xs)   = x*product(xs);
```

The first statement declares a *List* data type with two constructors *Nil* and *Cons*. This *List* type has a shorthand notation $[a_1, a_2, .., a_n]$ for a list of *n*-elements where $n \geq 0$, and an infix notation *(x:xs)* for *Cons(x,xs)*. The next two statements are pattern-matching equations of *product* for a base case and a recursive case. In the latter case, the result is computed sequentially via *x*product(xs)*. By using the fact that the operator * is associative, the parallel synthesis method proposed in [Chi95] can derive the following definition for *product*.

```
product(Nil)     = 1;
product([x])     = x;
product(xr++xs)  = product(xr) * product(xs);
```

Note the *++* operator is to split an input list into two halves. Details of our proposed parallelization methodology is described in [Chi95]. It comprises four main stages, which shall be briefly summarised below using *product* function as an example.

Procedure 1: *Four-Stage Parallel Synthesis Method*

Stage 1: Determine a desired *pre-parallel form* for the initial recursive equation.
Stage 2: Obtain a second recursive equation with the *same* pre-parallel form.
Stage 3: Use *second-order generalisation* to obtain an *equation template* from the two pre-parallel equations. This template may have one or more unknown functional variable(s).
Stage 4: *Derive* the unknown functional variable(s).

Stage 1 tries to obtain a desired pre-parallel form for the initial recursive equation. This is assisted by two simple heuristics:

Definition 1: *Heuristics for Desired Pre-Parallel Form*

(H1) All function calls (e.g. *, *product*), leading to the recursion variables (e.g. xs) should be either *associative* or *distributive*.
(H2) The depth of recursion variables (e.g. xs) from the root of its equation should be as *shallow* as possible.

The first heuristic is needed to match the associativity of the * operator in the LHS. The second heuristic is to reduce the number of function operators leading to the recursion variables. As these have to be associative (or distributive), the fewer the better. Guided by the two heuristics, we can obtain:

product($\underline{[x]}$++xs)   = $\underline{x}$ + product(xs)

The underlined sub-expressions are known as *holes* of the pre-parallel equation. Formally, each maximal sub-term, not including any recursion variables, shall be known as a *hole* of the pre-parallel equation. Given an expression e, we can decompose it via $\widehat{e^P}\langle h_i \rangle_{i \in 1..n}$ where $\widehat{e^P}\langle \rangle$ is a *pre-parallel context* (or *form*), and $\{h_i\}_{i \in 1..n}$ are the *pre-parallel holes*.

Stage 2 is to obtain a second recursive equation with *identical* pre-parallel context as the first equation. We unfold the self-recursive call followed by transformations (guided by the earlier two heuristics), to obtain:

product(($\underline{[x]++[y]}$)++xs)   = $\underline{(x+y)}$+ product(xs)

With two similar pre-parallel equations, Stage 3 will attempt a matching cum generalisation procedure to obtain a *parallel template equation*. The matching process will apply (possibly second-order) generalisations to those pre-parallel holes which *mismatch*. In the LHS, a pair of mismatched holes, *[x]* (first equation) and *[x]++[y]* (second equation), is resolved by replacing the two sub-terms by a new variable, xr, generalising the LHS to *product(xr++xs,c)*. The new LHS parameter, xr++xs, has two variables, xr and xs, called *active* and *passive* recursion variables, respectively. In the RHS, a mismatch occurs for the expressions: x (first equation) and (x+y) (second equation). This mismatch is resolved by supplying a generalised expression *H(xr)* where *H* is a function-type variable and xr is the active recursion variable. This generalisation is second-order since a function variable, *H*, is used. The resulting parallel template is:

product(xr++xs) = H(xr) + product(xs);

Lastly, Stage 4 is used to derive inductive definitions for the unknown functions. This stage involves *instantiation* of the active recursion variable; followed by *simplification* and *induction* to obtain identical pre-parallel forms (for both the LHS and RHS), before *unification* to obtain new definitions for the unknown functions. In the case of *product*, we can obtain:

H(Nil)   = 1;
H(x:xr)   = x + H(xr);

This definition of *H* is syntactically identical to the definition of *product*. Hence, we can assert that $H \equiv product$ holds.

This paper proposes an extension, to handle conditional recurrences, for our parallelization methodology. We introduce some notations and a key idea on conditional normalisation (in Sec 2). This is followed by a collection of techniques to handle different sub-classes of conditional recurrences (in Sec 3, 4 and 5). Lastly, related work are described before a conclusion (in Sec 6).

2 Conditional Recurrence and its Normalisation

We consider a first-order functional language. Valid expressions include variables (v), data constructors (C), functions (f), *let* and *if* statements. We shall abbreviate a list of terms $t_1, \ldots, t_n$ by $\vec{t}$, so that a function call $f(t_1, \ldots, t_n)$ could be abbreviated as $f(\vec{t}\,)$, or more precisely as $f(t_i)_{i \in 1..n}$. We introduce a special context notation with multiple holes.

Definition 2: *Context with Multiple Holes*
A *hole*, $\langle\rangle_m$, is a special variable labelled with a number, m.

A *context*, $\widehat{e}\langle\rangle$, is an expression with a finite number of holes, defined by the grammar:
$$\widehat{e}\langle\rangle \;::=\; v \;\mid\; \langle\rangle_m \;\mid\; C(\widehat{e}\langle\rangle^*) \;\mid\; f(\widehat{e}\langle\rangle^*) \;\mid\; let \; \{p = \widehat{e}\langle\rangle\}^+ \; in \; \widehat{e}\langle\rangle$$
$$\mid\; if \; \widehat{e}\langle\rangle \; then \; \widehat{e}\langle\rangle \; else \; \widehat{e}\langle\rangle \;;$$
Note that S^* denotes zero or more occurrences of S, while S^+ denotes one or more occurrences.

Each expression e can be decomposed into a context $\widehat{e}\langle\rangle$ and a sequence of sub-terms $[t_i]_{i \in 1..n}$ using the notation $e \equiv \widehat{e}\langle t_i \rangle_{i \in 1..n}$. This stands for the substitutions of sub-terms, $t_1, \ldots, t_n$, into their respective holes, $\langle\rangle_1, \ldots, \langle\rangle_n$, for context $\widehat{e}\langle\rangle$. We define self-recursive function calls and variables, as follows:

Definition 3: *Self-Recursive Function Calls and Variables*
A function, f_i, is said to be *self-recursive* if its (mutual) recursive set of functions consists of just $\{f_i\}$. Each occurrence of a self-recursive call, $f_i(\vec{t}\,)$, in the RHS of f_i shall be annotated as $f_{i_\#}(\vec{t}\,)$.
Consider *let* $(\vec{v})=f_{i_\#}(\vec{t}\,)$ *in* e. The variables $\vec{v}$ will be called *self-recursive variables*, and shall be annotated as $\vec{v}_\#$.

We mark a sub-expression, e, of an equation as either e^R (*recursive*) or $e^{\overline{R}}$ (*non-recursive*), depending on whether it has self-recursive functions/variables or not. To facilitate the parallelization of conditional recurrences, we propose a guarded conditional that is to be used as intermediate representation during transformation. This conditional has the form:

$gc \quad ::= \; cond \; \{b_i \rightarrow e_i\}_{i \in 1..n}$

where *exactly* one of the tests $b_1, .., b_n$ can be true; while $e_1, .., e_n$ are their respective branches.

To obtain suitable pre-parallel form, we shall transform each recursive expression with conditional(s) into a *canonical form* during Stages 1 and 2. In this form, each self-recursive call has at most one outer conditional (and one outer *let* construct), as specified next.

Definition 4: *Normalised Conditional Form*
An expression, without any outer conditional (or *let* construct) prior to its self-recursive calls, is defined by:
$$e^F ::= e^{\overline{R}} \mid v \mid v_\# \mid c(e^F{}^*) \mid f(e^F{}^*) \mid f_\#(e^{\overline{R}}{}^*)$$
Correspondingly, an expression with at most one outermost conditional and *let* for each self-recursive call is defined by:
$$e^N ::= let\ \{p = e^F\}^+\ in\ e^C \mid e^C$$
$$e^C ::= e^F \mid cond\ \{e^F \to e^F\}^*$$
An expression is said to be a *normalised conditional expression* if (i) it satisfies grammar term e^N, and (ii) the outermost conditional has at most one non-recursive branch.

Though only one outer guarded conditional is allowed prior to self-recursive calls of e^C, inner conditionals are still permitted in non-recursive expressions via $e^{\overline{R}}$ in the definition of e^F. To obtain normalised conditional expressions, we have the following procedure.

Procedure 2: *Normalisation of Conditional Expressions*

Step 1 Convert each simple (recursive) conditional to guarded conditional.
Step 2 Float out conditional (and let) constructs, where possible.
Step 3 Flatten nested conditional (and let) constructs.
Step 4 Combine all non-recursive branches of outermost conditional.
Step 5 Simplify conditionals, where possible.

The conditional normalisation rules are given in Figure 1. (Note that $\widehat{e^I}\langle\rangle$ is a special context which must not contain any conditional/*let* constructs, while $\widehat{e^L}\langle\rangle^V$ must not contain *let* constructs with bound variables from V.) To ensure that these rules are applied deterministically, we always transform the leftmost, outermost sub-term first.

The normalisation steps will be used to help parallelize three sub-classes of conditional recurrences, codenamed **LC1** with a *single recursive branch*, **LC2** with *multiple recursive branches*, and **LC3** with *recursive test(s)*.

3 Single Recursive Branch (LC1)

A normalised conditional recurrence is said to have a *single recursive branch* if its outermost conditional has exactly one recursive branch. An example is:

$ssc(Nil,p) = Nil$;
$ssc(x:xs,p) = if\ p(x)\ then\ p:ssc(xs,p)\ else\ Nil$;

Using the conditional normalisation procedure, we can obtain the following pre-parallel equations in Stages 1 & 2, respectively.

$ssc(\underline{[x]}$++$xs,\underline{p}) \quad = cond\ \{\ \underline{p(x)} \to \underline{[x]}$++$ssc(xs,\underline{p});$
$\qquad\qquad\qquad\qquad\qquad \underline{\neg\ p(x)} \to \underline{Nil}\ \}$
$ssc((\underline{[x]}$++$\underline{[y]})$++$xs,\underline{p}) = cond\ \{\ \underline{p(x) \wedge p(y)} \to (\underline{[x]}$++$\underline{[y]})$++$ssc(xs,\underline{p});$
$\qquad\qquad\qquad\qquad\qquad \underline{\neg\ (p(x) \wedge p(y))} \to cond\ \{\underline{p(x)} \to \underline{[x]}$++$Nil; \underline{\neg\ p(x)} \to \underline{Nil}\}\}$

$$
\begin{aligned}
(\mathcal{N}1) \quad & \text{if } b \text{ then } e_1 \text{ else } e_2 \equiv cond \ \{b \to e_1; \neg b \to e_2\} \\
& \text{IF } (b \in e^R) \vee (e_1 \in e^R) \vee (e_2 \in e^R)
\end{aligned}
$$

$$
\begin{aligned}
(\mathcal{N}2a) \quad & \widehat{e^I}\langle cond \ \{b_i \to e_i\}_{i \in N} \rangle \\
& \equiv cond \ \{b_i \to \widehat{e^I}\langle e_i \rangle\}_{i \in N} \\
& \text{IF } \exists i \in N. \ (b_i \in e^R) \vee (e_i \in e^R)
\end{aligned}
$$

$$
\begin{aligned}
(\mathcal{N}2b) \quad & \widehat{e^L}\langle let \ \{v_i \to e_i\}_{i \in N} \ in \ e \rangle^V \\
& \equiv let \ \{v_i \to e_i\}_{i \in N} \ in \ \widehat{e^L}\langle e \rangle^V \\
& \text{IF } (\exists i \in N. \ e_i \in e^R) \vee (e \in e^R) \ \& \ V = \bigcup_{i \in N} vars(e_i))
\end{aligned}
$$

$(\mathcal{N}3a) \ cond \ B \cup \{b_i \to cond \ \{b_{i_j} \to e_{i_j}\}_{j \in M}\} \equiv cond \ B \cup \{b_i \wedge b_{i_j} \to e_{i_j}\}_{j \in M}$

$(\mathcal{N}3b) \ cond \ B \cup \{cond \ \{b_{i_j} \to e_{i_j}\}_{j \in M} \to e_i\} \equiv cond \ B \cup \{\bigvee_{j \in M}(b_{i_j} \wedge e_{i_j}) \to e_i\}$

$(\mathcal{N}3c) \ let \ \{p_i = e_i\}_{i \in N} \ in \ (let \ \{p_i = e_i\}_{i \in M} \ in \ e) \equiv let \ \{p_i = e_i\}_{i \in (N \cup M)} \ in \ e$

$$
\begin{aligned}
(\mathcal{N}3d) \quad & let \ \{p_i = e_i\}_{i \in N} \cup \{v = cond \ \{b_j \to er_j\}_{j \in M}\} \ in \ e \\
& \equiv let \ \{p_i = e_i\}_{i \in N} \ in \ cond \ \{b_j \to [er_j/v]e\}_{j \in M} \\
& \text{IF } \exists j \in M. \ (b_j \in e^R \vee er_j \in e^R)
\end{aligned}
$$

$$
\begin{aligned}
(\mathcal{N}4) \quad & cond \ \{b_i \to e_i\}_{i \in N \cup M} \equiv cond \ \{b_i \to e_i\}_{i \in N} \cup \{\bigvee_{j \in M} b_j \to \\
& \quad cond \ \{b_i \to e_i\}_{i \in M}\} \ \text{IF } \forall i \in M. \ (b_i \in e^{\overline{R}} \wedge e_i \in e^{\overline{R}})
\end{aligned}
$$

$(\mathcal{N}5a) \ cond \ \{b_i \to e\}_{i \in N} \equiv e$

$(\mathcal{N}5b) \ cond \ B \cup \{True \to e\} \equiv e$

$(\mathcal{N}5c) \ cond \ B \cup \{False \to e\} \equiv cond \ B$

Fig. 1. Rules for Conditional Normalisation

In Stage 3, second-order generalisation can be applied to obtain the following parallel equation template with unknown functions H and G.

$ssc(xr\!+\!+xs,p) \qquad = cond \ \{H(xr,p) \to xr\!+\!+ssc(xs,p); \ \neg H(xr,p) \to G(xr,p)\}$;

In Stage 4, we can infer that $G \equiv ssc$, while H has the definition:

$H(Nil,p) \qquad = True$;
$H(x\!:\!xs,p) \qquad = p(x) \wedge H(xs,p)$;

The new H function can also be subjected to the parallel synthesis method in order to obtain:

$H(xr\!+\!+xs,p) \qquad = H(xr,p) \wedge H(xs,p)$;

The above equations are parallel but not efficient due to the presence of redundant H calls. This is indicated by the presence of calls with common recursion argument, e.g $H(xr,p)$ and $ssc(xr,p)$. To eliminate this redundancy, we can use the automatic tupling method of [Chi93] to introduce a new definition:

$tup(xs,p) \qquad = (H(xs,p), ssc(xs,p))$;

before it is transformed to:

$tup([x],p) \qquad = (p(x), \text{if } p(x) \text{ then } [x] \text{ else } Nil)$;
$tup(xr\!+\!+xs,p) \qquad = let \ \{(u,v)=tup(xr,p); (a,b)=tup(xs,p)\} \ in$
$\qquad\qquad\qquad\qquad (u \wedge a, \text{if } u \text{ then } xr\!+\!+b \text{ else } v)$;

Such functions are both parallel and efficient. Using a tuple of values, the results of function calls have been memoised to avoid redundant computation.

4 Multiple Recursive Branches (LC2)

The next sub-class of conditional recurrence we consider are those with *multiple recursive branches*. An example (with two recursive branches) is:

filter(Nil,p)   = Nil ;
filter(x:xs,p)   = if p(x) then x:filter(xs,p) else filter(xs,p) ;

To handle conditional recurrence with multiple branches, we shall use a preprocessing technique to convert each such recurrence to one with at most a single recursive branch. The steps are outlined below:

Procedure 3: *Combining Multiple Recursive Branches*

Step 1 Find a common pre-parallel form for the different recursive branches.
Step 2 Combine and unify the multiple recursive branches.

Step 1 is to find a common pre-parallel form for all the recursive branches. This is achieved by obtaining a desired pre-parallel form for each of the branches, before attempting to unify them via suitable transformations. In the case of filter function, the desired pre-parallel form for the first recursive branch is [x]++filter(xs,p̲), while that of the second branch is filter(xs,p̲). To make these two pre-parallel expressions unifiable, we would transform the second recursive branch to Nil++filter(xs,p̲).

With a common pre-parallel context, Step 2 would now combine all recursive branches into a single conditional. This is achieved by rule $\mathcal{N}4$ but with M being the set of recursive branches. After that, the multiple recursive branches are unified into a single branch by pushing the outer conditional into each hole of the common pre-parallel context. Assuming that $\widehat{e^P}\langle\rangle$ is a common pre-parallel context for multiple branches, the following rule is needed.

$(\mathcal{N}6) \quad \text{cond } \{b_i \rightarrow \widehat{e^P}\langle ti_j \rangle_{j \in M}\}_{i \in N} \equiv \widehat{e^P}\langle \text{cond}\{b_i \rightarrow ti_j\}_{i \in N}\rangle_{j \in M}$
$$\text{IF } \forall i \in N. \, j \in M. \, (ti_j \in e^{\overline{R}})$$

Applying it to *filter*, we obtain:

filter([x]++xs,p) = cond $\{p(x) \rightarrow$ [x]++filter(xs,p); $\neg p(x) \rightarrow$ Nil++filter(xs,p)$\}$; $\mathcal{N}6$
              = (cond $\{p(x) \rightarrow$ [x]; $\neg p(x) \rightarrow$ Nil$\}$)
                ++filter(xs,cond $\{p(x) \rightarrow p; \neg p(x) \rightarrow p\}$) ; $\mathcal{N}5a$
              = (cond $\{p(x) \rightarrow$ [x]; $\neg p(x) \rightarrow$ Nil$\}$)++filter(xs,p) ;

Once we have unified the multiple recursive branches into a single recursive branch, we can proceed with the earlier technique for the **LC1** sub-class.

5 Recursive Tests (LC3)

Another scenario for conditional recurrence involves multiple occurrences of a recursive call in the conditional's test, and its branches. An example is:

max([x])   = x ;
max(x:xs)   = if x>max(xs) then x else max(xs) ;

The recursive call *max(xs)* appears in both the test and a branch of the conditional. To make this equation into a linear recurrence, we use a *let* statement to abstract out the recursive call, as follows:

$max([x]++xs)$ $\quad = let\ z_\# = max(xs)\ in\ if\ \underline{x} > z_\#\ then\ \underline{x}\ else\ z_\#$;

Note that variable z is a proxy for the self-recursive call *max(xs)*, and is thus marked with the recursive annotation, $\#$. At the moment, there are two pre-parallel holes (shown underlined). However, since these holes represent the same variable x, we could reduce the number of pre-parallel holes (also an effect of heuristic H2), by abstracting it via a *let* variable, as follows:

$max([\underline{x}]++xs)$ $\quad = let\ \{z=max(xs);\ v=\underline{x}\}\ cond\ \{v>z \rightarrow v;\ \neg\ v>z \rightarrow z\}$;

Subsequently, in Stage 2, we perform a further unfold of *max* before applying the conditional normalisation procedure to obtain:

$max([x]++([y]++xs)) = let\ \{z'_\# = max(xs)\}\ in$
$\qquad\qquad\qquad\qquad\quad cond\ \{(y>z'_\#) \wedge (x>y) \rightarrow x\ ;$
$\qquad\qquad\qquad\qquad\qquad\quad\ (y>z'_\#) \wedge \neg(x>y) \rightarrow y\ ;$
$\qquad\qquad\qquad\qquad\qquad\quad\ \neg(y>z'_\#) \wedge (x>z'_\#) \rightarrow x\ ;$
$\qquad\qquad\qquad\qquad\qquad\quad\ \neg(y>z'_\#) \wedge \neg(x>z'_\#) \rightarrow z'_\#\}$;

At this point, the second pre-parallel equation has four conditional branches, while the original pre-parallel equation has only two. To simplify this more complex conditional, we use another procedure with three steps: (Step 1) Choose a non-recursive test to float out, (Step 2) Simplify the inner guarded conditionals, and (Step 3) Unify the inner guarded conditionals.

The three tests in the outermost conditional are $\{y>z'_\#,\ x>y,\ x>z'_\#\}$. Two of the tests are *recursive*, but $x>y$ is *non-recursive* since it did not depend on the self-recursive variable, $z'_\#$. Step 1 will choose this non-recursive test to float out, using the rule: $e \equiv cond\ \{(x > y) \rightarrow e; \neg(x > y) \rightarrow e\}$.

This step pushes all recursive tests and branches to inner conditionals. After this, the non-recursive assertions, $x>y$ and $\neg(x>y)$, can be used to simplify the inner conditionals in Step 2. Applying appropriate simplification results in:

$max(([x]++[y])++xs) = let\ z'=max(xs)\ in\ cond$
$\qquad\qquad\qquad\qquad\quad \{\ (x>y) \rightarrow\ let\ v=\underline{x}\ in\ cond\ \{\ (v>z') \rightarrow v\ ;\ (v \leq z') \rightarrow z'\ \}\ ;$
$\qquad\qquad\qquad\qquad\quad\ \ (x \leq y) \rightarrow\ let\ v=\underline{y}\ in\ cond\ \{\ (v>z') \rightarrow v\ ;\ (v \leq z') \rightarrow z'\ \}\ \}$;

The two inner recursive conditionals now have the same pre-parallel form. In Step 3, we unify these multiple recursive branches to obtain:

$max(([x]++[y])++xs)\ = let\ z'=max(xs)\ in$
$\qquad\qquad\qquad\qquad\qquad let\ v=\underline{cond\ \{x>y \rightarrow x\ ;\ x \leq y \rightarrow y\}}\ in$
$\qquad\qquad\qquad\qquad\qquad \overline{cond\ \{(v>z') \rightarrow v\ ;\ (v \leq z') \rightarrow z'\}}$;

This last equation has the same pre-parallel form as the first equation. Proceeding with Stages 3 & 4 can yield the expected parallel equation.

6 Related Work & Conclusion

A well-known approach for synthesizing parallel functional programs is through Bird-Meertens formalism [Ski90]. The emphasis there is to construct programs using a small set of higher-order functions (such as *map, reduce*) from which it is often possible to directly detect/derive the divide-and-conquer homomorphism *without induction*. However, the homomorphism sub-class is rather limiting since many programs (e.g. *ssc*) lie outside it. Cole realised this limitation and attempted to identify a larger class of so-called *near-homomorphism* [Col93]. His attempt is somewhat informal and appears difficult to automate.

In traditional imperative languages (e.g. Fortran), there are also many ongoing efforts at developing sophisticated techniques for parallelizing iterative loops. Lately, we became aware of a more systematic method for parallelizing complex scans and reductions [FG94]. This method is based on a parallel reduction of *function composition* where function-type values are propagated. This is allowed because function composition is associative. However, the complexity of the functions propagated could get progressively worse unless they match a certain *template* form. The steps needed for finding such template form is similar to our technique for finding a parallel template equation. Our technique was discovered independently with the initial methodology presented in [Chi90].

As illustrated in this paper, a more systematic approach is possible by augmenting the unfold/fold rules [BD77] with induction/generalisation steps. There is no need to restrict user programs to a closed set of higher-order functions. Our method is capable of generating suitable parallel equations, even in the presence of conditional expressions. The systematic formulation presented is an initial step towards an automation of the proposed parallelization methodology.

Acknowledgement: Work supported by EEC KIT143 grant for international scientific cooperation, and NUS research grant RP920614.

References

[BD77] R.M. Burstall and J. Darlington. A transformation system for developing recursive programs. *Journal of ACM*, 24(1):44–67, January 1977.

[Chi90] Wei-Ngan Chin. *Automatic Methods for Program Transformation*. PhD thesis, Imperial College, University of London, March 1990.

[Chi93] Wei-Ngan Chin. Towards an automated tupling strategy. In *3rd ACM Symposium on Partial Evaluation and Semantics-Based Program Manipulation, ACM Press*, pages 119–132, Copenhagen, Denmark, ACM Press, June 1993.

[Chi95] Wei-Ngan Chin. Synthesizing efficient parallel programs from sequential specifications. Technical report, (TRA9/95) Dept of IS/CS, NUS, September 1995.

[Col93] M Cole. Parallel programming, list homomorphism and the maximum segment sum problem. CSR 25-93, University of Edinburgh, 1993.

[FG94] AL. Fischer and AM. Ghuloum. Parallelizing complex scans and reductions. In *ACM PLDI*, pages 135–136, Orlando, Florida, ACM Press, 1994.

[Ski90] D. Skillicorn. Architecture-independent parallel computation. *IEEE Computer*, 23(12):38–50, December 1990.

Adaptive Data Parallel Computation in the Parallel Object-Oriented Language *OCore*

Hiroki Konaka, Yoshiaki Itoh, Takashi Tomokiyo,
Munenori Maeda, Yutaka Ishikawa, Atsushi Hori

Real World Computing Partnership,
1-6-1 Takezono, Tsukuba, Ibaraki 305, JAPAN

Phone: +81-298-53-1662

Fax: +81-298-53-1652

E-mail: {konaka,itoh,tomokiyo,m-maeda,ishikawa,hori}@rwcp.or.jp

Workshop: W05 Parallel Languages and Programming

Abstract

Data parallel computation is a typical parallel computing paradigm. It covers a wide range of application areas, some of which require the data set to vary dynamically to adapt to changes in the target system. In this paper, we will show how to implement adaptive data parallel computation in the parallel object-oriented language *OCore*. After giving an overview of *OCore*, we will present adaptive word recognition as a realistic example of adaptive data parallel computation.

1 Introduction

Data parallel computation, where many data items are subject to identical processing, is a typical parallel computing paradigm. It covers a wide range of application areas, some of which require the data set to vary dynamically to adapt to changes in the target system. However, such adaptive data parallel computation has not received much attention from existing parallel languages.

We are developing a parallel object-oriented language, called *OCore* [6]. *OCore* introduces the notion of *communities* to reduce the complexity in writing efficient parallel programs for multi computers. Communities support data parallel computation as well as multi-access data. In this paper, we will show how to implement adaptive data parallel computation in *OCore* using communities.

In Section 2, we will introduce *OCore*. In Section 3, we will present adaptive word recognition as an example of adaptive data parallel computation. After discussing some related work in Section 4, we offer our conclusions in Section 5.

2 OCore

2.1 Overview of *OCore*

OCore introduces the notion of *communities*, a *meta-level architecture* [11], and a *distributed garbage collection mechanism* [8], on top of a fundamental concurrent object-oriented layer, to support the programming of efficient parallel applications for multi computers. A prototype language processing system for the Intel Paragon XP/S [3], the Thinking Machines Corp. CM-5 [10], and Sun SPARC station clusters, is currently available [5]. However, parts of the meta-level architecture as well as garbage collection have not yet been implemented.

The rest of this section briefly describes the notion of objects and communities in *OCore*. See [7] and [6] for more details.

2.2 Objects

Objects in *OCore* perform operations such as message passing. The behavior of an object is described in a class, which defines slots, methods, broadcast handlers, local functions, and other meta-level definitions. Broadcast handlers are explained in Section 2.3.3. Classes can be defined using single inheritance and/or parametric types. These improve the reusability of definitions.

There is no intra-object concurrency in *OCore*. Messages are handled sequentially in an object. Message passing between objects is either synchronous or asynchronous according to the corresponding method. When an object sends a message asynchronously, it continues with its processing. When it sends a message synchronously, its processing is suspended until a reply is received.

2.3 Communities

A community structures a set of objects in a multi-dimensional logical space. Each member object in a community can handle messages independently. Communities support data parallel computations as well as multi-access data abstractions. Since community implementation depends heavily on whether the set of member objects varies dynamically, we have two kinds of communities: *static communities* and *dynamic communities*. Their difference will be explained later.

2.3.1 Community Templates

The behavior of a community is determined by the classes of its member objects and by what we call a *community template*. A community template specifies a member class, the dimensions of the logical space where member objects are structured, whether the community instances are static (this is the default) or dynamic, and *community procedures*. Community procedures are used mainly to describe the logical structure of a community. The mapping of the logical

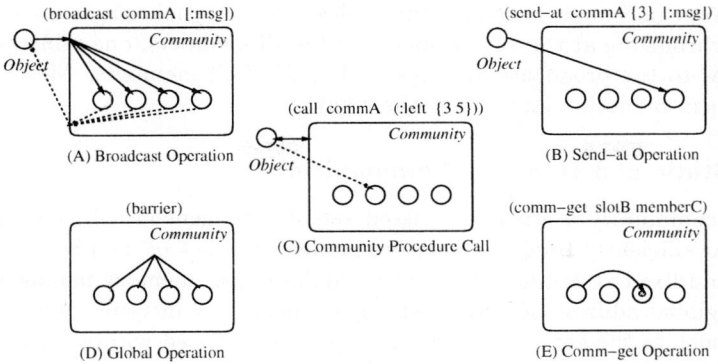

Figure 1: Operations related to Communities

space to the processor space can also be specified. Community templates can be defined parametrically but inheritance is not available.

2.3.2 Operations related to Communities

Figure 1 shows the images and usages of some operations related to communities. Any object that knows a community, whether it is a member of the community or not, can i) broadcast a message to all member objects in the community using the **broadcast** operator (Figure 1-A), ii) send a message to a member object having given indices in the community using the **send-at** operator (Figure 1-B), and call a community procedure using the **call** operator (Figure 1-C). The **broadcast** operation supports data parallel computation, while the **send-at** operation supports multi-access data. Community procedures make it easy to describe communication according to the logical structure of the community.

Each member object can obtain information about the community it belongs to, such as the global address of another member object in the same community. Barrier synchronization and reduction for all members in a community may also be performed (Figure 1-D). The **comm-get** operation allows a member object to obtain the value of the slot of another member object in the same community without explicit message passing(Figure 1-E).

2.3.3 Broadcast Handlers

After being delivered to each node, a broadcast message is shared among the member objects of a community on each node. It is then handled by a *broadcast handler* rather than by a method at each member object. Broadcasting a message is also either synchronous or asynchronous according to the corresponding broadcast handler. The main difference between a broadcast handler and a method appears in synchronous communication. In synchronous broadcasting, after synchronizing at the `reply` operation for all members, one arbitrary member replies to the broadcast message on behalf of all members. Members may also perform reduction and return the result.

2.3.4 Static and Dynamic Communities

A static community is used for a fixed set of homogeneous objects. It is implemented efficiently based on this restriction. The set of member objects are created and fixed at community creation under global memory management, so that the global address of a member object, necessary in some community operations such as the `send-at` operation, can be calculated without using remote operations.

In the case of a dynamic community, only the logical space is created at community creation. Then objects of a class or a subclass thereof may be added to or removed from the community using the `put-member` or `remove-member` operation. However, they are not actually added to or removed from the community until the `reorganize` operation is invoked. The `reorganize` operation determines a globally consistent set of member objects on which community operations are guaranteed to be performed correctly in a dynamic community. Each node manages the member objects corresponding to the logical subspace mapped on the node. Operations that need the global address of a member object require additional remote memory access overhead. Each node also keeps track of the local member objects registered in the remote logical subspace, so that broadcast and global operations may be performed almost as efficiently as in static communities. Dynamic communities provide more flexibility than static communities; they can be used for reconfigurable and/or sparse collections of heterogeneous objects.

2.3.5 Community Programming

We can distribute data to a community, then start data parallel computation by broadcasting a message. Member objects perform global operations when necessary. Data can be exchanged between member objects by message passing or the `comm-get` operation.

A data parallel computation can be encapsulated in a community. Multiple data parallel computations can run in parallel using multiple communities. A dynamic community allows member objects to be heterogeneous and changed dynamically. We showed an example of heterogeneous data parallel computation [7]. In Section 3, we present another type of data parallel computation, where the set of processing data changes dynamically.

A community may also be used as a communication medium where each member object is used to pool data. Objects may access member objects according to their status as well as other conditions; some objects put data into a member object while others take data out. Data are exchanged between objects accessing the same member object, thus realizing a kind of group communication where grouping may be changed dynamically. Each member object may propagate data according to the logical structure of the community, or pre-process data using the results of global or local operations [6].

3 Adaptive Data Parallel Computation

Although data parallel computation is applied to various areas, research efforts have been concentrated on cases where the set of processing data is fixed during execution. In some applications, however, the set of processing data should vary dynamically to adapt to changes in the target system. We present adaptive word recognition as an example of such adaptive data parallel computation.

3.1 Continuous Dynamic Programming

Speech recognition is important for realizing a flexible man-machine interface [4]. Word spotting, which extracts and understands words continuously from spontaneous speech, plays a basic role in speech recognition.

Continuous Dynamic Programming (CDP) is one way to realize segmentation-free word spotting [9]. Following is a brief explanation of the CDP algorithm.

Let P_i be a standard pattern, a sequence of phonemes or demi-phonemes, consisting of T_i frames. Standard patterns correspond to words recognized in word spotting. Let $d_i(t, \tau)$ be the distance between the input frame at time t and the τ'th frame of the standard pattern P_i. Then the accumulated distance $S_i(t, \tau)$ is obtained as follows:

$$S_i(-1, \tau) = S_i(0, \tau) = \infty$$
$$S_i(t, 1) = 3d_i(t, 1)$$
$$S_i(t, 2) = \min \begin{cases} S_i(t-2, 1) + 2d_i(t-1, 2) + d_i(t, 2) \\ S_i(t-1, 1) + 3d_i(t, 2) \\ S_i(t, 1) + 3d_i(t, 2) \end{cases}$$
$$S_i(t, \tau) = \min \begin{cases} S_i(t-2, \tau-1) + 2d_i(t-1, \tau) + d_i(t, \tau) \\ S_i(t-1, \tau-1) + 3d_i(t, \tau) \\ S_i(t, \tau-2) + 3d_i(t, \tau-1) + 3d_i(t, \tau) \end{cases} \quad (3 \leq \tau \leq T)$$

The accumulated distance is normalized by the length of P_i as follows:

$$A_i(t) = S_i(t, T_i)/3T_i$$

The normalized value $A_i(t)$ corresponds to the unlikelihood of completing the i'th pattern at time t. Values corresponding to all standard patterns can be obtained independently at each time step. When a normalized value corresponding to a standard pattern falls below some threshold and becomes minimal as to time, the pattern is output as the result of recognition.

3.2 Word Recognition in *OCore*

Word spotting speech recognition described above can be implemented in *OCore* as follows. First we construct a community wordcomm consisting of the member objects of class Word that correspond to standard patterns. The definition of the class Word includes the broadcast handler :cdp, which calculates the normalized CDP value between the pattern and the input frame, replying with the minimum value using reduction at each time step. To distribute input frames, we use another static community auxcomm which has one member object on each node.

The main loop of the word recognition program does the following. At each time step, an input frame is broadcast synchronously to auxcomm and stored in a local variable, so that the member objects of wordcomm on each node can share the frame. Then a :cdp message is broadcast synchronously to wordcomm. A replied minimum CDP value is kept in a ring buffer and checked to see if the value corresponding to some pattern satisfies the output condition described above. When the condition holds, the pattern is output.

3.3 Adaptive Word Recognition

As to word recognition using CDP, the recognition performance degrades as the number of words to be recognized increases. On the other hand, there are some cases in real-world application where the recognizing vocabulary varies according to the subject or situation. For example, in an interactive system for house design [4], the recognizing vocabulary could be narrowed down according to the place being designed; interior design terms would be necessary while designing a kitchen or a dining room, but unnecessary while designing a garden. Narrowing the recognizing vocabulary according to the context is expected to improve not only the recognition ratio but also the recognition rate.

Thus we extend the above program so that the recognizing vocabulary can be changed dynamically. Here we assume that the words to recognize are classified into several categories, each of which becomes active or inactive according to the context. Such adaptive word recognition can be realized in three different ways in *OCore*, as shown in Figure 2:

MS uses multiple static communities each of which corresponds to a category. :cdp messages are broadcast only to communities corresponding to active categories iteratively.

SS uses a single static community for wordcomm where each member object has a category ID. :cdp message is broadcast to all member objects with active category information. The member objects belonging to active categories calculate the CDP values, while those belonging to inactive categories do nothing but reply.

SD uses a single dynamic community for wordcomm. Word objects are created and registered in auxcomm in a distributed manner for each category. When

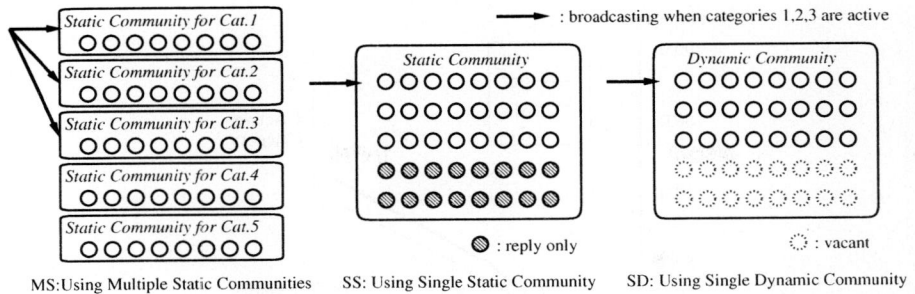

Figure 2: Adaptive Word Recognition Programs

the recognizing vocabulary changes, a message is broadcast to `auxcomm` to perform `put-member` and `remove-member` accordingly, and then the community `wordcomm` is reorganized so that it consists of `Word` objects belonging to active categories. While the recognizing vocabulary remains the same, `:cdp` message is broadcast to `wordcomm`.

In **MS**, the broadcast and reduction overhead increases as the number of active categories increases. In **SS**, wasteful processing increases as the number of inactive categories increases. In **SD**, `:cdp` messages are processed only by the `Word` objects belonging to active categories, while the recognizing vocabulary remains unchanged. However, vocabulary changes incur some overhead.

Additionally, if we support the dynamic addition of new words, we have to reserve extra member objects in **MS** and **SS**, which leads to an increase in the recognition overhead. In **SD**, however, only the logical space of the dynamic community needs to be large enough, which results in little increase in overhead.

3.4 Execution Results

The three programs described above were written in *OCore*, and executed on a Paragon XP/S running the operating system OSF/1 R1.3. The recognizing vocabulary consisted of 8 categories, each of which contained 320 words. The distribution of standard patterns for each category affects performance; however, we used 8×320 words of the same frame length to simplify the distribution.

Figure 3 shows the processing rates of these programs with one, three, and eight categories active, for varying numbers of nodes. I/O times are excluded. For convenience of comparison, it also shows the ideal speedup lines of the sequential execution of a corresponding C++ program [1].

---

[1] To avoid paging, the sequential performance is calculated based upon execution with the size of one category reduced to 64.

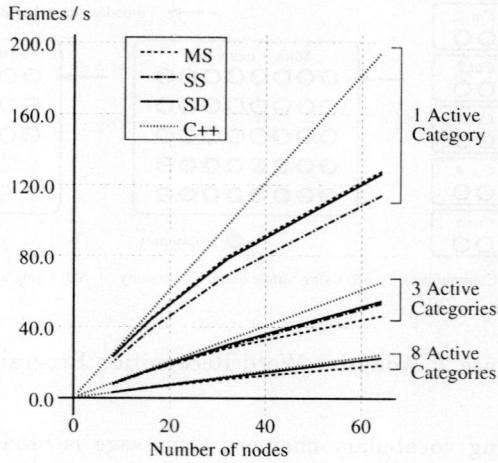

Figure 3: Speedup of Word Recognition Programs

For one active category, **MS** performs just one broadcast per input frame, like **SS** and **SD**, and shows the best performance. **SD** is slightly outperformed by **MS** because of the extra overhead for broadcasting messages. **SS** shows poor performance due to wasteful processing for seven inactive categories. On the whole, speedup is considerably reduced by the communication overhead.

With three categories active, the performance of **MS** is lower due to the broadcast and reduction overhead which is proportional to the number of active categories. **SS** shows a better performance because of relatively less wasteful processing; however, **SD** still outperforms **SS**.

With all categories active, the performances of **SD** and **SS** are now much the same, with only a small difference due to the extra broadcast overhead in **SD**. Both programs show relatively high speedup.

When the recognizing vocabulary changes, **SD** requires additional overhead to add or remove member objects, unlike **MS** and **SS**. Times for changing categories in **SD** on 64 nodes were around 5.5 ms, as shown in Table 1. They include the broadcast and reduction overhead in `auxcomm`, time for local operations such as `put-member` and `remove-member` on each node, and time for the `reorganize` operation in `wordcomm`. Above all, time for the `reorganize` operation, around 4.3 ms, was dominant. Since the recognizing vocabulary is unlikely to vary frequently, such a small overhead for changing categories is not significant.

As shown above, using a dynamic community for a reconfigurable set of objects makes it easy to program adaptive data parallel computations. It is

Table 1: Times for Changing Recognizing Categories

Num. of Active Categories	4 → 8	8 → 4	4 → 4'
Time (ms)	5.65	5.46	5.55

also shown that in some cases such an adaptive data parallel program using a dynamic community can outperform those using static communities.

4 Related Work

Some languages introduce the notion of abstract parallelism.

For example, CA (Concurrent Aggregates) [1] introduces the notion of *aggregate* for multi-access data. An aggregate is a set of homogeneous objects, called *representatives*. A message sent to an aggregate is actually sent to an arbitrarily chosen representative. Then, the receiver forwards the message to an appropriate representative, if necessary. Without optimization, such a message forwarding overhead is incurred, even if the appropriate representative is known to the sender. However, by replacing bottleneck objects with aggregates, a program can be made more efficient. CA supports neither broadcast nor global operations in its language specification.

pC++ [2], based on C++, has the class *collection*. Objects in a collection may be handled in a data parallel manner. The mapping of objects in a collection to real processors is specified by templates and alignments, as in HPF. pC++ supports the SPMD model; however, control parallelism is not supported.

Communities in *OCore* support both multi-access data abstractions and data parallel computations. In addition, a dynamic community allows the set of member objects to be heterogeneous and to be changed dynamically, which leads to flexible parallel programming.

5 Conclusions

We have presented an overview of *OCore* and the notion of communities. We have also shown adaptive word recognition as an example of adaptive data parallel computation using communities.

A static community, consisting of a fixed set of homogeneous objects, is adequate for regular parallel computation. On the other hand, a dynamic community, where the set of member objects can be changed dynamically as well as heterogeneous, supports adaptive and/or heterogeneous parallel processing. Preliminary evaluation of word recognition programs shows that in some cases an adaptive data parallel program using a dynamic community can outperform those using static communities without too frequent `reorganize` operations.

Further research includes the refinement of the *OCore* language specifications, and the exploration of community programming.

References

[1] A. A. Chien. *Concurrent Aggregates*. The MIT Press, 1993.

[2] D. Gannon. Libraries and Tools for Object Parallel Programming. In *Proc. CNRS-NSF Workshop on Environments and Tools For Parallel Scientific Computing*, 1992.

[3] Intel Supercomputer Systems Division. *Paragon OSF/1 C System Calls Reference Manual*, 1993.

[4] Y. Itoh, J. Kiyama, and R. Oka. Novel Interface System by Real-time Integration of Conversational Speech Understanding and Gesture Understanding by Multiple Users. *Proc. of SIGSLP, IPS Japan*, 7(3):17–22, 1995.

[5] H. Konaka, Y. Ishikawa, M. Maeda, T. Tomokiyo, and A. Hori. An Implementation of the Massively Parallel Object-based Language *OCore* for Multi Computers. *Trans.IPS.Japan*, 36(7):1520–1528, 1995. (in Japanese).

[6] H. Konaka, T. Tomokiyo, M. Maeda, Y. Ishikawa, and A. Hori. A Parallel Object-Oriented Language *OCore*. In T. Ito and Y. A., editors, *Lecture Notes in Computer Science*, volume 907, pages 167–186. Springer-Verlag, 1995. (Proc. of Intl. Workshop TPPP'94).

[7] H. Konaka, T. Tomokiyo, M. Maeda, Y. Ishikawa, and A. Hori. Data Parallel Programming in the Parallel Object-Oriented Language *OCore*. In *Pre-proceedings of French-Japanese Workshop on Object-Based Parallel and Distributed Computation*, 1995.

[8] M. Maeda, H. Konaka, Y. Ishikawa, T. Tomokiyo, A. Hori, and J. Nolte. On-the-fly global garbage collection based on partly mark-sweep. In H. Baker, editor, *Lecture Notes in Computer Science*, volume 986, pages 283–296. Springer-Verlag, Sept. 1995. (Proc. IWMM'95).

[9] R. Oka. Continuous Words Recognition by Use of Continuous Dynamic Programming for Pattern Matching. *Report of the Acoustic Society of Japan*, S78(20), 1978.

[10] Thinking Machines Corp. *CM5 Technical Summary*.

[11] T. Tomokiyo, H. Konaka, Y. Ishikawa, M. Maeda, and A. Hori. Meta-Level Programming in the Massively-Parallel Object-Based Language *OCore*. In *Proc. JSSST 11th Annual Conf.*, 1994. (in Japanese).

The Use of Interpreted Languages for Implementing Parallel Algorithms on Distributed Systems[*]

Noemi Rodriguez, Cristina Ururahy, Roberto Ierusalimschy, Renato Cerqueira

Departamento de Informática — PUC/Rio
R. M. S. Vicente 255, Rio de Janeiro — 22453-900, Brazil
noemi,ururahy,roberto,rcerq@inf.puc-rio.br

1 Introduction

Parallel programs have traditionally been implemented on distributed systems using conventional programming languages, such as Fortran or C, plus a library with support for communication, such as, for instance, PVM [11]. Also, some effort has been put into developing parallel programming languages [13].

Usually, the different development platforms are compared as to application execution time or parallel efficiency. However, as pointed out in [4], when measuring the performance of a parallel application or programming tool, different metrics may need to be taken into consideration. The ease of construction of an application and the time a programmer spends to implement it may in some cases be more important than the final execution time.

Working under this assumption, this work argues that the use of interpreted languages as a basis for parallel programming tools should be further investigated. As support for our argument, we show an example of use of an interpreted language in a classical application in parallel distributed programming, namely, that of detecting program termination. Two different algorithms for termination detection are implemented in Lua, an extension language developed at PUC-Rio [5].

2 Lua

Lua is a language specifically designed to be used as an extension language [6, 5]. Lua is small, portable (it is being used in platforms ranging from PC-DOS to CRAY), has a simple syntax and a simple semantics. And it is flexible.

Such flexibility has been achieved through some unusual mechanisms that make the language highly extensible. Among these mechanisms, we emphasize associative arrays, which act as a strong unifying data constructor, reflexive facilities, which allow facilities for "self manipulation" [8], and ease of incorporation of C libraries, which allows adding new functionalities to the language.

---

[*] This work has been partially supported by CNPq (The Brazilian Research Council) and by CENPES/PETROBRAS (the Research Center of the Brazilian Oil Company).

This last facility has been used to create the *Tche* environment [2, 3], which is of special consequence to this work. Tche is an application which creates an interface between Lua and the operating system, through an event-driven approach. It consists of two components: a basic library and a main loop. The basic library adds to the language a plethora of graphical functions (which are of no concern in this work), plus only one function for communication, called **send**. This function has a conventional behaviour, asynchronously sending a string to a specified recipient. The system has no symmetrical function to receive a message; instead, messages are received as special *events*.

The event loop is responsible for dispatching the OS events to appropriate handlers, written in Lua. The handler function for the communication event is called every time a process receives a message, with the message and the sender as its arguments. The default implementation for this handler function is to *execute* the message, assuming that the message is a piece of Lua code. This behaviour gives great flexibility to the communication mechanism, allowing the sender to modify global variables, to call functions, and even to define new functions in the receiver's environment.

An important property of the above protocol is that it is non blocking. The function **send** is, as mentioned above, asynchronous, and since receiving is carried on through an event driven mechanism, a process is never blocked. Another important property is that each event is handled to completion before the system starts handling the next event.

3 The Implementation of Termination Detection Algorithms Using Lua

3.1 The Problem of Termination Detection

Several algorithms for termination detection are available in the literature [12]. For the purpose of this work, two of these algorithms were chosen to be implemented. The chosen algorithms were the *four counter algorithm* [9] and the *coloured ring* algorithm [10]. The choice did not take into account the efficiency or number of messages exchanged, but focused instead on the diversity of the solutions and their relative simplicity.

In most termination detection algorithms, a detection procedure is started by one process (which may or not be one of the worker processes), which after receiving the results of this procedure, is able to determine whether termination has occurred. This process will be called the *initiator*. The messages exchanged between processes to carry out the detection procedure are called *control* messages, as opposed to messages exchanged due to the execution of the parallel program itself, called *basic* messages.

The algorithms used in this work assume that the processes involved in the parallel computation execute each at a different network node, and that the communication links between these nodes create a connected graph. Each process is essentially passive, becoming active only when it receives a basic message. At this moment, it may send a number of messages to its neighbours. The receipt of a basic message and the resulting transmissions is an atomic operation, ie, no other messages are received between the receipt of a message and the transmission of messages sent as its consequence.

It is interesting to notice how well the event driven paradigm fits into this model. As pointed out in the previous section, Lua works exactly this way, always processing an event to completion before receiving the next one.

3.2 Implementation

The implementation of the algorithms is structured in the following way. Each network node is represented by a Lua/Tche process running on a different physical machine. Communication links are simulated by providing each node with information about which machines are its neighbours. One particular machine runs the initiator code. This code, besides controlling the termination detection algorithm as described above, is also responsible for sending to all other nodes part of the code they should execute.

The code executed on all other nodes is structured in two parts. One of these parts contains the communication skeleton which is the same for any detection algorithm used. A second part of the code comprehends the functions which are dependent on the chosen termination detection procedure. This is the part of the code sent by the initiator. It has been called the *server* code, to emphasize the idea that it implements a specific service, the detection. With this structure, it has been possible to implement a function, called *SetAlgorithm*, which allows the user to interactively chose the detection algorithm to be used in the next execution round.

The server code consists of a set of functions which are called by the basic communication skeleton when sending or receiving messages, at "key" moments of operation. This set comprises the following functions:

1. InfoS(sender), called when a process receives a basic message from sender;
2. SendS(targetprocs), called to signal that each process in the targetprocs list will receive a basic message;

besides some other auxiliary functions, not worth mentioning here. Besides these, the server also defines a function Message, which each process invokes in another processes in order to implement the termination algorithm. Different implementations of these functions generate implementations of the different termination detection algorithms.

4 Final Remarks

The algorithms described above have also been implemented in C and SR [1], as an experience for the comparison of different programming platforms. Besides the algorithms described in this paper, we have also used Lua to implement a number of other distributed algorithms, such as heartbeat and filter algorithms.

This experience is described in [3], which also describes the extensions made to Lua to implement inter-process communication. The results of the experiment described in this previous work and in the present one show the importance of studying interpreted languages as a tool for parallel programming in distributed systems.

Doubtless, the ability to transmit code along with messages was one of the most important features introduced by using Lua. This ability is already being explored in the context of WWW-like applications [7]. It is interesting to note that one of the major problems of using "imported" code in these applications, which is the threat presented to security, is much simpler to deal with in the context of distributed parallel programming.

One interesting use of this ability has been shown by the architecture used in our implementation. A framework based on a chosen communication model for a parallel program was built, allowing for experimenting different implementations, with no need to test and debug the basic communication pattern for each variant.

References

1. G. Andrews and R. Olsson. *The SR Programming Language.* Benjamin Cummings, 1993.
2. A. Carregal and R. Ierusalimschy. Tche — a visual environment for the Lua language. In *VIII Simpósio Brasileiro de Computação Gráfica*, pages 227–232, São Carlos, 1995.
3. R. Cerqueira, N. Rodriguez, and R. Ierusalimschy. Uma experiência em programação distribuída dirigida por eventos. In *PANEL95 — XXI Conferência Latino Americana de Informática*, pages 225–236, Canela, 1995.
4. I. Foster. *Designing and Building Parallel Programs.* Addison Wesley, 1995.
5. R. Ierusalimschy, L. Figueiredo, and W. Celes. Lua - an extensible extension language. *Software: Practice and Experience*, 26(6), 1996.
6. R. Ierusalimschy, L. H. Figueiredo, and W. Celes. Reference manual of the programming language Lua version 2.1. Monografias em Ciência da Computação 08/95, PUC-Rio, Rio de Janeiro, Brazil, 1995. (available by ftp at ftp.inf.puc-rio.br/pub/docs/techreports).
7. Javabook, 1995. available through ftp at ftp://java.sun.com in directory /docs/JavaBook.ps.tar.Z.
8. G. N. Kirby. *Reflection and Hyper-Programming in Persistent Programming Systems.* PhD thesis, University of St. Andrews, St. Andrews, UK, 1992.
9. F. Mattern. Algorithms for ditributed termination detection. *Distributed Computing*, 2:161–175, 1987.
10. Michel Raynal. *Distributed Algorithms and Protocolos.* Wiley, 1988.
11. V. Sunderman. PVM: A framework for parallel distributed computing. *Concurrency: Practice and Experience*, December 1990.
12. Gerard Tel. *Introduction to Distributed Algorithms.* Cambridge University Press, 1994.
13. Barbara Wyatt, Krishna Kavi, and Steve Hufuangel. Parallelism in object-oriented languages: a survey. *IEEE Computer*, 9(6), November 1992.

Co-ordinating Heterogeneous Parallel Computation

Peter Au   John Darlington   Moustafa Ghanem
Yi-ke Guo   Hing Wing To   Jin Yang

Department of Computing,
Imperial College, London SW7 2BZ, U.K.
E-mail: {aktp, jd, mmg, yg, hwt, jy}@doc.ic.ac.uk

Abstract. There is a growing interest in heterogeneous high performance computing environments. These systems are difficult to program owing to the complexity of choosing the appropriate resource allocations and the difficulties in expressing these choices in traditional parallel languages. In this paper we propose that functional skeletons are used to express these resource allocation strategies. By associating performance models with each skeleton it is possible to predict and optimise the performance of different resource allocation strategies, thus providing a tool for guiding the choice of resource allocation. Through a case study of a parallel conjugate gradient algorithm on a mixed vector and scalar parallel machine we demonstrate these features of the SPP(X) approach.

1 Introduction

Parallel computation platforms often now have a heterogeneous structure arising from either exploiting clusters of workstations or from using parallel computers with specialist hardware such as vector units. The recent development of techniques for integrating high speed networks with high performance computers, such as the I-way project [5], provide strong evidence of the growing importance of heterogeneous parallel computing.

To fully exploit the power of a heterogeneous parallel computing system sophisticated resource allocation strategies are often required. This is a complex task as it is difficult to predict the performance of any particular resource allocation strategy on a system composed out of components which have widely differing functionalities and performance. To maximise the overall performance of an application on such systems, a more structured mechanism is needed where the resource requirement can be explicitly specified, and the performance of a resource allocation strategy can be quantified and calculated. This situation often occurs with homogeneous parallel machines, because of the diversities between different machine architectures. However, the problem is exacerbated in a heterogeneous system where the number of possible options is greatly increased and the differences in potential performance more marked.

Traditional approaches to parallel programming, such as the use of lower level communication libraries [7], are often unsatisfactory for this task. The main drawback of these approaches lies with the low-level nature of the mechanisms used for co-ordinating parallel computation. In these approaches it is difficult to both specify and predict the cost of a particular resource allocation strategy. Furthermore considerable effort is needed to change between resource allocation strategies owing to the low-level nature of the resulting programs.

In this paper, we explore the application of a more structured approach to the problem of programming heterogeneous systems. At Imperial College, we have developed a Structured Parallel Programming language, SPP(X), based on the idea of using a set of higher-order, pre-defined functions, known as *skeletons*, to co-ordinate the parallel activities of tasks defined using standard imperative languages [4]. This high-level approach provides the programmer with powerful control over the allocation of resources, which is necessary for the efficient use of heterogeneous machines. Furthermore this approach to parallel programming provides a uniform and simple method for guiding the complex resource allocation in a heterogeneous system by associating each skeleton with *performance models*. These performance models provide quantitative predictions of the use of the skeleton and can thus be used for predicting the cost of a particular resource allocation strategy.

In this paper, we present a pilot study in using SPP(X) to program a combined vector and scalar parallel computation. In Section 2, we survey the SPP(X) language. In Section 3, we describe performance guided resource organisation. Section 4 presents a set of experiments which implement a parallel Conjugate Gradient algorithm using different resource allocation strategies. Conclusions and further work are described in Section 5.

2 Parallel Programming with Structured Co-ordination

In this section we briefly outline the SPP(X) language. A more detailed description of the language can be found in [4]. In the SPP(X) model a Structured Co-ordination Language, SCL, is used to co-ordinate fragments of sequential code which are written in a base sequential language (X) such as Fortran and C. An application is therefore constructed in two layers: a higher co-ordination level and a lower base language level.

2.1 SCL: A Structured Co-ordination Language

SCL consists of skeletons (or co-ordination forms) that abstract parallel behaviour. Being functions, the skeletons can be easily composed and therefore, modularity and extensibility is naturally supported. Since all parallel behaviour arises from the known behaviour of these skeletons, they can be implemented by pre-defined libraries or code templates in the desired imperative language together with standard message passing libraries providing both efficiency and program portability. SCL has three main sets of constructs: configuration skeletons, data parallel skeletons and computational skeletons.

Configurations and Configuration Skeletons A *configuration* models the logical division and distribution of data objects. To program configurations a set of skeletons known as *configuration skeletons* are provided. For example the `distribution` skeleton defines the configuration of two arrays `A` and `B`:

distribution (f,p) (g,q) A B = align (p o partition f A)(q o partition g B)

where o is functional composition. The `partition` skeleton divides the initial arrays `A` and `B` into parallel arrays of sequential sub-arrays. The two functions `f` and `g` are the partitioning strategies used to divide the arrays. In SCL, some commonly occurring partitioning strategies, such as `row_block`, are provided as built-in functions. The functions `p` and `q` are data-movement skeletons which are described in more detail below. The `align` skeleton pairs the corresponding sub-arrays of two distributed arrays together. The `distribution` skeleton is usually generalised to a list of arrays rather than just two arrays.

The result of applying `distribution` is a parallel array of tuples (a configuration). Each element of the parallel array is a tuple of the sub-arrays that have been allocated to the same processor. As a short hand, rather than writing a configuration as an array of tuples, we can also regard a configuration as a tuple of (distributed and aligned) arrays and write it as <DA_1,...,DA_n> where DA_j represents the distributed version of array A_j. In particular we can pattern match to this notation to extract a particular distributed array from a configuration.

It is necessary in some applications to control the placement of configurations onto specific processors. This physical data placement is modelled in SCL by a *processor reference function* `@procSet` which explicitly specifies a mapping from a data configuration to a set of processors defined by `procSet`.

Data Parallel Operators SCL provides a set of skeletons which abstract the basic operations found in the data parallel computation model. For example, the function `map` abstracts the behaviour of executing the same task on all the components of a distributed array. While the function `fold` abstracts a parallel reduction computation over a distributed data structure, such as summing together the components of a distributed array. The skeleton `brdcast` broadcasts a data item over a configuration. Thus `brdcast d C` will associated a copy of the data item `d` with each component of `C`. To simplify the notation for broadcasting, the configuration syntax has been extended. Given a configuration <dA, dB> and a value `v` to be broadcast, this can be written as <dA, dB, v>.

Computational Skeletons for Abstracting Control Flow Commonly used parallel control flow patterns are abstracted in SCL as *computational skeletons*. For example, the skeleton `SPMD` abstracts the features of Single Program Multiple Data (SPMD) computation. The definition of this skeleton is given as:

SPMD [] = id
SPMD (globalFun, localFun):fs = (SPMD fs) o (globalFun o (map localFun))

where `h:tl` is a list whose head is `h` and tail is `tl`. SPMD takes a list of global-local operation pairs, which are applied over configurations of distributed data objects.

The local operations, `localFun`, are farmed to each processor and computed in parallel. The global operations, `globalFun`, operate over the whole configuration. These global operations are parallel operations that require synchronisation and communication.

Another computational skeleton `MPMD` abstracts the Multiple Program Multiple Data (MPMD) model. This can be defined in a similar way:

`MPMD [f1, f2, ... fn]  [c1, c2, ... cn] = [f1 c1, f2 c2, ... fn cn]`

The first argument is a list of parallel tasks, whilst the second argument is a list of configurations. MPMD provides a simple means to specify the concurrent execution of independent tasks over different groups of distributed data objects. SCL also provides other computational skeletons including `iterUntil s t c x` which iteratively applies `s` to `x` until condition `c` is satisfied whereupon `t` is applied to it.

2.2 Examples of using SCL

To illustrate the basic features of SPP(X) and SCL some short examples are described. Given the configuration of two distributed vectors `dv1` and `dv2`:

`< dv1, dv2 > = distribution [(block n, id), (block n , id)] [V1, V2]`

the following program computes the inner product of two vectors by `n` processors:

`innerProduct < dv1, dv2 >  = SPMD( fold(+), S_innerProduct ) < dv1, dv2 >`

where `S_innerProduct` is the sequential code in Fortran or C for performing a sequential inner product. The function begins by performing a local, sequential inner product on each pair of the distributed segments of the two vectors, and then performs the global operation of summing all the results of the local inner products. Similarly, suppose a matrix `A` is distributed row wise as `n` blocks (e.g. by the expression `dA = partition (row_block n) A`) and a vector `x` has been distributed as for an inner product then a parallel matrix-vector product can be defined as:

`matrixVectorProduct <dA,dx> = map S_matrixVectorProduct < dA, gather dx >`

where `S_matrixVectorProduct` is the sequential code for performing a sequential matrix-vector product. The parallel algorithm chosen for implementing a matrix-vector product begins by duplicating the *entire* argument vector on each of the processors. The result can then be computed locally with the result distributed in the same manner as the original matrix.

Currently we are building a prototype compiler/translator which translates SPP(X) programs into Fortran or C plus MPI [7] targeted at the AP1000. In this paper, SPP(X) with underlying imperative language as C is used to program the different approaches to the Conjugate Gradient algorithm.

3 Performance Guided Resource Organisation

Organising the computational resources in a heterogeneous parallel machine is often a difficult task owing to the diversity of ways in which an algorithm can exploit these resources. The usual solution is for the programmer to repeatedly execute the program under different resource strategies, observe the results and adjust the resource allocation decisions in an attempt to improve the performance of the program.

An alternative approach is to guide the choice of resource usage by predicting the performance of particular implementation strategies. This approach is very effective when the essential aspects of parallel behaviour are systematically abstracted as known program forms, such as those found in the SPP(X) approach, since each form naturally contains sufficient information for determining the performance under a given resource strategy. Thus, the process of producing and verifying performance models for an SPP(X) skeleton and machine pair need only be performed once. This approach has been adopted by several authors for homogeneous architectures [1, 2, 3].

The performance model associated with a SPP(X) skeleton is usually a function parameterised by the problem and machine characteristics, and returns a prediction of the total time to execute a specific instantiation of a skeleton. A performance model can be developed through a combination of analysing and benchmarking the implementation. In this paper the performance of the primitive SPP(X) skeletons are developed through benchmarking. Techniques for developing performance models through analysis are described in an earlier paper [3].

3.1 A Heterogeneous Architecture

The case study uses the Fujitsu AP1000 located at Imperial College. The basic architecture consists of 128 scalar Sparc processors connected by a two-dimensional torus for general point-to-point message passing and a dedicated network for broadcasting data [8]. Each scalar processor has a theoretical peak performance of approximately 5.6MFLOP/s. Interestingly, 16 of the scalar nodes have Numerical Computational Accelerators (NCA) attached to them. Each NCA consists of an implementation of Fujitsu's μ-VP vector processor each of which has a theoretical peak performance of 100MFLOP/s [6]. Communication between the two units on a single board is through a dedicated shared 16MB DRAM. This results in a heterogeneous architecture which can be exploited in many different ways.

3.2 Performance of Matrix and Vector Operations

In this section we highlight the details of performance modelling through several simple examples of performance models for some skeletons and their compositions. These will be later used in the case study described in Section 4. As discussed in Section 2, skeletons can be easily composed to produce higher-level co-ordination forms. Two examples `innerProduct` and `matrixVectorProduct` have already been defined. Two further higher-level co-ordination forms which will be useful for the case study are:

`scalarVectorProduct < s, v > = map S_scalarVectorProduct < s, v >`

where `S_scalarVectorProduct` is the sequential code for performing a sequential scalar-vector product. The other useful operator is a generalisation of vector addition. Given two vector v_1 and v_2, and a scalar value α, the operation computes $v_1 + \alpha v_2$. It is assumed that the scalar has been duplicated across all the processors, for example by using the `brdcast` operator. The definition of the function is:

`vectorAdd < v1, s, v2 > = map S_vectorAdd < v1, s, v2 >`

where `S_vectorAdd` is the fragment of sequential code for performing a generalised vector addition on the local segments of the vectors.

From a simple study of the definitions of the four operations, `innerProduct`, `matrixVectorProduct`, `scalarVectorProduct` and `vectorAdd`, performance models based on the performance of the primitive skeletons can be derived:

`innerProduct`	$t_{ip} = t_{sip}(N/P) + t_{fold+}(P)$
`matrixVectorProduct`	$t_{mvp} = t_{gather}(P,N) + t_{brdcst}(N) + t_{smvp}(N/P,N)$
`vectorAdd`	$t_{va} = t_{sva}(N/P)$
`scalarVectorProduct`	$t_{svp} = t_{ssvp}(N/P)$

The performance models are parameterised by the size of the vector N and the number of processors used P. It is assumed that the matrix is square and distributed row-blockwise. The components t_{fold+}, t_{gather} and t_{brdcst} reflect the cost of performing the primitive skeletons `fold (+)`, `gather` and `brdcast` (explicitly or using shorthand notation) respectively. Notice that there is no overhead involved in performing a `map` therefore the cost of `vectorAdd` and `scalarVectorProduct` is simply the cost of the local computation. The costs of the local computation performed by each operation are represented by t_{sip}, t_{mvp}, t_{sva} and t_{ssva}.

The performance of implementations of the primitive skeletons and the sequential code fragments used in the matrix-vector operations were benchmarked on the AP1000. For each of the basic components there are two models, reflecting the choice of executing the component on either the scalar or the vector units. Note that where a fragment of sequential code is needed either scalar or vectorised code is used depending on the target processing unit. The resulting performance models from the benchmarking exercise are:

Component	Scalar model (μs)	Vector model (μs)
$t_{sip}(N)$	$1.2 + 1.26N$	$2.1 + 0.028N$
$t_{smv}(M,N)$	$1.2 + M(1.56N + 1.2)$	$5.9 + 0.0028(M*N)$
$t_{sva}(N)$	$1.2 + 0.56N$	$2.1 + 0.022N$
$t_{ssvp}(N)$	$1.2 + 0.89N$	$2.1 + 0.022N$
$t_{fold+}(P)$	$130 + 30\log_2 P$	$175 + 270\log_2 P$
$t_{gather}(P,N)$	$150P + 0.72N$	$150P + 1.28N$
$t_{brdcst}(N)$	$260 + 1.45N$	$160 + 2.2N$

These are average figures, and do not involve extensive modelling of the cache which would give more accurate models. Notice that the cost of the `fold` function for the vector units is higher than for the scalar units. This is as a result of the irregular distribution of the vector units. The lower bandwidth seen by the vector processors for the `gather` and `broadcast` operations is due to cost of transferring data to and from the vector processor's memory.

4 Case Study: Parallel Conjugate Gradient Solver

The SPP(X) approach to co-ordinating heterogeneous parallel computation is illustrated through a case study of the Conjugate Gradient (CG) method for solving systems of linear equations. In particular we focus on the CG algorithm described by Quinn [9]. We are primarily concerned with the problems of co-ordinating parallel computation and therefore simplify the program by applying it to dense matrices rather than the sparse systems to which it is usually applied. Pseudo code for the CG algorithm for solving the system $Ax = b$ is given below:

$k = 0;\ d_0 = 0; x_0 = 0;\ g_0 = -b;\ \alpha_0 = \beta_0 = g_0^T g_0;$
while $\beta_k > \epsilon$ **do**
$\quad k = k + 1;$
$\quad d_k = -g_{k-1} + (\beta_{k-1}/\alpha_{k-1})d_{k-1};$
$\quad \rho_k = d_k^T g_{k-1};$
$\quad w_k = A d_k;$
$\quad \gamma_k = d_k^T w_k;$
$\quad x_k = x_{k-1} - (\rho_k/\gamma_k)d_k;$
$\quad \alpha_k = g_{k-1}^T g_{k-1};$
$\quad g_k = A x_k - b;$
$\quad \beta_k = g_k^T g_k;$
endwhile;
$x = x_k;$

where vectors are represented using roman letters (except k) and scalars are represented using greek letters. This algorithm can be parallelised by using parallel versions for the inner products, matrix-vector products and vector additions. Parallel versions of these operations written in SCL were described in Section 3. Since each of these operations can be executed on either the scalar or the vector units, the implementation of the algorithm can exploit only the scalar units of the AP1000, only the vector units of the AP1000, or some combination of the two. This leads to a difficult decision over how to use the resources effectively. The following experiments demonstrate how the performance models developed in Section 3 can be used to accurately predict the cost of various implementations of this algorithm in SCL. It is thus possible to use the performance models to aid in making the appropriate resource allocation. The experiments also emphasise the ease with which these complex resource strategies can be expressed using the SPP(X) approach.

4.1 Experiment 1: Scalar Processors Only

The first implementation only exploits the scalar processors of the AP1000. In this implementation all the vectors are distributed block-wise across the scalar processors and the matrix is distributed row-block-wise across the scalar processors. The following SPP(X) program expresses the CG algorithm:

```
CG A b e = iterUntil iterStep finalResult isConverge
                    (ipG0, < zeroVector, zeroVector, negb, ipG0, ipG0 >)
   where
      <dA,db>@SPG = distribution [(row-block nP, id), (block nP, id)] [A, b]
      ipG0@ROOT   = innerProduct < b, b >
      negb        = scalarVectorProduct < -1, db >
      isConverge (beta, < dx, dd, dg, dalpha, dbeta >) = beta < e
      finalResult (beta, < dx, dd, dg, dalpha, dbeta >) = gather dx
      iterStep (beta, < dx, dd, dg, dalpha, dbeta >)
          = (beta', < dx', dd', dg', alpha', beta' >)
      where negG        = scalarVectorProduct < -1, dg >
            dd'         = vectorAdd < negG, dbeta/dalpha, dd >
            rho@ROOT    = innerProduct < dd', dg >
            w           = matrixVectorProduct < dA, dd' >
            gamma@ROOT  = innerProduct < dd', w >
            dx'         = vectorAdd < dx, -(rho/gamma), dd' >
            alpha'@ROOT = innerProduct < dg, dg >
            u           = matrixVectorProduct < dA, dx' >
            dg'         = vectorAdd < u, -1, db >
            beta'@ROOT  = innerProduct < dg', dg' >
```

where `zeroVector` is a constant vector of zeros of size b distributed in the same manner as db, and nP returns the number of processors used. The distribution of the data onto the scalar processors is specified by the notation @SPG where SPG is the scalar parallel group of processors. The result of the inner products is placed on a unique processor specified by ROOT.

From the structure of the program it is possible to derive the following performance model for the main loop of the program:

$$t_{cg1} = i_{iter}(4t_{ip} + 4t_{brdcst} + 2t_{mvp} + 3t_{va} + t_{svp})$$

where i_{iter} is the number of iterations. All the experiments are conducted for a fixed number of iterations in order to exclude the effect on the number of iterations caused by differences in the accuracy of the arithmetic operations between the scalar and vector units. This enables comparisons to be made between alternative runs and across the two processing units. The reported results are thus standardised at 100 iterations for all experiments and exclude the time required to initially distribute and finally to collect the data.

The results of the experiments are shown in Fig. 1. The first graph, 1(a), shows the execution time vs. the number of processors for several problem sizes. The second graph, 1(b), shows the execution time vs. the problem size for several different processors numbers. The plotted dots represent the measured times, whilst the lines are the predicted times. The predictions are within 10% of the measured times and follow the trend of the measured times.

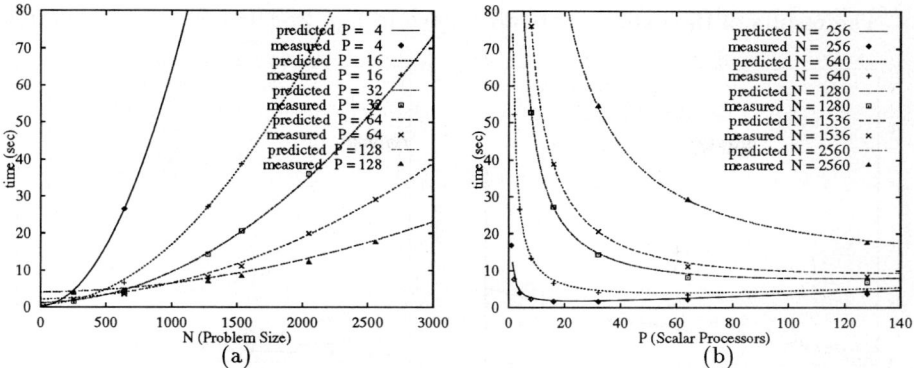

Fig. 1. Elapsed time for the parallel scalar case for various processor numbers and problem sizes.

4.2 Experiment 2: Vector Processors Only

The second implementation only utilises the vector units of the AP1000. Here the data is only distributed to the cells that have vector units attached to them. The SPP(X) program for this is:

```
CG A b e = iterUntil iterStep finalResult isConverge
                    (ipG0, < zeroVector, zeroVector, negb, ipG0, ipG0 >)
   where
      <dA,db>@VPG = distribution [(row-block nP, id), (block nP, id)] [A, b]
      ipG0@ROOT    = innerProduct < b, b >
      negb         = scalarVectorProduct < -1, db >
      isConverge (beta, < dx, dd, dg, dalpha, dbeta >) = beta < e
      finalResult (beta, < dx, dd, dg, dalpha, dbeta >) = gather dx
      iterStep (beta, < dx, dd, dg, dalpha, dbeta >)
            = (beta', < dx', dd', dg', alpha', beta' >)
        where  negG       = scalarVectorProduct < -1, dg >
               dd'        = vectorAdd < negG, dbeta/dalpha, dd >
               rho@ROOT   = innerProduct < dd', dg >
               w          = matrixVectorProduct < dA, dd' >
               gamma@ROOT = innerProduct < dd', w >
               dx'        = vectorAdd < dx, -(rho/gamma), dd' >
               alpha'@ROOT= innerProduct < dg, dg >
               u          = matrixVectorProduct < dA, dx' >
               dg'        = vectorAdd < u, -1, db >
               beta'@ROOT = innerProduct < dg', dg' >
```

The difference between this code and the code in experiment 1 is the change in the placement of the data from the scalar processor group to the vector processor group, **VPG**. Naturally the vector versions of the matrix-vector operations must be used. Owing to the similar nature of the code, the performance model of the program is the same as that of experiment 1, with but different costs associated with the components. Again the timing was performed over the main loop for 100 iterations.

The results of the experiments are shown in Fig. 2. The layout of the graphs

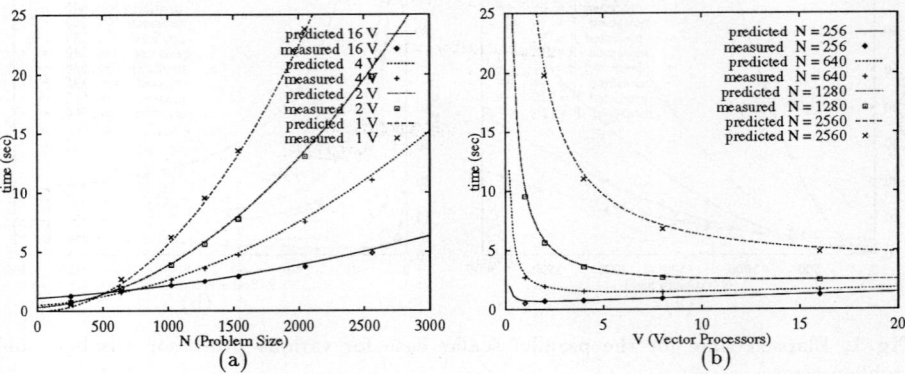

Fig. 2. Elapsed time for the parallel vector case for various processor numbers and problem sizes.

is the same as for experiment 1. The predictions are also within 10% of the measured times and again follow the trend of the measured results.

The results of this experiment reflect the superior performance of the vector units over the scalar units. In general the performance of the program for different processor numbers is as expected. However, notice that for very small problem sizes, a smaller number of vector units performs better owing to the high communication overheads of using more processors.

4.3 Experiment 3: Mixed Scalar and Vector Processors

The third implementation explores the use of both the scalar and the vector units in this algorithm. A study of the algorithm shows that the computation of the matrix-vector products can be overlapped with some of inner products as there are no data dependencies between these operations. A sketch of the data dependencies in the algorithm is shown in Fig. 3. It is thus possible to execute the

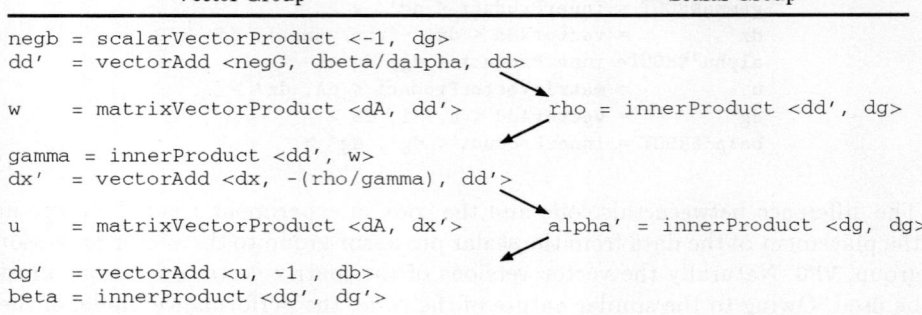

Fig. 3. Data dependencies in the CG algorithm.

more expensive matrix-vector product on the vector units whilst concurrently executing inner products on the scalar units. This can be implemented in SPP(X) as:

```
CG A b e = iterUntil iterStep finalResult isConverge
                    (ipG0, < zeroVector, zeroVector, negb, ipG0, ipG0 >)
   where
      <dA,db>@VPG = distribution [(row-block nP, id), (block nP, id)] [A, b]
      ipG0@ROOT     = innerProduct < b, b >
      negb          = scalarVectorProduct < -1, db >
      isConverge (beta, < dx, dd, dg, dalpha, dbeta >) = beta < e
      finalResult (beta, < dx, dd, dg, dalpha, dbeta >) = gather dx
      iterStep (beta, < dx, dd, dg, dalpha, dbeta >)
            = (beta', < dx', dd', dg', alpha', beta' >)
      where  negG       = scalarVectorProduct < -1, dg >
             dd'        = vectorAdd < negG, dbeta/dalpha, dd >
             [ rho@ROOT, w ] = MPMD [ innerProduct, matrixVectorProduct ]
                                    [ < dd', dg >@SPG, < dA, dd' > ]
             gamma@ROOT = innerProduct < dd', w >
             dx'        = vectorAdd < dx, -(rho/gamma), dd' >
             [ alpha'@ROOT, u ] = MPMD [ innerProduct, matrixVectorProduct ]
                                       [ < dg, dg >@SPG, < dA, dx' > ]
             dg'        = vectorAdd < u, -1, db >
             beta'@ROOT = innerProduct < dg', dg' >
```

The overlapping of the vector and scalar processing is expressed using the MPMD skeleton. Notice that the vectors used in the inner product must be marked as being mapped to the scalar processor group. In cases where more or fewer scalar processors are being used than vector processors, this re-allocation of data will have an associated cost for redistributing the data.

Owing to the change in the implementation of the algorithm there will be a corresponding change in the performance model for the program. By analysing the program it is possible to arrive at the following performance model:

$$t_{cg3} = i_{iter}(2t_{ip} + 4t_{brdcst} + t_{mpmd}(2t_{mvp}, 2t_{ip} + t_{redist}) + 3t_{va} + t_{svp})$$

where the performance for the MPMD skeleton is the maximum of time taken by any of its concurrent tasks plus some overhead for setting up the concurrent tasks. In this particular case there are no overhead cost, therefore for simplicity this cost has been omitted from the model.

$$t_{mpmd}(t_1, t_2) = \max(t_1, t_2)$$

The benchmarked cost of t_{redist} for V vector units, P scalar processors and a problem size N is $(P/V - 1)(3.12N + 192)\mu s$. Notice that each vector unit only communicates with $(P/V-1)$ other processors rather than P/V processors, since it can use the scalar processor attached to itself as one of the scalar units for performing the overlapped computation.

By comparing the cost of a vectorised matrix-vector product t_{smvp} with the cost of redistribution t_{redist}, it is possible to see that using a different number of scalar processors to vector processor has a prohibitive overhead. When a matching number of scalar processors are used there is no need to redistribute the data as the scalar units attached to the vector units can be used. Therefore the results shown in Fig. 4 are for a matching number of vector and scalar processors, where there is no redistribution. If a larger unit of work were being performed by the vector units, there may be a benefit in using a mismatching number of vector and scalar processor as the overhead of redistribution would then be hidden.

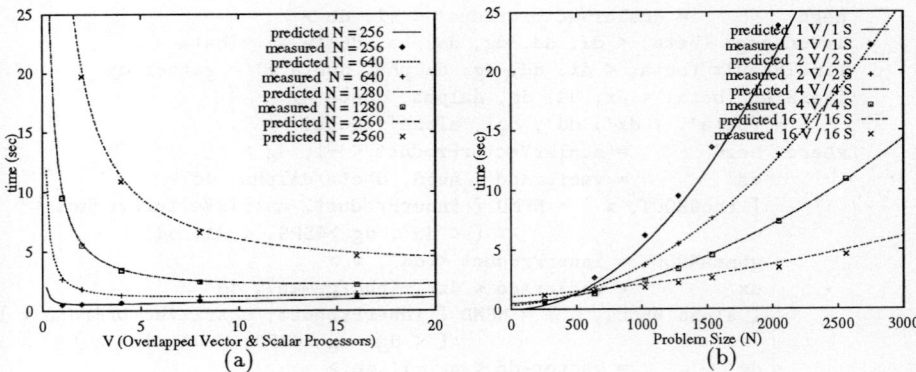

Fig. 4. Elapsed time for the mixed parallel vector and scalar case for various processor numbers and problem sizes.

The layout of the graphs are as for the previous experiments. The predicted results are within 10% of the measured times and again follow the trend of the measured results. For the majority of the given problem sizes 16 vector and scalar units perform best, although, as in the pure vector case, this is not true for very small problems.

To demonstrate the extra cost of redistribution, Fig. 5 shows the execution time vs. the problem size for 16 vector units and 128 scalar units in comparison with 16 vector units and 16 scalar processors.

4.4 Analysis of Results

As shown in Fig. 6 the best performance was achieved by using a mixture of 16 vector and 16 scalar processors. This combination gives better performance than using only 16 vector processors, or only 128 scalar processors. The performance models predict a similar trend. The ability to predict the performance of programs enables the appropriate resource decision to be made without recourse to implementing all the different strategies and executing them. This pilot study indicates the potential for exploiting heterogeneous environments. By using higher-level co-ordination forms different configurations of the machine can be easily organised and their performances systematically predicted.

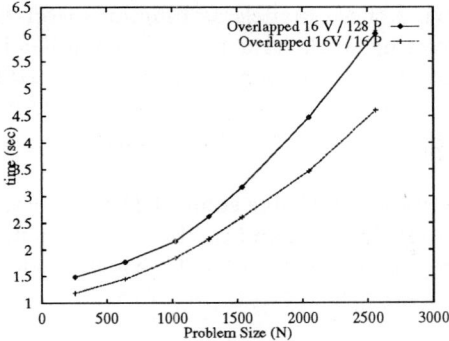

Fig. 5. Elapsed time for the mixed parallel vector and scalar case for 16 vector units and 16 scalar processors, and 16 vector units and 128 scalar processors.

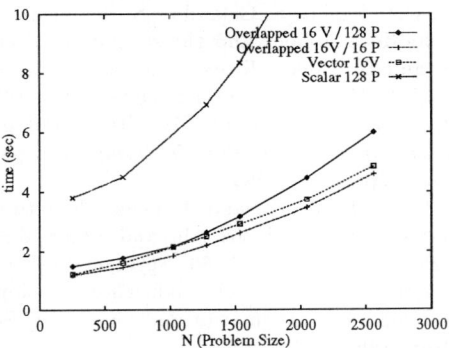

Fig. 6. Comparison of execution time for all four experiments vs. problem size.

5 Conclusions and Further Work

In this paper we have presented a methodology for co-ordinating and organising resources in heterogeneous parallel machines. By using higher-level co-ordination forms, different configuration structures of the machine can easily be expressed and their performances can be systematically predicted. The pilot study reported in this paper, which implemented a parallel conjugate gradient algorithm using different configurations of the vector and scalar processors of an AP1000, has demonstrated these features in the SPP(X) approach. The experiments indicate that the structured method of programming by using co-ordination forms allows sufficient control over the allocation of resources to make efficient use of heterogeneous machines. To further validate the approach, further work includes testing the approach on more sophisticated problems such as a multigrid solver and a parallel climate model where the overlapping of the vector computation and the scalar computation will become even more crucial to the overall performance.

The models presented in this paper were generated by hand. However, an analysis of these models will show that the models presented can be derived from the syntax of the program. Current work includes the development of a system for automatically deriving these models as functions over the basic skeletons. We are currently completing work on a portable compiler for the SPP(X) system together with an interactive tool for performance modelling.

Acknowledgements

The authors gratefully acknowledge support from Fujitsu Laboratories, the EPSRC funded project GR/K69988 and the British Council. We would like to thank Fujitsu for providing the facilities at IFPC, which made this work possible.

References

1. Tore Andreas Bratvold. *Skeleton-Based Parallelisation of Functional Programs*. PhD thesis, Heriot-Watt University, November 1994.
2. M. Danelutto, R. Di Meglio, S. Orlando, S. Pelagatti, and M. Vanneschi. A methodology for the development and the support of massively parallel programs. In D.B. Skillicorn and D. Talia, editors, *Programming Languages for Parallel Processing*, pages 205–220. IEEE Computer Society Press, 1994.
3. J. Darlington, M. Ghanem, and H. W. To. Structured parallel programming. In *Programming Models for Massively Parallel Computers*, pages 160–169. IEEE Computer Society Press, September 1993.
4. J. Darlington, Y. Guo, H. W. To, and J. Yang. Functional skeletons for parallel coordination. In Seif Haridi, Khayri Ali, and Peter Magnussin, editors, *EUROPAR'95 Parallel Processing*, pages 55–69. Springer-Verlag, August 1995.
5. Ian Foster and Warren Smith. I-WAY application developers and users guide version 1.2. Technical report, Mathematics and Computer Science Division, Argonne National Laboratory, 1995.
6. Fujitsu Ltd. *VPU (MB92831) Functional Specifications*, 1.20 edition, March 1992.
7. William Gropp, Ewing Lusk, and Anthony Skjellum. *Using MPI: Portable Parallel Programming with the Message-Passing Interface*. MIT Press, 1994.
8. Hiroaki Ishihata, Takeshi Horie, Satoshi Inano, Toshiyuki Shimizu, Sadayuki Kato, and Morio Ikesaka. Third generation message passing computer AP1000. In *International Symposium on Supercomputing*, pages 46–55, 1991.
9. Michael J. Quinn. *Parallel Computing: Theory and Practice*. McGraw-Hill, second edition, 1994.

Correctness of a Distributed-Memory Model for Scheme

Luc Moreau[*]

University of Southampton
L.Moreau@ecs.soton.ac.uk

Abstract. We propose a high-level approach to program distributed applications; it is based on the annotation **future** by which the programmer specifies which expressions may be evaluated remotely in parallel. We present the CEKDS-Machine, an abstract machine with a distributed memory, able to evaluate Scheme-like future-based programs. In this paper, we focus on the issue of task migration and prove that task migration is transparent to the user, i.e. task migration does not change the observable behaviour of programs.

1 Introduction

Distributed systems are omnipresent: local area networks and the success of the Internet in the past years are particular illustrations of the ubiquity of distributed computing. A major research focus in this area has been the design of new languages or programming paradigms to develop distributed applications, like e.g. PVM [10], MPI [8], Nexus [9], Cilk [1]. We argue that those systems were designed to build high-performance distributed applications, and that they favour efficiency over ease of programming. Therefore, these languages or paradigms overwhelm the programmer with the burden of dealing with the complexity of distribution. Some approaches even impose programming styles, with which the programmer may not be familiar; e.g. Cilk [1] demands programs written in continuation-passing style.

Mostly-functional languages like Scheme and SML have traditionally provided the programmer with abstraction, expressiveness, first-class citizenship of objects, and automatic garbage collection. We believe that there is a niche for a high-level approach to distributed computing. Following Halstead's work on MultiLisp [11], we extend a Scheme-like language with an annotation **future** by which the programmer specifies which expressions may be evaluated in parallel, possibly remotely. By definition, annotations must be *transparent*, i.e. annotated programs return the same result as in the absence of annotations. This approach is abstract because it hides the intricacies of distribution by giving the programmer the illusion that a distributed system is programmable as a sequential one.

We consider the idealised Scheme-like language defined in Figure 1. It is a purely functional language, extended with a primitive **makeref** to create boxes, with primitives **deref**, **setref!** to read and modify them, and with a primitive **callcc** to capture first-class continuations. In addition, there is a construct (**future** M) to create a producer-consumer type of parallelism [11]. Intuitively, the evaluation of (**future** M) immediately returns an object called placeholder, while another task evaluates the argument M in parallel. The purpose of the latter task, called the *producer*, is to compute and then store the value of M in the placeholder. The task using the placeholder is called the *consumer*.

For a long time, this approach has been characterised by a lack of formal semantics due to the difficulty of providing transparent annotations for parallelism in the presence of first-class continuations and side-effects. Recently, the author [20] defined the semantics of **future** for the language of Figure 1. The goal of this paper is to extend this semantics to a distributed framework (the proof of its correctness is available in a technical report [21]).

More specifically, the contributions of this paper are the following. *i)* We define a distributed architecture able to evaluate **future**-based programs; *ii)* We prove that

---

[*] This research was supported in part by the Engineering and Physical Sciences Research Council, grant GR/K30773. Author's address: Department of Electronics and Computer Science, University of Southampton, Southampton SO17 1BJ, United Kingdom.

$$
\begin{aligned}
&P \in \Lambda_f^0 && \text{(Program)} \\
&M \in \Lambda_f &&::= V \mid (M\ M) \mid (\text{if } M\ M\ M) \mid (\text{future } M) && \text{(Term)} \\
&V \in Value_f &&::= p \mid f \mid x \mid (\lambda x.M) && \text{(Syntactic Value)} \\
&p \in BConst &&= \{\text{true}, \text{false}, \text{nil}, 0, 1, \ldots\} && \text{(Basic Constant)} \\
&f \in FConst &&= \{\text{cons}, \text{car}, \text{cdr}, \text{makeref}, \text{deref}, \text{setref!}, \text{callcc}\} && \text{(Functional Constant)} \\
&x \in Vars &&= \{x, y, z \ldots\} && \text{(Variable)}
\end{aligned}
$$

Fig. 1. Syntax of Λ_f

task migration is transparent to the programmer, i.e. task migration does not change the observable behaviour of programs; *iii)* This result is the reference semantics that we can use to design and prove the correctness of program optimisations. The architecture is described in Section 2. Section 3 discusses related work and is followed by concluding remarks.

2 The Architecture

In this section, we present the CEKDS-Machine, an abstract machine with a distributed store, which extends Felleisen and Friedman's CEK and CEKS machines [4, 5], and the F-PCKS-machine [20]; its state space is formally described in Figure 2.

Some operations like deref, i.e. reading the content of a box, are rather complex. Indeed, as deref is *strict*[2], it touches its argument, checks whether it is legitimate to access the content of the box received in argument, and finally, reads the box content. In order to distinguish these three operations, we add two primitives touch and sync to Λ_{cekds}, the language accepted by the CEKDS-machine; besides, we translate every program of Λ_f into a term of Λ_{cekds} by the function $\mathcal{X}$ of Figure 2, which makes the touch and sync operations explicit.

In our distributed architecture, computational resources are called *sites* and are uniquely identified by *site names*. A site has its own memory and can run several tasks that share the site memory. A *world* is the set of sites that can be used to evaluate a program; sites in a world communicate by exchanging messages. More specifically, in a site, we distinguish active tasks, i.e. tasks that can be run, from suspended tasks, i.e. tasks that wait for a message or a synchronisation. As far as communications are concerned, a site is equipped with two spools of messages: the input spool contains pending input messages, while the output spool contains the messages that remain to be transmitted.

A store is a finite function, also represented as a set of pairs, associating locations with store contents. In our distributed architecture, each site has its own procedure of memory allocation, and its proper task naming mechanism. Hence, we use the notion of *qualified* location or task name, to unambiguously refer to a location or a task in the world. We now appreciate how site-names uniquess is important to define qualified locations or task names.

We abstract a task by a triple composed of a computational state, a *legitimacy* [15, 20] used to implement first-class continuations and side-effects, and a name. A computational state is a configuration of the CEK-machine [4], which can be either $\mathsf{Ev}\langle M, \rho, \kappa\rangle$ representing the evaluation of a term M in an environment ρ with a continuation κ, or $\mathsf{Ret}\langle V, \kappa\rangle$ meaning the return of a value to a continuation. The continuation, implemented as a data structure called *continuation code*, represents what remains to do after evaluating M in $\mathsf{Ev}\langle M, \rho, \kappa\rangle$. In conventional languages, the continuation is nothing else but the evaluation stack. The environment is a finite function mapping variables to values.

In Figure 3, transitions between computational states specify how to evaluate the purely functional and sequential subset of the language extended with first-class continuations; details can be found in [4, 20, 22, 19]. Figure 4 shows the transitions that involve a collaboration of a task with its site. According to *(fork)*, the evaluation of a future allocates a new placeholder ph and creates a new task, with a name τ_1, which speculatively evaluates the

---

[2] A strict function applied to a placeholder, accesses the value that the producer task has stored in the placehoder; this operation, called *touching* the placeholder, can suspend the current task when the placeholder has not received a value yet.

$$
\begin{aligned}
\mathcal{W} &::= \{m_1,\ldots,m_n\} & \text{(World)} \\
m \in \mathcal{M} &::= \langle T,\theta,s,S,I,O\rangle & \text{(Site)} \\
t \in Task &::= \langle C,\ell,\tau\rangle & \text{(Task)} \\
\theta \in Store &::= \{(\alpha_1\ X_1)\ldots(\alpha_n\ X_n)\} & \text{(Store)} \\
X &::= V \mid \bot \mid \ell & \text{(Store Content)} \\
s \in \mathcal{S} &= \{s_1,s_2,\ldots\} & \text{(Site Name)} \\
\tau \in \mathcal{T} &= \{\tau_1,\tau_2,\ldots\} & \text{(Task Name)} \\
u &::= \langle \tau,s\rangle & \text{(Qualified Task Name)} \\
a &::= \langle \alpha,s\rangle & \text{(Qualified Location)}
\end{aligned}
$$

Explicit translation: $\mathcal{X}: \Lambda_f \to \Lambda_{cekds}$

$$
\begin{aligned}
\mathcal{X}[\![x]\!] &= x \text{ if } x \in Vars \cup BConst \cup NStP \\
\mathcal{X}[\![\mathbf{car}]\!] &= \lambda x.(\mathbf{car}\ (\mathbf{touch}\ x)) \\
\mathcal{X}[\![\mathbf{cdr}]\!] &= \lambda x.(\mathbf{cdr}\ (\mathbf{touch}\ x)) \\
\mathcal{X}[\![\mathbf{deref}]\!] &= \lambda x.(\mathbf{deref}\ (\mathbf{sync}\ (\mathbf{touch}\ x))) \\
\mathcal{X}[\![\mathbf{setref!}]\!] &= \lambda x_1 x_2.(\mathbf{setref!}\ (\mathbf{sync}\ (\mathbf{touch}\ x_1))\ x_2) \\
\mathcal{X}[\![(\mathbf{future}\ M)]\!] &= (\mathbf{future}\ \mathcal{X}[\![M]\!]) \\
\mathcal{X}[\![(M_1\ M_2)]\!] &= (\lambda m_1 m_2.(\mathbf{touch}\ m_1)m_2)\mathcal{X}[\![M_1]\!]\mathcal{X}[\![M_2]\!] \\
\mathcal{X}[\![(\lambda x.M)]\!] &= (\lambda x.\mathcal{X}[\![M]\!]) \\
\mathcal{X}[\![(\mathbf{if}\ M_1\ M_2\ M_3)]\!] &= (\mathbf{if}\ (\mathbf{touch}\ \mathcal{X}[\![M_1]\!])\ \mathcal{X}[\![M_2]\!]\ \mathcal{X}[\![M_3]\!])
\end{aligned}
$$

$$
\begin{aligned}
C \in CoSt &::= \mathsf{Ev}\langle M,\rho,\kappa\rangle & \text{(Computational State)} \\
&\mid \mathsf{Ret}\langle V,\kappa\rangle \\
\rho \in Env &::= \{(x_1\ V_1)\ldots(x_n\ V_n)\} & \text{(Environment)} \\
I &= \{q_1,\ldots,q_n\} & \text{(Input Spool)} \\
O &= \{q_1,\ldots,q_n\} & \text{(Output Spool)} \\
q &::= R \mid A & \text{(Message)} \\
T &= \{t_1,\ldots,t_n\} & \text{(Active Tasks Set)} \\
S &= \{t_1,\ldots,t_n\} & \text{(Suspend Set)} \\
&\cup \{R_1,\ldots,R_m\}
\end{aligned}
$$

$$
\begin{aligned}
Unload[c,\mathcal{W}] &= c \\
Unload[(\mathbf{cons}\ V_1\ V_2),\mathcal{W}] &= (\mathbf{cons}\ Unload[V_1,\mathcal{W}]\ Unload[V_2,\mathcal{W}]) \\
Unload[\lambda x.M,\mathcal{W}] &= \mathbf{procedure} \\
Unload[b,\mathcal{W}] &= \mathbf{box} \\
Unload[f_c,\mathcal{W}] &= \mathbf{procedure} \\
Unload[\langle \mathbf{co}\ \kappa\rangle,\mathcal{W}] &= \mathbf{cont} \\
Unload[\langle \mathbf{ph}\ \alpha,s\rangle,\mathcal{W}] &= Unload[\mathcal{W}[\langle \alpha,s\rangle],\mathcal{W}]
\end{aligned}
$$

$$
\begin{aligned}
M \in \Lambda_{cekds} &::= V_s \mid (M\ M) & \text{(Term)} \\
&\mid (\mathbf{if}\ M\ M\ M) \mid (\mathbf{future}\ M) \\
V_s \in SValue &::= c \mid x \mid (\lambda x.M) & \text{(Syntactic Value)} \\
W \in PValue &::= c \mid \langle \mathbf{cl}\ \lambda x.M,\rho\rangle \mid f_c & \text{(Proper Value)} \\
&\mid (\mathbf{cons}\ V\ V) \mid \langle \mathbf{co}\ \kappa\rangle \mid b \\
V \in Value &::= W \mid ph & \text{(Runtime Value)} \\
ph &::= \langle \mathbf{ph}\ \alpha,s\rangle & \text{(Placeholder)} \\
b \in Box &::= \langle \mathbf{bx}\ \alpha,s,\ell\rangle & \text{(Box)} \\
\ell \in Leg &::= \langle \mathbf{leg}\ \alpha,s\rangle & \text{(Legitimacy)} \\
c \in Const &::= p \mid f & \text{(Constant)} \\
f_c \in PApp &::= (\mathbf{cons}\ V) \mid (\mathbf{setref!}\ V) & \text{(Partial Application)} \\
g \in AValue &::= \langle \mathbf{cl}\ \lambda x.M,\rho\rangle \mid f \mid f_c \mid \langle \mathbf{co}\ \kappa\rangle & \text{(Applicable Val.)} \\
p \in BConst &= \{\mathbf{true,false,nil},0,1,\ldots,\mathbf{void}\} & \text{(Basic Constant)} \\
f \in FConst &= \{\mathbf{cons,car,cdr,makeref,deref,} & \text{(Func. Cstnt)} \\
&\quad \mathbf{setref!,callcc,sync,touch}\} \\
f_n \in NStP &= \{\mathbf{cons,makeref,callcc}\} & \text{(Non Strict Primitives)} \\
x \in Vars &= \{x,y,z\ldots\} & \text{(User Variable)} \\
\kappa \in CCode &::= (\mathbf{init}) \mid (\kappa\ \mathbf{fun}\ V) & \text{(Continuation code)} \\
&\mid (\kappa\ \mathbf{arg}\ M\ \rho) \mid (\kappa\ \mathbf{cond}(M,M,\rho)) \\
&\mid (\kappa\ \mathbf{det}\ (ph,\ell)) \mid (\kappa\ \mathbf{leg}\ (\ell,\ell)) \\
R \in Req &::= \mathsf{Req}\langle s,u,rc\rangle & \text{(Request)} \\
rc \in RC &::= \mathsf{rtouch}(\alpha) \mid \mathsf{rdet}(\alpha,V,\alpha,\ell) & \text{(Request Contents)} \\
&\mid \mathsf{deref}(\alpha) \mid \mathsf{rset}(\alpha,V) \mid \mathsf{rleg}(\ell,\ell) \\
A \in Ans &::= \mathsf{Ans}\langle u,ac\rangle & \text{(Answer)} \\
ac \in AC &::= \mathsf{rtouch}(V) \mid \mathsf{rdet}(X) & \text{(Answer Contents)} \\
&\mid \mathsf{rset}(X) \mid \mathsf{deref}(V) \mid \mathsf{rleg}
\end{aligned}
$$

Store Operations:
$$\theta \uplus \{(\alpha\ V)\} = \theta \cup \{(\alpha\ V)\} \text{ with } \alpha \notin DOM(\theta)$$
$$\theta(\alpha) = V \text{ if } (\alpha\ V) \in \theta$$
$$\theta[\alpha := V] = (\theta \setminus \{(\alpha\ \theta(\alpha))\}) \cup \{(\alpha\ V)\}$$

Global reference:
$$\mathcal{W}[\langle \alpha,s\rangle] = \theta(s) \quad \text{if } \exists \langle T,s,\theta,S,I,O\rangle \in \mathcal{W}$$

Environment Operations:
$$\rho(x) = V \text{ if } (x\ V) \in \rho$$
$$\rho[x \leftarrow V] = (\rho \setminus \{(x\ V')\}) \cup \{(x\ V)\} \text{ if } (x\ V') \in \rho$$
$$\rho[x \leftarrow V] = \rho \cup \{(x\ V)\} \text{ if } x \notin DOM(\rho)$$

$$
\begin{aligned}
\mathsf{ltouch}(W,\theta,s) &= W \\
\mathsf{ltouch}(\langle \mathbf{ph}\ \alpha,s\rangle,\theta,s) &= \mathsf{ltouch}(\theta(\alpha),\theta,s) \text{ if } \theta(\alpha) \neq \bot \\
\mathsf{ltouch}(\langle \mathbf{ph}\ \alpha,s\rangle,\theta,s) &= \langle \mathbf{ph}\ \alpha,s\rangle \text{ if } \theta(\alpha) = \bot \\
\mathsf{ltouch}(\langle \mathbf{ph}\ \alpha,s_1\rangle,\theta,s) &= \langle \mathbf{ph}\ \alpha,s_1\rangle \text{ if } s_1 \neq s
\end{aligned}
$$

$$\ell \leadsto^s_\theta \ell$$
$$\langle \mathbf{leg}\ \alpha,s\rangle \leadsto^s_\theta \ell \text{ if } \theta(\alpha) \neq \bot \text{ and } \theta(\alpha) \leadsto^s_\theta \ell$$

$$\ell \leadsto_\mathcal{W} \ell$$
$$\langle \mathbf{leg}\ \alpha,s_1\rangle \leadsto_\mathcal{W} \ell \text{ if } \theta_{s_1}(\alpha) \neq \bot$$
$$\text{and } \mathcal{W}[\langle \alpha,s_1\rangle] \leadsto_\mathcal{W} \ell$$

Fig. 2. State Space of the CEKDS-machine

$$
\begin{array}{rll}
\mathsf{Ev}\langle (M\ N),\rho,\kappa\rangle &\to_{cek} \mathsf{Ev}\langle M,\rho,(\kappa\ \mathbf{arg}\ N,\rho)\rangle & (\textit{operator}) \\
\mathsf{Ev}\langle \lambda x.M,\rho,\kappa\rangle &\to_{cek} \mathsf{Ret}\langle\langle\mathsf{cl}\ \lambda x.M,\rho\rangle,\kappa\rangle & (\textit{lambda}) \\
\mathsf{Ev}\langle c,\rho,\kappa\rangle &\to_{cek} \mathsf{Ret}\langle c,\kappa\rangle & (\textit{constant}) \\
\mathsf{Ev}\langle x,\rho,\kappa\rangle &\to_{cek} \mathsf{Ret}\langle\rho(x),\kappa\rangle & (\textit{variable}) \\
\mathsf{Ret}\langle V,(\kappa\ \mathbf{arg}\ N,\rho)\rangle &\to_{cek} \mathsf{Ev}\langle N,\rho,(\kappa\ \mathbf{fun}\ V)\rangle & (\textit{operand}) \\
\mathsf{Ret}\langle V,(\kappa\ \mathbf{fun}\ \langle\mathsf{cl}\ \lambda x.M,\rho\rangle)\rangle &\to_{cek} \mathsf{Ev}\langle M,\rho[x\leftarrow V],\kappa\rangle & (\textit{apply}) \\
\mathsf{Ev}\langle(\mathsf{if}\ M\ M_1\ M_2),\rho,\kappa\rangle &\to_{cek} \mathsf{Ev}\langle M,\rho,(\kappa\ \mathbf{cond}\ (M_1,M_2,\rho))\rangle & (\textit{predicate}) \\
\mathsf{Ret}\langle V,(\kappa\ \mathbf{cond}\ (M_1,M_2,\rho))\rangle &\to_{cek} \mathsf{Ev}\langle M_2,\rho,\kappa\rangle \quad \text{if}\ V=\mathsf{false} & (\textit{if else}) \\
&\to_{cek} \mathsf{Ev}\langle M_1,\rho,\kappa\rangle \quad \text{if}\ V\neq\mathsf{false} & (\textit{if then}) \\
\mathsf{Ret}\langle V,(\kappa\ \mathbf{fun}\ \mathsf{callcc})\rangle &\to_{cek} \mathsf{Ret}\langle\langle\mathsf{co}\ \kappa\rangle,(\kappa\ \mathbf{fun}\ V)\rangle & (\textit{capture}) \\
\mathsf{Ret}\langle V,(\kappa'\ \mathbf{fun}\ \langle\mathsf{co}\ \kappa\rangle)\rangle &\to_{cek} \mathsf{Ret}\langle V,\kappa\rangle & (\textit{invoke}) \\
\mathsf{Ret}\langle V,(\kappa\ \mathbf{fun}\ V_1)\rangle &\to_{cek} \mathsf{Ret}\langle(V_1\ V),\kappa\rangle \quad \text{if}\ V_1\in PApp & (\textit{partial apply}) \\
\mathsf{Ret}\langle V,(\kappa\ \mathbf{fun}\ (\mathsf{cons}\ V_1))\rangle &\to_{cek} \mathsf{Ret}\langle(\mathsf{cons}\ V_1\ V),\kappa\rangle & (\textit{cons}) \\
\mathsf{Ret}\langle(\mathsf{cons}\ V_1\ V_2),(\kappa\ \mathbf{fun}\ \mathsf{car})\rangle &\to_{cek} \mathsf{Ret}\langle V_1,\kappa\rangle & (\textit{car}) \\
\mathsf{Ret}\langle(\mathsf{cons}\ V_1\ V_2),(\kappa\ \mathbf{fun}\ \mathsf{cdr})\rangle &\to_{cek} \mathsf{Ret}\langle V_2,\kappa\rangle & (\textit{cdr}) \\
\mathsf{Ret}\langle V,(\kappa\ \mathbf{fun}\ f)\rangle &\to_{cek} \mathsf{Ret}\langle\delta(f,V),\kappa\rangle & (\delta) \\
\mathsf{Ret}\langle V,(\kappa\ \mathbf{fun}\ V_1)\rangle &\to_{cek} \mathsf{Ret}\langle\mathsf{error},(\mathbf{init})\rangle\ \text{if}\ V_1\notin AValue & (\textit{apply error})
\end{array}
$$

Fig. 3. Transitions between computational states

continuation of future with the placeholder ph. After transition, the initial task τ evaluates the argument of future. We shall explain later the purpose of the legitimacies ℓ and ℓ_1.

A new box can be created by applying the functional constant makeref on a value. As a result, a new location is allocated in the local store, and a new box object, which refers to the new location and the site name, i.e. the qualified location, is returned.

In order to illustrate the behaviour of the machine, we consider the operation of reading a box, which will lead us to explain some rules of Figures 4, 5, 6, and 7; similar comments apply to box modification. A task considers that a box is local if the site name held in the box is the name of the current site. According to (*deref local*) in Figure 4, if the box is local, the value contained in the local store at the given location can be returned. Otherwise, rule (*deref remote*) adds to the output spool a *request* addressed to the site that allocated the box; in addition, the task is suspended, which is modelled by its transfer from the set T of runnable tasks to the set S of suspended tasks. We represent requests as triples composed of the destination site, the qualified name of the requesting task, and the message itself describing the type of the request. In the present case, the message deref(α) means that the distant site is asked to supply the content of location α.

Sites communicate according to the rules of Figure 5. In rule (*migrate request*), two sites exchange requests by moving them from the output spool of the source site to the input spool of the destination site. Figure 6 shows how a site handles incoming requests. In the case of rule (*request deref*), the content of the location is packaged up into an *answer* which must return back to the task that initiated the request; we again can see the interest of the qualified task name which indicates the name of the site that emitted the request. The answer is entered in the output spool and is migrated, like a request, by rule (*migrate answer*). Figure 7 shows the rules that handle incoming answers. The arrival of the answer deref(V) awakens the task waiting for this answer by transferring it back to the set of runnable tasks, with the value V as the content of the box.

Rule (*fork*) allows us to create new tasks on the current site. In Figure 5, rule (*migrate task*) shows that a task may be migrated from a site 1 running more that one active task to a site 2 without any active task. We see that a task is migrated by transferring its computational state, i.e. among others its continuation, and its legitimacy.

$\langle \{ \langle C, \ell, \tau \rangle \} \cup T, \theta, s, S, I, O \rangle$
  $\rightarrow_s \langle \{ \langle C_1, \ell, \tau \rangle \} \cup T, \theta, s, S, I, O \rangle$ if $C \rightarrow_{cek} C_1$ (*sequential*)

$\langle \{ \langle \mathsf{Ev}\langle (\mathsf{future}\ M), \rho, \kappa \rangle, \ell, \tau \rangle \} \cup T, \theta, s, S, I, O \rangle$
  $\rightarrow_s \langle \{ \langle \mathsf{Ev}\langle M, \rho, (\kappa\ \mathbf{det}\ ph, \ell_1) \rangle, \ell, \tau \rangle, \ \langle \mathsf{Ret}\langle ph, \kappa \rangle, \ell_1, \tau_1 \rangle \} \cup T, \theta_1, s, S, I, O \rangle$ (*fork*)
  with $ph = \langle \mathsf{ph}\ \alpha, s \rangle, \ell_1 = \langle \mathsf{leg}\ \alpha_1, s \rangle, \theta_1 = \theta \uplus \{ (\alpha\ \bot)\ (\alpha_1\ \bot) \}, \tau_1 \notin FN(T \cup S) \cup \{\tau\}$

$\langle \{ \langle \mathsf{Ret}\langle V, (\kappa\ \mathbf{fun}\ \mathsf{makeref}) \rangle, \ell, \tau \rangle \} \cup T, \theta, s, S, I, O \rangle$
  $\rightarrow_s \langle \{ \langle \mathsf{Ret}\langle b, \kappa \rangle, \ell, \tau \rangle \} \cup T, \theta_1, s, S, I, O \rangle$ with $b = \langle \mathsf{bx}\ \alpha, s, \ell \rangle, \theta_1 = \theta \uplus \{(\alpha\ V)\}$ (*makeref*)

$\langle \{ \langle \mathsf{Ret}\langle \langle \mathsf{bx}\ \alpha, s, \ell \rangle, (\kappa\ \mathbf{fun}\ \mathsf{deref}) \rangle, \ell_1, \tau \rangle \} \cup T, \theta, s_1, S, I, O \rangle$
  $\rightarrow_s \langle \{ \langle \mathsf{Ret}\langle \theta(\alpha), \kappa \rangle, \ell_1, \tau \rangle \} \cup T, \theta, s_1, S, I, O \rangle$ if $s_1 = s$ (*deref local*)
  $\rightarrow_s \langle T, \theta, s_1, \{ \langle \mathsf{Ret}\langle \langle \mathsf{bx}\ \alpha, s, \ell \rangle, (\kappa\ \mathbf{fun}\ \mathsf{deref}) \rangle, \ell_1, \tau \rangle \} \cup S, I, O_1 \rangle$ if $s_1 \neq s$ (*deref remote*)
  with $O_1 = \{\ \mathsf{Req}\langle s, \langle \tau, s_1 \rangle, \mathsf{deref}(\alpha) \rangle\ \} \cup O$

$\langle \{ \langle \mathsf{Ret}\langle V, (\kappa\ \mathbf{fun}\ (\mathsf{setref}\ \langle \mathsf{bx}\ \alpha, s, \ell \rangle))) \rangle, \ell_1, \tau \rangle \} \cup T, \theta, s_1, S, I, O \rangle$
  $\rightarrow_s \langle \{ \langle \mathsf{Ret}\langle \mathsf{void}, \kappa \rangle, \ell_1, \tau \rangle \} \cup T, \theta[\alpha := V], s_1, S, I, O \rangle$ if $s_1 = s$ (*setref local*)
  $\rightarrow_s \langle T, \theta, s_1, \{ \langle \mathsf{Ret}\langle V, (\kappa\ \mathbf{fun}\ (\mathsf{setref}\ \langle \mathsf{bx}\ \alpha, s, \ell \rangle))) \rangle, \ell_1, \tau \rangle \} \cup S, I, O_1 \rangle$ (*setref remote*)
  if $s_1 \neq s$, with $O_1 = \{\ \mathsf{Req}\langle s, \langle \tau, s_1 \rangle, \mathsf{rset}(\alpha, V) \rangle\ \} \cup O$

$\langle \{ \langle \mathsf{Ret}\langle V, (\kappa\ \mathbf{det}\ \langle \mathsf{ph}\ \alpha, s \rangle, \langle \mathsf{leg}\ \alpha_1, s \rangle) \rangle, \ell_2, \tau \rangle \} \cup T, \theta, s_2, S, I, O \rangle$
  $\rightarrow_s \langle T_1, \theta_1, s_2, S_1, I_1, O \rangle$ if $s_2 = s,\ \theta(\alpha) = \bot$ (*determine local*)
  with $I_1 = (I \cup I_2), T_1 = (T \cup T_2), S_1 = (S \setminus (I_2 \cup T_2)), \theta_1 = \theta[\alpha_1 := \ell_2][\alpha := V]$
  with $I_2 = \{\mathsf{Req}\langle s_3, \langle \tau_1, s_4 \rangle, \mathsf{rtouch}(\alpha) \rangle, \mathsf{Req}\langle s_3, \langle \tau_1, s_4 \rangle, \mathsf{rleg}(\langle \mathsf{leg}\ \alpha_1, s \rangle, \ell_4) \rangle \in S\}$
  with $T_2 = \{\langle \mathsf{Ret}\langle \langle \mathsf{ph}\ \alpha, s \rangle, (\kappa'\ \mathbf{fun}\ \mathsf{touch}) \rangle, \ell_3, \tau_1 \rangle \in S\} \cup$
              $\{\langle \mathsf{Ret}\langle V_1, (\kappa'\ \mathsf{leg}\ (\langle \mathsf{leg}\ \alpha_1, s \rangle, \ell_4)) \rangle, \ell_5, \tau_1 \rangle \in S\}$
  $\rightarrow_s \langle \{ \langle \mathsf{Ret}\langle V, \kappa \rangle, \ell_2, \tau \rangle \} \cup T, \theta, s_2, S, I, O \rangle$ if $s_2 = s,\ \theta(\alpha) \neq \bot$ (*determine localn*)
  $\rightarrow_s \langle T, \theta, s_2, \{ \langle \mathsf{Ret}\langle V, (\kappa\ \mathbf{det}\ \langle \mathsf{ph}\ \alpha, s \rangle, \langle \mathsf{leg}\ \alpha_1, s \rangle) \rangle, \ell_2, \tau \rangle \} \cup S, I, O_1 \rangle$ (*determine remote*)
  if $s_2 \neq s$ with $O_1 = \{\ \mathsf{Req}\langle s, \langle \tau, s_2 \rangle, \mathsf{rdet}(\alpha, V, \alpha_1, \ell_2) \rangle\ \} \cup O$

$\langle \{ \langle \mathsf{Ret}\langle V, (\kappa\ \mathbf{fun}\ \mathsf{touch}) \rangle, \ell, \tau \rangle \} \cup T, \theta, s, S, I, O \rangle$
  $\rightarrow_s \langle \{ \langle \mathsf{Ret}\langle \mathsf{ltouch}(V, \theta, s), \kappa \rangle, \ell, \tau \rangle \} \cup T, \theta, s, S, I, O \rangle$ (*touch local*)
    if $\mathsf{ltouch}(V, \theta, s) \in PValue$
  $\rightarrow_s \langle T, \theta, s, \{ \langle \mathsf{Ret}\langle \langle \mathsf{ph}\ \alpha, s_1 \rangle, (\kappa\ \mathbf{fun}\ \mathsf{touch}) \rangle, \ell, \tau \rangle \} \cup S, I, O_1 \rangle$ (*touch remote*)
    if $\mathsf{ltouch}(V, \theta, s) = \langle \mathsf{ph}\ \alpha, s_1 \rangle,\ s \neq s_1$, with $O_1 = \{\mathsf{Req}\langle s_1, \langle \tau, s \rangle, \mathsf{rtouch}(\alpha) \rangle\} \cup O$
  $\rightarrow_s \langle T, \theta, s, \{ \langle \mathsf{Ret}\langle \langle \mathsf{ph}\ \alpha, s_1 \rangle, (\kappa\ \mathbf{fun}\ \mathsf{touch}) \rangle, \ell, \tau \rangle \} \cup S, I, O \rangle$ (*touch suspend*)
    if $\mathsf{ltouch}(V, \theta, s) = \langle \mathsf{ph}\ \alpha, s_1 \rangle,\ s = s_1,\ \theta(\alpha) = \bot$

$\langle \{ \langle \mathsf{Ret}\langle V, (\kappa\ \mathbf{fun}\ \mathsf{sync}) \rangle, \ell, \tau \rangle \} \cup T, \theta, s, S, I, O \rangle$
  $\rightarrow_s \langle \{ \langle \mathsf{Ret}\langle V, (\kappa\ \mathsf{leg}\ (\ell, \ell_1)) \rangle, \ell, \tau \rangle \} \cup T, \theta, s, S, I, O \rangle$ if $V = \langle \mathsf{bx}\ \alpha, s_1, \ell_1 \rangle$ (*synchronise*)
  $\rightarrow_s \langle \{ \langle \mathsf{Ret}\langle \mathsf{error}, (\mathsf{init}) \rangle, \ell, \tau \rangle \} \cup T, \theta, s, S, I, O \rangle$ if $V \notin Box$ (*synchronise error*)

$\langle \{ \langle \mathsf{Ret}\langle V, (\kappa\ \mathsf{leg}\ (\ell, \ell_1)) \rangle, \ell_2, \tau \rangle \} \cup T, \theta, s, S, I, O \rangle$
  $\rightarrow_s \langle \{ \langle \mathsf{Ret}\langle V, \kappa \rangle, \ell_2, \tau \rangle \} \cup T, \theta, s, S, I, O \rangle$ if $\ell \leadsto_\theta^s \ell_1$ (*leg local*)
  $\rightarrow_s \langle T, \theta, s, \{ \langle \mathsf{Ret}\langle V, (\kappa\ \mathsf{leg}\ (\langle \mathsf{leg}\ \alpha, s_1 \rangle, \ell_1)) \rangle, \ell_2, \tau \rangle \} \cup S, I, O_1 \rangle$ (*leg remote*)
    if $\ell \leadsto_\theta^s \langle \mathsf{leg}\ \alpha, s_1 \rangle,\ s_1 \neq s$, with $O_1 = \{\ \mathsf{Req}\langle s_1, \langle \tau, s \rangle, \mathsf{rleg}(\langle \mathsf{leg}\ \alpha, s_1 \rangle, \ell_1) \rangle\ \} \cup O$
  $\rightarrow_s \langle T, \theta, s, \{ \langle \mathsf{Ret}\langle V, (\kappa\ \mathsf{leg}\ (\langle \mathsf{leg}\ \alpha, s \rangle, \ell_1)) \rangle, \ell_2, \tau \rangle \} \cup S, I, O \rangle$ (*leg suspend*)
    if $\ell \leadsto_\theta^s \langle \mathsf{leg}\ \alpha, s \rangle,\ \ell \not\leadsto_\theta^s \ell_1,\ \theta(\alpha) = \bot$

Fig. 4. Site Transitions

$\{\langle\{\langle C_1, \ell_1, \tau_1\rangle\}\cup T_1, \theta, s_1, S_1, I_1, O_1\rangle, \langle\emptyset, \theta_2, s_2, S_2, I_2, O_2\rangle\} \cup \mathcal{W}$ (*migrate task*)
$\rightarrow_c \{\langle T_1, \theta, s_1, S_1, I_1, O_1\rangle, \langle\{\langle C_1, \ell_1, \tau_2\rangle\}, \theta_2, s_2, S_2, I_2, O_2\rangle\} \cup \mathcal{W}$
     if $T_1 \neq \emptyset$, with $\tau_2 \notin FN(S_2)$
$\{\langle T_1, \theta_1, s_1, S_1, I_1, \{\mathsf{Req}\langle s_2, \langle \tau_3, s_3\rangle, rc\rangle\} \cup O_1\rangle, \langle T_2, \theta_2, s_2, S_2, I_2, O_2\rangle\} \cup \mathcal{W}$ (*migrate request*)
$\rightarrow_c \{\langle T_1, \theta_1, s_1, S_1, I_1, O_1\rangle, \langle T_2, \theta_2, s_2, S_2, \{\mathsf{Req}\langle s_2, \langle \tau_3, s_3\rangle, rc\rangle\} \cup I_2, O_2\rangle\} \cup \mathcal{W}$
$\{\langle T_1, \theta_1, s_1, S_1, I_1, \{\mathsf{Ans}\langle \tau_2, s_2, ac\rangle\} \cup O_1\rangle, \langle T_2, \theta_2, s_2, S_2, I_2, O_2\rangle\} \cup \mathcal{W}$ (*migrate answer*)
$\rightarrow_c \{\langle T_1, \theta_1, s_1, S_1, I_1, O_1\rangle, \langle T_2, \theta_2, s_2, S_2, \{\mathsf{Ans}\langle \tau_2, s_2, ac\rangle\} \cup I_2, O_2\rangle\} \cup \mathcal{W}$

Fig. 5. Communications between sites

$\langle T, \theta, s, S, \{\mathsf{Req}\langle s, \langle \tau, s_2\rangle, \mathsf{deref}(\alpha)\rangle\} \cup I, O\rangle$
   $\rightarrow_s \langle T, \theta, s, S, I, \{\mathsf{Ans}\langle \tau, s_2, \mathsf{deref}(\theta(\alpha))\rangle\} \cup O\rangle$ (*request deref*)
$\langle T, \theta, s, S, \{\mathsf{Req}\langle s, \langle \tau, s_2\rangle, \mathsf{rset}(\alpha, V)\rangle\} \cup I, O\rangle$
   $\rightarrow_s \langle T, \theta_1, s, S, I, \{\mathsf{Ans}\langle \tau, s_2, \mathsf{rset}\rangle\} \cup O\rangle$ with $\theta_1 = \theta[\alpha := V]$ (*request set*)
$\langle T, \theta, s, S, \{\mathsf{Req}\langle s, \langle \tau, s_2\rangle, \mathsf{rdet}(\alpha, V, \alpha_1, \ell)\rangle\} \cup I, O\rangle$
   $\rightarrow_s \langle T_1, \theta_1, s, S_1, I_1, \{\mathsf{Ans}\langle \tau, s_2, \mathsf{rdet}(\theta(\alpha))\rangle\} \cup O\rangle$ if $\theta(\alpha){=}\bot$ (*request det first*)
   with $I_1 = (I \cup I_2), T_1 = (T \cup T_2), S_1 = (S \setminus (I_2 \cup T_2)), \theta_1 = \theta[\alpha := V][\alpha_1 := \ell]$
   with $I_2 = \{\mathsf{Req}\langle s_3, \langle \tau_1, s_4\rangle, \mathsf{rtouch}(\alpha)\rangle, \mathsf{Req}\langle s_3, \langle \tau_1, s_4\rangle, \mathsf{rleg}(\langle\mathsf{leg}\ \alpha_1, s\rangle, \ell_1)\rangle \in S\}$
   with $T_2 = \{\langle\mathsf{Ret}\langle\langle\mathsf{ph}\ \alpha, s\rangle, (\kappa\ \mathbf{fun}\ \mathsf{touch})\rangle, \ell_1, \tau_1\rangle \in S\} \cup$
               $\{\langle\mathsf{Ret}\langle V_2, (\kappa\ \mathsf{leg}\ (\langle\mathsf{leg}\ \alpha_1, s\rangle, \ell_1))\rangle, \ell_2, \tau_1\rangle \in S\}$
   $\rightarrow_s \langle T, \theta, s, S, I, \{\mathsf{Ans}\langle \tau, s_2, \mathsf{rdet}(\theta(\alpha))\rangle\} \cup O\rangle$ if $\theta(\alpha){\neq}\bot$ (*request det mult*)
$\langle T, \theta, s, S, \{\mathsf{Req}\langle s, \langle \tau, s_2\rangle, \mathsf{rtouch}(\alpha)\rangle\} \cup I, O\rangle$
   $\rightarrow_s \langle T, \theta, s, S, I, \{\mathsf{Ans}\langle \tau, s_2, \mathsf{rtouch}(\mathsf{ltouch}(\langle\mathsf{ph}\ \alpha, s\rangle, \theta, s))\rangle\} \cup O\rangle$ (*request touch local*)
   if $\mathsf{ltouch}(\langle\mathsf{ph}\ \alpha, s\rangle, \theta, s) \in PValue, s_2 \neq s$
   $\rightarrow_s \langle T, \theta, s, S, \{\mathsf{Ans}\langle \tau, s_2, \mathsf{rtouch}(\mathsf{ltouch}(\langle\mathsf{ph}\ \alpha, s\rangle, \theta, s))\rangle\} \cup I, O\rangle$ (*request touch local'*)
   if $\mathsf{ltouch}(\langle\mathsf{ph}\ \alpha, s\rangle, \theta, s) \in PValue, s_2 = s$
   $\rightarrow_s \langle T, \theta, s, S, I, \{\mathsf{Req}\langle s_3, \langle \tau, s_2\rangle, \mathsf{rtouch}(\alpha_1)\rangle\} \cup O\rangle$ (*request touch remote*)
   if $\mathsf{ltouch}(\langle\mathsf{ph}\ \alpha, s\rangle, \theta, s) = \langle\mathsf{ph}\ \alpha_1, s_3\rangle,\ s \neq s_3$
   $\rightarrow_s \langle T, \theta, s, \{\mathsf{Req}\langle s, \langle \tau, s_2\rangle, \mathsf{rtouch}(\alpha_1)\rangle\} \cup S, I, O\rangle$ (*request touch suspend*)
   if $\mathsf{ltouch}(\langle\mathsf{ph}\ \alpha, s\rangle, \theta, s) = \langle\mathsf{ph}\ \alpha_1, s\rangle,\ \theta(\alpha_1){=}\bot$
$\langle T, \theta, s, S, \{\mathsf{Req}\langle s, \langle \tau, s_2\rangle, \mathsf{rleg}(\ell, \ell_1)\rangle\} \cup I, O\rangle$
   $\rightarrow_s \langle T, \theta, s, S, I, \{\mathsf{Ans}\langle \tau, s_2, \mathsf{rleg}\rangle\} \cup O\rangle$ if $\ell \leadsto_\theta^s \ell_1, s_2 \neq s$ (*request leg local*)
   $\rightarrow_s \langle T, \theta, s, S, \{\mathsf{Ans}\langle \tau, s_2, \mathsf{rleg}\rangle\} \cup I, O\rangle$ if $\ell \leadsto_\theta^s \ell_1, s_2 = s$ (*request leg local'*)
   $\rightarrow_s \langle T, \theta, s, S, I, \{\mathsf{Req}\langle s_3, \langle \tau, s_2\rangle, \mathsf{rleg}(\langle\mathsf{leg}\ \alpha, s_3\rangle, \ell_1)\rangle\} \cup O\rangle$ (*request leg remote*)
   if $\ell \leadsto_\theta^s \langle\mathsf{leg}\ \alpha, s_3\rangle,\ s \neq s_3$
   $\rightarrow_s \langle T, \theta, s, \{\mathsf{Req}\langle s, \langle \tau, s_2\rangle, \mathsf{rleg}(\langle\mathsf{leg}\ \alpha, s_3\rangle, \ell_1)\rangle\} \cup S, I, O\rangle$ (*request leg suspend*)
   if $\ell \leadsto_\theta^s \langle\mathsf{leg}\ \alpha, s\rangle,\ \theta(\alpha){=}\bot,\ \ell \not\leadsto_\theta^s \ell_1$

Fig. 6. Handling of Requests

According to (*fork*), the effect of evaluating (**future** M) is to allocate a placeholder that a new task speculatively passes to its continuation. The original task, which evaluates M, acts as a producer for the placeholder value, while the new task acts as a consumer.

A producer task has obtained the value V of a **future** argument, when V is returned to a continuation code of the form $(\kappa\ \mathbf{det}\ ph, \ell)$; the producer task is then expected to store V

$\langle T, \theta, s, \{\ \langle \mathsf{Ret}\langle V, (\kappa\ \mathbf{fun}\ \mathsf{deref})\rangle, \ell, \tau\rangle\ \} \ \cup\ S, \{\ \mathsf{Ans}\langle \tau, s, \mathsf{deref}(V_1)\rangle\ \}\ \cup\ I, O\rangle$
$\to_s \langle \{\ \langle \mathsf{Ret}\langle V_1, \kappa\rangle, \ell, \tau\rangle\ \} \ \cup\ T, \theta, s, S, I, O\rangle$ (*answer deref*)

$\langle T, \theta, s, \{\ \langle \mathsf{Ret}\langle V, (\kappa\ \mathbf{fun}\ (\mathsf{setref}\ b))\rangle, \ell, \tau\rangle\ \} \ \cup\ S, \{\ \mathsf{Ans}\langle \tau, s, \mathsf{rset}\rangle\ \}\ \cup\ I, O\rangle$
$\to_s \langle \{\ \langle \mathsf{Ret}\langle \mathsf{void}, \kappa\rangle, \ell, \tau\rangle\ \} \ \cup\ T, \theta, s, S, I, O\rangle$ (*answer rset*)

$\langle T, \theta, s, \{\ \langle \mathsf{Ret}\langle V, (\kappa\ \mathbf{det}\ ph, \ell)\rangle, \ell_1, \tau\rangle\ \} \ \cup\ S, \{\ \mathsf{Ans}\langle \tau, s, \mathsf{rdet}(X)\rangle\ \}\ \cup\ I, O\rangle$
$\to_s \langle T, \theta, s, S, I, O\rangle$ if $X=\bot$ (*answer det first*)
$\to_s \langle \{\ \langle \mathsf{Ret}\langle V, \kappa\rangle, \ell_1, \tau\rangle\ \} \ \cup\ T, \theta, s, S, I, O\rangle$ if $X\neq\bot$ (*answer det mult*)

$\langle T, \theta, s, \{\ \langle \mathsf{Ret}\langle ph, (\kappa\ \mathbf{fun}\ \mathsf{touch})\rangle, \ell, \tau\rangle\ \} \ \cup\ S, \{\ \mathsf{Ans}\langle \tau, s, \mathsf{rtouch}(V)\rangle\ \}\ \cup\ I, O\rangle$
$\to_s \langle \{\ \langle \mathsf{Ret}\langle V, \kappa\rangle, \ell, \tau\rangle\ \} \ \cup\ T, \theta, s, S, I, O\rangle$ (*answer touch*)

$\langle T, \theta, s, \{\ \langle \mathsf{Ret}\langle V, (\kappa\ \mathsf{leg}\ (\ell, \ell_1))\rangle, \ell_2, \tau\rangle\ \} \ \cup\ S, \{\ \mathsf{Ans}\langle \tau, s, \mathsf{rleg}\rangle\ \}\ \cup\ I, O\rangle$
$\to_s \langle \{\ \langle \mathsf{Ret}\langle \mathsf{V}, \kappa\rangle, \ell_2, \tau\rangle\ \} \ \cup\ T, \theta, s, S, I, O\rangle$ (*answer leg*)

Fig. 7. Handling of Answers

in the placeholder *ph*; this operation is called *determining* the placeholder. Depending on whether the current task is running on the site where the placeholder was allocated, rules (*determine local*) or (*determine remote*) take care of assigning the value V to the placeholder *ph*. However, placeholders are not boxes because they are defined as datastructures that can receive one value *at most* [11]; placeholders are like single-assignment variables in CC+ [2] and PCN [7]. As opposed to conventional languages, the language Λ_f has firstclass continuations which allow the programmer to write expressions that "return" multiple values; in other words, in Λ_f, different values can be passed to the same continuation. As a result, we distinguish the case where a placeholder is not assigned, in rules (*determine local*) and (*request det first*), from the case where it is already assigned, rules (*determine localn*) and (*request det mult*). In the former case, the placeholder is updated and the producer task ends its evaluation. In the latter case, the value is returned to the continuation of **future**, as if no **future** had existed, following Katz and Weise's implementation [15, 3, 20]. Let us observe that transitions are atomically executed in order to ensure a sound behaviour of (*determine local*) and (*determine localn*).

Strict primitives introduce synchronisations between the consumer task of a placeholder and its producer task: they require their arguments to be *proper values*, i.e. values different from placeholders; strict primitives are said to *touch* their argument. The translation $\mathcal{X}$ makes the touch action explicit by the call to **touch**, whose purpose is to return a *proper value*. We use an auxiliary function ltouch, displayed in Figure 2, which touches a value with respect to a local store θ of a site s. The function ltouch can return three results: a proper value, an undetermined placeholder that was allocated on site s, or a placeholder that was allocated on a different site. In the first case, the **touch** operation succeeds (*touch local*); in the second case, the task is suspended as long as the placeholder remains undetermined (*touch suspend*); in the third case, a request rtouch(α) is sent to the remote site (*touch remote*). The remote site behaves similarly: it can return a proper value, suspend the request, or pass it to another site. Tasks or requests that are suspended when touching a placeholder are reactivated when this placeholder gets determined, cfr. (*determine local*) or (*request det first*). Let us observe that the **touch** operation can initiate exchanges of messages between sites; as soon as a proper value is found, it is directly returned back to the site that started the operation, thanks to the qualified task name.

So far, our explanations have ignored legitimacies. Following Katz and Weise [15], we use a notion of legitimacy to keep track of the control flow that would exist if evaluation was sequential. An initial legitimacy is allocated when we start to evaluate a program, and each new task is given a new legitimacy. Legitimacies, like placeholders, are datastructures whose only slot can receive one value at most; unlike placeholders, legitimacies are not firstclass values. When a placeholder gets determined, the consumer task becomes dependent on the value of the placeholder; hence, the legitimacy of the producer task, recorded in the continuation (κ **det** $ph\ \ell$), is stored into the legitimacy of the consumer task. As evaluation proceeds, chains of legitimacies get formed into memory. The relation $\ell_1 \leadsto_\theta^s \ell_2$ states that there is a path from legitimacy ℓ_1 to legitimacy ℓ_2 in the local store θ of site s, which means that control has flowed from a task with legitimacy ℓ_2 to a task with legitimacy ℓ_1.

As we want **future** to be an annotation, every program should return the result that it would produce when evaluated sequentially in the absence of **future**. The solution adopted in our semantics is to perform causally-dependent [24] box accesses in the same order as in a sequential implementation; the solution relies on legitimacies. The translation of the primitive **deref**, $\lambda x.(\textsf{deref}\ (\textsf{sync}\ (\textsf{touch}\ x)))$, touches and then applies **sync** on its argument. The primitive **sync** behaves as the identity function if the legitimacy of the current task leads to the legitimacy associated with the box. In other words, **sync** acts as a synchronisation barrier by ensuring that all accesses to the box (read or write) that a sequential implementation would have performed before the current access are actually done in the parallel machine, and all accesses that a sequential implementation would perform after the current one remain to be performed by the parallel machine. The primitive **sync** suspends a task that illegitimately tries to access a box; it will be reactivated by (*determine local*) or (*request det first*).

In order to determine when a computation ends, the initial configuration contains a box aimed at receiving the final value. Consistent box accesses guarantee that the box will receive the legitimate final value (if there exists one).

It should be observed that using legitimacies to synchronise box accesses does not impose a *total* order on those operations, but a *partial* order. This property ensures that parallelism can exist for programs written in a mostly-functional style, where one generally considers that side-effects are performed locally in different modules or functions.

$\mathcal{W}_1 \to_{ds}^{1,m} \mathcal{W}_2$ if $\quad \mathcal{W}_1 \to_c \mathcal{W}_2$, or $\mathcal{W}_1 = \{m_1\} \cup \mathcal{W},\ \mathcal{W}_2 = \{m_2\} \cup \mathcal{W},\ m_1 \to_s m_2$,
$\quad\quad\quad\quad\quad$ with $m = 1$ if $\ell \leadsto_{w_1} \ell_0$, with ℓ the legitimacy of the task related to the transition
$\quad\quad\quad\quad\quad$ $m = 0$ otherwise

$\mathcal{W}_1 \to_{ds}^{n+n',m+m'} \mathcal{W}_2$ if $\mathcal{W}_1 \to_{ds}^{n,m} \mathcal{W}_3$ and $\mathcal{W}_3 \to_{ds}^{n',m'} \mathcal{W}_2$ $\hfill$ (transitive)

Conventions: $\mathcal{W}_1 \to_{ds} \mathcal{W}_2$ if $\mathcal{W}_1 \to_{ds}^{1,m} \mathcal{W}_2$; $\mathcal{W}_1 \to_{ds}^* \mathcal{W}_2$ if $\mathcal{W}_1 \to_{ds}^{n,m} \mathcal{W}_2, n \geq 0$;
$\quad\quad\quad\quad\quad$ $\mathcal{W}_1 \to_{ds}^+ \mathcal{W}_2$ if $\mathcal{W}_1 \to_{ds}^{n,m} \mathcal{W}_2, n > 0$.

Initial world for P, $InitWorld[P] = \{m_0, m_1, \ldots, m_n\}$, with

$\quad\quad$ Initial Store: $\theta_0 = \{(\alpha_0\ \bot)\ (\alpha_1\ \bot)\}$ $\quad\quad$ Initial Legitimacy: $\ell_0 = \langle \textsf{leg}\ \alpha_1, s_0 \rangle$
$\quad\quad$ Initial Box: $b_0 = \langle \textsf{bx}\ \alpha_0, s_0, \ell_0 \rangle$ $\quad\quad$ Initial Environment: $\rho_0 = \{(x\ b_0)\}$
$\quad\quad\quad\quad$ $m_0 = \langle \{\langle \textsf{Ev}\langle((\lambda v.(\textsf{setref!}\ (\textsf{sync}\ x)\ v))\ \mathcal{X}[\![P]\!]), \rho_0, (\textsf{init})\rangle, \ell_0, \tau_0\rangle\}, s_0, \theta_0, \emptyset, \emptyset, \emptyset \rangle$
$\quad\quad$ Empty Sites: $m_i = \langle \emptyset, s_i, \emptyset, \emptyset, \emptyset \rangle$ $\quad i = 1, \ldots, n$

Final World: $Final[\mathcal{W}_f]$ if $\mathcal{W}_f[\langle \alpha_0, s_0 \rangle] \neq \bot$.

$\text{eval}_{ds}(P) = \begin{cases} W & \text{if there exists } \mathcal{W}_0, \mathcal{W}_f, \text{ such that } \mathcal{W}_0 = InitWorld[P],\ \mathcal{W}_0 \to_{ds}^* \mathcal{W}_f, \\ & Final[\mathcal{W}_f] \text{ with } W = Unload(\mathcal{W}_f[\langle \alpha_0, s_0 \rangle], \mathcal{W}_f) \\ \bot & \text{if } \forall i \in \mathbf{N}, \exists \mathcal{W}_i \in World, n_i, m_i \in \mathbf{N}, \mathcal{W}_i \to_{ds}^{n_i, m_i} \mathcal{W}_{i+1} \text{ such that } m_i > 0 \\ & \mathcal{W}_0 = InitWorld[P] \end{cases}$

Fig. 8. Evaluation Relation

We now have all the components to define an evaluation relation that associates programs with their observable behaviour. The function *Unload* replaces each function, box, or continuation by a tag; in addition, *Unload* touches every placeholder appearing in the result. As values can be spread over different sites, *Unload* takes the world in argument.

Divergence should be defined with the greatest care, because **future** has the ability to create new tasks, but the scheduler may elect to evaluate any of them. One task only is *mandatory*; all the others are *speculative*. A task with legitimacy ℓ is mandatory if ℓ leads to the initial legitimacy ℓ_0 in the current world $\mathcal{W}$, which is written $\ell \leadsto_\mathcal{W} \ell_0$. Figure 8 defines the relation $\to_{ds}^{n,m}$ [6] to denote reductions that involve n steps among which m are mandatory. According to the evaluation relation $eval_{ds}$, a program is said to be divergent, i.e. its value is $\bot$, if it leads to an infinite transition sequence that regularly often contains mandatory transitions.

The soundness of the CEKDS-machine is estalished by proving that its evaluation relation $eval_{ds}$ is equal to the sequetial evaluation function of the CEK-Machine.

Theorem 1 (Soundness) $eval_{ds} = eval_{ss}$ □

3 Discussion and Related Work

This paper builds upon previous work about annotations for parallelism in functional languages. For a long time, research has focused on implementation issues and efficient designs [15, 3, 28, 14, 13, 11, 18]. Parallelism by annotations has been formalised recently only. Flanagan and Felleisen [6] have defined the semantics of future in a purely functional language. The author [19, 20] has proposed a semantic framework for continuations and side-effects in a language with the annotations pcall, fork, and future. This paper is the first to present a formal semantics for futures, side-effect, first-class continuations, and distribution. Our research is part of a project which aims at building a Virtual Multicomputer [23], which provides a soft architecture to support distributed applications, transcending the details of hardware architecture. The "distribution by annotation" paradigm is our contribution to this virtual multiprocessor, which provides the user with the view that a distributed network of computers is programmable as a sequential processor.

Both compile-time and runtime improvements could boost the performance of our architecture. Our semantics is a dynamic semantics, and there are opportunities to improve it using static analysis. Flanagan and Felleisen's [6] *touch analysis* remove provably redundant touch operations in purely functional future-based programs, using a set-based analysis [12]; extending their analysis to side-effects and first-class continuations would be desirable for the CEKDS-machine. Similarly, an analysis removing unnecessary sync operations would greatly reduce the cost of synchronisations associated with side-effects.

As far as the runtime system is concerned, a realistic implementation needs a distributed garbage collection; the approach "garbage collecting the world" [16] appears to be a suitable candidate. Similarly, we have to address the issue of process collection. Miller's MultiScheme [17] task collection is done during garbage collection: a task can be reclaimed if the placeholder that it determines is accessible from the gc roots.

According to Figure 4, every future creates a new task on the current site. This process creation strategy is referred to as *eager task creation* [11, 18, 3]. However, future-based programs can generate far more tasks than the number of sites in a CEKDS-world. In order to avoid the expensive cost of task creation, a *lazy task creation* [18, 3] strategy can be used: it postpones the creation of a task until a processor is ready to run it. A simple modification of our rules could make this strategy explicit. Though rule (*migrate task*) does not enforce any migration strategy, we think that task stealing [1, 3] would be appropriate; according to this strategy, a processor that becomes idle steals a task from a heavily loaded processor.

Queinnec's ICSLA [27, 24, 26, 25] is a dialect of Lisp offering primitives for parallelism, transparent migration of objects, and maintenance of their cache coherence over the network. Queinnec's purpose is different from ours: as he does not rely on transparent annotations, he does not preserve the sequential meaning of programs. However, he proposes a caching mechanism which is certainly lacking in our CEKDS machine. Although his notion of coherency is not suitable to the CEKDS machine because it is not relative to the sequential evaluation, the protocol that he proposes with lazy propagation of updated values is an interesting technique that would be worth investigating for our semantics.

4 Conclusion

Traditional approaches to distributed computing favour high-performance over ease of programming. We believe that there is a need for a high-level paradigm to program distributed systems. By supplying transparent annotations to create remote computations, we provide the programmer with the illusion that a distributed system is programmable like a sequential one, because the runtime system itself takes care of task and data migrations, race conditions, or critical sections.

This paper is the first step in this direction: we propose an abstract machine with a distributed store and prove that task migration is transparent to the user. Work is under way to refine this semantics and to precisely investigate the issue of data migration and distributed garbage collection within this framework.

References

1. Robert D. Blumofe, Christopher F. Joerg, Bradley C. Kuszmaul, C. E. Leiserson, Keith H. Randall, and Y. Zhou. Cilk: an Efficient Multithreaded Runtime System. In *PPOPP'95*, 1995.
2. K. M. Chandy and Kesselman C. CC++: A Declarative, Concurrent, Object Oriented Programming Notation. Technical Report CS-92-01, California Institute of Technology, 1992.
3. Marc Feeley. *An Efficient and General Implementation of Futures on Large Scale Shared-Memory Multiprocessors*. PhD thesis, Brandeis University, 1993.
4. M. Felleisen and D. Friedman. Control Operators, the SECD-Machine and the λ-Calculus. In *Formal Description of Programming Concepts III*, pages 193–217, 1986. Elsevier Pub.
5. Matthias Felleisen and Daniel P. Friedman. A Reduction Semantics for Imperative Higher-Order Languages. In *PARLE'87*, LNCS 259, pages 206–223. Springer-Verlag, 1987.
6. Cormac Flanagan and Matthias Felleisen. The Semantics of Future and Its Use in Program Optimization. In *POPL'95*. Also in Technical Reports 238, 239, Rice University, 1994.
7. I. Foster, R. Olson, and S. Tuecke. Productive Parallel Programming: The PCN Approach. *Scientific Programming*, 1(1):51–66, 1992.
8. Message Passing Interface Forum. A Message-Passing Interface Standard. Technical report, University of Tennessee, Knoxville, Tennessee, June 1995.
9. I. Foster, C. Kesselman, and S. Tuecke. The Nexus Apporach to Integrating Multithreading and Communications. Math. and Comp. Sci. Division, Argonne National Laboratory, 1995.
10. Al. Geist and al. PVM 3 User's Guide and Reference Manual. Technical report, Oak Ridge National Laboratory, Knoxville, Tennessee, May 1993.
11. Robert H. Halstead, Jr. New Ideas in Parallel Lisp : Language Design, Implementation. In *Parallel Lisp : Languages and Systems*, LNCS 441, pages 2–57, 1990.
12. Nevin Heintze. Set-Based Analysis of ML Programs. In *Proceedings of the 1994 ACM Conference on Lisp and Functional Programming*, pages 306–317, Orlando, Florida, June 1994.
13. Takayasu Ito and Manabu Matsui. A Parallel Lisp Language Pailisp and its Kernel Specification. In *Parallel Lisp : Languages and Systems*, LNCS 441, pages 58–100, 1990.
14. Takayasu Ito and Tomohiro Seino. On Pailisp Continuation and its Implementation. In *Proceedings of the ACM SIGPLAN workshop on Continuations CW92*, pages 73–90, 1992.
15. Morry Katz and Daniel Weise. Continuing Into the Future: On the Interaction of Futures and First-Class Continuations. In *LFP'90*, pages 176–184, June 1990.
16. Bernard Lang, Christian Queinnec, and José Piquer. Garbage Collecting the World. In *POPL'92*, pages 39–50, Albuquerque, New Mexico, 1992.
17. James S. Miller. *MultiScheme : A Parallel Processing System Based on MIT Scheme*. PhD thesis, MIT, 1987.
18. Eric Mohr, David A. Kranz, and Robert H. Halstead. Lazy Task Creation : a Technique for Increasing the Granularity of Parallel Programs. In *LFP'90*, pages 185–197, June 1990.
19. Luc Moreau. *Sound Evaluation of Parallel Functional Programs with First-Class Continuations*. PhD thesis, University of Liège, June 1994. Also available by anonymous ftp from ftp.montefiore.ulg.ac.be in directory pub/moreau.
20. Luc Moreau. The Semantics of Scheme with Future. In *In ACM SIGPLAN International Conference on Functional Programming (ICFP'96)*, Philadelphia, May 1996.
21. Luc Moreau. Correctness of a Distributed-Memory Model for Scheme. Technical report M96/3, University of Southampton, 1996.
22. Luc Moreau and Daniel Ribbens. The Semantics of pcall and fork. In *PSLS 95 – Parallel Symbolic Langages and Systems*, LNCS 1068, Beaune, France, October 1995.
23. Julian Padget. Controlling (Virtual) Multicomputers. In *Massively Parallel Computer Systems (MPCS'94)*, pages 102–112. IEEE Computer Society Press, 1994.
24. Christian Queinnec. Locality, Causality and Continuations. In *Proceedings of the 1994 ACM Conference on Lisp and Functional Programming*, Orlando, Florida, June 1994.
25. Christian Queinnec. Sharing mutable objects and controlling groups of tasks in a concurrent and distributed language. In *TPPP'94*, LNCS 700, pages 70–93, Sendai (Japan), 1994.
26. Christian Queinnec. DMEROON: a Distributed Class-based Causally-coherent Data Model: Preliminary Report. In *Parallel Symbolic Languages and Systems.*, LNCS 1068, 1995.
27. C. Queinnec and D. De Roure. Design of a Concurrent and Distributed Language. In *Parallel Symbolic Computing: Languages, Systems and Applications*, LNCS'748, p. 234–259, 1992.
28. Pete Tinker and Morry Katz. Parallel Execution of Sequential Scheme with ParaTran. In *LFP'88*, pages 28–39, 1988.

Partial Evaluation Scheme for Concurrent Languages and Its Correctness

Haruo Hosoya, Naoki Kobayashi and Akinori Yonezawa

Department of Information Science, University of Tokyo,
7–3–1 Hongo, Bunkyo–ku, Tokyo 113, Japan

Abstract. A simple, general, and well-formalized partial evaluation method for concurrent languages is proposed. In spite of many potential benefits, there are few partial evaluation techniques for concurrent languages. We choose a process calculus for the target language because it has theoretical clarity, and yet has expressive power enough to represent various high-level constructs in concurrent object-oriented languages. We realize effective optimization by allowing elimination of even nondeterministic interprocess communications. Furthermore, we prove correctness of our method with respect to *barb-agreed simulation*.

1 Introduction

Partial evaluation is a program transformation scheme to improve efficiency by specializing a program with respect to its known part of inputs. Many partial evaluation techniques have been proposed for sequential languages [6, 14]. They have been applied to programs in many fields such as compilers for reflective languages [10], logic circuit simulation [14], and numerical computation [2], speeding up programs by orders of magnitude. In concurrent languages, partial evaluation may optimize code for data distribution or load balancing if it depends heavily on the known inputs, or partial evaluation may enhance low-level optimizations by expanding sequences of message sending explicitly in output code. In spite of such potential benefits, there are few proposals of partial evaluation methods for concurrent languages.

Our goal is to develop a partial evaluation method that (1) is powerful enough to handle multiple processes and their communications, which are not treated in conventional partial evaluation methods for sequential languages, (2) is general enough to easily apply to various constructs in high-level concurrent (especially object-oriented) languages, and (3) is clear enough to formalize and prove its correctness since partial evaluation is non-trivial global program transformation. For these purposes, we make the following two fundamental decisions.

First, rather than doing complicated and specific partial evaluation directly on high-level languages, we first translate a program in a high-level language into a simple language based on *process calculi* [12, 9] and then apply partial evaluation to the translated program. It is because process calculi have theoretical clarity, and yet significant expressive power enough to represent high-level concurrent (object-oriented) languages, so that they have been foundations of static analyses and optimization techniques for concurrent languages [7, 8]. It is therefore easy to formalize our partial evaluation method and discuss its correctness. In this paper, we chose a subset of HACL [9], one of such process calculi, as our target language.

The second decision is that we allow our partial evaluation to eliminate communications between processes. This approach can not only reduce highcost communications themselves, but can also make it easy to propagate values sent by communications for further optimization of the subsequent code. Even when there are several *nondeterministic* choices in a communication, our method chooses one and discards the rest. In this sense, our approach *resolves nondeterminism*. Some previous work already takes similar approaches. For example, some existing optimizing compilers achieve high performance by generating code for statically fixed scheduling order [15]. In practice, results of most of concurrent applications do not depend on nondeterminism.

We formalize our partial evaluation method and prove its correctness. Because our approach resolves nondeterminism, a resultant program is not always equivalent to the original program with respect to usual process equivalence relations. We therefore introduce *barb-agreed simulation*, which is a relaxed version of barbed bisimulation [11]. We show that our method is correct with respect to barb-agreed simulation. We believe that it sufficiently expresses correctness criteria for most program transformation schemes as well as partial evaluation. As far as the authors know, this is the first study that develops a partial evaluation method for a concurrent language with enough expressive power, and also proves correctness of the method.

According to our preliminary experiments using a prototype system for our partial evaluation scheme with an application of logic expression interpreter, our method has power to eliminate a considerable number of communications in practice and improve efficiency dramatically.

The rest of this paper is organized as follows. Section 2 introduces the syntax and semantics of our target language. Section 3 gives our formal partial evaluation method. In section 4, we introduce barb-agreed simulation and prove our partial evaluation correct. In section 5, we discuss our approach to elimination of communications. In section 6, we remark on related work. Finally section 7 concludes this paper and touches upon future work. Full formalization, full proofs, and experiments, which are omitted here because of the lack of space, are found in our accompanying technical report [5].

2 Syntax and Semantics of the Target Language

This section introduces our target language, a subset of HACL. From the full HACL [9], we excluded static polymorphic types, choices, functions, and firstclass processes.

In HACL, computation is performed by concurrent processes communicating each other *asynchronously* via channels. m(v) is a process that sends a value v to a channel m. m(x)=>P is a process that receives a values v from a channel m,

Syntax

$$(\text{processes}) \; \mathcal{P} \ni P ::= P_1|P_2 \; | \; \$\texttt{x}.P \; | \; e_1(e_2) \; | \; e(\texttt{x})\texttt{=>}P \; | \; \texttt{T}$$
$$| \; \texttt{if } e \texttt{ then } P_1 \texttt{ else } P_2$$
$$(\text{procedure defs.}) \quad \Gamma ::= \{\texttt{f}_1(\texttt{x})\texttt{=}P_1, \ldots, \texttt{f}_n(\texttt{x})\texttt{=}P_n\}$$
$$(\text{programs}) \; \mathcal{G} \ni \Pi ::= \Gamma \triangleright P$$
$$(\text{arithmetic exprs.}) \; \mathcal{E} \ni e ::= \texttt{x} \; | \; \textit{const} \; | \; op(e) \; | \; (e_1,\ldots,e_n) \; | \; \#i(e)$$
$$(\text{values}) \quad v ::= \texttt{x} \; | \; \textit{const} \; | \; (v_1,\ldots,v_n)$$

Structural congruence
(1) $P_1|P_2 \cong P_2|P_1$ \qquad\qquad (2) $(P_1|P_2)|P_3 \cong P_1|(P_2|P_3)$
(3) $\$\texttt{x}.(P_1|P_2) \cong P_1|\$\texttt{x}.P_2$ if $\texttt{x} \notin Fv(P_1)$.

Reduction Rules

$$\frac{P \cong Q \quad \Gamma \triangleright P \to \Gamma \triangleright P' \quad P' \cong Q'}{\Gamma \triangleright Q \to \Gamma \triangleright Q'} \; (\text{SCong}) \qquad \frac{\Gamma \triangleright P \to \Gamma \triangleright P'}{\Gamma \triangleright P|Q \to \Gamma \triangleright P'|Q} \; (\text{Par})$$

$$\frac{\Gamma \triangleright P \to \Gamma \triangleright P'}{\Gamma \triangleright \$\texttt{x}.P \to \Gamma \triangleright \$\texttt{x}.P'} \; (\text{New}) \qquad \frac{}{\Gamma \triangleright \texttt{m}(v)|\texttt{m}(\texttt{x})\texttt{=>}P \to \Gamma \triangleright P[v/\texttt{x}]} \; (\text{Com})$$

$$\frac{}{\Gamma \cup \{\texttt{f}(\texttt{x})\texttt{=}P\} \triangleright \texttt{f}(v) \to \Gamma \cup \{\texttt{f}(\texttt{x})\texttt{=}P\} \triangleright P[v/\texttt{x}]} \; (\text{App})$$

Fig. 1. Syntax and operational semantics (selected)

and executes $P[v/\texttt{x}]$. Semantically, a channel is a bag of values, rather than a (FIFO) queue, and each of them is consumed by a receiver. If there are multiple senders and receivers, it is nondeterministic which pair of sender and receiver will communicate. Processes are spawned by parallel composition: $P_1|P_2$. $\$\texttt{x}.P$ creates a channel $\texttt{x}$ and executes P. Channels can be carried as first-class data.

We give the syntax and operational semantics in Figure 1. op is a primitive operator such as integer addition. Note that $e_1(e_2)$ can be either a sender process or a procedure (described below) call according to whether e_1 is evaluated to a channel or a procedure. T denotes a run-time error. A *context* $C[\cdot]$ is a process with a single occurrence of a hole $[\cdot]$ in it. $C[P]$ denotes a process obtained by replacing $[\cdot]$ in $C[\cdot]$ with P.

The operational semantics is defined via three reduction relations: an (applicative-order) reduction relation $\to_\mathcal{E}$ over arithmetic expressions $\mathcal{E}$; a structural congruence $\cong$ over processes $\mathcal{P}$; and a reduction relation $\to$ over programs $\mathcal{G}$, i.e., pairs of a set of *procedure* (parameterized recursive process) definitions and a process. Structural congruence $\cong$ is the smallest reflexive and transitive congruence relation closed under the rules in Figure 1. We define it in order to identify "equal" processes in terms of the structure and simplify the reduction rules. In the figure, we present part of the inference rules for $\to$. **SCong** rule allows structurally congruent processes to make the same reduction.

Example We give an example of simplified parallel "logic expression interpreter", which inputs a logic expression (exp) and the value for the variable X (valofx), and outputs the value for the expression. Arguments of each subexpression are evaluated in parallel. We describe below this example in our language, where a natural extension to continuation passing style (CPS) [1] is used. (ML-style `case` statement is a syntax sugar.) The procedure `logev` takes an extra argument r, which is a "reply" channel to which the procedure will pass a return value. On the other hand, each caller of the procedure first creates a new channel for reply, call the procedure with the new channel as an extra argument, and, in parallel to this, waits for a return value on the channel.

```
logev(exp,valofx,r) = case exp of
   TRUE or FALSE => r(exp)
 | X => r(valofx)
 | (AND,e1,e2) => $s1.$s2.( logev(e1,valofx,s1) | logev(e2,valofx,s2)
                          | s1(v1)=>s2(v2)=>r(v1 and v2))
```

3 Partial Evaluation Method

We represent each step of program transformation in partial evaluation by a reduction relation $\leadsto$ over programs. Its usage is as follows. Suppose we are given a program $\Pi_0 \equiv \Gamma \triangleright P_0$ and a context $C[\cdot] \equiv C_d[C_s[\cdot]]$ where $C_s[\cdot]$ and $C_d[\cdot]$ denote a statically known context and a dynamic (unknown) context, respectively. We first transform $\Gamma \triangleright C_s[P]$ in arbitrary number of steps by $\leadsto$: $\Gamma \triangleright C_s[P] \leadsto^* \Gamma \triangleright P_1$. We then execute $\Gamma \triangleright C_d[P_1]$ by $\rightarrow$.

In Figure 2, we present part of inference rules for $\leadsto_\mathcal{E}$ over $\mathcal{E}$ and $\leadsto$ over $\mathcal{G}$. Most of these rules are naturally derived from the operational semantics rules. In partial evaluation, any subexpression in an expression is allowed to reduce. For arithmetic expressions, we allow destructions of tuples containing even irreducible expressions (**PEArSel** rule), which is known as *partially static data structures* [6]. **PECom** rule eliminates communications between pairs of senders and receivers apparently in parallel; we say that P_1 and P_2 are *apparently in parallel* in Q if $Q \cong \$x_1.\ldots.\$x_n.(P_1 | P_2 | R)$ for some R, $x_1, \ldots, x_n$. **PEApp** rule inlines procedure calls. Our treatment of procedures makes it easy to allow specialization of procedures [5] (making specialized versions of recursive procedures w.r.t. particular arguments.)

We show some examples of reductions to illustrate our rules. Let Γ be the definition of the procedure `logev` shown in Section 2. $\Gamma \triangleright \$r.(\mathtt{logev(X,b,r)} | R)$, where $R \equiv \mathtt{r(y)=>s(y\ and\ TRUE)}$, invokes `logev`, waits for a result on r, and replies a value to s. We have the reductions

$\Gamma \triangleright \$\mathtt{r.(logev(X,b,r)} | R) \leadsto \Gamma \triangleright \$\mathtt{r.(case\ X\ of\ \ldots\ } | R)$ (**PEApp**)
$\leadsto \Gamma \triangleright \$\mathtt{r.(r(b)} | R)$ (**PEIf**)
$\leadsto \Gamma \triangleright \$\mathtt{r.(s(b\ and\ TRUE))}$ (**PECom**)

where the right-most column shows main rules applied in each reduction step. We present below the inference of the last reduction by **PECom**.

$$\frac{\dfrac{\Gamma \triangleright \mathtt{r(b)} | R \leadsto \Gamma \triangleright \mathtt{s(b\ and\ TRUE)}}{\Gamma \triangleright \$\mathtt{r.(r(b)} | R) \leadsto \Gamma \triangleright \$\mathtt{r.(s(b\ and\ TRUE))}}\ (\mathbf{PECom})}{}\ (\mathbf{PENew})$$

$$\frac{e_i \leadsto_{\mathcal{E}} e'_i}{(\cdots, e_i, \cdots) \leadsto_{\mathcal{E}} (\cdots, e'_i, \cdots)} \text{ (PEAr1)} \quad \frac{1 \le i \le n}{\#i((e_1, \cdots, e_n)) \leadsto_{\mathcal{E}} e_i} \text{ (PEArSel)}$$

$$\frac{\Gamma \triangleright P \leadsto \Gamma \triangleright P'}{\Gamma \triangleright P | Q \leadsto \Gamma \triangleright P' | Q} \text{ (PEPar)} \quad \frac{\Gamma \triangleright P \leadsto \Gamma \triangleright P'}{\Gamma \triangleright e(x) => P \leadsto \Gamma \triangleright e(x) => P'} \text{ (PERecv)}$$

$$\frac{}{\Gamma \triangleright \texttt{m}(e) | \texttt{m}(\texttt{x}) => P \leadsto \Gamma \triangleright P[e/\texttt{x}]} \text{ (PECom)}$$

$$\frac{}{\Gamma \cup \{\texttt{f}(\texttt{x}) = P\} \triangleright \texttt{f}(e) \leadsto \Gamma \cup \{\texttt{f}(\texttt{x}) = P\} \triangleright P[e/\texttt{x}]} \text{ (PEApp)}$$

Fig. 2. The partial evaluation rules (selected)

Nondeterminism **PECom** rule can *resolve nondeterminism*. Specifically, when there are multiple pairs of sender and receiver apparently in parallel on a channel, the rule chooses a pair and eliminates the communication between them. For example, among two choices of communications in the following program Π, our rule chooses one and may transform the program as follows.

$$\Pi \equiv \Gamma \triangleright \texttt{\$c.(c(a) | c(x)=>c(x+b1) | c(y)=>c(y+b2))}$$
$$\leadsto \Gamma \triangleright \texttt{\$c.(c(a+b1) | c(y)=>c(y+b2))} \leadsto \Gamma \triangleright \texttt{\$c.(c((a+b1)+b2))}$$

Nondeterminism in the original program is thus resolved by partial evaluation. As mentioned in the introduction, in our principle, any optimization scheme as well as partial evaluation can be allowed to resolve nondeterminism, for efficiency.

4 Correctness of Partial Evaluation

This section shows correctness of our partial evaluation method. For correctness criteria, we cannot use bisimulation equivalences because of our treatment of nondeterminism. On the other hand, the simulation relation [11] is too weak because it allows m(1)|n(2) to be replaced with m(1), for example. We therefore introduce *barb-agreed simulation*.

Barb-agreed simulation involves a binary relation $\mathcal{R}$ on programs. We require that for each $(\Pi, \Phi) \in \mathcal{R}$, (1) each possible one-step reduction from Φ corresponds to some possible multiple-step reduction from Π, and (2) Π and Φ have the same "actions observable from external processes." We require the latter condition because what *always* happen in Π should always happen in Φ. We allow to observe in Π sender processes via free channels, and the error, if any; we call them *barbs* of Π.

Definition 1 Barb. A program $\Pi \equiv \Gamma \triangleright P$ has a *barb* a, written $\Pi \Downarrow_a$, if (1) $P \cong \texttt{\$x}_1 \ldots \texttt{\$x}_n.(a(v)|Q)$ where $a \ne \texttt{x}_i$ and $a \notin Dom(\Gamma)$, for some Q; or (2) $a = \texttt{T}$ and $P \cong \texttt{T}|Q$, for some Q.

Definition 2 Barb-agreed Simulation. A relation $\mathcal{R}$ is a *barb-agreed simulation* if $(\Pi, \Phi) \in \mathcal{R}$ implies (1) $\Pi\Downarrow_a$ iff $\Phi\Downarrow_a$; and (2) if $\Phi \to \Phi'$, then $\Pi \to^* \Pi'$ and $(\Pi', \Phi') \in \mathcal{R}$, for some Π'. We write $\Pi \mathrel{\dot{>}} \Phi$ if $(\Pi, \Phi) \in \mathcal{R}$, for some barb-agreed simulation $\mathcal{R}$.

Because the barb-agreed simulation itself is too weak a relation (for example, it allows m(1) to be replaced with m(2)), in order to obtain a reasonable relation, we should further require that the relation be closed under any context, just as barbed congruence is obtained from barbed bisimulation by closing it under any context. We call the resulting relation a *barb-agreed quasi-congruence*.

In addition, we have to take the following into consideration: (1) in some rules such as **PECom**, Π need not to have the same barbs of Φ at the starting point, but to reduce in some steps to get the barbs. (2) some rules such as **PECom** and **PEApp** assume that Π never cause errors at execution time. We then obtain the following relation.

Definition 3 Barb-agreed Quasi-congruence. $\Gamma \triangleright P \succ^c \Delta \triangleright Q$ if, for each context $C[\cdot]$, $\Gamma \triangleright C[P] \not\to^* \mathtt{T}$ implies $\Gamma \triangleright C[P] \to^* \Pi'$ and $\Pi' \mathrel{\dot{>}} \Delta \triangleright C[Q]$ for some Π'.

We can finally prove soundness of partial evaluation by induction on the height of inferences of $\Pi \leadsto \Phi$, and the reflexivity and transitivity of $\succ^c$.

Theorem 4 Soundness of Partial Evaluation. *If $\Pi \leadsto^* \Phi$, then $\Pi \succ^c \Phi$.*

5 Discussions on Elimination of Communications

Power of a partial evaluation method mostly depends on how it can propagate constants to the whole program. Our approach propagates values to be sent by (even nondeterministic) communications by just eliminating them. It is analogous to conventional approaches that propagate argument values of function calls by inlining them. However, we have the following issues to discuss.

Which communications to eliminate? Although our **PECom** rule is sound as already shown, it is conservative in the sense that it eliminates a communication only if the sender and the receiver are apparently in parallel. It is far from trivial to judge whether the elimination is sound or not if they are not apparently in parallel. For example, in the process $P \equiv$ (n(5) | m(x)=>n(y)=>f(x+y)), we can show that it is not sound to eliminate the communication between the sender n(5) and the receiver n(y)=>f(x+y), while it is sound if the channel n is bound as \$n.P, intuitively because no external process can receive from n. Developing an analysis to detect such cases is one of our future research issues.

Possibility of loss of concurrency Eliminating communications has potential of decreasing concurrency. For example, consider the following transformation,

$$\Gamma \triangleright \mathtt{m1}(e_1) \mid \mathtt{m2}(e_2) \mid \mathtt{m1(x)=>m2(y)=>n(x+y)} \leadsto^* \Gamma \triangleright \mathtt{n}(e_1 + e_2)$$

At execution time, the expressions e_1 and e_2 will be computed concurrently in the original program, while they will be computed sequentially in the resultant program. It is a trade-off whether we should eliminate the communication and decrease concurrency; or leave it uneliminated and retain concurrency. A possible decision strategy may analyze the size of the expressions and whether they are serialized in the receivers. Another strategy may use annotations specifying location of each process, and leave remote communications uneliminated. Note that this issue exists even if we eliminate only deterministic communications.

6 Related Work

One of our motivations is to aggressively optimize high-cost communication and synchronization constructs in concurrent object-oriented languages. The Concert compiler [13] stands on a setting similar to ours and proposes several optimization techniques. However, their approach is rather ad-hoc and handles locks and communication constructs separately, while our framework can handle them uniformly.

In the reflective concurrent object-oriented language ABCL/R3 [10], partial evaluation is used for compiling away meta-level interpreter code. However, although their meta-level uses in principle concurrent objects, the target of their partial evaluation is restricted to functional part of the meta-level programs, and every inter-object communication is residualized as an I/O operation. It is one of our goals to develop a framework that can optimize such communications.

Some partial evaluation methods for concurrent logic languages have been proposed [4]. These methods, in contrast to ours, take approaches that preserve nondeterminism, though they have not clearly described it. It seems to us that such approaches are difficult to propagate values between processes effectively. Moreover, their correctness has not been proven yet to the best of our knowledge.

7 Conclusion and Future Work

In this paper, we have proposed a simple, general, and well-formalized partial evaluation method for a simple language based on process calculi. By taking an approach that eliminates even nondeterministic communications, we realized effective partial evaluation. We introduced barb-agreed simulation as correctness criteria for program transformation schemes that are allowed to resolve nondeterminism, and proved correctness of our partial evaluation method with respect to barb-agreed simulation.

Finally, to make our partial evaluation method more powerful, we are planning to utilize the following analyses: linear channel analysis [8], which analyzes channels used only "once", for elimination of more communications, and set-based analysis [3], which analyzes conservatively values to be sent, for specializing receivers.

Acknowledgements: We deeply appreciate Hidehiko Masuhara, Kenjiro Taura and Tatsurou Sekiguchi for helpful comments and advice.

References

1. A. W. Appel. *Compiling with Continuation.* Cambridge University Press, 1992.
2. R. Baier, R. Glük, and R. Zöchling. Partial evaluation of numerical programs in Fortran. In *Proceedings of Partial Evaluation and Semantics-Based Program Manipulation*, pages 119–132, 1994.
3. N. Heintze. Set-based analysis of ML programs. In *proceedings of the 1994 Conference on Lisp and Functional Programming*, pages 306–317, 1994.
4. H.Fujita, A.Okamura, and K.Furukawa. Partial evaluation of GHC programs based on the UR-set with constraints. In *proceedings of Logic Programming: Fifth International Conference and Symposium*, pages 924–941, 1988.
5. H. Hosoya, N. Kobayashi, and A. Yonezawa. Partial evaluation for concurrent languages and its correctness. Technical report of the Department of Information Science, the University of Tokyo, 1996. to appear.
6. N. D. Jones, C. K. Gomard, and P. Sestoft. *Partial Evaluation and Automatic Program Generation.* Prentice Hall, 1993.
7. N. Kobayashi, M. Nakade, and A. Yonezawa. Static analysis of communication for asynchronous concurrent programming languages. In *Second International Static Analysis Symposium (SAS'95)*, volume 983 of *Lecture Notes in Computer Science*, pages 225–242. Springer-Verlag, 1995.
8. N. Kobayashi, B. C. Pierce, and D. N. Turner. Linearity and the pi-calculus. In *Proceedings of ACM SIGACT/SIGPLAN Symposium on Principles of Programming Languages*, pages 358–371, 1996.
9. N. Kobayashi and A. Yonezawa. Higher-order concurrent linear logic programming. In *Theory and Practice of Parallel Programming*, volume 907 of *Lecture Notes in Computer Science*, pages 137–166. Springer-Verlag, 1995.
10. H. Masuhara, S. Matsuoka, K. Asai, and A. Yonezawa. Compiling away the metalevel in object-oriented concurrent reflective languages using partial evaluation. In *Proceedings of ACM SIGPLAN Conference on Object-Oriented Programming Systems, Languages, and Applications (OOPSLA'95)*, pages 300–315, 1995.
11. R. Milner and D. Sangiorgi. Barbed bisimulation. In *19th ICALP*, volume 623 of *Lecture Notes in Computer Science*, pages 685–695, 1992.
12. B. C. Pierce and D. N. Turner. Concurrent objects in a process calculus. In *Theory and Practice of Parallel Programming (TPPP)*, volume 907 of *Lecture Notes in Computer Science*, pages 187–215. Springer-Verlag, 1995.
13. J. Plevyak, X. Zhang, and A. A.Chien. Obtaining sequential efficiency for concurrent object-oriented languages. In *Proceedings of ACM SIGACT/SIGPLAN Symposium on Principles of Programming Languages*, pages 311–321, 1995.
14. E. Ruf. *Topics in Online Partial Evaluation.* PhD thesis, Stanford University, 1993. (Technical Reprt CSL-TR-93-563).
15. K. Taura, S. Matsuoka, and A. Yonezawa. StackThreads: An abstract machine for scheduling fine-grain threads on stock cpus. In *Proceedings of Workshop on Theory and Practice of Parallel Programming*, number 907 in Lecture Notes on Computer Science, pages 121–136. Springer Verlag, 1994.

Support for Implementation of Evolutionary Concurrent Systems in Concurrent Programming Languages

Raju Pandey[1] and J. C. Browne[2]

[1] Computer Science Department, University of California, Davis, CA 95616
[2] Department of Computer Sciences, The University of Texas, Austin, TX 78712

Abstract. In many concurrent programming languages, concurrent programs are difficult to extend and modify: small changes in a concurrent program may require re-implementations of a large number of its components. In this paper a novel concurrent program composition mechanism is presented in which implementations of computations and synchronizations are completely separated. Separation of implementations facilitates extensions and modifications of programs by allowing one to change implementations of both computations and synchronizations. The paper also describes a concurrent programming model and a programming language that support the proposed approach.

1 Introduction

Complex software systems are evolutionary in general. They change during the initial development stage, and often after they have been deployed. These changes may occur due to changes in the requirements, in the hardware configuration, and/or in the execution environment. Programming languages must support methodologies that allow implementations of evolutionary systems. Specifically, small changes in the implementations of such systems should be localized, and should require modifications of a small number of components.

In this paper we show that many concurrent programming languages do not adequately support implementation of evolutionary concurrent systems: changes in the implementations of a small number of components may affect the implementations of a *disproportionately* large number of components. More importantly, concurrent program abstractions cannot be composed easily with existing program abstractions. This has implications on the re-usability of program abstractions and on concurrent programming language design. Specifically, the inability to compose concurrent program abstractions causes breakdowns in many of the programming language composition mechanisms.

A novel structuring scheme for concurrent programs is presented in this paper. In this scheme, *implementations of computations and synchronizations are completely separated*. A concurrent program is, thus, composed from *separate* implementations of computations and synchronizations. This is unlike most existing approaches where implementations of computations and synchronizations are embedded within the implementations of components.

Separation of implementations of computations and synchronizations has direct implications on the extensibility and modifiability of programs. Concurrent programs can be easily extended and modified by adding and modifying implementations of either computations, synchronizations, or both. Further, the approach advocates a programming design methodology where concurrent programs can be quickly constructed from existing implementations of computations and synchronizations. We briefly describe a concurrent programming model and a concurrent programming language that supports this programming methodology. The model defines general mechanisms for representing computations, interactions, and program compositions. The object-oriented programming language, CYES-C++, supports extensibility and modifiability of concurrent programs as well as re-usability of specifications of computations and interactions.

This paper is organized as follows: In Section 2, we show that there is poor support for implementation of evolutionary concurrent systems in many existing approaches. In Section 3, we analyze the reasons for the problems, and show how some of these problems can be resolved. In Section 4, the details of a concurrent programming model and a language that support the programming methodology are presented. A brief survey of the related work is presented in Section 5. Section 6 contains concluding remarks and the status of the research.

2 Modifications of Concurrent Programs

In this section we show that it is difficult to change the implementation of a concurrent system implemented using traditional approaches to concurrent programming. In a majority of concurrent programming languages, the approach to implementing a concurrent program involves partitioning a problem into a set of components, each implemented as a process, task, or thread. An implementation of a component contains operations that implement its computations, synchronization with other components, data decomposition and distributions and task scheduling algorithms. We show that concurrent programs specified in this manner are difficult to change and modify: extensions and modifications in a concurrent program may require that a large number of its components be modified. We illustrate this by showing that extensions and modifications of a simple concurrent program require re-implementation of some or all of its components. Note that the conclusions of this exercise are independent of the example.

Example 2.1. *(Extensibility and Modifiability of Concurrent Programs).* Below we show a concurrent program, `examprog1`, that is composed from two components: `producer` and `consumer`. The `producer` component repeatedly produces data, which are consumed by the `consumer` component. The components interact through the `send` and `receive` primitives over a mailbox [1] in which programs can deposit and retrieve information in a FIFO manner. Primitives `send` and `receive` respectively are non-blocking and blocking.

```
examprog1() {
    channel buf;
    producer(buf) || consumer(buf);
}
producer(channel buf){                consumer(channel buf){
    while (TRUE) {                        while (TRUE) {
        info = produce();                     info = receive(buf);
        send(buf, info);                      consume(info);
}}                                    }}
```

A simple extension of examprog1 involves adding another consumer component, for instance because consumer is slow relative to producer, such that data are now shared between the two consumer components *alternately*. There are many possible implementations of the extended program. However, in all implementations, producer, consumer, or both must be re-implemented in order to implement the altered interaction among the producer and the two consumer components.

Similarly, a modification of examprog1 may involve defining additional synchronization constraints — for instance, producer must wait after N un-consumed data — between producer and consumer. Again, as in the case of the extension, either or both components must be re-implemented in order to implement the altered interaction. ∎

Even though the above program contains two simple components, implications of simple changes in the program are widespread. *Simple extensions and modifications in a concurrent program may therefore affect implementations of a large number of its components.* Implementations of component are not encapsulated from each other. Changes in a concurrent program may be visible in some or possibly all components.

Also, specifications of components cannot be reused easily. For instance, in three versions of the example program, much of the behavior of producer and consumer remains unchanged. However, different versions of the components are created by duplicating much of the code from one version to another. In addition, synchronization, task scheduling, data mapping, and data distribution algorithms cannot be reused easily because they are embedded *procedurally* inside the implementations of components.

Further, modifications in components often involve making modifications in existing source code. Such modifications in source programs are error prone. Indeed, they are one of the major sources of errors in concurrent programs.

More importantly, the example underlines the problem associated with constructing new concurrent program abstractions in terms of existing program abstractions.

Definition 2.1. *(Program Composition Anomaly).* The program composition anomaly denotes the phenomenon in which the concurrent program composition of program abstractions requires changes and modifications in some of all of the program abstractions. ∎

Example 2.1 shows an occurrence of the program composition anomaly. The program composition anomaly highlights the inability to compose concurrent program abstractions from existing program abstractions. Since programming languages use many composition mechanisms for defining abstractions in terms of other abstractions, the presence of the program composition anomaly causes breakdowns in many of these composition mechanisms. We enumerate two such cases below.

Object-oriented programming languages support two fundamental composition mechanisms: *aggregation* and *inheritance*. Aggregation is used to define the structure of an object in terms of its component objects. Inheritance, on the other hand, is used to extend the structure of an object. In a concurrent object-oriented programming languages, we can think of a concurrent object as a concurrent program, whose composition is defined in terms of its methods and interactions among the methods. Both aggregation and inheritance can be viewed as implicit concurrent program composition mechanisms: aggregation as defining the concurrent program associated with an object as a composition of programs associated with its component objects, and inheritance as a means for extending the program composition of concurrent objects. We show that instances of the program composition anomaly occur when defining the two composition mechanisms.

Aggregation anomaly: The aggregation anomaly occurs when an object defines additional interaction behavior for methods of its component objects.

Example 2.2. *(Aggregation anomaly).* Assume that an object of class TwoBufs contains two objects: LarBuf and SmBuf of a concurrent class AtBuf. Class AtBuf defines two methods: Read and Write. The two methods synchronize with each other while accessing common data structures of AtBuf. Let Class TwoBufs define addition constraints on invocations of Read and Write over LarBuf and SmBuf objects: Write invocations on LarBuf have higher priority than Write invocations on SmBuf. Since the synchronization operations of Write are embedded inside the implementation of Write, the new synchronization behavior can be specified only by re-implementing the methods in AtBuf, thereby requiring redefinition of AtBuf. ■

In this example, class TwoBufs is used to compose two instances of abstraction AtBuf along with additional synchronization constraint. However, such a composition requires changes in the abstraction (AtBuf).

Inheritance anomaly: The second problem, termed the *inheritance anomaly* [16], arises due to the diverse synchronization requirements of a class and its subclasses.

Example 2.3. *(Inheritance anomaly).* Let class NBuf extend class AtBuf by defining a new method GetLst. Method GetLst interacts with Read and Write of AtBuf. This implies that synchronization properties of Read and Write change. Since the implementations of Read and Write include synchronization operations, the interaction behaviors of the methods can be implemented only by re-implementing the methods. This can be achieved either by re-implementing AtBuf or by re-implementing Read and Write in NBuf. In the latter case, implementations of Read and Write cannot be inherited in NBuf. ∎

The inheritance anomaly is another instance of the program composition anomaly. Here, a subclass extends the program composition associated with a concurrent object either by adding new methods or by modifying inherited methods. Such extensions require changes in the composition, which, in this case, means redefinition of methods.

3 Support for Extensibility and Modifiability

We first examine the reason for occurrence of the program composition anomaly. There are two distinct behaviors of a component: *computational behavior* and *interaction behavior*. The computational behavior of a component specifies the operations performed during an execution of the component. For instance, computational behavior of the producer component is to produce data. The interaction behavior of a component determines the manner in which the component affects or is affected by other components. It represents a semantic relationship among components. For instance, the interaction behavior of consumer (example 2.1) specifies that every invocation of consume depends on a preceding invocation of produce, representing a data dependency relationship among the operations.

The program composition anomaly arises because implementations of both — computational and interaction — behaviors of a component are embedded within an implementation of the component. Any changes (either through extension or modification) in a concurrent program tend to change the existing interaction relationships among the components. Since implementations of the relationships are distributed in the implementation of the components, changes in an interaction relationship can be effected only by re-implementing all components that implement the relationship.

3.1 Concurrent Program Composition

Our approach, which we call *evolution through separation*, is based on a novel structuring technique for concurrent programs. It advocates a programming methodology in which *implementations of computational and interaction behaviors are completely separated*. A concurrent program is, thus, composed from *separate* implementations of computational and interaction behaviors.

Definition 3.1. *(Constrained concurrent program composition).* The expression

$$C = (C_1 \parallel C_2 \parallel \ldots \parallel C_n) \text{ where } \phi$$

specifies a concurrent program C. Program C is composed from components $C_1, C_2, \ldots,$ and C_n and expression ϕ that represents relationship among the operations of the components. ∎

The semantics of the composition is that during an execution of C, operations of components $C_1, C_2, \ldots,$ and C_n occur in parallel by default. However, there are invocations of operations that interact. Executions of these invocations must satisfy all interaction relationships specified by ϕ. Concurrent program `examprog1` is thus defined as:

```
examprog1 = (producer || consumer) where consexp1
```

In this definition, components `producer` and `consumer` define only their computational behavior. Expression `consexp1` defines interaction among operations of `producer` and `consumer`.

3.2 Implications of separation

Separation of implementations of computational and interaction behaviors have direct implications on extensibility and modifiability of concurrent programs, as well as re-usability of components.

Concurrent programs can be extended easily. Additions of components may require definition of new interaction behaviors, and possible modifications of existing ones. For instance, `examprog1` can be extended easily:

```
examprog2 = (producer || consumer || consumer) where consexp2
```

Expression `consexp2` represents the new interaction relationship among the three components. Implementations of either `producer` or `consumer` do not change.

A concurrent program can be modified easily either by modifying computational behavior of its components or their interaction behaviors. For instance, the following program

```
examprog3 = (producer || consumer) where consexp3
```

is composed from the same components as `examprog1` except that `consexp3` implements a different interaction behavior among the components. The approach supports encapsulation of implementations of both computational and interaction behaviors. For instance, `producer` can be re-implemented, in isolation, from the implementations of `consumer` and the interaction behavior. Even if this implementation implies changes in concurrent program, only the implementations of interaction behaviors needs to be changed. The computational behavior of `consumer` remains unaffected. Separation of implementations therefore *localizes* the effects of changes in a concurrent program. Further, it supports re-usability of implementations of both computational and interaction behaviors. For instance, different versions of `examprog1` can be constructed by combining `producer` and `consumer` in many different ways. Indeed, it advocates a programming design methodology in which concurrent programs can be quickly constructed from existing implementations of computational and interaction behaviors.

Verification of concurrent programs is also facilitated by the separation of the implementations. The approach allows one to verify properties of the system by looking at the implementations of computational and interaction behaviors in isolation.

Separation also forms the basis for the resolution of the aggregation and inheritance anomalies. In the case of inheritance anomaly, interaction behavior of inherited methods can be extended and/or modified by defining interaction behaviors in a subclass [21]. The inheritance anomaly has been studied in great detail and many solutions [14, 26, 23, 25] have been proposed. Most of these solutions are based on the separation of synchronization constraints from the method specifications as well.

Separation of implementations facilitates programming language design as well. By supporting mechanisms for defining abstractions for computational and interaction behaviors, a concurrent programming language can provide support for constructing powerful concurrent program abstractions by simply extending the existing composition mechanisms. The design of CYES-C++(Section 4.3) clearly benefited from this approach.

4 Support for Concurrent Programming

We now describe a concurrent programming model and a programming language that support the proposed programming methodology. We first present a model of concurrent computation, called the C-YES model [20]. The C-YES model defines representation mechanisms for computational and interaction behaviors. It has been used to define a compositional model for concurrent object-oriented languages [21], and a concurrent object-oriented programming language, CYES-C++ [22]. Due to the lack of space, we outline only the fundamental aspects of the model and the language. The details can be found in [19].

4.1 Representation of computational behavior

Given that implementations of components do not include implementations of interaction behaviors, the question is: how are component programs implemented so that their interaction behaviors can be specified in a concurrent program?

The execution behavior of a component is to repeatedly execute operations, and occasionally interact with its environment (other components) during the execution of certain operations. For instance, producer interacts with its environment during executions of produce operations. We call such operations *interaction points*. An interaction point denotes a set of possible invocations of operations where interaction may occur. A component in the C-YES model is therefore represented by its computations and interaction points. We call each invocation of an operation an *event*. An interaction point therefore denotes a set of possible events.

We represent an event by Operation[Selector]. Here, the term Selector is used to uniquely identify an occurrence of Operation. We use the notion of *event occurrence number* as a selector. An event occurrence number, i, of an event specifies that the event is the ith invocation of an operation in a computation. For instance, term `produce[0]` denotes the first invocation of `produce`.

Components are represented by extending the interfaces of procedures to incorporate the notion of interaction points. In CYES-C++, interaction points of a component are derived from the parameter variables: all methods on objects denoted by the variables are the interaction points of the component. (We assume that the parameters represent objects). For instance, the implementations of `producer` and `consumer` are shown below:

```
producer(buffer info){              consumer(buffer info) {
    while (TRUE) {                      while (TRUE) {
        info.produce();                     info.consume();
    }                                   }
}                                   }
```

Interaction points of `producer` are represented by the term `info.produce()`, which denotes the set of all possible invocations of `produce` during an execution of `producer`. Interaction behaviors of components are defined in terms of their interaction points.

4.2 Interaction specification

Interaction among programs is specified by an expression, called the *event ordering constraint expression*. An event ordering constraint expression is used to represent semantic dependencies among events of component programs by specifying execution orderings — deterministic or nondeterministic — among the events. An event ordering constraint expression is constructed from a set of *primitive ordering constraint expressions* and a set of *interaction composition operators*.

Primitive event ordering constraint expression: A primitive event ordering constraint expression (e1 < e2) specifies the constraint that event e1 must occur before event e2

Interaction composition operators: There are four operators for composing event ordering constraint expressions:
i) And constraint operator (&&): An execution of a program satisfies event ordering constraint expression (E1 && E2) containing && if it satisfies *both* E1 and E2.
ii) Or constraint operator(| |): An execution of a program satisfies event ordering constraint expression (E1 || E2) if it satisfies *at least one* of event ordering constraint expressions E1 or E2.
iii) forall operator: The `forall` operator extends && in order to specify ordering constraints over sets of events. There are two ways in which the `forall` operator can be specified. The first

```
forall var v in S { E(v) }
```

specifies that event ordering constraint expression E(v) holds true for all events v in event set S. In this expression, variable v iterates over the events of S. The second

```
forall occ i in S { E(S[exp(i)]) }
```

specifies that event ordering constraint expression E(S[exp(i)]) holds true for all events S[exp(i)] of S. In this expression, variable i ranges over the occurrence numbers of events of S. Expression exp(i) determines the occurrence number of the event for which E must hold.

iv) Exists operator: The exists operator is similar to forall in that it extends the || constraint operator over a set of events.

The interaction specification mechanism is declarative in nature. Its power stems from the ability to decompose global interactions among programs into a set of local interactions, each represented by event ordering constraint expressions, and combined with suitable interaction composition operators. One of the implications of the modularity property of event ordering constraint expressions is that interaction behaviors of programs can be changed by modifying only the relevant and local interaction specifications. Further, the interaction specification mechanism is is not based on the semantic properties of any synchronization primitive. It can be used to specify any interaction behavior for any invocation of any operation.

Example 4.1. *(Interaction specification).* We now present an example that illustrates the manner in which event ordering constraint expressions can be used for specifying interaction relationships. In this example, we show different instances of event ordering constraint expressions for the producer/consumer example.

Simple data dependency: In example 2.1, the synchronization constraint specifies that the ith invocation of consume cannot execute until the ith invocation of produce has occurred. Let the terms produce and consume respectively denote the interaction points of producer and consumer. The following expression implements the data dependency relationship between producer and consumer:

```
ConsExp1 =    forall occ i in produce { (produce[i] < consume[i]) }
```

Extended concurrent program: In this example, we consider interaction between a single producer and two consumers. Assume that the data produced by producer are shared between the two consumer components *alternately*. Also, assume that consume1 and consume2 denote the interaction points of the two consumer components. The interaction relationship between the components is derived by implementing two relationships: one between *odd* events of produce and events of one consumer, and the other between *even* events of produce and events of the other consumer. The two relationships are implemented by the following expression:

```
TwoRel = (produce[2*i-1] < consume1[i])&&(produce[2*i] < consume2[i])
```

The above relationship is true for all events of `produce`, which represents the interaction relationship among the components:

$$\text{ConsExp2} = \text{forall occ } i \text{ in produce } \{ \text{ TwoRel } \}$$

Modification of concurrent program: In this example, the interaction relationship between `producer` and `consumer` of example 2.1 is modified by defining an additional constraint: there are at most N unconsumed values. Component `producer` therefore must wait for `consumer` if there are N unconsumed values. The modified interaction relationship among the events of `producer` and `consumer` can be implemented by simply extending the existing interaction relationship (as implemented by `ConsExp1`) with suitable event ordering constraint expression that represents the additional constraint:

$$\text{ConsExp3} = \text{ConsExp1 \&\&} \\ \text{forall occ } i \text{ in consume } \{ \text{ (consume[i] < produce[i+N]) } \}$$

∎

4.3 Design of a Programming Language

The C-YES model is a general model of concurrent computation in that it can be applied to define many concurrent programming languages. In our research, we combined it with the object-oriented model [27] in order to design a concurrent extension of C++ [24], called CYES-C++ [22]. The design of CYES-C++ is facilitated, and in parts driven, by the notion of separation. In CYES-C++, both computations and interactions are defined as abstractions. CYES-C++ supports powerful concurrent programming abstractions by extending existing C++ abstractions that combine computational and interaction behavior abstractions in different ways. We briefly enumerate them below (See [22] for detail):

Concurrent class: CYES-C++extends the notion of a C++ class in order to define concurrent objects. In CYES-C++, a concurrent object is represented as a composition of a set of methods and a set of event ordering constraint expressions. The event ordering constraint expressions represent interaction relationships such as semantic dependencies, data consistency, and priority among the methods. Concurrent classes allow one to model concurrent objects that permit multiple concurrent activities to occur at the same time.

Inheritance: In CYES-C++, inheritance is a mechanism for extending the program composition of concurrent objects. Separation of implementations of computational and interaction behaviors allows one to extend and modify either components of a concurrent class. CYES-C++ supports inheritance of implementations of both computational and interaction behaviors.

Genericity: C++ provides the template mechanism for implementing generic data structures. CYES-C++ extends the notion of template classes in order define generic concurrent classes. Generic concurrent classes capture common computational and interaction behavior specifications of methods of concurrent classes. They can be instantiated with user classes to associate computational and interaction behaviors with user defined abstractions. Separation of implementations of computational and interaction behaviors allows either or both behaviors to be instantiated with a class.

Coordination Structure: Open software systems are often characterized by sets of autonomous and distributed objects whose execution behaviors must be coordinated. We have developed a coordination structure, called *object space*. An object space is a composition of a set of objects and a set of event ordering constraint expressions that define coordinate constraints among invocations of methods on the objects of an object space.

5 Related Work

In most approaches to concurrent programming, implementations of computations and synchronization are embedded within the implementation of components. Separation of implementation of computational and interaction behaviors has been proposed for the resolution of the inheritance anomaly [16]. However, focus here has mostly been on resolving a specific instance of the program composition anomaly. It has not been studied within the general context of concurrent program composition. Svend and Agha [10] also use the notion of separation of implementations of object and coordination constraints in order to define a distributed coordination structure. However, the focus here is on re-usability of object and coordination constraints, and not on the modifiability and extensibility of concurrent programs in general. Foster [8] also introduces the notion of separation of implementations of architectural elements from task implementations in order to support re-usability of implementations of the architectural specifications, and portability of concurrent programs. However, in the proposed approach, specifications of synchronization is not separated from computations.

There has been extensive work done in the area of concurrent programming. Most of this work has focussed on developing methodologies, languages, and tools for implementing concurrent programs. Most languages have added constructs for specifying concurrency and synchronization in a base languages. An extensive survey of these constructs is given in [19]. Examples of synchronization mechanisms are: semaphores [5, 3], write-once-read-many variables [6], data flow based data dependencies [13], signal variables, enable-based approaches [11, 18, 25, 17, 7, 12], disable based approaches [9], and behavior abstraction based approaches [14, 15].

Our proposed interaction specification mechanism differs from most approaches in that it is declarative, and compositional. It supports abstractions for defining interaction behaviors. The abstractions can be modified and extended in isolation from other abstractions. Further, they can composed with other computational abstractions in many different ways to construct powerful program abstractions. An example of a declarative mechanism is Path Expression [4]. Event ordering constraint expressions differ from Path Expressions in that they are used to specify the ordering constraints that must be satisfied. Path Expressions, on the other hand, are used to specify the valid sequences of operations through a regular expression. Further, Bloom [2] shows that path expressions do not adequately support modular development of interaction specifications because path expressions do not contain general mechanisms for directly representing states of objects, and for specifying interactions that depend on the states. States in event ordering constraints expressions can be easily captured through event sets [22].

6 Conclusion and Status

Concurrent programs can be easily modified and extended if implementations of both computational and interaction behaviors are separated. Separation supports encapsulation of implementations of both computational and interaction behaviors. It *localizes* the effects of changes in a concurrent program to specific implementations of computational and interaction behaviors. Further, implementations of both computational and interaction behaviors can be reused. In addition, implementations of computational and interaction behaviors can each be represented as separate abstractions. These abstractions can be combined with other programming language composition mechanisms such as aggregation, inheritance, and genericity to construct new and powerful concurrent programming abstractions.

A prototype implementation for CYES-C++ currently runs on a network of RS/6000 workstations.

References

1. Gregory R. Andrews. *Concurrent Programming*. The Benjamin/Cummings Publishing Company, Redwood City, CA, 1991.
2. Toby Bloom. Evaluating Synchronization Schemes. In *Proc. 7th Symposium on Operating Systems Principles*, pages 24–32. ACM, 1979.
3. Peter A. Buhr and Richard A. Strossbosscher. μC++ Annotated Reference Manual. Technical Report Version 3.7, University of Waterloo, Waterloo, Ontario, Canada, N2L 3G1, June 1993.
4. R. H. Campbell and A. N. Habermann. The Specification of Process Synchronization by Path Expressions. In *Lecture Notes on Computer Sciences*, volume 16, pages 89–102. Springer Verlag, 1974.
5. R. Chandra, A. Gupta, and J. L. Hennessy. COOL: A Language for Parallel Programming. In *Languages and Compilers for Parallel Computing Conference*, pages 126–147. Springer Verlag, 1992.
6. K. Mani Chandy and Carl Kesselman. Compositional C++: Compositional Parallel Programming. Technical Report Caltech-CS-TR-92-13, Cal Tech, 1992.
7. D. Dechouchant, S. Krakowiak, M. Meyesmbourg, M. Riveill, and X. Rousset de Pina. A Synchronization Mechanism for Typed Objects in a Distributed Systems. In *Workshop on Object-based Concurrent Programming*, pages 105–107. ACM SIGPLAN, ACM, Sept. 1989.
8. Ian T. Foster. Information Hiding in Parallel Programs. Technical Report MCS-P290-0292, Argonne National laboratory, 1992.

9. Svend Frolund. Inheritance of Synchronization Constraints in Concurrent Object-Oriented Programming Languages. In *ECOOP '92, LNCS 615*, pages 185–196. Springer Verlag, 1992.
10. Svend Frolund and Gul Agha. A Language Framework for Multi-Object Coordination. In *Proceedings of the ECOOP'93*, pages 346–360, 1993.
11. Narain H. Gehani. Capsules: A Shared Memory Access Mechanism for Concurrent C/C++. *IEEE Transactions on Parallel and Distributed Systems*, 4(7):795–810, July 1993.
12. J. E. Grass and R. H. Campbell. Mediators: A Synchronization Mechanism. In *Sixth International Conference on Distributed Computing Systems*, pages 468–477, 1986.
13. Andrew S. Grimshaw. Easy-to-Use Object-Oriented Parallel Processing with Mentat. *IEEE Computer*, 26(6):39–51, 1993.
14. Dennis Kafura and Keung Lee. Inheritance in Actor based Concurrent Object-Oriented Languages. In *Proceedings ECOOP'89*, pages 131–145. Cambridge University Press, 1989.
15. Satoshi Matsuoka. *Language Features for Re-use and Extensibility in Concurrent Object-Oriented Programming*. PhD thesis, The University of Tokyo, Japan, June 1993.
16. Satoshi Matsuoka, Keniro Taura, and Akinori Yonezawa. Highly Efficient and Encapsulated Re-use of Synchronization Code in Concurrent Object-Oriented Languages. In *OOPSLA'93*, pages 109–126. ACM SIGPLAN, ACM Press, 1993.
17. Ciaran McHale, Bridget Walsh, Seán Baker, and Alexis Donnelly. Scheduling Predicates. In *Object-Based Concurrent Computing Workshop, ECOOP'91, LNCS 612*, pages 177–193. Springer Verlag, 1991.
18. Christian Neusius. Synchronizing Actions. In *ECOOP '91*, pages 118–132. Springer Verlag, 1991.
19. Raju Pandey. *A Compositional Approach to Concurrent Programming*. PhD thesis, The University of Texas at Austin, August 1995.
20. Raju Pandey and James C. Browne. Event-based Composition of Concurrent Programs. In *Workshop on Languages and Compilers for Parallel Computation, Lecture Notes in Computer Science 768*. Springer Verlag, 1993.
21. Raju Pandey and James C. Browne. A Compositional Approach to Concurrent Object-Oriented Programming. In *IEEE International Conference on Computer Languages*. IEEE Press, May 1994.
22. Raju Pandey and James C. Browne. Support for Extensibility and Reusability in Concurrent Object-Oriented Programming Languages. In *Proceedings of the International Parallel Processing Symposium*, pages 241–248. IEEE, 1996.
23. S. Crespi Reghizzi and G. Galli de Paratesi. Definition of Reusable Concurrent Software Components. In *ECOOP '91*, pages 148–165. Springer–Verlag, 1991.
24. Bjarne Stroustrup. *The C++ Programming Language*. Addison Wesley, Second Edition edition, 1991.
25. Laurent Thomas. Extensibility and Reuse of Object-Oriented Synchronization Components. In *Parallel Architecture and Languages Europe, LNCS 605*, pages 261–275. Springer Verlag, 1992.
26. Chris Tomlinson and Vineet Singh. Inheritance and Synchronization with Enabled Sets. In *OOPSLA '89 Conference on Object-Oriented Programming*, pages 103–112. ACM Press, 1989.
27. Peter Wegner. Dimensions of Object–Based Language Design. In *OOPSLA'87*, page 168. ACM Press, 1987.

Structured Dagger: A Coordination Language for Message-Driven Programming

Laxmikant V. Kalé and Milind A. Bhandarkar

Department of Computer Science
University of Illinois, Urbana IL
{kale,milind}@cs.uiuc.edu

Abstract. Message-Driven Programming style, used in languages such as Charm, avoids the use of blocking receives and allows adaptive overlap of computation and communication by scheduling objects depending on availability of messages. Charm supports objects whose methods can be triggered by remote objects asynchronously, which enables Charm programs to tolerate communication latencies in an adaptive manner. However, many parallel object-based applications require the object to coordinate the sequencing of the execution of their methods. Structured Dagger is a coordination language built on top of Charm that supports such applications by facilitating a clear expression of the flow of control within the object without losing the performance benefits of adaptive message-driven execution.

1 Introduction

One of the daunting tasks for parallel programmers is to tolerate message latency and unpredictable delays in remote response. Message-driven style of parallel programming attempts to tolerate such latencies by disallowing any process to block the processor when trying to receive messages and scheduling computation depending upon availability of messages. Message-driven parallel programming languages provide constructs for attaching code blocks to availability of specific messages. In object-oriented systems these blocks correspond to methods of parallel objects. These blocks are scheduled for execution by the run-time system when the specified messages arrive. This scheme minimizes the performance impact of communication latency by scheduling a ready process for execution while other processes are waiting for data.

Charm [6] is one of the first object-based portable parallel programming languages that embodies message-driven execution and promotes modularity while exhibiting latency tolerance. The order of execution of processes is determined by the order of messages received. Due to unpredictable delays in remote response times, the messages may arrive in any order and the programmer must deal with all possible message orderings. However, imposing an order on the arrival of messages, as is done in the traditional message-passing systems, tends to make the parallel program inefficient by letting the communication latency affect its performance.

To solve this problem, a coordination language called Dagger [3] was developed on top of the Charm programming system. However, the structure of Dagger programs still does not clearly express the flow of control in certain situations. We propose a new coordination language called Structured Dagger, which reduces the complexity of message-driven objects further by providing constructs to express control flow as a series-parallel graph.

2 Charm

Charm is a machine independent parallel programming system [6]. Charm programs are written in C with a few syntactic extensions. Charm currently runs on many distributed and shared memory parallel machines, as well as workstation networks. Charm programs consist of potentially medium grained objects (*chares*), and a special type of replicated objects, called branch-office chares. Charm supports dynamic creation of chares, by providing dynamic (as well as static) load balancing strategies. Chares interact by sending messages to each other and via specific information sharing modes.

The runtime system is message-driven. It repeatedly selects an available message from a pool of messages, switches to the context of the chare to which it is directed, and initiates execution of the method specified by the message.

A Charm program consists of chare definitions, message definitions, and declarations of specifically shared objects in addition to regular C language constructs (except global variables). A chare definition consists of local variable declarations, entry-point definitions and private function definitions. Local variables of a chare are shared among the chare's entry-points and private functions. Calls are provided to create chares and send messages to existing chares.

A branch office chare (BOC) represents a group of chares. An instance of a BOC has a branch chare on every processor. A BOC definition is similar to a chare definition. All the branches of a single BOC instance share a global ID. One can send a message to a specific branch chare of a BOC, on a particular processor, or broadcast it to all its branches. BOC's are useful for some computations such as reduction operations, expressing static load balancing, and SPMD style programs.

In addition to messages, Charm provides other ways in which objects share information. The information sharing abstractions supported include readonly variables, monotonic variables, accumulators and distributed tables. Charm also provides a sophisticated module system that facilitates reuse, and large-scale programming. Details about these features can be found in [9].

Consider an algorithm for matrix multiplication that is dynamically load balanced. Matrix A is stored as a collection of entries where each entry is a block of contiguous rows. Similarly, the matrix B is stored as a collection of columns. The `mult_chare` used in this algorithm (Figure 1a) is responsible for multiplying a block of rows of A, and a block of columns of B. The entry `init` is executed when an instance of the chare is created. The message `msg` contains indices of the row and column blocks that are to be multiplied. First, the chare requests the row and columns from the tables `Atable` and `Btable` (these tables store the matrices A and B) by calling `Find` which is supported by the distributed tables mechanism in Charm. Note that the `Find` call is non-blocking, and it immediately returns. Eventually, the row (and column) data will be sent in a message to the entry-point `recv_row` (`recv_column`), and these messages may arrive in any order.

The multiplication depends on availability of both rows and columns. The dependence (i.e. the flow of control within mult_chare) must therefore be enforced using mechanisms such as counters and message buffers. Here, a chare-private variable, `count`, is initially set to 2, and is decremented with arrival of each message. When `count` becomes zero, the buffered messages are fetched and multiplication is performed. This example has been chosen to be a simple one in order to demonstrate the necessity of counters and buffers. In general, a parallel algorithm may have more interactions leading to the use of many counters, flags, and message buffers, which complicates the program development significantly.

```
chare mult_chare {
  int count, *row, *col;
  ChareIDType chareid;
  entry init: (message MSG *msg) {
     count = 2; MyChareID(&chareid);
     Find(Atable, msg->row_index, recv_row, &chareid,NOWAIT);
     Find(Btable, msg->col_index,recv_col,&chareid,NOWAIT);}
  entry recv_row: (message TBL_MSG *msg) {
     row = msg->data; if (--count == 0 ) multiply(row,col);}
  entry recv_col:(message TBL_MSG *msg){
     col = msg->data; if (--count == 0) multiply(row,col);}
}
```

Fig. 1(a) Matrix multiplication chare in Charm

```
chare mult_chare {
   structentry init : (message MSG *msg){
      atomic {
         Find(Atable, msg->row_index,...);
         Find(Btable, msg->col_index,...); }
      when recv_row(TBL_MSG *row), recv_col(TBL_MSG *col) {
         atomic{ multiply(row->data,col->data) }}
   }
}
```

Fig. 1(b) Matrix multiplication chare in Structured Dagger

3 Structured Dagger : The Language

Structured Dagger hides the details of counters, buffers, and tests mentioned in the last section from the programmer while clarifying the flow of control by providing structured constructs discussed below.

Structured Entry-Methods: The Structured Dagger language is defined by augmenting Charm with structured entry-methods, which specify pieces of computations (*when-blocks*) and dependences among computations and messages. A when-block is guarded by dependences that must be satisfied before it can be scheduled for execution. These dependences include arrival of messages or completion of other constructs. Before describing the language in detail, let us consider the matrix multiplication example once again.

Figure 1b shows the matrix multiplication written using Structured Dagger. Whenever the entries `recv_row` and `recv_column` receive messages, the `multiply` function is called with the rows and columns that have been received. Structured Dagger takes care of the bookkeeping functions such as incrementing counters, flags and buffering the messages. Therefore, the resulting code is more readable (and easy to program).

When-Blocks: When-blocks specify dependence between computation and message arrival at an entry-point. In general, a when-block may specify its dependence on more than one entry-point. When all constituent entry-points receive messages, computation corresponding to the when-block may be triggered.

When-blocks combined with the ordering constructs are adequate for specifying computations where multiple iterations of the same computations may not overlap. However, in many practical problems, such as Jacobi Relaxation in numerical methods, such overlap may occur. Then messages for different iterations must be matched separately. In order to handle this problem, Structured Dagger provides *reference numbers* attached to messages to distinguish between messages belonging to different phases of computation. A when-block optionally specifies the reference numbers for the messages triggering its constituent entry-points. Messages that belong to the same phase of the computation are given specific reference numbers by the user. Structured Dagger matches the messages with those reference numbers to activate a when-block.

Atomic Construct: The `atomic` construct is a wrapper around C statements and specifies that no Structured Dagger constructs appear inside it. further, it does not contain code executed depending on the arrival of remote messages and is therefore executed atomically.

Ordering Constructs: Receiving a message at an entry-point is not sufficient to trigger a computation. The computation must be in a state where it is ready to process the message. Even if all the entry-points specified in a when-block have received messages, the computation specified in the when-block is not triggered until other constructs occurring previously in the program order may not have completed. The program order may be specified in Structured Dagger using the ordering constructs, `seq` and `overlap`.

The `seq` construct is written as `seq{construct-list}` and ensures that each of the constructs in the list is enabled only after its predecessor completes. Note

```
BranchOffice Harlow_Welch{
//chare-local variables declarations
structentry init:(MSGINIT *msg){
  seq {
    atomic { initialize();for(i=0; i<Z; i++) convdone[i] = FALSE; }
    forall(i=0,Z-1,1){
      while(!convdone[i]){
        atomic { for (dir=0; dir<4; dir++){
                   m[i][dir] = copy_boundary(i,dir);
                   SendMsgBranch(entry_no[dir],m[i][dir],nbr[i][dir]);}}
        when North(Bdry *n),South(Bdry *s),East(Bdry *e),West(Bdry *w){
          atomic { update(i, n, s, e, w);
                   reduction(my_conv, i, Converge, &mycid);}}
        when[i] Converge(Conv *c) {atomic{convdone[i] = c->done;}}
      }
      atomic { print_results(); }
    }
  }
}}
```

Fig. 1. Harlow-Welch Program

that, `seq` construct is not the same as `atomic` construct because it may contain other Structured Dagger constructs. The `seq` construct completes when the last of its component constructs reaches completion.

The `overlap` construct enables all its component constructs concurrently and can execute these constructs in any order. Actual execution of these component constructs may be dependent on arrival of messages that they use. An `overlap` construct reaches its completion only after each of its component constructs has completed.

Conditional and Looping Constructs: In many situations, one may need to conditionally enable the Structured Dagger constructs, or to iterate over a set of constructs. Since `atomic` construct cannot include any Structured Dagger constructs, the C statements such as `if`, `while`, and `for` cannot be used for this purpose. Therefore, Structured Dagger provides the equivalent constructs. If more than one component constructs appear inside such a construct, they are implicitly enclosed by a `seq` construct. The constructs supported include:

    if (*condition*) {construct-list} **else** {construct-list}
    while (*condition*) {construct-list}
    for (*stmt; condition; stmt*) {construct-list}
    forall (var=*const, const, const*) {construct-list}

A `forall` construct enables its component constructs for the entire iteration space as opposed to the `while` and `for` constructs, which enable their component constructs for each element of the iteration space in strict sequence.

Example Program: We present an example Structured Dagger program that implements the Harlow-Welch scheme in Computational Fluid Dynamics. The control flow is expressed in Figure 1. Each iteration in this scheme consists of communicating the boundary elements with neighbors in the 2-D grid followed by a global reduction to check whether the scheme has converged. (The reduction is carried out asynchronously by a separate object and is not shown here.) This is done concurrently for all the planes and each of the planes could converge independently of each other.

4 Structured Dagger : Implementation

Structure Dagger is implemented on top of Charm as a translator and a run-time library. The translator transforms the program to an equivalent Charm program, by splitting a structured entry-point into a number of Charm entry-methods and chare-private functions, inserting counters and flags to specify dependences between different component constructs of the structured entry-point.

For each construct, the translator generates code for enabling the construct and for the completion of the construct. Code generated for completion of the construct contains code to free the message buffers occupied by the messages arrived during its execution as well as to enable the constructs that may be dependent on its completion.

The runtime library maintains one message queue for each object. Whenever any when-block is enabled, it checks for the messages intended for its component entries. If all of these are available, it enables its component constructs and if possible executes them (In particular, it executes the code in atomic constructs, which do not have dependence on message arrival.) The entry-method generated corresponding to each of the entries within when-blocks contains code to buffer the message, set the appropriate flags and awaken any when-blocks that may be waiting. By doing a careful analysis of this dependence, the translator avoids repeated and redundant checking for all enabled when-blocks.

For assessing the performance impact of our translation scheme, we ran a simple program on a single node of CM-5. This program creates two objects, which then start sending messages to each other in a loop for a specified number of times. We compared the performance of our Structured Dagger program with a Charm program and also with a multi-threaded program written using thread-objects in Converse [7]. The results for 10000 round-trip messages (each of size 4 bytes) are in table 1. As can be seen from these results, Structured Dagger program does not add significant overhead to the native Charm code, while it reduces the program complexity. The cost of context-switching in a multi-threaded program is very high, which justifies our use of message-driven execution in Structured Dagger.

Table 1. Performance Results

Program	Charm	Multi-Threaded	Structured Dagger
Time(seconds)	1.390	5.654	1.890

5 Related Work

Dagger [3] is an earlier attempt to build a coordination language on top of Charm. The concept and structure of when-blocks in Structured Dagger is borrowed from Dagger. Dagger permits a more general class of control flow graphs than Structured Dagger, using when-blocks, **expect** and **ready** statements, and condition variables. A when-block specifies dependences as a list of entries and condition variables. A Dagger program enables a when-block by issuing an **expect** statement. If the arrived message is not expected, it is buffered for later retrieval. A condition variable is used to signal the end of a when-block with a **ready** statement. Thus control-dependences among when-blocks belonging to the same chare can be expressed using condition variables. However, the structure of Dagger programs is not as perspicuous as Structured Dagger because a Dagger program is a flat collection of when-blocks. This perspicuity is obtained at the cost of sacrificing the generality that Dagger provides.

CC++ [4] is an object-parallel language that bears some similarities to Structured Dagger. CC++ is a thread-based system. A computation consists of one or more processor objects each with its own address space. Objects within these processor objects can be accessed by remote objects using global pointers. Within individual processor objects, new threads can be spawned using the structured constructs *par*, and *parfor*, and the unstructured construct *spawn*, which creates a new parallel thread. Multiple threads created by these statements may be executed by different processors, or interleaved on the same processor, and they may share variables. The *par* and *parfor* constructs of CC++ are analogous to the *overlap*, and *forall* constructs in Structured Dagger. However, they are different in a fundamental sense: two statements in a *par* construct may actually be executed in parallel by two different processors, whereas two constructs in an *overlap* statement are always executed by the same processor. Also they can interleave only in a disciplined fashion: only entire when-blocks can be interleaved, based on the arrival of messages, and not the individual C statements.

The most important difference between Structured Dagger and CC++ (and other systems such as Chant [5]) has to do with threads. Using threads creates a flexibility, but at a cost: thread context switches are more expensive than message-driven invocations of methods in Charm or Structured Dagger(as illustrated in fig. 1); also, threads waste memory: creating hundreds or thousands of threads, each with its own stack, may not be possible, whereas a large number of parallel objects can easily be created without reaching memory limits.

ABC++ [2] is a thread-based object-parallel language. There is one thread associated with each parallel object. This thread receives method invocation message and decides when and whether to invoke methods. Primitives are provided to selectively enable execution of individual methods. Unlike Structured Dagger, no direct expression of control flow across method invocations is possible.

The *enable set* construct [8] addresses the issue of synchronization within *Actors* [1]. Using this, one may specify which messages may be processed in the new state. Other messages received are buffered until the current enable set includes them. The ordering constructs in Structured Dagger achieve this in a cleaner manner. Also, there is no analogue of a when-block, viz. a computation block, that can be executed only when a specific group of messages have arrived.

6 Conclusion

We presented a coordination language called Structured Dagger which is a notation for specifying intra-process control dependences in message-driven programs. This language combines efficiency of message-driven execution with the explicitness of control specification. Structured Dagger allows easy expression of dependences among messages and computations and also among computations within the same object using when-blocks similar to Dagger and various structured constructs. Structured Dagger has been developed on top of Charm and is portable across many MIMD machines, with or without shared memory.

References

1. G.Agha, *Actors: A Model of Concurrent Computation in Distributed Systems*. MIT Press. 1986.
2. E. Arjomandi et. al., "ABC++: Concurrency by inheritance in C++", *IBM Systems Journal*, Vol 34, No. 1, 1995.
3. A.Gursoy, *Message Driven Execution and its Impact on the Performance of CFD and other Applications*, Ph.D Thesis, University of Illinois at Urbana-Champaign, Jan 1993.
4. K.Mani Chandy and C. Kesselman, "Compositional C++: Compositional Parallel Programming", *Technical Report no. Caltech-CS-TR-92-13*, Department of Computer Science, California Institute of Technology, 1992.
5. M. Hainer, D. Cronk and P. Mehrotra, "On the Design of Chant: A Talking Threads Package", *Proceedings of Supercomputing '94*, Nov 1994.
6. L.V.Kale, "The Chare Kernel parallel programming language and system", *Proceedings of the International Conference on Parallel Processing*, Vol II, Aug 1990, pp17-25.
7. L.V.Kale et.al., "Converse: An Interoperable Framework for Parallel Programming", *Submitted to International Parallel Processing Symposium, 1996*.
8. C.Tomlinson, V.Singh, "Inheritance and Synchronization with Enabled-Sets", ACM OOPSLA 1989 , pp103-112.
9. The CHARM(4.0) programming language manual, Department of Computer Science, University of Illinois at Urbana-Champaign, Urbana, IL, 1993.

TPascal - A Language for Task Parallel Programming

Ansgar Brüll and Herbert Kuchen

Lehrstuhl für Informatik I/II, RWTH Aachen, D-52056 Aachen, Germany
email: {bruell,herbert}@i2.informatik.rwth-aachen.de

Abstract. The programming language *TPascal* is designed for the programming of MIMD computers with distributed memory in a task parallel way using explicit message passing. In contrast to traditional message passing libraries major problems like deadlocks are already avoided in the definition of the language. The message passing and the creation of processes is fully integrated into the language making compile time checking and optimization possible. *TPascal* enriches a sequential programming language similar to *PASCAL* with the concept of *topologies* which are sets of processes arranged in a specific way.

1 Introduction

Programming MIMD computers with distributed memory is still difficult, error prone and time consuming. Mostly, traditional sequential programming languages together with a message passing library like *P4*, *PVM* or *MPI* (see [2] for references) are used. Such library calls allow only very limited compile-time checking and optimization of the communication. Various of the mentioned libraries transmit data as strings making even run-time checking impossible.

In contrast to the explicit use of message passing routines, different languages have been developed achieving a higher degree of abstraction from the underlying system, e.g. *High Performance Fortran (HPF)* [5]. The programmer can distribute data over virtual SPMD processes and perform data parallel operations on arrays of arbitrary shape. All communication statements necessary to execute such programs on truly parallel computers are inserted by the compiler. Deadlocks cannot occur.

While the number of data parallel applications is quite large, many applications and algorithms can be better thought of as a number of independent processes connected and communicating in a specific way. This *task parallel* paradigm has been used as the basis of a number of languages. E.g. *SVM Fortran* [1] uses parallel sections and parallel loops where each section and each loop iteration is an independent process. Synchronization and data exchange between the processes can be achieved by global data that has to be shared between the processes and is realized by means of virtual memory. To regulate the data access the programmer has to make use of standard low level synchronization methods like semaphores and atomic updates making deadlocks possible. The approach taken by *Fortran M* [6] is based on explicit message passing. The programmer can declare processes which can exchange data through typed channels. These channels connect two arbitrary processes, again enabling deadlocks. A more static approach for the communication has been chosen within *FX* [7], where tasks can communicate only through arguments at the time of creation and termination. The advantage of all these languages over the traditional message passing libraries is that the data exchange protocol can be type checked by the compiler. However, programs may still run into deadlocks.

To avoid them, *TPascal* takes an approach quite similar to *skeletons* [4] in functional programming. A skeleton can be regarded as a parameterized template of a specific parallel operation. The implementation of a skeleton is hidden from the programmer and therefore the programmer does not need to tackle the 'low level problems' of parallelism. Within *TPascal* these templates are represented by *topologies* which are sets of processes being connected to each other in a particular way, e.g. pipes, rings etc. Topologies can be nested enabling the construction of large collections of processes. A similar scheme has been used in P^3L [3] where the result of a computation is transferred to a different process by variables that are shared between the processes. *TPascal* however, uses explicit message passing. Only processes being connected can exchange data in a well defined, topology dependent way. For each topology only certain ways for a data exchange exist, making the occurrence of deadlocks impossible. Through the integration of the primitives into the language the compiler can perform many optimizations. Additionally, the data exchange protocol between two connected processes can be verified by the compiler and the runtime system with respect to the types of the data being exchanged.

2 A Description of TPascal

TPascal consists of two parts. The first one, the *host language*, is made up of a sequential programming language containing usual constructs like loops, conditionals etc. The second component is the *coordination part* of the language consisting of constructs to set up parallel processes and to exchange data between them. Let us first consider an example.

The program in Fig. 1 computes the sum of an array of integers. It consists of a ring of five processes. Each process initializes its local sum by an argument received at its creation. Then, it stores the local sum into two auxiliary variables, SUM1 and SUM2, which are then used to exchange these values with the neighbor processes. The EXCHANGE_NB operation has for each neighbor an argument list telling what to send to (OUT) and what to receive from this neighbor (IN). The received values are added to the local sum, and then passed to the opposite neighbor in the next step. Thus, in the i-th iteration, the initial local value reaches the neighbors with distance i.

2.1 The Coordination Language

The constructs from the coordination language are used for expressing parallelism within a program by setting up parallel processes and for exchanging data

```
PROGRAM ADD
VAR Data : ARRAY [1..5] OF INTEGER;

PROCEDURE Calc(A: INTEGER)
VAR I,SUM,SUM1,SUM2:INTEGER;
BEGIN
   SUM:=A; SUM1 := SUM; SUM2 := SUM;
   FOR I:=1 TO 2 DO
   BEGIN
      EXCHANGE_NB([IN SUM1, OUT SUM2],[OUT SUM1, IN SUM2]);
      SUM := SUM + SUM1 + SUM2;
   END
END

BEGIN
   /* Start of the main program, Initialization of array Data, ...*/
   RING(Calc(Data[1]), Calc(Data[2]), Calc(Data[3]), Calc(Data[4]), Calc(Data[5]));
END
```

Fig. 1. A simple TPascal program.

between these processes.

Parallelism is expressed by explicitly creating *topologies*, which are sets of parallel processes called *tasks* connected in a specific way. An atomic task is a procedure call started as a separate process with its own address space. Topologies are established by constructs consisting of a keyword describing the connection and a corresponding list of tasks. The constructs are integrated into the host language as statements. The interconnections between the tasks within a topology determine the way of exchanging data. Only connected tasks can communicate. The programmer can choose between different predefined topologies.

Communication Primitives. Depending on the topology certain tasks are connected and can exchange data by two primitives. To exchange data with all connected tasks (called *neighbor tasks*), EXCHANGE_NB can be used. These tasks are referred to as. A non-empty list of expressions has to be supplied to the primitive for each connected task together with an attribute that specifies whether the data will be received from (IN attribute) or send (OUT attribute) to a connected task. Another operation EXCHANGE (only available for some topologies) allows to communicate with one selected neighbor. Implementation aspects are discussed in [2].

Creating Topologies. In *TPascal* constructs are available for the most convenient topologies like parallel independent tasks, rings of tasks, pipelines of tasks etc. (see Fig. 2). For each type of topology a construct is available that has as parameter a list of topologies together with their actual parameters. When a new topology is created by executing a task construct, the parts of the state of the starting task needed for the new tasks are duplicated. Within *TPascal* tasks will not necessarily be executed on the same physical processor in which case the runtime system is responsible for creating an initial state on the processors

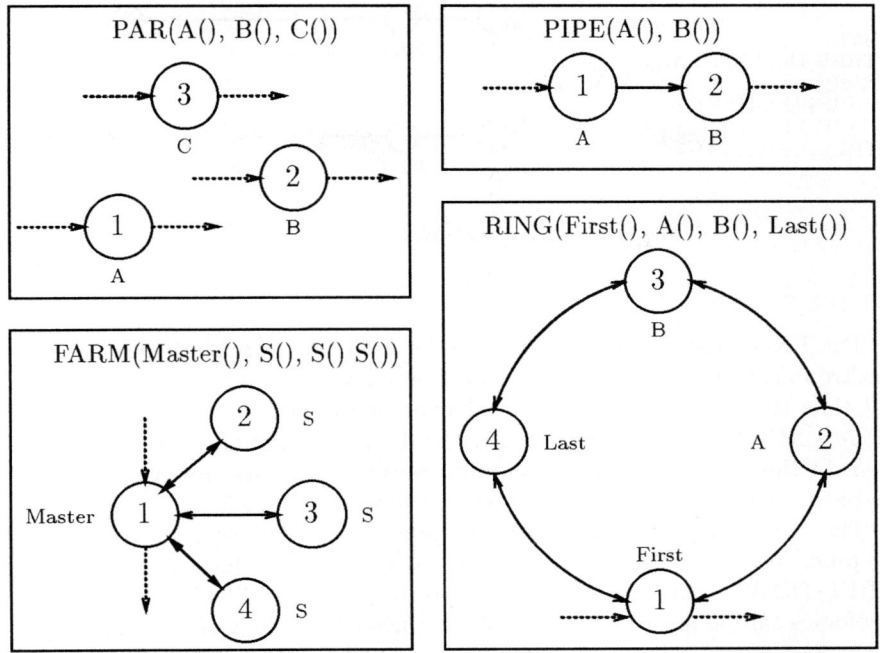

Fig. 2. Topologies: parallel independent processes, pipe, farm, and ring.

involved in that operation. Procedure calls started as tasks can, of course, contain further topology constructs. A topology construct terminates when all tasks within the topology have terminated.

The simplest topology consist of parallel independent tasks (created by the PAR construct), which cannot communicate with each other. In Fig. 2, the tasks are represented by circles labeled with some internal task number needed for the connection of different topologies. All tasks that have an incoming dotted arrow expect some input from previous topologies, while all tasks with an outgoing dotted arrow will send some output to subsequent topologies. The previous and subsequent topologies are determined by a topology construct at an outer level of nesting (explained below).

The PIPE construct enables the programmer to define a unidirectional sequence of not necessarily equal tasks. Ordinary SEND and RECEIVE instructions may be used, since a pipe is acyclic and there is no danger of deadlocks.

The ring topology allows a cyclic exchange of data in a bidirectional way. All communication that is performed between two tasks being connected by a RING construct has to be done by the EXCHANGE_NB primitive.

A farm allows one master to send jobs to a number of slaves which in turn work on these jobs and return intermediate results, final results and possibly new jobs. This process of distributing jobs to the slaves continues until no more jobs are available.

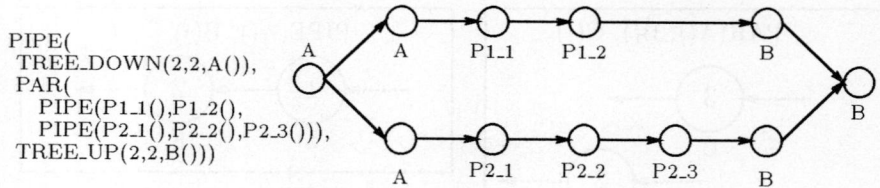

Fig. 3. Nesting topologies

The EXCHANGE_NB primitive can be used to ensure that the slaves work synchronously. The EXCHANGE operation allows to give jobs to selected slaves and thus to get an asynchronous behavior of the farm. The master may also use an EXCHANGE operation, where the desired communication partner is left open. In this case, the master communicates with the first ready slave. This is the only place, where non-determinism occurs.

The tree is a generalization of the pipe. It can be regarded as a branching pipe. We distinguish trees with a data flow from the root to the leaves (TREE_DOWN), and trees with the opposite data flow (TREE_UP). The above topologies can be nested. An example of a nested topology is shown in Fig. 3.

Combining Topologies with Constructs of the Host Language. The topologies described are static with respect to their communication structure. By using the topology primitives in combination with the control structures of the host language, tasks can be established dependent on the state of program execution e.g. within conditionals or loops. Tasks can be created recursively when used in connection with recursive procedures. Thus, in combination with the constructs of the host language a dynamic process structure can be established. Tasks started in different iterations of a loop or in different levels of a recursive procedure can only "communicate" via the host language. Certain features of the host language have to be handled in a restricted way, e.g. pointers and I/O. See [2] for details.

2.2 Deadlock Avoidance in TPascal

A necessary condition for a deadlock to occur is that a circular dependence has to exist between the processes. By induction on the nesting level, it can be shown that no deadlock can ever occur in a *TPascal* program. In the following we present a rough sketch of the induction step of this proof. For the induction hypothesis we assume the topologies $T_1, \ldots, T_n$, that are used for the construction of a new topology T, to be deadlock free. If T is a pipe or a tree topology, it is obvious that no deadlock can ever occur, as no cyclic dependences are introduced by these constructs. If T is a farm, the data exchange always takes place in a bidirectional way within a single primitive. Therefore no circular dependencies can occur. For a ring topology, a similar argument holds.

It should be noted that other forms of congestion besides deadlocks are still possible. This is mainly due to the fact that their avoidance within the language definition would impose large restrictions on the host language making it impractical. Among these problems are *livelocks* where a number of tasks wait for an event that is never going to happen. In particular, an infinite loop can prevent a process from reaching its next communication instruction. This will cause the communication partners to wait forever.

3 Summary and Discussion

We have presented *TPascal*, a language for programming distributed memory MIMD machines based on the task parallel paradigm using explicit message passing. Tasks are arranged in topologies. The topology determines which tasks can directly exchange data. *TPascal* simplifies programming as deadlocks within user programs are avoided by restricting the communication. Both the primitives for task creation and communication are part of the language enabling the compiler to perform necessary optimizations. Another feature of *TPascal* is its abstraction from the underlying hardware. The programmer can choose between a given set of topologies like trees, pipes, farms and parallel independent tasks. A mapping of the tasks that form a topology to the physical processors of the target machine is done by the runtime system in a way that the routing of messages between tasks is done on the shortest way possible and the load is equally distributed onto the physical processors available. This is possible as the way tasks exchange data is known in advance through the use of the topologies with only a fixed way of exchanging data. In the future, we want to extend *TPascal* by data parallelism. This will be done by some skeletons (topologies) operating on a distributed data structure.

References

1. R. Berrendorf, M. Gerndt, W. Nagel, and J. Prümmer. SVM-Fortran. Report KFA-ZAM-IB-9322, Forschungszentrum Jülich, 1993.
2. A. Brüll, H. Kuchen. TPascal - A Language for Task Parallel Programming. Report, http://www-i2.informatik.rwth-aachen.de/~herbert/TRGiBrKu.ps
3. M. Danelutto, R. DiMeglio, S. Orlando, S. Pelagatti, and M. Vanneschi. A methodology for the development and the support of massively parallel programs. *Future Generation Computer Systems*, 205 – 220. Elsevier, 1992.
4. J. Darlington et al. Parallel Programming Using Skeleton Functions. PARLE'93, LNCS 694, Springer, 1993.
5. The High Performance Fortran Forum. *High Performance Fortran Language Specification*, 1993.
6. I.T.Foster and K.M. Chandy. *Fortran M: A Language for Modular Parallel Programming*, June 1992.
7. J. Subhlok and T. Gross. Task parallel programming in FX. Report CS-94-112, Carnegie Mellon University, 1994.

OB(PN)²: An Object Based Petri Net Programming Notation
(Extended Abstract)

Johan Lilius[*]
Helsinki University of Technology
FINLAND

Abstract. In this paper we present a translation from the object-based language OB(PN)² to a class of high-level Petri nets. The OB(PN)² language is an extension of B(PN)² as defined by Best and Hopkins, and the semantics is inspired by the B(PN)² semantics defined in terms of M-nets. The translation relies on the CCS-like composition operators defined for M-nets. Each program construct is translated to a box (a special kind of net) or an operation for combining boxes. Thus in essence each program is translated into an expression in the algebra of boxes.

1 Introduction

Concurrency is a natural feature of objects. Given two objects, there is no reason why the execution order of their methods should be dependent on any other concept than synchronization. Indeed many object-oriented languages contain concurrency-primitives [YT87]. In this paper we present OB(PN)² (Object Based Petri Net Programming Notation), a parallel object-based high-level programming notation with semantics defined in terms of M-nets. The objects of OB(PN)² are inherently parallel, meaning that each object may services method requests in parallel, and the variables are accessed with a mutual exclusion semantic. This mutual exclusion semantics is enforced by the resources consciousness of Petri nets. The objects in OB(PN)² can communicate either synchronously through a hand-shake or asynchronously through a channel with an arbitrary capacity. The statements are atomic actions, where an atomic action

---

[*]This research has been partially supported by the DAAD through the Konrad Zuse Programm, while the author was at Institut für Informatik, Universität Hildesheim.

can be a variable access and/or a variable assignment or a procedure/method call. Currently only the scalar data-types integers and booleans are provided.

The intended application area of $OB(PN)^2$ is the analysis and verification of programs for safety-critical embedded systems. It is generally felt that direct analysis of programs is infeasible because of the state-space explosion problem. For example the work on static analysis of ADA [Tay83] aims at abstracting away details of the program so that reachability graph generation becomes feasible. However we feel that recent advances in new analysis techniques, as implemented in PROD [GTV93] and implemented in the PEP-tool [Gra95], make the full analysis of programs possible. Also we think that the area of safety-critical embedded systems, where the programs are usually relatively small, provides an application area where the full and automatic analysis of programs will not only be practical but also cost-effective.

In this work we view a kind of Petri nets, M-nets as an implementation formalism for an object-oriented language. Since M-nets are provided with process-algebra like operations, the semantics of $OB(PN)^2$ is compositional. The proposed formalism $OB(PN)^2$ is based on $B(PN)^2$, a Basic Programming Notation for Petri Nets [BH93], which was given a semantics in terms M-nets in [BHW+95]. In [LP] $B(PN)^2$ was extended with procedures that allowed parameters to pass by value and the corresponding procedure-box concept was developed. In this work we extend $B(PN)^2$ with objects and with reference variables. The semantics of $OB(PN)^2$ differs substantially from the original $B(PN)^2$ M-net semantics [BHW+95]. We have had to extend M-nets to be able to correctly treat procedure call instances. Also the treatment of reference parameters changes the translation of variables. This treatment of references variable is the key to the implementation of objects as *class boxes*.

We feel that the main advantages of our approach are twofold: (i) by giving semantics as Petri nets, instead of some new formalism, it is possible to analyze and verify the programs by using the powerful techniques available in existing tools (PEP, PROD), and (ii) the use of a syntactically simple but powerful programming notation makes the formalism accessible to the programmer, thus easing the acceptance of the approach in the intended application area. For a full description of the translation see [Lil].

2   The Class Box Concept

An object can trivially be seen essentially as an abstract data-type together with an internal state and an identity. It has a set of internal variables, and a set of methods for manipulating the state of the object. In terms of the box-algebra this means that we will obtain one *variable-box* for each internal variable, and one *method-box* for each method. The identity of the object is the stored in a *class-box*. These boxes are composed in parallel and synchronized and restricted over all internal labels.

The best way to explain the translation is through an example. Figure 1(b) is a part of the translation of the class declaration in figure 1(a). The net is

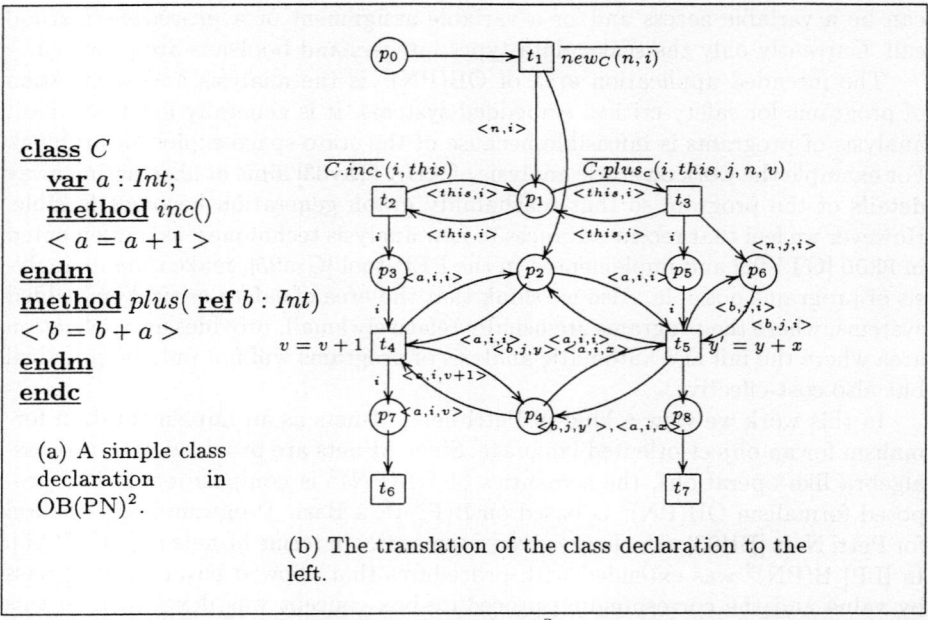

(a) A simple class declaration in OB(PN)^2.

(b) The translation of the class declaration to the left.

Fig. 1.: A simple class declaration in OB(PN)^2 and part of its translation into an M-net.

obtained as the synchronization and restriction of the parallel composition of the translations of the variable and method declarations together with a *class-box*. The places p_0, p_1 and the transition t_1 are generated by the class box. The method *inc* generates t_2, p_3, t_4, p_6, t_6, *plus* generates $t_3, p_5, p_6, t_5, p_8, t_7$, and place p_2 represents the variable a. The reason for place p_4 will become evident below. Transitions t_2 (labeled $\overline{C.inc_c}(i, this)$) and t_3 (labeled $\overline{C.plus_c}(i, this, j, n_1, v_1)$) are the *method-call* transitions. A call to *inc* is translated into a transition that is synchronized with t_2. As a result control passes to p_3, activating transition t_4. This transition is the result of the translation of the statement $< a = a + 1 >$. When firing the transition reads the reference to a, $< a, i, i >$ from place p_2 together with the corresponding value from p_4 $< a, i, v >$ and the writes the reference a back to p_2 and the value of a incremented by 1 $< a, i, v + 1 >$ to p_4. The reference $< x, k, l >$ consists of a *pointer* $< x, k >$ where x is the id of the variable, k is the creator of the variable, and l is the current owner of the variable, (the instance l of the procedure). We need this concept of a pointer to be able to pass variables by reference. The values are stored in global *type-boxes* (which has generated place p_4) and retrieved by the pair $< x, k >$. The current owner l (or the frame-index) is used to distinguish between the different instances of the procedures as will be explained below. Notice that after the call to *inc*, the object is ready to answer a new call to *inc* or a call to *plus*. Suppose thus, that right after the firing of transition t_2 the program makes a call *plus*(m), where m is some global variable. As you can see in the figure the label of the transition t_3 is more complicated than the corresponding label of t_2. The reason for this is the translation of the parameters to *plus*. The variables j, n, v are used to transfer the parameters, n, j are for reference and v for value parameters (since we only have one reference parameter v is not used as can be seen from the arc expressions). Through the call we store the reference $<n,j,i>$ to m in place p_6 "as" b. Transition t_5 representing the assignment $<b = b+a>$ is now enabled. As can be seen from the annotations, the value of the reference variable m aliased as b becomes $m + a$. Also notice that: (i) the semantics of places induces a mutual-exclusion semantics on the access of variables, and (ii) clearly the methods could have been composed with the other components using some other operation that parallel composition, eg. choice, indeed it would be easy to extend the class definition with some kind of synchronization constraint expressions.

3 Conclusions

In the paper we presented OB(PN)^2 and its semantics in terms of M-nets. The main points of the semantics are: (i) compositionality, which relies on the use of the algebra of boxes, (ii) the use of pairs of integers to simulate procedure call frames, (iii) the treatment of references, thus extending the procedure-box concept of [LP] to reference parameters, (iv) and the reliance on the symbolic firing rule instead of the unfolding as done in [BHW+95].

Acknowledgments I would like to thank Eike Best, Elisabeth Pelz and Hanna Klaudel for helpful discussions during the early stages of this work.

References

[BH93] E. Best and R.P. Hopkins. B(PN)^2 - A Basic Petri Net Programming Notation. In *Proc. of PARLE-93*, pp. 379–390.

[BHW+95] E. Best, H. Fleischhack, W. Fraczak, R.P. Hopkins, H. Klaudel, and E. Pelz. An M-net Semantics of B(PN)^2. In *Proc. of STRICT'95*, pp. 85–100.

[Gra95] B. Grahlmann. An introduction to the principles, the functionality and the usage of the **pep**-system. In *Proc. of PEP Workshop*, nr. 14/95 in Hildesheimer Inf. Ber., 1995.

[GTV93] P. Grönberg, M. Tiusanen, and K. Varpaaniemi. PROD - A Pr/T-Net reachability analysis tool. Technical Report B11, Digital Systems Laboratory, Helsinki University of Technology, Espoo, 1993.

[Lil] J. Lilius. OB(PN)^2: An Object Based Petri Net Programming Notation. Technical Report. Digital Systems Laboratory, Helsinki University of Technology, Espoo, to appear.

[LP] J. Lilius and E. Pelz. An M-net semantics for B(PN)^2 with procedures. Submitted to ISCIS-XI.

[Tay83] R. N. Taylor. A general purpose algorithm for analyzing concurrent programs. *Communications of the ACM*, 26(5):362–376, 1983.

[YT87] A .Yonezawa and M .Tokoro. *Object-Oriented Concurrent Programming*. Computer Systems Series. MIT Press, 1987.

Reusable Coordinator Modules for Massively Concurrent Applications

F. Arbab, C.L. Blom, F.J. Burger and C.T.H. Everaars

CWI, P.O. Box 94079, 1090 GB Amsterdam, The Netherlands

Abstract. Isolating computation and communication concerns into separate pure computation and pure coordination modules enhances modularity, understandability, and reusability of parallel and/or distributed software. **MANIFOLD** is a pure coordination language that encourages this separation. We use real, concrete, running **MANIFOLD** programs to demonstrate the concept of pure coordination modules and the advantage of their reuse in applications of different nature.

1 Introduction

Some of the shortcomings of the common approaches to the design and development of parallel and distributed applications stem from the fundamental properties of the various models of communication used to construct this software[4]. Without a proper programming paradigm for expressing the coordination of the cooperation of various active components that comprise a single concurrent application, programmers are forced to use low-level communication constructs, such as message passing, directly in their code. Because these primitives are generally scattered throughout the source code of the application and are typically intermixed with non-communication application code, the protocols of coordination generally never manifest themselves in a tangible form as easily identifiable pieces of source code. Thus, in spite of the fact that the coordination protocols are often the most complex and expensive-to-develop part of a non-trivial parallel or distributed application, they are not treated as a separate commodity that can be designed, developed, debugged, maintained, and reused, in isolation from the rest of the application code.

Intermixing communication concerns with computation decreases the comprehensibility, maintainability, and reusability of software modules. Moreover, the targeted send primitives used in message passing models of communication strengthen the dependence of individual processes on their environment. This too diminishes the reusability and maintainability of processes. It also complicates debugging and proving correctness of programs because a process that depends on the existence and certain expected "valid" behavior of some other processes for its own correctness, by itself is not a well encapsulated concept.

Coordination languages[1] ameliorate these problems to some extent, but existing coordination languages, such as Linda, still do not go all the way to support reusable, pure coordination program modules. The goal of this paper is to demonstrate the concept of pure coordination modules and their reusability in different applications through real, concrete, running examples using the coordination language **MANIFOLD**, which is based on the IWIM model of communication developed at CWI. Only a very brief summary of the IWIM model is presented here in §2. An introduction to the **MANIFOLD** language appears in §3 in this paper. More detailed descriptions of IWIM and **MANIFOLD** appear in [3] and [4], which also contain a review of some related work and a comparison with other coordination models and systems. For our purpose in this paper, the syntax and semantics of some of the relevant constructs of **MANIFOLD** are introduced in §4 through a trivial example program. In §5, we discuss a quite non-trivial example of sorting and show how **MANIFOLD** encourages isolating communication and computation concerns into separate modules. The reusability of the pure coordination module developed for sorting is then demonstrated in §6, where the same **MANIFOLD** program is applied to coordinate a parallel/distributed numerical optimization application using a domain decomposition algorithm. All examples presented in this paper run, without any change to any source code, on a variety of parallel and/or heterogeneous computing platforms. We close this paper with a short conclusion in §7.

2 The IWIM Model of Communication

IWIM stands for *Idealized Worker Idealized Manager* and is a generic, abstract model of communication that supports the separation of responsibilities and encourages a weak dependence of workers (processes) on their environment[3, 4].

The basic concepts in the IWIM model are *processes, events, ports*, and *channels*. The IWIM model supports *anonymous communication*: in general, a process does not, and need not, know the identity of the processes with which it exchanges information. This concept reduces the dependence of a process on its environment and makes processes more reusable.

A process in IWIM can be regarded as a worker process or a manager (or coordinator) process. The responsibility of a worker process is to perform a (computational) task. A worker process is not responsible for the communication that is necessary for it to obtain the proper input it requires to perform its task, nor is it responsible for the communication that is necessary to deliver the results it produces to their proper recipients. In general, *no process in* IWIM *is responsible for its own communication with other processes.* It is always the responsibility of a manager process to arrange for and to coordinate the necessary communications among a set of worker processes.

There is always a bottom layer of worker processes, called *atomic workers*, in an application. In the IWIM model, an application is built as a (dynamic) hierarchy of (worker and manager) processes on top of this layer. Note that a manager process may itself be considered as a worker process by another manager process.

3 The Manifold Coordination Language

In this section, we briefly introduce **MANIFOLD**: a coordination language for managing complex, dynamically changing interconnections among sets of independent, concurrent, cooperating processes, which is based on the IWIM model, described in §2.

A **MANIFOLD** application consists of a (potentially very large) number of (light- and/or heavy-weight) processes running on a network of heterogeneous hosts, some of which may be parallel systems. Processes in the same application may be written in different programming languages. Some of them may not know anything about **MANIFOLD**, nor the fact that they are cooperating with other processes through **MANIFOLD** in a concurrent application.

The **MANIFOLD** system consists of a compiler, a run-time system library, a number of utility programs, libraries of builtin and predefined processes[3], a link file generator called **MLINK** and a run-time configurator called **CONFIG**. The system has been ported to several different platforms (e.g., SGI 5.3, SUN 4, Solaris 5.2, IBM SP/1). **MLINK** uses the object files produced by the (**MANIFOLD** and other language) compilers to produce link files needed to compose the application executable files for each required platform. At run time of an application, **CONFIG** determines the actual host(s) where the processes which are created in the **MANIFOLD** application will run.

The library routines that comprise the interface between **MANIFOLD** and processes written in other languages (e.g. C), automatically perform the necessary data format conversions when data is routed between various different machines.

3.1 Processes

In **MANIFOLD**, the atomic workers of the IWIM model are called atomic processes. Any operating system-level process can be used as an atomic process in **MANIFOLD**. However, **MANIFOLD** also provides a library of functions that can be called from a regular C function running as an atomic process, to support a more appropriate interface between the atomic processes and the **MANIFOLD** world. Atomic processes can only produce and consume units through their ports, generate and receive events, and compute. In this way, the desired separation of computation and coordination is achieved.

Coordination processes are written in the **MANIFOLD** language and are called manifolds. The **MANIFOLD** language is a block-structured, declarative, event driven language. A manifold definition consists of a header and a body. The header of a manifold gives its name, the number and types of its parameters, and the names of its input and output ports. The body of a manifold definition is a block. A block consists of a finite number of states. Each state has a label and a body. The label of a state defines the condition under which a transition to that state is possible. It is an expression that can match observed event occurrences in the event memory of the manifold. The body of a simple state defines the set of actions that are to be performed upon transition to that state. The body of a compound state is either a (nested) block, or a call to a parameterized subprogram known as a *manner* in **MANIFOLD**. A manner consists of a header and a body. As for the subprograms in other languages, the header of a manner essentially defines its name and the types and the number of its parameters. A manner is either atomic or regular. The body of a regular manner is a block. The body of an atomic manner is a C function that can interface with the **MANIFOLD** world through the same interface library as for the compliant atomic processes.

3.2 Streams

All communication in **MANIFOLD** is asynchronous. In **MANIFOLD**, the asynchronous IWIM channels are called streams. A stream is a communication link that transports a sequence of bits, grouped into (variable length) *units*.

A stream represents a reliable and directed flow of information from its *source* to its *sink*. As in the IWIM model, the constructor of a stream between two processes is, in general, a third process. Once a stream is established between a producer process and a consumer process, it operates autonomously and transfers the units from its source to its sink. The sink of a stream requiring a unit is suspended only if no units are available in the stream. The suspended sink is resumed as soon as the next unit becomes available for its consumption. The source of a stream is never suspended because the infinite buffer capacity of a stream is never filled.

There are four basic stream types designated as BB, BK, KB, and KK, each behaving according to a slightly different protocol with regards to its automatic disconnection from its source or sink. Furthermore, in **MANIFOLD**, the BK and KB type streams can be declared to be *reconnectable*. See [3] or [4] for details.

3.3 Events and State Transitions

In **MANIFOLD**, once an event is *raised* by a process, it continues with its processing, while the event occurrence propagates through the environment independently. Any receiver process that is interested in such an event occurrence will automatically receive it in its *event memory*. The observed event occurrences in the event memory of a process can be examined and reacted on by this process at its own leisure. In reaction to such an event occurrence, the observer process can make a transition from one labeled state to another.

The only control structure in the **MANIFOLD** language is an event-driven state transition mechanism. More familiar control structures, such as the sequential flow of control represented by the connective ";" (as in Pascal and C), conditional (i.e., "if") constructs, and loop constructs can be built out of this event mechanism, and are also available in the **MANIFOLD** language as convenience features.

Upon transition to a state, the primitive actions specified in its body are performed atomically in some non-deterministic order. Then, the state becomes *preemptable*: if the conditions for transition to another state are satisfied, the current state is preempted, meaning that all streams that have been constructed are dismantled and a transition to a new state takes place. The most important primitive actions in a simple state body are (1) creating and activating processes, (2) generating event occurrences, and (3) connecting streams to the ports of various processes.

4 Hello World!

For our first example, consider a simple program to print a message such as "Hello World!" on the standard output. The MANIFOLD source file for this program contains the following:

```
1 manifold printunits import.
2
3 auto process print is printunits
4
5 manifold Main
6 {
7   begin: "Hello World!" -> print.
8 }
```

The first line of this code defines a manifold named `printunits` that takes no arguments, and states (through the keyword `import`) that the real definition of its body is contained in another source file. This defines the "interface" to a process type definition, whose actual "implementation" is given elsewhere. Whether the actual implementation of this process is an atomic process (e.g., a C function) or it is itself another manifold is indeed irrelevant in this source file. We assume that `printunits` waits to receive units through its standard input port and prints them. When `printunits` detects that there are no incoming streams left connected to its input port and it is done printing the units it has received, it terminates.

The second line of code defines a new instance of the manifold `printunits`, calls it `print`, and states (through the keyword `auto`) that this process instance is to be automatically activated upon creation, and deactivated upon departure from the scope wherein it is defined; in this case, this is the end of the application. Because the declaration of the process instance `print` appears outside of any blocks in this source file, it is a global process, known by every instance of every manifold whose body is defined in this source file.

The last lines of this code define a manifold named `Main` that takes no parameters. Every manifold definition (and therefore every process instance) always has at least three default ports: `input`, `output`, and `error`. The definition of these ports are not shown in this example, but the ports are defined for `Main` by default.

The body of this manifold is a block (enclosed in a pair of braces) and contains only a single state. The name `Main` is indeed special in MANIFOLD: there must be a manifold with that name in every MANIFOLD application and an automatically created instance of this manifold, called `main`, is the first process that is started up in an application. Activation of a manifold instance automatically posts an occurrence of the special event `begin` in the event memory of that process instance; in this case, `main`. This makes the initial state transition possible: `main` enters its only state – the `begin` state.

The `begin` state contains only a single primitive action, represented by the stream construction symbol, "→". Entering this state, `main` creates a stream instance (with the default BK-type) and connects the `output` port of the process instance on the left-hand side of the → to the `input` port of the process instance on its right-hand side. The process instance on the right-hand side of the → is, of course, `print`. What appears to be a character string constant on the left-hand side of the → is also a process instance: conceptually, a constant in **MANIFOLD** is a special process instance that produces its value as a unit on its `output` port and then dies.

Having made the stream connection between the two processes, `main` now waits for all stream connection made in this state to break up (on at least one of their ends). The stream breaks up, in this case, on its source end as soon as the string constant delivers its unit to the stream and dies. Since there are no other event occurrences in the event memory of `main`, the default transition for a state reaching its end (i.e., falling over its terminator period) now terminates the process `main`.

Meanwhile, `print` reads the unit and prints it. The stream type BK ensures that the connection between the stream and its sink is preserved even after a preemption, or its disconnection from its source. Once the stream is empty and it is disconnected from its source, it automatically disconnects from its sink. Now, `print` senses that it has no more incoming streams and dies. At this point, there are no other process instances left and the application terminates.

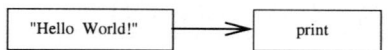

Fig. 1. The "Hello World" example in Manifold

Note that our simple example, here, consists of three process instances: two worker processes, a character string constant and `print`, and a coordinator process, `main`. Figure 1 shows the relationship between the constant and `print`, as established by `main`. Note also that the coordinator process `main` only establishes the connection between the two worker processes. It does *not* transfer the units through the stream(s) it creates, nor does it interfere with the activities of the worker processes in other ways.

5 Bucket Sort

The example in the previous section was simple enough to require only a static pattern of communication. In this section, we illustrate the dynamic capabilities of **MANIFOLD** through a program for sorting an unspecified number of input units. The particular algorithm used in this example is not necessarily the most effective one. However, it is simple to describe, and serves our purpose of demonstrating the dynamic aspects of the **MANIFOLD** language well. The sort algorithm is as follows.

There is a sufficiently large (theoretically, infinite) number of *atomic sorters* available, each of which is able to sort a bucket of $n > 0$ units very efficiently. (The number n may even vary from one atomic sorter to the next.) Each atomic sorter receives its input through its **input** port; raises a specific event it receives as a parameter to inform other processes that it has filled up its input bucket; sorts its units; produces the sorted sequence of the units through its **output** port; and terminates.

The parallel bucket sort program is supposed to feed as much of its own input units to an atomic sorter as the latter can take; feed the rest of its own input as the input to another copy of itself; merge the two output sequences (of the atomic sorter and its new copy); and produce the resulting sequence through its own **output** port. Merging of the two sorted sequences can be done by a separate merger process, or by a subprogram (i.e., a manner) called by the sorter.

We assume our application consists of several source files. The first source file contains our **Main** manifold, as shown below. We assume that the merger is a separate process. The merger and the atomic sorter can be written in the **MANIFOLD** language, but they will be more efficient if they are written in a computation language, such as C. We do not concern ourselves here with the details of the merger and the atomic sorter, and assume that each is defined in a separate source file.

```
1  manifold printunits import.
2  manifold Sorter import.
4  manifold ReadFile(process filename) atomic {internal.}.
5  manifold AtomicSorter(event) atomic {internal.}.
6  manifold AtomicIntMerger port in a, b. atomic {internal.}.
7  /************************************************************/
8  manifold Main
9  {
10    auto process read is ReadFile("unsorted").
11    auto process sort is Sorter.
12    auto process print is printunits.
13
14    begin:  read -> sort -> print.
15 }
```

The **main** manifold in this application creates **read**, **sort**, and **print** as instances of manifold definitions **ReadFile**, **Sorter**, and **printunits**, respectively. It then connects the **output** port of **read** to the **input** port of **sort**, and the **output** port of **sort** to the **input** port of **print**. The process **main** terminates when both of these connections are broken.

The process **read** is expected to read the contents of the file named **unsorted** and produce a unit for every sort item in this file through its **output** port. When it is through with producing its units, **read** simply terminates. The process **sort** is an instance of the manifold definition **Sorter**, which is expected to sort the units it receives through its **input** port. This process terminates when its **input** is disconnected and all of its output units are delivered through its **output** port.

The manifold definition **Sorter**, shown below, is our main interest. The keyword **export** on line 1 allows other separately compiled **MANIFOLD** source files to **import** and use this coordinator manifold. In its **begin** state, an instance of **Sorter** connects its own **input** to an instance of the **AtomicSorter**, it calls

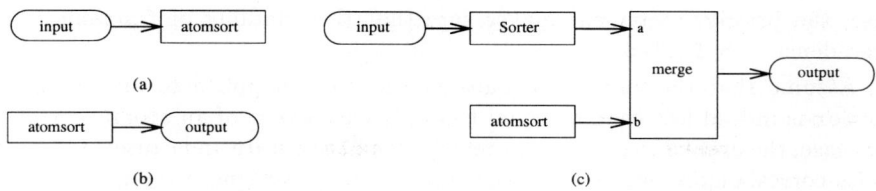

Fig. 2. Bucket sort

`atomsort`. It also installs a *guard* on of its `input` port. This guard posts the event `finished` if it has an empty stream connected to its departure side, after the arrival side of this port has no more stream connections, following a first connection. This means that the event `finished` is posted in an instance of `Sorter` after a first connection to the arrival side of its `input` is made, then all connections to the arrival side of its `input` are severed, and all units passed through this port are consumed. The connections in this state are shown in Figure 2.a.

```
 1  export manifold Sorter()
 2  {
 3    event filled, flushed, finished.
 4    process atomsort is AtomicSorter(filled).
 5    stream reconnect KB input -> *.
 6    priority filled < finished.
 7
 8    begin: (
 9      activate(atomsort), input -> atomsort,
10      guard(input, a_everdisconnected!empty, finished) // no more input
11    ).
12
13    finished: {
14      ignore filled. //possible event from atomsort
15
16      begin: atomsort -> output   //your output is only that of atomsort
17    }.
18
19    filled: (
20      process merge<a, b | output> is AtomicIntMerger.
21      stream KK * -> (merge.a, merge.b).
22      stream KK merge -> output.
23
24      begin: (
25        activate(merge),
26        input -> Sorter -> merge.a,
27        atomsort -> merge.b,
28        merge -> output
29      ).
30
31      end | finished:.
32    }.
33
34    end: (
35      begin: (
36        guard(output, a_disconnected, flushed), // ensure flushing
37        terminated(void) //wait for units to flush through output
38      ).
39
40      flushed: halt.
41    ).
42  }
```

Two events can preempt the `begin` state of an instance of `Sorter`: (1) if the incoming stream connected to `input` is disconnected (no more incoming units) and `atomsort` reads all units available in its incoming stream, the guard on `input` posts the event `finished`; and (2) the process `atomsort` can read its fill and raise the event `filled`. Normally, only one of these events occurs; however, when the number of input units is exactly equal to the bucket size, n, of `atomsort`, both `finished` and `filled` can occur simultaneously. In this case, the priority statement makes sure that the handling of `finished` takes precedence over `filled`.

Assume that the number of units in the input supplied to an instance of `Sorter` is indeed less than or equal to the bucket size n of an atomic sorter. In this case, the event `finished` will preempt the `begin` state and cause a transition to its corresponding state in `Sorter`. In this state, we ignore the occurrence of `filled` that may have been raised by `atomsort` (if the number of input units is equal to the bucket size n); and deliver the output of `atomsort` as the output of the `Sorter`. The connections in this state are shown in Figure 2.b.

Now suppose the number of units in the input supplied to an instance of `Sorter` is greater than the bucket size n of an atomic sorter. In this case, the event `filled` will preempt the `begin` state and cause a transition to its corresponding state in `Sorter`. In this state we create an instance of the merger process, called `merge`. A new instance of the `Sorter` is created in the `begin` state of the nested block. The rest of the input is passed on as the input to this new `Sorter`, and its output is merged with the output of the atomic sorter and the result is passed as the output of the `Sorter` itself. The connections in this state are shown in Figure 2.c. An occurrence of `finished` in this state preempts the connected streams and causes a transition to the local `finished` state in this block. This preemption is necessary to inform the new instance of `Sorter` (by breaking the stream that connects `input` to it) that it has no more input to receive, so that it can terminate. The empty body of the `finished` state means that it causes an exit from its containing block.

In the `end` state, a `Sorter` instance installs a guard on its `output` port, to post the event `flushed` after there is no stream connected to the arrival side of this port following its first connection. This means that the event `flushed` is posted in an instance of `Sorter` after a connection is made to its arrival side, and all units arriving at this port have passed through. The `Sorter` instance then waits for the termination of the special predefined process `void`, which will never happen (the special process `void` never terminates). This effectively causes the `Sorter` instance to hang indefinitely. The only event that can terminate this indefinite wait is an occurrence of `flushed` which indicates there are no more units pending to go through the `output` port of the `Sorter` instance.

An interesting aspect of the `Sorter` manifold is the dynamic way in which it switches connections among the process instances it creates. Perhaps more interesting, is the fact that, in spite of its name, `Sorter` knows nothing about sorting! If we change its name to X, and systematically change the names of the identifiers it uses to Y_1 through Y_k, we realize that all it knows is to divert its own input to an instance of some process it creates; when this instance raises a certain event, it is to divert the rest of its input to a new instance of itself; and to divert the output of these two processes to a third process, whose output is to be passed out as its own output.

What `Sorter` embodies is a protocol that describes how instances of two process definitions (e.g., `AtomicSorter` and `AtomicIntMerger` in our case) should communicate with each other. Our `Sorter` manifold can just as happily orchestrate the cooperation of any pair of processes that have the same input/output and event behavior as `AtomicSorter` and `AtomicIntMerger` do, regardless of what computation they perform. The cooperation protocol defined by `Sorter` simply doles out chunks of its input stream to instances of what it knows as `AtomicSorter` and diverts their output streams to instances of what it knows as `AtomicIntMerger`. What is called `AtomicSorter` needs not really sort its input units, the process called `AtomicIntMerger` needs not really merge them, and neither has to produce as many units through its `output` as it receives through its `input` port. They can do any computation they want.

By parameterizing the names of the manifolds used in `Sorter` and changing its name to `ProtocolX`, we obtain a more general program:

```
1  export manifold ProtocolX(manifold M1(event), manifold M2<a, b | output>)
2  {
3    event filled, flushed, finished.
4    process m1 is M1(filled).
5    stream reconnect KB input -> *.
6    priority filled < finished.
7
8    begin: (
9      activate(m1), input -> m1,
10     guard(input, a_everdisconnected!empty, finished) // no more input
11   ).
12
13   finished: (
14     ignore filled. //possible event from m1
15
16     begin: m1 -> output   //your output is only that of m1
17   ).
18
19   filled: (
20     process m2<a, b | output> is M2.
21     stream KK * -> (m2.a, m2.b).
22     stream KK m2 -> output.
23
24     begin: (
25       activate(m2),
26       input -> ProtocolX(M1, M2) -> m2.a,
27       m1 -> m2.b,
28       m2 -> output
29     ).
30
31     end | finished:.
32   ).
33
34   end: (
35     begin: (
36       guard(output, a_disconnected, flushed), // ensure flushing
37       terminated(void) //wait for units to flush through output
38     ).
39
40     flushed: halt.
41   ).
42 }
```

The new version of our bucket sort main program using `ProtocolX` is:

```
1  manifold printunits import.
2  manifold ProtocolX(manifold M1(event), manifold M2) import.
3  manifold ReadFile(process filename) atomic {internal.}.
4  manifold AtomicSorter(event) atomic {internal.}.
5  manifold AtomicIntMerger port in a, b. atomic {internal.}.
6
7  /****************************************************************/
8  manifold Main
9  {
10   auto process read is ReadFile("unsorted").
11   auto process sort is ProtocolX(AtomicSorter, AtomicIntMerger).
12   auto process print is printunits.
13
14   begin:   read -> sort -> print.
15 }
```

As a concrete demonstration of the reusability of coordinator modules, in the next section, we present an example that uses the coordinator `ProtocolX` in a numerical optimization problem.

6 Domain Decomposition

Consider the following optimization problem:

$$\max z = x^2 + y^2 - 0.5*\cos(18*x) - 0.5*\cos(18*y) \text{ with } (x, y) \in [-1.0, 1.0] \quad (1)$$

Figure 3 shows the landscape formed by this function on its domain.

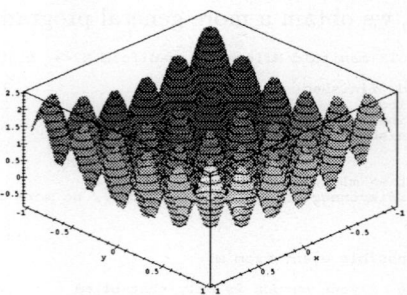

Fig. 3. The function $z = x^2 + y^2 - 0.5 * \cos(18 * x) - 0.5 * \cos(18 * y)$

Analytical solutions to such problems are, in general, non-existent and domain decomposition is a common search technique used to solve them through numerical methods. Domain decomposition imposes a grid on the domain of the function, splitting it into a number of sub-domains, as determined by the size of the grid. Next, we obtain a (number of) good rough estimate(s) for the highest value of z in each sub-domain. Then, we select the sub-domains with the most promising z values and decompose them into smaller sub-domains. New estimates for the highest value of z in each of these sub-domains, recursively narrow this search process further and further into smaller and smaller regions that (hopefully) tend towards the area with the real maximum z, while the estimates for the obtained maximum z values become more and more accurate. In single grid domain decomposition, the same grid is imposed on all successive sub-domains. Multi-grid domain decomposition techniques allow a different grid for each sub-domain, whose granularity and other properties may depend on the attributes of the sub-domain and those of the function within that region.

For our example, we consider a single grid method. We need four computation modules for this example: ap_printobjects, Split, AtomicEval, and AtomicObjMerger. An instance of ap_printobjects simply prints the units it reads from its input, each of which describes a (sub-)domain and the x, y, and z values for the estimated maximum z value in that (sub-)domain. An instance of Split receives as its parameters the specification of a grid (in our case, 6×6). Next, it reads from its input port a unit that describes a (sub-)domain, produces units on its `output` port that describe the sub-domains obtained by imposing the grid on this input domain, and terminates. An instance of *AtomicEval* reads a bucket of $n > 0$ sub-domains (for simplicity, let $n = 1$) from its `input` port and raises a specific event, which it receives as a parameter, to inform other processes that it has filled up its input bucket with some sub-domains descriptions. It then finds the best estimate for the optimum z value in each of its sub-domains, producing an ordered sequence of units describing the best solutions it has found through its `output` port, and terminates. In our example, we use sampling: we simply evaluate z for a number of (say 1000) sample points in each sub-domain and consider the sample point with the maximum z as the best estimate for that sub-domain. An instance of `AtomicObjMerger` reads from its ports a and b two ordered sequences of units describing sub-domains and their best estimates, and produces a sequence of one or more of its best sub-domains on its `output` port.

We need a **MANIFOLD** program, say *Eval*, to coordinate the cooperation of the instances of `AtomicEval` and `AtomicObjMerger` to solve our optimization problem in a parallel/distributed fashion. *Eval* receives through its `input` port units describing (sub-)domains. It is supposed to feed as much of its own input units to an atomic evaluator as the latter can take; feed the rest of its own input as the input to another copy of itself; merge the two output sequences (of the atomic evaluator and its new copy); and produce the resulting sequence through its own `output` port. The similarity between the description of *Eval* and that of `Sort` in §5 suggests that we can use the same coordination module for our optimization problem. Indeed, *Eval* is merely a version of `ProtocolX` with `AtomicEval` and `AtomicObjMerger` as its parameters. The following **MANIFOLD** program shows a single iteration of our domain decomposition application using the separately compiled `ProtocolX` of §5.

```
1 manifold ap_printobjects atomic {internal.}.
2 manifold ProtocolX(manifold M1(event), manifold M2) import.
3 manifold Split(port in, port in) atomic {internal.}.
4 manifold AtomicEval(event) atomic {internal.}.
5 manifold AtomicObjMerger port in a, b. atomic {internal.}.
6
7 /***************************************************************/
8 manifold Main
9 {
10    auto process split is Split(6, 6).
11    auto process eval is ProtocolX(AtomicEval, AtomicObjMerger).
12    auto process print is ap_printobjects.
13
14    begin: <<1, -1.0, -1.0, 1.0, 1.0>> -> split -> eval -> print.
15 }
```

The output of this program (of which only the first 8 lines appear below) shows the result produced by 36 instances of `AtomicEval`, each taking in the description of a single sub-domain. The top four lines show the best estimates to be in the neighborhoods of the four corners of the domain for our symmetric function in Figure 3.

```
domain = (-1.000, -1.000) (-0.667, -0.667) point = (-0.883, -0.880), z = 2.541
domain = ( 0.667,  0.667) ( 1.000,  1.000) point = ( 0.889,  0.884), z = 2.540
domain = ( 0.667, -1.000) ( 1.000, -0.667) point = ( 0.881, -0.889), z = 2.539
domain = (-1.000,  0.667) (-0.667,  1.000) point = (-0.878,  0.881), z = 2.539
domain = (-0.667,  0.667) (-0.333,  1.000) point = (-0.528,  0.884), z = 2.048
domain = ( 0.333, -1.000) ( 0.667, -0.667) point = ( 0.533, -0.882), z = 2.048
domain = ( 0.667, -0.667) ( 1.000, -0.333) point = ( 0.885, -0.527), z = 2.048
domain = (-0.667, -1.000) (-0.333, -0.667) point = (-0.534, -0.885), z = 2.047
```

A straight-forward generalization of this program repeats this single step until a termination criterion (such as a maximum number of iterations, or the diminishing of improvements below a threshold) is reached. Each iteration selects a (few of the) best sub-domain(s) found so far as input to another instance of Split and *Eval*. This would be yet another MANIFOLD program that coordinates the cooperation of different instances of *Eval* and Split. The following output is produced by such a program using a 2 × 2 grid. The first line in this output is our initial input unit representing the whole domain. Each succeeding group of four lines then represents one iteration. The best sub-domain found in each iteration is fed as input to the next iteration. The first line of the last group (representing the third iteration) shows the best solution found ($z = 2.542$) which is slightly better than the best solution we found using our single step 6 × 6 grid ($z = 2.541$).

```
domain = (-1.000, -1.000) ( 1.000,  1.000)

domain = ( 0.000,  0.000) ( 1.000,  1.000) point = ( 0.885,  0.879), z =  2.540
domain = (-1.000, -1.000) ( 0.000,  1.000) point = (-0.884, -0.890), z =  2.539
domain = ( 0.000, -1.000) ( 1.000,  0.000) point = ( 0.890, -0.893), z =  2.532
domain = (-1.000,  0.000) ( 0.000,  1.000) point = (-0.880,  0.911), z =  2.484

domain = ( 0.500,  0.500) ( 1.000,  1.000) point = ( 0.879,  0.892), z =  2.536
domain = ( 0.000,  0.500) ( 0.500,  1.000) point = ( 0.498,  0.866), z =  1.941
domain = ( 0.500,  0.000) ( 1.000,  0.500) point = ( 0.880,  0.490), z =  1.920
domain = ( 0.000,  0.000) ( 0.500,  0.500) point = ( 0.498,  0.499), z =  1.400

domain = ( 0.750,  0.750) ( 1.000,  1.000) point = ( 0.883,  0.883), z =  2.542
domain = ( 0.500,  0.750) ( 0.750,  1.000) point = ( 0.530,  0.883), z =  2.049
domain = ( 0.750,  0.500) ( 1.000,  0.750) point = ( 0.886,  0.529), z =  2.048
domain = ( 0.500,  0.500) ( 0.750,  0.750) point = ( 0.531,  0.530), z =  1.555
```

The highly modular structure of this application is remarkable. Its computation modules (C functions) are simple and have no idea of how they relate to or cooperate with one another. The coordination module *Eval* knows nothing about what these computation modules actually do; it is just as happy coordinating the sorter workers in §5 as it is managing these numerical optimization workers. The various processes comprising this application can run on parallel or distributed platforms without any change to their source code.

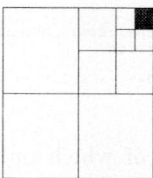

Fig. 4. Visualizer snap-shot of 2 × 2 distributed domain decomposition

The plumbing paradigm of MANIFOLD makes it easy to divert the flows of units, change the coordination structures, and dynamically modify the topology of communication links among (computation as well as coordination) modules to adapt an application to new requirements. We can plug in graphics modules to display an on-going computation. Indeed, we have a small computational steering environment built around this example, using **MANIFOLD** coordinators and a few generic graphics interaction modules. The user interface for this program graphically shows the on-going activity of various atomic evaluators (that may be running on different hosts) and allows the user to interactively direct the focus of the attention of the program onto one or more areas of interest, simply by drawing a box to designate a sub-domain. Figure 4 shows a snap-shot of this user interface as our 2×2 distributed domain decomposition optimization application moved on (in its third iteration) to the top-right corner sub-domain in the run that produced the above output.

7 Conclusion

IWIM is a model of communication that supports anonymous communication and separation of computation responsibilities from communication and coordination concerns. **MANIFOLD** is a coordination language that takes full advantage of these two key concepts of IWIM. Unlike other coordination languages, **MANIFOLD** encourages decomposition of a parallel and/or distributed application into a hierarchy of pure computation and pure coordination modules, none of which contain hard-coded dependencies on their environment. This leads to highly reusable computation modules, and more interestingly, also to highly reusable coordination modules.

The examples in this paper show a single coordination module used in two very different applications. We have also used **MANIFOLD** to reorganize existing Fortran 77 sequential code into a parallel and distributed application. The usefulness of the IWIM model and, in particular, the **MANIFOLD** language in these and other applications has been very encouraging. The plumbing paradigm inherent in IWIM makes it easy to compose and recompose a **MANIFOLD** application and adapt it to new requirements. To enhance the effectiveness of this coordination language, we are presently developing a visual programming environment around **MANIFOLD** which takes advantage of its underlying plumbing paradigm.

References

1. D. Gelernter and N. Carriero, "Coordination languages and their significance," *Communication of the ACM*, vol. 35, pp. 97–107, February 1992.
2. F. Arbab, "Coordination of massively concurrent activities," Tech. Rep. CS–R9565, Centrum voor Wiskunde en Informatica, Kruislaan 413, 1098 SJ Amsterdam, The Netherlands, 1995. Available on-line:
   http://www.cwi.nl/ftp/CWIreports/IS/CS-R9565.ps.Z
3. F. Arbab, "Manifold version 2: Language reference manual," Tech. Rep. in preparation, Centrum voor Wiskunde en Informatica, Kruislaan 413, 1098 SJ Amsterdam, The Netherlands, 1995.
4. F. Arbab, "The IWIM model for coordination of concurrent activities," in *Coordination '96*, Lecture Notes in Computer Science, Springer-Verlag, April 1996.

Introducing Dynamicity in the Data-Parallel Language 8१/२ *

Olivier Michel

LRI u.r.a. 410 du CNRS
Bâtiment 490, Université de Paris-Sud, F-91405 Orsay Cedex, France.
Tel: +33 (1) 69 41 76 01 email: michel@lri.fr

Abstract. The main motivation of 8१/२ is to develop a high-level language that supports the parallel simulation of dynamical processes [1, 2]. To achieve this goal, a new data-structure, that merges the concept of *stream* and *collection* is introduced in a declarative framework. After a brief description of 8१/२ basics, we describe the introduction of *dynamicity* and *symbolic values* in the language. We focus on the expressivity and issues brought by the new dynamic possibilities of the language and show, through several paradigmatic examples, that our computation model is able to support parallel symbolic processing.

1 The Declarative Data-Parallel Language 8१/२

1.1 Motivations: the Implicit Data-Parallel Approach to Parallel Symbolic Processing

8१/२ is an experimental language combining features of *collection* and *stream* oriented languages in a declarative framework. It tries to promote the construction of parallel programs by isolating the programmer from the complexities of parallel processing. To let the designer concentrate on the modeling aspects, we advocate the use of a high-level language, where the entities expressed are close to the concepts used in the target application [3, 4] and hiding implementation details.

The use of functions and lists to provide parallel symbolic processing capabilities has been advocated for a long time and largely demonstrated. However, from the point of view of parallelism exploitation, this approach naturally leads to control-parallelism with some drawbacks: a) lists are sequentially accessed even in a distributed implementation, inducing some unnecessary bottlenecks; b) there is an "impedance mismatch" problem between tasks and functions: b.1) function invocations are fine-grained entities while task activations are more heavy weight. Using tasks to implement functions is therefore too expensive, even when using light-weight threads [5]; b.2) mapping only some functions to tasks, while using a more standard sequential implementation for other functions, can be achieved on an explicit or implicit basis. The explicit approach

---

* This research is partially supported by the operation "Programmation parallèle et distribuée" of the french "GDR de programmation".

looses the benefits of the implicit expression of parallelism and comes close to the traditional task-oriented languages. The implicit approach encounters the difficulties of the dynamic load-balancing strategies [6, 7].

So, we propose to explore an alternative approach focussed on data-types rather than on control-structures, through the concept of *fabric*, embedded into a declarative programming style. This new structure, allows the programmer to write programs as mathematical expressions and to *implicitly* express *control* **and** *data* parallelism.

In the next section, we briefly detail the concepts of collection, stream and fabric needed to understand the concepts and examples appearing in the paper (see [2] for a complete description of the language).

1.2 A Brief Introduction to the $8_{1/2}$ Concepts

The concept of Collection in $8_{1/2}$. A collection is a data structure that represents a set of elements *as a whole* [8]. From the point of view of the parallel implementation, the elements of a collection are distributed over the processing elements (PEs).

Here we consider collections that are *ordered* sets of elements. An element of a collection, also called a *point* in $8_{1/2}$ can be accessed through an index (the $T.\text{`}n\text{'}$ operation gives the nth point of T) or a label. If necessary, the type system implicitly and automatically coerces a collection with one point into a scalar and vice-versa [1].

Geometric operators change the *geometry* of a collection, i.e. its *structure*. The geometry of the collection is the hierarchical structure of point values. Collection nesting allows multiple levels of parallelism and can be found, for example, in ParalationLisp and NESL. It is possible to *pack* fabrics together: the $\{a, b\}$ expression computes a nested collection from the collections a and b. Elements of a collection may also be named and the result is a *system*. Assuming $rectangle = \{height = 5, width = 3\}$ the elements of this collection can be reached through the *dot* construct using either their label, e.g. $rectangle.height$, or their index: $rectangle.\text{`}0\text{'}$.

The *concatenation* operator # (also called and "amalgam", see Sect. 2.2 for the use of this operator in symbolic computations) concatenates the values and merges the systems: $box = rectangle \# \{length = 3\} \Longrightarrow \{height = 5, width = 3, length = 3\}$.

Four kinds of function applications can be defined. The first one, the *application*: $f(c_1, \ldots, c_n)$ is the standard function application. The second one is the *extension*: $f\hat{\ }(c_1, \ldots, c_n)$ produces a collection whose elements are the "pointwise" application of the function to the elements of the arguments. For instance, using a scalar addition, we obtain an addition between collections. Extension is implicit for the basic operators $(+, *, \ldots)$ but is explicit for user-defined functions to avoid ambiguities between application and extension. The third type of function application is the *reduction* : $f\backslash c$. Reduction of a collection using the binary scalar addition, results in the summation of all the elements of the collection. The last function application is the *scan*: $f\backslash\backslash c$, which application mode is similar to the reduction but returns the collection of all partial results. For instance: $+\backslash\backslash\{1,1,1\} \Longrightarrow \{1,2,3\}$. Reductions and scans can be performed in $O(\log_2(n))$ steps on SIMD architecture, where n is the number of elements in the collection, if the number of PEs is greater than n.

The Concept of Stream in 8$_{1/2}$**.** Streams in 8$_{1/2}$ are infinite series of values as in LUCID [9]. Streams in 8$_{1/2}$ are computed in a strict ascending order, and at a given instant of the computation, there is always only one value (the "current" value) of the stream stored in the memory. No dynamic allocation of memory nor garbage-collector is required.

Two streams may have different *clocks*, that is, their elements are not computed at the "same speed"; it is nevertheless possible to perform operations between them. Here, we assume that all streams share the same clock (the operator X *when* Y is used to constraint the clock of the stream X to be that of Y). The concept of stream in 8$_{1/2}$ is close to the synchronous stream found in LUSTRE [10] and SIGNAL [11].

8$_{1/2}$ expresses relations between data, it does not describe how to produce them. For instance, the definition $C = A + B$ means that the stream C is always equal to the sum of values in the stream A and B (we assume that the changes of the values are propagated instantaneously). When A (or B) changes, so does C at the same logical instant.

Scalar operations are extended to denote elementwise application of the operation on the values of the streams. The delay operator, $, shifts the entire stream to give access, at the current time, to the previous stream value. This operator is the only operator that does not act in a pointwise fashion.

Fabrics: a New Data Structure for the Declarative Simulation of Time-Evolving Processes. A *fabric* is a *stream of collections* or a *collection of streams*. In fact, we have to distinguish between two kinds of fabrics: *static* and *dynamic*. A static fabric is a collection of streams where every element has the same clock. It is equivalent to say that, a static fabric is a stream of collections where every collection has the same geometry. Fabrics that are not static are called dynamic. The compiler detects the kind of the fabrics and accepts the static ones. At that time, programs involving dynamic fabrics are interpreted.

8$_{1/2}$ is a declarative language: a program is a set of equations representing a set of fabric definitions. A fabric definition has a syntax similar to $T = A + B$. This equation is an expression defining the fabric T from the fabric A and B (A and B are the parameters of T). This expression can be read as a *definition* (the naming of the expression $A + B$ by the identifier T) as well as a *relationship*, satisfied at each moment and for each collection element of T, A and B.

Running an 8$_{1/2}$ program consists in solving the fabric equations. Solving a fabric equation means "enumerating the values of the fabric". This set of values is structured by the stream and collection aspects of the fabric: let a fabric be a stream of collections; in accordance to the time interpretation of streams, the values constituting the fabric are enumerated in the stream's ascending order.

Therefore, running an 8₁/₂ program means enumerating, in sequential order, the values of the collections making the stream. The enumeration of the collection values is not subject to some predefined order and may be done in parallel.

1.3 Example: Three Ways of Computing a Factorial

The paradigmatic example of the computation of a factorial is used to illustrate the possibilities of 8₁/₂. Through the same example, we exhibit the expression of sequentiality, recursion and data-parallelism. It is also an example of three different programming styles.

The Iterative Way. The first way of computing a factorial is to enumerate the values of the function in time, that is:

$fact@0 = 1;$ $\quad fact = counter * \$fact;$
$counter@0 = 1;$ $\quad counter = \$counter + 1$ when $Clock;$

$counter$ is a stream that enumerates the integers at the speed of $Clock$. The quantified equation $counter@0 = 1$ gives the initial value of the counter. In this example, $n!$ is computed as the nth value of the fabric $fact$. This way of computing factorial is iterative. There is no parallelism to be exhibited because the stream elements are computed sequentially (and $fact$ cannot be computed in parallel with $counter$ because of data dependences).

The Space Mapping of Data: the Use of Collections. The second way of computing a factorial relies on collections: $iota[n] = +\backslash\backslash 1$ computes a vector of size n with element i equal to $(i+1)$: the scalar constant 1 is implicitly coerced into a vector of n elements of value 1 (see [1]) and then scanned using the $+$ operation. It is then possible to define $fact$ as: $fact = *\backslash\backslash iota$. The pth element of vector $fact$ is $p!$. This definition exhibits data-parallelism (in the scan operations) and has complexity of $log(n)$ in a SIMD implementation [12].

The Recursively Defined Collection. The third way of computing a factorial is also in space, using a recursively defined collection: $fact = 1 \# (fact : [n-1] * iota)$ where $: [n]$ is the *take* operator which truncates (or extends, if needed) its argument to size n. To convince ourselves that this expression really computes the factorial values, we can see that (using transparential referency):

$fact . \ `0´ \equiv 1$ \qquad (because of $\#$)
$fact . \ `i´ \equiv (1 \# (fact : [n-1] * iota)) . \ `i´$ \qquad (and subsequently, for $i > 0$)
$\equiv (fact : [n-1] * iota) . \ `(i-1)´$
$\equiv fact . \ `(i-1)´ * iota . \ `(i-1)´$ \qquad (extension of $*$)
$\equiv fact . \ `(i-1)´ * i$ \qquad (value of $iota$) □

Note that although computed as a collection, this definition of factorial has a linear complexity because there are dependencies between the elements which induce a sequential order of computation.

2 Introducing Dynamicity in $8_{1/2}$

The three previous examples involve static fabrics, that is, fabrics with collections of fixed geometry (see Sect. 1.2) defined before execution. The original restriction to static fabric was motivated by the effective description and implementation of a class of problems: the problems that have a static behaviour that could be known at compile-time [13].

Nevertheless, this restriction is too firm to describe a whole class of phenomena: the phenomena described by systems with a dynamical structure (modelisation of plant growing, morphogenesis, ...). To describe, manipulate and simulate those dynamical processes, we propose an extension to the static fabrics: dynamically shaped fabrics.

2.1 Dynamic Collections in $8_{1/2}$

Pascal's Triangle. In this example, we use a dynamically shaped fabric to accommodate a combinatorial data structure. The value of the point $(line, col)$ in the triangle is the sum of the point value $(line - 1, col)$ and point value $(line - 1, col - 1)$. If we decide to map the rows in time, the fabric representation of Pascal's triangle is a stream of growing collections. We can identify that the row l ($l > 0$) is the sum of row $(l - 1)$ concatenated with 0 and 0 concatenated with row $(l - 1)$. The $8_{1/2}$ program with its 4 first values is:

$t@0 = 1;$
$t = (\$t \# 0) + (0 \# \$t)$ when $Clock;$

Top : 0 : $\{1\}$: int[1]
Top : 1 : $\{1, 1\}$: int[2]
Top : 2 : $\{1, 2, 1\}$: int[3]
Top : 3 : $\{1, 3, 3, 1\}$: int[4]

Eratosthenes's Sieve. The *Eratosthenes's sieve* is a paradigmatic example of the use of dynamically created tasks in the concurrent programming style: a task is associated to each prime number and linked to the previous tasks, to increase a filter. We describe here an alternative solution, in the data-parallel style, using dynamically shaped collections.

The program used to compute prime numbers consists of a generator producing increasing integers and a collection of known primes numbers (starting with the single element 2). Whenever a new number is generated, we try to divide it with all previously computed prime numbers (a number that is not divisible by a prime number is a prime number itself and is added to the list of prime numbers). *generator* is a fabric that produces a stream of integers. *extend* is a vector with the same size as the collection of already computed prime numbers. *modulo* is a fabric where each element is the modulo of the produced number and the prime number in the same column. *zero* is the fabric containing boolean values that are **true** whenever the number generated is divisible by a prime number. Finally, *reduced* is a reduction with an *or* operation, which result is **true** if one of the computed prime numbers divides the generated number. The $x : |y|$ operator shrinks the fabric x to the rank specified by y. The rank of a collection x is a vector where the ith element represents the number of elements of x in the ith dimension. Table 1 presents the details of the computation of prime numbers following Eratosthene's method.

$generator@0 = 2;\qquad generator = \$generator + 1$ when $Clock$;
$extend \qquad\qquad = generator : |\$sieve|;$
$modulo \qquad\qquad = extend \,\%\, \$sieve;$
$zero \qquad\qquad\, = (modulo == (0 : |modulo|));$
$reduced \qquad\quad = or\backslash zero;$
$sieve@0 \qquad\quad\, = generator;\quad sieve = \$sieve \,\#\, generator$ when (not $reduced$);

In this example, data-parallelism is found in the extension of the $==$ operator, *modulo*, in reductions, etc. There is no control-parallelism because *sieve* depends on *reduced* which depends itself on *zero*, *modulo*, *extend* and finally *generator*. Note that in the data-parallel version, the amount of parallelism grows with the size of the collections. In the concurrent programming version, the speedup is due to the pipeline effect between the tasks associated to the primes; this pipeline effect also grows with the prime numbers.

	0	1	2	3	4	5
generator	{2}	{3}	{4}	{5}	{6}	{7}
extend		{3}	{4, 4}	{5, 5}	{6, 6, 6}	{7, 7, 7}
modulo		{1}	{0, 1}	{1, 2}	{0, 0, 1}	{1, 1, 2}
zero		{0}	{1, 0}	{0, 0}	{1, 1, 0}	{0, 0, 0}
reduced		{0}	{1}	{0}	{1}	{0}
sieve	{2}	{2, 3}	{2, 3}	{2, 3, 5}	{2, 3, 5}	{2, 3, 5, 7}

Table 1. The computation of the Eratosthene's sieve.

2.2 Symbolic Values in $8_{1/2}$

We have seen, in the two previous examples, the possibility brought by the dynamically shaped fabrics. These new possibilities have been made possible by the removal of the static constraint on the fabrics. Furthermore, in $8_{1/2}$, equations defining fabrics have to involve only *defined* identifiers. Equations like $T = a + 1$; or $U = \{a = b + c, b = 2\}$; are rejected because they involve identifiers (a in the first example and c in the second) with unknown values; these variables are usually referred to as *free variables* (the same would happen with more complex equations as long as identifiers appearing in the right hand-side of a definition do not appear in a left hand-side of another definition in an enclosing scope). We see that with little more work in the definition of the language, releasing the constraint of allowing only closed equations, could lead us to define equations with values of *symbolic* type. This extension, and its relevance to "classical" symbolic processing, is presented in the next section.

We only have seen numerical *systems* so far, that is, collections with elements of numerical value (possibly accessible through a label). We consider now that a free variable has a symbolic value: namely itself. A symbolic expression is an expression involving free identifiers or symbolic sub-expressions. Such a symbolic expression is a first citizen value although it is not a numerical. An expression

E involving a symbolic value evaluates to a symbolic value except when the expression E provides the missing definitions.

For example, assuming that S has no definition at top-level, equation $X = S+1$; defines a fabric X with a symbolic value. Nevertheless, equation $E = \{S = 33; X\}$; evaluates to $\{33, 34\}$ (a numeric value) because E provides the missing definition of S to X. Remark that the evaluation process in 8$_{1/2}$ always tries to *evaluate* all numerical values.

Factoring Computations: Building Once and Evaluating Several Times a Power Series. A wide range of series in mathematics require to compute a sequence of symbolic computation (e.g. a Taylor series) and then to instantiate the sequence with numerical values to get the desired result. We exemplify this through the computation of the exponential: $e^x = 1 + x + x^2/2! + x^3/3! + \ldots$ The 8$_{1/2}$ program computing the symbolic sequence is:

```
n@0    = 0.0;   n    = $n + 1.0 when Clock;
fact@0 = 1.0;   fact = n * $fact when Clock;
term@0 = 1.0;   term = ($term * x) when Clock;
exp@0  = 1.0;   exp  = ($exp + term/fact) when Clock;
```

The symbolic value *exp* corresponding to the series and computed only once, is completed in a local scope and accessed through the dot operator: $e = \{x = 1.0; val = exp\}$. *val*. This method factorizes the computation of the call-tree and can be used to a wide range of sequence of the same type. Once the initial computation of the symbolic "tree of computations" has been achieved, various results can be computed very easily through an "instantiation-like" mechanism.

3 Conclusions and Future Work

The examples in Sect. 2 have shown that the expressivity of dynamically shaped fabrics with symbolic values is fairly efficient to express some paradigmatic examples in symbolic processing. In addition, 8$_{1/2}$ is able to concisely express standard numerical processing problems, like numerical resolution of partial differential equations [14]. The general idea is to use more specific and sophisticated data-types to ease the programmer's life. Nevertheless, further experimentations have to be done to comfort the relevance of this approach.

A compiler for the static subset of 8$_{1/2}$ has already been implemented. All the compiler phases assume a full MIMD execution model and we are currently working on the MIMD code generation. The static examples of this article have been processed by the existing compiler whereas the dynamic ones have been interpreted by a sequential interpreter which triggers low-level vector operations (currently implemented in C as a virtual SIMD machine). Data-parallelism could be exploited by just adapting the low-level virtual machine. The current work on the 8$_{1/2}$ language concerns the extension of the notion of collection (towards a group structure [15]), the efficient treatment of dynamically shaped fabrics and their relations to symbolic computation.

References

1. J.-L. Giavitto. Typing geometries of homogeneous collection. In *2nd Int. workshop on array manipulation, (ATABLE)*, Montréal, 1992.
2. O. Michel. Design and implementation of 81/2, a declarative data-parallel language. *special issue on Parallel Logic Programming in Computer Languages*, 1996. (to appear).
3. E. V. Zima. Recurrent relations and speed-up of computations using computer algebra systems. In *International Symposium, DISO'92, Bath, U.K.*, number 721 in Lecture Notes in Computer Sciences. Springer Verlag, April 1992.
4. P. Fritzson and N. Andersson. Generating parallel code from the equations in the ObjectMath programming environment. In *Second international ACPC conference, Gmunden, Austria*, number 734 in Lecture Notes in Computer Sciences. Springer Verlag, October 1993.
5. J.-L. Giavitto, C. Germain, and J. Fowler. OAL: an implementation of an actor language on a massively parallel message-passing architecture. In *2nd European Distributed Memory Computing Conf. (EDMCC2)*, volume 492 of *LNCS*, Münich, 22-24 April 1991. Springer-Verlag.
6. M. Lemaitre, M. Castan, M.H. Durand, G. Durrieur, and B. Lecussan. Mechanisms for efficient multiprocessor combinator reduction. In *Proc. of the 1986 ACM Conference on LISP and Functionnal Programming*, pages 113–121, Cambridge, Ma., August 1986. ACM.
7. B. Hunberman and T. Hog. *The ecology of computation*, chapter The behavior of computationnal ecologies. Studies in computer science and artificial intelligence. North-Holland, 1988.
8. J. Sipelstein and G. E. Blelloch. Collection-oriented languages. *Proc. of the IEEE*, 79(4), April 1991.
9. E. Ashcroft, A. Faustini, R. Jagannatha, and W. Wadge. *Multidimensional Programming*. Oxford University Press, February 1995. ISBN 0-19-507597-8.
10. P. Caspi, D. Pilaud, Halbwachs N., and J. Plaice. Lustre: a declarative language for programming synchronous systems. In *Fourteenth annual symposium on principles of programming languages*. ACM, ACM Press, January 1987.
11. P. Le Guernic, A. Benveniste, P. Bournai, and T. Gautier. Signal, a dataflow oriented language for signal processing. *IEEE-ASSSP*, 34(2):362–374, 1986.
12. W. D. Hillis and G. L. Steele. Data parallel algorithms. *Communication of the ACM*, 29(12), December 1986.
13. F. Cappello, J.-L. Béchennec, and J.-L. Giavitto. PTAH: Introduction to a new parallel architecture for highly numeric processing. In *Conf. on Parallel Architectures and Languages Europe, Paris, LNCS 605*. Springer-Verlag, 1992.
14. O. Michel, J.-L. Giavitto, and J.-P. Sansonnet. A data-parallel declarative language for the simulation of large dynamical systems and its compilation. In *SMS-TPE'94: Software for Multiprocessors and Supercomputers*, Moscow, 21-23 September, 1994. Office of Naval Research USA & Russian Basic Research Foundation.
15. J.-L. Giavitto, O. Michel, and J.-P. Sansonnet. Group based fields. In I. Takayasu, R. H. Jr. Halstead, and C. Queinnec, editors, *Parallel Symbolic Languages and Systems (International Workshop PSLS'95)*, volume 1068 of *LNCS*, pages 209–215, Beaune (France), 2-4 October 1995. Springer-Verlag.

Astro-Gofer: Parallel Functional Programming with Co-ordinating Processes

Andrew Douglas, Niklas Röjemo, Colin Runciman and Alan Wood

Department of Computer Science*, University of York,
Heslington, York YO1 5DD

Abstract. This paper investigates the addition of operations for explicit parallelism to the lazy functional language, Gofer. For this purpose, we choose to use a co-ordination language, derived from Linda, based on logically shared associative memories, called *spaces*.

We describe the use of Gofer type classes in building an interface to spaces, which highlights and extends the traditional associative access mechanism. We then show how processes can be added to Gofer, and how we can combine spaces and processes to produce a parallel functional language, which we have called Astro-Gofer.

1 Introduction

In this paper, we describe a parallel lazy functional programming language based on the notion of co-ordination[4]. The functional language used is Gofer[8], a dialect of Haskell in which type classes may have multiple type parameters. The co-ordination language is based on shared associative memories called *spaces*, derived from Linda[4].

2 Simple Spaces in Gofer

We begin with a preliminary exploration of how we can include a kind of associative structure into Gofer. This associative structure will be used initially for sequential programming, but will motivate our parallel extension of Gofer in future sections.

We use Gofer type classes to enforce typed interaction with our associative structure. A type class associates a number of *methods* with one or more type parameters. An *instance* of a class declares a type to be a member of the class, and gives function definitions for the class methods. Although method names are overloaded and functions using methods can be polymorphic over the types declared as instances, correct method implementations are chosen for a particular type. The existence of a suitable method is guaranteed by compile time checks.

An association between two types of values, one of which (the *template* type) is used to retrieve the other (the *object* type), is represented by an instance of a

---

* For correspondence, wood@minster.york.ac.uk

class `Match`, with two type parameters. The method of the class, `match`, tells us how we can match objects of the required object-type and the template-type.

```
class Match o t where match :: o -> t -> Bool
```

One simple instance matches by equality between basic values of the same type:

```
instance Match Int Int where match a b = (a == b)
```

Generalising, templates of type `Maybe Int` can be used to specify *either* a specific integer, *or any integer*.

```
data Maybe a = Any | Just a
```

```
instance Match Int (Maybe Int) where
        match n Any = True
        match n (Just m) = (n == m)
```

As another example, we can match template-structures against object-structures component by component. For example, triples:

```
instance (Match a d, Match b e, Match c f) =>
        Match (a,b,c) (d,e,f) where

        match (a,b,c) (p,q,r) = (match a p) && (match b q)
                                            && (match c r)
```

This declaration states that `Match (a,b,c) (d,e,f)` is an instance of the class, for all types `a, b, c, d, e` and `f`, provided that (as specified to the left of the `=>`) `Match a d`, `Match b e` and `Match c f` are also instances of the class.

So now we can have templates such as `(Any :: Maybe Int, 10, 6)` which matches any integer triple whose second and third components are 10 and 6, but whose first component is unconstrained apart from its type.

Finally, we could introduce more flexibility into matching, through constraints:

```
instance Match a (a -> bool) where
        match n f = f n
```

This extends the traditional Linda style of matching to include templates such as `((<99), 10, 6)`, which matches any integer triples whose second and third components are 10 and 6, but whose first component is constrained to be less than 99.

2.1 Simple space simulation

The `Match` class can be used to define computation in the context of several named *spaces*. Spaces are multisets containing values of a variety of object-types, together with operations to add new values and to retrieve values matching a given template.

Consider first a sequential implementation of uniformly-typed spaces, using *continuation-passing style*, a standard means of expressing sequences of operations in a purely functional language.

```
type Space o  - multisets of values of object-type o
type Name     - names, eg integers
type Table o  - associations of Names to Spaces
type Cont o = Table o -> Table o
```

First, there must be a means of generating a new (empty) space with a new name. The function is given a continuation to which is passed the identifier of the new space.

```
newspace :: (Name -> Cont o) -> Cont o
```

The simplest continuation is `stop`, which does nothing.

```
stop :: Cont o
stop = id
```

We can place things into a space using `put`:

```
put :: Name -> o -> Cont o -> Cont o
```

We also want to retrieve items from a space. It is this function, `get`, that relies on the class `Match o t` (being polymorphic only over those types which are instances of `Match`). Given a template of type `t`, a matching item of object-type `o` is retrieved from a named space. If no matching item is found, then the expression evaluates to $\bot$.

```
get :: Match o t => Name -> t -> (o -> Cont o) -> Cont o
```

The $\bot$ semantics for a failed match might appear to be a little drastic. In section 4, when we use spaces shared between processes, the reason for our choice will become clear.

Finally, we include two operations which act *between* spaces. The first of these, `collect`, *moves objects matching a template from one space to another*. The continuation is passed the number of items moved – which may be zero.

```
collect :: Match o t =>
           Name -> Name -> t -> (Int -> Cont o) -> Cont o
```

The `copy_collect` operation behaves like `collect`, except that the items matching the template are *copied* rather than moved.

```
copy_collect :: Match o t =>
                Name -> Name -> t -> (Int -> Cont o) -> Cont o
```

The concept of *space*, together with the operations `put` and `get`, is derived from the *tuple space* concept in Linda [4]. The *collect* primitive is justified in [1], while the *copy_collect* primitive has been shown to be indispensable for maximising concurrency in certain algorithms [10]. Although no detailed semantics is given, it is probable that this sequential system will be deterministic.

```
single    :: Name -> Int -> Name -> Cont Int -> Cont Int
dolevels  :: Name -> Cont Int -> Int -> Name -> Cont Int
histogram :: Name -> Int -> (Name -> Cont (Int,Int,Int))
                         -> Cont (Int, Int, Int)

single img level hist cont
    = collect img img (Any :: Maybe Int, Any :: Maybe Int, level) $
       \occurs -> put hist (level, occurs, 0) $
                  cont

dolevels img cont 0     hist = cont ts
dolevels img cont (n+1) hist = single img n hist (dolevels n ts)

histogram img maxgrey cont = newspace $ \newts ->
                              dolevels img cont maxgrey newts
```

Fig. 1. Histogram program in Gofer extended with spaces.

2.2 An example space program

We give a brief example of how space operations can be used: producing a histogram of an image (see **Fig.1.**).

An image is represented by a space, containing triples; each triple represents a co-ordinate together with the grey level to which it is set in the image. For the moment, all spaces must hold objects of the same type, so a histogram of an image is also represented by a space of triples, where the first component represents a grey level, the second component represents the number of pixels set to that grey level in the image, and the third is always 0.

The `histogram` function takes the name of the image space, and an integer representing the number of grey levels allowed in the image. It also takes a continuation, to which is passed the name of the histogram space. The function creates a new space identifier, which it passes to `dolevels`.

The `dolevels` function creates a histogram entry for each grey level. Given an integer n, it performs `single` on the values 0 to n − 1, placing the results in space `hist`.

`single` is given a grey level value, the names of the image and the histogram spaces, and a continuation. The function will `collect` the pixels from the image space *into the image space*. This way, the pixels of the appropriate grey level value are counted, but not removed from the image. This value is then placed into the histogram space, tagged with the grey level value, and then the function behaves as the continuation.

An important deficiency of the above sequential system is that, although we can have multiple spaces, these all have to be of the same type. This will be remedied by hiding the implementation of spaces within the Gofer system. The result will be strongly and statically typed, through the use of Gofer type classes. In section 4.1 we will show that building a concurrent (parallel) function to create a histogram requires very little modification to the above program.

3 Processes and Gofer

We have considered spaces in a sequential programming context. However, spaces are intended for use in the context of concurrent processes, where they are used for communication and co-ordination.

Processes communicate through common access to named spaces. These named spaces constitute a *global state* which is hidden from the user by the Gofer system. Each *individual* process will have its own *world state*, which is acted upon by continuations – this enforces a single thread of control.

The type of a process is as follows:

```
type Process a = a -> a
```

A process is a state transformer, which, given a state of type a (the process's world state), returns a new state after performing some action. An action will affect the world state, and might also affect the global state.

Two functions are provided; the first provides a mechanism for spawning processes:

```
spawn :: Process a -> a -> Process b -> Process b
```

where spawn P s Q creates a new process which behaves as P, with initial state s, before continuing as the process Q.

The second function returns a process which terminates a computation:

```
stop :: Process ()
```

Operations for communication are provided in the next section.

4 Processes + Spaces = Astro-Gofer

This section combines spaces with processes to produce Astro-Gofer. With the introduction of processes, the *global state*, the named spaces, will be shared between *all* processes. So the implementation of spaces and their associated operations become hidden in the underlying Gofer system. This conveniently means that we can remove the typing restrictions on spaces, so that they can hold objects of a variety of types. However, Astro-Gofer is still strongly and statically typed, because of the Match type class.

In the place of the type Cont t, we use the Process a type to create a notion of threading. The implementation of spaces and the type Name is hidden from the user, as is the implementation of the space operations. The first of these is newspace, whose type becomes:

```
newspace :: (Name -> Process a) -> Process a
```

As before, the put function places an object into a space, and then continues. This implies that process communication is *asynchronous*.

```
put :: Name -> a -> Process b -> Process b
```

Objects are removed from a space by the operation get:

```
get :: Match o t =>
        Name -> t -> (o -> Process a) -> Process a
```

The operation get uses a template of type t to retrieve an object of type o from space. The operation will *block* (no result will be returned) until such an item can be found. This reflects the ⊥ semantics given to get in section 2.1. As a result of adding processes, the semantics of get is now non-deterministic (that is, given many matching objects, it is not determined which will be chosen, and given several processes competing for an object, it is not determined which will get it).

The types given to collect and copy_collect are as follows:

```
type Rep a = Int

collect :: Match t u => Space -> Space ->
           u -> (Rep t -> Process a) -> Process a

copy_collect :: Match t u => Space -> Space ->
                u -> (Rep t -> Process a) -> Process a
```

The type Rep a is necessary because Gofer requires that both types of a two-type class must appear in the type of the function.

4.1 Example space program - revisited

We return to our histogram example, and show how we can now take advantage of processes (see **Fig.2.**).

```
single    :: Name -> Int -> Name -> Process a  -> Process a
dolevels  :: Name -> Process a -> Int -> Name  -> Process a
histogram :: Name -> Int -> (Name -> Process a) -> Process a

dolevels img cont 0     hist = cont ts
dolevels img cont (n+1) hist = spawn (single img n hist stop)
                                     (dolevels n ts)
```

Fig. 2. Concurrent histogram program in Astro-Gofer

The types of the functions have changed to reflect the addition of processes. On each recurse of the function dolevels, a new process is created to calculate the histogram entry for the grey level value. So, for n grey level values, n processes are spawned.

The histogram is known to be complete when n triples have appeared in the histogram space, which can be checked easily. However, a consumer might not need to wait for the histogram to be complete before it begins to consume the histogram values.

5 Conclusion

We have described an approach to providing explicit parallelism in a lazy functional language, using Linda-like co-ordination. This is quite different to other attempts at providing explicit parallelism within a functional setting, which use message passing. Our approach is highly asynchronous and decoupled, and in this respect, it is closest to Concurrent Haskell [5].

No other attempt to provide *space-based* parallelism within a lazy functional language has been undertaken, and the extendibility of the language (in its matching capabilities) and the inclusion of laziness within spaces is novel.

Various Linda-based systems have been undertaken in strict functional languages, such as Lisp [7], which has untyped tuple spaces, and ML[11]. The latter takes an approach which is rather static, with fixed algebraic types representing tuples and templates. This provides little typing information other than *this is a tuple* or *this is a template*. None of these languages has capabilities for extending matching.

5.1 Semantics

A two-part semantics for Astro-Gofer can be given. One part provides a functional semantics for individual processes, another deals with the interaction of processes.

An individual process can be seen as an interactive function, which produces a list of requests, and receives a list of replies to these requests. We could, then, give a semantics to processes which is of the kind used for describing IO [6].

The second level of the semantics of Astro-Gofer is given by the semantics for spaces. The semantics of Linda, as given in [2], can be used as a foundation for spaces. Difficulty comes when describing the two new operations, collect and copy_collect – a concurrent semantics is quite straight-forward, but when we consider truly parallel implementations such as [3], an intuitive semantics is much more difficult to describe.

A consequence of using continuations is that processes are *single threads of control which cannot interact*. The only way of performing space operations is as part of one of these threads. Consequently, no value expressions can perform space operations, and no space operations can be performed by the process of matching, or by items evaluated while in space. We can pass functions through space, and hence we can pass objects of type Process a – these can only be evaluated as part of a process.

Items within space may be evaluated only once but used many times (true laziness in space), or they may be copied and distributed as part of the underlying system, and evaluated many times (loss of laziness).

Finally, we note that the language Astro-Gofer has a model of processes and asynchronous communication which results in the following properties characteristic of a co-ordination language.

Spatial decoupling: a process knows little or nothing about other processes.

Temporal decoupling: communicating processes need not overlap temporally.

5.2 Implementation

The current implementation of Astro-Gofer is concurrent, being based on a version of Gofer extended with processes[12]. The scheduling strategy is simple – evaluate a process until a space operation is performed, and then swap. A space mechanism can be built quite simply.

We intend to build a parallel implementation to exploit networks of workstations. Recent developments in implementation technology [9] show that parallel implementations can be very efficient.

References

1. P. Butcher, A. Wood, and M. Atkins. Global synchronisation in Linda. *Concurrency: Practice and Experience*, 6(6):505 – 516, 1994.
2. P. Ciancarini, K. Jenson, and D. Yankelevich. On the operational semantics of a co-ordination language. In *ECOOP'94*, LNCS 924. Springer Verlag, 1995.
3. A. Douglas, A. Rowstron, and A. Wood. Linda implementation revisited. In *Transputer and occam Developments*. IOS Press, 1995.
4. D. Gelernter. Generative communication in Linda. *ACM Transactions on Programming Languages*, 7(1), January 1985.
5. A. Gordon, S. L. Peyton-Jones, and S. Finne. Concurrent Haskell. In *ACM Symposium on Principles of Programming Languages*. ACM, 1996.
6. A. D. Gordon. *Functional Programming and Input/Output*. Distinguished Dissertations in Computer Science. Cambridge University Press, 1994.
7. S. Jagannathan. Customisation of first-class tuple spaces in a higher-order language. In *PARLE*, LNCS 506. Springer Verlag, 1991.
8. M. P. Jones. The implementation of the Gofer functional programming system. Technical Report YALEU/DCS/RR-1030, Yale University, 1994.
9. A. Rowstron and A. Wood. An efficient distributed tuple space implementation for networks of workstations. In *EuroPar'96*, LNCS. Springer Verlag, 1996.
10. A. Rowstron and A. Wood. Solving the Linda mutliple rd problem. In *Coordination Languages and Models*, LNCS 1061. Springer Verlag, 1996.
11. E. H. Siegel and E. C. Cooper. Implementing distributed Linda in Standard ML. Technical Report CMU-CS-91-151, Carnegie Mellon University, 1991.
12. M. Wallace. *Functional Programming and Embedded Systems*. PhD thesis, Department of Computer Science, 1995. Also Technical Report YCST95/04.

Multiple OR-Parallel Resolution: Meta-Level Control of Parallel Logic Programs

Petros Kefalas

Ioannis Vlahavas

Department of Computer Science
City Liberal Studies, 13 Tsimiski Str
54624 Thessaloniki, Greece
kefalas@hyper.gr

Department of Informatics
Aristotle University of Thessaloniki
54006 Thessaloniki, Greece
vlahavas@olymp.ccf.auth.gr

Abstract. Multiple OR-parallel Resolution (MORE) Prolog is a combination of a pure logic language and control directives expressed as a meta-program. The meta-program affects the default resolution strategy by suspending execution of particular predicates, ordering the suspended processes and selectively reactivating them, thus achieving the desired kind of resolution. In this paper, we formally define the computation process of MORE-Prolog and illustrate how a set of primitive directives could be combined, leading in effect to application of different parallel search algorithms over the same state space. Finally, the effectiveness of MORE-Prolog is demonstrated by presenting different meta-programs which result in different performance if applied on the same logic program.

1 Introduction

Logic programming languages are highly suited to the type of symbolic computation found in Artificial Intelligence applications. A logic program is a set of *Horn Clauses* of the form: $H \leftarrow B_1 \wedge B_2 \wedge ... \wedge B_n$ $(n \geq 0)$ where H and $B_1,...,B_n$ are predicates, H is the *head* and $B_1 \wedge B_2 \wedge ... \wedge B_n$ is the *body* of the clause.

A Horn clause has its *declarative* reading, i.e. *H is true if B_1 is true and ... and B_n is true*, but could also be interpreted in a *procedural* way: *in order to prove H, prove B_1 and ... and prove B_n*. A logic program does not impose any ordering of Horn clauses or any ordering of subgoals within the body of the clause. The execution requires an extra control mechanism applied to the program. In an ideal logic language, the *resolution strategy*, i.e. how the program is executed, is completely separated from the logic [1]. In practice, however, control primitives are introduced to the syntax of a logic language, thus altering the declarative reading of the program. For example, the sequential Prolog language uses a fixed depth-first left-to-right search to resolve the initially set goal (*query*).

Parallelism is inherent in logic programs. There are mainly two types of parallelism, namely *OR-parallelism* (a goal is resolved by many different program clauses) and *AND-parallelism* (a conjunction of goals in the body of a clause is resolved in parallel). Recent implementations show that parallel execution of logic programs can result into efficiently speeding up their execution [2, 3].

1.1 The Resolution Principle of Prolog

Prolog uses LUSH resolution [4], a kind of SLD-resolution [5], which involves chronological *backtracking* in order to explore alternatives. LUSH resolution inherits efficiency but also incompleteness from depth-first search.

A conjunction of subgoals $\leftarrow A_1, A_2, ..., A_q$ which remain to solve the initial goal (query) is called *resolvent*. A query is solved if the resolvent becomes empty ($\leftarrow \varepsilon$). The *Computation State (CS)* is defined by the sequence: $CS = <R_1, R_2, ..., R_z>$ where $R_1, ..., R_z$ are resolvents which are generated during the resolution.

At any stage of the resolution, given the computation state CS, a *resolution* step in Prolog is defined as a three stage operation:
i) Pop the first resolvent $R_1 = \leftarrow A_1, A_2, ..., A_q$ from CS, i.e. $R_1 = head\ CS$.
ii) Generate k new resolvents $R_{11}, ..., R_{1k}$, by finding k clauses whose head match with goal A_1.
iii) Push the k new resolvents in the order that they were generated in CS thus producing a new computation state $CS' = <R_{11}, ..., R_{1k}, R_2, ..., R_z>$

In parallel implementations, OR-Parallel LUSH resolution is used. The *Parallel Computation State (PCS)* is defined as: $PCS = <CS_1, CS_2, ..., CS_n>$, where n can be seen as the number of processors working cooperatively to solve the query. The overall resolution proceeds in an asynchronous way and each processor i resolves its current computation state CS_i as described above.

1.2 Coding Search Algorithms in OR-Parallel Prolog

It is relatively easy to exploit Prolog's built-in execution mechanism to declaratively write applications that perform OR-parallel depth-first search over a problem's state space. Even though parallel depth-first search improves on its sequential counterpart, it still may not find a solution even if one exists. In addition, depth-first search is a naive blind search algorithm and therefore it is not practical for applications with large state spaces that contain infinite branches. However, the same state space could be explored by a different blind search algorithm, such as breadth-first search, iterative deepening or by a heuristic algorithm, such as hill-climbing, best-first search, A* etc. [6] . The burden for the programmers is that they should write a completely different Prolog program for each different type of algorithm. It also turns to be rather procedural to write logic programs that perform any type of search except depth-first, since the programmers need to employ the control features of Prolog.

1.3 An Alternative Approach

This paper shows how control of resolution could be achieved as a separate programming activity. Multiple OR-Parallel REsolution (MORE) Prolog is basically pure Prolog without extra-logical features together with suitable syntax for meta-programs responsible for controlling the parallel execution strategy. The programmer needs only to combine the directives in the meta-program to implement a different search algorithm over a state space, leaving the logic program untouched.

Reasoning about control in the form of meta-rules initially appeared in [9], achieving pruning of the search space. In logic programming, the importance of control over logic programs has been demonstrated, either by altering the left-to-right execution of subgoals of the body of a clause [10], or by altering the ordering of program clauses if certain conditions are satisfied [11, 12]. In all the above cases, the aim was to reduce the expansion of the search tree rather than to facilitate the coding of search algorithms. Also, new built-in predicates were introduced to give priority to proof paths of Prolog execution [13, 14], but with rather unclear declarative semantics. A similar approach to ours, is presented in [15], but with a pre-fixed directive for best-first search only.

In the following sections we provide the syntax and the procedural semantics of MORE-Prolog programs and we present the results of applying different control on the same logic program, for simple search problems.

2 Multiple OR-Parallel Resolution Prolog

MORE-Prolog consists of two parts: an *object program* which is a pure logic program without annotations or control primitives, and a *meta-program*. The meta-program is a sequence of directives to control the OR-parallel execution. If the object program is left alone, it is executed in the default parallel depth-first strategy. In the meta-program, a directive causes suspension of specific calls to annotated predicates. Eventually, all calls may be suspended. Then, other directives impose the reactivation of some suspended processes, the order of reactivation as well as determine the pruning performed to the rest. Execution proceeds in such a way that the overall behaviour mimics parallel variations of breadth or best-first search.

2.1 The syntax of Meta-Program

The syntax of the meta-program is the following:
```
metaprogram(   suspend <predicate(...,X,...)>,     mode <scope>,
               order <ordering>,                    release <reactivation>,
               prune <pruning>    ).
```

<predicate(..,X,.)>	→ any predicate defined in the logic program
<scope>	→ local | global
<ordering>	→ ascending(X) | descending(X) | random
<reactivation>	→ all | best [<N>] | any <N>
<pruning>	→ none | rest | any <N>
<N>	→ any positive integer number | <#p>
<#p>	→ the number of processors

2.2 MORE-Prolog Rewrite Computation Rules

In MORE-Prolog the computation state of processor i (CS_i) is described by the tupple: $CS_i = (<AR_1,...,AR_n>_i, <SR_1,...,SR_m>_i)$, where $<AR_1,...,AR_n>$ and $<SR_1,...,SR_m>$ are sequences of active and suspended resolvents respectively generated during the execution. A resolvent may be transferred to the suspended sequence and vice versa under specific circumstances. Given a query $\leftarrow Q$, the parallel computation state (PCS) is initialised:
$$PCS = <CS_1, CS_2,...,CS_n> = < (<\leftarrow Q>_1, <>_1), (<>_2, <>_2), ..., (<>_n, <>_n)>$$

Suspend. In MORE-Prolog, all predicates not mentioned in the meta-program are executed with the default depth-first execution strategy. However, any call to the predicate annotated with the directive *suspend* in the meta-program should be treated differently. The call to the predicate suspends, and so does the proof path. Execution continues to the rest of the unexplored choices. This is in effect a forced backtracking without, however, undoing the variable bindings made by the suspended proof path. The following rewrite rule applies:
$$(< (\leftarrow predicate(...,X,...),...,A_q), AR_2,...,AR_n >_i, < SR_1,...,SR_m >_i) \Rightarrow$$
$$(< (AR_2,...,AR_n >_i, < NR_1,...,NR_k, SR_1,...,SR_m >_i) \text{ and}$$
$NR_1, ..., NR_k$ are the resolvents generated by $\leftarrow predicate(...,X,...)$

Mode. There are two modes of suspension, namely *local* and *global*. These refer to the situation which should be detected in order to take specific reactivation actions. After exploring all possible alternatives, a processor i suspends when there are only suspended resolvents in its computation state: $\exists CS_i \in PCS: CS_i = (<>_i, Suspended_i)$.

If the mode of suspension is specified as *local*, the processor should reactivate its resolvents in a way described later. If the mode of suspension is specified as *global*, the processor takes no action (except possibly from taking available work from another processor) until a global suspension appears: $\forall CS_i \in PCS: CS_i = (<>_i, Suspended_i)$. Then, a reactivation takes place as described below.

Order. In the case of a local suspension, the suspended calls are ordered according to the directive *order*. This specifies ordering according to argument X of the suspended call (which must be bound to a numeric value), i.e. $(<>_i, Suspended_i) \Rightarrow (<>_i, Suspended_i')$ where $Suspended'$ is the $Suspended$ sequence sorted on *ascending* or *descending* order of X or a *random* permutation (no ordering). In the case of global suspension, the sorting performed concerns all processors' suspended resolvents.

Release. The reactivation field specifies which of the suspended processes should be awakened after ordering. The flag *all* indicates that all suspended calls should be activated: $(<>_i, Resolvents_i) \Rightarrow (Resolvents_i, <>_i)$.
   The flag *best* indicates that the best call is reactivated:
$$(<>_i, <SR_1,...,SR_n>_i) \Rightarrow (<SR_1>_i, <R_2,...,R_n>_i)$$
The user could also specify a number of calls, e.g. the *best N* to be reactivated:

$(<>_i, <SR_1,...,SR_n,...,SR_k>_i) \Rightarrow (<SR_1,...,SR_n>_i, <SR_{n+1},...,SR_k>_i)$ if $k>n$
or $(<>_i, <SR_1,...,SR_n,...,SR_k>_i) \Rightarrow (<SR_1,...,SR_k>_i, <>_i)$ if $k \leq n$.

Similar re-write rule apply to reactivation of *any N*.

In global suspension the flags *best* and *any* have similar effect, but the selection of the *best* or *any* resolvents is made by considering all computation states.

Prune. The last directive of the meta-program specifies how the implementation should treat the calls that remain suspended after the reactivation stage. It should either discard *none*, i.e. $CS_i \Rightarrow CS_i$, or the *rest* of the remaining suspended calls, i.e. $(Resolvents_i, <SR_1,...,SR_n>_i) \Rightarrow (Resolvents_i, <>_i)$. Similar re-write rule apply to pruning *any N*.

In global suspension the flag *any* have similar effect, but the selection of the resolvents is made by considering all computation states.

Task Switching. Under global suspension mode, a processor may become idle when its *ActiveResolvent* list becomes empty, i.e. $IdleCS_i = (<>_i, Suspended_i)$. It could then look for work from another processor's j computation state, as the following rewrite rule indicates: $PCS \Rightarrow PCS'$, where

$PCS = <..., (<>_i, Suspended_i), ..., (Active_j, Suspended_j),...>$ and $\#Active > 1$
$PCS' = <...,(<AR>_i, Suspended_i), ..., (NewActive_j, Suspended_j), ... >$ and
$NewActive = Active - <AR>$ where '-' indicates set difference.

Under local suspension a processor becomes idle only when both lists become empty, i.e. $IdleCS_i = (<>_i, <>_i)$. A similar to the above rewrite rule applies.

The order in which the rewrite rules for suspension and task switching are applied is non determinate. The rewrite rules for ordering, reactivation and pruning are applied as listed.

3 Execution of MORE-Prolog

Consider the search space of Fig.1(a). A simple Prolog program that implements a depth-first search over the state space is the following:

    search(State, []) ← goal(State).   % Goal is reached
    search(State, [State|SolutionPath]) ←
        operator(State, NextState),   % Generate the next state
        search(NextState, SolutionPath). % Keep on searching for the solution.

where predicates *goal/1* and *operator/2* are domain specific logic programs and their definition depends on the problem being solved. For reasons of exposition, these predicate definitions are restricted to simple facts in Fig.1(a).

The solution is returned in form of a list of the visited states, from the initial to the goal state. Given the query: *←initial_state(IS), search(IS,Solution)* OR-parallel Prolog execution imposes any call to predicates *search/2* and *operator/2* to generate alternatives. If there are no available processors, then the normal Prolog execution is employed. If there exist available processors then they pick up alternatives and work in parallel. A part of the execution OR-tree is shown in Fig.1(b), where two processors are used for the execution of the above program.

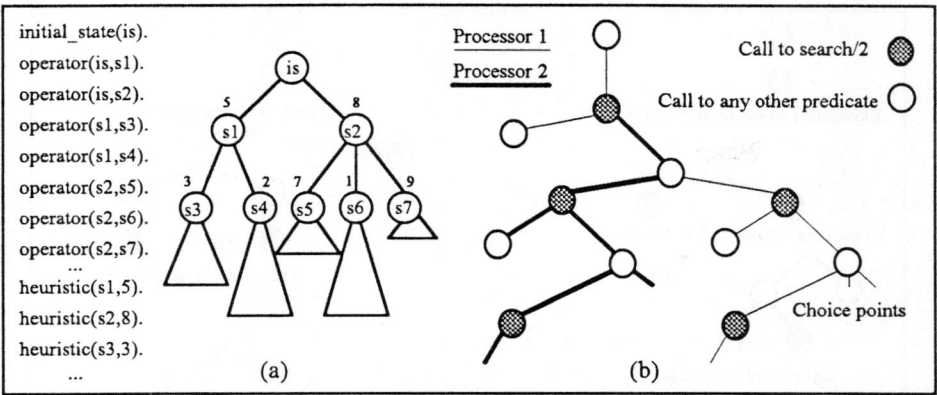

Fig.1. (a) A sample search space of a problem and its corresponding code. (b) A possible OR-parallel execution of the program by 2 processors.

With MORE-Prolog the resolution of the logic program is guided by the meta-program that determines the way in which the execution tree is generated and expanded. Combining the directives defined in the previous section, someone could apply different execution strategies to the above program.

Example 1. One possible strategy is expressed by the following meta-program:
  suspend search(_,_), mode global, order random, release all, prune none.
  Fig.2(a) shows that the execution proceeds according to the default OR-parallel Prolog strategy, suspending any call to *search/2*, until an overall suspension is detected. Then, all suspended calls are reactivated. This is in effect a breath-first search over the search space.

Example 2. In most problems, the state space grows exponentially and blind search is inefficient. There are more sophisticated algorithms which use heuristics to guide the search to the most promising paths. Such search algorithms are not easily coded neither in sequential nor in OR-Parallel Prolog. The following code implements a MORE-Prolog program in a search space where a heuristic function could be used (see heuristic values attached to nodes in Fig.1(a)):
search(State, _,[]) ←goal(State).   % Goal is reached
search(State, HV,[State|SolutionPath])←
    operator(State, NextState),   % Generate next state,
    heuristic(NextState,Value),   % find the heuristic value
    search(NextState, Value, SolutionPath). % Keep on searching.
If the program is left to be executed as listed, it would perform the default parallel depth-first search. However, if the following meta-program is added:
suspend search(_,V,_), mode global, order descending(V), release best, prune rest.
then the execution will perform a hill-climbing search (Fig.2(b)). The same program, but with *prune none* directive instead of *prune rest* will perform a best-first search over the same state space.

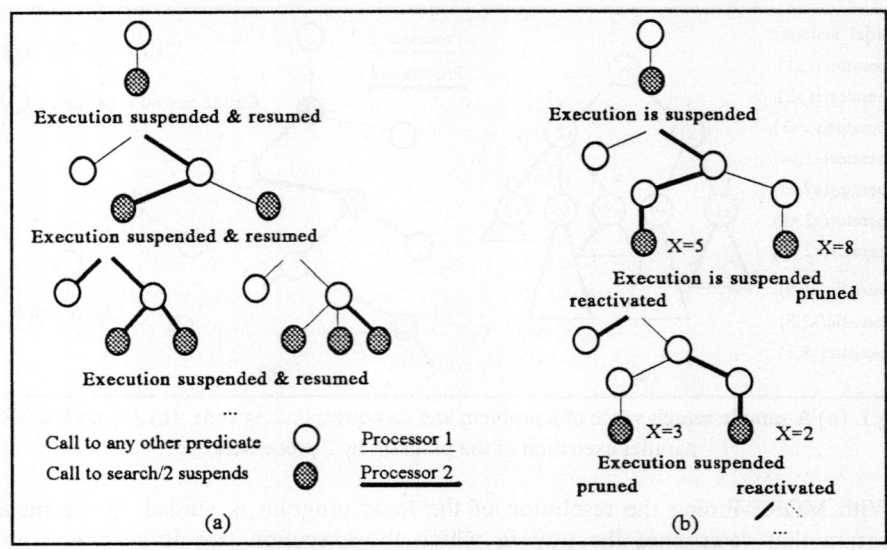

Fig.2: Possible Execution by 2 processors in MORE-Prolog:
(a) Breadth-First. (b) Hill-Climbing

The above meta-program codes have the same effect with the corresponding sequential algorithms. Thus, they do not take too much advantage of a parallel machine, since only one best suspended process is allowed to proceed whereas there could be as many best as the number of processors. On the other hand, since the suspension is *global*, there is an extra overhead due to synchronisation as well as due to global ordering. These problems could be faced by either making the scope of suspension to be *local* or by reactivating more than one best resolvents. This results in different kinds of search altogether, but demonstrates the flexibility of the combinations of directives of MORE-Prolog.

4 Results

An interpreter of MORE-Prolog has been developed as a prototype implementation in order to investigate the effectiveness of the language to run a series of example programs. Our aim is to integrate the interpreter with APIM [7] in a complete implementation which will be faster in terms of actual execution time. In a complete MORE-Prolog system, suspension of calls enforces the implementation to be able to handle multiple bindings of the same variables occurring in different paths of the execution tree. Suitable techniques for handling multiple binding environments in OR-parallel execution could be borrowed [8]. This, however, may introduce an extra overhead to memory management.

At the moment, only one predicate of the program may be annotated. In principle, there should be no restriction to the number of predicates that should be controlled by the meta-program. However, in such case the user should also define the scope of the directives as well as the effect of their interaction. This increases the overhead in both programming activity and execution of the program.

We have tried MORE-Prolog programs in a series of simple problems, like route finding on a maze, N-puzzle, N-queens etc., each with different search strategies. We measured the resolution steps required for 1 to 6 processors to find one solution. The search spaces of these problems are small and therefore more processors could not improve the performance any further.

Fig.3(a) shows the effect of global and local suspension in a parallel breadth-first search when applied to the N-queens problem. In global suspension the behaviour is equivalent to sequential breadth-first search, only with faster execution. In local suspension processors are allowed to proceed in breadth-first mode independently of the others and may produce a super-linear leap, as it happens in the example with 2 processors. This is because local suspension gives the opportunity to some processors to explore solution paths which under global suspension should have been left for later consideration. The speedup remains the same up to 5 processors since the execution time is only affected by the second processor which finds the solution.

The same remarks apply to Fig.3(b), although for different reasons. Parallel best-first is employed to solve the N-puzzle. In one case, only one best path is allowed to proceed. In the other case, the *best #p* resolvents are reactivated, giving the chance to explore the solution path faster, although in the early stages of the search another path seemed more promising. The speedup brutally grows for 6 processors because of the effective branching factor of the execution tree (three alternative children states on average by two alternative clauses of the logic program for each state). It is worth noticing that parallel depth-first search for that problem results in such performance that makes the algorithm practically incomplete.

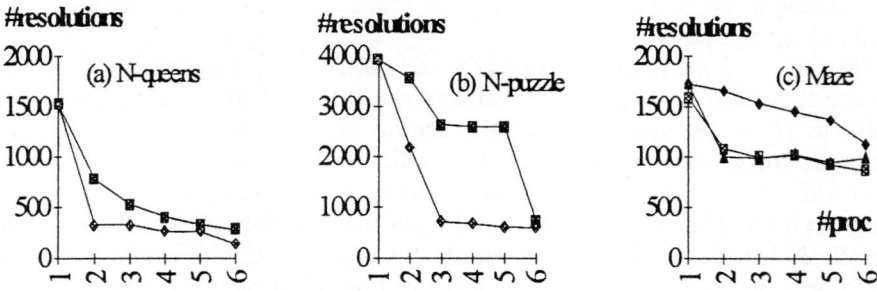

Fig.3: Resolutions required to solve three example problems:
(a) Breadth-first global or local suspension. (b) Best or Best #p reactivation.
(c) Best-first local or global suspension and Depth-first.

Fig.3(c) involves a parallel depth-first search. In this problem, it seems that parallel depth-first search has similar performance with parallel local best-first search. This, however, is true only for small search spaces as the one of the maze problem, where no infinite paths appear. It is worth noticing that the number of resolutions required for breadth-first search by one processor is way out of range in this problem (not showed in the figure), since it is approximately 8 times slower than depth-first.

The importance of the approach is that the results were obtained by minimum programming effort. Efficiency largely depends on the search algorithm used as well as on the specific problem. This was expected, since MORE-Prolog strategies inherit the advantages and disadvantages of the search algorithms which simulate.

Although these results refer to resolution cycles, we are expecting that the speedups obtained on a real implementation would be close to these figures since there is no much overhead of task switching in OR-parallel execution. On the other hand, synchronisation under global suspension might be a burden, but this comes as a consequence of retaining the behaviour of sequential search algorithm under parallel execution. Parallel execution should impose a completely different philosophy under which search algorithms are written and implemented. The results encourage us to proceed with this approach in two directions: to establish a new programming discipline in logic programming and to attempt to use this methodology in order to implement customised scheduling for parallel logic systems.

5 Conclusions and Further Work

We have presented a logic language, MORE-Prolog, which was designed to meet the specific requirements for parallel search algorithms. This language is pure Prolog enriched with control directives which are able to alter the overall default OR-parallel depth-first strategy by suspending calls to predicates, ordering and pruning suspended processes. The combination of object and meta-program introduces a new programming discipline, which results in more clear semantics. The coding of the underling logic program becomes easier while more attention is paid to control aspects.

There is room for further development of MORE-Prolog features. New primitive directives as well as additional constraints can be added to the meta-language in order to allow the user to extend the control mechanism in various ways. For example, suspension could depend on specific conditions being met, i.e. *suspend predicate when condition*, where *condition* is a separate and general control meta-program itself. This will facilitate encoding of more flexible control strategies, like branch & bound, iterative deepening etc. In addition, task-switching schedulers could be in principle coded with an augmented meta-language and used in parallel Prolog systems. Finally, further work is concerned with the development of a sophisticated self-customised control which could automatically decide which algorithm to apply while executing a program.

References

1. R.A. Kowalski: Algorithm=Logic+Control, CACM, 22, 424-431 (1979).
2. E.L. Lusk, D.H.D. Warren, S. Haridi, et al.: The Aurora OR-Parallel Prolog System, New Generation Computing, 7, 234-271 (1990).

3. R. Yang, A. Beaumont, I. Dutra, V. Santos-Costa and D.H.D. Warren: Performance of the Compiler Based Andorra-I System, In Proc. of the 10th Int. Conf. on Logic Programming, (1993).
4. R. Hill: LUSH Resolution and it Completeness, DCS Memo 78, Dept. of Artificial Intelligence, University of Edinburgh, (1974).
5. J.A. Robinson: A Machine-Oriented Logic Based on the Resolution Principle, Journal of the ACM, 12:1, 23-41 (1965).
6. E. Rich and K. Knight: Artificial Intelligence, Mc-Graw Hill, (1992).
7. I. Vlahavas and P. Kefalas: The AND/OR Parallel Prolog Machine APIM: Execution Model and Abstract Design, Journal of Programming Languages, 1, 245-261, Chapman & Hall (1993).
8. D.S. Warren: Efficient Prolog Memory Management for Flexible Control Strategies, In Proc. of the 2nd Int. Symp. on Logic Programming, 198-202 (1984)
9. R. Davis: Meta-rules: Reasoning about Control, Artificial Intelligence, 15:3, 179-222 (1980).
10. L. Naish: Prolog Control Rules, In Proc. of the Int. Conf. on Artificial Inteligence, IJCAI-85, 720-723 (1985).
11. M. Dincbas and J-P. LePape: Metacontrol of Logic Programs in Metalog:, In Proc. of the Int. Conf. on Fifth Generation Computer Systems, Elsevier/North-Holland, 361-370 (1984).
12. H. Gallaire and C. Lasserre: Metalevel control of Logic Programs, In Logic Programming: K.L.Clark, S.A.Tarnlund (eds.), Academic Press, 173-185 (1982).
13. P. Szeredi: Exploiting Or-parallelism in Optimisation Problems, In Proc. of the 9th Int. Symp. on Logic Programming, 703-716 (1992).
14. K. Nakamura: Heuristic Prolog: Logic Program Execution by Heuristic Search, In Proc. of the Int. Conf. on Logic Programming 1986, Springer, 149-155 (1986).
15. T.J. Reynolds and P. Kefalas: Or-Parallel Prolog and Search Problems in Artificial Intelligence Applications, In Proc. of the 7th Int. Conf. on Logic Programming, D.H.Warren and P.Szeredi (eds.), MIT Press, 340-453 (1990).

High Level Parallel Programming Based on Automatic Coordination

Jürgen Knopp

Siemens Corporate Research and Development
email: juergen.knopp@zfe.siemens.de

Abstract. This paper presents an architecture independent paradigm for explicit parallel programming. The focus is on the coordination (data) model: we introduce **coordinators**, an uniform passive object approach for distributed and shared variables programming. Programming with coordinators allows static definition of dynamic access properties (access patterns) in a declarative manner. These access specifications lead to automatic synchronization or remote communication, depending on the underlying architecture. Coordinators are very well suited for generalized data parallel programming thus enabling comfortable and efficient programming in application fields like scientific computing and image processing. Nevertheless they can be used for task parallelism, too.
While the distributed implementation is currently being designed the shared memory implementation already shows good results.

1 Overview

In conventional explicit parallel programming there are two aspects to be considered: the task (or process) model and the coordination (or memory) model. The task model describes the creation and destruction of parallelism and the work distribution. The interaction between tasks (i.e synchronization or remote communication, depending on the underlying architecture) is referred to as the coordination model.

1.1 A coordination model

A way to get rid of differences between coordination primitives suitable to specific architectures is to completely avoid explicit coordination. We do so by specifying access patterns for (logically) global objects. Coordination takes place implicitly when these objects are accessed. By mapping implicit coordination differently to shared resp. distributed architectures we can take advantage of the architectural peculiarities without sacrificing architectural independence. Coordinators allow for the specification of access patterns which constitute abstractions of parallel algorithms. These abstractions describe the parallel behavior w.r.t. coordination. By relating parallel programs not containing any explicit coordination to the specified access patterns the overall parallel semantics is well defined. In other words, the parallel algorithm is factorized by defining a parallel program which does not contain any coordination plus access patterns capturing the coordination aspects only. These patterns are statically specified in a way which is akin to type declarations.

1.2 A task model

Coordinators may be combined with any task model featuring shared and distributed architectures. We use a workpool model [KR96b] which allows programming with abstract tasks rather than with processes or threads. The workpool is architecture independent, too and thus nicely complements the coordinator paradigm.

2 Introducing coordinators by example

Consider a producer-consumer scenario: some tasks write to an object, some tasks read the object. Note that according to our focus on performance this is a non-buffering approach. The following code[1] shows two code fragments that write and read a global object, respectively. These functions access the global object glob in an alternating manner, ideally scheduled in a just-in-time order.

```
while(...){ ... someWritingFunction(glob,...);...};
```

```
while(...){ ... someReadingFunction(glob,...);...};
```

This code does not contain any well defined coordination. Fortunately, we can add it without changing the code. We just specify glob as coordinator. There are several variants of parallel behavior one might like to have. Consider for instance the following specification:

```
typedef coordinator<vector<double>> globVect;
globVect glob(write(each(producers,1)),read(each(consumers,1)));
```

It mirrors the fact that every producing task should be involved exactly once before some information is read. producers and consumers are task groups in this example. When obeying the specified access pattern the two while loops within the producer and consumer code fragments are performed in a coherent manner: glob is released for reading if and only if the writing tasks have completed the write access. Moreover, the next loop traversal for the write access proceeds if and only if all tasks have read the object glob (each of them once).

To show another example, consider the case that one task fills the object by a number - say writeTimes - of accesses and the identity of the reader is irrelevant.

```
producers = group(someTask);              // a single task
typedef coordinator<vector<double>> globVect;
globVect glob(write(each(producers,writeTimes)),read(arbitrary(1)));
```

Fully coded numeric examples may be found in [KR96a] and [KR96c].

---

[1] We use C++ syntax since our system is based on C++. Any other languages would be suitable too. However, for implementation reasons languages supporting overloading and generic programming are favorable.

3 On the semantics of coordinators

After this introduction we now describe coordinators more systematically. Due to lack of space we ignore some properties. A more complete description may be found in [KR96a]. An operational semantics is given in [KR96c].

3.1 Dynamic behavior of coordinators

Reading and writing The basic assumption about accesses to coordinators is that *writing* accesses are distinguished from *reading accesses*.

Phases and computational steps In imperative programs, switches between *writing* and *reading* typically occur repeatedly inside loops. All such activities within one iteration are referred to as a *computational step*. A *computational step* with respect to a coordinator is defined to consist of a *write phase* and a *read phase*[2]. A coordinator is initially in the *write* phase. Depending on the access patterns defined for each state, the switches may be triggered by different actions.

Blocking Depending on the specified access pattern and the access history accesses to coordinators may automatically block[3]. Note that the access pattern may (and must) be specified by the user. Blocking occurs

- when the access does not correspond to the phase (i.e. reading accesses in the write phase and writing accesses in the read phase) or
- the accessing task has already completed all accesses specified by the access pattern for the current phase.

Switching Once the access pattern for a phase is completed (i.e. all the specified accesses within the phase have occurred) the phase is switched.

3.2 Defining access patterns

Every coordinator reveals a specific access pattern with respect to both phases of a computational step. The access patterns must be specified by the user separately for both phases. Access patterns correspond to algorithmic patterns as they typically occur in "data-oriented" parallel algorithms (e.g. in numerical computation). We show the two main regular patterns here. More patterns may be found in [KR96a].

- each(<*task group*>,<*no of accesses*>)
- arbitrary(<*no of accesses*>)

---

[2] In functional programs there is exactly one computational step.
[3] Readers familiar with *futures* or *I-structures* might see some relationship to these.

The standard way to use coordinators is to declare how many times each involved task will access the coordinator within the phase. To do so the **each** pattern is used with a <*no of accesses*> specification (<*no of accesses*> will be 1 in many cases). Note that two accesses of a task are different from one access of two tasks. This gives us the opportunity to avoid semantically incorrect overtaking by some eager task.

In contrast, the **arbitrary** pattern is used if an arbitrary, possibly varying number of possibly inhomogeneous tasks share the coordinator. Within this co-operation mode, we can provide only global switch conditions: In the **arbitrary** pattern a <*no of accesses*> specification refers to the overall number of accesses, i.e. the state of the coordinator switches after exactly <*no of accesses*> accesses, no matter which tasks were involved.

3.3 Coordinators as path expressions

Regular access patterns can be defined using a variant of the well known formalism of path expressions. Path expressions have been introduced to describe abstract synchronization constraints. While there are some properties of coordinators [KR96a] which are not expressible with path expressions we believe that it is useful to show the relationship w.r.t. the abstract coordination aspects.

A dynamic trace semantics may be defined in terms of path expressions (i.e. regular expressions) when atoms with an adequate semantics are used.

Atoms are (reading or writing) accesses to coordinators. They are denoted by a. Accesses may be executed by tasks which are denoted by t. Tasks may be members of groups denoted by g. We denote accesses of tasks t with $t \in g$ by a_t. Given this notation we express the trace semantics by path expressions using $";"$, $"[]"$ and $"*"$ to denote sequencing, interleaving execution (i.e. execution in any order) and repetition, respectively.

The overall trace semantics is then defined by the regular expression

$$(writePhase; readPhase)*.$$

The used items are defined as follows:

- Given a pattern **phase(each(g,n))** we get
  $readOrWritePhase ::= (A_{t_1}[]...[]A_{t_m})$ where
  $g = \{t_1, ..., t_m\}$ and $A_{t_i} ::= (a_{t_i}[])^n$ where $a_{t_i} = r_{t_i}$ or $a_{t_i} = w_{t_i}$, respectively.
- Given a pattern definition **phase(arbitrary(n))** we get
  $readOrWritePhase ::= (a[])^n$ where $a = r$ or $a = w$, respectively.

We have used $(a_t[])^{count}$ in order to express a fixed number of accesses of the same task. It should be noted that the semantics of $[]$ is that accesses must (rather than may) be executed (in any order).

Given this trace semantics the operational behavior of coordinators is that accesses are blocked when the access history does not (yet) match the path expression.

4 Implementation

The programming model including the workpool and the coordinators is designed for both shared and distributed platforms - featuring a common semantics.
We have implemented a prototype for the shared case for both parallel *SUN* and SGI workstations. In figure 1 some benchmark results are given. The program is a conjugate gradient linear system solver for sparse linear systems using a fully abstract sparse C++ library.

All benchmarks (including those omitted here for brevity) show that there is a speed down of 20 to 30 percent when moving to a parallel version running a single task. Moreover, with respect to scalability memory contention is a limiting factor. The limitations of shared memory workstations w.r.t scalability are well known. As expected, this occurs for SUN workstations earlier than for Silicon Graphics machines. Low level versions of our benchmarks (without coordinators) show similar results. Note that these figures refer to non-academic applications.

In order to allow parallel debugging at the level of coordinators we have devel-

machine	sequential	1 task	5 tasks	10 tasks	20 tasks
SUN SS 10 with 4 processors	100	132	50	43	44
SGI Challenge with 8 processors	100	123	26	25	22

Fig. 1. Normalized sample runtimes for different numbers of tasks

oped the visualization tool *covis* which can be found as Java applet on the world wide web under
http://www.informatik.tu-muenchen.de/~weidendo/covis/index.html.

5 Related work

Coordinators are akin to *FUTURES* and to *LINDA*. They extend the single assignment approach of both to imperative programming introducing user defined access patterns. These could be expressed with modified path expressions too, see section 3.3. However, coordinators have some additional properties which fit less well to path expressions [KR96a].

In the context of object based parallel programming several languages and concepts featuring implicit coordination or synchronization have been defined. First, there are object based distributed languages such as Emerald and Orca. These languages perform automatic object migration on distributed platforms. The implementation techniques used within our forthcoming distributed implementation are along the lines of Emerald and Orca. Note that this does not hold for the access patterns as these are not part of Orca and Emerald.

Implicit coordination is omnipresent in the newer literature discussing object based concurrency and active objects. The names found there are synchronization constraints, delayed queues synchronizers, synchronization counters, concurrency constraints, and delay declarations.

The scope of these constraint mechanisms is to enable and disable methods in object based (and object oriented) languages.

There are several differences to coordinators. We consider access types rather than specific functions or methods. We focus on ease of implementation and speedup of real applications rather than on generality. Thus we do not consider objects to be active. Moreover, we do not support inter-object synchronization. Obviously, the above mentioned approaches are more general than coordinators. On the other hand, they do not deal with features needed in data parallel computations such as monitoring of different tasks operating on the same object: the difference between several accesses by a task and accesses by several tasks is crucial for high level parallel programming. To our knowledge, this has not been discussed at a level comparable to coordinators up to now.

6 Conclusions

Our main contribution is a mechanism which allows programming in a declarative fashion: Global objects (coordinators) are declared with statically specified access patterns which describe the dynamic behavior of accesses to these objects. The coordinators follow the access pattern rules, thus automatically performing (implicit) synchronization of the accessing processes in the shared case and remote communication in the distributed case. We believe that our declarative approach has a great impact on ease and safeness of explicit parallel programming without sacrificing efficiency.

7 Acknowledgments

Thomas Henties, Hermann Ilmberger, Leo Mandl, Matthias Reich and Christian Weiß have contributed to the design and implementation of the coordinators. Josef Weidendorfer has designed and implemented the *covis* tool (see section 4).

References

[KR96a] J. Knopp and M. Reich. A Data Model For Architecture Independent Parallel Programming. In *Workshop on High-Level Programming Models and Supportive Environments held in conjunction with the IEEE International Parallel Processing Symposium, Honolulu.* IEEE Computer Press, 1996.

[KR96b] J. Knopp and M. Reich. A Workpool Model for Parallel Computing. In *Workshop on High-Level Programming Models and Supportive Environments held in conjunction with the IEEE International Parallel Processing Symposium, Honolulu.* IEEE Computer Press, 1996.

[KR96c] J. Knopp and M. Reich. On the Semantics of Coordinators. Technical report, SIEMENS Corporate Research and Development, April 1996. In german.

Eden — The Paradise of Functional Concurrent Programming*

S. Breitinger[1], R. Loogen[1], Y. Ortega-Mallén[2], R. Peña-Marí[2]

[1] Philipps-Universität Marburg, Fachbereich Mathematik, Fachgebiet Informatik,
Hans Meerwein Straße, Lahnberge, D-35032 Marburg, Germany,
{breiting,loogen}@informatik.uni-marburg.de
[2] Universidad Complutense de Madrid, Sec. Dept. de Informática y Automática,
Facultad de C.C. Matemáticas, E-28040 Madrid, Spain,
{yolanda,ricardo}@dia.ucm.es

The functional concurrent language *Eden* [1] is an extension of the lazy functional language Haskell [4] by constructs for the explicit specification of dynamic process systems. It employs stream-based communication and is tailored for distributed memory systems. Eden supports and facilitates the task of parallel and concurrent programming.

Eden incorporates special concepts for the efficient treatment of general reactive systems, i.e. systems which maintain some interaction with the environment and which may be time-dependent. The *dynamic creation of reply channels* simplifies the generation of complex communication topologies and increases the flexibility of the language. *Predefined nondeterministic processes* MERGE and SPLIT are used to model many-to-one and one-to-many communication in process systems.

Eden incorporates a two level structure: the level of user-defined *processes* and the level of *process systems*. User-defined processes can be seen as deterministic mappings from input channels to output channels. Nondeterminism is only handled at the system level which consists of all (predefined and user-defined) processes interacting via communication on channels.

1 Eden in a nutshell

Haskell forms the *computation language* of Eden. This is extended by a *coordination model* that introduces processes in a functional style, embodying constructs which allow for the definition and creation of processes, communication and synchronization, and the specification of interconnections between processes.

Eden distinguishes between *process abstractions*, which specify process behaviour in a purely functional way, and *process instantiations* in which process abstractions are supplied with input values in order to create new processes.

A process abstraction defines a general parameterized process scheme. It specifies a process which maps (streams of) input values $in_1, \ldots, in_m$ to (streams of) output values $out_1, \ldots, out_n$:

---

* Supported by the DAAD (Deutscher Akademischer Austauschdienst) and the Spanish Ministry of Education and Science in the context of the German-Spanish Acción Integrada n.142B.

process abstraction:

$$p :: \tau_1 \to \ldots \to \tau_k \to Process\ (\tau_1', \ldots, \tau_m')\ (\tau_1'', \ldots, \tau_n'')$$
$$p\ par_1\ \ldots\ par_k = process\ (in_1, \ldots, in_m) \to (out_1, \ldots, out_n)$$
$$\text{where}\quad equation_1\ \ldots\ equation_r$$

Process creation takes place when a process abstraction without unspecified parameters[3], e.g. $(p\ e_1 \ldots e_k)$, is applied to a tuple of inport expressions. This is called *process instantiation* and defines the tuple of outports of the newly created process:

process instantiation: $(p\ e_1 \ldots e_k)\ \sharp\ (in\_exp_1, \ldots, in\_exp_m)$

Often, process instantiations occur in equations of the form

$$(out_1, \ldots, out_n) = (p\ e_1 \ldots e_k)\ \sharp\ (in\_exp_1, \ldots, in\_exp_m)$$

The left hand side of such an equation is the tuple of output channels of the created process.

It is important that a process abstraction must be closed, i.e. the expressions in the body may depend only on parameters, input values, local auxiliary definitions or functions contained in the standard Haskell prelude. This guarantees that a process is an independent unit of computation which communicates only via its ports and can be executed without any implicit need to access global information. This property is essential as Eden is a language for distributed memory systems.

We do not allow the duplication of running processes because they are dynamic entities with an internal state. Therefore it is essential to presuppose a lazy semantics, which guarantees that process instantiations passed as argument to functions, to other processes, or to abstractions will be shared.

Communication channels transmit completely evaluated values of arbitrary type. In order to model the transmission of a stream of values we introduce a new algebraic data type *Strm* a with data constructor $<.>:: [a] \to Strm\ a$. A channel of type *Strm* τ transmits values of type τ one by one. Streams correspond to lazy lists, but note that a channel of type *list* τ is assumed to transfer exactly one list of type τ, which is finite because its evaluation must be completed before the transmission, while a channel of type *Strm* τ transmits a potentially infinite list component-wise.

Example. A sorting network which transforms an input stream into a sorted output stream by subsequently merging sorted sublists with increasing length is specified by the following process abstraction:

```
mergesort :: Process (Strm a) (Strm a)
mergesort = process <s> -> <sort s>
            where
```

---

[3] Actual parameter expressions are copied into the body of process abstractions before the process is created.

```
              sort []     = []
              sort [x]    = [x]
              sort xs     = smerge (mergesort # <l1>) (mergesort # <l2>)
                            where (l1,l2) = unshuffle xs
              smerge [] l = l
              smerge l [] = l
              smerge (x:l) (y:t) = if x<=y then x:smerge l (y:t)
                                          else y:smerge (x:l) t
              unshuffle []      = ([],[])
              unshuffle [x]     = ([x],[])
              unshuffle (x:y:t) = (x:t1,y:t2)
                                  where (t1,t2) = unshuffle t
```

Streams with at least two elements are split into two substreams for which recursive instantiations of the mergesort process are generated. As a result a tree of mergesort processes is created.

2 Programming in Eden

Due to space limitations we present only the specification of a parallel matrix multiplication algorithm. Matrix multiplication is an important operation in many scientific and engineering problems. It has a good potential for parallelization and has become a standard example for parallel programming. We consider a parallel algorithm which uses a torus topology. The matrices are partitioned into submatrices which are distributed on the torus nodes. For simplicity we assume the partition size to be 1 and the matrix dimension to be $n \times n$.

Each element of the torus first gets the corresponding elements of the input matrices. The torus node with the position (i, j) then has to compute the element (i, j) of the result matrix, i.e. the scalar product of the ith row of the first input matrix and the jth column of the second input matrix. In order to place the elements to be multiplied on the same node, the rows of the first matrix and the columns of the second matrix are rotated by $(i - 1)$ and $(j - 1)$ positions respectively before the proper computation starts. Then all torus nodes perform an iteration of $n - 1$ steps in which they multiply corresponding elements of the matrices which they read subsequently from their input channels.

We presuppose the definition of a process abstraction torus which creates a torus topology according to the dimension of the given matrix (see [2] or [3]):

```
type Matrix a    = [[a]]
type NodeProc a b c d = (Int -> Int -> a ->
                         Process (Strm b, Strm c) (d, Strm b, Strm c)
torus :: Matrix a -> NodeProc a b c d -> Process () (Matrix d)
```

The node processes are connected row- and column-wise by stream channels from right to left and bottom to top. Their behaviour is specified by an abstraction which is passed as a parameter to the torus process abstraction.

```
matmult :: Int -> (Matrix Int) -> (Matrix Int) -> (Matrix Int)
```

```
matmult n ass bss
  = torus (map uncurry.zip (zip ass bss)) (scalarprod n)
scalarprod :: Int -> Int -> Int -> (Int,Int)
              -> Process (Strm Int, Strm Int) (Int, Strm Int, Strm Int)
scalarprod n i j (a,b) -- process abstraction for each torus node
  = process  (inrow, incol) -> (result, outrow, outcol)
    where
      outrow = a:inrow   -- rotate row of first matrix
      outcol = b:incol   -- rotate col of second matrix
      result = iterate n arow bcol (a0 * b0)
      iterate 0          row       col  val = val
      iterate (n+1) (r:row) (c:col) val = iterate n row col (val+r*c)
      arow = (drop (i-1) inrow)
      bcol = (drop (j-1) incol)
      a0 = (a:inrow)!!(i-1)
      b0 = (b:incol)!!(j-1)
```

3 Where to find more about Eden

The interested reader can find a more detailed discussion of Eden features in [2]. There, more programming examples are provided. A collection of skeletons for the instantiation of various process topologies can be found in [3].

The semantics of Eden is an extension of the standard operational semantics of Haskell. It reflects the two-layered nature of Eden: systems and processes. On the upper level global effects on process systems are described. The lower level handles local effects within processes. The interface between the two levels consists of so-called 'actions' by which the need for global events is communicated to the upper level.

In a prototype implementation of Eden the low-level primitives of Concurrent Haskell have been used to model communication and synchronization between Eden processes explicitly. In the future there will be a direct implementation of Eden on an IBM SP-2 using the MPI (Message Passing Interface) standard.

References

1. S. Breitinger, R. Loogen, Y. Ortega-Mallén: Towards a Declarative Language for Concurrent and Parallel Programming, Glasgow Workshop on Functional Programming, Springer Workshops in Computing 1995.
2. S. Breitinger, R. Loogen, Y. Ortega-Mallén, R. Peña-Marí: Eden — The Paradise of Functional Concurrent Programming, TR DIA-UCM-20/96, Universidad Complutense de Madrid 1996, also available as
http://www.mathematik.uni-marburg.de/~loogen/paper/paradise.ps.
3. L. A. Galán, C. Pareja, R. Peña: Functional Skeletons Generate Process Topologies in Eden, APPIA-GULP-PRODE'96 Joint Conference on Declarative Programming, San-Sebastian, Spain, July 1996.
4. P. Hudak, Ph. Wadler (eds.): Report on the Programming Language Haskell: a nonstrict, purely functional language, SIGPLAN Notices, 27(5):1–162,1992.

A Straightforward Translation of D0L Systems in the Declarative Data-Parallel Language $8_{1/2}$ *

Olivier Michel

LRI u.r.a. 410 du CNRS
Bâtiment 490, Université de Paris-Sud, F-91405 Orsay Cedex, France.
Tel: +33 (1) 69 41 76 01 email: michel@lri.fr

Abstract. L Systems are a mathematical formalism originally designed to model the biological growth of plants. They are based upon a parallel rewriting mechanism. We show in this paper that there exists an equivalence between the class of D0L Systems and a particular construction of programs in the $8_{1/2}$ declarative data-parallel language.

1 Introduction

$8_{1/2}$ is a declarative data-parallel language aimed at the simulation of dynamical systems. Its main structure is the *fabric* which is a *stream* [1, 2, 3] of *collections* [4]. The stream aspect of the fabric is used to describe the time axis of the simulation whereas the collection aspect of the fabric describe the space axis of the simulation. Because of space restrictions, we will not detail the $8_{1/2}$ language here. The reader should refer to the paper *"Introducing dynamicity in the data-parallel language $8_{1/2}$"* in these proceedings or to [5].

An $8_{1/2}$ program is a set of equations defining the relations between the variables of the dynamical system. $8_{1/2}$ has operators to handle the collection and streams aspect of the fabric but we will only detail the concatenation, the packing and the delay operator. The *concatenation* ($a \# b$) and the *packing* ($\{\dots\}$) are two spatial operators. The relationship between the concatenation and the packing is the same as between the cons and append operations for lists: $\{A, B, \dots\} = \{A\} \# \{B\} \# \dots$ The *delay* operator, which is a temporal operator, shifts an entire stream to give at the current instant of the simulation, the previous stream value. For example, if we have $S = > 0, 1, 2, 3, \dots >$ (that is, at time instant 0, the value of S is 0, at time instant 1, the value is 1, etc.), then the value of $\$S$ is $> \bot, 0, 1, 2, 3 >$. The definition of an $8_{1/2}$ fabric is usually made of two parts: 1) the definition at some specific instants expressed by quantified equations; 2) the definition for all other instants. For example, $T@0 = 1$ means that the value of the fabric at time instant 0 is 1 and $T = \$T + 1$ when Clock means that for all other instants, the current value of T is the previous one increased by one. The time instants where this increase occurs are specified by the "when Clock" expression.

---

* This research is partially supported by the operation "Programmation parallèle et distribuée" of the french "GDR de programmation".

2 L Systems and 8½: From Transformation Rules to Equations Between Fabrics

2.1 Lindenmayer Systems

An *L System* is a *parallel* rewriting system (every production rule that might be used at each derivation state is used simultaneously) developed by A. Lindenmayer in 1968 [6].Since their initial design for the specification of growing processes [6], L Systems have proved to be an effective paradigm (for example, in the field of image processing, symbolic processing [7], cellular development modelisation [8], parallel computation model [9], ...).

We will restrict ourselves to the simplest form of L Systems: *D0L Systems*. The D letter stands for deterministic, that is there exists at most a single production rule for each element of Σ. Therefore the derivation sequence is unique while in non-deterministic L Systems, since there can be more than one production rule applied at each derivation state, there exists more than one derivation sequence. The numerical argument of the L System gives the number of interactions in the rewriting process; therefore a 0L System is a context free L System (whereas an nL *System* is context sensitive with n interactions). A *D0L System* is a triple $G = (\Sigma, h, \omega)$ where Σ is an alphabet, h is a finite substitution on Σ (into the set of subsets of Σ^*) and ω, referred to as the axiom, is an element of Σ^+. The derivation $H(w)$ of a word $w \in \Sigma^*$ is defined in the following manner:

$$H(a) \;\; = h(a) \qquad \text{for } a \in \Sigma$$
$$H(u \,.\, v) = H(u) \,.\, H(v) \qquad \text{for } u, v \in \Sigma^+$$

that is, H is the homomorphism extending h.

Our goal is to show that D0L Systems can be mimicked in the 8½ language *without adding any new features* to the current definition of the language. We sketch in the next section the relation between L Systems and the 8½ language.

2.2 A Translation of L Systems into 8½

We are interested in describing the derivation mechanism involved in the translation of the L Systems into the 8½ language. Fabrics with a static structure cannot describe a phenomenon that grows in space (like plants). To describe those structures, we need dynamically structured fabrics. We do not need to introduce new operators, current fabrics definition already enables the construction of dynamically shaped fabrics. The translation of a D0L System in 8½ relies on the following: 1) a word is a *vector of elements* of Σ; 2) a fabric W is associated to the axiom ω, and the value $W(t)$ of W at time t is $H^t(\omega)$.

To implement this in 8½, each rule defining h is translated into the following fabric definition:

$$a \to a_1 \ldots a_n \text{ gives } \begin{cases} A@0 = {'a'} \\ A = \$A_1 \# \ldots \# \$A_n \text{ when } Clock \end{cases} \qquad (1)$$

and a word $w = b_1 \ldots b_n$ gives the fabric definition $W = B_1 \# \ldots \# B_n$.

We prove that for any word w, $W(t) = H^t(w)$. First, we prove by recurrence on t that $A(t) = H^t(a)$. This is obviously true for $t = 0$. Suppose the property true for $t' < t$. Then,

$$\begin{aligned}
A(t) &= (\$A_1)(t) \# \ldots \# (\$A_n)(t) \\
&= A_1(t-1) \# \ldots \# A_n(t-1) && \text{(definition of \$)} \\
&= H^{t-1}(a_1) \ldots H^{t-1}(a_n) && \text{(by hypothesis)} \\
&= H^{t-1}(a_1 \ldots a_n) && \text{(H homomorphism)} \\
&= H^{t-1}(H(a)) \\
&= H^t(a)
\end{aligned}$$

Secondly, $W(t) = B_1(t) \# \ldots \# B_n(t) = H^t(b_1) \ldots H^t(b_n) = H^t(b_1 \ldots b_n) = H^t(w)$. □

It is therefore possible, using only the $\#$ and delay operators to program in $8_{1/2}$ the whole class of D0L Systems. Note that it is possible to describe the whole class of D0L Systems in $8_{1/2}$ (even non propagating L Systems). We consider each transition rule as a recursive $8_{1/2}$ equation. Parallelism is present in this "rewriting" mechanism, in the form of:

- *data-parallelism:* actually no data-parallelism is directly involved in the recursive operations except that the data structure involved (collections) are distributed (so that multiple concatenations can be implemented in parallel). However, the mechanism presented here is very general. L Systems are a *model* used to describe phenomena. At each step, data-parallel (reductions, scans, etc.) operations can be performed locally (on each collection) to reflect a particular behaviour taking place after the rewriting.

- *control-parallelism:* control parallelism, depends on the data dependencies between the equations involved in the axiom and in each rewriting rule. There are no dependencies because each fabric refers to past values of other fabrics. Therefore, the computation of the current value of each fabric can be fully processed in parallel.

An implementation based solely on the exploitation of control-parallelism will map the computations of a fabric to some processors. Assume $A = \$A \# \$B$; the "owner computes rule" (as in HPF) assigns the computation of the concatenations to the processor owning A and implies the communication of the past values of B and C from the processors owning B and C to the one of A. This can be avoided using locality, that is assigning B and/or C to the processor owning A. This is in contradiction with the maximal parallelism exploitation but minimizes the communications. A mapping and scheduling heuristic, taking into account data-communications, is presented in [10].

3 Conclusions and Future Work

We have described in this paper the translation of L Systems into the declarative data-parallel language $8_{1/2}$. Each transition rule of a D0L System can easily be rewritten into equations between streams and collections. Following this basic scheme, it is very simple to implement even more complex programs in $8_{1/2}$ (like the graph generation scheme using G0L Systems found in [11]). We are currently working on the implementations of more general L Systems and especially the class of parametric L Systems.

Acknowledgments. The author wishes to thank the members of the $8_{1/2}$ team: Jean-Louis Giavitto, Jean-Paul Sansonnet, Dominique De Vito and Abderrahmane Mahiout for many stimulating discussions. He also thanks Dan Truong for careful reading the paper.

References

1. E. Ashcroft, A. Faustini, R. Jagannatha, and W. Wadge. *Multidimensional Programming.* Oxford University Press, February 1995. ISBN 0-19-507597-8.
2. P. Caspi, D. Pilaud, Halbwachs N., and J. Plaice. Lustre: a declarative language for programming synchronous systems. In *Fourteenth annual symposium on principles of programming languages.* ACM, ACM Press, January 1987.
3. P. Le Guernic, A. Benveniste, P. Bournai, and T. Gautier. Signal, a dataflow oriented language for signal processing. *IEEE-ASSSP*, 34(2):362–374, 1986.
4. G. E. Blelloch and G. W. Sabot. Compiling collection-oriented languages onto massively parallel computers. *Journal of Parallel and Distributed Computing*, 8:119–134, 1990.
5. O. Michel. Design and implementation of $8_{1/2}$, a declarative data-parallel language. *special issue on Parallel Logic Programming in Computer Languages*, 1996. (to appear).
6. A. Lindenmayer. Mathematical models for cellular interactions in development, parts i and ii. *Journal of Theoretical Biology*, 18:280–315, 1968.
7. N. S. Goel and M. D. Goodwin. Symbolic computation using L systems. *Applied mathematics and computation*, 42:223–253, 1991.
8. M. J. M. de Boer, F. D. Fracchia, and P Prusinkiewicz. A model for cellular development in morphogenetic fields. In G. Ronzenberg and A. Salomaa, editors, *Lindenmayer Systems, Impacts on Theoretical Computer Science, Computer Graphics and Developmental Biology*, pages 351–370. Springer Verlag, February 1992.
9. P. Prusinkiewicz and J. Hanan. L systems: from formalism to programming languages. In G. Ronzenberg and A. Salomaa, editors, *Lindenmayer Systems, Impacts on Theoretical Computer Science, Computer Graphics and Developmental Biology*, pages 193–211. Springer Verlag, February 1992.
10. A. Mahiout, J.-L. Giavitto, and J.-P. Sansonnet. Distribution and scheduling dataparallel dataflow programs on massively parallel architectures. In *SMS-TPE'94: Software for Multiprocessors and Supercomputers*, Moscow, September, 1994. Office of Naval Research USA & Russian Basic Research Foundation.
11. E. J. W. Boers. Using L systems as graph grammars: G2L systems. Technical Report TR 95-30, Leiden University, 1995.

Efficient Parallel Programming with Algorithmic Skeletons

George Horatiu Botorog*   Herbert Kuchen

Aachen University of Technology, Lehrstuhl für Informatik II
Ahornstr. 55, D-52074 Aachen, Germany
{botorog,herbert}@i2.informatik.rwth-aachen.de

Abstract. Algorithmic skeletons are polymorphic higher-order functions representing common parallelization patterns and implemented in parallel. They can be used as the building blocks of parallel and distributed applications by integrating them into a sequential language. In this paper, we present a new approach to programming with skeletons. We integrate the skeletons into an imperative host language enhanced with higher-order functions and currying, as well as with a polymorphic type system. We thus obtain a high-level programming language which can be implemented very efficiently. After describing a series of skeletons which work with distributed arrays, we give two examples of parallel algorithms implemented in our language, namely matrix multiplication and a statistical numerical algorithm for solving partial differential equations. Run-time measurements show that we approach the efficiency of message-passing C up to a factor between 1 and 1.75.

1 Introduction

Algorithmic skeletons represent an approach to parallel programming which combines the advantages of high-level (declarative) languages and those of efficient (imperative) ones. The aim is to obtain languages that allow easy parallel programming, in which the user does not have to handle low-level features such as communication and synchronization, and does not face problems like deadlocks or non-deterministic program runs. At the same time, skeletons should be efficiently implemented, coming close to the performance of low-level approaches, such as message-passing on distributed memory machines.

A *skeleton* is an algorithmic abstraction common to a series of applications, which can be implemented in parallel [4]. Skeletons are embedded into a sequential host language, thus being the only source of parallelism in a program. Most skeletons, like for instance *map*, *farm* and *divide&conquer* [5] are polymorphic higher-order functions, and can thus be defined in functional languages in a straightforward way. This is why most languages with skeletons build upon a functional host [6, 9]. However, programs written in these languages are still 5 to 10 times slower than their low-level counterparts, e.g. parallel C [9].

---
* The work of this author is supported by the "Graduiertenkolleg Informatik und Technik" at the Aachen University of Technology.

A series of approaches use an imperative language as a host, like for instance P^{3}L [1], which builds on top of C++, and in which skeletons are internal language constructs. The main drawbacks are the difficulty to add new skeletons and the fact that only a restricted number of skeletons can be used. A related approach is PCN [8], where *templates* provide a means to define reusable parallel program structures. The language also contains lower-level features, such as *definitional variables*, by which processes can communicate, but which can also lead to problems such as deadlocks. In the language ILIAS [12], arithmetic and logic operators are extended by overloading to pointwise operators that can be applied to the elements of matrices. These operations are actually not skeletons in the sense of the former definition, but their functionality resembles that of some skeletons. SPP(X) [6] uses a 'two-layer' language: a high-level, functional language for the application and a low-level language (at present Fortran) for efficient sequential code that is coordinated by the skeletons.

In this paper, we present a new approach to programming with algorithmic skeletons. Firstly, we describe the host language, which contains higher-level features, but can be efficiently implemented. After that, we define the data structure "distributed array" and present some data-parallel skeletons operating on it. We then show how two parallel applications can be programmed in a sequential style by using these skeletons. Run-time measurements show that our results are better than those obtained by using pure functional languages with skeletons, approaching direct low-level (C) implementations.

2 The Approach

An important characteristic of skeletons is their *generality*, i.e. the possibility to use them in different applications. For this, most skeletons are parameterized by functions and have a polymorphic type. Since we wanted to obtain a high efficiency, we have used the imperative language C as the basis for our host. However, in order to allow skeletons to be integrated in their full generality, we have provided the language with the following features:

- *higher-order functions*, i.e. functions with functional arguments and/or result.
- *partial applications*, i.e. the application of an n-ary function to $k \leq n$ arguments. Partial applications are useful in generating new functions at runtime, as shown in Section 4.
- *conversion of operators to functions*, which allows passing operators as functional arguments, as well as partial applications of operators. This conversion is done by enclosing the operator between brackets, e.g. (+).
- a *polymorphic type system*, which allows skeletons to be reused in solving similar or related problems. Polymorphism is achieved by using *type variables*, which can be instantiated with arbitrary types. Syntactically, a type variable is an identifier which begins with a '$', e.g. $t.
- *distributed data structures* can be defined by means of the 'pardata' construct, which has the syntax

pardata *name* <t_1, ..., t_n> [*implem*] ;

where t_1, ..., t_n are type variables and *implem* is a (polymorphic) type, representing the data structure to be created on each processor. Distributed data structures may not be nested, in particular the type arguments of a pardata construct cannot be instantiated with other pardatas. Moreover, some additional problems appear if *dynamic* (pointer-based) data types are used for this instantiation. This issue is addressed in [3].

We have called our language *Skil*, as an acronym for *Sk*eleton *I*mperative *L*anguage. A detailed description of the language and its functional features is given in [2].

After this brief outline, we want to compare Skil with two related approaches. SISAL [7] is a functional language which achieves Fortran performance. However, it does not support partial applications and polymorphism, and thus cannot be used directly as a host for our skeletons. On the other hand, SPP(X)'s functional language [6] contains more features than are strictly needed to integrate the skeletons. We therefore expect our language to be more efficient. A comparison of Skil with further related approaches, like data-parallel languages and C++-extensions, can be found in [3].

3 Skeletons for Arrays

Depending on the kind of parallelism used, skeletons can be classified into *process parallel* and *data parallel* ones. In the first case, the skeleton creates a series of processes which run concurrently, whereas in the second case, it acts upon some distributed data structure. Although both categories can be integrated in Skil, we place the emphasis here on data parallelism.

We shall describe a series of skeletons which work on the distributed data structure "array". This data structure is defined as:

pardata array <$t> ... *implementation* ... ;

where the type parameter $t denotes the type of the elements of the array.

At present, arrays can be distributed only block-wise onto processors. Each processor thus gets one block (partition) of the array, which, apart from its elements, contains the local bounds of the partition. These bounds are accessible via the macro array_part_bounds:

Bounds array_part_bounds (array <$t> a) ;

where Bounds is a data structure comprising two Index structures (see Subsection 3.1) representing the lowest and highest indices of the current partition, respectively. Array elements can be accessed by using the macros:

$t array_get_elem (array <$t> a, Index ix) ;
void array_put_elem (array <$t> a, Index ix, $t newval) ;

which read, respectively overwrite a given array element. An important aspect is that these macros can only be used to access *local* elements, i.e. the index `ix` should be within the bounds of the array partition currently placed on each processor. The reason for this restriction is that remote accessing of single array elements easily leads to inefficient programs. Non-local element accessing is still possible, however only in a coordinated way by means of skeletons.

We shall now present some skeletons for the distributed data structure `array`. For each skeleton, its syntax, informal semantics and complexity are given. Note that for the higher-order skeletons, this complexity is given as a function of the complexities of its functional arguments, since the use of different customizing functions may lead to a different overall complexity. p denotes the number of processors (and, hence, of array partitions) and d the dimension of the array. We assume for simplicity that our arrays have the same size n in each dimension.

For each skeleton, we give the actual computation time, $t(n)$, whereas the overall complexity can be derived as $c(n) = t(n) \cdot p$. For simplicity, we consider that both a local operation on a processor and the sending of one array element from one processor to one of his neighbors equally take one time unit.

3.1 Constructor Skeletons

array_create creates a new, block-wise distributed array and initializes it using a given function. The skeleton has the following syntax:

```
array <$t> array_create (int dim, Size size, Size blocksize,
                        Index lowerbd, $t init_elem (Index), int distr);
```

where:
- `dim` is the number of dimensions of the array. At present, only one- and two-dimensional arrays are supported. The types `Size` and `Index` are (classical) arrays with `dim` components.
- `size` contains the global sizes of the array.
- `blocksize` contains the sizes of a partition. Passing a zero value for a component lets the skeleton fill in an appropriate value depending on the network topology.
- `lowerbd` is the lowest index of a partition. Passing a negative value for a component lets the skeleton derive the lower local bound for this dimension.
- `init_elem` is a user-defined function that initializes each element of the array depending on its index.
- `distr` gives the virtual (software) topology onto which the array should be mapped (where available). In our implementation based on Parix [13, 15], an array can be mapped
  - directly onto the hardware topology (`DISTR_DEFAULT`)
  - onto a virtual ring topology (`DISTR_RING`) – this can be useful for 1-dimensional arrays
  - onto a virtual 2-dimensional torus topology (`DISTR_TORUS2D`) – this can be useful for 2-dimensional arrays, as it is shown in the first example in Section 4.

If we notate the complexity of the initialization function with t_i, then the time complexity of the skeleton is $t(n) \in \mathcal{O}\left(t_i \cdot n^d/p\right)$, since we have to call `init_elem` for each element of a partition and all partitions are processed in parallel.

array_destroy deallocates an existing array. It has constant complexity $t(n) \in \mathcal{O}(1)$.

```
void array_destroy (array <$t> a) ;
```

3.2 Computational Skeletons

array_map applies a given function to all elements of an array, and puts the results into another array. However, the two arrays can be identical; in this case the skeleton makes an in-situ replacement. The syntax is:

```
void array_map ($t2 map_f ($t1, Index), array<$t1> from, array<$t2> to) ;
```

The source and the target arrays do not necessarily have the same element type. The result is placed in another array rather than returned, since the latter would lead to the creation of a temporary distributed array, whereas our solution avoids this additional memory consumption. The return-solution is however used in `array_create`, since this skeleton allocates the new array anyway, whereas `array_map` only 'fills in' new values in an existing array. Note that this efficiency improvement is not directly possible in a functional host language, where side-effects are not allowed[1].

The complexity of this skeleton is similar to that of `array_create`: $t(n) \in \mathcal{O}\left(t_m \cdot n^d/p\right)$, where t_m is the complexity of the applied function `map_f`.

array_fold composes ("folds together") all elements of an array.

```
$t2 array_fold ($t2 conv_f ($t1, Index), $t2 fold_f ($t2, $t2),
                array <$t1> a) ;
```

The skeleton first applies the conversion function `conv_f` to all array elements in a *map*-like way[2]. After that, each processor composes all elements of its partition using the folding function `fold_f`. In the next step, the results from all partitions are folded together. Since the order of composition is non-deterministic, the user should provide an associative and commutative folding function, otherwise the result is non-deterministic. In our implementation, this step is performed along the edges of a virtual tree topology, with the result finally collected at the root. In order to make the result known to all processors, it is broadcasted from the root along the tree edges to all other processors.

---

[1] A work-around might be here *deforestation*, however this is difficult for higher-order programs.

[2] This step could also be done by a preliminary `array_map`, but our solution is more efficient.

If t_c and t_f are the complexities of the conversion and folding function, respectively, then the complexity of the local computations is given by the initial *map* and the local (sequential) folding and amounts to $\mathcal{O}\left((t_c + t_f) \cdot n^d/p\right)$. The complexity of communication and non-local computation is given by the folding of the single results from each processor and by the broadcasting of the final results and is $\mathcal{O}\left((t_f + 1) \cdot \log_2 p\right)^3$. The overall complexity of the *fold* skeleton is thus: $t(n) \in \mathcal{O}\left(t_c \cdot n^d/p + t_f \cdot (n^d/p + \log_2 p)\right)$.

3.3 Communication Skeletons

array_rotate_parts_horiz works on a hardware mesh topology of $\sqrt{p} \times \sqrt{p}$ processors, on which a virtual torus topology has been defined, by cyclically shifting the partitions of an array in horizontal direction. The number of shifts is given by the return value of the argument function, which takes the row coordinate of the current processor as argument. If the result of this function is positive, then a shift to the right is performed, otherwise a shift to the left. A number of shifts s, with $|s| > \lfloor \sqrt{p}/2 \rfloor$ is reduced to $\sqrt{p} - (|s| \bmod \sqrt{p})$ shifts in the opposite direction. The skeleton *array_rotate_parts_vert* analogously rotates the partitions of an array vertically on the processor topology, whereas the positive direction is downwards. The syntax of the two skeletons is:

```
void array_rotate_parts_horiz (int f (int), array <$t> a, array <$t> b);
void array_rotate_parts_vert  (int f (int), array <$t> a, array <$t> b);
```

The complexity of the two skeletons depends on the result of the argument function f: $t(n) \in \mathcal{O}\left(n^2/p \cdot \min\left\{\frac{\sqrt{p}}{2}, \max_{i \in \{0,\ldots,\sqrt{p}-1\}}\{\mathtt{f}(i)\}\right\}\right)$.

4 Sample Applications

We shall now present two sample applications illustrating the way parallel programs can be written using algorithmic skeletons. The first program multiplies two matrices using Gentleman's algorithm [14], the second uses a statistical ("Monte Carlo") numerical method to solve a partial differential equation [11].

We have implemented the skeletons and applications on a Parsytec MC system with 64 T800 transputers connected as a 2-dimensional mesh and running at 20Mhz, under the Parix operating system [13]. Since only 1MB of memory was available per node, larger problem sizes could only be fitted into larger networks.

We have compared our results with those obtained for the same applications using the data-parallel functional language DPFL [9, 10] and the same skeletons. Our run-times are faster than those of DPFL, approaching in most cases those of hand-written C code. On the one hand, this is due to the efficiency of imperative languages, which is higher than that of their functional counterparts, on the other hand, it is due to the translation of the functional features done by the Skil compiler [2]. Detailed results are given throughout this section.

---

[3] Actually, if the hardware topology is not a tree, then this complexity becomes $\mathcal{O}\left((t_f + 1) \cdot \log_2 p \cdot \delta\right)$, where δ is the dilation of embedding the tree into the hardware topology [15].

4.1 Matrix Multiplication

The algorithm of Gentleman multiplies two matrices distributed block-wise onto a two-dimensional torus network. It firstly shifts the partitions of the first matrix placed on the ith row of processors cyclically i times to the left, and the partitions of the second matrix placed on the jth column of processors j times upwards. After this restructuring, the partitions of the two matrices placed on the same processor can be multiplied, since their pairwise corresponding inner indices are equal (i.e. a_{ik} and b_{kj}). After this multiplication, the partitions of the first matrix are shifted one step to the left and those of the second matrix one step upwards. This leads to new index combinations, but these combinations also have the 'appropriate' indices, so that the partitions can again be multiplied with one another and the result added to that of the previous multiplication. After repeating this combination of computation and communication $\sqrt{p}$ times, where $\sqrt{p} \times \sqrt{p}$ is the size of the network, we obtain the product matrix of the two initial matrices. The algorithm is presented in detail in [14].

The Straightforward Implementation with Skeletons. The straightforward solution is to use the *map* skeleton for the computation steps and the *rotate_parts* skeletons for the communication steps. The program is given below.

```
void matmult (int n) {
   array <int> a, b, c ;
   a = array_create (2, {n,n}⁴, {0,0}, {-1,-1}, init_f1, DISTR_TORUS2D) ;
   b = array_create (2, {n,n}, {0,0}, {-1,-1}, init_f2, DISTR_TORUS2D) ;
   c = array_create (2, {n,n}, {0,0}, {-1,-1}, zero, DISTR_TORUS2D) ;

   array_rotate_parts_horiz ((-)(0), a, a) ;
   array_rotate_parts_vert  ((-)(0), b, b) ;
   for (i = 0 ; i < xPartDim ; i++) {
      array_map (local_scal_prod (a, b), c, c) ;
      array_rotate_parts_horiz (minone, a, a) ;
      array_rotate_parts_vert  (minone, b, b) ; }
   /* output array c */
   array_destroy (a) ;   array_destroy (b) ;   array_destroy (c) ; }
```

In the first line, the variables a, b and c are declared as having the type "distributed array with integer elements". The arrays are then created with dimension 2 and size total size n × n by the *create* skeleton. They are distributed onto a 2-dimensional virtual torus topology, since this optimizes the communication inside the *rotate_parts* skeletons[5]. The dimensions of the torus (here $\sqrt{p}$), which determine the number of partition rotations, are given by the predefined variables xPartDim and yPartDim. The elements of a and b are initialized by some user-defined functions init_f1 and init_f2, whereas those of the result

---
[4] In order to avoid excessive code details, we have used the pseudo-code notation {a,b} for the 'classic' array with elements a and b.

[5] These skeletons can also work on non-torus topologies, however less efficiently.

matrix c are set to zero. The first calls of the *rotate_parts* skeletons perform the initial restructuring. The partial operator application (-)(0) implements in an efficient way the negation function[6], whereas **minone** is the constant function -1 [7]. The local partition multiplications are done by mapping the function local_scal_prod to every element of the result matrix c. This function computes the result of the scalar product of the corresponding row of the a-partition and the corresponding column of the b-partition, and adds it to the current element of c (denoted here by **v**):

```
$t local_scal_prod (array <$t> a, array <$t> b, <$t> v, Index ix) {
    int k, xLowerBd, xUpperBd ;
    Bounds bds ;

    bds = array_part_bounds (a) ;
    xLowerBd = bds->lowerBd[0] ;
    xUpperBd = bds->upperBd[0] ;

    for (k = xLowerBd ; k < xUpperBd ; k++)
        v += array_get_elem (a, {ix[0],k}) * array_get_elem (b, {k,ix[1]});
    return (v) ; }
```

Note that this function is called from within array_map and that it gets the current array element **v** and its index **ix** from this skeleton (see Subsection 3.2). However, it needs as further arguments the arrays **a** and **b**. This can be done without altering the type of the functional argument of the *map* skeleton by *partially applying* local_scal_prod to these two arguments in the main procedure, thus creating a function which has the type expected by *map*.

We have measured the run-times of the Skil program for matrix sizes between 100×100 and 800×800 on 4 to 64 transputers. The results are given in Table 1, where bold entries stand for absolute run-times, roman font entries denote speed-ups relative to the DPFL implementation, and entries in italics remaining slow-down factors with respect to the Parix-C[8] implementation. Note that the Skil program is on the average 4.5 times faster that the DPFL one, while being about 2 times slower than its Parix-C counterpart. This is already a good result, but we shall show in the next subsection how it can be improved further.

The Optimized Version. Upon analyzing the performance of the above matrix multiplication program, we have found out, that the main cause for the remaining overhead lies in the way array elements are accessed. In order to keep the mapping of the array onto processors transparent to the user, the macro array_get_elem gets the global index of an array element and converts it internally to the local index used for the actual access, by subtracting from it the lower index ('offset') of the current partition. In the above algorithm, this

---

[6] Since $(-)(0)(x)$ is the explicitly curried form of $(-)(0, x)$ which is equivalent to $0 - x = -x$. This is more efficient than using an new function neg, because it avoids the additional function call.

[7] The values of these functions are negative, since the rotations are done to the left and upwards, respectively (see Subsection 3.3).

[8] Parix-C is a parallel C dialect based on message-passing [13].

$\sqrt{p} \times \sqrt{p}$ \ n	100	200	300	400	500	600	700	800
2 × 2	**4.22** 4.53 *2.11*	**33.46** 4.51 *2.10*	**113.26** 4.46 *2.08*	–	–	–	–	–
4 × 4	**1.12** 4.44 *2.04*	**8.57** 4.48 *2.07*	**28.67** 4.47 *2.07*	**67.54** 4.47 *2.07*	**131.52** 4.48 *2.07*	**228.00** 4.46 *2.07*	–	–
6 × 6	**0.56** 4.34 *2.00*	**4.16** 4.47 *2.03*	**12.92** 4.48 *2.05*	**30.92** 4.46 *2.06*	**60.55** 4.47 *2.06*	**101.73** 4.47 *2.06*	**162.38** - *2.06*	–
8 × 8	**0.36** 4.33 *1.89*	**2.26** 4.47 *2.00*	**7.69** 4.46 *2.03*	**17.26** 4.47 *2.04*	**34.20** 4.49 *2.05*	**57.74** 4.46 *2.05*	**92.82** 4.47 *2.05*	**135.93** 4.46 *2.05*

Table 1. Run-time results for matrix multiplication (straightforward version) [**bold**: absolute times; roman: DPFL/Skil; *italics*: Skil/Parix-C]

conversion is done in the innermost loop, consequently its overhead is amplified by a factor of $\mathcal{O}(n^3)$, where n is the size of the array. We have therefore defined a more efficient macro for element accessing, which gets directly the *local* index, and additionally, for access efficiency, the size of the current partition:

`$t array_get_local_elem (array <$t> a, Index local_ix, Size part_size) ;`

where both `local_ix` and `part_size` can be derived from the local bounds of the current array partition. Using this feature (which does not collide with the transparency of the array mapping onto the processors), we can perform the conversion from global to local coordinates outside the innermost loop. The function `local_scal_prod` can thus be rewritten as follows.

```
                    /* optimized version with local array element access */
$t local_scal_prod (array <$t> a, array <$t> b, <$t> v, Index ix) {
  Index a_lowerix, b_lowerix, a_localix, b_localix ;
  Size a_psize, b_psize ;

  /* derive lower indices (*_lowerix) and sizes (*_psize) for the
     partitions of the matrices a and b from their lower bounds      */
  a_localix[0] = ix[0] - a_lowerix[0] ;
  b_localix[1] = ix[1] - b_lowerix[1] ;
  for (k = 0 ; k < a_psize ; k++) {
     a_localix[1] = b_localix[0] = k ;
     v += array_get_local_elem (a, a_localix, a_psize) *
          array_get_local_elem (b, b_localix, b_psize)  ; }
  return (v) ; }
```

The run-time results for this optimized version, given in Table 2, show that we have removed the main cause for the remaining efficiency gap. We have thus approached the performance of the C version up to about 20%, while becoming 7 to 8 times faster than DPFL.

$\sqrt{p} \times \sqrt{p}$ \ n	100	200	300	400	500	600	700	800
2×2	**2.40** 7.97 *1.20*	**18.79** 8.03 *1.18*	**63.57** 7.94 *1.17*	–	–	–	–	–
4×4	**0.67** 7.42 *1.22*	**4.93** 7.79 *1.19*	**16.35** 7.84 *1.18*	**38.24** 7.90 *1.17*	**74.13** 7.95 *1.16*	**128.80** 7.89 *1.17*	–	–
6×6	**0.35** 6.94 *1.25*	**2.45** 7.57 *1.20*	**7.47** 7.75 *1.19*	**17.74** 7.80 *1.18*	**34.51** 7.84 *1.17*	**57.70** 7.88 *1.17*	**91.78** - *1.16*	–
8×8	**0.23** 6.78 *1.21*	**1.36** 7.39 *1.20*	**4.52** 7.58 *1.19*	**9.99** 7.72 *1.18*	**19.62** 7.82 *1.17*	**33.04** 7.80 *1.17*	**52.86** 7.85 *1.17*	**77.21** 7.85 *1.17*

Table 2. Run-time results for matrix multiplication (optimized version) [**bold**: absolute times; roman: DPFL/Skil; *italics*: Skil/Parix-C]

4.2 A Monte Carlo Algorithm for Solving PDEs

The second application we consider is a statistical numerical method for the solution of a partial differential equation (PDE). The idea is to find a random variable θ, such that its expectation $E\theta$ coincides with the solution of the problem. Having found such a θ, we need to generate n realizations of it $(\theta_1, \ldots, \theta_n)$ and obtain the value of the solution as their arithmetic mean $E\theta \approx \frac{1}{n} \sum_{i=1}^{n} \theta_i$. An important advantage of this method is that the realizations θ_i are independent from one another, and can thus be computed in parallel with no communication in between. Communication is only needed in the final phase, for collecting the single realizations.

We consider the following partial differential equation:

$$\frac{\partial^2 u(x,y)}{\partial x^2} + \frac{\partial^2 u(x,y)}{\partial y^2} = 0$$

defined on the domain $[0,1] \times [0,1]$ with the boundary conditions

$$u(0,y) = y \text{ and } u(1,y) = 1+y, \; \forall y \in [0,1]$$
$$u(x,0) = x \text{ and } u(x,1) = x+1, \; \forall x \in [0,1]$$

The domain is discretized to a rectangular mesh with step h, and the PDE is discretized by finite differences to

$$u(x,y) = \frac{u(x+h,y) + u(x-h,y) + u(x,y+h) + u(x,y-h)}{4}$$

If (x_0, y_0) is the point where we want to know the solution of the PDE, then we must find a random variable θ, such that $u(x_0, y_0) = E\theta$. For that, we construct a random trajectory from our initial point to the boundary of the domain using the following Markov process:

$(x, y) := (x_0, y_0)$
while (x, y) is not on the boundary of the domain
$\quad (x, y) := (x', y')$
$\quad$ where (x', y') is selected from the set of neighboring points
$\quad\quad \{(x - h, y), (x + h, y), (x, y - h), (x, y + h)\}$
$\quad\quad$ with the same probability $p = 1/4$

Let (x^*, y^*) be the final point on the boundary. Then, our random variable is $\theta := u(x^*, y^*)$. It can be proved that $u(x_0, y_0) = E\theta$, which validates the result. The algorithm is described in detail in [10, 11].

The Skil implementation of this Monte Carlo algorithm is very simple, compared to its Parix-C counterpart presented in [11]. Each of the p processors computes n/p random trajectories[9] and places the results in its elements of a 1-dimensional distributed array of total size n. The trajectories are computed during the initialization phase of the array by the argument function rand_traj. The additional parameters needed by rand_traj (the initial point (x_0, y_0), the discretization step h and the number of trajectories n) are supplied via *partial application*. After the random trajectories have been computed and the results placed in the distributed array, the single results must be collected and combined, in order to yield the actual result $E\theta$. This is done by the skeleton array_fold, which computes the sum of all elements of the array. Finally, the arithmetic mean is computed by dividing the result of the *fold* operation by n. The procedure is given below.

```
double monte_carlo (double x0, double y0, double h, int n) {
   double res ;
   a = array_create (1, {n,0}, {0,0}, {-1,-1}, rand_traj (x0, y0, h,
                    n/netSize), DISTR_DEFAULT) ;
   res = array_fold (ident, (+), a) / n ;
   array_destroy (a) ;
   return (res) ; }
```

The run-times obtained for the Skil implementation of this algorithm are given in Table 3. We have compared these results with those obtained for the Parix-C and DPFL programs, respectively. For more clarity, we have plotted the speedups relative to DPFL and the slow-downs relative to Parix-C against the number of processors, for different values of h and n and obtained the graphics in Fig. 1. Again, the Skil implementation is faster than the DPFL one, with most of the relative speedups grouped around 5-6 (left graphic). On the other hand, the performance of Skil is comparable to that of C, most of the relative slow-downs being clustered around 1 (right graphic).

---

[9] The predefined variable netSize contains the number of processors p.

h		0.05			0.02			0.005		
p	n	1000	5000	10000	1000	5000	10000	1000	5000	10000
3		0.98	4.67	9.31	5.31	26.51	52.56	87.05	423.02	844.96
7		0.46	2.06	4.08	2.38	11.64	22.99	38.10	186.29	371.88
15		0.25	1.02	1.96	1.20	5.54	11.07	19.23	88.80	172.56
31		0.17	0.55	1.04	0.72	2.92	5.51	10.15	45.26	86.95
63		0.11	0.30	0.55	0.39	1.52	2.84	5.83	23.30	44.91

Table 3. Run-time results for the Monte Carlo algorithm

Notice that in some cases the relative slow-downs are slightly below 1, i.e. the Skil run-times even beat the C run-times. The reason is that the C implementation referred to here is an older version, which does not use virtual topologies or asynchronous communication, as our skeleton implementation does. Of course, a Skil program could never beat an equally well optimized C version of that program, since Skil is translated to message-passing C.

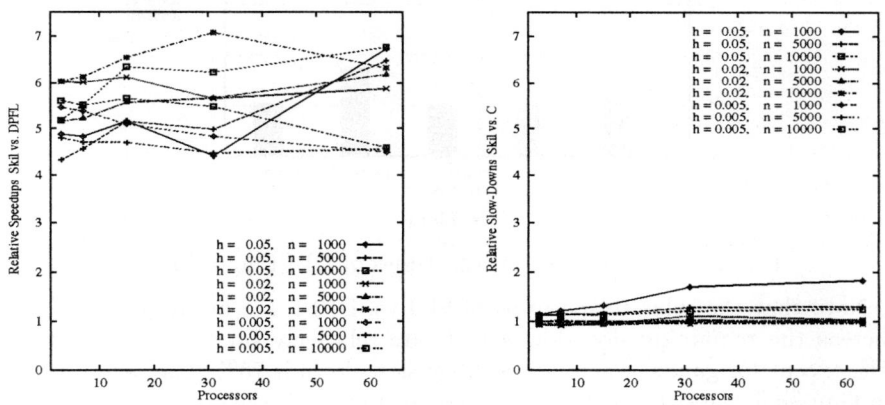

Fig. 1. Comparison: Skil vs. DPFL (left) and Skil vs. Parix-C (right) for the Monte Carlo algorithm

4.3 Further Run-Time Results

We have implemented further applications using both the skeletons presented above, and some additional skeletons, like for instance partition broadcasting, row permutation and generic matrix multiplication [2]. The latter encapsulates the above Gentleman algorithm, however parameterized with the functions used to compute the "scalar product" of a row and a column. If the actual multiplication and addition are supplied, then we obtain the classical matrix multiplication. However, by using other functions, different algorithms with this communication pattern can be implemented, like for instance shortest paths in graphs[10].

---

[10] Here, we 'multiply' the distance matrix of a graph with itself using addition in the role of the generic multiplication and the minimum function in the role of the generic addition [2].

We have considered the following applications: `matrix mult1` is the optimized matrix multiplication procedure (Subsection 4.1.2); `matrix mult2` is the implementation of this procedure based on the generic skeleton outlined above; `shortest paths` is also based on this skeleton; `gauss` is the Gaussian elimination procedure, and finally `monte carlo` is the implementation of the algorithm given in Subsection 4.2. For each of these programs, we have compared the average of all run-time measurements for the Skil version with the averages for the DPFL and C versions. The results are depicted in Fig. 2.

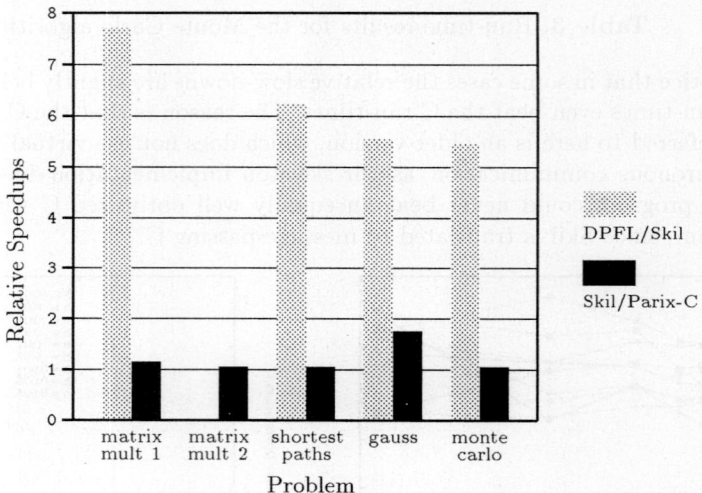

Fig. 2. Skil run-times vs. run-times of DPFL and Parix-C

Note that the relative speedups of Skil vs. DPFL are between 5.5 and 7.5, whereas the remaining slow-downs to C are relatively small, between 5% and 20%, except for `gauss`, where the average slow-down is 75%. The reason is that the Gaussian algorithm cannot be implemented with skeletons as straightforwardly as the other applications, so that this overhead has a harder impact on the overall performance. However, a worst-case slow-down factor of 1.75 is still a good result for practical applications.

5   Conclusions and Future Work

We have presented a new approach to parallel programming with algorithmic skeletons. We have designed a language that allows high-level parallel programming and at the same time can be efficiently implemented, namely an imperative language enhanced with functional features and a polymorphic type system.

We have then described a series of skeletons for the work with distributed arrays and showed how two matrix applications can be implemented on the basis of skeletons on a parallel system. Run-time results have shown that our implementation was faster than that of a functional language with skeletons, while approaching the performance of hand-written C code based on message-passing.

Future work is necessary in several directions. Firstly, the distributed data type `array` together with its skeletons are only at a prototype level. It would be interesting to support other distributions onto processors, apart from block-wise, like cyclic, block-cyclic etc. Further, in case of block distributions, it should be possible to define overlapping areas for the single partitions, in order to reduce communication in operations which require more than one element at a time. Such operations are used for instance in solving partial differential equations [10] or in image processing. In order to be able to cope with 'real world' applications, new skeletons, for instance for (parallel) I/O must be designed and implemented. Moreover, the implementation of skeletons can be optimized, for instance by allowing the application of argument functions only to certain elements of a distributed data structure, since this would be more efficient than testing the condition in the customizing function itself.

References

1. B. Bacci, M. Danelutto, S. Orlando, S. Pelagatti, M. Vanneschi: P^3L : *a Structured High-level Parallel Language and its Structured Support*, Technical Report HPL-PSC-93-55, Hewlett-Packard Laboratories, Pisa Science Center, 1993.
2. G. H. Botorog, H. Kuchen: Skil: An Imperative Language with Algorithmic Skeletons for Efficient Distributed Programming, to appear in *Proceedings of the Fifth International Symposium on High Performance Distributed Computing (HPDC-5)*, IEEE Computer Society Press, 1996.
3. G. H. Botorog, H. Kuchen: Using Algorithmic Skeletons with Dynamic Data Structures, to appear in *Proceedings of IRREGULAR '96*, LNCS, Springer, 1996.
4. M. I. Cole: *Algorithmic Skeletons: Structured Management of Parallel Computation*, MIT Press, 1989.
5. J. Darlington, A. J. Field, P. G. Harrison et al: Parallel Programming Using Skeleton Functions, in *Proceedings of PARLE '93*, LNCS 694, Springer, 1993.
6. J. Darlington, Y. Guo, H. W. To, J. Yang: Functional Skeletons for Parallel Coordination, in *Proceedings of EURO-PAR '95*, LNCS 966, Springer, 1995.
7. J. T. Feo, D. C. Cann: A Report on the Sisal Language Project, in *Journal of Parallel and Distributed Computing*, Vol. 10, No. 4, 1990.
8. I. Foster, R. Olson, S. Tuecke: Productive Parallel Programming: The PCN Approach, in *Scientific Programming*, Vol. 1, No. 1, 1992.
9. H. Kuchen, R. Plasmeijer, H. Stoltze: Efficient Distributed Memory Implementation of a Data Parallel Functional Language, in *Proceedings of PARLE '94*, LNCS 817, Springer, 1994.
10. H. Kuchen: *Datenparallele Programmierung von MIMD-Rechnern mit verteiltem Speicher*, Thesis (in German), Shaker-Verlag, Aachen, 1996.
11. H. Kuchen, H. Stoltze, I. Dimov, A. Karaivanova: Distributed Memory Implementation of Elliptic Partial Differential Equations in a Dataparallel Functional Language, in *Proceedings of MPPM '95*, IEEE Computer Society Press, 1995.
12. L. D. J. C. Loyens, J. R. Moonen: ILIAS, a Sequential Language for Parallel Matrix Computations, in *Proceedings of PARLE '94*, LNCS 817, Springer, 1994.
13. Parsytec Computer GmbH: PARIX 1.2, Software Documentation, Aachen, 1993.
14. M. J. Quinn: *Parallel Computing: Theory and Practice*, McGraw Hill, 1994.
15. M. Röttger, U. P. Schroeder, J. Simon: *Virtual Topology Library for PARIX*, Technical Report 148, University of Paderborn, 1994.

A Loosely Synchronized Execution Model for a Simple Data-Parallel Language
(Extended Abstract)

Yann Le Guyadec[2], Emmanuel Melin[1], Bruno Raffin[1]
Xavier Rebeuf[1] and Bernard Virot[1][*]

[1] LIFO - IIIA Université d'Orléans
4, rue Léonard De Vinci - BP 6759 F-45067 Orléans Cedex 02 - FRANCE
[2] Currently affiliated to VALORIA 8 rue Montaigne - BP 1104
F-56014 Vannes - FRANCE

Abstract. Data-parallel languages offer a programming model structured and easy to understand. The challenge consists in taking advantage of the power of present parallel architectures by a compilation process allowing to reduce the number and the complexity of synchronizations. In this paper, we clearly separate the synchronous programming model from the asynchronous execution model by the way of a translation from a synchronous data-parallel programming language into an asynchronous target language. The synchronous data-parallel programming language allows to temporarily mask local computations. The asynchronous target language handles explicit and partial synchronizations through the use of structural clocks.

In the area of parallel programming, a crucial step has been performed with the emergence of the data-parallel programming model. Parallelism is expressed by means of data-types which are promoted from scalars to vectors. Its leads to a distribution mechanism which maps components of data-types over a network of *virtual processors*. From the programmer's point of view, a program is a sequential composition of operations (local computations or global rearrangements) which are applied to restricted parts of large data-structures. The distribution mechanism is easily scalable and the model is adapted to many scientific applications. The challenge consists in taking advantage of the power of present parallel architectures by a compilation process allowing to reduce the number and the complexity of synchronizations. The effort is then transferred to the compiler which has to fill the gap between the abstract *synchronous* and *centralized* programming model and an *asynchronous* and *distributed* execution model that depends on the target architecture. Data-parallel compilers have to perform complex data-flow analysis to manage data distribution and desynchronizations between processors, especially in loop structures. The efficiency of the produced

---

[*] Authors contact: Bernard Virot (Bernard.Virot@lifo.univ-orleans.fr). This work has been partly supported by the French CNRS Coordinated Research Program on Parallelism PRS.

code heavily depends on static informations relating to data dependences which are available at compile time. Automatic parallelizers can detect regular dependences especially in nested loops. When complex and irregular data-structures must be handled, explicit expression of dependences is mandatory in order to obtain automatic and efficient desynchronizations.

In existing data-parallel languages data-dependence is the default and only independence is made explicit (cf. the **FORALL** and **INDEPENDENT** constructs in HPF [2]). We propose a kernel data-parallel language called $\mathcal{TML}$ (which stands for Twin Memory Language) which purpose is to offer both a synchronous programming model and an asynchronous execution model. In this language independence is the default and data dependence is made explicit in the syntax. To achieve this aim, we reuse the twin memory model, previously introduced for the language $\mathcal{D}$ [10], to temporarily mask local computations. The $\mathcal{TML}$ language yields a synchronous programming model, similar to classical data-parallel languages such as C^* [12], HYPERC [9], Data–Parallel C [8] or $\mathcal{L}$ [1]. Taking advantage of the absence of implicit dependence in the language, our main contribution consists in showing that it is possible to perform an automatic desynchronization of $\mathcal{TML}$ programs. We propose an execution model relying on the translation of $\mathcal{TML}$ programs into the asynchronous and non-deterministic data-parallel language $\mathcal{SCL}$, previously introduced in [5, 7]. The desynchronization is based on a partial synchronization algorithm relying on structural clocks. We extend the $\mathcal{SCL}$ language with a wait instruction performing a point to point synchronization. Determinism of translated $\mathcal{SCL}$ programs yields the correctness of the translation function.

In this paper, we first introduce the data-parallel model. We give an informal description of the $\mathcal{TML}$ language. Next, we briefly recall the semantics of the $\mathcal{SCL}$ language. The translation of $\mathcal{TML}$ towards $\mathcal{SCL}$ is presented in the third section. Finally we illustrate the benefit of our approach with an example.

1 The $\mathcal{TML}$ language

We turn now to the description of the $\mathcal{TML}$ language (which stands for Twin Memory Language) which aim is to offer both a synchronous programming model and an asynchronous execution model.

1.1 The data-parallel programming model

In the data-parallel programming model, basic objects are arrays with parallel access, called vectors. They are denoted with uppercase initial letters. The component of the parallel variable X located at index u is denoted by $X|_u$. Expressions are evaluated by applying operators componentwise to parallel values. Each action is associated with the set of array indices to which it is applied. This set is called the *activity context* or *extent of parallelism*. Indices to which an operation is applied are called *active*, whereas others are *idle*. Legal expressions are usual *pure* expressions, i.e. without side effect, like the definition of *pure* functions in HPF [2]. The value of a pure expression at each index u only depends on the values of variable components at index u. We make use of a special vector constant called *This*. The value of its component at each index u is the value u itself: $This|_u = u$.

1.2 The twin memory model

$\mathcal{TML}$ reuses the twin memory management previously introduced for the language $\mathcal{D}$ [10]. The model relies on an abstract machine where each index u owns a *private memory* containing the components of all vector variables at index u, and a *public memory* containing a copy of the private memory of the same index. Private memories are dedicated to local computations (evaluation of *pure* expressions or assignments) whereas public memories are referred by communications. Public memory updates are explicitly executed through a specific instruction (dump). As only the dump instruction makes a local transition visible for other indices, this mechanism suppresses all implicit dependences.

1.3 Informal description of the $\mathcal{TML}$ language

We turn now to an informal description of the $\mathcal{TML}$ language.

No action: skip. This instruction does nothing and just terminates.

Memory copy: dump. All *active* indices copy their private memory into their public one.

Assignment: $X := E$. For each active index u, the component $X|_u$ of the *private* memory is updated with the local value $E|_u$. As E is a *pure* expression, $E|_u$ is evaluated in the *private* memory of the index u.

Communication: get X from A into Y. The address A is a *pure* expression. Each active index u evaluates $A|_u$ in its private memory. Then, u fetches the remote value of X in the *public* memory of the index $A|_u$. The result is assigned to the component $Y|_u$, in the *private* memory of the index u.

Sequence: $S;T$. At the end of the execution of the last instruction of S, the execution of the instructions of T starts.

Conditioning: where B do S elsewhere T end. This construct splits the set of currently active indices into two disjoint subsets depending on the value of the *pure* expression B. The active indices evaluating B to *true* execute S. Next, the other active indices execute T. This is the instruction if/else of MPL [3] or where/elsewhere of HYPERC [9].

Iteration: loopwhere B do S end. Iteration is expressed by classical loop unfolding. Only the active indices evaluating the *pure* expression B to *true* execute S with the new extent of parallelism. The other ones do not execute the loop. Note that the set of active indices may decrease at each loop iteration. An execution of a loop terminates if all active indices evaluate B to *false*. This is the instruction while of MPL [3] or the whilesomewhere of HYPERC [9].

Example. In this $\mathcal{TML}$ program (Fig. 1), after the first instruction, all the indices copy their private memory in their public one. Then, the indices executing the first branch of the where/elsewhere construct execute a dump again. Hence, at the end of the execution, the indices for which $This > 3$ store the value 0 in their component of Y and the index 3 stores 1. The data dependences between indices occur only through the dump and the get instructions.

```
X := 0;
dump;
where (This < 3) do
    X := 1;
    dump
elsewhere
    get X from This − 1 into Y
end
```

Fig. 1. A $\mathcal{TML}$ program.

2 The target language $\mathcal{SCL}$

For the sake of completeness, in this section we briefly present the target language $\mathcal{SCL}$ previously introduced in [6, 7]. It yields an asynchronous and non-deterministic execution model.

2.1 Informal description

Let us turn now to an informal description of the $\mathcal{SCL}$ language. Like $\mathcal{TML}$, this language uses a twin memory management. From $\mathcal{TML}$, it reuses some instructions with the same semantics: assignment, skip, get, dump. We only present the other $\mathcal{SCL}$ instructions concerning sequence, loop, conditioning and synchronization.

Asynchronous sequence: $S; T$. Each index independently executes S and then T. Hence, the sequence yields only local synchronizations: an index may execute T before the other indices terminate the execution of S.

Concurrent conditioning: where B do S elsewhere T end. This statement is an asynchronous double-branched conditioning construct as found in some optimizations of HYPERC [11]. It was also introduced in the $\mathcal{D}$ language [10] to express parallel execution of disjoint programs. The set of active indices is divided into two sets depending on the value of the pure expression B, which is evaluated in the private memories. An active index u which evaluates the expression $B|_u$ to $true$ (respectively $false$) executes the program S (respectively T). In contrast to the where/elsewhere statement of $\mathcal{TML}$, the two branches are executed concurrently.

Asynchronous loop: loopwhere B do S end. This is an asynchronous version of the $\mathcal{TML}$ instruction loopwhere. Each currently active index u repeats the computation of the block S while the pure expression $B|_u$ evaluates to $true$ in its private memory. Note that the set of active indices may decrease after each loop iteration. When an index terminates the loop, it goes on executing instructions that directly follow the loop without any implicit synchronization.

Waiting: wait A. The wait A statement is used to manage the consistency of public memory references by serializing instructions which interfere with this memory. The expression A is pure. The instruction wait A delays the index u which is executing it, until the target index $A|_u$ has stepped over all the instructions referencing public memory and preceding the wait instruction. Since where and loopwhere govern the activity, if the wait point is nested in such control structures, it becomes possible that v could never reach some preceding instruction referencing the public memory. In this case u can step over the wait point as soon as v is engaged in a different branch from u in a where/elsewhere structure, or as soon as v has terminated the current loopwhere executed by u. We call *meaningful* the instructions that play a part in the wait management, namely dump, get, where and loopwhere. To sum up, the instruction wait A can terminate if and only if the two following conditions **C1** and **C2** are satisfied.

- **C1:** The index $A|_u$ has evaluated the boolean conditions of all enclosing where and loopwhere constructs.
- **C2:** If the index $A|_u$ enters the innermost instruction block inside which u is waiting, it must have passed all *meaningful* statements coming before the wait instruction.

Remark. Note that an index may step over a waiting point even if the waited index needs to execute non-meaningful statements before reaching it.

Synchronization: wait_all. This instruction is a generalization of the wait instruction. An index terminates this instruction when it has performed a wait towards all other indices. Note that this instruction is a partial synchronization since some indices may not execute it.

2.2 Structural Clocks presentation

In order to formalize index positions we introduce *Structural Clocks* [7]. Each index owns a clock that encodes its current position, during program execution, with regard to the meaningful statements. An index position is defined by the meaningful control structures it is nested in, and by the last meaningful instruction executed in the innermost one. The *Structural Clock* t_u of an index u is expressed by a list of pairs. Each term of the list corresponds to a nesting level. Each pair (l,c) is composed of a label l and an instruction counter c. The counter c represents the number of meaningful instructions already executed in the corresponding instruction block. The label l is used to distinguish which branch of a where/elsewhere statement an index is inside. Lists are built up by popping or pushing pairs on their right hand side.

- When the index u executes a get or a dump instruction its clock $t_u = t(l,c)$ becomes $t_u = t(l, c+1)$.
- After evaluating the condition B of a where/elsewhere statement, if $B|_u$ is $true$ (resp. $false$) its clock $t_u = t$ becomes $t_u = t(1,0)$ (resp. $t_u = t(2,0)$). When the index u exits a where or a elsewhere branch, its clock $t_u = t(l,c)(m,d)$ becomes $t_u = t(l, c+1)$.

In a loopwhere, an index which does not compute an iteration directly exits the loop. Therefore, it turns out that we do not have to consider each iteration as a new nesting level for the structural clock list. We only push a new term on the list at the first loop iteration and next, we count instructions already executed in the loopwhere body.

- When the index u enters for the first time a loopwhere instruction, if the condition $B|_u$ evaluates to $true$, then its clock $t_u = t$ becomes $t_u = t(1,0)$. Otherwise, its clock $t_u = t(l,c)$ becomes $t_u = t(l, c+1)$.
- If it has already executed at least one loop iteration, then if $B|_u = true$ its clock $t_u = t$ remains unchanged, otherwise $t_u = t(l,c)(1,d)$ becomes $t_u = t(l, c+1)$.

Remark. At the beginning of the execution, each index initializes its structural clock to $(0,0)$.

Structural clock ordering. To order two structural clocks, we compare the local counters corresponding to the common instruction block of the innermost nested level. We so define this partial ordering as a lexicographical order based on partially ordered pairs.

Definition. An index u is said to be *later* than an other index v (denoted by $t_u \prec t_v$) if one of the following conditions holds:
- there exists t not empty such that $t_v = t_u\, t$;
- there exists t^1, t^2, t^3 (possibly empty) and c_u, c_v, l such that $t_u = t^1(l, c_u)t^2$, $t_v = t^1(l, c_v)t^3$ and $c_u < c_v$.

Handling wait statements using structural clocks. Structural clocks yield a general mechanism to handle wait statements. The conditions **C1** and **C2** above are formalized by the following theorem.

Theorem 1. *Consider an index u with structural clock t_u that executes an instruction* wait *A. This index can terminate the wait statement if and only if the index $v = A|_u$ holds a structural clock t_v which satisfies $\neg(t_v \prec t_u)$.*

3 Translation from $\mathcal{TML}$ into $\mathcal{SCL}$

In this section, we show that it is possible to perform an automatic desynchronization of $\mathcal{TML}$ programs by the way of a translation function $\mathcal{F}$ from $\mathcal{TML}$ programs to $\mathcal{SCL}$ ones.

We introduce some necessary notations to explain the translation function. The predicate $Hasdump$ allows to know if a program S contains an instruction dump: $Hasdump(S) = true$ if S contains at least one dump, $Hasdump(S) = false$ otherwise. In the same way, the predicate $Hasget(S)$ denotes the presence of get instructions in the program S.

The function $\mathcal{F}$ (Fig. 2) must translate a program while preserving its semantics. We guarantee determinism using several mechanisms:

- To prevent wrong references, an index can perform a dump instruction only if there is no index later than it. Since all the indices can refer a dump, the function $\mathcal{F}$ introduces a partial synchronization wait_all before each instruction dump (cf. 1, Fig. 2).
- An index can perform a communication only if the referred index has updated its public memory. Moreover, the instruction get points out accurate index dependences. Therefore the function $\mathcal{F}$ introduces an instruction wait before each instruction get (cf. 2, Fig. 2).
- Since the two branches of a $\mathcal{SCL}$ conditioning structure are concurrent, a $\mathcal{TML}$ where/elsewhere can be directly replaced by a $\mathcal{SCL}$ where/elsewhere only if the two branches are independent. The translation is based on a simple syntactic analysis. Independence between the two branches is ensured if a get instruction does not occur in a branch whereas a dump instruction occurs in the other one (cf. 4, Fig. 2). If the analysis does not prove the independence, the $\mathcal{TML}$ where/elsewhere is replaced by a sequence of two $\mathcal{SCL}$ where/elsewhere. The former triggers the execution of the first branch of the $\mathcal{TML}$ where/elsewhere, whereas the latter is concerned by the second branch (cf. 3, Fig. 2).

We turn now to the correctness proof of the translation function. The intuitive idea is the following. The unique (synchronous) computation of a $\mathcal{TML}$ program S can be seen as a particular computation of the translated program $\mathcal{F}(S)$. Therefore, to ensure the correctness of the translation function, it is sufficient to prove that all computations of the translated program are equivalent, i.e. translated programs are deterministic.

Definition. A synchronous execution of a $\mathcal{SCL}$ program is a computation performing a global synchronization after the execution of each instruction.

The next theorem states that for the execution of a $\mathcal{TML}$ program S and the synchronous execution of $\mathcal{F}(S)$ are equivalent.

Theorem 2. *The execution of a $\mathcal{TML}$ program terminates if and only if the synchronous execution of $\mathcal{F}(S)$ terminates. If they both terminate, they yield the same final state for the twin memories.*

$\mathcal{TML}$	$\longrightarrow$ $\mathcal{SCL}$	
$\mathcal{F}(\text{skip})$	skip	
$\mathcal{F}(X := E)$	$X := E$	
$\mathcal{F}(\text{dump})$	wait_all; dump	(1)
$\mathcal{F}(\text{get } X \text{ from } A \text{ into } Y)$	wait A; get X from A into Y	(2)
$\mathcal{F}(S;T)$	$\mathcal{F}(S); \mathcal{F}(T)$	
If $(Hasdump(S) \wedge Haseget(T)) \vee (Hasget(S) \wedge Hasdump(T))$ then $\mathcal{F}\begin{pmatrix} \text{where } B \text{ do } S \\ \text{elsewhere } T \\ \text{end} \end{pmatrix}$ otherwise	$\begin{cases} Tmp := B; \\ \text{where } Tmp \text{ do } \mathcal{F}(S) \\ \text{elsewhere skip} \\ \text{end}; \\ \text{where } Tmp \text{ do skip} \\ \text{elsewhere } \mathcal{F}(T) \\ \text{end} \end{cases}$	(3)
$\mathcal{F}\begin{pmatrix} \text{where } B \text{ do } S \\ \text{elsewhere } T \\ \text{end} \end{pmatrix}$	$\begin{cases} \text{where } B \text{ do } \mathcal{F}(S) \\ \text{elsewhere } \mathcal{F}(T) \\ \text{end} \end{cases}$	(4)
$\mathcal{F}(\text{loopwhere } B \text{ do } S \text{ end})$	loopwhere B do $\mathcal{F}(S)$ end	

Fig. 2. The translation function $\mathcal{F}$.

Theorem 3. *All $\mathcal{SCL}$ programs produced by the translation function $\mathcal{F}$ are deterministic.*

From the two previous theorems, we deduce that all the executions of a translated $\mathcal{TML}$ program always compute the expected result. The correctness of the translation function $\mathcal{F}$ is thus proven.

3.1 Example

The example (Fig. 3) illustrates the $\mathcal{TML}$ and $\mathcal{SCL}$ potentialities. Thanks to the synchronous programming model, the behavior of the $\mathcal{TML}$ program is simple. Initially each index u holds a value $X|_u$. First, it computes the maximum of the values belonging to the indices on its left side, by using a scan. Then, it computes an average between its result and its neighbor ones.

The translated $\mathcal{SCL}$ program is loosely synchronized. Overlapping of communications by computations can occur. For instance, the index 1 can execute the instruction of the line 15 (Fig. 3) while the index 2 refers its public memory through the instruction get of the line 7. But the index 1 must wait the index 2

line	instruction	clock
1	$Val := X$;	$(0,0)$
2	wait_all;	$(0,0)$
3	dump;	$(0,1)$
4	$I := 1$;	$(0,1)$
5	loopwhere $This - I > 0$ do	$(0,1)(1,3j)$
6	wait $This - I$;	$(0,1)(1,3j)$
7	get Val from $This - I$ into Y;	$(0,1)(1,3j+1)$
8	$Val := max(Val, Y)$;	$(0,1)(1,3j+1)$
9	wait_all;	$(0,1)(1,3j+1)$
10	dump;	$(0,1)(1,3j+2)$
11	$I := I * 2$	$(0,1)(1,3j+2)$
12	end;	$(0,2)$
13	wait $This - 1$;	$(0,2)$
15	get Val from $This - 1$ into Aux_1;	$(0,3)$
16	wait $This + 1$;	$(0,3)$
17	get Val from $This + 1$ into Aux_2;	$(0,4)$
18	$Val := (Val + Aux_1 + Aux_2)/3$	$(0,4)$

Fig. 3. The second column displays a $\mathcal{TML}$ program and its $\mathcal{SCL}$ translation. For the sake of consistency, the new $\mathcal{SCL}$ instructions added by the translation are displayed on the right hand side. For each instruction, the structural clock an index owns, *when it has finished to execute it*, appears in the third column. The variable j denotes the number of loop iterations already executed. We consider an array of 2^n indices numbered from 1 to 2^n. The get instructions addressing a non existing index return the default value 0.

before it executes the instruction of the line 17. The structural clock mechanism forbids the index 1 to step over the wait of the line 16 since its structural clock $(0,3)$ is greater than the structural clock $(0,1)(1,0)$ of the index 2.

Remark. The partial synchronization wait_all (line 2, Fig. 3) introduced by the translation function is useless since in the program there is no communication before. A more precise syntactic analysis can permit to detect such configurations.

4 Conclusion

In this paper, we have presented a kernel data-parallel language $\mathcal{TML}$ relying on a twin memory management. It offers a synchronous data-parallel programming model where dependences between indices are explicit in the syntax and simple to express. The execution model based on a translation into $\mathcal{SCL}$ is asynchronous, and thereby adapted to present MIMD architectures.

The translation suppresses some useless synchronizations and preserves the initial semantics of $\mathcal{TML}$ programs by only imposing waiting points. It is possible to take advantage of more syntactic information to reduce synchronization requirements [5]. We also provide formal semantics of $\mathcal{TML}$ and $\mathcal{SCL}$ languages in [5]. This theoretical framework allows us to validate the correctness of the translation function. The extension of the $\mathcal{SCL}$ language to other structures, especially functions and escape operators, is a current research direction.

An implementation of the $\mathcal{TML}$ language is under progress. This is a meaning step to tackle the problem of comparison with other approaches of compilation and optimization of data-parallel programs [4, 8].

References

1. L. Bougé and J.-L. Levaire. Control Structures for Data–Parallel SIMD Languages: Semantics and Implementation. In *Future Generation Computer Systems*, pages 363–378. North Holland, 1992.
2. C.H.Koelbel, D.B.Loveman, R.S.Schreiber, G. Jr., and M.E.Zosel. *The High Performance Fortran Handbook*. The MIT Press, 1994.
3. Digital Equipment Corporation. *DECmpp Programming Langage, Reference Manual*, 1992.
4. F.Coelho, C.Germain, and J. Pazat. State of the Art in Compiling HPF. In *Spring School on Data Parallelism*. Springer-Verlag, 1996. *(To appear)*.
5. Y. L. Guyadec, E. Melin, B. Raffin, X. Rebeuf, and B. Virot. A Loosely Synchronized Execution Model For a Simple Data-Parallel Language. Technical Report RR96-5, LIFO, Orléans, France, February 1996.
6. Y. L. Guyadec, E. Melin, B. Raffin, X. Rebeuf, and B. Virot. Horloges structurelles pour la désynchronisation de programmes data-parallèles. In *Actes de RenPar'8*, pages 77–80. Université de Bordeaux, France, may 1996.
7. Y. L. Guyadec, E. Melin, B. Raffin, X. Rebeuf, and B. Virot. Structural Clocks for a Loosely Synchronized Data-Parallel Language. In *MPCS'96* . IEEE, may 1996. *(To appear)*.
8. P. J. Hatcher and M. J. Quinn. *Data–Parallel Programming on MIMD Computers*. The MIT Press, 1991.
9. Hyperparallel Technologies. *HyperC Documentation*, 1993.
10. Y. Le Guyadec. Désynchronisation des programmes data-parallèles: une approche sémantique. *TSI*, 14(5):619–638, 1995.
11. N. Paris. Compilation du flot de contrôle pour le parallélisme de données. *TSI*, 12(6):745–773, 1993.
12. Thinking Machines Corporation, Cambridge MA. *C* Programming Guide*, 1990.

A Nonannotative Approach to Distributed Data-Parallel Computing

A Shafarenko

Department of Electronic and Electrical Engineering, University of Surrey, GU2 5XH, England.
e-mail: a.shafarenko@ee.surrey.ac.uk

Abstract. An approach to data-parallel computing is presented which avoids annotation by introducing a type system with symmetric subtyping. The properties that are usually specified in annotations in a machine-dependent way become deducible from type signatures of data objects. The chief advantage of the method is that it caters for portability by presenting a data description in terms of algorithmic properties (most importantly symmetry of data and of access to it) rather than any machine-specific terms.

1 Introduction

Significant effort invested into the development of HPF and similar languages shows the importance of array properties, such as distribution, alignment, etc for efficient generation of code. However, the use of a virtual machine for data parallelism with a (very specific) rectangular array processor structure and equally specific data distribution modes poses the questions of exactly how much must the user know about the construction of the machine to produce efficient code and whether this code could survive a change of the hardware platform.

The central issue here is one of describing a hierarchy of array properties in a sufficiently abstract way, but not too abstract in order to keep the bearing on real hardware. This is precisely the main difficulty too: what is "too abstract" depends on the assumptions the code designer has to make and what if they are wrong. Should we have to move on to a Fortran 2000 because nonrectangular arrays become rife and the most efficient distribution mode is finally pseudo-random?

The characteristic feature of von Neumann computing is its impressive ability to *approximate* the hardware. We have expressions in languages and the machine has commands. So expressions are turned into commands by a compiler and still the programmer can think expressions and loosely identify commands as "operations" — whence comes most of the usable part of complexity theory. The hardware instructions may change after porting the code to a different machine, but the operations will not, nor will the cost intuition based on them. Why can we not expect the same in the parallel domain?

The answer lies with the nature of approximation in the DP paradigm. It is extremely difficult to regard the translation process as one which simply brings granularity down to singular atomic actions of the hardware (by compiling down the chain "functions → statements → operations → instructions"). Because of the complex spatial structure of distributed array objects and high communication costs the operational decomposition of an algorithm is inferior to a spatial decomposition of a data structure. As we decompose a von Neumann algorithm into instructions of decreasing granularity, so we should be able to decompose a data-parallel (DP) object's type into a system of subtypes of decreasing abstraction. It should be noted, however, that while operational decomposition is implicit and is done by a compiler, the breaking-down of a type into subtypes has got to be explicit and known to the programmer. The compiler can only choose which subtypes to recognise and efficiently translate; others will not be distinguished from their supertypes.

To summarise, here is the proposed translation scheme. In the source program, data objects are typed using a type system rich enough to reflect any access feature or other particularity of a distributed object following from the algorithm and so, by definition, machine-independent. Some of these properties will be important for implementation on a particular machine, some will not. If a feature is not important for a given implementation, the compiler will ignore the subtype and use the supertype (i.e. approximate). It will then use the operators defined in the supertype to support computing with the subtype. The result of such computation should be consistent with the approximation used.

By itself, this is very similar to the agenda of OOP with its subclasses and inheritance. There is, however, an important distinction. The OOP approach makes types rigid; the use of implicit coercions is very limited indeed even when operators naturally extend to a variety of nested types. With OOP, it is believed to be a virtue of the method that the programmer has to be constantly aware of precisely the data types being used, so that no unexpected ambiguity occurs. Since the intention of our type system is approximation, we have to require that all versions of an operator are homomorphic, so an automatic coercion not anticipated by the programmer would not ruin the consistency of the program.

The nature of approximation "in space" is nothing new either. Consider what we do with the shapes of three-dimensional objects, say pieces of wood, in real life. If a craftsman needs to process such an object, he may have a tool for the simplest particular case, for example, a cubic shape. Such a shape is characterised by the fewest parameters and can be processed very efficiently. What if the same processing needs to be done on a more complex shape, e.g. a parallelepiped? There may be a tool for that, too, albeit less efficient. Such a tool would work on a cube (since a cube is an instance of parallelepiped). Finally there may be a tool that does the job on an arbitrary prism, perhaps even less efficiently: it would still have to be applicable to the previous two cases of the object as they are instances of prism, too. This example shows that the classification principle in 3d can be identified with the spatial symmetry of objects. The less symmetry an object has the greater generality and the more senior its type in the subtyping hierarchy.

Symmetry is indeed the key to any spatial approximation. We can easily iden-

```
DIMENSION Q(IL,JL,KL), V(KL), W(JL,IL)
...
DO I=1,IL
    DO J=1,JL
        DO K=1,KL
            Q(I,J,K)=V(K)*W(J,I)
        END DO
    END DO
END DO
```

Fig. 1. Example of a vectorisable loop nest

tify certain symmetries that are important for data parallelism, not all of them being purely geometric. The following sections will introduce translational symmetry (which is essentially one of replication), affine symmetry (which comes from analysis of constant-strided integer objects) and the symmetry of distributed access. We shall define and exemplify the type lattices corresponding to our classification and shall introduce appropriate coercion rules.

Finally, the general model limitation, which we shall assume hereinafter, is that nested DP computing (in the sense of [Ble93]) is not supported, although it appears nesting would not change the nature of fundamental DP symmetries too much. It nevertheless requires a separate study.

2 Translational symmetry

Fig 1 shows an example of a vectorisable loop nest as it appears in Fortran. This example uses 3 arrays of different ranks in a treble loop nest. Consequently, some of the indexed variables will not depend on some of the loop indices, for example V(K) is not affected by I- or J-iterations and W(J,I) does not change with iterations in K. According to f-code[MSS93] this symmetry should be interpreted as *orientation* of operands to a DP operation and should be defined statically using constant Boolean masks.

Definition 1. *An m-orientation of a rank-R array A is an array object that, if indexed with $[i_0, i_1, \ldots, i_r]$, where r equals the length of the mask $|m|$, selects the element $[j_0, j_1, \ldots, j_R]$ of the array A, with the indices $[j_0, j_1, \ldots, j_R]$ drawn from $[i_0, i_1, \ldots, i_r]$ according to the Boolean mask m in order. The number of ones in a Boolean mask is called the* character *of the mask and is denoted $\Box m$. For any valid orientation of A, $\Box m = R$.*

Note that orientation introduces translational symmetry in each result dimension corresponding to a zero in the mask.

Using orientations instead of the original arrays one can bring the example in fig 1 to a common dimensionality and then drop the explicit iteration space altogether:

```
Q=[001]V * [110]W'
```

where the prime denotes matrix transposition. The notation here is syntactically similar to the "numbers in brackets" of APL 2[BPP88].

In a complete DP world, a function can be applied to a nonscalar argument. Although we can limit our analysis to a function of a single argument (and use currying), it has to have a certain rank: since the rank is a component of the multi-type, a function should have to have a static type signature in the rank component. If a function is applied to an object of a rank higher than the one the function assumes for the argument, this can only be interpreted as a DP application of the function in the co-space of the argument.

Definition 2. *An m-orientation of a function of a rank-r argument is a function that accepts an array argument of a higher rank $R = |m| > r$. It uses a subset of the argument indices, according to the mask m, with the rest of the indices appended to the index list of the function result.*

The type inclusion relation for types of translational symmetry follows from the fact that the lack of symmetry along an axis is a more general case than its presence, taking into account that symmetries associated with different axes are independent.

Definition 3. *Let two objects x and y have different rank masks $\rho(x) \neq \rho(y)$, with the actual ranks being the same: $|\rho(x)| = |\rho(y)| = r$. Then the type inclusion relation $\rho(x) \subset \rho(y)$ is defined by the partial order $(\forall i : 1..r)\rho(x)_i \leq \rho(y)_i$, according to the standard subsumption.*

The orientation symbol used above for array orientation (a mask in square brackets) has the following "rank-mask signature":

$$(\forall a : |a| = \Box m)[m_1..m_n] : \{a_i\} \rightarrow \{b_j\}, \text{ where } b_j = \begin{cases} a_{\omega(m,j)}, & \text{if } m_j = 1 \\ 0, & \text{otherwise} \end{cases}.$$

Here $\omega(m,j) = \sum_{k=1}^{j} m_k$. Finally, let us join all rank lattices together at the bottom, by making every scalar type a member of all ranks since this only introduces unambiguous upgrading coercions.

3 Individual access symmetry

Abstract parallelism of data can be described as the lack of interference between different elements of a nonscalar assignment so that the hardware *may* perform all elemental assignments at once. In practice, however, a distributed implementation would perform DP assignment in a certain fuzzy order to minimise the communication and scheduling costs. In the simplest case of a rectangular processor array, data objects participating in the same DP operation will be co-mapped onto the array with a certain block size. Although scheduling of different blocks may be totally independent, within a block computing would have to be strictly serial. Such an arrangement is less symmetric than the source DP model, but the user does not see much *operational* manifestation of that (i.e., in terms of what can and what can not be done with a given data object). To separate out objects with different symmetries we shall introduce an *a priori* access cost, which is an asymptotic ($N \gg 1$, with N being the object size) measure that guides the user in the choice of the correct access type.

Definition 4. *The a priori access cost is a triple* $(c_\tau, c_\alpha, c_\rho)$, *where* c_τ *is the cost of* **total access**, *i.e. retrieval of all items of the arrangement, but not necessarily in order;* c_α *is the maximum cost of* **affine access**, *i.e. an arrangement of array elements with the indices forming an arithmetic progression, and* c_ρ *is the maximum cost of* **random access**.

Now we are in a position to introduce access subtypes, initially for a single dimension of an array, by giving upper bounds to the corresponding access costs.

Subtype	c_τ	c_α	c_ρ
locator	$O(N)$	$O(N)$	$O(N)$
collector	$O(1)$	$O(\log N)$	$O(\log N)$
sequencer	$O(1)$	$O(1)$	$O(\log N)$
director	$O(1)$	$O(1)$	$O(1)$
replicator	0	0	0

The above definitions also define the chain of type inclusions

$$replicator \subset director \subset sequencer \subset collector \subset locator,$$

which is a linear order on types.

Definition 5. *The access type of a multidimensional object is the Cartesian product of per-axis types.*

The type inclusion relation between multidimensional access types is one of partial order: a subtype has to be junior in *all* dimensions of a supertype. Objects of different ranks have incommensurable access types. All access types with a common rank form a lattice.

4 Data-parallel skeletons

In the framework of the skeleton approach[Col89], the DP operators can be regarded as instances of a few high-order functions that depend on functional parameters or introduce appropriate data structures. We shall consider some of them below. Since we need to use product types as well as array types in type signatures, it is important that we use some unambiguous notation. We shall denote as $^r x$ the type of an array which has rank r and el-type x. When a superscript follows a type variable, as in x^n, this should be interpreted as a product type, i.e. the type of all n-tuples of objects of type x. When we use both preceding and succeeding superscripts, an ambiguity may result as that can be interpreted either as an array of tuples or a tuple of arrays. In all such cases we shall bracket the type expression explicitly. Finally, wherever the access component of type has to be specified, we shall use a preceding subscript, so $^2_{sl}t$ denotes the type of any 2d array with el-type t whose access types in the first and second dimensions are s and l, respectively. Note that sl in this example is, in fact, the Cartesian product of per-axis types (see def. 5), which makes it legal to use power as well, e.g. $c^3 = ccc$.

We have fully defined 4 skeletons for DP computing: Map, $Juxtapose$, $Select$ and $Concatenate$. In this brief overview we shall not, for lack of space, describe all of them, but only the most interesting ones: Map and $Select$.

5 Map

This is the fundamental skeleton of DP computing. It applies a pure function to an array element-wise and has the following type signature:

$$(\forall r > 0, a, b)\,(^0a \to {}^0b) \to {}^ra \to {}^rb,$$

which introduces overloading in rank. For any function f, $Map\ f$ is indifferent to the access type of the argument: the access part of the signature is therefore fully decoupled from the rest and is given by

$$(\forall r > 0, x \subseteq \mathbf{1}^r)\,() \to x \to x,$$

where $\mathbf{1}$ is the locator access type and $()$ is the access type of a scalar.

A generic Map skeleton must also allow the function argument to accept arrays of any rank not exceeding the rank of the second argument of the Map. Therefore there has to be a family of skeletons $\{Map_m\}$ parametrised with an orientation mask m, with the following signature:

$$(\forall k = |m|;\ \forall a, b)\,(^{\square m}a \to {}^0b) \to {}^ka \to {}^{k-\square m}b\,.$$

Now let us define the (still disjoint) $access$ type signature of Map_m:

$$(\forall y : (\widehat{m}y) \subseteq x)\,(x \to ()) \to y \to (\widehat{\overline{m}}y)\,,$$

where the bar above m is the standard denotation of bit compliment, and the hat over the mask denotes the projection operator defined earlier. Note that the first argument, a function returning a scalar, is antimonotonic in the access type of its argument.

The functional argument (call it functional parameter to avoid confusion) can be any function taking an object of rank $\square m$ into an object of rank 0 (the latter guarantees non-nesting). However, three important cases below structure the functional parameter further, down to the level of scalar user-defined functions, which can be regarded as parameter-operators, and hence be treated algebraically.

Computation. This is a case of applying the functional parameter to the nonscalar argument to compute a new array. If the rank of the functional parameter argument is 0 then it defines an ordinary unary operator, such as $(-)$; if the rank is 1 or higher, the meaning of the Map is one of a reduction. Define three subskeletons:

$$\Gamma^l : (\forall a, b)^0 (a \to b \to a) \to {}^0 a \to {}^1_\mathbf{l} b \to {}^0 a \,,$$
$$\Gamma^s : (\forall a)^0 (a \to a \to a) \to {}^0 a \to {}^1_\mathbf{s} a \to {}^0 a \,,$$
$$\Gamma^c : (\forall a, r)^0 (a \to a \to a) \to {}^0 a \to {}^r_\mathbf{c} a \to {}^0 a \,.$$

The reader familiar with high-order functions will easily recognise the `foldr` type signature of Γ^s, which has the meaning of a reduction with any associative (but not necessarily commutative) operator typed $a \to b \to a$ and its identity value typed a. Due to noncommutativity, the access signature requires type sequencer for the last argument. If the reduction operator is commutative as well, Γ^c should be used instead, generally with an increase in parallelism. Γ^c is polymorphic in the rank of the last argument as its semantics is not sensitive to the array structure (it uses the array argument as a bag).

Selection. This is a case of using the nonscalar argument of Map to provide some location information that the functional parameter can use to select a specific element from another array: such a function can always be represented as $\lambda x.(\Xi\, S\, f(x))$, with some numerical function f, some array S and the constant Ξ being the element selection function which returns the element of its first argument selected using the second argument as index tuple. Unfortunately such a selection primitive turns out to be insensitive to the access type of its argument and so a separate primitive is required which is not based on an instance of Map.

6 Select

There are two reasons for treating selections separately from the Map skeleton. Firstly, as was mentioned in section 5, they should be sensitive to the access type of the array source. Secondly, a more complex subtyping structure is required for the nonscalar index argument, which combines the already encountered translational with yet another, affine, symmetry, which occurs in integer objects.

The type signature of the Select skeleton is as follows:

$$Sel :: (\forall r, d, x)^r x \to ({}^d I)^r \to {}^d x \,,$$

where ${}^d I$ is some rank-d index type defined below, which we shall assume to be a subtype of ${}^d int$. (Remember the notation t^n is used for the nth power of type t in the Cartesian product sense, i.e. the type of n-tuples of type-x components.)

7 Affine integer type

In this section we shall use the translational symmetry notation introduced in the end of section 2.

Definition 6. *The purely affine type ^dA is the type of all d-dimensional, integer arrays v whose elements satisfy the following formula*

$$v_{i_1 i_2 \ldots i_d} = \sum_{k=1}^{d} a^{[k]} i_k + b$$

with some integer $a^{[k]}$ and b. (We enclose the superscript in square brackets to avoid any confusion with Cartesian powers of types)

An array may not have a purely affine type, with some of the dimensions still being purely affine. The most general case is described by the following expression:

$$v_{\mathbf{j}} = \sum_{k=1}^{n} a_{\mathbf{p}}^{[k]} i_k + b_{\mathbf{p}} ,$$

where $\mathbf{p} = \widehat{m}\mathbf{j}$, $i_k = \{\widehat{\overline{m}}\mathbf{j}\}_k$, for some mask m, and all the coefficients are of the same rank $\Box m = |m| - n$.

Definition 7. *The index type ^dI is a type of a d-dimensional, integer array all elements of which satisfy the above formula with some mask m, $|m| = d$ and rank-l coefficients (where $l = \Box m$) $a^{[k]}$ and b. The general index type is fully defined by two Boolean masks:*

$$\tau = [m] \overline{\left(\bigvee_{k=1}^{\Box m} \rho(a^{[k]}) \right) \vee \rho(b)} ,$$

which indicates by 1's which dimensions have translational symmetry, and $\alpha = \overline{m}$, showing which dimensions have affine symmetry.

The data constructor Υ for the general index type is parametrised with the mask α and accepts as the argument an $(l+1)$-tuple (where $l = \Box \alpha$) of affine form coefficients of equal rank: $\Upsilon \alpha [a^{[1]} : e_1, a^{[2]} : e_2, \ldots, a^{[l]} : e_l, b]$, where the integer scalars $e_1..e_l$ define the dimensions of the result along the affine axes.

Now we are well-equipped to define *affine subtyping* on type $^n int$. For a single dimension the type inclusion relation is as follows: ts $\subset$ as $\subset$ ns, where "ts" stands for translational symmetry, "as" for affine symmetry and "ns" for no symmetry. As before, a multidimensional subtype must be junior or equal to a supertype in all dimensions. We exemplify the type lattice in fig 2, where the case $d = 2$ is displayed.

The access type of an axis of affine symmetry is replicator. The implementation may choose to introduce "smart" upgrading coercions from the replicator type, which modify the way the coercee is produced rather than moving it about when the production is completed.

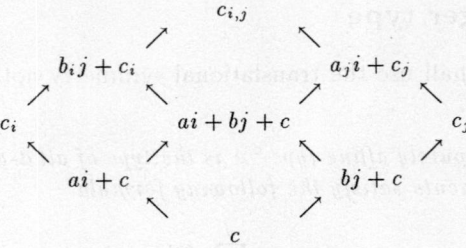

Fig. 2. 2d affine type classification.

7.1 *Sel* skeleton

This function takes as many other arguments as the rank of the first one, the source. The reason they are not juxtaposed in the sense of the general DP paradigm, see [MS96] is because we do not wish to coerce the nonscalar index tuple to a single affine type, which would cause a loss of type information and therefore an excessive generalisation. Nevertheless, as far as the result contents are concerned, these can be defined element-wise as follows:

$$(Sel\ Z\ X_0\ X_1\ \ldots X_n)_{\mathbf{k}} = (Map\ (\Xi Z)\ [[X_0, X_1, \ldots, X_n]])_{\mathbf{k}}$$

for any valid multi-index $\mathbf{k}$.[1]

However, function *Sel*, unlike *Map*, can use the information about affine symmetries of the indices as well as the source argument access type to choose the most efficient *particular* selection. This is achieved by overloading *Sel* for any combination of τ and α of each index argument.

The access type requirements for the source of the *Sel* function are very easy to establish. Indeed, if the index corresponding to an axis of the source has an affine dimension, the axis type can be as high as sequencer. Otherwise the source axis is required to be a director. *Sel* is obviously polymorphic in the access type of all indices. How is the access type of the result defined? Denote as $\{w_k\}$ the access type tuple of $[[X_0, X_1, \ldots, X_n]]$. For any k, consider the following cases:

1. w_k is senior to type replicator. The respective result axis has the same type and alignment.
2. w_k is of type replicator. If the kth axis of each of the $X_0, X_1, \ldots, X_n$ is translationally symmetric, so is the result axis, and it has the same type and alignment. Else if all but one axis are such, with the remaining axis of an X_m being affine, then the result axis is aligned with the mth axis of the source. Otherwise, same as case 1.

---

[1] This is *not* how Sel should be implemented, see section 5; we only use Ξ to define the value of the elements of the result.

How does *Sel* act on an affine integer object as the source? If τ_i^s and α_i^s are the parameters of the affine symmetry of the source, τ_i^r and α_i^r the respective parameters of the result, and $\tau_i^{[k]}$ and $\alpha_i^{[k]}$ of the kth index object,

$$\tau_i^r = \bigwedge_{k=1}^{d} \tau_k^s \vee \tau_i^{[k]}$$

$$\alpha_i^r = \bigwedge_{k=1}^{d} \tau_i^{[k]} \vee \left(\alpha_i^{[k]} \wedge (\tau_k^s \vee \alpha_k^s) \right) \ .$$

It should be noted that the power of *Sel* surpasses all known non-nested DP selections so that they can be expressed via it straight away. For example, a SLICE of a vector V is given by $Sel\ V\ (\Upsilon 1[k:l,m])$, where m is the start, k is the increment, and l is the new horizontal dimension and the transposition of a matrix R can be encoded as

$$Sel\ R\ (\Upsilon 10[[1]1 : \dim_2(R), [1]0])\ (\Upsilon 01[[1]1 : \dim_1(R), [1]0])\ ,$$

which clearly shows the 1d-affine, 1d-translational symmetry of the operation. ($\dim_k$ stands for the kth component of the object shape).

8 Conclusions

A type system based on analysis of symmetries inherent in distributed DP computing has been introduced and the fundamental DP skeletons have been typed accordingly. Programming with those could be free from HPF style annotation while conveying similar information to the compiler.

The support from EPSRC under grant GR/H78993, as well as funding from the Nuffield Foundation under their research grant scheme is acknowledged.

References

[Ble93] Guy E Blelloch. Nesl: A nested data-parallel language (version 2.6). Technical Report CMU-CS-93-129, School of Computer Science, Karnegie Mellon University, 1993.

[BPP88] J A Brown, S Pakin, and R P Polivka. *APL2 at a glance*. Prentice Hall, Englewood Cliffs, N.J. 07632, 1988.

[Col89] M I Cole. *Algorithmic Skeletons: Structured Management of Parallel Computation*. Pitman, 1989.

[MS96] V B Muchnick and A V Shafarenko. *Data-Parallel Computing: the Language Dimension*. Thompson Publishers, 1996.

[MSS93] V B Muchnick, A V Shafarenko, and C D Sutton. F-code and its implementation: a portable software platform for data parallelism. *The Computer Journal*, 36(8):712–721, 1993.

Petri Net Modelling of PARSE Designs*

Stefano Russo[1], Carlo Savy[1], Innes Jelly[2] and Peter Collingwood[2]

[1] Dipartimento di Informatica e Sistemistica, Università di Napoli
Via Claudio 21, 80125 Napoli, Italy
[2] Computing Research Centre, Sheffield Hallam University
Napier Street, Sheffield, S11 8HD, UK

Abstract. PARSE is a staged object-based design methodology for parallel and distributed software systems. In the highest stage, a graphical notation is used to describe the system components (objects) and their interconnections. In the subsequent stage, the designer introduces the behaviour of the objects, by means of a textual notation. This paper shows a mechanical transformation of the textual representation of a PARSE design into a complete Petri net model. This supports the integration of formal analysis techniques into the early stage of the software development process, and provides a formal semantics for the design notation.

1 Introduction and motivations

PARSE is a design methodology for parallel software systems, which provides a coherent framework for parallel software production from top level design analysis to implementation [1]. PARSE is based on a graphical design notation, known as *process graphs*. These enable the developer to describe a parallel system in terms of a collection of concurrent, hierarchically structured, interconnected components known as *process objects*.

The PARSE methodology proceeds iteratively through four stages, until a proper refinement level has been achieved: identification of the system structural properties; definition of the system dynamic behaviour; design verification; implementation. For the production of reliable software, the design verification stage is crucial. This is a recognised engineering practice, which can be performed in PARSE by means of sound formal techniques such as Petri nets [2] or CSP. While formal analysis can be optional for some applications, it is of fundamental importance for many classes of systems, such as safety critical ones.

The focus of this paper is the transformation of the design specification into a complete Petri net model of the system, to provide the basis for qualitative and quantitative design analysis. Previous work in the PARSE Project has shown how a process graph can be semi-mechanically transformed in two steps into a Petri net [3]. First, the interconnections among the lowest level objects are mechanically transformed into a skeleton Petri net, which captures the structural aspects of the interaction between processes. Then, the designer has to complete manually the net, modelling the internal behaviour of each process.

---

* This work has been supported in part by grants from BC and CNR, and by the University of Naples under the International Collaborations Programme.

This paper describes a new method by which Petri nets are generated fully automatically from PARSE BSL design descriptions, and shows its usage with GreatSPN [4], a generalised stochastic Petri net toolset. The BSL (Behavioural Specification Language) is a textual design language which complements the graphical notation [5].

The transformation of PARSE BSL designs into Petri net models relieves the developer from the need to manually enhance the skeleton net obtainable from only the process graph. Indeed, this activity requires a non trivial modelling skill, is tedious and error-prone. This approach also provides the basis for the provision of a fully integrated toolset which supports formal verification of system behaviour, both at the design and coding stages [6]. This allows the software engineer to employ formal mathematical techniques in an accessible manner, thus giving increased confidence in the behaviour and performance of the system under construction. Finally, this work defines formally the semantics of the PARSE design notation. This is an important issue for any design method, needed to ensure that an ambiguous meaning is given to any specific design [7].

2 Graphical parallel software design within PARSE

The process graph notation [1] categorizes a parallel system's components in three general classes (*process objects*): *data servers, function servers* and *control processes*. Data servers and function servers are passive objects, while control processes are responsible for initiating and coordinating processing. Differently from function servers, data server can have a state that persists between two requests. Finally, processes interfacing external equipments or devices are named *external interfaces*. They perform the interaction (input/output) with the external world, without any processing. A parallel design is expressed in a hierarchical way, since every process object can decompose into a set of lower-lever parallel processes. Decomposible objects are called *classes*, while really executable processes are named *primitive objects*.

Interactions between processes takes place over *communication paths*, which are categorised in *synchronous, asynchronous, bi-directional* (client-server) and *broadcast* (one-to-many). A non-primitive object has internal paths between lower-level objects, but of course its external interface must remain unchanged.

An example PARSE design is shown at the top-level in Fig. 1, which represents the parallel graphics processing system described in Reference [1]. There are five process objects and two external interfaces. The Capture class converts raw sensory data into input for subsequent processing, decomposing it into a pipeline of lower-level primitive processes. Process object Pool packetizes data in appropriate chunks, which it sends to the Master when requested. Master is a control process, which handles the data processing done by a farm of replicated Worker processes. Sending and receiving data to/from workers is done in parallel by lower-level processes. Finally, output to the external display device is filtered by an appropriate Filter object. Fig. 2 depicts the the internal design of the Master and the Capture classes.

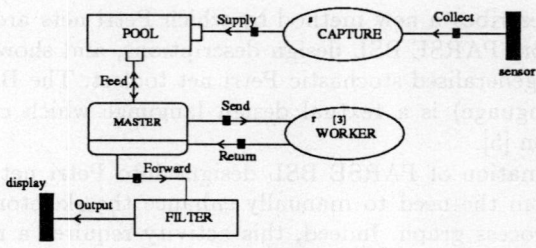

Fig. 1. Top-level PARSE design of the parallel graphics processing system.

3 Defining process objects behaviour

After having designed the process graph, the developer has to specify the low-level behaviour of each component. The Behavioural Specification Language [5] has been defined to support this task. The BSL is a pseudo-language meant to capture the essential dynamic behaviour of the software, which is not expressed in the process graph. For this purpose, it provides basic flow control and communication constructs, which should be mapped easily into a variety of message-passing languages and environments. The BSL features are summarized in the following.

BSL Behavioural components

Variables and flow control. Like most procedural languages, the BSL provides data types, variable declarations and the common sequential, conditional (IF .. THEN, IF .. THEN .. ELSE, CASE) and loop (WHILE .. DO) flow control constructs. In the BSL to PN transformation procedure, we do not model the value of the variables and their assignments. It is known that the modelling of the variables in the PN reduces the number of *non-faults* detected by the net analysis steps. However it increases considerably the complexity of the net, in terms of number of places, transitions, and arcs.

Communication constructs. The BSL offers the `sy-send`, `as-send` and `br-send` primitives for the three types of send operations, synchronous, asynchronous or broadcast respectively. There is only one `receive` primitive for all three cases. Synchronous and asynchronous operations have the usual semantics. In a broadcast operation the sender blocks until the message has been delivered to all the destination ports.

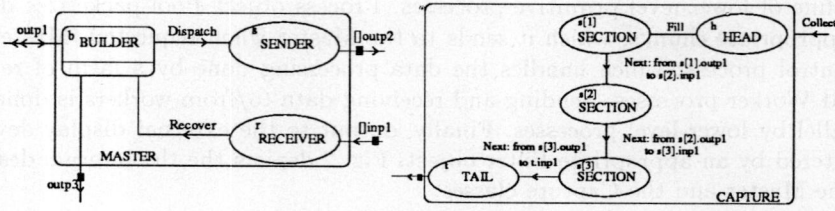

Fig. 2. PARSE design of the MASTER and CAPTURE classes.

On a bi-directional communication link, a client process starts the communication by issuing a **request** to the partner, which acts as server. The server receives the **request** with an **accept**, performs the necessary computation, and then answers with a **reply**. A server cannot accept any further input until it has replied to a pending request.

Communication path constructors. These constructs are used to specify which policy is used when receiving from several input ports. In the graphical PARSE design notation, there are four *communication path constructors*, namely *undefined*, *concurrent*, *non-deterministic* and *deterministic*.

Undefined input handling is the most general case, with no implied order of input response. This is specified by the user at the BSL code level. By definition, concurrent input handling is not allowed for the non-decomposable objects. Non-deterministic input handling has the same semantics as in many CSP-based languages (like ALT in OCCAM), that is, the receiving process selects ready inputs randomly. Deterministic input handling means that the receiving process selection among ready paths is prioritized.

BSL Structural components

Design. The textual representation of a PARSE design is made up by a single unit (module), although file inclusion facilities can be used to split it into several source files. This lists all process objects' definitions. At the topmost level there is the DESIGN, which has the following structure.

```
DESIGN design-name
PROTOCOLS
   ... // data types to be associated to paths and ports
PATHS
   ... // top level paths
PRIMITIVES
   ... // sequential process objects descriptions
CLASSES
   ... // decomposable process objects descriptions
EXTERNALS
   ... // external interface objects
EXECUTION
   ... // top level parallel process objects
CONNECT PATHS
   ... // inter-object path connections
END
```

The EXECUTION section defines the instances of executing objects visible at the topmost level. They may be either primitive objects, external objects, or instances of class objects. The CONNECT PATHS section is used to associate paths to input and output ports of executing objects, that is, to specify actual connections among them.

Primitives. Each primitive process definition has an *interface* and a *body*. The interface specifies the structural components, namely input ports, output ports and constructors. The body defines the variables used, and the actual behaviour by means of the communication and flow control constructs. Fig. 3 (left) shows the BSL code for a primitive process component of a PARSE design, namely the RECEIVER process, which is part of the MASTER CLASS of our case study.

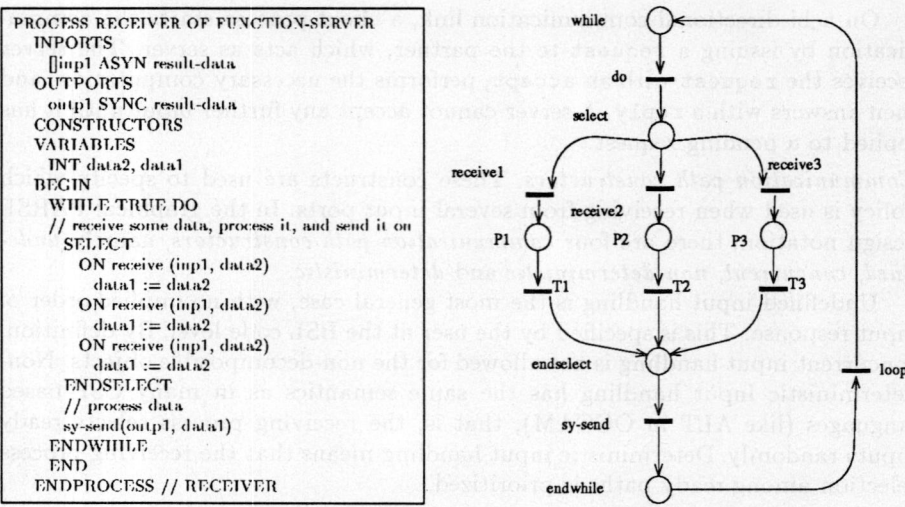

Fig. 3. BSL description and Petri net model of the Receiver process.

Classes. A class declaration is syntactically a mixing of DESIGN and sequential PROCESS definitions. Similarly to a PROCESS, it has INPORTS and OUTPORTS sections, which represent its interface with the external world. However, similarly to the DESIGN definition, it comprises a PATHS, an EXECUTION and a CONNECT PATHS section, since a class instance in turns specifies concurrent activities within itself. There is also a CONNECT PORTS section, that specifies which of the input/output ports of processes in the EXECUTION section is attached to the input/output ports of the class instance itself.

4 Modelling designs with Petri nets

The transformation of the BSL code into a Petri net is accomplished in two steps. The first one consists into applying proper translation rules for the various BSL statements. The complete set of translation rules for the BSL is defined in [8]. Basically, in this first translation step the communication constructs *sy-send, as-send, receive, request, accept, reply* are represented by a proper sequence place-transition-place, with the firing of the transition corresponding to the statement execution. The translation rules for the IF and WHILE flow control constructs and the SELECT statement exploit the well-known Petri net concept of free choice. The user is referred to Ref. [6] for the modelling of flow control constructs and that of synchronous and asynchronous message passing. As an example of the application of the rules to a complete BSL process, Fig. 3 (right) shows the Petri net model of the Receiver process of our case study.

The subnets modelling the processes of a PARSE design must be properly linked in order to obtain the Petri net which models the design as a whole.

The integration points of isolated subnets are all the potential interaction points between any two pair of processes in the BSL program, that is any correspondent pair of inter-process communication statements.

Fig. 4 (left) shows the Petri net model of a broadcast operation. As many buffer output places from transition $br - send$ have to be added, as there are destination processes. Moreover, since the semantics of broadcast is such that the sender cannot continue execution until the message has been received by all destination processes, an inhibitor arc connects each buffer process to the "synchronization" transition t. Finally, Fig. 4 (left) shows the Petri net model for the bi-directional path. The composition of the subnets of the client and the server in this kind of communication requires to remove the *request* transition, which is in a certain sense fused with both the *accept* and *reply* corresponding transitions. Transition T models the activity of the server to prepare the reply.

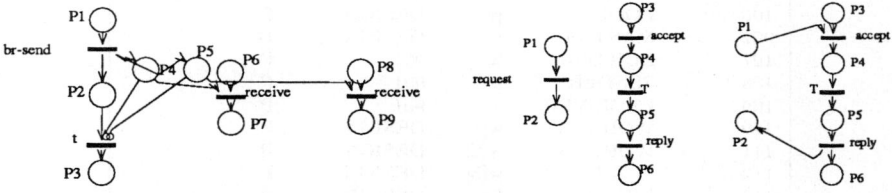

Fig. 4. Composition rule for broadcast (left) and client-server communication (right).

5 Transformation procedure

The BSL to PN transformation algorithm consists, at a macroscopic level, of the following steps:

1. Construction of the list of executing objects.
2. Construction of the list of actual connections among them.
3. Translation of the behaviour of each primitive object into a Petri net.
4. Composition of these isolated nets, to produce the final Petri net model of the overall PARSE design.

The lists built in the first two steps are produced by parsing the BSL program, examining the structural components of the design. These lists represent a set of cross-reference information about the program components. The third step is based on the behavioural components, i.e. on the sequential part of each process. The parsing tool produces a set of intermediate net descriptions for the various executable components. The fourth step is performed by a proper tool, which links the intermediate isolated component nets, on the basis of the cross-reference information. Each step is now described in more detail.

Step 1. The text of the BSL program is parsed, and a complete list is constructed of all executables and class instances that are declared in the EXECUTION section of the DESIGN and of each DCLASS. A unique (integer) identifier is assigned to each of them. Class instances are decomposable objects and are marked as such, whereas the lowest-level objects are those which actually execute in parallel. This list is called *Process Identifiers Table* (PIT). The PIT for the graphics system is shown in Fig. 5.

DESIGN Graphics — No. of non-decomposible processes: 15				
Process instance	Behaviour		Parent process	Process type (Prim/Decomp)
100	SENSOR	s	DESIGN	P
900	CAPTURE	c	DESIGN	D
101	HEAD	h	900	P
102	SECTION	s[1]	900	P
103	SECTION	s[2]	900	P
104	SECTION	s[3]	900	P
105	TAIL	t	900	P
106	POOL	p	DESIGN	P
902	MASTER	m	DESIGN	D
107	BUILDER	b	901	P
108	SENDER	s	901	P
109	RECEIVER	r	901	P
110	WORKER	w[1]	DESIGN	P
111	WORKER	w[2]	DESIGN	P
112	WORKER	w[3]	DESIGN	P
113	FILTER	f	DESIGN	P
114	DISPLAY	d	DESIGN	P

Fig. 5. Process Identifier Table for the graphics processing system.

Step 2. This step works on the INPORTS, OUTPORTS of each PROCESS and DCLASS, and on the CONNECT PORTS sections of the DESIGN and of DCLASSes. A *Communication Paths Table* (CPT) is built, which lists the actual paths in the BSL program and, for each of them, the process instances connected to it. This table is needed to establish the potential interaction points between processes, that is, the potential corresponding pairs of input/output statements. The CPT for the graphics system is shown in Fig. 6. It tells, for instance, that "Feed" is a bi-directional path connecting formal port "outp1" of the process whose ID is 107 (which is BUILDER, see the PIT) to formal port "inp2" of the process whose ID is 106 (POOL). "Send" is an asynchronous "one to many" path connecting the output port of process "108" to the input ports of three processes (110, 111, 112).

Step 3. This step consists in the generation of a subnet for each non-decomposible object, and it is straightforward. For each PROCESS definition in the BSL program, the sequential code is parsed, and the statement translation rules are applied. In the intermediate net representation produced in this step, places and transitions must be properly "tagged", in some way resembling the original statement, so that the subsequent composition step can properly merge or connect corresponding communication transitions.

Step 4. Based on the cross-reference information provided by the first two steps, in this final step the isolated nets are composed into a single global net, according to the rules defined in section 4. In our approach, only instances of executable objects appear explicitly in the final net. These are the lowest level, non-decomposable objects, i.e. primitive processes and external objects. All these objects execute in parallel. Class instances are not explicitly represented. The Petri net "flattens" the hierarchical structure of the design.

DESIGN Graphics — No. of paths: 17						
Path name	Path type	From Proc. ID	OutPort	To Proc. ID	InPort	Protocol
Feed	BIDI	107	outp1	106	inp2	input-data
Collect	ASYN	100	outp1	101	inp1	raw-data
Supply	ASYN	105	outp1	106	inp1	input-data
Send	ASYN	108	outp1	110	inp1	work-data
Send	ASYN	108	outp1	111	inp1	work-data
Send	ASYN	108	outp1	112	inp1	work-data
Return	ASYN	110	outp1	109	inp1	result-data
Return	ASYN	111	outp1	109	inp1	result-data
Return	ASYN	112	outp1	109	inp1	result-data
Forward	ASYN	107	outp3	113	inp1	display-data
Output	SYN	113	outp1	114	inp1	display-data
Dispatch	SYN	107	outp2	108	inp1	work-data
Recover	SYN	109	outp1	107	inp1	result-data
Fill	SYN	101	outp1	102	inp1	input-data
Next[0]	SYN	102	outp1	103	inp1	input-data
Next[1]	SYN	103	outp1	104	inp1	input-data
Next[2]	SYN	104	outp1	105	inp1	input-data

Fig. 6. Communication Paths Table for the graphics processing system.

The final net model of the whole parallel graphics system is shown in Fig. 7. As we can see, such a net is difficult to generate manually, particularly because there are many potential interaction points between processes (corresponding to "diagonal" arcs). Moreover, an automatic tool can generate a graphical layout, such that the primitive PARSE processes are easily recognizable. This helps in understanding the final model, and also in moving from the PARSE design domain to the Petri net model domain and viceversa when reasoning on the system behaviour. In Fig. 7 we recognize, for instance, from the left to the right the fifteen executable processes making up our graphics processing system, from the input Sensor external interface to the output Display process. Such CASE tools that associate a graphical layout to a net automatically generated are already incorporated in the EPOCA system [6].

It is possible to apply to the complete model the consolidated Petri net structural and reachability analysis techniques for qualitative design verification (eg deadlock detection), as well as techniques for predictive performance evaluation, provided that a class of timed nets is used. Proper CASE tools such as GreatSPN [4] can be exploited for these purposes. We have used GreatSPN to simulate the behaviour of the system, executing the Petri net by playing the so colled "token game". However, we point out explicitly that net visualization is not mandatory to perform the analysis steps. This is important for net models of complex systems, for which the visualization reveals useless.

6 Conclusions

In this paper we have described a method to derive Petri net models from designs of parallel software systems, specified according to the PARSE notation. In particular, we have shown that a complete PARSE design contains all the information needed for the automated generation of a net model.

The transformation from the design domain to the Petri net domain provides two fundamental advantages: (a) it enables the developer to formally verify the correctness of the parallel system at an early stage of its development, that is during the design phase, before the implementation; (b) it allows to simulate the system behaviour, by executing the net.

Being based on an algorithmic transformation, the method is fully automatable. This relieves the developer from the need to manually construct the net model. The method forms the basis for the integration of existing software engineering techniques and tools, in order to provide proper support in the design verification stage of parallel software development.

References

1. I. Gorton, P. Gray and I. Jelly: Object-Based Modelling of Parallel Programs. IEEE Parallel and Distributed Technology, Vol 3, No. 2, 1995.
2. Murata, T.: Petri Nets: Properties, Analysis and Applications. Proc. of the IEEE, Vol. 77, No. 4, April 1989
3. Gorton, I., Jelly, I., Soon Chan, T.: Engineering High Quality Parallel Software Using PARSE. Proc. of CONPAR94, Linz, Austria, (1994) LNCS, Springer Verlag
4. Chiola, G.: GreatSPN 1.5 software architecture. Proc. 5th Int. Conf. Modelling Techniques and Tools for Computer Performance Evaluation, Torino, Italy, 1991
5. Bradney, W., Gray, J. (supervisor): Parse code translator. M.Sc. Project, Dept of Computer Science, University of Wollongong, Nov. 1994
6. Donatelli, S., Mazzocca, N., Russo, S.: A CASE System for Petri Net Modelling of CSP-like Programs. Transputer Communications, Vol.3(1), (1996) 15–22
7. Nixon, P.A., Shi, L.: Concurrent semantics for structured design methods. in: Jelly, Gorton, Croll (eds.), Software Engineering for Parallel and Distributed Systems, Berlin, (1996) Chapman&Hall
8. Jelly, I., Collingwood, P., Russo, S., Savy, C.: From textual representation of PARSE design to Petri Nets. Joint Technical Report No 7/95, April 1995, Computing Research Centre, Sheffield Hallam University and University of Naples

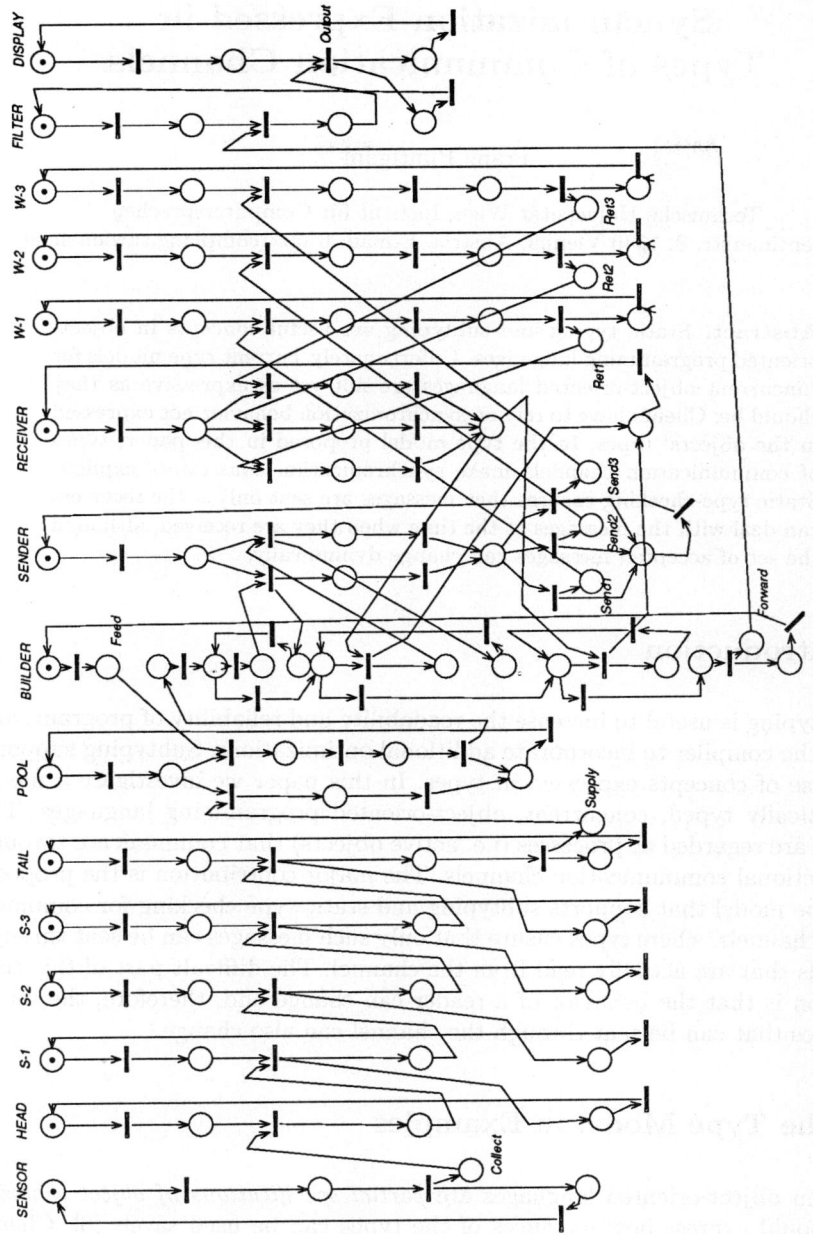

Fig. 7. Final Petri net model of the parallel graphics processing system.

Synchronization Expressed in Types of Communication Channels

Franz Puntigam

Technische Universität Wien, Institut für Computersprachen
Argentinierstr. 8, 1040 Vienna, Austria. E-mail: franz@complang.tuwien.ac.at

Abstract. Static typing and subtyping are useful concepts in object-oriented programming languages. Unfortunately, current type models for concurrent object-oriented languages are not yet as expressive as they should be: Clients have to rely on synchronization behavior not expressed in the objects' types. In the type model proposed in this paper, types of communication channels make synchronization constraints explicit. Static type-checking ensures that messages are sent only if the receivers can deal with the messages at the time when they are received, although the set of accepted messages can change dynamically.

1 Introduction

Static typing is useful to increase the readability and reliability of programs and allows the compiler to incorporate additional optimizations. Subtyping supports the reuse of concepts expressed in types. In this paper we investigate a model for statically typed, concurrent, object-oriented programming languages. The objects are regarded as processes (i.e. active objects) that communicate through unidirectional communication channels. The major contribution is the proposal of a type model that supports subtyping and static type checking for communication channels, where types ensure that only such messages can be sent through channels that are actually read from the channel. The difficult part of this contribution is that the behavior of a reader can change and, therefore, the set of messages that can be sent through the channel can also change.

2 The Type Model in Examples

Types in object-oriented languages are *partial specifications of object behavior* that should express how instances of the types can be used savely [6]. Clients must not assume anything about an object's behavior except what is expressed in the object's type. Static type checking ensures that each object actually behaves as specified. A supertype is a less complete specification of the behavior specified by a subtype; all properties that hold for the supertype also hold for the subtype. The *principle of substitutability* says that an instance of a subtype can always be used in any context in which an instance of a supertype is expected [16].

Syntax and semantics of the language/model used in this paper resemble that of concurrent logic languages [14]; relationships of these languages to the object-based actor model [4] and to process algebra are widely known. An example of a process definition for a simple buffer with a capacity of one element shows the basic constructions of our language/model:

$$\text{Buffer1}(q) := q \triangleright \text{put}(e) \cdot \text{get}(a) \cdot r, \, a \triangleleft \text{got}(e), \text{Buffer1}(r).$$

The parameter q is the read-port of a channel transporting commands to an object behaving according to Buffer1. The read-expression $q \triangleright \text{put}(e) \cdot \text{get}(a) \cdot r$ specifies that, first, a message put(e) is read from q, then a message get(a); r denotes the remaining read-port. The variable e stands for the element put into the buffer, and a for a write-port of a channel through which the element is returned. The write-expression $a \triangleleft \text{got}(e)$ writes the message got(e) to a. The dot $(\cdot)$ is the sequential composition operator: $c \cdot c'$ means that a channel first transports messages according to c and then according to c'. In a read-expression $v \triangleright c \cdot c'$, variables occurring in c (and c') are initialized when a message corresponding to c (and c') is received; if c' is a variable, c' is initialized to the read-port v after receiving c (and removing c from the read-port). An expression in the body of a process definition is executable when all variables in this expression have been initialized. Hence, the recursive call Buffer1(r) becomes executable in parallel with $a \triangleleft \text{got}(e)$ when the messages put(e) and get(a) have been received.

The next example shows a process definition for a buffer of capacity n:

$$\text{Buffer}[n](q) := n = 1 \rightarrow \text{Buffer1}(q);$$
$$n > 1 \rightarrow q \triangleright r \,\|\, s, \text{Buffer1}(r), \text{Buffer}[n-1](s).$$

For $n = 1$ the buffer behaves as Buffer1; for $n > 1$ the expression $q \triangleright r \,\|\, s$ splits the read-port q into two concurrent parts r and s. $\|$ stands for parallel composition: A channel behaving according to $c \,\|\, c'$ can transport messages according to c and c' in arbitrary interleaving. Because of parallel composition, r and s are initialized immediately and the calls of Buffer1 and Buffer$[n-1]$ are not delayed.

Our language also supports alternative composition expressed by $+$. A channel of behavior $c + c'$ transports messages according to either c or c'. In a read-expression $v \triangleright c + c'$, the variables occurring in either c or c' are initialized.

Read- and write-expressions determine the messages transported through channels. A channel's type describes expected orderings of messages by using sequential, parallel and alternative composition operators. For each channel type τ, $!\tau$ denotes the write-port of a channel of this type, and $?\tau$ the read-port. The next example shows the type definition of a communication channel that transports commands to a buffer:

$$\text{BC\_t}[A, n] :: n = 1 \rightarrow \text{put}(!A) \cdot \text{get}(!\text{got}(!A)) \cdot \text{BC\_t}[A, 1];$$
$$n > 1 \rightarrow \text{BC\_t}[A, 1] \,\|\, \text{BC\_t}[A, n-1].$$

A stands for the type of the channels through which elements in the buffer can be accessed. An instance of BC_t$[A, 1]$ transports first a message put(e), where e is the write-port of a channel of type A; the next message is of the form get(a), where a is the write-port of a channel that transports a message got(e'), and e' is the write-port of a channel of type A. A sequence of such messages (put and get in alternation) is transported through the channel. For $n > 1$, n arbitrarily interleaved sequences of such messages are transported.

We annotate parameters of process definitions with type information:

Buffer$[A,n](q{:}?\mathrm{BC\_t}[A,n])\ :=$
$\quad n = 1\ \to\ q \triangleright \mathrm{put}(e) \cdot \mathrm{get}(a) \cdot r,\ a \triangleleft \mathrm{got}(e),\ \mathrm{Buffer}[A,1](r);$
$\quad n > 1\ \to\ q \triangleright r \,\|\, s,\ \mathrm{Buffer}[A,n-1](r),\ \mathrm{Buffer}[A,1](s).$

3 Typed Channels and Ports

Types of channels are denoted by σ, τ, φ and ψ, the type of the write-port of a channel of type τ by !τ, and the type of the read-port by ?τ. We write $\hat{\tau}$ for the type of a port if the kind (read or write) does not matter.

A channel transports a sequence of messages from its write-port to its read-port so that the ordering is preserved. The channel's type specifies the allowed sequences of messages. The type of a message $m(v_1,\ldots,v_n)$ is the *message prototype* $m(\hat{\tau}_1,\ldots,\hat{\tau}_n)$, where $\hat{\tau}_i$ is the type of v_i (for $1 \leq i \leq n$). A channel's type is specified by a *trace set*, i.e. the set of all sequences of message prototypes which correspond to all message sequences that can be transported through the channel [13]. Let P denote the set of all message prototypes usable in a programming system, P^* the set of all words over the alphabet P, and ε the empty word in P^*. Then, $S \subseteq P^*$ is a trace set if $S \neq \emptyset$ and $ww' \in S$ implies $w \in S$ for all $w,w' \in P^*$. Each type τ is defined by the trace set $\mathrm{tr}(\tau)$ of τ.

Read- and write-ports express different concepts to their users. A write-port means that the user (writer) can send messages according to *any* trace in the trace set. A read-port means that the user (reader) must be able to deal with messages received according to *all* traces in this set. The writer selects the written messages, while the reader cannot influence the received messages. This difference becomes evident when dealing with subtyping.

Let a process write messages to a write-port of type !τ. These messages depend only on the process' behavior and input, but not on the write-port's type. It is possible to replace this write-port with another write-port of type !σ without affecting the process. We have to define !$\sigma \leq$!τ (!σ is a subtype of !τ) so that a process that assumes a write-port of type !τ also works with a write-port of type !σ. A sufficient condition for !$\sigma \leq$!τ is $\mathrm{tr}(\tau) \subseteq \mathrm{tr}(\sigma)$, i.e. a channel of type σ transports all messages transported through a channel of type τ.

Let a process receive messages at a read-port of type ?τ. The process must be able to deal with all messages specified by $\mathrm{tr}(\tau)$. We have to define ?$\sigma \leq$?τ so that a process that assumes a read-port of type ?τ can deal with all messages received from a read-port of type ?σ. A sufficient condition for ?$\sigma \leq$?τ is $\mathrm{tr}(\sigma) \subseteq \mathrm{tr}(\tau)$, i.e. a channel of type σ transports only those messages transported through a channel of type τ. Surprisingly, ?$\sigma \leq$?τ is equivalent to !$\tau \leq$!σ.

We extend the definition of subtyping for ports by considering variant argument types. Let a process send messages to a channel v of a type !τ, and let

τ describe a message with an argument of type $!\varphi$. For each σ with $!\sigma \leq !\tau$, a reader may expect messages from v according to σ. Let $!\psi$ be the argument type of the message in σ that corresponds to $!\varphi$ in τ. The reader of v may send messages to the received write-port according to ψ, although the writer of v provides a write-port of type $!\varphi$. Hence, type errors are prevented if $\mathrm{tr}(\psi) \subseteq \mathrm{tr}(\varphi)$ or (equivalently) $!\varphi \leq !\psi$. By applying similar argumentations to other kinds of ports we get these relationships:

Relation	Argument in σ	Argument in τ	Trace sets	Subtyping
$!\sigma \leq !\tau$	$!\varphi$	$!\psi$	$\mathrm{tr}(\psi) \subseteq \mathrm{tr}(\varphi)$	$!\varphi \leq !\psi$
$!\sigma \leq !\tau$	$?\varphi$	$?\psi$	$\mathrm{tr}(\varphi) \subseteq \mathrm{tr}(\psi)$	$?\varphi \leq ?\psi$
$?\sigma \leq ?\tau$	$!\varphi$	$!\psi$	$\mathrm{tr}(\varphi) \subseteq \mathrm{tr}(\psi)$	$!\psi \leq !\varphi$
$?\sigma \leq ?\tau$	$?\varphi$	$?\psi$	$\mathrm{tr}(\psi) \subseteq \mathrm{tr}(\varphi)$	$?\psi \leq ?\varphi$

From these relationships we can derive a less restrictive definition of subtyping:

Definition 1. $!\sigma \leq !\tau$ holds if for each $p_1 \cdots p_n \in \mathrm{tr}(\tau)$ there is a $q_1 \cdots q_n \in \mathrm{tr}(\sigma)$ so that $!\varphi_{i,j} \leq !\psi_{i,j}$, where (for $1 \leq i \leq n; 1 \leq j \leq l_i$) $p_i = m_i(\hat{\varphi}_{i,1}, \ldots, \hat{\varphi}_{i,l_i})$ and $q_i = m_i(\hat{\psi}_{i,1}, \ldots, \hat{\psi}_{i,l_i})$. Furthermore, $?\sigma \leq ?\tau$ iff $!\tau \leq !\sigma$.

The syntax of prototypes $p \in P$ and type expressions $\tau \in T$ is:

$$\begin{array}{rcl}
\tau & ::= & \mathrm{nil} \mid p \mid \tau \cdot \tau' \mid \tau \,\|\, \tau' \mid \tau + \tau' \\
p & ::= & m(\hat{\tau}_1, \ldots, \hat{\tau}_n) \qquad (0 \leq n) \\
\hat{\tau} & ::= & !\tau \mid ?\tau
\end{array}$$

where m is a message name. $T^!$ denotes the set $\{!\tau : \tau \in T\}$ of write-port types, $T^?$ the set $\{?\tau : \tau \in T\}$ of read-port types. Elements of T are denoted by σ, τ, φ and ψ, elements of P by p and q.

A set of algebraic laws is defined on T (as well as $T^!$ and $T^?$):

$$\begin{array}{ll}
(\sigma \cdot \tau) \cdot \varphi = \sigma \cdot (\tau \cdot \varphi) & \sigma \cdot (\tau + \varphi) = \sigma \cdot \tau + \sigma \cdot \varphi \\
\tau \cdot \mathrm{nil} = \tau & (\sigma + \tau) \cdot \varphi = \sigma \cdot \varphi + \tau \cdot \varphi \\
\mathrm{nil} \cdot \tau = \tau & \sigma \,\|\, \tau = \tau \,\|\, \sigma \\
(\sigma + \tau) + \varphi = \sigma + (\tau + \varphi) & \tau \,\|\, \mathrm{nil} = \tau \\
\sigma + \tau = \tau + \sigma & p \cdot \sigma \,\|\, q \cdot \tau = p \cdot (\sigma \,\|\, q \cdot \tau) + q \cdot (p \cdot \sigma \,\|\, \tau) \\
\sigma + \tau = \sigma \quad \text{for } !\sigma \leq !\tau & (\sigma + \tau) \,\|\, \varphi = \sigma \,\|\, \varphi + \tau \,\|\, \varphi
\end{array}$$

Further general laws are derivable from the given equations. For example, $\sigma + \tau = \sigma$ for $!\sigma \leq !\tau$ implies $\sigma + \mathrm{nil} = \sigma$ and $\sigma + \sigma = \sigma$. The parallel composition operator is in fact syntactic sugar; each type expression containing $\|$ is equal to a type expression not containing $\|$.

We define the function tr already used in computing type trace sets:

$$\begin{array}{l}
\mathrm{tr}(\mathrm{nil}) = \{\varepsilon\} \\
\mathrm{tr}(p \cdot \tau) = \{\varepsilon\} \cup \{pw : w \in \mathrm{tr}(\tau)\} \\
\mathrm{tr}(\sigma + \tau) = \mathrm{tr}(\sigma) \cup \mathrm{tr}(\tau)
\end{array}$$

The definition of tr for parallel composition follows implicitly from the equality relation on type expressions. There is an implicit condition for the usefulness of the above definitions: $\sigma = \tau \Leftrightarrow \text{tr}(\sigma) = \text{tr}(\tau)$. This condition is fulfilled as can be seen easily by adapting a proof of a similar sentence [1, 13].

We provide a sound and complete axiomatization of subtyping as given by Definition 1. The following rule set defines $\leq$ as a binary relation on $T^!$ and the auxiliary relation $\sqsubseteq$ on $T^!$. These relations are easily extended to $T^! \cup T^?$ using the equivalences $!\sigma \leq !\tau \Leftrightarrow ?\tau \leq ?\sigma$ and $!\sigma \sqsubseteq !\tau \Leftrightarrow ?\tau \sqsubseteq ?\sigma$.

$$!\tau \leq !\tau \qquad \text{S1} \qquad !\tau \sqsubseteq !\tau \qquad \text{R1}$$

$$!\tau \leq !\text{nil} \qquad \text{S2} \qquad \frac{!\sigma \sqsubseteq !\tau \quad !\varphi \sqsubseteq !\psi}{!(\sigma \cdot \varphi) \sqsubseteq !(\tau \cdot \psi)} \qquad \text{R2}$$

$$\frac{!\sigma \sqsubseteq !\tau \quad !\varphi \leq !\psi}{!(\sigma \cdot \varphi) \leq !(\tau \cdot \psi)} \qquad \text{S3} \qquad \frac{!\sigma \sqsubseteq !\tau \quad !\varphi \sqsubseteq !\psi}{!(\sigma + \varphi) \sqsubseteq !(\tau + \psi)} \qquad \text{R3}$$

$$\frac{!\sigma \leq !\tau \quad !\varphi \leq !\psi}{!(\sigma + \varphi) \leq !(\tau + \psi)} \qquad \text{S4} \qquad \frac{!\sigma \sqsubseteq !\tau}{!(\sigma + \varphi) \sqsubseteq !\tau} \qquad \text{R4}$$

$$\frac{!\sigma_1 \leq !\tau_1 \; \cdots \; !\sigma_n \leq !\tau_n}{!m(\hat{\tau}_1,\ldots,\hat{\tau}_n) \leq !m(\hat{\sigma}_1,\ldots,\hat{\sigma}_n)} \qquad \text{S5} \qquad \frac{!\sigma_1 \leq !\tau_1 \; \cdots \; !\sigma_n \leq !\tau_n}{!m(\hat{\tau}_1,\ldots,\hat{\tau}_n) \sqsubseteq !m(\hat{\sigma}_1,\ldots,\hat{\sigma}_n)} \qquad \text{R5}$$

As shown in Sect. 2, functions over type expressions allow us to easily specify large or infinite type expressions of regular structures. Because of space limitations we refrain from formally defining syntax and semantics of type functions. Type equivalence and subtyping shall be extended to cover type functions. Complete axiomatizations of these notions are not feasible because known algorithms for determining the equivalence of two expressions do not terminate in some cases. However, there are sound and feasible axiomatizations of type equivalence and subtyping that relate only structurally similar type functions.

4 Typed Processes

The syntax of channel expressions $c \in C$, process expressions $b \in B$ and process definitions d is given by:

$$
\begin{array}{lll}
c & ::= & v \;|\; m(v_1,\ldots,v_n) \;|\; c_1 \cdot c_2 \;|\; c_1 \| c_2 \;|\; c_1 + c_2 \\
b & ::= & \text{skip} \;|\; v \triangleright c \;|\; v \triangleleft c \;|\; b,b \;|\; x[s_1,\ldots,s_m](v_1,\ldots,v_n) \;|\; \text{cond} \\
d & ::= & x[s_1,\ldots,s_m](v_1{:}\hat{\tau}_1,\ldots,v_n{:}\hat{\tau}_n) := b
\end{array}
$$

where m is a message name, $v, v_1, \ldots, v_n$ are variables, x is the name of a process definition, $s_1, \ldots, s_m$ are structural parameters, and *cond* stands for conditional expressions not dealt with here. The semantics should be clear from the explanations of the examples in Sect. 2. The actions are read-expressions ($v \triangleright c$), write-expressions ($v \triangleleft c$) and "skip" (which does nothing). Although actions in process expressions are separated by comma and can be rearranged arbitrarily, the rule about using only initialized variables imposes dependences on the actions. We assume that some equality relation on process expressions is defined so that a compiler can determine whether $b = b'$ holds for each $b, b' \in B$.

Read- and write-expressions can be split into basic components. Each $v \triangleright c$ can be split into actions of the forms $v \triangleright m(v_1, \ldots, v_n) \cdot v'$, $v_1 \triangleright v_2 \parallel v_3$ and $v_1 \triangleright v_2 + v_3$; and $v \triangleleft c$ can be split into $v \triangleleft m(v_1, \ldots, v_n) \cdot v'$, $v_1 \triangleleft v_2 \parallel v_3$ and $v_1 \triangleleft v_2 + v_3$. There are some special cases: An action $v \triangleright v'$ copies messages from a read-port v to a write-port v'; $v \triangleleft v'$ creates a new channel with read-port v and write-port v'.

Variables occurring in the head of a process definition are annotated explicitly with a type. The type of other variables can be derived from the process definition. (The derivation of this type information is not shown in this paper.) $\text{pt}(v)$ denotes a port type associated with a variable v. Often there are several ways to associate types with variables. A compiler has to check whether there is a type assignment so that a process definition is well-typed. There are only few conditions that must be fulfilled:

Argument consistency: A call $x[s_1, \ldots, s_m](v_1, \ldots, v_n)$ is consistent with a process definition $x[s'_1, \ldots, s'_m](v'_1{:}\hat{\tau}_1, \ldots, v'_n{:}\hat{\tau}_n) := b$ if s_i and s'_i are compatible and $\text{pt}(v_j) \leq \hat{\tau}_j$ (for $1 \leq i \leq m$ and $1 \leq j \leq n$).

Variable occurrence: Each variable occurs either twice or not at all in a process definition. A single action contains a variable at most once at the top level (not as an argument). If a variable occurs in two actions at the top level, one occurrence is at the left of $\triangleright$ or $\triangleleft$, the other at the right of $\triangleright$ or $\triangleleft$.

Well-typed actions: An action is well-typed if its condition is satisfied:

Action	Condition
$v \triangleleft m(v_1, \ldots, v_n) \cdot v'$	$\text{pt}(v) \leq !(m(\text{pt}(v_1), \ldots, \text{pt}(v_1))) \cdot \tau)$ where $!\tau = \text{pt}(v')$
$v_1 \triangleleft v_2 \parallel v_3$	$\text{pt}(v_1) \leq !(\sigma \parallel \tau)$ where $!\sigma = \text{pt}(v_2)$ and $!\tau = \text{pt}(v_3)$
$v_1 \triangleleft v_2 + v_3$	$\text{pt}(v_1) \leq !(\sigma + \tau)$ where $!\sigma = \text{pt}(v_2)$ and $!\tau = \text{pt}(v_3)$
$v \triangleleft v'$	$\text{pt}(v)?\tau$ and $\text{pt}(v') = !\tau$
$v \triangleright m(v_1, \ldots, v_n) \cdot v'$	$\text{pt}(v) \leq ?(m(\text{pt}(v_1), \ldots, \text{pt}(v_1))) \cdot \tau)$ where $?\tau = \text{pt}(v')$
$v_1 \triangleright v_2 \parallel v_3$	$\text{pt}(v_1) \leq ?(\sigma \parallel \tau)$ where $?\sigma = \text{pt}(v_2)$ and $?\tau = \text{pt}(v_3)$
$v_1 \triangleright v_2 + v_3$	$\text{pt}(v_1) \leq ?(\sigma + \tau)$ where $?\sigma = \text{pt}(v_2)$ and $?\tau = \text{pt}(v_3)$
$v \triangleright v'$	$!\sigma \leq !\tau$ where $?\tau = \text{pt}(v)$ and $!\sigma = \text{pt}(v')$

Well-typed process expressions ensure that for each message written to a write-port there is an action reading this message from a corresponding read-port. A message can be written to a port v only if $\text{pt}(v)$ specifies that the channel associated with v can transport the message. This port can be combined with other parallel or alternative write-ports; the conditions on well-typed actions ensure that all messages from these ports are transported in the channel. If the write-port has been received as an argument of a message or as a head variable, the channel may be able to transport more messages, but not less. If the port was created by an action $v' \triangleleft v$, exactly the messages specified by $\text{pt}(v)$ are transported. Each channel was created by an action of this form so that there is an appropriate read-port. Conditions on variable occurrences ensure that no variable can be read or written several times and that there are actions reading from each read-port and splitting each read-port at parallel and alternative compositions. Conditions on types associated with read-ports ensure that all messages written to a channel are read from the channel.

5 Related Work

Much of the work on types for concurrent languages and models is based on the π-calculus [7]; especially the problem of inferring most general types was considered [3, 15]. Literature on subtyping in such calculi is also available [5, 8, 9, 10, 15]. Each of these type models differs in important aspects from the one presented in this paper.

Nierstrasz [9] proposes "regular types" and "request substitutability"—a specific form of the principle of substitutability—as foundations of subtyping. However, his very general results are not concrete enough to develop a static type system from them.

The proposal of Kobayashi and Yonezawa [5] is more practical; subtyping is defined in a similar way as in sequential object-oriented languages based on a typed λ-calculus [2]: A type of an active object specifies the set of messages that will be accepted by all instances; a subtype specifies an extended set of messages. In fact, their approach is a subset of our approach if types of write-ports are regarded as types of the objects reading from the corresponding ports. Each type definition in the subset is of the form $x :: (p_1 + \cdots + p_n) \cdot x$, where $p_1, \ldots, p_n$ are the prototypes of all messages understood by instances of the type.

Kobayashi and Yonezawa assume that messages in a mail queue are processed in arbitrary ordering; but their type system cannot ensure that each received message will be processed eventually. The models of Vasconcelos [15], Pierce and Sangiorgi [10] and Nielson and Nielson [8] also cannot ensure that the receiver reacts to all received messages. However, if an object's type is understood as a contract between the object and its clients, the object must be able to deal properly with all messages received from its clients; the clients can expect that the object responses their type-conforming requests.

In the model proposed in the present paper as well as in the process type model [13] objects react to all received messages. This property was a primary goal for this work because we regard it as a precondition for the inclusion of more expressive (partial) behavior specifications into types. The most important difference to the work presented in [11, 12, 13] is that process types are types of active objects, not types of ports. Process types can be regarded as equivalent to write-ports; but there is nothing comparable with read-ports.

Although the model proposed by Nielson and Nielson [8] does not ensure that all messages are understood, it ensures that instances of subtypes preserve the properties expressed in supertypes: If instances of a supertype understand some message sequences, instances of subtypes also understand them. Because types specify the complete behavior of processes as far as the communication is concerned, subtyping is rather restricted; a subtype cannot specify additional message orderings and messages.

6 Conclusion

We defined the type of a communication channel as the set of all sequences of messages that can be transported through the channel. Types of channel ports are not symmetrical: A write-port allows a process to write arbitrary messages specified in the type; but the process must be prepared for reading all messages specified in the type of a read-port. For write-ports a subtype is essentially a superset of message sequences, for read-ports a subset. This definition is compatible with the principle of substitutability so that extensions of object behavior are expressible in types. Process expressions can be type-checked statically.

References

1. J. C. M. Baeten and W. P. Weijland. *Process Algebra*. Cambridge Tracts in Theoretical Computer Science 18. Cambridge University Press, 1990.
2. Luca Cardelli and Peter Wegner. On understanding types, data abstraction, and polymorphism. *ACM Computing Surveys* **17** (1985) 471–522.
3. Simon J. Gay. A sort inference algorithm for the polyadic π-calculus. In *Proceedings POPL'93*, January 1993.
4. Kenneth M. Kahn and Vijay A. Saraswat. Actors as a special case of concurrent constraint programming. *ACM SIGPLAN Notices* **25**:10 (October 1990) 57–65. Proceedings OOPSLA/ECOOP'90.
5. Naoki Kobayashi and Akinori Yonezawa. Type-theoretic foundations for concurrent object-oriented programming. *ACM SIGPLAN Notices* **29**:10 (October 1994) 31–45. Proceedings OOPSLA'94.
6. Barbara Liskov and Jeannette M. Wing. Specifications and their use in defining subtypes. *ACM SIGPLAN Notices* **28**:10 (October 1993) 16–28. Proceedings OOPSLA'93.
7. R. Milner, J. Parrow, and D. Walker. A calculus of mobile processes (parts I and II). *Information and Computation* **100** (1992) 1–77.
8. Flemming Nielson and Hanne Riis Nielson. From CML to process algebras. In *Proceedings CONCUR'93*, LNCS 715, Springer-Verlag, 1993, 493–508.
9. Oscar Nierstrasz. Regular types for active objects. *ACM SIGPLAN Notices* **28**:10 (October 1993) 1–15. Proceedings OOPSLA'93.
10. Benjamin Pierce and Davide Sangiorgi. Typing and subtyping for mobile processes. In *Proceedings LICS'93*, 1993.
11. Franz Puntigam. Flexible types for a concurrent model. In *Proceedings of the Workshop on Object-Oriented Programming and Models of Concurrency*, Torino, Italy, June 1995.
12. Franz Puntigam. Type specifications with processes. In *Proceedings FORTE'95*, IFIP, October 1995.
13. Franz Puntigam. Types for active objects based on trace semantics. In *Proceedings of the IFIP Workshop on Formal Methods for Open Object-based Distributed Systems*, Paris, France, March 1996.
14. Ehud Shapiro. The family of concurrent logic programming languages. *ACM Computing Surveys* **21** (1989) 412–510.
15. Vasco T. Vasconcelos. Typed concurrent objects. In *Proceedings ECOOP'94*, LNCS 821, Springer-Verlag, 1994, 100–117.
16. Peter Wegner and Stanley B. Zdonik. Inheritance as an incremental modification mechanism or what like is and isn't like. In S. Gjessing and K. Nygaard (eds.), *Proceedings ECOOP'88*, LNCS 322, Springer-Verlag, 1988, 55–77.

Laws of Data Parallel Assignment

J.P. Wray

Department of Computer Science
The Queen's University of Belfast
Belfast BT7 1NN, UK
jp.wray@qub.ac.uk

Abstract. A set of laws for data parallel assignment is outlined. The laws illustrate the mathematical tractability of this programming construct and provide a means of correctly transforming a complex assignment into a sequence of simpler assignments which may then be interpreted on a variety of parallel architectures.

1 Introduction

One means of exploiting the potential of supercomputers for the efficient execution of scientific programs is to specify a set of "fine grained" order independent operations which may be applied to a data structure. The concept of applying order independent updates to data structures [1, 8, 7, 12, 10] may be viewed as an extension of multiple assignment [3, 4]. Data parallel assignment [9, 10] captures the concept of independence of a set of operations; it may be executed using any combination (parallel or sequential) of its atomic constituents (individual updates) and, consequently, is suitable for implementation on a wide range of parallel architectures.

Previous work on data parallel assignment [9, 10, 13] has resulted in the development of complementary denotational and axiomatic definitions of its semantics. These definitions exhibit mathematical regularity and facilitate the development and verification of correct data parallel programs. In further work a formal array processor based operational semantics of the construct has been specified [15] and data parallel assignment has been compared with apparently similar constructs in Fortran 90 and High Performance Fortran [11].

In this paper a set of laws for data parallel assignment is sketched. The motivation for studying the algebraic properties of data parallel assignment stems from the observation [5] that

   ... conventional programs are mathematical expressions, and are subject to a set of laws as rich and elegant as those of any other branch of mathematics, engineering, or natural science.

That is, it is possible to define laws which allow a programmer to manipulate and reason about programs in much the same way that mathematicians reason about, for example, logic and calculus. These manipulations may be carried out without reference to underlying domains or assertions. In particular, algebraic laws may be used to define correctness-preserving transformations that can facilitate the interpretation of data parallel assignment on various parallel computer architectures. Furthermore, another consistent complementary description of the construct will provide more information about its mathematical regularity—as observed by Hoare and Lauer [6], a good intuitive criterion for such regularity of a language is that it is easy to describe formally *in more than one way*.

Section 2 provides a brief overview of data parallel assignment. The laws of data parallel assignment are outlined in Section 3. A structural normal form for data parallel assignment, the validity of which is established using some of the laws, is proposed in Section 4.

2   Data Parallel Assignment

Data parallel assignment is a means of specifying a set of *independent* updates that are to be applied to a fine grained data structure (an array). For example, if a, b, and c are $n \times n$ matrices, $\{(1,1),\ldots,(n,n)\}$ denotes the set of index pairs (i,j) such that $1 \leq i,j \leq n$, and $\{(k,l) \in S \mid P(k,l)\}$ denotes the set of index pairs which belong to S and satisfy the predicate P, then the data parallel assignment

$$\forall (i,j) \in \{(1,1),\ldots,(n,n)\}.a(i,j) := b(i,j) + c(i,j)$$

assigns to a the sum of b and c, and

$$\forall (i,j) \in \{(k,l) \in \{(1,1),\ldots,(n,n)\} \mid k = l\}.a(i,j) := 1$$

assigns the value "1" to each element in the diagonal of a.

The examples above involve scalars and structures of uniform dimension. A structure of one dimension may be assigned values from a structure of a different dimension using a data parallel assignment incorporating an *index function*, i.e., a mapping from the indices of the structure on the left-hand side of the assignment to the indices of a structure on the right-hand side. Data parallel assignments may also involve conditional expressions. In general, a data parallel assignment statement has the form

$$\forall ilist \in S.id(ilist) := exp$$

where *ilist* is a list of index variables, S is a set expression defining a set of actual indices (integer tuples), *id* is an array variable, and *exp* is an expression which may contain *ilist* as well as index functions. The effect of the assignment is to "simultaneously" assign, for all indices in the set S, the values of *exp* to the corresponding elements of *id*. Full details of the the syntax and denotational semantics of data parallel assignment may be found in [13].

3 The Laws

Each law is an equivalence of the form "$A1 \equiv A2$", where $A1$ and $A2$ are data parallel assignment schemas, and signifies that $A1$ may be rewritten as $A2$, and vice-versa, whilst preserving meaning. The laws fall into three main categories:

1. Substitution laws, which define how bound variables may be renamed whilst preserving meaning;
2. Laws derived from standard set and function laws such as the associative and distributive laws for set union;
3. Laws for simplifying assignments, for example, laws which define absorption of assignments and laws which define the removal of conditional expressions from assignments.

Lack of space precludes listing the laws in full here. For complete details see [14]. The following is an example of a law in the third of the above categories:

$$a \notin free(S, B, e2) \models$$
$$\forall i \in S.a(i) := If\ B\ Then\ e1\ Else\ e2$$
$$\equiv \forall i \in \{j \in S \mid B\langle j/i\rangle\}.a(i) := e1; \forall i \in \{j \in S \mid \neg B\langle j/i\rangle\}.a(i) := e2$$

This law defines how a data parallel assignment involving a conditional expression may be replaced by a pair of unconditional assignments. It is valid only if the array variable a does not occur free in any of the sub-expressions S, B, or $e2$. The set expressions in each of the two unconditional assignments could then be further simplified through the application of substitution and set laws.

The validity of each of the laws can be established [14] using a straightforward application of the denotational definitions. One application of the laws is in establishing the equivalence of a data parallel assignment and a sequence of such assignments having a special form—structural normal form—discussed in the next section.

4 A Structural Normal Form

The definition of an operational interpretation of data parallel assignment is simplified if the number of possible syntactic forms which needs to be considered is reduced. This can be achieved by defining a structural form into which any particular data parallel assignment (involving arithmetic operators of maximum arity 2) may be transformed without changing its meaning:

Definition 1. A data parallel assignment of the form

$$\forall i \in S.a(i) := L\ OP\ NL$$

where

1. **L** is an expression in which the only term (if any) involving free tuple variables is **a(i)**;
2. **NL** is an expression in which the only term (if any) involving free tuple variables is **b(F(i))**, where b is an array variable and **F** is an index function;
3. **OP** is a binary operator

is in **structural normal form**.

Data parallel assignment is interpreted on a model distributed array processor in [15]. To this end, the structural normal form ensures that each "atomic" assignment involves at most one "non-local" data element (that which occurs in the sub-expressions NL).

It is necessary to prove that any data parallel assignment can be transformed into an equivalent sequence of structural normal form assignments. The strict denotational equivalence of two sequences of data parallel assignments is inappropriate for this purpose since, for example, the final values of temporary variables (such as "dummy" variables used in swapping the values of two program variables) are not significant from the programmer's point of view. Therefore the notion of "relative equivalence", as defined below, is used.

Definition 2. Let **SS1** and **SS2** be finite sequences of data parallel assignments, and let $a \in Id$. We say that **SS1** and **SS2** are **equivalent with respect to** a iff
$$\forall \sigma \in St. \forall i \in iddomain(\sigma, a). \mathcal{C}[SS1]\sigma(a,i) = \mathcal{C}[SS2]\sigma(a,i)$$
where C is the semantic function for commands, and $iddomain(\sigma, a)$ denotes the index range of the array a in state σ.

Informally, $SS1$ and $SS2$ are equivalent with respect to a iff they always effect the same transformation on a, regardless of their effect on other program variables. The following theorem states that the structural normal form defined above possesses the required properties.

Theorem 3. *For any data parallel assignment* **A** *to a variable* **a**, *there exists a finite sequence of structural normal form data parallel assignments which is equivalent, with respect to* **a**, *to* **A**.

Proof. By induction on the structure of A, appealing to the laws discussed earlier. Details are omitted. □

5 Conclusion

Data parallelism continues to be a significant feature of parallel computing. There is a need for a model of data parallel computation that is tractable and efficiently implementable. It has been shown [11] that correctness proofs of programs based on data parallel assignment are more tractable than their equivalents expressed in Fortran 90 and HPF; it remains to be shown how data parallel assignment may be efficiently implemented on a range of parallel architectures.

One possible approach is to derive an efficient implementation of a program expressed using data parallel assignment through the automated application of a sequence of correctness preserving transformations. This approach has been successfully applied in the derivation of efficient parallel implementations from functional specifications [2]. The laws presented in this paper provide the basis for many of the necessary transformation rules. It would, however, be necessary to define a set of special purpose transformations for carrying out optimizations for particular architectures. Alternatively, it may be possible to transform programs expressed using data parallel assignment into equivalent Fortran 90 or HPF versions and thereby exploit the sophisticated compilers that have been developed for these languages. This is a topic for further investigation.

References

1. K.M. Chandy and J. Misra. *Parallel Program Design: A Foundation*. Addison-Wesley, 1988.
2. M. Clint, S. Fitzpatrick, T.J. Harmer, P.L. Kilpatrick, and J.M. Boyle. A family of data-parallel derviations. In W. Gentzsch and U. Harms, editors, *Proceedings of High Performance Computing and Networking, Volume II, LNCS 797*, pages 457–462. Springer-Verlag, 1994.
3. E.W. Dijkstra. *A Discipline of Programming*. Prentice-Hall, 1976.
4. D. Gries. *The Science of Programming*. Prentice-Hall International, 1981.
5. C.A.R. Hoare, I.J. Hayes, He Jifeng, C.C. Morgan, A.W. Roscoe, J.W. Sanders, I.H. Sorensen, J.M. Spivey, and B.A. Sufrin. Laws of programming. *Communications of the ACM*, 30(8):672–686, 1987.
6. C.A.R. Hoare and P.E. Lauer. Consistent and complementary formal theories of the semantics of programming languages. *Acta Informatica*, 3:135–153, 1974.
7. M. Metcalf and J. Reid. *Fortran 90 explained*. Oxford University Press, 1990.
8. R.H. Perrott. A language for array and vector processors. *ACM Transactions on Programming Languages and Systems*, 2:266–287, 1979.
9. A. Stewart. SIMD language design using prescriptive semantics. *BIT*, 28:639–650, 1988.
10. A. Stewart. An axiomatic treatment of SIMD assignment. *BIT*, 30:70–82, 1990.
11. A. Stewart. Reasoning about data-parallel array assignment. *Journal of Parallel and Distributed Computing*, 27:79–85, 1995.
12. P.J.L. Wallis. Some primitives for the portable programming of array and vector processors. *BIT*, 21:436–448, 1981.
13. J.P. Wray. *The Semantics of Synchronised Assignment*. PhD thesis, The Queen's University of Belfast, July 1992.
14. J.P. Wray. Algebraic laws and a normal form for data parallel assignment. Technical report, Department of Computer Science, The Queen's University of Belfast, May 1996.
15. J.P. Wray and A. Stewart. Correct translation of data parallel assignment onto array processors. *Formal Aspects of Computing*, 6:417–439, 1994.

Proving Progress Properties of non Terminating Programs under Fairness Assumptions

Ricardo Peña*   Luis A. Galán*

It is commonly accepted that, in order to prove progress properties of a concurrent system, some fairness assumptions about the underlying scheduling of actions are needed. An important liveness property is *eventuality*, i.e., to be able to prove that "state P eventually leads to state Q". In the language UNITY [CM88], eventuality properties are proved under unconditional fairness assumptions about the language. The differences here are that we have an explicit notion of process and rely on strong fairness assumptions. The work of Francez [Fra86] and that of K.R. Apt and E-R. Olderog [AO91] are mainly concerned with proving termination, under strong guard fairness assumptions. In this paper, we start from the work of Apt and Olderog, and prove other progress properties, different from termination, under strong channel fairness assumptions. We deviate from their approach in two main aspects: our programs may terminate or not and the semantics given to our language is a denotational one. We define a simple CSP-like language. A program denotes a set of finite and infinite traces corresponding to all its possible computations. We restrict the semantics of a program to consist only of fair traces. Proof rules concerning the notion of progress properties are given. Finally, we verify a non-trivial case study. Interested readers can find more details in [PG95].

1 Programming Language Syntax

A distributed program is $P = [\|_{i=1}^{n} P_i]$ where $P_i = [S_{i0}; *[[]_{j=1}^{m_i} B_{ij}; \alpha_{ij} \rightarrow S_{ij}]]$ for $i \in \{1..n\}$, being each B_{ij} a boolean expression, each α_{ij} either an output or an input communication, and each S_{ij} either a multiple assignment or the **skip** sentence. Additionally, we require every channel to communicate exactly two processes and to appear at most once in the guards of a process.

Definition 1. Given a distributed program P, the predicate $match(i, j, k, l)$ holds if $i \neq k$, and either $\alpha_{ij} = c\,!\,E$ and $\alpha_{kl} = c\,?\,x$, or $\alpha_{ij} = c\,!$ and $\alpha_{kl} = c\,?$. We will call c the *matched communication* of α_{ij} and α_{kl}, and we will write $matched\_comm(\alpha_{ij}, \alpha_{kl}) = c$. We will denote by $Matches(P)$ the set $\{(i, j, k, l) \mid match(i, j, k, l)\}$.

2 Semantics

A *global state*, in short a *state*, is an application $\sigma : Var(P) \longrightarrow \mathcal{D}$, where $Var(P)$ are all the variables of the distributed program P (possibly renamed to avoid

---

* Departamento de Informática y Automática, Universidad Complutense de Madrid, 28040 Madrid, Spain. e-mails: {ricardo,lagalan}@dia.ucm.es

name conflicts), an $\mathcal{D}$ is the value domain. We denote by $\sigma_\perp$ the *undefined state*, by $\sigma \models B$ the satisfaction of a boolean expression B in state σ, by $\sigma[x_1, \ldots, x_n \leftarrow v_1, \ldots, v_n]$ the *simultaneous updating* in σ of a set of variables by a set of values. We use $\sigma[\leftarrow S]$, S being the assignment statement $x_1, \ldots, x_n := E_1, \ldots, E_n$, as an abbreviation of $\sigma[x_1, \ldots, x_n \leftarrow \sigma(E_1), \ldots, \sigma(E_n)]$.

Definition 2. Given a distributed program P, and a state σ, we say that there exists a *transition* from σ to σ' with communication c, denoted $\sigma \xrightarrow{c} \sigma'$, if there exists $(i, j, k, l) \in Matches(P)$ such that the following conditions hold:

1. $c = matched\_comm(\alpha_{ij}, \alpha_{kl})$, $\alpha_{ij} = c\,!\,E$ and $\alpha_{kl} = c\,?\,x$.
2. $\sigma \models B_{ij} \wedge B_{kl}$.
3. $S_{ij} = x_1, \ldots, x_r := E_1, \ldots, E_r$ and $S_{kl} = y_1, \ldots, y_s := H_1, \ldots, H_s$ and
   $\sigma' = \sigma[x_1, \ldots, x_r, x, y_1, \ldots, y_s \leftarrow \sigma(E_1), \ldots, \sigma(E_r), \sigma(E), \sigma''(H_1), \ldots, \sigma''(H_s)]$
   where $\sigma'' = \sigma[x \leftarrow \sigma(E)]$.

We will denote by $\sigma \xrightarrow{t}{}^* \sigma'$, $t \in Comm^*$, the reflexive, transitive closure of the $\xrightarrow{c}$ relation.

Definition 3. The set of *reachable states* of a distributed program P, denoted Σ_P, is the following inductively generated set:

- $\sigma_0 \in \Sigma_P$, where $\sigma_0 \stackrel{\text{def}}{=} \sigma_\perp [\leftarrow \cup_{i=1}^n S_{i0}]$.
- $\sigma \in \Sigma_P \wedge \sigma \xrightarrow{c} \sigma' \Rightarrow \sigma' \in \Sigma_P$

Definition 4. A *computation* of a distributed program P is a (finite or infinite) sequence $\sigma_0 \xrightarrow{c_1} \sigma_1 \xrightarrow{c_2} \sigma_2 \xrightarrow{c_3} \ldots$. The *trace* of a computation is $t \stackrel{\text{def}}{=} c_1 ^\frown c_2 ^\frown c_3 \ldots$. If the computation is finite and successful (all processes terminate) we will write $\sigma_0 \xrightarrow{t^\frown \sqrt{}}{}^* \sigma_f$; if it is finite and ends in a deadlock (at least one process does not terminate) we will write $\sigma_0 \xrightarrow{t^\frown \triangle}{}^* \sigma_f$. In general, we will write $\sigma_0 \xrightarrow{t}{}^\infty$ meaning that t is either a finite (i.e. ending in $\sqrt{}$ or $\triangle$) or an infinite trace of P.

Definition 5. The *semantics* of a distributed program P, denoted $[\![P]\!]$, is defined as the set of its traces: $[\![P]\!] \stackrel{\text{def}}{=} \{t \in (Comm \cup \{\sqrt{}, \triangle\})^\infty \mid \sigma_0 \xrightarrow{t}{}^\infty\}$

Definition 6. Given a program P and $t \in [\![P]\!]$, $t = c_0 ^\frown c_1 ^\frown c_2 \ldots$, we define the *run* of t as the following sequence of pairs: $(E_0, c_0), (E_1, c_1), (E_2, c_2), \ldots$ where, for all i, $E_i \stackrel{\text{def}}{=} enabled\,(t_i, P)$ and $t_0 \stackrel{\text{def}}{=} \epsilon$, $t_{i+1} \stackrel{\text{def}}{=} t_i ^\frown c_i$, being $enabled\,(t, P) = \{c \in Comm \mid \exists t'. t ^\frown c ^\frown t' \in [\![P]\!]\}$

Definition 7. Given a program P and $t \in [\![P]\!]$, let $(E_0, c_0), (E_1, c_1), (E_2, c_2), \ldots$ be the run of t. Then, t is called a *fair* trace, denoted $fair(t)$, if its run satisfies the following predicate $\forall c \in Comm.((\stackrel{\infty}{\exists} i \in \mathbf{N}.c \in E_i) \Rightarrow (\stackrel{\infty}{\exists} j \in \mathbf{N}.c = c_j))$ where the predicate $\stackrel{\infty}{\exists} i \in C.P(i)$ holds when there exists an infinite number of values of i satisfying $P(i)$. Let us note that every finite trace is fair.

Definition 8. The *fair semantics* of a program P, denoted $[\![P]\!]_{fair}$, is defined as the set of its fair traces: $[\![P]\!]_{fair} \stackrel{\text{def}}{=} \{t \mid t \in [\![P]\!] \wedge fair\,(t)\}$

3 Proof Rules

Definition 9. Given a distributed program P, a *global invariant*, in short an *invariant*, is a predicate I having variables of P as free variables, such that:

1. $\{True\} \cup_{i=1}^{n} S_{i0}\{I\}$
2. $\forall (i,j,k,l) \in Matches(P).\{I \wedge B_{ij} \wedge B_{kl}\} S_{ijkl}\{I\}$, where
   $S_{ijkl} \stackrel{def}{=} x_1,\ldots,x_r,x,y_1,\ldots,y_s := E_1,\ldots,E_r,E,H_1\mid_x^E,\ldots,H_s\mid_x^E$ being S_{ij}
   and S_{kl} the assignments $x_1,\ldots,x_r := E_1,\ldots,E_r$ and $y_1,\ldots,y_s := H_1,\ldots,H_s$.

In what follows, we will denote by B_c and S_c respectively the combined guard and assignment corresponding to the communication $c = matched\_comm(i,j,k,l)$.

Proposition 10. *Given a distributed program P, and an invariant I, then $\forall \sigma \in \Sigma_P.\sigma \models I$. Obviously, if $I \Rightarrow SAFE_P$ then all states of P satisfy $SAFE_P$.*

Definition 11. Given a distributed program P and an invariant I, a communication c is: (1) *enabled* in states satisfying Q, denoted $Q \stackrel{c}{\hookrightarrow}$, if $I \wedge Q \neq false$ and $I \wedge Q \Rightarrow B_c$. (2) *forbidden* in states satisfying Q, denoted $Q \stackrel{c}{\not\hookrightarrow}$, if $I \wedge Q \neq false$ and $I \wedge Q \Rightarrow \neg B_c$.

Definition 12. Given a distributed program P and an invariant I, a transition *may happen* from states satisfying Q to states satisfying R, denoted $Q \stackrel{c}{\hookrightarrow} R$, if $Q \stackrel{c}{\hookrightarrow} \wedge \{Q\} S_c \{R\}$.

Definition 13. Given a distributed program P and an invariant I, we say that states Q *eventually lead* to states R, denoted $Q \Longmapsto^* R$, if one of the following conditions applies:

1. $Q \Rightarrow R$
2. $\exists Q'. Q \Longmapsto^* Q' \wedge Q' \Longmapsto^* R$
3. $Q \Longmapsto_F R$ where $Q \Longmapsto_F R$ is defined as: $(\exists c \in Comm. Q \stackrel{c}{\hookrightarrow} R)$
   $\wedge (\forall c' \in Comm, c' \neq c. Q \stackrel{c'}{\not\hookrightarrow} \vee (\exists Q'. Q \stackrel{c'}{\hookrightarrow} Q' \wedge Q' \Longmapsto^* Q \vee R))$

Theorem 14. *Given a distributed program P and an invariant I,*

$Q \Longmapsto^* R$ iff $(\forall \sigma \in \Sigma_P, \forall t \in [\![P]\!]_{fair}.$
$\sigma \models Q \wedge t = t_1 \frown t_2 \wedge \sigma_0 \stackrel{t_1}{\longrightarrow}{}^* \sigma \Rightarrow \exists \sigma', \exists t_{21}. t_2 = t_{21} \frown t_{22} \wedge \sigma \stackrel{t_{21}}{\longrightarrow}{}^* \sigma' \wedge \sigma' \models R)$

4 Case Study

In this section we apply the theory developed in the previous sections to a token passing ring of processes achieving mutual exclusion between a set of n distributed user-processes $Ring :: [(\|_{i=1}^{n} U_i) \| (\|_{i=1}^{n} A_i)]$, where $\|_{i=1}^{n} U_i$ stands for the users and $\|_{i=1}^{n} A_i$ stands for the distributed arbiter. Only the arbiter is shown:

$A_i :: [st_i, neighb_i, priv_i := thinking, false, f(i);$
*[$st_i.t; \mathbf{req}_i?$ → $st_i := hungry$
    $st_i.h \wedge priv_i; \mathbf{ack}_i!$ → $st_i := eating$
    $st_i.e; \mathbf{rel}_i?$ → $st_i := thinking$
    $(neighb_i \vee (st_i.h)) \wedge \neg priv_i; \mathbf{req}_{ii\ominus 1}!$ →skip
    $\neg neighb_i; \mathbf{req}_{i\oplus 1i}?$ → $neighb_i := true$
    $priv_i \wedge \neg st_i.e \wedge neighb_i; \mathbf{ack}_{ii\oplus 1}!$ → $priv_i, neighb_i := false, false$
    $\mathbf{ack}_{i\ominus 1i}?$ → $priv_i := true$
]
]

where we abbreviate $st_i = thinking$ by $st_i.t$, $st_i = hungry$ by $st_i.h$, and $st_i = eating$ by $st_i.e$ and use a function $f(i)$ evaluating to $true$ for only one $i \in \{1..n\}$. The aim is to initially assign the token to exactly one ring process. Variable st_i records the current state of user i, $priv_i$ is true for the only process i holding the token, $neighb_i$ is used to propagate requests for the token. One invariant is: $I \stackrel{\text{def}}{=} ((\mathcal{N}i.priv_i) = 1) \wedge (\forall i.st_i.e \Rightarrow priv_i) \wedge (\forall i.neighb_i \Rightarrow ((st_{i\oplus 1}.h \vee neighb_{i\oplus 1}) \wedge \neg priv_{i\oplus 1}))$, where $\mathcal{N}$ is the counting quantifier. Mutual exclusion (i.e. $I \Rightarrow (\mathcal{N}i.st_i.e) \leq 1$) is trivial from the invariant. The program does not terminate and does not get blocked, i.e. the following implication holds:

$$I \Rightarrow \bigvee_{(i,j,k,l) \in Matches(Ring)} B_{ij} \wedge B_{kl} \equiv (\bigvee_{i=1}^{n} st_i.t) \vee (\bigvee_{i=1}^{n}(st_i.h \wedge priv_i)) \vee (\bigvee_{i=1}^{n} st_i.e) \vee ...$$

In case all users were hungry, by I, at least one of them holds the token and can progress. Now, we use definition 13 to prove that every hungry user will eventually be served.

Proposition 15. $st_i.h \Longmapsto^* st_i.h \wedge priv_i$ (the proof uses lemmas 16, 17, 18 and 19)

Lemma 16. If $i \neq j$ then $st_i.h \wedge priv_j \Longmapsto^* st_i.h \wedge priv_j \wedge neighb_j$

Lemma 17. If $k \neq j$ then $neighb_k \wedge priv_j \Longmapsto^* neighb_k \wedge priv_j \wedge neighb_j$

Lemma 18. If $i \neq j$ then $st_i.h \wedge priv_j \wedge neighb_j \Longmapsto^* st_i.h \wedge priv_j \wedge neighb_j \wedge \neg st_j.e$

Lemma 19. If $i \neq j$ then $st_i.h \wedge priv_j \wedge neighb_j \wedge \neg st_j.e \Longmapsto^* st_i.h \wedge priv_i$

Proposition 20. $st_i.h \wedge priv_i \Longmapsto^* st_i.e$

Corollary 21. $st_i.h \Longmapsto^* st_i.e$

References

[AO91] K. R. Apt and E.-R. Olderog. *Verification of Sequential and Concurrent Programs.* Springer-Verlag, 1991.

[CM88] K. M. Chandy and J. Misra. *Parallel Program Design: A Foundation.* Addison Wesley, Reading, Massachusetts, 1988.

[Fra86] N. Francez. *Fairness.* Springer-Verlag, 1986.

[PG95] R. Peña and L.A. Galán. Proving Progress Properties of non Terminating Programs under Fairness Assumptions. Technical Report DIA-UCM 95/6, Universidad Complutense de Madrid, 1995.

Workshop 06

Parallel Discrete Algorithms

A Simple Parallel Dictionary Matching Algorithm*

Paolo Ferragina

Dipartimento di Informatica
Università di Pisa, Italy. E-mail: ferragin@di.unipi.it

Abstract. In the Parallel Dictionary Matching problem a set of patterns $\mathcal{D}$ is fixed at the beginning, and the following Query(T) operation has to be quickly supported: given an arbitrary text $T[1:t]$, for each position i retrieve the longest pattern in $\mathcal{D}$ that is prefix of text suffix $T[i:t]$. In this paper, we present a simple CRCW PRAM algorithm achieving optimal work for answering Query(T) in the case of a constant-sized alphabet.

1 Introduction

The classical *pattern matching* problem on strings consists of finding all occurrences of a single pattern $P[1:p]$ as a substring of a text $T[1:t]$, where P and T are drawn from an ordered alphabet Σ. The goal is to preprocess P such that all the succeeding queries on an arbitrary text T can be answered quickly, that is, in optimal $O(t)$ time [7, 15]. The "dual" version of this problem, in which P is given on-line and T is fixed, has been studied as well attaining optimal solutions [16, 19].

One generalization of the pattern matching problem is the *multiple pattern matching* problem, commonly called *dictionary matching* (shortly, DM) problem. Here, instead of a single pattern, a *set* of patterns $\mathcal{D} = \{P_1, \ldots, P_k\}$, called the *dictionary*, is given to be preprocessed and an arbitrary text T is provided on-line with the intention of finding all the occurrences of the patterns in $\mathcal{D}$ that appear in T (let *tocc* be their number). In addition to its theoretical importance, DM problem has many practical applications. For example, in molecular biology, one is often concerned with determining the sequence of DNA, and then compare that sequence against all the known strings to find the ones that are related to it. Also, in computer virus detection applications, a dictionary of computer viruses is given and new programs are queried on-line to find out if they are *infected*.

Any pattern matching algorithm can be trivially extended to a set of patterns by matching each pattern separately, thus requiring $O(d + tk)$ time, where $d = \sum_{i=1}^{k} |P_i|$ is the dictionary size (i.e., brute-force method). However, one may clearly hope that, once $\mathcal{D}$ has been preprocessed, the cost of finding all the occurrences of $\mathcal{D}$'s patterns in T be proportional only to the length t of T and to the number *tocc*, independent of the length d of the (usually) much larger dictionary. Aho and Corasick [1] were the first to solve optimally the DM problem in $O(d \log \sigma)$ sequential time for the preprocessing of $\mathcal{D}$, and $O(t \log \sigma + tocc)$ time for answering a query on text T, where $\sigma = \min\{d, |\Sigma|\}$. This result is perhaps surprising because the text scanning time is *independent* of the dictionary size (for a constant-sized Σ). Since then, a dynamic formulation of this problem has been also well studied achieving very interesting results (e.g., see [2, 3, 4, 11]).

The DM problem has been also deeply investigated in the widely used Parallel Random Access Machine (shortly PRAM [13]). In particular, the powerful Concurrent-Read-Concurrent-Write variant of this model has been employed to describe various parallel solu-

---
* Work supported in part by MURST of Italy.

tions. We remark that in the parallel context, Query(T) operation is defined as follows: For each position of T, retrieve the *longest pattern* in $\mathcal{D}$ that occurs in T starting at that position. Notice that the whole information about shorter patterns is contained implicitly in this representation. Moreover, the output size does not prevent the algorithm to have polylogarithmic time complexity when using $O(t)$ processors. Amir and Farach [2] were the first to provide an efficient parallel algorithm for the DM problem requiring $O(\log m \log d)$ time and $O(t \log m \log d)$ work [2] for answering Query(T), and $O(\log m \log d)$ time and $O(d \log d)$ work for preprocessing $\mathcal{D}$, where m denotes the length of the longest pattern in $\mathcal{D}$. Then, Muthukrishnan and Palem [17] presented an algorithm requiring $O(\log m)$ time and $O(t \log m)$ work for answering Query(T), and $O(\log m)$ time and $O(d)$ optimal work for preprocessing $\mathcal{D}$. They also presented an improved algorithm, in the case of a constant-sized Σ, which requires $O(\log m)$ time and $O(d \log m)$ work for preprocessing $\mathcal{D}$, and $O(\log m)$ time and $O(t)$ optimal work for answering Query(T). Both two solutions use a large amount of space, i.e. $O(md^{1+\epsilon})$, for any given $\epsilon > 0$. Using *randomization*, first Amir and Farach and Matias [5], and later Farach and Muthukrishnan [9], have reported very efficient algorithms with expected work optimal bounds both for answering Query(T) and preprocessing $\mathcal{D}$.

In this paper we provide a *simple* CRCW PRAM algorithm achieving optimal work for answering Query(T) and requiring small space, in the case of a constant-sized alphabet. The dictionary can be preprocessed in $O(\log d)$ time and $O(d \log m)$ total work. Answering Query(T) requires $O(\log m)$ time and $O(t)$ optimal work. The total required space is $O(d^2 \log m)$. It is worth noting that our solution achieves the same time and work bounds as in [17], but it is simpler and also requires less space for $d^{1-\epsilon} = O(\frac{m}{\log m})$.

2 Preliminaries

In what follows we will use the classical naming technique [14] and the suffix tree data structure [16, 19] as basic tools to develop our parallel solution (see the corresponding literature for more details).

Let X be a string of x characters and assume \$ $\notin \Sigma$.[3] The suffix tree ST_X built on $X\$$ is a digital search tree containing all the suffixes of $X\$$ and occupying optimal $O(x)$ space [16]. The character \$ is used to prevent that a suffix $X[i:x]\$$ is a prefix of another suffix $X[j:x]\$$; thus there exists a unique leaf in ST_X for each suffix of $X\$$. Each arc of ST_X is labeled with a substring $X[i:j]$, which is represented as a triple (X,i,j). Given the suffix tree ST_X and a node u, we denote by $W(u)$ the concatenation of the labels on the path from the root to node u. Clearly, $W(u)$ is a substring of X and thus every ancestor w of u in ST_X denotes a string $W(w)$ which is a proper prefix of $W(u)$. In general, given a substring V of $X\$$, we define the (exact) *locus* of V as the node v in ST_X such that $V = W(v)$. Moreover, we define the *extended locus* of a string U as the node u in ST_X such that U is a prefix of $W(u)$ and $W(p(u))$ is a *proper* prefix of U, where $p(u)$ is the parent of u in ST_X.

The parallel construction of the suffix tree works on the arbitrary CRCW PRAM and requires two phases [6] (see [12] for a work-optimal algorithm). In the first phase, called *naming*, we label all of X's substrings of power-of-two length. Labels are integers between 1 and $x + 1$, and equal substrings get the same label (this is called *consistent naming*). In the second phase, called *refining*, a sequence of refinement trees $RT^{(r)}$ is produced for $r = \lceil \log x \rceil, \ldots, 0$. The final tree $RT^{(0)}$ is basically the suffix tree ST_X, except for some minor adjustments. For each intermediate value r, $RT^{(r)}$ is a better and better approximation of suffix tree ST_X.

---

[2] By *work* of a parallel algorithm, we mean the total number of operations performed to solve a problem [13]. A parallel algorithm is called *work optimal* if its work is of the order of time of the best possible sequential algorithm for the same problem.

[3] From now on we assume that alphabet Σ has constant size.

Theorem 1. *[6] Given a string $X[1:x]$, the names of all its substrings of power-of-two length and its set of refinement trees can be computed in $O(\log x)$ time and $O(x \log x)$ work. The total required space is $O(x^2 \log x)$.*

To search for the *longest prefix* of a string $Y[1:y]$ which occurs in string X, we maintain all of X's refinement trees and partition Y in substrings $\sigma_1, \sigma_2, \ldots, \sigma_k$, where $k \leq \lfloor \log y \rfloor, |\sigma_i| = 2^{r_i}, r_i > r_{i+1}$. Then, we label these substrings consistently with X and search for them in X's refinement trees, thus obtaining:

Lemma 2. *[6] Given an arbitrary string $Y[1:y]$, Y's longest prefix occurring in X (and its extended locus in ST_X) can be found in $O(\log y)$ time and $O(y)$ work.*

Before concluding this section, let us recall a simple result which will be used later.

Lemma 3. *[10] Let T be a tree in which the root and some nodes are marked. The pointer to the deepest marked ancestor of each node in T can be computed in $O(\log |T|)$ time and $O(|T|)$ optimal work on the EREW PRAM.*

3 Preprocessing $\mathcal{D}$

Let $\mathcal{D} = \{P_1, \ldots, P_k\}$ be a dictionary of *patterns* of total size $d = \sum_{i=1}^{k} |P_i|$ and maximal pattern length $m = \max\{|P_i| : 1 \leq i \leq k\}$ (w.l.o.g. assume m is a power of two). Since the dictionary is fixed at the beginning, our goal is to preprocess it by building a proper set of data structures to support work-optimal queries on arbitrary texts that are provided on-line.

Preprocessing $\mathcal{D}$ consists of two main steps. In Step (1), all the patterns in $\mathcal{D}$ are labeled and the corresponding set of refinement trees is built. In Step (2), the suffix tree ST_D, built on the patterns of $\mathcal{D}$, is augmented with some additional information.

Step (1): Consider the string $D = P_1\$_1 P_2\$_2 \ldots P_k\$_k$, where $\$_i \neq \$_j$ for each $i \neq j$, and $\$_i \notin \Sigma$ for all $i = 1, \ldots, k$. Notice that D's total length is still $O(d)$. Apply the naming technique to consistently label all of D's substrings having power-of-two length at most m (Theorem 1). Then, build the suffix tree ST_D and its set of refinement trees $RT^{(i)}$ only for $i = \log m, \ldots, 1, 0$, by exploiting the fact that all of D's substrings longer than m are distinct, so taking $O(\log m)$ time and $O(d \log m)$ total work. The total space required by the set of refinement trees is therefore $O(d^2 \log m)$. It is worth noting that the final suffix tree ST_D has a distinct leaf for each suffix of a pattern in $\mathcal{D}$. Hence Lemma 2 can be easily extended as follows:

Lemma 4. *Given a string $Y[1:y]$, Y's longest prefix occurring in a pattern of $\mathcal{D}$ (and its extended locus in ST_D) can be found in $O(\log y)$ time and $O(y)$ work.*

Step (2): Augment the suffix tree ST_D computing for each node $u \in ST_D$ the deepest ancestor of u, called $lp(u)$, which is the locus of a pattern in $\mathcal{D}$ (i.e., $W(lp(u)) \in \mathcal{D}$). To do this, we mark the root of ST_D and all of its leaves that are locus of some $P_i\$_i$. If the leaf storing a string $P_i\$_i$ is connected to its parent by an arc whose first labeling character is $\$_i$, then we delete the mark from the leaf and mark its parent (i.e., this node is the exact locus of the pattern P_i). From the properties of suffix trees [16], it immediately follows that $lp(u)$ is u's deepest marked ancestor. Therefore, we can use Lemma 3 (with $|T| = O(d)$) and compute $lp(u)$ in $O(\log d)$ time and $O(d)$ work.

We further augment ST_D computing a set of pointers $ext(c, u)$, for all $c \in \Sigma$ and $u \in ST_D$, defined as follows:

Definition 5. For each node $u \in ST_D$ and for each character $c \in \Sigma$, we define $ext(c, u) = v$ if and only if v is the extended locus in ST_D of the longest prefix of $cW(u)$ occurring in D (possibly $cW(u)$ itself).

Notice that $ext(c, u)$ is different from the pointer defined in [8], because in that case a pointer is defined for a character c and a node u only if the string $cW(u)$ occurs in D and thus its extended locus is defined. In our case, instead, it may be $|W(ext(c, u))| < W(u) + 1$, because $cW(u)$ might not occur in any pattern of $\mathcal{D}$. We remark also that the augmented ST_D still requires $O(d)$ space, because $|\Sigma| = O(1)$ (by the hypothesis), and thus we have a constant number of ext-pointers leaving from each node in ST_D. We prove the following result:

Lemma 6. *For each node $u \in ST_D$ and for each character $c \in \Sigma$, the pointer $ext(c, u)$ can be computed in $O(\log m)$ time and $O(d \log m)$ total work on the CRCW PRAM.*

Proof. In Step (1), all the substrings of the patterns in $\mathcal{D}$ have been consistently labeled, and the corresponding set of refinement trees has been built accordingly. Given a node $u \in ST_D$ and a character $c \in \Sigma$, let us consider the string $\alpha = cW(u)$, and its substrings of length 2^q, for $0 \leq q \leq \log |\alpha|$. The substrings $\alpha[i : i + 2^q - 1]$, with $i > 1$, are actually substrings of $W(u)$, and thus they have been labeled in Step (1). Conversely, the names of substrings $\alpha[1 : 2^q]$, for all $0 \leq q \leq \log |\alpha|$, are not directly available (because we do not know even if α occurs in a pattern of $\mathcal{D}$). They are computed inductively by observing that $\alpha[1 : 2^q] = \alpha[1 : 2^{q-1}] \alpha[2^{q-1} + 1 : 2^q]$, where the substring $\alpha[2^{q-1} + 1 : 2^q]$ is entirely contained in $W(u)$ and thus its name is already known. Hence, we can label all of α's prefixes having power-of-two length in $O(\log |\alpha|) = O(\log m)$ sequential time by using the BB matrices previously adopted to label $\mathcal{D}$'s patterns. Finally, using Lemma 4, we search for α in the set of refinement trees built on D, thus finding the extended locus $ext(c, u)$ of the longest prefix of α that occurs in some pattern of $\mathcal{D}$. □

Furthermore, ST_D is preprocessed in $O(\log d)$ time and $O(d)$ total work to support constant-time LCA queries [18]. This way, given two arbitrary leaves $\ell, \ell' \in ST_D$, the longest common prefix between the two suffixes $W(\ell)$ and $W(\ell')$ can be computed in $O(1)$ sequential time by means of $LCA(\ell, \ell')$. Therefore we have:

Theorem 7. *Preprocessing phase requires $O(\log d)$ time and $O(d \log m)$ work on the CRCW PRAM. The total required space is $O(d^2 \log m)$.*

4 Answering Query(T)

We describe an approach that answers **Query**(T) based upon the information computed in Section 3 and available in the augmented ST_D. We first introduce a problem which arises in answering **Query**(T) and whose solution is used as a key tool in our parallel algorithm.

4.1 Left Extension problem

Let X be a substring of a pattern in $\mathcal{D}$. Clearly, X is consistently labeled and $|X| \leq m$. Furthermore, let $c_h, \ldots, c_1$ be a sequence of characters drawn from Σ. For $1 \leq i \leq h$, we define lcp_i as the longest prefix of the string $c_i \ldots c_1 X$ that occurs in some pattern of $\mathcal{D}$ (i.e., it occurs in D). The following proposition is easily provable:

Proposition 8. lcp_i *is the longest prefix of* $c_i lcp_{i-1}$ *that occurs in some pattern of* $\mathcal{D}$.

Proposition 8 highlights that lcp_{i-1} contains the whole information that suffices for computing lcp_i. The next step consists of solving efficiently the *Left Extension* problem defined as follows: For all $i = 1, \ldots, h$, retrieve the extended locus u_i of lcp_i in $ST_\mathcal{D}$ (notice that u_i exists since lcp_i occurs in D, by definition). This problem was studied in [9]. We propose below a simpler solution based upon *ext*-pointers in $ST_\mathcal{D}$.

Algorithm-LEP($X, c_h \ldots c_1$)

Step 1: Let u_0 be the extended locus of X in $ST_\mathcal{D}$. Retrieve u_0 by searching for X in the refinement trees built on D. Since X is consistently labeled, u_0 can be retrieved in $O(\log |X|) = O(\log m)$ sequential time (by Lemma 4).

Step 2: For $i = 1, \ldots, h$, set $u_i := ext(u_{i-1}, c_i)$.

Before showing the correctness of Algorithm-LEP, we state an intermediate result.

Lemma 9. *Let Z be a string and node z be its* extended locus *in $ST_\mathcal{D}$. For each character $c \in \Sigma$, the node $ext(c,z)$ is the extended locus of the longest prefix of cZ occurring in some pattern of $\mathcal{D}$.*

Proof. Let v be the *extended locus* of the longest prefix of cZ occurring in some pattern of $\mathcal{D}$. Recall that *ext*-pointers are computed only for the substrings of D that have exact locus in $ST_\mathcal{D}$. Therefore, if $Z = W(z)$, then the lemma clearly follows by Definition 5. Otherwise (i.e., Z is a proper prefix of $W(z)$, and $W(p(z))$ is a proper prefix of Z), we do not have directly the *ext*-pointer for Z. Nevertheless, we can show that $v = ext(c,z)$.

Since Z is a proper prefix of $W(z)$, we have that cZ is a proper prefix of $cW(z)$. Thus v is an ancestor of $ext(c,z)$ in $ST_\mathcal{D}$ (by suffix tree's structure). By contradiction, assume that v is a *proper* ancestor of $ext(c,z)$, that is, $v \neq ext(c,z)$. Thus, $W(v)$ is a *proper* prefix of $W(ext(c,z))$. Now, since v is a node in $ST_\mathcal{D}$, it has at least two outgoing arcs that have, for example, as first labeling characters c' and c'', with $c' \neq c''$. Thus, $W(v)c'$ and $W(v)c''$ occur in D. Let $W(v) = c\beta$, for some $\beta \in \Sigma^*$, then we may conclude that $\beta c'$ and $\beta c''$ also occur in D. By suffix tree's properties, β must therefore have locus u_β in $ST_\mathcal{D}$. Moreover, β is a proper prefix of $W(z)$ (recall that $W(v)$ is a proper prefix of $W(ext(c,z))$), so that u_β is a *proper* ancestor of z, and $ext(c, u_\beta) = v$ (exact locus). From the definition of v, the longest prefix of cZ occurring in some pattern of $\mathcal{D}$ must be $c\beta$ and, by the hypothesis, $|\beta| < |Z| < |W(z)|$. Hence, the longest prefix of $cW(z)$ occurring in some pattern of $\mathcal{D}$ should be $c\beta$, contradicting the hypothesis that $v \neq ext(c,z)$! □

The correctness of Algorithm-LEP is proved by induction. The basis holds since u_0 is the extended locus of X (by definition). Let us set $lcp_0 := X$. By the inductive hypothesis, u_{i-1} is the extended locus of lcp_{i-1}, thus lcp_{i-1} is a prefix of $W(u_{i-1})$. From Proposition 8, it immediately follows that lcp_i is the longest prefix of $c_i lcp_{i-1}$ occurring in some pattern of $\mathcal{D}$. Hence, applying Lemma 9 (with $c = c_i$, $Z = lcp_{i-1}$ and $z = u_{i-1}$), we derive that $ext(c_i, u_{i-1})$ is the extended locus of lcp_i.

Theorem 10. *Left Extension problem defined on a sequence of characters $c_h, \ldots, c_1 \in \Sigma$ and on a substring X of some pattern in $\mathcal{D}$, can be solved in $O(h)$ sequential time once the extended locus of X in $ST_\mathcal{D}$ is given.*

4.2 The algorithm

It consists of four steps, called Preprocessing, Sampling, Left Extension, and Retrieval. Let us describe first their main features and then proceed to their detailed discussion.

In the Preprocessing step, we label only the text substrings of length 2^q which start at positions $(h2^q + 1)$, for all $0 \leq q \leq \log m$ and $0 \leq h \leq \lfloor \frac{t}{2^q} \rfloor$. These substrings are $O(t)$ in total. In the Sampling step, we consider a subset $\mathcal{S}$ of text positions which are $O(\log m)$ positions apart each other (note that $|\mathcal{S}| = O(\frac{t}{\log m})$). For each $i \in \mathcal{S}$, we compute the longest prefix $T[i : i + L_i - 1]$ of $T[i : t]$, for a proper value L_i, occurring in some pattern of $\mathcal{D}$ by using the refinement trees built on D and the text consistent labeling (see [9, 17] for a different approach). In the Left Extension step, we compute $T[j : j + L_j - 1]$ for all the other positions $j \in [1 : t]$, by exploiting the *ext*-pointers stored in the nodes of ST_D (see Theorem 10). In the Retrieval step, we use *lp*-pointers and substring $T[j : j + L_j - 1]$, for all $j = 1, \ldots t$, to retrieve the longest pattern in $\mathcal{D}$, say $P_{long(j)}$, that is prefix of $T[j : t]$.

PREPROCESSING STEP. Recall that m is the length of the longest pattern in $\mathcal{D}$. Let us assume to append to the end of T, t special symbols $\$ \notin \Sigma$, and $\$ \neq \$_j$, for $j = 1, \ldots, k$. This way $T = T[1 : 2t]$. We perform a "partial naming" of T by applying the labeling procedure of [6] to all text substrings $T[h2^q + 1 : h2^q + 2^q]$, where $0 \leq h \leq \lfloor \frac{t}{2^q} \rfloor$ and $0 \leq q \leq \log m$. For a fixed q, these substrings cover entirely T without any overlapping and thus their number is $O(\frac{t}{2^q})$. Hence, their total number is $O(t)$. Therefore, by using the BB matrices employed to label the patterns in $\mathcal{D}$, Preprocessing step takes $O(\log m)$ time and $O(t)$ work. We point out that, we are saving space by avoiding the labeling of those text substrings that do not occur in any pattern of $\mathcal{D}$ (i.e., whose entries in the BB matrices are not initialized). Moreover, we are saving time and work by performing only a partial naming.

SAMPLING STEP. Let us consider the subset $\mathcal{S}$ of text positions defined as follows: $\mathcal{S} = \{h 2^{\log \log m} + 1 : h = 0, 1, 2, \ldots, \lfloor \frac{t}{2^{\log \log m}} \rfloor\}$. That is, $\mathcal{S}$ is the set of every other $2^{\log \log m}$ positions in T, so that $|\mathcal{S}| = O(t/\log m)$. For each suffix $T[i : t]$, with $i \in \mathcal{S}$, the longest prefix $T[i : i + L_i - 1]$, for a proper value $L_i \geq 0$, occurring as a substring of some pattern in $\mathcal{D}$ satisfies the following property:

Proposition 11. $|P_{long(i)}| \leq L_i \leq m$.

Notice that, even if $T[i : t]$ could have been not completely labeled in the Preprocessing step, each substring $T[h2^q + 1 : h2^q + 2^q]$ that occurs in $T[i : i + L_i - 1]$ has been consistently labeled because $T[i : i + L_i - 1]$ is a substring of some pattern in $\mathcal{D}$ (by its definition) and $i = h' 2^{\log \log m} + 1$, for some $h' \geq 0$. Hence, we can find $T[i : i + L_i - 1]$, searching for $T[i : t]$ in the refinement trees built on D (Lemma 2). Sampling step requires $O(\log m)$ time and $O(t)$ total work, since we are searching for $O(t/\log m)$ suffixes in total. We remark that, this step determines also the exact length L_i, for all $i \in \mathcal{S}$.

EXTENSION STEP. We compute the longest prefix $T[j : j + L_j - 1]$ occurring in some pattern of $\mathcal{D}$, for all the other positions $j \in [1 : t]$. Indeed, we exploit the *ext*-pointers stored in each node of ST_D, the length L_i (for each $i \in \mathcal{S}$), and the result proved in Theorem 10.

We map one processor to each position $i \in \mathcal{S}$ (hence we need $O(\frac{t}{\log m})$ processors in total). Each processor executes the following algorithm:

- Let u_i be the extended locus of $T[i : i + L_i - 1]$, where $i \in \mathcal{S}$ (determined in the Sampling step).
- For $s := 1$ to $2^{\log \log m} - 1$ do
  - $u_{i-s} := ext(T[i-s], u_{i-s+1})$.
  - Compute the length L_{i-s} of the longest prefix of $T[i - s : t]$ occurring in some pattern of $\mathcal{D}$, by using L_{i-s+1} and an LCA query. That is, if u_{i-s} is the root of ST_D then set $L_{i-s} := 0$; otherwise proceed as follows:
    * Let ℓ_{i-s+1} and ℓ_{i-s} be any two leaves of ST_D descending from u_{i-s+1} and u_{i-s}, respectively, and assume that $W(\ell_{i-s}) = P[r : |P|]$, for some pattern $P \in \mathcal{D}$.
    * Determine the new leaf ℓ associated with the second suffix of $P[r : |P|]$, i.e., $W(\ell) = P[r + 1 : |P|]$.
    * Set $L_{i-s} := 1 + \min\{L_{i-s+1}, |W(LCA(\ell_{i-s+1}, \ell))|\}$.

From Proposition 8, we have $L_{i-s+1} \geq L_{i-s} - 1$ and thus $T[i - s : i - s + L_{i-s} - 1]$ is a prefix of the string $T[i - s : i - s + L_{i-s+1}]$. The computation of the node u_{i-s} is correct, as immediately derives from Theorem 10 because node u_{i-s} is the extended locus (maybe exact locus) of $T[i - s : i - s + L_{i-s} - 1]$ (with $lcp_0 = T[i : i + L_i - 1]$ and $lcp_s = T[i - s : i - s + L_{i-s} - 1]$).

It remains to be shown that L_{i-s} is correctly computed by the LCA query executed in the above algorithm. Indeed, let us assume that u_{i-s} is not the root of ST_D, thus $L_{i-s} > 0$. Let $W(\ell_{i-s+1})$ and $W(\ell_{i-s})$ be two suffixes of some patterns in $\mathcal{D}$ associated with the two leaves ℓ_{i-s+1} and ℓ_{i-s}, respectively. From suffix tree's properties, we clearly have that $T[i - s + 1 : i - s + L_{i-s+1}] = W(\ell_{i-s+1})[1 : L_{i-s+1}]$ and $T[i - s : i - s + L_{i-s} - 1] = W(\ell_{i-s})[1 : L_{i-s}]$. From the observations above and since $L_{i-s+1} \geq L_{i-s} - 1 \geq 0$, we can state the following result:

Proposition 12. $W(\ell)[1 : L_{i-s} - 1]$ *is a (maybe proper) prefix of* $W(\ell_{i-s+1})[1 : L_{i-s+1}]$, *where ℓ is the leaf storing the second suffix of $W(\ell_{i-s})$.*

Hence, two cases may arise in the computation of L_{i-s}, either $L_{i-s} = L_{i-s+1} + 1$, or $L_{i-s} < L_{i-s+1} + 1$. From Proposition 12, it is clear that both two cases are managed correctly by $LCA(\ell_{i-s+1}, \ell)$. For the time complexity of the Left Extension step, we observe that all of the $O(\frac{t}{\log m})$ processors can execute simultaneously the loop in $O(\log m)$ time (by Theorem 10). After that, for each suffix $T[j : t]$ we have the extended locus u_j of its longest prefix $T[j : j + L_j - 1]$ occurring in some pattern of $\mathcal{D}$ and its length L_j.

RETRIEVAL STEP. We retrieve the longest pattern, say $P_{long(j)}$, that is prefix of $T[j : t]$. If u_j is the exact locus of $T[j : j + L_j - 1]$ (i.e., $|W(u_j)| = L_j$) and u_j is the locus of some pattern in $\mathcal{D}$ (i.e., $W(u_j) \in \mathcal{D}$), then we set $P_{long(j)} := W(u_j)$. Otherwise, we set $P_{long(j)} := W(lp(u_j))$. The correctness derives from Proposition 11 and from the definitions of L_j and u_j. As far as for the time complexity, this step requires $O(1)$ sequential time for each position in T. Hence, using $O(t/\log m)$ processors, this step takes $O(\log m)$ time in total.

Summing up the time and work bounds required by the four steps above, we immediately derive:

Theorem 13. *Given an arbitrary text $T[1 : t]$, Query(T) can be answered in $O(\log m)$ time and $O(t)$ total work on the CRCW PRAM.*

Acknowledgments I thank R. Grossi and the anonymous referees for their helpful comments and suggestions on the early version of this paper.

References

1. A. V. Aho, and M. J. Corasick. Efficient string matching: an aid to bibliographic search. *Communications of the ACM*, 333–340, 1975.
2. A. Amir and M. Farach. Adaptive dictionary matching. In *IEEE Symposium on Foundations of Computer Science*, 760–766, 1991.
3. A. Amir, M. Farach, Z. Galil, R. Giancarlo, and K. Park. Dynamic dictionary matching. *Journal of Computer and System Science*, 208–222, 1994.
4. A. Amir, M. Farach, R. M. Idury, H. La Poutré, and A. A. Schäffer. Improved dictionary matching. *Information and Computation*, 258–282, 1995.
5. A. Amir, M. Farach, and Y. Matias. Efficient randomized dictionary matching algorithms. In *Combinatorial Pattern Matching*, 259–272, 1992.
6. A. Apostolico, C. Iliopolus, G. M. Landau, B. Schieber, and U. Vishkin. Parallel construction of a suffix tree with applications. *Algorithmica*, 347–365, 1988.

7. R. S. Boyer and J. S. Moore. A fast string searching algorithm. *Communications of the ACM*, 762-772, 1977.
8. M. T. Chen and J. Seiferas. Efficient and elegant subword tree construction. In *Combinatorial Algorithms on Words*, 97-107, 1985.
9. M. Farach and S. Muthukrishnan. Optimal parallel dictionary matching and compression. In *ACM Symposium on Parallel Algorithms and Architectures*, 244-253, 1995.
10. P. Ferragina. Incremental Text Editing: a new data structure. In *European Symposium on Algorithms*, LNCS 855, 495-507, 1994.
11. P. Ferragina and F. Luccio. On the parallel Dictionary Matching problem: new results with applications. In *European Symposium on Algorithms*, 1996 (to appear).
12. R. Hariharan. Optimal parallel suffix tree construction. In *ACM Symposium on Theory of Computing*, 290-299, 1994.
13. J. Já Já. *An introduction to parallel algorithms*. Addison-Wesley, 1992.
14. R. Karp, R. Miller, and A. Rosenberg. Rapid identification of repeated patterns in strings, arrays and trees. In *ACM Symposium on Theory of Computing*, 125-136, 1972.
15. D. E. Knuth, J. H. Morris, and V. R. Pratt. Fast pattern matching in strings. *SIAM Journal on Computing*, 63-78, 1977.
16. E. M. McCreight. A space-economical suffix tree construction algorithm. *Journal of the ACM*, 262-272, 1976.
17. S. Muthukrishnan and K. Palem. Highly efficient dictionary matching in parallel. In *ACM Symposium on Parallel Algorithms and Architectures*, 69-78, 1993.
18. B. Schieber and U. Vishkin. On finding lowest common ancestor: simplification and parallelization. *SIAM Journal on Computing*, 1253-1262, 1988.
19. P. Weiner. Linear pattern matching algorithm. In *IEEE Switch. Aut. Theory*, 1-11, 1973.

Scalability and Granularity Issues of the Hierarchical Radiosity Method

AXEL PODEHL   THOMAS RAUBER   GUDULA RÜNGER *

Computer Science Dep., Universität des Saarlandes, 66041 Saarbrücken, Germany

Abstract. The radiosity method is a global illumination method from computer graphics to visualize light in scenes of diffuse objects within an enclosure. The hierarchical radiosity method reduces the problem size considerably but results in a highly irregular algorithm which makes a parallel implementation more difficult. We investigate a task-oriented shared memory implementation and present optimizations with different effects concerning locality and granularity properties. As execution platform we use the SB-PRAM, a shared-memory machine with uniform memory access time, which allows us to concentrate on load balancing and scalability issues.

1 Introduction

The radiosity method is a simulation method from computer graphics to generate photo-realistic images of computer-generated three-dimensional environments with objects which diffusely reflect or emit light [4]. The method is based on the energy radiation between surfaces of objects and accounts for direct illumination and multiple reflections between the surfaces. First, all intensity values of an environment are determined in a view-independent stage, and then an image is computed using conventional visible-surface and interpolative shading algorithms.

The radiosity method uses a decomposition of the surfaces of objects in the scene into small elements, for each of which a radiosity value has to be computed. The radiosity values are the radiant energy per unit time and per unit area which are determined by solving a transport equation of energy, a system of linear equations describing the mutual interactions between radiosity energies of different surfaces with the help of geometric configuration factors. Building up and solving the transport equation causes high computational costs. For a reduction of the computational costs, a variety of methods have been proposed, including an adaptive refinement technique [7], hierarchical methods [6], or progressive methods [2]. Adaptive and hierarchical methods reduce the number configuration factors to be computed by combining several mutual dependencies. The progressive method reduces the costs of solving the linear equation system. A further reduction of computation time can be achieved by parallel implementations [9].

The efficient computational technique of the hierarchical radiosity method is achieved by computing the mutual illumination of surfaces more precisely only for short distances and less precisely for far surfaces. The mutual influence decays with the square of the distance and, thus, a uniform accuracy is achieved. This

---

* author supported by DFG

results in a smaller number of radiosity values to be computed and a reduction of interactions which can be computed efficiently on the hierarchical data structure supporting the hierarchical method.

In this article, we investigate the implementation of the hierarchical radiosity method on shared memory machines. The starting point of the investigation is the SPLASH-2 benchmark implementation [11] described in [9] which is tuned towards an execution on a cache-based virtual shared-memory machine with a physically distributed memory (Stanford DASH). In this implementation load balance is realized by using distributed task queues with task stealing. This competes with the locality of accesses to the task queues, because each attempt to steal a task includes an access to a remote task queue.

Usually, the competition between load balance (and granularity) and data locality hinders a concise study for scalability. This limitation vanishes when using an execution platform like the SB-PRAM providing a large number of processors and a global shared memory with unit access time [1]. Thus, the implementation can concentrate on the efficient exploitation of the task granularity and can neglect effects of locality. The original implementation is optimized on the algorithmic level, on the design level for tasks (towards a finer granularity), and on the task administration level. The optimized version is derived from the SPLASH-2 implementation by several optimization steps improving the exploitation of the degree of parallelism. Both implementations have good speedup values on the SB-PRAM for a small number of processors, exceeding the speedup values on the DASH. The optimized version exhibits good speedup values also for large numbers of processors (up to 2048).

The remainder of the paper is organized as follows. Section 2 summarizes the classical and the hierarchical radiosity methods. Section 3 describes the shared memory implementation. Section 4 discusses the experiments and Section 5 concludes.

2 The hierarchical radiosity method

The classical radiosity method starts with the subdivision of the input polygons (representing the surfaces of the objects in the scene) into a number of small patches with area A_j, $j = 1, \ldots, n$. For each of the patches, a *radiosity value* B_j (of dimension [watt/m^2]) is computed which describes the specific radiant energy per unit time and per unit area of A_j. Because of the mutual illumination of diffuse objects, a radiosity value B_j is composed of two parts: the emission energy per unit area E_j and the reflections of light that is incident on patch j from all other visible patches not occluded by patches in between. The light incident from patch i on patch j is a portion $H_i = B_i F_{ij}$ of the radiosity B_i; the dimensionless *configuration factor* or *form factor*) F_{ij} describes the fraction of lightening from patch i incident to patch j. Each form factor is a double integral depending only on the geometric constellation of the two elements i and j. Using the symmetry relation $F_{ij} A_i = F_{ji} A_j$, and diffuse reflectivity factors ρ_j, the unknown radiosity values B_j can be specified by a linear system of equations (see [3]):

$$B_j = E_j + \rho_j \sum_{i=1}^{n} F_{ji} B_i, \quad j = 1, \ldots, n. \tag{1}$$

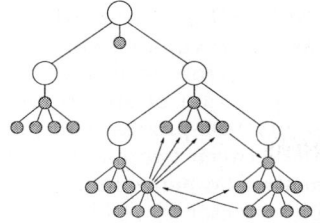

Fig. 1. Mixed BSP-quad-tree data structure: the large circles represent nodes of the BSP tree, the small shaded circles represent nodes of the quadtrees. The arrows show interactions between nodes of different quadtrees.

System (1) is solved by an iterative solution method for linear equation systems like the Jacobi method. But the main computational effort of the radiosity method consists in computing the $n(n-1)/2$ configuration factors.

The *hierarchical* radiosity method [6] reduces the computational effort by adaptively subdividing polygons into a hierarchy of smaller pieces of the surfaces. For each input polygon the subdivision into the hierarchy is organized in a quadtree such that the four children of a node represent a partition of the patch attached to the parents node. The input polygons themselves are organized in a BSP tree (binary space partitioning tree) where each node of the BSP tree represents the root of the corresponding quad-tree (see Figure 1). The union of the patches attached to leaves of all quad-trees of this data structure now represent the patches A_i, $i = 1, \ldots, n$, of the scene.

The algorithm starts by building up the BSP-tree in an initialization step. During the computation of the form factors, the quad-trees are built up adaptively in order to guarantee the computations to be of sufficient precision.

For each patch, the decision about the set of interacting elements with which the energy exchange is computed depends on an a-priori estimation of the influence of the radiosity. The method computes the energy transport (i.e., the configuration factor) between two patches only if it is not too large; otherwise the patches are subdivided. The subdivision of patches is performed if the energy transport between two patches is not small enough, i.e., $V_{ij}F_{ij}B_j > \mathrm{BF}_\epsilon$. When the radiant surface A_j is larger (i.e., $A_j > A_i$) then this area A_j is divided and the interaction list are adapted appropriately. The patch A_i is deleted from the interaction list of A_j and is inserted in the interaction lists of the children of A_j. If $A_j < A_i$ then A_i is divided and the children of A_i are inserted into the interaction list of A_j. The division of patches is stopped such that the patches are not smaller than A_ϵ. This procedure is performed for all pairs of input polygons and because all interactions use the same quadtree, each patch in the tree has its individual set of interaction patches for which the configuration factors have to be computed. Thus, the computation of radiosity for a leaf element may use configuration factors with internal elements of the quadtree unifying all the interaction with nodes in the corresponding subtree (representing a further division of the node element).

The a-priori-estimations of the energy transport take into account the configuration factors, the radiosity values and the visibility values. The use of radiosity values is possible because the hierarchical method alternates iteration steps of the Jacobi method to solve the energy system (1) with a re-computation of the quad-tree and the interaction sets based on estimations of $V_{ji}F_{ji}B_i$ with radiosity values B_i from the last step. The values $V_{ji}F_{ji}B_i$ are estimated and must not be larger than a bound BF_ϵ. The iteration stops when the difference of the total radiosity of two successive iterations is small enough.

In this paper, we consider a version of the radiosity method described in

[6] and adopted in the implementation of the SPLASH-2 benchmark suite [11] with the following methods: The form factors are approximated numerically by a ray tracing method proposed in [10] using form factors from four points of A_i (which are points from pieces which form a partition of A_i) to four small disks A_{D_l}, $l = 1, \ldots, 4$, covering A_j. The visibility test proposed in [6] uses a ray tracing method with a fixed number of rays between two patches. Here we use rays from 16 sub-elements to 16 sub-elements The percentage of rays not blocked by intervening surfaces is used as visibility factor $V_{ij} \in [0, 1]$.

One iteration step of the Jacobian method consists in a top-down and a bottom-up traversal for each quadtree: In the top-down traversal, a radiosity value for each node is computed by computing the influence of all interaction partner $j \in I(patch)$ and adding the radiosities of the parents node. The leaves of the quadtrees also take into account the emission E_j. In the bottom-up traversal, each node computes area-weighted radiosity of the children in order to make new radiosity values available on every level of the tree for the estimations of energy exchange and the next iteration step.

Figure 2 shows the hierarchical radiosity methods already expressing explicitly the potential parallelism. Portions of the program that can be executed independently from each other in parallel are separated by a || sign. Loops with independent iterations are described by forall.

---

(1) do recursively: insert-polygon(p_1)
with procedure insert-polygon(p_i) =
{insert p_i into BSP tree;
do in parallel {insert-polygon(p_{i+1}) || forall $p_j, j < i$, compute $F_{p_j p_i}, F_{p_i p_j}$}}
(2) Compute visibility factors $V_{p_j p_i}$;
do { forall polygons $p \in \{p_1, \ldots, p_k\}$
do recursively: compute-interaction(root(p));
with procedure compute-interaction(i) =
{ forall interactions $j \in I(i)$ do {
Compute visibility factor V_{ij};
if $V_{ij} F_{ij} B_j > \text{BF}_\epsilon$ and $A_i, A_j > A_\epsilon$ then {
Divide (A_i, A_j);
compute configuration factors of new interactions; }
Compute $B(i) = \rho_i \sum_{j \in I(i)} V_{ij} F_{ij} B(j) + B(parent(i))$;
if i is leaf then {
$B(i) = B(i) + E(i)$;
while i is last child of $parent(i)$ do {
$i = parent(i)$;
$B(i) = 1/4 \sum_{j=child(i)} B(j)$; }
else { forall children k of i do compute-interaction(k); } } } }
while (ERROR $> \epsilon$)
(3) forall elements in all quadtrees do { bilinear interpolation; }

Fig. 2. Hierarchical radiosity method with maximum degree of parallelism.

3 Parallel implementation

For the parallel implementation of the hierarchical radiosity method we used a task-oriented shared memory model and the SB-PRAM as execution platform. The SB-PRAM is a realization of a modified fluent machine [1]. A number p of physical processors has access to p memory modules each consisting of m memory cells. The processors are connected to the memory modules via a butterfly interconnection network. Thus, the memory is accessed as a virtual linear shared memory distributed among the modules. Besides the usual load and store operations to access memory cells, the SB-PRAM also offers *multiprefix* instructions which enable several processors to perform simple operations on a memory cell in parallel. The multiprefix instruction with addition MPADD starts with one local value for each processor and produces the sequence of prefix sums; initially one input value resides on one processors and at the end each processor holds one sum. A multiprefix operation is performed in two time units, independently of the number of participating processors. It is even possible that different groups of processors perform separate multiprefix operations in parallel. Multiprefix operations can be used for an efficient realization of access coordination to the global memory (*locking*) or parallel data structures like parallel task queues [5]. They can also be used for the implementation of a parallel loop which is controlled by a shared counter. The access to the counter by a multiprefix operation allows a dynamic execution of the loop iterations without sequentializations.

SPLASH implementation: The SPLASH2 benchmark suite comprises several parallel example programs realizing irregular applications, which are mainly intended for the Standorf DASH multiprocessor [8]. The multiprocessor is a cache coherent shared address space processor with physically distributed memory. The software simulator of the DASH (described in [9], [9]) simulates the non-uniform memory access with a constant access time for each level, i.e., cache, local memory and non-local memory or cache.

The parallel implementation of the hierarchical radiosity method realizes parallelism that occurs across input polygons, across the patches that a polygon is divided into, and across the interactions computed for a patch. This parallelism is reflected in the choice of tasks: The B-tasks and F-tasks realize parallelism in the first phase. The parallel computation of interactions and parallel BF-refinement are realized by R-tasks and V-tasks. The last phase uses A-tasks.

TASK NAME	COMPUTATIONS PERFORMED BY THE TASK
B-task(p)	insert input polygon p into BSP-tree and create B- and F-tasks
F-task(ij)	compute form factors F_{ij} and F_{ji} for input polygons i and j
R-task(p)	compute phase (2) for patch/element p except visibility factor
V-task(ij)	compute visibility factor V_{ij} between patches i and j
A-task(p)	create A-tasks for children of patch p and perform bilinear interpolation if p is an element

In the original SPLASH-2 implementation, each processor has its own task queue and inserts new tasks created by local tasks into this queue to maintain locality. But because of the dynamically changing hierarchical data structure, load balance cannot be achieved by a static assignment of tasks. To avoid a bad load balance, a processor is allowed to access task queues of the other processors, if its own queue is empty (*task stealing*) [11]. Concurrent accesses to the same data are avoided by the locking mechanism, e.g. when interactions between patches/elements are computed.

```
(1) do recursively: B-task($p_1$)
    with task B-task($p_i$) =
      { insert polygon $p_i$ into BSP-tree;
        do in parallel { B-task($p_{i+1}$); || forall $p_j, j < i$ do F-task(ij)}}
(2) do { forall polygons $p \in \{p_1, \ldots p_k\}$
      do recursively: R-task(root(p));
      with task R-task(i) = {
        forall interactions $j \in I(i)$ do {
          V-task(ij);
          hierarchical subdivision and computation of radiosity values;}
        forall children $k$ of $i$ do {
          compute configuration factors of new interactions;
          R-task($k$);}}
      while (ERROR $> \epsilon$ )
(3) forall polygons $p \in \{p_1, \ldots p_k\}$ do recursively: A-task(root(p));
    with task A-task(q) = { forall children $r$ of $q$ do A-task($r$) }
```

Fig. 3. Task organization of the Splash implementation.

The SPLASH implementation is illustrated in a pseudo-code task-program in Figure 3. The pseudo language reflects the fact that tasks perform both computations and initiations of other tasks (similar to procedures in sequential programming calling other procedures). The B-tasks have to be executed sequentially. The initiation of data dependent F-tasks is expressed by a recursive call-structure using the keyword do recursively. The corresponding task is defined by a with task statement having a recursive structure due to the hierarchical tree structure used in the algorithm. A possible schedule of B-tasks and F-tasks on 4 processors is depicted in Figure 5 on the left. The independence of computations on different quad-trees in phase (2) is expressed by a forall construct. The interactions within each quadtree are performed recursively according to the tree structure determining data dependencies between R-tasks. The visibility V-tasks for children are initiated by the parents R-task. The last phase creates a hierarchy of A-tasks of which only the leaf-tasks perform the bilinear interpolation.

SB-PRAM implementation: The SB-PRAM implementation uses the SPLASH implementation as starting point. Optimizations of the parallel implementation include a parallel construction of the BSP tree, the use of a parallel task queue, and the use of parallel loops where locality can be ignored. Table ?? summarize the modifications.

The BSP tree is constructed by a parallel search over the polygon tree in contrast to a sequential construction in order to reduce the sequential parts of the computation. In the second phase, the tasks to compute interactions (one task for each input element) do not offer enough parallelism for a large number of processors. In the first iteration step the number of tasks cannot be increased. But because these interactions all take place on the same level, this phase is separated from the rest of the iteration and the mutual configuration factors are solved with a parallel loop. Moreover, the symmetry of the configuration factors is exploited. In all following iterations, the computation of configuration

PHASE	MODIFICATION	SPECIFIC MODIFICATION	ADVANTAGES
1	algorithmic	redefined B'-tasks for building BSP-tree	larger potential parallelism; worthwhile for large numbers of processors;
1, 2	algorithmic and task design	parallel loop for combined initial interactions V_{pq}, F_{pq}	creation of regular forall-loop realized by *parallel loop*; exploitation of form-factor symmetries is possible
2	task design	modified V'-tasks for computing V_{pq}, F_{pq}	refinement of granularity
3	implementation	*parallel loop* for bilinear interpolation	task-administration for forall-loop is avoided
1-3	software support	task allocation	parallel queue avoids failures in task stealing

Table 1. Optimizations for an efficient SB-PRAM implementation

factors and the visibility values are completely moved to lower levels in the quadtrees thus creating a high degree of parallelism. The tasks for the final bilinear interpolation for smoothing the solution are also executed in a parallel loop over the leaf elements. This strategy replaces the version where all internal patch nodes were involved in creating tasks for their child nodes for locality reasons. The modified tasks are:

TASK NAME	COMPUTATIONS PERFORMED BY THE TASK
B'-task(p)	insert polygon p into BSP-tree and build sublists of elements
F'-comp.(ij)	compute form factors F_{ij}, F_{ji} and visibility factors V_{ij}, V_{ji} for input polygons i and j
R'-task(p)	compute phase (2) for element/patch except form and visibility factors
V'-task(ij)	compute visibility factor V_{ij} and form factor F_{ij} between patches/elements i and j
A'-comp.(p)	compute bilinear interpolation for element p

The corresponding task program is given in Figure 4 showing four instead of three phases which are still separated by synchronization points. The implementation supports the use of parallel loops and increases the granularity. A possible schedule on 4 processors is given in Figure 5 on the right.

4 Experiments

Figure 6 (left) shows the speedup values of the original SPLASH implementation on the DASH (as reported in [9]) and on the SB-PRAM simulators for the SPLASH test scene (with 346 input polygons). Due to the task stealing mechanism the DASH reaches the best efficiency with a coarser granularity where the V-tasks are chosen to compute four visibility values instead of one; this causes locality advantages on the BSP-tree data structure (DASH(default) in Figure 6 left). In contrast, the SB-PRAM achieves a better speedup when a finer granularity with one visibility computation per V-task is chosen (SB-PRAM(finest) in Figure 6 left). The original SPLASH implementation performs better on the SB-PRAM (SB-PRAM(default) in Figure 6 left) than on the DASH because of unit memory access times and the redundance of locality.

```
(1')  do recursively: B'-task($p_1, ..p_k$);
        with task B'-task($p_1, ..p_k$) = {
          insert $p_1$ in BSP-tree;
          built lists of polygons ($q_1, ..q_l$) visible and ($r_1, ..r_m$) invisible from $p_1$;
          do in parallel {B'-task($q_1, ..q_l$) || B'-task($r_1, ..r_m$) } }
(1")  forall $i, j = 1, ..., k$ with $i < j$ do F'-computation(ij);
(2')  do { forall polygons $p \in \{p_1, ... p_k\}$
        do recursively: R'-task(root(p));
          with task R'-task(i) = {
            forall interactions $j \in I(i)$ do {
              V'-task(i,j);
              hierarchical subdivision and computation of radiosity values;}
            forall children $k$ of $i$ do R'-task($k$);}
        while (ERROR $> \epsilon$ )
(3')  forall elements $p$ in all quadtrees do A'-computations(p)
```

Fig. 4. Task organization for the SB-PRAM implementation

The large number of processors and the additional software support (see Section 3) makes the SB-PRAM to an ideal platform to study the scalability properties inherent in an algorithms. But the massive parallelism and the uniform access time require a different implementation strategy for the study of the scalability of the hierarchical radiosity method. The optimizations described in Section 3 take this modified concept into account by replacing the expensive task concept by *parallel loops* when loops exhibit a regular, independent parallel structure (F'-computations, A'-computations), decreasing the granularity (new V'-tasks and R'-tasks with smaller runtime), increasing the degree of potential parallelism by destroying data locality (B'tasks), and exploiting a new task-administration concept by using unit access time task-queues.

Figure 6 (right) shows the speedup values for the original SPLASH implementation and the optimized implementation on the SB-PRAM simulator for up to 2048 processors. The SPLASH implementation does not scale well for more than 256 processors. For 1024 processors the optimized version still has an efficiency of 0.742. Table 7 reports the speedup values S_{1024} and the absolute runtimes T_{1024} for both implementations of the SPLAH test scene. The values for the SPLASH implementation with phases (1), (2), (3) are on the left; the values for the optimized version with phases (1'), (1"), (2'), (3') are on the right. The phases overlap according to the algorithmic structure. The timings in columns T_1 left and right show that the optimized version performs even better in the sequential case (5.8 %).

The efficient parallelization of the B-tasks (BSP-tree) is more important for massively parallel implementations than for a relatively small number of pro-

Fig. 5. Task scheduling for SPLASH tasks and SB-PRAM tasks

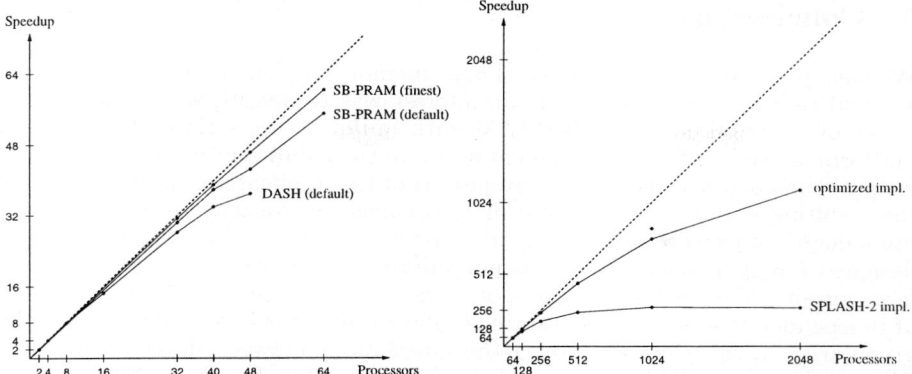

Fig. 6. Speedup values on the SB-PRAM and on the DASH. Left diagram: original SPLASH-2 implementation on the DASH and the SB-PRAM with default granularity of the tasks and the finest granularity that is possible. Right diagram: default granularity finest granularity on the SB-PRAM for larger number of processors.

cessors where the fraction $\frac{1}{14000}$ of the computation for building the BSP-tree is neglectable. Moreover, the prephase for sorting the input polygons is done before building the BSP-tree. Experiments have shown that the global execution time of the entire algorithms depends significantly on the order of the input polygons; the time varies by a factor of up to 10. This is a general phenomenon which should be separated from issues of parallelization.

A different phenomenon concern the convergence of the iteration steps solving the system of radiosity values (1). The iteration may converge faster in the parallel implementation than in the sequential one thus saving a full iteration step. The reason is the use of updated values within one iteration step if the number of patches is greater than the number of processors. The faster convergence corresponds to the faster convergence in the Gauss-seidel method for solving linear systems. This effect is exploited in the (non-hierarchical) progressive radiosity method [2]. Besides the improvements of the efficiency our investigations show that the optimized version leads to a much simpler source code. The main reasons are the lack of locality, the use of *parallel loops* which corresponds to the loops in the pseudocode algorithms, and the simplified task administration.

	SPLASH2 implementation				optimized implementation				
dimension	Phase	T_1 sec	T_{1024} sec	S_{1024}	S_{1024}	T_{1024} sec	T_1 sec	Phase	
BSP and form factor	1	6898.7	104.6	66.0	1.44	5.409	7.805	1'	BSP
					876.3	7.413	6496.2	1"	root inter.
iteration	2	90726	244.1	371.7	791.5	108.0	85487	2'	iteration.
bil.interp.	3	33.782	4.225	8.0	814.9	0.036	29.336	3'	bil.interp.
total		97658	353.2	276.5	760.5	121.0	92021		total

Fig. 7. Timings of the SPLASH2 and the optimized implementation on the SB-PRAM. The input scene is the SPLASH2 test scene.

5 Conclusions

We have presented a task oriented shared memory implementation for the hierarchical radiosity method. The main interest was to investigate the scalability issues of the method. The SB-PRAM with uniform access time offers a good platform to study efficient implementations and scalability properties for irregular problems because the locality properties of the applications do not influence the resulting performance and the investigations can concentrate on the maximum degree of parallelism. The experiments have shown that an implementation designed for up to 64 processors is not suitable to achieve good speedups on a large number of processors. A redesign of the algorithms provides a large number of independent tasks. The means are regular forall-loops and an decrease of the granularity in the phases realizing the interactions between different surfaces. The resulting parallel algorithms shows good speedup for up to 2048 processors. Thus, the hierarchical radiosity methods can be implemented efficiently although it has highly irregular computation and access patterns. Moreover, the investigations have shown that parallel data structures provided by the underlying machine can support massively parallel, efficient implementations of highly irregular algorithms.

References

1. F. Abolhassan, J. Keller, and W.J. Paul. On the Cost–Effectiveness of PRAMs. In *Proceeding of the 3rd IEEE Symposium on Parallel and Distributed Processing*, pages 2–9, 1991.
2. Micheal Cohen, Shenchang Chen, John Wallace, and Donald Greenberg. A progressive refinement approach to fast radiosity image generation. *Computer Graphics*, 22(4):75–84, 1988. Proceedings of the SIGGRAPH '88.
3. James Foley, Adries van Dam, Steven Feiner, and John Hughes. *Computer Graphics: Principles and Practice*. Addison–Wesley, Reading, USA, 1990.
4. Cindy M. Goral, Kenneth E. Torrance, Donald P. Greenberg, and Bennet Battaile. Modeling the interaction of light between diffuse surfaces. *Computer Graphics*, 18((3):212–222, 1984. Proceedings of the SIGGRAPH '84.
5. Th. Grün, Th. Rauber, and J. Röhrig. The programming environment of the SB-PRAM. In *Proc. 7th IASTED/ISMM Int.l Conf. on Parallel and Distributed Computing and Systems, Washington DC*, pages 504–509, October 1995.
6. Pat Hanrahan, David Salzman, and Larry Aupperle. A rapid hierarchical radiosity algorithm. *Computer Graphics*, 1991.
7. Paul S. Heckbert. *Simulating Global Illumination using Adaptive Meshing*. PhD thesis, university of California, Berkeley, 1991.
8. Daniel Lenoski, James Laudon, Truman Joe David Nakahira, Luis Stevens, Anoop Gupta, and John Hennessy. The dash prototype: Logic overhead and performance. *IEEE Transactions on Parallel and Distributed Systems*, 4(1):41–61, 1993.
9. J.P. Singh, C. Holt, T. Totsuka, A. Gupta, and J. Hennessy. Load balancing and data locality in adaptive hierarchical N-body methods: Barnes-hut, fast multipole, and radiosity. *Journal of Parallel and Distributed Computing*, 27:118–141, 1995.
10. John Wallace, Kells Elmquist, and Eric Haines. A ray tracing algorithm for progressive radiosity. *Computer Graphics*, 23(3):315–324, 1989. Proceedings of the SIGGRAPH '89.
11. S.C. Woo, M. Ohara, E. Torrie, J.P. Singh, and A. Gupta. The SPLASH-2 Programs: Characterization and Methodological Considerations. In *ISCA '95*, pages 24–36, 1995.

List Ranking on Interconnection Networks

Jop F. Sibeyn

Max-Planck-Institut für Informatik,
Im Stadtwald, 66123 Saarbrücken, Germany.
E-mail: jopsi@mpi-sb.mpg.de, Fax: +49-681-3025401.

Abstract

The list-ranking problem is considered for parallel computers which communicate through an interconnection network. Each PU holds k nodes of a set of singly linked lists. An easy randomized algorithm gives a considerable improvement over earlier ones.

For a large class of networks, the algorithm takes only twice the number of steps required by a k-k routing. The only conditions are that: (1) $k = \omega(k^*)$, where k^* is so large that the time consumption of k^*-k^* routing is determined by the bisection bound, and (2) the routing time slightly increases with the number of PUs in the network.

For special networks we can prove stronger results. Particularly, for $n \times \cdots \times n$ meshes, the list ranking problem is solved in $(1/2 + o(1)) \cdot k \cdot n$ steps, if $k = \omega(1)$. For hypercubes with N PUs, assuming all-port communication, the algorithm requires only $(2 + o(1)) \cdot k$ steps, if $k = \omega(\log^2 N)$.

We show that list ranking requires at least the time required for k-k routing. So, the results are within a factor two from optimal. For meshes we even match the lower bound up to lower-order terms.

Keywords: parallel algorithms, interconnection networks, list ranking, randomization.

1 Introduction

Lists. A *linked list*, hereafter just *list*, is a basic data structure: it consists of nodes which are linked together, such that every node has precisely one predecessor and one successor, except for the *initial node*, which has no predecessor, and the *final node*, which has no successor. Lists play a central role, both in sequential and in parallel algorithms. For example, they are used in the Euler-tour technique (see [8]), which is of outstanding importance in the theory of parallel computation: it is applied as a subroutine in many parallel algorithms for problems on trees.

List Ranking. An important problem connected to the use of lists is the *list ranking* problem: the determination of the rank of a node within its list. Once the nodes have been ranked, the lists can be transformed into arrays, on which many parallel operations can be performed more efficiently. For example, the parallel computation of 'prefix sums' is much easier on arrays than on lists. Also, the Euler-tour technique involves ranking lists.

In sequential computation the list ranking problem goes unnoticed, because it is trivial to solve: following the links, a list of length N is ranked in $\mathcal{O}(N)$ time. In particular, list-ranking is easier than sorting. In parallel computation it is much more important, and often the time for the list ranking determines the time of the algorithm in which it is applied.

Related Work. List-ranking has been studied intensively on various parallel computer models. Here we only mention the most relevant works, a more complete description is given in the full version of this paper (there one can also find the omitted proofs and alternative algorithms) [16].

On PRAMs optimal deterministic list ranking is achieved in [4]: time $\mathcal{O}(\log N)$ and work $\mathcal{O}(N)$. More realistic than PRAMs are parallel computers consisting of N processing units, *PUs*, that communicate through an interconnection network. On such parallel computers, it is hard to achieve anything worth mentioning: by its nature, list ranking is an extremely non-local problem. For hypercubes Ryü and JáJá [15] have shown that linear speed-up can be achieved if every PU holds at least $k = N^\epsilon$ nodes, for $\epsilon > 0$ a constant. Though the time *order* is optimal, $\mathcal{O}(k)$ in the all-port model, the leading constant strongly increases with $1/\epsilon$, limiting its practicality. Randomizedly, the problem has been solved in $(7 + o(1)) \cdot k$, if $k = \omega(\log^2 N)$ [16]. The list-ranking problem on meshes has been considered in [2, 5, 16]. The fastest algorithm for $n \times n$ meshes (randomized) takes $(3^{1/2} + o(1)) \cdot k \cdot n$ steps [16].

Reid-Miller [12] analyzes list-ranking algorithms for a 'vector processor', the CRAY C-90. She advocates the use of an algorithm, which follows the same basic approach as ours. Both algorithms can be viewed as simplified, randomized versions of the earlier algorithm by Anderson and Miller [1]. We have added various novel ideas, and for the first time we

- precisely formulate the communication complexity of list ranking for a large class of networks, allowing for general application and prediction of its cost;

- allow for a set of lists, rather than one long list, without needing extra routing steps;

- give nearly optimal results for list ranking on meshes and hypercubes, thereby solving two important open problems.

The second point is important, because in most applications, e.g. in parallel connected components algorithms the ranking has to be performed on a set of intermixed lists.

This Paper. We develop efficient randomized list-ranking algorithms. The main algorithm is formulated for arbitrary networks. Its versatility is demonstrated by giving modifications which perform very well for meshes and hypercubes. We also show that list ranking requires at least as many steps as routing. Because this paper aims to contribute theory with a practical impact, we concentrate on the case that every PU holds $k \gg 1$ nodes of a set of lists.

For hypercubes under the all-port assumption we obtain a running time of $(2 + o(1)) \cdot k$ steps for all $k = \omega(\log^2 N)$. On $n \times \cdots \times n$ meshes, for $k = \omega(1)$, a refined algorithm only requires $(1/2 + o(1)) \cdot k \cdot n$ steps. This equals the routing time, and hence the lower bound, up to lower-order terms.

We do not formulate our results that way, but the algorithm fits well into a BSP-like approach [17]. In fact, our algorithm can be viewed as a work-optimal PRAM algorithm with sub-optimal running time, which is very well suited for implementation on networks.

2 Preliminaries

Networks. There are various models for parallel computers. We consider distributed memory machines. That is, every PU has its own local memory, and data are exchanged over a fixed interconnection network. The number of PUs is N. The *degree* of a network is the maximum degree of any PU; its *diameter* is the maximal distance between any pair of PUs. The *bisection width*, $BW(N)$, is the minimum number of connections that has to be removed to obtain two disjoint networks with $\lfloor N/2 \rfloor$ and $\lceil N/2 \rceil$ PUs, respectively.

Meshes and Hypercubes. are among the best studied and most constructed machines with a fixed interconnection network, because of their regularity, programmability and other positive features. In a d-dimensional *mesh*, the N PUs are laid out in a d-dimensional $n \times \cdots \times n$ grid. Here and in the remainder, we use $n = N^{1/d}$. A PU is connected to its (at most) $2 \cdot d$ immediate neighbors. A *hypercube* with N PUs is a $\log N$-dimensional $2 \times \cdots \times 2$ grid.

Communication. We assume that the connections allow bi-directional communication. The PUs can communicate with all their neighbors at the same time: the *all-port model*. If on a real mesh or hypercube this assumption does not hold, then one step can trivially be split into several more elementary routing operations.

Performance Measure. The performance of an algorithm running on an interconnection network is measured by the maximum number of routing steps T it may take. This 'store-and-forward model' is very common in theoretical papers. It reflects the features of routing on SIMD machines (CM-2, MasPar). However, modern parallel computers tend rather to be MIMD machines consisting of larger and more powerful PUs (GCel, AP1000, iWarp, Touchstone Delta, Intel Paragon, J-Machine). With the applied 'wormhole routing' the distance between nodes plays a minor role (see [7, Sec. 7.3.2]). Fortunately, our algorithms are robust relative to such details of the model. In the general algorithm, the time consumption is expressed in terms of the time consumption of a corresponding routing operation.

Problem Definition. Initially, every PU holds k nodes. Every node has a unique index: node p, residing in memory position j, $0 \le j < k$, of PU i, $0 \le i < N$, has index $ind(p) = i \cdot k + j$. This index has nothing to do with its rank, but is used to compute the PU in which a node resides. The successor (predecessor) of a node p in the list is denoted $suc(p)$ ($pred(p)$). p knows the index of its successor, but, it does not know the index of its predecessor. The first and last nodes of every list know their status. The rank of p is denoted $r(p)$. It gives the 'distance' from the initial nodes. The rank of the initial nodes is fixed on 0. The described problem will be denoted $lrp(k, N)$ and the required number of steps by $T_{rank}(k, N)$.

Routing and Sorting. k-k *routing* is the routing problem, in which every PU is the source and destination of at most k packets. In k-k sorting, the packets must be sorted on a key from a totally ordered set with respect to a given indexing of the PUs. For a given network, $T_{route}(k, N)$ denotes the number of steps to perform k-k routing. $T_{sort}(k, N)$ is defined analogously. Let k^* be the smallest number such that for all $k \ge k^*$ the routing time increases linearly with k and equals the sorting time:

$$T_{route}(k, N) = k/k^* \cdot T_{route}(k^*, N), \quad (1)$$
$$T_{sort}(k, N) = T_{route}(k, N). \quad (2)$$

The existence of a k satisfying (1) is a basic assumption of the BSP model [17]. The existence of a k satisfying (2) is a consequence of (1): applying sample sort [14, 13], sorting can be reduced to routing. Generally, (1) and (2) are accurate only up to lower order-terms. To get a pleasant notation, *these* lower-order terms are omitted. The lower-order terms resulting from the list-ranking algorithm are taken into account.

Mathematical Issues. Let f and g be positive functions on **R**. f **is of larger order than** g, denoted $f = \omega(g)$, if for all $c > 0$, there is a $C(c)$, such that for all $x > C(c)$, $f(x) > c \cdot g(x)$. For estimates on the binomial distribution, we use

Lemma 1 (Chernoff Bounds) [6] *Let $X_1, \ldots, X_m$ be independent Bernoulli trials, with $P[X_i = 1] = p$. Let $Z = \sum_i X_i$. Then for any $t > 0$,*

$$P[Z \ge p \cdot m + t] \le e^{-t^2/(3 \cdot p \cdot m)}.$$

For more general estimates the following inequality by McDiarmid is very useful:

Lemma 2 (Azuma Inequality) [11] *Let $X_1, \ldots, X_m$ be independent random variables. For each i, X_i takes values in a set A_i. Let $f : \prod_i A_i \to \mathbf{R}$ be a measurable function satisfying $|f(x) - f(y)| \le c$, when x and y differ only in a single coordinate. Let Z be the random variable $f(X_1, \ldots, X_m)$. Then for any $t > 0$,*

$$P[|Z - E[Z]| \ge t] \le 2 \cdot e^{-2 \cdot t^2/(c^2 \cdot m)}.$$

All results in this paper hold with *high probability*: their probability on failure is bounded by n^{-c}, for some constant $c > 0$.

Pointer Jumping. Suppose we have a set of lists of total length S. Suppose that for a final node p, $suc(p) = p$. The following process of repeatedly doubling is called *pointer jumping*:

> **repeat** $\lceil \log S \rceil$ **times**
> **for** all p **do** $suc(p) := suc(suc(p))$.

Hereafter, for all p, $suc(p)$ gives the final node of the list of p (see [8] for a proof). The algorithm can easily be modified to compute functions like the rank. Implementing each of the $\lceil \log S \rceil$ concurrent reads with a constant number of routing and sorting operations, gives

Lemma 3 *Pointer jumping can be performed in $\mathcal{O}(\log S \cdot T_{sort}(k, N))$ steps.*

3 Lower Bounds

For routing or sorting it is clear which packets have to go where: distance and bisection arguments give strong lower bounds. For list-ranking, there is no trivial lower bound: a priori it is not clear what information actually has to be exchanged. By counting the 'information content' of the list-ranking problem, we prove that it is not easier than routing.

Lemma 4 *$lrp(k, N)$ takes at least as many steps as transferring an unknown number $m \in \{1, 2, \ldots, (k \cdot N/2 - 1)!\}$ over the bisection.*

Proof: Consider a single list, of which all nodes with even rank are stored on the 'left' of the bisection of the network. These nodes are stored in some order in $k \cdot N/2$ memory positions. Suppose that the initial node, whose rank is already known, is stored in position 0. The remaining arrangement is one out of $(k \cdot N/2 - 1)!$ permutations: there is a one-one correspondence between the numbers in $\{1, 2, \ldots, (k \cdot N/2 - 1)!\}$, and arrangements. Suppose that on the right side the encoded number is known. Communication and computation within each half are free. That is, we assume a two-processor system, with infinitely powerful PUs that communicate through a connection with capacity equal to the sum of the capacities of the connections through the bisection. Clearly, solving the actual list-ranking problem takes at least as long as solving it on this two-processor system. Solving the list-ranking problem in T steps gives a method to transfer the encoded number in T steps as well: if the ranks of all nodes on the left are known, then the number can be computed instantaneously. □

Lemma 5 *If at most l bits go over the bisection in a step, then any protocol for transferring numbers $m \in \{1, 2, \ldots, M\}$ over the bisection takes at least $\lceil \log M / \log(2^l + 1) \rceil$ steps.*

Proof: Consider the two-processor system from the proof of Lemma 4. Let A be a protocol for transferring any number $m \in \{1, 2, \ldots, M\}$, from one PU to the other. A is known by each of them. The action of A can be described by a $(2^l + 1)$-ary tree. The protocol starts at the root, at level 0. With every packet sent, it goes one level deeper. The additional term 1 is due to the case 'no-packet-transferred'. The nodes represent the set of remaining possibilities for the number m. At the root we find the whole set of cardinality M. For some inputs, the protocol may have a fast strategy, expressed by a leaf close to the root. On the other hand, it is easy to prove by induction, that at level i, there is a set with cardinality at least $\lceil M/(2^l + 1)^i \rceil$. Hence, on any level i with $i < \lceil 2^{l+1} \log M \rceil = \lceil \log M / \log(2^l + 1) \rceil$ there is a set with more than one element: there are still numbers among which A cannot distinguish. □

Combining these lemmas and applying Stirling's formula gives an interesting result:

Theorem 1 *If at most $\log(k \cdot N) + \mathcal{O}(1)$ bits go over any connection in a step, then*

$$T_{rank}(k, N) \geq (1 - o(1)) \cdot k \cdot N / (2 \cdot BW(N)).$$

4 General Algorithm

The only conditions for efficient list ranking with the algorithm of this section are

$$k = \omega(k^*), \tag{3}$$
$$T_{\text{route}}(k, \ln N) = o(T_{\text{route}}(k, N)). \tag{4}$$

Let $f = f(N)$ be the largest function such that

$$f^6 \cdot \ln f \leq k/k^*, \tag{5}$$
$$f^2 \cdot \ln f \leq T_{\text{route}}(k, N)/T_{\text{route}}(k, \ln N). \tag{6}$$

Algorithm. The algorithm, is described and analyzed step by step. Some readers may prefer to read through the clearly marked steps, before reading the remarks and lemmas. Due to a lack of space all proofs had to be omitted (see [16]). $T_i(k, N)$ denotes the number of routing steps taken by Step i.

Step 1. Every non-initial node is selected uniformly and independently as *ruler* with probability f^{-1}. Every initial node p sets $r(p) := 0$.

Lemma 6 $T_1(k, N) = 0$.

Step 2. Each ruler p initiates a 'sending wave': It sends a message $(ind(p), 1)$ to $suc(p)$. Upon reception of a packet $(ind(p), r)$, a node p' assigns $r(p') := r$. If p' is not a final node or a ruler, and if $r < f \cdot \ln f$, then it sends a packet $(ind(p), r + 1)$ to $suc(p')$.

Notice that we do *not* perform pointer jumping: every node sends at most one packet in Step 2. The sending of the packets $(*, r)$, $1 \leq r \leq f \cdot \ln f$, is called *sending round r*. The sending rounds should have barrier synchronizations between them: no packet $(*, r + 1)$ should be send before all packets $(*, r)$ have arrived. The following operations are added to Step 2, to increase the efficiency of the algorithm:

Step 2$^+$. At the beginning of each sending round r, every initial node p that was not selected before, is designated as *pseudo-ruler* with probability $f^{-1} \cdot (1 - f^{-1})^{r-1}$. Upon selection, p behaves as a ruler, and initiates a sending wave.

At the beginning of each sending round r, the packets $(*, r)$ are spread within blocks of $\ln N$ PUs. In general, this can be done easily. Similarly, at the end of it, the packets are first sent to preliminary destinations within their $\ln N$ block.

Lemma 7 *In total over all rounds, the spreadings in Step 2 can be implemented such that the required number of steps is at most $4/f \cdot T_{route}(k, N)$.*

We analyze the sending rounds. The following lemma plays a central role in our analysis. It essentially goes back to the special way the initial nodes are designated as pseudo-rulers. Let $P_r(p)$ be the probability that a node p sends a packet in round r.

Lemma 8 *For all p and r, $1 \leq r \leq f \cdot \ln f$, $P_r(p) = f^{-1} \cdot (1 - f^{-1})^{r-1}$.*

Let $k_r = k \cdot P_r$, be the expected number of packets sent by a PU in round r.

Lemma 9 *In round r, $1 \leq r \leq f \cdot \ln f$, the number of packets sent by a PU is bounded by $k'_r = k_r + \mathcal{O}(k_r/f)$.*

Corollary 1 *The number of steps required for the routing in round r is bounded by $T_{route}(k_r + \mathcal{O}(k_r/f), N)$.*

Lemma 10 $T_2(k, N) = (1 + \mathcal{O}(f^{-1})) \cdot T_{route}(k, N)$.

Some nodes have not been reached during Step 2: nodes at the end of a section between two rulers which lie more than $f \cdot \ln f$ out of each other, and some nodes at the beginning of the lists. Instead of expensively walking step by step, we proceed with

Step 3. Perform a list-ranking algorithm on all nodes that were not reached in Step 2. The nodes that were reached in round $f \cdot \ln f$ and the initial nodes that were not designated as pseudo-rulers act as initial list elements. The unreached rulers act as final elements. If k is still large enough, recursion may be applied. Pass along the data related to the ruler or initial node, and add appropriate values to the computed ranks.

First the participating nodes are spread within $\ln N$ blocks.

The participating nodes in Step 3 are precisely those which did not send during Step 2. By Lemma 8, the probability that a node did not send during Step 2 can be estimated by $(1 - f^{-1})^{f \cdot \ln f} \leq f^{-1}$. Hence, the expected number of nodes participating in Step 3 is less than k/f in every PU.

Lemma 11 $T_3(k, N) = (1 + o(f^{-1})) \cdot T_{rank}(k/f, N)$.

Initial nodes and nodes reached by waves from them know their ranks. They do not participate further. A non-ruler node, reached by a wave coming from a ruler p, is said to be a *subject* of p. Rulers and their subjects remain active.

Step 4. If a ruler p has been reached by a wave coming from a ruler p', then p sends a packet holding $ind(p)$ to p'. If p has been reached by a wave from an initial node, it marks itself as an initial node in the new set of lists.

First the packets are spread within $\ln N$ blocks.

A ruler p, receiving a packet holding $ind(p')$, marks p' as its new successor. If p does not receive a packet, it marks itself as a final node.

Lemma 12 $T_4(k, N) = (f^{-1} + o(f^{-1})) \cdot T_{route}(k, N)$.

Step 5. Perform a weighted list ranking (prefix sum) on the rulers. The weight of a ruler is the number r from the packet $(*, r)$ by which it was reached during Step 2 or Step 3. If k is still large enough, recursion may be applied.

First the rulers are spread within $\ln N$ blocks.

Lemma 13 *Rulers know their ranks after Step 5.* $T_5(k, N) = (1 + o(f^{-1})) \cdot T_{rank}(k/f, N)$.

Step 6.a, 6.b. Spread the ranks of the rulers back to their subjects, by performing obvious modifications of Step 2 and 3.

Lemma 14 $T_6(k, N) = (1 + \mathcal{O}(f^{-1})) \cdot T_{route}(k, N) + (1 + o(f^{-1})) \cdot T_{rank}(k/f, N)$.

Summing over all steps, we obtain:

Theorem 2 *If (5) and (6) hold, then*

$$T_{rank}(k, N) \leq (2 + \mathcal{O}(f^{-1})) \cdot T_{route}(k, N) + (3 + o(f^{-1})) \cdot T_{rank}(k/f, N).$$

The result is very general. The recurrency cannot be solved without knowing more about the network, or without imposing additional conditions on k.

Consequences. For sufficiently large k, we now achieve within a factor two from optimal:

Theorem 3 *There is a constant x such that, if (4) holds, and $k \geq \log\log^x N \cdot \log N \cdot k^*$, then*
$$T_{rank}(k, N) \leq (2 + o(1)) \cdot T_{route}(k, N).$$

The value of the x in Theorem 3 strongly depends on the analysis and on the precise formulation of Theorem 2. The term $(3 + o(f^{-1})) \cdot T_{rank}(k/f, N)$ in Theorem 2 comes from two sources: ranking the rulers, and ranking the tails. It can be reduced to $(1 + o(f^{-1})) \cdot T_{rank}(k/f, N)$, if the 'walking' along the paths is carried out for f^2 instead of $f \cdot \ln f$ steps: the number of nodes in a tail becomes extremely small. A refined analysis shows that then $k = \log\log\log^x N \cdot \log N \cdot k^*$ is sufficient for list ranking in $(2 + o(1)) \cdot T_{route}(k, N)$ steps.

k can be taken smaller, if a known algorithm achieves the optimal time order:

Theorem 4 *Let $T_{rank}(k, N) = \mathcal{O}(T_{route}(k, N))$, for all $k \geq k^{**}$. If (4) holds, then for all $k = \omega(\max\{k^*, k^{**}\})$,*
$$T_{rank}(k, N) \leq (2 + o(1)) \cdot T_{route}(k, N).$$

Theorem 4 is very general: for $k \geq \log N \cdot k^*$, the only condition for an optimal time-order list-ranking algorithm is a parallel prefix algorithm running in $\mathcal{O}(T_{route}(k, N))$ steps [16]. This will be granted on any reasonably structured network.

5 Large-Bisection Networks

For hypercubes with all-port communication, and other networks with a large bisection width, (4) does not hold: for $k \geq k^*$, routing or sorting in a subcube is as expensive as on the whole hypercube. Thus, the given algorithm cannot be applied: the spreadings in the $\ln N$ blocks would dominate the total time consumption. What determined the condition on k, (3), and the definition of f, (5)?

- In order to apply (1) and (2), the number of packets sent by each PU should be at least k^* in all routing operations.

- The maximum number of packets sent by a PU should never exceed the expected number by more than a factor $1 + \mathcal{O}(f^{-1})$.

Both conditions become most critical at the end of Step 2, were each PU sends only k/f^2 packets. Now we want to apply the same algorithm, but without spreadings. Our new condition and choice of $f = f(N)$ are inspired by the above two demands:

$$\begin{aligned} k &= \omega(\max\{\ln N, k^*\}), \\ k &= f^6 \cdot \ln f \cdot \max\{\ln N, k^*\}. \end{aligned} \quad (7)$$

Here k^* only needs to satisfy (1): no sorting subroutines are applied. For such k the number of packets in a single PU is as large as in an $\ln N$ block before (we never used $k^* > 1$). This gives

Theorem 5 *If (7) holds, then*
$$T_{rank}(k, N) \leq (2 + \mathcal{O}(f^{-1})) \cdot T_{route}(k, N) + (3 + o(f^{-1})) \cdot T_{rank}(k/f, N).$$

Theorem 5 has important consequences:

Theorem 6 *There is a constant x such that, if $k \geq \log\log^x N \cdot \log N \cdot \max\{\ln N, k^*\}$, then*
$$T_{rank}(k, N) \leq (2 + o(1)) \cdot T_{route}(k, N).$$

Theorem 7 *Let $T_{rank}(k, N) = \mathcal{O}(T_{route}(k, N))$, for all $k \geq k^{**}$. For all $k = \omega(\max\{\ln N, k^*, k^{**}\})$,*
$$T_{rank}(k, N) \leq (2 + o(1)) \cdot T_{route}(k, N).$$

From [3] the following result can be derived:

Lemma 15 *For a hypercube with all-port communication, if $k = \omega(\log N)$,*
$$T_{route}(k, N) = (1 + o(1)) \cdot k.$$

Thus, $k^* = \omega(\log N)$. This implies $T_{rank}(k, N) = \mathcal{O}(T_{route}(k, N))$, for $k \geq \log^2 N$, and thus

Theorem 8 *For a hypercube with all-port communication, if $k = \omega(\log^2 N)$,*
$$T_{rank}(k, N) \leq (2 + o(1)) \cdot k.$$

6 Meshes

Meshes are slow in comparison with other architectures, but often the loss is partially gained back because 'local' and 'sparse' operations can be performed almost for free. The spreadings are local operations: for meshes (4) holds. The operation needed at the end of the algorithm to distribute the computed ranks of the rulers back to their subjects is sparse: only a small amount of information has to be made available. Generally, it is a problem that the subjects of a ruler may be scattered over the network. On meshes, however, the ranks of the rulers can be broadcast to suitable submeshes, and then a subject can 'look-up' its rank locally. In this way the total routing time is reduced by a factor of two and becomes optimal: $(1 + o(1)) \cdot T_{route}$.

For meshes, [10, 9] offer the first optimal deterministic k-k routing and sorting algorithms:

Lemma 16 *For a d-dimensional mesh, if $k \geq 4 \cdot d$,*
$$T_{route}(k, N), T_{sort}(k, N) = (1/2 + o(1)) \cdot k \cdot n.$$

Choice of Parameters. The algorithm as given is correct. But, for d-dimensional $n \times \cdots \times n$ meshes the spreadings can be performed in much larger blocks. This enables us to optimize the parameter choices. For k and f we assume

$$\begin{aligned} k &= 4 \cdot d \cdot f^{d+1}, \\ f &= \omega(1). \end{aligned} \tag{8}$$

In Step 1, every non-initial node is selected as ruler with probability f^{-d}. In Step 2, the probability of selecting an initial node as pseudo-ruler is modified accordingly. The number of sending rounds becomes $f^d \cdot \ln f$. We assume that

$$\begin{aligned} k &\leq n^{1/2 - 2 \cdot \epsilon}, & (9) \\ \epsilon &= 1/(2 \cdot d + 2). & (10) \end{aligned}$$

For larger k, good performance can be achieved with different choices. At the start and end of each sending round, the participating packets are spread in $n^a \times \cdots \times n^a$ submeshes, with

$$a = 1/2 + \epsilon. \tag{11}$$

Step 1, ..., Step 5.

Lemma 17 *In total the spreadings take less than $n^{-\epsilon} \cdot T_{route}(k, N)$ steps.*

Lemma 18 *Let k_r be the expected number of packets sent by a PU in sending round r, $1 \leq r \leq f^d \cdot \ln f$, of Step 2. The number of packets sent by a PU in round r is bounded by*
$$k'_r = (1 + \ln^{1/2} n \cdot n^{-\epsilon}) \cdot k_r.$$

Lemma 19
$$T_2(k, N) = (1 + \ln^{1/2} n \cdot n^{-\epsilon}) \cdot T_{route}(k, N),$$
$$T_3(k, N) = (1 + \ln^{1/2} n \cdot n^{-\epsilon}) \cdot T_{rank}(k/f, N),$$
$$T_4(k, N) = T_{route}(k/f^d + 1, N),$$
$$T_5(k, N) = T_{rank}(k/f^d + 1, N).$$

Alternative Step 6. We describe an alternative to Step 6, which runs in $\mathcal{O}(T_{route}(k/f, N))$.

The mesh is divided in $n/f \times \cdots \times n/f$ submeshes. After spreading, there are at most $k/f^d + 1$ rulers in a PU. The rulers travel along all dimensions, and drop a copy in all positions that are shifted from their original positions by multiples of n/f.

Lemma 20 *Broadcasting the rulers to all submeshes takes $2 \cdot k \cdot n/(f \cdot d)$ steps. Afterwards every PU holds at most $k + f^d$ rulers.*

Now, in all submeshes the packets are sorted in a 'snake-like' order. Rulers get their index as key, subjects the index of their ruler $+ 1/2$.

Lemma 21 *In every submesh, rulers and subjects can be sorted together in $(2 + f^{-1}) \cdot T_{sort}(k, N/f)$ steps. Afterwards, each ruler stands at the head of its subjects.*

The rulers now send their ranks 'along the snake' to their subjects. Those add their off-set, computed in Step 2 or 3, and are done.

Lemma 22 *Distributing the rank of a ruler to its subjects can be performed in $\log n$ steps.*

Summing the time consumptions of the substeps,

Lemma 23 $T_6(k, N) = (2 + 4/d + o(1)) \cdot T_{route}(k/f, N).$

With Lemma 19, this gives our last main result:

Theorem 9 *For list ranking on d-dimensional meshes, with k satisfying (8),*
$$T_{rank}(k, N) \leq (1 + \mathcal{O}(f^{-1})) \cdot T_{route}(k, N) = (1 + \mathcal{O}(f^{-1})) \cdot k \cdot n/2.$$

7 Conclusion

We have given a complete analysis of the list ranking problem on networks. First we showed that the problem requires at least as many steps as routing. Then we gave a general algorithm, which uses only twice as many steps. A modification also achieves this for list ranking on hypercubes. For meshes we even match the lower bound up to a lower order term. We consider our algorithm to be an excellent candidate for actual implementation.

Two questions remain unanswered:

(1) Can the algorithm be made deterministic? We notice that the randomization in our algorithm is fundamentally different from the randomization in sorting algorithms. For sorting, a good sample can also be obtained by sorting subsets and taking regularly interspaced subsets thereof. The problem with list ranking is that there is no total order on the elements, and hence it is hard to deterministically select a small subset that more or less regularly subdivides the lists. Deterministic coin tossing [4] can be used to select an r-ruling set, but this mere selection is more expensive than our entire algorithm. In practice, taking the local minima of a simple deterministic coloring with $\log(k \cdot N)$ or with $\log \log(k \cdot N)$ colors (as it is applied in [1]) will work fine. But, there is no guarantee that the selected set of rulers is small, so along these lines no strong claims can be made.

(2) Can we get rid of the condition that the initial elements of the lists are known? This condition is required only if the average list length is constant. Even then it is not hard to modify the main algorithm such that the theorems in Section 4 and Section 5 still hold. The real problem is to achieve $T_{rank}(k, N) = (1 + o(1)) \cdot T_{route}(k, N)$ for meshes. This implies that we should walk along most links only once. The trick with the rulers works fine in the case of long lists, but they are not useful for lists of length $\mathcal{O}(1)$.

References

[1] Anderson, R.J., G.L. Miller, 'Deterministic Parallel List Ranking,' *Algorithmica*, 6, pp. 859–868, 1991.

[2] Atallah, M.J., S.E. Hambrusch, 'Solving Tree Problems on a Mesh-Connected Processor Array,' *Information and Control*, 69, pp. 168–187, 1986.

[3] Chang, Y., J. Simon, 'Continuous Routing and Batch Routing on the Hypercube,' *Proc. 18th Symp. on Theory of Computing*, pp. 272–281, ACM, 1986.

[4] Cole, R., U. Vishkin, 'Deterministic Coin Tossing and Accelerated Cascades: Micro and Macro Techniques for Designing Parallel Algorithms,' *Proc. 18th Symp. on Theory of Computing*, pp. 206–219, ACM, 1986.

[5] Gibbons, A.M., Y. N. Srikant, 'A Class of Problems Efficiently Solvable on Mesh-Connected Computers Including Dynamic Expression Evaluation,' *Information Processing Letters*, 32, pp. 305–311, 1989.

[6] Hagerup, T., C. Rüb, 'A Guided Tour of Chernoff Bounds,' *Information Processing Letters*, 33, 305–308, 1990.

[7] Hwang, K., *Advanced Computer Architecture; Parallelism, Scalability, Programmability*, McGraw-Hill, Inc., 1993.

[8] JáJá, J., *An Introduction to Parallel Algorithms*, Addison-Wesley Publishing Company, Inc., 1992.

[9] Kaufmann, M., J.F. Sibeyn, T. Suel, 'Derandomizing Routing and Sorting Algorithms for Meshes,' *Proc. 5th Symp. on Discrete Algorithms*, pp. 669–679, ACM-SIAM, 1994.

[10] Kunde, M., 'Block Gossiping on Grids and Tori: Deterministic Sorting and Routing Match the Bisection Bound,' *Proc. European Symp. on Algorithms*, LNCS 726, pp. 272–283, Springer-Verlag, 1993.

[11] McDiarmid, C., 'On the Method of Bounded Differences,' in *Surveys in Combinatorics*, J. Siemons, editor, 1989 London Mathematical Society Lecture Note Series 141, pp. 148–188, Cambridge University Press, 1989.

[12] Reid-Miller, M., 'List Ranking and List Scan on the Cray C-90,' *Proc. 6th Symp. on Parallel Algorithms and Architectures*, pp. 104–113, ACM, 1994.

[13] Reif, J., L.G. Valiant, 'A logarithmic time sort for linear size networks,' *Journal of the ACM*, 34(1), pp. 68–76, 1987.

[14] Reischuk, R., 'Probabilistic Parallel Algorithms for Sorting and Selection,' *SIAM Journal of Computing*, 14, pp. 396–411, 1985.

[15] Ryu, K.W., J. JáJá, 'Efficient Algorithms for List Ranking and for Solving Graph Problems on the Hypercube,' *IEEE Transactions on Parallel and Distributed Systems*, Vol. 1, No. 1, pp. 83–90, 1990.

[16] Sibeyn, J.F., 'List Ranking on Interconnection Networks,' *Techn. Rep. 11/1995, SFB 124-D6*, Universität Saarbrücken, Saarbrücken, Germany, 1995. Preliminary version in *Proc. Computing Science in the Netherlands*, pp. 271–280, SION, Amsterdam, 1994. Submitted to *Acta Informatica*.

[17] Valiant, L.G., 'A Bridging Model for Parallel Computation,' *Communications of the ACM*, 33(8), pp. 103–111, 1990.

Parallel Algorithm for Computing the Fragment Vector in Steiner Triple Systems

Erik Urland

Centre Universitaire d' Informatique
Université de Genève, 24 rue Général Dufour
1211 Genève 4, Switzerland
urland@cui.unige.ch

Abstract. In this paper we describe a linear time algorithm using $O(n^2)$ processors for computing the fragment vector in Steiner triple systems. The algorithm is designed for $SIMD$ machines having a grid interconnection network. We discuss an implementation and some experimental results obtained on the Connection Machine CM-2.

1 Introduction

Let n be a positive integer. By a *Steiner triple system* of order n, denoted by $STS(n)$, we understand a pair (V, B), where V is a set of elements called *points* (or *vertices*) such that $|V| = n$, and B is a set of such 3-subsets of V, called *lines* (*blocks* or *triples*), that every unordered pair of distinct points of V occurs exactly once among the lines of B. It is well known that $STS(n)$ exists if and only if $n \equiv 1 \pmod 6$ or $n \equiv 3 \pmod 6$. For example, an $STS(7)$ over $V = \{1, 2, ..., 7\}$ can be formed with the following set of lines $B = \{[1, 2, 3]; [4, 1, 5]; [1, 6, 7]; [4, 6, 2]; [2, 5, 7]; [3, 4, 7]; [3, 5, 6]\}$. Two $STS(n)$ (V_1, B_1) and (V_2, B_2) are said to be *isomorphic* if there is a bijection $\phi : V_1 \to V_2$ such that $[\alpha, \beta, \gamma] \in B_1$ if and only if $[\phi(\alpha), \phi(\beta), \phi(\gamma)] \in B_2$. A $k - line\ configuration$, $k \geq 1$, is defined as any collection of k lines of an $STS(n)$. An Erdös configuration of order k is a $k-$line configuration on $k + 2$ points which contains no subconfiguration of m lines on $m + 2$ points for $1 < m < k$. Two $k-$line configurations C_1 and C_2 are considered to be *isomorphic* if there is a bijection between the vertices of the configurations mapping lines to lines. If $C_1 = C_2$ then such an isomorphism is called an *automorphism*. By $frequency$ (or $number\ of\ occurrences$) of a configuration C in a given $STS(n)$ we understand the number of all different representations of the configuration C in the $STS(n)$. Let C be a configuration and let S be a subset of vertices of C. Then by a *partial* configuration P_S of C we understand a subconfiguration of C which consists only from lines having at least one vertex in S. Let C be a configuration and let π be a vertex of C. Then the *number of symmetries* of C according to the vertex π, denoted by $\Upsilon(C, \pi)$, is the number of vertices v of C such that there is an automorphism of C mapping π to v. Let C be a configuration. Let S be a subset of vertices of C and let P_S be the partial configuration of C, which is formed by a set of lines

$L \subseteq B$. Then by $R(C, P_S)$ we denote the number of all different representations of C in $STS(n) = (V, B)$ containing L.

Non-isomorphic STSs are frequently used as source data for various kinds of statistical experiments. Similarly, we need different STSs in order to determine a linear basis for k-line configurations [2, 3, 5]. The classical approach is to randomly generate STSs on a computer using the hill-climbing technique [4]. This technique appears to be extremely fast but unfortunately cannot guarantee that STSs constructed in this way are non-isomorphic. As there is no known polynomial time algorithm to test isomorphism of STSs, in practice one can use invariants as a proof that two given STSs are non-isomorphic.

In this paper, we concentrate on one invariant called *fragment vector*, originally introduced by Gibbons in [1]. Consider an Erdös configuration of order 4 called the *Pasch* configuration. Let π be a point of an $STS(n)$ and let $f(\pi)$

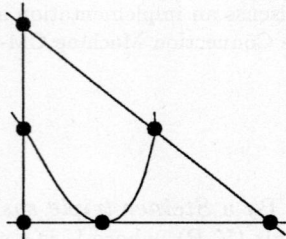

Fig. 1. Pasch configuration

denote the number of Pasch configurations containing π. Then the *fragment vector* of an $STS(n)$ is a sequence of integers $f(\pi_1), f(\pi_2), ..., f(\pi_n)$ for $\pi_i \in V$, $1 \leq i \leq n$, sorted in non-decreasing order. It is easy to see that determining the number of occurrences of the Pasch configuration in an $STS(n)$ forms the main part of the algorithm for computing the fragment vector. In the next section we shall deal with a parallel algorithm for counting the frequency of the Pasch configuration designed for $SIMD$ machines having a grid interconnection network.

2 Parallel algorithm

Before we describe the algorithm itself we shall present the following two lemmas, which will be used later in the design of the algorithm.

Lemma 1. *Let α be an arbitrary vertex of the Pasch configuration. Then $\Upsilon(Pasch, \alpha) = 6$.*

It follows from Lemma 1 that if α is a vertex of the Pasch configuration and if $S = \{\alpha\}$ then up to isomorphism, every vertex of the Pasch configuration induces the same partial configuration P_S. The following lemma shows the number of different representations of the Pasch configuration in an $STS(n)$ which can be obtained from a partial configuration P_S.

Lemma 2. *Let $STS(n)$ be a Steiner triple system of order n. Let $S = \{\alpha\}$, where α is an arbitrary vertex of the Pasch configuration and P_S denotes the partial configuration of the Pasch configuration. Then $R(Pasch, P_S) \leq 2$.*

As a next step we present an algorithm for massively parallel machine with a set of processors running in $SIMD$ mode and having a grid interconnection network. One can intuitively see that a matrix representation of STS is well suited for $SIMD - MC^2$ parallel machine. According to this representation the proposed algorithm consists of the following three levels.

Algorithm 1.

Level 1. Initialization

- For each point α of $STS(n)$, $0 \leq \alpha < n$, let $f(\alpha) := 0$. Let $N := 0$, where N denotes the number of occurrences of the Pasch configuration.
- Let $T_{n \times n}$ be a matrix and let $t_{(i,j)} := -1$ for $0 \leq i, j < n$.
- Transform the input list of triples forming $STS(n)$ to matrix $T_{n \times n}$ in such a way that the entry $t_{(i,j)} := x$, $0 \leq i, j < n$, where x is a point of the triple $[i, j, x]$.
- Let $B_{n \times n}$ be an index matrix such that $b_{(i,j)} := j$ for $0 \leq i, j < n$.

Level 2. Precomputation and reduction

- As each triple of $STS(n)$ is represented six times in T, we compress redundant representations. Let $A_{n \times n}$ be a matrix and let $a_{(i,j)} := t_{(i,j)}$, $0 \leq i, j < n$. For all rows i, $0 \leq i < n$, and each column j, $0 \leq j < n$, if $a_{(i,j)} \neq -1$ then $a_{(i, a_{(i,j)})} := -1$.
- Delete the entries $a_{(i,j)} < 0$, $0 \leq i, j < n$, from the matrix A and the corresponding entries $b_{(i,j)}$ from B. Note that after this reduction the matrices A and B are of the size $n \times \varphi$, where $\varphi = \frac{n-1}{2}$.
- Let $C_{n \times \varphi}$ and $D_{n \times \varphi}$ be two matrices and let $c_{(i,j)} := a_{(i,j)}$ and $d_{(i,j)} := b_{(i,j)}$ for $0 \leq i < n$ and $0 \leq j < \varphi$.

Level 3. Main computation

- Let us shift the entries $c_{(i,j)}$ and $d_{(i,j)}$ of the matrices C and D in such a way that $c_{(i,j)} := c_{(i,j+1)}$ and $d_{(i,j)} := d_{(i,j+1)}$ for $0 \leq i < n$ and $0 \leq j < \varphi - 1$.
- Delete the last column of the matrices A, B, C and D, and set $\varphi := \varphi - 1$. Assume that vertex $\alpha \in P_{\{\alpha\}}$. Then the triple $h_1 = [i, b_{(i,j)}, a_{(i,j)}]$ corresponds to one line of $P_{\{\alpha\}}$, for $\alpha = i$ and for some j-th triple from the list of triples containing α. Similarly, the triple $h_2 = [i, d_{(i,j)}, c_{(i,j)}]$ corresponds to the second line of $P_{\{\alpha\}}$. Thus the pairs of entries $\{a_{(i,0)}, b_{(i,0)}\}$, $\{a_{(i,1)}, b_{(i,1)}\}$,

..., $\{a_{(i,\varphi-1)}, b_{(i,\varphi-1)}\}$ of the matrices A and B, together with some vertex $\alpha = i$, form the triples which represent one line of the partial configuration $P_{\{\alpha\}}$. The second line of $P_{\{\alpha\}}$ is represented in a similar way, by pairs of corresponding entries of the rows i in C and D. Note that the lines h_1 and h_2, containing the j-th pair of entries of the matrices A, B and C, D, forming $P_{\{\alpha\}}$ for $\alpha = i$, $0 \le \alpha < n$, cannot be the same.

- For all $a_{(i,j)}$, $b_{(i,j)}$, $c_{(i,j)}$ and $d_{(i,j)}$, $0 \le i < n$ and $0 \le j < \varphi$, check if $t_{(a_{(i,j)}, c_{(i,j)})} = t_{(b_{(i,j)}, d_{(i,j)})}$. If yes, then without loss of generality the two other lines, not forming $P_{\{\alpha\}}$ for $\alpha = i$, have a common point. Thus we obtain one representation of the Pasch configuration from Lemma 2. In this case let $N:=N+1$ and $f(\pi):=f(\pi)+1$ for all points π forming the triples which represent the Pasch configuration.
- Repeat the previous step for all $a_{(i,j)}$, $b_{(i,j)}$, $c_{(i,j)}$ and $d_{(i,j)}$, $0 \le i < n$ and $0 \le j < \varphi$, under the condition $t_{(a_{(i,j)}, d_{(i,j)})} = t_{(b_{(i,j)}, c_{(i,j)})}$, which checks the second representation of the Pasch configuration from Lemma 2.
- Repeat the last four steps until $\varphi = 1$.
- By Lemma 1, let $N:=N/6$ and for each point α of $STS(n)$, $0 \le \alpha < n$, let $f(\alpha):=f(\alpha)/6$. The computation is performed and the algorithm terminates.

Theorem 3. *Algorithm 1 computes the frequency of the Pasch configuration in an $STS(n)$ in linear time using $O(n^2)$ processors.*

3 Implementation and experimental results

As the processors of the Connection Machine CM-2 can be configured as a k-dimensional grid, we use this computational model for implementing our algorithm. The speedup of the parallel approach becomes significant with the order of STS greater than 100. For $STS(249)$ we have achieved speedup approximately 30 compared to the best known $O(n^3)$ sequential algorithm running on a Sun SPARCstation 4 computer. Note that the original algorithm can be optimized by assuming wrap-around connections among processors in the grid, but this modification gives only slightly better results.

References

1. Gibbons, P. D.: *Computing techniques for the construction and analysis of block designs*, Ph.D. Thesis, University of Toronto, 1976.
2. Grannell, M. J.-Griggs, T. S. and Mendelsohn, E.: *A small basis for four-line configurations in Steiner triple systems*, Journal of Combinatorial Designs, Vol. 3, No. 1 (1995), p. 51-59.
3. Horak, P.-Phillips, N.-Wallis, W. D. and Yucas, J.: *Counting frequencies of configurations in Steiner triple systems*, to appear in Journal of Combinatorial Designs.
4. Stinson, D. R.: *Hill-climbing algorithms for the construction of combinatorial designs*, Annals of Discrete Math. **26**, (1985), p. 321-334.
5. Urland, E.: *A linear basis for the 7-line configurations*, submitted for publication, Journal of Comb. Math. and Comb. Computing.

Representation of the Gabow Algorithm for Finding Smallest Spanning Trees with a Degree Constraint on Associative Parallel Processors*

A. S. Nepomniaschaya

Supercomputer Software Department, Computing Center,
Siberian Division of Russian Academy of Sciences,
pr. Lavrentieva, 6, Novosibirsk, 630090, Russia
E–mail: anep@comcen.nsk.su

Abstract. In this paper by means of an abstract model (the STAR–machine) we describe an associative version of the Gabow algorithm for finding smallest spanning trees with a degree constraint. We have obtained that on an associative parallel processor this algorithm takes the same time as a minimal spanning tree algorithm.

1 Model of Associative Parallel Machine

We will analyze algorithms for associative parallel processors belonging to fine–grained SIMD systems with bit–serial (vertical) processing and simple single–bit processing elements (PEs). Such an architecture provides massively parallel search by contents and processing of unordered tabular data.

Our model is defined as an abstract STAR–machine of the SIMD type with vertical data processing. It consists of the following components:
 – a sequential control unit where programs and scalar constants are stored;
 – an associative processing unit consisting of m single–bit PEs;
 – a matrix memory for the associative processing unit.

Binary data are loaded in the matrix memory in the form of the two–dimensional table in which each datum occupies an individual row. A row (word) or a column (slice) may be accessed equally easy. The associative processing unit is represented as h vertical registers each consisting of m bits. The bit columns of the tabular data are stored in the registers which perform the necessary Boolean operations and record the search results.

The STAR–machine run is described by means of the language STAR [1] being an extension of Pascal. Consider briefly the STAR constructions needed for the paper. To simulate data processing in the matrix memory we use three new data types **word**, **slice** and **table**. We employ the types **slice** and **word** for bit column access and bit row access, respectively, and the type **table** for defining the tabular data. Assume that any variable of the type **slice** or **word** consists of m components which belong to $\{0,1\}$.

---

* This work was supported in part by the Russian Foundation for Basic Research under Grant N 96-01-01704

Consider operations and predicates for slices.

Let X, Y be variables of the type **slice** and i be a variable of the type **integer**. We define the following operations: SET(Y) sets all components of Y to '1'; CLR(Y) sets all components of Y to '0'; $Y(i)$ selects the i-th component of Y; FND(Y) returns the ordinal number i of the first component '1' of Y, $i \geq 0$; STEP(Y) returns the same result as FND(Y) and then resets the first component '1'.

We utilize the bitwise Boolean operations and predicates ZERO(Y) and SOME(Y) which are introduced in the obvious way.

For a variable T of the type **table** we will use the following operations: $T(i)$ returns the i-th row in the matrix T ($1 \leq i \leq m$); col(i, T) returns the i-th column in the matrix T; with(Y, T) attaches to the matrix T left the contents of the slice Y.

In the STAR-machine matrix memory we represent an *undirected weighted graph* $G=(V, E, w)$ in the form of association of the matrices *left*, *right* and *weight* in which each edge $e = (p, q)$ corresponds to the triple $< left(e), right(e), weight(e) >$. Note that $V = \{1, 2, \ldots, n\}$.

2 Associative Algorithm

Let t be a degree constraint for the vertex r in G and $\deg(r) = k$. Without loss of generality we assume that in the matrix memory any edge incident to r has the form $< r, s, w(r, s) >$ and all the edges incident to r occupy the first k rows in the graph representation.

The Gabow algorithm [2] constructs the smallest spanning tree of G containing only t edges incident to r. Denote by P a set of edges incident to r. The basic idea of the algorithm is to find a smallest spanning tree containing P. Then elementary exchanges are executed until the degree of r decreases to t. For speeding up the computation at first a minimal spanning forest U of the graph $G - r$ is constructed. Then only edges from $P \cup U$ will be used.

For representing this algorithm on the STAR-machine we will employ a group of basic procedures and constructions.

At first, enumerate a group of procedures which employ a given global slice X for indicating by '1' the row positions being used in the corresponding procedure.

The procedure MERGE(T, X, F) writes into the resulting matrix F those rows of the given matrix T which correspond to positions '1' in the slice X. The procedure MATCH(T, X, w, Z) defines the row positions in the given matrix T coinciding with the given w. Its result is the slice Z in which $Z(i) ='1'$ if $T(i)=w$. The procedure MIN(T, X, Z) defines the row positions in the given matrix T where the minimal element is located. It returns the slice Z in which $Z(i) =' 1'$ if $T(i)$ is the minimal matrix element. The procedure MAX(T, X, Z) is defined by analogy with MIN. The procedure ADDC(T, X, w, F) adds the binary word w to the rows of the matrix T and writes the result into the matrix F. The procedure SUBTV(T, Q, X, F) writes into the matrix F the result of subtraction of the matrix Q from the matrix T.

For finding a minimal spanning tree of a connected component we employ the procedure MSTC($left, right, weight, S1, S, R$) which is obtained by analogy with MST1 [1]. It uses the slice $S1$ for indicating the positions of the graph edges among which a connected component is constructed. The procedure returns the slices S and R. In the slice R we store positions of the edges belonging to the minimal spanning tree of the connected component. In the slice S we save positions of those edges from $S1$ which are not incident to the vertices included into the minimal spanning tree.

We construct a minimal spanning forest of $G-r$ by means of the procedure FOREST($left, right, weight, Z, U$). It uses the slice Z for indicating positions of the edges among which connected components are looked for. The procedure returns the slice U in which positions of edges belonging to the minimal spanning forest are indicated by $'1'$.

The procedure EXCHANGE($left, right, j, Z, R, U$) returns the slice U in which positions of all the replacing edges for an edge $e = (r, s)$ from the j-th position are indicated by $'1'$. It uses the slice Z for indicating positions of edges deleted from the minimal spanning forest of $G-r$ and the slice R for indicating positions of edges which belong to the smallest spanning tree containing all the edges incident to r.

Now, consider the main constructions.

The first construction allows one by means of the Prim–Dijkstra algorithm to obtain the smallest spanning tree including all the edges incident to r. For performing it, we will use the procedures MAX, ADDC, MERGE and MSTC.

Construction 1. At first, we define the maximal weight ($w1$) of the edges incident to r. Then we construct a new matrix $cost$ from the matrix $weight$ as follows. Weights of the edges, not incident to r, are increased by the value of $w1$. Now, we perform the procedure MSTC using the matrix $cost$ instead of the matrix $weight$.

In the second construction for any edge e belonging to the j-th position in the graph representation we define the position of its replacing edge f.

Construction 2. For any edge $e=(r, s)$ from the j-th position by means of the procedure EXCHANGE($left, right, j, Z, R, U$) we define in the slice U positions of all the edges which can replace it. If there are such positions, among them using the procedure MIN we define the position i of the replacing edge f having the minimal weight. The position j of the edge e is saved in the slice W and the position i of its replacing edge f is stored in the j-th component of the array A. At last, we carry out the statement $cost(j) := cost(i)$.

The third construction is employed for selecting the current exchange.

Construction 3. After performing the procedure SUBTV($cost, weight, W, cost1$) in the matrix $cost1$ we obtain the augments of the rows corresponding to positions $'1'$ in the slice W. For defining the current exchange position we carry out the procedure MIN($cost1, W, X$) and the statement $m :=$FND(X). Thus, the pair ($m, A[m]$) is the current exchange. For finding the replacing edge position we fulfil the statement $i := A[m]$. Now, we perform the exchange as follows. We delete edge position m both from the slice W and from the resulting slice R. Then we add the edge position i to the slice R.

In the fourth construction we define whether there are at least two edges incident to r having the same replacing edge.

Construction 4. Let (m, i) be a current exchange. Knowing the position m of the deleted edge e we define the column number p in the matrix *change* where positions of its replacing edges are stored. We delete the i-th position from the p-th column. Now, if there exists another edge incident to r with the same replacing edge, then in the i-th row of the matrix *change* a bit '1' must be necessary. Knowing the i-th row by means of the operation STEP we define the column number (say l) in the matrix *change* whose i-th bit is '1' and replace this bit with '0'. Then, we perform the disjunction between the p-th column and the l-th column and we write the result into the l-th column of the matrix *change*. In the Gabow algorithm this is equivalent to removing the occurrence of the edge f from $F(e')$ and to merging $F(e)$ and $F(e')$. Finally, we set all the components of the p-th column to '0' using the operation CLR.

The associative algorithm consists of the following three stages.

At **the first stage**, using the procedure FOREST, we construct a minimal spanning forest of $G-r$. Using Construction 1 we obtain the matrix *cost*. Now, knowing positions of edges incident to r and positions of edges included in the minimal spanning forest by means of the procedure MSTC we construct the smallest spanning tree T containing all the edges incident to r. Finally, positions of edges from the minimal spanning forest not belonging to T are stored into a slice Z.

At **the second stage**, using Construction 2, we get the slice W and the array A forming the list of pairs $(j, A[j])$. Moreover, for any deleted edge from the j-th position in the current p-th column of the matrix *change* we save positions of its replacing edges. Finally, by means of the statements $B[j] := p$ and $C[p] := j$ we determine a one–to–one correspondence between j and p. Note that the matrix *change* will be used at the next stage.

At **the third stage**, using Construction 3, we carry out the current exchange (m, i). If the added edge from the i-th position is a replacing edge for another edge e' incident to r, then by means of Construction 4 we update the l-th column of the matrix *change*, where positions of new replacing edges for e' will be written. Now, knowing the l-th column with the use of the procedure MIN we define the position (say q) of a new replacing edge for e' having the minimal weight. For defining the position in the graph representation, where the edge e' is located, we perform the statement $h := C[l]$. Finally, we fulfil the statement $cost(h) := cost(q)$. Note that this stage is repeated until $deg(r) > t$.

Remark. Let us briefly explain a construction realized in the procedure EXCHANGE. For any edge $e = (r, s)$ belonging to the j-th position we construct a path from its right vertex (s) to a terminal vertex by means of the procedure MATCH. Knowing the slice R, for every new vertex w added to this path we verify whether there are such edges, incident to w, whose positions are indicated by '1' in the slice Z. Positions of such edges are accumulated in U.

We represent the Gabow algorithm on the STAR–machine using the procedure SSTEQ whose correctness is established by the following theorem.

Theorem 1. Let a graph G be represented by the association of the matrices *left*, *right* and *weight*. For a given vertex r let *node* be its binary code, k be its degree and t be a constraint degree. Then, using the procedure SSTEQ($left, right, weight, node, k, t, R$), we construct a minimal spanning tree T_S whose edge positions are indicated by '1' in the slice R and T_S satisfies the degree constraint.

By means of the STAR–machine we simulate the vertical data processing *at micro–level*. Therefore assume that any elementary operation performed in the matrix memory of our model takes one unit of time.

Time complexity of an algorithm will be measured by counting all the elementary operations performed in the worst case.

We have obtained that the procedure SSTEQ takes $0(n \log n)$ time since its first stage takes $0(n \log n)$ time, the second stage requires $0((n - deg(r)) \log n)$ time and the third stage takes $0(deg(r) - t)$ time.

3 Conclusions

In this paper we have presented the associative version of the Gabow algorithm for the degree–restricted problem. We have obtained that on the STAR–machine both this algorithm and the MST algorithm [1] take $0(n \log n)$ time.

In [3] it has been shown that on sequential computers the degree–restricted problem is time–equivalent to the MST–problem. In [4], the best sequential algorithm for the MST–problem runs in time $0(m \log \beta(m, n))$, where $\beta(m, n) = min\{i \mid \log^{(i)} n \leq m/n\}$, n is the number of vertices and m is the number of edges. To this end a special data structure F-heap is used. However, the model considered in the paper can employ only tabular data structures.

Acknowledgements. I wish to thank the anonymous referees for their helpful comments.

References

1. A. Sh. Nepomniaschaya, Comparison of two MST Algorithms for Associative Parallel Processors, in: *Proc. of the 3-d Intern. Conf. "Parallel Computing Technologies"*, PaCT–95, (St. Petersburg, Russia), *Lecture Notes in Computer Science*, **964**, (1995) 85–93.
2. H. N. Gabow, A Good Algorithm for Smallest Spanning Trees with a Degree Constraint, in: *Networks*, **8**, No 3, (1978) 201–208.
3. H. N. Gabow, R. E. Tarjan, Efficient Algorithms for a Family of Matroid Intersection Problems, in: *J. Algorithms*, **5**, (1984) 80–131.
4. H. N. Gabow, Z. Galil, T. Spencer, R. E. Tarjan, Efficient Algorithms for Finding Minimum Spanning Trees in Undirected and Directed Graphs, in: *Combinatorica*, **6**, No 2, (1986) 109–122.

Runtime Support for Replicated Parallel Simulators of an ATM Network on Workstation Clusters

Kam Hong Shum[1] and Shuo-Yen Robert Li[2]

[1] Computer Laboratory, University of Cambridge, Cambridge CB2 3QG, UK
[2] Dept. of Information Engineering, Chinese University of Hong Kong, Hong Kong

Abstract. An effective approach of speeding up the simulation of an ATM network on workstation clusters is presented. In this approach, multiple simulation runs are performed by replicated parallel simulators (RPSs) concurrently. Since the execution platform of the simulation is in a shared-network environment, the RPSs must compete with other applications for resources. The RPSs support adaptive execution by reconfiguring the grain-size of their logical processes dynamically. In addition, scheduling policies are proposed to facilitate efficient allocation of workstations to the RPSs. Experiments are conducted to evaluate the performance of three proposed scheduling policies in the scenarios of homogeneous and heterogeneous workstation clusters.

1 Replicated Parallel Simulators

Many discrete event simulations are regarded as computationally intensive. To reduce the turn-around time of simulation, parallelism is introduced to perform simulation operations in multi-processor or multi-computer machines. In a parallel simulation, a model is decomposed into logical processes and then the logical processes execute in distributed processors. Another approach of parallelism applied to simulation is to run multiple serial simulation programs on multiple processors in parallel and average the results at the end of the runs. This approach is referred to as replicated serial simulation (RSS) [3]. The major advantage of this approach is providing a simple implementation to reduce the overall turnaround time of multiple simulation runs. However, the RSS approach may not be adequate if the time and computational complexity of the simulation model is too demanding to be executed serially.

With the advent of recent technology, workstations have become a powerful and yet inexpensive computational resource. In this paper, a parallelism in which simulation runs are executed by replicated copies of parallel simulation on workstation clusters in parallel is proposed. This approach aims at combining the benefits of the parallel simulation approach and the RSS approach, which are reduction in the turnaround time of each simulation run and the overall turnaround time of all the runs. Although the proposed approach can be applied to different simulation applications, the RPSs described in this paper are designed for the purpose of evaluating the performance of the traffic flow control and call set-up algorithms for an ATM network proposed in [2].

2 The Runtime System and Scheduling Policies

An RPS adapts to the dynamics of resource availability such as changing workloads on individual workstations by reconfiguring the agglomeration of tasks, called *grains*, at runtime to improve workload distribution. An RPS can reconfigure its grains in two different ways [4]: first, the grains and their neighboring grains race one another until all their tasks are accomplished. If the speed of processing the tasks between the grains are different, the workloads of the slow grains will be shared with the faster grains through task relocation. Second, the tasks can be grouped into predefined partitioning levels and an RPS can switch from one partitioning level to another partitioning level. Although these partitioning levels restrict the number of possible task groupings, they reduce the time in searching for suitable grain-to-workstation mappings at runtime.

In addition, scheduling is required to allocate workstations fairly and efficiently to parallel applications in a shared-network environment. A system called *Comedians* (Competitive Environment for Distributed and Adaptive Applications) [5] is developed to tackle the problem of workstation allocation and, at the same time, to maximize the speedup of individual parallel applications. In this paper, the Comedians system is used to support the scheduling and adaptive execution of RPSs. The Comedians system coordinates the execution of parallel applications on workstation clusters. In the system, workstations are partitioned into clusters according to their processing speed. An application can run on more than one workstation clusters, but one of the clusters is specified as local cluster and the others as remote clusters.

Like the RSS, the *first N replications initiated* (FNI) scheduling method [3] is applied by the RPSs to obtain statistically accurate simulation results. In this scheduling policy, N results of simulation runs are recorded from the first N replications initiated. The value of N is determined by the termination condition of simulation, for example when a desirable confidence interval of output results is obtained. Since the FNI does not specify any control over workstation allocation, an RPS will compete with all the other applications on the Comedians system based on its runtime performance. To harness the dynamic computational resource of workstation clusters, special scheduling policies are proposed to build on the top of the FNI. These policies are enforced by the Comedians system. Three scheduling policies are proposed and compared in this paper; they are FNI-SA *(Static Assignment)*, FNI-DA *(Dynamic Assignment)*, and FNI-DAC *(Dynamic Assignment with Coalition)*.

In the FNI-SA policy, an RPS can only be executed in the workstations of its local cluster, but different RPSs can be assigned to different local clusters. Whereas, the RPSs in the FNI-DA policy can be executed in the workstations of multiple clusters dynamically. The FNI-DA and the FNI-DAC policies are basically the same except that the FNI-DAC policy allows the RPSs from the same user to form coalition partners. Since RPSs from the same user have a single objective – the completion of all simulation runs, the RPSs should not compete with one another for workstations. Instead the RPSs with earlier initiated simulation run should have higher priority than the other to get resources because the simulation terminates as soon as the first N replications finishes their runs. This speciality allows the RPSs to be executed more efficiently on workstation clusters. If a coalition is formed, other parallel applications that are running on the Comedians system are regarded as alien applications to the coalition. However, the alien applications regard the coalition as independent parallel applications.

3 Experiments and Results

The experiments are carried out on DEC3100 and DEC Alpha workstation clusters in which all the workstations are connected by Ethernet in different subnets. Unexpected interference from other users is minimized throughout the experiments so as to ensure a controlled environment. All RPSs are written in a single program multiple data (SPMD) model meaning that all workstations run the same piece of program. They can be partitioned into 1 grain (partitioning level 0), 3 grains (partitioning level 1), and 9 grains (partitioning level 2). The RPSs and the Comedians system use PVM [1] as their message passing interface.

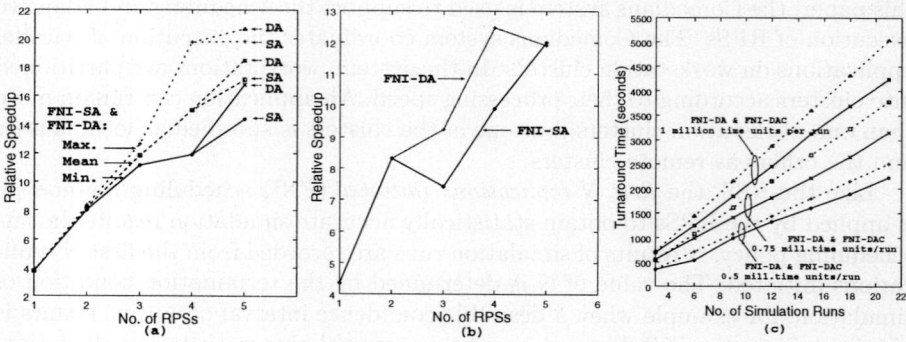

Fig. 1. Comparison of (a) FNI-SA and FNI-DA policies (Scenario One); (b) same as (a) but with an alien application; (c) FNI-DA and FNI-DAC policies (Scenario Two)

The scheduling policies are tested in two scenarios. In the first scenario, simulation runs are running on two homogeneous DEC3100 workstation clusters; each cluster consists of twenty-four workstations. Figure 1a shows the relative speedup of the FNI-SA and the FNI-DA policies when different number of RPSs are running. The figure records the maximum, the mean, and the minimum relative speedup of the RPSs for finishing ten simulation runs, each run performs half a million of simulation time units. The performance of the FNI-SA and the FNI-DA policies is similar until the number of RPS becomes five. In this case, there are not enough workstations for all the RPSs to split to the highest partitioning level. As indicated in figure 1a, the speedup is improved by applying the FNI-DA policy because the RPSs can relocate their grains to the remote cluster for execution. The discrepancy of the speedup between the maximum and the minimum is great when four RPSs are running because two of the RPSs perform one more simulation run than the other RPSs. These two policies are also tested in the situation when an alien application is running on one of the clusters. The alien application is a block-matrix multiplication program which has the same partitioning levels as the RPSs. Figure 1b depicts the relative speedup of the simulation when different numbers of RPSs are running with the alien application. The results show that the FNI-DA policy can sustain performance gain by allowing one of the RPSs to migrate its grains to the remote cluster when there are not enough workstations available in its local cluster. When the number of RPSs is greater than or equal to five, both clusters become overcrowded.

In the second scenario, the simulation is performed on two heterogeneous workstation clusters – one cluster of thirty DEC3100 workstations and the other cluster of six DEC Alpha workstations. Since the speedup of an RPS at partitioning level one in DEC Alpha workstations is higher than the speedup at partitioning level two in DEC3100 workstations, the RPSs will compete for DEC Alpha machines even at the expense of merging to a lower partitioning level. Due to the competition of workstations in the Comedians system, it is possible that some RPSs are constantly excluded from running in DEC Alpha machines. As a result, the overall turnaround time will be restricted by the slowest RPS. Experiments are conducted to compare this special case of the FNI-DA policy with the FNI-DAC policy. Figure 1c depicts the performance of the policies when three RPSs are running in this scenario. The figure suggests that the difference in overall turnaround time between these two policies rises slowly when the number of simulation runs increases.

In summary, the experimental results demonstrate that the overall performance of the simulation can be enhanced significantly by combining the parallelism and the replication of modelling. The results also indicate that the overall performance can further be improved if (i) the RPSs can relocate their grains to remote clusters dynamically; (ii) the RPSs can form a coalition so that the priority of workstation allocation decreases with the order of initiation.

References

1. A. Giest, A. Beguelin, J. Dongarra, W. Jiang, R. Manchek, and V. Sunderam. *PVM 3.0 User's Guide and Reference Manual*, Feb. 1993.
2. S.-Y.R. Li and K.H. Shum. Distributed algorithms for flow control & call set-up in a broadband packet network. In *Proc. of 12th EFOC&N*, pages 182–186, 1994.
3. Y.-B. Lin. Parallel independent replicated simulation on a network of workstations. In *Proc. of 8th Workshop on Parallel and Distributed Simulation*, pages 73–80, 1994.
4. K.H. Shum. Adaptive distributed computing through competition. In *Proc. of the International Conference on Configurable Distributed Systems*, pages 200–207, Maryland, May 1996. IEEE Computer Society.
5. K.H. Shum and K. Moody. A competitive environment for parallel applications on heterogeneous workstation clusters. In *Proc. of the Heterogeneous Computing Workshop (HCW'96), IPPS'96, presented in the joint session with the 2nd Workshop on Job Scheduling Strategies for Parallel Processing*, Hawaii, April 1996.

Shared-Memory Implementation of an Irregular Particle Simulation Method

THOMAS RAUBER   GUDULA RÜNGER *   CARSTEN SCHOLTES

Computer Science Dep., Universität des Saarlandes, 66041 Saarbrücken, Germany

Abstract. We investigate a parallel implementation of an irregular particle simulation algorithm. We concentrate on the issue which programming and system support is needed to yield an efficient implementation for a large number of processors. As execution platform we use the SB-PRAM, a shared memory machine with up to 4096 processors.

1 Introduction

Most investigations of irregular applications concentrate on implementations on shared-memory machines with a small or medium number of processors (usually not more than 64) [5, 6]. In this article, we consider the question whether a typical irregular applications can efficiently be executed on larger machines. As example application, we consider the particle simulator MP3D [7]. We investigate which issues are important for large speedup values, which system support should be be available, and which programming effort must be invested. For hundreds or thousands of processors, it is essential to investigate the exploitable degree of parallelism which represents an upper bound for the achievable speedup. Moreover, sequential portions of the parallel implementations, the granularity of the computations, and load balancing issues have a much larger influence on the efficiency than for a smaller number of processors.

The investigations are executed on a simulator of the SB-PRAM. The SB-PRAM is designed for 128 physical processors which provide a total number of 4096 virtual processors seen by the programmer [1]. The machine provides a global shared memory with uniform access time, i.e., from a virtual processor's point of view, an access to the global memory takes the same time as two arithmetic operations, independently from the memory location that is addressed. Because of this memory organization, locality properties can be neglected and the investigations can concentrate on the exploitation of the maximum degree of parallelism and the usage of efficient parallel data structures. Thus, we can obtain an upper bound on the attainable speedup, which also takes into consideration the overhead for shared data structures and the task management overhead. The avoidance of sequentializations and a good load balance are supported by the powerful multiprefix operation provided by the SB-PRAM[3].

2 The Particle Simulation MP3D

MP3D is a particle simulator for fluid dynamics problems, which uses a uniform subdivision into cubic cells [2, 7, 5]. MP3D is used to study rarefied fluid flow

---
* author supported by DFG

around objects as they pass through the upper atmosphere at hypersonic speed. The significance of the method is the reduction of the interaction phase from $O(n^2)$ to $O(n)$. The program simulates a set of particles by computing their position and velocity in each time step. The space vehicle is represented as a flat object in a rectangular wind tunnel with openings at each end and reflecting walls on the remaining sides. Particles generally flow through the tunnel in positive x-direction. A particle may collide with the boundaries of the wind tunnel, the object, and other particles. Particles can collide with each other only if they reside in the same space cell. The program starts the simulation with a fixed number of particles that are moving with a constant wind velocity in x-direction. The density of the particles represents the resulting air pressure. Particles exiting the wind tunnel are replaced by new particles from a small reservoir, which enter the tunnel at random positions near the entrance.

The main data structures of the MP3D program are two large arrays. One array is used to store the particle information (position, velocity). A second array is used to store the cell information (collision probability, collision information, number and velocity statistics for particles entering the cell). Each time step consists of five phases. The *initialization phase* resets the collision counters and the population counters of the space cells and recomputes the collision probabilities. The *move phase* moves all particles to new positions and computes collisions. In the *add phase*, new particles are entered into the tunnel to simulate a constant current of air. Particles in the reservoir are moved in the *move reservoir* phase. In the *collide reservoir* phase, the reservoir particles are paired and a collision is computed for each of the particle pairs. Tests with a sequential version of the program show that more than 96% of the total runtime is used for the move phase. This percentage is increasing for increasing numbers of processors.

The move phase computes a new position and velocity for each particle ρ in the wind tunnel. Moving a particle may lead to a collision with a surface of an object or a boundary of the tunnel or with another particle. The program treats the different kinds of collisions one after another, i.e., for each particle, first a collision with a boundary is tested and then a collision with another particle. A collision with a surface may happen, if the cell $cell(\rho)$ in which particle ρ is positioned contains a surface. If the particle hits the surface, its position and velocity are changed according to the laws of elastic impact. The collision between particles is computed by using cell tags of particles located in each cell and formulas expressing the collision probability. A particle ρ sets the cell tag of $cell(\rho)$ if the tag information is empty. If $cell(\rho)$ already contains a tag for another particle ρ', a collision between ρ and ρ' may happen. The collision in $cell(\rho)$ is computed according to a collision probability. If a collision takes place, the velocity of ρ and ρ' are changed accordingly and the tag of $cell(\rho)$ is cleared.

3 Implementation for small numbers of processors

The original implementation of MP3D from the SPLASH-1 benchmark suite was developed for a small or medium number p of processors ($p \leq 64$) [5]. This is reflected in the implementation of the different phases. In the initialization phase, the 60 rows of inner cells, each one containing 14 cells, are statically distributed round-robin among all processors. The relatively large grain size of 14 cells results in a load imbalance if $60/p \notin N$. In the move phase, the set of particles to be simulated is split up into packets of 64 particles each, mainly to improve the memory alignment and cache behavior on some architectures. The packets are distributed statically round-robin among all processors. Each processor simulates the particles of the packets assigned to it in turn. Since the particles assigned to a specific processor q move around unpredictably in the wind tunnel, q may have to access different cells' data during the simulation of different particles and even during the simulation of one particle. Basically, the simulation of one particle ρ comprises the following steps: First, some particle related statistics of cell(ρ) are updated. The access to the statistical data of cell(ρ) is protected by a lock to avoid simultaneous updates by other processors. After this, the particle is moved, possibly into another cell. Second, the cell in which the particle now resides must be checked, whether it contains a boundary or not. If cell(ρ) contains a boundary, an interaction with it is simulated making an additional lock necessary. The computation of the interaction may place ρ into a third cell. This possibly new cell(ρ) is then locked for a possible particle-particle collision: it must be checked whether it contains a collision particle ρ' or not. If ρ' exists, a collision between ρ and ρ' is computed. The lock of cell(ρ) is released after the computation of the collision. The simulation of a particle-particle collision only affects the velocities of the particles ρ and ρ', not their positions, i.e., no further space cell can be entered. Thus, up to three different space cells may be accessed for the simulation of one particle in one simulation step and between two or three locks are used. The time to simulate a particle may differ considerably depending on whether collisions have to be computed. Thus, the statical assignment of particles to processors may lead to large differences in the work load, especially if each processor has to simulate only a few particles. Additionally, the distribution in packets of 64 particles may lead to an additional load imbalance if $\lfloor n/64 \rfloor /p \notin N$. In the remaining three phases the particles to be added, the reservoir particles to be moved, and the pairs of reservoir particles to be collided, respectively, are distributed statically in a round-robin way among all processors.

Figure 1 (left) shows the resulting speedup values for different shared-memory multiprocessor platforms, the cache based KSR-1 and Stanford DASH, a bus-based Silicon Graphics 4D/380S (SGI), and the SB-PRAM. The values for the KSR-1 are taken from [7] and show a simulation run with 50000 particles, the values for the SGI are taken from [2] and show a simulation run with 8000 particles, the values for the DASH are taken from [4]. Both the DASH and the SB-PRAM runs show a simulation with 40000 particles. For the KSR-1, the speedup is decreasing for the KSR-1 with increasing numbers of processors, mainly because of contention for accesses to the shared space array. For the DASH, the speedups are poor because of frequent sharing and updates of space cells. As the number of processors increases, it becomes more and more likely that the newest version of a space cell being referenced resides in the cache of another processor. For the same reason, the speedup for the SGI is not increasing with increasing numbers of processors. The efficiency of the original implementation on the SB-PRAM is quite large because of the efficient realization of the locking mechanism with the help of the multiprefix operations. As shown in [3], the execution of a lock or unlock operation is independent from the number of participating processors, as long as no sequentializations occur. Moreover, the lack of spatial locality does not have a negative influence as it is the case for the KSR-1 or the DASH. For larger number of processors, the efficiency on the SB-PRAM is slightly decreasing because the probability for sequentializations through concurrent accesses to the same space cell by different processors is increasing. The efficiency on

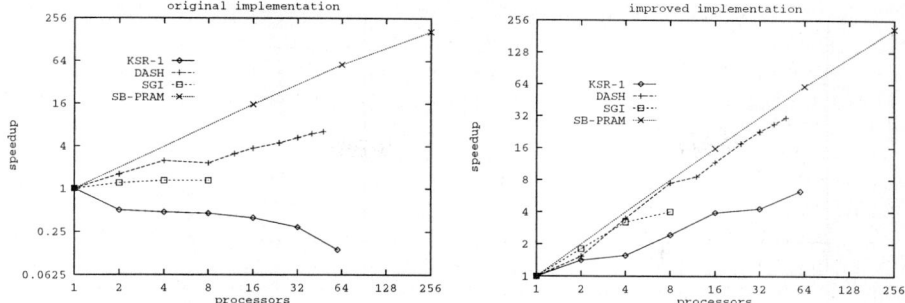

Fig. 1. Speedup values for the original implementation (left) and of improved or restructured implementations (right) of MP3D on different multiprocessor platforms. For an SB-PRAM-simulation with 256 virtual processors, the speedup of the original implementation is 166, the speedup of the improved implementation is 212.

the SB-PRAM is not optimal because of the overhead of the locking mechanism and because of the load imbalance caused by the static scheduling. Figure 1 (right) shows the runtime of restructured versions of the MP3D program. The versions on the KSR-1, SGI, and DASH result from major changes in the program to enhance the locality of MP3D, including a spatial decomposition among the processors, see [7, 2, 4] for a detailed description. The changes for the SB-PRAM are restricted to the use of efficient parallel data structures with dynamic scheduling and the use of multiprefix operations.

4 Implementation for large number of processors

Although the original implementation yields a good performance on the SB-PRAM, it can be considerably increased by reorganizations that increase the degree of parallelism and by using dynamic scheduling. The degree of parallelism is increased by using a fine grain size for the different tasks: In the initialization phase, the degree of parallelism is increased to 840 by assigning single cells to processors. Because the computational effort per space cell is the same, a static assignment of cells to processors can still be used. In the move phase, the degree of parallelism is increased to the number of particles to be simulated by removing the building of packets of 64 particles.

As discussed in the previous subsection, the static assignment of particles to processors may lead to a load imbalance. This can be removed by a dynamic assignment of particles to processors which can be realized by the use of a parallel loop. The loop is controlled by a shared counter c which is accessed by the processors with a multiprefix operation to get the next particle to be simulated. Because of the use of the multiprefix operation, no sequentializations occur when different processors access the shared counter concurrently. The same effect can be obtained by the use of a central *parallel* task queue which can be accessed by all processors in parallel without sequentializations, see [3]. A further improvement is obtained when replacing locks to shared data structures by multiprefix operations if this results in the same behavior. This can be applied for the update of the cell statistics at the beginning of the move phase, because each processor contributes only informations of the particle that is currently simulated and because the statistical data is not used before the next simulation round. Locks

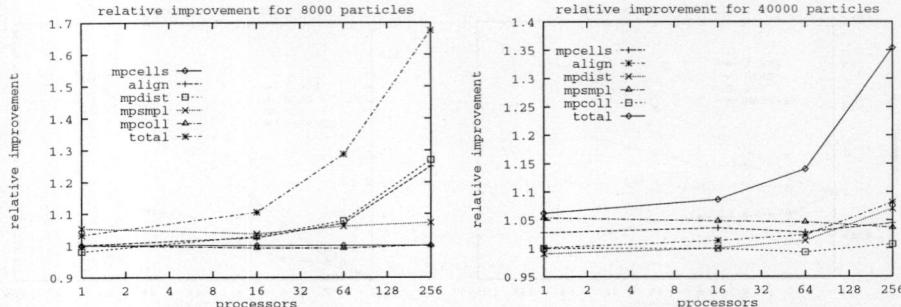

Fig. 2. Stepwise improvement of the original implementation on the SB-PRAM for 8000 (left) and 40000 (right) particles.

can also be replaced by multiprefix operations when particles are added near the entrance to get a constant air pressure and when particles are moved to the entrance if they leave the tunnel through the exit.

Figure 2 shows the percentage of the improvements on the MP3D program for runs with 8000 and 40000 particles, respectively. The figure shows the effects of the *single* changes compared to the original implementation. The following changes are investigated: **mpcells** shows the effect of increasing the degree of parallelism in the initialization phase by assigning single cells to the processors; a dynamic scheduling is applied by using a parallel loop; **align** shows the effect of increasing the degree of parallelism in the move phase by assigning single particles to the processors instead of packets of 64 particles; a static scheduling is used for the distribution; **mpdist** shows the effect of using the **align** variant with a dynamic scheduling which is obtained by a parallel loop; **mpsmpl** uses a multiprefix operation instead of a lock for the update of the cell statistics at the beginning of the move phase of each particle; **mpcoll** uses a multiprefix operation instead of a lock for the statistics at the end of each move phase.

The use of the **mpcells** option leads to a smaller runtime because of a better load balance. The improvement in percentage is increasing for increasing number of processors until the maximum degree of parallelism (840) is reached. The use of the **align** option leads to the largest reduction in the runtime because of a better load balance. The effect is larger than the effect of the **mpcells** option because the move phase contributes a much larger fraction to the global runtime than the initialization phase. The improvement in percentage is large if the load balance of the original implementation can be improved by a better distribution of the particles among the processors. This is, for example, the case for 8000 particles and 256 processors. In the original implementation, 125 processors get 64 particles each, the rest of the processors is idle. With the **align** option, each processor simulates 31 or 32 particles, respectively. The additional use of the **mpdist** option can lead to an additional improvement because the load is balanced even better by the dynamic scheduling. This effect has a considerable impact mainly in the case when each processor has to simulate only a few particles. If the number of particles per processor is large, the advantage of the dynamic scheduling diminishs, because the static scheduling already establishs a quite even distribution among the processors. For large numbers of particles per processor, the overhead of the dynamic scheduling even leads to a small increase of the runtime compared to the **align** variant. The **mpsmpl** option leads to a slight improvement because the direct use of a multiprefix operation is faster than the use of a lock operation. The improvement is not increasing for increasing numbers of processors because the execution time of a lock or unlock operation is constant, independent from the number of participating processors. Moreover for up to 256 processors, there are usually only a few sequentializations through concurrent accesses to the same space cell when using the locking variant. For larger numbers of processors ($p > 256$), the percentage of improvement of the **mpsmpl** option is getting larger because the possibility for sequentializations increases (not shown in the figure). The **mpcoll** option has only a small influence on the runtime because the update of the statistics at the end of each move phase only contributes a small fraction to the global runtime. The use of a multiprefix operation instead of a lock operation for the update of the number of particles in the wind tunnel and the number of particles in the reservoir at the end of each simulation round does not have a significant influence on the runtime and, therefore, is not shown in the figure.

5 Conclusions

For most irregular applications, it is difficult to achieve good speedup values for large number of processors. One of the main reasons lies in the lack of locality of these applications. The negative influence of missing locality is usually getting worse for increasing number of processors. To enhance the spatial locality and the resulting speedup values, major reorganizations are required for most applications. We use the MP3D program as example and show that much better results can be obtained, if the underlying hardware provides appropriate operations to implement shared data structures and to support dynamic task scheduling. Another advantage of this approach is that the resulting programs are much easier to write and to maintain.

References

1. F. Abolhassan, J. Keller, and W.J. Paul. On the Cost–Effectiveness of PRAMs. In *Proceeding of the 3rd IEEE Symposium on Parallel and Distributed Processing*, pages 2–9, 1991.
2. D.R. Cheriton, H.A. Goosen, and P. Machanick. Restructuring a parallel simulation to improve cache behavior in a shared-memory multiprocessor: A first experience. In *Proc. International Symposium on Shared Memory Multiprocessing*, 1991.
3. Th. Grün, Th. Rauber, and J. Röhrig. The programming environment of the SB-PRAM. In *Proc. 7th IASTED/ISMM Int.l Conf. on Parallel and Distributed Computing and Systems, Washington DC*, pages 504–509, October 1995.
4. D. Lenoski, J. Laudon, T. Joe, D. Nakahira, L. Stevens, A. Gupta, and J. Hennessy. The DASH prototype: Logic overhead and performance. *IEEE Trans. on Parallel and Distributed Systems*, 4 (1):41–61, 1993.
5. J.P. Singh, W.D. Weber, and A. Gupta. SPLASH: Stanford Parallel Applications for Shared-Memory. *Computer Architecture News*, 20(1):5–44, 1992.
6. S.C. Woo, M. Ohara, E. Torrie, J.P. Singh, and A. Gupta. The SPLASH-2 Programs: Characterization and Methodological Considerations. In *ISCA '95*, pages 24–36, 1995.
7. X. Zhang, K. He, and G. Butchee. Performance botleneck identification and application program improvement on network-based shared-memory architectures. Technical report, University of Texas at San Antonio, 1993.

A Parallel Algorithm for the Technology Mapping of LUT-Based FPGAs

Vamsi Boppana[1]* and Prashant Saxena[2]** and Prithviraj Banerjee[1]*
W. Kent Fuchs[1]* and C. L. Liu[2]**

[1] Coordinated Science Lab, 1308 W. Main St.
[2] Department of Computer Science, Digital Computer Lab, 1304 W. Springfield Ave.
University of Illinois at Urbana-Champaign, Urbana, IL 61801, USA.

Abstract. FLOWMAP ([1]) was the first delay-optimal algorithm for the technology mapping of LUT-based FPGAs. However, even though this algorithm is polynomial, rapid prototyping using FPGAs requires faster solutions. This paper provides an efficient parallelization of FLOWMAP that minimizes locking on shared memory architectures. The influence of scheduling strategies and technology-specific parameters on speedups is studied. The expected running time is also analyzed. The parallel algorithms yield speedups of around 4 to 5 on 8 processors.

1 Introduction

Parallel processing has extensive applications in all phases of VLSI design ([2]). A key step in the design cycle that does not yet have a parallel solution is the field programmable gate array (FPGA) technology mapping problem. Fast technology mapping is vital in maintaining one of the main advantages offered by FPGAs, viz., a small design turn-around time for rapid, interactive prototyping.

An LUT-based FPGA consists of an array of programmable logic blocks containing *lookup tables* (LUTs). A K-input LUT can realize any arbitrary boolean function of upto K variables. The process of realizing a general boolean network using LUTs is called FPGA technology mapping. Cong & Ding ([1]) developed a polynomial algorithm, FLOWMAP, for optimal-delay technology mapping for LUT-based FPGAs under the unit delay model. FLOWMAP yields minimum depth FPGA realizations for combinational circuits. It first computes labels for all the nodes in the circuit in topological order. A node's label represents the depth of the optimal realization of its fanin cone. The second phase uses the labels to select gates to be realized by each LUT. Since the second phase runs in linear time, most of the run-time is spent in the labeling phase.

FLOWMAP is an efficient polynomial algorithm; therefore, sequential algorithmic advances are unlikely to yield significantly faster solutions to the problem. We provide an alternative by exploiting the parallelism available in the labeling phase. Our parallel algorithms are designed for shared memory architectures and guarantee a final mapping identical to the corresponding sequential solutions.

---

* Supported in part by the Semiconductor Research Corporation and the Joint Services Electronics Program under grants 95-DP-109 and N00014-96-1-0129 respectively.
** Supported in part by the National Science Foundation under grant MIP-9222408.

2 Parallelization of FLOWMAP

The computation of a node label involves solving a flow problem on a network that depends only on the labels in the fanin cone of the node. A simple parallel strategy to compute node labels uses a centralized task queue. This event-driven scheme starts with the queue containing only the primary inputs (PIs), and places a node into the queue as soon as the labels of all its fanins are computed. However, this requires excessive synchronization. First, whenever the label computation of a node is completed, each of the fanouts of that node must be locked (one at a time) so that their counts of remaining fanin computations may be decremented and any node with count 0 may be placed in the task queue. Second, the task queue is a hot spot, with every idle processor competing for nodes in it.

In contrast, our approach works on the levelized circuit graph; level i containing all nodes whose maximum distance from any PI is i. Computation of the labels proceeds *level by level*, starting with level 0 (the PIs). Correct label computation across levels can be ensured by a barrier at the end of each level. The scheduling of nodes within a level may be **static** or **dynamic**. In the case of static scheduling, *no locking is required*, since the only time a node's label is written is when the level containing the node is being processed. But this node cannot be in any fanin cone currently being processed. Therefore, it is safe to read and write its label without locking. With dynamic scheduling, although labels can still be written without any locking, we need to lock the variable pointing to the next available node. However, our experiments show that this overhead is not significant. The memory requirements for the level-based schemes are small. The dynamic memory required by a processor is bounded by the size of a fanin cone. The static memory requirements are comparable to those of the sequential algorithm because the circuit data structure is read-only.

A potential disadvantage of our approach is the possibility of processors being idle even when work is available in subsequent levels. However, the probability of this occurring becomes small with sufficiently large circuits. Circuits for which this parallelization would be most beneficial, would have near-uniform level sizes with a large number of nodes at each level, and similar-sized fanin cones within each level. In general, speedups on non-ideal circuit topologies degrade less with dynamic scheduling than with static scheduling.

Our implementation of the level-based scheme uses an array *levellist[i][1 .. levelsize[i]]* for the *levelsize[i]* nodes in the graph at level i. The pseudocode for static-interleaved scheduling, shown in Figure 1, uses no locking. The pseudocode for dynamic scheduling (Figure 2) requires a lock on the pointer to the next available node. However, we do not need to lock the global *label* and *cut_nodes* arrays that store the optimal labels and cuts.

3 Analysis of Expected Running Time

Let n, m and p respectively represent the total number of nodes and edges in the circuit graph and the number of processors in use. Let $L_0, L_1, \ldots, L_{\nu_l}$ denote the levels in the levelized graph, with s_i being the number of nodes in

```
procedure STATIC-FLOWMAP
    j ← my_processor_ID;
    while (j ≤ levelsize[i])
        curr_node ← levellist[i][j];
        compute the label and optimum cut for node curr_node;
        write this label to label[curr_node] and the cut to cut_nodes[curr_node];
        j ← j+num_processors;
```

Fig. 1. Pseudocode for parallel label computation using static-interleaved scheduling

```
procedure DYNAMIC-FLOWMAP
/* Start with curr_node ← 0 on each processor, and the global next_node ← 1 */
    while (curr_node ≥ 0)
        lock next_node;
        if (next_node ≤ levelsize[i]) then
            curr_node ← levellist[i][next_node];
            next_node ← next_node +1;
        else curr_node ← −1;
        unlock next_node;
        if (curr_node ≥ 0)
            compute the label and optimum cut for node curr_node;
            update label[curr_node] and cut_nodes[curr_node];
```

Fig. 2. Pseudocode for parallel label computation iterations using dynamic scheduling

L_i. If $t_i^{\max}$ is the maximum of the label computation times for the nodes in L_i, the barrier synchronization implies that the total label computation time for the graph is at most $\sum_{i=0}^{\nu_l}(t_i^{\max}\cdot\lceil s_i/p\rceil)$ in the static-interleaved case. With dynamic scheduling, each level L_i also requires s_i lock-unlock operation pairs that cannot be pairwise concurrent. If the average time for one such pair is t_l, the label computation time increases by nt_l units. However, the dynamic balancing reduces the $\sum_{i=0}^{\nu_l}(t_i^{\max}\cdot\lceil s_i/p\rceil)$ term to $\sum_{i=0}^{\nu_l}(t_i^{avg}\cdot\lceil s_i/p\rceil)$.

The initial topological sorting and level computation uses $3m$ time units. Another m units are required for the mapping phase. Let C and c' be the times required to respectively create and reactivate a thread (typically, $c' \ll C$). Then, the total overhead associated with thread management is $p(C + \nu_l c')$.

Thus, the run-times for the static-interleaved and dynamic schemes are given by $4m+p(C+\nu_l c')+\sum_{i=0}^{\nu_l}(t_i^{\max}\cdot\lceil s_i/p\rceil)$ and $4m+p(C+\nu_l c')+\sum_{i=0}^{\nu_l}(t_i^{avg}\cdot\lceil s_i/p\rceil)+nt_l$ respectively. In contrast, the sequential run-time is $3m + \sum_{i=0}^{\nu_l} s_i t_i^{avg}$, since the m units required for level computation and the $p(C + \nu_l c')$ units required for thread management in the parallel algorithms are not needed here.

4 Experimental Results

We tested our parallel implementations of FLOWMAP on several large circuits from the MCNC '91 CML suite on a SGI Power Challenge multiprocessor. All circuits were first decomposed into a 2-input network of simple gates. This was followed by the actual technology mapping on 2, 4, 6 and 8 processors under both static-interleaved and dynamic scheduling policies.

Table 1 presents the run-times and speedup figures obtained with $K = 5$

(5 being typical for many industrial FPGAs). The speedups obtained for the smaller circuits indicate that the parallelism inherent in them is limited by the small number of gates in each level. However, larger circuits, towards which the parallelization is targeted, yield substantially better speedups and scalability. Dynamic scheduling results in slightly better scalability despite the overhead of the task queue. Running times are similar under static-interleaved and dynamic schedules. This implies that the overhead involved in the task queue synchronization is quite small, and that the times required for the processing of different nodes within a level are comparable.

Table 1. Sequential run-times in seconds and parallel speedups (with $K = 5$)

Circuit	Optimal depth	Sequential run-time	Static speedup # processors				Dynamic speedup # processors			
			2	4	6	8	2	4	6	8
apex7	4	12.43	1.4	1.9	2.4	2.5	1.4	1.9	2.4	2.1
i7	2	57.87	1.5	2.1	2.6	2.9	1.4	2.1	1.9	3.3
apex6	5	10.97	1.2	2.3	3.3	3.9	1.2	2.4	3.3	3.7
c2670	7	89.26	1.6	2.6	3.3	4.3	1.7	2.9	3.6	4.7
pair	5	50.05	1.6	2.6	3.5	3.9	1.5	2.8	3.5	4.1
c5315	8	276.84	1.5	2.3	3.1	4.1	1.4	2.1	2.7	3.3
i10	13	389.34	1.4	2.3	3.5	4.0	1.5	2.4	3.5	4.1

We also studied the speedups obtained with different values of K. Increasing the value of K increases the grain size of the available parallelism, since each network flow sub-problem is solved in $O(Km)$ time. However, this effect is often offset by the increased imbalance created in small-sized levels due to the larger grain size. Thus, the speedups obtained depend primarily on the circuit topology. As an illustration, although $K = 7$ resulted in the best speedup for i7 (6.9 on 8 processors), it was the worst choice for c5315 (2.8 on 8 processors).

5 Conclusions

In this paper, we presented a strategy for the parallelization of FLOWMAP that yields good speedups and scalability by addressing grain size and load balance.

FLOWMAP has been extended to the general delay model by Yang & Wong ([3]) by generalizing the labeling phase. Our parallel approach can also be extended to their algorithm in a manner analogous to that described in this paper.

References

1. Cong, J., and Y. Ding, "FlowMap: An Optimal Technology Mapping Algorithm for Delay Optimization in Lookup-Table Based FPGA Designs," *IEEE Trans. Computer-aided Design* **13** (1), January 1994, 1–13.
2. Banerjee, P., *Parallel Algorithms for VLSI Computer-Aided Design*, Prentice Hall, Inc., Englewood Cliffs, NJ, 1995.
3. Yang, H., and D. F. Wong, "Edge-Map: Optimal Performance Driven Technology Mapping for Iterative LUT Based FPGA Designs," *Digest IEEE Intl. Conf. on CAD*, November 1994, 150–155.

Distributed String Matching Algorithm on the N-cube

Fouzia Moussouni   Christian Lavault
LIPN, *CNRS URA 1507*
Université Paris-XIII,
Av. J.-B. Clément 93430 Villetaneuse. France
email: mouf@lipn.univ-paris13.fr

Abstract. In this paper, we present a distributed algorithm which runs on the N-cube and solves the string matching problem. A basic prefix-suffix matching technique is used as a building block for the construction of the algorithm. As opposed to the parallel algorithms so far designed on shared-memory PRAM models, our algorithm runs distributively on fixed hypercube networks and especially applies to long texts. Given a pattern x and a text t of size n, the occurrences of x in t are found in time $O(n/N + \log N)$ on the N-cube, including the preprocessing computation time.

1 Introduction

Let $t[1..n]$ be the text and $x[1..m]$ the pattern, with the two strings taken over an alphabet Σ. In the **string matching problem**, we want to know whether x is a substring of t, and if so wherever x matches t, i.e where the starting positions of all the occurrences of x are in t.

Since the 80's, the string matching problem was extensively studied on the Parallel RAM computation model. The first work optimal parallel string matching algorithm was given in [6], for the CRCW/PRAM. By introducing the idea of sampling in [7], Vishkin achieved an $O(\log^* m)$ time (work optimal) algorithm for the problem. Galil improved this result to a constant, though with a high time overcost due to a slow preprocessing.
Recently, Gąsieniec *et al.* speeded up the preprocessing to $O(\log \log m)$ time, and designed a new recursive parallel divide and conquer algorithm based on the pseudo-period technique (see [2]).

On more realistic models with fixed interconnection networks and non-shared memory, the routing must be taken into account. As an example, a parallel pattern matching algorithm is considered in [1] on a mesh connected array of processors. It is also based on the pseudo-period technique and runs in time $O(\sqrt{n})$ on a $\sqrt{n} \times \sqrt{n}$ mesh computation model. Although non optimal, current algorithms on such models have a fine-granularity. One can use the same techniques with fewer number of processors. Each processor deals with s successive symbols of the text by simulating s virtual processors in its local memory. But it turns out to be unfeasible when having quite a long text per processor. More exactly, on a parallel computer with N fixed processors and a reasonably big ratio $\frac{n}{N}$, it is more efficient to use sequential computations in parallel and perform limited communication. Without loss of generality we may assume that N divides n (otherwise, the text t may be padded on the right with extra symbols). Each processor deals with a text segment T_i of size $|T_i| = s = \lceil \frac{n}{m} \rceil \geq m$. Some of the N-cube's links connect the end of each segment T_i to the beginning of the next segment T_{i+1}. By contrast with the classical string matching problem, there exist pattern occurrences that span two consecutive segments of the text. We use a prefix-suffix matching technique as a basic building block of the algorithm for finding all overlapping occurrences caused by the text segmentation. We show that the overall string matching problem can be solved in time $O(s)$, in constant space, and linear total work on a MIMD hypercube with no more than N processors.

2 The parallel computation model

Many computation models may be thought of to deal with string processing, e.g. the hypercube network was chosen herein for its versatility. The d-dimensional hypercube Q_d has $N = 2^d$ processors which communicate by exchanging messages through its $d \cdot 2^{d-1}$ links. An important property is that the hypercube can simulate many known networks ($O(N)$-node arrays, rings, meshes, binary trees, etc.). This way of looking at the hypercube makes it possible to implement simple parallel algorithms on any such networks while using the attractive properties of Q_d. To have a good task mapping, string matching algorithms use general simulations on lines, rings (e.g, pipelining or simple linear string matching algorithms), and similarly on meshes with a snake-like ordering of the pattern.

3 N-cube String Matching with Long Texts

The text t is first broken into non-overlapping s-blocks and then distributed over the PEs such that each PE_i contains subtext $T_i = t_{si+1}...t_{s(i+1)}$. We say that position j in the text is at position j' in T_i whenever $j = i \cdot s + j$', where $s = \lceil \frac{n}{N} \rceil$. By using a spanning tree embedding technique for broadcasting $O(m)$ items to all PEs, the pattern is broadcasted to all other PEs. The overall time needed for this initialization phase is bounded by $O(m + \log N)$. The allocation scheme places adjacent segments of the text on adjacent processors. Thus, what we want to do is to find all the occurrences of the pattern across multiple segments by serial computations in each processor.

The main difficulty is that the pattern may occur in some subtext T_i as a substring overlapping two adjacent segments, viz. either T_i and T_{i+1}, or T_i and T_{i-1}, caused by the text segmentation.

To avoid text block transfers, our algorithm operates according to whether the pattern has some periods or none. The possible periodicity of the pattern x is the key part of the algorithm.

We search for occurrences that are within the subtexts by running the fastest sequential string matching algorithm on each processor. Now, to find occurrences in the overlapping endpoints of a segment T_i, we need to compute some prefix-suffix information. Recall that a subword u of the pattern is said to be a prefix of x if $x = uz$. Similarly, a subword v of the pattern is a suffix of x if $x = zv$.

To find occurrences that occur in two segments T_i and T_{i+1}, each processor PE_i sends to PE_{i+1} one constant size message carrying the prefix length $|u|$ (if any). Because of the periodicity property, we also consider the longest suffix v located at the beginning of T_{i+1}. Upon receipt of the prefix length value, the processor PE_{i+1} checks which of the following property holds:

1. $|u| + |v| = m$ and the pattern x is strictly nonperiodic (i.e., without periodic prefixes and suffixes).

2. $|u| + |v| \equiv m \pmod{p}$ and x is periodic with shortest period p.

3. $p_x \equiv |u| + |v| - m \pmod{p}$ and x is periodic with shortest period p, p_x being a greater period of x and nonmultiple of p.

4. $|u| + |v| \equiv m \pmod{pp \ (resp. \ sp)}$ and x is partially periodic. Which means that x is not periodic while some prefix (resp. suffix) is periodic with period pp (resp. sp).

In whichever case, there is at least one pattern occurrence across the sections T_i and T_{i+1}. The corresponding position in the text t clearly depends on which of the above properties holds. More details can be found in [5].

Note that only one constant size message carrying the prefix size $|u|$ is sent to processor PE_{i+1}. An additional linear time (in the pattern size) is used for searching the words u and v, which is negligible with respect to the global communication time. Finally, an upper bound on local time complexity is $O(s)$ since we run a serial string matching algorithm to match both the word x, the prefix u and the suffix v. The space complexity related to this string matching algorithm is either linear in the pattern size (e.g using KMP) or a constant. Since $m = O(\frac{n}{N})$, this provides a sketchproof of the following proposition:

Proposition *Taking the preprocessing into account, the time complexity of the above N-cube string matching algorithm is $O(\frac{n}{N} + \log(N))$.*

4 Concluding Remarks

We proposed a new flexible distributed algorithm which solves the string matching problem on a fixed N-cube with long texts stored on processors. Compared to classical parallel string matching algorithms, the text has a broken structure caused by the distribution of computations in non-shared memory models. So, we consider that pattern occurrences may overlap adjacent segments.

For finding all pattern occurrences, we presented a more involved linear time algorithm, with pattern periods assumptions. Note that, a kind of trade-off between the size of the hypercube and the time complexity may be achieved to give the first optimal algorithm with the smallest possible size of the cube.

Open questions. Design a more general algorithm to search for long texts as well as small texts allocated to processors. Kim and Park address a similar problem in [4] in the case of an hypertext which is also a nonlinear structure of a text.

Extend the relevant prefix-suffix technique (mainly used to deal with overlapping occurrences), to sequential and incremental string matching problems (see [3]). All pattern occurrences may be derived by matching maximal prefixes and suffixes on independent blocks of the text. Using this technique, a linear algorithm in the number of comparisons is expected.

References

1. B.S. Chlebus and L. Gąsieniec. Optimal pattern matching on meshes. In *11th Ann. Symp. on Theoretical Aspects of Computer Science*, pages 213–224, 1994.
2. A. Czumaj, Z. Galil, L. Gąsieniec, K. Park, and W. Plandowski. Work-time optimal parallel algorithms for strings problems. Manuscript, 1994.
3. Z.M. Kedem, G.M. Landau, and K.V. Palem. Optimal parallel suffix-prefix matching algorithm and applications. In *ACM Symposium on Parallel Algorithms and Architectures*, pages 388–398, 1989.
4. D.K. Kim and K. Park. String matching in hypertext. In *6th Annuel Symp. on Combinatorial Pattern Matching*, Lecture Notes in Computer Science, pages 318–329, 1995.
5. C. Lavault and F. Moussouni. N-cube string matching algorithm with long texts. In *The Proceeding of CCS'95, Combinatorics and Computer Science*, To appear in Lecture Notes in Computer Science, 1995.
6. U. Vishkin. Optimal parallel pattern matching in strings. *Information and Control*, 67:91–113, 1985.
7. U. Vishkin. Deterministic sampling - a new technique for fast pattern matching. *SIAM J.Comput*, 20:22–40, 1991.

Index of Authors

Aarts (E. H. L.)   II: 226
Abdallah (Ali E.)   II: 911
Addison (C.)   I: 178
Albers (Patrick)   II: 266
Alexandrov (Vassil N.)   II: 72
Alonso (Luis)   I: 522
Altman (Erik R.)   II: 833
Antelo (Elisardo)   II: 155
Anuta (Michael A.)   II: 193
Appiani (Enrico)   II: 139
Arbab (F.)   I: 664
Arbenz (Peter)   II: 11
Asselin de Beauville (Jean-Pierre)   II: 377
Attali (Isabelle)   I: 136
Au (Peter)   I: 601
Avellana (Narcís)   II: 470
Ayguade (Eduard)   II: 644

Backschat (Martin)   II: 631
Badouel (Didier)   I: 98
Baeumker (Armin)   II: 369
Bagherzadeh (Nader)   I: 247
Bagherzadeh (Nader)   II: 747
Balakrishnan (Shobana)   I: 287
Banerjee (Prithviraj)   I: 828
Barcaccia (Piera)   II: 594
Barrena (Manuel)   II: 866
Barreteau (Michel)   I: 463
Barth (Dominique)   I: 243
Barthou (Denis)   I: 424
Basu (Anindya)   I: 187
Bellone (Jacques)   II: 266
Benaini (Abdelhamid)   II: 535
Bennett (Andrew J.)   I: 106
Bennett (Andrew J.)   II: 445
Bermond (Jean-Claude)   I: 313
Bernabéu-Aubán (José M.)   I: 526
Besch (Matthias)   I: 455
Betzos (George A.)   II: 457
Bhandarkar (Milind A.)   I: 646
Blom (C. L.)   I: 664
Boier Martin (Ioana M.)   II: 255
Bonuccelli (Maurizio A.)   II: 594

Boppana (Vamsi)   I: 828
Bordawekar (Rajesh)   I: 541
Botorog (George Horatiu)   I: 718
Bourdin (H.)   I: 218
Bozas (Giannis)   II: 881
Brandes (Thomas)   I: 459
Breitinger (S.)   I: 710
Brilman (Matthieu)   II: 734
Brown (T. J.)   II: 213
Browne (J. C.)   I: 633
Bruguera (Javier D.)   II: 155
Brüll (Ansgar)   I: 654
Brunie (Lionel)   I: 74
Brunie (Lionel)   II: 887
Burger (F. J.)   I: 664
Burgess (David A.)   II: 697

Cabillic (Gilbert)   I: 114
Calinescu (Radu)   II: 555
Caromel (Denis)   I: 136
Carretero (Jesus)   I: 522
Carro (Manuel)   II: 724
Catthoor (Francky)   II: 103, 217
Cerqueira (Renato)   I: 597
Chan (Sun)   II: 801
Chang (Chialin)   II: 109
Chen (Chien-Ming)   II: 757
Chen (Huey-Ling)   II: 611
Chen (Lei)   II: 359
Chin (Wei-Ngan)   I: 579
Cho (Cheng-Hong)   II: 843
Cho (Sangyeun)   II: 492
Cholvi-Juan (Vicente)   I: 526
Choudhary (Alok N.)   II: 486
Choudhary (Alok)   I: 541
Chung (Yongwha)   II: 123
Clémençon (Christian)   I: 64
Clint (Maurice)   II: 22, 26
Coelho (C.)   II: 139
Coelho (Fabien)   I: 571
Collard (Jean-François)   I: 406, 424
Collingwood (Peter)   I: 752
Colombet (Laurent)   II: 653
Cornu (Thierry)   II: 689

Cortadella (Jordi) II: 824
Cortes (Toni) I: 477
Cortes (Toni) II: 665
Corvi (Marco) II: 139
Crivelli (Silvia A.) I: 151
Crookes (D.) II: 213
Crumpton (Paul I.) II: 697

Dai (H. K.) I: 234
Damm (Werner) II: 453, 461
Danckaert (Koen) II: 217
Darlington (John) I: 579, 601
Darte (Alain) I: 379
Davis (Larry) II: 103, 109
Davy (John R.) II: 319
Daydé (Michel J.) II: 34
De Coster (Luc) II: 236
De Man (Hugo) II: 217
Dehnert (Jim) II: 801
Dekeyser (Jean-Luc) I: 173
Delgado-Frias (José G.) I: 213
Della Vecchia (Gennaro) II: 147
Delmas (Olivier) I: 370
Delorme (Charles) I: 283
Delosme (Jean-Marc) II: 103
Demian (Vladimir) II: 243
Desbat (Laurent) II: 653
Desprez (Frédéric) I: 165, 459
Desprez (Frédéric) II: 3, 243
Devirmiş (Timuçin) II: 862
Dew (Peter M.) II: 319
Di Ianni (Miriam) I: 258
Di Ianni (Miriam) II: 594
Díaz de Cerio (Luis) I: 253
Diderich (Claude G.) I: 451
Dikenelli (Oduz) II: 892
Dimakopoulos (Vassilios V.) I: 341, I: 347
Dimopoulos (Nikitas J.) I: 341, 347
Diniz (Pedro C.) I: 414
Distasi (Riccardo) II: 147
Dittrich (Wolfgang) II: 369
Domas (Stéphane) II: 3
Dong (Y.) II: 213
Dongarra (Jack J.) II: 251
Douglas (Andrew) I: 686
Duato (José) I: 205

Dyer (Martin E.) II: 319

Eberhart (Andreas) I: 428
Eicken (Thorsten von) I: 187
Eisenbeis (Christine) II: 745
Eisenbiegler (Jörn) II: 602
Endo (Akiyoshi) I: 64
Engels (Marc) II: 236
Everaars (C. T. H.) I: 664

Feautrier (Paul) I: 424, 463
Fernandes (Edil S. T.) II: 773
Fernández (Agustin) I: 402
Fernández (Maria José) II: 724
Ferragina (Paolo) I: 781
Ferreira (A.) I: 218
Fey (Dietmar) II: 478
Filho (Eliseu M. C.) II: 773
Flynn (Michael J.) II: 183
Ford (Rupert W.) I: 432
Formella (Arno) II: 425
Foster (Ian) I: 3
Freytag (Johann Christoph) II: 872
Friedman (Roy) I: 84
Fritscher (Josef) I: 64
Frougny (Christiane) II: 175
Fuchs (W. Kent) I: 828
Fujita (Satoshi) I: 353
Fukuda (Akira) II: 706
Fulgham (Melanie L.) I: 195

Galán (Luis A.) I: 775
Gao (Guang R.) II: 745, 833
García (Felix) I: 522
Garibotto (Giovanni) II: 139
Gasperoni (Franco) II: 515
Gaudiot (Jean-Luc) I: 506
Geist (Al) I: 128
Gemund (Arjan J. C. van) II: 397
Gengler (Marc) I: 451
Gerbessiotis (Alexandros V.) II: 348
Gesmann (Michael) II: 852
Ghanem (Moustafa) I: 601
Giannakos (Aristotelis) II: 578
Gibbons (Phillip B.) II: 279
Girona (Sergi) I: 477
Girona (Sergi) II: 665

Goldberg (Allen) I: 145
Goldin (Maxim) I: 84
González (Antonio) I: 253
Goossens (Bernard) II: 789
Gorlatch (Sergei) II: 401
Gould (Nicholas I. M.) II: 34
Goward (Samuel) II: 109
Graeb (Robert) II: 91
Gregoris (Luis) II: 665
Griebl (Martin) I: 406, 467
Gropp (William) I: 128
Guenther (Michael) II: 91
Guerdoux-Jamet (Pascale) I: 11
Guider (Romain) I: 136
Guil (Nicolas) II: 131
Guo (Yi-ke) I: 601
Guo (Yike) I: 579

Ha (Soonhoi) II: 573
Hackstadt (Steven T.) I: 55
Hains (Gaétan) II: 409
He (Jifeng) II: 359
Hermenegildo (Manuel) II: 635, 724
Hernández (Juan) II: 866
Heun (Volker) I: 222
Hey (Tony) II: 251
Hill (Jonathan M. D.) II: 697
Hollingsworth (Jeffrey K.) I: 88
Hoppe (Hans-Christian) II: 899
Hori (Atsushi) I: 587
Hosoya (Haruo) I: 625
Hu (Zhenjiang) I: 553
Huber (Walter) II: 62
Huss-Lederman (Steve) I: 128

Ierusalimschy (Roberto) I: 597
Ikenaga (Takeshi) II: 203
Ishikawa (Yutaka) I: 587
Itoh (Yoshiaki) I: 587
Itzkovitz (Ayal) I: 84
Iwasaki (Hideya) I: 553

Jaedicke (Michael) II: 881
Jégou (Yvon) I: 563
Jelly (Innes) I: 752
Jessup (Elizabeth R.) I: 151
Jézéquel (Fabienne) II: 97

Jiménez (Marta) I: 402
Joe (Kazuki) II: 706
Juurlink (Ben H. H.) I: 361
Juurlink (Ben H. H.) II: 339

Kaklamanis (Christos) I: 270
Kalé (Laxmikant V.) I: 646
Kautonen (Anssi) II: 307
Kefalas (Petros) I: 694
Keller (Joerg) II: 425
Kelly (Paul H. J.) I: 106
Kelly (Paul H. J.) II: 445
Kessler (Christoph W.) II: 66
Keyes (David) II: 387
Kim (Myung-Kyun) I: 278
King (Chung-Ta) II: 611, 757
Kirkham (Chris C.) II: 563
Knoop (Jens) I: 441
Knopp (Juergen) I: 704
Kobayashi (Naoki) I: 625
Konaka (Hiroki) I: 587
Kong (Jinseok) II: 435
König (Jean-Claude) II: 578
Konuru (Ravi B.) II: 621
Kosch (Harald) II: 887
Kraemer-Fuhrmann (Ottmar) I: 120
Krizanc (Danny) I: 270
Krotz-Vogel (Werner) II: 899
Kruskal (Clyde P.) II: 352
Kshemkalyani (Ajay D.) I: 496
Kuchen (Herbert) I: 654, 718
Kuo (Sy-Yen) I: 514

Labarta (Jesús) II: 665
Labarta (Jesús) I: 477
Labarta (Jesús) II: 644
Laforest (Christian) I: 333, 353
Lahlou (Chams) II: 539
Laiymani (David) II: 535
Lakka (Spyridoula) II: 72
Lanet (Jean-Louis) II: 640
Lang (Tomas) II: 155
Langauer (Christian) I: 467
Langendoerfer (Horst) II: 615
Latifi (Shahram) I: 247
Lauwereins (Rudy) II: 236
Lavault (Christian) I: 832

Lavenier (Dominique)   I: 11
Le Gouëslier d'Argence (Patrick)   II: 501
Le Guyadec (Yann)   I: 732
Lee (Gyungho)   II: 435, 492
Lengauer (Christian)   I: 377
Leppänen (Ville)   II: 303, 307
Letteron (Jean-Marc)   II: 259
Levrouw (L. J.)   I: 70
L'Excellent (Jean-Yves)   II: 34
Li (Jingke)   I: 428
Li (Shuo-Yen Robert)   I: 818
Li (T. Y.)   II: 44
Li (Zhonghua)   II: 563
Lilius (Johan)   I: 660
Lin (Jenn-Wei)   I: 514
Lippens (P. E. R.)   II: 226
Listl (Andreas)   II: 881
Liu (C. L.)   I: 828
Llabería (José M.)   I: 402
Lo (Raymond)   II: 801
Loogen (R.)   I: 710
Löwe (Welf)   II: 602
Lozier (Daniel W.)   II: 193
Ludwig (Thomas)   I: 78
Lumsdaine (Andrew)   I: 128
Lusk (Ewing)   I: 128

MacKenzie (Philip D.)   II: 293
Maeda (Munenori)   I: 587
Maeng (Seung-Ryoul)   I: 278
Malony (Allen D.)   I: 55
Malumbres (M. P.)   I: 205
Mano (António)   II: 259
Marchetti-Spaccamela (Alberto)   I: 313
Marcus (K.)   I: 218
Marinescu (Dan C.)   II: 255
Marrakchi (Mounir)   II: 907
Martínez (José M.)   II: 866
Martorell (Xavier)   II: 644
Matias (Yossi)   II: 279
Mayr (Ernst W.)   I: 222
Mayr (Ernst W.)   II: 543
McAleese (G.)   II: 213
McColl (W. F.)   I: 25
Meerbergen (J. L. van)   II: 226

Melin (Emmanuel)   I: 732
Meyer auf der Heide (Friedhelm)   I: 299
Meyer auf der Heide (Friedhelm)   II: 369
Michel (Olivier)   I: 678, 714
Miguel (Pedro de)   I: 522
Miguel (Pedro de)   II: 866
Mikschl (Alfred)   II: 453, 461
Miller (Barton P.)   I: 88
Miller (Quentin)   II: 359
Mills (Peter)   I: 145
Mitkas (Pericles A.)   II: 457
Mitschang (Bernhard)   II: 881
Moraes de Azevedo (Marcelo)   I: 247
Morancho (Enric)   I: 402
Moreau (Luc)   I: 615
Moreira (José E.)   II: 621
Morrow (P. J.)   II: 213
Moussouni (Fouzia)   I: 832
Müller (Andreas)   I: 64
Muller (Jean-Michel)   II: 165
Mullins (John)   II: 409
Munier (Alix)   II: 578
Munk Nielsen (Asger)   II: 165
Muñoz (Xavier)   I: 313
Murphy (Kieran)   II: 22, 26

Naik (Vijay K.)   II: 621
Nappi (Michele)   II: 147
Naroska (Edwin)   II: 582
Nash (Jonathan M.)   II: 319
Navarro (Nacho)   II: 644
Nepomniaschaya (Ann S.)   I: 813
Nibhanupudi (Mohan V.)   II: 311
Nieto (M.)   II: 866
Nisbet (Andrew P.)   I: 432
Nyland (Lars)   I: 145

Oberhuber (Michael)   I: 78
Oberman (Stuart F.)   II: 183
O'Boyle (Michael F. P.)   I: 432
Ogura (Takeshi)   II: 203
Oh (Hyunok)   II: 573
Okuda (Kunio)   I: 398
Oliver (T.)   I: 178
Openshaw (Stan)   II: 270

Ortega-Mallén (Y.)   I: 710
Özgüner (Füsun)   I: 287
Ozkarahan (Esen)   II: 892

Pahud (Michel)   II: 689
Panaite (Petrişor)   I: 283
Pandey (Raju)   I: 633
Paugam-Moisy (Hélène)   II: 243
Peña (Ricardo)   I: 775
Peña-Marí (R.)   I: 710
Penttonen (Martti)   II: 307
Peperstraete (J. A.)   II: 236
Perennes (Stéphane)   I: 325, 353, 370
Pérez (Fernando)   I: 522
Petri (Stefan)   II: 615
Petrini (Fabrizio)   I: 307
Pfaffinger (Alexander)   II: 631
Pillet (Vincent)   II: 665
Piquer (José M.)   I: 532
Podehl (Axel)   I: 789
Pohl (Hans Werner)   I: 455
Polo (Antonio)   II: 866
Potter (Richard)   II: 779
Pourzandi (Makan)   II: 243
Prasanna (Viktor K.)   II: 123
Prata dos Santos (Sérgio)   II: 259
Prins (Jan)   I: 145
Priol (Thierry)   I: 98
Prylli (Loïc)   I: 155
Puaut (Isabelle)   I: 114
Puntigam (Franz)   I: 762

Quinton (Patrice)   I: 11

Raffin (Bruno)   I: 732
Rahm (Erhard)   I: 37
Rahola (Jussi)   II: 81
Rajopadhye (Sanjay)   I: 389
Ramachandran (Vijaya)   II: 279, 293
Ramanujam (J.)   I: 541
Ramet (Pierre)   I: 165
Rangaswami (Roopa)   II: 417
Rao (P. S.)   I: 361
Rapine (Christophe)   II: 527
Rauber (Thomas)   I: 789, 822
Rauber (Thomas)   II: 52
Rebeuf (Xavier)   I: 732

Refstrup (Jacob G.)   II: 445
Reif (John)   I: 145
Reiser (Angelika)   II: 881
Renambot (Luc)   I: 98
Reymann (Olivier)   I: 74
Rice (Daniel G.)   I: 213
Rieping (Ingo)   II: 369
Rinard (Martin C.)   I: 414
Risau (Juergen)   II: 453
Roantree (D. K.)   II: 213
Rodriguez (Noemi)   I: 597
Röjemo (Niklas)   I: 686
Roman (Jean)   I: 165
Ronsse (M. A.)   I: 70
Rowstron (Antony)   I: 510
Runciman (Colin)   I: 686
Rünger (Gudula)   I: 789, 822
Rünger (Gudula)   II: 52
Russo (Stefano)   I: 752

Saltz (Joel)   II: 109
Sánchez (Fermín)   II: 824
Saphir (William)   I: 128
Savy (Carlo)   I: 752
Saxena (Prashant)   I: 828
Schabanel (Nicolas)   II: 193
Schmidt (Sebastian)   II: 815
Schnor (Bettina)   II: 615
Scholtes (Carsten)   I: 822
Schroeder (Klaus)   I: 299
Schuster (Assaf)   I: 84
Schwarze (Frank)   I: 299
Schwiegelshohn (Uwe)   II: 515, 582
Shafarenko (A.)   I: 742
Sharma (Vishal)   II: 738
Shen (Kish)   II: 635
Shock (Carter T.)   II: 109
Shrivastava (S. K.)   I: 487
Shum (Kam Hong)   I: 818
Sibeyn (Jop F.)   I: 361, 799
Sigmund (Ulrich)   II: 797
Simon (Jens)   II: 675
Siniolakis (Constantinos J.)   II: 348
Skjellum (Tony)   I: 128
Smith (J. A.)   I: 487
Snir (Marc)   I: 128
Snyder (Lawrence)   I: 195

Spence (I. T. A.)   II: 213
Spiliopoulou (Myra)   II: 872
Stadtherr (Hans)   II: 543
Steven (Gordon)   II: 779
Stoica (Ion)   II: 387
Strey (Alfred)   II: 470
Strohmaier (Erich)   II: 251
Sueur (Dominique)   I: 173
Sultan (Florin)   II: 387
Summerville (Douglas H.)   I: 213
Sunderland (A.)   I: 178
Sussman (Alan)   II: 109
Szularz (Marek)   II: 22, 26
Szymanski (Boleslaw K.)   II: 311

Takeichi (Masato)   I: 553
Talbot (Sarah A. M.)   I: 106
Talbot (Sarah A. M.)   II: 445
Temam (Olivier)   II: 765
Thorelli (Lars-Erik)   II: 714
Tiskin (Alexandre)   II: 327
To (Hing Wing)   I: 601
Tomokiyo (Takashi)   I: 587
Torre (Pilar de la)   II: 352
Tourancheau (Bernard)   I: 155
Tourancheau (Bernard)   II: 3
Towle (Ross)   II: 801
Trystram (Denis)   II: 527
Tumuluri (Chaitanya)   II: 486
Turner (Peter R.)   II: 193
Turton (Ian)   II: 270

Uhl (Andreas)   II: 151
Ulusoy (Özgür)   II: 862
Ünalýr (M. Osman)   II: 892
Ungerer (Theo)   II: 797
Urland (Erik)   I: 809
Ururahy (Cristina)   I: 597

Valero-García (Miguel)   I: 253
Vanneschi (Marco)   I: 307
Varvarigos (Emmanouel A.)   II: 738
Verhaegh (W. F. J.)   II: 226
Verley (Gilles)   II: 377
Villalba (Julio)   II: 155
Vincent (Jean-Marc)   II: 734
Virot (Bernard)   I: 732

Vitulano (Domenico)   II: 147
Vivien (Frédéric)   I: 379
Vlahavas (Ioannis)   I: 694
Vlassov (Vladimir)   II: 714
Vu (Duc Thang)   II: 789

Wagner (Charles)   I: 11
Wagner (Robert)   I: 145
Wallace (Steven)   II: 747
Walle (Thomas)   II: 425
Wang (Jer-Tsang)   II: 843
Wang (Jian)   II: 745
Welsh (Matt)   I: 187
Wendelborn (Andrew L.)   I: 136
Werf (A. van der)   II: 226
Weston (James)   II: 26
Weston (Jim)   II: 22
Wever (Utz)   II: 91
Wierum (Jens-Michael)   II: 675
Wijshoff (Harry A. G.)   II: 339
Wilde (Doran)   I: 389
Winckel (Laurent)   II: 819
Wismueller (Roland)   I: 78
Wolf (Klaus)   I: 120
Wolf (Klaus)   II: 259
Wolfe (Andrew)   II: 773
Wood (Alan)   I: 510, 686
Wray (J. P.)   I: 770
Wylie (Brian J. N.)   I: 64

Yang (Jin)   I: 601
Yonezawa (Akinori)   I: 625
Yoon (Dae-Kyun)   I: 506
Yoon (Hyunsoo)   I: 278

Zapata (Emilio L.)   II: 131, 155
Zenger (Christoph)   II: 631
Zheng (Qinghua)   II: 91
Zimmermann (Stephan)   II: 881
Zimmermann (Wolf)   II: 602
Zou (Xiulin)   II: 44

Springer-Verlag and the Environment

We at Springer-Verlag firmly believe that an international science publisher has a special obligation to the environment, and our corporate policies consistently reflect this conviction.

We also expect our business partners – paper mills, printers, packaging manufacturers, etc. – to commit themselves to using environmentally friendly materials and production processes.

The paper in this book is made from low- or no-chlorine pulp and is acid free, in conformance with international standards for paper permanency.

Lecture Notes in Computer Science

For information about Vols. 1–1054

please contact your bookseller or Springer-Verlag

Vol. 1055: T. Margaria, B. Steffen (Eds.), Tools and Algorithms for the Construction and Analysis of Systems. Proceedings, 1996. XI, 435 pages. 1996.

Vol. 1056: A. Haddadi, Communication and Cooperation in Agent Systems. XIII, 148 pages. 1996. (Subseries LNAI).

Vol. 1057: P. Apers, M. Bouzeghoub, G. Gardarin (Eds.), Advances in Database Technology — EDBT '96. Proceedings, 1996. XII, 636 pages. 1996.

Vol. 1058: H. R. Nielson (Ed.), Programming Languages and Systems – ESOP '96. Proceedings, 1996. X, 405 pages. 1996.

Vol. 1059: H. Kirchner (Ed.), Trees in Algebra and Programming – CAAP '96. Proceedings, 1996. VIII, 331 pages. 1996.

Vol. 1060: T. Gyimóthy (Ed.), Compiler Construction. Proceedings, 1996. X, 355 pages. 1996.

Vol. 1061: P. Ciancarini, C. Hankin (Eds.), Coordination Languages and Models. Proceedings, 1996. XI, 443 pages. 1996.

Vol. 1062: E. Sanchez, M. Tomassini (Eds.), Towards Evolvable Hardware. IX, 265 pages. 1996.

Vol. 1063: J.-M. Alliot, E. Lutton, E. Ronald, M. Schoenauer, D. Snyers (Eds.), Artificial Evolution. Proceedings, 1995. XIII, 396 pages. 1996.

Vol. 1064: B. Buxton, R. Cipolla (Eds.), Computer Vision – ECCV '96. Volume I. Proceedings, 1996. XXI, 725 pages. 1996.

Vol. 1065: B. Buxton, R. Cipolla (Eds.), Computer Vision – ECCV '96. Volume II. Proceedings, 1996. XXI, 723 pages. 1996.

Vol. 1066: R. Alur, T.A. Henzinger, E.D. Sontag (Eds.), Hybrid Systems III. IX, 618 pages. 1996.

Vol. 1067: H. Liddell, A. Colbrook, B. Hertzberger, P. Sloot (Eds.), High-Performance Computing and Networking. Proceedings, 1996. XXV, 1040 pages. 1996.

Vol. 1068: T. Ito, R.H. Halstead, Jr., C. Queinnec (Eds.), Parallel Symbolic Languages and Systems. Proceedings, 1995. X, 363 pages. 1996.

Vol. 1069: J.W. Perram, J.-P. Müller (Eds.), Distributed Software Agents and Applications. Proceedings, 1994. VIII, 219 pages. 1996. (Subseries LNAI).

Vol. 1070: U. Maurer (Ed.), Advances in Cryptology – EUROCRYPT '96. Proceedings, 1996. XII, 417 pages. 1996.

Vol. 1071: P. Miglioli, U. Moscato, D. Mundici, M. Ornaghi (Eds.), Theorem Proving with Analytic Tableaux and Related Methods. Proceedings, 1996. X, 330 pages. 1996. (Subseries LNAI).

Vol. 1072: R. Kasturi, K. Tombre (Eds.), Graphics Recognition. Proceedings, 1995. X, 308 pages. 1996.

Vol. 1073: J. Cuny, H. Ehrig, G. Engels, G. Rozenberg (Eds.), Graph Grammars and Their Application to Computer Science. Proceedings, 1994. X, 565 pages. 1996.

Vol. 1074: G. Dowek, J. Heering, K. Meinke, B. Möller (Eds.), Higher-Order Algebra, Logic, and Term Rewriting. Proceedings, 1995. VII, 287 pages. 1996.

Vol. 1075: D. Hirschberg, G. Myers (Eds.), Combinatorial Pattern Matching. Proceedings, 1996. VIII, 392 pages. 1996.

Vol. 1076: N. Shadbolt, K. O'Hara, G. Schreiber (Eds.), Advances in Knowledge Acquisition. Proceedings, 1996. XII, 371 pages. 1996. (Subseries LNAI).

Vol. 1077: P. Brusilovsky, P. Kommers, N. Streitz (Eds.), Mulimedia, Hypermedia, and Virtual Reality. Proceedings, 1994. IX, 311 pages. 1996.

Vol. 1078: D.A. Lamb (Ed.), Studies of Software Design. Proceedings, 1993. VI, 188 pages. 1996.

Vol. 1079: Z.W. Raś, M. Michalewicz (Eds.), Foundations of Intelligent Systems. Proceedings, 1996. XI, 664 pages. 1996. (Subseries LNAI).

Vol. 1080: P. Constantopoulos, J. Mylopoulos, Y. Vassiliou (Eds.), Advanced Information Systems Engineering. Proceedings, 1996. XI, 582 pages. 1996.

Vol. 1081: G. McCalla (Ed.), Advances in Artificial Intelligence. Proceedings, 1996. XII, 459 pages. 1996. (Subseries LNAI).

Vol. 1082: N.R. Adam, B.K. Bhargava, M. Halem, Y. Yesha (Eds.), Digital Libraries. Proceedings, 1995. Approx. 310 pages. 1996.

Vol. 1083: K. Sparck Jones, J.R. Galliers, Evaluating Natural Language Processing Systems. XV, 228 pages. 1996. (Subseries LNAI).

Vol. 1084: W.H. Cunningham, S.T. McCormick, M. Queyranne (Eds.), Integer Programming and Combinatorial Optimization. Proceedings, 1996. X, 505 pages. 1996.

Vol. 1085: D.M. Gabbay, H.J. Ohlbach (Eds.), Practical Reasoning. Proceedings, 1996. XV, 721 pages. 1996. (Subseries LNAI).

Vol. 1086: C. Frasson, G. Gauthier, A. Lesgold (Eds.), Intelligent Tutoring Systems. Proceedings, 1996. XVII, 688 pages. 1996.

Vol. 1087: C. Zhang, D. Lukose (Eds.), Distributed Artificial Intelliegence. Proceedings, 1995. VIII, 232 pages. 1996. (Subseries LNAI).

Vol. 1088: A. Strohmeier (Ed.), Reliable Software Technologies – Ada-Europe '96. Proceedings, 1996. XI, 513 pages. 1996.

Vol. 1089: G. Ramalingam, Bounded Incremental Computation. XI, 190 pages. 1996.

Vol. 1090: J.-Y. Cai, C.K. Wong (Eds.), Computing and Combinatorics. Proceedings, 1996. X, 421 pages. 1996.

Vol. 1091: J. Billington, W. Reisig (Eds.), Application and Theory of Petri Nets 1996. Proceedings, 1996. VIII, 549 pages. 1996.

Vol. 1092: H. Kleine Büning (Ed.), Computer Science Logic. Proceedings, 1995. VIII, 487 pages. 1996.

Vol. 1093: L. Dorst, M. van Lambalgen, F. Voorbraak (Eds.), Reasoning with Uncertainty in Robotics. Proceedings, 1995. VIII, 387 pages. 1996. (Subseries LNAI).

Vol. 1094: R. Morrison, J. Kennedy (Eds.), Advances in Databases. Proceedings, 1996. XI, 234 pages. 1996.

Vol. 1095: W. McCune, R. Padmanabhan, Automated Deduction in Equational Logic and Cubic Curves. X, 231 pages. 1996. (Subseries LNAI).

Vol. 1096: T. Schäl, Workflow Management Systems for Process Organisations. XII, 200 pages. 1996.

Vol. 1097: R. Karlsson, A. Lingas (Eds.), Algorithm Theory – SWAT '96. Proceedings, 1996. IX, 453 pages. 1996.

Vol. 1098: P. Cointe (Ed.), ECOOP '96 – Object-Oriented Programming. Proceedings, 1996. XI, 502 pages. 1996.

Vol. 1099: F. Meyer auf der Heide, B. Monien (Eds.), Automata, Languages and Programming. Proceedings, 1996. XII, 681 pages. 1996.

Vol. 1100: B. Pfitzmann, Digital Signature Schemes. XVI, 396 pages. 1996.

Vol. 1101: M. Wirsing, M. Nivat (Eds.), Algebraic Methodology and Software Technology. Proceedings, 1996. XII, 641 pages. 1996.

Vol. 1102: R. Alur, T.A. Henzinger (Eds.), Computer Aided Verification. Proceedings, 1996. XII, 472 pages. 1996.

Vol. 1103: H. Ganzinger (Ed.), Rewriting Techniques and Applications. Proceedings, 1996. XI, 437 pages. 1996.

Vol. 1104: M.A. McRobbie, J.K. Slaney (Eds.), Automated Deduction – CADE-13. Proceedings, 1996. XV, 764 pages. 1996. (Subseries LNAI).

Vol. 1105: T.I. Ören, G.J. Klir (Eds.), Computer Aided Systems Theory – CAST '94. Proceedings, 1994. IX, 439 pages. 1996.

Vol. 1106: M. Jampel, E. Freuder, M. Maher (Eds.), Over-Constrained Systems. X, 309 pages. 1996.

Vol. 1107: J.-P. Briot, J.-M. Geib, A. Yonezawa (Eds.), Object-Based Parallel and Distributed Computation. Proceedings, 1995. X, 349 pages. 1996.

Vol. 1108: A. Díaz de Ilarraza Sánchez, I. Fernández de Castro (Eds.), Computer Aided Learning and Instruction in Science and Engineering. Proceedings, 1996. XIV, 480 pages. 1996.

Vol. 1109: N. Koblitz (Ed.), Advances in Cryptology – Crypto '96. Proceedings, 1996. XII, 417 pages. 1996.

Vol. 1110: O. Danvy, R. Glück, P. Thiemann (Eds.), Partial Evaluation. Proceedings, 1996. XII, 514 pages. 1996.

Vol. 1111: J.J. Alferes, L. Moniz Pereira, Reasoning with Logic Programming. XXI, 326 pages. 1996. (Subseries LNAI).

Vol. 1112: C. von der Malsburg, W. von Seelen, J.C. Vorbrüggen, B. Sendhoff (Eds.), Artificial Neural Networks – ICANN 96. Proceedings, 1996. XXV, 922 pages. 1996.

Vol. 1113: W. Penczek, A. Szałas (Eds.), Mathematical Foundations of Computer Science 1996. Proceedings, 1996. X, 592 pages. 1996.

Vol. 1114: N. Foo, R. Goebel (Eds.), PRICAI'96: Topics in Artificial Intelligence. Proceedings, 1996. XXI, 658 pages. 1996. (Subseries LNAI).

Vol. 1115: P.W. Eklund, G. Ellis, G. Mann (Eds.), Conceptual Structures: Knowledge Representation as Interlingua. Proceedings, 1996. XIII, 321 pages. 1996. (Subseries LNAI).

Vol. 1116: J. Hall (Ed.), Management of Telecommunication Systems and Services. XXI, 229 pages. 1996.

Vol. 1117: A. Ferreira, J. Rolim, Y. Saad, T. Yang (Eds.), Parallel Algorithms for Irregularly Structured Problems. Proceedings, 1996. IX, 358 pages. 1996.

Vol. 1118: E.C. Freuder (Ed.), Principles and Practice of Constraint Programming — CP 96. Proceedings, 1996. XIX, 574 pages. 1996.

Vol. 1119: U. Montanari, V. Sassone (Eds.), CONCUR '96: Concurrency Theory. Proceedings, 1996. XII, 751 pages. 1996.

Vol. 1120: M. Deza. R. Euler, I. Manoussakis (Eds.), Combinatorics and Computer Science. Proceedings, 1995. IX, 415 pages. 1996.

Vol. 1121: P. Perner, P. Wang, A. Rosenfeld (Eds.), Advances in Structural and Syntactical Pattern Recognition. Proceedings, 1996. X, 393 pages. 1996.

Vol. 1122: H. Cohen (Ed.), Algorithmic Number Theory. Proceedings, 1996. IX, 405 pages. 1996.

Vol. 1123: L. Bougé, P. Fraigniaud, A. Mignotte, Y. Robert (Eds.), Euro-Par'96. Parallel Processing. Proceedings, 1996, Vol. I. XXXIII, 842 pages. 1996.

Vol. 1124: L. Bougé, P. Fraigniaud, A. Mignotte, Y. Robert (Eds.), Euro-Par'96. Parallel Processing. Proceedings, 1996, Vol. II. XXXIII, 926 pages. 1996.

Vol. 1125: J. von Wright, J. Grundy, J. Harrison (Eds.), Theorem Proving in Higher Order Logics. Proceedings, 1996. VIII, 447 pages. 1996.

Vol. 1126: J.J. Alferes, L. Moniz Pereira, E. Orlowska (Eds.), Logics in Artificial Intelligence. Proceedings, 1996. IX, 417 pages. 1996. (Subseries LNAI).

Vol. 1127: L. Böszörményi (Ed.), Parallel Computation. Proceedings, 1996. XI, 235 pages. 1996.

Vol. 1128: J. Calmet, C. Limongelli (Eds.), Design and Implementation of Symbolic Computation Systems. Proceedings, 1996. IX, 356 pages. 1996.

Vol. 1129: J. Launchbury, E. Meijer, T. Sheard (Eds.), Advanced Functional Programming. Proceedings, 1996. VII, 238 pages. 1996.